ZENECA LIBRARY
WILMINGTON, DE 19897

65647

Y0-AGC-140

SA RS 55 N4   2001

Organic-Chemical Drugs A
nd Their Synonyms: An...

Martin Negwer
Hans-Georg Scharnow

**Organic-Chemical Drugs
and Their Synonyms**

Volume 3

Martin Negwer
Hans-Georg Scharnow

Organic-Chemical Drugs and Their Synonyms

(An International Survey)

Eighth Extensively
Enlarged Edition

Volume 3

$C_{19}H_{18}BrF_3N_2O_4 -$
$C_{24}H_{20}O_8$
$(8409 - 12213)$

FIZ CHEMIE BERLIN
Fachinformationszentrum Chemie GmbH

Weinheim · New York · Chichester
Brisbane · Singapore · Toronto

Dr. Martin Negwer, Berlin
Dr. Hans-Georg Scharnow, Berlin

This book was carefully produced. Nevertheless authors and publisher do not warrant the information contained therein to be free of errors. Readers are advised to keep in mind that statements, data, illustrations, procedural details or other items may inadvertently be inaccurate.

1st edition 1959
2nd edition 1962
3rd edition 1966
4th edition 1971
5th edition 1978
6th edition 1987
7th edition 1994
8th edition 2001

Library of Congress Card No.: applied for

A catalogue record for this book is available from the British Library.

Die Deutsche Bibliothek - CIP Cataloguing-in-Publication-Data
A catalogue record for this publication is available from Die Deutsche Bibliothek

ISBN 3-527-30247-6

© WILEY-VCH Verlag GmbH, D-69469 Weinheim (Federal Republic of Germany), 2001
Printed on acid-free paper.
All rights reserved (including those of translation in other languages). No part of this book may be reproduced in any form - by photoprinting, microfilm, or any other means - nor transmitted or translated into machine language without written permission from the publishers. Registered names, trademarks, etc. used in this book, even when not specifically marked as such, are not to be considered unprotected by law.
Electronic documents and indexes: Fachinformationszentrum Chemie GmbH, D-10587 Berlin
Printing: betz-druck gmbh, D-64291 Darmstadt.
Bookbinding: J. Schäffer GmbH & Co. KG, D-67269 Grünstadt.
Printed in the Federal Republic of Germany.

8409   (6891) $C_{19}H_{18}BrF_3N_2O_4$
41113-86-4

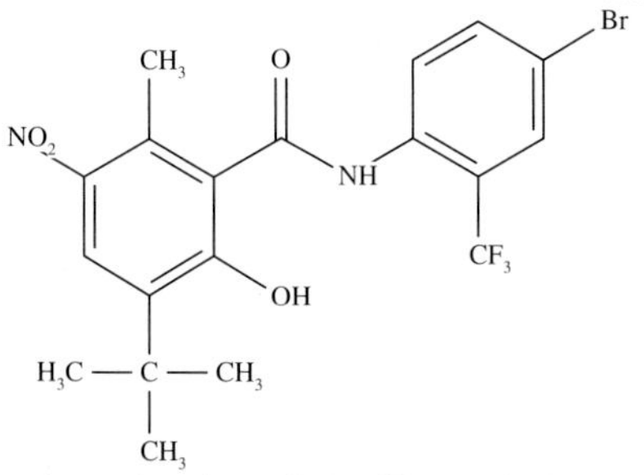

4'-Bromo-3-*tert*-butyl-α',α',α'-trifluoro-5-nitro-2,6-cre-
soto-*o*-toluidide = 4'-Bromo-3-*tert*-butyl-6-methyl-5-ni-
tro-2'-(trifluoromethyl)salicylanilide = *N*-[4-Bromo-2-
(trifluoromethyl)phenyl]-3-(1,1-dimethylethyl)-2-hydr-
oxy-6-methyl-5-nitrobenzamide (●)
S   *Bromoxanide**, SKF 61636
U   Anthelmintic

8410   (6892) $C_{19}H_{18}ClFN_2O_3$
27060-91-9

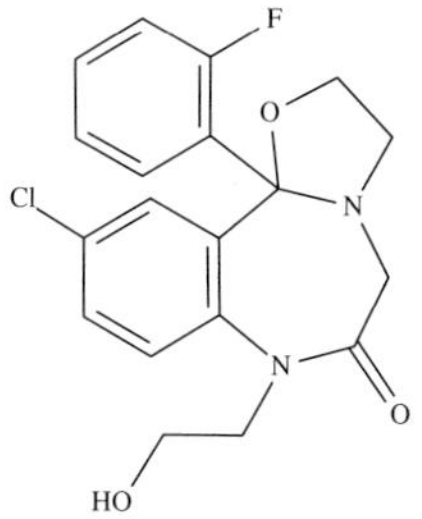

10-Chloro-11b-(2-fluorophenyl)-2,3,7,11b-tetrahydro-7-
(2-hydroxyethyl)oxazolo[3,2-*d*][1,4]benzodiazepin-
6(5*H*)-one (●)
S   Coreminal, *Flutazolam**, MS 4101, Ro 7-6102
U   Tranquilizer

8411   (6893) $C_{19}H_{18}ClFN_2O_3S$
52042-01-0

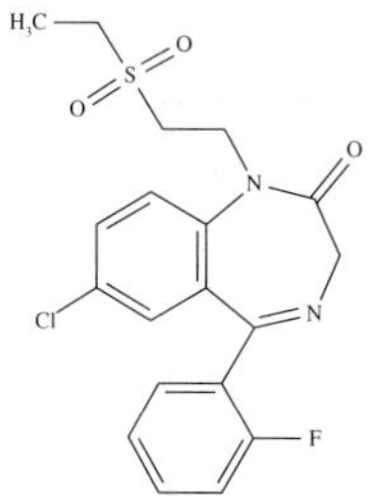

7-Chloro-1-[2-(ethylsulfonyl)ethyl]-5-(2-fluorophenyl)-
1,3-dihydro-2*H*-1,4-benzodiazepin-2-one (●)
S   *Elfazepam**, SKF 72517
U   Appetite stimulant (veterinary)

8412   (6894) $C_{19}H_{18}ClF_2N_3O_3$
127254-12-0

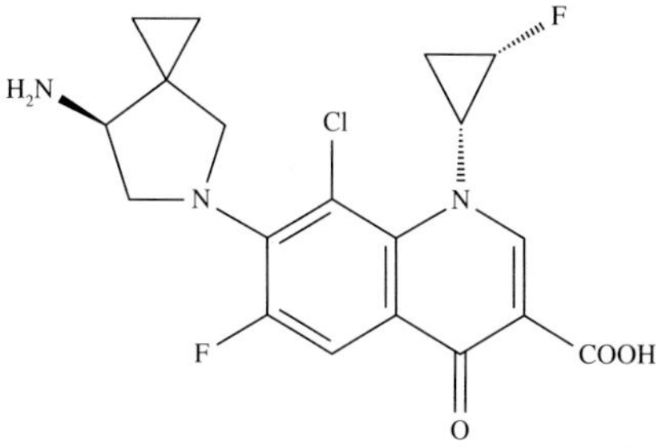

(−)-7-[(7*S*)-7-Amino-5-azaspiro[2.4]hept-5-yl]-8-chloro-
6-fluoro-1-[(1*R*,2*S*)-fluorocyclopropyl]-1,4-dihydro-4-
oxo-3-quinolinecarboxylic acid = [1*R*-[1α(*S**),2α]]-7-(7-
Amino-5-azaspiro[2.4]hept-5-yl)-8-chloro-6-fluoro-1-(2-
fluorocyclopropyl)-1,4-dihydro-4-oxo-3-quinolinecarb-
oxylic acid (●)
R   also hydrate
S   DU 6859, *Sitafloxacin**, *Spifloxacin*
U   Antibacterial

8413   (6895) $C_{19}H_{18}ClNO$
60085-78-1

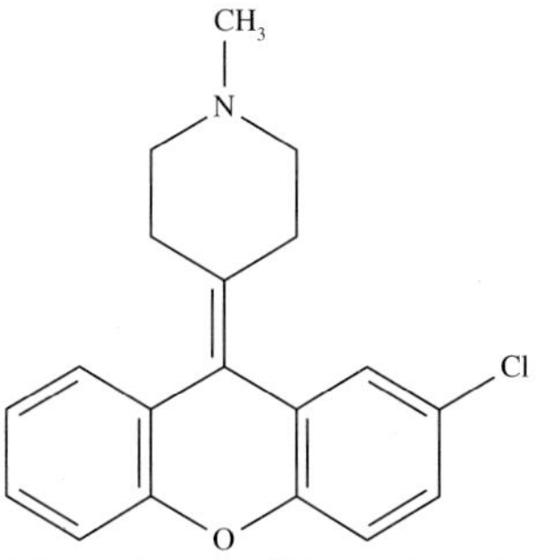

4-(2-Chloro-9*H*-xanthen-9-ylidene)-1-methylpiperidine
(●)
R   Methanesulfonate (60086-22-8)
S   *Clopipazan mesilate**, SKF 69634
U   Antipsychotic

8414　(6896)

$C_{19}H_{18}ClN_3O$
15687-07-7

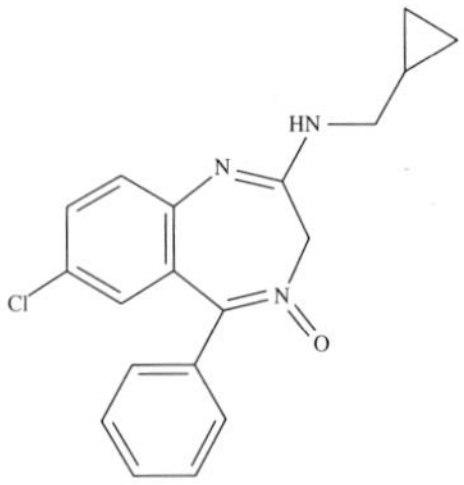

7-Chloro-2-[(cyclopropylmethyl)amino]-5-phenyl-3*H*-1,4-benzodiazepine 4-oxide = 7-Chloro-*N*-(cyclopropyl-methyl)-5-phenyl-3*H*-1,4-benzodiazepin-2-amine 4-oxide (●)

S　*Ciprazepam, Cyprazepam***, W 3623
U　Tranquilizer

8415

$C_{19}H_{18}ClN_3O_2$
138752-33-7

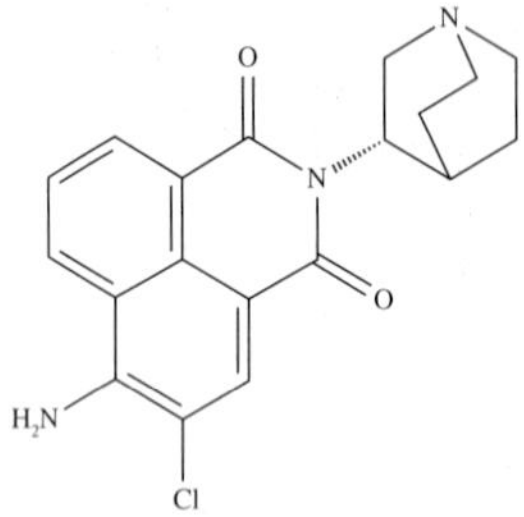

(*S*)-6-Amino-5-chloro-2,3-dihydro-2-(3-quinuclidinyl)-1*H*-benz[*de*]isoquinoline-1,3-dione = (*S*)-6-Amino-2-(1-azabicyclo[2.2.2]oct-3-yl)-5-chloro-1*H*-benz[*de*]isoqui-noline-1,3(2*H*)-dione (●)

R　Monohydrochloride (138752-34-8)
S　RS-56532
U　Anti-arrhythmic (positive inotropic), anxiolytic

8416

$C_{19}H_{18}ClN_3O_2$
128672-84-4

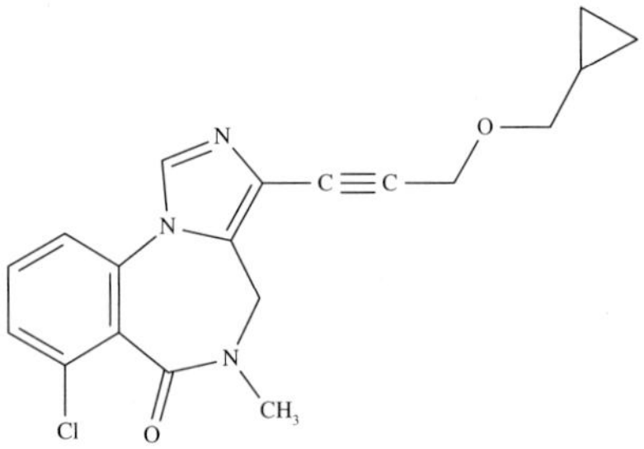

7-Chloro-3-[3-(cyclopropylmethoxy)-1-propynyl]-4,5-dihydro-5-methyl-6*H*-imidazo[1,5-*a*][1,4]benzodiazepin-6-one (●)

S　Ro 42-8773
U　Benzodiazepine receptor agonist

8417　(6897)

$C_{19}H_{18}ClN_3O_3$
36104-80-0

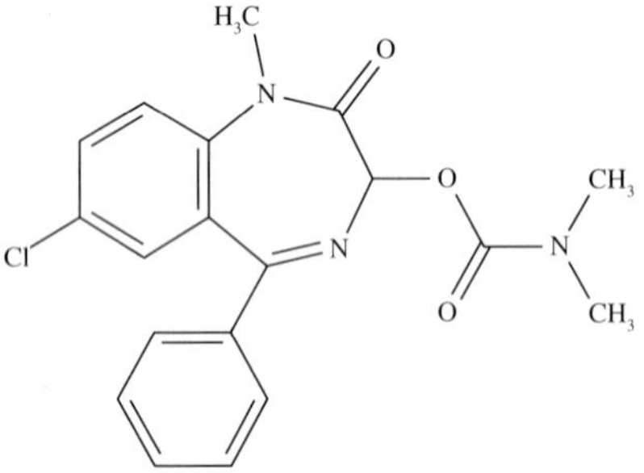

7-Chloro-1,3-dihydro-3-hydroxy-1-methyl-5-phenyl-2*H*-1,4-benzodiazepin-2-one dimethylcarbamate (ester) = Di-methylcarbamic acid 7-chloro-2,3-dihydro-1-methyl-2-oxo-5-phenyl-1*H*-1,4-benzodiazepin-3-yl ester (●)

S　Albego, Amotril "Vianex", *Camacepán, Camaze-pam***, KTH 497, Limpidon, Nebolan, Panevril, Pa-xaure, Paxor, SB 5833, *Temazepam dimethylcarba-mate*
U　Tranquilizer

8418 (6898) $C_{19}H_{18}ClN_3O_5S$
61-72-3

(2S,5R,6R)-6-[3-(o-Chlorophenyl)-5-methyl-4-isoxazole-carboxamido]-3,3-dimethyl-7-oxo-4-thia-1-azabicy-clo[3.2.0]heptane-2-carboxylic acid = (6R)-6-(3-o-Chlo-rophenyl-5-methyl-4-isoxazolecarboxamido)penicillanic acid = 3-(o-Chlorophenyl)-5-methyl-4-isoxazolylpenicil-lin = [2S-(2α,5α,6β)]-6-[[[3-(2-Chlorophenyl)-5-methyl-4-isoxazolyl]carbonyl]amino]-3,3-dimethyl-7-oxo-4-thia-1-azabicyclo[3.2.0]heptane-2-carboxylic acid (●)

R Sodium salt (642-78-4) [or sodium salt monohydrate (7081-44-9)]

S Alclox, Anaclosil, Ankerbin "Ankerfarm", Apo-Clo-xi, Austrastaph, Bactopen, Benecil, Benicil, Bi-ject-Cloxa, Bioclox, Biocloxil, Bristapen, BRL 1621, cf-Cloxamin, Clobex, Clocillin "Asia Pharm.", Cloxacap, *Cloxacillin-Natrium***, *Cloxacil-lin Sodium***, Cloxalene, Cloxapen, Cloxilean, Clo-xillin, Cloxin "Labethica", Cloxipen, Cloxipenil, Cloxypen, Coclox, Dariclox, Diner, Ekvacillin, Elle-cid, Equacillin, Eumacid, Gelstaph, Gertemin, Isoxa-cillin, Kloksacillin, Klox, Kloxerate-QR, Lactaclox, Lactocillin, Landerclox, Loxavit, Lutrexin "Proel", Meiclox, Meixam, Methocillin S., Monoclox, No-roclox, Novocloxin, Nu-Cloxi, Opticlox, Orbenil, Orbenin, Orbénine, P-25, Penivet, Penstapho N (Pen-staphon), Prevencilina P, Pro-Cloxa, Prostaphilin A, Prostaphlin A, Rivoclox, Semicillina, Serviclox, *So-dium Cloxacillin*, Staflocil, Staphobristol, Staphybio-tic, Staphyclox, Syntarpen, Tabroclox, Tegopen, Ubracillin

U Antibiotic

8418-01 (6898-01) 23736-58-5

R Compd. with N,N'-dibenzylethylenediamine (no. 6346) (2:1)

S Belaclox, *Benzathine Cloxacillin*, Boviclox, Centra-clox HL, *Cloxacillin Benzathine*, Cloxalene asciutta, Cloxaman, Cloxar, Cloxastop, Cloxataryl, Clo-xavan, Cloxine H.L., Coxalin-TS, Cuxavet TS, Dic-lomam, Dry-Clox, Embaclox, Gelstaph T.S., Izoclo-

xacillina, Klateclox, Kloxerate-DC, Mammin TS, Noroclox-DC, Orbenin retard, Orbenor, Orbicilline, Penivet-TS, Tariclone, Tarigermel, Tarilac, Trait-mam HL, Trokacilon, Vetoscon, Vetriclox, Wede-clox

U Depot-antibiotic

8419 (6899) $C_{19}H_{18}ClN_5$
37115-32-5

8-Chloro-1-[(dimethylamino)methyl]-6-phenyl-4H-s-tri-azolo[4,3-a][1,4]benzodiazepine = 8-Chloro-N,N-dime-thyl-6-phenyl-4H-[1,2,4]triazolo[4,3-a][1,4]benzo-diaze-pine-1-methanamine (●)

R also monomethanesulfonate (57938-82-6)

S *Adinazolam***, Deracyn, U-41123

U Antidepressant, sedative

8420 (6900) $C_{19}H_{18}Cl_2N_2O_2$
75616-03-4

4'-Chloro-2'-(o-chlorobenzoyl)-2-(cyclopropylamino)-N-methylacetanilide = N-[4-Chloro-2-(2-chloroben-zoyl)phenyl]-2-(cyclopropylamino)-N-methylacetamide (●)

S *Ciprazafone***

U Anxiolytic

8421 (6901)
$C_{19}H_{18}Cl_2N_4O$
114606-98-3

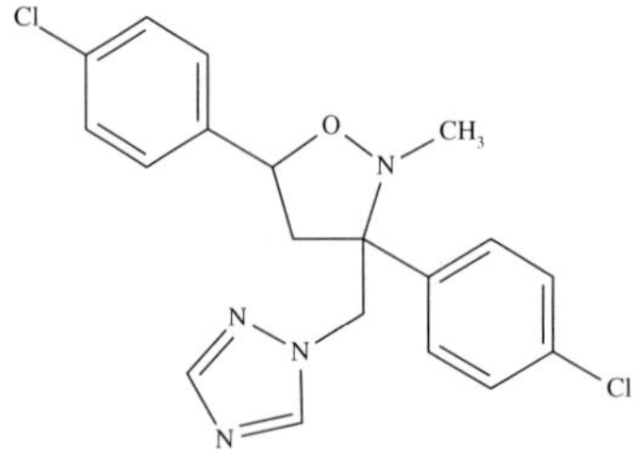

1-[[3,5-Bis(4-chlorophenyl)-2-methyl-3-isoxazolidi-
nyl]methyl]-1*H*-1,2,4-triazole (●)
S PR 988-399 FB
U Antifungal

8422 (6902)
$C_{19}H_{18}Cl_2N_4O_4S$
15949-72-1

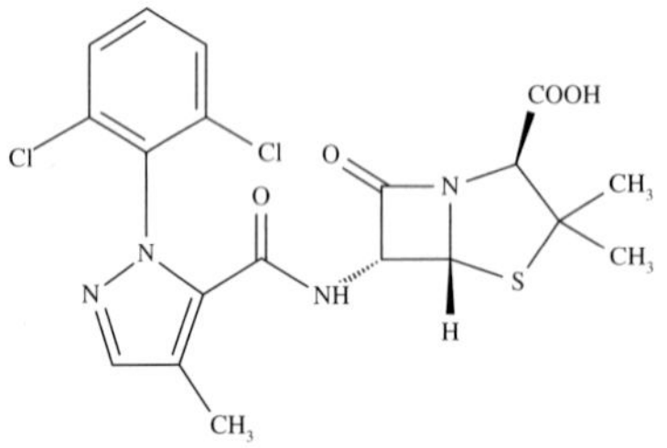

(2*S*,5*R*,6*R*)-6-[1-(2,6-Dichlorophenyl)-4-methyl-5-pyra-
zolecarboxamido]-3,3-dimethyl-7-oxo-4-thia-1-azabicy-
clo[3.2.0]heptane-2-carboxylic acid = (6*R*)-6-[1-(2,6-Di-
chlorophenyl)-4-methyl-5-pyrazolecarboxamido]penicil-
lanic acid = 1-(2,6-Dichlorophenyl)-4-methyl-5-pyrazo-
lylpenicillin = [2*S*-(2α,5α,6β)]-6-[[[1-(2,6-Dichlorophe-
nyl)-4-methyl-1*H*-pyrazol-5-yl]carbonyl]amino]-3,3-di-
methyl-7-oxo-4-thia-1-azabicyclo[3.2.0]heptane-2-carb-
oxylic acid (●)
R also sodium salt (15949-69-6)
S F-75, *Pirazocillin, Prazocillin**, Pyrazocillin*
U Antibiotic

8423 (6903)
$C_{19}H_{18}FN_3$
87642-38-4

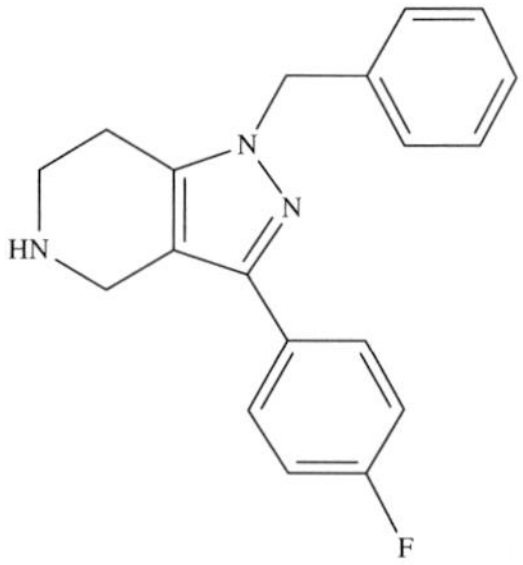

1-Benzyl-3-(*p*-fluorophenyl)-4,5,6,7-tetrahydropyrazo-
lo[4,3-*c*]pyridine = 3-(4-Fluorophenyl)-4,5,6,7-tetrahy-
dro-1-(phenylmethyl)-1*H*-pyrazolo[4,3-*c*]pyridine (●)
S L 16052
U Antihypertensive

8424
$C_{19}H_{18}F_2N_8OS$
151856-47-2

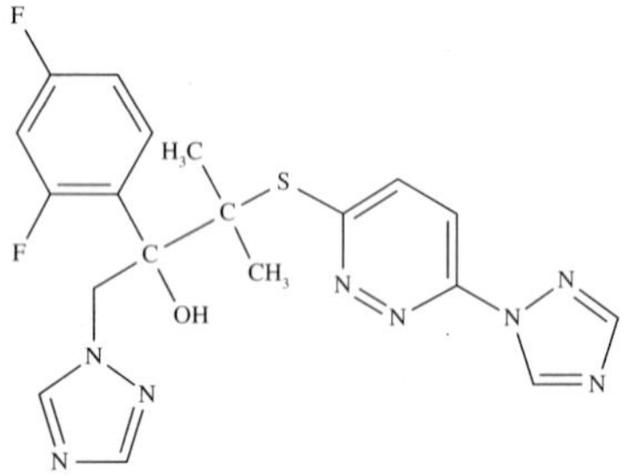

(+)-α-(2,4-Difluorophenyl)-α-[1-methyl-1-[[6-(1*H*-
1,2,4-triazol-1-yl)-3-pyridazinyl]thio]ethyl]-1*H*-1,2,4-tri-
azole-1-ethanol (●)
S MFB 1041
U Antifungal

8425 (6904)
$C_{19}H_{18}F_3NO_2$
57775-28-7

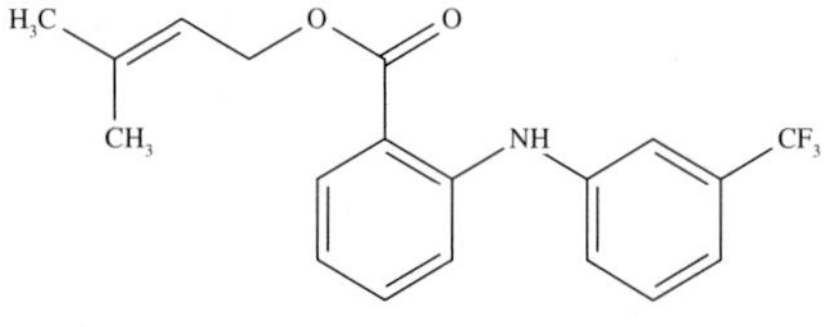

3-Methyl-2-butenyl *N*-(α,α,α-trifluoro-*m*-tolyl)anthrani-
late = *N*-(α,α,α-Trifluoro-*m*-tolyl)anthranilic acid 3-me-
thyl-2-butenyl ester = 3-[3-(Trifluoromethyl)anilino]
benzoic acid isopentyl ester = 2-[[3-(Trifluoro-

methyl)phenyl]amino]benzoic acid 3-methyl-2-butenyl ester (●)

S   Fexamin, *Prefenamate***

U   Anti-inflammatory, analgesic

8426 $C_{19}H_{18}F_3N_5O$
                                         167298-74-0

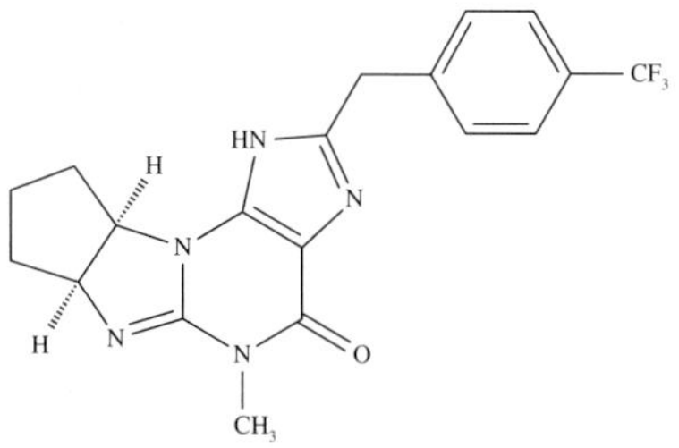

(6a*R*,9a*S*)-*rel*-5,6a,7,8,9,9a-Hexahydro-5-methyl-2-[[4-(trifluoromethyl)phenyl]methyl]cyclopent[4,5]imidazo[2,1-*b*]purin-4(1*H*)-one (●)

S   Sch 51866

U   Phosphodiesterase inhibitor

8427   (6905) $C_{19}H_{18}F_6N_2O$
                                         66364-73-6

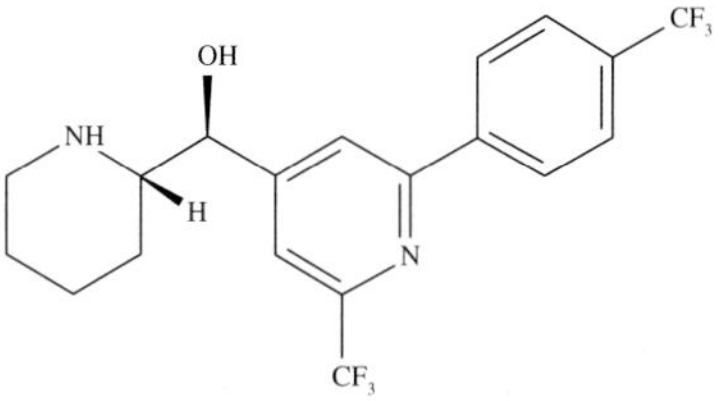

(±)-(*R**,*R**)-α-[2-(Trifluoromethyl)-6-(α,α,α-trifluoro-*p*-tolyl)-4-pyridyl]-2-piperidinemethanol = (*R**,*R**)-(±)-α-(2-Piperidinyl)-2-(trifluomethyl)-6-[4-(trifluoromethyl)phenyl]-4-pyridinemethanol (●)

R   Phosphate (1:1) (66364-74-7)

S   *Enpiroline phosphate***, WR 180409

U   Antimalarial

8428   (6906) $C_{19}H_{18}F_6O_3$
                                         123407-36-3

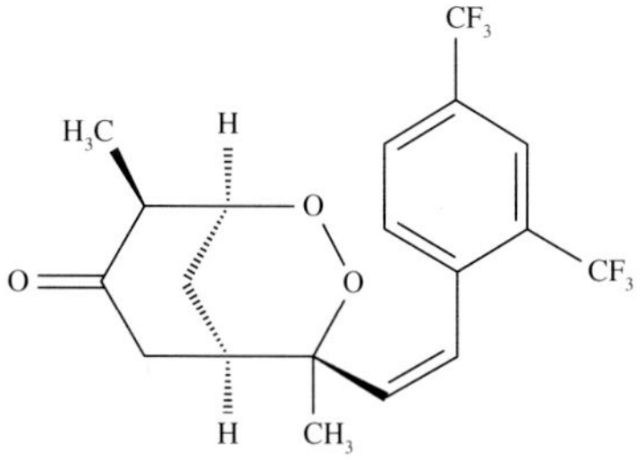

(1*S*,4*R*,5*R*,8*S*)-4-[(*Z*)-2,4-Bis(trifluoromethyl)styryl]-4,8-dimethyl-2,3-dioxabicyclo[3.3.1]nonan-7-one (●) = [1*S*-[1α,4β(*Z*),5α,8β]]-4-[2-[2,4-Bis(trifluoromethyl)phenyl]ethenyl]-4,8-dimethyl-2,3-dioxabicyclo[3.3.1]nonan-7-one (●)

S   *Arteflene***, Ro 42-1611

U   Antimalarial

8429   (6907) $C_{19}H_{18}N_2$
                                         100999-26-6

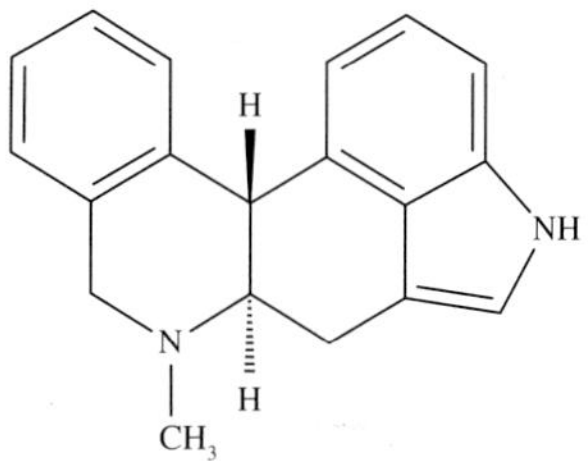

(–)-(6a*R*,12b*R*)-4,6,6a,7,8,12b-Hexahydro-7-methylindolo[4,3-*ab*]phenanthridine = *trans*-(–)-4,6,6a,7,8,12b-Hexahydro-7-methylindolo[4,3-*ab*]phenanthridine (●)

S   CY 208-243

U   Analgesic (dopaminergic)

8430   (6908) $C_{19}H_{18}N_2O_2$
                                         54318-59-1

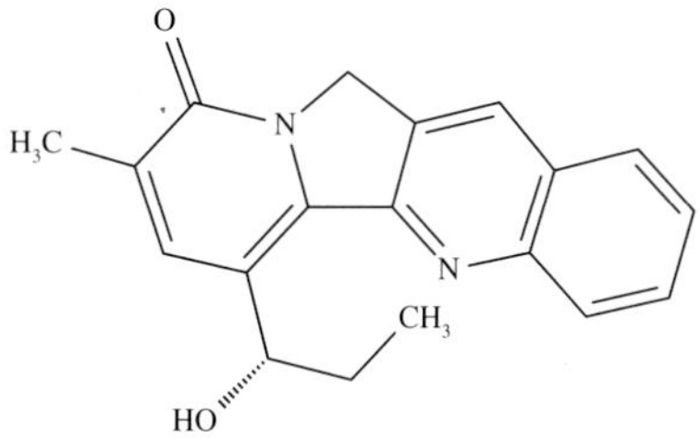

(*S*)-7-(1-Hydroxypropyl)-8-methylindolizino[1,2-*c*]quinolin-9(11*H*)-one (●)

S *Mappicine*
U Antiviral (alkaloid from *Mappia foetida*)

8431 (6909) $C_{19}H_{18}N_2O_2$
33453-23-5

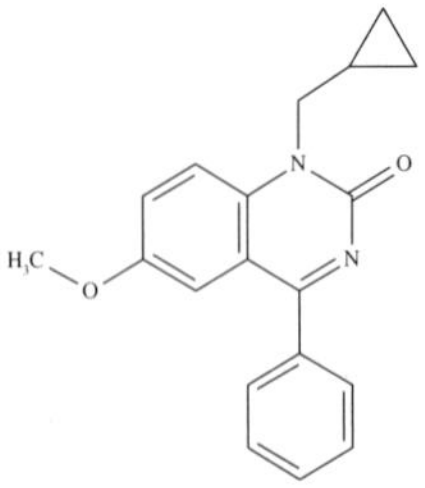

1-(Cyclopropylmethyl)-6-methoxy-4-phenyl-2-(1*H*)-qui-
nazolinone (●)
S *Ciproquazone***, *Cyproquazone*, SL-573
U Anti-inflammatory, antirheumatic

8432 (6910) $C_{19}H_{18}N_2O_2$
33016-12-5

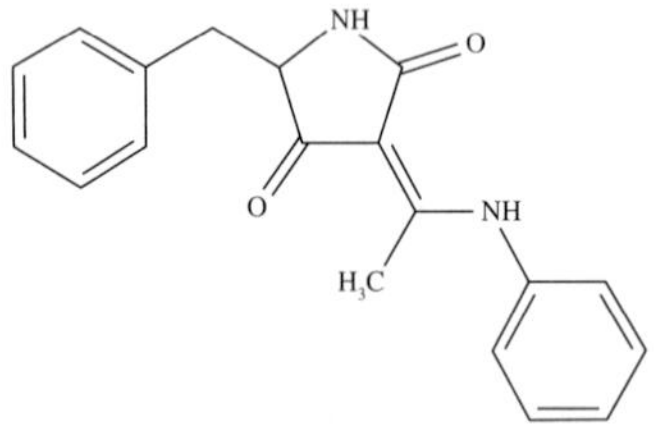

3-(1-Anilinoethylidene)-5-benzyl-2,4-pyrrolidinedione =
3-[1-(Phenylamino)ethylidene]-5-(phenylmethyl)-2,4-
pyrrolidinedione (●)
S TN-16
U Antineoplastic

8433 (6911) $C_{19}H_{18}N_2O_3$
75357-55-0

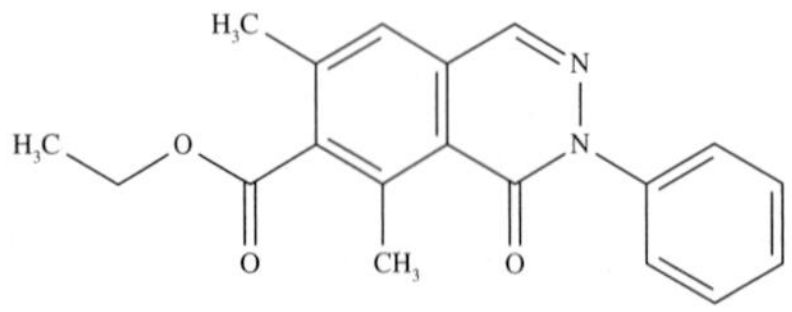

3,4-Dihydro-5,7-dimethyl-4-oxo-3-phenyl-6-phthala-
zinecarboxylic acid ethyl ester (●)
S TH 016
U Anti-atherosclerotic

8434 $C_{19}H_{18}N_2O_3$
133550-34-2

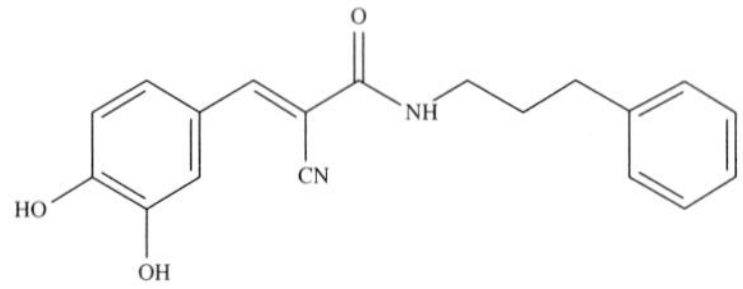

(2*E*)-2-Cyano-3-(3,4-dihydroxyphenyl)-*N*-(3-phenylpro-
pyl)-2-propenamide (●)
S AG-555
U Tyrosin kinase inhibitor

8435 (6912) $C_{19}H_{18}N_2O_3$
853-34-9

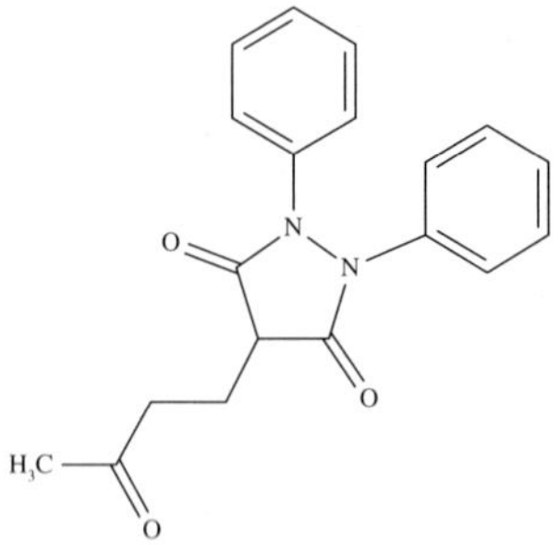

4-(3-Oxobutyl)-1,2-diphenyl-3,5-pyrazolidinedione (●)
S Aprhazon, Audominen, Chebutan, Chepirol, Cheta-
zol, Chetazolidin, Chetil, *Chetofenilbutazone*, Cheto-
pir, Chetosol, Copirene, Ejor, Gammachetone, Gluta-
zon, Hichillos, Isolidin, *Kebuzone**, Keltagon, Ken-
tan-S, Ketazon(e), Ketobutane, Ketobutazon(e), Ketofen (old form), Ketonizon, *Ketophenylbutazone*,
Ketophezon, Nemec, Neo-Panalgyl, Neufenil, Neu-
phenyl, Pecnon, Phloguron, Recheton, Reuchetal,
Reumo-Campil, Rheumalgin, Ryumapirin K, Telos-
min, Trixidol, Vintop, Zenmasal
U Anti-inflammatory, antirheumatic

8436 (6913) $C_{19}H_{18}N_2O_3S$
107761-24-0

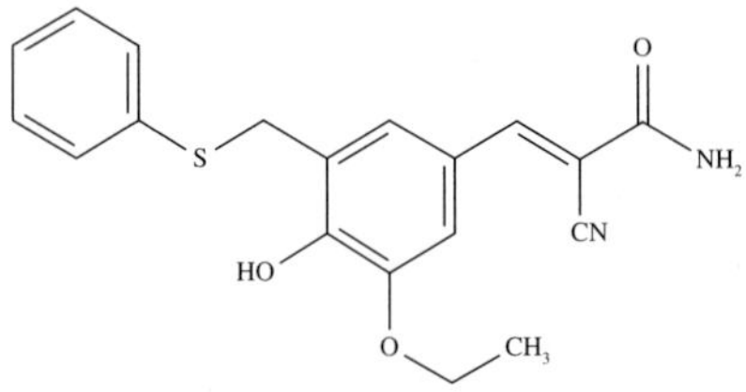

2-Cyano-3-[3-ethoxy-4-hydroxy-5-[(phenylthio)me-thyl]]-2-propenamide (●)
S ST 638
U Proteinkinase inhibitor (tyrosine spezific)

8437 (6914) $C_{19}H_{18}N_2O_4$
73815-11-9

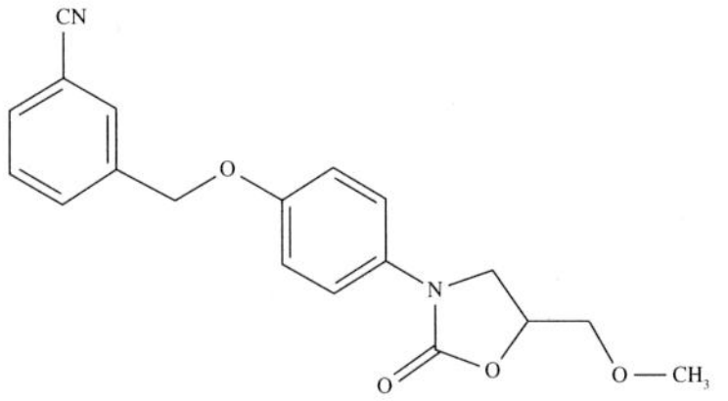

α-[p-[(5-Methoxymethyl)-2-oxo-3-oxazolidinyl]phen-oxy]-m-toluonitrile = 3-[4-[(3-Cyanophenyl)methoxy]phe-nyl]-5-(methoxymethyl)-2-oxazolidinone = 3-[[4-[5-(Me-thoxymethyl)-2-oxo-3-oxazolidinyl]phenoxy]me-thyl]benzonitrile (●)
S *Cimoxatone**, MD 780515
U Antidepressant (monoamine oxidase inhibitor)

8438 $C_{19}H_{18}N_2O_4$
130736-65-1

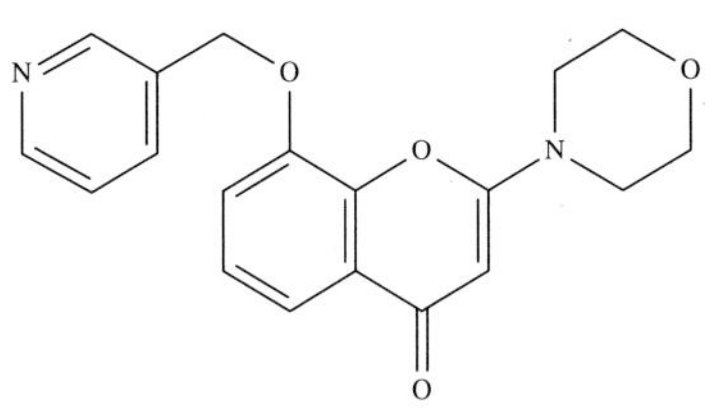

2-(4-Morpholinyl)-8-(3-pyridinylmethoxy)-4H-1-benzo-pyran-4-one (●)
S U-86983
U Antiproliferative

8439 $C_{19}H_{18}N_2O_4S$
198470-84-7

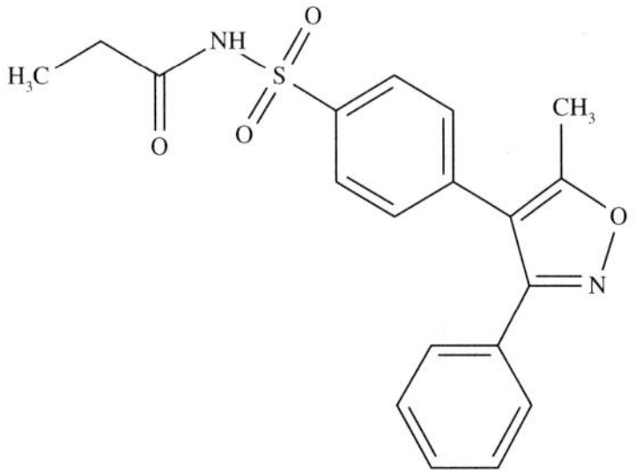

N-[[4-(5-Methyl-3-phenyl-4-isoxazolyl)phenyl]sulfo-nyl]propanamide (●)
R Sodium salt (198470-85-8)
S *Parecoxib sodium***, SC-69124A
U Anti-inflammatory, analgesic (cyclooxygenase-2 in-hibitor)

8440 $C_{19}H_{18}N_2O_4S$
145739-56-6

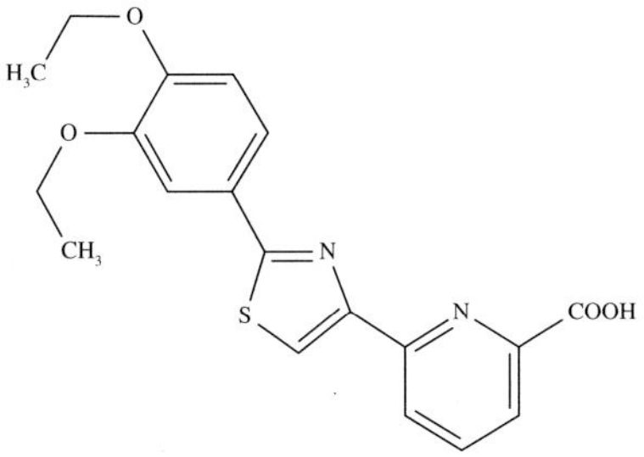

6-[2-(3,4-Diethoxyphenyl)-4-thiazolyl]-2-pyridinecarb-oxylic acid (●)
S OPC-6535
U Superoxide production inhibitor

8441 (6915) $C_{19}H_{18}N_2O_6$
115666-98-3

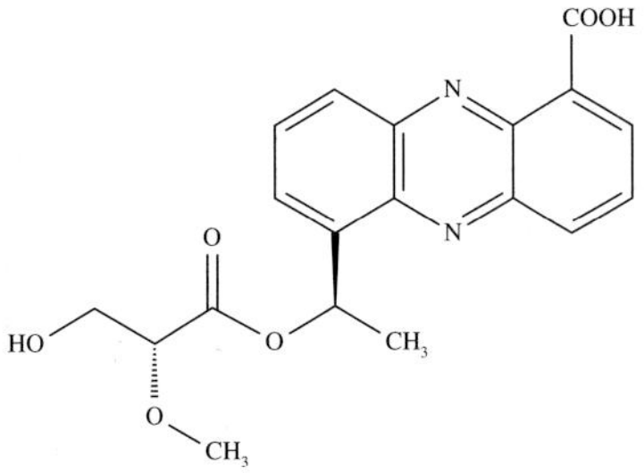

6-(1-Hydroxyethyl)-1-phenazinecarboxylic acid 3-hydr-oxy-2-methoxypropionate = (R*,R*)-6-[1-(3-Hydroxy-2-

methoxy-1-oxopropoxy)ethyl]-1-phenazinecarboxylic acid (●)
S DOB 41
U Antineoplastic, gram-positive antibiotic from a *Pseudomonas* strain

---

8442 (6916) $C_{19}H_{18}N_4O_2$
74150-27-9

4,5-Dihydro-6-[2-(p-methoxyphenyl)-5-benzimidazolyl]-5-methyl-3(2H)-pyridazinone = 4,5-Dihydro-6-[2-(4-methoxyphenyl)-1H-benzimidazol-5-yl]-5-methyl-3(2H)-pyridazinone (●)
R also monohydrochloride (77469-98-8)
S Acardi, *Pimobendan***, UD-CG 115-BS, Vetmedin
U Cardiotonic (phosphodiesterase inhibitor)

8443 $C_{19}H_{18}N_4O_2S$
172152-36-2

2-[[(4-Methoxy-3-methyl-2-pyridinyl)methyl]sulfinyl]-5-(1H-pyrrol-1-yl)-1H-benzimidazole (●)
S *Gilaprazole*, IY-81149
U gastric antisecretory (proton pump inhibitor)

8444 (6917) $C_{19}H_{18}N_4O_5S_3$
41952-52-7

(6R,7R)-7-[(R)-Mandelamido]-3-[[(5-methyl-1,3,4-thiadiazol-2-yl)thio]methyl]-8-oxo-5-thia-1-azabicyclo[4.2.0]oct-2-ene-2-carboxylic acid = (7R)-7-(D-Mandelamido)-3-[[(5-methyl-1,3,4-thiadiazol-2-yl)thio]methyl]-3-cephem-4-carboxylic acid = [6R-[6α,7β(R*)]]-7-[(Hydroxyphenylacetyl)amino]-3-[[(5-methyl-1,3,4-thiadiazol-2-yl)thio]methyl]-8-oxo-5-thia-1-azabicyclo[4.2.0]oct-2-ene-2-carboxylic acid (●)

R see also no. 13425
S *Cefcanel***
U Antibiotic

8445 (6918) $C_{19}H_{18}N_6O_5S_3$
104145-95-1

(+)-(6R,7R)-7-[2-(2-Amino-4-thiazolyl)glyoxylamido]-3-[(Z)-2-(4-methyl-5-thiazolyl)vinyl]-8-oxo-5-thia-1-azabicyclo[4.2.0]oct-2-ene-2-carboxylic acid 7²-(Z)-(O-methyloxime) = [6R-[3(Z),6α,7β(Z)]]-7-[[(2-Amino-4-thiazolyl)(methoxyimino)acetyl]amino]-3-[2-(4-methyl-5-thiazolyl)ethenyl]-8-oxo-5-thia-1-azabicyclo[4.2.0]oct-2-ene-2-carboxylic acid (●)
R see also no. 12783
S *Cefditoren***, *Cefoviten*
U Antibiotic

8446 (6919) $C_{19}H_{18}N_6O_6S_2$
57268-80-1

(6R,7R)-7-[(R)-O-Formylmandelamido]-3-[[(1-methyl-1H-tetrazol-5-yl)thio]methyl]-8-oxo-5-thia-1-azabicyclo[4.2.0]oct-2-ene-2-carboxylic acid = (7R)-7-[(αR)-α-Formyloxyphenylacetamido]-3-[[(1-methyl-1H-tetrazol-5-yl)thio]methyl]-3-cephem-4-carboxylic acid = [6R-[6α,7β(R*)]]-7-[[(Formyloxy)phenylacetyl]amino]-3-[[(1-methyl-1H-tetrazol-5-yl)thio]methyl]-8-oxo-5-thia-1-azabicyclo[4.2.0]oct-2-ene-2-carboxylic acid (●)
R Monosodium salt (42540-40-9)
S Acemycin, Bergacef, Cedol, Cefadol "Atlantic", Cefaformil, Cefam, Cefamandole "Dong Shin", Cefamandole Derly, *Cefamandole nafate*, Céfamandole Panpharma, Cefamen, Cefaseptolo, Cefiran, Cefman, Cemado, Cemandil, *Cephamandole nafate*, Dardokef, Fado, Forcef, *Formylcefamandole*

sodium, Ibiman, Kefabiotic, Kefadol, Kefandol, Kefdol(e), Lampomandol, Lilly 106223, Mancef, Mandokef, Mandol, Mandolsan, Neocefal, Pavacef, Septomandolo

U Antibiotic

8447 (6920)

$C_{19}H_{18}N_6O_7S$
93129-06-7

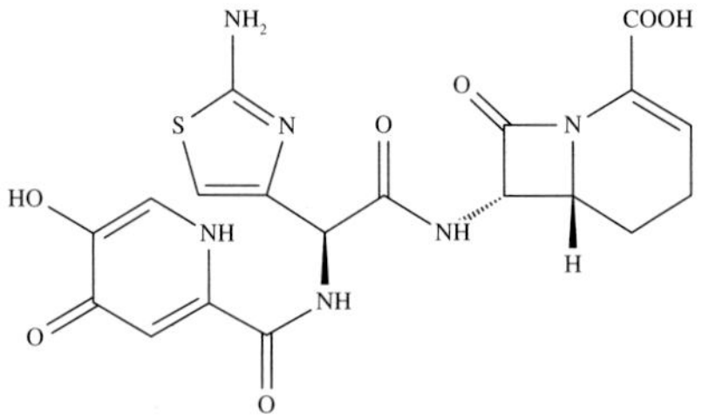

7-[2-(2-Amino-4-thiazolyl)-2-[(1,4-dihydro-5-hydroxy-4-oxo-2-pyridyl)carboxamido]acetyl]carbacephem = [6R-[6α,7β(S*)]]-7-[[(2-Amino-4-thiazolyl)[[(1,4-dihydro-5-hydroxy-4-oxo-2-pyridinyl)carbonyl]amino]acetyl]amino]-8-oxo-1-azabicyclo[4.2.0]oct-2-ene-2-carboxylic acid (●)

S KT 4697

U Antibiotic

8448 (6921)

$C_{19}H_{18}O_4$
64761-48-4

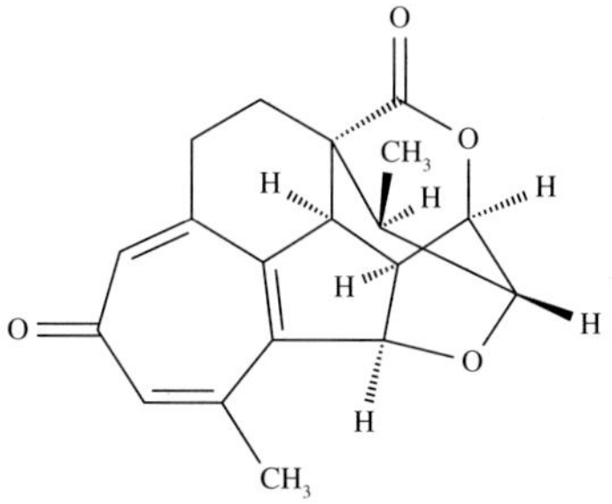

[2aR-(2aα,2bα,3β,4aα,10aβ,10bα,11S*)]-2a,3,4a,9,10,10b-Hexahydro-5,11-dimethyl-1H-3,10a-methano-2,4-dioxacyclohepta[bc]cyclopent[jk]acenaphthylene-1,7(2bH)-dione (●)

S *Hainanolide, Harringtonolide*

U Antineoplastic from *Chephalotaxus* sp.

8449

$C_{19}H_{18}O_5$
36413-91-9

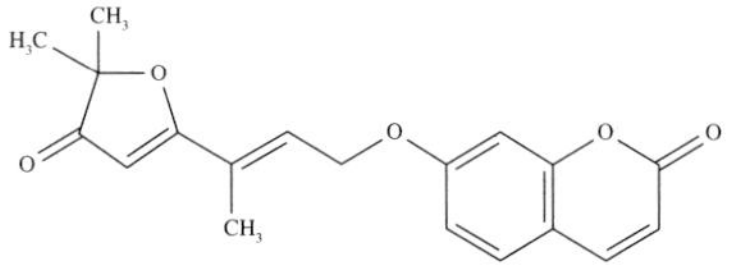

7-[[(2E)-3-(4,5-Dihydro-5,5-dimethyl-4-oxo-2-furanyl)-2-butenyl]oxy]-2H-1-benzopyran-2-one (●)

S *Geiparvarin*

U Antiproliferative from *Geijera parviflora*

8450 (6922)

$C_{19}H_{18}O_5$
79181-48-9

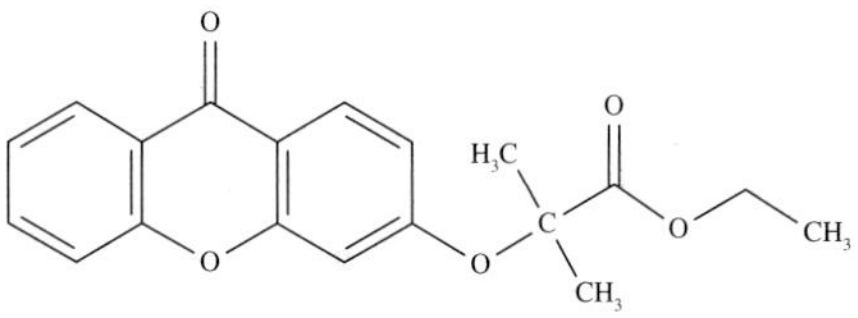

Ethyl 2-methyl-2-[(9-oxo-9H-xanthen-3-yl)oxy]propionate = 2-Methyl-2-[(9-oxo-9H-xanthen-3-yl)oxy]propanoic acid ethyl ester (●)

S MR 981

U Antihyperlipidemic

8451 (6923)

$C_{19}H_{18}O_6S$
105937-56-2

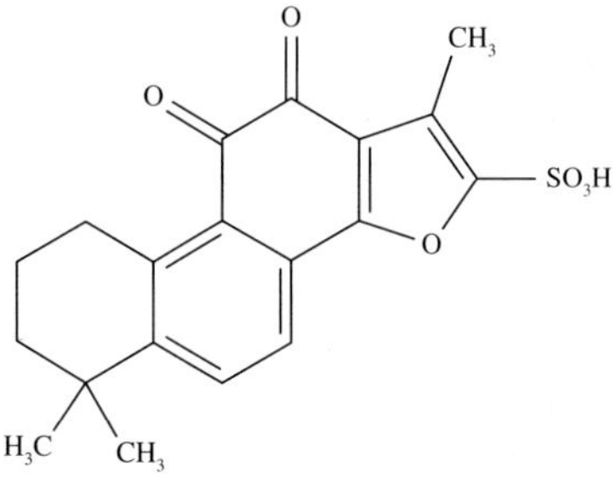

6,7,8,9,10,11-Hexahydro-1,6,6-trimethyl-10,11-dioxophenanthro[1,2-b]furan-2-sulfonic acid (●)

R Sodium salt (69659-80-9)

S Danshen-201, DS 201, *Sodium tanshinone II-A sulfonate*

U Antihemolytic, coronary vasodilator

8452 (6925) $C_{19}H_{18}O_7$
118-27-4

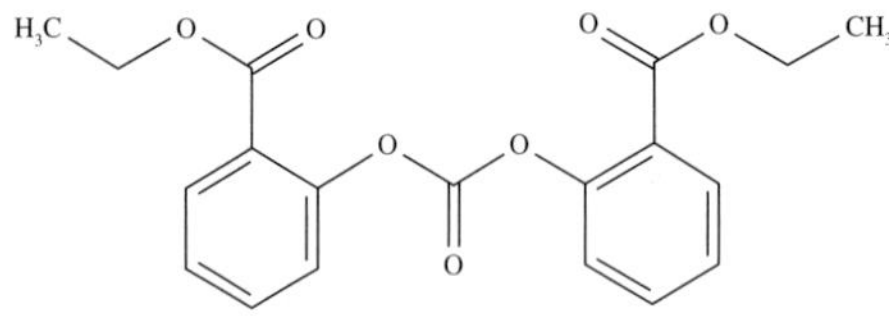

O,O'-Carbonylbis[ethyl salicylate] = Carbonic acid di-
ester with ethyl salicylate = 2,2'-[Carbonyl-
bis(oxy)]bis[benzoic acid] diethyl ester (●)
R see also no. 4125-19
S *Carbethyl Salicylate*, Sal-Ethyl Carbonate
U Analgesic, antipyretic

8453 (6924) $C_{19}H_{18}O_7$
3447-95-8

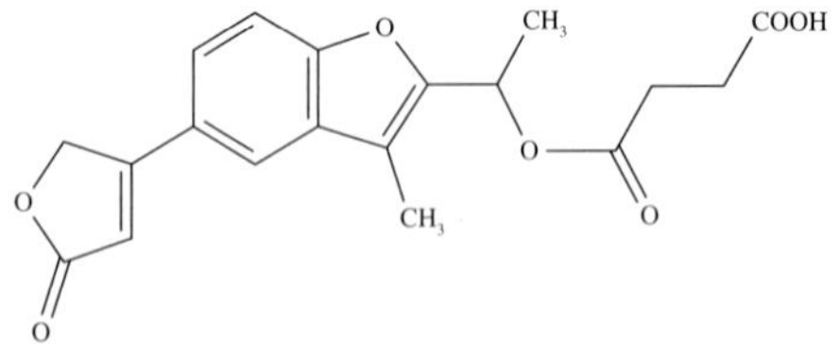

2-(1-Hydroxyethyl)-β-(hydroxymethyl)-3-methyl-5-ben-
zofuranacrylic acid γ-lactone hydrogen succinate = 1-[3-
Methyl-5-(2,5-dihydro-5-oxo-3-furyl)-2-benzofura-
nyl]ethyl hydrogen succinate = Butanedioic acid mono[1-
[5-(2,5-dihydro-5-oxo-3-furanyl)-3-methyl-2-benzofura-
nyl]ethyl] ester (●)
R also morpholinium salt
S *Benfurodil hemisuccinate***, *Benfurodili succinas*,
4091 C.B., Clinodilat, Eucilat, Eudilat
U Vasodilator, cardiotonic

8454 (6927) $C_{19}H_{18}O_8$
479-20-9

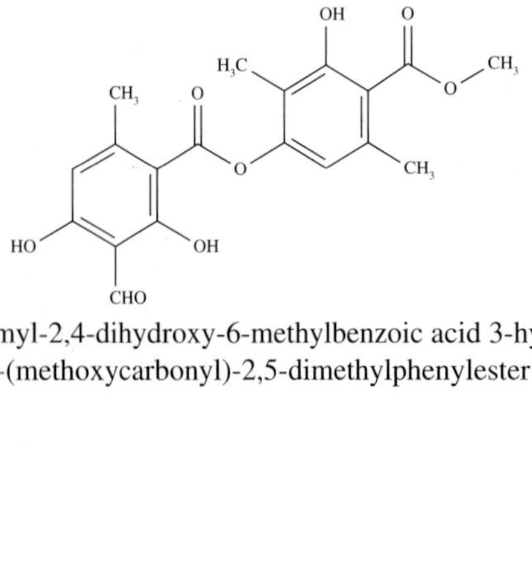

3-Formyl-2,4-dihydroxy-6-methylbenzoic acid 3-hydr-
oxy-4-(methoxycarbonyl)-2,5-dimethylphenylester (●)

S *Atranoric acid, Atranorin, Atranorsäure*, Parmelin,
Usnarin
U Antibiotic from number of lichens

8455 (6928) $C_{19}H_{18}O_8$
56003-01-1

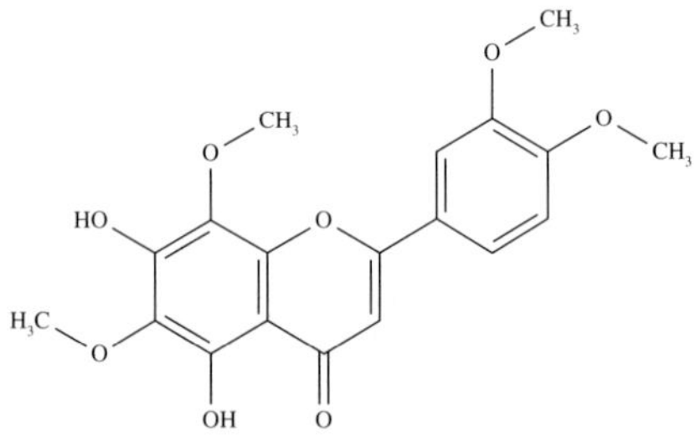

5,7-Dihydroxy-3',4',6,8-tetramethoxyflavone = 2-(3,4-Di-
methoxyphenyl)-5,7-dihydroxy-6,8-dimethoxy-4*H*-1-
benzopyran-4-one (●)
S *Hymenoxin*
U Cytotoxic from *Scoparia dulcis*

8456 (6929) $C_{19}H_{18}O_8$
80140-31-4

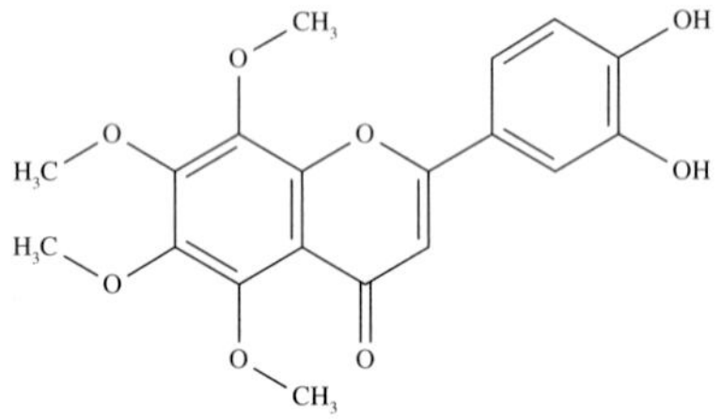

3',4'-Dihydroxy-5,6,7,8-tetramethoxyflavone = 2-(3,4-Di-
hydroxyphenyl)-5,6,7,8-tetramethoxy-4*H*-1-benzopyran-
4-one (●)
S LARI 4
U Aldose reductase inhibitor

8457 $C_{19}H_{18}O_8$
479-91-4

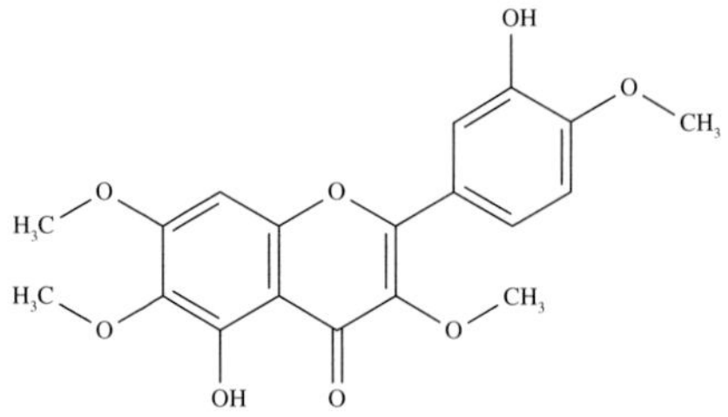

3',5-Dihydroxy-3',4',6,7-tetramethoxyflavone = 5-Hy-
droxy-2-(3-hydroxy-4-methoxyphenyl)-3,6,7-trimeth-
oxy-4*H*-1-benzopyran-4-one (●)
S   *Casticin, Vitexicarpin*
U   Anti-inflammatory, antimalarial from many plants

8458   (6926) $C_{19}H_{18}O_8$
60539-13-1

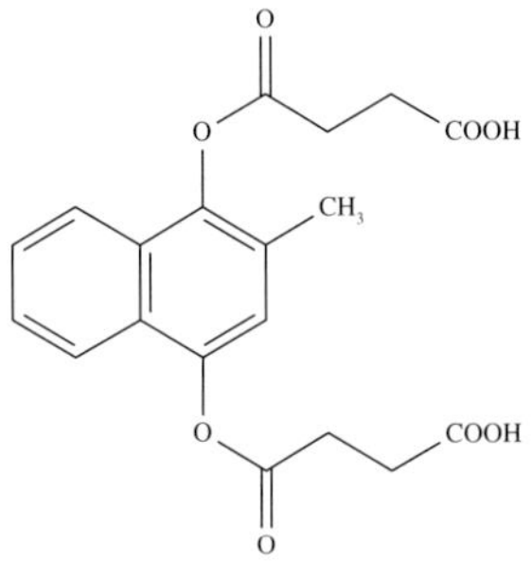

2-Methyl-1,4-naphthylene bis(hydrogen succinate) = 2-
Methyl-1,4-disuccinylnaphthohydroquinone = Butane-
dioic acid 2-methyl-1,4-naphthalenediyl ester (●)
S   Kapathrom, *Menadiol disuccinate*
U   Prothrombogenic

8459   (6930) $C_{19}H_{18}O_9$
85338-85-8

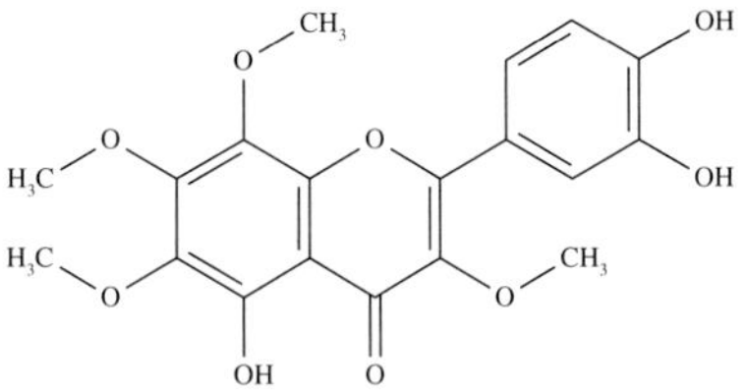

3',4',5-Trihydroxy-3,6,7,8-tetramethoxyflavone = 2-(3,4-
Dihydroxyphenyl)-5-hydroxy-3,6,7,8-tetramethoxy-4*H*-
1-benzopyran-4-one (●)
S   THTMF
U   Antineoplastic

8460   (6931) $C_{19}H_{18}O_{11}$
4773-96-0

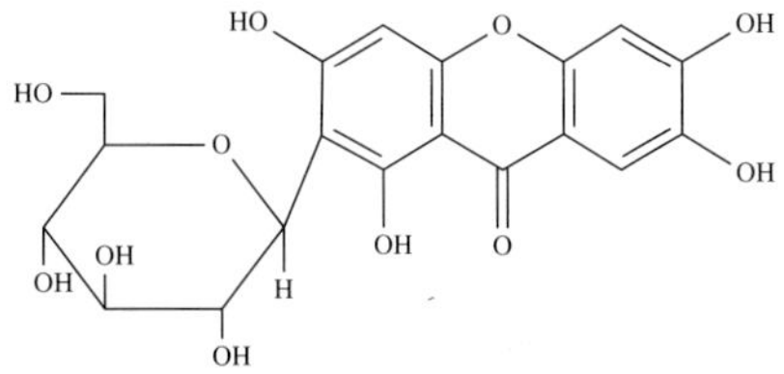

2-β-D-Glucopyranosyl-1,3,6,7-tetrahydroxy-9*H*-xanthen-
9-one (●)
S   *Alpisarin, Alpizarin, Aphloiol, Chinomin, Chinonin,*
       Hedysarid, Mangiferin
U   Anti-inflammatory, antiviral, antidiabetic from many
       plants, antineoplastic

8461   (6932) $C_{19}H_{18}O_{11}$
24699-16-9

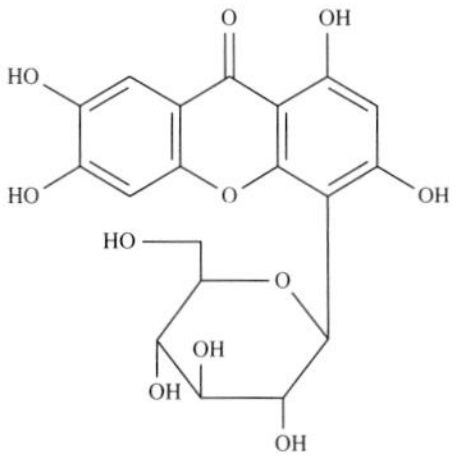

4-β-D-Glucopyranosyl-1,3,6,7-tetrahydroxy-9*H*-xanthen-
9-one (●)
S   *Isomangiferin*
U   Antiviral (antiherpetic)

8462   (6933) $C_{19}H_{19}BrClNO_2$
150490-85-0

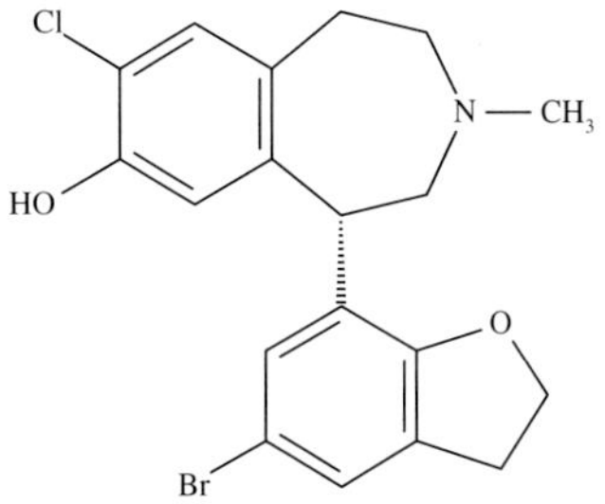

(*S*)-5-(5-Bromo-2,3-dihydro-7-benzofuranyl)-8-chloro-
2,3,4,5-tetrahydro-3-methyl-1*H*-3-benzazepin-7-ol (●)
S   *Berupipam**, NNC 22-0010

1715

U Antipsychotic (dopamine D_1 antagonist)

8463 $C_{19}H_{19}ClN_2$
                                                                  100643-71-8

8-Chloro-6,11-dihydro-11-(4-piperidinylidene)-5H-ben-
zo[5,6]cyclohepta[1,2-b]pyridine (●)
S Aerius, Allex, Azomyr, DCL, *Desloratadine***,
      Neoclarityn, Opulis, Sch 34117
U Anti-allergic (H_1 receptor antagonist)

8464   (6934) $C_{19}H_{19}ClN_2$
                                                                  39051-50-8

2-Chloro-11-[3-(dimethylamino)propylidene]-11H-di-
benz[b,e]azepine = 2-Chloro-11-[3-(dimethylamino)pro-
pylidene]morphanthridine = 3-(2-Chloro-11H-di-
benz[b,e]azepin-11-ylidene)-N,N-dimethyl-1-propan-
amine (●)
S EX 11528 A, RMI 81582
U Antihypertensive, CNS depressant

8465   (6935) $C_{19}H_{19}ClN_2O_2S$
                                                                  3576-51-0

2-Chloro-10-(3-morpholinopropionyl)phenothiazine = 2-
Chloro-10-[3-(4-morpholinyl)-1-oxopropyl]-10H-pheno-
thiazine (●)
S Morfafen
U Diuretic, saluretic

8466   (6936) $C_{19}H_{19}ClN_2S$
                                                                  22431-21-6

8-Chloro-10-(4-methyl-1-piperazinyl)dibenzo[b,f]thi-
epin = 1-(8-Chlorodibenzo[b,f]thiepin-10-yl)-4-methylpi-
perazine (●)
S *Dehydroclothepine*
U Neuroleptic

8467   (6937) $C_{19}H_{19}ClN_4O_2$
                                                                  125652-47-3

1-[4-[1-(4-Chlorophenyl)-1-methylethyl]-2-methoxyphe-
nyl]-1*H*-1,2,4-triazole-3-carboxamide (●)
S   CGP 31358
U   Anticonvulsant

8468   (6938) $C_{19}H_{19}ClO_3$
13878-10-9

L-4-(*o*-Chlorobenzyl)-5-oxo-4-phenylcaproic acid = (–)-
γ-Acetyl-2-chloro-γ-phenylbenzenepentanoic acid (●)
R   Sodium salt
S   *Caprochlorone*
U   Antiviral

8469 $C_{19}H_{19}Cl_2N_3O_2$
136122-46-8

6-Chloro-3-(*p*-chlorobenzyl)-β,β-dimethyl-3*H*-imida-
zo[4,5-*b*]pyridine-2-butyric acid = 4-[6-Chloro-3-(4-chlo-
robenzyl)-3*H*-imidazo[4,5-*b*]pyridin-2-yl]-3,3-dimethyl-
butanoic acid = 6-Chloro-3-[(4-chlorophenyl)methyl]-
β,β-dimethyl-3*H*-imidazo[4,5-*b*]pyridine-2-butanoic acid
(●)
S   *Mipitroban***, Pimitroban, UP 116-77
U   Antithrombotic

8470   (6939) $C_{19}H_{19}Cl_2N_3O_3$
38955-22-5

1-[(8-Chlorodibenz[*b,f*][1,4]oxazepin-10(11*H*)-yl)carbo-
nyl]-2-(5-chlorovaleryl)hydrazine = 8-Chlorodi-
benz[*b,f*][1,4]oxazepine-10(11*H*)-carboxylic acid 2-(5-
chloro-1-oxopentyl)hydrazide (●)
S   *Pinadoline**, SC-25469
U   Analgesic

8471   (6940) $C_{19}H_{19}Cl_2N_3O_3$
103946-15-2

Isopropyl (–)-(*S*)-4-(2,3-dichlorophenyl)-1,4-dihydro-
2,6-dimethyl-5-(1,3,4-oxadiazol-2-yl)nicotinate = (–)-4-
(2,3-Dichlorophenyl)-1,4-dihydro-2,6-dimethyl-5-(1,3,4-
oxadiazol-2-yl)-3-pyridinecarboxylic acid 1-methylethyl
ester (●)
S   *Elnadipine***
U   Calcium antagonist

8472 $C_{19}H_{19}Cl_2N_5O_4$
188804-07-1

β-[[[[[3-[(Aminoiminomethyl)amino]phenyl]carbo-nyl]amino]acetyl]amino]-3,5-dichlorobenzenepropanoic acid = N-[3-[(Aminoiminomethyl)amino]benzoyl]glycyl-3-(3,5-dichlorophenyl)-β-alanine (●)
S SC-68448
U Antineoplastic (integrin $\alpha_v\beta_3$ antagonist)

8473 $C_{19}H_{19}FN_2O$
153249-83-3

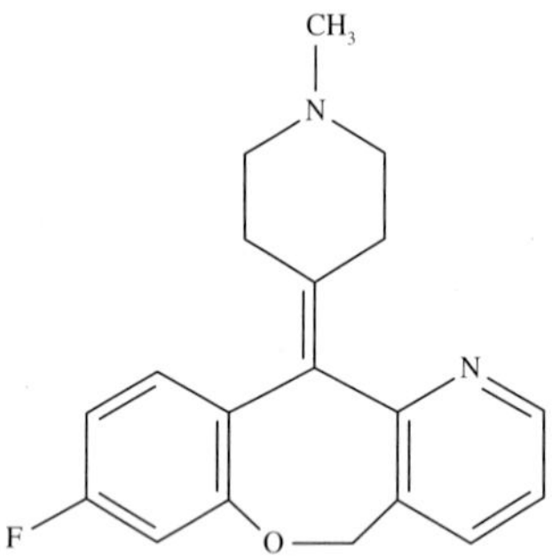

8-Fluoro-5,11-dihydro-11-(1-methyl-4-piperidinyli-dene)-[1]benzoxepino[4,3-b]pyridine (●)
S PY-608
U Anti-allergic

8474   (6941) $C_{19}H_{19}FO_4$
52995-24-1

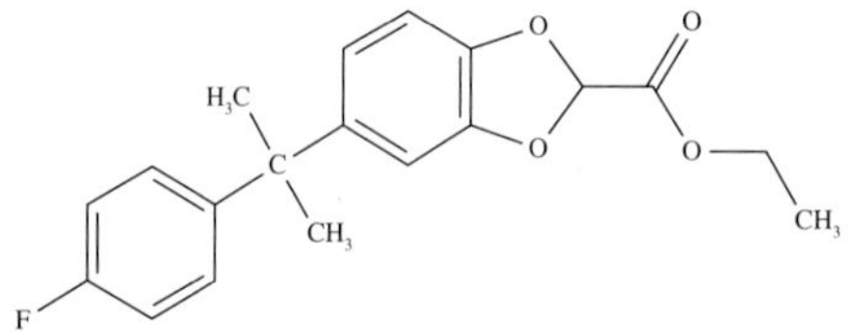

5-(4-Fluoro-α,α-dimethylbenzyl)-1,3-benzodioxole-2-carboxylic acid ethyl ester = 5-[1-(4-Fluorophenyl)-1-me-thylethyl]-1,3-benzodioxole-2-carboxylic acid ethyl ester (●)
S RMI 14676
U Antihyperlipidemic

8475   (6942) $C_{19}H_{19}FO_4$
91503-79-6

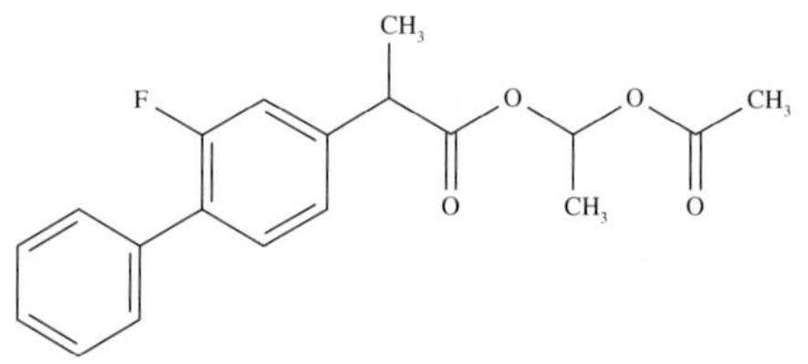

1-Acetoxyethyl 2-(2-fluoro-4-biphenylyl)propionate = 2-Fluoro-α-methyl[1,1'-biphenyl]-4-acetic acid 1-(acetyl-oxy)ethyl ester (●)
S *Flurbiprofen axetil**, FP 83, Lipfen, Ropion
U Anti-inflammatory, analgesic, antipyretic

8476   (6943) $C_{19}H_{19}F_2NO_2$
21221-18-1

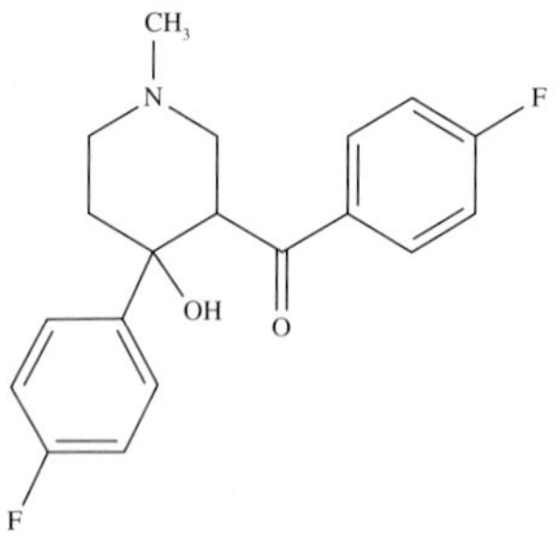

p-Fluorophenyl 4-(p-fluorophenyl)-4-hydroxy-1-methyl-3-piperidyl ketone = 3-(4-Fluorobenzoyl)-4-(4-fluorophe-nyl)-1-methyl-4-piperidinol = (4-Fluorophenyl)[4-(4-fluorophenyl)-4-hydroxy-1-methyl-3-piperidinyl]metha-none (●)
R Hydrochloride (34039-01-5)
S *Flazalone hydrochloride*, *Flumefenine*, NSC-102629, R 760, Riker 760
U Anti-inflammatory

8477   (6944) $C_{19}H_{19}HgN_5O_5$
                                                               20619-89-0

N-[3-Methoxy-2-(7-theophyllinylmercuri)propyl]phthal-
imide = (3,7-Dihydro-1,3-dimethyl-1*H*-purine-2,6-diona-
to-*N*[7])[2-(1,3-dihydro-1,3-dioxo-2*H*-isoindol-2-yl)-1-
(methoxymethyl)ethyl]mercury (●)
R   Sodium salt
S   Merphtallyl, Poliurene, Polyurene
U   Mercurial diuretic

8478   (6945) $C_{19}H_{19}N$
                                                               57262-94-9

2,3,4,9-Tetrahydro-2-methyl-1*H*-dibenzo[3,4:6,7]cyclo-
hepta[1,2-*c*]pyridine (●)
R   also maleate (1:1) (85650-57-3)
S   *Carbamianserin*, MO-8282, MOD-20, Org 8282,
        Setiptiline**, *Teciptiline*, Tecipul, Tesipool, Tesipul,
        Tesolon
U   Antidepressant, anticonvulsant

8479   (6946) $C_{19}H_{19}N$
                                                               82-88-2

2-Methyl-9-phenyl-2,3,4,9-tetrahydro-1*H*-pyridindene =
1,2,3,4-Tetrahydro-2-methyl-9-phenyl-2-azafluorene =
2,3,4,9-Tetrahydro-2-methyl-9-phenyl-1*H*-inde-
no[2,1-*c*]pyridine (●)
R   Tartrate (1:1) (569-59-5)
S   Nolahist, Nu 1504, Pernovin, *Phenindamine tar-*
        trate**, PM 254, Teforin(a), Theophorin, Thephorin
U   Anti-allergic, antihistaminic

8480   (6947) $C_{19}H_{19}NO$
                                                               65509-24-2

2,3,4,5-Tetrahydro-3-methyl-1*H*-dibenz[2,3:6,7]oxepi-
no[4,5-*d*]azepine (●)
S   CGP 13442, *Maroxepin***
U   Neuroleptic

8481   (6948) $C_{19}H_{19}NOS$
                                                                 34580-13-7

4,9-Dihydro-4-(1-methyl-4-piperidinylidene)-10*H*-ben-
zo[4,5]cyclohepta[1,2-*b*]thiophen-10-one (●)
R   also fumarate (1:1) (34580-14-8)
S   Airvitess, Allerkif, Anatifen, Asdron, Asmafen, As-
malergin, Asmanoc, Asmasedil, Asmax, Asmen,
Astafen, Asthafen, Asthotifen, Astifat, Astifen, Azu-
tifen, Barosin, Broncoten, Broniten, Catifen, *Cetoti-
feno*, *Chetotifene*, Cipanfeno, Cosolve, Demetofrin,
Denerel, Dezuwart, Difen "Biogen", Dihalar, Durati-
fen, Eucycline, Firmapol, Frenasma "Faran", Fu-
mast, Galitifen, Globofil, Harelun, HC 20-511, Hi-
statec, Histaten, Irifen, Jomen, Kaler, K-Asmal, Kel-
me, Kepiten, Kerofu, Ketarfen, Ketasma, Keten, Ke-
tifen, Ketobron, Ketof, Ketofar, Ketofen "Higea; Li-
far", Ketofex, Ketonal "Roemmers", Ketonil "Roemmers", Ketotif, *Ketotifen***, Ketotilon, Ketotisin, Ke-
tox, Kleovistamin, Klevistamin, Labelphen, Licof-
ten, Matigen, Mediket, Musibon, Nazalen, Nemesil,
Nostimex, Novo-Ketotifen, Orbidix, Orpidix, Pädia-
tifen, Pellexeme, Positan, Pozitan, Privent, Profilar,
Profilas, Profilasmin, Profiten, Quefeno, Respimex,
Sadifen, Santopen, Saratin "Nippon Y. Kog", Seki-
ton, Stafen, Structum "Armstrong", Sudaved, Ta-
mon, Tanodin, Tifen "Acromax", Tophen "Bexim-
co", Totifen, Tritophen, Vemelan, Ventisol, Xeblen,
Xidanef, Xinanef, Zadec, Zadetin, Zadiken, Zadin,
Zaditen, Zaditor, Zasten, Zatofug, Zatotiten, Zeros-
ma, Zetifen, Zidox, Zikilion, Ziroten, Zolfen
U   Antihistaminic

8482   (6949) $C_{19}H_{19}NO_2$
                                                                 22108-99-2

N,N-Dimethylphenanthro[3,4-*d*]-1,3-dioxole-5-ethan-
amine (●)
S   *Stephenanthrine*
U   ACE inhibitor from *Stephania tetrandra*

8483   (6950) $C_{19}H_{19}NO_2$
                                                                 30011-11-1

2,3-Bis(*p*-methoxyphenyl)-5-methylpyrrole = 2,3-Bis-(4-
methoxyphenyl)-5-methyl-1*H*-pyrrole (●)
S   *Bimetopyrol*
U   Anti-inflammatory

8484   (6951) $C_{19}H_{19}NO_3$
                                                                 10400-01-8

7-(β-Dimethylaminoethoxy)flavone = 7-[2-(Dimethyl-
amino)ethoxy]-2-phenyl-4*H*-1-benzopyran-4-one (●)
R   7-Theophyllineacetate (no. 1619) (1715-55-5)
S   Aperflavon, Laoflavon, Perflavon
U   Coronary vasodilator

8485   (6953) $C_{19}H_{19}NO_4$
298-45-3

10-Methoxy-1,2-(methylenedioxy)-6aα-aporphin-11-ol =
3-Methoxy-5,6-methylenedioxyaporphin-4-ol = (S)-
6,7,7a,8-Tetrahydro-11-methoxy-7-methyl-5H-benzo[g]-
1,3-benzodioxolo[6,5,4-de]quinolin-12-ol (●)
S   *Bulbocapnine*
U   Treatment of muscular tremors (alkaloid from *Corydalis cava*)

8486   (6952) $C_{19}H_{19}NO_4$
15547-50-9

1-(3,4-Dimethoxyphenyl)-6,7-dimethoxyisoquinoline (●)
S   Chinoparin, Quinoparine
U   Spasmolytic

8487 $C_{19}H_{19}NO_5$
190726-45-5

4-(3,4-Dimethoxyphenoxy)-6,7-dimethoxyquinoline (●)
S   Ki 6783

U   PDGF receptor inhibitor

8488   (6954) $C_{19}H_{19}NO_6$
62992-61-4

Salicylic acid acetate ester with β-hydroxy-*p*-acetophene-
tidide = 2-(*p*-Acetamidophenoxy)ethyl salicylate acetate
(ester) = 2-(*p*-Acetamidophenoxy)ethyl *o*-acetoxyben-
zoate = 2-(Acetyloxy)benzoic acid 2-[4-(acetylamino)phenoxy]ethyl ester (●)
S   Daital, *Eterilate*, *Etersalate***, *Eterylate*
U   Analgesic, anti-inflammatory, platelet aggregation inhibitor

8489   (6955) $C_{19}H_{19}NS$
314-03-4

1-Methyl-4-(thioxanthen-9-ylidene)piperidine = 9-(1-
Methyl-4-piperidinylidene)thioxanthene = 1-Methyl-4-
(9H-thioxanthen-9-ylidene)piperidine (●)
S   Audicalm, Bovix simple, BP 400, Calmixen(e),
Haustin, LC 115, Metixopan, Muricalm(a), Neocal-
metta, *Pimethixene***, *Pimétixène*, Salcemetic, Sedo-
sil, Sonin "Labofarma; Wesley", Zaditen (old form)
U   Antihistaminic, bronchospasmolytic, sedative

8489-01   (6955-01)   59729-34-9
R   7-Theophyllineacetate (1:1) (no. 1619)
S   *Acefylline Pimethixene*, Restoren
U   Appetite stimulant

8490 $C_{19}H_{19}N_2O$
70173-20-5

9-Methoxy-N^2-methylellipticinium = 9-Methoxy-2,5,11-trimethyl-6H-pyrido[4,3-b]carbazolium (●)
R Acetate (98510-80-6)
S MMEA
U Antineoplastic

8491 (6956) $C_{19}H_{19}N_3$
60539-14-2

N^1-Phenyl-5,6,7,8-tetrahydro-3-carbazolecarboxamidine = 6-(N^1-Phenylamidino)-1,2,3,4-tetrahydrocarbazole = 2,3,4,9-Tetrahydro-N-phenyl-1H-carbazole-6-carbox-imidamide (●)
S Pathcole
U Serotonin inhibitor

8492 (6957) $C_{19}H_{19}N_3$
72444-62-3

1-Phenyl-3-(1-piperazinyl)isoquinoline (●)
R also monohydrochloride (79779-44-5)
S HR 459, *Perafensine***
U Antidepressant

8493 $C_{19}H_{19}N_3O$
139191-80-3

(R)-2-[[[2-(1H-Imidazol-4-yl)-1-methylethyl]imino]phenylmethyl]phenol (●)
S BP 2-94
U Histamine H_3 receptor agonist

8494 (6958) $C_{19}H_{19}N_3O_2$
32061-14-6

4-(Anisylideneamino)antipyrine = 1,2-Dihydro-4-[[(4-methoxyphenyl)methylene]amino]-1,5-dimethyl-2-phenyl-3H-pyrazol-3-one (●)
S Compound IS IV, IS_4
U Analgesic, antipyretic, anti-inflammatory

8495 $C_{19}H_{19}N_3O_2S$
155106-73-3

2-(4-Piperonyl-1-piperazinyl)benzothiazole = 2-[4-(1,3-Benzodioxol-5-ylmethyl)-1-piperazinyl]benzothiazole (●)
S VB 20B7

U Gastrokinetic (serotonin 5-HT$_3$ receptor antagonist and 5-HT$_4$ receptor agonist)

8496 $C_{19}H_{19}N_3O_3$
161832-65-1

(4R)-3-Acetyl-1-(4-aminophenyl)-3,4-dihydro-4-methyl-7,8-(methylenedioxy)-5H-2,3-benzodiazepine = (8R)-7-Acetyl-5-(4-aminophenyl)-8,9-dihydro-8-methyl-7H-1,3-dioxolo[4,5-h][2,3]benzodiazepine (●)
S GYKI 53773, IDR-53773, LY 300164, *Talampanel***
U Neuroprotector (AMPA receptor antagonist)

8497   (6959) $C_{19}H_{19}N_3O_4$
40915-84-2

2-Methyl-2-[p-[5-(3-pyridyl)-1,2,4-oxadiazol-3-yl]phenoxy]propionic acid ethyl ester = 2-Methyl-2-[4-[5-(3-pyridinyl)-1,2,4-oxadiazol-3-yl]phenoxy]propanoic acid ethyl ester (●)
S AT-308
U Antihyperlipidemic

8498   (6960) $C_{19}H_{19}N_3O_5S$
66-79-5

(2S,5R,6R)-3,3-Dimethyl-6-(5-methyl-3-phenyl-4-isoxazolecarboxamido)-7-oxo-4-thia-1-azabicyclo[3.2.0]hep-

tane-2-carboxylic acid = (6R)-6-(5-Methyl-3-phenyl-4-isoxazolecarboxamido)penicillanic acid = 5-Methyl-3-phenyl-4-isoxazolylpenicillin = [2S-(2α,5α,6β)]-3,3-Dimethyl-6-[[(5-methyl-3-phenyl-4-isoxazolyl)carbonyl]amino]-7-oxo-4-thia-1-azabicyclo[3.2.0]heptane-2-carboxylic acid (●)
R see also no. 13440;
Sodium salt monohydrate (7240-38-2)
S AB 1400, Bactocill, Bristopen, BRL 1400, Cercantal NF-Inj., Cryptocillin, Estafilin, Micropenin, Oksin, Oxabel, Oxacilin, *Oxacillin-Natrium***, *Oxacillin Sodium***, Oxazillin, Oxazina "Made-Lisboa", *Oxazocillin(-Sodium)*, Oxilin, Oxin "Mustafa Nevzat", P-12, Pebenal, Penicillin P-12, Penistafil, Penstapho, Penstaphocid, Prostafilina, Prostaphcillin, Prostaphlin, Resistopen "Bristol; Squibb", *Sodium Oxacillin*, SQ 16423, Stafcilin "V", Staficilin-N, Stapenor, Staphcillin V, Viostatin
U Antibiotic

8498-01   18303-77-0
R Compd. with N,N'-dibenzylethylenediamine (no. 4743)
S Oxaclen
U Antibiotic

8499   (6961) $C_{19}H_{19}N_3O_6$
75530-68-6

5-Isopropyl 3-methyl 2-cyano-1,4-dihydro-6-methyl-4-(m-nitrophenyl)-3,5-pyridinedicarboxylate = 2-Cyano-1,4-dihydro-6-methyl-4-(3-nitrophenyl)-3,5-pyridinedicarboxylic acid 3-methyl 5-(1-methylethyl) ester (●)
S Arcadipin, CL 287389, Escor, FK 235, FR 34235, *Nilvadipine***, *Niprodipine*, Nivadil, *Nivadipine*, SKF 102362, Tensan "Klinge"
U Calcium antagonist, antihypertensive

8500 (6962) $C_{19}H_{19}N_5$
90685-01-1

3-(1-Piperazinyl)-9H-dibenzo[c,f]-1,2,4-triazolo[4,3-a]azepine (●)
S *Pitrazepin*
U GABA antagonist

8501 (6963) $C_{19}H_{19}N_5O_3$
13236-40-3

N-(2,3-Dimethyl-5-oxo-1-phenyl-3-pyrazolin-4-yl)-α-nicotinamidoacetamide = 2,3-Dimethyl-4-(nicotinamidoacetamido)-1-phenyl-5-pyrazolone = N-[2-[[(2,3-Dihydro-1,5-dimethyl-3-oxo-2-phenyl-1H-pyrazol-4-yl)amino]-2-oxoethyl]-3-pyridinecarboxamide (●)
R Hydrochloride
S *Acenifenazone hydrochloride*, Nicopyron ad injectionum, Nicozone-I
U Antineuralgic, antirheumatic

8502 $C_{19}H_{19}N_5O_4S_2$
150094-84-1

6-Methoxy-4-(1-methylethyl)-2-[[(1-phenyl-1H-tetrazol-5-yl)thio]methyl]-1,2-benzisothiazol-3(2H)-one 1,1-dioxide (●)
S Win 633395
U Elastase inhibitor

8503 (6964) $C_{19}H_{19}N_5O_4S_2$
120635-27-0

7-[D-2-Amino-2-(2-amino-6-benzothiazolyl)acetamido]-3(Z)-(1-propenyl)-3-cephem-4-carboxylic acid = [6R-[3(Z),6α,7β(R*)]]-7-[[Amino(2-amino-6-benzothiazolyl)acetyl]amino]-8-oxo-3-(1-propenyl)-5-thia-1-azabicyclo[4.2.0]oct-2-ene-2-carboxylic acid (●)
R Monohydrate
S Bay v 3522
U Antibacterial

8504 (6965) $C_{19}H_{19}N_5O_5S_3$
51627-20-4

(6R,7R)-7-[(R)-2-Amino-2-(p-hydroxyphenyl)acetamido]-3-[[(5-methyl-1,3,4-thiadiazol-2-yl)thio]methyl]-8-oxo-5-thia-1-azabicyclo[4.2.0]oct-2-ene-2-carboxylic acid = (7R)-7-[(R)-2-Amino-2-(p-hydroxyphenyl)acetamido]-3-[[(5-methyl-1,3,4-thiadiazol-2-yl)thio]methyl]-3-cephem-4-carboxylic acid = [6R-[6α,7β(R*)]]-7-[[Amino(4-hydroxyphenyl)acetyl]amino]-3-[[(5-methyl-1,3,4-thiadiazol-2-yl)thio]methyl]-8-oxo-5-thia-1-azabicyclo[4.2.0]oct-2-ene-2-carboxylic acid (●)
S *Cefaparole**, Lilly 110264
U Antibiotic

8505 (6966) $C_{19}H_{19}N_7O_6$
59-30-3

N-[p-[[(2-Amino-4-hydroxy-6-pteridinyl)methyl]amino]benzoyl]-L-glutamic acid = N-[4-[[(2-Amino-1,4-di-

hydro-4-oxo-6-pteridinyl)methyl]amino]benzoyl]-L-glu-
tamic acid (●)
R   also sodium salt (6484-89-5)
S   Acfol, *Acide folique***, *Acidum folicum***, *Acidum*
pteroylglutaminicum, Acifol "Ameripharma", Acifo-
lic, AFI-folsyre, Afolic, A.f. Valdecasas,
Aleukon N, Andreafol, Apo-Folic, Bio-Fol, Bipifol,
Callanish, Cantassium, Clonfolic, Cytofol, Dreisa-
Fol, Elfolin, Endofolin, Facid "Scruggs", Femisanit,
Filicine, fol-5(15)-biomo, Folac, Folacid, Folacin(e),
Folämin, Folan, Folarell, Folasic, Fol-ASmedic,
Folate sodium, Folatine, Folavit, Folbal, Folbiol,
Folcidin, Folcur, Foldine, Folecina, Folettes, Fol-
gamma, Foliamin, *Folic acid***, Folical, Folicare,
Folicet, Folicil, Folicindon, Folico "Ecobi; Taricco",
Folico-Cidan, Folicum, Folijuste, Folillon, Folin(a),
Foliner, Folinor, Folinsyra ACO, Folinsyre, Folipac,
Foli-Rivo, Folisamin, Folisyx, Folitab, *Foliumzuur*,
Foliumzuurtabl., Folivin, Fol Lichtenstein, Folmac,
Folsäure**, Folsan, Folsav, Folverlan, Folvite, Iber-
folico, Incafolic, Innovafolat, *Lactobacillus casei*
Factor, Lafol, Lanes preconceive, Lexpec, Megafol,
Millafol, Mission Prenatal, Nifolin, Novofolacid,
NSC-3073, PGA, Piofolin, Preconceive, *Pteroylglut-*
amic acid, *Pteroylglutaminsäure*, Recuvolin,
Ren-O-fol, RubieFol, SAF, Speciafoldine, Tecnovo-
rin, Terofolin, Vibefol, *Vitamin B_{11}*, *Vitamin B_c*,
Vitamin M, Vitanatal, Witamina B_c, Witamina M,
Yesunan
U   Hematopoietic vitamin

8506 $C_{19}H_{20}BrN_2O_4P$
                                                     133208-93-2

[[4-[[(4-Bromo-2-cyanophenyl)amino]carbonyl]phe-
nyl]methyl]phosphonic acid diethyl ester (●)
S   NO-1886
U   Antihyperlipidemic

8507   (6967) $C_{19}H_{20}BrN_3O_3$
                                                     84379-13-5

tert-Butyl (*S*)-8-bromo-11,12,13,13a-tetrahydro-9-oxo-
9*H*-imidazo[1,5-*a*]pyrrolo[2,1-*c*][1,4]benzodiazepine-1-
carboxylate = (*S*)-8-Bromo-11,12,13,13a-tetrahydro-9-
oxo-9*H*-imidazo[1,5-*a*]pyrrolo[2,1-*c*][1,4]benzodiaze-
pine-1-carboxylic acid 1,1-dimethylethyl ester (●)
S   *Bretazenil***, Ro 16-6028
U   Anxiolytic (partial benzodiazepine receptor agonist)

8508   (6968) $C_{19}H_{20}ClN$
                                                     69175-77-5

(±)-(3aα,4α,9aα)-6-Chloro-3a,4,9,9a-tetrahydro-2-me-
thyl-4-phenylbenz[*f*]isoindoline = (3aα,4α,9aα)-(±)-6-
Chloro-2,3,3a,4,9,9a-hexahydro-2-methyl-4-phenyl-1*H*-
benz[*f*]isoindole (●)
S   BI 27-062, *Losindole***
U   Antidepressant

8509   (6969) $C_{19}H_{20}ClNO$
                                                   112108-01-7

(6aS-*trans*)-11-Chloro-6,6a,7,8,9,13b-hexahydro-7-me-
thyl-5H-benzo[d]naphth[2,1-b]azepin-12-ol (●)
R   Hydrochloride (190133-94-9)
S   *Ecopipam hydrochloride***, Sch 39166
U   Antipsychotic (dopamine D_1 receptor antagonist)

8510   (6970) $C_{19}H_{20}ClNOS$
                                                   83986-02-1

2-Chloro-11-(1-methyl-4-piperidyl)-11H-dibenz[b,e]
[1,4]oxathiepin = 4-(2-Chloro-11H-dibenz[b,e][1,4]oxa-
thiepin-11-yl)-1-methylpiperidine (●)
R   Maleate (1:1) (83986-04-3)
S   *Cloxathiepin*, VÚFB-14107
U   Neuroleptic

8511 $C_{19}H_{20}ClNO_2$
                                                   80751-65-1

3-Allyl-6-chloro-7,8-dihydroxy-1-phenyl-2,3,4,5-tetra-
hydro-1H-3-benzazepine = 6-Chloro-2,3,4,5-tetrahydro-
1-phenyl-3-(2-propenyl)-1H-3-benzazepine-7,8-diol (●)
R   also hydrobromide (74115-01-8)
S   SKF 82958
U   Dopamine D_1 receptor agonist

8512   (6971) $C_{19}H_{20}ClNO_2$
                                                   131796-63-9

(S)-8-Chloro-5-(2,3-dihydro-7-benzofuranyl)-2,3,4,5-te-
trahydro-3-methyl-1H-3-benzazepin-7-ol (●)
S   NNC 01-0756, NNC-756, *Odapipam***
U   Antipsychotic (dopamine D_1 receptor antagonist)

8513   (6972) $C_{19}H_{20}ClNO_3$
                                                   77386-12-0

3-Allyl-6-chloro-2,3,4,5-tetrahydro-1-(*p*-hydroxyphe-
nyl)-1*H*-3-benzazepine-7,8-diol = 6-Chloro-2,3,4,5-tetra-
hydro-1-(4-hydroxyphenyl)-3-(2-propenyl)-1*H*-3-benz-
azepine-7,8-diol (●)
S   SKF 85174
U   Antihypertensive

8514   (6973) $C_{19}H_{20}ClNO_4$
                                                           41859-67-0

2-[*p*-[2-(*p*-Chlorobenzamido)ethyl]phenoxy]-2-methyl-
propionic acid = 2-[4-[2-[(4-Chlorobenzoyl)ami-
no]ethyl]phenoxy]-2-methylpropanoic acid (●)
S   Autasynt, Azufibrat, Becait, Befibrat, Befizal,
        Beza 1A Pharma, Beza 200 von ct, Beza AbZ, Bez-
        abeta, Bezacur, *Bezafibrate**,* Bezafisal, Bezagam-
        ma, Beza-Lande, Bezalin, Bezalip, Bezamerck, Be-
        zamidin, Bezapham, Beza-Puren, Bezarip, Bezatol,
        Bezifal, BF 759, BM 15075, Cedur, Demia, Detrex
        "Vargas", Difaterol, Durabezur, Elpilip, Eulitop, Ge-
        tur, Hadiel, Lacromid, Lipocor, Lipox, LO 44, Nor-
        lip, Oralipin, Reducterol, Regadrin B, Sklerofibrat,
        Solibay, Verbital, Zalipid
U   Antihyperlipidemic

8515   (6974) $C_{19}H_{20}ClNO_5$
                                                           42597-57-9

3-Hydroxypropyl nicotinate 2-(*p*-chlorophenoxy)-2-me-
thylpropionate (ester) = 3-(Nicotinoyloxy)propyl α-(*p*-
chlorophenoxy)isobutyrate = 3-Pyridinecarboxylic acid
3-[2-(4-chlorophenoxy)-2-methyl-1-oxopropoxy]propyl
ester (●)
S   Cloprane, I 612, *Ronifibrate***
U   Antihyperlipidemic

8516   (6975) $C_{19}H_{20}ClN_3$
                                                           442-52-4

1-*p*-Chlorobenzyl-2-(1-pyrrolidinylmethyl)benzimida-
zole = 1-[(4-Chlorophenyl)methyl]-2-(1-pyrrolidinylme-
thyl)-1*H*-benzimidazole (●)
R   Monohydrochloride (1163-36-6);
        see also nos. 6247-09, 6249-04
S   AL 20, Alercur, Allercur, Allercurol, Allerpant, Ben-
        midina, Betistin, *Clemizole hydrochloride**,* Desen-
        sibil, Hérol, Histacur, Histacuran, Klemidox,
        Pan-Allerg, Reactrol, Sanall-Vi
U   Antihistaminic, anti-allergic

8517   (6976) $C_{19}H_{20}ClN_3O$
                                                          103844-86-6

N-[[2-(*p*-Chlorophenyl)imidazo[1,2-*a*]pyridin-3-yl]me-
thyl]-*N*-methylbutyramide = *N*-[[2-(4-Chlorophenyl)imi-
dazo[1,2-*a*]pyridin-3-yl]methyl]-*N*-methylbutanamide
(●)
S   *Saripidem**,* SL 850274
U   Anxiolytic

8518   (6977) $C_{19}H_{20}ClN_3O_4$
                                                        98374-54-0

5-[*N,N*-bis(2-hydroxyethyl)glycyl]-8-chloro-5,10-dihy-
dro-11*H*-dibenzo[*b,e*][1,4]diazepin-11-one = 5-[[Bis(2-
hydroxyethyl)amino]acetyl]-8-chloro-5,10-dihydro-11*H*-
dibenzo[*b,e*][1,4]diazepin-11-one (●)
R   Monohydrochloride (98374-55-1)
S   AWD 26-06, *Siltenzepine hydrochloride***
U   Anti-ulcer agent

8519 $C_{19}H_{20}ClN_3O_5S$
                                                        102251-91-2

8-Chlorodibenz[*b,f*][1,4]oxazepine-10(11*H*)-carboxylic
acid 2-[3-(ethylsulfonyl)-1-oxopropyl]hydrazide (●)
S   SC-42867
U   PGE2 antagonist

8520   (6978) $C_{19}H_{20}Cl_2N_2O_3$
                                                        53498-76-3

4-[[[4-[Bis(2-chloroethyl)amino]phenyl]acetyl]ami-
no]benzoic acid (●)
R   see also no. 8586

S   Pafencil, Paphencyl
U   Antineoplastic

8521   (6979) $C_{19}H_{20}Cl_2N_2O_5$
                                                        25287-60-9

2,2-Dichloro-*N*-(2-ethoxyethyl)-*N*-[(*p*-nitrophenoxy)ben-
zyl]acetamide = *N*-(2-Ethoxyethyl)-*N*-[*p*-(*p*-nitrophen-
oxy)benzyl]dichloroacetamide = 2,2-Dichloro-*N*-(2-eth-
oxyethyl)-*N*-[[4-(4-nitrophenoxy)phenyl]methyl]acet-
amide (●)
S   *Ethychlordiphene, Etichlordifene, Etofamide***,
        K-430, Kitnos, Kitnosil
U   Anti-amebic

8522   (6980) $C_{19}H_{20}FNO_3$
                                                        61869-08-7

(–)-*trans*-4-(*p*-Fluorophenyl)-3-[[3,4-(methylenedioxy)phenoxy]methyl]piperidine = (–)-*trans*-5-[(4-*p*-
Fluorophenyl-3-piperidyl)methoxy]-1,3-benzodioxole =
(3*S*-*trans*)-3-[(1,3-Benzodioxol-5-yloxy)methyl]-4-(4-
fluorophenyl)piperidine (●)
R   also hydrochloride (78246-49-8) or hydrochloride hy-
        drate (2:1) (110429-35-1)
S   Aropax, Aroxat, BRL 29060, Casbol, Cebrilin, De-
        roxat, Eutimil, FG 7051, Frosinor, Motivan "Faes",
        *Paroxetine***, Paxil "SK Beecham", Paxyl "SK Bee-
        cham", Pondera, Sereupin, Seroxat, Tagonis
U   Antidepressant

8523 $C_{19}H_{20}FNO_5$
158836-71-6

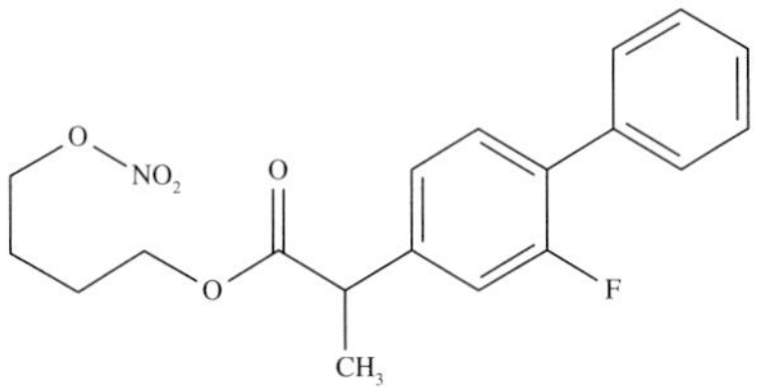

2-Fluoro-α-methyl-[1,1'-biphenyl]-4-acetic acid 4-(nitro-oxy)butyl ester (●)
S *Flurbiprofen nitroxybutyl ester*, HCT-1026, *Nitro-flurbiprofen*
U Nitric oxide donor, anti-inflammatory

8524 (6981) $C_{19}H_{20}FN_3$
67121-76-0

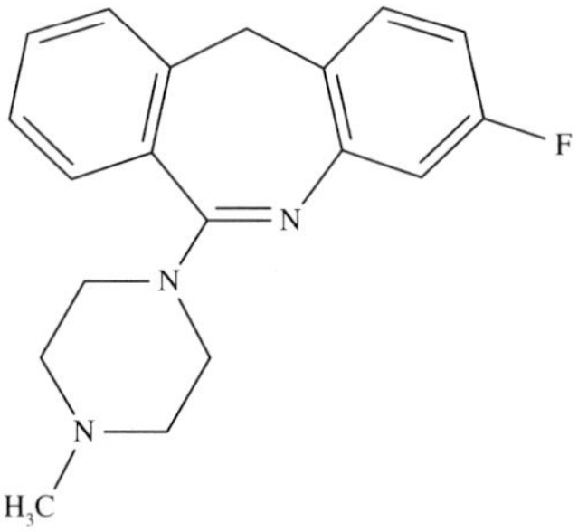

3-Fluoro-6-(4-methyl-1-piperazinyl)morphanthridine = 3-Fluoro-6-(4-methyl-1-piperazinyl)-11*H*-dibenz[*b,e*]azepine (●)
S *Fluoroperlapine, Fluperlapine***, NB 106-689, SNS 106-689
U Neuroleptic, antidepressant

8525 (6982) $C_{19}H_{20}FN_3$
107266-06-8

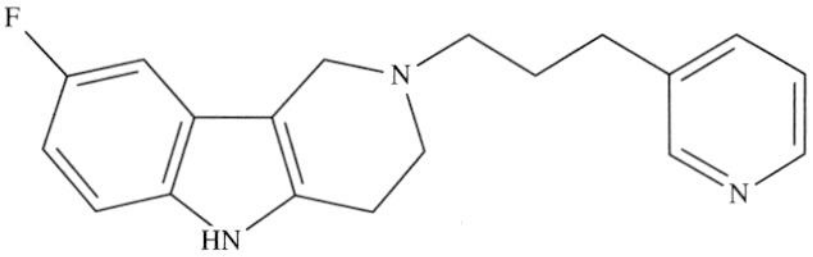

8-Fluoro-2,3,4,5-tetrahydro-2-[3-(3-pyridinyl)propyl]-1*H*-pyrido[4,3-*b*]indole (●)
R Monohydrochloride (112243-58-0)
S *Gevotroline hydrochloride***, Wy-47384
U Antipsychotic

8526 (6983) $C_{19}H_{20}FN_3O_3$
112398-08-0

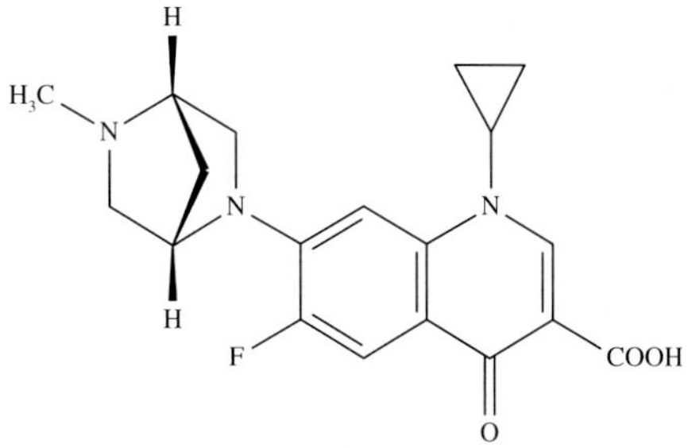

1-Cyclopropyl-6-fluoro-1,4-dihydro-7-[(1*S*,4*S*)-5-methyl-2,5-diazabicyclo[2.2.1]hept-2-yl]-4-oxo-3-quinolinecarboxylic acid = (1*S*)-1-Cyclopropyl-6-fluoro-1,4-dihydro-7-(5-methyl-2,5-diazabicyclo[2.2.1]hept-2-yl)-4-oxo-3-quinolinecarboxylic acid (●)
R Monomethanesulfonate (119478-55-6)
S Advocid, Advocin, CP-76136-27, *Danofloxacin mesilate***
U Antibacterial (veterinary)

8527 (6984) $C_{19}H_{20}FN_3O_4$
108405-24-9

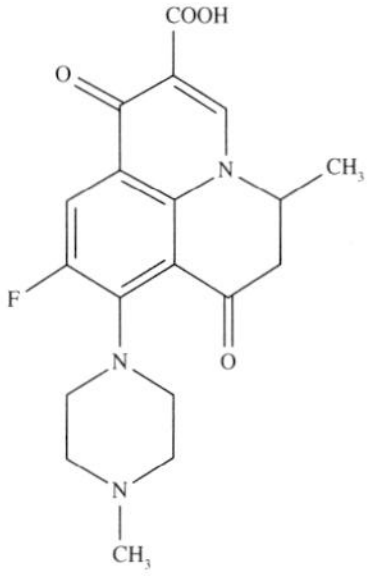

9-Fluoro-6,7-dihydro-5-methyl-8-(4-methyl-1-piperazinyl)-1,7-dioxo-1*H*,5*H*-benzo[*ij*]quinolizine-2-carboxylic acid (●)
R Monohydrochloride (108405-58-9)
S QA-241
U Antibacterial

8528 $C_{19}H_{20}F_2N_2O_3S$
                                                           141699-86-7

1-Cyclopropyl-6,8-difluoro-1,4-dihydro-7-[3-[(methylthio)methyl]-1-pyrrolidinyl]-4-oxo-3-quinolinecarb-
oxylic acid (●)
S DWQ-013
U Antibacterial

8529 $C_{19}H_{20}F_2N_4O_4$
                                                           154187-75-4

5-Amino-1-cyclopropyl-6,8-difluoro-7-(hexahydro-4-
methyl-5-oxo-1H-1,4-diazepin-1-yl)-1,4-dihydro-4-oxo-
3-quinolinecarboxylic acid (●)
S FA 107
U Antibacterial

8530 (6985) $C_{19}H_{20}F_3NO$
                                                           10355-14-3

1-[2-[[4'-(Trifluoromethyl)-4-biphenylyl]oxy]ethyl]pyr-
rolidine = 1-[2-[[4'-(Trifluoromethyl)[1,1'-biphenyl]-4-
yl]oxy]ethyl]pyrrolidine (●)
S Boxidine*, CL 65205
U Antihyperlipidemic

8531 (6986) $C_{19}H_{20}F_3NO_2$
                                                           23602-78-0

2-[[α-Methyl-m-(trifluoromethyl)phenethyl]amino]etha-
nol benzoate (ester) = N-(2-Benzoyloxyethyl)norfenflura-
mine = 2-[[1-Methyl-2-[3-(trifluoromethyl)phenyl]ethyl]
amino]ethanol benzoate (ester) (●)
R also hydrochloride (23642-66-2)
S Balans, Benfluorex*, Benfluramate, JP 992, Liparex,
      Lipascor, Lipophoral, Mediator, Mediaxal, Minolip,
      Modulator, Palameda, Proplatone, Redrax, S 780,
      SE 780
U Antihyperlipidemic, anorexic

8532 $C_{19}H_{20}F_3NO_4$
                                                           137887-41-3

Dimethyl 1,4-dihydro-1,2,6-trimethyl-4-(α,α,α-trifluoro-
p-tolyl)pyridine-3,5-dicarboxylate = 1,4-Dihydro-1,2,6-
trimethyl-4-[4-(trifluoromethyl)phenyl]-3,5-pyridinedi-
carboxylic acid dimethyl ester (●)
S BAY w 9798, Reoflusine
U Treatment of peripheral arterial occlusive disease

8533 (6987) $C_{19}H_{20}F_3N_3O_3$
65847-85-0

2-Morpholinoethyl 2-(α,α,α-trifluoro-*m*-toluidino)nicoti-nate = 2-(α,α,α-Trifluoro-*m*-toluidino)nicotinic acid 2-morpholinoethyl ester = 2-[[3-(Trifluoromethyl)phe-nyl]amino]-3-pyridinecarboxylic acid 2-(4-morpholi-nyl)ethyl ester (●)

R see also no. 3793-03
S Actol supos., Flomax "Chiesi", Flumarin "Promedi-ca", Morniflu, *Morniflumate***, Niflactol, Niflam "Upsa"-Suppos., Nifluril-Suppos., Niflux supos., UP 164
U Anti-inflammatory

8534 (6988) $C_{19}H_{20}F_3N_3O_3$
113617-63-3

cis-1-Cyclopropyl-7-(3,5-dimethyl-1-piperazinyl)-5,6,8-trifluoro-1,4-dihydro-4-oxo-3-quinolinecarboxylic acid (●)

S *Marufloxacin*, OBFX, Orbax, *Orbifloxacin***
U Antibacterial

8535 $C_{19}H_{20}F_3N_3O_4$
153808-85-6

(*S*)-1-Cyclopropyl-8-(difluoromethoxy)-6-fluoro-1,4-di-hydro-7-(3-methyl-1-piperazinyl)-4-oxo-3-quinoline-carboxylic acid (●)

R Monohydrochloride (128427-55-4)
S *Caderofloxacin hydrochloride, Cadrofloxacin hydro-chloride***, CS 940
U Antibacterial

8536 (6989) $C_{19}H_{20}N_2$
524-81-2

5-Benzyl-2,3,4,5-tetrahydro-2-methyl-1*H*-pyri-do[4,3-*b*]indole = 9-Benzyl-1,2,3,4-tetrahydro-3-methyl-γ-carboline = 2,3,4,5-Tetrahydro-2-methyl-5-(phenylme-thyl)-1*H*-pyrido[4,3-*b*]indole (●)

R also 1,5-naphthalenedisulfonate (2:1) (6153-33-9)
S Diazolin, *Diazolin(um)*, Fabahistin, Incidal, Incidan, Insidal, Mebcidal, *Mebhydrolin***, Mebolin, Omeril
U Antihistaminic

8537 (6990) $C_{19}H_{20}N_2O$
117946-91-5

N-Acetyl-2-benzyltryptamine = N-[2-[2-(Phenylmethyl)-
1H-indol-3-yl]ethyl]acetamide (●)
S   Luzindole, N-0774
U   Melatonin receptor antagonist

8538 (6991) $C_{19}H_{20}N_2O$
6880-54-2

(19E)-2,16,19,20-Tetrahydrocuran-17-al (●)
R   also monohydrochloride (17682-63-2)
S   Barvincan, Vincanine
U   Analeptic (alkaloid from Vinca erecta)

8539 (6992) $C_{19}H_{20}N_2O$
73931-96-1

(±)-α-(4-Phenethylphenyl)-1-imidazoleethanol = N-[2-
(4-Phenethylphenyl)-2-hydroxyethyl]imidazole = α-[4-
(2-Phenylethyl)phenyl]-1H-imidazole-1-ethanol (●)
R   Monohydrochloride (77234-90-3)
S   Denzimol hydrochloride**, Rec 15/1533
U   Anticonvulsant, anti-epileptic

8540 $C_{19}H_{20}N_2O_2$
148372-04-7

3-(2,4-Dimethoxybenzylidene)anabaseine = (3E)-3-[(2,4-
Dimethoxyphenyl)methylene]-3,4,5,6-tetrahydro-2,3'-bi-
pyridine (●)
R   Dihydrochloride (156223-05-1)
S   DMBX, DMXBA, GTS-21
U   Nootropic (nicotinic agonist)

8541 (6993) $C_{19}H_{20}N_2O_2$
50-33-9

3,5-Dioxo-1,2-diphenyl-4-butylpyrazolidine = 4-Butyl-
1,2-diphenyl-3,5-pyrazolidinedione (●)
R   see also nos. 4919-01, 6457-01, 14039;
    also sodium enolate (129-18-0)
S   Acrizeal, Algoverine, Alindor, Alka Butazolidin, Al-
    kabutazona, Ambene, Amrazoladin, Antadol, Anu-
    spiramin, Apo-Phenylbutazone, Arkazolidine, Ar-
    thrazone, Arthril, Arthrizon, Artrizin, Artrizone,
    Artropan, Atrofen "MD Farm.", Azobutil, Azolid,
    Basireuma, Benzone, Betazed, Bipizolidin, Bizolin,
    B.T.Z., Bufenil, Buffazone (active substance), Bu-
    lentan, Busone, Butacal, Butacompren, Butacote,
    Butadin, Butadion(a), Butadion(um), Butadyne, Bu-
    tafenil, Butagen, Butagesic, Butalan "Lancet", Butal-
    gin, Butalgina "Esteve", Butalidon, Butalona, Bu-
    taluy, Buta-Phen, Butaphen "Marois; Prosana; Wie-
    denmann"-Salbe, Butapirazol, Butaran, Butarecbon,
    Butarex, Butartril, Butartrina, Butasan, Butaval, Bu-

tazina, Butazolidin(e), Butazona, Butazone, Bute, Butidiona, Butina, Butiwas-simple, Butofar, Butone, Butoroid, Butosal, Butoz, Butozone, Butrex, Butylpyrin, Buvetzone, Buzon, Chembutazone, Chillos-P, Chrobutazon, Clonbute, Colbutan, Companazone, Curozoladin, Delbutan, Deltabutanyl, Delta-Pyrazol, Demoplas-Amp., Dibutone, Dibuzon, Digibutina, Diossidone, Diozol, Diphebuzol, *Diphenylbutazone*, Ecobutazone, Elcazolidine, Elmedal, Equi-Bute, Equipalazone, Equiphen, Eributazone, Erobutal, Exrheudon OPT-Dragees, Febuzina, *Fenbutazone*, Fenibutasan, Fenibutol, Fenilblan, *Fenilbutazona*, Fenilbutina, Fenilidina, Fenotone, *Fenylbutazone*, Fezona, Flexazone, Folirozon, Fylafen, G 13871, Ia-but, Imphabutazone, Inflazone, Intrabutazone, Intrazone, Ipsoflame, Jagozolidine, Kadol, Larbutazone, Lingel, Malgesic, Mephabutazon, Merizone, Nadozone, Neo-Algolaz, Neo-Zoline, Neuplus, Novobutazone, Novophenyl, Panazone, Phebutan, Phebuzine, Phenatone, Phen-Buta-Vet, Phenbutazol, Phenbutazone, Phenogel, Phenopyrine, Phenycare, Phénylarthrite, Phenylbetazone, *Phenylbutazone***, Phenyl-Mobuzon, Phenylone, Phenyzone, Pilazon, Pirarreumol-B, Praecirheumin-Drag., Prodynam, Rectofasa, Reudo(x), Reumaphen, Reumasyl, Reumazin, Reumazol, Reumune, Reumuzol, Reupolar, Rheumal, Rheumaphen, Rhomatoryl, Robizone-V, Rubatone, Salyren, Scanbutazone, Schmergen, Scriptozone, Sedazole, Sepvadol Inj., Servizolidin, Shigrodin, Silalidin, Spondyril, Tazone, Tetnor, Tevcodyne, Therazone, Thilozone, Tibutazone, Ticinil "de Angeli", Tisatin, Todalgil-Supposit., Tokugen, Umezolina, Unifalgan, Uzone, Wescozone, Zolandin, Zolaphen, Zolidinum

U Antirheumatic, anti-inflammatory

8541-01 (6993-01)   36298-23-4
R Calcium salt (2:1)
S Butazona Cálcica, Neo Ticinil Calcium, Peralgin, Pyrazon "Un.Am.Ph.", Rhumifen, Siobutazine, Ticinil Calcico
U Antirheumatic, anti-inflammatory

8541-02 (6993-02)
R Gentisate
S Azogen
U Antirheumatic, anti-arthritic

8541-03 (6993-03)   4985-25-5
R Compd. with piperazine (1:1)
S Aflamina, Carudol, Carudolin, Catalgin, Clavezona, Crucium, Damixa (old form), Dartranol, DB 139, Doriol, Ischialzid, Moritanin, Niflamin, *Phénylbutazone-Pipérazine*, Piribute, Pirzon, *Pyrasanonum*, *Pyrazinobutazone*, Ranoroc
U Anti-inflammatory

8541-04 (6993-04)   34427-79-7
R Compd. with dextropropoxyphene (no. 11375) (1:1)
S *Fenoxipirazofenona*, MI 1310, *Phenoxypyrazophenone*, *Proxifezone***, *Prozodiona*
U Anti-inflammatory, analgesic

8541-05 (6993-05)   34214-49-8
R Compd. of the sodium enolate with glycerol (1:1)
S G 26872, *Phenbutazone Sodium Glycerate*
U Anti-inflammatory

8541-06 (6993-06)   54749-86-9
R Compd. with 2-thiazolin-2-amine (no. 140) (1:1)
S Deflogix "UCB", Fordonal (old form), LAS 11871, *Thiazolinobutazone*
U Anti-inflammatory, antirheumatic

8541-07 (6993-07)   37598-92-8
R Compd. of the 5-enolate with benzydamine (no. 8780) (1:1)
S *Butazidamina*, Butial, LS-701
U Anti-inflammatory, antirheumatic

8541-08 (6993-08)   62025-40-5
R Compd. with fenyramidol (no. 3955) (1:1)
S Decabutin rectal, Fenirabutol
U Anti-inflammatory, anti-arthritic

8541-09 (6993-09)   71789-17-8
R Mixture with quercetin (no. 5119)
S Butaquertin
U Anti-inflammatory

8542   (6994)

$C_{19}H_{20}N_2O_2S$
126769-16-2

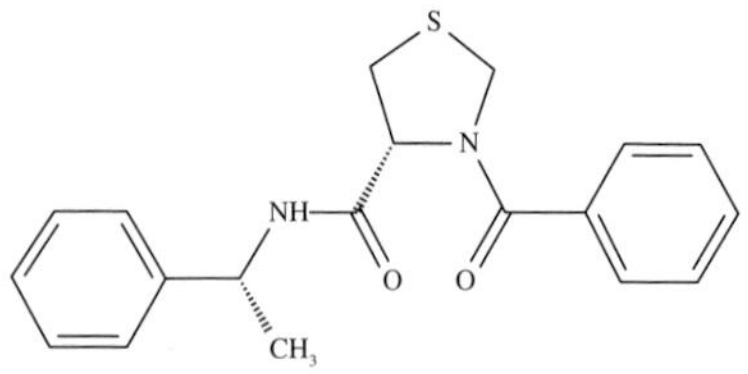

(4R)-3-Benzoyl-N-[(1R)-1-phenylethyl]-4-thiazolidine-
carboxamide = (R*,R*)-3-Benzoyl-N-(1-phenylethyl)-4-
thiazolidinecarboxamide (●)
S   RS-0481
U   Antineoplastic, immunopotentiator

8543   (6997)

$C_{19}H_{20}N_2O_3$
1673-06-9

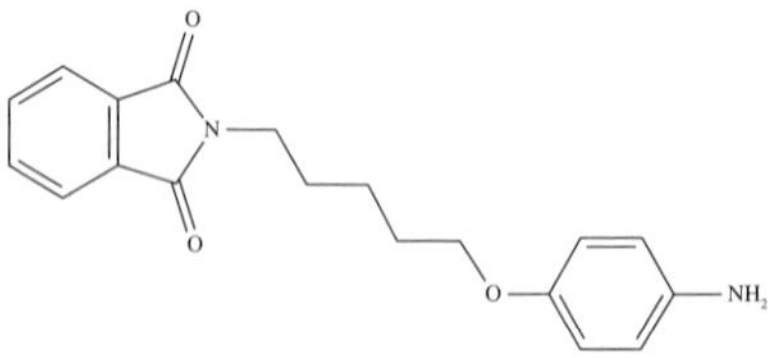

N-[5-(p-Aminophenoxy)pentyl]phthalimide = 2-[5-(4-
Aminophenoxy)pentyl]-1H-isoindole-1,3-(2H)-dione (●)
S   Amphotalide**, M. & B. 1948 A, 6171 R.P., Schi-
stomide
U   Antischistosomal

8544   (6998)

$C_{19}H_{20}N_2O_3$
115956-12-2

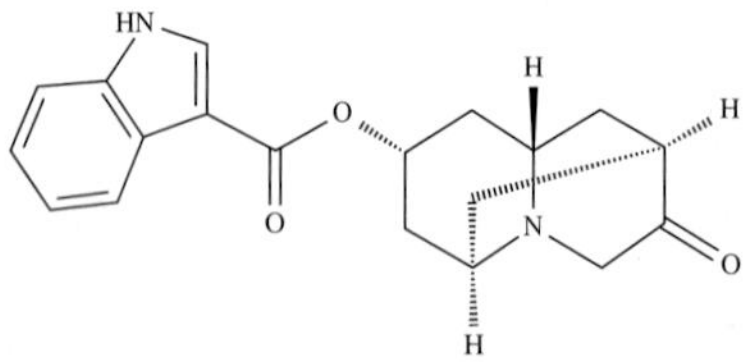

Indole-3-carboxylic acid, ester with (8r)-hexahydro-8-
hydroxy-2,6-methano-2H-quinolizin-3(4H)-one =
(2α,6α,8α,9aβ)-Octahydro-3-oxo-2,6-methano-2H-qui-
nolizin-8-yl 1H-indole-3-carboxylate = (2α,6α,8α,9aβ)-
1H-Indole-3-carboxylic acid octahydro-3-oxo-2,6-metha-
no-2H-quinolizin-8-yl ester (●)
R   Monomethanesulfonate (115956-13-3)
S   Anemet, Anzemet, Dolasetron mesilate**,
MDL 73147 EF

U   Serotonin antagonist (antimigraine, anti-emetic, antipsychotic)

8545   (6996)

$C_{19}H_{20}N_2O_3$
18471-20-0

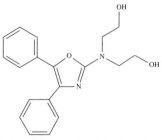

2,2'-[(4,5-Diphenyl-2-oxazolyl)imino]diethanol = 4,5-Di-
phenyl-2-(diethanolamino)oxazole = 2,2'-[(4,5-Diphenyl-
2-oxazolyl)imino]bis[ethanol] (●)
S   Ageroplas, APT-574, Diethamphenazole, Ditazo-
le**, Fendazol, S 222
U   Anti-inflammatory, analgesic, platelet aggregation inhibitor, vasoprotectant

8546   (6999)

$C_{19}H_{20}N_2O_3$
130641-38-2

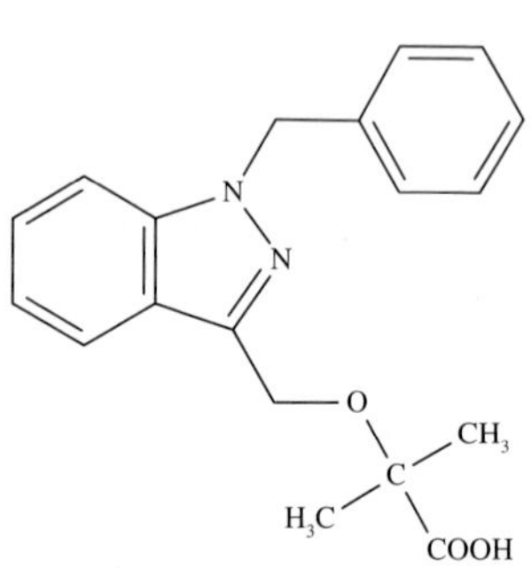

2-[(1-Benzyl-1H-indazol-3-yl)methoxy]-2-methylpro-
pionic acid = 2-Methyl-2-[[1-(phenylmethyl)-1H-inda-
zol-3-yl]methoxy]propanoic acid (●)
S   AF 2838, Bindarit**
U   Anti-inflammatory, antirheumatic

8547 (6995) $C_{19}H_{20}N_2O_3$
129-20-4

4-Butyl-1-(4-hydroxyphenyl)-2-phenyl-3,5-pyrazolidine-
dione (●)
R also monohydrate (7081-38-1)
S Aflamox, Altodol, Amsafenil, Aradinum, Arkadil,
Artroflog, Artzone, Azopirin "Dogu", Breis, Buta-
fen, Butaflogin, Butalgon-Drag., Butanova, Buta-
phen "Mulda", Butapirone, Butazolon, Butazonic,
Buteril, Butilene, Buxifen, Californit, Cinophan-N,
Crovaril (old form), Deflogin "Valeas", Delilidin,
Diflamil Suppos., Edefan, Eroxyzone, Etrozolidina,
Eugaril, Famaril, Fedezona, Fendril, Fibutrox, Fina-
ril, Flamaril, Flammryl, Flegmostam, Flogal, Floghe-
ne, Flogistin, Flogitolo, Flogodin "Firma", Flogoril,
Flogostop, Flopirina, Frabel, G 27202, Genal, *Hy-
droxyphenylbutazone*, Idrobutazina, Iltazon, Imbun
(old form), Inflamil, Inflammil, Inflanil, Ipebutona,
Iridil, Isobutil, Jagril, Ladryl, Maderil, Medoxyl
"Medoz", Mepharil, Miorelase, Miyadril, Mysite,
Naleran, Neo-Farmadol, Neofen, Nevraspis, Octiro-
na, Offitril, *Oksifenbutazon*, Ophen, Optimal "Do-
jin", *Ossifenbutazone*, Otone, Oxalid, Oxazoli-
din-Geigy, Oxiblan, Oxibutil, Oxibutol, *Oxifenbuta-
zona*, Oxi-Fenibutol, *Oxiphenbutazonum*, Oxi-Pirabu-
tal, Oxi-Reumopha, Oxitazona, Oxyacidon, Oxybu-
tazone, Oxybuton, Oxybuzon, Oxyfenibutol, Oxy-
fentamin, Oxygesic, Oxylone "Kobayashi Kako",
Oxyperol, Oxyphen, *Oxyphenbutazone***, Oxyphen-
tamin, Oxyron, Oxzan, Phenabid, Phenderil, Phen-
macalm, Phlogase-Suppos., Phlogistol, Phlogont,
Phloguran, Pirabutina, Piraflogin, Poliflogil, Port-
oril, Prolin "Mercury", Rapostan "Mepha", Reducin,
Remazin, Reumabutal, Reumox, Rheumapax, Rheu-
mazone, Ronderil, Rumapax, Rumazolidin, Rumo-
xil, Sederil, Seskazon, Sioril, Suganril, Tabazone,
Tandacote, Tandearil, Tandelut, Tanderal, Tanderil,
Tandrex, Tandron, Tantal, Telidal, Teneral, Trifazin,
Validil, Visubutina, Zatine

U Anti-inflammatory, antirheumatic

8547-01 (6995-01) 25146-18-3
R Compd. with fenyramidol (no. 3955) (1:1)
S Anabutol, Butofen, Butolfen, CG-21, Clandilon (old
form), Febutol(o), *Fenbutamidolum, Fenhidroxibu-
tol*, Fepirina, Litial, *Oxifenidol*, Reumatex
U Anti-inflammatory, analgesic

8547-02 (6995-02) 57148-78-4
R Compd. with piperazine (1:1)
S Diflamil, Difmedol, *Oxipizone, Oxyphenbutazonum
piperazinum*
U Anti-inflammatory

8548 (7001) $C_{19}H_{20}N_2O_3S$
34316-48-8

2',3-Bis(*p*-ethoxyphenyl)pseudothiohydantoin = 3-(4-
Ethoxyphenyl)-2-[(4-ethoxyphenyl)imino]-4-thiazolidi-
none (●)
S Ditofen, Ditophen
U Tuberculostatic

8549 (7000) $C_{19}H_{20}N_2O_3S$
111025-46-8

(±)-5-[*p*-[2-(5-Ethyl-2-pyridyl)ethoxy]benzyl]-2,4-thia-
zolidinedione = (±)-5-[[4-[2-(5-Ethyl-2-pyridinyl)eth-
oxy]phenyl]methyl]-2,4-thiazolidinedione (●)
R Monohydrochloride (112529-15-4)
S Actos, AD 4833, *Pioglitazone hydrochloride***,
U-72107 (A), Zactos
U Antidiabetic

8550　(7002)　　　　　　　　　　　$C_{19}H_{20}N_2O_4$
　　　　　　　　　　　　　　　　　128022-68-4

(+)-5-(2,3-Dihydro-7-benzofuranyl)-2,3,4,5-tetrahydro-
3-methyl-8-nitro-1*H*-3-benzazepin-7-ol (●)
S　CEE 03-310, NNC 01-0687, NNC-687
U　Antipsychotic (dopamine D_1 receptor antagonist)

8551　(7003)　　　　　　　　　　　$C_{19}H_{20}N_2O_7$
　　　　　　　　　　　　　　　　　86780-90-7

Acetonyl methyl 1,4-dihydro-2,6-dimethyl-4-(*o*-nitrophe-
nyl)-3,5-pyridinedicarboxylate = 1,4-Dihydro-2,6-dime-
thyl-4-(2-nitrophenyl)-3,5-pyridinedicarboxylic acid methyl 2-oxopropyl ester (●)
S　*Aranidipine***, *Asanidipine*, Bec, MPC-1304, Sapre-
sta
U　Antihypertensive (calcium antagonist)

8552　(7004)　　　　　　　　　　　$C_{19}H_{20}N_2O_8$
　　　　　　　　　　　　　　　　　128429-19-6

α-Formamido-5'-(2-formamido-1-hydroxyethyl)-β,2',6-
trihydroxy-3-biphenylpropionic acid = α-(Formylamino)-
5'-[2-(formylamino)-1-hydroxyethyl]-β,2',6-trihydroxy-
[1,1'-biphenyl]-3-propanoic acid (●)

R　Sodium salt (128524-51-6)
S　FR 900280, WF-2421
U　Aldose reductase inhibitor from the fungus *Humicola
grisea*

8553　(7005)　　　　　　　　　　　$C_{19}H_{20}N_2S$
　　　　　　　　　　　　　　　　　85273-95-6

2-[[2-(Dimethylamino)ethyl]thio]-3-phenylquinoline =
N,*N*-Dimethyl-2-[(3-phenyl-2-quinolinyl)thio]ethan-
amine (●)
R　Monohydrochloride (85273-96-7)
S　ICI 169369
U　Antidepressant

8554　　　　　　　　　　　　　　　$C_{19}H_{20}N_4O$
　　　　　　　　　　　　　　　　　167172-77-2

N-[2-(2-Methoxyphenyl)ethyl]-*N*'-[4-(1*H*-imidazol-4-
yl)phenyl]formamidine = *N*-[4-(1*H*-Imidazol-4-yl)phe-
nyl]-*N*'-[2-(2-methoxyphenyl)ethyl]methanimidamide
(●)
S　IY-80843
U　Histamine H_2 receptor antagonist (anti-ulcer)

8555　(7006)　　　　　　　　　　　$C_{19}H_{20}N_4O_2$
　　　　　　　　　　　　　　　　　96449-05-7

(±)-6,11-Dihydro-1-(1-methylnipecotoyl)-5*H*-pyri-
do[2,3-*b*][1,5]benzodiazepin-5-one = (±)-6,11-Dihydro-

11-[(1-methyl-3-piperidinyl)carbonyl]-5H-pyrido[2,3-b]
[1,5]benzodiazepin 5-one (●)
R Monohydrochloride (121798-89-8)
S DF 594, *Rispenzepine hydrochloride**, Ulvenzepine
U Antispasmodic

8556   (7007) $C_{19}H_{20}N_4O_2$
                                                        96487-37-5

6,11-Dihydro-11-(1-methylisonipecotoyl)-5H-pyri-
do[2,3-b][1,5]benzodiazepin-5-one = 6,11-Dihydro-11-
[(1-methyl-4-piperidinyl)carbonyl]-5H-pyri-
do[2,3-b][1,5]benzodiazepin-5-one (●)
R Monohydrochloride (128292-56-8)
S DF 545, K 545, *Nipenzepine hydrochloride, Nuven-
zepine hydrochloride**
U Anti-ulcer agent, gastric antisecretory

8557 $C_{19}H_{20}N_4O_3$
                                                        143692-18-6

1-(4-Aminophenyl)-3,4-dihydro-4-methyl-3-(methylcarbamoyl)-7,8-(methylenedioxy)-5H-2,3-benzodiaze-
pine = 5-(4-Aminophenyl)-8,9-dihydro-N,8-dimethyl-
7H-1,3-dioxolo[4,5-h][2,3]benzodiazepine-7-carbox-
amide (●)
R Monohydrochloride (143692-48-2)
S GYKI 53655, LY 300168
U Anticonvulsant, muscle relaxant

8558 $C_{19}H_{20}N_4O_3$
                                                        161832-71-9

(–)-1-(4-Aminophenyl)-3-methylcarbamoyl-4-methyl-
7,8-methylenedioxy-3,4-dihydro-5H-2,3-benzodiazepine
= (8R)-5-(4-Aminophenyl)-8,9-dihydro-N,8-dimethyl-
7H-1,3-dioxolo[4,5-h][2,3]benzodiazepine-7-carbox-
amide (●)
S GYKI 53784, LY 303070
U Neuroprotector (AMPA receptor antagonist)

8559   (7008) $C_{19}H_{20}N_4O_3S$
                                                        116091-80-6

5-(4,5-Dihydro-2-oxazolyl)-2-[[(4-methoxy-3,5-dime-
thyl-2-pyridinyl)methyl]sulfinyl]-1H-benzimidazole (●)
S HN-11203
U Gastric antisecretory, anti-ulcer agent

8560   (7009) $C_{19}H_{20}N_4O_4$
                                                        79690-61-2

10-(2,3-Dimethylpentanamido)-4-oxo-4H-pyrimi-
do[1,2-c]quinazoline-3-carboxylic acid = 10-[(2,3-Di-
methyl-1-oxopentyl)amino]-4-oxo-4H-pyrimido[1,2-c]
quinazoline-3-carboxylic acid (●)
R Monosodium salt (100508-89-2)
S FR 50948
U Anti-allergic

8561   (7010) $C_{19}H_{20}N_6O$
27885-92-3

3,3'-Di-2-imidazolin-2-ylcarbanilide = 1,3-Bis[3-(2-imi-
dazolin-2-yl)phenyl]urea = N,N'-Bis[3-(4,5-dihydro-1H-
imidazol-2-yl)phenyl]urea (●)
R   Dihydrochloride (5318-76-3)
S   4 A 65, *Imidocarb hydrochloride**, Imizocarb
U   Antiprotozoal (*Babesia*)

8561-01   (7010-01)   55750-06-6
R   Dipropionate
S   Carbésia, *Imidocarb dipropionate**,
Imizad Equine Inj., Imizol
U   Antiprotozoal (*Babesia*)

8562 $C_{19}H_{20}N_6O_3$
93522-20-4

[5-[[4-(2-Pyridinyl)-1-piperazinyl]carbonyl]-1H-benz-
imidazol-2-yl]carbamic acid methyl ester (●)
S   CDRI 81/470
U   Antischistosomal

8563   (7011) $C_{19}H_{20}N_8O_5$
54-62-6

N-[p-[[(2,4-Diamino-6-pteridinyl)methyl]amino]ben-
zoyl]-L-glutamic acid = 4-Aminopteroylglutamic acid =
N-[4-[[(2,4-Diamino-6-pteridinyl)methyl]amino]ben-
zoyl]-L-glutamic acid (●)
R   also disodium salt (58602-66-7)
S   *4-Amino-4-deoxyfolic acid, 4-Aminofolsäure, 4-Ami-
no-PGA, Aminopterin**, 4-Aminopteroylglutamic*

*acid, 4-Aminopteroylglutaminsäure, AMT, Antifolic
acid, Antifolsäure, APGA, NSC-739, Pteramina,
SK 1072*
U   Antineoplastic (antimetabolite)

8564   (7012) $C_{19}H_{20}O_2$
6033-98-3

p-[1,2-Diethylidene-2-(p-methoxyphenyl)ethyl]phenol =
3,4-Bis(p-hydroxyphenyl)-2,4-hexadiene monomethyl
ether = 4-[1-Ethylidene-2-(4-methoxyphenyl)-2-bute-
nyl]phenol (●)
S   Depot-Dienol forte, *Dienestrol monomethyl ether,
Dienöstrolmonomethyläther*
U   Synthetic estrogen

8565   (7013) $C_{19}H_{20}O_3$
35825-57-1

(R)-1,2,6,7,8,9-Hexahydro-1,6,6-trimethylphenan-
thro[1,2-b]furan-10,11-dione (●)
S   *Cryptotanshinone*
U   Antibacterial from *Salvia miltiorrhiza* (Dan-Shen)

8566   (7014) $C_{19}H_{20}O_5$
60539-15-3

4-(5-Benzyl-2,4-dimethoxyphenyl)-4-oxobutyric acid =
3-(5-Benzyl-2,4-dimethoxybenzoyl)propionic acid = 2,4-
Dimethoxy-γ-oxo-5-(phenylmethyl)benzenebutanoic acid
(●)
S Diphébyl
U Choleretic

8567 (7015)

$C_{19}H_{20}O_5$
131420-91-2

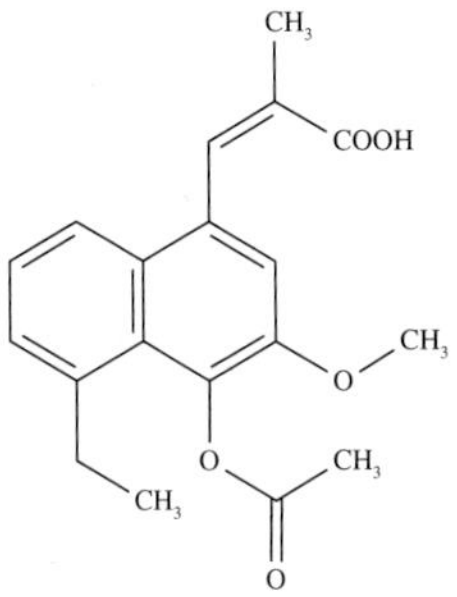

(Z)-3-[4-(Acetyloxy)-5-ethyl-3-methoxy-1-naphthale-
nyl]-2-methylpropenoic acid (●)
S E-5090
U Anti-inflammatory

8568 (7016)

$C_{19}H_{20}O_5$
76554-66-0

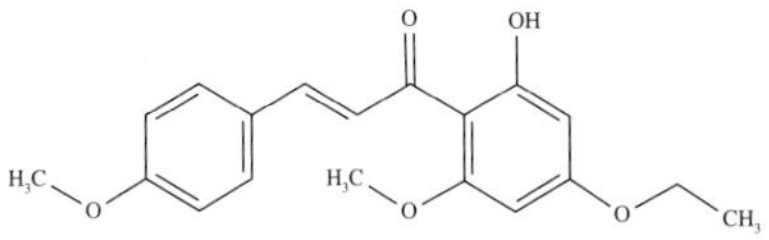

4'-Ethoxy-2'-hydroxy-4,6'-dimethoxychalcone = 1-(4-
Ethoxy-2-hydroxy-6-methoxyphenyl)-3-(4-methoxyphe-
nyl)-2-propen-1-one (●)
S Ro 9-0410
U Antiviral

8569

$C_{19}H_{20}O_7$
210419-07-1

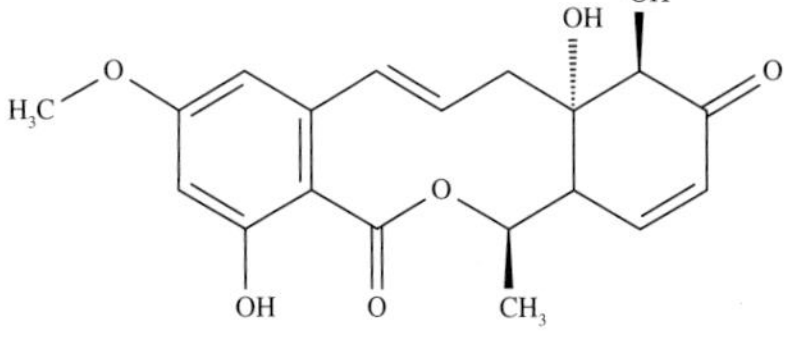

(1R,5R,12E,14aS)-rel-4a,5,14,14a-Tetrahydro-1,8,14a-
trihydroxy-10-methoxy-5-methyl-2H-dibenz[c,h]oxecin-
2,7(1H)-dione (●)
S Ro 09-2210
U MEK inhibitor from *Curvularia* sp.

8570 (7017)

$C_{19}H_{20}O_7$
13017-11-3

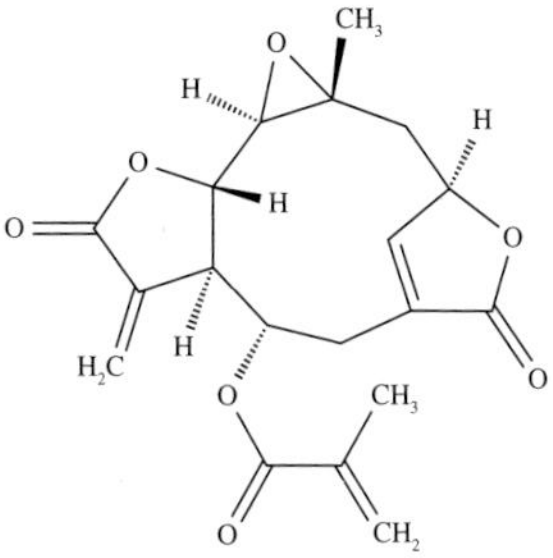

2-Methyl-2-propenoic acid [1aR-
(1aR*,3R*,8S*,8aR*,11aS*,11bS*)]-
1a,2,3,7,8,8a,9,10,11a,11b-decahydro-1a-methyl-9-me-
thylene-5,10-dioxo-5H-3,6-metheno[2,3-f]oxireno[d]oxa-
cycloundecin-8-yl ester (●)
S *Elephantopin*, NSC-100046
U Cytostatic (sesquiterpene)

8571 (7018)

$C_{19}H_{20}O_8$
4106-97-2

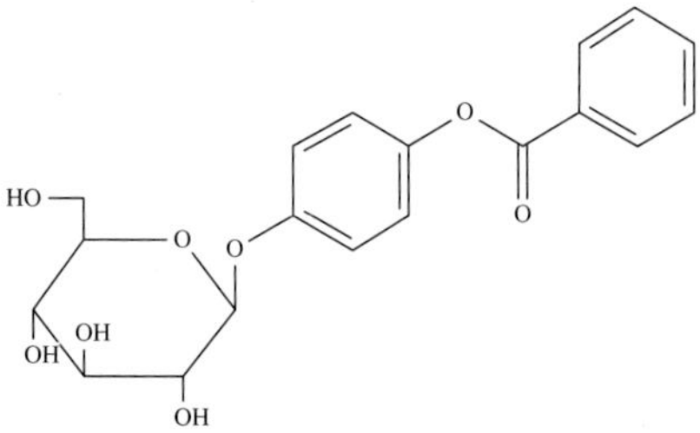

p-Glucosyloxyphenyl benzoate = Hydroquinone β-D-glucopyranoside benzoate = Monobenzoylarbutin = 4-(Benzoyloxy)phenyl β-D-glucopyranoside (●)
S Cellotropin
U Anti-infective

8572   (7019) $C_{19}H_{20}O_9$
                                                 18700-78-2

4a,9a-Epoxy-3-(2,3-epoxybutyryl)-1,2,3,4,4a,9a-hexahy-
dro-1,3,4,5,10-pentahydroxy-2-methylanthrone = 1,2,3,4-
Tetrahydro-1,3,4,5,10-pentahydroxy-2-methyl-3-[(3-me-
thyloxiranyl)carbonyl]-4a,9a-epoxyanthracen-9(10*H*)-
one (●)
S   *Cervicarcin*, NSC-65380
U   Antineoplastic antibiotic from *Streptomyces ogaensis*

8573   (7020) $C_{19}H_{20}O_{10}$
                                                 17226-75-4

7-Hydroxymethyl-4-methoxy-5*H*-furo[3,2-*g*][1]benzopy-
ran-5-one glucoside = 5-Methoxy-2-hydroxymethylfu-
ro[3,2-*g*]chromone β-D-glucoside = 7-[(β-D-Glucopyranosyloxy)methyl]-4-methoxy-5*H*-furo[3,2-*g*][1]benzopy-
ran-5-one (●)
S   *Khellinin, Khellol glucoside, Khelloside**, Kille
U   Coronary vasodilator

8574 $C_{19}H_{21}ClFNO$
                                                 64169-45-5

5-Chloro-1-[3-(dimethylamino)propyl]-1-(4-fluorophe-
nyl)phthalan = 5-Chloro-1-(4-fluorophenyl)-1,3-dihydro-
N,N-dimethyl-1-isobenzofuranpropanamine (●)
S   Lu 10-134C

U   Serotonin reuptake inhibitor

8575   (7021) $C_{19}H_{21}ClN_2$
                                                 6196-08-3

3-Chloro-11-[3-(dimethylamino)propylidene]-5,6-dihy-
dro-11*H*-dibenz[*b,e*]azepine = 3-Chloro-11-[3-(dimethyl-
amino)propylidene]-5,6-dihydromorphanthridine = 3-(3-
Chloro-5,6-dihydro-11*H*-dibenz[*b,e*]azepin-11-ylidene)-
N,N-dimethyl-1-propanamine (●)
S   *Elanzepine***, W-A 363
U   Antidepressant

8576   (7022) $C_{19}H_{21}ClN_2O$
                                                 86187-86-2

5-Amino-4'-chloro-2-(4-methylpiperidino)benzophe-
none = [5-Amino-2-(4-methyl-1-piperidinyl)phenyl](4-
chlorophenyl)methanone (●)
S   LF 1695, Modulim, SF 1695
U   Immunomodulator

8577   (7023) $C_{19}H_{21}ClN_2OS$
                                                 15599-36-7

5-Chloro-2-[*p*-(β-diethylaminoethoxy)phenyl]benzothia-
zole = 2-[4-(5-Chlorobenzothiazol-2-yl)phenoxy]ethyl-
diethylamine = 2-[4-(5-Chloro-2-benzothiazolyl)phen-
oxy]-*N,N*-diethylethanamine (●)
R also hydrochloride
S Episol "Ceookes", *Haletazole***, *Halethazole*
U Antiseptic, antifungal

8578 (7024)   C$_{19}$H$_{21}$ClN$_2$OS
800-22-6

2-Chloro-10-[3-(diethylamino)propionyl]phenothiazine =
2-Chloro-10-(*N,N*-diethyl-β-alanyl)phenothiazine = 2-
Chloro-10-[3-(diethylamino)-1-oxopropyl]-10*H*-pheno-
thiazine (●)
R also monohydrochloride (1045-82-5)
S *Chloracizine*, *Chloracyzine***, *Chlorazisin*, G-020
U Coronary vasodilator, spasmolytic

8579 (7025)   C$_{19}$H$_{21}$ClN$_2$O$_2$
89845-16-9

Ethyl [4-(*p*-chlorophenyl)-1,2,3,4-tetrahydro-2-methyl-8-
isoquinolyl]carbamate = [4-(4-Chlorophenyl)-1,2,3,4-te-
trahydro-2-methyl-8-isoquinolinyl]carbamic acid ethyl
ester (●)
R Monohydrochloride (89845-17-0)
S AN$_5$, DZO-200, Gastrophenzine
U Anti-ulcer agent

8580 (7026)   C$_{19}$H$_{21}$ClN$_2$O$_2$S
138977-28-3

N-[2-(4-Chlorophenyl)ethyl]-1,3,4,5-tetrahydro-7,8-di-
hydroxy-2*H*-benzazepine-2-carbothioamide (●)
S *Capsazepine*
U Capsaicin antagonist

8581 (7027)   C$_{19}$H$_{21}$ClN$_2$O$_3$
70541-17-2

2-Benzoyl-4'-chloro-2-[(2-hydroxyethyl)methylamino]-
N-methylacetanilide = 2'-Benzoyl-4'-chloro-*N*'-(2-hydr-
oxyethyl)-*N,N*-dimethylglycinanilide = *N*-(2-Benzoyl-4-
chlorophenyl)-2-[(2-hydroxyethyl)methylamino]-*N*-me-
thylacetamide (●)
S *Oxazafone***
U Anxiolytic

8582 (7028)   C$_{19}$H$_{21}$ClN$_2$S
4789-63-3

2-Chloro-10-(4-methyl-1-piperazinyl)-10,11-dihydrodi-benzo[b,f]thiepin = 1-(2-Chloro-10,11-dihydrodiben-zo[b,f]thiepin-10-yl)-4-methylpiperazine (●)
S Doclothepin, VÚFB-10030
U Neuroleptic

8583 (7029) $C_{19}H_{21}ClN_2S$
13448-22-1

8-Chloro-10-(4-methyl-1-piperazinyl)-10,11-dihydrodi-benzo[b,f]thiepine = 1-(8-Chloro-10,11-dihydrodiben-zo[b,f]thiepin-10-yl)-4-methylpiperazine (●)
R also maleate (1:1) (4789-68-8) or dimethanesulfonate (51327-13-0)
S Clopiben, *Clorotepine**, Clorothepin, Clotepin, Clothepin, *Octoclothepine*, VÚFB-6281
U Neuroleptic

8584 (7030) $C_{19}H_{21}ClN_4O_5$
54504-70-0

2-(p-Chlorophenoxy)-2-methylpropionic acid ester with 7-(2-hydroxyethyl)theophylline = β-(7-Theophylli-nyl)ethyl 2-(p-chlorophenoxy)-2-methylpropionate = 2-(4-Chlorophenoxy)-2-methylpropanoic acid 2-(1,2,3,6-tetrahydro-1,3-dimethyl-2,6-dioxo-7H-purin-7-yl)ethyl ester (●)
S Duolip, *Etofylline clofibrate**, Etolip "Spofa", ML 1024, *Theofibrate*
U Antihyperlipidemic

8585 (7031) $C_{19}H_{21}ClN_4O_5$
70788-27-1

2-(p-Chlorophenoxy)-2-methylpropyl 1,2,3,6-tetrahydro-1,3-dimethyl-2,6-dioxopurine-7-acetate = 7-Theophylli-neacetic acid 2-(p-chlorophenoxy)-2-methylpropyl ester = 1,2,3,6-Tetrahydro-1,3-dimethyl-2,6-dioxo-7H-purine-7-acetic acid 2-(4-chlorophenoxy)-2-methylpropyl ester (●)
S *Acefylline clofibrol**, Fibrafyllin*
U Antihyperlipidemic

8586 (7032) $C_{19}H_{21}Cl_2N_3O_2$
17173-85-2

2-[p-[Bis(2-chloroethyl)amino]phenyl]-4'-carbamoyl-acetanilide = N-[4-(Aminocarbonyl)phenyl]-4-[bis(2-chloroethyl)amino]benzeneacetamide (●)
R see also no. 8520
S Pafencil, Paphencyl, Paraphenacyl
U Antineoplastic

8587 (7033) $C_{19}H_{21}FN_2O_2$
130579-75-8

(E)-2'-Fluoro-4-hydroxychalcone (Z)-O-[2-(dimethyl-amino)ethyl]oxime = (Z,E)-1-(2-Fluorophenyl)-3-(4-hydroxyphenyl)-2-propen-1-one O-[2-(dimethylami-no)ethyl]oxime (●)
R Fumarate (2:1) (salt) (130580-02-8)
S *Eplivanserin fumarate**, SR 46349 B
U 5HT$_2$-receptor antagonist

8588   (7034)

$C_{19}H_{21}FN_2O_4$
124858-35-1

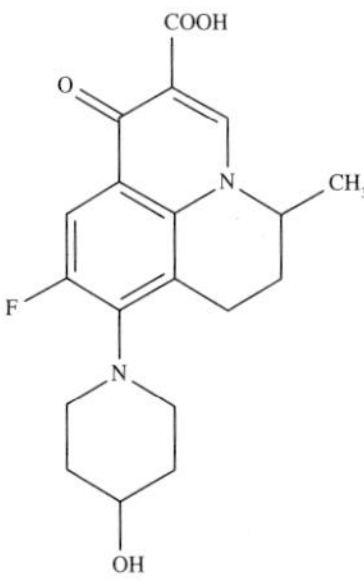

(±)-9-Fluoro-6,7-dihydro-8-(4-hydroxypiperidino)-5-me-
thyl-1-oxo-1H,5H-benzo[ij]quinolizine-2-carboxylic acid
= (±)-9-Fluoro-6,7-dihydro-8-(4-hydroxy-1-piperidinyl)-
5-methyl-1-oxo-1H,5H-benzo[ij]quinolizine-2-carb-
oxylic acid (●)
R   also without (±)-definition (81962-84-7)
S   Acuatim, Aquatim, *Nadifloxacin***, NDFX,
        OPC-7251, *Zinofloxacin*
U   Antibacterial

8589   (7035)

$C_{19}H_{21}FN_2S$
19905-05-6

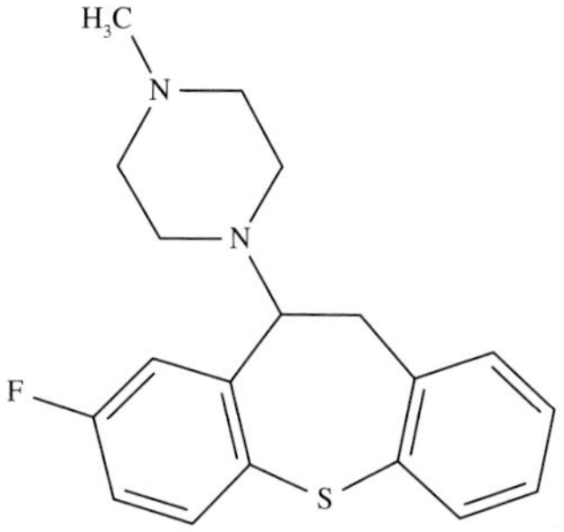

1-(8-Fluoro-10,11-dihydrodibenzo[b,f]thiepin-10-yl)-4-
methylpiperazine (●)
S   *Fluothepine*
U   Neuroleptic

8590

$C_{19}H_{21}FN_4$
75970-99-9

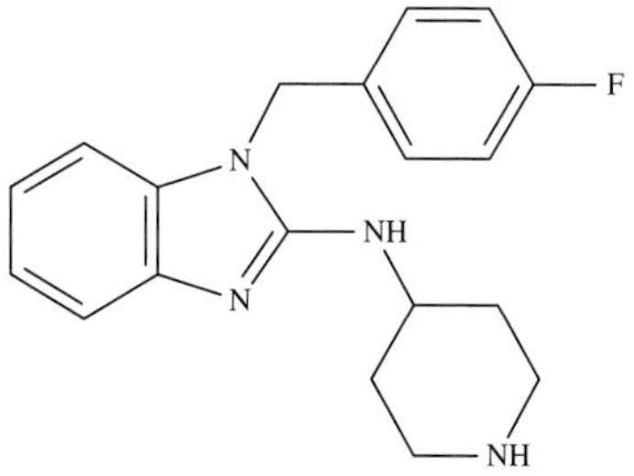

1-[(4-Fluorophenyl)methyl]-N-4-piperidinyl-1H-benz-
imidazol-2-amine (●)
R   Dihydrobromide (75970-64-8)
S   *Norastemizole hydrobromide*, R 41232
U   Anti-allergic

8591

$C_{19}H_{21}FN_4O_3$
162301-05-5

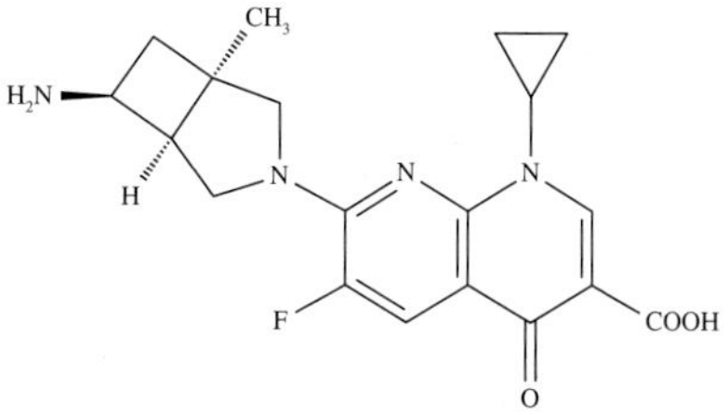

(1α,5α,6β)-(+)-7-(6-Amino-1-methyl-3-azabicy-
clo[3.2.0]hept-3-yl)-1-cyclopropyl-6-fluoro-1,4-dihydro-
4-oxo-1,8-naphthyridine-3-carboxylic acid (●)
R   Monohydrochloride (162424-67-1)
S   CFC 222, *Ecenofloxacin hydrochloride***
U   Antibacterial

8592   (7036)

$C_{19}H_{21}F_3N_2S$
2622-37-9

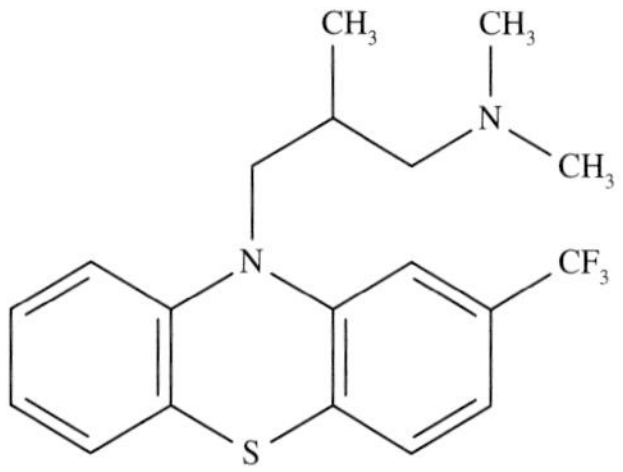

10-[3-(Dimethylamino)-2-methylpropyl]-2-(trifluoromethyl)phenothiazine = Dimethyl[2-methyl-3-[2-(trifluoromethyl)-10-phenothiazinyl]propyl]amine = N,N,β-Tri-

methyl-2-(trifluoromethyl)-10*H*-phenothiazine-10-pro-
panamine (●)
R also maleate (1:1) (71609-19-3)
S Nortan "Norden", Nortran, SKF 5354-A, *Trifluome-*
*prazine***, Triflutrimeprazine*
U Neuroleptic

8593 $C_{19}H_{21}F_3N_6O_2$
                                         152939-42-9

4-Amino-2-(4,4-dimethyl-2-oxo-1-imidazolidinyl)-*N*-
ethyl-α,α,α-trifluoro-5-pyrimidinecarboxy-*m*-toluidide =
4-Amino-2-(4,4-dimethyl-2-oxo-1-imidazolidinyl)-*N*-
ethyl-*N*-[3-(trifluoromethyl)phenyl]-5-pyrimidinecarbox-
amide (●)
S *Opanixil***
U Antihyperlipidemic

8594 (7037) $C_{19}H_{21}N$
                                         7395-90-6

N,N-Dimethyl-1-phenylindene-1-ethylamine = 1-[2-(Di-
methylamino)ethyl]-1-phenylindene = *N,N*-Dimethyl-1-
phenyl-1*H*-indene-1-ethanamine (●)
R Hydrochloride (2988-32-1)
S *Indriline hydrochloride***, Lu 3-083, MJ 1986
U Stimulant (central)

8595 (7038) $C_{19}H_{21}N$
                                         72-69-5

10,11-Dihydro-*N*-methyl-5*H*-dibenzo[*a,d*]cycloheptene-
$\Delta^{5,\gamma}$-propylamine = 10,11-Dihydro-5-(3-methylamino-
propylidene)-5*H*-dibenzo[*a,d*]cycloheptene = *N*-Methyl-
3-(dibenzo[*a,d*]-1,4-cycloheptadien-5-ylidene)propyl-
amine = 3-(10,11-Dihydro-5*H*-dibenzo[*a,d*]cyclohepten-
5-ylidene)propyl(methyl)amine = 3-(10,11-Dihydro-5*H*-
dibenzo[*a,d*]cyclohepten-5-ylidene)-*N*-methyl-1-propan-
amine (●)
S *Desitriptyline, Desmethylamitriptyline,* ELF 101,
*Nortriptilina, Nortriptyline***
U Antidepressant

8595-01 (7038-01)   894-71-3
R Hydrochloride
S Acetexa, Allegron, Altilev, Apo-Nortriptyline, Ate-
ben, Avantyl, Aventyl, E.L.F. 101, G 38048, Kareon
"Ima", Lilly 38489, Martimil, Motipres, N 7048, No-
ridyl "Gepepharm", Noritren, Norpress "Pacific",
Nortab(s), Nortrilen, Nortrilin, Nortrip "Protea",
Nortrix, Nortyl, Nortylin, Norzepine, NSC-169453,
Pamelor, Paxtibi, Psychostyl, Sensaval, Sensival,
Sintyben, Vividyl
U Antidepressant

8596 (7039) $C_{19}H_{21}N$
                                         438-60-8

N-Methyl-5*H*-dibenzo[*a,d*]cycloheptene-5-propylamine
= 5-(3-Methylaminopropyl)-5*H*-dibenzo[*a,d*]cyclohep-
tene = 3-(5*H*-Dibenzo[*a,d*]cyclohepten-5-yl)propyl(methyl)amine = *N*-Methyl-5*H*-dibenzo[*a,d*]cycloheptene-5-
propanamine (●)
R Hydrochloride (1225-55-4)

S *Amimethyline hydrochloride*, Anelun, Concordin, Maximed, MK-240, *Protriptyline hydrochloride***, Triptil, Vivactil "M.S.D."

U Antidepressant

8597 (7040) $C_{19}H_{21}N$
                                                                    5370-41-2

3-(Diphenylmethylene)-1-ethylpyrrolidine (●)

R Hydrochloride (23239-78-3)

S AHR-1118, *Pridefine hydrochloride***

U Antidepressant

8598 (7041) $C_{19}H_{21}NO$
                                                                    1668-19-5

N,N-Dimethyldibenz[b,e]oxepin-$\Delta^{11(6H),\gamma}$-propylamine = 11-(3-Dimethylaminopropylidene)dibenz[b,e]oxepin = 3-(Dibenz[b,e]oxepin-11-ylidene)propyldimethylamine = 3-Dibenz[b,e]oxepin-11(6H)-ylidene-N,N-dimethyl-1-propanamine (●)

R Hydrochloride [*cis-trans*-mixture of approximately (1:5) (1229-29-4)]

S Adapin, Alti-Doxepin, Anten, Apo-Doxepin, Aponal, Co-Dox, Curatin "Pfizer", Deptran "Alphapharm", Desidox, Desidox epin, Doksapan, Dolat, Doneurin, Doxal "Orion", Doxecan, Doxederm, Doxedyn, *Doxepin hydrochloride***, Doxetar, Gilex, Kenral-Doxepin, Mareen, MF 10, Novodoxepin, Novoxapin, NSC-108160, P-3693 A, Poldoxin, Quita-

xon, Rho-Doxepin, Serecan, Sinequan, Sinquan, Spectra, Sumikang, Toruan, Triadapin, Xepin, Zonalon

U Antidepressant, antipruritic

8599 (7042) $C_{19}H_{21}NO$
                                                                    3607-18-9

(Z)-N,N-Dimethyldibenz[b,e]oxepin-$\Delta^{11(6H),\gamma}$-propylamine = (Z)-3-Dibenz[b,e]oxepin-11(6H)-ylidene-N,N-dimethyl-1-propanamine (●)

R Hydrochloride (25127-31-5)

S *Cidoxepin hydrochloride***, P-4599

U Antidepressant

8600 (7043) $C_{19}H_{21}NO$
                                                                    59143-05-4

3-p-Tolylspiro[isobenzofuran-1,4'-piperidine] = 3-(4-Methylphenyl)spiro[isobenzofuran-1(3H),4'-piperidine] (●)

S HRP 197

U Antidepressant

8601 $C_{19}H_{21}NO_2$
                                                           104422-04-0

S   *Beloxepin***, Org 4428
U   Antidepressant

(±)-3-Allyl-2,3,4,5-tetrahydro-7,8-dihydroxy-1-phenyl-
1*H*-3-benzazepine = 2,3,4,5-Tetrahydro-1-phenyl-3-(2-
propenyl)-1*H*-3-benzazepine-7,8-diol (●)
R   Hydrochloride (62751-58-0)
S   SKF 77434
U   Dopamine D$_1$ agonist

8602   (7045) $C_{19}H_{21}NO_2$
                                                           29541-85-3

2-[(10,11-Dihydro-5*H*-dibenzo[*a,d*]cyclohepten-5-
yl)oxy]-*N,N*-dimethylacetamide (●)
S   B.S. 7679, *Oxitriptyline***
U   Anticonvulsant

8603 $C_{19}H_{21}NO_2$
                                                           135928-30-2

(4a*R*,13b*R*)-*rel*-1,3,4,13b-Tetrahydro-2,10-dimethyldi-
benz[2,3:6,7]oxepino[4,5-*c*]pyridin-4a(2*H*)-ol (●)

8604   (7048) $C_{19}H_{21}NO_2$
                                                           23623-36-1

3-Cyanatoestra-1,3,5(10)-trien-17-one (●)
S   *Estrone cyanate*, Östroncyanat
U   Estrogen

8605   (7046) $C_{19}H_{21}NO_2$
                                                           39787-47-8

cis-2-(Dimethylamino)-1,2,3,4-tetrahydro-1-naphthyl
benzoate = *cis*-1-(Benzoyloxy)-2-(dimethylamino)-
1,2,3,4-tetrahydronaphthalene = *cis*-2-(Dimethylamino)-
1,2,3,4-tetrahydro-1-naphthalenol benzoate (ester) (●)
S   *Benamocaine*, YAU-17
U   Gastricmucosa anesthetic

8606   (7047) $C_{19}H_{21}NO_2$
                                                           40077-13-2

2-[β-(4-Hydroxyphenyl)ethylaminomethyl]tetralone =
3,4-Dihydro-2-[[[2-(4-hydroxyphenyl)ethyl]amino]me-
thyl]-1(2*H*)-naphthalenone (●)
R   Hydrochloride (30007-39-7)

S   BE 2254, HEAT
U   α-Adrenoceptor blocker, hypothermic

8607   (7044)   C₁₉H₂₁NO₂
                                                            6538-22-3

2-Methoxy-N,N-dimethylxanthene-Δ⁹,ᵞ-propylamine = 9-
(3-Dimethylaminopropylidene)-2-methoxyxanthene = 3-
(2-Methoxy-9H-xanthen-9-ylidene)-N,N-dimethyl-1-pro-
panamine (●)
S   *Dimeprozan***, *Dimeprozinum*
U   Tranquilizer

8608   C₁₉H₂₁NO₃
                                                            72527-29-8

3,4-Dihydro-6,7-dimethoxy-1-[(4-methoxyphenyl)me-
thyl]isoquinoline (●)
S   GS 386
U   Calcium antagonist

8609   (7051)   C₁₉H₂₁NO₃
                                                            62-67-9

17-Allyl-7,8-didehydro-4,5α-epoxymorphinan-3,6α-diol
= 17-Allyl-4,5α-epoxymorphin-7-ene-3,6-diol = N-Allyl-
normorphine = (–)-(5R,6S)-9a-Allyl-4,5-epoxymorphin-
7-ene-3,6-diol = (5α,6α)-7,8-Didehydro-4,5-epoxy-17-
(2-propenyl)morphinan-3,6-diol (●)
R   also hydrochloride (57-29-4) or hydrobromide
        (1041-90-3);
        see also nos. 11766, 14441
S   Acetorphin, Allorphin, *Allylnormorphine*, Anarcon,
        Antorfin, Antorphin, Cloridrato de Nalorfina, Lethid-
        rone, Lithidrone, Miromorfalil, Nallin(e), *Nalor-
        fin(a)*, *Nalorphine***, Norfin "Lusofarmaco"
U   Antagonist to narcotics

8610   (7050)   C₁₉H₂₁NO₃
                                                            115-37-7

4,5α-Epoxy-3,6-dimethoxy-N-methyl-6,8-morphina-
diene = (5α)-6,7,8,14-Tetradehydro-4,5-epoxy-3,6-di-
methoxy-17-methylmorphinan (●)
S   *Paramorphine, Tebaina, Thebaine*
U   Opium alkaloid

8611   (7052)   C₁₉H₂₁NO₃
                                                            136100-14-6

4-Phenoxyphenyl 4-methyl-1-piperidinecarboxylate = 4-
Methyl-1-piperidinecarboxylic acid 4-phenoxyphenyl es-
ter (●)
S   WAY-121898
U   Coronary heart disease risk inhibitor

8612 (7053)

$C_{19}H_{21}NO_3$
119518-25-1

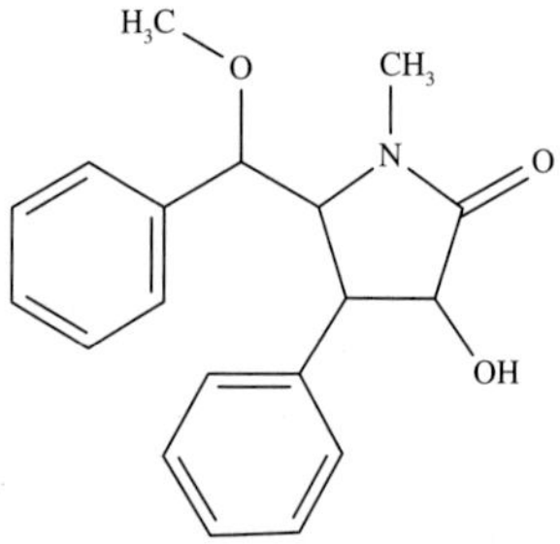

3-Hydroxy-5-(α-methoxybenzyl)-1-methyl-4-phenyl-2-pyrrolidinone = 3-Hydroxy-5-(methoxyphenylmethyl)-1-methyl-4-phenyl-2-pyrrolidinone (●)
S *Lansimide*
U Spasmolytic from *Clausena lansium*

8613 (7049)

$C_{19}H_{21}NO_3$
47254-05-7

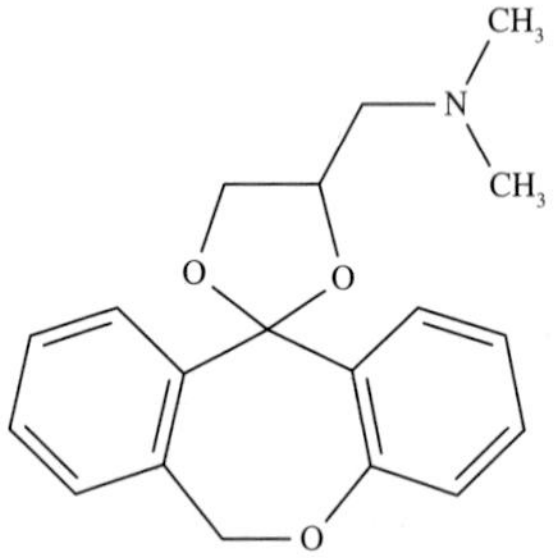

N,N-Dimethylspiro[dibenz[*b,e*]oxepin-11(6*H*),2'-[1,3]di-oxolane]-4'-methylamine = *N,N*-Dimethylspiro[di-benz[*b,e*]oxepin-11(6*H*),2'-[1,3]dioxolane]-4'-methan-amine (●)
S *Spiroxepin**
U Antidepressant

8614 (7054)

$C_{19}H_{21}NO_3S$
96335-13-6

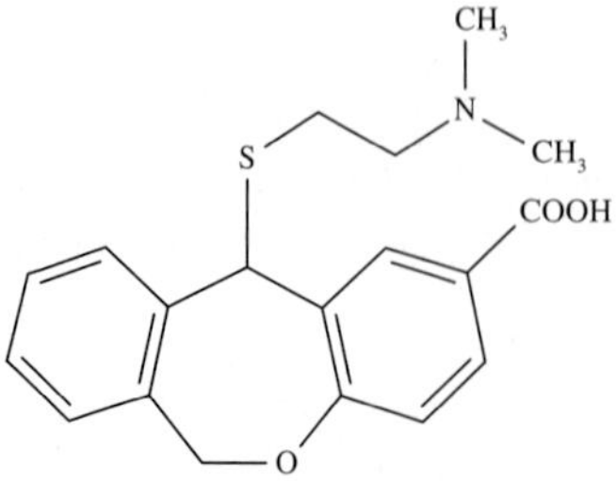

8615 (7055)

$C_{19}H_{21}NO_4$
113079-40-6

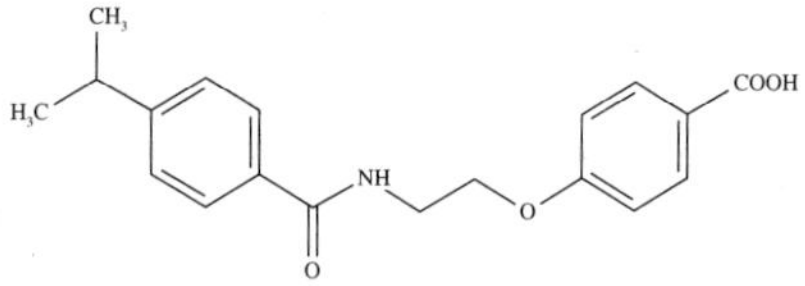

11-[[2-(Dimethylamino)ethyl]thio]-6,11-dihydrodi-benz[*b,e*]oxepine-2-carboxylic acid (●)
S KW 4994
U Anti-allergic, antihistaminic

p-[2-(*p*-Isopropylbenzamido)ethoxy]benzoic acid = 4-[2-[[4-(1-Methylethyl)benzoyl]amino]ethoxy]benzoic acid (●)
S NS-1, OCO-1112
U Hypolipidemic

8616 (7059)

$C_{19}H_{21}NO_4$
3019-51-0

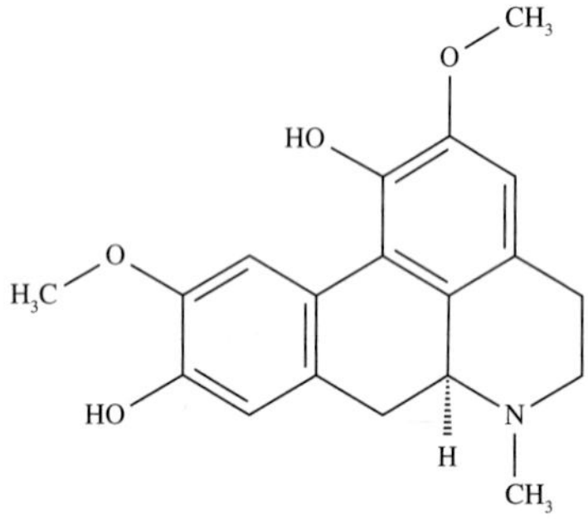

2,10-Dimethoxy-6aα-aporphine-1,9-diol = (*S*)-5,6,6a,7-Tetrahydro-2,10-dimethoxy-6-methyl-4*H*-diben-zo[*de,g*]quinoline-1,9-diol (●)
S *Isoboldine*
U Alkaloid from *Nandina domestica* and other plants

8616-01 (7059-01)

79979-14-9

R Mixture with bracteoline (no. 8617)
S *Isotheoline, Izoteolin*
U Cerebral vasodilator, antihypertensive

8617 (7058) $C_{19}H_{21}NO_4$
25651-04-1

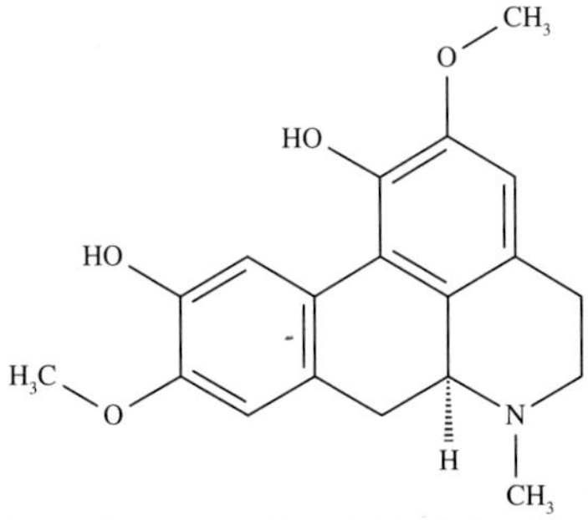

2,9-Dimethoxy-6aα-aporphine-1,10-diol = (S)-5,6,6a,7-
Tetrahydro-2,9-dimethoxy-6-methyl-4H-diben-
zo[de,g]quinoline-1,10-diol (●)
R see also no. 8616-01
S *Bracteoline*
U Alkaloid from *Papaver bracteatum* and *P. orientale*

8618 (7057) $C_{19}H_{21}NO_4$
476-70-0

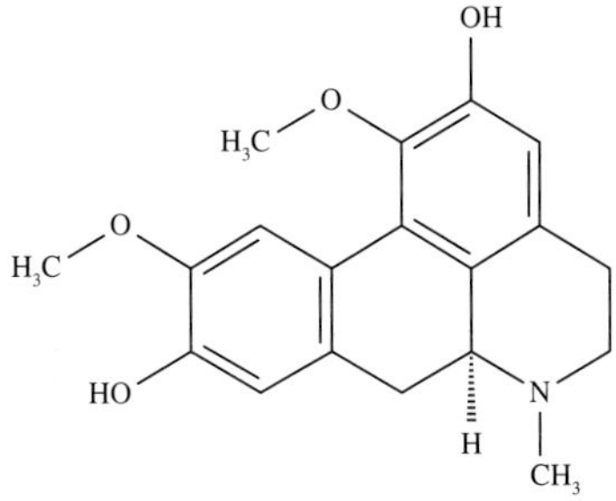

1,10-Dimethoxy-6aα-aporphine-2,9-diol = 3,5-Dime-
thoxyaporphin-2,6-diol = (S)-5,6,6a,7-Tetrahydro-1,10-
dimethoxy-6-methyl-4H-dibenzo[de,g]quinoline-2,9-diol
(●)
R see also no. 12722-01
S Boldicina, *Boldine*, Uniboldina
U Choleric, laxative (alkaloid from *Peumus boldus*)

8619 (7060) $C_{19}H_{21}NO_4$
16562-13-3

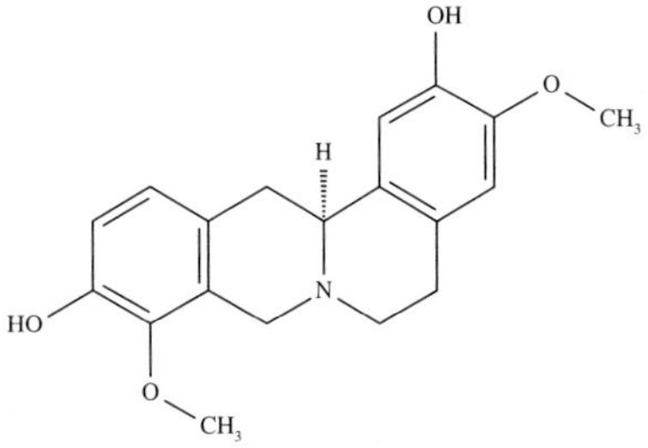

3,9-Dimethoxy-13aα-berbine-2,10-diol = (S)-5,8,13,13a-
Tetrahydro-3,9-dimethoxy-6H-dibenzo[a,g]quinolizine-
2,10-diol (●)
S *l-Stepholidine*
U Analgesic, antidyskinetic

8620 $C_{19}H_{21}NO_4$
1356-73-6

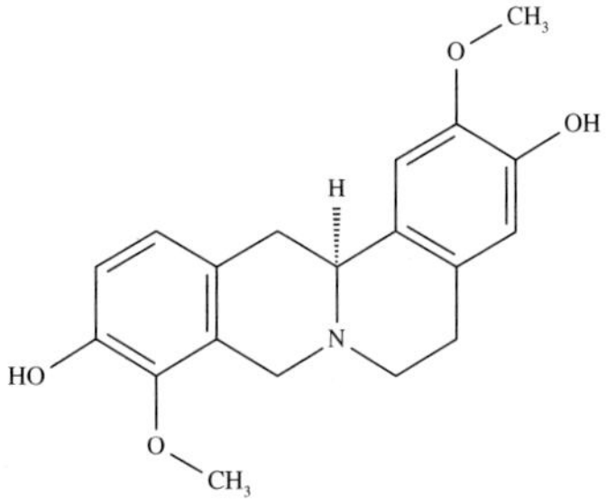

(13aS)-5,8,13,13a-Tetrahydro-2,9-dimethoxy-6H-diben-
zo[a,g]quinolizine-3,10-diol (●)
S *Aequaline, Discretamine*
U α-Adrenoceptor, 5-HT receptor antagonist from *Fis-
sistigma glaucescens*

8621 (7056) $C_{19}H_{21}NO_4$
465-65-6

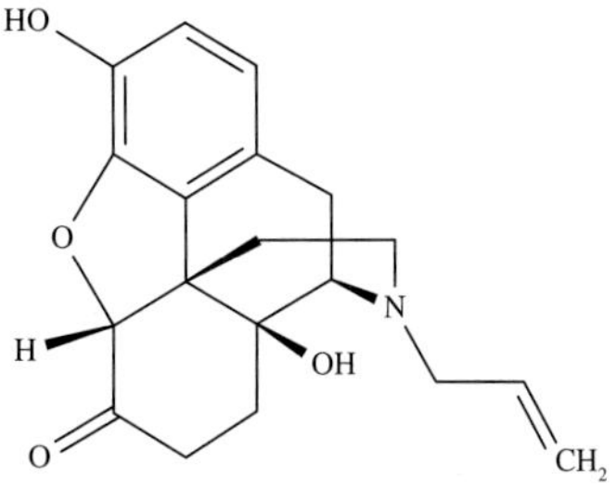

(–)-17-Allyl-4,5α-epoxy-3,14-dihydroxymorphinan-6-
one = (–)-N-Allyl-14-hydroxynordihydromorphinone =

(−)-12-Allyl-7,7a,8,9-tetrahydro-3,7a-dihydroxy-4aH-
8,9c-iminoethanophenanthro[4,5-bcd]furan-5(6H)-one =
(−)-(5R,14S)-9a-Allyl-4,5-epoxy-3,14-dihydroxymorphi-
nan-6-one = (5α)-4,5-Epoxy-3,14-dihydroxy-17-(2-pro-
penyl)morphinan-6-one (●)
R Hydrochloride (357-08-4) [or hydrochloride dihydrate (51481-60-8)]
S EN-1530, EN-15304, Intrenon, Min-I-Jet Naloxone,
Nalokson, Nalone(e), Naloselect, *Naloxone cloridrato*, *Naloxone hydrochloride***, Narcan, Narcanti, Na-
xan, Zynox
U Antagonist to narcotics

8622   C₁₉H₂₁NO₄
                                               135204-83-0

(5R)-3-[6-(Cyclopropylmethoxy)-2-naphthalenyl]-5-
(methoxymethyl)-2-oxazolidinone (●)
S T-794
U Antidepressant (MAO inhibitor)

8623   C₁₉H₂₁NO₄
                                               37933-99-6

3-[2-Hydroxy-3-[(1-methylethyl)amino]propoxy]-9H-
xanthen-9-one (●)
S Xanthonolol
U β-Adrenergic blocker, calcium antagonist

8624   C₁₉H₂₁NO₅
                                               137275-81-1

5-[3-[[(2S)-1,4-Benzodioxan-2-ylmethyl]amino]prop-
oxy]-1,3-benzodioxole = (2S)-N-[3-(1,3-Benzodioxol-5-
yloxy)propyl]-2,3-dihydro-1,4-benzodioxin-2-methan-
amine (●)
R Hydrochloride (137275-80-0)
S Esmozotan hydrochloride, MCI-242, MKC-242
U Anxiolytic, antidepressant (5-HT₁A receptor agonist)

8625   (7061)   C₁₉H₂₁NO₅S
                                               89163-44-0

N-Acetyl-S-[2-(6-methoxy-2-naphthyl)propionyl]cys-
teine = (S)-N-Acetyl-L-cysteine 6-methoxy-α-methyl-2-
naphthaleneacetate (ester) (●)
S Cinaproxen**
U Anti-inflammatory, analgesic, antipyretic, mucolytic

8626   C₁₉H₂₁NO₅S
                                               96108-55-3

[R-(R*,S*)]-N-[2-[[2-(6-Methoxy-2-naphthalenyl)-1-
oxopropyl]thio]-1-oxopropyl]glycine (●)
S Naproxen tiopronin ester, Naproxen tioproninate
U Anti-inflammatory

8627   (7062) $C_{19}H_{21}NO_6$
90729-41-2

Ethyl methyl 1,4-dihydro-2,6-dimethyl-4-(2,3-methy-lenedioxyphenyl)-3,5-pyridinedicarboxylate = 4-(1,3-Benzodioxol-4-yl)-1,4-dihydro-2,6-dimethyl-3,5-pyri-dinedicarboxylic acid ethyl methyl ester (●)
S   IQB-837-v, *Oxodipine***
U   Calcium antagonist (antihypertensive)

8628   (7063) $C_{19}H_{21}NS$
113-53-1

N,N-Dimethyldibenzo[*b,e*]thiepin-$\Delta^{11(6H),\gamma}$-propyl-amine = 11-(3-Dimethylaminopropylidene)diben-zo[*b,e*]thiepin = 5-(3-Dimethylaminopropylidene)-10,11-dihydro-11-thiadibenzo[*a,c*]cycloheptene = 3-(Diben-zo[*b,e*]thiepin-11-ylidene)propyldimethylamine = 3-Di-benzo[*b,e*]thiepin-11(6*H*)-ylidene-*N,N*-dimethyl-1-pro-panamine (●)
R   Hydrochloride (897-15-4)
S   Altapin, Depresym, *Dosulepin hydrochloride***, Dothapax, Dothep, *Dothiepin hydrochloride*, D-Press, Idom, IZ 914, Jardin, KS 1596, Prepadine, Prothiaden, Protiaden(e), Thaden, Tihilor, VÚFB-10615, Xerenal
U   Antidepressant

8629   (7064) $C_{19}H_{21}NS$
15574-96-6

9,10-Dihydro-4-(1-methyl-4-piperidinylidene)-4*H*-ben-zo[4,5]cyclohepta[1,2-*b*]thiophene = 4-(9,10-Dihydro-4*H*-benzo[4,5]cyclohepta[1,2-*b*]thien-4-ylidene)-1-me-thylpiperidine (●)
R   also malate (1:1) (5189-11-7)
S   BC 105, Litec, Lysagor, Mosegor, *Pizotifen***, *Pizo-tyline*, Polomigran, Sandolitec, Sandomigran(e), San-domigrin, Sandomiran, Sanmigran, Sanomigran, Tri-litec
U   Serotonin inhibitor (specific in migraine), appetite sti-mulant, antidepressant

---

8630 $C_{19}H_{21}NS$
96795-89-0

(+)-5,6-Dihydro-6α-[4-(methylthio)phenyl]pyrrolidi-no[2,1-*a*]isoquinoline = *trans*-1,2,3,5,6,10b-Hexahydro-6-[4-(methylthio)phenyl]pyrrolo[2,1-*a*]isoquinoline (●)
S   McN 5652Z
U   Serotonin uptake inhibitor

8631 (7065) $C_{19}H_{21}NS$
                                                      76865-42-4

trans-(±)-2,3,4,4a,7,8,12b,13-Octahydro-1*H*-6-thia-13a-azabenzo[*f*]naphth[1,2,3-*cd*]azulene (●)
S QM-7184
U Neuroleptic

8632 (7066) $C_{19}H_{21}N_3$
                                                      1977-11-3

6-(4-Methyl-1-piperazinyl)morphanthridine = 6-(4-Methyl-1-piperazinyl)-11*H*-dibenz[*b,e*]azepine (●)
S AW-142333, HF-2333, Hypnodin(e), MP-11, *Perlapine***
U Hypnotic

8633 (7070) $C_{19}H_{21}N_3O$
                                                      82626-48-0

N,N,6-Trimethyl-2-*p*-tolylimidazo[1,2-*a*]pyridine-3-acetamide = *N,N*,6-Trimethyl-2-(4-methylphenyl)imidazo[1,2-*a*]pyridine-3-acetamide (●)
R Tartrate (2:1) (99294-93-6)

S Ambien, Bikalm, Cedrol, Cymerion, Dalparan, Durnit, Eudorm "Rontag", Ivadal, Lioram, Myslee, Niotal, Nottem, Ridaxil, Sanval, SL 800750-23 N, Somit, Sovigen, Stilnoct, Stilnox, Sumenan, *Zolpidem tartrate***
U Hypnotic

8634 (7067) $C_{19}H_{21}N_3O$
                                                      78279-88-6

2-[(5,6-Dihydro-3-phenyl-2*H*-1,2,4-oxadiazin-6-yl)methyl]-1,2,3,4-tetrahydroisoquinoline (●)
S CH-141
U Peripheral vasodilator

8635 (7068) $C_{19}H_{21}N_3O$

5-[2-(α-Methylphenethylamino)ethyl]-3-phenyl-1,2,4-oxadiazole = *N*-(1-Methyl-2-phenylethyl)-3-phenyl-1,2,4-oxadiazole-5-ethanamine (●)
R Monohydrochloride (49561-54-8)
S Wd 67/2
U Anorexic

8636 (7069) $C_{19}H_{21}N_3O$
                                                      16188-61-7

4-Benzyl-2-[2-(dimethylamino)ethyl]-1(2H)-phthalazi-none = 2-[2-(Dimethylamino)ethyl]-4-(phenylmethyl)-1(2H)-phthalazinone (●)
R Monohydrochloride (16188-76-4)
S Ahanon, *Benzylphthalazone hydrochloride*, HL 2186, *Talastine hydrochloride***
U Antihistaminic, anti-allergic

8637 (7071) $C_{19}H_{21}N_3O_2S$
                                                             80883-55-2

(E)-2-Amino-1-(isopropylsulfonyl)-6-(1-phenyl-1-prope-nyl)benzimidazole = (E)-1-[(1-Methylethyl)sulfonyl]-6-(1-phenyl-1-propenyl)-1H-benzimidazol-2-amine (●)
S *Enviradene**, LY 127123
U Antiviral

8638 (7072) $C_{19}H_{21}N_3O_3$
                                                             36179-23-4

5,6-Dimethoxy-N-(3-phenylpropyl)-1H-indazole-3-carb-oxamide (●)
S Egyt-1331
U Analgesic

8639 $C_{19}H_{21}N_3O_4$
                                                             150693-65-5

(–)-6-Diazo-3-methyl-4-[(1E)-1,3,5-trimethyl-1-hexe-nyl]-2,5,7,8(1H,6H)-quinolinetetrone (●)
S *Lagunamycin*
U Anti-inflammatory from *Streptomyces* sp.

8640 (7073) $C_{19}H_{21}N_3O_5$
                                                             72803-02-2

Diethyl 4-(4-benzofurazanyl)-1,4-dihydro-2,6-dimethyl-3,5-pyridinedicarboxylate = Diethyl 4-(2,1,3-benzoxadia-zol-4-yl)-1,4-dihydro-2,6-dimethyl-3,5-pyridinedicarb-oxylate = 4-(4-Benzofurazanyl)-1,4-dihydro-2,6-dime-thyl-3,5-pyridinedicarboxylic acid diethyl ester (●)
S *Darodipine**, Dazodipine*, PY-108, PY 108-068
U Coronary and peripheral vasodilator (calcium antago-nist)

8641 (7074) $C_{19}H_{21}N_3O_5$
                                                             75695-93-1

Isopropyl methyl 4-(4-benzofurazanyl)-1,4-dihydro-2,6-dimethyl-3,5-pyridinedicarboxylate = 4-(2,1,3-Ben-zoxadiazol-4-yl)-1,4-dihydro-2,6-dimethyl-3,5-pyridine-dicarboxylic acid methyl 1-methylethyl ester (●)
S Clivoten, Dilatol "Jaba", Dynacirc, Dynacrine, Esra-din, Icaz, *Isradipine**, Isra SRO, *Isrodipine*, Lomir, PN 200-110, Prescal, Rebriden "Sandoz-USA", Ten-zipin, Vascal, Vaslan
U Calcium antagonist (coronary and peripheral vasodi-lator)

8642 $C_{19}H_{21}N_3O_8$
102409-92-7

Acetic acid 11-acetyl-8-(carbamoyloxymethyl)-4-formyl-6-methoxy-14-oxa-1,11-diazatetracyclo[7.4.1.0²,⁷.0¹⁰,¹²]tetradeca-2,4,6-trien-9-yl ester = 1-Acetyl-9-(acetyloxy)-8-[[(aminocarbonyl)oxy]methyl]-1,1a,2,8,9,9a-hexahydro-7-methoxy-3,9-epoxy-3*H*-azirino[2,3-*c*][1]benzazocine-5-carboxaldehyde (●)
S FK 317
U Antineoplastic antibiotic from *Streptomyces sandaensis* sp. 6897

8643 (7076) $C_{19}H_{21}N_3S$
5800-19-1

2-Methyl-11-(4-methyl-1-piperazinyl)dibenzo[*b,f*][1,4]thiazepine (●)
S *Metiapine***
U Neuroleptic

8644 (7075) $C_{19}H_{21}N_3S$
3546-03-0

10-[3-(Dimethylamino)-2-methylpropyl]phenothiazine-2-carbonitrile = 2-Cyano-10-(3-dimethylamino-2-methyl-propyl)phenothiazine = 10-[3-(Dimethylamino)-2-methylpropyl]-10*H*-phenothiazine-2-carbonitrile (●)
R also tartrate (98537-28-1)
S Cianatil, *Cyamemazine***, *Cyamepromazine, Cyanotrimeprazine*, F.I. 6229, *Kyamepromazin*, 7204 R.P., SKF 6477, Tercian, 2602 TH
U Neuroleptic, antihistaminic

8645 $C_{19}H_{21}N_5O_2$
187985-20-2

1,2,3,4-Tetrahydro-*N*-[3-(2-nitro-1*H*-imidazol-1-yl)propyl]-9-acridinamine (●)
R Monohydrochloride (163714-83-8)
S THNLA-1
U Cytotoxic, radiosensitizer

8646 $C_{19}H_{21}N_5O_2$
151272-90-1

3-[[1-Methyl-2(*R*)-pyrrolidinyl]methyl]-5-[(3-nitro-2-pyridinyl)amino]-1*H*-indole = 3-[[(2*R*)-1-Methyl-2-pyrrolidinyl]methyl]-*N*-(3-nitro-2-pyridinyl)-1*H*-indol-5-amine (●)

S CP-135807

U 5-HT$_{1D}$ receptor agonist (migraine therapeutic)

8647 (7077) $C_{19}H_{21}N_5O_2$
28797-61-7

5,11-Dihydro-11-[(4-methyl-1-piperazinyl)acetyl]-6*H*-pyrido[2,3-*b*][1,4]benzodiazepin-6-one (●)

R Dihydrochloride (29868-97-1) [or dihydrochloride monohydrate]

S Abrinac, Acilec, Bisvanil, Cevanil, Droxol "Microsules Bernabo", Duogastral, Durapirenz, Executiv, Folinzepin, Frazim, Gardenopine, Gaspirene, Gasteril "Ripari-Gero", Gastricur, Gastril "Torrent", Gastri-P-Tablinen, Gastrizin, Gastrol, Gastromen, Gastropin "Boehringer-Ingelheim", Gastropiren, Gastrosed "Amsa", Gastrozem, Gastrozepin(a), Gastrozépine, Gastsion, Gatanple, Gipatinil, Indone 50, Kakumine, Karoderin, Kiccalzin, Leblon, Leblun, Ligeral, L-S 519, Lulcus, Maghen, Norsecretol, Novogastrina, Pilenzel, Pin, Pirahexal, Pirefar, PireHexal, Piren, Pirenbasan, Pirengast, Piren von ct, *Pirenzepine hydrochloride***, Pirenzet, Pirezan, Pirigast, Pirodeine, Ranclic, Renzepin, Stomazepin, Tabe, Threptin, Ugaston, Ulcepin, Ulcin, Ulcmor, Ulcopir, Ulcoprotect, Ulcosafe, Ulcosan "Dompé", Ulcosyntex, Ulcozepin, Ulcuforton, Ulgescum, Ulpir "IBP", Ulzepin

U Treatment of peptic ulcer, anti-emetic (M$_1$-antagonist)

8648 (7078) $C_{19}H_{21}N_5O_3$
99009-20-8

2-[3-(Dimethylamino)propyl]-2,6-dihydro-9-methoxy-5-nitropyrazolo[3,4,5-*kl*]acridine = 9-Methoxy-*N,N*-dimethyl-5-nitropyrazolo[3,4,5-*kl*]acridine-2(6*H*)-propanamine (●)

S NCS-366140, PD 115934, PZA

U Antineoplastic

8649 (7079) $C_{19}H_{21}N_5O_3S$
72830-39-8

2-[[2-[[(5-Methyl-5-imidazolyl)methyl]thio]ethyl]amino]-5-piperonyl-4(1*H*)-pyrimidinone = 5-(1,3-Benzodioxol-5-ylmethyl)-2-[[2-[[(5-methyl-1*H*-imidazol-4-yl)methyl]thio]ethyl]amino]-4(1*H*)-pyrimidinone (●)

R Dihydrochloride (63204-23-9)

S *Oxmetidine hydrochloride***, SKF 92994-A$_2$

U Histamine H$_2$-receptor antagonist (anti-ulcer, antihistaminic)

8649-01 (7079-01) 84455-52-7

R Methanesulfonate (1:2)

S *Oxmetidine mesylate***, SKF 92994-J$_2$

U Histamine H$_2$-receptor antagonist (anti-ulcer, antihistaminic)

8650   (7080) $C_{19}H_{21}N_5O_4$
                                                        96392-15-3

(R)-N-(2,3-Dihydro-1H-inden-1-yl)adenosine (●)
S   CI-947, PD 117519
U   Peripheral vasodilator, diuretic

8651   (7081) $C_{19}H_{21}N_5O_4$
                                                        19216-56-9

1-(4-Amino-6,7-dimethoxy-2-quinazolinyl)-4-(2-fu-
royl)piperazine = 2-[4-(2-Furoyl)piperazin-1-yl]-6,7-di-
methoxyquinazolin-4-ylamine = 1-(4-Amino-6,7-dimeth-
oxy-2-quinazolinyl)-4-(2-furanylcarbonyl)piperazine (●)
R   Monohydrochloride (19237-84-4)
S   Adversuten, Adverzuten, Alphavase, Alpresin, Al-
        press, Alti-Prazosin, Apo-Prazo, Coltock,
        CP-12299-1, Daldanon, Decliten, Deprazolin,
        Dexazosin, Downat, Downpress, Duramipress, Enzo-
        sine, Eurex, Fellhye, Flaboido, *Furazosin hydrochlo-
        ride*, Hexapress, Hypotens, Hypovase, Isepress, Ita-
        ka, Kachilet, Kenral-Prazosin, Lentopress, Lopress
        "Siam", Metrasen, Minebar, Minipres, Minipress,
        Mizpiron, Mysial, Novo-Prazin, Nu-Prazo, Orbisan,
        Patsolin, Peripress, Polpressin, Polypress, Prasig,
        Pratsiol, Prazac, Prazoberag, Prazocor, Prazopress,
        *Prazosin hydrochloride***, Prazozinbene, Pressin,
        Queenpress, Rexibet, Rho-Prazosin, Sedaxin, Sine-
        tens, Sinozzard, Stroken, Trabuzon, Vasoflex "Alka-
        loid"
U   Antihypertensive (α-blocker)

8652   (7083) $C_{19}H_{22}ClNO_2$
                                                        494-14-4

4-[3-[(3-Chloro-4-biphenylyl)oxy]propyl]morpholine =
4-(γ-Morpholinopropoxy)-3-chlorobiphenyl = 4-[3-[(3-
Chloro[1,1'-biphenyl]-4-yl)oxy]propyl]morpholine (●)
S   *Chlordimorine***, *Chlordimorphine*
U   Antifungal

8653   (7084) $C_{19}H_{22}ClNO_2$
                                                        5617-26-5

4-[2-[(p-Chloro-α-phenylbenzyl)oxy]ethyl]morpholine =
4-[2-(p-Chlorodiphenylmethoxy)ethyl]morpholine = 1-
Phenyl-1-(p-chlorophenyl)-1-(β-morpholinoethoxy)me-
thane = 4-[2-[(4-Chlorophenyl)phenylmeth-
oxy]ethyl]morpholine (●)
R   most hydrochloride (1798-49-8)
S   *Difencloxazine***, *Diphenchloxazine*, L.D. 2630,
        Olympax
U   Tranquilizer

8654 $C_{19}H_{22}ClNO_2$
                                                        210757-90-7

N-[[6-Hydroxy-1,2,3,4-tetrahydronaphthalen-2(R)-yl]
methyl]-2(R)-hydroxy-2-(3-chlorophenyl)ethanamine =

(6R)-6-[[[(2R)-2-(3-Chlorophenyl)-2-hydroxyethyl]amino]methyl]-5,6,7,8-tetrahydro-2-naphthalenol (●)
R Hydrochloride (136758-90-2)
S SR 59104A
U β_3-Adrenoceptor agonist

8655 $C_{19}H_{22}ClNO_4$
90730-96-4

rel-[4-[(2R)-2-[[(2R)-2-(3-Chlorophenyl)-2-hydroxyethyl]amino]propyl]phenoxy]acetic acid (●)
S BRL 37344
U β_3-Adrenoceptor agonist

8656 (7085) $C_{19}H_{22}ClNS$
87906-31-8

(E)-3-Chloro-N-(6,6-dimethyl-2-hepten-4-ynyl)-N-methylbenzo[b]thiophene-7-methanamine (●)
R Hydrochloride (87906-33-0)
S SDZ 87-469
U Antifungal

8657 $C_{19}H_{22}ClN_3O_5S_2$
158751-64-5

1-[[5-[2-(5-chloro-o-anisamido)ethyl]-2-methoxyphenyl]sulfonyl]-3-methylthiourea = 5-Chloro-2-methoxy-N-[2-[4-methoxy-3-[[[(methylamino)thioxomethyl]amino]sulfonyl]phenyl]ethyl]benzamide (●)
S *Clamikalant**, HMR 1883

U Cardioselective K_{ATP}-channel blocker

8658 $C_{19}H_{22}ClN_5O$
200484-11-3

N-[6-(4-Chlorophenoxy)hexyl]-N'-cyano-N''-4-pyridinylguanidine (●)
S CHS 828
U Antineoplastic

8659 (7086) $C_{19}H_{22}ClN_5O$
19794-93-5

2-[3-[4-(3-Chlorophenyl)-1-piperazinyl]propyl]-1,2,4-triazolo[4,3-a]pyridin-3(2H)-one (●)
R Monohydrochloride (25332-39-2)
S AF 1161, Alti-Trazodone, Apo-Trazodone, Azona, Beneficat, Bimaran, Deprax "Lepori", Deprel, Depresil, Depyrel, Desirel, Desyrel, Devidon "Lek", Geripax, Manegan, Molipaxin, Novo-Trazodone, Nu-Trazodone, PMS-Trazodone, Pragmarel, Pragmazone, Reslin(e), Sideril, Syn-Trazodone, Taxagon, Taxagonad, Thombran, Tombran, Tramensan, Trandin, Trazodil, *Trazodone hydrochloride***, Trazolan, Trazon(e), Tresin, Trialodine, Triticum, Trittico
U Antidepressant, anxiolytic, adjuvant agent in preanesthesia

---

8660 (7087)　　　　　　　　　　$C_{19}H_{22}Cl_2N_2O_3$
857-95-4

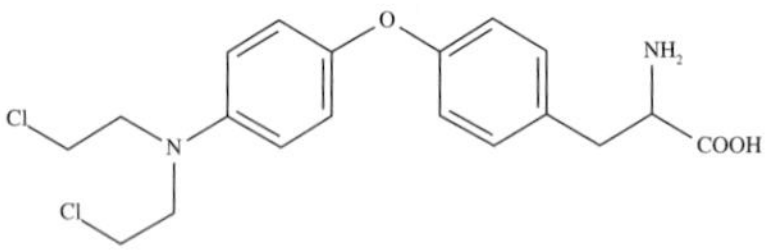

3-[p-[p-[Bis(2-chloroethyl)amino]phenoxy]phenyl]-DL-alanine = DL-2-Amino-3-[p-[p-[bis(β-chloroethyl)amino]phenoxy]phenyl]propionic acid = O-[4-[Bis-(2-chloroethyl)amino]phenyl]-DL-tyrosine (●)
R Dihydrochloride (56348-18-6)
S Fentirin, Phentyrin
U Antineoplastic, immunosuppressive

8661　　　　　　　　　　　　　　$C_{19}H_{22}Cl_2N_4O$
145544-79-2

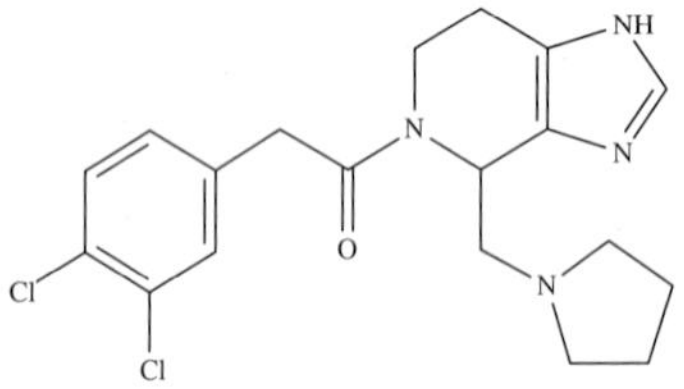

5-[(3,4-Dichlorophenyl)acetyl]-4,5,6,7-tetrahydro-4-(1-pyrrolidinylmethyl)-1H-imidazo[4,5-c]pyridine (●)
S BRL 52974
U κ-Opioid receptor agonist

8662 (7088)　　　　　　　　　　　$C_{19}H_{22}FN_3O$
1649-18-9

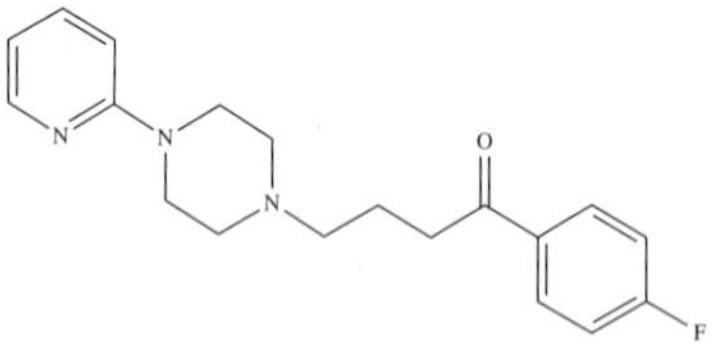

4'-Fluoro-4-[4-(2-pyridyl)-1-piperazinyl]butyrophenone = 1-(4-Fluorophenyl)-4-[4-(2-pyridinyl)-1-piperazinyl]-1-butanone (●)
S Afiperon, *Azaperone***, Porcador, Porcirelax, R 1929, Sedaperone, Stresnil, Suicalm
U Neuroleptic, hypnotic

8663 (7089)　　　　　　　　　　　$C_{19}H_{22}FN_3O_3$
79644-90-9

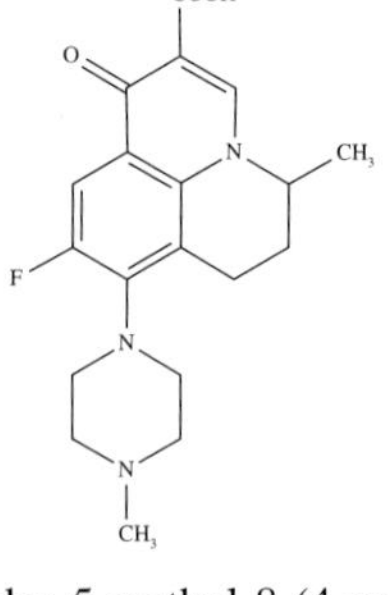

9-Fluoro-6,7-dihydro-5-methyl-8-(4-methyl-1-piperazinyl)-1-oxo-1H,5H-benzo[ij]quinolizine-2-carboxylic acid (●)
S *Benofloxacin*, OPC-19A, *Vebufloxacin***
U Antibacterial

8664 (7090)　　　　　　　　　　　$C_{19}H_{22}FN_3O_3$
93106-60-6

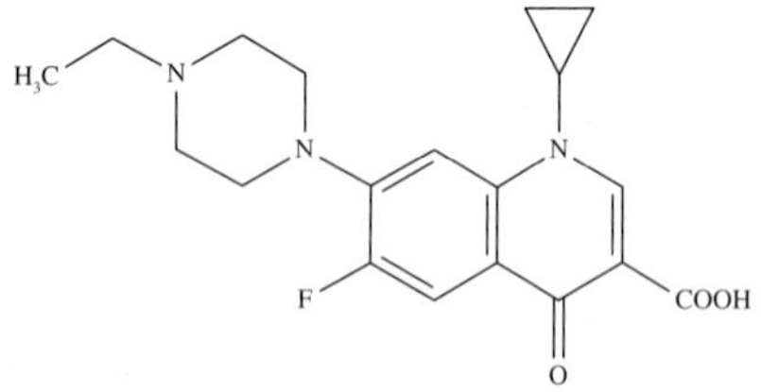

1-Cyclopropyl-7-(4-ethyl-1-piperazinyl)-6-fluoro-1,4-dihydro-4-oxo-3-quinolinecarboxylic acid (●)
S Baytril, Bay Vp 2674, CPFQ, *Enrofloxacin***
U Antibacterial (veterinary)

8665 (7091)　　　　　　　　　　　$C_{19}H_{22}FN_3O_3$
119914-60-2

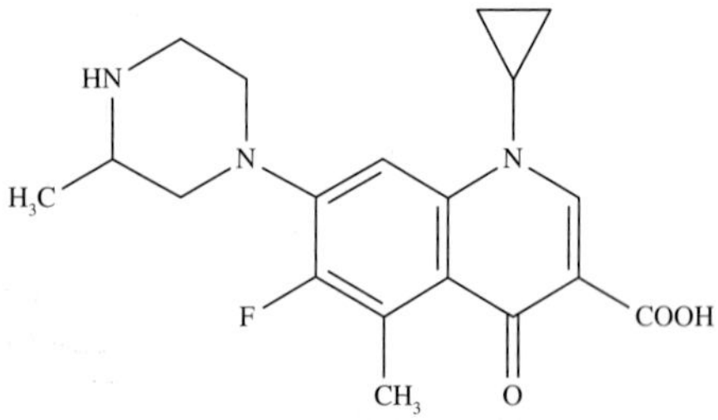

1-Cyclopropyl-6-fluoro-1,4-dihydro-5-methyl-7-(3-methyl-1-piperazinyl)-4-oxo-3-quinolinecarboxylic acid (●)
R also monohydrochloride (161967-81-3)

S Cerobin, GPFX, *Grepafloxacin***, Grepax, Lunga-skin, OPC-17116, Raxar, *Tomefloxacin*, Vaxar, Vorzan

U Antibacterial

8666   (7092) $C_{19}H_{22}FN_3O_3$
                                                           108437-28-1

7-(1,4-Diazabicyclo[3.2.2]non-4-yl)-1-ethyl-6-fluoro-1,4-dihydro-4-oxo-3-quinolinecarboxylic acid (●)

S *Binfloxacin***, CP-73049

U Antibacterial (veterinary)

8667   (7093) $C_{19}H_{22}FN_3O_4$
                                                           112811-59-3

1-Cyclopropyl-6-fluoro-1,4-dihydro-8-methoxy-7-(3-methyl-1-piperazinyl)-4-oxo-3-quinolinecarboxylic acid (●)

R also sesquihydrate (180200-66-2)

S AM-1155, BMS 206584-01, *Gafloxacin*, Gatiflo, *Gatifloxacin***, GFLX, Tequin

U Antibacterial

8668   (7094) $C_{19}H_{22}FN_3O_8$
                                                           124012-42-6

N-[1-(5-Deoxy-β-D-ribofuranosyl)-5-fluoro-1,2-dihydro-2-oxo-4-pyrimidinyl]-3,4,5-trimethoxybenzamide = 5'-Deoxy-5-fluoro-*N*-(3,4,5-trimethoxybenzoyl)cytidine (●)

S *Galocitabine***, Ro 9-1390

U Antineoplastic, immunosuppressive

8669 $C_{19}H_{22}FN_7O_6S_2$
                                                           162081-75-6

7β-[(*Z*)-2-(5-Amino-1,2,4-thiadiazol-3-yl)-2-(fluoro-methoxyimino)acetylamino]-3-[(*E*)-2-[(*S*)-2,2-dimethyl-5-isoxazolidinio]ethenyl]-3-cephem-4-carboxylate = [6*R*-[3[*E*(*S**)],6α,7β(*Z*)]]-5-[2-[7-[[(5-Amino-1,2,4-thiadi-azol-3-yl)][(fluoromethoxy)imino]acetyl]amino]-2-carb-oxy-8-oxo-5-thia-1-azabicyclo[4.2.0]oct-2-en-3-yl]ethe-nyl]-2,2-dimethylisoxazolidinium inner salt (●)

S YM 40220

U Antibiotic

8670 $C_{19}H_{22}F_2N_4O$

7α-[(4-Fluorophenoxy)methyl]-2-(5-fluoro-2-pyrimidi-nyl)perhydropyrido[1,2-*a*]pyrazine = (7*R*,9a*S*)-7-[(4-Fluorophenoxy)methyl]-2-(5-fluoro-2-pyrimidinyl)octa-hydro-2*H*-pyrido[1,2-*a*]pyrazine (●)

R Monohydrochloride (178930-30-8)
S CP-293019
U Antipsychotic (dopamine D_4 receptor antagonist)

8671 (7095)

$C_{19}H_{22}F_2N_4O_3$
110871-86-8

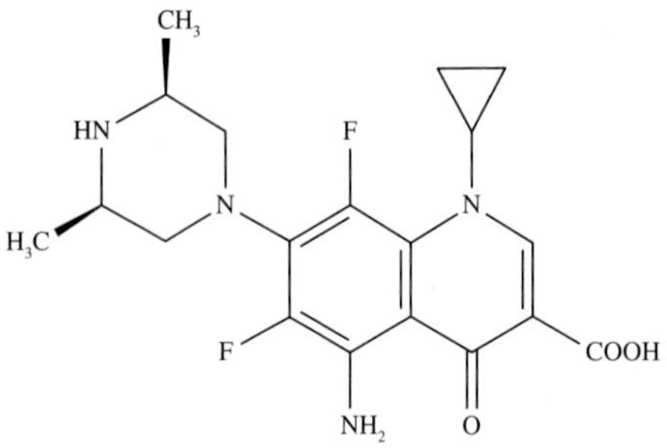

cis-5-Amino-1-cyclopropyl-7-(3,5-dimethyl-1-piperazi-nyl)-6,8-difluoro-1,4-dihydro-4-oxo-3-quinolinecarb-oxylic acid (●)
S AT-4140, CI-978, *Esparfloxacino*, PD 131501, Rexpar, RP 64206, Spara, Sparaxin, Sparfloxac, *Sparflo-xacin***, Sparlox, SPFX, Torospar, Zagam, Zegam
U Antibacterial, tuberculostatic

8672

$C_{19}H_{22}F_3N_3$
1814-64-8

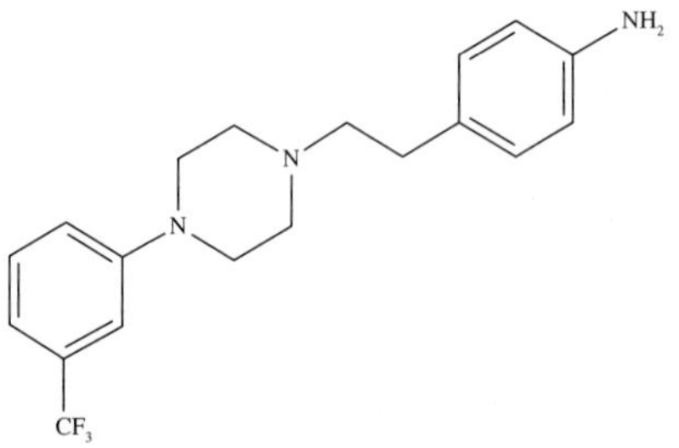

1-(*p*-Aminophenethyl)-4-(α,α,α-trifluoro-*m*-tolyl)pipe-razine = 4-[2-[4-[3-(Trifluoromethyl)phenyl]-1-piperazi-nyl]ethyl]benzenamine (●)
S LY 165163
U $5-HT_{1A}$ receptor agonist

8673 (7096)

$C_{19}H_{22}NO_4S_2$
186691-13-4

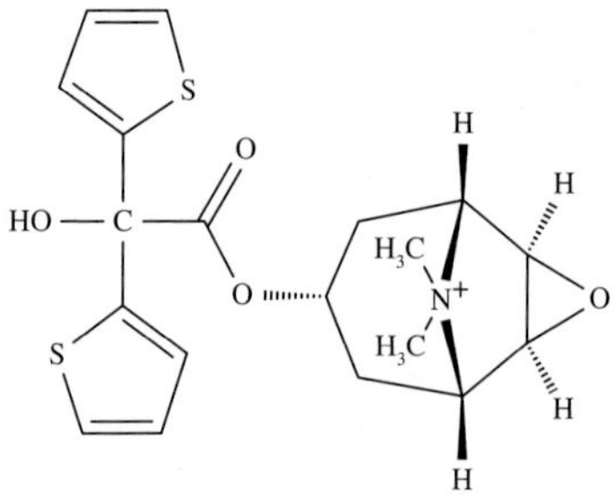

6β,7β-Epoxy-3β-hydroxy-8-methyl-1α*H*,5α*H*-tropa-nium di-2-thienylglycolate = (1α,2β,4β,5α,7β)-7-[[Hy-droxydi-2-thienylacetyl]oxy]-9,9-dimethyl-3-oxa-9-azoniatricyclo[3.3.1.0^{2,4}]nonane (●)
R Bromide (136310-93-5) or bromide hydrate (139404-48-1)
S Ba 679 BR, Spiriva, *Tiotropium bromide***
U Anticholinergic

8674 (7097)

$C_{19}H_{22}N_2$
6293-01-2

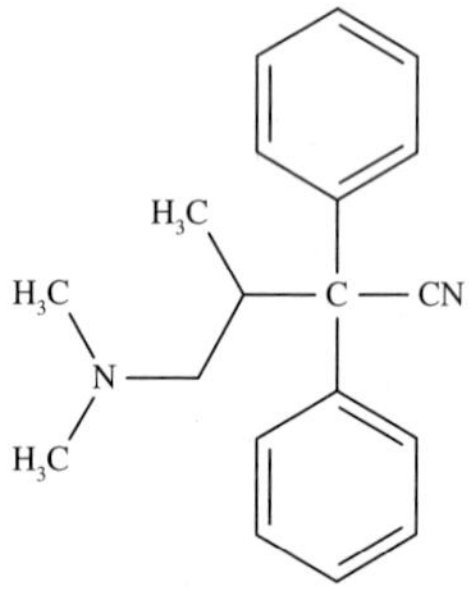

4-(Dimethylamino)-3-methyl-2,2-diphenylbutyronitrile = α-[2-(Dimethylamino)-1-methylethyl]-α-phenylbenzene-acetonitrile (●)
R Salicylate
S ATE-15, Bronco-Blandin, Tuso-Blandin
U Antitussive

8675   (7098) $C_{19}H_{22}N_2$
                                                   125-79-1

4-(Dimethylamino)-2,2-diphenylvaleronitrile = 3-Di-
methylamino-1,1-diphenyl-1-cyanobutane = α-[2-(Dimethylamino)propyl]-α-phenylbenzeneacetonitrile (●)
S   Methadone intermediate, Methadon-Zwischenpro-
        dukt, Pre-Methadone
U   Analgesic

8676   (7099) $C_{19}H_{22}N_2$
                                                   303-54-8

5-[3-(Dimethylamino)propyl]-5H-dibenz[b,f]azepine = 3-
(5H-Dibenz[b,f]azepin-5-yl)propyldimethylamine = N,N-
Dimethyl-5H-dibenz[b,f]azepine-5-propanamine (●)
S   Balipramine, Dehydroimipramine, Depramine**,
        GP 31406
U   Antiparkinsonian

8677   (7100) $C_{19}H_{22}N_2$
                                                   486-12-4

trans-2-[3-(1-Pyrrolidinyl)-1-p-tolylpropenyl]pyridine =
trans-1-(2-Pyridyl)-3-pyrrolidino-1-p-tolyl-1-propene =
(E)-2-[1-(4-Methylphenyl)-3-(1-pyrrolidinyl)-1-prope-
nyl]pyridine (●)
R   Monohydrochloride (550-70-9) [or monohydrochlori-
        de monohydrate (6138-79-0)]
S   Actidil, Actidilon, Actiphyll, Alleract, Bayidyl,
        B.W. 51-295, 295 C 51, Entra, Histradil, Jupick,
        Myidyl, Pro-Actidil, Pro-Actidilon, Pro-Entra, Tri-
        prolidine hydrochloride**, Venen "Tanabe"
U   Antihistaminic

8678 $C_{19}H_{22}N_2O$
                                                   135729-56-5

2,4,5,6-Tetrahydro-2-[(3S)-3-quinuclidinyl]-1H-
benz[de]isoquinolin-1-one = (S)-2-(1-Azabicy-
clo[2.2.2]oct-3-yl)-2,4,5,6-tetrahydro-1H-benz[de]iso-
quinolin-1-one (●)
R   Monohydrochloride (135729-55-4);
        corrected definition see no. 8818
S   Palonosetron hydrochloride**, RS-25259-197
U   Anti-emetic, antinauseant (selective 5-HT₃ receptor
        antagonist)

8679   (7105) $C_{19}H_{22}N_2O$
                                                   118-10-5

(5-Vinyl-2-quinuclidinyl)(4-quinolyl)methanol = (9S)-
Cinchonan-9-ol (●)
S   Cinchonine

U Antimalarial, tonic, reagent (alkaloid from *Cinchona* bark)

8680 (7103) $C_{19}H_{22}N_2O$
                                                796-29-2

5-[3-(Dimethylamino)propyl]-5,11-dihydro-10*H*-dibenz[*b,f*]azepin-10-one (●)
R also fumarate (1:1) (17243-32-2)
S G 35259, *Ketimipramine**, Ketipramine, Ketoimipramine*
U Antidepressant

8681 (7104) $C_{19}H_{22}N_2O$
                                                3362-45-6

5-[2-(Dimethylamino)ethyloximino]-5*H*-dibenzo[*a,d*]cyclohepta-1,4-diene = 2-[[(10,11-Dihydro-5*H*-dibenzo[*a,d*]cyclohepten-5-ylidene)amino]oxy]ethyldimethylamine = 10,11-Dihydro-5*H*-dibenzo[*a,d*]cyclohepten-5-one *O*-[2-(dimethylamino)ethyl]oxime (●)
R Monohydrochloride (4985-15-3)
S Agedal, Bayer 1521, CD 37 B, *Dibenzoxin*, Elronon, Nogédal, *Noxiptiline hydrochloride**, Noxiptyline hydrochloride*, Sipcar (old form)
U Antidepressant

8682 (7106) $C_{19}H_{22}N_2O$
                                                4880-88-0

(–)-Eburnamenin-14(15*H*)-one = 3α,16α-Eburnamonine = (3α,16α)-Eburnamenin-14(15*H*)-one (●)
R also phosphate (77117-62-5)
S Atrican "Montpellier", Cervoxan, CH-846, Eburnal, (–)-*Eburnamonine*, Eburnoxin, Eubornoxin, Eubran, Luvenil, Monil, Scleramin, Tensiplex, *Vinburnine**, Vincamon*
U Cerebrotonic

8683 (7101) $C_{19}H_{22}N_2O$
                                                441-91-8

3-(2-Aminoethyl)-1-benzyl-5-methoxy-2-methylindole = 5-Methoxy-2-methyl-1-(phenylmethyl)-1*H*-indole-3-ethanamine (●)
R Monohydrochloride (525-02-0)
S BAS, *Benanserin hydrochloride, Benzylantiserotonin*, Woolley's Antiserotonin
U Serotonin antagonist

8684   (7102) $C_{19}H_{22}N_2O$
                                                                   22136-26-1

3-Methyl-3-[3-(methylamino)propyl]-1-phenyl-2-indoli-
none = 1,3-Dihydro-3-methyl-3-[3-(methylamino)pro-
pyl]-1-phenyl-2*H*-indol-2-one (●)
R   Monohydrochloride (22232-73-1)
S   *Amedalin hydrochloride***, UK 3540-1
U   Antidepressant

8685   (7109) $C_{19}H_{22}N_2OS$
                                                                   5845-26-1

5-[2-(Dimethylamino)ethyl]-2,3-dihydro-2-phenyl-1,5-
benzothiazepin-4(5*H*)-one (●)
R   Monohydrochloride (3122-01-8)
S   Altinil, SQ 10496, *Thiazenone, Thiazesim hydrochlo-
ride, Tiazesim hydrochloride***
U   Antidepressant

8686   (7108) $C_{19}H_{22}N_2OS$
                                                                   13461-01-3

10-[2-(Dimethylamino)propyl]phenothiazin-2-yl methyl
ketone = 2-Acetyl-10-[2-(dimethylamino)propyl]pheno-
thiazine = 1-[10-[2-(Dimethylamino)propyl]-10*H*-pheno-
thiazin-2-yl]ethanone (●)
S   *Aceprometazine***, 1664 C.B.
U   Neuroleptic, antitussive

8687   (7107) $C_{19}H_{22}N_2OS$
                                                                   61-00-7

10-[3-(Dimethylamino)propyl]phenothiazin-2-yl methyl
ketone = 2-Acetyl-10-[3-(dimethylamino)propyl]pheno-
thiazine = 1-[10-[3-(Dimethylamino)propyl]-10*H*-pheno-
thiazin-2-yl]ethanone (●)
R   also hydrochloride (53048-64-9) or maleate (1:1)
       (3598-37-6)
S   *Acepromazine***, Acetacin, Acetazin, *Acetopromazi-
ne, Acetylpromazine*, ACP C-Vet, Anatran "Ayerst",
Atravet, AY-57062, Bayer 1212, Berkace, BK Ace,
Calmivet, 1522 C.B., Concentrat VO 34, Kavmos,
Killitam, Neurotranq, Notensil, Oralject Sedazine,
Plegicil, Plegicin, Plegicyl, Plivaphen, Prequillan,
7214 R.P., Sedalin "Chassot", Soprontin, Vétran-
quil, Wy-1172
U   Neuroleptic, anti-emetic

8688   (7110) $C_{19}H_{22}N_2O_2$
                                                                   65792-35-0

4-[2-(Dimethylamino)ethyl]-6-methyl-2-phenyl-2*H*-1,4-
benzoxazin-3(4*H*)-one (●)
S   AR 17048

U Anti-inflammatory

8689 (7111) $C_{19}H_{22}N_2O_2$
60607-25-2

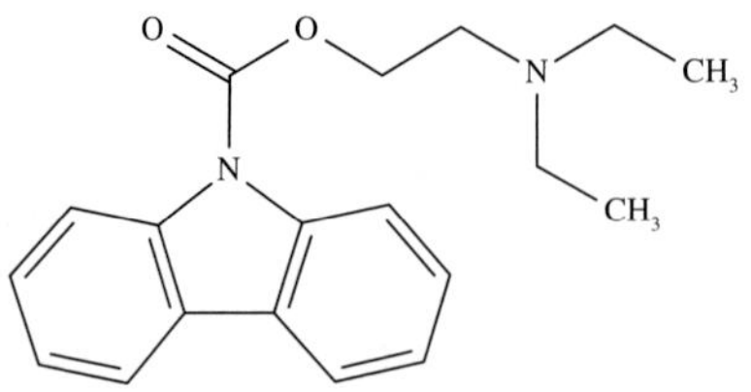

9-Carbazolecarboxylic acid β-(diethylamino)ethyl ester =
9H-Carbazole-9-carboxylic acid 2-(diethylamino)ethyl
ester (●)
S Carbacain
U Local anesthetic

8690 (7112) $C_{19}H_{22}N_2O_2$
70877-75-7

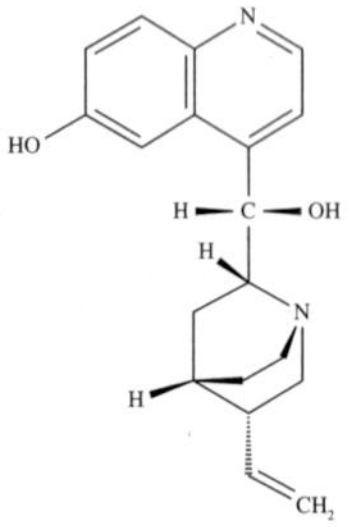

6'-Hydroxycinchonine = (9S)-Cinchonan-6',9-diol (●)
S *Cupreidine*
U Anti-arrhythmic

8691 (7113) $C_{19}H_{22}N_2O_2$
7008-15-3

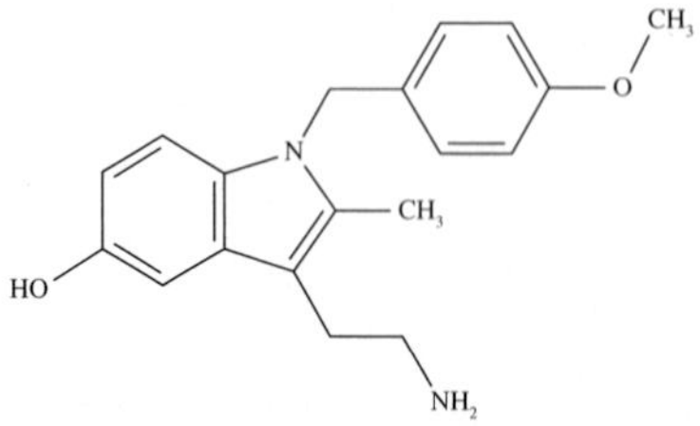

3-(2-Aminoethyl)-1-(p-methoxybenzyl)-2-methylindol-
5-ol = 5-Hydroxy-1-(p-methoxybenzyl)-2-methyltrypta-
mine = 3-(2-Aminoethyl)-1-[(4-methoxyphenyl)methyl]-
2-methyl-1H-indol-5-ol (●)
R also monohydrochloride (24353-32-0)

S *Hydroxindasol**, Oxindasolum*, Presid
U Stimulant, serotonin antagonist

8692 (7114) $C_{19}H_{22}N_2O_2$
482-68-8

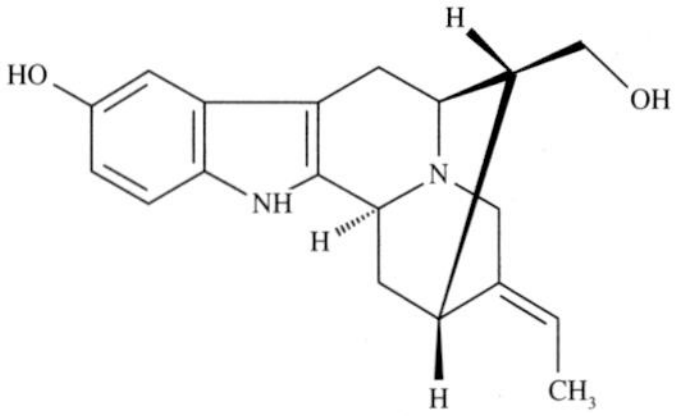

Sarpagan-10,17-diol (●)
S *Raupine, Sarpagine*
U Antihypertensive (rauwolfia alkaloid)

8693 (7115) $C_{19}H_{22}N_2O_2S$
82-00-8

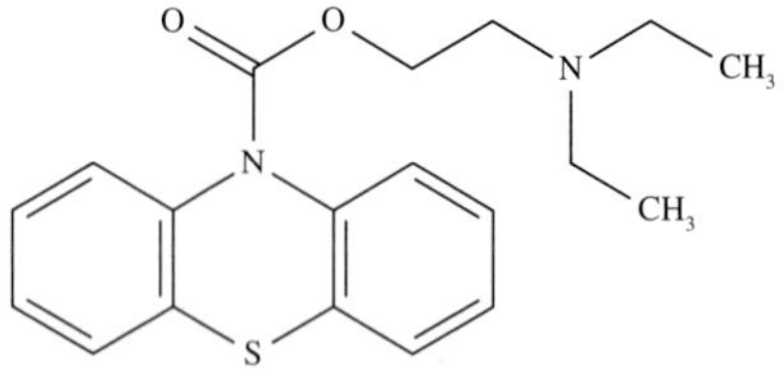

10-Phenothiazinecarboxylic acid β-(diethylamino)ethyl
ester = 10H-Phenothiazine-10-carboxylic acid 2-(diethyl-
amino)ethyl ester (●)
R Monohydrochloride (298-51-1)
S Novazin, Transergan
U Spasmolytic

8694 (7116) $C_{19}H_{22}N_2O_3$
69365-73-7

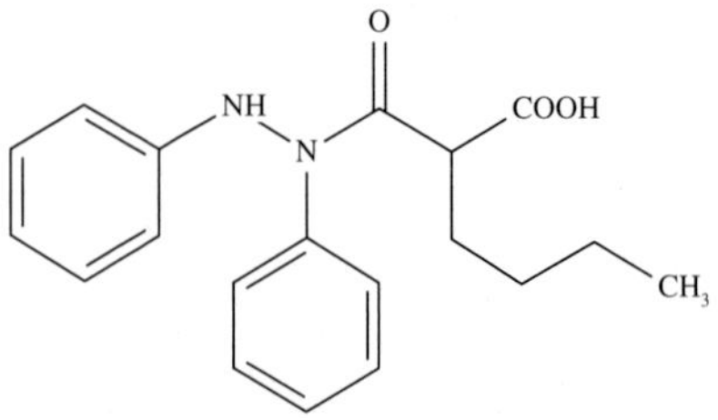

Butylmalonic acid mono(1,2-diphenylhydrazide) = Buty-
lpropanedioic acid mono(1,2-diphenylhydrazide) (●)
R Calcium salt hemihydrate (34461-73-9)

S B 64114-Ca, *Bumadizone Calcium***, Bumaflex (old form), Desflam, Dibilan, Eumotol, Exflam, Imotol, Rheumatol "Tosse"

U Anti-inflammatory, antipyretic

8695 (7117) $C_{19}H_{22}N_2O_3S$
                                                      49858-38-0

1-[4-(Methylsulfonyl)phenacyl]-4-phenylpiperazine = 1-[4-(Methylsulfonyl)phenyl]-2-(4-phenyl-1-piperazinyl)ethanone (●)

R Maleate (1:1) (50648-51-6)

S *Mesylphenacyrazine*, VÚFB-8752

U Tranquilizer

8696 (7118) $C_{19}H_{22}N_2O_3S$
                                                      477-93-0

2-[2-(Dimethylamino)ethoxy]ethyl phenothiazine-10-carboxylate = 10-Phenothiazinecarboxylic acid 2-[2-(dimethylamino)ethoxy]ethyl ester = 10H-Phenothiazine-10-carboxylic acid 2-[2-(dimethylamino)ethoxy]ethyl ester (●)

R Monohydrochloride (518-63-8)

S Atuss, Cothera, Cotrane, *Dimethoxanate hydrochloride***, Egotux, Metaxan "Faro", Perlatos, Pulmoll-Sir., Toa, Toxanon, Tussidin, Tussizid

U Antitussive

8697 (7119) $C_{19}H_{22}N_2O_4$
                                                      101071-43-6

Ethyl 5-isopropoxy-4-(methoxymethyl)-β-carboline-3-carboxylate = 4-(Methoxymethyl)-5-(1-methylethoxy)-9H-pyrido[3,4-b]indole-3-carboxylic acid ethyl ester (●)

S ZK 95962

U Anxiolytic (benzodiazepine partial agonist)

8698 (7120) $C_{19}H_{22}N_2O_6S$
                                                      983-85-7

(2S,5R,6R)-3,3-Dimethyl-7-oxo-6-(phenylacetamido)-4-thia-1-azabicyclo[3.2.0]heptane-2-carboxylic acid (acetyloxy)methyl ester = Acetoxymethyl (6R)-6-(2-phenylacetamido)penicillanate = Acetate ester of the hydroxymethyl ester of benzylpenicillin = Benzylpenicillin acetoxymethyl ester = [2S-(2α,5α,6β)]-3,3-Dimethyl-7-oxo-6-[(phenylacetyl)amino]-4-thia-1-azabicyclo[3.2.0]heptane-2-carboxylic acid (acetyloxy)methyl ester (●)

S Havapen, Maripen, *Penamecillin***, Penclen, Wy-20788

U Antibiotic

8699 (7121) $C_{19}H_{22}N_2O_7$
                                                      62760-70-7

Diethyl 1,4-dihydro-2-(hydroxymethyl)-6-methyl-4-(*m*-nitrophenyl)-3,5-pyridinedicarboxylate = 1,4-Dihydro-2-(hydroxymethyl)-6-methyl-4-(3-nitrophenyl)-3,5-pyridinedicarboxylic acid diethyl ester (●)
S FR 7534
U Coronary vasodilator (calcium antagonist)

8700 (7124) $C_{19}H_{22}N_2S$
                                                   1526-83-6

10-(4-Methyl-1-piperazinyl)-10,11-dihydrodibenzo[*b,f*]thiepine = 1-(10,11-Dihydrodibenzo[*b,f*]thiepin-10-yl)-4-methylpiperazine (●)
R Maleate (1:1) (4774-32-7)
S Perathiepine
U Neuroleptic

8701 (7123) $C_{19}H_{22}N_2S$
                                                   15302-12-2

10-[(1,3-Dimethyl-3-pyrrolidinyl)methyl]phenothiazine = 10-[(1,3-Dimethyl-3-pyrrolidinyl)methyl]-10*H*-phenothiazine (●)
R also hydrochloride
S B 1250, Centrophène (old form), *Dimelazine***, *Diprothazinum*, *Mepyrrotazinum*
U Psychosedative

8702 (7122) $C_{19}H_{22}N_2S$
                                                   60-89-9

10-[(1-Methyl-3-piperidyl)methyl]phenothiazine = 10-[(1-Methyl-3-piperidinyl)methyl]-10*H*-phenothiazine (●)
R also monohydrochloride (2975-36-2) or acetate (24360-97-2)
S III-2318, Lacumin, *Mepasin*, *Mepazine*, MPMP, Nothiazine, P 391, Pacatal, Pacatol, Pactal, Papital, Paxital, *Pecazine**, *Pekazin*, Ravenil (old form), Seral "Cresci", W 1224
U Neuroleptic

8703 $C_{19}H_{22}N_4$
                                                   115178-28-4

(8α)-8-(1*H*-Imidazol-1-ylmethyl)-6-methylergoline (●)
S BAM-2101
U Antihypertensive, dopaminergic

8704 (7125) $C_{19}H_{22}N_4O$
                                                   74627-35-3

(α-*RS*)-α-Cyano-6-methylergoline-8β-propionamide =
(8β)-α-Cyano-6-methylergoline-8-propanamide (●)
S   355/1057, *Cianergoline***
U   Antihypertensive

8705 $C_{19}H_{22}N_4O_2$
                                 158364-59-1

Methyl 2-[[(2,3-dimethylimidazo[1,2-*a*]pyridin-8-
yl)amino]methyl]-3-methylcarbanilate = 2,3-Dimethyl-8-
[[[2-(methoxycarbonylamino)-6-methylphenyl]me-
thyl]amino]imidazo[1,2-*a*]pyridine = [2-[[(2,3-Dimethy-
limidazo[1,2-*a*]pyridin-8-yl)amino]methyl]-3-methyl-
phenyl]carbamic acid methyl ester (●)
S   B 9208-041, BY 841, *Pumaprazole***
U   Anti-ulcer agent (proton pump inhibitor)

8706   (7126) $C_{19}H_{22}N_4O_2$
                                 95688-34-9

1-[[(8β)-6-Methylergolin-8-yl]methyl]-2,4-imidazoli-
dinedione (●)
S   FCE 22716
U   Antihypertensive

8707   (7127) $C_{19}H_{22}N_4O_2S$
                                 80880-90-6

4,9-Dihydro-3-methyl-4-[(4-methyl-1-piperazinyl)ace-
tyl]-10*H*-thieno[3,4-*b*][1,5]benzodiazepin-10-one (●)
S   B 818-03, By 803, *Telenzepine***
U   Anti-ulcer agent (M_1-antagonist)

8708 $C_{19}H_{22}N_4O_3$
                                 53671-71-9

N-[2-[[3-(2-Cyanophenoxy)-2-hydroxypropyl]ami-
no]ethyl]-*N'*-phenylurea (●)
S   ICI 89406
U   β-Adrenoceptor antagonist

8709   (7128) $C_{19}H_{22}N_4O_3$
                                 118989-56-3

4-Methyl-*N*-(2-nitrophenyl)-*N*-phenyl-1-piperazineacet-
amide (●)
R   Maleate [(*Z*)-2-butenedioate] (1:2) (118989-82-5)
S   VÚFB-17104
U   Anti-ulcer agent

8710 (7129)　　　　　　　　　　　$C_{19}H_{22}N_4O_3$
145194-32-7

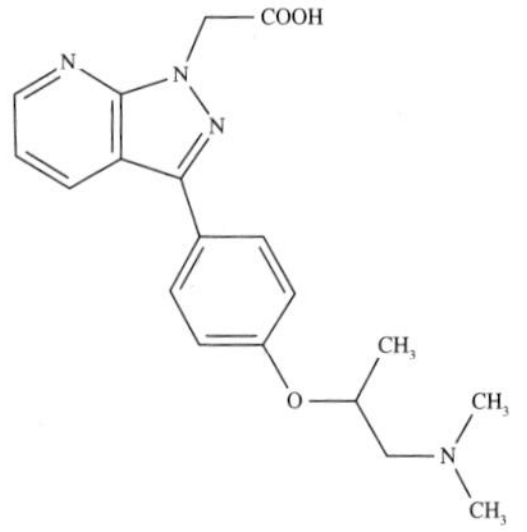

(±)-3-[4-[2-(Dimethylamino)-1-methylethoxy]phenyl]-1H-pyrazolo[3,4-b]pyridine-1-acetic acid (●)
S　Y-25510
U　Immunostimulant

8711　　　　　　　　　　　　　　　$C_{19}H_{22}N_4O_5$
170148-29-5

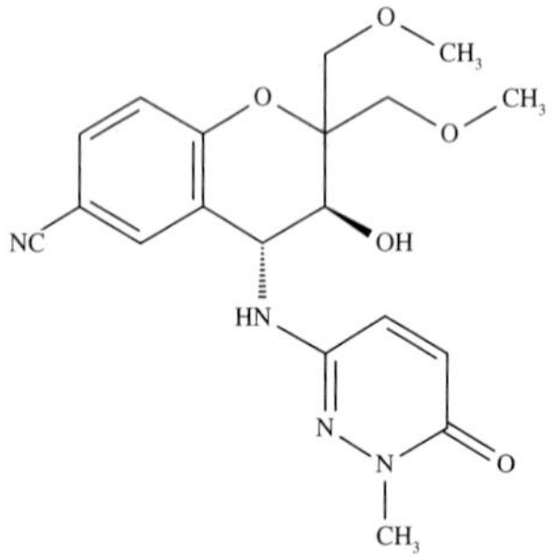

(−)-(3S,4R)-6-Cyano-3,4-dihydro-4-[(1,6-dihydro-1-me-thyl-6-oxo-3-pyridazinyl)amino]-2,2-bis(methoxyme-thyl)-2H-1-benzopyran-3-ol = (3S,4R)-4-[(1,6-Dihydro-1-methyl-6-oxo-3-pyridazinyl)amino]-3,4-dihydro-3-hydroxy-2,2-bis(methoxymethyl)-2H-1-benzopyran-6-carbonitrile (●)
S　JTV-506
U　Coronary vasodilator (potassium channel opener)

8712 (7130)　　　　　　　　　　　$C_{19}H_{22}N_4O_6S_2$
2603-23-8

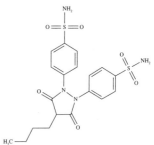

4-Butyl-1,2-bis(p-sulfamoylphenyl)-3,5-pyrazolidine-dione = 4,4'-(4-Butyl-3,5-dioxo-1,2-pyrazolidinedi-yl)bis[benzenesulfonamide] (●)
S　*Butaglionamide*, DSB
U　Anti-inflammatory

8713 (7131)　　　　　　　　　　　$C_{19}H_{22}N_4O_7$
6599-84-4

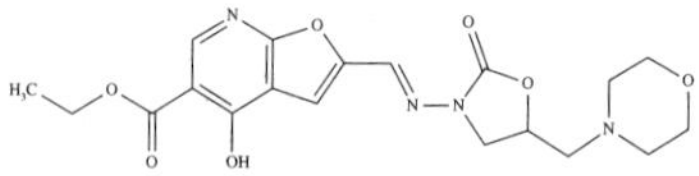

Ethyl 4-hydroxy-2-[[[5-(morpholinomethyl)-2-oxo-3-oxazolidinyl]imino]methyl]furo[2,3-b]pyridine-5-carb-oxylate = 4-Hydroxy-2-[[[5-(4-morpholinylmethyl)-2-oxo-3-oxazolidinyl]imino]methyl]furo[2,3-b]pyridine-5-carboxylic acid ethyl ester (●)
S　*Ebifuramin*, NSC-201047
U　Cytotoxic

8714 (7132)　　　　　　　　　　　$C_{19}H_{22}N_8O_6S_2$
122841-10-5

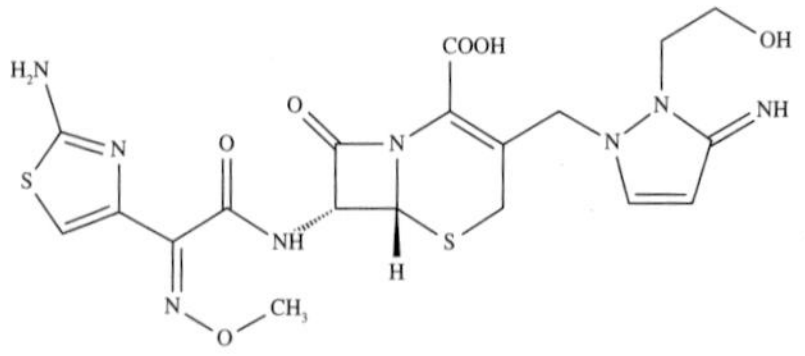

(−)-5-Amino-2-[[[(6R,7R)-7-[2-(2-amino-4-thiazolyl)gly-oxylamido]-2-carboxy-8-oxo-5-thia-1-azabicy-clo[4.2.0]oct-2-en-3-yl]methyl]-1-(2-hydroxyethyl)-1H-pyrazolium hydroxide inner salt 7^2-(Z)-(O-methyloxime) = [6R-[6α,7β(Z)]]-7-[[(2-Amino-4-thiazolyl)(methoxy-imino)acetyl]amino]-3-[[2,3-dihydro-2-(2-hydroxy-

ethyl)-3-imino-1*H*-pyrazol-1-yl]methyl]-8-oxo-5-thia-1-
azabicyclo[4.2.0]oct-2-ene-2-carboxylic acid (●)
R Sulfate (1:1) (122841-12-7)
S *Cefolis sulfate, Cefoselis sulfate***, CFSL sulfate,
      FK 037, Wincef
U Antibiotic

8715 $C_{19}H_{22}O_2$
                                                 157583-18-1

14α,15α-Methyleneestra-1,3,5(10),8-tetraene-3,17α-diol
= (14*S*,15β,17α)-3',15-Dihydrocycloprop[14,15]estra-
1,3,5(10),8-tetraene-3,17-diol (●)
S J 861
U Cytoprotective, anti-oxidant

8716 $C_{19}H_{22}O_2$
                                                 71109-09-6

2-(4-Cyclohexyl-1-naphthyl)propionic acid = 4-Cyclo-
hexyl-α-methyl-1-naphthaleneacetic acid (●)
S CERM 10202, PM 150, Quadrisol, *Vedaprofen***
U Anti-inflammatory (veterinary)

8717 (7133) $C_{19}H_{22}O_2$
                                                 18839-90-2

trans-p-[1,2-Diethyl-2-(*p*-methoxyphenyl)vinyl]phenol =
trans-p,p'-(1,2-Diethylvinylene)diphenol monomethyl
ether = *trans*-4-Hydroxy-4'-methoxy-α,β-diethylstilbene
= (*E*)-4-[1-Ethyl-2-(4-methoxyphenyl)-1-butenyl]phenol
(●)
S *Diäthylstilböstrolmonomethyläther, Diethylstil-
      bestrol monomethyl ether*, Mestilbol, Monomestrol
U Synthetic estrogen

8718 (7134) $C_{19}H_{22}O_3$
                                                 15372-34-6

DL-*cis*-1-Ethyl-7-methoxy-2-methyl-1,2,3,4-tetrahydro-
phenanthrene-2-carboxylic acid = 7-Methylbisdehydro-
doisynolic acid = *cis*-(±)-1-Ethyl-1,2,3,4-tetrahydro-7-
methoxy-2-methyl-2-phenanthrenecarboxylic acid (●)
S BDDA-ME, *Doisynestrol, Doisynoestrol*, Fenocicli-
      na, Fenocyclin, Fenocylin, Phenocyclin, RS-2874,
      Surestrine, Surestryl
U Estrogen

8719 (7135) $C_{19}H_{22}O_3$
                                                 2012-73-9

2-[4-(α,α-Dimethylbenzyl)phenoxy]-2-methylpropionic
acid = 2-Methyl-2-[4-(1-methyl-1-phenylethyl)phen-
oxy]propanoic acid (●)
S ICI 53072, Su-13314
U Antihyperlipidemic

8720   (7136) $C_{19}H_{22}O_4$
                                                 53370-44-8

2-Methyl-1,4-naphthohydroquinone dibutyrate = Butano-
ic acid 2-methyl-1,4-naphthalenediyl ester (●)
S   Karan, Karanum, *Menadiol dibutyrate*
U   Prothrombogenic

8721   (7137) $C_{19}H_{22}O_4$
                                                 13164-04-0

6-(1,1-Dimethyl-2-propenyl)-2,3-dihydro-2-(1-hydroxy-
1-methylethyl)-7H-furo[3,2-g][1]benzopyran-7-one (●)
S   *Chalepin*
U   Anticoagulant from *Clausena anisata*

8722 $C_{19}H_{22}O_5$
                                                 165561-14-8

(1E,5R*,7R*,8aS*,10aS*,12R*,13aR*,13bS*)-
5,6,7,8,8a,10a,12,13,13a,13b-Decahydro-7-hydroxy-12-
methyl-2,5-ethanoindeno[4,5-e]oxecin-3,11,15-trione (●)
S   *Macquarimicin A*
U   Anti-inflammatory (neutral sphingomyelinase inhibi-
        tor)

8723   (7138) $C_{19}H_{22}O_5$
                                                 35413-70-8

α,α-Dimethyl-6,7-dihydrofuro[4',5':6,7]coumarin-7-yl-
methyl valerate = Pentanoic acid 1-(2,3-dihydro-7-oxo-
7H-furo[3,2-g][1]benzopyran-2-yl)-1-methylethyl ester
(●)
S   *Marmesin valerate*
U   Spasmolytic

8724 $C_{19}H_{22}O_7$
                                                 66018-38-0

(3S,5E,8S,9S,11E)-3,4,9,10-Tetrahydro-8,9,16-trihydr-
oxy-14-methoxy-3-methyl-1H-2-benzoxacyclotetrade-
cin-1,7(8H)-dione (●)
S   LL-Z 1640-2
U   Protein and cytokine release inhibitor

8725 $C_{19}H_{22}O_8$
                                                 76958-67-3

(1aR,3S,4S,6Z,9S,15bR)-1a,8,9,15b-Tetrahydro-3,4,12-
trihydroxy-14-methoxy-9-methyl-3H-oxireno[k][2]benz-
oxacyclotetradecin-5,11(2H,4H)-dione (●)
S   *Hypothemycin*
U   Antineoplastic (rassignaling inhibitor)

8726 (7139)

$C_{19}H_{22}O_8S$
57150-72-8

1-[2-Hydroxy-4-(3-sulfopropoxy)phenyl]-3-(3-hydroxy-4-methoxyphenyl)-1-propanone = 3-[3-Hydroxy-4-[3-(3-hydroxy-4-methoxyphenyl)-1-oxopropyl]phenoxy]-1-propanesulfonic acid (●)
R Sodium salt (59881-19-5)
S Chinoin-401
U Sweetener

8727 (7140)

$C_{19}H_{23}ClIN_3O_2$
125141-02-8

5-Chloro-N-[2-[(p-iodobenzyl)methylamino]ethyl]-4-(methylamino)-o-anisamide = 5-Chloro-N-[2-[[(4-iodo-phenyl)methyl]methylamino]ethyl]-2-methoxy-4-(me-thylamino)benzamide (●)
S Spectramide
U Dopamine D_2-receptor antagonist

8728 (7141)

$C_{19}H_{23}ClNO_2$
51489-68-0

[2-(2-Acetyl-4-chlorophenoxy)ethyl]benzyldimethylam-monium = N-[2 (2-Acetyl-4-chlorophenoxy)ethyl]-N,N-dimethylbenzenemethanaminium (●)
R p-Chlorobenzenesulfonate (closilate) (51489-69-1)
S Bemosat, G 526
U Anthelmintic

8729 (7142)

$C_{19}H_{23}ClNO_3$
47364-20-5

N-(3-Acetyl-5-chloro-2-hydroxybenzyl)-N,N-dimethyl-N-(β-phenoxyethyl)ammonium = 3-Acetyl-5-chloro-2-hydroxy-N,N-dimethyl-N-(2-phenoxyethyl)benzeneme-thanaminium (●)
R 3-Hydroxy-2-naphthoate (no. 2544) (34987-38-7)
S Difesyl, Difezil, Diphezyl, G-472-b
U Anthelmintic

8730 (7143)

$C_{19}H_{23}ClNO_3S$
139995-67-8

5-[(2-Chlorophenyl)methyl]-2-(2,2-dimethyl-1-oxoprop-oxy)-4,5,6,7-tetrahydro-5-hydroxythieno[3,2-c]pyridi-nium (●)
S SR 26831
U Elastase inhibitor

8731 (7144)

$C_{19}H_{23}ClN_2$
848-53-3

1-(p-Chloro-α-phenylbenzyl)hexahydro-4-methyl-1H-1,4-diazepine = 1-(p-Chlorodiphenylmethyl)-4-methyl-perhydro-1,4-diazepine = 1-(p-Chlorodiphenylmethyl)-4-methyl-1,4-diazacycloheptane = 1-(4-Chlorobenzhy-

dryl)perhydro-4-methyl-1,4-diazepine = 1-[(4-Chloro-phenyl)phenylmethyl]hexahydro-4-methyl-1H-1,4-diaze-pine (●)
R also dihydrochloride (1982-36-1)
S Aller, Antihismine, Attackmin, Berahalten, Clomon-S, Curosajin, Echlizin, Hischro, Histalizine, Homadamon, Homochlo, Homochlogyl, *Homochlorcyclizine***, Homoclicin, Homoclomin, Homocolzine, Homodyki, Homoginin, Homojin, Homolesmin, Homomallermin, Homoradin, Homorestar, Homotolin, Lysilan, Neochlophin, Neohista, Neukohis, Palphard, Puradenin, Rimskin, Rolemin, SA-97, Sacronal, Sankumin, Wagmalin, Wicron, Zenchlomin
U Antihistaminic, anti-allergic (serotonin antagonist)

8732 (7145) $C_{19}H_{23}ClN_2$
303-49-1

3-Chloro-5-[3-(dimethylamino)propyl]-10,11-dihydro-5H-dibenz[b,f]azepine = 3-(3-Chloro-10,11-dihydro-5H-dibenz[b,f]azepin-5-yl)propyldimethylamine = 3-Chloro-10,11-dihydro-N,N-dimethyl-5H-dibenz[b,f]azepine-5-propanamine (●)
R Monohydrochloride (17321-77-6)
S Anafranil, Anafril, Apo-Clomipramine, *Chlorimipramine hydrochloride*, Clofranil, Clomicalm, Clomifril, *Clomipramine hydrochloride***, Clopress, Deprelin "Astorga", G 34586, Gen-Clomipramine, Hydiphen, *Klomipraminhydrochlorid*, Maronil, *Monochlorimipramine hydrochloride*, Neoprex, Novo-Clopamine, Placil, Tranquax
U Antidepressant

8733 (7146) $C_{19}H_{23}ClN_2O_2$
103624-59-5

(±)-2-Chloro-12-[3-(dimethylamino)-2-methylpropyl]-12H-dibenzo[d,g][1,3,6]dioxazocine = 2-Chloro-N,N,β-trimethyl-12H-dibenzo[d,g][1,3,6]dioxazocine-12-propanamine (●)
R Monohydrochloride (70133-85-6)
S Dimopridin, Egyt-2509, *Traboxopine hydrochloride***
U Neuroleptic, antipsychotic

8734 (7147) $C_{19}H_{23}ClN_2S$
84-01-5

2-Chloro-10-[3-(diethylamino)propyl]phenothiazine = 2-Chloro-N,N-diethyl-10H-phenothiazine-10-propanamine (●)
R also monohydrochloride (4611-02-3)
S *Chlorproethazine***, Neuriplège, 4909 R.P.
U Neuroleptic, spasmolytic

8735 (7148) $C_{19}H_{23}ClN_2S$
67643-14-5

1-[1-[5-Chloro-2-(phenylthio)phenyl]ethyl]-4-methylpi-
perazine (●)
R   Maleate (1:2) (67729-52-6)
S   *Fethiozine maleate*, VÚFB-12257
U   Local anesthetic, anthelmintic

8736 (7149) $C_{19}H_{23}ClO_2$
4091-75-2

16α-Chloro-3-methoxyestra-1,3,5(10)-trien-17-one =
16α-Chloroestrone 3-methyl ether = (16α)-16-Chloro-3-
methoxyestra-1,3,5(10)-trien-17-one (●)
S   Arteriosterol, Arterolo, Athéran-N, *Chloroestrone
methyl ether, Chloröstronmethyläther*, Clomestro-
num, Colesterel, Iposclerone, Lesterol "Scharper",
Liprotene, *Methoxychloroestrone*, Metilalfa, Ormo-
vas, Persclerol, SC-8246, Sclerokin, Sincolestrol,
Steroclazina
U   Anticholesteremic

8737 (7150) $C_{19}H_{23}Cl_2NO_4$
104970-08-3

(±)-1-(2,4-Dichlorophenoxy)-3-[[2-(3,4-dimethoxyphe-
nyl)ethyl]amino]-2-propanol (●)
R   also without (±)-definition (83335-62-0)

S   B 24/76
Y   β-Adrenergic blocker

8738 (7151) $C_{19}H_{23}FN_2O_3$
2804-00-4

8-[3-(*p*-Fluorobenzoyl)propyl]-2-methyl-2,8-diazaspi-
ro[4.5]decane-1,3-dione = 8-[4-(4-Fluorophenyl)-4-oxo-
butyl]-2-methyl-1,8-diazaspiro[4.5]decane-1,3-dione (●)
S   FR-33, R 7158, *Roxoperone***
U   Neuroleptic

8739 (7152) $C_{19}H_{23}F_2N_3O_3$
91188-00-0

1-Ethyl-7-[3-[(ethylamino)methyl]-1-pyrrolidinyl]-6,8-
difluoro-1,4-dihydro-4-oxo-3-quinolinecarboxylic acid
(●)
R   also with (±)-definition (110013-21-3)
S   CI-934, FL-17, *Merafloxacin***, PD 114843
U   Antibacterial (veterinary)

8740 (7153) $C_{19}H_{23}I_2NO_2$
22103-14-6

4-Hydroxy-3,5-diiodo-α-[1-[(1-methyl-3-phenylpro-
pyl)amino]ethyl]benzyl alcohol = 1-(4-Hydroxy-3,5-di-
iodophenyl)-2-[(1-methyl-3-phenylpropyl)amino]-1-pro-
panol = 4-Hydroxy-3,5-diiodo-α-[1-[(1-methyl-3-phenyl-
propyl)amino]ethyl]benzenemethanol (●)

S *Bufeniode***, Diastal "Bayropharm", *Diiodobupheni-ne*, 241-HF, Proclival
U Antihypertensive, vasodilator

8741 (7154) $C_{19}H_{23}N$
60607-24-1

N-Methyl-*N*-(2-phenylpropyl)cinnamylamine = 1-(*N*-Cinnamylmethylamino)-2-phenylpropane = *N*,β-Dimethyl-*N*-(3-phenyl-2-propenyl)benzeneethanamine (●)
S Filocor, Vasoflex "Merrell"
U Coronary vasodilator

8742 (7155) $C_{19}H_{23}NO$
90-86-8

α-[1-(Cinnamylmethylamino)ethyl]benzyl alcohol = 2-(Cinnamylmethylamino)-1-phenyl-1-propanol = α-[1-[Methyl(3-phenyl-2-propenyl)amino]ethyl]benzenemethanol (●)
S *Cinnamedrine***, *Cinnamylephedrine*
U Smooth muscle relaxant

8743 (7156) $C_{19}H_{23}NO$
52742-40-2

N-(3-Methoxy-3,3-diphenylpropyl)allylamine = *N*-Allyl-3-methoxy-3,3-diphenylpropylamine = 1,1-Diphenyl-1-methoxy-3-(allylamino)propane = γ-Methoxy-γ-phenyl-*N*-2-propenylbenzenepropanamine (●)

S *Alimadol***
U Analgesic

8744 (7157) $C_{19}H_{23}NO$
147-20-6

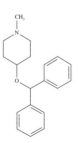

1-Methyl-4-piperidyl benzhydryl ether = 4-(Benzhydryloxy)-1-methylpiperidine = 4-(Diphenylmethoxy)-1-methylpiperidine (●)
R Hydrochloride (132-18-3)
S Aerogastol, Alergipan, Allergen, Allergie S, Allerzine, 1041 A.N., Anginosan, Anti-H 10, Antihistamin "Linz", Antihist Retard, Antivom "Ritsert", Arbid N, Atinal, Belfene, Cupertin, Dayfen, Diafen "3M", Difenpir, Difrin, Diphenine "Maney", *Diphenylpyraline hydrochloride***, *Diphenylpyrilene hydrochloride*, Dorahist, Eskayol, Hispril, Histalert, Histargan, Histrol, Histryl, Histyn, Hystamin, Hystryl, Kolten-Gelee, Lergoban, Lergobine, Lyssipoll, Mepiben, Neargal, Neo-Kohis, Neo-Lergic, P 253, Phenypyra, Pyrahist, Sumadil
U Antihistaminic, anti-allergic

8744-01 (7157-01) 606-90-6
R 8-Chlorotheophyllinate (1:1)
S Agiell, Colton, *Diphenylpyraline teoclate*, Kolton, Koltonal, Mepedyl, P 284, *Piprinhydrinate***, Plokon
U Antihistaminic, anti-emetic

8744-02 (7157-02)
R 7-Theophyllineacetate (no. 1619)
S Disergan
U Antihistaminic, anti-emetic

8745 (7158) $C_{19}H_{23}NO_2$
                                      105051-87-4

4-Aminoandrosta-1,4,6-triene-3,17-dione (●)
S FCE 24928, *Minamestane***
U Antineoplastic (aromatase inhibitor)

8746 (7159) $C_{19}H_{23}NO_2$
                                      56030-50-3

(+)-2,3,4,5-Tetrahydro-7,8-dimethoxy-3-methyl-1-phe-
nyl-1*H*-3-benzazepine (●)
R Maleate (1:1) (39624-66-3)
S Sch 12679, *Trepipam maleate***, *Trimopam maleate*
U Antipsychotic

8747 (7160) $C_{19}H_{23}NO_2$
                                      64622-45-3

2-Pyridylmethyl 2-(*p*-isobutylphenyl)propionate = Ibu-
profen 2-pyridylmethyl ester = α-Methyl-4-(2-methylpro-
pyl)benzeneacetic acid 2-pyridinylmethyl ester (●)
R also with (±)-definition (112017-99-9)
S BE-100, Freshing, *Ibuprofen piconol*, Pair-Acne, *Pi-
       meprofen*, Pymeprofen, Staderm, U-75630, Vesicum
U Topical anti-inflammatory, antipsoriatic

8748 $C_{19}H_{23}NO_2$
                                      3423-13-0

1,2,3,4-Tetrahydro-6,7-dimethoxy-2-methyl-1-(phenyl-
methyl)isoquinoline (●)
R Hydrochloride (6130-52-5)
S CSH 72
U PAF antagonist

8749 (7161) $C_{19}H_{23}NO_2$
                                      25314-87-8

α-[(Diethylamino)methyl]benzyl alcohol benzoate (ester)
= Benzoic acid α-[(diethylamino)methyl]benzyl ester =
α-[(Diethylamino)methyl]benzenemethanol benzoate
(ester) (●)
S *Elucaine**
U Local anesthetic, gastric anticholinergic

8750 (7162) $C_{19}H_{23}NO_2$
                                      525-01-9

β-Morpholinoethyl benzhydryl ether = 4-[2-(Diphenyl-methoxy)ethyl]morpholine (●)
S   A-446, Linadryl
U   Antihistaminic

8751   (7163) $C_{19}H_{23}NO_2S$
                                                      4295-63-0

9-[3-(Dimethylamino)propyl]-2-methoxythioxanthen-9-ol = 9-[3-(Dimethylamino)propyl]-2-methoxy-9*H*-thioxanthen-9-ol (●)
S   *Meprothixol, Meprotixol**, N 7020
U   Antitussive, neuroleptic, anti-inflammatory, analgesic

8752   (7164) $C_{19}H_{23}NO_3$
                                                      77955-41-0

(±)-Methyl *p*-[2-[(β-hydroxyphenethyl)amino]pro-pyl]benzoate = (*R**,*R**)-(±)-4-[2-[(2-Hydroxy-2-phenyl-ethyl)amino]propyl]benzoic acid methyl ester (●)
R   Fumarate (2:1) (87857-42-9)
S   BRL 26830 A
U   Antihyperglycemic (thermogenic β-adrenoceptor agonist)

8753   (7171) $C_{19}H_{23}NO_3$
                                                      3572-52-9

2-(Diethylamino)ethyl 3-phenylsalicylate = 3-Phenylsali-cylic acid β-(diethylamino)ethyl ester = 2-Hydroxy-[1,1'-biphenyl]-3-carboxylic acid 2-(diethylamino)ethyl ester (●)
R   Hydrochloride (5560-62-3)
S   Alvinine (active substance), *Biphenamine hydrochlo-ride*, Melsaphine, Sebaclen, Sébaklen, *Xenysalate hy-drochloride**
U   Topical anesthetic, antibacterial, antifungal (treatment of seborrhea)

8754   (7165) $C_{19}H_{23}NO_3$
                                                      69174-75-0

Succinic acid mono-3-guaiazulenamide = 4-[[3,8-Di-methyl-5-(1-methylethyl)-1-azulenyl]amino]-4-oxobuta-noic acid (●)
S   TPH-3
U   Gastric antisecretory, anti-ulcer agent

8755   (7172) $C_{19}H_{23}NO_3$
76-58-4

4,5α-Epoxy-3-ethoxy-17-methylmorphin-7-en-6α-ol =
3-Ethoxy-6-hydroxy-N-methyl-4,5-epoxymorphin-7-ene
= Ethylmorphine = (5α,6α)-7,8-Didehydro-4,5-epoxy-3-
ethoxy-17-methylmorphinan-6-ol (●)
R   Hydrochloride (125-30-4) [or hydrochloride dihydra-
te (6746-59-4)]
S   *Aethomorphinum*, *Äthylmorphinhydrochlorid*, Chlo-
romyl, Cocillana, *Codethyline*, *Codetilina*,
Collins Elixier, Cosylan, Diocil, Diolan, Dionin,
Dionina, Dionjuste, Diosan, *Ethomorphine*, *Ethyl-
morphine hydrochloride*, Etyfin, Pneumax, Syropon,
Tionidel
U   Antitussive, analgesic, mydriatic

8756   (7169) $C_{19}H_{23}NO_3$
98769-81-4

(±)-(2R*)-2-[(αR*)-α-(o-Ethoxyphenoxy)benzyl]mor-
pholine = (R*,R*)-(±)-2-[(2-Ethoxyphenoxy)phenylme-
thyl]morpholine (●)
R   Monomethanesulfonate (98769-82-5)
S   Davedax, Edronax, FCE 20124, Lonol "Promeco",
Norebox, PNU-155950 E, Prolift, *Reboxetine mesila-
te**, Vestra
U   Antidepressant

8757   (7170) $C_{19}H_{23}NO_3$
6037-21-4

α-Methyl-α-1-naphthyl-4-morpholinebutyric acid = α-
Methyl-α-(2-morpholinoethyl)-1-naphthaleneacetic acid
= α-Methyl-α-1-naphthalenyl-4-morpholinebutanoic acid
(●)
R   Hydrochloride (6680-44-0)
S   DA 1627
U   Choleretic

8758   (7166) $C_{19}H_{23}NO_3$
94915-04-5

3',4'-Dihydroxy-α-[(1-methyl-3-phenylpropyl)ami-
no]propiophenone = 1-(3,4-Dihydroxyphenyl)-2-[(1-me-
thyl-3-phenylpropyl)amino]-1-propanone (●)
R   Hydrochloride (3810-85-3)
S   Asthma-Keton 11, Asthma-Tropon
U   Anti-asthmatic

8759   (7167) $C_{19}H_{23}NO_3$
15687-41-9

L-3-[(β-Hydroxy-α-methylphenethyl)amino]-3'-meth-
oxypropiophenone = L-3-Methoxy-ω-(1-hydroxy-1-phe-
nylisopropylamino)propiophenone = [R-(R*,S*)]-3-[(2-
Hydroxy-1-methyl-2-phenylethyl)amino]-1-(3-methoxy-
phenyl)-1-propanone (●)
R   Hydrochloride (16777-42-7)

S CGG 6 E, D 563, Ildamen, Modacor "I.S.H.", Myofedrin (new form), *Oxifedrini chloridum*, *Oxyfedrine hydrochloride***

U Coronary vasodilator

8760 (7168)

$C_{19}H_{23}NO_3$
21071-51-2

DL-3-[(β-Hydroxy-α-methylphenethyl)amino]-3'-methoxypropiophenone = (R*,S*)-(±)-3-[(2-Hydroxy-1-methyl-2-phenylethyl)amino]-1-(3-methoxyphenyl)-1-propanone (●)

R Hydrochloride (14223-94-0)

S Myofedrin, DL-*Oxyfedrine hydrochloride***, DL-*Oxyphedrinum hydrochloricum*

U Coronary vasodilator

8761 (7174)

$C_{19}H_{23}NO_4$

3β-Benzoyloxytropane-2β-carboxylic acid allyl ester = Benzoylecgonine allyl ester = 3β-Benzoyloxy-2β-allyloxycarbonyltropane = [1R-(exo,exo)]-3-(Benzoyloxy)-8-methyl-8-azabicyclo[3.2.1]octane-2-carboxylic acid 2-propenyl ester (●)

R Hydrochloride (60595-63-3)

S *Benzoylallylecgonine*, Nycain

U Anesthetic

8762 (7175)

$C_{19}H_{23}NO_4$
67950-87-2

(±)-1,2,3,4-Tetrahydro-1-(3,4,5-trimethoxybenzyl)-6-isoquinolinol = (±)-1,2,3,4-Tetrahydro-1-[(3,4,5-trimethoxyphenyl)methyl]-6-isoquinolinol (●)

R Hydrochloride (65718-00-5)

S CV 705

U Vasodilator

8763 (7173)

$C_{19}H_{23}NO_4$
115-53-7

4-Hydroxy-3,7-dimethoxy-17-methylmorphin-7-en-6-one = (9α,13α,14α)-7,8-Didehydro-4-hydroxy-3,7-dimethoxy-17-methylmorphinan-6-one (●)

S *Coculine, Kukoline, Sinomenine*

U Antirheumatic, antineuralgic (alkaloid from *Sinomenium acutum*), antidote

---

8764 (7176)

$C_{19}H_{23}NO_4$
75305-17-8

6-[[2-(3,4-Dihydroxyphenyl)-1-methylethyl]amino]-5,6,7,8-tetrahydro-2,3-naphthalenediol (●)

S ASL-7022

U Coronary vasodilator

8765 (7177) $C_{19}H_{23}NO_4S$
92642-94-9

2-[[[2-(2,6-Dimethoxyphenoxy)ethyl]amino]methyl]-
1,4-benzoxathiane = *N*-[2-(2,6-Dimethoxyphen-
oxy)ethyl]-2,3-dihydro-1,4-benzoxathiin-2-methan-
amine (●)
R also hydrochloride (92642-97-2)
S *Benoxathian*
U α_1-Adrenoceptor antagonist

8766 (7178) $C_{19}H_{23}NO_4S$
123875-01-4

2-(Phenylthio)ethyl methyl 1,4-dihydro-2,4,6-trimethyl-
3,5-pyridinedicarboxylate = 1,4-Dihydro-2,4,6-trimethyl-
3,5-pyridinedicarboxylic acid methyl 2-(phenylthio)ethyl
ester (●)
S PCA-4248
U PAF receptor antagonist, antihypertensive

8767 (7179) $C_{19}H_{23}NO_5$
613-67-2

N-[2-(2,6-Dimethoxyphenoxy)ethyl]-1,4-benzodioxan-2-
methylamine = *N*-[2-(2,6-Dimethoxyphenoxy)ethyl]-2,3-
dihydro-1,4-benzodioxine-2-methanamine (●)
S WB 4101
U α_1-Adrenoceptor antagonist

8768 (7180) $C_{19}H_{23}NO_5$
30418-38-3

(−)-1,2,3,4-Tetrahydro-1-(3,4,5-trimethoxybenzyl)-6,7-
isoquinolinediol = (*S*)-1,2,3,4-Tetrahydro-1-[(3,4,5-tri-
methoxyphenyl)methyl]-6,7-isoquinolinediol (●)
R Hydrochloride (18559-59-6)
S Antene, AQ-110, AQL 208, Caluyon, Coughjust,
Darob (old form), Erumeril, Expansolin "Parke-Da-
vis", Festrin, Inalon, Inolin, Izanamin, Methocough,
Pneumobenzil, RO 7-5965, Taisoquinol, TMQ, Tos-
merian, *Tretoquinol hydrochloride**, Trikvinol, *Tri-
methoquinol hydrochloride*, Triquinol, Veberon,
Vems, Youmetoquinol
U Bronchodilator, anti-asthmatic

8769 (7181) $C_{19}H_{23}NO_5$
88939-40-6

(−)-4,5α-Epoxy-3,14-dihydroxy-17-(2-methoxy-
ethyl)morphinan-6-one = (5α)-4,5-Epoxy-3,14-dihydr-
oxy-17-(2-methoxyethyl)morphinan-6-one (●)
R Hydrochloride (88939-41-7)
S Mr 2264 Cl, *Semorphone hydrochloride***
U Narcotic analgesic

8770 $C_{19}H_{23}NO_6$
                                                  141269-99-0

[4-[2-[[(2S)-2-Hydroxy-3-phenoxypropyl]amino]eth-
oxy]phenoxy]acetic acid (●)
S   ZM 215001
U   Lipolysis partial agonist

8771   (7182) $C_{19}H_{23}NO_6$
                                                 55165-22-5

9-[3-(*tert*-Butylamino)-2-hydroxypropoxy]-4-hydroxy-7-
methyl-5*H*-furo[3,2-g][1]benzopyran-5-one = 9-[3-(*tert*-
Butylamino)-2-hydroxypropoxy]-4-hydroxy-7-methylfu-
ro[3,2-g]chromone = 9-[3-[(1,1-Dimethylethyl)amino]-2-
hydroxypropoxy]-4-hydroxy-7-methyl-5*H*-fu-
ro[3,2-g][1]benzopyran-5-one (●)
S   Betanitran, *Butocrolol***, P 18
U   β-Adrenergic blocker

8772 $C_{19}H_{23}NO_7$
                                                 113485-69-1

Oxolinic acid carbitol ester = 5-Ethyl-5,8-dihydro-8-oxo-
1,3-dioxolo[4,5-g]quinoline-7-carboxylic acid 2-(2-eth-
oxyethoxy)ethyl ester (●)
S   JL 93V, Vetoquinol (new form)
U   Antibacterial

8773   (7183) $C_{19}H_{23}NS$
                                                 1886-45-9

11-[3-(Dimethylamino)propyl]-6,11-dihydrodiben-
zo[b,e]thiepine = 6,11-Dihydro-N,N-dimethyldiben-
zo[b,e]thiepine-11-propanamine (●)
R   Hydrochloride (846-54-8)
S   Hydrothiaden
U   Thymoleptic

8774   (7184) $C_{19}H_{23}N_2OS$
                                                 47268-75-7

N,N-Diethyl-N-methyl(10-phenothiazinylcarbonyl)me-
thylammonium = N,N-Diethyl-N-methyl-β-oxo-10H-phe-
nothiazine-10-ethanaminium (●)
R   Methyl sulfate (6249-68-9)
S   Mefazine, Mephasine, Mephazin
U   Spasmolytic

8775　　　　　　　　　　　　　　　　$C_{19}H_{23}N_2O_6P$
147923-04-4

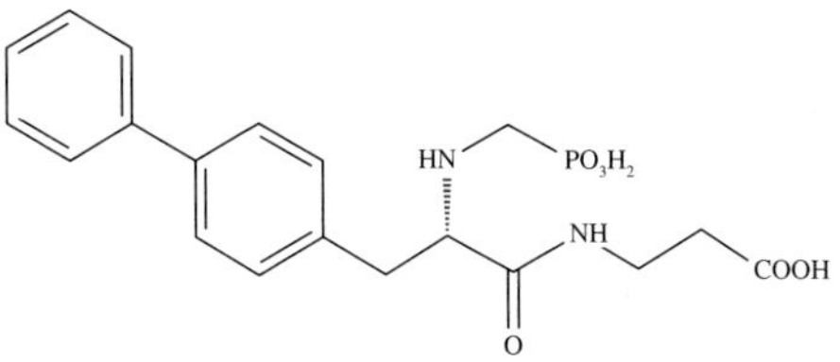

3-[1,1'-Biphenyl]-4-yl-N-(phosphonomethyl)-L-alanyl-β-alanine (●)
S CGS 24592
U Neutral endopeptidase-24,11 inhibitor

8776　(7185)　　　　　　　　　　　　　$C_{19}H_{23}N_3$
56548-50-6

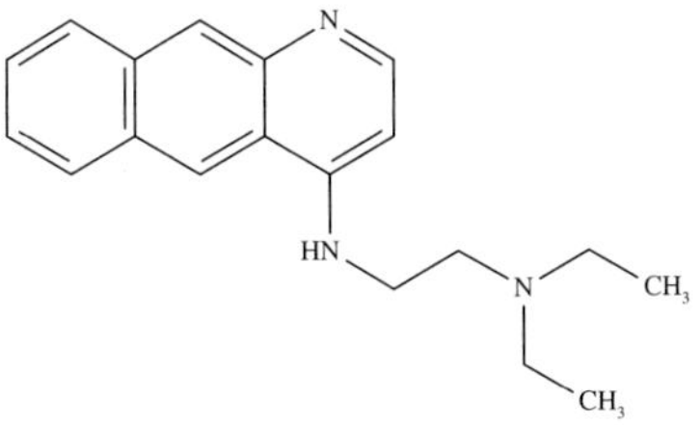

4-[[2-(Diethylamino)ethyl]amino]benzo[g]quinoline = N'-Benzo[g]quinolin-4-yl-N,N-diethyl-1,2-ethanediamine (●)
R Phosphate (1:2) (56548-51-7)
S Dabekhin, Dabequine, G 800
U Antimalarial

8777　(7186)　　　　　　　　　　　　　$C_{19}H_{23}N_3$
60662-16-0

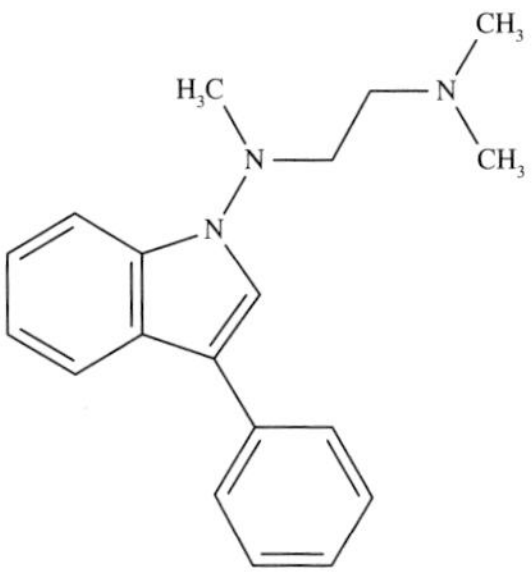

1-[[2-(Dimethylamino)ethyl]methylamino]-3-phenylindole = N,N,N'-Trimethyl-N'-(3-phenyl-1H-indol-1-yl)-1,2-ethanediamine (●)
R Monohydrochloride (57647-35-5)

S *Binedaline hydrochloride**, Binodaline hydrochloride*, Ixprim, RU 780, Sgd Scha 1059
U Antidepressant

8778　(7187)　　　　　　　　　　　　　$C_{19}H_{23}N_3$
57614-23-0

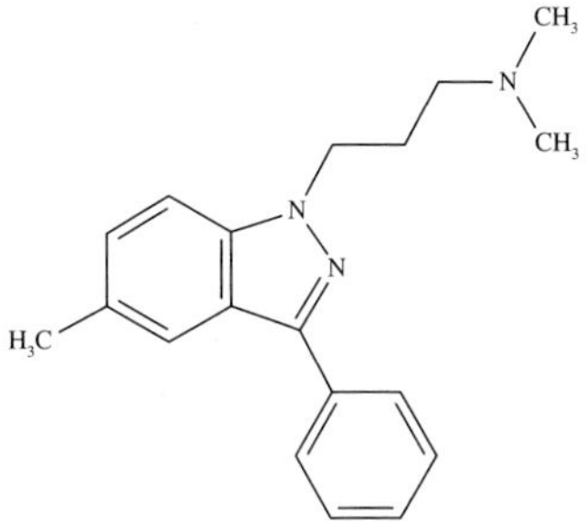

1-[3-(Dimethylamino)propyl]-5-methyl-3-phenyl-1H-indazole = N,N,5-Trimethyl-3-phenyl-1H-indazol-1-propanamine (●)
S FS-32
U Antidepressant

8779　(7188)　　　　　　　　　　　　　$C_{19}H_{23}N_3$
33089-61-1

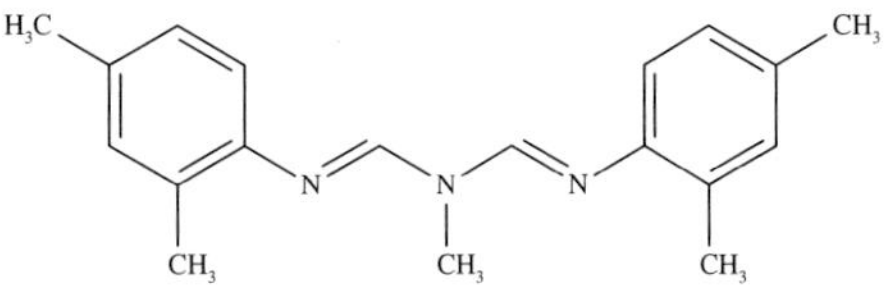

N-Methyl-N'-2,4-xylyl-N-(N-2,4-xylylformimidoyl)formamidine = Methylbis(2,4-xylyliminomethyl)amine = 1,5-Bis(2,4-dimethylphenyl)-3-methyl-1,3,5-triaza-1,4-pentadiene = N'-(2,4-Dimethylphenyl)-N-[[(2,4-dimethylphenyl)imino]methyl]-N-methylmethanimidamide (●)
S Aludex, *Amitraz**, Apivar, BAAM, Biocani-Tique, BTS 27419, Demotick, Ectodex, ENT-27967, Kenaz, Mitaban, Mitac, OMS 1820, Préventic, Taktic, Topline, Triatix, Triatox, U-36059
U Acaricide, scabicide, tickicide, appetite stimulant

8780   (7189) $C_{19}H_{23}N_3O$
642-72-8

1-Benzyl-3-[3-(dimethylamino)propoxy]-1*H*-indazole = 3-[(1-Benzylindazol-3-yl)oxy]propyldimethylamine = *N*,*N*-Dimethyl-3-[[1-(phenylmethyl)-1*H*-indazol-3-yl]oxy]-1-propanamine (●)

R   Monohydrochloride (132-69-4);
see also no. 8541-07

S   AF-864, Afloben, Alcidol, Algiflog, Algiflogue, Alscott, Andolex, Antol, A-Termadol, Azutan, Benalgin "Polfa", Benchixin, Benciflam, Bendacillin, Bendaminol, Bendazol "Kobayashi", Benflogin "Labofarma", Bentazon, Benzaflax, Benzidam, *Benzidammina (cloridrate di)*, Benzidan, *Benzindamina*, Benziplex, Benzirin "Fater; Fixalia", Benzitrat, *Benzydamine hydrochloride***, Benzyrin, Benzyta, Biozendi, Bucco-Tantum, Bucodrin, Bucofaringe, C 1523, Ciclinalgin, Ciflogex, Difflam, Dorinamin, DP 84, Easy Gel, Enzamin, Epirotin, Ernex, Flodont, Flogaton, Flogi Ped, Flogoral, Flogo-Rosa, Ginesal, I.M. 25, Imotryl, Impryl, Indolin (old form), Lagin, Lilizin, Limbron, Lonol "Promero", Maitonin S, Milten, Multum "Lampugnani", Multum rosa, Mytonin, Neo-Barmidon, Omniflam, Opalgyne, Oxinazin, Panflogin, Panxin, Petiflog, Radicalin, Reutissome, Rhino-Tantum, Riripen, Rosalgin, Salyzoron, Sanal, Saniflor, Sawapen, Sidamin, Sun-Benz, Sursum "York", Tamas, Tanflex, Tantum, Tantumar, Vantal, Verax "Tosi", Zydan

U   Analgesic, antipyretic, anti-inflammatory

8780-01   (7189-01)   59831-61-7
R   Salicylate
S   *Benzasal*, Fulgium
U   Anti-inflammatory, analgesic

8781   (7190) $C_{19}H_{23}N_3OS$
104340-86-5

(±)-2-[[2-(Isobutylmethylamino)benzyl]sulfinyl]-1*H*-benzimidazole = 2-[(1-*H*-Benzimidazol-2-ylsulfinyl)methyl]-*N*-methyl-*N*-(2-methylpropyl)benzenamine (●)
S   Leminon, *Leminoprazole***, NC 1300-0-3
U   Anti-ulcer agent (proton pump inhibitor)

8782 $C_{19}H_{23}N_3O_2$
177476-74-3

4-[[3,4-Dioxo-2-[[(1*R*)-1,2,2-trimethylpropyl]amino]-1-cyclobuten-1-yl]amino]-3-ethylbenzonitrile (●)
S   WAY-133537
U   Potassium channel opener

8783   (7191) $C_{19}H_{23}N_3O_2$
60-79-7

N-(2-Hydroxy-1-methylethyl)-D(+)-lysergamide = D-Lysergic acid 1-(hydroxymethyl)ethylamide = D(+)-Lysergic acid β-hydroxyisopropylamide = [8β(*S*)]-9,10-Didehydro-*N*-(2-hydroxy-1-methylethyl)-6-methylergoline-8-carboxamide (●)
S   *Ergobasine*, Ergoklinine, *Ergometrine***, *Ergonovine*, Ergostetrin, *Ergotocine*
U   Oxytocic (ergot alkaloid)

8783-01   (7191-01)   129-51-1
R   Maleate (1:1)

S Arconovina, Cornocentin, Cryovinal, Ergo-Bio-quim, Ergofar, Ergomal, Ergomed "Promed", Ergomet, Ergometine, Ergometron, Ergomine, Ergonovine maleate, Ergostabil, Ergoton-B, Ergotrate Maleate, Ermalate, Ermeton, Ermetrin, Hemogen, Margonovine, Metrergina, Metriclavin, Metrisanol, Novergo, Panergal, Secalysat-EM, Secometrin, Takimetrin, Uteron

U Oxytocic

8783-02 (7191-02)   129-50-0

R Tartrate (1:1)

S Basergin, Ergomar "Nordson", Neofemergen

U Oxytocic

8784 (7192) $C_{19}H_{23}N_3O_2S$
118788-41-3

3-[3,5-Bis(*tert*-butyl)-4-hydroxyphenyl]-7*H*-thiazolo[3,2-*b*][1,2,4]triazin-7-one = 3-[3,5-Bis(1,1-dimethylethyl)-4-hydroxyphenyl]-7*H*-thiazolo[3,2-*b*][1,2,4]triazin-7-one (●)

S HWA 131

U Anti-inflammatory, immunomodulator

8785 (7193) $C_{19}H_{23}N_3O_3$
13424-56-1

2-(Diethylamino)ethyl *p*-(nicotinamido)benzoate = *p*-(Nicotinoylamino)benzoic acid β-(diethylamino)ethyl ester = 4-[(3-Pyridinylcarbonyl)amino]benzoic acid 2-(diethylamino)ethyl ester (●)

S Atheroplex (active substance), D 384, *Nicotinoyl-Procaine*

U Local anesthetic, respiration catalyst

8786 (7194) $C_{19}H_{23}N_3O_3$
132829-83-5

(*R*)-3-Quinuclidinyl (*R*)-α-(hydroxymethyl)-α-phenylimidazole-1-acetate = (*R*)-3-Quinuclidinyl (*R*)-3-hydroxy-2-imidazol-1-yl-2-phenylpropionate = [*R*-(*R**,*R**)]-α-(Hydroxymethyl)-α-phenyl-1*H*-imidazole-1-acetic acid 1-azabicyclo[2.2.2]oct-3-yl ester (●)

S *Espatropate***, UK-88060

U Bronchodilator, anticholinergic

8787 (7195) $C_{19}H_{23}N_3O_3$
73674-85-8

17-Allyl-4,5α-epoxy-3,14-dihydroxymorphinan-6-one hydrazone = (5α)-4,5-Epoxy-3,14-dihydroxy-17-(2-propenyl)morphinan-6-one hydrazone (●)

S *Naloxazone, Naloxone hydrazone*

U Opiate antagonist

8788 $C_{19}H_{23}N_3O_3$
                                                      135463-81-9

2-Oxo-N-(5,6,7,8-tetrahydro-2,3-dimethylfuro[2,3-b]qui-
nolin-4-yl)-1-pyrrolidineacetamide (●)
S *Colupracetam*, MKC-231
U Acetylcholinesterase inhibitor

8789 (7196) $C_{19}H_{23}N_3O_4S$
                                                      3511-16-8

(2S,5R,6R)-6-[(R)-2,2-Dimethyl-5-oxo-4-phenyl-1-imi-
dazolinyl]-3,3-dimethyl-7-oxo-4-thia-1-azabicy-
clo[3.2.0]heptane-2-carboxylic acid = (6R)-6-(2,2-Di-
methyl-5-oxo-4-phenyl-1-imidazolidinyl)penicillanic
acid = [2S-[2α,5α,6β(S^*)]]-6-(2,2-Dimethyl-5-oxo-4-
phenyl-1-imidazolidinyl)-3,3-dimethyl-7-oxo-4-thia-1-
azabicyclo[3.2.0]heptane-2-carboxylic acid (●)
R also potassium salt (5321-32-4);
      see also no. 10590
S Ambiopen, Amfogram, Ampliben "Beta", Anfalin,
      Asklemycin, BL-P 804, BRL 804, Crystallin, Deme-
      tocipen, Emyfarcilina, Etabus "Genefarm", Etaci-
      land, *Etacillina*, Etaciphenol, Etasepti, Fismicin,
      Heciline, Hertin, Hetabiotic, *Hetacillin**, Hetacin
      (-K), Hetacycline, Hetadoce, Hetamcisina, Hetan-
      cinato, H-K Mastitis, Hystra, Keicin, Kyligram, Ma-
      zolica, Natacillin, Neotynen, Nepopen, Penplenum,
      Pensabina, Pentalina, *Phenazacillin*, Procilin, Reclo-
      cidin, Reclomycetin, Relivoxil, Relyosiptic, Stato-
      gram, Tasmacran, Toxedin "Inco", Uropen, Vekfar-
      lin, Versapen (-K), Versatrex, Versuline, Viderbio-
      tic, Vilcobiotic, Vimocilin, Viobretine

U Antibiotic

8790 (7197) $C_{19}H_{23}N_3O_4S$
                                                      18715-92-9

(2S,5R,6S)-6-(2,2-Dimethyl-5-oxo-4-phenyl-1-imidazoli-
dinyl)-3,3-dimethyl-7-oxo-4-thia-1-azabicy-
clo[3.2.0]heptane-2-carboxylic acid = (6S)-6-(2,2-Di-
methyl-5-oxo-4-phenyl-1-imidazolidinyl)penicillanic
acid = [2S-[2α,5α,6α(S^*)]]-6-(2,2-Dimethyl-5-oxo-4-
phenyl-1-imidazolidinyl)-3,3-dimethyl-7-oxo-4-thia-1-
azabicyclo[3.2.0]heptane-2-carboxylic acid (●)
S *Epihetacillin*
U Antibiotic

8791 (7198) $C_{19}H_{23}N_3O_5S$
                                                      53861-02-2

(2S,5R,6R)-6-[(R)-4-(p-Hydroxyphenyl)-2,2-dimethyl-5-
oxo-1-imidazolidinyl]-3,3-dimethyl-7-oxo-4-thia-1-aza-
bicyclo[3.2.0]heptane-2-carboxylic acid = [2S-
[2α,5α,6β(S^*)]]-6-[4-(4-Hydroxyphenyl)-2,2-dimethyl-
5-oxo-1-imidazolidinyl]-3,3-dimethyl-7-oxo-4-thia-1-
azabicyclo[3.2.0]heptane-2-carboxylic acid (●)
R see also no. 7604
S LH-380, *Oxetacillin***
U Antibiotic

8792   (7199) $C_{19}H_{23}N_3S$
                                                        75522-73-5

2-Benzyl-3-[[3-(dimethylamino)propyl]thio]-2H-inda-
zole = 3-[(2-Benzyl-2H-indazol-3-yl)thio]-N,N-dimethyl-
propylamine = N,N-Dimethyl-3-[[2-(phenylmethyl)-2H-
indazol-3-yl]thio]-1-propanamine (●)
S   Dazidamine**
U   Anti-inflammatory

8793   (7200) $C_{19}H_{23}N_4O_6PS$
                                                        22457-89-2

N-[(4-Amino-2-methyl-5-pyrimidinyl)methyl]-N-(4-
hydroxy-2-mercapto-1-methyl-1-butenyl)formamide S-
benzoate O-phosphate = Benzenecarbothioic acid S-[2-
[[(4-amino-2-methyl-5-pyrimidinyl)methyl]formylami-
no]-1-[2-(phosphonooxy)ethyl]-1-propenyl] ester (●)
S   Anevrex, Benfogamma, Benfotial, Benfotiamine**,
    Benphothiamin, Benzoylthiamine monophosphate,
    Berdi, Betivina, Bietamine, Biotamin, Bio-Towa,
    BTMP, 8088 C.B., Milgamma-mono, Neuroluy,
    Neurostop, Tabiomin, Tabiomyl, Vitanévril
U   Analgesic (Vitamin B_1 source)

8794 $C_{19}H_{23}N_5O_2$
                                                        158020-82-7

1-Cyclopentyl-6-(3-ethoxy-4-pyridinyl)-3-ethyl-1,7-di-
hydro-4H-pyrazolo[3,4-d]pyrimidin-4-one (●)
S   SR 265579, Win 65579
U   cGMP phosphodiesterase 5 inhibitor

8795   (7201) $C_{19}H_{23}N_5O_2$
                                                        73090-70-7

2,4-Diamino-5-(3,5-diethoxy-4-pyrrol-1-ylbenzyl)pyri-
midine = 5-[[3,5-Diethoxy-4-(1H-pyrrol-1-yl)phenyl]me-
thyl]-2,4-pyrimidinediamine (●)
S   Epiroprim**, RO 11-8958
U   Antibacterial, sulfa-potentiator

8796   (7202) $C_{19}H_{23}N_5O_3$
                                                        52128-35-5

2,4-Diamino-5-methyl-6-[(3,4,5-trimethoxyanilino)me-
thyl]quinazoline = 5-Methyl-6-[[(3,4,5-trimethoxyphe-
nyl)amino]methyl]-2,4-quinazolinediamine (●)
S   CI-898, JB 11, NSC-249008, TMQ, TMTX, Trime-
    trexate**
U   Antineoplastic (dihydrofolate reductase inhibitor)

8796-01 (7202-01) 82952-64-5
R Mono-D-glucuronate
S Neutrexin, NSC-352122, Oncotrex, *Trimetrexate glucuronate*
U Antineoplastic

N-[[5-[2-[(6*R*)-2-Amino-1,4,5,6,7,8-hexahydro-4-oxopyrido[2,3-*d*]pyrimidin-6-yl]ethyl]-2-thienyl]carbonyl]-L-glutamic acid (●)
S LY 309887
U Antineoplastic (GARFT inhibitor)

8797 (7203) $C_{19}H_{23}N_5O_3S$
95847-70-4

2-[4-[4-(2-Pyrimidinyl)-1-piperazinyl]butyl]-1,2-benzisothiazolin-3-one 1,1-dioxide = 2-[4-[4-(2-Pyrimidinyl)-1-piperazinyl]butyl]-1,2-benzisothiazol-3(2*H*)-one 1,1-dioxide (●)
R Monohydrochloride (92589-98-5)
S Bay q 7821, *Ipsapirone hydrochloride***, TVX Q 7821
U Anxiolytic

8798 $C_{19}H_{23}N_5O_4$
38594-96-6

(*R*)-(−)-*N*[6]-(β-Phenylisopropyl)adenosine = *N*-[(1*R*)-1-Methyl-2-phenylethyl]adenosine (●)
S PIA, *R*-PIA, Th 162
U Adenosine A_2 receptor agonist

8799 $C_{19}H_{23}N_5O_6S$
127228-54-0

8800 (7204) $C_{19}H_{24}ClNO$
5668-06-4

2-[(*p*-Chloro-α-methyl-α-phenylbenzyl)oxy]-*N,N*-dimethylpropylamine = β-(Dimethylamino)isopropyl 4-chloro-α-methylbenzylhydryl ether = 2-[1-(4-Chlorophenyl)-1-phenylethoxy]-*N,N*-dimethyl-1-propanamine (●)
R Citrate (1:1) (56050-03-4)
S Isopropyl-Systral, *Mecloxamine citrate***, *Prophenoxamine citrate*
U Anticholinergic

8801 $C_{19}H_{24}ClNO_2$
199460-44-1

Benzyl(2-chloroethyl)[2-(2-methoxyphenoxy)-1-methylethyl]amine = *N*-(2-Chloroethyl)-*N*-[2-(2-methoxyphenoxy)-1-methylethyl]benzenemethanamine (●)
R Hydrochloride (164857-88-9)
S CM 18
U α-Adrenoceptor blocker

8802

$C_{19}H_{24}Cl_2N_2OS$
157824-29-8

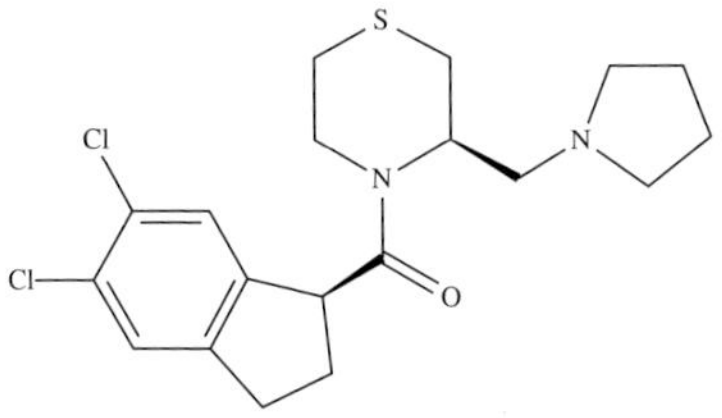

(3*R*)-3-(1-Pyrrolidinylmethyl)-4-[(1*S*)-5,6-dichloro-1-indanylcarbonyl]tetrahydro-1,4-thiazine = [*R*-(*R**,*S**)]-4-[(5,6-Dichloro-2,3-dihydro-1*H*-inden-1-yl)carbonyl]-3-(1-pyrrolidinylmethyl)thiomorpholine (●)
R Monohydrochloride (157824-23-2)
S R 84760
U κ-Opioid receptor agonist

8803

$C_{19}H_{24}Cl_2N_2O_4S$
133229-23-9

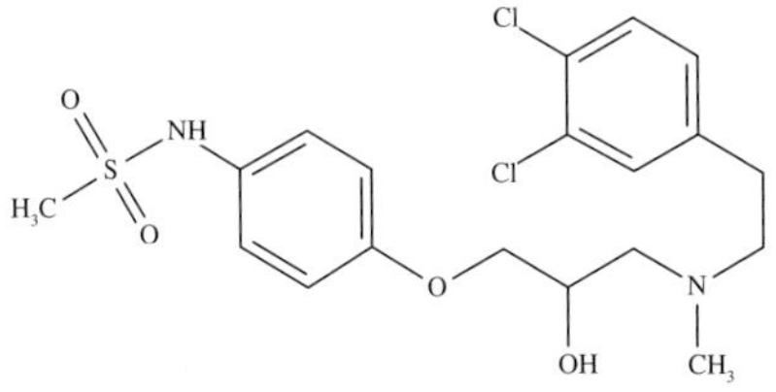

1-[4-(Methanesulfonamido)phenoxy]-3-(*N*-methyl-3,4-dichlorophenethylamino)-2-propanol = *N*-[4-[3-[[2-(3,4-Dichlorophenyl)ethyl]methylamino]-2-hydroxyprop-oxy]phenyl]methanesulfonamide (●)
R Monobenzoate (salt) (178894-81-0)
S AM 92016
U Anti-arrhythmic

8804 (7205)

$C_{19}H_{24}Cl_2N_4O_3$
2764-56-9

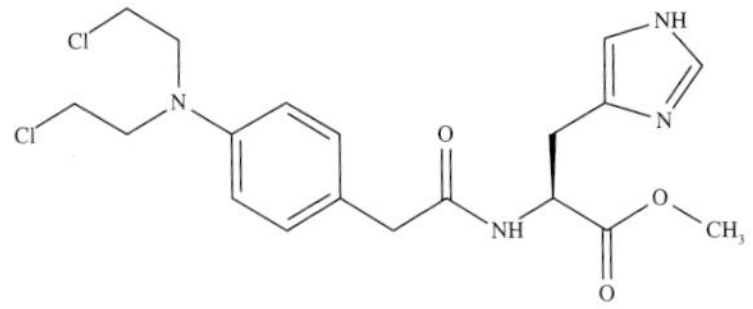

N-[*p*-[Bis(2-chloroethyl)amino]phenylacetyl]-3-(4-imi-dazolyl)alanine methyl ester = *N*-[[4-[Bis(2-chloro-ethyl)amino]phenyl]acetyl]-L-histidine methyl ester (●)
R also dihydrochloride (2792-04-3)
S Hisfen, Hisphen, MD 2

U Antineoplastic

8805

$C_{19}H_{24}FN_3O$
2804-05-9

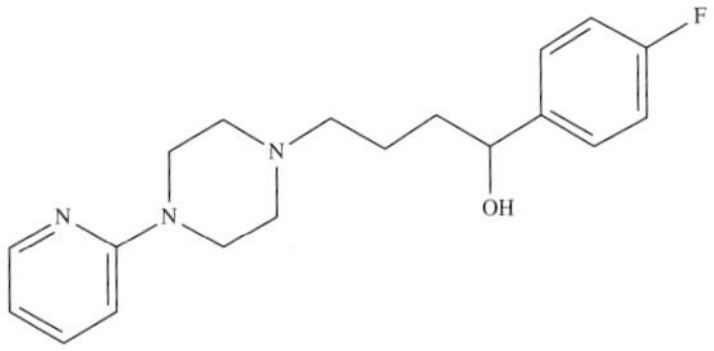

α-(4-Fluorophenyl)-4-(2-pyridinyl)-1-piperazinebutanol (●)
S *Azaperol*
U Neuroleptic

8806 (7206)

$C_{19}H_{24}NO_3$
1553-33-9

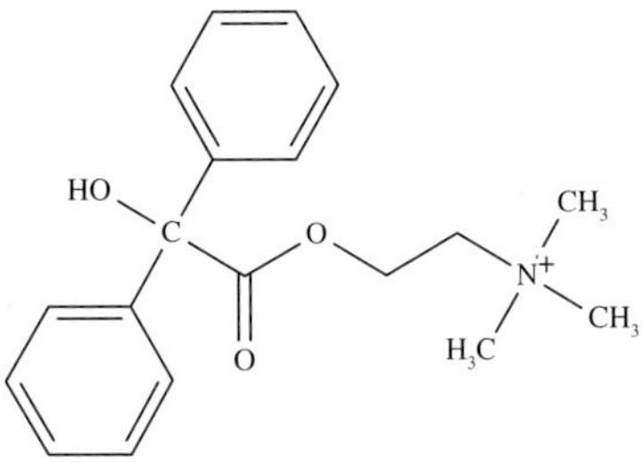

(2-Hydroxyethyl)trimethylammonium benzilate = *N,N,N*-Trimethyl-2-(benziloyloxy)ethylammonium = 2-[(Hydro-xydiphenylacetyl)oxy]-*N,N,N*-trimethylethanaminium (●)
R Bromide (55019-64-2)
S *Benzilylcholine bromide*, Hypaventral
U Anticholinergic, spasmolytic

8806-01 (7206-01)   2424-71-7
R Iodide
S Metacin, *Methacin(um)*, Methoxin, *Metocinium iodi-de**
U Anticholinergic, spasmolytic

8807 $C_{19}H_{24}NO_3$
6871-67-6

(1*R*)-1,2,3,4-Tetrahydro-6-hydroxy-1-[(4-hydroxyphe-nyl)methyl]-7-methoxy-2,2-dimethylisoquinolinium (●)
S *Lotusine*
U Antihypertensive (alkaloid from *Plumula nelumbinis*)

8808 (7207) $C_{19}H_{24}NO_3$
51730-07-5

4,5α-Epoxy-6α-hydroxy-3-methoxy-*N,N*-dimethylmor-phin-7-enium = (5α,6α)-7,8,Didehydro-4,5-epoxy-6-hydroxy-3-methoxy-17,17-dimethylmorphinanium (●)
R Bromide (125-27-9)
S *Codeine methyl bromide*, Eucodin
U Antitussive, sedative, narcotic analgesic

8809 (7208) $C_{19}H_{24}NS_2$

trans-3-(Di-2-thienylmethylene)octahydro-5-methyl-2*H*-quinolizinium (●)
R Bromide (71731-58-3)

S HSR-902, Thiaton, *Tiquizium bromide***
U Anticholinergic, spasmolytic

8810 (7212) $C_{19}H_{24}N_2$
4757-49-7

9,9-Dimethyl-10-[3-(methylamino)propyl]acridan = *N*,9,9-Trimethyl-10(9*H*)-acridinepropanamine (●)
S *Monometacrine***, SD 735
U Antidepressant

8811 (7213) $C_{19}H_{24}N_2$
73-07-4

5,6-Dihydro-*N*-[3-(dimethylamino)propyl]-11*H*-di-benz[*b,e*]azepine = 6,11-Dihydro-*N,N*-dimethyl-5*H*-di-benz[*b,e*]azepine-5-propanamine (●)
S NSC-172129, *Prazepine**, Propazepine
U Antidepressant

8812 (7214) $C_{19}H_{24}N_2$
50-49-7

5-[3-(Dimethylamino)propyl]-10,11-dihydro-5*H*-di-benz[*b,f*]azepine = *N*-(3-Dimethylaminopropyl)iminodibenzyl = 3-(10,11-Dihydro-5H-dibenz[b,f]azepin-5-yl)propyldimethylamine = 10,11-Dihydro-N,N-dimethyl-5H-dibenz[b,f]azepine-5-propanamine (●)

R most monohydrochloride (113-52-0)

S Antidep, Antideprin, Antipress "Lemmon", Apo-Imipramine, Berkomine, Censtim, Chemipramine, Chimoreptin, Chrystemin, Chrytemin, Clonimip, Daypress, Depramine "Adelco", Depranil, Deprenil "Polfa", Depress, Deprimin, Deprinol, Depsonil, Dimipressin, Dynaprin, Dyna-Zina, Efuranol, Emprime, Ethipramine, Eupramin, Feinalmin, G 22355, Ia-pram, Imavate "Robins", *Imidobenzyle*, Imidol, Imilanyle, Iminil, Imipra, Imipramina, *Imipramine***, Imipran, Imipranil, Imiprin, *Imizin(um)*, Impamin(e), Impram, Impramin, Impranil, Imprasine, Impratab, Impratone, Impril, Imp-Tab, Intalpram, Iprogen, Iramil, Irmin, Janimine, Medipramine, Melipramin, Mepramin, Meripramin, Nixapress, Norfranil, Norpramine "Norton", Novopramine, NSC-169866, Panpramine, Pramin "Unique", Praminil, Presamine, Priloigan, Primonil, Prodepress, Promiben, Propep, Pryleugan, Psicopramin, Psychoforin, Sedacoroxen, Servipramine, Sipramine, Skat, SK-Pramine, Smeralium, Somipra, Surplix, Talpramin, Thymopramin(e), Timolet, Tipramine, Tizipramine, Tofranil, Venefon, Vionyl, W.D.D. Tabs

U Antidepressant

8812-01   (7214-01)   10075-24-8
R Embonate (pamoate) (2:1)
S Tofranil pamoato, Tofranil-PM
U Antidepressant

8813   (7211)   C$_{19}$H$_{24}$N$_2$
                                                      22136-27-2

3-Methyl-3-[3-(methylamino)propyl]-1-phenylindoline = 2,3-Dihydro-N,3-dimethyl-1-phenyl-1H-indole-3-propanamine (●)

R p-Toluenesulfonate (1:1) (23226-37-1)

S *Daledalin tosilate***, UK 3557-15
U Antidepressant

8814   (7210)   C$_{19}$H$_{24}$N$_2$
                                                      4945-47-5

4-(N-Benzylanilino)-1-methylpiperidine = N-Benzyl-N-phenyl-4-amino-1-methylpiperidine = N-Benzyl-N-(1-methyl-4-piperidyl)aniline = 1-Methyl-N-phenyl-N-(phenylmethyl)-4-piperidinamine (●)

R also monohydrochloride (1229-69-2) or lactate (1:1) (61670-09-5)
S *Bamipine***, *Piperamine*, Soventelo, Soventilo, Soventol, Taumidrine, Tomindrina
U Antihistaminic, anti-allergic

8815   (7209)   C$_{19}$H$_{24}$N$_2$
                                                      493-80-1

1-(2-N-Benzylanilinoethyl)pyrrolidine = N-Phenyl-N-(2-pyrrolidinoethyl)benzylamine = N-Phenyl-N-(phenylmethyl)-1-pyrrolidineethanamine (●)

R Monohydrochloride (6113-17-3)
S Domistan, *Histapyrrodine hydrochloride***, Luvistin
U Antihistaminic, anti-allergic

8816   C$_{19}$H$_{24}$N$_2$O
178929-75-4

(1R,2S,3S,5S)-8-Methyl-2-(3-methyl-5-isoxazolyl)-3-(4-methylphenyl)-8-azabicyclo[3.2.1]octane (●)
R Monohydrochloride (179089-36-2)
S RTI 171
U Narcotic antagonist

8817   (7218)   C$_{19}$H$_{24}$N$_2$O
60-46-8

4-(Dimethylamino)-2,2-diphenylvaleramide = α-[2-(Di-methylamino)propyl]-α-phenylbenzeneacetamide (●)
R Sulfate (1:1) (29620-30-2)
S *Aminopentamide sulfate*, BL 139, Centrin(e), Compound 3 K 1358, *Dimevamide sulfate***, Estomida, Pentamyl, Valeramide-OM, Valerylamide
U Anticholinergic

8818   C$_{19}$H$_{24}$N$_2$O
135729-61-2

(3aS)-2,3,3a,4,5,6-Hexahydro-2-[(3S)-3-quinuclidinyl]-1H-benz[de]isoquinolin-1-one = (3aS)-2-(3S)-1-Azabicy-clo[2.2.2]oct-3-yl-2,3,3a,4,5,6-hexahydro-1H-benz[de]isoquinolin-1-one (●)
R Monohydrochloride (135729-62-3);
see also no. 8678
S *Palonosetron hydrochloride***, RS 25259-197
U Anti-emetic, antinauseant (5-HT$_3$ receptor antagonist)

8819   (7215)   C$_{19}$H$_{24}$N$_2$O
6829-98-7

5-[3 (Dimethylamino)propyl]-10,11-dihydro-5H-di-benz[b,f]azepine N-oxide = 10,11-Dihydro-N,N-dime-thyl-5H-dibenz[b,f]azepine-5-propanamine N-oxide (●)
R also monohydrochloride (19864-71-2)
S Diamifan, Ditisan, Elepsin, Iminox, *Imipramine N-oxide, Imipraminoxide***, Imiprex, Nediol, NSC-169424
U Antidepressant

8820   (7216)   C$_{19}$H$_{24}$N$_2$O
19877-89-5

(3α,14β,16α)-14,15-Dihydroeburnamenin-14-ol (●)
S RGH-4406, Torondel, *Vincanol**
U Cerebrotonic

8821 $C_{19}H_{24}N_2O$
                                                                136982-36-0

(+)-(2S,3S)-3-[(2-Methoxybenzyl)amino]-2-phenylpipe-
ridine = (2S,3S)-N-[(2-Methoxyphenyl)methyl]-2-phe-
nyl-3-piperidinamine (●)
S CP-99994
U Anti-arthritic, anti-asthmatic (tachykinin NK_1 receptor antagonist)

8822   (7217) $C_{19}H_{24}N_2O$
                                                               13931-73-2

N-(1-Phenylheptyl)nicotinamide = N-(1-Phenylheptyl)-3-
pyridinecarboxamide (●)
S Lytosin
U Spasmolytic

8823   (7219) $C_{19}H_{24}N_2OS$
                                                               3735-90-8

S-[2-(Diethylamino)ethyl] diphenylthiocarbamate =
Diphenylcarbamic acid 2-(diethylamino)ethyl thioester =
Diphenylcarbamothioic acid S-[2-(diethylamino)ethyl]
ester (●)

S Bayer 1355, *Diphencarbamide*, Escorpal, *Fencarba-*
mide**, *Fenocarbamida*, *Phencarbamide*, Rispa-
sulf, Wh-3363
U Anticholinergic, spasmolytic

8824   (7220) $C_{19}H_{24}N_2OS$
                                                               60-99-1

(−)-10-[3-(Dimethylamino)-2-methylpropyl]-2-methoxy-
phenothiazine = (−)-3-(2-Methoxyphenothiazin-10-yl)-2-
methylpropyldimethylamine = (R)-2-Methoxy-N,N,β-tri-
methyl-10H-phenothiazine-10-propanamine (●)
R most maleate (1:1) (7104-38-3) or monohydrochloride (1236-99-3)
S Bayer 1213, CL 36467, CL 39743, Dedoran, Hirn-
amin, Laevomazine, *Laevomepromazine*, Lepurin,
Lerium, Levamin, Levaru, Levium "Hexal", Levoci-
na, Levohalte, *Levomeprazine*, *Levomepromazine***,
Levomezine, Levonormal, Levopromazin "Spofa",
Levopromazina, Levoprome, Levotomin, Levozan,
Levozin(e), *Methotrimeprazine*, Milezin, Minozi-
nan, Neozine "Rhodia", Neuractil, Neurocil,
NIH 7788, Nirvan, No-calm, Novo-Meprazine, Nozi-
nan, Procrazine, Ronexine, 7044 R.P., Sinogan,
SKF 5116, Sofmin, Tisercin, Tisercinetta, Tiserzin,
Togrel, Veractil
U Neuroleptic, analgesic

8824-01   (7220-01)
R Embonate (pamoate)
S Nozinan Embonate, Specia 246
U Neuroleptic

8825   (7222) $C_{19}H_{24}N_2O_2$
6376-26-7

2-[2-(Diethylamino)ethoxy]benzanilide = O-[2-(Diethylamino)ethyl]salicylanilide = Salicylanilide diethylaminoethyl ether = 2-[2-(Diethylamino)ethoxy]-N-phenylbenz-
amide (●)
R   also monohydrochloride
S   Ajucin simplex, Ajurac, Biospal, M 811, *Salverine***
U   Spasmolytic, analgesic

8826   (7223) $C_{19}H_{24}N_2O_2$
74249-07-3

[R-(R*,S*)]-4-[3-[(2-Hydroxy-2-phenylethyl)amino]bu-
tyl]benzamide (●)
R   Monohydrochloride (72332-32-2)
S   LY 104119
U   Anorexic (β-adrenergic agonist)

8827   (7224) $C_{19}H_{24}N_2O_2$
6443-50-1

N-[2-(m-Methoxyphenoxy)propyl]-2-m-tolylacetamidine
= N-[2-(3-Methoxyphenoxy)propyl]-N-[α-(m-tolyl)acet-
imidoyl]ammonium = N-[2-(3-Methoxyphenoxy)propyl]-
3-methylbenzeneethanimidamide (●)
R   p-Toluenesulfonate (1:1) (6443-40-9)
S   B.W. 64-545, B.W. 545-C-64, *Xylamide tosylate*,
*Xylamidine tosilate**

U   Serotonin antagonist

8828   (7221) $C_{19}H_{24}N_2O_2$
15585-88-3

2-(Diethylamino)ethyl diphenylcarbamate = Diphenyl-
carbamic acid 2-(diethylamino)ethyl ester (●)
R   Monohydrochloride (668-37-1)
S   *Dicarfen hydrochloride***, SD 25, Sipasyl
U   Antiparkinsonian, local anesthetic, anticholinergic

8829   (7225) $C_{19}H_{24}N_2O_2$
55268-74-1

2-(Cyclohexylcarbonyl)-1,2,3,6,7,11b-hexahydro-4H-py-
razino[2,1-a]isoquinolin-4-one (●)
S   Azinox, Biltricide, Biltride, Caniquantel, Cenaride,
Cesol, Cestocur, Cestodur, Cestox, Cisticid, Cutter
Tape-Tabs, Cysticide, Cystiside, Distocide, Docatel,
Droncit, Dronsit, Drontal, EMBay 8440,
EMD 29810, Hemintel, Parkevesmin, Plativers, *Pra-
ziquantel***, Pyquiton, Seroquantel, Stopharin, Ver-
com, Vermaqpharma Vet
U   Anthelmintic (teniacide, antischistosomal)

8830　(7226)　　　　　　　　　　　　$C_{19}H_{24}N_2O_3$
99450-03-0

3-[3-[[2-(3,4-Dihydroxyphenyl)ethyl]amino]butyl]benz-
amide (●)
R　Monohydrochloride (99450-04-1)
S　KM-13
U　Cardiotonic

8831　(7227)　　　　　　　　　　　　$C_{19}H_{24}N_2O_3$
36894-69-6

5-[1-Hydroxy-2-[(1-methyl-3-phenylpropyl)ami-
no]ethyl]salicylamide = 2-Hydroxy-5-[1-hydroxy-2-[(1-
methyl-3-phenylpropyl)amino]ethyl]benzamide (●)
R　also monohydrochloride (32780-64-6)
S　Abetol, AH 5158 A, Albetal, Albetol, Alfabetal,
Amipress, Amitalol, Antimisol, Ascool, Betadren
"Crosara", Blocan, Coreton "Spofa", Frenocor, Ftali-
ned, *Ibidomide*, Imavate "Doctum", Ipolab, Labelol,
Labestonin, *Labetalol***, Labeten, Labetol, Labro-
col, Lamitol, Liondox "Roemmers-Argentina", Lir-
capil, Lobecor, Lolum, Mitalolo, Normadate, Nor-
modyne, Opercol, Presdate, Presolol "Alphapharm;
Generics", Pressalolo, Pressocard, Resporito, Salma-
gne, Sch 15719 W, SN 202, Trandate, Trandol
U　Antihypertensive, vasodilator (α- and β-adrenergic
blocker)

8832　(7228)　　　　　　　　　　　　$C_{19}H_{24}N_2O_3$
75659-07-3

(−)-5-[(1R)-1-Hydroxy-2-[[(1R)-1-methyl-3-phenylpro-
pyl]amino]ethyl]salicylamide = [R-(R*,R*)]-2-Hydroxy-
5-[1-hydroxy-2-[(1-methyl-3-phenylpropyl)ami-
no]ethyl]benzamide (●)
R　Monohydrochloride (75659-08-4)
S　AH 19501, *Dilevalol hydrochloride**, Dilevalon, Le-
vadil, *(R.R)-Labetalol hydrochloride*, Sch 19927,
Unicard(e)
U　Antihypertensive, vasodilator (β-adrenergic blocker)

8833　(7229)　　　　　　　　　　　　$C_{19}H_{24}N_2O_4$
38103-61-6

p-[2-[[2-Hydroxy-3-(o-tolyloxy)propyl]amino]eth-
oxy]benzamide = 1-[2-(4-Carbamoylphenoxy)ethylami-
no]-3-(2-methylphenoxy)-2-propanol = 4-[2-[[2-Hy-
droxy-3-(2-methylphenoxy)propyl]amino]ethoxy]benz-
amide (●)
R　Monohydrochloride (51599-37-2)
S　Coptin "Pfizer", *Tolamolol hydrochloride***,
UK 6558-01
U　β-Adrenergic blocker

8834   (7230)   C$_{19}$H$_{24}$N$_2$O$_4$
15534-05-1

α-(3,4-Dihydroxyphenyl)-4-(2-methoxyphenyl)-1-pipe-
razineethanol = 3,4-Dihydroxy-α-[[4-(2-methoxyphe-
nyl)-1-piperazinyl]methyl]benzyl alcohol = 4-[1-Hy-
droxy-2-[4-(2-methoxyphenyl)-1-piperazinyl]ethyl]-1,2-
benzenediol (●)
S   JP 711, *Pipratecol***, S 4216, SE 711
U   Vasodilator

8835   (7231)   C$_{19}$H$_{24}$N$_2$O$_4$
73573-87-2

(±)-2'-Hydroxy-5'-[(RS)-1-hydroxy-2-[[(RS)-p-methoxy-
α-methylphenethyl]amino]ethyl]formanilide = 3-Form-
amido-4-hydroxy-α-[[N-(p-methoxy-α-methylphenethyl)amino]methyl]benzyl alcohol = (R*,R*)-(±)-N-[2-
Hydroxy-5-[1-hydroxy-2-[[2-(4-methoxyphenyl)-1-me-
thylethyl]amino]ethyl]phenyl]formamide (●)
R   Fumarate (2:1) (43229-80-7)
S   Asmatec, Atock, BD 40 A, CGP 25827 A-F, *Efor-
moterol fumarate*, Eolus, Foradil, *Formoterol fumarate***, Neblik, Oxis, Vanetril "Padro", YM-08316
U   Bronchodilator

8836   (7232)   C$_{19}$H$_{24}$N$_2$O$_4$S
62658-88-2

Diethyl 1',4'-dihydro-2',6'-dimethyl-2-(methylthio)[3,4'-
bipyridine]-3',5'-dicarboxylate = 1,4-Dihydro-2,6-dime-
thyl-4-[2-(methylthio)-3-pyridinyl]-3,5-pyridinedicarb-
oxylic acid diethyl ester (●)
S   *Mesudipine***
U   Coronary vasodilator

8837   C$_{19}$H$_{24}$N$_2$O$_4$S$_2$
167305-00-2

(4S,7S,10aS)-Octahydro-4-[(S)-α-mercaptohydrocinnam-
amido]-5-oxo-7H-pyrido[2,1-b][1,3]thiazepine-7-carb-
oxylic acid = (4S,7S,10aS)-Octahydro-4-[[(2S)-2-mercap-
to-1-oxo-3-phenylpropyl]amino]-5-oxo-7H-pyri-
do[2,1-b][1,3]thiazepine-7-carboxylic acid (●)
S   BMS 186716, BMS 186716-01, *Omapatrilat***,
Vanlef, Vanlev
U   Antihypertensive, cardiotonic (vasopeptidase inhibitor)

8838   (7233)   C$_{19}$H$_{24}$N$_2$O$_7$S
54340-65-7

(2S,5R,6R)-6-[(R)-2-Hydroxy-4-methylvaleramido]-3,3-
dimethyl-7-oxo-4-thia-1-azabicyclo[3.2.0]heptane-2-
carboxylic acid 2-furoate (ester) = (6R)-6-[(R)-2-Furoyl-
oxy-4-methylvaleramido]penicillanic acid = [2S-
[2α,5α,6β(S*)]]-6-[[2-[(2-Furanylcarbonyl)oxy]-4-me-
thyl-1-oxopentyl]amino]-3,3-dimethyl-7-oxo-4-thia-1-
azabicyclo[3.2.0]heptane-2-carboxylic acid (●)
S   *Furbucillin***
U   Antibiotic

8839   (7234) $C_{19}H_{24}N_2S$
522-00-9

10-[2-(Diethylamino)propyl]phenothiazine = 1-Methyl-2-phenothiazin-10-ylethyldiethylamine = N,N-Diethyl-α-methyl-10H-phenothiazine-10-ethanamine (●)
R   Monohydrochloride (1094-08-2)
S   *Äthopropazinhydrochlorid*, Dibutil, *Ethopropazine hydrochloride*, *Isothazine*, *Isothiazine*, Lysivane, Parcidol, Pardisol, Parfezin, Parkin, Parkisol, Parphezein, Parsidol, Parsitan, Parsotil, Phenopropazin, Preparten, *Profenamine hydrochloride**, Prophenamine hydrochloride*, *Prophenamonium*, Rochipel, Rodipal, 3356 R.P., SC-2538, W-483
U   Anticholinergic, antiparkinsonian

8839-01
R   2-(4-Hydroxybenzoyl)benzoate (hibenzate)
S   *Parkin-Powder*, *Profenamine hibenzate***
U   Antiparkinsonian

8840   (7235) $C_{19}H_{24}N_2S_2$
7009-43-0

(±)-10-[3-(Dimethylamino)-2-methylpropyl]-2-(methyl-thio)phenothiazine = (±)-$N,N,β$-Trimethyl-2-(methyl-thio)-10H-phenothiazine-10-propanamine (●)
S   *Methiomeprazine**, 7238 R.P., SKF 6270
U   Anti-emetic, neuroleptic

8841   (7236) $C_{19}H_{24}N_2S_2$
1759-09-7

(–)-10-[3-(Dimethylamino)-2-methylpropyl]-2-(methyl-thio)phenothiazine = (–)-$N,N,β$-Trimethyl-2-(methyl-thio)-10H-phenothiazine-10-propanamine (●)
S   *Levomethioprazinum*, *Levometiomeprazine***, Phenaceda
U   Anti-emetic, neuroleptic

8842   (7237) $C_{19}H_{24}N_4O$
115768-29-1

6,7-Dihydro-N-[2-(4-morpholinyl)ethyl]-5H-benzo[6,7]cyclohepta[1,2-c]pyridazin-3-amine (●)
R   Dihydrochloride (115767-94-7)
S   SR 95639A
U M_1-Muscarinic agonist

8843   (7238) $C_{19}H_{24}N_4O_2$
100-33-4

4,4'-(Pentamethylenedioxy)dibenzamidine = p,p'-Diami-dino-1,5-diphenoxypentane = 4,4'-[1,5-Pentanediyl-bis(oxy)]bis[benzenecarboximidamide] (●)
S   *Pentamidine**, WR 245720
U   Antiprotozoal

8843-01　(7238-01)　　　　　　　　　　　140-64-7
R　Isethionate (2-hydroxyethanesulfonate) (1:2)
S　Aeropent, Benambax, Diamidine, M. & B. 800, Ne-
　　　bupent, Pentacarinat(e), Pentam 300, *Pentamidine is-*
　　　*ethionate***, Pneumopent, 2512 R.P., Zethionat
U　Antiprotozoal

8843-02　(7238-02)　　　　　　　　　　　6823-79-6
R　Dimethanesulfonate
S　Lomidine, *Pentamidine mesilate***
U　Antiprotozoal

8843-03　(7238-03)
R　Salt with suramin (no. 15661)
S　4891 R.P.
U　Anti-infective

8844　　　　　　　　　　　　　　　　$C_{19}H_{24}N_4O_2$
　　　　　　　　　　　　　　　　　　147676-85-5

1,4-Dihydro-1-methyl-5-(5-methyl-2-propoxyphenyl)-3-
propyl-7H-pyrazolo[4,3-d]pyrimidin-7-one (●)
S　UK-122764
U　Vasodilator (PDE 5 inhibitor)

8845　(7239)　　　　　　　　　　　　$C_{19}H_{24}N_4O_3$
　　　　　　　　　　　　　　　　　　103922-33-4

3-Amino-4-[[(Z)-4-[[4-(piperidinomethyl)-2-pyri-
dyl]oxy]-2-butenyl]amino]-3-cyclobutene-1,2-dione =
(Z)-3-Amino-4-[[4-[[4-(1-piperidinylmethyl)-2-pyridi-
nyl]oxy]-2-butenyl]amino]-3-cyclobutene-1,2-dione (●)
R　Monohydrochloride (126463-66-9)
S　IT-066, *Pibutidine hydrochloride***
U　Gastric antisecretory (histamine H_2-receptor antago-
　　　nist)

8846　(7240)　　　　　　　　　　　　$C_{19}H_{24}N_4O_4$
　　　　　　　　　　　　　　　　　　106260-91-7

4-[4-(4-Furo[3,2-c]pyridin-4-yl-1-piperazinyl)butyl]-3,5-
morpholinedione (●)
S　BMY-20661
U　Antipsychotic

8847　(7241)　　　　　　　　　　　　$C_{19}H_{24}N_6O_2$
　　　　　　　　　　　　　　　　　　78208-13-6

4,9-Dihydro-1,3-dimethyl-4-[(4-methyl-1-piperazi-
nyl)acetyl]pyrazolo[4,3-b][1,5]benzodiazepin-10-(1H)-
one (●)
S　*Zolenzepine***
U　Anti-ulcer agent

8848　(7242)　　　　　　　　　　　　$C_{19}H_{24}N_6O_2S$
　　　　　　　　　　　　　　　　　　78463-83-9

N-(2-Azidoethyl)-N-[[8,9-didehydro-6-methylergolin-8-
yl]methyl]methanesulfonamide (●)
R　Maleate (1:1) (78463-86-2)
S　GYKI 32887, RGH-7825
U　Antiparkinsonian (dopaminergic agonist)

8849 (7243) $C_{19}H_{24}N_6O_5S_2$
88040-23-7

1-[[(6R,7R)-7-[2-(2-Amino-4-thiazolyl)glyoxylamido]-2-carboxy-8-oxo-5-thia-1-azabicyclo[4.2.0]oct-2-en-3-yl]methyl]-1-methylpyrrolidinium hydroxide inner salt 7^2-(Z)-(O-methyloxime) = (Z)-(7R)-7-[2-(2-Amino-4-thiazolyl)-2-(methoxyimino)acetamido]-3-[(1-methyl-1-pyrrolidinio)methyl]-3-cephem-4-carboxylate = [6R-[6α,7β(Z)]]-1-[[7-[[(2-Amino-4-thiazolyl)(methoxyimino)acetyl]amino]-2-carboxy-8-oxo-5-thia-1-azabicyclo[4.2.0]oct-2-en-3-yl]methyl]-1-methylpyrrolidinium hydroxide inner salt (●)
R also dihydrochloride monohydrate (123171-59-5)
S Axepim, Axépim(e), Axepime, BMY-28142, Cefelim, *Cefepime***, Cefeplus, Cefquattro, Cepim, Cepimex, CFPM, Maxcef, Maxipim(e)
U Antibiotic

8850 (7244) $C_{19}H_{24}N_6O_6$
118252-44-1

N-[4-[[3-(2,6-Diamino-1,4-dihydro-4-oxo-5-pyrimidinyl)propyl]amino]benzoyl]-L-glutamic acid (●)
S 5-DACTHF, 543 U 76
U Antileukemic

8851 $C_{19}H_{24}O_2$
105455-76-3

14α,15α-Methyleneestradiol = (15β,17β)-3',15-Dihydrocycloprop[14,15]estra-1,3,5(10)-triene-3,17-diol (●)
S J 824
U Estrogenic

8852 (7245) $C_{19}H_{24}O_2$
10448-96-1

3-Hydroxy-7α-methylestra-1,3,5(10)-trien-17-one = (7α)-3-Hydroxy-7-methylestra-3,5(10)-trien-17-one (●)
S *Almestrone***, Ba-38372, Ciba 38372, *7α-Methylestrone*, *7α-Methylöstron*
U Estrogen

8853 (7246) $C_{19}H_{24}O_2$
965-93-5

17β-Hydroxy-17-methylestra-4,9,11-trien-3-one = (17β)-17-Hydroxy-17-methylestra-4,9,11-trien-3-one (●)
S *Methyltrienolone, Metribolone***, NSC-92858, R 1881, RU 1881
U Anabolic

8854 (7247) $C_{19}H_{24}O_2S$
69425-13-4

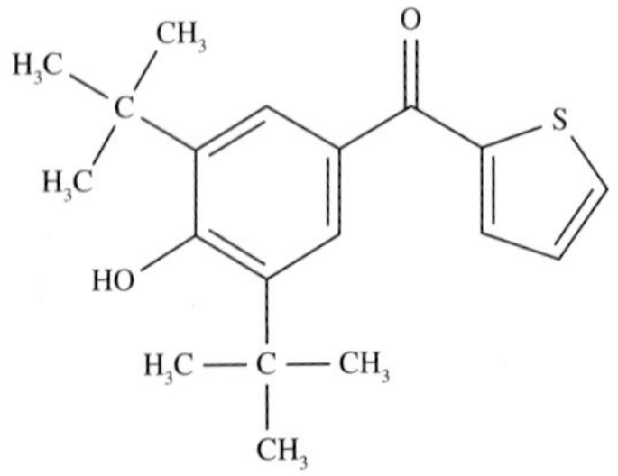

3,5-Di-*tert*-butyl-4-hydroxyphenyl 2-thienyl ketone =
[3,5-Bis(1,1-dimethylethyl)-4-hydroxyphenyl]-2-thienyl-
methanone (●)
S *Prifelone***, R 830, S 16820
U Anti-inflammatory, anti-oxidant

8855 (7248) $C_{19}H_{24}O_3$
19825-63-9

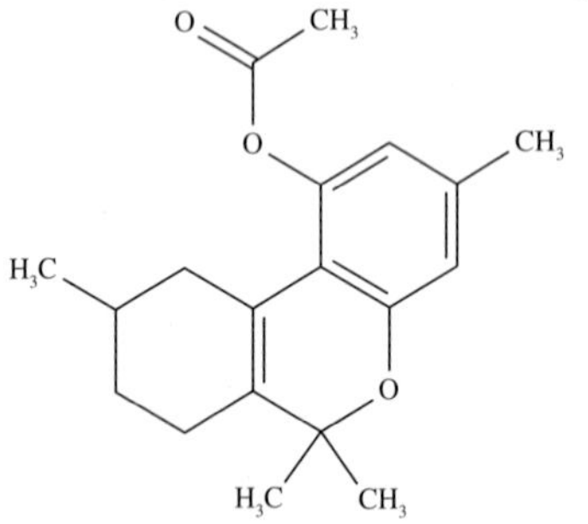

7,8,9,10-Tetrahydro-3,6,6,9-tetramethyl-6*H*-diben-
zo[*b,d*]pyran-1-ol acetate (●)
R also with (±)-definition (68298-00-0)
S *Pirnabin(e)***, SP-304
U Antiglaucoma agent

8856 (7249) $C_{19}H_{24}O_3$
968-93-4

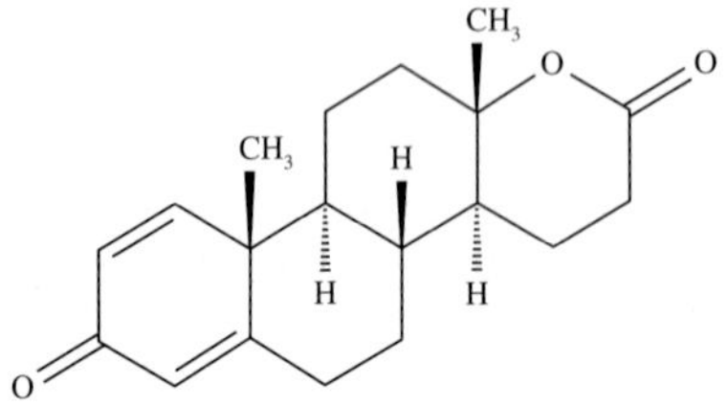

13-Hydroxy-3-oxo-13,17-secoandrosta-1,4-dien-17-oic
acid δ-lactone = Δ^1-Testololactone =
1,2,3,4,4a,4b,7,9,10,10a-Decahydro-2-hydroxy-2,4b-di-

methyl-7-oxo-1-phenanthrenepropionic acid δ-lactone =
D-Homo-17a-oxaandrosta-1,4-diene-3,17-dione (●)
S Fludestrin, NSC-23759, SQ 9538, Teolit, Teslac,
Teslak, *Testolactone***
U Antineoplastic

8857 $C_{19}H_{24}O_3$
72944-19-5

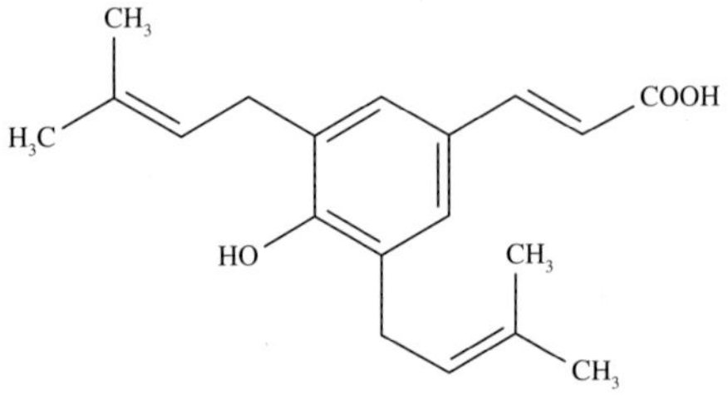

3,5-Diprenyl-4-hydroxycinnamic acid = (2*E*)-3-[4-Hy-
droxy-3,5-bis(3-methyl-2-butenyl)phenyl]-2-propenoic
acid (●)
S *Artepillin C*
U Antineoplastic from propolis

8858 $C_{19}H_{24}O_4$
120051-39-0

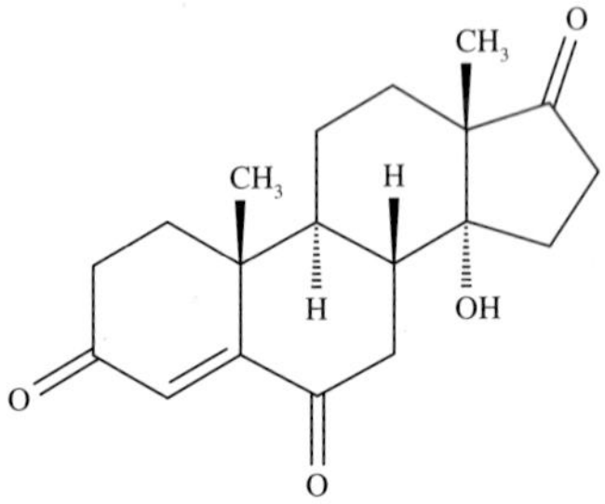

14-Hydroxyandrost-4-ene-3,6,17-trione (●)
S NKS-01
U Antineoplastic (aromatase inhibitor)

8859 (7250) $C_{19}H_{24}O_5$
6379-69-7

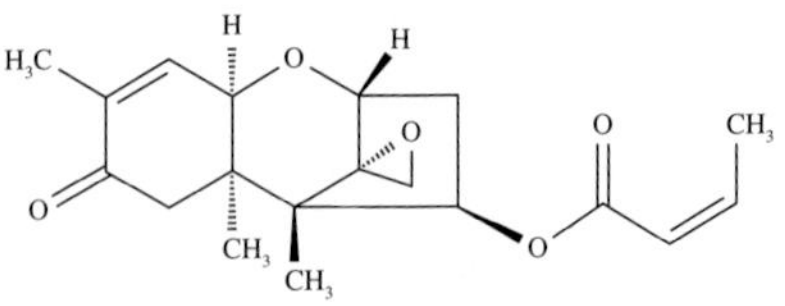

[4β(*Z*)]-12,13-Epoxy-4-[(1-oxo-2-butenyl)oxy]tricho-
thec-9-en-8-one (●)

S *Trichothecin*
U Antibiotic from *Trichothecium roseum*

8860 $C_{19}H_{25}BN_4O_4$
179324-69-7

[(1*R*)-3-Methyl-1-[[(2*S*)-1-oxo-3-phenyl-2-[(pyrazinyl-carbonyl)amino]propyl]amino]butyl]boronic acid (●)
S LDP-341, PS-341
U Antineoplastic (proteasome inhibitor)

8861 (7251) $C_{19}H_{25}ClN_2O_2$
123482-22-4

5-Chloro-2,3-dihydro-2,2-dimethyl-*N*-1α*H*,5α*H*-tropan-3α-yl-7-benzofurancarboxamide = *endo*-5-Chloro-2,3-di-hydro-2,2-dimethyl-*N*-(8-methyl-8-azabicyclo[3.2.1]oct-3-yl)-7-benzofurancarboxamide (●)
R Maleate (1:1) (123482-23-5)
S LY 277359 maleate, *Zatosetron maleate***
U Serotonin antagonist (antimigraine, anti-arrhythmic, gastroprokinetic)

8862 $C_{19}H_{25}Cl_2N_3O_3$
126766-31-2

4-[(3,4-Dichlorophenyl)acetyl]-3-(1-pyrrolidinylmethyl)-1-piperazinecarboxylic acid methyl ester (●)
R (2*E*)-2-Butenedioate (fumarate) (1:1) (126766-32-3)
S GR 89696
U κ$_2$-Opioid agonist

8863 (7252) $C_{19}H_{25}Cl_2N_3O_3$
126766-42-5

(*R*)-4-[(3,4-Dichlorophenyl)acetyl]-3-(1-pyrrolidinylme-thyl)-1-piperazinecarboxylic acid methyl ester (●)
R Maleate (1:1) (126766-43-6)
S GR 103545
U Analgesic (κ-opioid), sedative, diuretic

8864 (7253) $C_{19}H_{25}FN_2O_5S$
125961-37-7

(*R*)-5-[2-[[2-(5-Fluoro-2-methoxyphenoxy)ethyl]ami-no]propyl]-2-methoxybenzenesulfonamide (●)
R Monohydrochloride (125961-36-6)
S HSR-175
U α$_1$-Adrenoceptor antagonist

8865 (7254) $C_{19}H_{25}F_3N_2O$

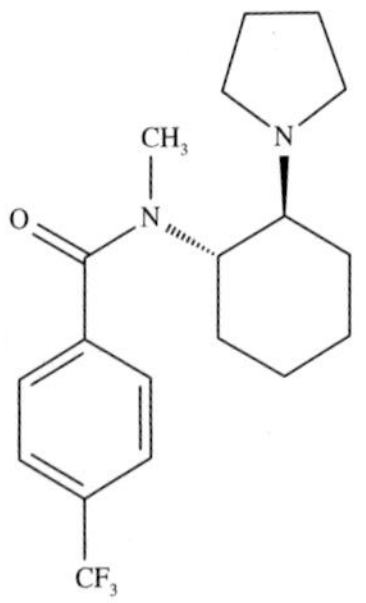

trans-N-Methyl-*N*-[2-(1-pyrrolidinyl)cyclohexyl]-4-(tri-
fluoromethyl)benzamide (●)
R Mono(4-methylbenzenesulfonate) (monotosylate)
     (142013-44-3)
S U-49524 E
U Anticonvulsant

8866 (7255) $C_{19}H_{25}F_3N_2O$
                                   130497-40-4

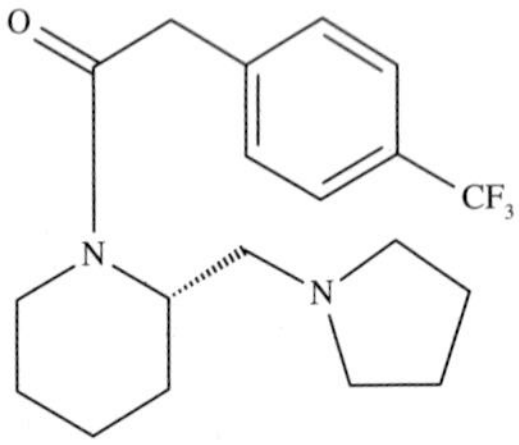

(*S*)-2-(1-Pyrrolidinylmethyl)-1-[[4-(trifluoromethyl)phe-
nyl]acetyl]piperidine (●)
R also monohydrochloride (115642-84-7)
S BRL 52656, ZT 52656 A
U κ-Opioid receptor agonist

8867 (7256) $C_{19}H_{25}NO$
                                   15599-37-8

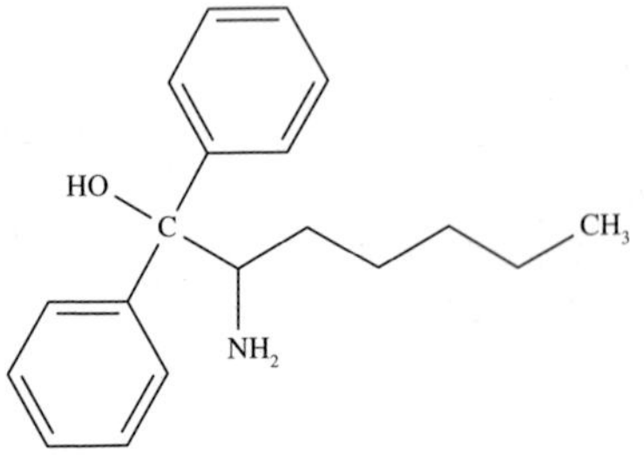

α-(1-Aminohexyl)benzhydrol = 2-Amino-1,1-diphenyl-
1-heptanol = α-(1-Aminohexyl)-α-phenylbenzenemetha-
nol (●)
S *Hexapradol***
U Psycho-analeptic stimulant

8868 (7257) $C_{19}H_{25}NO$
                                   58473-73-7

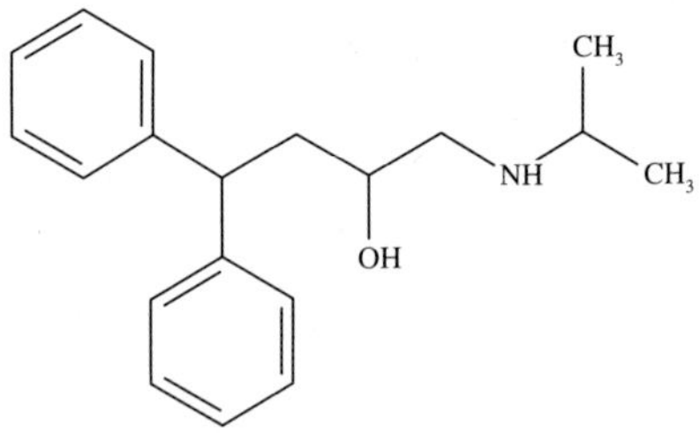

(±)-1-(Isopropylamino)-4,4-diphenyl-2-butanol = (±)-α-
[[(1-Methylethyl)amino]methyl]-γ-phenylbenzenepropa-
nol (●)
S *Drobuline***, Lilly 122587
U Anti-arrhythmic

8869 (7260) $C_{19}H_{25}NO$
                                   1157-87-5

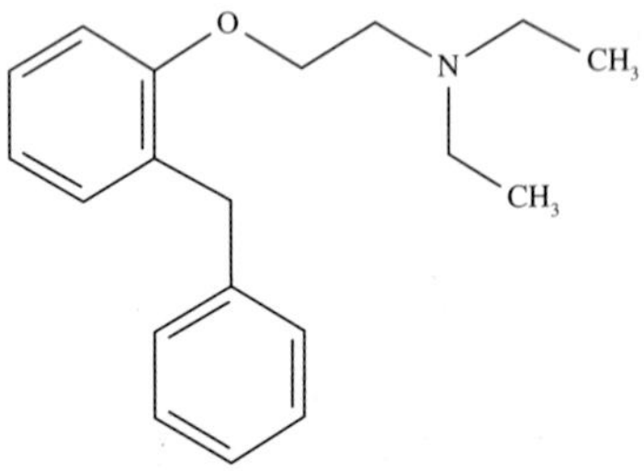

2-[(α-Phenyl-*o*-tolyl)oxy]triethylamine = *N,N*-Diethyl-2-
(*o*-benzylphenoxy)ethylamine = β-(Diethylamino)ethyl
2-benzylphenyl ether = *N,N*-Diethyl-2-[2-(phenylmethyl)phenoxy]ethanamine (●)
R Hydrochloride (2087-37-8)
S *Ätholoxaminhydrochlorid*, AH 3, CI-072, *Diaethyl-
     aminoaethoxydiphenylmethanum hydrochloricum,
     Etoloxamine hydrochloride***
U Antihistaminic, anti-allergic

8870

$C_{19}H_{25}NO$
98774-23-3

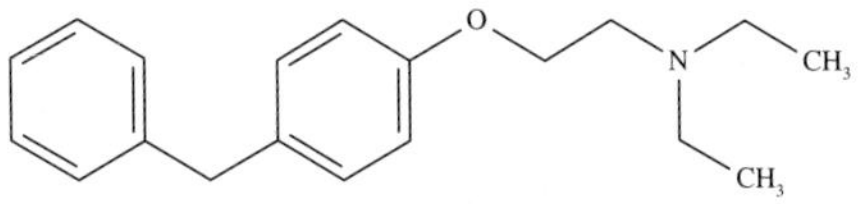

2-[(α-Phenyl-*p*-tolyl)oxy]triethylamine = *N,N*-Diethyl-2-[4-(phenylmethyl)phenoxy]ethanamine (●)
R Hydrochloride (92981-78-7)
S BMS 217380-01, BMY-33419, DPPE, *Tesmilifene hydrochloride***
U Anti-estrogen

8871 (7259)

$C_{19}H_{25}NO$
642-58-0

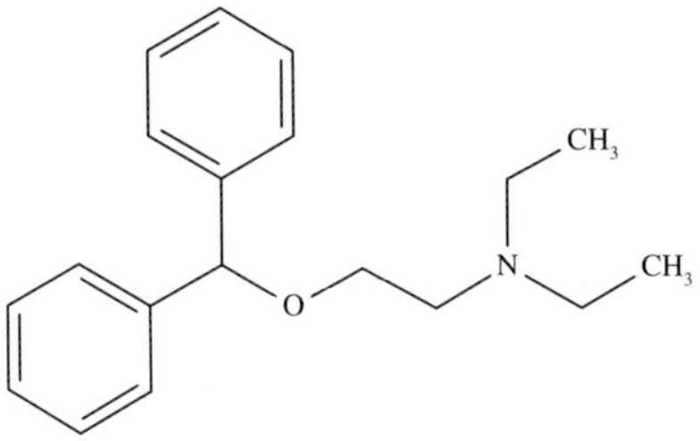

2-(Benzhydryloxy)triethylamine = *N,N*-Diethyl-2-(diphenylmethoxy)ethylamine = β-(Diethylamino)ethyl benzhydryl ether = 2-(Diphenylmethoxy)-*N,N*-diethylethanamine (●)
R Hydrochloride (86-24-8)
S *Äthylbenzhydraminhydrochlorid*, Antiparkin (old form), *Ethylbenzhydramine hydrochloride*, PKM-Tabletten, Rigidyl, S-45
U Antiparkinsonian, anti-allergic

8872 (7261)

$C_{19}H_{25}NO$
152-02-3

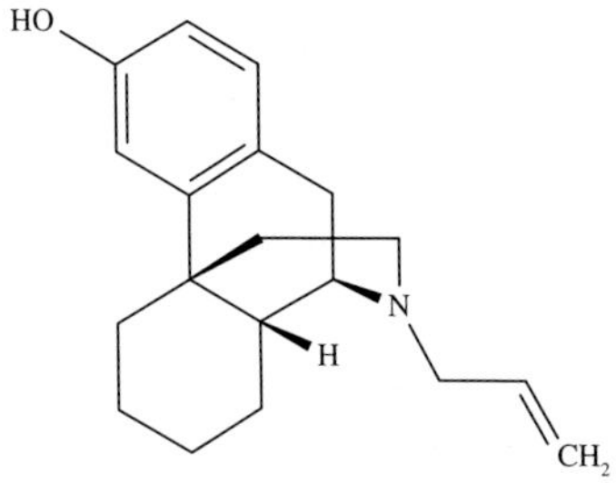

L-17-Allylmorphinan-3-ol = (–)-3-Hydroxy-*N*-allylmorphinan = 17-(2-Propenyl)morphinan-3-ol (●)
R see also no. 11411

S *Lävallorphan, Levallorphan**, Naloxifane, Naloxiphane*
U Narcotic antagonist

8872-01 (7261-01)

71-82-9
R Tartrate (1:1)
S Lorfan, Ro-1-7700
U Narcotic antagonist

8873 (7262)

$C_{19}H_{25}NO$
112891-97-1

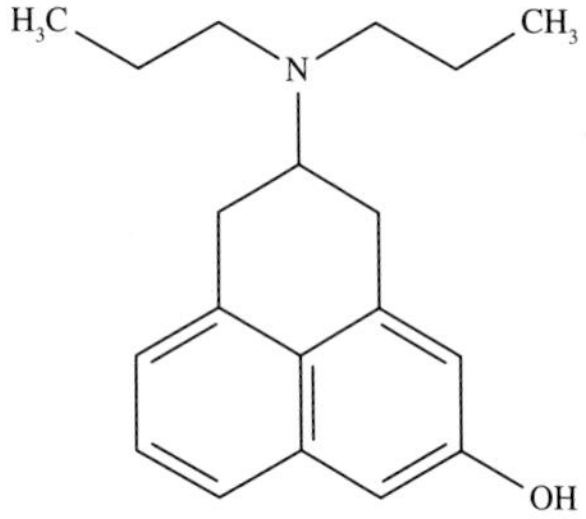

(+)-2-(Dipropylamino)-2,3-dihydro-1*H*-phenalen-5-ol (●)
R Hydrobromide (112892-81-6)
S *Alentemol hydrobromide**, U-68553 B
U Antipsychotic

8874

$C_{19}H_{25}NO$
315209-09-7

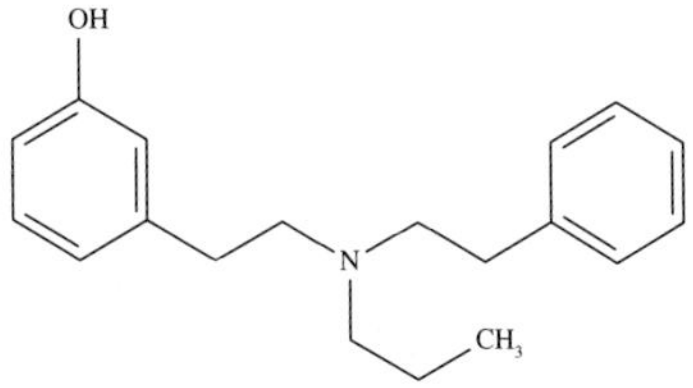

3-[2-[(2-Phenylethyl)propylamino]ethyl]phenol (●)
R Hydrochloride (67383-44-2)
S RU 24213
U Dopamine D_2 receptor agonist

8875   (7258) $C_{19}H_{25}NO$
                                                                     29775-46-0

3-[(3,3-Diphenylpropyl)methylamino]-1-propanol (●)
R   Hydrochloride (23903-11-9)
S   PF-82
U   Antidepressant

8876   (7263) $C_{19}H_{25}NOS$
                                                                     92206-54-7

2-[Propyl(2-thienylethyl)amino]-5-hydroxytetralin =
5,6,7,8-Tetrahydro-6-[propyl[2-(2-thienyl)ethyl]amino]-
1-naphthalenol (●)
R   also with (±)-definition (113349-24-9)
S   N-0437
U   Antiparkinsonian, antiglaucoma agent (dopamine D$_2$-
         agonist)

8877 $C_{19}H_{25}NOS$
                                                                     99755-59-6

(S)-2-[Propyl[2-(2-thienyl)ethyl]amino]-5-hydroxytetra-
lin = (6S)-5,6,7,8-Tetrahydro-6-[propyl[2-(2-thie-
nyl)ethyl]amino]-1-naphthalenol (●)
S   Botagolide, N-0923, Rotigotine*, SPM-962
U   Antiparkinsonian (dopamine D$_2$ receptor agonist)

8878   (7265) $C_{19}H_{25}NO_2$
                                                                     109525-44-2

(±)-(αR*)-3-Hydroxy-4-methyl-α-[(1S*)-1-[(3-phenyl-
propyl)amino]ethyl]benzyl alcohol = (±)-erythro-1-(3-
Hydroxy-4-methylphenyl)-2-[(3-phenylpropyl)amino]-1-
propanol = (R*,S*)-3-Hydroxy-4-methyl-α-[1-[(3-phe-
nylpropyl)amino]ethyl]benzenemethanol (●)
R   Hydrochloride (99203-37-9)
S   Clipoxamine hydrochloride, Cliropamine hydrochlo-
         ride**, D-16427
U   Cardiotonic (positive inotropic)

8879   (7264) $C_{19}H_{25}NO_2$
                                                                     447-41-6

1-(p-Hydroxyphenyl)-2-[(1-methyl-3-phenylpropyl)ami-
no]-1-propanol = p-Hydroxy-α-[1-[(1-methyl-3-phenyl-
propyl)amino]ethyl]benzyl alcohol = 4-Hydroxy-α-[1-
[(1-methyl-3-phenylpropyl)amino]ethyl]benzenemetha-
nol (●)
R   Hydrochloride (849-55-8)
S   Adrin, Agyrax, Arlibide, Arlidin, Atemnen, Belid-
         rin, Bufedon, Buphedrin, Buphenine hydrochlori-
         de**, Certadyn, Circovenil, CS 6712, Diatolil, Dila-
         tal, Dilatol, Dilatropin, Dilatropon, Dilatyl, Dilaver,
         Dilydrin, Flumil, Mefeprol, Nalide, Nilvitam, Ny-
         deral, Nylidrin hydrochloride, Nylin, Opino "Bayro-
         pharm", Penitardon, Perdilat, Perdilatal, Pervadil
         "Empire", Pharmadil, PMS-Nylidrin, RHC 3432-A,
         Rinard, Rolidrin, Rudilin, Rydrin, Shatorn,
         SKF 1700-A, Tocodilydrin, Tocodrin, Vasiten, Veri-
         na "Daiichi"
U   Peripheral vasodilator

8880 (7266)

$C_{19}H_{25}NO_2$
69815-38-9

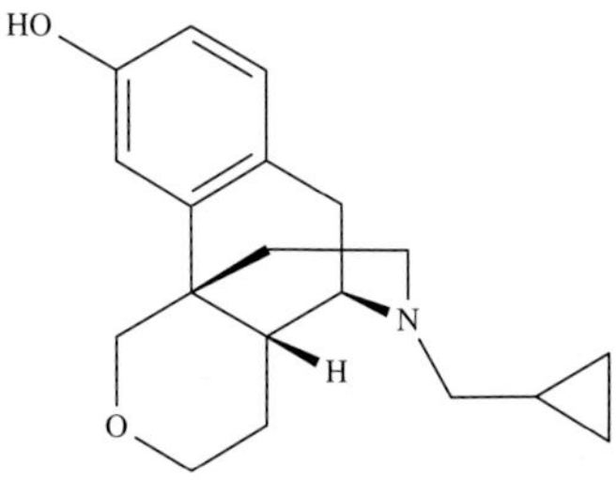

(–)-(4aR,5R,10bS)-13-(Cyclopropylmethyl)-4,4a,5,6-te-
trahydro-3H-5,10b-(iminoethano)-1H-naphtho[1,2-e]py-
ran-9-ol = 17-(Cyclopropylmethyl)-6-oxamorphinan-3-ol
(●)

R D(–)-Tartrate (2:1) (69815-39-0)
S BL-5572 M, *Proxorphan tartrate***
U Analgesic, antitussive

8881 (7267)

$C_{19}H_{25}NO_2$
65934-61-4

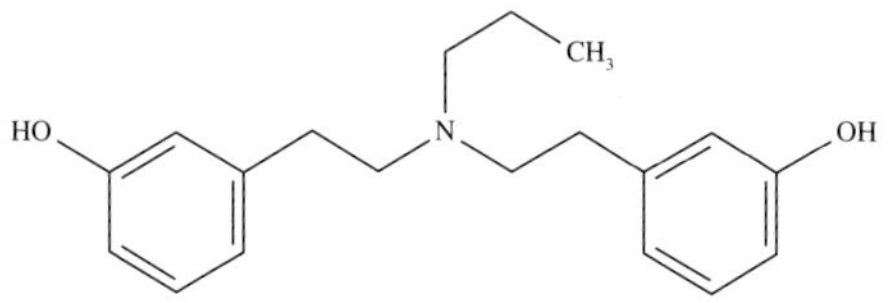

3,3'-[(Propylimino)diethylene]diphenol = 3,3'-[(Propylimino)di-2,1-ethanediyl]bis[phenol] (●)

R Hydrochloride (74515-04-1)
S RU 24926
U Dopamine D_2 agonist, κ-opioid receptor antagonist

8882 (7268)

$C_{19}H_{25}NO_2$
3811-53-8

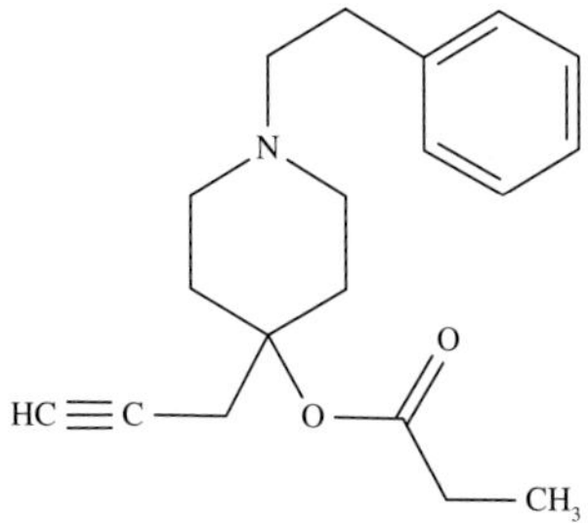

1-Phenethyl-4-(2-propynyl)-4-piperidinol propionate = 1-
(2-Phenylethyl)-4-propargyl-4-propionyloxypiperidine =

1-(2-Phenylethyl)-4-(2-propynyl)-4-piperidinol propanoate (ester) (●)

S L. 2909, Labazyl (new form), *Propinethidine, Propi-
netidine***
U Antitussive

8883 (7269)

$C_{19}H_{25}NO_2$
18965-97-4

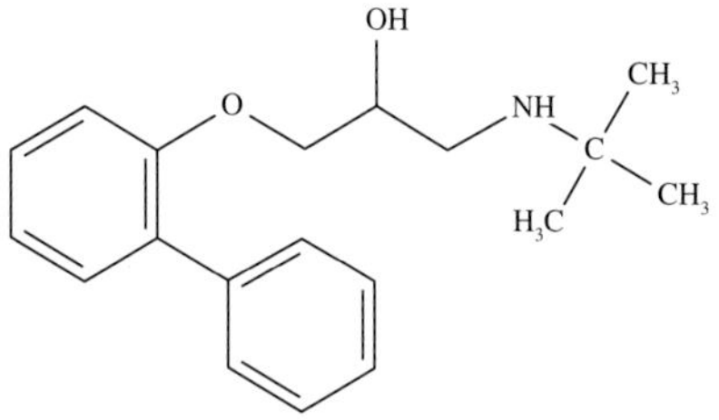

(±)-1-(2-Biphenylyloxy)-3-(*tert*-butylamino)-2-propanol
= (±)-1-([1,1'-Biphenyl]-2-yloxy)-3-[(1,1-dimethyl-
ethyl)amino]-2-propanol (●)

R Hydrochloride (18965-98-5)
S *Berlafenone hydrochloride***, Bipranol hydrochlori-
de, GK 23-G
U Anti-arrhythmic

8884

$C_{19}H_{25}NO_2$
162382-62-9

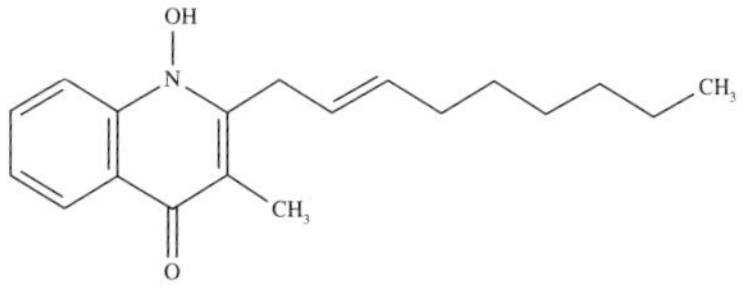

(E)-1-Hydroxy-3-methyl-2-(2-nonenyl)-4(1H)-quinoli-
none (●)

S YM-30059
U Antibacterial (gram-positive) from *Arthrobacter* sp.
YL-02729S

8885 (7270) $C_{19}H_{25}NO_3$
73-54-1

8-Allyl-1αH,5αH-nortropan-3α-ol (±)-tropate = Tropic acid 8-allyl-3α-nortropanyl ester = N-Allylnoratropine = endo-α-(Hydroxymethyl)benzeneacetic acid 8-(2-prope-nyl)-8-azabicyclo[3.2.1]oct-3-yl ester (●)
S *Allylnoratropine*, N-728, *Naltropine*
U Anticholinergic

8886 $C_{19}H_{25}NO_3$
145375-43-5

(−)-(2S,3a,7a-*cis*)-α-Benzylhexahydro-γ-oxo-2-isoindo-linebutyric acid = (αS,3aR,7aS)-Octahydro-γ-oxo-α-(phe-nylmethyl)-2H-isoindole-2-butanoic acid (●)
R Calcium salt (2:1) (145525-41-3) or calcium salt (2:1) dihydrate (207844-01-7)
S KAD 1229, *Mitiglinide calcium***, *Mituglinide calci-um*, S 21403
U Hypoglycemic (insulotropic)

8887 (7271) $C_{19}H_{25}NO_3$
39907-68-1

N-(3,4-Dihydroxyphenethyl)-1-adamantanecarboxamide = N-[2-(3,4-Dihydroxyphenyl)ethyl]tricy-clo[3.3.1.1^{3,7}]decane-1-carboxamide (●)
S *Dopamantine***, Sch 15507
U Antiparkinsonian

8888 (7272) $C_{19}H_{25}NO_4$
55608-72-5

O-Benzoyl-D-pseudotropine-2-carboxylic acid propyl es-ter = [1R-(2-*endo*,3-*exo*)]-3-(Benzoyloxy)-8-methyl-8-azabicyclo[3.2.1]octane-2-carboxylic acid propyl ester (●)
S Neopsicaine, Psicain-Neu
U Local anesthetic

8889 $C_{19}H_{25}NO_4$
64091-46-9

(1R,2R,3S,5S)-3-(Benzoyloxy)-8-methyl-8-azabicy-clo[3.2.1]octane-2-carboxylic acid propyl ester (●)
S *Cocapropylene*, *Propylcocaine*
U Local anesthetic (narcotic)

8890 (7273) $C_{19}H_{25}NO_4$
18910-65-1

4-Hydroxy-α-[(*p*-methoxy-α-methylphenethylami-
no)methyl]-*m*-xylene-α,α'-diol = 1-[4-Hydroxy-3-(hydroxymethyl)phenyl]-2-[(4-methoxy-α-methylphenethyl)
amino]ethanol = 4-Hydroxy-α^1-[[[2-(4-methoxyphenyl)-
1-methylethyl]amino]methyl]-1,3-benzenedimethanol
(●)
S   AH 3923, *Salmefamol***
U   Bronchodilator

8891   (7274) $C_{19}H_{25}NO_4$
7696-12-0

2,2-Dimethyl-3-(2-methylpropenyl)cyclopropanecarb-
oxylic acid ester with *N*-(hydroxymethyl)-1-cyclohexene-
1,2-dicarboximide = 1-Cyclohexene-1,2-dicarboximido-
methyl 2,2-dimethyl-3-(2-methylpropenyl)cyclopropane-
carboxylate = (4,5,6,7-Tetrahydro-1,3-dioxo-2-isoindoli-
nyl)methyl 2,2-dimethyl-3-(2-methyl-1-propenyl)cyclo-
propanecarboxylate = 2,2-Dimethyl-3-(2-methyl-1-pro-
penyl)cyclopropanecarboxylic acid (1,3,4,5,6,7-hexahy-
dro-1,3-dioxo-2*H*-isoindol-2-yl)methyl ester (●)
S   FMC-9260, Neo-Pynamin, OMS 1011, *Phthalthrin*,
SP 1103, *Tetramethrin***
U   Pediculicide, insecticide

8892   (7275) $C_{19}H_{25}NO_5$
84957-89-1

2-(2,6-Dimethoxyphenoxy)-2'-(2-methoxyphenoxy)di-
ethylamine = *N*-[2-(2,6-Dimethoxyphenoxy)ethyl]-2-(2-
methoxyphenoxy)ethanamine (●)
R   Hydrochloride (95110-09-1)
S   ACC-7513
U   α-Adrenoceptor and 5-HT-receptor blocker

8893 $C_{19}H_{25}NO_6$
74513-77-2

4-[3-[[2-(3,4-Dimethoxyphenyl)ethyl]amino]-2-hydroxy-
propoxy]-1,2-benzenediol (●)
S   Ro 363
U   β$_1$- and β$_3$-adrenoceptor partial agonist

8894   (7276) $C_{19}H_{25}NS$
112726-66-6

1-(1-Benzo[*b*]thien-2-ylcyclohexyl)piperidine (●)
S   BTCP, GK 13
U   Dopaminergic agonist

8895   (7277) $C_{19}H_{25}N_2OS$
7647-63-4

N-(2-Hydroxyethyl)-*N*,*N*-dimethyl-*N*-[1-methyl-2-(10-
phenothiazinyl)ethyl]ammonium = *N*-(2-Hydroxyethyl)-
N,*N*,α-trimethyl-10*H*-phenothiazine-10-ethanaminium
(●)
R   Chloride (2090-54-2)
S   Aprobit, *Hydroxyäthylpromethazin*, Lergobit, *Pro-
methazine hydroxyethyl chloride*
U   Antihistaminic, anticholinergic

8896　　　　　　　　　　　　　　　　　$C_{19}H_{25}N_3$
　　　　　　　　　　　　　　　　　136924-88-4

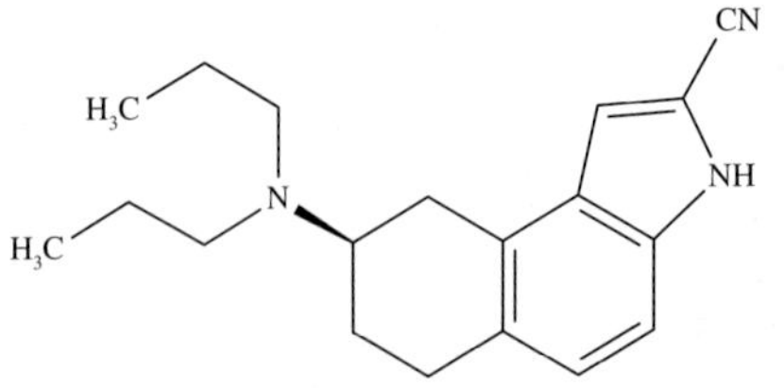

(*R*)-2-Cyano-8-(dipropylamino)cyclohexeno[*e*]indole = (*R*)-8-(Dipropylamino)-6,7,8,9-tetrahydro-3*H*-benz[*e*]indole-2-carbonitrile (●)
S　U-92016A
U　Antihypertensive (5-HT$_{1A}$ agonist)

8897　(7278)　　　　　　　　　　　　　$C_{19}H_{25}N_3$
　　　　　　　　　　　　　　　　　21755-66-8

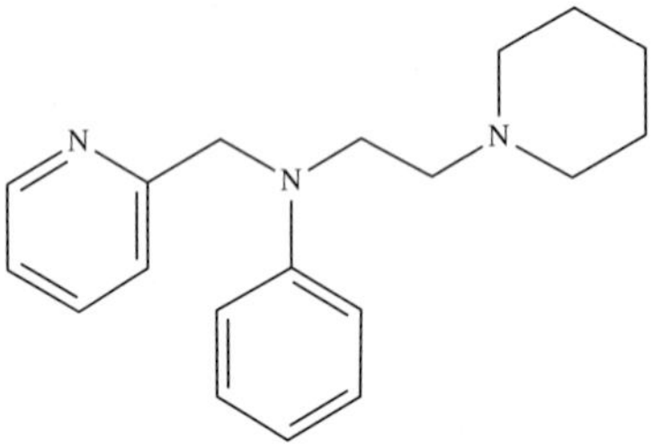

1-[2-[*N*-(2-Pyridylmethyl)anilino]ethyl]piperidine = *N*-(2-Picolyl)-*N*-phenyl-*N*-(2-piperidinoethyl)amine = *N*-Phenyl-*N*-[2-(1-piperidinyl)ethyl]-2-pyridinemethanamine (●)
R　also monohydrochloride (24699-40-9) or tripalmitate (24656-22-2)
S　Coben, *Picoperidamine*, *Picoperine***, TAT-3
U　Antitussive

8898　(7282)　　　　　　　　　　　　　$C_{19}H_{25}N_3O$
　　　　　　　　　　　　　　　　　67658-45-1

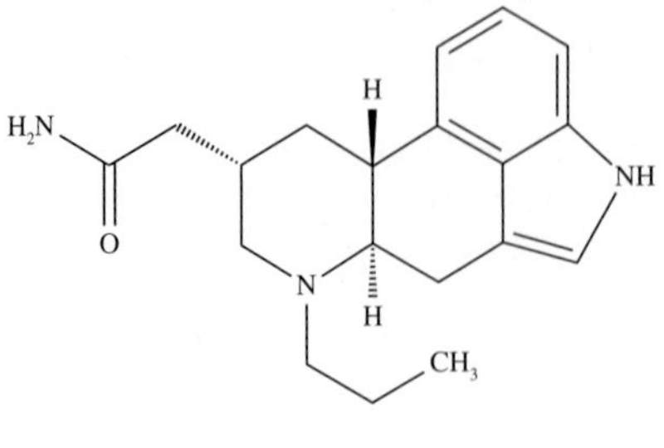

D-6-Propyl-8-ergolineacetamide = D-6-Nor-6-propyldihydrohomolysergamide = (8β)-6-Propylergoline-8-acetamide (●)
R　Tartrate (2:1) (67658-46-2)
S　Propyldeprenon, VÚFB-10915
U　Prolactin inhibitor

8899　(7279)　　　　　　　　　　　　　$C_{19}H_{25}N_3O$
　　　　　　　　　　　　　　　　　95968-63-1

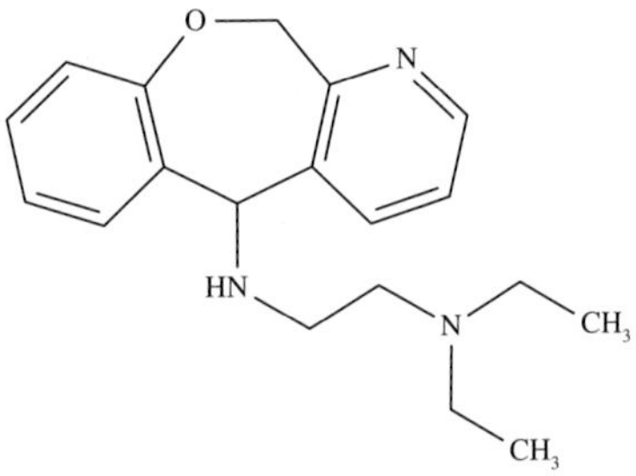

5-[[2-(Diethylamino)ethyl]amino]-5,11-dihydro[1]benzoxepino[3,4-*b*]pyridine = *N*'-(5,11-Dihydro[1]benzoxepino[3,4-*b*]pyridin-5-yl)-*N*,*N*-diethyl-1,2-ethanediamine (●)
R　Trihydrochloride (95968-62-0)
S　KW 5805
U　Gastroprotective, anti-ulcer agent

8900　(7280)　　　　　　　　　　　　　$C_{19}H_{25}N_3O$
　　　　　　　　　　　　　　　　　54063-58-C

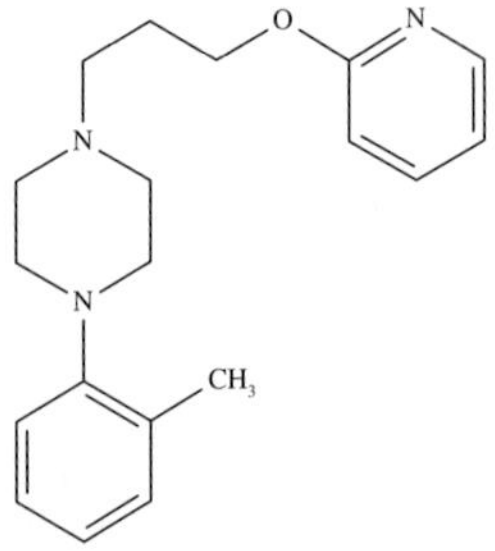

1-[3-(2-Pyridyloxy)propyl]-4-*o*-tolylpiperazine = 1-(2-Methylphenyl)-4-[3-(2-pyridinyloxy)propyl]piperazine (●)
R　Hydrochloride (37816-49-2)
S　Hoechst 757, *Toprilidine hydrochloride***
U　Peripheral vasodilator, sedative

8901　　　　　　　　　　　　　　　　$C_{19}H_{25}N_3O$
141725-09-9

(*R*)-2-Amino-*N*-[3-(3-pyridyl)propyl]-propiono-2',6'-xylidide = (*R*)-2-Amino-*N*-(2,6-dimethylphenyl)-*N*-[3-(3-pyridinyl)propyl]propanamide (●)
R　Tartrate (1:1) (141725-10-2)
S　*Milacainide tartrate**, Ro 22-9194, Xilacainide tartrate*
U　Anti-arrhythmic, cardiac depressant

8902　(7283)　　　　　　　　　　　$C_{19}H_{25}N_3OS$
500-89-0

1-(*p*-Butoxyphenyl)-3-[*p*-(dimethylamino)phenyl]-2-thiourea = 4-Butoxy-4'-(dimethylamino)thiocarbanilide = *N*-(4-Butoxyphenyl)-*N*'-[4-(dimethylamino)phenyl]thiourea (●)
S　Ciba 1906, Su-1906, *Thiambutosine***
U　Leprostatic

8903　(7284)　　　　　　　　　　$C_{19}H_{25}N_3O_2S_2$
7456-24-8

10-[2-(Dimethylamino)propyl]-*N*,*N*-dimethylphenothiazine-2-sulfonamide = 10-(2-Dimethylaminopropyl)-2-(dimethylsulfamoyl)phenothiazine = 10-[2-(Dimethylamino)propyl]-*N*,*N*-dimethyl-10*H*-phenothiazine-2-sulfonamide (●)

R　also monomethanesulfonate (mesilate) (7455-39-2)
S　Alius "Roussel; Scharper", Arumejisten, Banigren, Banistyl, Bayer 1483, Bisbermin, Bonpac, Calsekin, Cylmitiazine, *Dimethothiazine, Dimétiotazine, Dimetotiazine**,* Flue, *Fonazine mesylate,* Fusaban, IL-6302, Migrethiazin, Migristène, Neomestin, Normelin, Promaquid, Rorantoin, 8599 R.P., Serevirol, Yoristen
U　Serotonin inhibitor, antihistaminic, anti-anaphylactic, analgesic

---

8904　(7285)　　　　　　　　　　$C_{19}H_{25}N_3O_3$
86134-80-7

3-Amino-4-[[3-[3-(1-piperidinylmethyl)phenoxy]propyl]amino]-3-cyclobutene-1,2-dione (●)
S　BMY-25368, SKF 94482
U　Histamine H_2-receptor antagonist

8905　(7286)　　　　　　　　　　$C_{19}H_{25}N_3O_3$
79848-61-6

3'-Cyano-4'-[3-[(1,1-dimethyl-2-propynyl)amino]-2-hydroxypropoxy]isobutyranilide = *N*-[3-Cyano-4-[3-[(1,1-dimethyl-2-propynyl)amino]-2-hydroxypropoxy]phenyl]-2-methylpropanamide (●)
S　Koe 3290
U　Anti-arrhythmic

8906　(7287)　　　　　　　　　　$C_{19}H_{25}N_3O_3S$
70144-16-0

[3a*R*-[2(2*S**,5*R**,6*R**),3aα,4α,7α,7aα]]-6-[[(1,3,3a,4,7,7a-Hexahydro-4,7-ethano-2*H*-isoindol-2-

yl)methylene]amino]-3,3-dimethyl-7-oxo-4-thia-1-azabi-
cyclo[3.2.0]heptane-2-carboxylic acid (●)
S   CGP 12553
U   Antiviral, antibacterial

8907 $C_{19}H_{25}N_3O_4$
                                               162203-65-8

N-Benzylcarbamoyl-L-prolyl-L-prolylmethanol = (2S)-2-
[[(2S)-2-(Hydroxyacetyl)-1-pyrrolidinyl]carbonyl]-N-
(phenylmethyl)-1-pyrrolidinecarboxamide (●)
S   JHP-4819, JTP-4819
U   Endopeptidase inhibitor, cognition enhancer

8908   (7288) $C_{19}H_{25}N_3S$
                                               58-37-7

10-[2,3-Bis(dimethylamino)propyl]phenothiazine = Tet-
ramethyl(1-phenothiazin-10-ylmethylethylene)diamine =
N,N,N′,N′-Tetramethyl-3-(10H-phenothiazin-10-yl)-1,2-
propanediamine (●)
R   Fumarate (2:1) (3688-62-8)
S   *Aminopromazine fumarate***, Bayer A 124, Jenoto-
       ne, Lispamol, Lorusil, Myspamol, NSC-169468,
       Proquamezine fumarate, 3828 R.P., Sedofarmolo,
       Spamol, *Tetraméprozine*
U   Spasmolytic

8909   (7289) $C_{19}H_{25}N_4O_6PS_2$
                                               52406-01-6

Diethyl [thio[o-[3-(p-tolylsulfonyl)ureido]phenyl]carb-
amoyl]phosphoramidate = Diethyl 3-[2-(3-tosylureido)-
phenyl]-2-thioureidophosphonate = [[[2-[[[[(4-Methyl-
phenyl)sulfonyl]amino]carbonyl]amino]phenyl]ami-
no]thioxomethyl]phosphoramidic acid diethyl ester (●)
S   RH-565, RH-32565, Sansalid, *Uredofos***
U   Anthelmintic (veterinary)

8910   (7290) $C_{19}H_{25}N_4O_7PS$
                                               52406-09-4

Diethyl [[o-[3-(p-tolylsulfonyl)ureido]phenyl]carb-
amoyl]phosphoramidate = [[[2-[[[[(4-Methylphenyl)sul-
fonyl]amino]carbonyl]amino]phenyl]amino]carbonyl]
phosphoramidic acid diethyl ester (●)
S   Diuredosan
U   Anthelmintic (veterinary)

8911 $C_{19}H_{25}N_5O_2$
                                               134017-78-0

rel-N-Cyano-N'-[(3R,4S)-6-cyano-3,4-dihydro-3-hydroxy-2,2-dimethyl-2H-1-benzopyran-4-yl]-N"-(1,1-dimethylpropyl)guanidine (●)
S BMS 189365, U-89232
U Potassium channel opener

8912 (7291) $C_{19}H_{25}N_5O_4$
                                                      63590-64-7

1-(4-Amino-6,7-dimethoxy-2-quinazolinyl)-4-(tetrahydro-2-furoyl)piperazine = 1-(4-Amino-6,7-diethoxy-2-quinazolinyl)-4-[(tetrahydro-2-furanyl)carbonyl]piperazine (●)
R also monohydrochloride (63074-08-8) or monohydrochloride dihydrate (70024-40-7)
S Abbott-45975, Adecur, Banmet, Blavin, Deflox "Abbott", Dysalfa, Dysalpha, Ezosina, Flotrin, Flumarc, Fosfomic, Heitrin, Hitrin, Hyron, Hytracin, Hytrin(e), Hytrinex, Isontyn, Itrin, Magnurol, My 208, Sinalfa, Teralfa, Teraprost, *Terazosin hydrochloride***, Unoprost "Guidotti", Urodie, Uroflo, Vasocard "Abbott", Vasomet, Vicard
U Antihypertensive, in benign prostatic hypertrophy

8913 (7292) $C_{19}H_{25}N_5O_4S_2$
                                                    138490-53-6

N-Methyl-N-[2-[methyl(1-methyl-1H-benzimidazol-2-yl)amino]ethyl]-4-[(methylsulfonyl)amino]benzenesulfonamide (●)
S WAY-123398
U Anti-arrhythmic

8914 (7293) $C_{19}H_{26}ClNO_3$
                                                      68163-03-1

(2α,4α,4aβ,5α,7α,7aβ,8R*)-(−)-N-[2-(2-Chlorophenyl)ethyl]hexahydro-4-methoxy-7a,8-dimethyl-2,5-methanocyclopenta-1,3-dioxin-7-amine (●)
S *Valorphin*
U Opioid analgesic

8915 $C_{19}H_{26}ClN_3O_2$
                                                      166743-07-3

(−)-4-Amino-N-[2-(1-azabicyclo[3.3.0]octan-5-yl)ethyl]-5-chloro-2,3-dihydro-2-methylbenzo[b]furan-7-carboxamide = (2S)-4-Amino-5-chloro-2,3-dihydro-2-methyl-N-[2-(tetrahydro-1H-pyrrolizin-7a(5H)-yl)ethyl]-7-benzofurancarboxamide (●)
R Fumarate (2:1) (166743-12-0)
S SK-951
U 5-HT$_4$ receptor agonist

8916 (7294) $C_{19}H_{26}Cl_2N_2O$
                                                      83913-05-7

trans-(±)-3,4-Dichloro-N-methyl-N-[2-(1-pyrrolidinyl)cyclohexyl]benzeneacetamide (●)

R Monomethanesulfonate (83913-06-2)
S U-50488 H, UM 1382
U Analgesic (selective κ-opioid agonist)

8917 $C_{19}H_{26}FN_3O_5S$
144625-67-2

5-Fluoro-2-methoxy-1H-indole-3-carboxylic acid [1-[2-[(methylsulfonyl)amino]ethyl]-4-piperidinyl]methyl ester (●)
S GR 125487
U 5-HT$_4$ receptor antagonist

8918 (7295) $C_{19}H_{26}FeO$
65606-61-3

(3,5,5-Trimethylhexanoyl)ferrocene = (3,5,5-Trimethyl-1-oxohexyl)ferrocene (●)
S Diciferron*
U Hematinic

8919 (7296) $C_{19}H_{26}I_3N_3O_9$
66108-95-0

N,N'-Bis(2,3-dihydroxypropyl)-5-[N-(2,3-dihydroxypropyl)acetamido]-2,4,6-triiodoisophthalamide = 5-[Acetyl(2,3-dihydroxypropyl)amino]-N,N'-bis(2,3-dihydroxypropyl)-2,4,6-triiodo-1,3-benzenedicarboxamide (●)

S Accupaque, C 545, Exypaque, Iohexol**, Myelo-Kit, Omnigraf, Omnipaque, Omnitrast, Win 39424
U Diagnostic aid (radiopaque medium for injection)

8920 (7297) $C_{19}H_{26}NO$
59866-76-1

2-(1,2-Diphenylethoxy)ethyltrimethylammonium = 2-(1,2-Diphenylethoxy)-N,N,N-trimethylethanaminium (●)
R Bromide (15585-70-3)
S Bibenzonium bromide**, Bractos, ES 132, LIM-tussis, Lisibex, Lysbex, Lysibex, Lysobex, Medipectol, OM-tussic, OM-tussis, Rea-tos, Sedobex "Continental Pharma", Thoragol, Tussibex, Tussis-OM
U Antitussive

8921 (7298) $C_{19}H_{26}NO_3S$
17989-37-6

Diethyl(2-hydroxyethyl)methylammonium α-phenyl-2-thiopheneglycolate (ester) = N,N-Diethyl-N-methyl-N-[2-(α-phenyl-2-thiopheneglycoloyloxy)ethyl]ammonium = N,N-Diethyl-2-[(hydroxyphenyl-2-thienylacetyl)oxy]-N-methylethanaminium (●)
R Bromide (17692-63-6)
S F-70, Nibitor, Oxitefonium bromide**, Oxytéfonium bromide
U Anticholinergic, spasmolytic

8922 (7299) $C_{19}H_{26}NO_4$
                                              99571-64-9

(8r)-6β,7β-Epoxy-8-ethyl-3α-hydroxy-1αH,5αH-tropa-
nium (−)-tropate = N-Ethyl-O-tropoylscopinium =
(3s,6R,7S,8r)-8-Ethyl-3-[(S)-tropoyloxy]-6,7-epoxytro-
panium = [7(S)-(1α,2β,4β,5α,7β)]-9-Ethyl-7-(3-hydr-
oxy-1-oxo-2-phenylpropoxy)-9-methyl-3-oxa-9-azonia-
tricyclo[3.3.1.0^{2,4}]-nonane (●)
R   Bromide (30286-75-0)
S   Äthylhyoscinbromid, Ba 253-BR-L, Balsigan, Ethyl-
    hyoscine bromide, Oxitropium bromide**, Oxivent,
    Oxytropium bromide, Pulsigan, Scopolamine
    ethobromide, Tersigan, Tersigat, Ventilat, Ventox
U   Anticholinergic, bronchospasmolytic

8923 (7300) $C_{19}H_{26}N_2$
                                              10019-18-8

N-Benzyl-N-[β-(diethylamino)ethyl]aniline = N-Benzyl-
N',N'-diethyl-N-phenylethylenediamine = N,N-Diethyl-
N'-phenyl-N'-(phenylmethyl)-1,2-ethanediamine (●)
R   also dihydrochloride or maleate
S   Rodismin
U   Antihistaminic, anti-allergic

8924 $C_{19}H_{26}N_2O$
                                              163562-15-0

8-(Dipropylamino)-6,7,8,9-tetrahydro-3H-benz[e]indole-
1-carboxaldehyde (●)
S   OSU 191
U   5-HT$_{1A}$ receptor agonist

8925 $C_{19}H_{26}N_2O$
                                              141318-62-9

1-[(4R)-4-(Dipropylamino)-1,3,4,5-tetrahydro-
benz[cd]indol-6-yl]ethanone (●)
S   LY 293284
U   5-HT$_{1A}$ receptor agonist

8926 (7301) $C_{19}H_{26}N_2O$
                                              1505-95-9

α-[2-(Dimethylamino)ethyl]-α-isopropyl-1-naphthalene-
acetamide = 4-Dimethylamino-2-isopropyl-2-(1-naph-
thyl)butyramide = α-[2-(Dimethylamino)ethyl]-α-(1-me-
thylethyl)-1-naphthaleneacetamide (●)
S   DA 992, Naftipramide, Naftypramide**, Naph-
    thypramide
U   Anti-inflammatory

8927 (7302) $C_{19}H_{26}N_2O_2$
2519-77-9

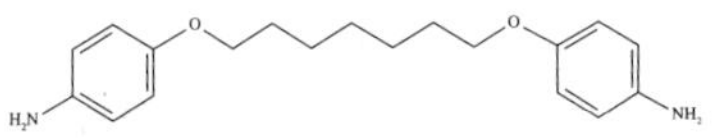

4,4'-(Heptamethylenedioxy)dianiline = 1,7-Bis(*p*-amino-phenoxy)heptane = 4,4'-[1,7-Heptanediyl-bis(oxy)]bis[benzenamine] (●)
S 153 C 51
U Antischistosomal

8928 (7303) $C_{19}H_{26}N_2O_2$
111676-78-9

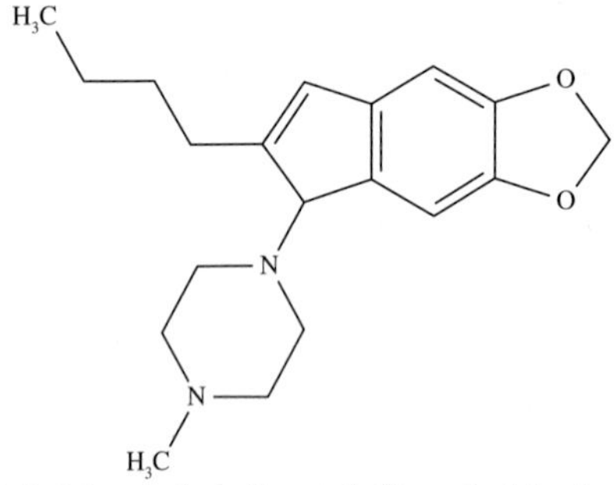

2-Butyl-5,6-(methylenedioxy)-1-(4-methyl-1-piperazi-nyl)indene = 1-(6-Butyl-5*H*-indeno[5,6-*d*]-1,3-dioxol-5-yl)-4-methylpiperazine (●)
R Dihydrochloride (118848-33-2)
S TN 871
U Cerebral vasodilator (potassium and calcium antagonist)

8929 $C_{19}H_{26}N_2O_2$
112603-82-4

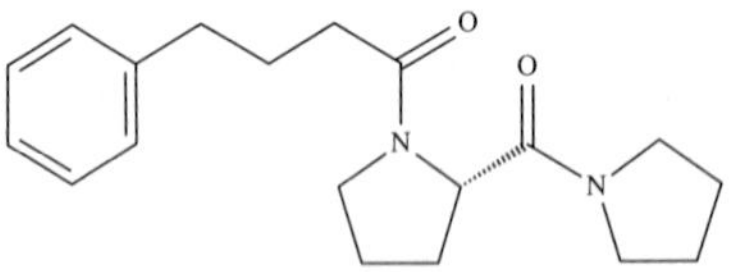

1-[1-(4-Phenylbutanoyl)-L-prolyl]pyrrolidine = (*S*)-1-(1-Oxo-4-phenylbutyl)-2-(1-pyrrolidinylcarbonyl)pyrroli-dine (●)
S SUAM 1221
U Prolyl endopeptidase inhibitor

8930 (7306) $C_{19}H_{26}N_2O_3$
93047-40-6

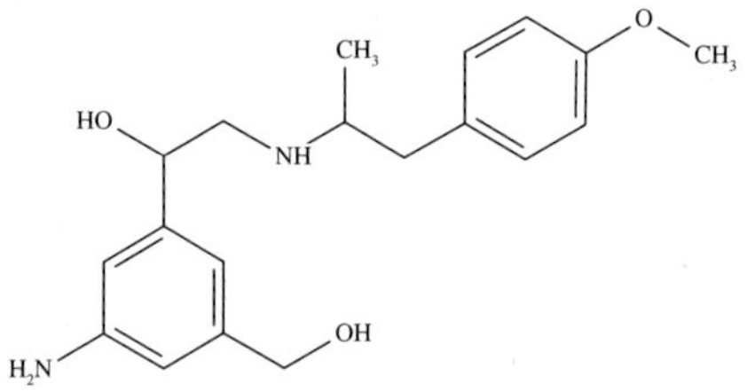

5-Amino-α-[[(*p*-methoxy-α-methylphenethyl)amino]me-thyl]-*m*-xylene-α,α'-diol = 3-Amino-5-(hydroxymethyl)-α-[[(*p*-methoxy-α-methylphenethyl)amino]methyl]ben-zyl alcohol = 1-[3-Amino-5-(hydroxymethyl)phenyl]-2-[[2-(4-methoxyphenyl)-1-methylethyl]amino]ethanol = 5-Amino-α-[[[2-(4-methoxyphenyl)-1-methylethyl]ami-no]methyl]-1,3-benzenedimethanol (●)
S *Naminterol***
U Bronchodilator

8931 (7304) $C_{19}H_{26}N_2O_3$
61864-30-0

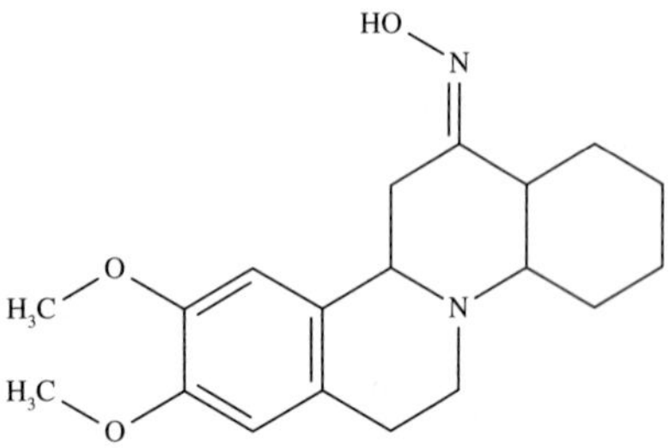

2 (Hydroxyimino)-3,4-cyclohexano-9,10-dimethoxy-1,2,3,4,6,7-hexahydro-11b*H*-benzo[*a*]quinolizine = 1,2,3,4,4a,6,7,11b,12,13a-Decahydro-9,10-dimethoxy-13*H*-dibenzo[*a,f*]quinolizin-13-one oxime (●)
S *Benolizime***
U Tranquilizer

8932 (7305) $C_{19}H_{26}N_2O_3$
75949-60-9

(±)-(*E*)-1-(*tert*-Butylamino)-3-[*o*-[2-(3-methyl-5-isoxa-
zolyl)vinyl]phenoxy]-2-propanol = (±)-(*E*)-3-Methyl-5-
[2-[2-hydroxy-3-*tert*-butylaminopropoxy]styryl]isoxa-
zole = (±)-(*E*)-1-[(1,1-Dimethylethyl)amino]-3-[2-[2-(3-
methyl-5-isoxazolyl)ethenyl]phenoxy]-2-propanol (●)
S *Isoxaprolol***, LU 27937
U α-and β-Adrenergic blocker, antihypertensive

8933 $C_{19}H_{26}N_2O_4$
205242-61-1

(*R*)-(+)-2-[3-(Morpholinomethyl)-2*H*-chromen-8-yloxy-
methyl]morpholine = (2*R*)-2-[[[3-(4-Morpholinylme-
thyl)-2*H*-1-benzopyran-8-yl]oxy]methyl]morpholine (●)
R Monomethanesulfonate (205242-62-2)
S NAS-181
U 5-HT$_{1B}$ receptor antagonist

8934 $C_{19}H_{26}N_2O_4S$
160135-92-2

(6*S*)-Hexahydro-6-[(α*S*)-α-mercaptohydrocinnamami-
do]-2,2-dimethyl-7-oxo-1*H*-azepine-1-acetic acid = (6*S*)-
Hexahydro-6-[[(2*S*)-2-mercapto-1-oxo-3-phenylpro-

pyl]amino]-2,2-dimethyl-7-oxo-1*H*-azepine-1-acetic acid
(●)
S BMS-189921, *Gemopatrilat*
U Antihypertensive, cardiotonic (vasopeptidase inhibi-
tor)

8935 (7307) $C_{19}H_{26}N_2O_4S$
37000-20-7

5-[2-[(α,α-Dimethylphenethyl)amino]-1-hydroxyethyl]-
2'-hydroxymethanesulfonanilide = *N*-[5-[2-[(1,1-Di-
methyl-2-phenylethyl)amino]-1-hydroxyethyl]-2-hydr-
oxyphenyl]methanesulfonamide (●)
R Monohydrochloride (38241-28-0)
S MJ 9184-1, *Zinterol hydrochloride***
U Bronchodilator

8936 $C_{19}H_{26}N_2O_4S$
182967-43-7

(*R**,*S**)-2-[2-[[(2-Hydroxy-3-phenoxypropyl)amino]pro-
pyl]-5-thiazolebutanoic acid (●)
S BMS 187257
U β$_3$-Adrenergic receptor agonist

8937 $C_{19}H_{26}N_2O_5$
132875-68-4

4-(Methoxycarbonyl)-4-[(1-oxopropyl)phenylamino]-1-piperidinepropanoic acid (●)
S GR 90291
U Anesthetic, analgesic (remifentanil metabolite)

8938 (7308) $C_{19}H_{26}N_2O_5S$
7541-30-2

2'-Hydroxy-5'-[1-hydroxy-2-[(p-methoxyphenethyl)amino]propyl]methanesulfonanilide = N-[2-Hydroxy-5-[1-hydroxy-2-[[2-(4-methoxyphenyl)ethyl]amino]propyl]phenyl]methanesulfonamide (●)
R Monohydrochloride (7660-71-1)
S *Mesuprine hydrochloride**, MJ 1987
U Vasodilator, smooth muscle relaxant

8939 $C_{19}H_{26}N_2O_9$
92236-41-4

(S)-2-[2,6-Dimethyl-3,5-bis(ethoxycarbonyl)-1,4-dihydropyridine-4-carboxamido]glutaric acid = 4-[[[(1S)-1,3-Dicarboxypropyl]amino]carbonyl]-1,4-dihydro-2,6-dimethyl-3,5-pyridinedicarboxylic acid 3,5-diethyl ester (●)
R Disodium salt (92236-42-5)
S *Glutapyrone*
U Calcium antagonist

8940 (7309) $C_{19}H_{26}N_2S$
66104-22-1

8β-[(Methylthio)methyl]-6-propylergoline = Methyl (8R,10R)-(6-propyl-8-ergolinyl)methyl sulfide = (8β)-8-[(Methylthio)methyl]-6-propylergoline (●)
R Monomethanesulfonate (66104-23-2)
S Celance, Celeance, Cellance, Lilly 127809, LY 127809, Nopar "Lilly", Parkotil, Parlide, *Pergolide mesilate***, Permax, Pharken
U Dopamine agonist (antiparkinsonian)

8941 $C_{19}H_{26}N_4O_2$

2,3-Dihydro-N-[(3-endo)-8-methyl-8-azabicyclo[3.2.1]oct-3-yl]-3-(1-methylethyl)-2-oxo-1H-benzimidazole-1-carboxamide (●)
R Monohydrochloride (134296-40-5)
S BIMU 8
U Gastroprokinetic (5-HT$_4$ receptor agonist)

8942 $C_{19}H_{26}N_4O_3$
                                                        176219-00-4

2-[[4-(o-Methoxyphenyl)-1-piperazinyl]methyl]-1,3-di-
oxoperhydoimidazo[1,5-a]pyridine = Tetrahydro-2-[[4-
(2-methoxyphenyl)-1-piperazinyl]methyl]imida-
zo[1,5-a]pyridine-1,3(2H,5H)-dione (●)
S   B-20991
U   5-HT$_{1A}$ receptor agonist

8943   (7310) $C_{19}H_{26}N_4O_4$
                                                        21560-58-7

Isobutyl 4-(6,7-dimethoxy-4-quinazolinyl)-1-piperazine-
carboxylate = 4-(6,7-Dimethoxy-4-quinazolinyl)-1-pipe-
razinecarboxylic acid 2-methylpropyl ester (●)
R   Monohydrochloride (23256-26-0)
S   CP 12521-1, *Piquizil hydrochloride***
U   Bronchodilator

8944   (7311) $C_{19}H_{26}N_4O_5$
                                                        21560-59-8

2-Hydroxy-2-methylpropyl 4-(6,7-dimethoxy-4-quinazo-
linyl)-1-piperazinecarboxylate = 4-(6,7-Dimethoxy-4-
quinazolinyl)-1-piperazinecarboxylic acid 2-hydroxy-2-
methylpropyl ester (●)
R   Monohydrochloride (23256-28-2)
S   CP 14185-1, *Hoquizil hydrochloride***
U   Bronchodilator

8945 $C_{19}H_{26}N_6O$
                                                        186692-46-6

(R)-2-[(1-Ethyl-2-hydroxyethyl)amino]-6-(benzylamino)-9-isopropyl-9H-purine = (2R)-2-[[9-(1-Methylethyl)-
6-[(phenylmethyl)amino]-9H-purin-2-yl]amino]-1-buta-
nol (●)
S   *Roscovitine*
U   Antimitotic, antineoplastic

8946 $C_{19}H_{26}N_6O$
                                                        135017-30-0

[1-[2-[[[5-(1-Piperidinylmethyl)-2-furanyl]methyl]amino]ethyl]-2-imidazolidinylidene]propanedinitrile (●)
R　Fumarate (1:1) 135017-85-5
S　KW 5092
U　Gastroprokinetic

8947　(7313)　　　　　　　　　　　　$C_{19}H_{26}O_2$
302-76-1

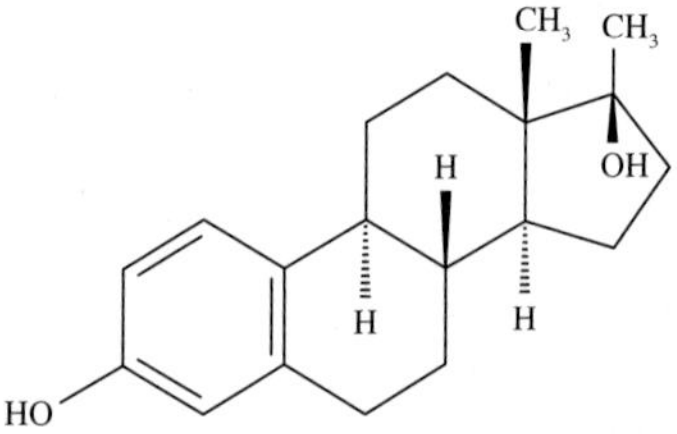

17-Methylestra-1,3,5(10)triene-3,17β-diol = 17α-Methylestradiol = (17β)-17-Methylestra-1,3,5(10)-triene-3,17-diol (●)
S　Follikosid-Tabl., *Methylestradiol, Methylöstradiol*
U　Estrogen

8948　(7312)　　　　　　　　　　　　$C_{19}H_{26}O_2$
846-48-0

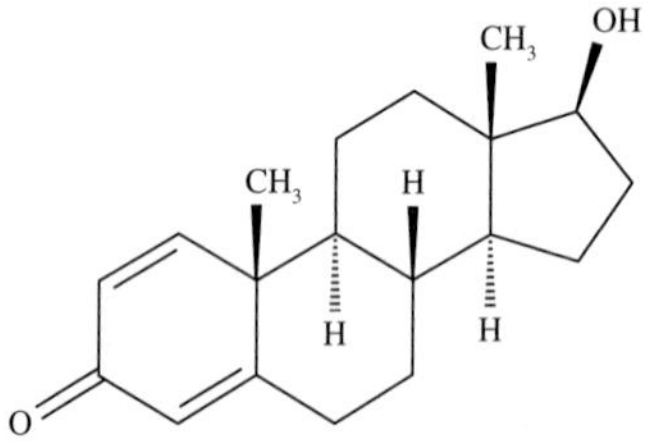

17β-Hydroxyandrosta-1,4-dien-3-one = (17β)-17-Hydroxyandrosta-1,4-dien-3-one (●)
R　see also no. 14391
S　*Boldenone***, *1-Dehydrotestosterone*
U　Anabolic

8949　(7314)　　　　　　　　　　　　$C_{19}H_{26}O_3$
566-48-3

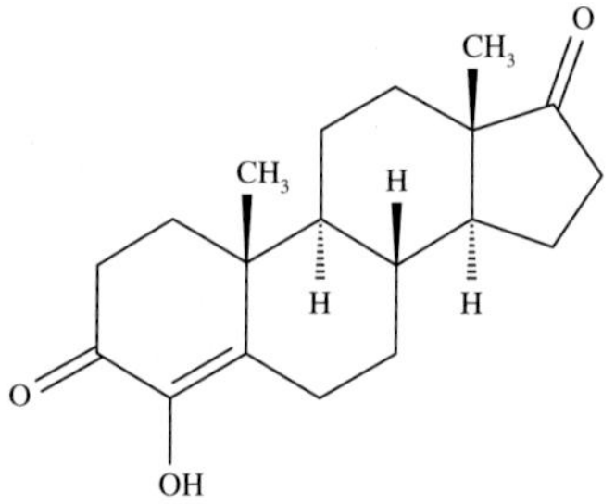

4-Hydroxyandrost-4-ene-3,17-dione (●)
S　CGP 32349, *Formestane***, *Formustine*, Lentare, Lentaron, 4-OHA
U　Aromatase inhibitor

8950　(7315)　　　　　　　　　　　　$C_{19}H_{26}O_3$
584-79-2

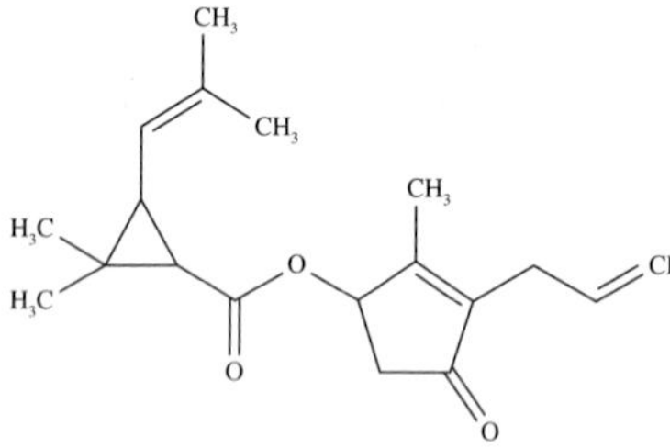

3-Allyl-2-methyl-4-oxo-2-cyclopenten-1-yl 2,2-dimethyl-3-(2-methyl-1-propenyl)cyclopropanecarboxylate = 2,2-Dimethyl-3-(2-methyl-1-propenyl)cyclopropanecarboxylic acid 2-methyl-4-oxo-3-(2-propenyl)-2-cyclopenten-1-yl ester (●)
S　Actomite, *Allethrin I, Allylcinerin I, Bioallethrin*, ENT-17510, Pynamin, Pyresin
U　Insecticide

8951　　　　　　　　　　　　　　　　$C_{19}H_{26}O_3$
362-07-2

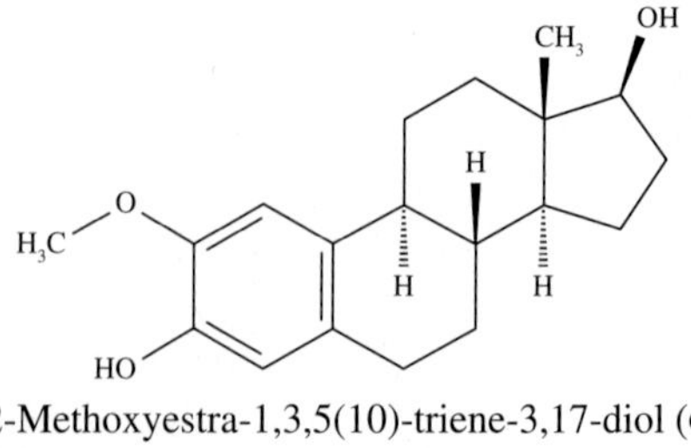

(17β)-2-Methoxyestra-1,3,5(10)-triene-3,17-diol (●)
S　2 ME2, *2-Methoxyestradiol*

U Antineoplastic, angiogenesis inhibitor

8952 (7316) $C_{19}H_{26}O_3$
7004-98-0

3-Methoxyestra-1,3,5(10)-triene-16α,17α-diol =
(16α,17α)-3-Methoxyestra-1,3,5(10)-triene-16,17-diol
(●)
S Alene, *Epimestrol***, NSC-55975, Org 817, Stimo-
vul
U Anterior pituitary activator (ovulation stimulant)

8953 (7317) $C_{19}H_{26}O_3$
83677-24-1

α-(3,5-Di-*tert*-butyl-4-hydroxybenzylidene)-γ-butyrolac-
tone = 3-[[3,5-Bis(1,1-dimethylethyl)-4-hydroxyphe-
nyl]methylene]dihydro-2(3*H*)-furanone (●)
S KME-4
U Anti-inflammatory

8954 (7319) $C_{19}H_{26}O_4$
60595-57-5

3-*p*-Menthyl acetylsalicylate = Acetylsalicylic acid men-
thol ester = (1α,2β,5α)-2-(Acetyloxy)benzoic acid 5-me-
thyl-2-(1-methylethyl)cyclohexyl ester (●)

S Menthospirin
U Therapeutic for laryngitis, pharyngitis and rhinitis

8955 (7318) $C_{19}H_{26}O_4$
115-66-2

1-(*p*,α-Dimethylbenzyl) hydrogen camphorate = Camphoric acid mono-*p*,α-dimethylbenzyl ester = Mono-D-
camphoric acid ester of *p*-tolylmethylcarbinol = 1,2,2-Tri-
methyl-1,3-cyclopentanedicarboxylic acid 1-[1-(4-me-
thylphenyl)ethyl] ester (●)
R see also no. 1729;
compd. with 2,2'-iminodiethanol (1:1) (5634-42-4)
S Bilagen, Biliphorin, Gallogen, Hepasynthyl, Hépa-
toxane, Licarbin, Lymethol, Syncuma, Syntabil, Syn-
thobilin, *Tocamphyl***
U Choleretic

8956 (7320) $C_{19}H_{26}O_5$
105846-98-8

[4α,5α(Z)]-7-[4-(2-Hydroxyphenyl)-2,2-dimethyl-1,3-
dioxan-5-yl]-5-heptenoic acid (●)
S ICI 180080
U Thromboxane receptor antagonist

8957 (7321) $C_{19}H_{26}O_7$
2270-40-8

5a-[(Acetyloxy)methyl]-2,3,4,5,5a,6,7,9a-octahydro-5,8-
dimethylspiro[2,5-methano-1-benzoxepin-10,2'-oxirane]-
3,4-diol 4-acetate = (3α,4β)-12,13-Epoxytrichothec-9-
ene-3,4,15-triol 4,15-diacetate (●)
S ANG-66, *Anguidin, Diacetoxyscirpenol,*
NSC-141537
U Antiviral, immunosuppressive, cytostatic from
Fusarium sp.

8958 $C_{19}H_{27}ClN_2O_4$
148702-58-3

(1-Butyl-4-piperidinyl)methyl 8-amino-7-chloro-1,4-ben-
zodioxan-5-carboxylate = 8-Amino-7-chloro-2,3-dihy-
dro-1,4-benzodioxin-5-carboxylic acid (1-butyl-4-piperi-
dinyl)methyl ester (●)
S SB 204070
U Smooth muscle relaxant (5-HT$_4$ receptor antagonist)

8959 (7322) $C_{19}H_{27}ClO_2$
1093-58-9

8960 (7323) $C_{19}H_{27}ClO_2$
3415-90-5

4-Chloro-17β-hydroxyandrost-4-ene-3-one = 4-Chlorote-
stosterone = (17β)-4-Chloro-17-hydroxyandrost-4-en-3-
one (●)
R see also nos. 10655, 11456, 12961
S *Chlortestosterone, Clostebol***
U Anabolic

4-Chloro-17β-hydroxy-17-methylestr-4-en-3-one = 4-
Chloro-17α-methyl-19-nortestosterone = (17β)-4-
Chloro-17-hydroxy-17-methylestr-4-en-3-one (●)
S *Chlordrolone, Clordrolone,* SKF 6612
U Anabolic

8961 (7324) $C_{19}H_{27}FN_2O_3$
20977-50-8

Isopropylcarbamic acid ester with 4'-fluoro-4-(4-hydr-
oxypiperidino)butyrophenone = 1-[3-(4-Fluoroben-
zoyl)propyl]-4-piperidyl isopropylcarbamate = (1-Meth-
ylethyl)carbamic acid 1-[4-(4-fluorophenyl)-4-oxobutyl]-
4-piperidinyl ester (●)
S AL-1021, *Carperone***
U Antipsychotic, anti-emetic

8962

$C_{19}H_{27}FO_2$
154604-49-6

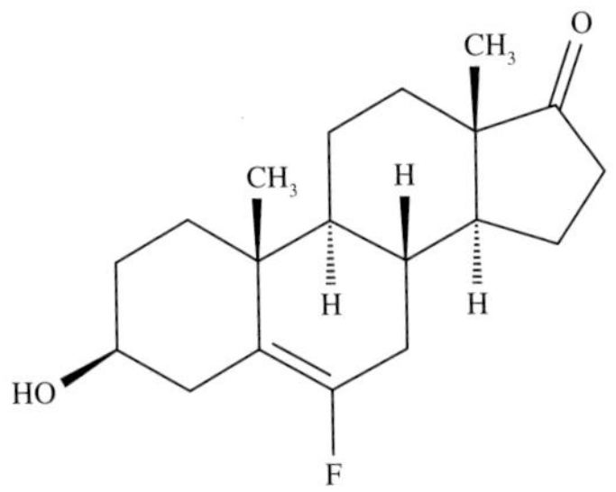

(3β)-6-Fluoro-3-hydroxyandrost-5-en-17-one (●)
S *Fluasteone, Fluoroprasterone*
U Antineoplastic

8963

$C_{19}H_{27}IN_2O_4$
148703-08-6

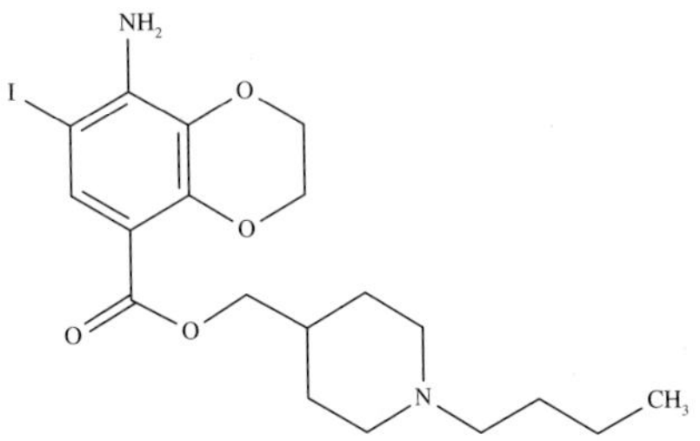

8-Amino-2,3-dihydro-7-iodo-1,4-benzodioxin-5-carb-oxylic acid (1-butyl-4-piperidinyl)methyl ester (●)
S SB 207710
U 5-HT$_4$ receptor antagonist

8964 (7325)

$C_{19}H_{27}NO$
359-83-1

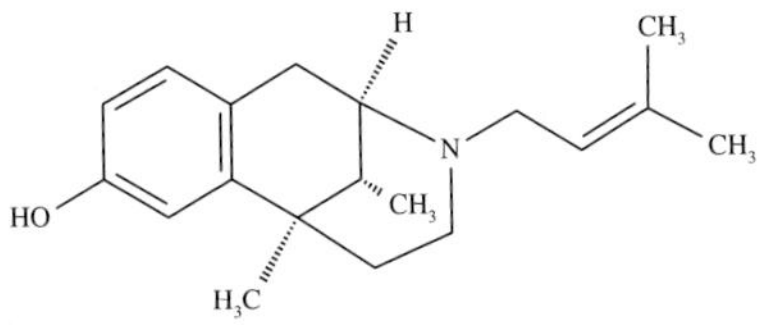

2-(3,3-Dimethylallyl)-2'-hydroxy-5,9-dimethyl-6,7-ben-zomorphan = (2α,6α,11R*)-1,2,3,4,5,6-Hexahydro-6,11-dimethyl-3-(3-methyl-2-butenyl)-2,6-methano-3-benz-azocin-8-ol (●)
R also hydrochloride (64024-15-3) or lactate (17146-95-1)
S Algopent, Basta, CS 350, Dolapent, Dolofortin, Fortal, Fortalgesic, Fortral, Fortralin, Fortwin, He-

xat, Lexir, Liticon, Magadolin, NIH 7958, NSC-107430, Ospronim, Peltazon, Pentacina, Penta-fen "Zoja", Pentafort "Coli", Pentagin, Pentajin, Pen-talgina, Pentavon, Pentawin, *Pentazocine**, Penta-zonicum, Penzin, Perutagin, Silin, Sosegon, Sose-nol, Sosenyl, Sosigon, Sossegon, Susevin, Talwin, Talwinsup, Talwintab, Win 20228
U Analgesic

8965 (7326)

$C_{19}H_{27}NO$
59889-36-0

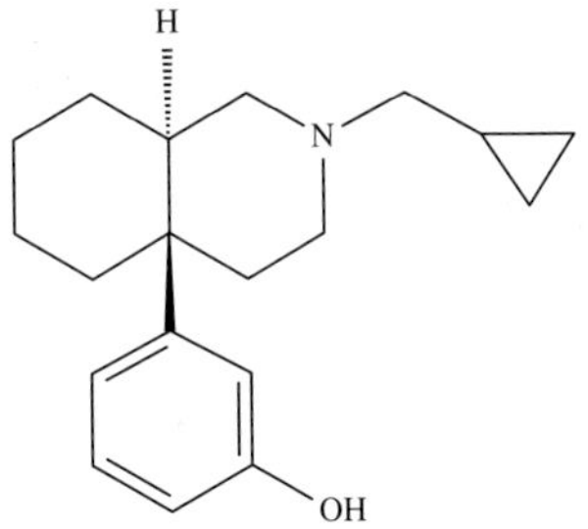

(−)-*m*-[2-(Cyclopropylmethyl)-1,3,4,5,6,7,8,8aα-octahy-dro-4aβ(2H)-isoquinolyl]phenol = (−)-*trans*-3-[2-(Cyclo-propylmethyl)octahydro-4a(2H)-isoquinolinyl]phenol (●)
R Succinate (1:1) (60719-85-9)
S *Ciprefadol succinate**, Lilly 113878
U Analgesic

8966

$C_{19}H_{27}NO_2$
146145-19-9

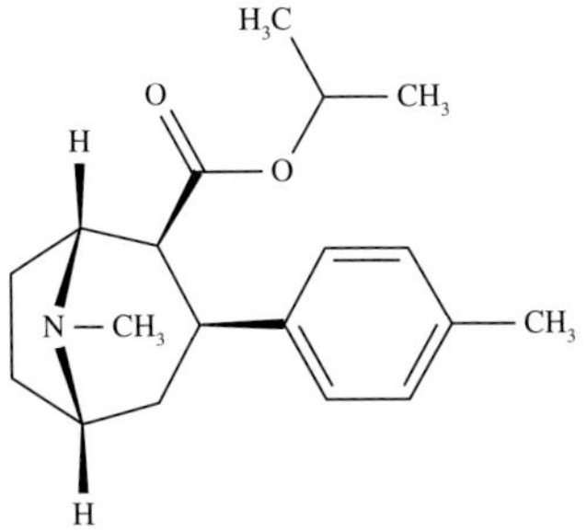

(1R,2S,3S,5S)-8-Methyl-3-(4-methylphenyl)-8-azabicy-clo[3.2.1]octane-2-carboxylic acid 1-methylethyl ester (●)
R Hydrochloride (141807-61-6)
S RTI 117
U Dopamine uptake inhibitor

8967   (7327) $C_{19}H_{27}NO_2$
                                        80158-12-9

17β-(Methylamino)estra-1,3,5(10)-triene-3,16α-diol =
(16α,17β)-17-(Methylamino)estra-1,3,5(10)triene-3,16-
diol (●)
R   Maleate (1:1) (80177-51-1)
S   Org 7797
U   Anti-arrhythmic

8968   (7328) $C_{19}H_{27}NO_2$
                                        57236-89-2

(±)-N-(Tetrahydrofurfuryl)normetazocine =
[2α,3(R*),6α,11R*]-(±)-1,2,3,4,5,6-Hexahydro-6,11-di-
methyl-3-[(tetrahydro-2-furanyl)methyl]-2,6-methano-3-
benzazocin-8-ol (●)
S   Mr 2033, NIH 9102, UM 1072
U   Analgesic (non-morphine-like)

8969   (7329) $C_{19}H_{27}NO_3$
                                        58-46-8

1,3,4,6,7,11b-Hexahydro-3-isobutyl-9,10-dimethoxy-
2H-benzo[a]quinolizin-2-one = 1,3,4,6,7,11b-Hexahy-
dro-9,10-dimethoxy-3-(2-methylpropyl)-2H-ben-
zo[a]quinolizin-2-one (●)
R   Methanesulfonate (804-53-5)

S   Nitoman, Regulin "Takeda", Ro 1-9569, Rubigen,
       *Tetrabenazine mesilate***
U   Antipsychotic, neuroleptic

8970 $C_{19}H_{27}NO_3$
                                        138847-85-5

1-Phenylcyclohexanecarboxylic acid 2-(4-morpholi-
nyl)ethyl ester (●)
S   PRE-084
U σ_1-Receptor agonist

8971   (7330) $C_{19}H_{27}NO_3$
                                        105816-04-4

trans-N-(4-Isopropylcyclohexanecarbonyl)-D-phenylala-
nine = trans-N-[[4-(1-Methylethyl)cyclohexyl]carbonyl]-
D-phenylalanine (●)
S   A-4166, AY-4166, DJN-608, Fastic, *Nateglinide***,
       SDZ-DJN 608, *Senaglinide*, Starlix, Starsis, YM-026
U   Antidiabetic

8972 $C_{19}H_{27}NO_3S$
                                        121048-14-4

(±)-1-[[(3,4-Dimethoxybenzyl)oxy]methyl]-N,N-dime-
thyl-1-(2-thienyl)propylamine = α-[[(3,4-Dimethoxyphe-
nyl)methoxy]methyl]-α-ethyl-N,N-dimethyl-2-thio-
phenemethanamine (●)
R   2-(4-Hydroxybenzoyl)benzoate (149998-21-0)

S　T-1815
U　Colony prokinetic

8973　(7331)　　　　　　　　　　　　　　$C_{19}H_{27}NO_4$
　　　　　　　　　　　　　　　　　　　　28598-08-5

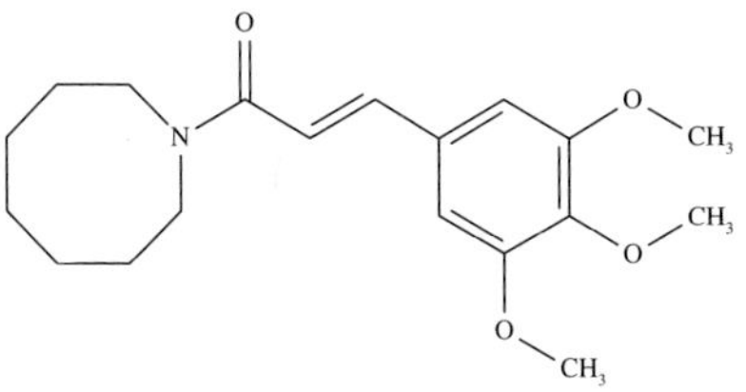

Octahydro-1-(3,4,5-trimethoxycinnamoyl)azocine = Oc-tahydro-1-[1-oxo-3-(3,4,5-trimethoxyphenyl)-2-prope-nyl]azocine (●)
S　*Cinoctramide***
U　Tranquilizer

8974　(7332)　　　　　　　　　　　　　　$C_{19}H_{27}NO_4$
　　　　　　　　　　　　　　　　　　　　3176-03-2

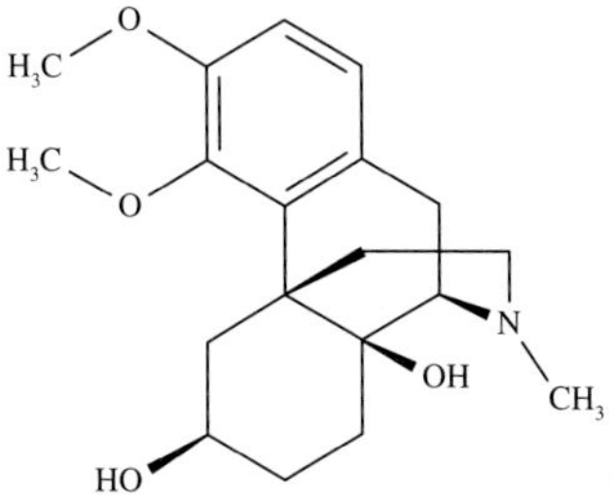

3,4-Dimethoxy-17-methylmorphinan-6β,14-diol = 14-Hydroxydihydro-6β-thebainol 4-methyl ether = (6β)-3,4-Dimethoxy-17-methylmorphinan-6,14-diol (●)
S　*Drotebanol***, Metebanyl, Methebanyl, *Oxymethe-banol*, RAM 327
U　Antitussive, analgesic

8975　(7333)　　　　　　　　　　　　　　$C_{19}H_{27}NO_4$
　　　　　　　　　　　　　　　　　　　　470-68-8

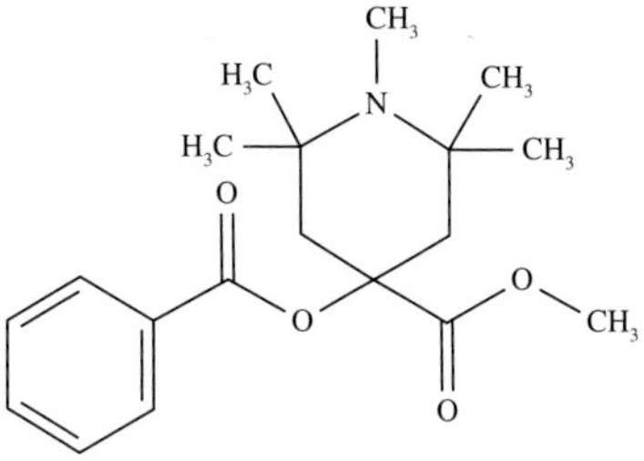

4-(Benzoyloxy)-1,2,2,6,6-pentamethyl-4-piperidinecarb-oxylic acid methyl ester (●)
R　also hydrochloride
S　Alphacaine, α-Eucaine, Eucaine (A)
U　Local anesthetic

8976　　　　　　　　　　　　　　　　　$C_{19}H_{27}NO_4$
　　　　　　　　　　　　　　　　　　　　90060-42-7

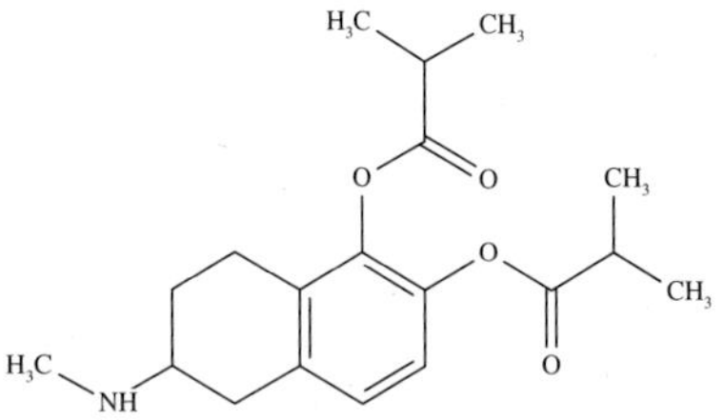

5,6,7,8-Tetrahydro-6-(methylamino)-1,2-naphthalene di-isobutyrate = 2-Methylpropanoic acid 5,6,7,8-tetrahydro-6-(methylamino)-1,2-naphthalenediyl ester (●)
S　*Nolomirole**
U　Dopamine receptor agonist

8977　　　　　　　　　　　　　　　　　$C_{19}H_{27}NO_4S$
　　　　　　　　　　　　　　　　　　　　149926-91-0

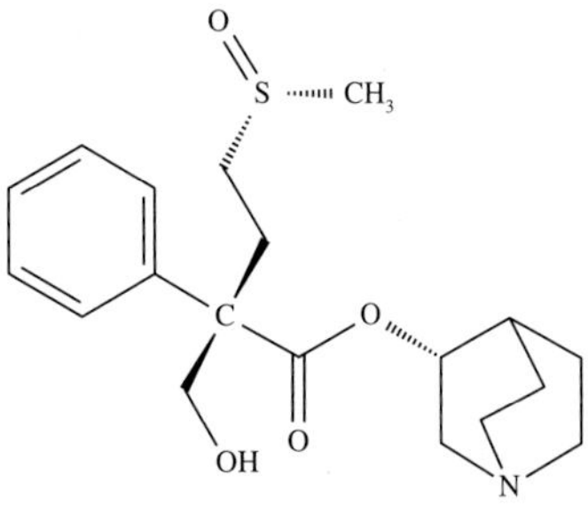

(R)-3-Quinuclidinyl (S)-β-hydroxy-α-[2-[(R)-methylsul-finyl]ethyl]hydratropate = (R)-3-Quinuclidinyl (2S,R_S)-2-(hydroxymethyl)-4-(methylsulfinyl)-2-phenylbutanoate = [3R-[3R*[S*(R*)]]]-α-(Hydroxymethyl)-α-[2-(methyl-sulfinyl)ethyl]benzeneacetic acid 1-azabicyclo[2.2.2]oct-3-yl ester (●)
S　*Revatropate***, UK-112166
U　Antimuscarinic, bronchodilator

8978   (7334) $C_{19}H_{27}N_2$

1-(2-Methoxyphenyl)-4-(4-succinimidobutyl)piperazine
= 1-[4-[4-(2-Methoxyphenyl)-1-piperazinyl]butyl]-2,5-
pyrrolidinedione (●)
R   Dihydrochloride (159187-70-9)
S   MM-77
U   5-HT$_{1A}$ receptor antagonist

2-(N-Benzylanilino)diethyldimethylammonium = *N*-Eth-
yl-*N,N*-dimethyl-2-(*N*-benzylanilino)ethylammonium =
N-Ethyl-*N,N*-dimethyl-2-[phenyl(phenylmethyl)ami-
no]ethanaminium (●)
R   Bromide (13928-80-8)
S   Dispasmol, *Phenbenzamine ethyl bromide*
U   Spasmolytic

8981 $C_{19}H_{27}N_3O_4S$
                                                        144625-51-4

1-Methyl-1*H*-indole-3-carboxylic acid [1-[2-[(methylsul-
fonyl)amino]ethyl]-4-piperidinyl]methyl ester (●)
S   GR 113808
U   5-HT$_4$ receptor antagonist

8979   (7335) $C_{19}H_{27}N_3O$
                                                        117086-68-7

8982   (7336) $C_{19}H_{27}N_3O_4S$
                                                        10004-67-8

3,3-Dimethyl-*N*-1α*H*,5α*H*-tropan-3α-yl-1-indolinecarb-
oxamide = *endo*-2,3-Dihydro-3,3-dimethyl-*N*-(8-methyl-
8-azabicyclo[3.2.1]oct-3-yl)-1*H*-indole-1-carboxamide
(●)
R   Monohydrochloride (140865-88-9)
S   BRL 46470A, *Carisetron hydrochloride*, *Ricasetron
        hydrochloride***
U   5HT$_3$-receptor antagonist (anxiolytic, anti-emetic)

(2*S*,5*R*,6*R*)-6-(3-Amino-1-adamantanecarboxamido)-3,3-
dimethyl-7-oxo-4-thia-1-azabicyclo[3.2.0]heptane-2-
carboxylic acid = (6*R*)-6-(3-Amino-1-adamantanecarbox-
amido)penicillanic acid = [2*S*-(2α,5α,6β)]-6-[[(3-Amino-
tricyclo[3.3.1.1^{3,7}]dec-1-yl)carbonyl]amino]-3,3-dime-
thyl-7-oxo-4-thia-1-azabicyclo[3.2.0]heptane-2-carb-
oxylic acid (●)
S   *Amantocillin***
U   Antibiotic

8980 $C_{19}H_{27}N_3O_3$
                                                        159311-94-1

8983 (7337) $C_{19}H_{27}N_3O_5$
66203-94-9

1-[4,7-Dimethoxy-6-(2-piperidinoethoxy)-5-benzofura-nyl]-3-methylurea = N-[4,7-Dimethoxy-6-[2-(1-piperidi-nyl)ethoxy]-5-benzofuranyl]-N'-methylurea (●)
R Monohydrochloride (84272-16-2)
S MD 750819, *Murocainide hydrochloride***
U Anti-arrhythmic

8984 (7338) $C_{19}H_{27}N_3O_5S$
88519-56-6

1-[4-Hydroxy-3-[(dimethylamino)sulfonamido]phenyl]-2-[(1-phenoxyisopropyl)amino]ethanol = N-[2-Hydroxy-5-[1-hydroxy-2-[(1-methyl-2-phenoxyethyl)ami-no]ethyl]phenyl]-N',N'-dimethylsulfamide (●)
R Monohydrochloride (88543-00-4)
S Me 693
U Anorexic

8985 (7339) $C_{19}H_{27}N_3O_5S_2$
115256-11-6

β-[[p-(Methanesulfonamido)phenethyl]methylamino]me-thanesulfono-p-phenetidide = 4'-[2-[Methyl[4-[(methyl-sulfonyl)amino]phenethyl]amino]ethoxy]methanesulfon-anilide = N-[4-[2-[Methyl[2-[4-[(methylsulfonyl)ami-no]phenoxy]ethyl]amino]ethyl]phenyl]methanesulfon-amide (●)
S *Dofetilide***, Tikosyn, UK-68798, Xelide
U Anti-arrhythmic

8986 (7340) $C_{19}H_{27}N_5$
72822-12-9

5,6,7,8-Tetrahydro-3-[2-(4-o-tolyl-1-piperazinyl)ethyl]-s-triazolo[4,3-a]pyridine = 5,6,7,8-Tetrahydro-3-[2-[4-(2-methylphenyl)-1-piperazinyl]ethyl]-1,2,4-tri-azolo[4,3-a]pyridine (●)
R Monohydrochloride (72822-13-0)
S AF 2139, Bengglau, Benglau, *Dapiprazole hydro-chloride***, Glamidol, Glamidolo, Remydrial, Rever-sil, Rev-Eyes
U Tranquilizer, antiglaucoma agent

8987 (7341) $C_{19}H_{27}N_5O_3$
80755-51-7

1-(4-Amino-6,7-dimethoxy-2-quinazolinyl)-4-butyrylhe-xahydro-1H-1,4-diazepine = 4-Amino-2-(N-butyryl-N'-homopiperazinyl)-6,7-dimethoxyquinazoline = 1-(4-Amino-6,7-dimethoxy-2-quinazolinyl)hexahydro-4-(1-oxobutyl)-1H-1,4-diazepine (●)
R Monohydrochloride (52712-76-2)
S Andante, Andante "Boehringer-Ingelh.", Bunatenon, *Bunazosin hydrochloride***, Dentol, Detantol, E 643, E 1025, EA 0643

U   Antihypertensive (α-blocker)

8988   (7342) $C_{19}H_{27}N_5O_4$
                                                    81403-80-7

(±)-N-[3-[(4-Amino-6,7-dimethoxy-2-quinazolinyl)me-
thylamino]propyl]tetrahydro-2-furancarboxamide (●)
R   also monohydrochloride (81403-68-1)
S   Alfetim, Alfoten, Alfuzol, *Alfuzosin**, Benestan,
    Dalfaz, Dilotex "Beta", Mittoval, SL 7749-10, Uri-
    on, Uroxatral, Xatral (uno), Zatral
U   Antihypertensive (α-blocker), treatment of benign
    prostatic hyperplasia

8989   (7343) $C_{19}H_{27}N_5O_5$
                                                    130636-43-0

6-[[2-[(2-Hydroxyethyl)[3-(p-nitrophenyl)propyl]ami-
no]ethyl]amino]-1,3-dimethyluracil = 6-[[2-[(2-Hydroxy-
ethyl)[3-(4-nitrophenyl)propyl]amino]ethyl]amino]-1,3-
dimethyl-2,4(1H,3H)-pyrimidinedione (●)
R   Monohydrochloride (130656-51-8)
S   MS-551, *Nifekalant hydrochloride**, Shinbit
U   Anti-arrhythmic (potassium channel blocker)

8990   (7344) $C_{19}H_{27}O_9P$
                                                    87810-56-8

5,6-Dihydro-6-[3,4,6,13-tetrahydroxy-3-methyl-1,7,9,11-
tridecatetraenyl]-2H-pyran-2-one 4-(dihydrogen phosphate) = 5,6-Dihydro-6-[3,6,13-trihydroxy-3-methyl-4-

(phosphonooxy)-1,7,9,11-tridecatetraenyl]-2H-pyran-2-
one (●)
R   Monosodium salt (87860-39-7)
S   CI-920, CL 1565 A, *Fostriecin sodium**, Fostrie-
    nin*, NSC 339638, PD 110161
U   Antineoplastic, antileukemic (antibiotic from *Strepto-
    myces* ATCC 31906)

8991   (7345) $C_{19}H_{28}ClNO_6$
                                                    129298-91-5

O-(Chloroacetylcarbamoyl)fumagillol = [3R-
[3α,4α(2R*,3R*),5β,6β]]Chloroacetylcarbamic acid 5-
methoxy-4-[2-methyl-3-(3-methyl-2-butenyl)oxiranyl]-
1-oxaspiro[2.5]oct-6-yl ester (●)
S   AGM-1470, TNP-470
U   Anti-angiogenic from *Aspergillus fumigatus*

8992   (7346) $C_{19}H_{28}ClN_3$
                                                    85-10-9

7-Chloro-4-[[4-(diethylamino)-1-methylbutyl]amino]-3-
methylquinoline = N^4-(7-Chloro-3-methyl-4-quinolinyl)-
N^1,N^1-diethyl-1,4-pentanediamine (●)
S   *Methylchloroquine*, Nivaquine C, 3038 R.P., San-
    tochin, Santoquine, S.N. 6911, Sontochin, Sontoqine
U   Antimalarial

8993　　　　　　　　　　　　　　$C_{19}H_{28}ClN_3O_5S$
219757-90-1

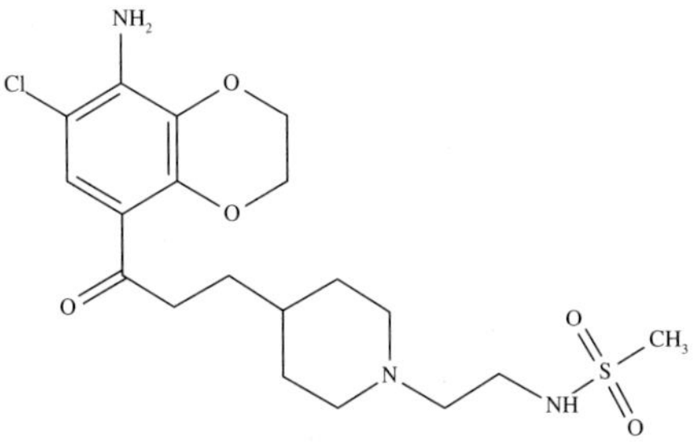

N-[2-[4-[2-[(8-Amino-7-chloro-1,4-benzodioxan-5-yl)carbonyl]ethyl]piperidino]ethyl]methanesulfonamide
= N-[2-[4-[3-(8-Amino-7-chloro-2,3-dihydro-1,4-benzodioxin-5-yl)-3-oxopropyl]-1-piperidinyl]ethyl]methanesulfonamide (●)
R　Monohydrochloride (184159-40-8)
S　RS-100302-190, *Sulamserod hydrochloride***
U　5-HT$_4$ receptor antagonist

8994　(7347)　　　　　　　　　　$C_{19}H_{28}ClN_5O$
52942-31-1

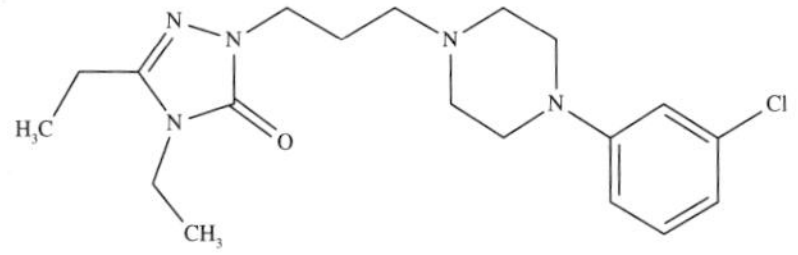

1-[3-[4-(*m*-Chlorophenyl)-1-piperazinyl]propyl]-3,4-diethyl-Δ^2-1,2,4-triazolin-5-one = 2-[3-[4-(3-Chlorophenyl)-1-piperazinyl]propyl]-4,5-diethyl-2,4-dihydro-3*H*-1,2,4-triazol-3-one (●)
R　Monohydrochloride (57775-22-1)
S　Axiomin, Centren, Clopadrone, Depracer, Depraser, Etonin, *Etoperidone hydrochloride***, Etoran, Eupsy-Mite, McN-A 2673-11, Praxium, ST 1191, Staff, Triazolinone, Tropene
U　Antidepressant, anxiolytic

8995　　　　　　　　　　　　　　$C_{19}H_{28}Cl_2N_2$
130609-93-7

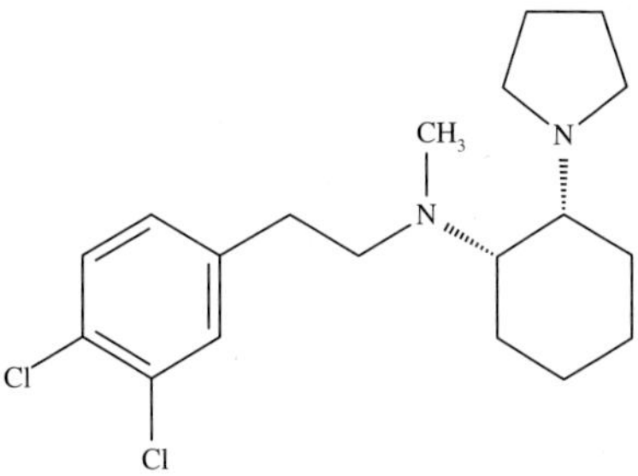

3,4-Dichloro-N-methyl-N-[(1S,2R)-2-(1-pyrrolidinyl)cyclohexyl]benzeneethanamine (●)
S　BD 737
U　Psychotropic (σ opiate receptor antagonist)

8996　(7348)　　　　　　　　　　$C_{19}H_{28}Cl_2N_2O_3$
3568-16-9

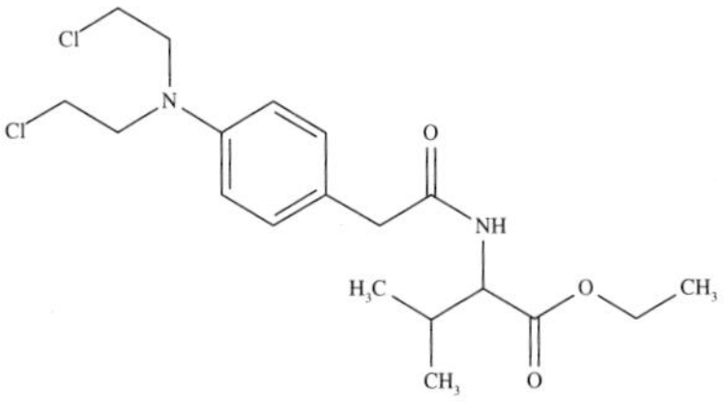

N-[*p*-Bis(β-chloroethyl)aminophenylacetyl]valine ethyl ester = N-[[4-[Bis(2-chloroethyl)amino]phenyl]acetyl]valine ethyl ester (●)
S　Phenaline
U　Antineoplastic

8997　(7349)　　　　　　　　　　$C_{19}H_{28}Cl_2N_2O_3S$
3819-34-9

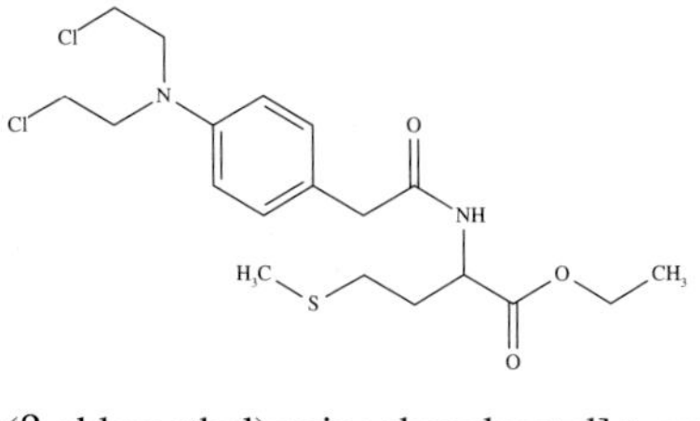

N-[*p*-Bis(β-chloroethyl)aminophenylacetyl]-L-methionine ethyl ester = N-[[4-[Bis(2-chloroethyl)amino]phenyl]acetyl]-L-methionine ethyl ester (●)
S　Fenamet, Phenamet
U　Antineoplastic

8998   (7350) $C_{19}H_{28}NO_3$
                                                   13283-82-4

3-Hydroxy-1,1-dimethylpyrrolidinum α-cyclopentyl-
mandelate = 3-(α-Cyclopentylmandeloyloxy)-1,1-dime-
thylpyrrolidinium = 3-[(Cyclopentylhydroxyphenylace-
tyl)oxy]-1,1-dimethylpyrrolidinium (●)
R   Bromide (596-51-0)
S   Acpan, AHR-504, Asécryl, Contrem, Distnart, Ga-
        strodyn, *Glycopyrrolate*, *Glycopyrronium bromi-
        de**, Nodapton, Pirokonal, Robanul, Robinal, Ro-
        binul, Sroton, Taional, Tarodyl, Tarodyn
U   Anticholinergic

8999   (7351) $C_{19}H_{28}NO_3$

erythro-3-Hydroxy-1,1-dimethylpyrrolidinium α-cyclo-
pentylmandelate = (*R**,*R**)-3-[(Cyclopentylhydroxyphe-
nylacetyl)oxy]-1,1-dimethylpyrrolidinum (●)
R   Bromide (53808-86-9)
S   *Ritropirronium bromide***
U   Anticholinergic

9000 $C_{19}H_{28}NO_5P$

3-[(3*S*,6*R*)-6-[[(Cyclohexylmethyl)hydroxyphosphi-
nyl]methyl]-3-morpholinyl]benzoic acid (●)
S   CGP 76290A
U   GABA$_B$ receptor antagonist

9001   (7353) $C_{19}H_{28}N_2$
                                                   5560-72-5

5-[3-(Dimethylamino)propyl]-6,7,8,9,10,11-hexahydro-
5*H*-cyclooct[*b*]indole = 1-(3-Dimethylaminopropyl)-2,3-
hexamethyleneindole = 6,7,8,9,10,11-Hexahydro-*N*,*N*-di-
methyl-5*H*-cyclooct[*b*]indole-5-propanamine (●)
R   Monohydrochloride (20432-64-8)
S   E-330, Galatur, *Iprindole hydrochloride***, *Pramin-
        dole hydrochloride*, Prindol, Prondol, Sprindol, Ter-
        tran, Wy-3263-HCl
U   Antidepressant

9002   (7352) $C_{19}H_{28}N_2$
                                                   1228-02-0

3-Methyl-2-phenyl-2-(2-piperidinoethyl)valeronitrile = α-*sec*-Butyl-α-phenyl-4-piperidinobutyronitrile = α-(1-Methylpropyl)-α-phenyl-1-piperidinebutanenitrile (●)

R Monohydrochloride (53198-87-1)

S Eurazyl, R 154

U Analgesic, spasmolytic

9003 (7356)

C$_{19}$H$_{28}$N$_2$O$_3$
64779-98-2

(±)-5-Butyryl-*N*-[(1-ethyl-2-pyrrolidinyl)methyl]-*o*-anisamide = 5-Butyryl-*N*-[(1-ethyl-2-pyrrolidinyl)methyl]-2-methoxybenzamide = *N*-[(1-Ethyl-2-pyrrolidinyl)methyl]-2-methoxy-5-(1-oxobutyl)benzamide (●)

S *Irolapride***

U Anti-emetic

9004 (7354)

C$_{19}$H$_{28}$N$_2$O$_3$
25331-92-4

3-Ethyl-6,7-dihydro-2-methyl-5-(4,4-ethylenedioxypiperidinomethyl)indol-4(5*H*)-one = 5-[(1,4-Dioxa-8-azaspiro[4.5]dec-8-yl)methyl]-3-ethyl-6,7-dihydro-2-methyl-1*H*-indol-4(5*H*)-one (●)

S AL-1612

U Antipsychotic, anti-emetic

9005 (7355)

C$_{19}$H$_{28}$N$_2$O$_3$
117539-16-9

(±)-2-[3-(Dimethylamino)propyl]-2-(*m*-methoxyphenyl)-3,3-dimethylglutarimide = (±)-3-[3-(Dimethylamino)propyl]-3-(3-methoxyphenyl)-4,4-dimethyl-2,6-piperidinedione (●)

R Monohydrochloride (103353-87-3)

S AGN 2979, BTG 1501, SC-48274

U Anxiolytic

9006

C$_{19}$H$_{28}$N$_2$O$_3$S

(8a*R*,12a*S*,13a*S*)-5,8,8a,9,10,11,12,12a,13,13a-Decahydro-3-methoxy-12-(ethylsulfonyl)-6*H*-isoquino[2,1-*g*][1,6]naphthyridine = [8a*R*-(8aα,12aα,13aα)]-12-(Ethylsulfonyl)-5,8,8a,9,10,11,12,12a,13,13a-decahydro-3-methoxy-6*H*-isoquino[2,1-*g*][1,6]naphthyridine (●)

R Monohydrochloride (186002-54-0)

S RS-79948-197

U α$_2$-Adrenoceptor antagonist

9007 (7357)

C$_{19}$H$_{28}$N$_2$O$_4$
78628-28-1

N-[3-[(α-Piperidino-*m*-tolyl)oxy]propyl]glycolamide acetate (ester) = 2-Acetoxy-*N*-[3-[*m*-(piperidinomethyl)phenoxy]propyl]acetamide = [[3-[(α-Piperidino-*m*-tolyl)oxy]propyl]carbamoyl]methyl acetate = 2-(Acetyloxy)-*N*-[3-[3-(1-piperidinylmethyl)phenoxy]propyl]acetamide (●)

R   Monohydrochloride (93793-83-0)
S   *Aceroxatidine hydrochloride, Acetylroxatidine hydrochloride,* Altat, Gastralgin, Hoechst 760, Meluston, NeoH₂, *Pifatidine hydrochloride,* Rotane, Roxane, *Roxatidine acetate hydrochloride,* Roxid "Hoechst", Roxin "Hoechst", Roxit, Roxiwas, Sarilen, TZU-0460, Xarcin, Zarocs
U   Histamine H₂-receptor antagonist (anti-ulcer agent)

9008   (7358) $C_{19}H_{28}N_2O_4$
                                                       39731-05-0

Isopropyl (±)-4-[3-(*tert*-butylamine)-2-hydroxypropoxy]indole-2-carboxylate = (±)-4-[3-[(1,1-Dimethylethyl)amino]-2-hydroxypropoxy]-1*H*-indole-2-carboxylic acid 1-methylethyl ester (●)

S   *Carpindolol***, LM 21009
U   β-Adrenergic blocker

9009 $C_{19}H_{28}N_2O_4$
                                                       192214-20-3

(3*Z*)-2,8-Dimethyl-1-oxa-8-azaspiro[4.5]decan-3-one *O*-[2-(3-methoxyphenoxy)ethyl]oxime (●)

S   ARL 17444
U   Neuroprotective (muscarinic agonist)

9010   (7359) $C_{19}H_{28}N_3O_3$

N,N-Diethyl-*N*-methyl-2-(5-ethyl-2,4,6-trioxo-5-phenyl-hexahydro-1-pyrimidinyl)ethylammonium = *N,N*,5-Triethylhexahydro-*N*-methyl-2,4,6-trioxo-5-phenyl-1-pyrimidineethanaminium (●)

R   Iodide (6191-48-6)
S   *Barbetonii iodidum,* CS 11, Defenale
U   Anticholinergic (anti-ulcer)

9011 $C_{19}H_{28}N_4$

N-(1-Adamantyl)-*N'*,*N'*-[3-[4(5)-1*H*-imidazolyl]pentamethylene]formamidine = 4-(1*H*-Imidazol-4-yl)-1-[(tricyclo[3.3.1.1^{3,7}]dec-1-ylimino)methyl]piperidine (●)

R   Dihydrochloride (175033-29-1)
S   AQ-0145
U   Histamine H₃ receptor antagonist

9012   (7360) $C_{19}H_{28}N_4O$
                                                       50565-83-8

p,p'-Ureylenebis(phenyltrimethylammonium) = 4,4'-(Carbonyldiimino)bis[*N,N,N*-trimethylbenzenaminium] (●)

R   Bis(metyl sulfate) (26271-84-1)
S   Hämosporidin, Hemosporidin, LP 2
U   Antibacterial

9013   (7361) $C_{19}H_{28}N_4O_2S$
64795-23-9

N,N-Diethyl-*N'*-(6-methylergolin-8α-yl)sulfamide = *N,N*-
Diethyl-*N'*-[(8α)-6-methylergolin-8-yl]sulfamide (●)
S   CQ 32-084, *Etisulergine***
U   Antiparkinsonian

9014   (7362) $C_{19}H_{28}N_4O_2S$
104317-90-0

N,N-Dimethyl-*N'*-[(8α)-6-propylergolin-8-yl]sulfamide
(●)
S   CQP 201-403
U   Dopamine agonist

9015   (7363) $C_{19}H_{28}N_4O_3$
117690-79-6

5'-Ethyl-2'-hydroxy-4'-[[6-methyl-6-(1*H*-tetrazol-5-
yl)heptyl]oxy]acetophenone = 1-[5-Ethyl-2-hydroxy-4-
[[6-methyl-6-(1*H*-tetrazol-5-yl)heptyl]oxy]phenyl]etha-
none (●)
S   LY 255283
U   Antipsoriatic, anti-asthmatic

9016 $C_{19}H_{28}N_4O_5S_2$
140695-21-2

(±)-1-[2-Hydroxy-2-(4-hydroxyphenyl)ethyl]-2-(methylsulfonyl)-3-[2-[5-(methylaminomethyl)-2-furanylme-
thylthio]ethyl]guanidine = (±)-*N*-[(*E*)-[(*p*,β-Dihydroxy-
phenethyl)amino][[2-[[5-[(methylamino)methyl]furfu-
ryl]thio]ethyl]amino]methylene]methanesulfonamide =
[*N*(*E*)]-*N*-[[[2-Hydroxy-2-(4-hydroxyphenyl)ethyl]ami-
no][[2-[[[5-[(methylamino)methyl]-2-furanyl]methyl]
thio]ethyl]amino]methylene]methanesulfonamide (●)
S   *Osutidine***, T-593, *Tomitidine*
U   Histamine H_2 receptor antagonist (anti-ulcer)

9017   (7364) $C_{19}H_{28}N_4O_6S_2$
33075-00-2

(6*R*,7*R*)-7-[2-[(*N,N'*-Diisopropylamidino)thio]acetami-
do]-3-(hydroxymethyl)-8-oxo-5-thia-1-azabicy-
clo[4.2.0]oct-2-ene-2-carboxylic acid acetate (ester) =
(7*R*)-7-[2-[(*N,N'*-Diisopropylamidino)thio]acetamido]ce-
phalosporanic acid = (6*R-trans*)-3-[(Acetyloxy)methyl]-
7-[[[[[(1-methylethyl)amino][(1-methylethyl)imino]me-
thyl]thio]acetyl]amino]-8-oxo-5-thia-1-azabicy-
clo[4.2.0]oct-2-ene-2-carboxylic acid (●)
S   *Cefathiamidine*
U   Antibiotic

9018 $C_{19}H_{28}O_2$
481-30-1

(17α)-17-Hydroxyandrost-4-en-3-one (●)
S   *Epitestosterone, Isotestosterone,* 17α-*Testosterone*

U Anti-androgen

9019 (7365)

$C_{19}H_{28}O_2$
58-22-0

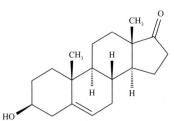

17β-Hydroxyandrost-4-en-3-one = Δ^4-Androsten-17β-ol-3-one = (17β)-17-Hydroxyandrost-4-en-3-one (●)

R see also nos. 9055, 9967, 10720, 11506, 12015, 12113, 12434, 12849, 13299, 13321, 13552, 13631, 13635, 13879, 13956, 14400, 14411, 14562, 14885, 15280;

S Andro 100, Androderm, Androderm Pflaster, Androflex, Androgel, Androgen, Android-T, Androlan aqueous, Androlin, Andronaq, Andropatch, Androsorb, Androtab-SL, Androtest, Andrusol, Aqua-Testerone, Aquaviron, Atmos "Astra; TheraTech", Cristaviron, Dep-Androle, Enarmon, Femaderm, Framviron, Harvogen, Histerone, Homogene-S, Homosteron, Hydrotest, Malerone, Malestrone, Malogen aqueous inj., Malotrone, Mertestate, Naga-Hormo-M, Nendron, Neo-Hombreol F, Neotestis, NSC-9700, Omnadren, Omnadren Ampullen, Oreton F, Orquisteron, Pan-Test, Perandren, Percutacrine androgénique forte, Primoniat T, Primoteston T, Primotest T, Proviron T, Rektandron, ReLibra, Stérandryl, Sterotate, Strong-Lar, Synandrets, Synandrol F, Synandrulin, Telipex-Lösung, Tesamone, Tes-Hol "Z", Tesionate 200, Teslen, Tesone, Testadenos, Testahomen T, Testalong, Testamone, Testandro, Testandrone, Testaqua, Testeplex Aq. inj., *Testikelhormon*, Testisan, Testoban, Testobase Aqueous, Testoderm, Testodrin-lösning, Testoject 50, Testolin, Testonex, Testop, Testopel, Testopropon, Testoral "Organon", Testorona-Tropfen, Testosir, Testosteroid, *Testosterone***, Testostop TTS, Testotop, Testovena, Testoviron T, Testraq, Testro-Med, Testrone, Testryl, Tetrasterone, TheraDerm, Tostrex, TTI-103, Virex, Virormone, Virosterone, Virotest Mariatherma

U Androgen

9020 (7366)

$C_{19}H_{28}O_2$
53-43-0

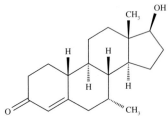

3β-Hydroxyandrost-5-en-17-one = Δ^5-Androsten-3β-ol-17-one = *trans*-Dehydroandrosterone = (3β)-3-Hydroxyandrost-5-en-17-one (●)

R also esters;
    see also nos. 9025, 13322

S Aslera, Cetovister, 17-Chetovis, Dastonil S, Deandros, *Dehydroandrosterone*, *Dehydroepiandrosterone*, *Dehydroisoandrosterone*, DHA, DHEA, Diandron, Dinistenile, EL-10, GL-701, Hormobagó, 17-Hormoforin, 17-Ketovis, Mentalormon, *Prasterone***, Psicosterone

U Andrenocortical hormone with psychotonic and antidepressant action

9021 (7367)

$C_{19}H_{28}O_2$
3764-87-2

17β-Hydroxy-7α-methylestr-4-en-3-one = (7α,17β)-17-Hydroxy-7-methylestr-4-en-3-one (●)

R see also no. 10721

S MENT, *Trestolone**

U Androgen, anabolic

9022 (7368)

C$_{19}$H$_{28}$O$_2$
514-61-4

17β-Hydroxy-17-methylestr-4-en-3-one = 17α-Methyl-19-nortestosterone = (17β)-17-Hydroxy-17-methylestr-4-en-3-one (●)

S   Lutenin, Matronal, Metalutin, Methalutin, *Methylestrenolone, Methylnortestosterone, Methyloestrenolon, Normethandrolone, Normethandrone, Normethisterone*, NSC-10039, Orgasteron, Orosteron, P-6051

U   Progestin

9023 (7369)

C$_{19}$H$_{28}$O$_2$
3570-10-3

17β-Hydroxy-17-methyl-*B*-norandrost-4-en-3-one = 17α-Methyl-*B*-nortestosterone = (17β)-17-Hydroxy-17-methyl-*B*-norandrost-4-en-3-one (●)

S   *Benorterone***, FC-612, SKF 7690

U   Anti-androgen

9024 (7370)

C$_{19}$H$_{28}$O$_4$
85702-60-9

L-Menthol guaiacolglycolate = L-*p*-Menthan-3-yl-*o*-methoxyphenoxyacetate = [1*R*-(1α,2β,5α)]-(2-Methoxy-phenoxy)acetic acid 5-methyl-2-(1-methylethyl)cyclohexyl ester (●)

S   *Mengualate*

U   Expectorant, antipyretic

9025 (7371)

C$_{19}$H$_{28}$O$_5$S
651-48-9

3β-Hydroxyandrost-5-en-17-one hydrogen sulfate = (3β)-3-(Sulfooxy)androst-5-en-17-one (●)

R   Sodium salt (1099-87-2)

S   Astenile, *Dehydroepiandrosterone sulfate sodium*, DHA-S, DHEAS, KYH 3102, Mylis, PB 005, *Prasterone sodium sulfate*, Teloin

U   Psychotropic, anti-asthmatic

9026 (7372)

C$_{19}$H$_{28}$O$_8$
88495-63-0

(3*R*,5a*S*,6*R*,8a*S*,9*R*,10*S*,12*R*,12a*R*)-Decahydro-3,6,9-trimethyl-3,12-epoxy-12*H*-pyrano[4,3-*j*]-1,2-benzodioxepin-10-ol hydrogen succinate = [3*R*-(3α,5aβ,6β,8aβ,9α,10α,12β,12a*R**)]-Butanedioic acid mono(decahydro-3,6,9-trimethyl-3,12-epoxy-12*H*-pyrano[4,3-*j*]-1,2-benzodioxepin-10-yl) ester (●)

S   Arnate, *Artesunate***, *Artesunic acid*, ARTS

U   Antischistosomal, antimalarial

9027 (7373) $C_{19}H_{29}BClN_6O_6{}^{99m}Tc$
104716-22-5

[Bis[(1,2-cyclohexanedione dioximato)(1–)-*O*][(1,2-cy-clohexanedione dioximato)(2–)-*O*]methylborato(2–)-*N,N',N'',N''',N'''',N'''''*]chloro[^{99m}Tc]technetium(III) = (*TPS*-7-1-232'4'54)-[Bis[(1,2-cyclohexanedione dioxima-to)(1–)-*O*][(1,2-cyclohexanedione dioximato)(2–)-*O*]me-thylborato(2–)-*N,N',N'',N''',N'''',N'''''*]chlorotechnetium-^{99m}Tc (●)
S Cardiotec, CDO-MeB, SQ 30217, TEBO, *Techne-tium (^{99m}Tc) teboroxime***
U Radioactive diagnostic (myocardial perfusion ima-ging agent)

9028 $C_{19}H_{29}ClN_2O_2$
160845-95-4

4-[3-(1-Butyl-4-piperidinyl)-1-oxopropyl]-2-chloro-5-methoxyphenylamine = 1-(4-Amino-5-chloro-2-meth-oxyphenyl)-3-(1-butyl-4-piperidinyl)-1-propanone (●)
R Monohydrochloride (168986-60-5)
S RS-67333
U Cognition enhancer (5-HT$_4$ receptor agonist)

9029 (7374) $C_{19}H_{29}IO_2$
99-79-6

Ethyl 10-(*p*-iodophenyl)undecanoate = 10-(*p*-Iodophe-nyl)undecanoic acid ethyl ester = 4-Iodo-τ-methylben-zenedecanoic acid ethyl ester (●)
R also mixture with isomers (1320-11-2)
S Aethyotrastum, Ethiodan, Etiodrast, *Iofendylate***, *Iophendylate*, *Jofendylatum*, Mulsopaque, Myelodil, Myodil, Neurotrast, Pantopaque
U Diagnostic aid (radiopaque medium-myelographic)

9030 (7375) $C_{19}H_{29}NO$
77-39-4

1-Cyclopentyl-1-phenyl-3-piperidino-1-propanol = α-Cyclopentyl-α-phenyl-1-piperidinepropanol (●)
R Hydrochloride (126-02-3)
S *Cicrimin*, Compound 08958, *Cycrimine hydrochlori-de***, Lilly 08958, Pagitane
U Anticholinergic, antiparkinsonian

9031 (7376) $C_{19}H_{29}NO$
77-37-2

1-Cyclohexyl-1-phenyl-3-(1-pyrrolidinyl)-1-propanol = α-Cyclohexyl-α-phenyl-1-pyrrolidinepropanol (●)
R Hydrochloride (1508-76-5)
S Arpicolin, Arpicolon, Kemadren, Kemadrin(e), Me-tanin, Osnervan, PMS-Procyclidine, *Prociclidina*, Procipar, Procyclid, *Procyclidine hydrochloride***
U Anticholinergic, antiparkinsonian

9032 (7377)

$C_{19}H_{29}NO_2$
53716-48-6

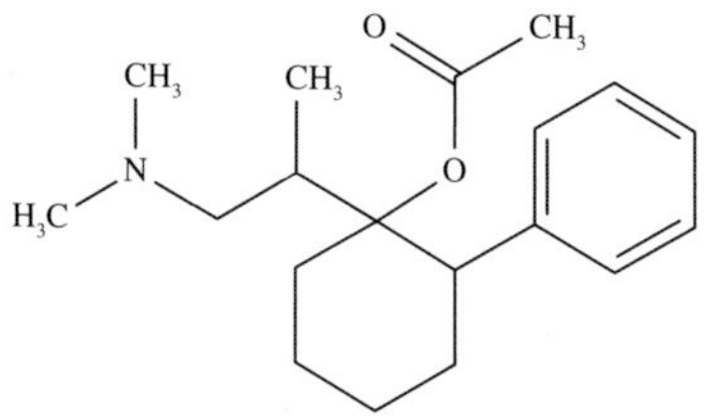

1-[2-(Dimethylamino)-1-methylethyl]-2-phenylcyclohe-
xanol acetate (ester) (●)
R Hydrochloride (53716-47-5)
S Compound 673-082, *Nexeridine hydrochloride***
U Analgesic

9033 (7378)

$C_{19}H_{29}NO_2$
5662-89-5

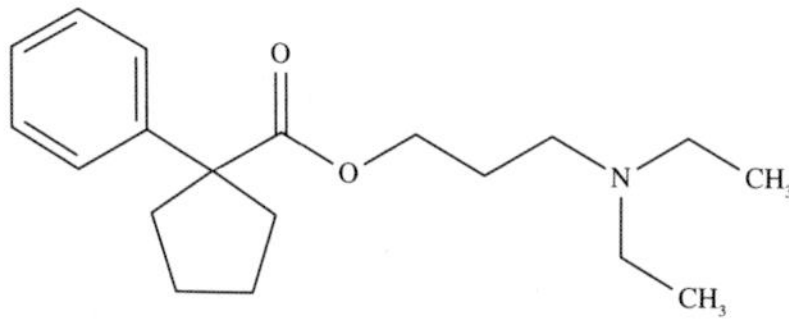

1-Phenylcyclopentanecarboxylic acid 3-(diethylami-
no)propyl ester (●)
R Hydrochloride (3613-69-2)
S Cipenam, Cypenam, ITOX 3808, Tsipenam
U Anticonvulsant

9034

$C_{19}H_{29}NO_2$
193278-47-6

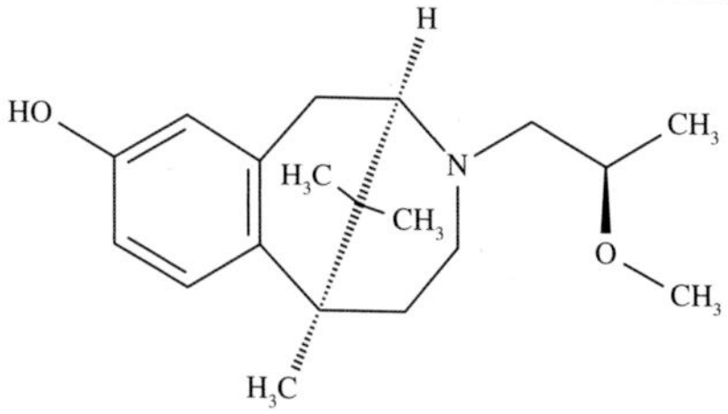

(2R,6S)-1,2,3,4,5,6-Hexahydro-3-[(2R)-2-methoxypro-
pyl]-6,11,11-trimethyl-2,6-methano-3-benzazocin-9-ol
(●)
R Hydrochloride (147759-53-3)
S BI II 277CL
U Neuroprotective (NMDA receptor antagonist)

9035 (7379)

$C_{19}H_{29}NO_2$
66451-06-7

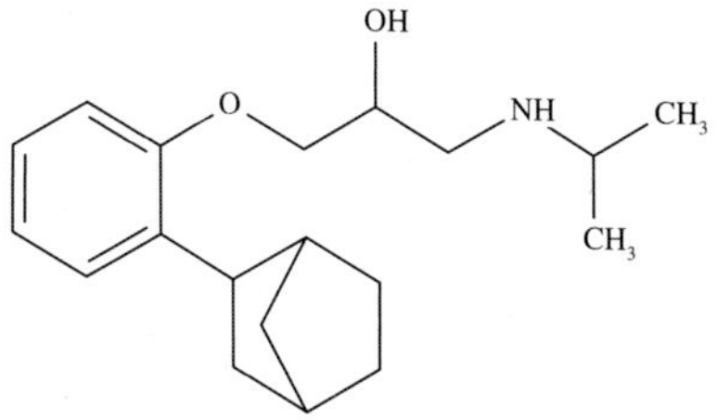

1-(Isopropylamino)-3-(o-2-exo-norbornylphenoxy)-2-
propanol = 1-(2-Bicyclo[2.2.1]hept-2-ylphenoxy)-3-[(1-
methylethyl)amino]-2-propanol (●)
R Hydrochloride (69319-47-7)
S *Bornaprolol hydrochloride***, FM 24
U β-Adrenergic blocker

9036 (7380)

$C_{19}H_{29}NO_2$
3670-69-7

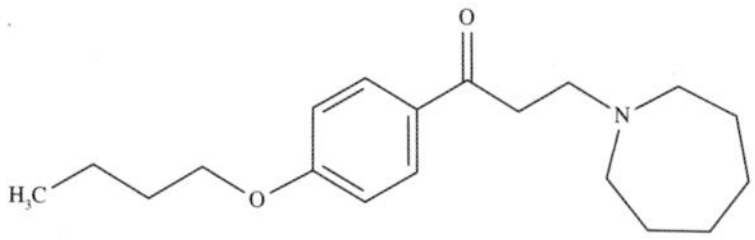

3-(Perhydroazepin-1-yl)-4'-butoxypropiophenone = β-N-
Hexamethylenimino-p-butoxypropiophenone = 1-(4-Bu-
toxyphenyl)-3-(hexahydro-1H-azepin-1-yl)-1-propanone
(●)
S TG-17
U Anti-arrhythmic

9037 (7381)

$C_{19}H_{29}NO_3$
52109-93-0

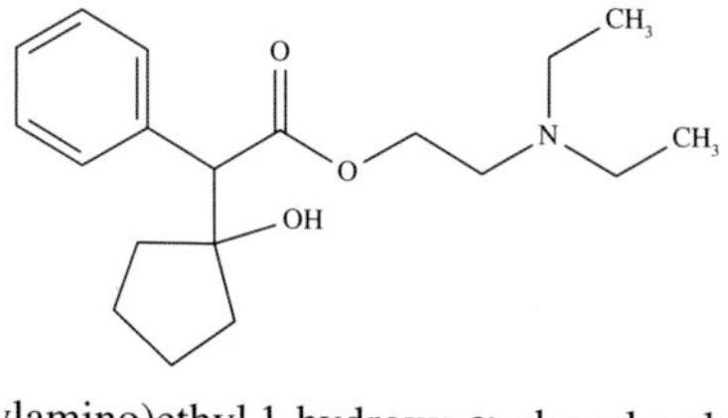

2-(Diethylamino)ethyl 1-hydroxy-α-phenylcyclopentane-
acetate = α-(1-Hydroxycyclopentyl)phenylacetic acid β-
(diethylamino)ethyl ester = α-(1-Hydroxycyclopen-
tyl)benzeneacetic acid 2-(diethylamino)ethyl ester (●)
R Hydrochloride (78853-39-1)
S *Cyclodrine hydrochloride*, Cyclopent, G.T. 92, *Zy-
klodrinhydrochlorid*

U Anticholinergic, spasmolytic, mydriatic

9038 (7382)

$C_{19}H_{29}NO_3$
42050-23-7

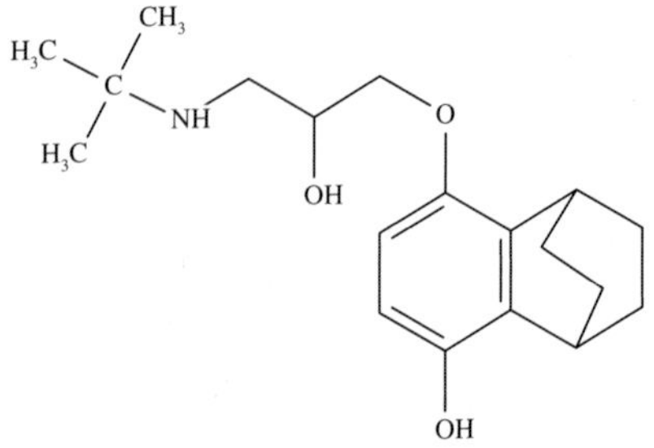

1-(*tert*-Butylamino)-3-[(1,2,3,4-tetrahydro-8-hydroxy-1,4-ethanonaphthalen-5-yl)oxy]-2-propanol = 8-[3-[(1,1-Dimethylethyl)amino]-2-hydroxypropoxy]-1,2,3,4-tetra-hydro-1,4-ethanonaphthalen-5-ol (●)
S K 5407, *Nafetolol***
U β-Adrenergic blocker

9039 (7383)

$C_{19}H_{29}NO_3$
55837-15-5

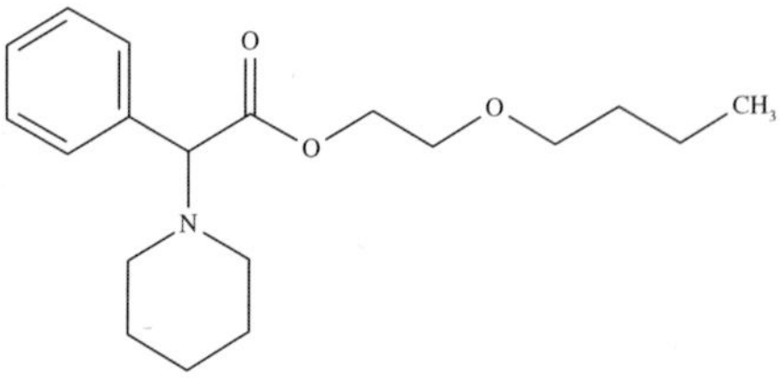

2-Butoxyethyl α-phenyl-1-piperidineacetate = 2-Phenyl-2-piperidinoacetic acid β-butoxyethyl ester = α-Phenyl-1-piperidineacetic acid 2-butoxyethyl ester (●)
R Hydrobromide (60595-56-4)
S *Butopiprine hydrobromide***, Felitussyl, Laucalon, LD 2351, Rutacel, Soditoux, Taci-Bex
U Antitussive

9040 (7384)

$C_{19}H_{29}NO_4$
35507-03-0

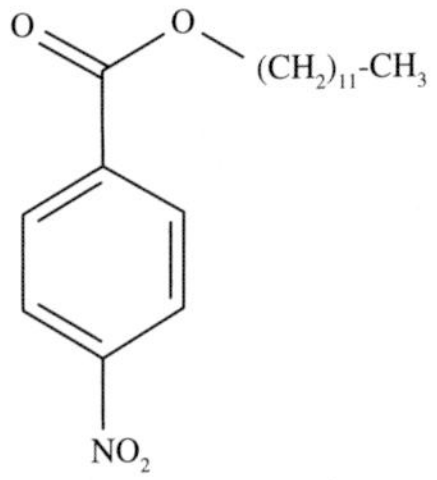

Dodecyl *p*-nitrobenzoate = 4-Nitrobenzoic acid dodecyl ester (●)
S Amonal
U Antibacterial

9041 (7385)

$C_{19}H_{29}NO_4$
92268-40-1

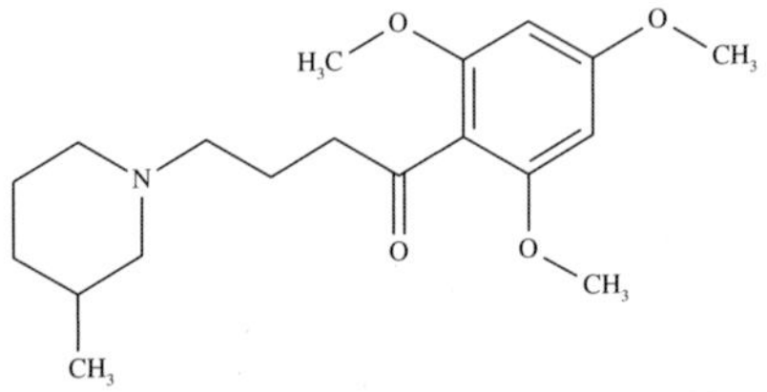

(±)-2',4',6'-Trimethoxy-4-(3-methylpiperidino)butyro-phenone = 4-(3-Methyl-1-piperidinyl)-1-(2,4,6-trimeth-oxyphenyl)-1-butanone (●)
R Succinate
S CRL 41034, *Perfomedil succinate***
U Peripheral vasodilator

9042 (7386)

$C_{19}H_{29}NO_4$
107811-55-2

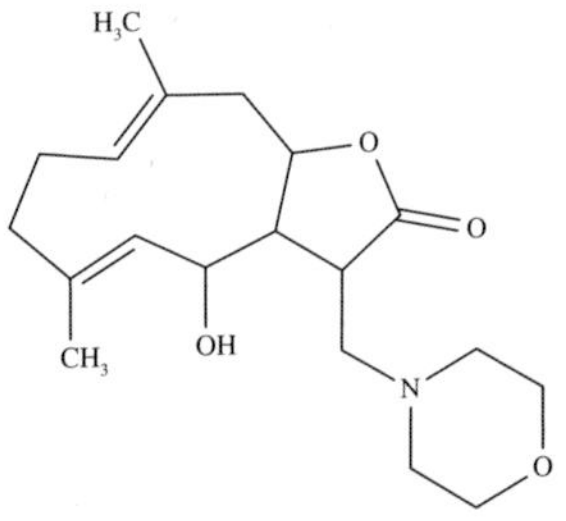

3a,4,7,8,11,11a-Hexahydro-4-hydroxy-6,10-dimethyl-3-(4-morpholinylmethyl)cyclodeca[*b*]furan-2(3*H*)-one (●)
S *Lamorolid*
U Antineoplastic

9043 (7387) $C_{19}H_{29}NO_4$
14436-50-1

Ethyl 2-(diethylamino)ethyl 2-ethyl-2-phenylmalonate =
2-Ethyl-2-phenylmalonic acid ethyl 2-(diethylamino)ethyl ester = Ethylphenylpropanedioic acid 2-(diethylamino)ethyl ethyl ester (●)
S PHEM
U Tremorine inhibitor

9044 (7388) $C_{19}H_{29}NO_5$
36121-13-8

1-Pyrrolidineethanol 4-butoxy-3,5-dimethoxybenzoate (ester) = 4-Butoxy-3,5-dimethoxybenzoic acid 2-(1-pyrrolidinyl)ethyl ester (●)
S *Burodiline**, Vasopentol
U Spasmolytic

9045 (7389) $C_{19}H_{29}NO_5$
52365-63-6

(±)-3,4-Dihydroxy-α-[(methylamino)methyl]benzyl alcohol 3,4-dipivalate = (±)-4-(1-Hydroxy-2-methylaminoethyl)-o-phenylene dipivalate = (±)-2,2-Dimethylpropanoic acid 4-[1-hydroxy-2-(methylamino)ethyl]-1,2-phenylene ester (●)
R also hydrochloride (64019-93-8)
S *Adrenaline dipivalate*, AkPro, DE-016, D-Epifrin (Depifrin), Diopin(e), Diphemin "Alcon-Thilo", *Dipivalylepinephrine*, *Dipivefrin(e)***, *Dipivéphrine*, Dipoquin, Diprin "Alcon", DPE, *Epinephrine dipivalate*, Glaucothil, Glaudrops, K 30.081, Oftanex, Oftapinex, Ophtho-Dipiverin, Pivalephrine, Pivepol, *Pro-Epinephrine*, Propine,
Propine Sterile Ophthalmic, Thilodrin, Vistapin
U Ophthalmic sympathomimetic

9046 $C_{19}H_{29}NO_5$
208465-34-3

3-[4-[3-[(1,1-Dimethylethyl)amino]-2-hydroxypropoxy]-3-methoxyphenyl]-2-propenoic acid ethyl ester (●)
S *Ferulinolol*
U β_1-Adrenoceptor antagonist with partial β_2-agonist activity

9047 (7390) $C_{19}H_{29}N_3O$
491-92-9

8-[(4-Dimethylamino-1-methylbutyl)amino]-6-methoxyquinoline = N^1,N^1-Diethyl-N^4-(6-methoxy-8-quinolinyl)-1,4-pentanediamine (●)
R Salt of 4,4'-methylenebis(3-hydroxy-2-naphthoic acid) = embonate (1:1) (635-05-2)
S Aminoquin, Beprochin, F 710, Gamefar, *Gametocid*, Oxypentyl simplex, *Pamachin*, *Pamaquine***, Plasmochin, Plasmokin, Plasmoquin, Praequine, Prequine, Proechin, Quipenyl, 4516 R.P., S.N. 971, Trobochine

U Antimalarial

9048 (7391)

$C_{19}H_{29}N_3O$
128505-56-6

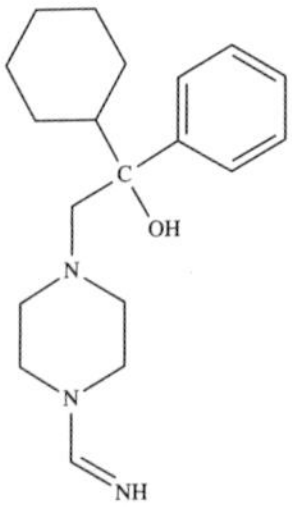

N-(Iminomethyl)-*N'*-(2-hydroxy-2-phenyl-2-cyclohexyl-ethyl)piperazine = α-Cyclohexyl-4-(iminomethyl)-α-phenyl-1-piperazineethanol (●)
R Monohydrochloride (124065-13-0)
S DAC 5945
U Muscarinic antagonist

9049 (7393)

$C_{19}H_{29}N_3O_2$
81907-78-0

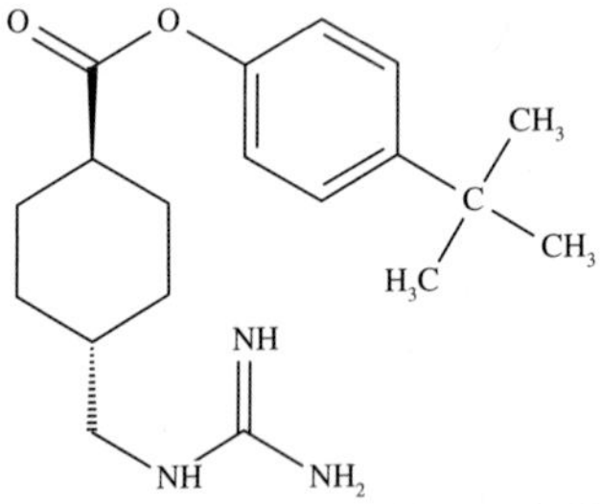

p-tert-Butylphenyl *trans*-4-(guanidinomethyl)cyclohe-xanecarboxylate = *trans*-4-[[(Aminoiminomethyl)ami-no]methyl]cyclohexanecarboxylic acid 4-(1,1-dimethyl-ethyl)phenyl ester (●)
R Monohydrochloride (83373-31-3)
S *Batebulast hydrochloride***, NCO-650, *Telbulast hy-drochloride*
U Anti-asthmatic, anti-allergic

9050 (7392)

$C_{19}H_{29}N_3O_2$
31386-24-0

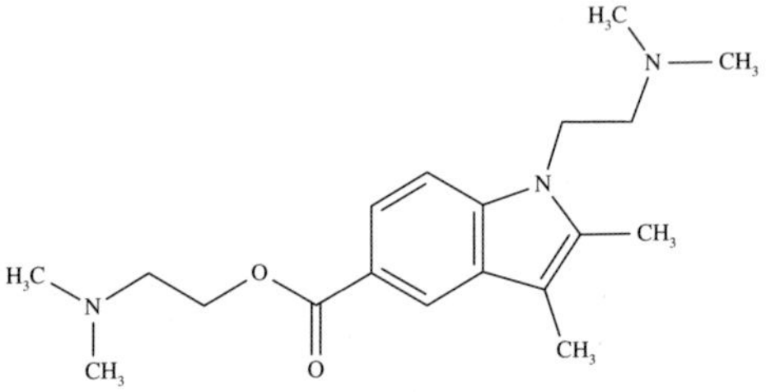

2-(Dimethylamino)ethyl 1-[2-(dimethylamino)ethyl]-2,3-dimethylindole-5-carboxylate = 1-[2-(Dimethylami-no)ethyl]-2,3-dimethyl-1*H*-indole-5-carboxylic acid 2-(dimethylamino)ethyl ester (●)
S *Amindocate***, Diamind
U Serotonin inhibitor

9051

$C_{19}H_{29}N_3O_5S$
312774-20-2

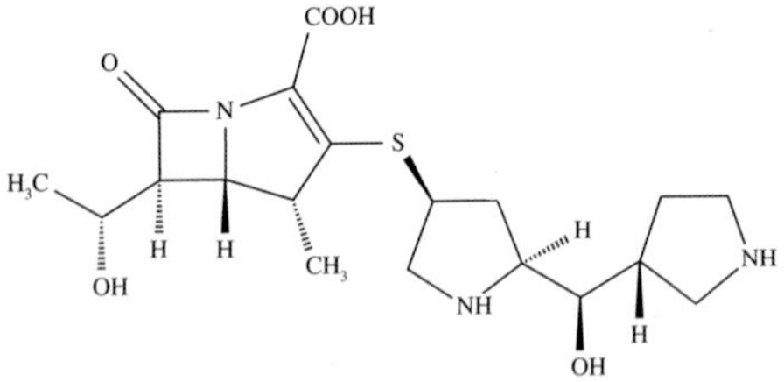

(+)-(1*R*,5*S*,6*S*)-6-[1(*R*)-Hydroxyethyl]-2-[2(*S*)-[1(*R*)-hydroxy-1-[3(*R*)-pyrrolidinyl]methyl]-4(*S*)-pyrrolidinyl-sulfanyl]-1-methyl-5-oxo-1-carba-2-penem-3-carboxylic acid = (4*R*,5*S*,6*S*)-6-[(1*R*)-1-Hydroxyethyl]-3-[[(3*S*,5*S*)-5-[(*R*)-hydroxy(3*R*)-3-pyrrolidinylmethyl]-3-pyrrolidi-nyl]thio]-4-methyl-7-oxo-1-azabicyclo[3.2.0]hept-2-ene-2-carboxylic acid (●)
R Monohydrochloride (186319-97-1)
S E-1010, ER-35786
U Carbapenem antibiotic

9052 (7395)

$C_{19}H_{29}N_5O_2$
83928-76-1

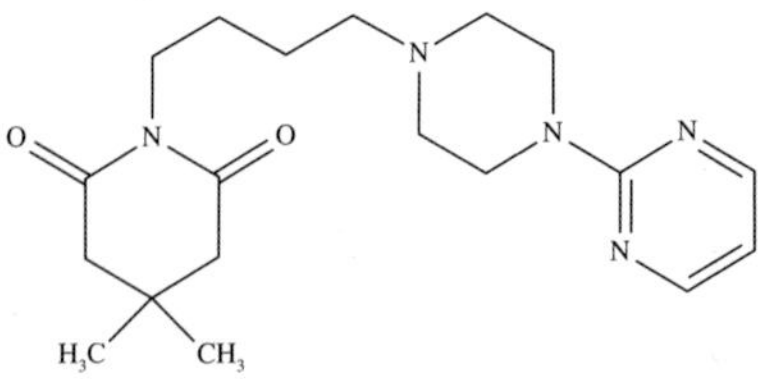

3,3-Dimethyl-*N*-[4-[4-(2-pyrimidinyl)-1-piperazinyl]bu-
tyl]glutarimide = 4,4-Dimethyl-1-[4-[4-(2-pyrimidinyl)-
1-piperazinyl]butyl]-2,6-piperidinedione (●)
R Monohydrochloride (83928-66-9)
S Ariza, BMY-13805-1, *Gepirone hydrochloride***,
MJ 13805-1
U Anxiolytic (5HT$_{1A}$ partial agonist)

9053 (7394) $C_{19}H_{29}N_5O_2$
76956-02-0

1-Methyl-5-[[3-[(α-piperidino-*m*-tolyl)oxy]propyl]ami-
no]-1*H*-1,2,4-triazole-3-methanol = 1-Methyl-5-[[3-[3-
(1-piperidinylmethyl)phenoxy]propyl]amino]-1*H*-1,2,4-
triazole-3-methanol (●)
S AH 23844, *Lavoltidine***, Loxotidine, Loxtidine
U Histamine H$_2$-receptor antagonist

9053-01 (7394-01) 86160-82-9
R Succinate (1:1)
S AH 23844 A, *Lavoltidine succinate***
U Histamine H$_2$-receptor antagonist

9054 $C_{19}H_{29}N_5O_6$
163259-37-8

1,3-Dibutylxanthine-7-riboside-5'-*N*-methylcarboxamide
= 1-Deoxy-1-(1,3-dibutyl-1,2,3,6-tetrahydro-2,6-dioxo-
7*H*-purin-7-yl)-*N*-methyl-β-D-ribofuranuronamide (●)
S DBXRM
U Adenosine A$_3$ receptor partial agonist

9055 (7396) $C_{19}H_{29}O_5P$
1242-14-4

3-Oxoandrost-4-en-17β-yl dihydrogen phosphate = 17β-
Hydroxyandrost-4-en-3-one 17-(dihydrogen phosphate) =
(17β)-17-(Phosphonooxy)androst-4-en-3-one (●)
S Telipex aquosum, *Testosterone phosphate*
U Androgen

9056 (7397) $C_{19}H_{30}NO_2$

(4-Cyclopentyl-4-hydroxy-3-oxo-4-phenylbutyl)ethyldi-
methylammonium = δ-Cyclopentyl-*N*-ethyl-δ-hydroxy-
N,*N*-dimethyl-γ-oxobenzenebutanaminium (●)
R Chloride (77214-85-8)
S Chlorosyl, Chlorozil, Khlorozil
U Anticholinergic, spasmolytic

9057 (7398) $C_{19}H_{30}NO_2$
7591-00-6

Diethylmethyl-2-(1-phenylcyclopentanecarbonyloxy)-
ethylammonium = *N*,*N*-Diethyl-*N*-methyl-2-[[(1-phenyl-
cyclopentyl)carbonyl]oxy]ethanaminium (●)
R Methyl sulfate (3690-66-2)
S *Caramiphen methylsulfomethylate*, Mepanit, Merpa-
nit
U Anticholinergic

9058 (7399) $C_{19}H_{30}NO_2$
26372-86-1

4-Hydroxy-1,1-dimethylpiperidinium 3-methyl-2-phenylvalerate = 1,1-Dimethyl-4-(3-methyl-2-phenylvaleroyloxy)piperidinium = 1,1-Dimethyl-4-[(3-methyl-1-oxo-2-phenylpentyl)oxy]piperidinium (●)
R Methyl sulfate (7681-80-3)
S AY-5810, Crilin, Crylène, Hycholin, *Pentapiperide metilsulfate*, *Pentapiperium metilsulfate***, Perium, Quilene, *Valpipamate methylsulfate*
U Anticholinergic, spasmolytic

9059 $C_{19}H_{30}NO_5P$
153994-93-5

3-[(1S)-1-[[(2S)-3-[(Cyclohexylmethyl)hydroxyphosphinyl]-2-hydroxypropyl]amino]ethyl]benzoic acid (●)
S CGP 56433
U GABA$_B$ receptor antagonist

9060 $C_{19}H_{30}NO_5P$
153994-97-9

3-[(1R)-1-[[(2S)-3-[(Cyclohexylmethyl)hydroxyphosphinyl]-2-hydroxypropyl]amino]ethyl]benzoic acid (●)
R Monolithium salt (153994-81-1)
S CGP 56999A

U GABA$_B$ receptor antagonist

9061 (7402) $C_{19}H_{30}N_2O_2$
54063-28-4

2-Phenyl-N-[2-(1-pyrrolidinyl)ethyl]glycine isopentyl ester = α-Phenyl-α-[2-(1-pyrrolidinyl)ethylamino]acetic acid isopentyl ester = α-[[2-(1-Pyrrolidinyl)ethyl]amino]benzeneacetic acid 3-methylbutyl ester (●)
S *Camiverine***, *Piramilofine*, Sanaspasmina
U Spasmolytic

9062 $C_{19}H_{30}N_2O_2$
153490-72-3

2,3-Dihydro-2,2,4,6,7-pentamethyl-3-[(4-methyl-1-piperazinyl)methyl]-5-benzofuranol (●)
R Dihydrochloride (153490-71-2)
S MDL 74180
U Neuroprotective (radical scavenger)

9063 $C_{19}H_{30}N_2O$
183142-37-2

(3R)-2,3-Dihydro-2,2,4,6,7-pentamethyl-3-[(4-methyl-1-piperazinyl)methyl]-5-benzofuranol (●)
S MDL 74722
U Lipid peroxidation inhibitor

9064 (7403) $C_{19}H_{30}N_2O_2$
149494-37-1

(*R*)-3-(Isopropylpropylamino)-5-(isopropylcarb-
amoyl)chroman = (*R*)-*N*-Isopropyl-3-(isopropylpropylamino)-5-chromancarboxamide = (3*R*)-3,4-Dihydro-*N*-
(1-methylethyl)-3-[(1-methylethyl)propylamino]-2*H*-1-
benzopyran-5-carboxamide (●)
S *Ebalzotan***
U Antidepressant (serotonin receptor agonist)

9065 (7400) $C_{19}H_{30}N_2O_2$
479-81-2

2-(Diethylamino)ethyl α-phenyl-1-piperidineacetate = α-
Phenylpiperidinoacetic acid β-(diethylamino)ethyl ester =
α-Phenyl-1-piperidineacetic acid 2-(diethylamino)ethyl
ester (●)
R also dihydrochloride (2691-46-5)
S *Bietamiverine***, *Diétamiverine*, Fine-Dol, Novospa-
rol, Paparid, Sparine "Tokyo Tanabe", Spasmaparid,
Spasmisolvina, Spasmo-Paparid, Tripaverin
U Spasmolytic

9066 (7401) $C_{19}H_{30}N_2O_2$
24671-26-9

4-Benzyl-1-piperidinecarboxylic acid 2-(diethylamino)ethyl ester = 4-(Phenylmethyl)-1-piperidinecarboxylic
acid 2-(diethylamino)ethyl ester (●)
S *Benrixate***
U Anti-arrhythmic

9067 (7404) $C_{19}H_{30}N_2O_3$
4551-59-1

Ethyl *N*-[2-(diethylamino)ethyl]-2-ethyl-2-phenylmalon-
amate = *N*-[2-(Diethylamino)ethyl]-2-ethyl-2-phenylma-
lonamic acid ethyl ester = α-[[[2-(Diethylamino)ethyl]
amino]carbonyl]-α-ethylbenzeneacetic acid ethyl ester
(●)
S *Fenalamide**, *Phemamide*, Sch 5706, SH 30858,
Spasmamide (semplice)
U Spasmolytic, smooth muscle relaxant

9068 $C_{19}H_{30}N_2O_3S$
127966-78-3

2-[(2-Hydroxyethyl)thio]-*N*-[3-[3-(1-piperidinylme-
thyl)phenoxy]propyl]acetamide (●)
R Mono[2-(4-hydroxybenzoyl)benzoate]
(128289-57-6)
S Z-300 "Zeria"
U Histamine H_2 receptor antagonist

9069 (7405) $C_{19}H_{30}N_2O_4S$
50656-94-5

(2*S*,5*R*,6*R*)-3,3-Dimethyl-7-oxo-6-(10-undecenamido)-4-
thia-1-azabicyclo[3.2.0]heptane-2-carboxylic acid =
(6*R*)-6-(10-Undecenamido)penicillanic acid = [2*S*-
(2α,5α,6β)]-3,3-Dimethyl-7-oxo-6-[(1-oxo-10-undece-
nyl)amino]-4-thia-1-azabicyclo[3.2.0]heptane-2-carb-
oxylic acid (●)

S *Undecillin*
U Antibiotic

9070 (7406) $C_{19}H_{30}N_2O_5S$
85856-54-8

(−)-1-[(2*S*)-3-Mercapto-2-methylpropionyl]-L-proline ester with *N*-(cyclohexylcarbonyl)thio-D-alanine = (−)-*N*-[(*S*)-3-[[*N*-(Cyclohexylcarbonyl)-D-alanyl]thio]-2-methylpropionyl]-L-proline = [*R*-(*R**,*S**)]-1-[3-[[2-[(Cyclohexylcarbonyl)amino]-1-oxopropyl]thio]-2-methyl-1-oxopropyl]-L-proline (●)
R Calcium salt (2:1) (85921-53-5)
S *Altiopril calcium*, Lowpres, MC 838, *Moveltipril calcium***
U Antihypertensive (ACE inhibitor)

9071 (7407) $C_{19}H_{30}N_2O_6S$
35423-51-9

N-[3-(Dimethylamino)-1,3-dimethylbutyl]-6,7-dimethoxy-2,1-benzoxathiane-3-carboxamide 1,1-dioxide = *N*-[3-(Dimethylamino)-1,3-dimethylbutyl]-3,4-dihydro-6,7-dimethoxy-2,1-benzoxathiin-3-carboxamide 1,1-dioxide (●)
S 16-244 AWD, DMAH, Impavido, *Tisocromide***
U Antidepressant

9072 $C_{19}H_{30}N_5O_{10}P$
201341-05-1

(*R*)-[[2-(6-Amino-9*H*-purin-9-yl)-1-methylethoxy]methyl]phosphonic acid bis(isopropoxycarbonyloxymethyl) ester = 9-[(*R*)-2-[[Bis[[(isopropoxycarbonyl)oxy]methoxy]phosphinyl]methoxy]propyl]adenine = 5-[[(1*R*)-2-(6-Amino-9*H*-purin-9-yl)-1-methylethoxy]methyl]-2,4,6,8-tetraoxa-5-phosphanonanedioic acid bis(1-methylethyl) ester 5-oxide (●)
R Fumarate (1:1) (202138-50-9)
S Bis(POC)PMPPA fumarate, GS-4331-05, oral PMPA, PMPA Prodrug, *Tenofovir DF, Tenofovir disoproxil fumarate*
U Antiretroviral

9073 (7408) $C_{19}H_{30}OS$
2363-58-8

2α,3α-Epithio-5α-androstan-17β-ol = (2α,3α,5α,17β)-2,3-Epithioandrostan-17-ol (●)
S *Epithioandrostanol, Epitiostanol***, NSC-194684, 10275-S, Thiodrol
U Anti-estrogen, androgen, cytostatic

9074   (7410) $C_{19}H_{30}O_2$
53-41-8

3α-Hydroxy-5α-androstan-17-one = 5α-Androstan-3α-ol-17-one = (3α,5α)-3-Hydroxyandrostan-17-one (●)
S   Androkinine, *Androsterone*, Androtine, NSC-9898
U   Androgen

9075   (7411) $C_{19}H_{30}O_2$
53-42-9

3α-Hydroxy-5β-androstan-17-one = (3α,5β)-3-Hydroxyandrostan-17-one (●)
S   *Etiocholanolone*, Etiolone
U   Diagnostic aid

9076   (7412) $C_{19}H_{30}O_2$
521-18-6

17β-Hydroxy-5α-androstan-3-one = 5α-Androstan-17β-ol-3-one = (5α,17β)-17-Hydroxyandrostan-3-one (●)
R   see also nos. 11546, 12622, 13224, 13333
S   Anaboleen, Anabolex "Lloyd-Hamol", Anabol-Tablinen, Anaprotin, Andractim, Androgel-DWT, *Androstanolone***, Apeton, Cristerona-MB, *Dihydro-*

testosterone, Gelovit, L.G. 152, Neodrol, Pesomax, Protéina, Protona, Stanaprol, *Stanolone*, Stanorone
U   Androgen, anabolic

9077   (7409) $C_{19}H_{30}O_2$
3642-89-5

Androst-5-ene-3β,16α-diol = (3β,16α)-Androst-5-ene-3,16-diol (●)
S   Cetadiol
U   Tranquilizer (treatment of alcoholism)

9078   (7413) $C_{19}H_{30}O_3$
53-39-4

17β-Hydroxy-17-methyl-2-oxa-5α-androstan-3-one = Dodecahydro-3-hydroxy-6-(hydroxymethyl)-3,3a,6-trimethyl-1*H*-benz[*e*]indene-7-acetic acid δ-lactone = (5α,17β)-17-Hydroxy-17-methyl-2-oxaandrostan-3-one (●)
S   Anavar, Antitriol, 8075 C.B., Hepandrin, Lipidex, Lonavar, Lonovar, NSC-67068, Oxandrin, *Oxandrolone***, Protivar, SC-11585, Vasorome-Kowa
U   Anabolic, antihyperlipidemic, hepatoprotectant (alcoholic hepatitis)

9079   (7414) $C_{19}H_{30}O_5$
51-03-6

α-[2-(2-Butoxyethoxy)ethoxy]-4,5-methylenedioxy-2-propyltoluene = [4,5-Methylenedioxy-2-propylbenzyl]

butyl diethyleneglycol ether = 5-[[2-(2-Butoxye-
thoxy)ethoxy]methyl]-6-propyl-1,3-benzodioxole (●)
S Butacide, Buzpel, ENT 14,250, NSC-8401, Para pio,
Piperonyl Butoxide, Prevent, Quitoso
U Acaricide, insecticide

9080 (7415) $C_{19}H_{30}O_5$
58186-27-9

2-(10-Hydroxydecyl)-5,6-dimethoxy-3-methyl-*p*-benzo-
quinone = 2-(10-Hydroxydecyl)-5,6-dimethoxy-3-me-
thyl-2,5-cyclohexadiene-1,4-dione (●)
S Aban, Arzkat, Avan, Cerestabon, CV-2619, Daru-
ma, Esanic, Geniceral, *Idebenone***, Idesole, Lucen-
abol, Mnesis, Nemocebral
U Cardiovascular agent, nootropic

9081 (7416) $C_{19}H_{31}NO$
2179-37-5

3-[(1-Benzylcycloheptyl)oxy]-*N,N*-dimethylpropyl-
amine = 1-Benzyl-1-[3-(dimethylamino)propoxy]cyclo-
heptane = *N,N*-Dimethyl-3-[[1-(phenylmethyl)cyclo-hep-
tyl]oxy]-1-propanamine (●)
R Fumarate (1:1) (14286-84-1)
S Angiociclan, Angiodel, Benciclan 200, Bencycdor,
Bencyclane fumarate**, Bencycrate, Bioarterol,
DAN-991, Dantifar, Dantrium "Marxer", Deaselon,
Desoblit, Diacyclan, Dilangio inyect., Divear,
Egyt-201, Fludilat, Flussema, Fluxema,

Fumagiol EC, Gerbatz, Halidor, Hemoflux
"Uni-Pharma", Iloramine, Inphos, Isamic, Ludilat,
Norilate, Ronben, Rospon, Tardilat, Vasodarkey, Va-
sorelax
U Spasmolytic, vasodilator

9081-01 (7416-01)   53521-46-3
R Cyclamate (no. 717)
S DAN-992, Dilangio
U Spasmolytic, vasodilator

9082 (7417) $C_{19}H_{31}NO_2$
50588-47-1

3α-Amino-2β-hydroxy-5α-androstan-17-one =
(2β,3α,5α)-3-Amino-2-hydroxyandrostan-17-one (●)
R Hydrochloride (51740-76-2)
S *Amafolone hydrochloride***, Org 6001
U Anti-arrhythmic

9083 $C_{19}H_{31}NO_2$
50652-76-1

2'-Hydroxy-5'-methyllaurophenone oxime = 1-(2-Hy-
droxy-5-methylphenyl)-1-dodecanone oxime (●)
S FLM 5011
U Lipoxygenase inhibitor (topical anti-inflammatory)

9084 $C_{19}H_{31}NO_2S$
138254-82-7

(R*,R*)-2-Amino-1-[5-(1-dodecynyl)-2-thienyl]-1,3-propanediol (●)
S   P-9645
U   Anti-inflammatory (protein kinase C inhibitor)

9085   (7418) $C_{19}H_{31}NO_4$
                                                     14817-09-5

4-(Decyloxy)-3,5-dimethoxybenzamide (●)
S   *Decimemide***, Denegyt, Egyt-1050, V-285
U   Anti-epileptic, spasmolytic

9086   (7419) $C_{19}H_{31}NO_4$
                                                     135357-96-9

Glycine [2S-[2α(1E,3E,5E),3β,4α,5α]]-tetrahydro-4-methoxy-5-methyl-2-(1-methyl-1,3,5-nonatrienyl)-2H-pyran-3-yl ester (●)
S   *Restricticin*, Ro 9-1470
U   Antifungal antibiotic from *Penicillium restrictum*

9087   (7420) $C_{19}H_{31}NO_5$
                                                     95368-75-5

5-(Diethylamino)pentyl 3,4,5-trimethoxybenzoate =
3,4,5-Trimethoxybenzoic acid 5-(diethylamino)pentyl ester (●)
R   Hydrochloride (38243-83-3)
S   TMB 5
U   Anti-arrhythmic, skeletal muscle relaxant

9088 $C_{19}H_{31}N_3O$
                                                     152241-24-2

4-[1-(5-Cyclohexylpentanoyl)-4-piperidinyl]-1H-imidazole = 1-(5-Cyclohexyl-1-oxopentyl)-4-(1H-imidazol-4-yl)piperidine (●)
S   GT-2016
U   Histamine H_3 receptor antagonist

9089 $C_{19}H_{31}N_7OS_2$
                                                     88022-86-0

2-[2-[[[2-[(Diaminomethylene)amino]-4-thiazolyl]methyl]thio]ethyl]-5-[3-(diethylamino)propyl]-6-methyl-4(1H)-pyrimidinone = [4-[[[2-[5-[3-(Diethylamino)propyl]-1,4-dihydro-6-methyl-4-oxo-2-pyrimidinyl]ethyl]thio]methyl]-2-thiazolyl]guanidine (●)
R   Trihydrochloride (88022-87-1)
S   YM-14471
U   Histamine H_2 receptor antagonist

9090   (7421) $C_{19}H_{31}N_7O_4$
                                                     13665-88-8

2,2',2'',2'''-[(4-Piperidinopyrimido[5,4-*d*]pyrimidine-2,6-diyl)dinitrilo]tetraethanol = 2,2',2'',2'''-[[4-(1-Piperidinyl)pyrimido[5,4-*d*]pyrimidine-2,6-diyl]dinitrilo]tetrakis[ethanol] (●)

S *Mopidamol***, OLX-102, R-A 233-BS, Rapenton, Repentone

U Platelet aggregation inhibitor (prevention of metastases after surgery for sarkoma and malignant lymphoma)

9091 (7422)

$C_{19}H_{32}NO$
13004-75-6

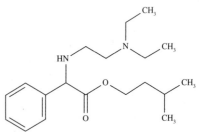

1-(3-Hydroxy-5-methyl-4-phenylhexyl)-1-methylpiperidinium (●)

R Bromide (520-20-7)

S Compound 1575, Darstin(e), *Mepiperphenidol (bromide)*, Mepiperphenidyl, *Piperphenamine*, *Piperphenidol methyl bromide*

U Anticholinergic

9092 (7423)

$C_{19}H_{32}NO_2$
16376-74-2

Diethyl(2-hydroxyethyl)methylammonium 3-methyl-2-phenylvalerate = *N,N*-Diethyl-*N*-methyl-2-(3-methyl-2-phenylvaleroyloxy)ethylammonium = *N,N*-Diethyl-*N*-methyl-2-[(3-methyl-1-oxo-2-phenylpentyl)oxy]ethanaminium (●)

R Bromide (90-22-2)

S Barespan, Baretaval, Barumaido, Beruhgen, Bisechilon, Brothamate, Celvan, Cranfupan, Daimate, Donopon, Elist, Epidosan, Epidosin, Faibunin-B, Five-

nin B, Frenant, Funapan V, Gosper C, Kaichyl, Letamate, Murel, Narest, Panmate, Pastan, Peridomin, Redimeton, Release-V, Resitan, S 78, Satotase, Shikitan "Shiki", Shinmetane, Spantrin, Study, Ulban-Q, Valate, Valemate, Valemeton, Valethalin, *Valethamate bromide*, Valethamin, Vameita (old form), Velamate

U Anticholinergic, spasmolytic

9093 (7424)

$C_{19}H_{32}N_2$
90961-53-8

3,7-Bis(cyclopropylmethyl)-9,9-tetramethylene-3,7-diazabicyclo[3.3.1]nonane = 3',7'-Bis(cyclopropylmethyl)spiro[cyclopentane-1,9'-[3,7]diazabicyclo[3.3.1]nonane] (●)

R Dihydrochloride (132523-84-3)

S KC-8857, *Tedisamil hydrochloride***

U Bradycardic, anti-ischemic

9094 (7425)

$C_{19}H_{32}N_2O_2$
54-30-8

N-[2-(Diethylamino)ethyl]-2-phenylglycine isopentyl ester = α-[*N*-(β-Diethylaminoethyl)amino]phenylacetic acid isoamyl ester = α-[[2-(Diethylamino)ethyl]amino]benzeneacetic acid 3-methylbutyl ester (●)

R Dihydrochloride (5892-41-1)

S Acamil, *Acamylophenin*, Adister, Adopon, Apasmo, Avacan, *Avacina*, Avadyl, Belosin, *Camylofin hydrochloride***, Fisiozim, Intrifor, Licosin, Navadyl, Novospasmina, Rugo, Sintespasmil, Sionol, Spasmocan, Tospasmol

U Anticholinergic, spasmolytic

9095 $C_{19}H_{32}N_2O_2$
                                                         72456-63-4

N-Ethyl-*N*-heptyl-4-nitrobenzenebutanamine (●)
S LY 97241
U Potassium channel activator

9096   (7426) $C_{19}H_{32}N_2O_2S$
                                                         13957-60-3

β-(Diethylamino)ethyl 4-amino-2-(hexyloxy)thioben-
zoate = 4-Amino-2-(hexyloxy)thiobenzoic acid β-(diethylamino)ethyl ester = 4-Amino-2-(hexyloxy)benzene-
carbothioic acid *O*-[2-(diethylamino)ethyl] ester (●)
S *Hexothiocaine*, Largacaine, Mocaton, Onocaine,
       Win 4510
U Local anesthetic, antipruritic

9097   (7427) $C_{19}H_{32}N_2O_3$
                                                         53352-75-3

2-(Diethylamino)ethyl 4-(butylamino)-2-ethoxybenzoate
= 4-(Butylamino)-2-ethoxybenzoic acid 2-(diethylamino)ethyl ester (●)

R Monohydrochloride (60319-18-8)
S Gynodal, Prusocain
U Local anesthetic

9098   (7428) $C_{19}H_{32}N_2O_3$

2-Butoxy-4,6-dimethylcarbanilic acid β-(diethylami-
no)ethyl ester = (2-Butoxy-4,6-dimethylphenyl)carbamic
acid 2-(diethylamino)ethyl ester (●)
R Hydrochloride
S LAC-18
U Local anesthetic

9099   (7429) $C_{19}H_{32}N_2O_4$
                                                         3818-62-0

2-[2-(Diethylamino)ethoxy]ethyl 3-amino-4-butoxyben-
zoate = 3-Amino-4-butoxybenzoic acid 2-[2-(diethylami-
no)ethoxy]ethyl ester (●)
R Hydrochloride (5003-47-4)
S *Betoxicaini chloridum, Betoxycaine hydrochlori-
       de**, Millicaine, Posicaine
U Local anesthetic

9100   (7430) $C_{19}H_{32}N_2O_5$
72880-75-2

[R-(R*,S*)]-3-(3-Cyclohexyl-3-hydroxypropyl)-2,5-di-
oxo-4-imidazolidineheptanoic acid (●)
S   B.W. 245 C
U   Platelet aggregation inhibitor (prostaglandin analog)

9101   (7431) $C_{19}H_{32}N_2O_5$
82834-16-0

(2S,3aS,7aS)-1-[(S)-N-[(S)-1-Carboxybutyl]alanyl]hexa-
hydro-2-indolinecarboxylic acid 1-ethyl ester =
(2S,3aS,7aS)-1-[(S)-2-[[(S)-1-(Ethoxycarbonyl)bu-
tyl]amino]propionyl]octahydro-2-indolecarboxylic acid =
[2S-[1[R*(R*)],2α,3aβ,7aβ]]-1-[2-[[1-(Ethoxycarbonyl)butyl]amino]-1-oxopropyl]octahydro-1H-indole-2-
carboxylic acid (●)
S   McN-A 2833, *Perindopril***, S 9490
U   Antihypertensive (ACE inhibitor)

9101-01   (7431-01)   107133-36-8
R   Compd. (1:1) with 2-methyl-2-propanamine (*tert*-bu-
tylamine)
S   Aceon, Acertil, Coverax, Coverene, Coverex, Cover-
sil, Coversum, Coversum Cor, Coversyl, Electan,
McN-A 2833-109, *Perindopril erbumine***, Prestari-
um, Prestimin, Prexanil, Prexum, Procaptan,
S 9490-3
U   Antihypertensive (ACE inhibitor)

9102   (7432) $C_{19}H_{32}N_6O_4$

1-[1-(1-Cyano-1-methylethoxy)-2,2,6,6-tetramethyl-4-
piperidinylamino]-3-(2-nitro-1-imidazolyl)-2-propanol =
2-[[4-[[2-Hydroxy-3-(2-nitro-1H-imidazol-1-yl)pro-
pyl]amino]-2,2,6,6-tetramethyl-1-piperidinyl]oxy]-2-me-
thylpropanenitrile (●)
R   Dihydrochloride (79820-31-8)
S   RSU 4013
U   Radiosensitizer

9103   (7433) $C_{19}H_{32}O_2$
1852-53-5

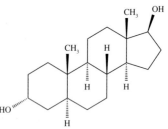

5α-Adrostane-3α,17β-diol = (3α,5α,17β)-Androstane-
3,17-diol (●)
S   Adiol, *Androstanediol, Dihydroandrosterone*, Hom-
breol
U   Androgen

9104   (7434) $C_{19}H_{32}O_4$
54857-86-2

5-(Tetradecyloxy)-2-furoic acid = 5-(Tetradecyloxy)-2-
furancarboxylic acid (●)
S   RMI 14514, TOFA

U Antihyperlipidemic

9105 $C_{19}H_{32}O_4$
                                                                    151519-02-7

(5R,6R)-5-Ethyl-5,6-dihydro-6-[(2R,3S,4R,5S,7E)-2-hydroxy-4-methoxy-3,5-dimethyl-7-nonenyl]-2H-pyran-2-one (●)
S *Pironetin*
U Antineoplastic

9106   (7435) $C_{19}H_{32}O_5$
                                                                    120755-15-9

Acetic acid 1-(2,5-dihydro-2-hydroxy-5-oxo-3-furanyl)tridecyl ester = 4-[1-(Acetyloxy)tridecyl]-5-hydroxy-2(5H)-furanone (●)
S AGN 190383
U Anti-inflammatory (topical)

9107   (7436) $C_{19}H_{33}N$
                                                                    3626-67-3

2-(2,2-Dicyclohexylvinyl)piperidine = 2-(2,2-Dicyclohexylethenyl)piperidine (●)

S *Hexadiline**, Hexadylamine*, MRL-38
U Coronary vasodilator

9108   (7437) $C_{19}H_{33}NO_2$
                                                                    18966-39-7

N-(3-Methylbutyl)-N-[(3-methyl-2-norbornyl)methyl]levulinamide = N-[(3-Methylbicyclo[2.2.1]hept-2-yl)methyl]-N-(3-methylbutyl)-4-oxopentanamide (●)
S Bornamid, *Pentanobornamide*, SIR 93
U Spasmolytic

9109 $C_{19}H_{33}NO_2$
                                                                    162359-55-9

2-Amino-2-[2-(4-octylphenyl)ethyl]-1,3-propanediol (●)
R Hydrochloride (162359-56-0)
S FTY 720
U Immunosuppressive

9110   (7438) $C_{19}H_{33}NO_3$
                                                                    799-25-7

4-[N-Bis(β-hydroxyethyl)aminomethyl]-2,6-di-*tert*-butylphenol = 2,2'-(3,5-Di-*tert*-butyl-4-hydroxybenzylimino)diethanol = 4-[[Bis(2-hydroxyethyl)amino]methyl]-2,6-bis(1,1-dimethylethyl)phenol (●)
R Hydrochloride (2226-97-3)
S Ambunol

U Antineoplastic, antiseptic

9111 (7439) $C_{19}H_{33}N_3O$
105630-62-4

3-[[2-(Diisopropylamino)ethyl]amino]propiono-2,6-xyli-
dide = 3-[[2-[Bis(1-methylethyl)amino]ethyl]amino]-*N*-
(2,6-dimethylphenyl)propanamide (●)
R Phosphate (1:2) (105668-70-0)
S AN 132
U Anti-arrhythmic

9112 (7440) $C_{19}H_{33}N_3O$
78833-03-1

α-[2-(Diisopropylamino)ethyl]-α-isobutyl-2-pyridine-
acetamide = 2-[2-(Diisopropylamino)ethyl]-4-methyl-2-
(2-pyridyl)valeramide = α-[2-[Bis(1-methylethyl)ami-
no]ethyl]-α-(2-methylpropyl)-2-pyridineacetamide (●)
R also with (±)-definition (96513-83-6)
S CM 7857, ME 3202, *Penticainide, Pentisomide***,
Propisomide
U Anti-arrhythmic

9113 (7441) $C_{19}H_{33}N_4S$

4-[1-[(Aminothioxomethyl)hydrazono]octyl]-*N*-ethyl-
N,N'-dimethylbenzenaminium (●)

R Iodide (26672-76-4)
S M. & B. 15944
U Muscle relaxant

9114 (7442) $C_{19}H_{34}ClN_3O_{12}$
70189-62-7

1-(2-Chloroethyl)-3-isobutyl-3-(β-maltosyl)-1-nitroso-
urea = *N*-(2-Chloroethyl)-*N*'-(4-*O*-α-D-glucopyranosyl-β-
D-glucopyranosyl)-*N*'-(2-methylpropyl)-*N*-nitrosourea
(●)
S TA-077
U Antineoplastic

9115 (7443) $C_{19}H_{34}N_2O_3S$
133267-19-3

(+)-4'-[(*R*)-4-(Dibutylamino)-1-hydroxybutyl]methane-
sulfonanilide = (*R*)-*N*-[4-[4-(Dibutylamino)-1-hydroxy-
butyl]phenyl]methanesulfonamide (●)
R Fumarate (2:1) (salt) (133267-20-6)
S *Artilide fumarate***, U-88943 E
U Anti-arrhythmic

9116 (7444) $C_{19}H_{34}N_8O_8$
3808-42-2

2-[4-O-Carbamoyl-2-deoxy-2-(3,6-diaminohexanami-do)-β-D-gulopyranosylamino]-3,3a,5,6,7,7a-hexahydro-5-hydroxy-4H-imidazo[4,5-c]pyridin-4-one = [3aS-[2(R*),3aα,7aβ]]-2-[[4-O-(Aminocarbonyl)-2-deoxy-2-[(3,6-diamino-1-oxohexyl)amino]-β-D-gulopyrano-syl]amino]-3,3a,5,6,7,7a-hexahydro-7-hydroxy-4H-imidazo[4,5-c]pyridin-4-one (●)
S   AY-24546, *Racemomycin A*, S 15-1 A,
       Streptothricin F, Streptothricin VI, Yazumycin A
U   Antibiotic from *Actinomyces lavendulae*

9117   (7445) $C_{19}H_{34}O_3$
                                                     40596-69-8

Isopropyl (2E,4E)-11-methoxy-3,7,11-trimethyl-2,4-do-decadienoate = (E,E)-11-Methoxy-3,7,11-trimethyl-2,4-dodecadienoic acid 1-methylethyl ester (●)
S   Altosid, Cevrin, ENT-70460, Manta, *Methoprene**,*
       Ovitrol "Novartis", Precor, ZR-515
U   Insecticide, uterine relaxant

9118 $C_{19}H_{34}O_3$
                                                     147317-12-2

2-Hydroxy-2,4-dimethyl-5-(1,3,5,7-tetramethylnonyl)-3(2H)-furanone (●)
S   AS 183
U   Anti-arteriosclerotic from *Scedosporium* sp. SPC-15549

9119   (7446) $C_{19}H_{35}ClN_2O_5S$
                                                     31101-25-4

cis-trans mixture of Methyl 7-chloro-6,7,8-trideoxy-6-(4-pentyl-L-2-pyrrolidinecarboxamido)-1-thio-L-*threo*-α-D-*galacto*-octopyranoside = 7-Chloro-N-demethyl-7-deoxy-3'-depropyl-3'-pentyllincomycin = Methyl 7-chloro-6,7,8-trideoxy-6-[[(4-pentyl-2-pyrrolidinyl)carbonyl]amino]-1-thio-L-*threo*-α-D-*galacto*-octopyranoside (●)
R   Monohydrochloride (8063-91-0)
S   *Mirincamycin hydrochloride**, U-24729 A
U   Antibiotic

9120   (7447) $C_{19}H_{35}N$
                                                     6621-47-2

2-(2,2-Dicyclohexylethyl)piperidine (●)
R   Maleate (1:1) (6724-53-4)
S   Corzepin, Daprin, Nodixil, *Perhexiline maleate**,*
       Pexid, WSM 3978 G
U   Coronary vasodilator

9121   (7448) $C_{19}H_{35}NO_2$
                                                     77-19-0

2-(Diethylamino)ethyl (bicyclohexyl)-1-carboxylate = β-(Diethylamino)ethyl 1-cyclohexylcyclohexanecarboxy-late = 1-Cyclohexylcyclohexanecarboxylic acid β-(diethylamino)ethyl ester = [1,1'-Bicyclohexyl]-1-carb-oxylic acid 2-(diethylamino)ethyl ester (●)
R   Hydrochloride (67-92-5)
S   Ametil, Antispas, A-Spas, Atumin, Babyspasmil, Ba-lacone, Baycyclomine, Benacol, Bentomine, Bentyl, Bentylol, Byclomine, Ciclan, Cicloima, Clomin "SCS Pharmalab", Coloplex "Laquifa", Cyclobec, Cyclogen "Central", Cyclomine, Cyclominol, Cyclo-nil, Cyclopam "Indoco", DCC, Dibent, Dicen, Dicic-

lomina, Dici-Gotas, Diclomin, *Dicyclomine hydro-chloride*, Di-Cyclonex, Dicyclon No.1, *Dicycloverine hydrochloride**, Dicycol, Dilomine, Diocyl, Di-Spaz, Di-Syntramine, Dycimina, Dyspas, Efison, Esentil, Formulex, Icramin, Incron, Isospasmex, J.L. 998, Kolantyl, Lomine, M 33536, Mabex, Magesan P, Mamiesan, Medicyclomine, Menospasm, Merbantal, Merbentyl, Mydocalm "Lennon", Neoquess, Nomocramp, Nospaz, Notensyl, Optimal "Inibsa", OR-Tyl, Pamoterap, Panakiron, Procyclomin, Protylol, Reparal, Respolimin, Rocyclo, Rotyl, Sawamin, Spascol, Spasmoban, Spasmoject, Spasmo-Rhoival-N, Spastyl, Stannitol-Inj., Tarestin, Viscerol, Wyovin

U Anticholinergic, spasmolytic

9122 (7449) $C_{19}H_{35}N_3O_5$
13434-13-4

3-[[1-[[2-(Hydroxymethyl)-1-pyrrolidinyl]carbonyl]-2-methylpropyl]carbamoyl]octanohydroxamic acid = stereoisomer of N^4-Hydroxy-N^1-[1-[[2-(hydroxymethyl)-1-pyrrolidinyl]carbonyl]-2-methylpropyl]-2-pentylbutanediamide (●)

S *Actinonin*
U Antibacterial (aminopeptidase inhibitor)

9123 (7450) $C_{19}H_{36}NO_2$

(8r)-3α-Hydroxy-8-isopropyl-1αH,5αH-tropanium 2-propylvalerate = *endo*-8-Methyl-*syn*-8-(1-methylethyl)-8-azoniabicyclo[3.2.1]octan-3-ol 2-propylvalerate = (*endo,syn*)-8-Methyl-8-(1-methylethyl)-3-[(1-oxo-2-propylpentyl)oxy]-8-azoniabicyclo[3.2.1]octane (●)

R Bromide (79467-19-9)
S *Sintropium bromide***
U Anticholinergic, spasmolytic

9124 (7451) $C_{19}H_{36}N_2$
126825-36-3

3'-Isobutyl-7'-isopropylspiro[cyclohexane-1,9'-[3,7]diazabicyclo[3.3.1]nonane] = 3-Isobutyl-7-isopropyl-9,9-pentamethylene-3,7-diazabicyclo[3.3.1]nonane = 3'-(1-Methylethyl)-7'-(2-methylpropyl)spiro[cyclohexane-1,9'-[3,7]diazabicyclo[3.3.1]nonane] (●)

R also difumarate ethanolate
S *Bertosamil**, KC-8851
U Anti-ischemic

9125 $C_{19}H_{37}N_2$
176388-52-6

N,N,N-Trimethyl-5-[(tricyclo[3.3.1.1^{3,7}]dec-1-ylmethyl)amino]-1-pentanaminium (●)

R Bromide hydrobromide (121034-89-7)
S IEM-1460
U AMPA/Kainate receptor blocker

9126 (7452) $C_{19}H_{37}N_5O_7$
55870-64-9

O-3-Deoxy-4-C-methyl-3-(methylamino)-β-L-arabinopyranosyl-(1→1)-O-[2,6-diamino-2,3,4,6-tetradeoxy-α-D-*glycero*-hex-4-enopyranosyl-(1→3)]-4,6-diamino-4,5,6-trideoxy-D-*myo*-inositol (●)

S 5-*Episisomicin*, Mu 6, *Mutamicin 6*, *Pentisomicin***, Sch 22591
U Antibiotic

9127 (7453) $C_{19}H_{37}N_5O_7$
32385-11-8

(2S-cis)-4-O-[3-Amino-6-(aminoethyl)-3,4-dihydro-2H-pyran-2-yl]-2-deoxy-6-O-[3-deoxy-4-C-methyl-3-(methylamino)-β-L-arabinopyranosyl]-D-streptamine = O-3-Deoxy-4-C-methyl-3-(methylamino)-β-L-arabinopyranosyl-(1→6)-O-[2,6-diamino-2,3,4,6-tetradeoxy-α-D-glycero-hex-4-enopyranosyl-(1→4)]-2-deoxy-D-streptamine (●)
R also sulfate (2:5) (53179-09-2)
S *Antibioticum 6640*, Bay c 8990, Baymicin(a), Biracin, Ensamycin, Exacin "Lucky Ph.", Extramycin, Geonyn, Mensiso, Milsido, Pathomycin, *Rickamycin*, Salvamina, Sch 13475, Sch Bay 13475, Siomycin, Siseptin, Sismine, Sisobiotic, Sisolline, Sisoloa, *Sisomicin***, Sisomin(a), Sisoptin, *Sissomicin*, Somazinal, Udolin
U Antibiotic from *Micromonospora inyoensis*

9128 (7454) $C_{19}H_{38}N_4O_{10}$
36889-15-3

O-6-Amino-6-deoxy-α-D-glucopyranosyl-(1→4)-O-[3-deoxy-4-C-methyl-3-(methylamino)-β-L-arabinopyranosyl-(1→6)]-2-deoxy-D-streptamine (●)

R Sulfate (43169-50-2)
S *Betamicin sulfate***, *Gentamicin B sulfate*, Sch 14342
U Antibiotic

9129 (7455) $C_{19}H_{39}NO_3$
6582-30-5

N-(Lauryloxypropyl)-β-aminobutyric acid = 3-[[3-(Dodecyloxy)propyl]amino]butanoic acid (●)
S *Acidum 3-(lauryloxypropylamino)butyricum*, Fongibactyl (main active substance), *Lopobutan***, *Lopobutate*, T 59
U Antiseptic, antifungal

9130 $C_{19}H_{40}NO_2$
87297-88-9

Tetradecyl betainate = N,N,N-Trimethyl-2-oxo-2-(tetradecyloxy)ethanaminium (●)
R Chloride (54514-50-0)
S B 14
U Antifungal

9131 (7456) $C_{19}H_{40}N_2$

(Methylenedi-1,4-cyclohexylene)bis(trimethylammonium) = 4,4'-Methylenebis(cyclohexyltrimethylammonium) = 4,4'-Methylenebis[N,N,N-trimethylcyclo-hexanaminium] (●)
R Diiodide (7681-78-9)
S Compound 12038, *Mebezonium iodide***, T 61
U Muscle relaxant

9132 (7457) $C_{19}H_{40}N_2O_2$
6582-31-6

3-[3-(Dodecylamino)propylamino]butyric acid = 3-[[3-
(Dodecylamino)propyl]amino]butanoic acid (●)
S *Dapabutan***, PS 620
U Antiseptic

9133 (7458) $C_{19}H_{40}O_3$
53584-29-5

3-(Hexadecyloxy)-1,2-propanediol (●)
S LY 217749
U Phospholipase A_2-inhibitor

9134 (7459) $C_{19}H_{42}N$
6899-10-1

Hexadecyltrimethylammonium = Cetyltrimethylam-
monium = *N,N,N*-Trimethyl-1-hexadecanaminium (●)
R Bromide (the commercial product also contains other
alkyltrimethylammonium bromides) (57-09-0);
see also nos. 4603-06, 7449-01
S Aseptiderm, Biocetab, Bromat, *Cemetrimoni bromidum*, Cetab, Cetabrom, Cetamium, Cetril, Cetrimex,
Cetriminium, *Cetrimonium bromide***, Cetriseptin,
Cetyl "Akpa", Cetylamine, Cirrasol-OD, CTAB,
Dermanatal, Desitur, Gomaxine, Lissolamine, Mi-
col, Quamonium, Savlex PCD, Savlon-Cream, Sebo-
derm, Senol, Septol "Hemofarm", Turisan
U Antiseptic, cationic detergent

9134-01 (7459-01) 112-02-7
R Chloride
S *Cetrimonium chloride***, Scadan "Dome", Surfak-
tivo
U Antiseptic, cation detergent

9134-02 (7459-02) 138-32-9
R *p*-Toluenesulfonate
S Aflogine, *Cetrimonium tosilate***, Golaval, Intima,
Intol, Neo-Intol
U Antiseptic

9134-03 (7459-03) 87-76-3
R Pentachlorophenolate (pentachlorophenoxide)
S Neo-Tachizol, TCAP, *Trimethylcetylammonium pen-
tachlorophenate*, T.S.P.
U Topical antiseptic, fungicide

9135 (7460) $C_{20}H_4Cl_4{}^{131}I_4O_5$
42352-53-4

4,5,6,7-Tetrachloro-2',4',5',7'-tetra-^{131}iodofluorescein =
4,5,6,7-Tetrachloro-3',6'-dihydroxy-2',4',5',7'-tetra(iodo-
^{131}I)spiro[isobenzofuran-1(3*H*),9'-[9*H*]xanthen]-3-one
(●)
R Disodium salt (15251-14-6) (with spiroisobenzofuran
structure: (50291-21-9))
S *Bengalrosa-*^{131}I, Radio-Rose-Bengal-I-131,
Robengatope I-131, *Rosebengal I 131*, Rose bengal
sodium (^{131}I) **, Roseum bengalense natricum
(^{131}I)**
U Diagnostic aid (hepatic function determination)

9136 (7461)

$C_{20}H_8Br_4O_5$
152-75-0

2',4',5',7'-Tetrabromofluorescein = 2',4',5',7'-Tetrabromo-3',6'-dihydroxyspiro[isobenzofuran-1(3H),9'-[9H]xanthen]-3-one (●)

R Disodium salt (17372-87-1)

S *Bromeosine*, C.I. 45380, C.I. Acid Red 87, Eosinal, *Eosine yellowish*, *Eosin gelblich*

U Dye, reagent, dermatic

9137 (7462)

$C_{20}H_8I_4O_5$
15905-32-5

2',4',5',7'-Tetraiodofluorescein = 3',6'-Dihydroxy-2',4',5',7'-tetraiodospiro[isobenzoforan-1(3H),9'-[9H]xanthen]-3-one (●)

R Disodium salt (16423-68-0) [also monohydrate (49746-10-3)]

S Blendax Anti-Belag Färbetabl., Butler Red-Cote, Ceplac, Diaplac, Disclo-Gel, Disclo-Tabs, *Erythrosine (sodium)*, FD & C Red No. 3, Felumin, *Iodeosine*, *Jodeosin*, Plaquefärbetabletten, Prevident Disclosing, Trace

U Diagnostic aid (radiopaque medium-cholecystographic), dye (indicator, orthochromatic sensitizer)

9138 (7463)

$C_{20}H_{10}Br_2HgO_6$
55728-51-3

2,7-Dibromo-4-(hydroxymercuri)fluorescein = [2,7-Dibromo-9-(2-carboxyphenyl)-6-hydroxy-3-oxo-3H-xanthen-4-yl]hydroxymercury = (2',7'-Dibromo-3',6'-dihydroxy-3-oxospiro[isobenzofuran-1(3H),9'-[9H]xanthen]-4'-yl)hydroxymercury (●)

R Disodium salt (129-16-8)

S Antiseptine, Aromer, Asept'Aqua, Aseptichrome, Aseptochrome, Badyl a la Mercuresceine, Biocrom, Brocasept, Bromochrom(e), Chibromercurobrome, Chromargyre, Chromine, Cinfacromin, Colluchromine, Cromer Orto, Cromo-Utin, Curichrome, Curocromo, Cynochrome, Dakermina, Davurcrom, D.O.M.F., Ekachrome, Emcerol, Flavurol, Flurochrom, Gallochrome, Glubelcromo, Gynochrome, Histochrome, Lafircron, Lapicromo, Logacram, Logacron, Lupurol, Medichrom, *Merbromin**, Merbromine sodique*, Mercromina, Mercromsal, Mercrotona, Mercroverk, Mercuchrom, Mercuranine, Mercuranum DAK., Mercurasept, *Mercurescéine*, Mercurin "Monik", Mercurio rojo "Ideal Tide", *Mercurobromfluorescein*, Mercurobromo, Mercurocéine, *Mercurochrome*, Mercurocil, Mercuro Clinico, Mercurocol, *Mercurocromo*, Mercuro-Fluoren, Mercurome, Mercurophage, Mercutina brota, Merkromol, Mersol "Merkez", Neostyl a la Mercuresceine, No. 220 soluble, Osmobrom, Pharmadose, Pharmadose mercuresceine, Pintacrom, Planochrome, Septichrome, Soluchrom(e), Stellachrome, Stick-Cytochrome, Stickrome, Stylochrome, Sulfacromo, Super-Cromer-orto, Tube Chauvin Dermichrome, Tube P.O.S. Chromercure, Veriscrom, Yocrom

U Antiseptic

9139 (7464) $C_{20}H_{10}Br_4O_4$
76-62-0

3,3-Bis(3,5-dibromo-4-hydroxyphenyl)phthalide = 3',3",5',5"-Tetrabromophenolphthalein = 3,3-Bis(3,5-dibromo-4-hydroxyphenyl)-1(3*H*)-isobenzofuranone (●)
R Disodium salt
S *Bromophthalein Sodium*, Brom-Tetragnost, *Tetrothalein Sodium*, Tetrabrom
U Diagnostic aid (radiopaque medium-cholecystographic)

9140 (7465) $C_{20}H_{10}Br_4O_{10}S_2$
297-83-6

5,5'-(4,5,6,7-Tetrabromophthalidylidene)bis(2-hydroxy-benzenesulfonic acid) = 4,5,6,7-Tetrabromophenolphthalein-3',3"-disulfonic acid = 3,3'-(4,5,6,7-Tetrabromo-3-oxo-1(3*H*)-isobenzofuranylidene)bis[6-hydroxybenzenesulfonic acid] (●)
R Disodium salt (71-67-0)
S Bromotaleina, *Bromsulfalein*, Bromsulfan, *Bromsulfophalein sodium*, *Bromsulphalein*, *Bromsulphthalein*, Bromthalein, B.S.F. Simes, BSP, Hepartest "Cassella", Hepatestabrome, Hepatosulfalein, Hepatosulphalein, Hepatotestbrom, *Sulfobromophthalein Sodium*, *Sulfobromphthalein*, *Sulphobromophthalein sodium*
U Diagnostic aid (hepatic function determination)

9141 (7466) $C_{20}H_{10}Cl_2F_5N_3O_3$
86811-58-7

1-[4-Chloro-3-[[3-chloro-5-(trifluoromethyl)-2-pyridyl]oxy]phenyl]-3-(2,6-difluorobenzoyl)urea = *N*-[[[4-Chloro-3-[[3-chloro-5-(trifluoromethyl)-2-pyridinyl]oxy]phenyl]amino]carbonyl]-2,6-difluorobenzamide (●)
S Acatik, CGA-157419, *Fluazuron***
U Antiparasitic (veterinary)

9142 (7467) $C_{20}H_{10}Cl_4O_4$
81-89-0

3,3-Bis(3,5-dichloro-4-hydroxyphenyl)phthalide = 3',3",5',5"-Tetrachlorophenolphthalein = 3,3-Bis-(3,5-dichloro-4-hydroxyphenyl)-1(3*H*)-isobenzofuranone (●)
R Disodium salt
S *Chlorophthalein Sodium*, Chlor-Tetragnost, Cholegnostyl
U Diagnostic aid (radiopaque medium)

9143 (7469) $C_{20}H_{10}I_4O_4$
386-17-4

3,3-Bis(4-hydroxy-3,5-diiodophenyl)phthalide =
3',3",5',5"-Tetraiodophenolphthalein = 3,3-Bis(4-hydr-
oxy-3,5-diiodophenyl)-1(3H)-isobenzofuranone (●)
R most disodium salt (2217-44-9)
S Antinosine, Bilagnost, Bilicontrast, Bilitrast, Chola-
gnost, Cholepulvis, Cholotrast, Cholumbral, Cho-
lumbrin, Cistopac, Cystopac, Foriod, Galisol, Iodei-
kon, *Iodognost(um)*, Iodophene, *Iodophthalein Sodium***, *Iodophthalein soluble*, Iodo-Ray, Iodo-
ray-oral, Jodafen, Jod-Cholumbral, Jodognost, Jodo-
phen, *Jodophthalein*, *Jodphthalein*, Jod-Tetragnost,
Keraphen, Kontrastol G, Nosophen, Opacin(e), Opa-
col, Oral-Tetragnost, Oro-Bilopac, Photobiline, Pi-
liophen, Protistène, Radiotétrane, Shadocol, Sombra-
chol, Stipolac, Syntagnost, Tetiothalein, Tetra-Con-
trast, Tetraiode, *Tetraiodophthalein Sodium*, Tetra-
jod, Tetrajod-Contrast, Tetrothalein, TIP, T.I.P.P.S.,
Videophel
U Diagnostic aid (radiopaque medium-cholecystographic), antiseptic

9143-01 (7469-01)
R Mercury salt
S Apallagin
U Antiseptic

9143-02 (7469-02)
R Bismuth salt
S Eudoxin
U Antiseptic (intestinal)

9144 (7468) $C_{20}H_{10}I_4O_4$
27458-03-3

3,3-Bis(4-hydroxyphenyl)-4,5,6,7-tetraiodophthalide =
4,5,6,7-Tetraiodophenolphthalein = 3,3-Bis(4-hydroxy-
phenyl)-4,5,6,8-tetraiodo-1(3H)-isobenzofuranone (●)
R Disodium salt (128-72-3)
S Iso-Iodeikon, *Phentetiothalein Sodium*
U Diagnostic aid (radiopaque medium)

9145 $C_{20}H_{10}O_5$
62417-80-5

4,6,9-Trihydroxynaphth[1,2-*a*]acenaphthylene-5,10-
dione = 4,7,9-Trihydroxybenzo[*j*]fluoranthene-3,8-dione
(●)
S *Bulgarein*
U Antineoplastic from *Bulgaria inquinans*

9146 $C_{20}H_{11}Br_2Cl_2NO_3$
67358-44-5

3,5-Dibromo-4'-chloro-3'-(4-chlorobenzoyl)salicylani-
lide = 3,5-Dibromo-*N*-[4-chloro-3-(4-chlorobenzoyl)phe-
nyl]-2-hydroxybenzamide (●)
S Tegalid

U Anthelmintic

9150 (7473) $C_{20}H_{12}HgO_6$
                                                    26836-06-6

U Coccidiostatic

9147 (7470) $C_{20}H_{11}Cl_2I_2NO_3$
                                                    36093-47-7

3'-Chloro-4'-(*p*-chlorobenzoyl)-3,5-diiodosalicylanilide =
N-[3-Chloro-4-(4-chlorobenzoyl)phenyl]-2-hydroxy-3,5-
diiodobenzamide (●)
S R 23050, *Salantel***
U Anthelmintic (fasciolicide)

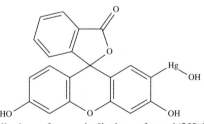

(3',6'-Dihydroxy-3-oxospiro[isobenzofuran-1(3*H*),9'-
[9*H*]xanthen]-2'-yl)hydroxymercury (●)
S Mercurasan, Mercurascan
U Cardioprotectant, anti-arrhythmic

9148 (7471) $C_{20}H_{11}N_3O_4$
                                                    133805-03-5

12,13-Dihydro-1,11-dihydroxy-5*H*-indolo[2,3-*a*]pyrro-
lo[3,4-*c*]carbazole-5,7(6*H*)-dione (●)
S BE-13793 C
U Antineoplastic, antileukemic antibiotic from *Strepto-
verticillium* BA-13793

9151 (7474) $C_{20}H_{12}O_5$
                                                    2321-07-5

o-(6-Hydroxy-3-oxoisoxanthen-9-yl)benzoic acid = 4"-
Hydroxy-2,2'-epoxyfuchsone-2-carboxylic acid = 3',6'-
Fluorandiol = 3',6'-Dihydroxyspiro[isobenzofuran-
1(3*H*),9'-[9*H*]xanthen]-3-one (●)
R see also no. 15476;
       Disodium salt (518-47-8)
S AK-Fluor, Broxotest, C.I. 45350, C.I. Acid
       Yellow 73, *Dihydroxyfluorane sodium*, Diofluor, *Di-
       oxyfluran sodium*, Disclo-Plaque, Fluo 10 (20),
       Fluoftal, Fluor "Tobishi-Santen", Fluoralfa, Flu-
       or-AMPS, Fluo-rectal, *Fluorescein Sodium, Fluore-
       scein soluble*, Fluorescite, Fluoreseptic, Fluores
       (IV), Fluorescein SE Thilo, Fluorets, Fluorin, Flu-
       or-I-Strip, Fluorovets, Flurenat, Ful-Glo, Fun-
       duscein, *Natrii fluoresceinas*, Obiturine, Ophthiflu-
       or, Optifluor Diba, Opulets, Plak-Lite, *Resorcin-
       phthalein Sodium*, Uranin(e), Vectidan inyectable
U Diagnostic aid (corneal trauma indicator), therapeutic
       for erythema, indicator

9149 (7472) $C_{20}H_{12}F_{12}N_2O_2$
                                                    60131-74-0

N,N'-Bis[3,4-di(trifluoromethyl)phenyl]methylmalon-
amide = *N,N'*-Bis[3,4-bis(trifluoromethyl)phenyl]-2-me-
thylpropanediamide (●)
S Sch 18545

9151-01   (7474-01)   25931-86-6
R   Silver salt
S   *Argenti fluoresceinas*, Collogène d'Argent fort
U   Antibacterial

9151-02   (7474-02)
R   basic bismuth salt
S   *Bismuti subfluoresceinas*, Bis-Oralin, Bi-Steril "Bouty", Collogène de Bismuth
U   Antibacterial

9152   (7475) $C_{20}H_{13}ClF_3NO_3$
                                                   53966-34-0

7-Chloro-3,4-dihydro-10-hydroxy-3-(α,α,α-trifluoro-*p*-tolyl)-1,9(2*H*)-acridandione = 7-Chloro-3,4-dihydro-10-hydroxy-3-[4-(trifluoromethyl)phenyl]-1,9(2*H*,10*H*)-acridinedione (●)
S   *Floxacrine**, Hoechst 991
U   Antimalarial

9153 $C_{20}H_{13}ClN_2O_3$
                                                   181122-94-1

7-Chloro-4-hydroxy-3-(3-phenoxyphenyl)-1,5-naphthyridin-2(1*H*)-one (●)
S   ACEA-762
U   NMDA/glycine site antagonist

9154 $C_{20}H_{13}F_2N_3O_2$
                                                   145915-60-2

4,5-Bis(4-fluoranilino)phthalimide = 5,6-Bis[(4-fluorophenyl)amino]-1*H*-isoindole-1,3(2*H*)-dione (●)
S   CGP 53353
U   Antineoplastic (tyrosine kinase receptor inhibitor)

9155 $C_{20}H_{13}IN_6$
                                                   172747-51-2

4-[2-(4-Azidophenyl)-5-(3-iodophenyl)-1*H*-imidazol-4-yl]pyridine (●)
S   SB 206718
U   Immunosuppressive, cytokine inhibitor

9156   (7476) $C_{20}H_{13}NO_2$
                                                   91753-07-0

5,14-Dihydrobenz[5,6]isoindolo[2,1-*b*]isoquinoline-8,13-dione (●)
S   GR 30921, *Mitoquidone**, NSC-382057
U   Antineoplastic

9157   (7477) $C_{20}H_{13}N_3$
                                                      95360-17-1

1-(2-Quinolinyl)-9H-pyrido[3,4-b]indole (●)
S   *Nitramarine*
U   Antihypertensive, spasmolytic (alkaloid from *Nitraria komarovii*)

9158 $C_{20}H_{13}N_3O_3$
                                                      149550-36-7

2-Amino-4-(3-nitrophenyl)-4H-naphtho[1,2-b]pyran-3-carbonitrile (●)
S   LY 290181
U   Antiproliferative

9159   (7478) $C_{20}H_{14}ClN_3$
                                                      78351-75-4

1-(3-Chloroanilino)-4-phenylphthalazine = N-(3-Chlorophenyl)-4-phenyl-1-phthalazinamine (●)
S   MY-5445

U   Platelet aggregation inhibitor

9160 $C_{20}H_{14}ClN_3O_3S$
                                                      136381-85-6

2-[[4-(o-Chlorophenyl)-2-thiazolyl]carbamoyl]indole-1-acetic acid = 2-[[[4-(2-Chlorophenyl)-2-thiazolyl]amino]carbonyl]-1H-indole-1-acetic acid (●)
S   *Lintitript***, SR 27897
U   Cholecystokinin CCK$_A$ receptor antagonist

9161 $C_{20}H_{14}Cl_2N_2O_3$
                                                      166974-22-7

4,6-Dichloro-3-[(E)-(2-oxo-1-phenyl-3-pyrrolidinylidene)methyl]-1H-indole-2-carboxylic acid (●)
R   Monosodium salt (166974-23-8)
S   GV 196771A
U   Analgesic (NMDA glycine site receptor antagonist)

9162 $C_{20}H_{14}FN_3O$
                                                      152121-30-7

4-[4-(4-Fluorophenyl)-5-(4-pyridinyl)-1H-imidazol-2-yl]phenol (●)
S   SB 202190
U   Cytokine suppressive, anti-inflammatory

9163 $C_{20}H_{14}F_3N_3O_3$
199789-36-1

7-(3-Azabicyclo[3.1.0]hex-3-yl)-1-(2,4-difluorophenyl)-6-fluoro-1,4-dihydro-4-oxo-1,8-naphthyridine-3-carboxylic acid (●)
S CP-415145
U Antibacterial

9164 (7479) $C_{20}H_{14}I_6N_2O_6$
606-17-7

3,3'-(Adipoyldiimino)bis(2,4,6-triiodobenzoic acid) = N,N'-Adipoylbis(3-amino-2,4,6-triiodobenzoic acid) = Adipic acid bis(3-carboxy-2,4,6-triiodoanilide) = 3,3'-[(1,6-Dioxo-1,6-hexanediyl)diimino]bis[2,4,6-triiodobenzoic acid] (●)
R also disodium salt (2618-26-0) or methylglucamine salt (1:2) (3521-84-4)
S Adipiodone**, BE 426, Bilgrafin, Bilignost(um), Biligrafin(a), Biligran, Bilipolinum, Bilispect, Cavumbren, Cholografin, Cholospekt, Endocistobil, Endografin, Intrabilix, Iodipamide, Iodipamide meglumine, Jodipamid, Kontrastmittel SH 216, Pobilan, Radiosélectan biliaire fort, RG 45, S 307, SH 216, Sodium iodipamide, Transbilix, Ultrabil, ZK 1
U Diagnostic aid (radiopaque medium)

9164-01 (7479-01)
R Lithium salt
S SH 243
U Diagnostic aid (radiopaque medium)

9165 (7480) $C_{20}H_{14}{}^{131}I_6N_2O_6$
42028-09-1

^{131}I-Adipiodone = 3,3'-[(1,6-Dioxo-1,6-hexanediyl)diimino]bis[2,4,6-tri(iodo-^{131}I)benzoic acid] (●)
R Disodium salt (24360-85-8)
S Iodipamide Sodium I 131, Radio-Cholografin
U Diagnostic aid (radiopaque medium)

9166 (7481) $C_{20}H_{14}NO_4$
2447-54-3

13-Methyl-[1,3]benzodioxolo[5,6-c]-1,3-dioxolo[4,5-i]phenanthridinium (●)
S Pseudochelerythrine, Sanguinarine
U Antibacterial alkaloid from Sanguinaria canadensis and other Papaveraceae

9166-01 (7481-01)   5578-73-4
R Chloride
S Sanguinarium chloride**
U Antibacterial, antifungal, anti-inflammatory

9167 (7482) $C_{20}H_{14}N_4O_2$
112515-43-2

6-Hydroxy-1H-indol-3-yl 5-(1H-indol-3-yl)-1H-imidazol-2-yl ketone = (6-Hydroxy-1H-indol-3-yl)[4-(1H-indol-3-yl)-1H-imidazol-2-yl]methanone (●)
S Topsentine B1
U Anti-infective (marine natural product)

9168   (7483)   C$_{20}$H$_{14}$N$_4$O$_3$
138304-90-2

1,2-Dihydro-4-hydroxy-2-oxo-1-phenyl-*N*-3-pyridinyl-
1,8-naphthyridine-3-carboxamide (●)
S   KF 17515
U   Anti-inflammatory

9169   (7484)   C$_{20}$H$_{14}$O$_3$
519-95-9

γ-Oxo-8-fluoranthenebutyric acid = 4-(8-Fluoranthenyl)-
4-oxobutyric acid = β-(8-Fluoranthoyl)propionic acid = γ-
Oxo-8-fluoranthenebutanoic acid (●)
S   Bilyn, Cistoplex, *Florantiron, Florantyrone***,
Fluochol, Idrobil, Idroepar, SC-1674, Zanchol
U   Hydrocholeretic

9170   (7485)   C$_{20}$H$_{14}$O$_4$
97743-96-9

(*S*)-2,3-Dihydro-12b-methyl-1*H*-benzo[6,7]phenan-
thro[10,1-*bc*]furan-6,8,11(12b*H*)-trione (●)
S   *Xestoquinone*, XQN
U   Cardiotonic from *Xestospongia sapra*

9171   (7486)   C$_{20}$H$_{14}$O$_4$
77-09-8

3,3-Bis(*p*-hydroxyphenyl)phthalide = 3-Oxo-1,1-bis(4-
hydroxyphenyl)phthalan = 3,3-Bis(4-hydroxyphenyl)-
1(3*H*)-isobenzofuranone (●)
S   Agafan, Alin "Biofarma", Alophen, Anfagoran,
Ap-La-Day, Arkalax, Becalax, Bitalax, Bom-bon,
Bonomint, Brooklax, Caolax, Castivlax, Castoline,
Castroline Laxative, Chocolax, Chocolaxine, Cioco-
lax, Cirulaxia, Cocolax, Darmol, Darolax, *Dihydro-
xyphthalophenone*, Dilsuave, Easylax, Egmolax, Egmol (new form), Espotabs, Euchessina, Evac-Q-Tabs
(old form), Evac-u-gen, Evac-U-lax, Evasof,
Ex-Lax, Fedolax, Feen-A-Mint (old form), Fenolax,
Fenolftalein, Fenotyl, Fentalax, Figsen, Flatax-Oran-
ge, Fructines-Vichy (Canada; Switzerland), Fructi-
ne-Vichy, Fructosan, Gastol, Healax, Huxol, Impha-
lax, Jaglax, Julabin, Kolagol, Koprol, Lacto-Purga,
Laksatin-Bodo, Laxal, Laxane, Laxante "Pyre", Laxante Wallace Carters, Laxante Yer, Laxa-Tabs, La-
xatina, Laxatone, Laxeine-N, Laxen-Busto, Laxettes
Chocolate, Laxicaps, Laxil, Laxiline, Laxin, Laxo-
gen, Laxol, Lax-Pills "G + W", Lecodol laxante,
Lilo, Medilax, Merilax, Metalax "Bombay Drug
House", Modane (old form), Musilaks, Musilax, Nat-
lex, Nectalax, Neopurghes, Novopuren, Opilax,
Phenaloin, Phenolax, Phenolaxan, Phenolaxol, *Phenolphthalein***, Phthalimetten, Prifinol, Prulet, Pru-
netta, Pürjen Sahab, Purex, Purga, Purgalon, Purga-
nol-Daguin, Purgante Falqui, Purgante Orravan, Pur-
garol, Purgativ, Purgatol, Purgax-Zazzera, Purgen,
Purgestol, Purgophen, Purgyl, Reguletts, Rilaxan,
Rowol-lax, Spulmakolax, Stanlax, Superlax, Su-
re-Lax, Syntholax, Thalinol, Thiolax, Trilax "Teva",
Vaculax, Wa-fer-lax
U   Laxative, indicator

9171-01   (7486-01)   37721-39-4
R   Mixture of an equal quantity of acetic acid ester and isovaleric acid ester (acetate + isovalerate)
S   Eval, Laxatol, *Phenovalin*
U   Laxative

9172   (7487) $C_{20}H_{14}O_6$
1821-16-5

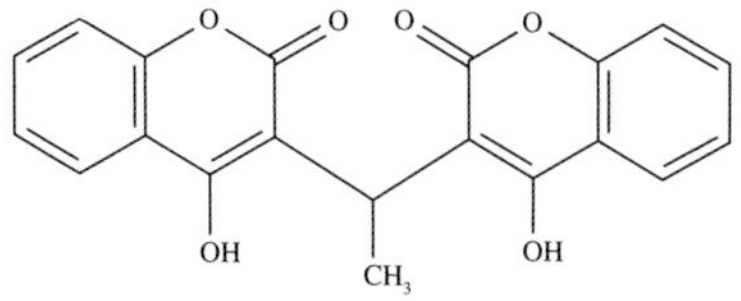

3,3'-Ethylidenebis(4-hydroxycoumarin) = 3,3'-Ethylidenebis[4-hydroxy-2*H*-1-benzopyran-2-one] (●)
S   E.D.C., *Ethylidene Dicoumarin, Ethylidenedicoumarol, Eticoumarolum*, Pertrombon, Thrombaton
U   Anticoagulant

9173 $C_{20}H_{14}O_7$
105037-92-1

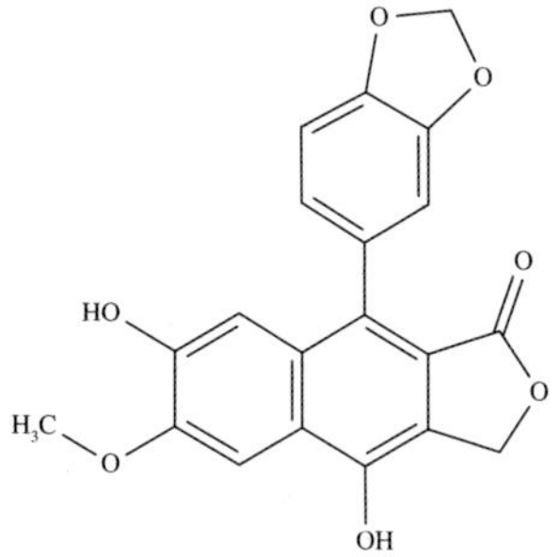

)-(1,3-Benzodioxol-5-yl)-4,7-dihydroxy-6-methoxy-naphtho[2,3-*c*]furan-1(3*H*)-one (●)
Haplomyrtin
U   Cytotoxic

9174 $C_{20}H_{14}O_7$
152607-03-9

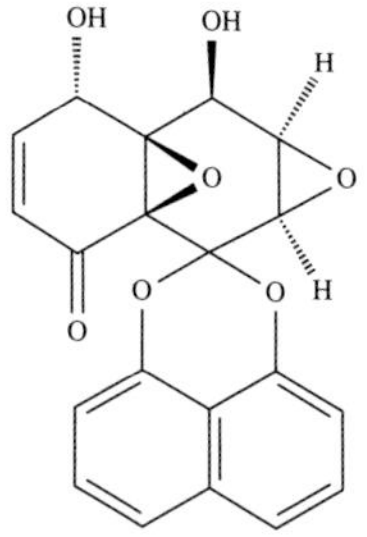

(1a*R*,2a*R*,6*S*,6a*S*,7*S*,7a*R*)-7,7a-Dihydro-6,7-dihydroxy-spiro[2a,6a-epoxynaphth[2,3-*b*]oxirene-2(1a*H*),2'-naphtho[1,8-*de*][1,3]dioxin]-3(6*H*)-one (●)
S   *Cladospirone bisepoxide*
U   Antibacterial, antifungal from a coelomycete

9175   (7488) $C_{20}H_{15}Br$
1607-57-4

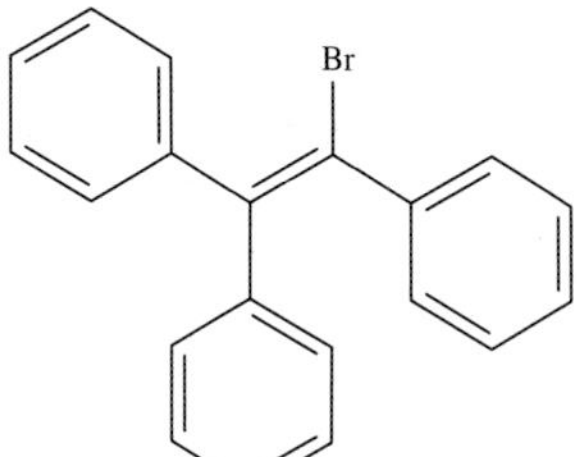

Bromotriphenylethene = Triphenylbromoethylene = Phenylstilbene bromide = 1,1',1''-(1-Bromo-1-ethenyl-2-ylidene)tris[benzene] (●)
S   Bromylène, *Fenbrostilbenum*, Oestronyl, *Phenylstilbene bromide*, Prostilban, Tribenorm
U   Synthetic estrogen

9176   (7489) $C_{20}H_{15}Cl$
18084-97-4

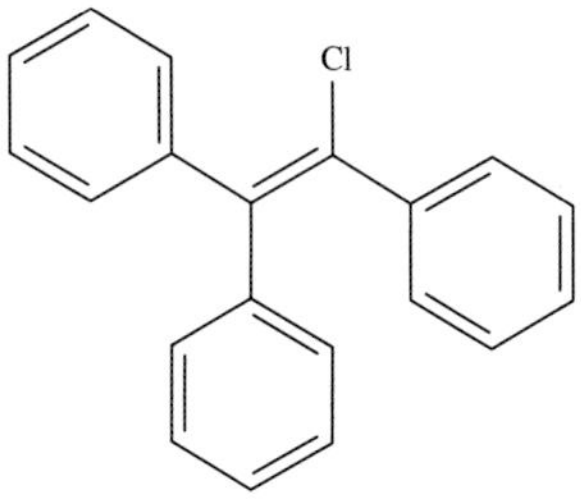

Chlorotriphenylethene = Triphenylchloroethylene = Phenylstilbene chloride = 1,1',1''-(1-Chloro-1-ethenyl-2-ylidene)tris[benzene] (●)
S Gynosone, Oestrogyl, *Phenylstilbene chloride*
U Synthetic estrogen

9177 $C_{20}H_{15}ClN_4$
212141-54-3

(4-Chlorophenyl)-[4-(4-pyridinylmethyl)-1-phthalazinyl]amine = 1-(*p*-Chloroanilino)-4-(4-pyridylmethyl)phthalazine = *N*-(4-Chlorophenyl)-4-(4-pyridinylmethyl)-1-phthalazinamine (●)
R also succinate (1:1) (212142-18-2)
S CGP 79787D, PTK 787, *Vatalanib**, ZK 222584
U Antineoplastic (VEGF receptor inhibitor)

9178   (7490) $C_{20}H_{15}Cl_3N_2OS$
99592-32-2

(±)-1-[2,4-Dichloro-β-[(7-chlorobenzo[*b*]thien-3-yl)methoxy]phenethyl]imidazole = 1-[2-[(7-Chlorobenzo[*b*]thien-3-yl)methoxy]-2-(2,4-dichlorophenyl)ethyl]-1*H*-imidazole (●)
R Mononitrate (99592-39-9)
S Dermofix, Dermoseptic, Extens, FI-7056, Fisderm, Monazol, Mykosert, *Sertaconazole nitrate***, Zalain
U Antifungal

9179   (7491) $C_{20}H_{15}Cl_3O_3$
6642-07-5

4-Chloro-2,6-bis(5-chloro-2-hydroxybenzyl)phenol = 4-Chloro-2,6-bis[(5-chloro-2-hydroxyphenyl)methyl]phenol (●)
S G-610, Trichlorophen
U Anthelmintic

9180 $C_{20}H_{15}F_3N_4O_3$
147059-72-1

7-[(1*R*,5*S*,6*s*)-6-Amino-3-azabicyclo[3.1.0]hex-3-yl]-1-(2,4-difluorophenyl)-6-fluoro-1,4-dihydro-4-oxo-1,8-naphthyridine-3-carboxylic acid = (1α,5α,6α)-7-(6-Amino-3-azabicyclo[3.1.0]hex-3-yl)-1-(2,4-difluorophenyl)-6-fluoro-1,4-dihydro-4-oxo-1,8-naphthyridine-3-carboxylic acid (●)
R Monomethansulfonate (147059-75-4); see also no. 13063
S CP-99219-27, *Trovafloxacin mesilate***, Trovan, Turvel
U Antibacterial

9181   (7492) $C_{20}H_{15}IN_2O_5S$

p'-(4-Iodophenylsulfamoyl)phthalanilic acid = *N*¹-(4-Iodophenyl)-*N*⁴-phthaloylsulfanilamide = 2-[[[4-[[(4-Iodo-

phenyl)amino]sulfonyl]phenyl]amino]carbonyl]benzoic
acid (●)
S Ftalil-Medeyol
U Antibacterial (intestinal)

9182 (7493)

$C_{20}H_{15}NO$
110033-17-5

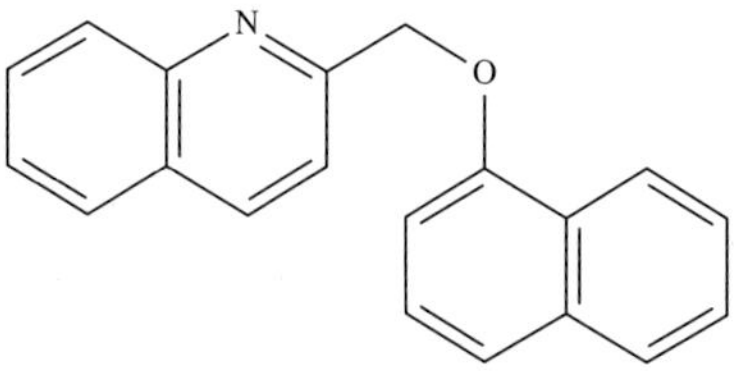

2-[(1-Naphthalenyloxy)methyl]quinoline (●)
S Wy-47288
U Anti-inflammatory (topical)

9183 (7494)

$C_{20}H_{15}NO_3$
125-13-3

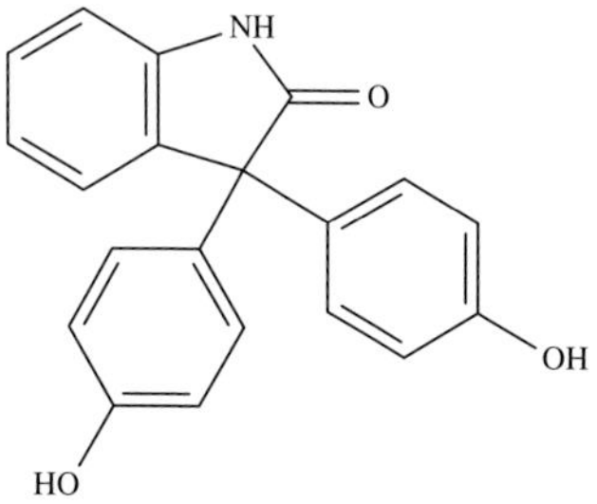

3,3-Bis(p-hydroxyphenyl)-2-indolinone = 3,3-Bis(p-
hydroxyphenyl)indol-2(3H)-one = 3,3-Bis(4-hydroxy-
phenyl)oxindole = 3,3-Bis-p-phenolisatin = 1,3-Dihydro-
3,3-bis(4-hydroxyphenyl)-2H-indol-2-one (●)
R see also nos. 12198, 13038, 14582, 15142, 15843
S Critex (old form), *Dioxyphenylisatin*, Hoscolax, Iso-
lax, Lavema, Med-Laxan-Suppos. (old form), Neodrast (old form), Normalax "Wampole", *Oxiphenisa-
tinum, Oxyphenisatin(e)***, PCL 243, *Phenolisatin*,
Propellax-R, Recolon, Veripaque
U Laxative

9184 (7495)

$C_{20}H_{15}NO_4$
17692-24-9

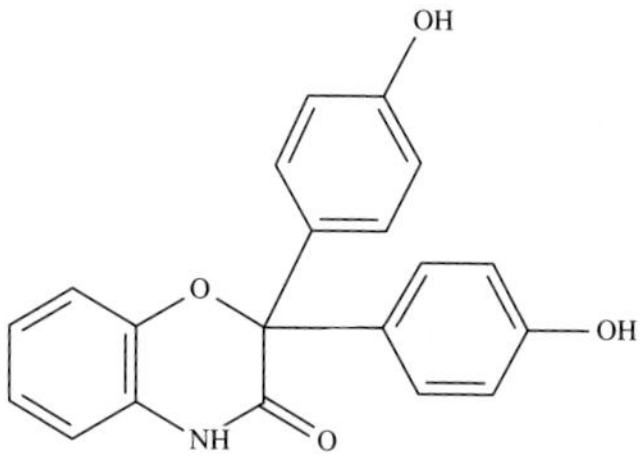

2,2-Bis(p-hydroxyphenyl)-2H-1,4-benzoxazin-3(4H)-
one = 2,2-Bis(4-hydroxyphenyl)-2H-1,4-benzoxazin-3-
(4H)-one (●)
R see also no. 12199
S *Bisoxatin***
U Laxative

9185

$C_{20}H_{15}N_3O_2$
157168-02-0

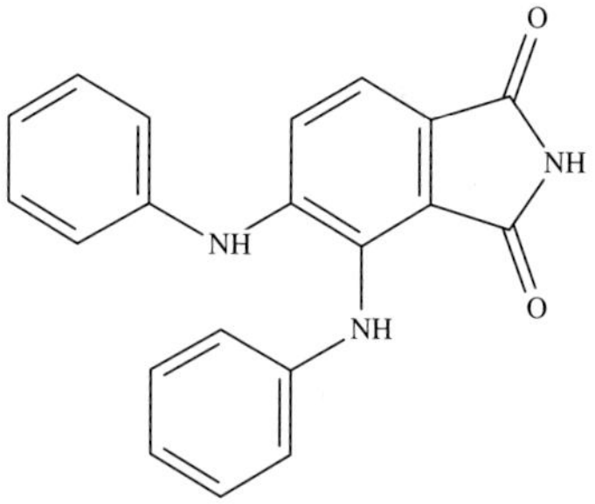

3,4-Dianilinophthalimide = 4,5-Bis(phenylamino)-1H-
isoindole-1,3(2H)-dione (●)
S CGP 52411
U Tyrosine kinase inhibitor

9186 (7496)

$C_{20}H_{15}N_3O_2$
60662-69-3

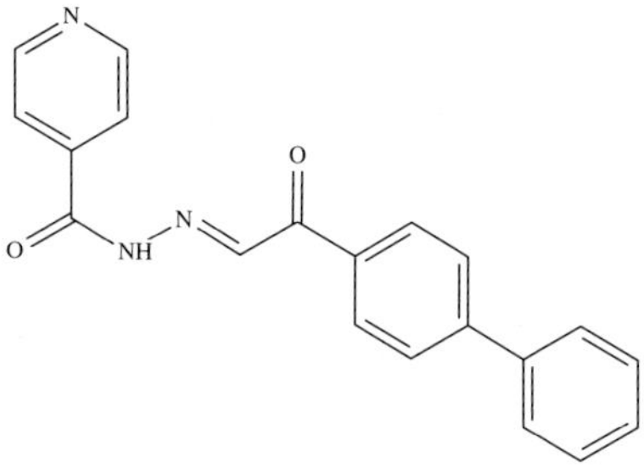

N^2-(p-Phenylphenacylidene)isonicotinic acid hydrazide = 4-Biphenylylglyoxal isonicotinoylhydrazone = 4-Pyridinecarboxylic acid (2-[1,1'-biphenyl]-4-yl-2-oxoethylidene)hydrazide (●)
S C.V. 58858, *Xenalazone*
U Antiviral

9187 (7497)

$C_{20}H_{15}N_3O_2S$
117038-09-2

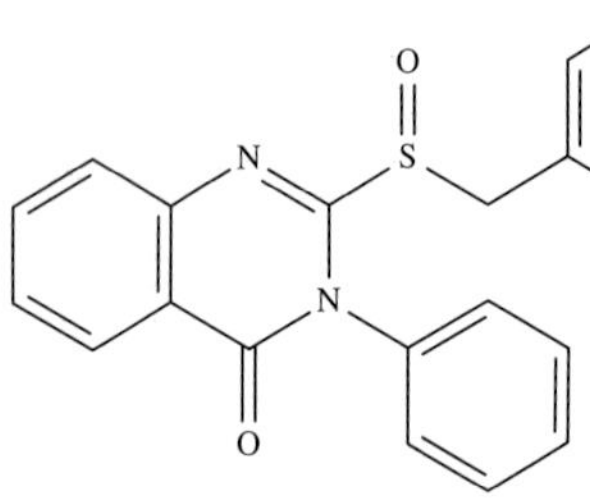

3-Phenyl-2-[(4-pyridinylmethyl)sulfinyl]-4(3H)-quinazolinone (●)
S N-2220
U Anti-ulcer agent

9188

$C_{20}H_{15}N_3O_6$
91421-42-0

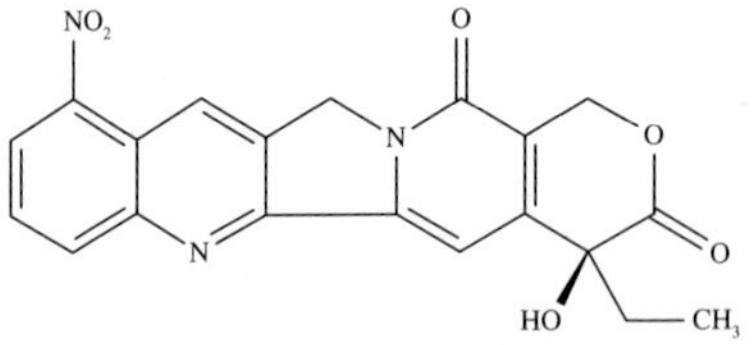

(4S)-4-Ethyl-4-hydroxy-10-nitro-1H-pyrano[3',4':6,7]indolizino[1,2-b]quinoline-3,14(4H,12H)-dione (●)
S Camptogen, 9 NC, *9-Nitrocamptothecin*, RFS 2000, *Rubitecan***
U Antineoplastic (topoisomerase inhibitor)

9189 (7498)

$C_{20}H_{15}N_5O_5$
53736-51-9

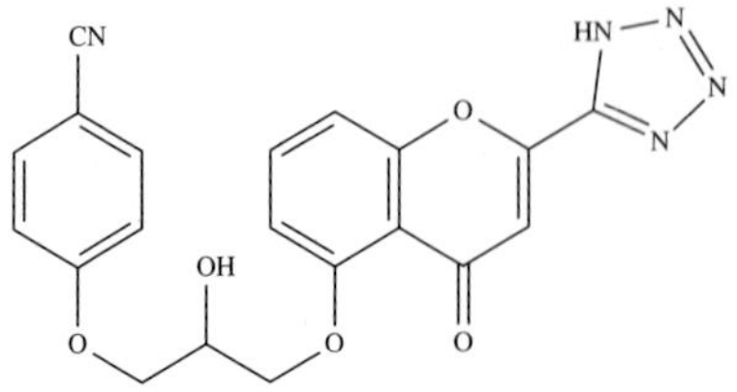

(±)-4-[2-Hydroxy-3-[[4-oxo-2-(1H-tetrazol-5-yl)-4H-1-benzopyran-5-yl]oxy]propoxy]benzonitrile (●)
R Monosodium salt (53736-52-0)
S *Cromitrile sodium***, TR-2855
U Anti-asthmatic

9190

$C_{20}H_{16}ClF_2N_5O_2$
187949-02-6

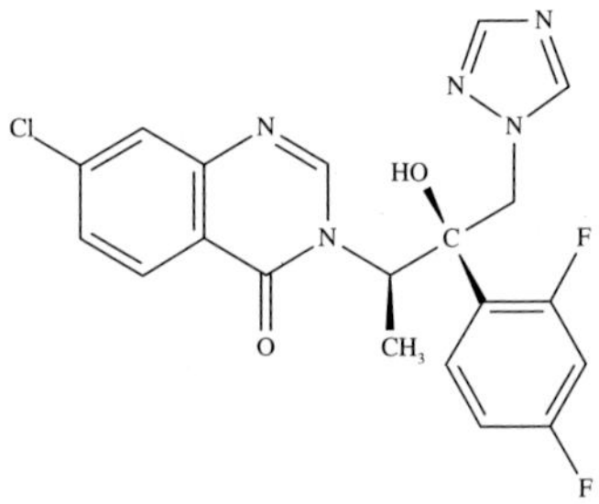

(1R,2R)-7-Chloro-3-[2-(2,4-difluorophenyl)-2-hydroxy-1-methyl-3-(1H-1,2,4-triazol-1-yl)propyl]-4(3H)-quinazolinone (●)
S *Albaconazole*, UR-9825
U Antifungal

9191 (7499)

$C_{20}H_{16}ClNO_3$
63608-11-7

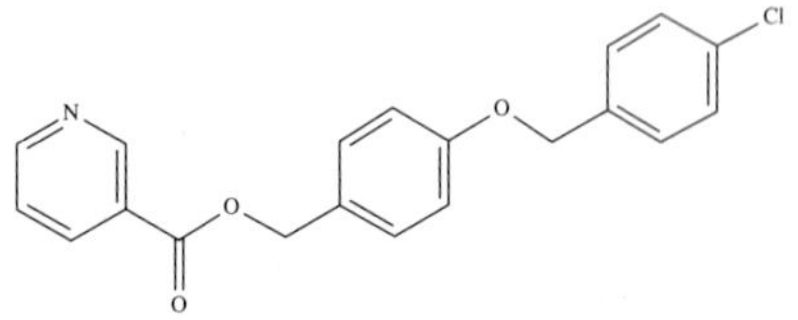

4-[(4-Chlorobenzyl)oxy]benzyl nicotinate = 3-Pyridinecarboxylic acid [4-[(4-chlorophenyl)methoxy]phenyl]methyl ester (●)
S KCD-232
U Antihyperlipidemic

9192　　　　　　　　　　　　　　　　$C_{20}H_{16}FN_3O_4$
　　　　　　　　　　　　　　　　　　111783-55-2

6-(3-Amino-1-pyrrolidinyl)-5-fluoro-3-oxo-3*H*-pyri-
do[3,2,1-*kl*]phenoxazine-2-carboxylic acid (●)
R　Monohydrochloride (111783-54-1)
S　A-62176
U　Antineoplastic (topoisomerase II inhibitor)

9193　(7500)　　　　　　　　　　　$C_{20}H_{16}F_3NS$
　　　　　　　　　　　　　　　　　　64301-63-9

N,N-Dimethyl-3-[8-(trifluoromethyl)dibenzo[*b,f*]thiepin-
10-yl]-2-propyn-1-amine (●)
R　Hydrochloride (64301-64-0)
S　Ro 11-7330
U　Antipsychotic

9194　(7501)　　　　　　　　　　　$C_{20}H_{16}F_3N_3O_3$
　　　　　　　　　　　　　　　　　　141725-88-4

(−)-7-[(2*S*,3*R*)-3-Amino-2-methyl-1-azetidinyl]-1-(2,4-
difluorophenyl)-6-fluoro-1,4-dihydro-4-oxo-3-quinoline-
carboxylic acid (●)
R　Tosylate (*p*-toluenesulfonate) (141725-89-5)
S　Ceflox, *Cetefloxacin tosilate***, E-4868
U　Antibacterial

9195　　　　　　　　　　　　　　　　$C_{20}H_{16}NO_4$
　　　　　　　　　　　　　　　　　　149998-48-1

7-Hydroxy-8-methoxy-5-methyl-2,3-(methylenedi-
oxy)benzo[*c*]phenanthridinium = 1-Hydroxy-2-methoxy-
12-methyl-[1,3]benzodioxolo[5,6-*c*]phenanthridinium
(●)
R　Sulfate (1:1) (143201-31-4)
S　NK-109
U　Antineoplastic (topoisomerase II inhibitor)

9196 $C_{20}H_{16}N_2O_2$
204717-44-2

1-(4-Methylphenyl)-9H-pyrido[3,4-b]indole-3-carb-
oxylic acid methyl ester (●)
S Compound 92/596
U Antifilarial

9197 (7502) $C_{20}H_{16}N_2O_4$
7689-03-4

(S)-4-Ethyl-4-hydroxy-1H-pyrano[3',4':6,7]indolizi-
no[1,2-b]quinoline-3,14(4H,12H)-dione (●)
R see also nos. 9249, 10927, 14755
S *Camptothecin*, CPT, NSC-94600, Prothecan
U Antineoplastic (alkaloid from *Camptotheca acumina-
ta*)

9198 (7503) $C_{20}H_{16}N_6O_4$
89224-56-6

3,7-Dimethoxy-4-phenyl-N-1H-tetrazol-5-yl-4H-fu-
ro[3,2-b]indole-2-carboxamide (●)

R Compd. with L-arginine (97958-08-2)
S CI-922, PD 106549
U Anti-allergic

9199 $C_{20}H_{16}O_5$
130548-09-3

(3R,4S)-3,4-Dihydro-4-hydroxy-8-methoxy-3-methyl-
benz[a]anthracene-1,7,12(2H)-trione (●)
S *Fujianmycin B, Rubiginone A$_2$*, SNA-8073-A
U Antibiotic, antileukemic

9200 $C_{20}H_{16}O_5$
157110-17-3

(S)-1,5,8-Trihydroxy-1,2,3,4-tetrahydronaphth-4-ylspiro-
1,8-dioxynaphthalene = (4S)-3,4-Dihydrospiro[naphtha-
lene-1(2H),2'-naphtho[1,8-de][1,3]dioxin]-4,5,8-triol (●)
S CJ 12372, *Palmarumycin CP1*
U Antibaterial from a fungus N 983-46

9201 (7504) $C_{20}H_{16}O_6$
16203-97-7

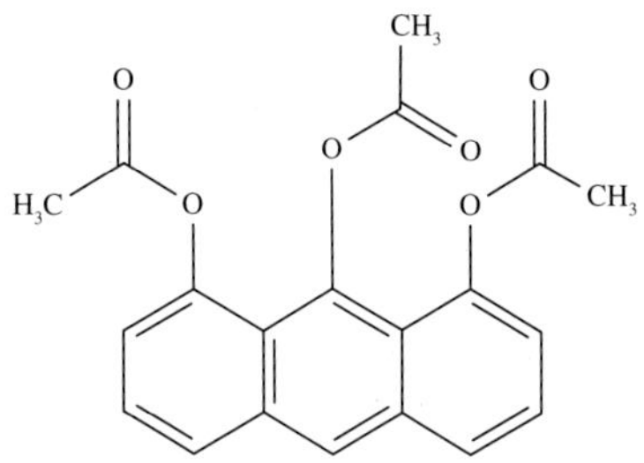

1,8,9-Anthratriyl triacetate = 1,8,9-Triacetoxyanthracene
= 1,8,9-Anthracenetriol triacetate (●)

S *Dithranol triacetate*, Exolan, Psoralex, *Triacetoxyan-thracene*

U Antipsoriatic

9202 (7505) $C_{20}H_{16}O_7$
98015-54-4

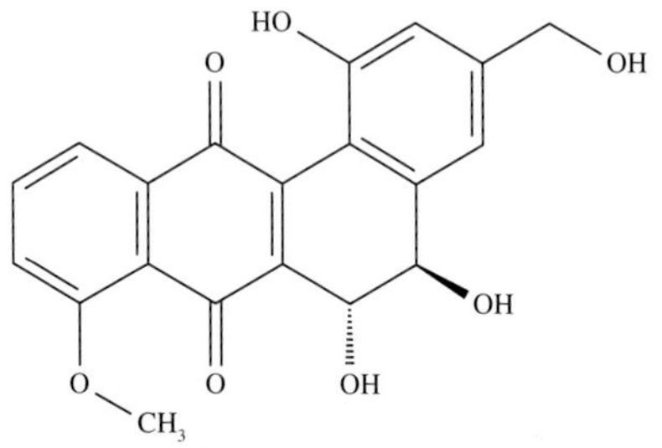

(+)-5,6-Dihydro-1,5β,6α-trihydroxy-3-(hydroxymethyl)-8-methoxybenz[a]anthraquinone = *trans*-(+)-5,6-Dihy-dro-1,5,6-trihydroxy-3-(hydroxymethyl)-8-methoxy-benz[a]anthracene-7,12-dione (●)

S PD 116740

U Antineoplastic

9203 $C_{20}H_{16}O_8$
186751-73-5

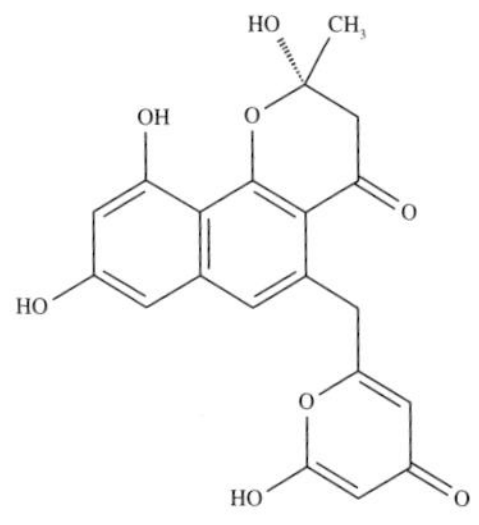

(2S)-2,3-Dihydro-2,8,10-trihydroxy-5-[(6-hydroxy-4-oxo-4H-pyran-2-yl)methyl]-2-methyl-4H-naph-tho[1,2-b]pyran-4-one (●)

S RM 80

U Antibacterial

9204 (7506) $C_{20}H_{17}Br_2N_3O_5$
105211-23-2

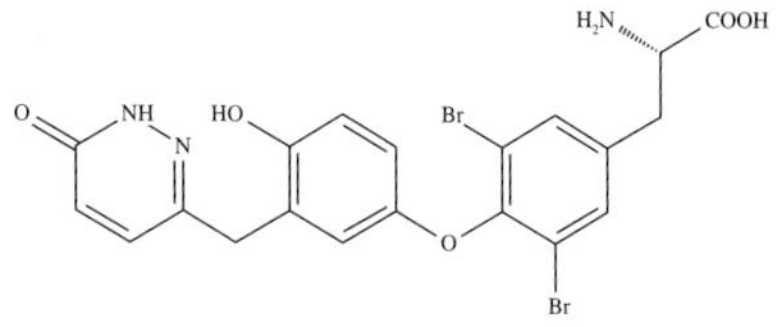

3,5-Dibromo-3'-[(1,6-dihydro-6-oxo-3-pyridazinyl)me-thyl]-L-thyronine = 3,5-Dibromo-O-[3-[(1,6-dihydro-6-oxo-3-pyridazinyl)methyl]-4-hydroxyphenyl]-L-tyrosine (●)

S L 94901, SKF L-94901

U Antihyperlipidemic, thyromimetic

9205 (7507) $C_{20}H_{17}ClFN_3O_4$
65400-85-3

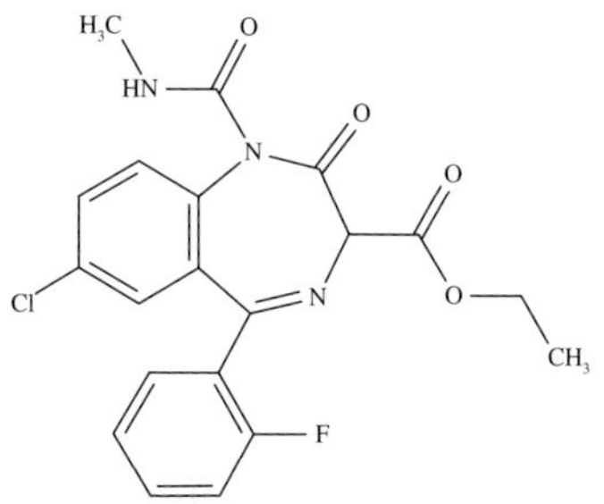

Ethyl 7-chloro-5-(o-chlorophenyl)-2,3-dihydro-1-(me-thylcarbamoyl)-2-oxo-1H-1,4-benzodiazepine-3-carb-oxylate = 7-Chloro-5-(2-fluorophenyl)-2,3-dihydro-1-[(methylamino)carbonyl]-2-oxo-1H-1,4-benzodiazepine-3-carboxylic acid ethyl ester (●)

S CM 7120, *Ethyl carfluzepate**, Ethylis carfluze-pas***

U Anxiolytic, sedative

9206   (7508) $C_{20}H_{17}ClN_2O_3$
                                           27223-35-4

11-Chloro-8,12b-dihydro-2,8-dimethyl-12b-phenyl-4*H*-
[1,3]oxazino[3,2-*d*][1,4]benzodiazepine-4,7(6*H*)-dione
(●)
S   Anseren, Ansieten, Ansietil, Anxon, Contamex
      "SmithKline Beecham; Wülfing", Grifoketam, *Keta-
      zolam***, Larpaz, Loftran, Marcen, Parcil, Sedatival
      "Pharmalab; Recalcine", Sedotime, Solatran,
      U-28774, Unakalm
U   Tranquilizer

9207   (7509) $C_{20}H_{17}Cl_3N_2O_2$
                                           74512-12-2

(*Z*)-1-[2,4-Dichloro-β-[2-(*p*-chlorophenoxy)ethoxy]-α-
methylstyryl]imidazole = (*Z*)-1-[2-[2-(4-Chlorophen-
oxy)ethoxy]-2-(2,4-dichlorophenyl)-1-methylvinyl]imi-
dazole = (*Z*)-1-(2,4-Dichlorophenyl)-1-[2-(4-chlorophen-
oxy)ethoxy]-2-(1-imidazolyl)propene = (*Z*)-1-[2-[2-(4-
Chlorophenoxy)ethoxy]-2-(2,4-dichlorophenyl)-1-me-
thylethenyl]-1*H*-imidazole (●)
R   Mononitrate (83621-06-1)
S   10 80 07, Afongan "Alcon; Galderma", Azameno,
      CM 8282, Fangorex, Fongamil, Fongarex, Fungisan
      "Galderma", Hoe 155, Melur "Siegfried", *Omocona-
      zole nitrate***, Omoderm, Omotrex, OMZ,
      Sgd 12878
U   Antifungal

9208   (7510) $C_{20}H_{17}Cl_3O_5$
                                           10089-10-8

(*E*)-2,4,7-Trichloro-3-hydroxy-8-methoxy-1,9-dimethyl-
6-(1-methyl-1-propenyl)-11*H*-dibenzo[*b,e*][1,4]dioxe-
pin-11-one (●)
S   Methylustin, *Nidulin*
U   Antifungal antibiotic from *Aspergillus nidulans*

9209   (7511) $C_{20}H_{17}FN_2O_3$
                                           123942-04-1

1-Cyclopropyl-7-(2,6-dimethyl-4-pyridinyl)-6-fluoro-
1,4-dihydro-4-oxo-3-quinolinecarboxylic acid (●)
S   Win 57273
U   Antibacterial (leprostatic)

9210 $C_{20}H_{17}FN_2O_3S$
                                           122033-48-1

(3*S*)-10-(2,6-Dimethyl-4-pyridinyl)-9-fluoro-2,3-dihy-
dro-3-methyl-7-oxo-7*H*-pyrido[1,2,3-*de*]-1,4-benzothia-
zine-6-carboxylic acid (●)

S Win 58161
U Antineoplastic (topoisomerase inhibitor)

9211 (7512) $C_{20}H_{17}FO_3S$
                                                 38194-50-2

(*Z*)-5-Fluoro-2-methyl-1-[*p*-(methylsulfinyl)benzyli-
dene]indene-3-acetic acid = (*Z*)-5-Fluoro-2-methyl-1-[[4-
(methylsulfinyl)phenyl]methylene]-1*H*-indene-3-acetic
acid (●)
R also sodium salt (63804-15-9)
S Aclin, Aflodac, Algocetil, Apo-Sulin, Arthridex, Ar-
throcine, Artribid, Artrobid, Biflace, Citireuma, Cli-
nax, Clinoril, Clisundac, Clusinol, Copal, Dorindac,
Etusapol, Flusoril, Imbaral, Kenalin, Klimacobal,
Klinoril, Leskosul, Lindac, Lyndak, MK-231, Mobi-
lin, MSD-943, Nodituss, Norilafin, Novosundac,
Nu-Sulindac, Overpon, Ratimazol, Reumofil, Reu-
myl "Lenza", R-Flex, Saldac, Skanoorin, Slidanil,
Sudac, Sukurinoru, Sulartrene, Sulen(e), Sulic, *Sulin-
dac***, Sulindal, Sulindor, Sulinid, Sulinol, Sulin-
pen, Sulreuma, Sutalan, Tinagnol, Udolac "Norma
Hellas", Vulbenol, Zirofalen
U Anti-inflammatory, antipyretic, analgesic

9212 $C_{20}H_{17}FO_4S$
                                               59973-80-7

5-Fluoro-2-methyl-1-[(*Z*)-*p*-(methylsulfonyl)benzyli-
dene]indene-3-acetic acid = (1*Z*)-5-Fluoro-2-methyl-1-

[[4-(methylsulfonyl)phenyl]methylene]-1*H*-indene-3-
acetic acid (●)
S *Aposulind*, Aptosyn, *Exisulind***, FGN-1, Prevatac,
Sulindac sulfone
U Antineoplastic

9213 (7513) $C_{20}H_{17}F_2N_3O_3$
                                               98105-99-8

6-Fluoro-1-(4-fluorophenyl)-1,4-dihydro-4-oxo-7-(1-pi-
perazinyl)-3-quinolinecarboxylic acid (●)
R Monohydrochloride (91296-87-6)
S A-56620, Abbott-56620, Saraflox, *Sarafloxacin hy-
drochloride***
U Antibacterial (veterinary)

9214 (7514) $C_{20}H_{17}F_3N_2O_4$
                                               23779-99-9

2,3-Dihydroxypropyl *N*-[8-(trifluoromethyl)-4-quino-
lyl]anthranilate = *N*-[8-(Trifluoromethyl)-4-quinolyl]an-
thranilic acid 2,3-dihydroxypropyl ester = 2-[[8-(Trifluoromethyl)-4-quinolinyl]amino]benzoic acid 2,3-dihydr-
oxypropyl ester (●)
S Diralgan, Duralgin, *Floctafenine***, Floktin, Glifa-
nil, Glifax, Idalon, Idarac, Novodolan, R 4318,
RU 15750
U Analgesic

9215 $C_{20}H_{17}F_3N_2O_4S$
213252-19-8

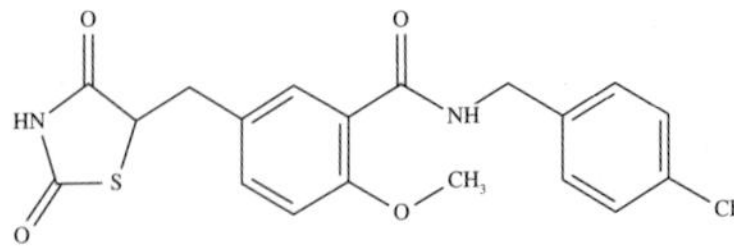

5-[[4-Methoxy-3-[[[[4-(trifluoromethyl)phenyl]me-
thyl]amino]carbonyl]phenyl]methyl]-2,4-thiazolidine-
dione = 5-[(2,4-Dioxo-5-thiazolidinyl)methyl]-2-meth-
oxy-N-[[4-(trifluoromethyl)phenyl]methyl]benzamide
(●)
S   KRP-297
U   Hypoglycemic

9216   (7515) $C_{20}H_{17}NO_2$
5443-63-0

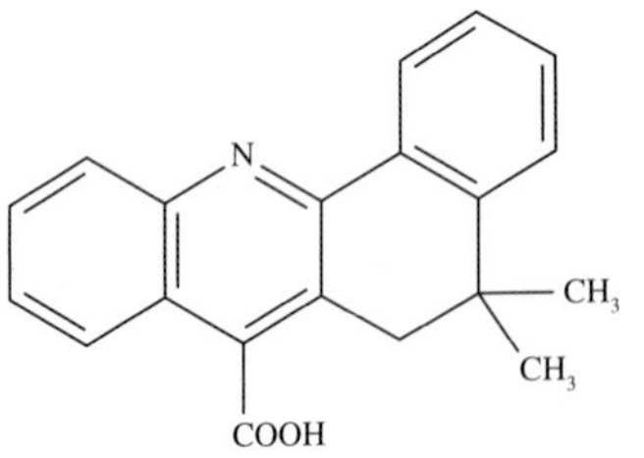

5,6-Dihydro-5,5-dimethylbenz[c]acridine-7-carboxylic
acid (●)
S   "Benzacridine"
U   Anti-arrhythmic

9217 $C_{20}H_{17}NO_5$
549-21-3

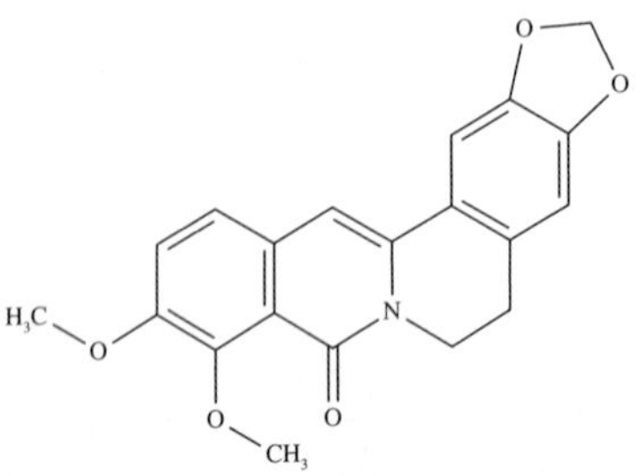

13,13a-Didehydro-9,10-dimethoxy-2,3-(methylenedioxy)berbin-8-one = 8-Oxoberberine = 5,6-Dihydro-9,10-
dimethoxy-8H-benzo[g]-1,3-benzodioxolo[5,6-a]quinoli-
zin-8-one (●)
S   JKL 1073A
U   Cardiotonic (positive inotropic)

9218 $C_{20}H_{17}NO_5$
5574-24-3

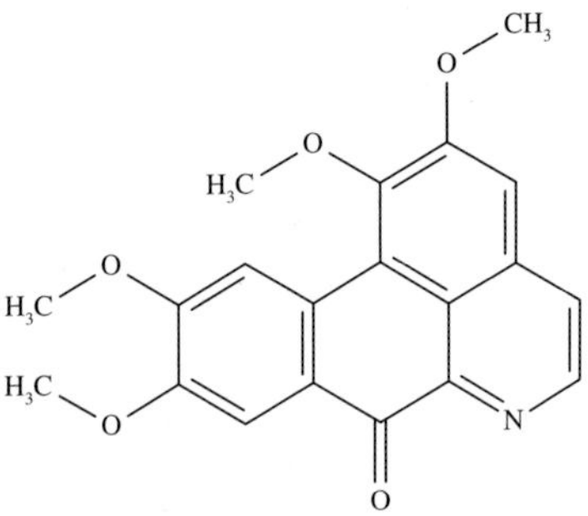

4,5,6,6a-Tetradehydro-1,2,9,10-tetramethoxynorapor-
phin-7-one = 1,2,9,10-Tetramethoxy-7H-diben-
zo[de,g]quinolin-7-one (●)
S   O-Methylatheroline, Oxoglaucine
U   Immunosuppressive

9219   (7517) $C_{20}H_{17}NO_6$
17298-36-1

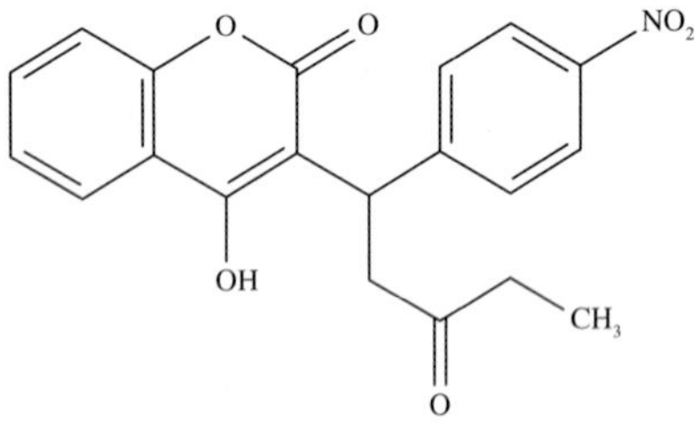

4-Hydroxy-3-[p-nitro-α-(2-oxobutyl)benzyl]coumarin =
4-Hydroxy-3-[1-(p-nitrophenyl)-3-oxopentyl]coumarin =
4-Hydroxy-3-[1-(4-nitrophenyl)-3-oxopentyl]-2H-1-ben-
zopyran-2-one (●)
S   Nitrofarin, Nitropharin
U   Anticoagulant

9220 $C_{20}H_{17}NO$
485-49-

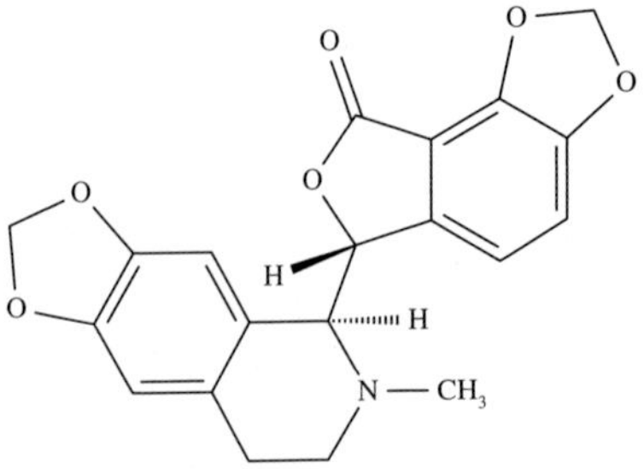

(6R)-6-[(5S)-5,6,7,8-Tetrahydro-6-methyl-1,3-di-oxolo[4,5-g]isoquinolin-5-yl]furo[3,4-e]-1,3-benzodi-oxol-8(6H)-one (●)
S *Bicuculline*
U $GABA_A$ antagonist (alkaloid from many plants)

9221 (7516)

$C_{20}H_{17}NO_6$
21722-66-7

5-Methoxy-2-methyl-1-(3,4-methylenedioxybenzoyl)-3-indolylacetic acid = 1-[(1,3-Benzodioxol-5-yl)carbonyl]-5-methoxy-2-methyl-1H-indole-3-acetic acid (●)
S ID-955
U Anti-inflammatory

9222 (7518)

$C_{20}H_{17}N_3O_2S$
53478-38-9

'-(9-Acridinylamino)methanesulfonanilide = N-[4-(9-Acridinylamino)phenyl]methanesulfonamide (●)
S AMSA
U Antineoplastic

9223

$C_{20}H_{17}N_3O_4$
91421-43-1

(4S)-10-Amino-4-ethyl-4-hydroxy-1H-pyra-no[3',4':6,7]indolizino[1,2-b]quinoline-3,14(4H,12H)-dione (●)
S 9-AC, *9-Aminocamptothecin*, IDEC-132, NSC-603071
U Antineoplastic

9224

$C_{20}H_{17}N_3O_4S$
199113-98-9

5-[[4-[(3,4-Dihydro-3-methyl-4-oxo-2-quinazoli-nyl)methoxy]phenyl]methyl]-2,4-thiazolidinedione (●)
R Potassium salt (199114-17-5)
S *Balaglitazone potassium**, DRF-2593
U Antidiabetic (insulin sensitizer)

9225 (7519)

$C_{20}H_{17}N_5O_9S_2$
117211-03-7

(6R,7R)-7-[2-(2-Amino-4-thiazolyl)glyoxylamido]-8-oxo-5-thia-1-azabicyclo[4.2.0]oct-2-ene-2-carboxylic acid 7^2-(Z)-[O-[(S)-α-carboxy-3,4-dihydroxyben-zyl]oxime] = (7R)-7-[2-(2-Amino-4-thiazolyl)-2-[[[(Z)-(S)-α-carboxy-3,4-dihydroxybenzyl]oxy]imino]acetami-do]-3-cephem-4-carboxylic acid = [6R-[6α,7β[Z(S*)]]]-7-[[(2-Amino-4-thiazolyl)[[carboxy(3,4-dihydroxyphe-nyl)methoxy]imino]acetyl]amino]-8-oxo-5-thia-1-azabi-cyclo[4.2.0]oct-2-ene-2-carboxylic acid (●)
R also tetrahydrate (127182-67-6)
S *Cefetecol***, GR 69153 X
U Antibiotic

9226　(7520)　　　　　　　　　　　$C_{20}H_{18}BrClN_4S$
58765-21-2

2-Bromo-4-(2-chlorophenyl)-9-cyclohexyl-6*H*-thie-
no[3,2-*f*][1,2,4]triazolo[4,3-*a*][1,4]diazepine (●)
S　*Ciclotizolam***, We 973-BS
U　Anxiolytic

9227　(7521)　　　　　　　　　　　$C_{20}H_{18}Br_2N_4$
113389-11-0

4,9-Dibromo-6-(4-methyl-1-piperazinyl)benzo[*b*]pyrro-
lo[3,2,1-*jk*][1,4]benzodiazepine (●)
S　HP 370
U　Neuroleptic, antipsychotic

9228　　　　　　　　　　　　　　　$C_{20}H_{18}ClN_5O_2$
144301-94-0

N-(4-Chlorophenyl)-*N*'-cyano-*N*''-[(3*S*,4*R*)-6-cyano-3,4-
dihydro-3-hydroxy-2,2-dimethyl-2*H*-1-benzopyran-4-
yl]guanidine (●)
S　BMS 180448

U　Potassium channel opener, cardioprotectant

9229　(7522)　　　　　　　　　　　$C_{20}H_{18}Cl_2N_2O_3$
57773-81-6

7-Chloro-5-(*o*-chlorophenyl)-1,3-dihydro-3-(pivaloyloxy)-2*H*-1,4-benzodiazepin-2-one = 2,2-Dimethylpropa-
noic acid 7-chloro-5-(2-chlorophenyl)-2,3-dihydro-2-
oxo-1*H*-1,4-benzodiazepin-3-yl ester (●)
S　Divial, Drupal, *Lorazepam pivalate*, Piralone, Pla-
cinoral
U　Tranquilizer

9230　(7523)　　　　　　　　　　　$C_{20}H_{18}Cl_2N_2O_6$
14399-14-5

(−)-*threo*-1-(*p*-Nitrophenyl)-2-(dichloroacetamido)-1,3-
propanediol 3-cinnamate = 3-Phenyl-3-propenoic acid [*R*
(*R**,*R**)]-2-[(dichloroacetyl)amino]-3-hydroxy-3-(4-ni-
trophenyl)propyl ester (●)
S　Biclomicil Jarabe, *Chloramphenicol monocinnama-
te*, Cloromilen, Farmicetina Sciroppo, Ismicetina
Cinnamato, Tramina liquida
U　Antibiotic

9231   (7524) $C_{20}H_{18}FN_3O_5$
110230-93-8

5-(4-Fluorophenyl)-2,3-dihydro-7-nitro-2-oxo-1*H*-1,4-
benzodiazepine-1-pentanoic acid (●)
S   RU 44502
U   Benzodiazepine receptor antagonist

9232 $C_{20}H_{18}FN_5O_2$
150867-88-2

3-(3-Cyclopropyl-5-isoxazolyl)-6-fluoro-5-(4-morpholi-
nyl)imidazo[1,5-*a*]quinazoline (●)
S   NNC 14-0185
U   Anticonvulsant (BDZ receptor partial agonist)

9233 $C_{20}H_{18}F_2N_4O_3$
164150-99-6

6-Fluoro-1-(5-fluoro-2-pyridinyl)-1,4-dihydro-7-(4-me-
thyl-1-piperazinyl)-4-oxo-3-quinolinecarboxylic acid (●)

R   Monohydrochloride (164150-85-0)
S   DW-116, *Fandofloxacin hydrochloride***
U   Antibacterial

9234   (7525) $C_{20}H_{18}F_3NO_4$
124916-54-7

(3*S-trans*)-2-[3,4-Dihydro-3-hydroxy-2,2-dimethyl-6-
(trifluoromethoxy)-2*H*-1-benzopyran-4-yl]-2,3-dihydro-
1*H*-isoindol-1-one (●)
S   *Celikalim*, WAY-120491
U   Smooth muscle relaxant (potassium channel activator)

9235   (7526) $C_{20}H_{18}F_3N_3O_4S$
105504-93-6

4-Hydroxy-2-[1-(1-oxopropoxy)propyl]-*N*-2-thiazolyl-8-
(trifluoromethyl)-3-quinolinecarboxamide (●)
R   also with (±)-definition (114351-00-7)
S   RU 43526
U   Anti-inflammatory

9236　　　　　　　　　　　　　$C_{20}H_{18}FeO_2$
75862-42-9

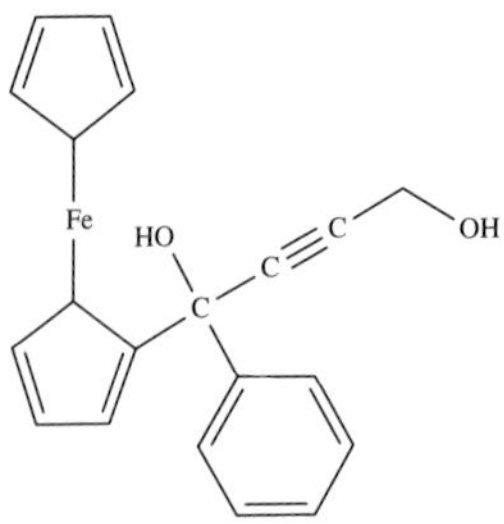

1-Ferrocenyl-1-phenyl-2-butyne-1,4-diol = (1,4-Dihydroxy-1-phenyl-2-butynyl)ferrocene (●)
S　Ferrocene-A
U　Antineoplastic, antibacterial

9237　(7527)　　　　　　　　　　$C_{20}H_{18}NO_4$
2086-83-1

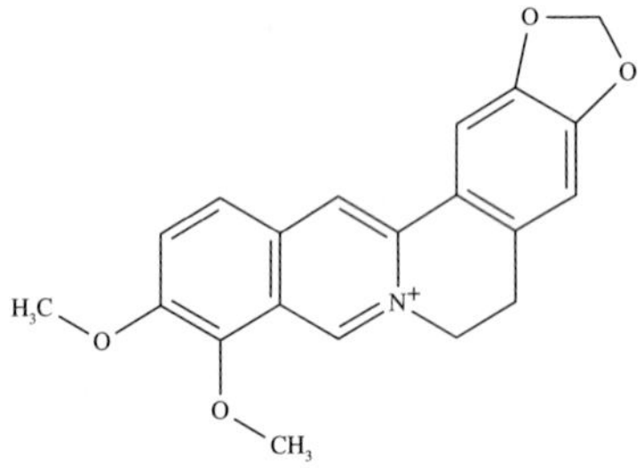

5,6-Dihydro-9,10-dimethoxybenzo[g]-1,3-benzodioxolo[5,6-a]quinolizinium (●)
S　*Berberine, Umbellatine*
U　Stomachic, antibacterial, tonic (alkaloid from *Hydrastis canadensis* and many other *Berberidaceae*)

9237-01　(7527-01)　　　　　　　　633-65-8
R　Chloride
S　Dirin, Kyoberin, Phelloberin, Pool Eyes, T-Up, Wakamatu
U　Antidiarrheal, antibacterial, stomachic

9237-02　(7527-02)　　　　　　　　316-41-6
R　Sulfate (1:1)
S　Berberal, Berisol, NSC-5355, Stogerin-Amp., Stopnin
U　Antidiarrheal, antibacterial, stomachic

9237-03　(7527-03)　　　　　　　　79236-58-1
R　Citrate (1:1)
S　Enterin "Kyoto"

U　Antidiarrheal, antibacterial, stomachic

9237-04　(7527-04)
R　Tannate
S　Erben-Powd., Tanberin
U　Antidiarrheal, antibacterial

9238　(7528)　　　　　　　　　　$C_{20}H_{18}N_2$

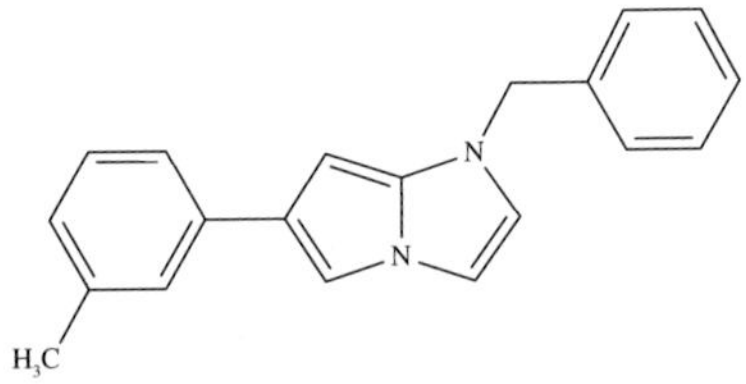

1-Benzyl-6-*m*-tolylpyrrolo[1,2-a]imidazole = 6-(3-Methylphenyl)-1-(phenylmethyl)-1*H*-pyrrolo[1,2-a]imidazole (●)
S　Tolbizol
U　Serotonin antagonist

9239　(7529)　　　　　　　　　　$C_{20}H_{18}N_2O$
553-06-0

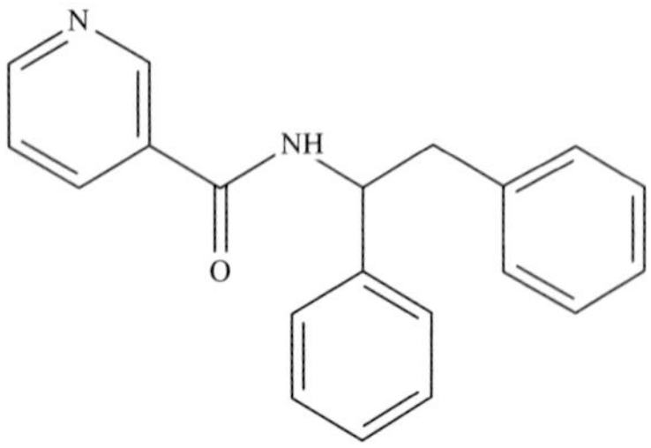

N-(1,2-Diphenylethyl)nicotinamide = Nicotinic acid 1,2-diphenylethylamide = N-(1,2-Diphenylethyl)-3-pyridinecarboxamide (●)
S　C 1065, *Diphénéthyl-nicotinamide*, Lispamina, Lyspamin, *Nicofetamide*
U　Spasmolytic

9240   (7530)

$C_{20}H_{18}N_2O$
17019-04-4

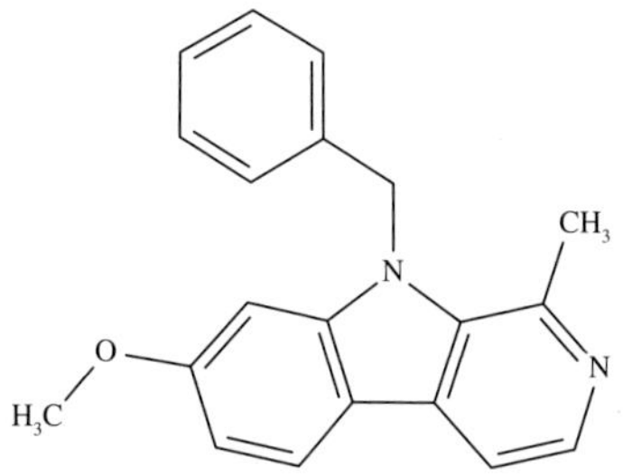

9-Benzyl-7-methoxy-1-methyl-β-carboline = 7-Methoxy-1-methyl-9-(phenylmethyl)-9H-pyrido[3,4-b]indole (●)
R   Monohydrochloride (1679-77-2)
S   *Benzylharmine*
U   Spasmolytic, vasodilator (alkaloid from *Peganum harmala*)

9241

$C_{20}H_{18}N_2O_2$
149398-59-4

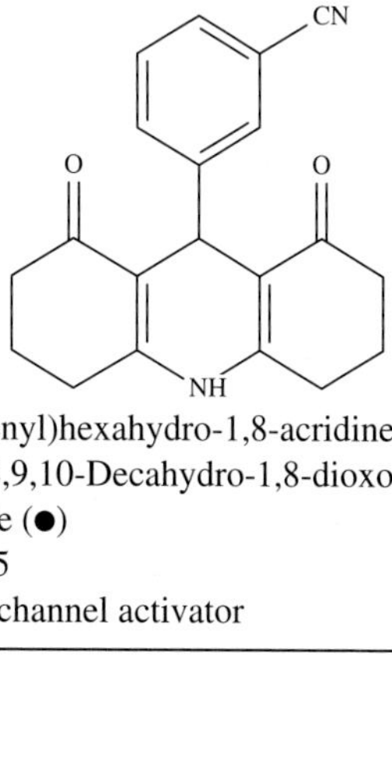

9-(3-Cyanophenyl)hexahydro-1,8-acridinedione = 3-(1,2,3,4,5,6,7,8,9,10-Decahydro-1,8-dioxo-9-acridinyl)benzonitrile (●)
S   ZM 244085
U   Potassium channel activator

9242

$C_{20}H_{18}N_2O_2$
140917-67-5

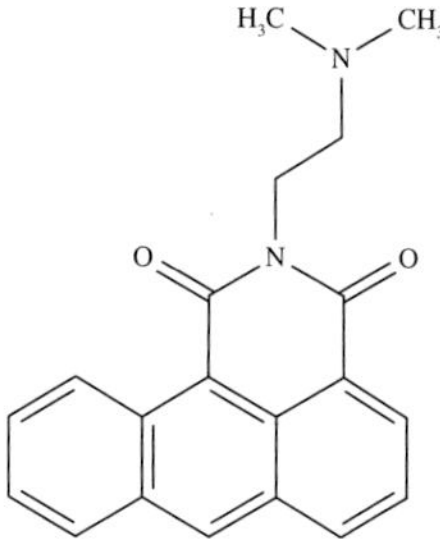

2-[2-(Dimethylamino)ethyl]-1H-dibenz[de,h]isoquinoline-1,3(2H)-dione (●)
S   *Azonafide*
U   Antineoplastic

9243   (7531)

$C_{20}H_{18}N_2O_3$
477-80-5

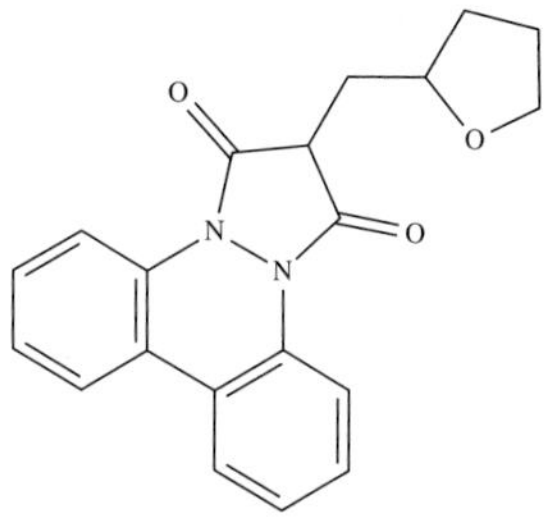

2-Tetrahydrofurfuryl-1H-benzo[c]pyrazolo[1,2-a]cinnoline-1,3(2H)-dione = 4-Tetrahydrofurfuryl-1,2-(benzo[c]cinnolino)-3,5-pyrazolidinedione = 1,2-Dihydro-1,2-(2-tetrahydrofurfuryl-1,3-dioxotrimethylene)benzo[c]cinnoline = 2-[(Tetrahydro-2-furanyl)methyl]-1H-benzo[c]pyrazolo[1,2-a]cinnoline-1,3(2H)-dione (●)
S   *Cinnofuradione***, *Cinnofuronum*
U   Analgesic

9244   (7532)

$C_{20}H_{18}N_2O_3$
148317-76-4

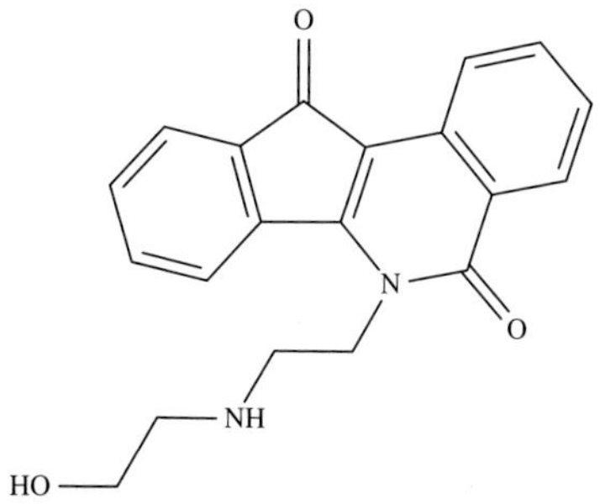

2-[2-[(5,11-Dioxo-5,6-dihydro-11H-indeno[1,2-c]isoquinolin-6-yl)ethyl]amino]ethanol = 6-[2-[(2-Hydroxyethyl)amino]ethyl]-5H-indeno[1,2-c]isoquinoline-5,11(6H)-dione (●)
R   Monohydrochloride (148296-83-7)
S   Oracin (czech. prep.), VÚFB-16956
U   Antineoplastic

9245 (7533) $C_{20}H_{18}N_2O_3$
112636-17-6

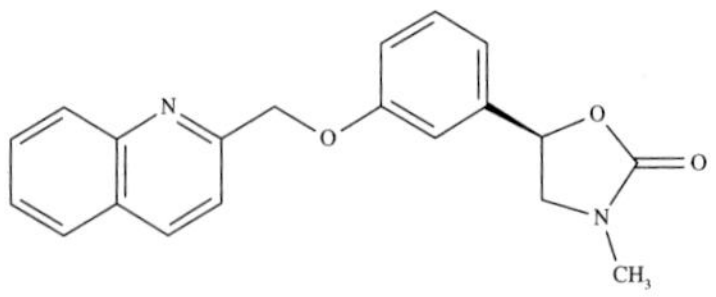

(*R*)-3-Methyl-5-[3-(2-quinolinylmethoxy)phenyl]-2-oxazolidinone (●)
S Wy-47674
U Anti-allergic

9246 $C_{20}H_{18}N_2O_3S$
195604-21-8

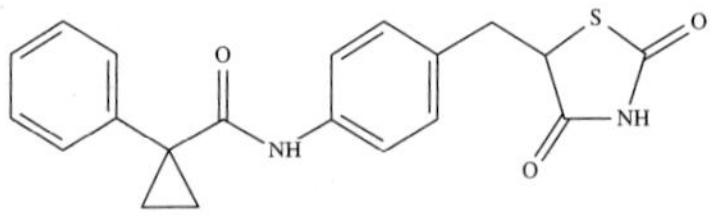

5-[4-(1-Phenyl-1-cyclopropanecarbonylamino)benzyl]-2,4-thiazolidinedione = *N*-[4-[(2,4-Dioxo-5-thiazolidinyl)methyl]phenyl]-1-phenylcyclopropanecarboxamide (●)
S DN-108
U Hypoglycemic

9247 $C_{20}H_{18}N_2O_3S$
172647-53-9

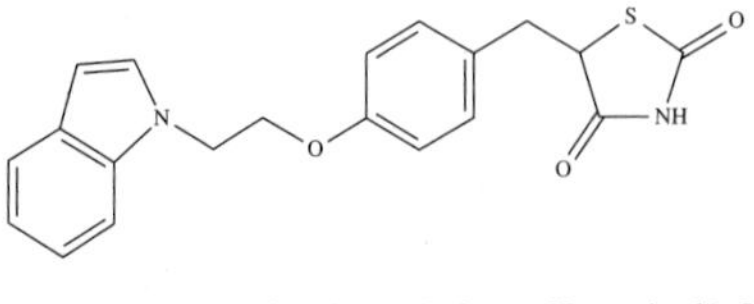

5-[[4-[2-(1*H*-Indol-1-yl)ethoxy]phenyl]methyl]-2,4-thiazolidinedione (●)
S DRF-2189
U Hypoglycemic (insulin sensitizer)

9248 (7534) $C_{20}H_{18}N_2O_4S$
107447-71-2

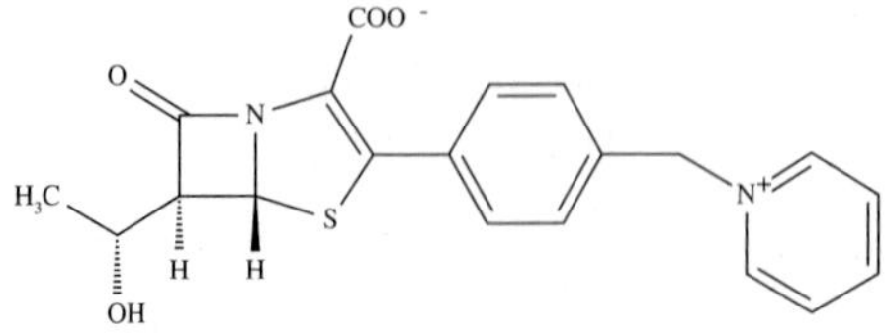

6-(1-Hydroxyethyl)-2-[4-(pyridiniomethyl)phenyl]-2-penem-3-carboxylate = [5*R*-[5α,6α(*R**)]]-1-[[4-[2-Carboxy-6-(1-hydroxyethyl)-7-oxo-4-thia-1-azabicyclo[3.2.0]hept-2-en-3-yl]phenyl]methyl]pyridinium hydroxide inner salt (●)
S FCE 24362
U Antibiotic

9249 (7535) $C_{20}H_{18}N_2O_5$
34079-22-6

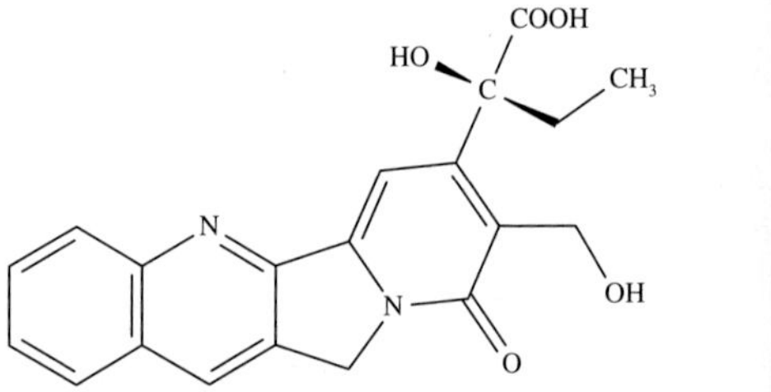

21,22-Secocamptothecin-21-oic acid = (*S*)-α-Ethyl-9,11-dihydro-α-hydroxy-8-(hydroxymethyl)-9-oxoindolizino[1,2-*b*]quinoline-7-acetic acid (●)
R Monosodium salt (25387-67-1)
S *Camptothecin Sodium*, NSC-100880
U Antineoplastic (alkaloid from *Camptotheca acuminata*)

9250 (7536) $C_{20}H_{18}N_2O_6$
94985-29-2

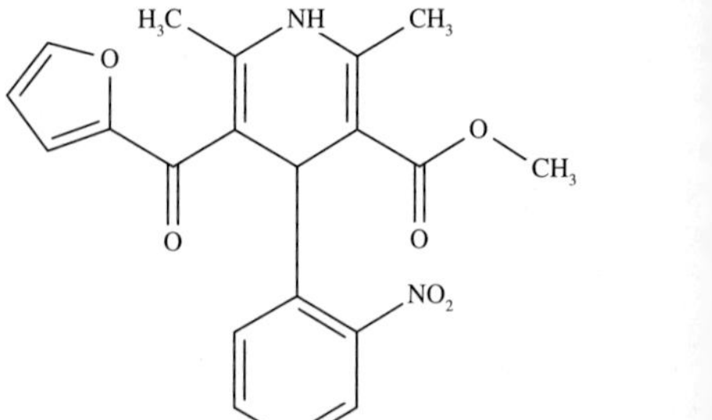

Methyl 5-(2-furoyl)-1,4-dihydro-2,6-dimethyl-4-(*o*-nitrophenyl)-3-pyridinecarboxylate = 5-(2-Furanylcarbonyl)-1,4-dihydro-2,6-dimethyl-4-(2-nitrophenyl)-3-pyridinecarboxylic acid methyl ester (●)
S MDL 72567
U Coronary vasodilator (calcium antagonist)

9251 (7537) $C_{20}H_{18}N_2O_7S_2$
                                                    19885-51-9

[5S-(5α,5aα,7aβ,13α,13aα,15aβ)]-5-(Acetyloxy)-
5,5a,13,13a-tetrahydro-13-hydroxy-8H,16H-7a,15a-epi-
dithio-7H,15H-bisoxepino[3',4':4,5]pyrrolo[1,2-a:1',2'-
d]pyrazine-7,15-dione (●)
S Aranotin**, Lilly 53183
U Antiviral from Arachniotus aureus

9252 (7538) $C_{20}H_{18}N_2S$
                                                    65509-66-2

2,3,4,5-Tetrahydro-3-methyl-1H-dibenzo[2,3:6,7]thiepi-
no[4,5-d]azepine-7-carbonitrile (●)
S CGP 14175, Citatepine**
U Neuroleptic

9253 (7539) $C_{20}H_{18}N_3$
                                                    20566-69-2

3,8-Diamino-5-methyl-6-phenylphenanthridinium (●)

R Bromide (518-67-2)
S Dimidium bromide, Phenanthridinium 1553, Trypa-
dine
U Trypanocide

9254 $C_{20}H_{18}N_4$
                                                    2683-78-5

7,8,17,18-Tetrahydro-21H,23H-porphine (●)
S Bacteriochlorin
U Photosensitizer

9255 (7540) $C_{20}H_{18}N_4O_5S_2$
                                                    5575-21-3

(6R,7R)-3-(4-Carbamoyl-1-pyridiniomethyl)-8-oxo-7-(2-
thienylacetamido)-5-thia-1-azabicyclo[4.2.0]oct-2-ene-2-
carboxylic acid = (7R)-3-(4-Carbamoyl-1-pyridiniome-
thyl)-7-(2-thienylacetamido)-3-cephem-4-carboxylate =
(6R-trans)-4-(Aminocarbonyl)-1-[[2-carboxy-8-oxo-7-
[(2-thienylacetyl)amino]-5-thia-1-azabicyclo[4.2.0]oct-
2-en-3-yl]methyl]pyridinium hydroxide inner salt (●)
S Cefalonium*, Cefalonum, Cepalonium, Cephaloni-
um, Cepravin, GL 87/90, Lilly 41071
U Antibiotic

9256 $C_{20}H_{18}N_6O_5S_3$
                                                    176302-54-8

(6R-trans)-4-[[2-[7-[[(2-Amino-4-thiazolyl)(hydroxyimino)acetyl]amino]-2-carboxy-8-oxo-5-thia-1-azabicy-

clo[4.2.0]oct-2-en-3-yl]ethenyl]thio]-1-methylpyridi-
nium inner salt (●)
S TOC-50
U Antibiotic

9257 (7541)

$C_{20}H_{18}N_8O_8S_3$
77360-52-2

(6R,7R)-7-[2-(2-Amino-4-thiazolyl)glyoxylamido]-3-
[(E)-2-[[4-(formylmethyl)-1,4,5,6-tetrahydro-5,6-dioxo-
as-triazin-3-yl]thio]vinyl]-8-oxo-5-thia-1-azabicy-
clo[4.2.0]oct-2-ene-2-carboxylic acid 7^2-(Z)-(O-methyloxime) = (7R)-7-[(Z)-2-(2-Amino-4-thiazolyl)-2-(methoxyimino)acetamido]-3[(E)-2-[[1,4,5,6-
dioxo-4-(2-oxoethyl)-1,2,4-triazin-3-yl]thio]vinyl]-3-ce-
phem-4-carboxylic acid = [6R-[3(E),6α,7β(Z)]]-7-[[(2-
Amino-4-thiazolyl)(methoxyimino)acetyl]amino]-8-oxo-
3-[2-[[1,4,5,6-tetrahydro-5,6-dioxo-4-(2-oxoethyl)-1,2,4-
triazin-3-yl]thio]ethenyl]-5-thia-1-azabicyclo[4.2.0]oct-
2-ene-2-carboxylic acid (●)
S *Ceftiolene***, 42980 R.P.
U Antibiotic

9258 (7542)

$C_{20}H_{18}O_3$
81-92-5

2-[Bis(p-hydroxyphenyl)methyl]benzyl alcohol = 2-
[Bis(4-hydroxyphenyl)methyl]benzenemethanol (●)
S Agarolax, Bical, Egmol, Gentiapol, Normolax, Osne-
roll, *Phenolphthalol*, Regmolax, Regolax, Velaxin
U Laxative

9259 (7543)

$C_{20}H_{18}O_4$
15620-34-5

4-Hydroxy-3-[α-(2-oxobutyl)benzyl]coumarin = 4-Hy-
droxy-3-(3-oxo-1-phenylpentyl)coumarin = 4-Hydroxy-
3-(3-oxo-1-phenylpentyl)-2H-1-benzopyran-2-one (●)
R also sodium salt (5058-39-9)
S Fepromaron, Nafarin, Napharin, Phepromaron
U Anticoagulant

9260 (7544)

$C_{20}H_{18}O_4$
518-20-7

3,4-Dihydro-(2-methyl-2-methoxy-4-phenyl)dihydropy-
ranocoumarin = 3,4-Dihydro-2-methoxy-2-methyl-4-phe-
nyl-2H,5H-pyrano[3,2-c][1]benzopyran-5-one (●)
S Anticoagulans 63, BL-5, Cumopyran, Cumopyrin,
Cyclocoumarol, *Cyclocumarol*,
Link's Compound No. 63, Methanopyranorin, Me-
thopyranorin
U Anticoagulant

9261 (7545)

$C_{20}H_{18}O_5$
135635-83-5

1,2,3,4-Tetrahydro-1,3-dihydroxy-8-methoxy-3-methyl-benz[a]anthracene-7,12-dione (●)
S SM 196 B
U Antiviral, antibacterial from *Streptomyces* sp. DSM 4769

9262 (7546) $C_{20}H_{18}O_5$
127984-70-7

1,3-Dioxano[4,5,6-*gh*]naphthalene-spiro(2,1)-(1,2,3,4,5,8,9,10-octahydro-4,8-dihydroxynaphthalen-5-one) = 2,3,4,4a,8,8a-Hexahydro-4,8-dihydroxyspi-ro[naphthalene-1(5*H*),2'-naphtho[1,8-*de*][1,3]dioxin]-5-one (●)
S MK 3018
U Antibacterial from *Tetraploa aristata*

9263 $C_{20}H_{18}O_6$
171927-44-9

3,4-Dihydro-3,7-dihydroxy-8-methoxy-3-methyl-6a,12a-epoxybenz[a]anthracen-1,12(2*H*,7*H*)-dione (●)
S EI 1507-1
U Anti-inflammatory from *Streptomyces* sp. E-1507

9264 $C_{20}H_{18}O_8$
203191-10-0

7-(Carboxymethoxy)-3',4',5-trimethoxyflavone = [[2-(3,4-Dimethoxyphenyl)-5-methoxy-4-oxo-4*H*-1-benzo-pyran-7-yl]oxy]acetic acid (●)
S DA 6034 "Dong A"
U Anti-inflammatory

9265 (7547) $C_{20}H_{18}O_{10}$
73536-69-3

7,7'-Dimethoxy[4,4'-bi-1,3-benzodioxole]-5,5'-dicarb-oxylic acid dimethyl ester (●)
S Bifendate, *"Biphenyl dicarboxylate"*, DDB
U Choleretic, hepatoprotectant

9266 $C_{20}H_{19}ClFNO_4$
175013-84-0

N-[(3*S*,4*S*)-6-Acetyl-3-hydroxy-2,2-dimethyl-4-chroma-nyl]-3-chloro-4-fluorobenzamide = (3*S-cis*)-*N*-(6-Acetyl-3,4-dihydro-3-hydroxy-2,2-dimethyl-2*H*-1-benzopyran-4-yl)-3-chloro-4-fluorobenzamide (●)
S SB 220453, *Tamabacet, Tonabersat***
U Anticonvulsant, antimigraine

9267 (7548) $C_{20}H_{19}ClF_5NO_4$
95445-79-7

5-Isopropyl 3-methyl 4-[3-chloro-6-fluoro-2-(trifluoro-methyl)phenyl]-2-(fluoromethyl)-1,4-dihydro-6-methyl-3,5-pyridinedicarboxylate = 4-[3-Chloro-6-fluoro-2-(tri-fluoromethyl)phenyl]-2-(fluoromethyl)-1,4-dihydro-6-methyl-3,5-pyridinedicarboxylic acid 3-methyl 5-(1-me-thylethyl) ester (●)
S FPL 62129
U Antihypertensive (calcium antagonist)

9268 (7549) $C_{20}H_{19}ClN_2OS$

2-Chloro-10-(6,7-epoxy-3-tropyl)phenothiazine = 3-(2-Chloro-10-phenothiazinyl)scopine = [7(*S*)-(1α,2β,4β,5α,7β)]-7-(2-chloro-10*H*-phenothiazin-10-yl)-9-methyl-3-oxa-9-azatricyclo[3.3.1.0^{2,4}]nonane (●)
S *Chlorophenothiazinylscopine, Clorofenotiazinylsco-pina*
U Anti-emetic, tranquilizer, spasmolytic

9269 (7551) $C_{20}H_{19}ClN_2O_3$
79046-96-1

4-Chloro-1-(4-methoxy-3-methylbenzyl)-2-phenyl-5-imidazoleacetic acid = 4-Chloro-1-[(4-methoxy-3-me-thylphenyl)methyl]-2-phenyl-1*H*-imidazole-5-acetic acid (●)
S CV-2973
U Antihypertensive, diuretic

9270 (7550) $C_{20}H_{19}ClN_2O_3$
55299-10-0

7-Chloro-1,3-dihydro-3-hydroxy-5-phenyl-2*H*-1-ben-zodiazepin-2-one pivalate (ester) = 2,2-Dimethylpropa-noic acid 7-chloro-2,3-dihydro-2-oxo-5-phenyl-1*H*-1,4-benzodiazepin-3-yl ester (●)
S *Oxazepam pivalate, Pivoxazepam***
U Tranquilizer

9271 (7552) $C_{20}H_{19}ClN_6O_5$
18921-73-8

N-[4-[[(2,4-Diamino-5-chloro-6-quinazolinyl)me-thyl]amino]benzoyl]-L-aspartic acid (●)

R Dihydrate
S *Chlorasquin*, NSC-529861, SK 29861
U Antineoplastic

9272 (7553) $C_{20}H_{19}Cl_2NO_3$
148-07-2

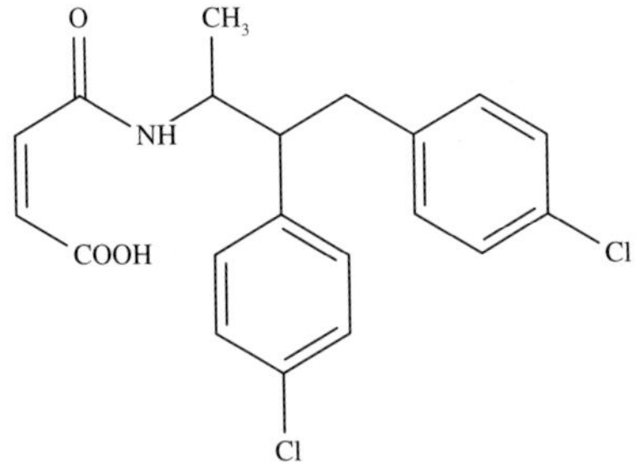

α-*N*-[2,3-Bis(*p*-chlorophenyl)-1-methylpropyl]maleamic
acid = 4-[[2,3-Bis(4-chlorophenyl)-1-methylpropyl]ami-
no]-4-oxo-2-butenoic acid (●)
S *Benzmalacene, Benzmalecene* **, MK-135
U Anticholesteremic, antihistaminic

9273 (7554) $C_{20}H_{19}Cl_3N_2O_2$
71097-23-9

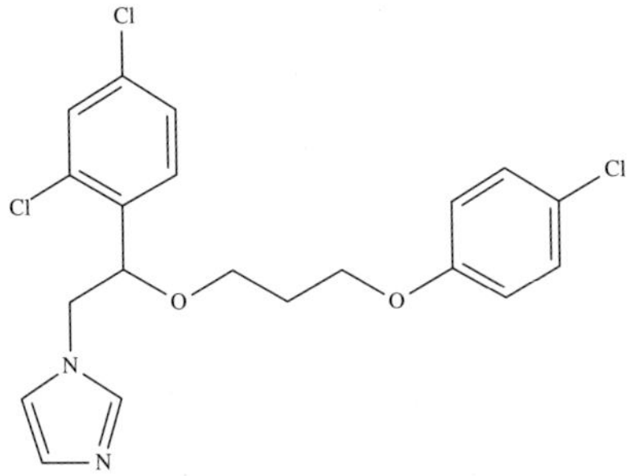

1-[2,4-Dichloro-β-[3-(*p*-chlorophenoxy)propoxy]phen-
ethyl]imidazole = 1-[2-[3-(4-Chlorophenoxy)propoxy]-2-
(2,4-dichlorophenyl)ethyl]-1*H*-imidazole (●)
S *Zoficonazole* **
U Antifungal

9274 $C_{20}H_{19}FN_4O_4$
209342-40-5

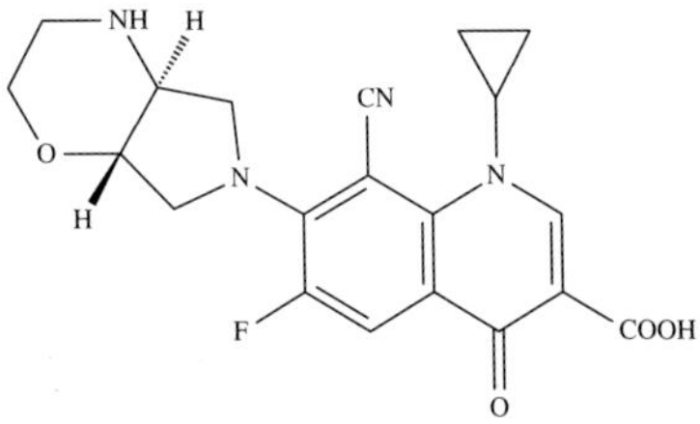

(–)-8-Cyano-1-cyclopropyl-6-fluoro-7-[(1*S*,6*S*)-2-oxa-
5,8-diazabicyclo[4.3.0]non-8-yl]-1,4-dihydro-4-oxo-3-
quinolinecarboxylic acid = 8-Cyano-1-cyclopropyl-6-
fluoro-7-[(4a*S*,7a*S*)-hexahydropyrrolo[3,4-*b*]-1,4-oxazin-
6(2*H*)-yl]-1,4-dihydro-4-oxo-3-quinolinecarboxylic acid
(●)
S BY 377, Byk 60621, *Finafloxacin*
U Antibacterial (topoisomerase inhibitor)

9275 (7555) $C_{20}H_{19}FN_6O_5$

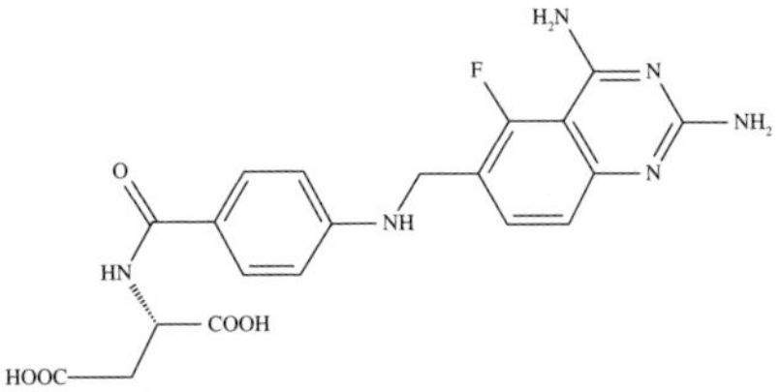

N-[4-[[(2,4-Diamino-5-fluoro-6-quinazolinyl)me-
thyl]amino]benzoyl]-L-aspartic acid (●)
S *Fluorasquin*
U Antineoplastic

9276 $C_{20}H_{19}F_2NO_4$
175013-73-7

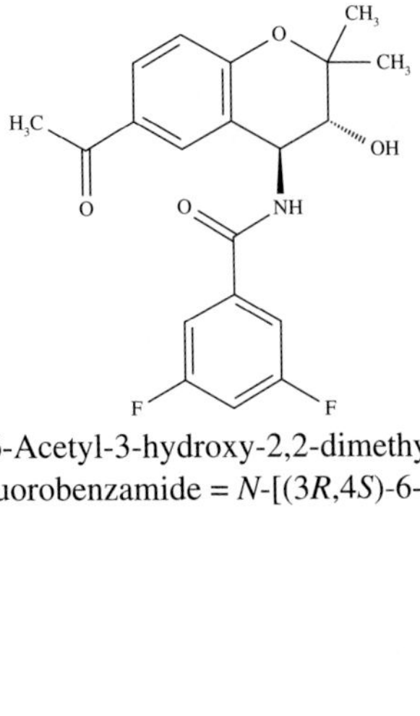

N-[(3*R*,4*S*)-6-Acetyl-3-hydroxy-2,2-dimethyl-4-chroma-
nyl]-3,5-difluorobenzamide = *N*-[(3*R*,4*S*)-6-Acetyl-3,4-

dihydro-3-hydroxy-2,2-dimethyl-2*H*-1-benzopyran-4-yl]-3,5-difluorobenzamide (●)
S　SB 218842, *Tidembersat**
U　Migraine therapeutic

9277　　　　　　　　　　　　　　　　$C_{20}H_{19}F_3N_2O_4$
　　　　　　　　　　　　　　　　　　148430-28-8

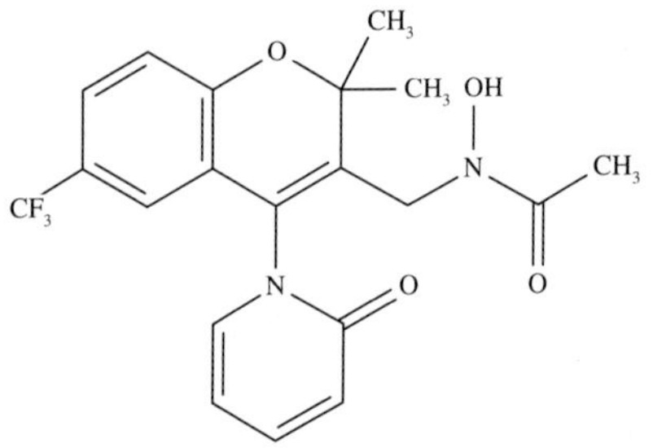

N-[[2,2-Dimethyl-4-(2-oxo-1(2*H*)-pyridyl)-6-(trifluoromethyl)-2*H*-1-benzopyran-3-yl]methyl]acetohydroxamic acid = *N*-[[2,2-Dimethyl-4-(2-oxo-1(2*H*)-pyridinyl)-6-(trifluoromethyl)-2*H*-1-benzopyran-3-yl]methyl]-*N*-hydroxyacetamide (●)
S　BIIX1XX, RS-91309, *Sarakalim***
U　Potassium channel activator

9278　(7556)　　　　　　　　　　　$C_{20}H_{19}F_3N_4O$

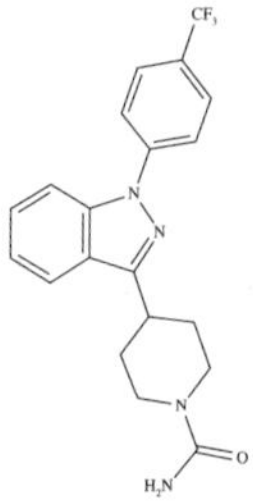

4-[1-[4-(Trifluoromethyl)phenyl]-1*H*-indazol-3-yl]-1-piperidinecarboxamide (●)
R　Monohydrobromide
S　HP 654
U　Anticonvulsant

9279　(7557)　　　　　　　　　　$C_{20}H_{19}F_3N_4O_3S$
　　　　　　　　　　　　　　　　　　90259-87-3

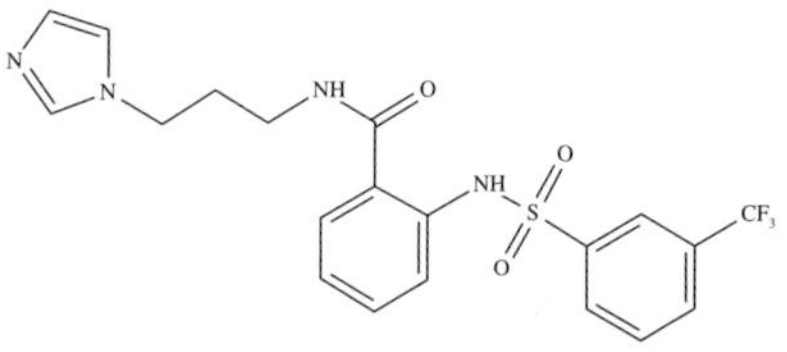

N-[3-(1-Imidazolyl)propyl]-2-[3-(trifluoromethyl)benzenesulfonamido]benzamide = *N*-[3-(1*H*-Imidazol-1-yl)propyl]-2-[[[3-(trifluoromethyl)phenyl]sulfonyl]amino]benzamide (●)
R　Monohydrochloride (90260-18-7)
S　M. & B. 39890 A
U　Hypoglycemic

9280　　　　　　　　　　　　　　　$C_{20}H_{19}F_6NO$
　　　　　　　　　　　　　　　　　　148700-85-0

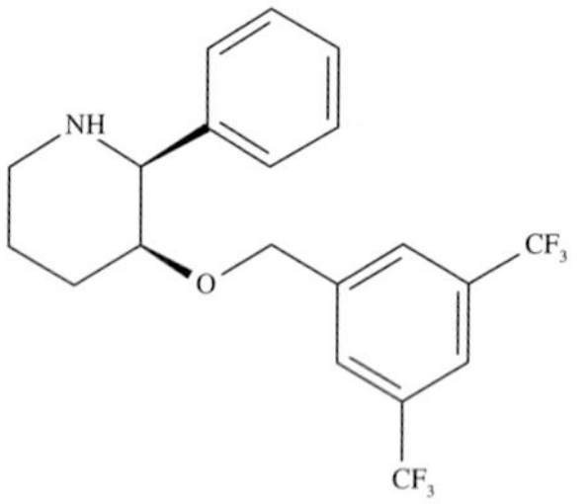

(2*S*,3*S*)-3-[[3,5-Bis(trifluoromethyl)phenyl]methoxy]-2-phenylpiperidine (●)
S　L 733060
U　Tachykinin NK$_1$ receptor antagonist (analgesic)

9281　　　　　　　　　　　　　　　$C_{20}H_{19}NO_2$
　　　　　　　　　　　　　　　　　　132836-11-4

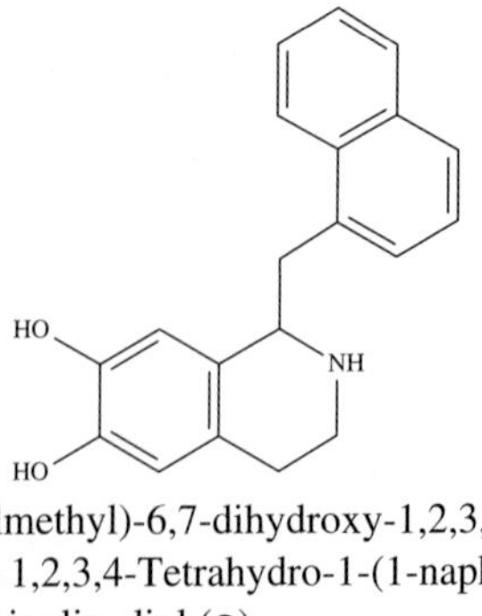

1-(α-Naphthylmethyl)-6,7-dihydroxy-1,2,3,4-tetrahydroisoquinoline = 1,2,3,4-Tetrahydro-1-(1-naphthalenylmethyl)-6,7-isoquinolinediol (●)

R Hydrobromide (213179-96-5)
S YS 49
U Cardiotonic (positive inotropic)

9282 (7558) $C_{20}H_{19}NO_3$
27591-42-0

2-(Morpholinomethyl)-2-phenyl-1,3-indanedione = 2-(4-Morpholinylmethyl)-2-phenyl-1H-indene-1,3(2H)-dione (●)
S Amplidione, L.D. 4610, *Mofedione*, *Oxazidione***,
Transidione
U Anticoagulant

9283 (7559) $C_{20}H_{19}NO_3$
7008-42-6

3,12-Dihydro-6-methoxy-3,3,12-trimethyl-7H-pyrano[2,3-c]acridin-7-one (●)
S *Acronine***, *Acronycine*, Compound 42339,
NSC-403169
U Antineoplastic (alkaloid from *Acronychia baueri*)

9284 $C_{20}H_{19}NO_3$
95737-68-1

4-Phenoxyphenyl (RS)-2-(2-pyridyloxy)propyl ether = 2-[1-Methyl-2-(4-phenoxyphenoxy)ethoxy]pyridine (●)
S Cyclio spot on, Nylar, *Pyriproxyfen*, S 9138
U Juvenile hormone analog

9285 (7560) $C_{20}H_{19}NO_3S$
109229-58-5

(−)-5-[[(2R)-2-Benzyl-6-chromanyl]methyl]-2,4-thiazolidinedione = 5-[[3,4-Dihydro-2-(phenylmethyl)-2H-1-benzopyran-6-yl]methyl]-2,4-thiazolidinedione (●)
R Sodium salt (109229-57-4)
S CP-72467-2, *Englitazone sodium***
U Antidiabetic

9286 (7561) $C_{20}H_{19}NO_4$
111753-73-2

4-Cyano-5,5-bis(4-methoxyphenyl)-4-pentenoic acid (●)
S E-5510, *Saigrel*, *Satigrel***
U Platelet aggregation inhibitor

9287 (7562) $C_{20}H_{19}NO_4S$
76631-44-2

1,2,3,4-Tetrahydro-1-(3-mercaptopropionyl)quinaldic acid benzoate = 1-[3-(Benzoylthio)-1-oxopropyl]-1,2,3,4-tetrahydro-2-quinolinecarboxylic acid (●)
R Sodium salt (75782-88-6)
S EU-4867
U Antihypertensive

9288 (7563) $C_{20}H_{19}NO_5$
476-32-4

[5bS-(5bα,6β,12bα)]-5b,6,7,12b,13,14-Hexahydro-13-methyl[1,3]benzodioxolo[5,6-c]-1,3-dioxolo[4,5-i]phenanthridin-6-ol (●)
S Chelidonine, Diphylline, Stylophorin
U Spasmolytic (alkaloid from Chelidonium majus and other Papaveraceae)

9289 $C_{20}H_{19}NO_5$
130-86-9

7-Methyl-2,3:9,10-bis(methylenedioxy)-7,13a-secoberbin-13a-one = 4,6,7,14-Tetrahydro-5-methylbis[1,3]benzodioxolo[4,5-c:5',6'-g]azecin-13(5H)-one (●)
S Biflorine, Corydinine, Fumarine, Macleyine, Protopine
U Antithrombotic, anti-inflammatory (alkaloid from many plants)

9290 (7564) $C_{20}H_{19}NO_5$
25771-23-7

3-(p-Anisoyl)-6-methoxy-2-methylindole-1-acetic acid = 6-Methoxy-3-(4-methoxybenzoyl)-2-methyl-1H-indole-1-acetic acid (●)
S Duometacin**, R 4444
U Anti-inflammatory, analgesic

9291 (7565) $C_{20}H_{19}NO_5$
119060-88-7

N-Methyl-N-[3,4-(methylenedioxy)phenethyl]-3,4-(methylenedioxy)cinnamamide = (E)-3-(1,3-Benzodioxol-5-yl)-N-[2-(1,3-benzodioxol-5-yl)ethyl]-N-methyl-2-propenamide (●)
S Dioxamide
U Antibacterial

9292 $C_{20}H_{19}NO_5$
201943-63-7

2(S)-Amino-3-[2(S)-carboxy-1(S)-cyclopropyl]-3-(9H-xanthen-9-yl)propionic acid = (αS)-α-Amino-α-[(1S,2S)-2-carboxycyclopropyl]-9H-xanthene-9-propanoic acid (●)
S LY 341495
U Cerebral protectant

9293 $C_{20}H_{19}NO_8$
88828-25-5

(4aR,5S,5aR,6R,12aS)-1,4,4a,5,5a,6,11,12a-Octahydro-3,5,10,12,12a-pentahydroxy-6-methyl-1,11-dioxo-2-naphthacenecarboxamide (●)
S CMT-8
U Antineoplastic (collagenase and matrix metalloproteinase inhibitor)

9294   (7566) $C_{20}H_{19}N_3$
                                                           3248-93-9

α^4-(p-Aminophenyl)-α^4-(4-imino-2,5-cyclohexadien-1-
ylidene)-2,4-xylidine = 4-[(4-Aminophenyl)(4-imino-2,5-
cyclohexadien-1-ylidene)methyl]-2-methylbenzenamine
(●)
R   Monohydrochloride (632-99-5); Definition also as
       mixture with the demethyl derivative (see no. 8399)
S   C.I. 42510, C.I. Basic Violet 14, *Diamantfuchsin,
       Fuchsin (basic)*, Magenta I, *Rosaniline (hydrochlori-
       de)*
U   Anti-infective (topical), dye

9295   (7567) $C_{20}H_{19}N_3$
                                                           75375-52-9

5,6-Dihydro-7,13-dimethyl-7H,13H-pyrido[4',3':6,5]aze-
pino[1,2,3-*lm*]-β-carboline = (*R*)-5,6,7,13-Tetrahydro-
7,13-dimethyl-7,11,13a-triazabenzo[5,6]cyclohep-
ta[1,2,3-*jk*]fluorene (●)
S   *Decussine*
U   Muscle relaxant from *Strychnos decussata*

9296 $C_{20}H_{19}N_3O$
                                                           183799-95-3

4-(Dimethylamino)-4'-(1-imidazolyl)chalcone = 3-[4-
(Dimethylamino)phenyl]-1-[4-(1H-imidazol-1-yl)phe-
nyl]-2-propen-1-one (●)
S   RL 3142
U   Antimalarial (cysteine protease inhibitor)

9297 $C_{20}H_{19}N_3O_2$
                                                           174634-08-3

6-[[2-(Dimethylamino)ethyl]amino]-3-hydroxy-7H-inde-
no[2,1-*c*]quinolin-7-one (●)
R   Dihydrochloride (174634-09-4)
S   TAS-103
U   Antineoplastic (topoisomerase I and II inhibitor)

9298 $C_{20}H_{19}N_3O_2$
                                                           186610-89-9

3-[[4-(4-Formyl-1-piperazinyl)phenyl]methylene]-2-in-
dolinone = 4-[4-[(1,2-Dihydro-2-oxo-3H-indol-3-yli-
dene)methyl]phenyl]-1-piperazinecarboxaldehyde (●)
S   Su-4984
U   Tyrosine kinase inhibitor

9299 $C_{20}H_{19}N_3O_2$
80279-24-9

(11S,11aS)-1,2,3,10,11,11a-Hexahydro-9-hydroxy-11-
(1H-indol-3-yl)-5H-pyrrolo[2,1-c][1,4]benzodiazepin-5-
one (●)
S   *Tilivalline*
U   Antineoplastic

9300   (7568) $C_{20}H_{19}N_3O_4S$
112859-12-8

1,3-Dihydro-7-[4-(phenylsulfonyl)butoxy]-2H-imida-
zo[4,5-b]quinolin-2-one (●)
S   BMY-21638
U   Antithrombotic

9301 $C_{20}H_{19}N_3O_4S$
185428-18-6

5-[[4-[(6-Methoxy-1-methyl-1H-benzimidazol-2-
yl)methoxy]phenyl]methyl]-2,4-thiazolidinedione (●)
S   CI-1037, CS 1011, *Rivoglitazone**
U   Antidiabetic

9302 $C_{20}H_{19}N_5$
244767-67-7

4-[[4-[(2,4,6-Trimethylphenyl)amino]-2-pyrimidi-
nyl]amino]benzonitrile (●)
S   R 147681
U   Antiviral (HIV)

9303   (7569) $C_{20}H_{19}N_5$
79152-85-5

N-Methyl-4'-[[7-methyl-1H-imidazo[4,5-f]quinolin-9-
yl]amino]acetanilide = N-Methyl-N-[4-[(7-methyl-1H-
imidazo[4,5-f]quinolin-9-yl)amino]phenyl]acetamide (●)
R   Monohydrochloride (55435-65-9)
S   *Acodazole hydrochloride***, EU-3120, NSC-305884
U   Antineoplastic

9304 $C_{20}H_{19}N_5O$
124294-43-5

4,5-Dihydro-5-methyl-6-[4-[(phenylmethyl)amino]-7-
quinazolinyl]-3(2H)-pyridazinone (●)
S   KF 15232
U   Cardiotonic (positive inotropic)

9305 (7570) $C_{20}H_{19}N_5O_2$
107813-63-8

2-[[3-[3-(1H-Tetrazol-5-yl)propoxy]phenoxy]me-
thyl]quinoline (●)
S   RG 7152
U   Anti-inflammatory, anti-allergic

9306 (7571) $C_{20}H_{19}N_8O_5S_2$

[(6R,7R)-7-[2-(2-Amino-4-thiazolyl)glyoxylamido]-2-
carboxy-8-oxo-5-thia-1-azabicyclo[4.2.0]oct-2-en-3-yl]-
3-methyl-3H-imidazo[4,5-c]pyridin-5-ium 7²-(Z)-(O-methyloxime) = [6R-[6α,7β(Z)]]-5-[7-[[(2-Amino-4-thiazo-
lyl)(methoxyimino)acetyl]-amino]-2-carboxy-8-oxo-5-
thia-1-azabicyclo[4.2.0]oct-2-en-3-yl]-3-methyl-3H-imi-
dazo[4,5-c]pyridinium (●)
R   Sulfate (1:1) (115681-28-2)
S   LY 217332
U   Antibiotic

9307 (7572) $C_{20}H_{20}BrNO_4$
51449-10-6

1-(2-Bromo-4,5-dimethoxybenzyl)-6,7-dimethoxyisoqui-
noline = 1-[(2-Bromo-4,5-dimethoxyphenyl)methyl]-6,7-
dimethoxyisoquinoline (●)
S   Brompapaverine
U   Spasmolytic

9308 $C_{20}H_{20}BrN_3O$
145204-79-1

(8β)-2-Bromo-9,10-didehydro-N,6-dimethyl-N-2-propy-
nylergoline-8-carboxamide (●)
S   LEK-8841
U   Antipsychotic

9309 $C_{20}H_{20}BrN_3O_2S$
127243-85-0

N-[2-[[3-(4-Bromophenyl)-2-propenyl]amino]ethyl]-5-
isoquinolinesulfonamide (●)
S   H-89
U   Protein kinase inhibitor

9310 (7573) $C_{20}H_{20}Br_2O_2$
32953-81-4

2,3-Dibromo-4,4-bis(4-ethylphenyl)isocrotonic acid =
(Z)-2,3-Dibromo-4,4-bis(4-ethylphenyl)-2-butenoic acid
(●)
S   Edikron, EFBK, VÚFB-6688
U   Antineoplastic

9311 (7574)

$C_{20}H_{20}CINO$
52758-02-8

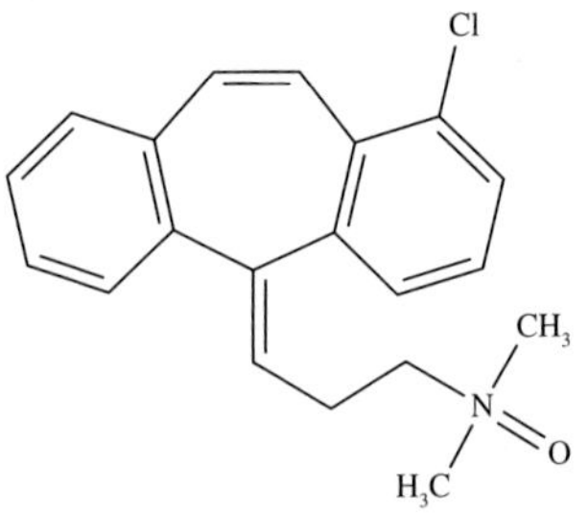

1-Chloro-N,N-dimethyl-5H-dibenzo[a,d]cycloheptene-$\Delta^{5,\gamma}$-propylamine N-oxide = 3-(1-Chloro-5H-dibenzo[a,d]cyclohepten-5-ylidene)-N,N-dimethyl-1-propanamine N-oxide (●)
S *Benzaprinoxide**, Ro 8-0254
U Antidepressant

9312

$C_{20}H_{20}CINO_3$
150206-14-7

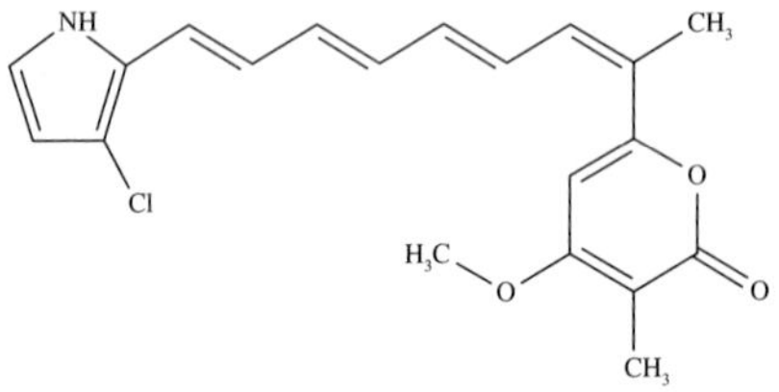

6-[(1Z,3E,5E,7E)-8-(3-Chloro-1H-pyrrol-2-yl)-1-methyl-1,3,5,7-octatetraenyl]-4-methoxy-3-methyl-2H-pyran-2-one (●)
S *Rumbrin*
U Cytoprotective

9313

$C_{20}H_{20}CINO_7$
183720-02-7

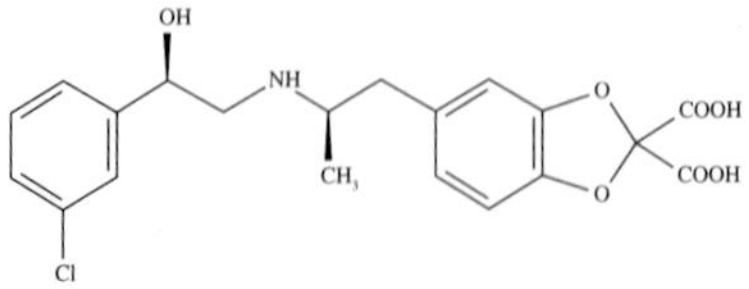

[R-(R^*,R^*)]-5-[2-[[2-(3-Chlorophenyl)-2-hydroxyethyl]amino]propyl]-1,3-benzodioxole-2,2-dicarboxylic acid (●)
R Disodium salt (138908-40-4)
S BTA-243, CL 316243
U β_3-Adrenergic agonist (antidiabetic, anti-obesity)

9314 (7575)

$C_{20}H_{20}CIN_3O$
550-81-2

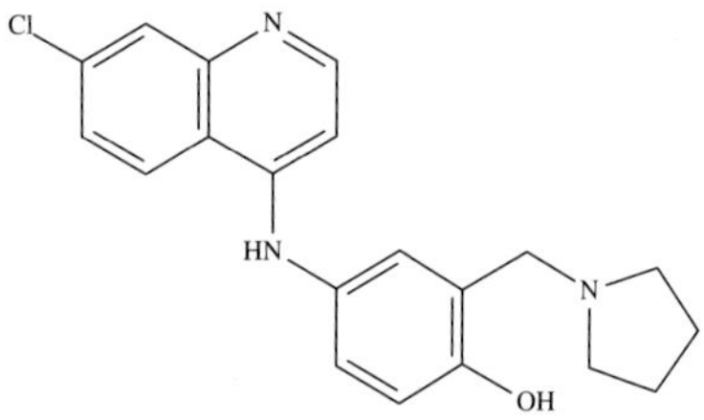

4-[(7-Chloro-4-quinolyl)amino]-α-1-pyrrolidinyl-o-cresol = 7-Chloro-4-[4-hydroxy-3-(pyrrolidinomethyl)anilino]quinoline = 4-[(7-Chloro-4-quinolinyl)amino]-2-(1-pyrrolidinylmethyl)phenol (●)
R also dihydrochloride (10350-81-9)
S *Amopyrochinum*, *Amopyroquine**, CI-356, PAM-780, Propoquin, WR 4835
U Antimalarial

9315

$C_{20}H_{20}CIN_3O_2$
75000-30-5

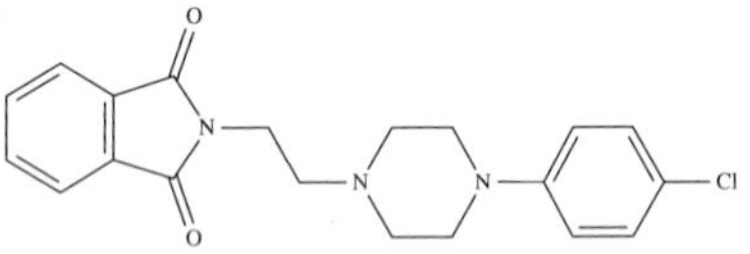

1-(p-Chlorophenyl)-4-(2-phthalimidoethyl)piperazine = 2-[2-[4-(4-Chlorophenyl)-1-piperazinyl]ethyl]-1H-isoindole-1,3(2H)-dione (●)
S CPPEP
U Anorexic, antipsychotic

9316 (7576)

$C_{20}H_{20}CIN_3O_3$
103475-41-8

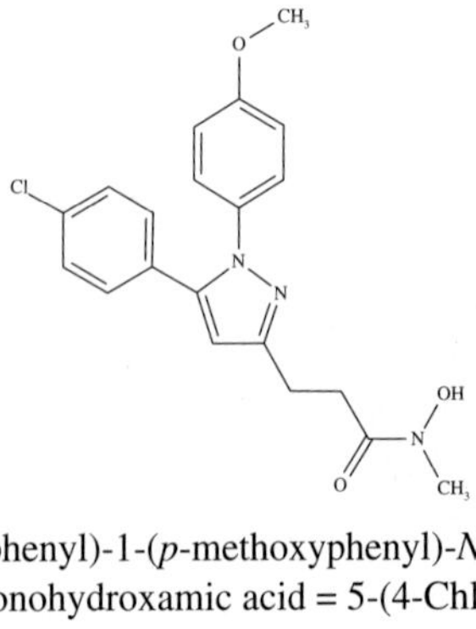

5-(p-Chlorophenyl)-1-(p-methoxyphenyl)-N-methylpyrazole-3-propionohydroxamic acid = 5-(4-Chlorophenyl)-

N-hydroxy-1-(4-methoxyphenyl)-*N*-methyl-1*H*-pyra-
zole-3-propanamide (●)
S ORF 20485, RWJ 20485, *Tepoxalin***
U Antipsoriatic, anti-inflammatory

9317 (7577) $C_{20}H_{20}ClN_5$
53257-71-9

8-Chloro-*N,N*-dimethyl-4*H*-[1,2,4]triazolo[4,3-*a*][1,4]
benzodiazepine-1-ethanamine (●)
R 4-Toluenesulfonate (tosylate) (83983-74-8)
S U-43465F
U Antidepressant

9318 (7578) $C_{20}H_{20}ClN_5O_3S_2$
53813-83-5

4-Methyl-1-piperazinecarboxylic acid ester with (±)-6-(7-
chloro-1,8-naphthyridin-2-yl)-2,3,6,7-tetrahydro-7-hydr-
oxy-5*H*-p-dithiino[2,3-*c*]pyrrol-5-one = (*RS*)-6-(7-
Chloro-1,8-naphthyridin-2-yl)-2,3,6,7-tetrahydro-7-oxo-
5*H*-1,4-dithiino[2,3-*c*]pyrrol-5-yl 4-methylpiperazine-1-
carboxylate = (±)-4-Methyl-1-piperazinecarboxylic acid
5-(7-chloro-1,8-naphthyridin-2-yl)-2,3,6,7-tetrahydro-7-
oxo-5*H*-1,4-dithiino[2,3-*c*]pyrrol-5-yl ester (●)
S Celexane, 31264 R.P., *Suriclone***, Suril "RPR"
U Tranquilizer, hypnotic

9319 $C_{20}H_{20}Cl_2N_4O_2S$
178979-85-6

5-[(3,5-Dichlorophenyl)thio]-4-isopropyl-1-(4-pyridyl-
methyl)imidazole-2-methanol carbamate (ester) = 5-
[(3,5-Dichlorophenyl)thio]-4-(1-methylethyl)-1-(4-pyri-
dinylmethyl)-1*H*-imidazole-2-methanol carbamate (ester)
(●)
S AG-1549, *Capravirine**, S-1153
U Antiviral (HIV-1)

9320 $C_{20}H_{20}FNO_3S$
150322-43-3

2-[2-(Acetyloxy)-6,7-dihydrothieno[3,2-*c*]pyridin-
5(4*H*)-yl]-1-cyclopropyl-2-(2-fluorophenyl)ethanone (●)
S CS 747
U Platelet aggregation inhibitor

9321 $C_{20}H_{20}FNO_4$
184653-84-7

N-[(3*R*,4*S*)-6-Acetyl-3-hydroxy-2,2-dimethyl-4-chroma-
nyl]-4-fluorobenzamide = *N*-[(3*R*,4*S*)-6-Acetyl-3,4-dihy-

dro-3-hydroxy-2,2-dimethyl-2H-1-benzopyran-4-yl]-4-fluorobenzamide (●)
R also hydrate (2:1) (185122-82-1)
S *Carabersat**, Elabacet*, SB 204269
U Anticonvulsant, antimigraine

9322 (7579)   C$_{20}$H$_{20}$FN$_3$O
                                                   98295-26-2

1-Benzoyl-6-fluoro-3-(1-methyl-4-piperidinyl)-1H-indazole (●)
S HP 818
U Analgesic, neuroleptic

9323   C$_{20}$H$_{20}$FN$_7$O
                                                   150867-91-7

3-(5-Cyclopropyl-1,2,4-oxadiazol-3-yl)-7-fluoro-5-(4-methyl-1-piperazinyl)imidazo[1,5-a]quinazoline (●)
S NNC 14-0189
U Anticonvulsant (BDZ receptor partial agonist)

9324 (7580)   C$_{20}$H$_{20}$F$_2$N$_4$
                                                   105102-20-3

2,2'-[(3,3'-Difluoro-4,4'-biphenylene)dinitrilo]dipyrrolidine = 3,3'-Difluoro-4,4'-bis(2-pyrrolidinylideneamino)biphenyl = 3,3'-Difluoro-N,N'-bis(2-pyrrolidinylidene)[1,1'-biphenyl]-4,4'-diamine (●)
S HL 707, *Liroldine***
U Anti-amebic

9325 (7581)   C$_{20}$H$_{20}$F$_2$N$_4$O$_3$
                                                   139294-87-4

5-Cyclopropyl-6,8-difluoro-1-hydroxy-7-(4-methyl-1-piperazinyl)benzo[b][1,6]naphthyridine-3,10(2H,5H)-dione (●)
S AT-5755
U Antibacterial

9326   C$_{20}$H$_{20}$F$_4$N$_4$OS

3-[2-[4-(4-Fluorophenyl)-1-piperazinyl]ethyl]-2-imino-6-(trifluoromethoxy)benzothiazoline = 3-[2-[4-(4-Fluorophenyl)-1-piperazinyl]ethyl]-6-(trifluoromethoxy)-2(3H)-benzothiazolimine (●)
R Dihydrochloride (133783-43-4)
S RP 66055
U Neuroprotective, anticonvulsant

9327　　　　　　　　　　　　　　　　　$C_{20}H_{20}NO_5P$
164173-10-8

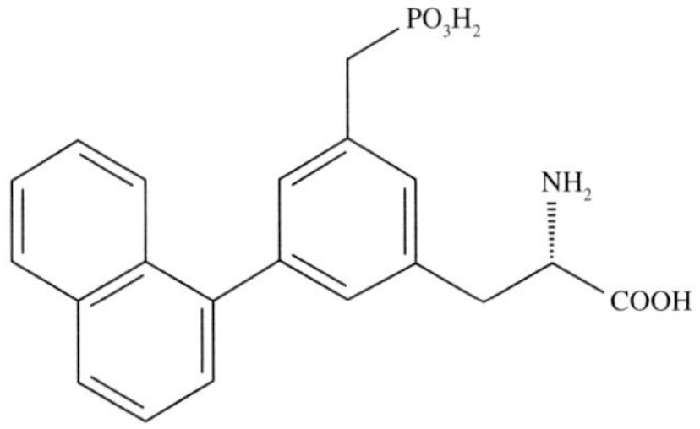

PO₃H₂ ... NH₂ ... COOH

(S)-2-Amino-3-[3-(1-naphthyl)-5-(phosphonomethyl)phenyl]propionic acid = 3-(1-Naphthalenyl)-5-
(phosphonomethyl)-L-phenylalanine (●)
S　PD 158473
U　Neuroprotective, anticonvulsant (NMDA receptor
　　antagonist)

9328　(7582)　　　　　　　　　　　　$C_{20}H_{20}N_2$
75437-14-8

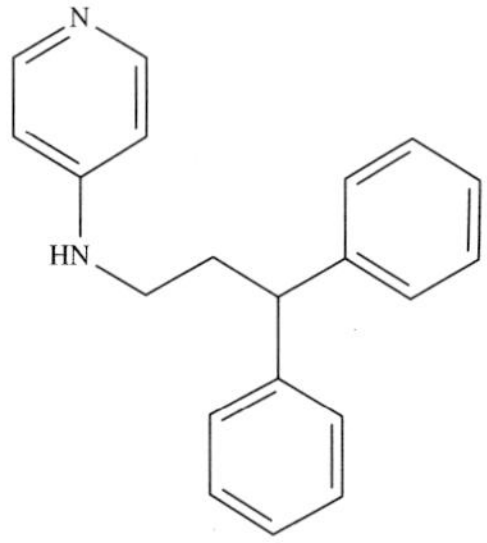

N ... HN

4-[(3,3-Diphenylpropyl)amino]pyridine = N-(4-Pyridyl)-
3,3-diphenylpropylamine = N-(3,3-Diphenylpropyl)-4-
pyridinamine (●)
R　Monohydrochloride (29769-70-8)
S　Fenprin, *Fenpyramine hydrochloride*, *Milverine hy-
　　drochloride**, Miospasm
U　Spasmolytic

9329　(7583)　　　　　　　　　　　　$C_{20}H_{20}N_2O$
121445-19-0

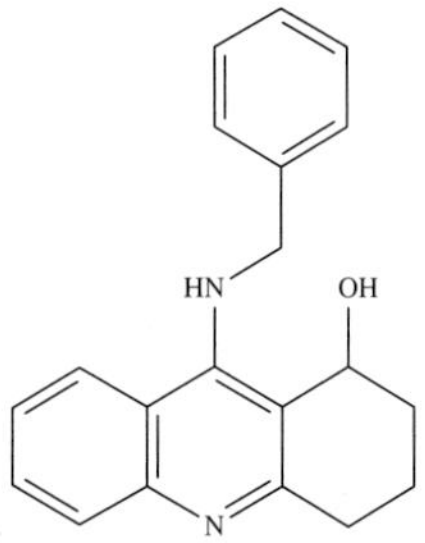

HN ... OH ... N

(±)-9-(Benzylamino)-1,2,3,4-tetrahydro-1-acridinol =
(±)-1,2,3,4-Tetrahydro-9-[(phenylmethyl)amino]-1-acri-
dinol (●)
R　Maleate (1:1) (121445-20-3)
S　HP 128, *Suronacrine maleate***
U　Cholinesterase inhibitor, cognition activator

9330　(7584)　　　　　　　　　　　　$C_{20}H_{20}N_2O_2$
137460-88-9

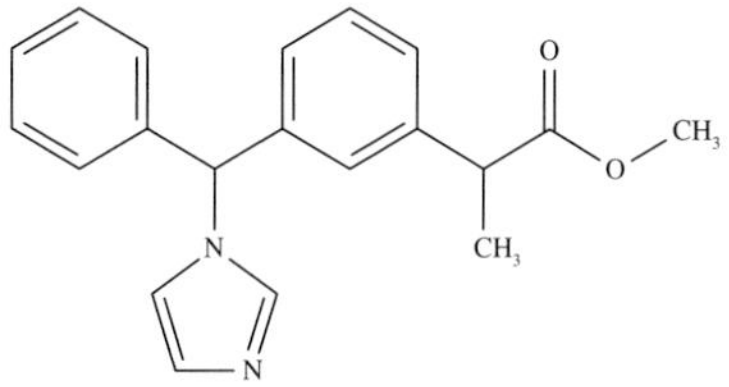

O ... O—CH₃ ... CH₃ ... N ... N

Methyl *m*-(α-imidazol-1-ylbenzyl)hydratropate = Methyl
2-[3-[(1*H*-imidazol-1-yl)phenylmethyl]phenyl]propio-
nate = 3-[(1*H*-Imidazol-1-yl)phenylmethyl]-α-methyl-
benzeneacetic acid methyl ester (●)
S　*Odalprofen***, WAS 5608
U　Anti-inflammatory, analgesic

9331　(7585)　　　　　　　　　　　　$C_{20}H_{20}N_2O_2$
41717-30-0

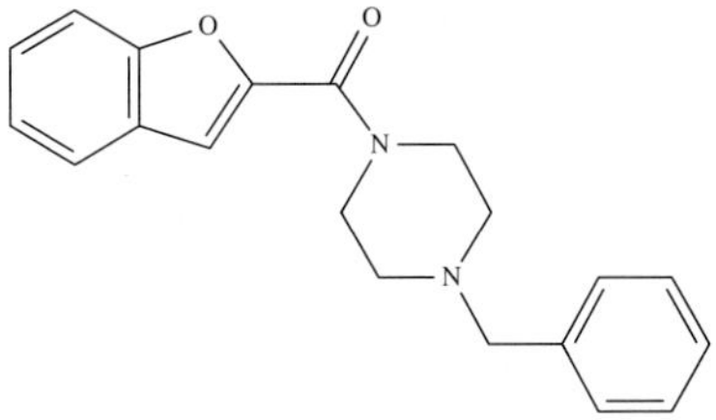

O ... O ... N ... N

1-(2-Benzofuranylcarbonyl)-4-benzylpiperazine = 1-(2-
Benzofuranylcarbonyl)-4-(phenylmethyl)piperazine (●)
R　also monohydrochloride (41716-84-1)

S *Befuraline***, DIV 154
U Antidepressant (phosphodiesterase inhibitor)

9332 (7586) $C_{20}H_{20}N_2O_2$
30748-29-9

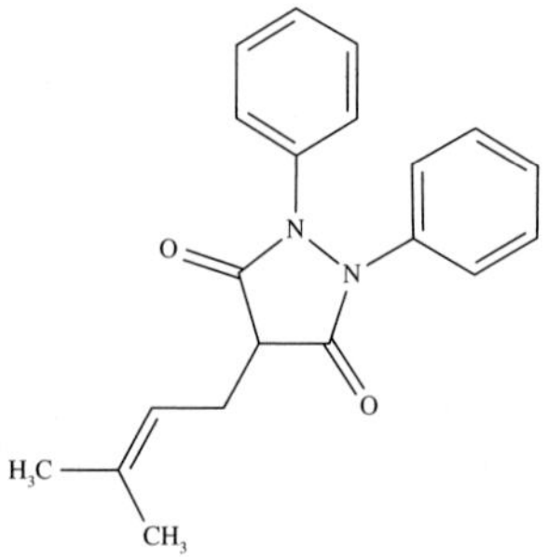

4-Prenyl-1,2-diphenyl-3,5-pyrazolidinedione = 4-(3-Methyl-2-butenyl)-1,2-diphenyl-3,5-pyrazolidinedione (●)

S Analud, Artrozone, Bentudor, Brotazona, Cocresol, DA 2370, Danfenona, Deflogen (old form), *Fenilprenazona, Feprazone***, Golaman, Grisona, Impremial, Medoprazone, Methrazone, Metrazone, Naloven, Nazona, Nessazona, Nilatin, *Phenylprenazone*, Prenakes, *Prenazone*, Rangozona, Represil, Rumax, Tabrien, Vapesin, Zepelan, Zepelin, Zepellin, Zontal
U Anti-inflammatory, analgesic

9332-01 (7586-01) 57148-60-4
R Salt with piperazine (1:1)
S Lubetrone, *Pinazone*, Reuflodol
U Anti-inflammatory

9333 $C_{20}H_{20}N_2O_2S$

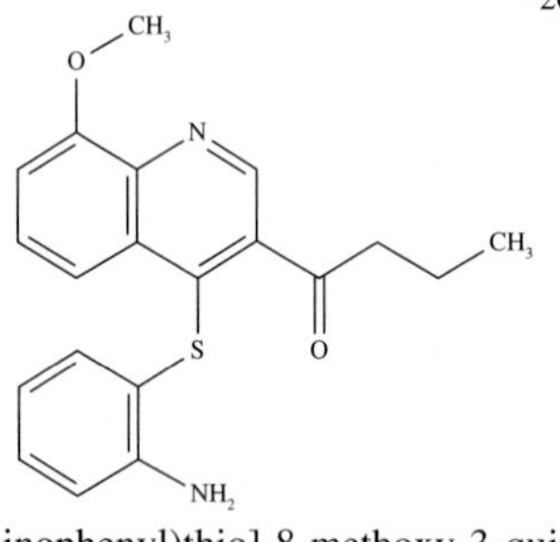

1-[4-[(2-Aminophenyl)thio]-8-methoxy-3-quinolinyl]-1-butanone (●)
R Monohydrochloride (209479-38-9)
S YJA 20379-6

U Proton-pump inhibitor

9334 (7587) $C_{20}H_{20}N_2O_3$
107433-19-2

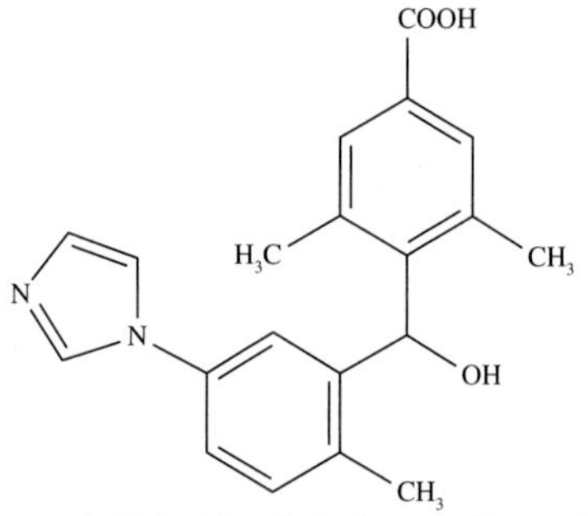

4-[α-Hydroxy-5-(1-imidazolyl)-2-methylbenzyl]-3,5-dimethylbenzoic acid = 4-[Hydroxy[5-(1H-imidazol-1-yl)-2-methylphenyl]methyl]-3,5-dimethylbenzoic acid (●)
R Monosodium salt (104363-98-6)
S Y-20811
U Thromboxane synthetase inhibitor

9335 $C_{20}H_{20}N_2O_3$
127654-03-9

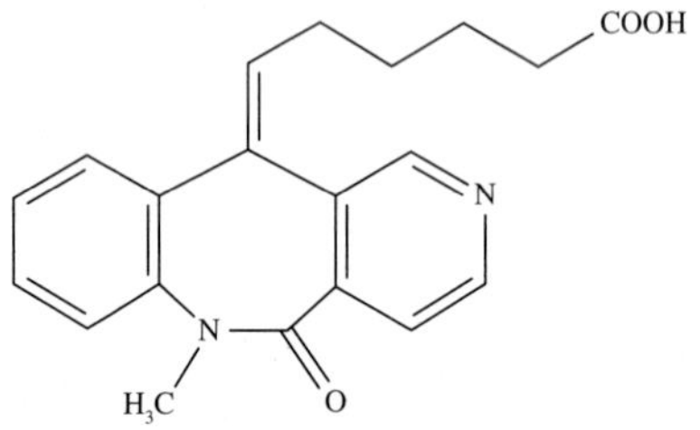

(6Z)-6-(5,6-Dihydro-6-methyl-5-oxo-11H-pyrido[4,3-c][1]benzazepin-11-ylidene)hexanoic acid (●)
S KF 13218
U Thromboxane synthase inhibitor

9336 (7588) $C_{20}H_{20}N_2O_3$
113206-32-9

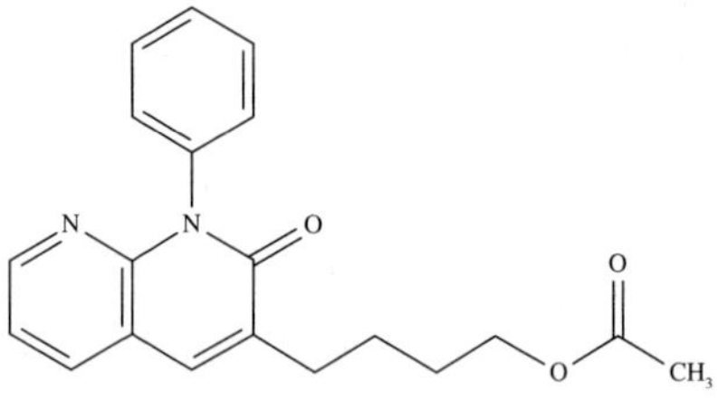

3-[4-(Acetyloxy)butyl]-1-phenyl-1,8-naphthyridin-2(1H)-one (●)

S UP 5145-52

U Anti-ulcer agent, gastric antisecretory

9337 (7589) $C_{20}H_{20}N_2O_3$
42438-73-3

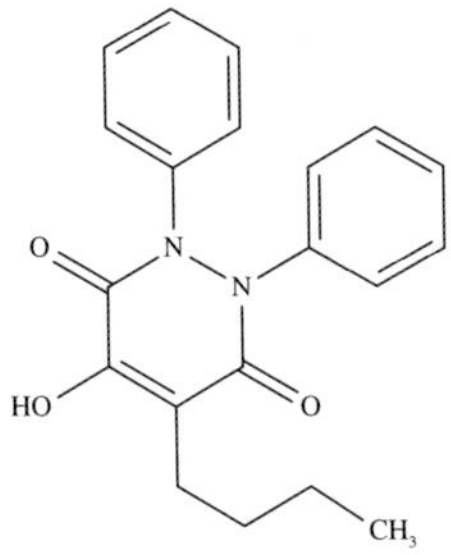

4-Butyl-1,2-dihydro-5-hydroxy-1,2-diphenyl-3,6-pyrida-zinedione (●)

S Denpidazone**

U Muscle relaxant

9338 (7590) $C_{20}H_{20}N_2O_4$
10351-50-5

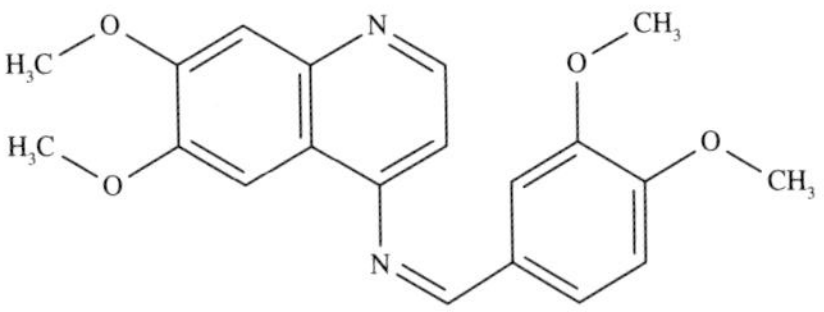

6,7-Dimethoxy-4-(veratrylideneamino)quinoline = N-[(3,4-Dimethoxyphenyl)methylene]-6,7-dimethoxy-4-quinolinamine (●)

S EU-1085, Leniquinsin**, U-1085

U Antihypertensive

9339 (7591) $C_{20}H_{20}N_2O_6$
73625-86-2

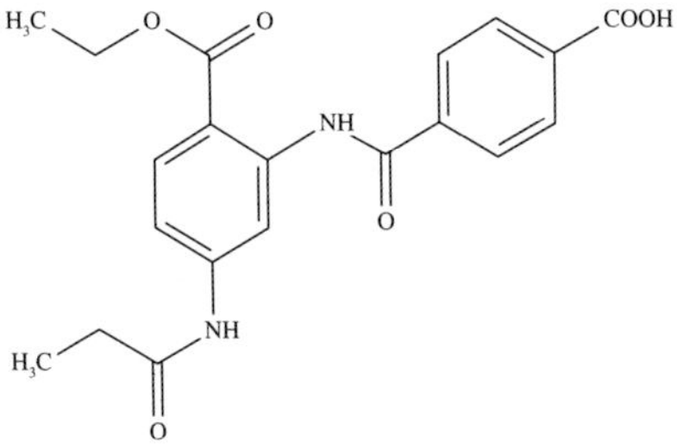

4-[[[2-(Ethoxycarbonyl)-5-[(1-oxopropyl)amino]phe-nyl]amino]carbonyl]benzoic acid = Ethyl 2-(4-carboxy-benzamido)-4-propionamidobenzoate = 2-[(4-Carboxy-benzoyl)amino]-4-[(1-oxopropyl)amino]benzoic acid 1-ethyl ester (●)

R Monosodium salt (124391-91-9)

S AM-682

U Anti-allergic

9340 (7592) $C_{20}H_{20}N_2O_7S$
59753-24-1

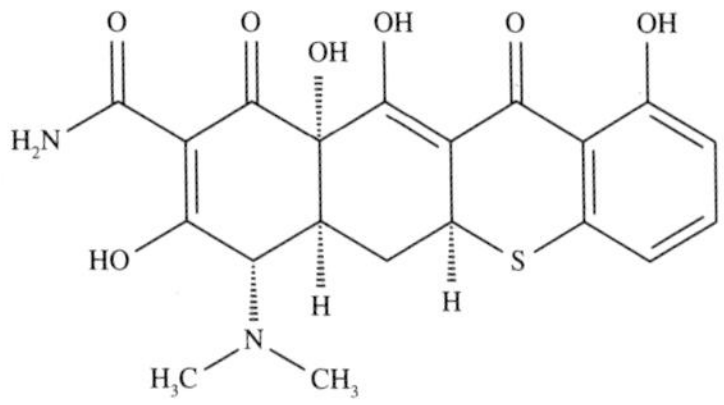

(4R*,4aR*,12aR*)-4-(Dimethylamino)-1,4,4a,5,5a,6,11,12a-octahydro-3,10,12,12a-tetrahydr-oxy-1,11-dioxo-6-thianaphthacene-2-carboxamide = [5aS-(5aα,6aα,7α,10aα)]-7-(Dimethylamino)-5a,6,6a,7,10a,12-hexahydro-1,8,10a,11-tetrahydroxy-10,12-dioxo-10H-benzo[b]thioxanthene-9-carboxamide (●)

S EMD 33330, Thiacycline, 6-Thiatetracycline

U Antibiotic, anthelmintic

9341 $C_{20}H_{20}N_2O_8S$
165948-72-1

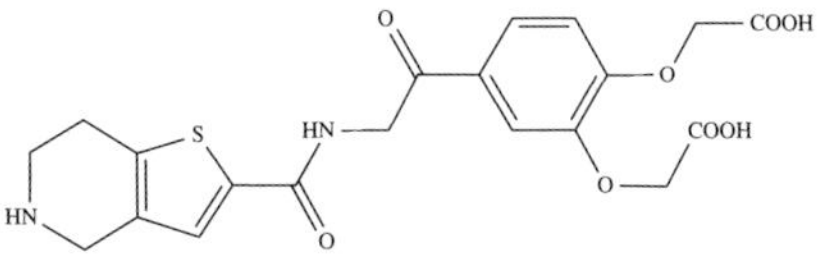

2,2'-[[4-[[[(4,5,6,7-Tetrahydrothieno[3,2-c]pyridin-2-yl)carbonyl]amino]acetyl]-1,2-phenylene]bis(oxy)]bis[acetic acid] (●)

R Monosodium salt

S ME-3277

U GP IIb/IIIa receptor antagonist

9342 (7593)   C$_{20}$H$_{20}$N$_2$O$_8$S$_2$
                                                83912-90-7

[1aS-(1α,4β,4aβ,8α,9β,11aα,12aR*)]-4,4a,8,9-Tetrahy-
dro-4-hydroxy-9-(2-hydroxy-3,4-dimethoxyphenyl)-
12H-8,11a-(iminomethano)-1aH,7H-[1,2,4]dithiazepi-
no[4,3-b]oxireno[e][1,2]benzoxazine-7,13-dione (●)
S *Gliovirin*
U Antifungal antibiotic from *Gliocladium virens*

9343   C$_{20}$H$_{20}$N$_2$S
                                                115475-38-2

(E)-11-[3-(Dimethylamino)propylidene]-6,11-dihydrodi-
benzo[b,e]thiepin-2-carbonitrile (●)
R Monohydrobromide (115475-40-6)
S BTS 56424, *Cyanodothiepin hydrobromide*
U Antidepressant

9344 (7594)   C$_{20}$H$_{20}$N$_4$O
                                                39038-32-9

2-(4-Methylpiperazinomethyl)-1,3-diazafluoranthene 1-
oxide = 2-[(4-Methyl-1-piperazinyl)methyl]inde-
no[1,2,3-de]quinazoline 1-oxide (●)
S AC 3579, NSC-170561
U Tranquilizer, antineoplastic

9345   C$_{20}$H$_{20}$N$_4$O$_3$
                                                181048-29-3

2-(2-Methylphenyl)-5,7-dimethoxy-4-quinolylcarbonyl-
guanidine = N-(Aminoiminomethyl)-5,7-dimethoxy-2-(2-
methylphenyl)-4-quinolinecarboxamide (●)
R Dihydrochloride
S MS-31-050
U Cardioprotective (Na$^+$/H$^+$ exchange inhibitor)

9346   C$_{20}$H$_{20}$N$_4$O$_2$
                                                181048-36-?

2-Phenyl-8-(2-methoxyethoxy)-4-quinolylcarbonylgua-
nidine = N-(Aminoiminomethyl)-8-(2-methoxyethoxy)-
2-phenyl-4-quinolinecarboxamide (●)
R Bis(methanesulfonate)
S MS-31-038
U Cardioprotective (Na$^+$/H$^+$ exchange inhibitor)

9347 (7595)   C$_{20}$H$_{20}$N$_4$O$_4$S
                                                114568-26-?

4-[(2-Succinimidoethyl)thio]phenyl 4-guanidinobenzoate
= 4-[(Aminoiminomethyl)amino]benzoic acid 4-[[2-(2,5-
dioxo-1-pyrrolidinyl)ethyl]thio]phenyl ester (●)
R Monomethanesulfonate (114568-32-0)
S E-3123, *Patamostat mesilate***

U   Trypsin inhibitor

9348 $C_{20}H_{20}N_4S$
                                    153628-86-5

5-[4-(Phenylmethyl)-1-piperazinyl]pyrrolo[1,2-a]thieno[3,2-e]pyrazine (●)
S   S 21007
U   5-HT$_3$ receptor agonist

9349   (7596) $C_{20}H_{20}N_6O_3$
                                    104961-19-5

5-Methoxy-3-(1-methylethoxy)-1-phenyl-N-1H-tetrazol-5-yl-1H-indole-2-carboxamide (●)
R   Compd. with L-arginine (1:1) (121530-58-3)
S   CI-949
U   Anti-inflammatory, anti-allergic

9350   (7597) $C_{20}H_{20}N_6O_5$
                                    18921-65-8

N-[4-[[(2,4-Diamino-6-quinazolinyl)methyl]amino]benzoyl]-L-aspartic acid (●)
NSC-112846, *Quinospar*, SK 29728

U   Antineoplastic

9351   (7598) $C_{20}H_{20}N_6O_7S_4$
                                      69739-16-8

(6R,7R)-7-[2-(2-Amino-4-thiazolyl)glyoxylamido]-3-[[[5-(carboxymethyl)-4-methyl-2-thiazolyl]thio]methyl]-8-oxo-5-thia-1-azabicyclo[4.2.0]oct-2-ene-2-carboxylic acid 7^2-(Z)-(O-methyloxime) = (Z)-(7R)-7-[2-(2-Amino-4-thiazolyl)-2-(methoxyimino)acetamido]-3-[[[5-(carboxymethyl)-4-methyl-2-thiazolyl]thio]methyl]-3-cephem-4-carboxylic acid = [6R-[6α,7β(Z)]]-7-[[(2-Amino-4-thiazolyl)(methoxyimino)acetyl]amino]-3-[[[5-(carboxymethyl)-4-methyl-2-thiazolyl]thio]methyl]-8-oxo-5-thia-1-azabicyclo[4.2.0]oct-2-ene-2-carboxylic acid (●)
R   Disodium salt (86329-79-5)
S   CDZM, *Cefodizime sodium**, Cefolactam, Diezime, HR 221, Kenicef, Modivid, Neucef, Noycef, Opticef, S 771221 B, THR 221, Timecef, Timicef, Unizime
U   Antibiotic

9352   (7599) $C_{20}H_{20}N_6O_9S$
                                      64952-97-2

N-[(6R,7R)-2-Carboxy-7-methoxy-3-[[(1-methyl-1H-tetrazol-5-yl)thio]methyl]-8-oxo-5-oxa-1-azabicyclo[4.2.0]oct-2-en-7-yl]-2-(p-hydroxyphenyl)malonamic acid = (7R)-7-[2-Carboxy-2-[p-hydroxyphenyl]acetamido]-7-methoxy-3-[[(1-methyl-5-tetrazolyl)thio]methyl]-1-oxa-3-cephem-4-carboxylic acid = [6R-[6α,7α,7(R*)]]-7-[[Carboxy-(4-hydroxyphenyl)acetyl]amino]-7-methoxy-3-[[(1-methyl-1H-tetrazol-5-yl)thio]methyl]-8-oxo-5-oxa-1-azabicyclo[4.2.0]oct-2-ene-2-carboxylic acid (●)
R   Disodium salt (64953-12-4)

S Betalactam, Festamoxin, *Lamoxactam disodium*, *Latamoxef disodium***, Latoxacef, LMOX, LY 127935, Mactam, Moxa "Ital Suisse", Moxacef "Pulitzer", Moxalactam, *Moxalactam disodium*, Moxam, Moxatres, Oxacef, Oxalactam, Oxalam, Oxcef, Polimoxal, Priolatt, 6059-S, Sectam, Shiomarin

U Antibiotic

9353 $C_{20}H_{20}N_6S$
162192-95-2

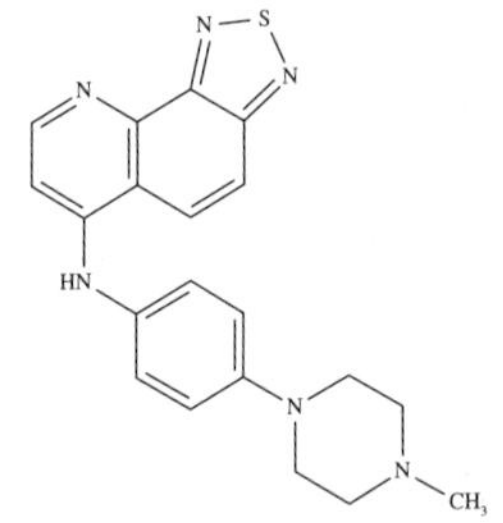

6-[[4-(4-Methyl-1-piperazinyl)phenyl]amino]-1,2,5-thiadiazolo[3,4-*h*]quinoline = *N*-[4-(4-Methyl-1-piperazinyl)phenyl]-[1,2,5]thiadiazolo[3,4-*h*]quinolin-6-amine (●)

S Drug G-1574

U Antiparasitic

9354 $C_{20}H_{20}O_4$
59870-68-7

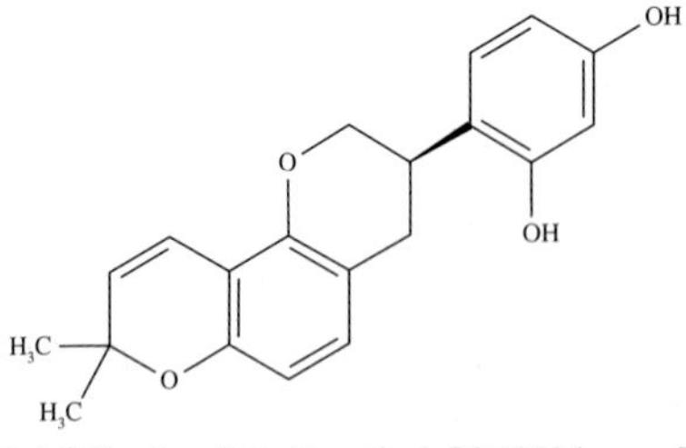

4-[(3*R*)-3,4-Dihydro-8,8-dimethyl-2*H*,8*H*-benzo[1,2-*b*:3,4-*b*']dipyran-3-yl]-1,3-benzenediol (●)

S *Glabridin*

U Anti-inflammatory, melanin synthesis inhibitor from licorice extract

---

9355 $C_{20}H_{20}O_5$
153890-03-0

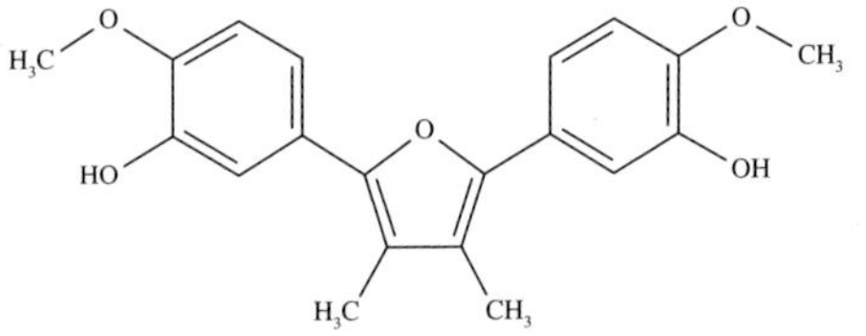

3,3'-(3,4-Dimethyl-2,5-furandiyl)bis[6-methoxyphenol] (●)

S GS 01

U Antineoplastic (anti-angiogenic)

9356 $C_{20}H_{20}O_5$
99217-68-2

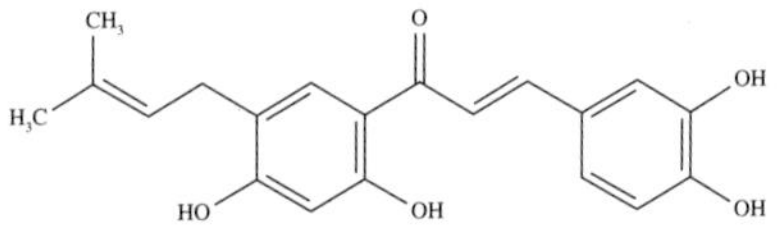

(*E*)-1-[2,4-Dihydroxy-5-(3-methyl-2-butenyl)phenyl]-3-(3,4-dihydroxyphenyl)-2-propen-1-one (●)

S *Broussochalcone A*

U Protein kinase C inhibitor from *Broussonetia papyrifera*

9357 $C_{20}H_{21}ClFN_3O_3$
151213-16-8

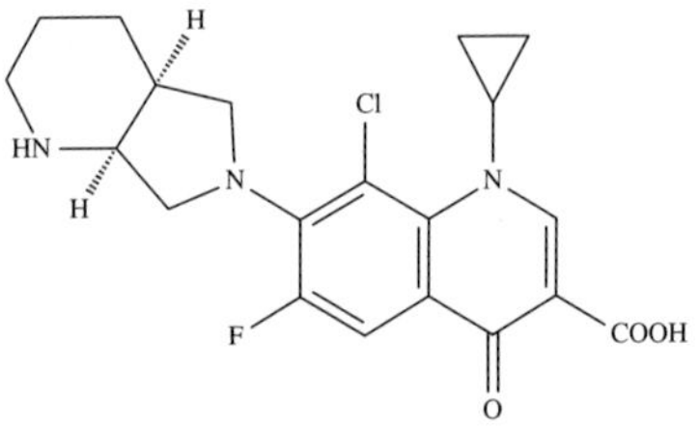

(4a*S*-*cis*)-8-Chloro-1-cyclopropyl-6-fluoro-1,4-dihydro-7-(octahydro-6*H*-pyrrolo[3,4-*b*]pyridin-6-yl)-4-oxo-3-quinolinecarboxylic acid (●)

R Monohydrochloride (144194-96-7)

S Bay y 3118

U Antibacterial

9358 $C_{20}H_{21}ClN_2O$
81226-60-0

4-(4-Chlorophenyl)-1-(1*H*-indol-3-ylmethyl)-4-piperidi-
nol (●)
S L 741626
U Antipsychotic (dopamine D_4 receptor antagonist)

9359 (7600) $C_{20}H_{21}ClN_2O_2$
71119-12-5

2'-Benzoyl-4'-chloro-*N*-methyl-2-[(2-methylallyl)ami-
no]acetanilide = 2'-Benzoyl-4'-chloro-*N*-methyl-*N*'-(2-
methylallyl)glycinanilide = *N*-(2-Benzoyl-4-chlorophe-
nyl)-*N*-methyl-2-[(2-methyl-2-propenyl)amino]acet-
amide (●)
S *Dinazafone***, F 1797
U Anxiolytic

9360 (7601) $C_{20}H_{21}ClN_2O_2$
95853-92-2

5-Chloro-1-methyl-2-(3-pyridinyl)-1*H*-indole-3-hexa-
noic acid (●)
R Monohydrochloride (107190-22-7)
S CGS 15435 A
U Thromboxane synthetase inhibitor

9361 (7602) $C_{20}H_{21}ClN_2O_4$
34161-24-5

1-[(*p*-Chlorophenoxy)acetyl]-4-piperonylpiperazine = 1-
(1,3-Benzodioxol-5-ylmethyl)-4-[(4-chlorophenoxy)ace-
tyl]piperazine (●)
R Monohydrochloride (34161-23-4)
S Attentil, BP 662, *Fipexide hydrochloride***, Fi-
pexium, Vigalor, Vigilor, Vilor
U Antidepressant, psychotonic

9362 $C_{20}H_{21}ClN_6O_4S_2$
162684-35-7

(*R*)-*N*-[2-(2-Benzothiazolylthio)-1-methylethyl]-2-chlo-
roadenosine (●)
S NNC 21-0136
U Adenosine A_1 receptor agonist

9363 (7603) $C_{20}H_{21}ClO_3$
102612-16-8

7-Chloro-2-(*p*-methoxybenzyl)-3-methyl-5-propyl-4-
benzofuranol = 7-Chloro-2-[(4-methoxyphenyl)methyl]-
3-methyl-5-propyl-4-benzofuranol (●)
S L 656224
U Anti-allergic

9364   (7604) $C_{20}H_{21}ClO_4$
                                                49562-28-9

Isopropyl 2-[p-(p-chlorobenzoyl)phenoxy]-2-methylpro-
pionate = 2-Methyl-2-[4-(4-chlorobenzoyl)phenoxy]pro-
pionic acid isopropyl ester = 2-[4-(4-Chlorobenzoyl)-
phenoxy]-2-methylpropanoic acid 1-methylethyl ester (●)
S   Ankebin, *Benprofibratum*, Chlorosteran, Cil, Cli-
    mage, Controlip, Daquin, Durafenat, Elasterin, Fa-
    gran, Fagratyl, Feno 100, Feno-AbZ, Fenobeta,
    Fenobrate "Sanofi-Winthrop", Fenodur, Fenofanton,
    *Fenofibrate**, Fenolibs, Fenolip "Lannacher", Feno
    Sanorania, Fenotard, Fulcro, Gestefol, Grofibrat, Hi-
    peremex, Hyperchol, Hypolipid, Katalip, Letomode,
    LF 178, Lipanon, Lipanthyl, Lipantil, Liparison, Lip-
    cor, Lipidax, Lipidil, Lipidil-Ter, Lipidrina, Lipil,
    Lipoclar, Lipofen(e), Liponat, Lipoplasmin, Lipo
    red, Liposit, Lipotriad, Lipovas, Lipsin, Lisomyex,
    Lofat, Neo-Disterin, Nolipax, Norm 100, Normalip,
    Normalip pro, Normolip "Synthesis", Palisix, Panli-
    pal, *Phenofibrate*, Planitrix, Procetin, *Procetofene*,
    Procetoken, Promeral, Protolipan, Qualecon, Saba-
    vulton, Scleril, Sclerofin (Reductor), Secalip, Sed-
    ufen, Sigurtil, Sitronella, Supralip, Tenactan, Tilene,
    Tricor "Abbott", Versamid, Volutine, W 13635, Zer-
    lubron
U   Antihyperlipidemic

9364-01   (7604-01)
R   Compd. with cinnarizine (no. 13119)
S   Fenobrate, *Fenofibrate cinnarizine*
U   Antihyperlipidemic

9365   (7605) $C_{20}H_{21}Cl_2NO_4$
                                             22204-91-7

1-Methyl-4-piperidyl glyoxylate 2-[bis(p-chlorophenyl)acetal] = 1-Methyl-4-piperidyl bis(p-chlorophenoxy)acetal = Bis(4-chlorophenoxy)acetic acid 1-methyl-
4-piperidinyl ester (●)
S   *Lifibrate**, MPCA, SaH 42-348, SaH-2348
U   Antihyperlipidemic

9366   (7606) $C_{20}H_{21}Cl_2NO_4$
                                             54063-27-3

1-Methyl-2-pyrrolidinylmethyl glyoxylate 2-[bis(p-chlorophenyl)acetal] = (1-Methyl 2-pyrrolidinyl)methyl bis(p-
chlorophenoxy)acetal = Bis(4-chlorophenoxy)acetic acid
(1-methyl-2-pyrrolidinyl)methyl ester (●)
S   *Biclofibrate**
U   Antihyperlipidemic

9367   (7607) $C_{20}H_{21}FN_2O$
                                             59729-33-8

1-[3-(Dimethylamino)propyl]-1-(p-fluorophenyl)-5-
phthalancarbonitrile = 1-[3-(Dimethylamino)propyl]-1-
(4-fluorophenyl)-1,3-dihydro-5-isobenzofurancarboni-
trile (●)
R   also monohydrobromide (59729-32-7)
S   Apertia, Celexa, Cipram, Cipramil, Ciprex "Parano-
    va", *Citalopram**, Elopram, Lu 10-171, Lupram,
    Nitalapram, Prisdal, Sepram, Seralgan, Seropram
U   Antidepressant

9368 $C_{20}H_{21}FN_2O$
128196-01-0

(+)-(S)-1-[3-(Dimethylamino)propyl]-1-(p-fluorophenyl)-5-phthalancarbonitrile = (1S)-1-[3-(Dimethylamino)propyl]-1-(4-fluorophenyl)-1,3-dihydro-5-isobenzo-
furancarbonitrile (●)
R   Oxalate (1:1) (219861-08-2)
S   *Escitalopram oxalate**, Lu 26-054-0
U   Antidepressant

9369   (7608) $C_{20}H_{21}FN_8O_5$
95755-20-7

N-[p-[[(2,4-Diamino-6-pteridinyl)methyl]-methylami-
no]benzoyl]-4-fluoroglutamic acid = N-[4-[[(2,4-Diami-
no-6-pteridinyl)methyl]methylamino]benzoyl]-4-fluoro-
glutamic acid (●)
S   *Fluoromethotrexate*, FMTX
U   Antineoplastic

9370 $C_{20}H_{21}FN_8O_{10}S_2$

[2R,5R,6R]-6-[2(R)-(2-Fluoro-4-hydroxyphenyl)-2-sul-
foacetamido]-2-[3-(2-semicarbazonoethylideneamino)-2-
oxo-1-imidazolidinyl]penam-2-carboxylic acid = [3R-
[3α(1E,2E),5α,6β]]-3-[3-[[[(Aminocarbonyl)hydrazo-
no]ethylidene]amino]-2-oxo-1-imidazolidinyl]-6-[[(2-
fluoro-4-hydroxyphenyl)sulfoacetyl]amino]-7-oxo-4-
thia-1-azabicyclo[3.2.0]heptane-3-carboxylic acid (●)
R   Disodium salt (168434-93-3)

S   T 5575
U   Antibacterial

9371 $C_{20}H_{21}F_2N_3O_4$
209340-26-1

7-[1-(Aminomethyl)-2-oxa-7-azabicyclo[3.3.0]oct-7-yl]-
1-cyclopropyl-6,8-difluoro-1,4-dihydro-4-oxo-3-quino-
linecarboxylic acid = 7-[6a-(Aminomethyl)hexahydro-
5H-furo[2,3-c]pyrrol-5-yl]-1-cyclopropyl-6,8-difluoro-
1,4-dihydro-4-oxo-3-quinolinecarboxylic acid (●)
R   Monohydrochloride (155448-15-0)
S   Bay x 9181
U   Antibacterial, proconvulsant

9372   (7609) $C_{20}H_{21}F_3N_2OS$
30223-48-4

10-[3-(Diethylamino)propionyl]-2-(trifluoromethyl)phe-
nothiazine = 10-[3-(Diethylamino)-1-oxopropyl]-2-(trifluoromethyl)-10H-phenothiazine (●)
R   also monohydrochloride (27312-93-2)
S   *Fluacizine**, *Fluoracizine*, Ftoracizin, Phthoraci-
zin, Phtorazisin
U   Antidepressant

9373   (7610) $C_{20}H_{21}F_3N_2O_3$
40256-99-3

α-[[α-Methyl-*m*-(trifluoromethyl)phenethyl]carbamoyl]-
p-acetanisidide = 2-(*p*-Acetamidophenoxy)-*N*-[α-methyl-
m-(trifluoromethyl)phenethyl]acetamide = 2-[4-(Acetylamino)phenoxy]-*N*-[1-methyl-2-[3-(trifluoromethyl)phe-
nyl]ethyl]acetamide (●)
S   *Flucetorex***, PM 3944
U   Anorexic

9374   (7611) $C_{20}H_{21}F_3N_2S$
24495-62-3

10-(4-Methyl-1-piperazinyl)-8-(trifuoromethyl)-10,11-
dihydrodibenzo[*b,f*]thiepine = 1-[10,11-Dihydro-8-(trifluoromethyl)dibenzo[*b,f*]thiepin-10-yl]-4-methylpipera-
zine (●)
S   *Trifluthepin*
U   Neuroleptic

9375 $C_{20}H_{21}F_3N_4O$
167933-07-5

1-[2-[4-(α,α,α-Trifluoro-*m*-tolyl)-1-piperazinyl]ethyl]-2-
benzimidazolinone = 1,3-Dihydro-1-[2-[4-[3-(trifluoromethyl)phenyl]-1-piperazinyl]ethyl]-2*H*-benzimidazol-2-
one (●)
R   also monohydrochloride (147359-76-0)
S   BIMT 17BS, *Flibanserin***
U   Antidepressant (5-HT$_{1A}$ and 5-HT$_2$ receptor antagonist)

9376   (7613) $C_{20}H_{21}N$
303-53-7

3-(5*H*-Dibenzo[*a,d*]cyclohepten-5-ylidene)-*N,N*-dime-
thylpropylamine = *N,N*-Dimethyl-5*H*-dibenzo[*a,d*]cyclo-
heptene-$\Delta^{5,\gamma}$-propylamine = 5-(3-Dimethylaminopro-py-
lidene)dibenzo[*a,d*]cycloheptatriene = 3-(5*H*-Diben-
zo[*a,d*]cyclohepten-5-ylidene)-*N,N*-dimethyl-1-propan-
amine (●)
R   also hydrochloride (6202-23-9)
S   Alti-Cyclobenzaprine, Apo-Cyclobenzaprine,
Benzamin BDO, Ciclamil, *Cyclobenzaprine***, Cy-
cloflex, Flexeril, Flexiban, G 38116, Lisseril, Mio-
san, MK-130, Novo-Cycloprine, NSC-173379, *Pro-
heptatriene*, Ro 4-1577, 9715 R.P., Tensodox, Yure-
lax
U   Muscle relaxant, tranquilizer

9377   (7612) $C_{20}H_{21}N$
47166-67-

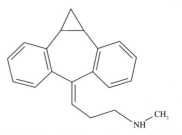

1a,10b-Dihydro-*N*-methyldibenzo[*a,e*]cyclopropa[*c*]cy-
cloheptene $\Delta^{6(1H),\gamma}$-propylamine = 3-(1a,10b-Dihydrodi-

benzo[a,e]cyclopropa[c]cyclohepten-6(1H)-ylidene)-N-methyl-1-propanamine (●)
R Phosphate (1:1) (51481-67-5)
S *Octriptyline phosphate**, SC-27123
U Antidepressant

9378 (7614)

$C_{20}H_{21}NO$
31232-26-5

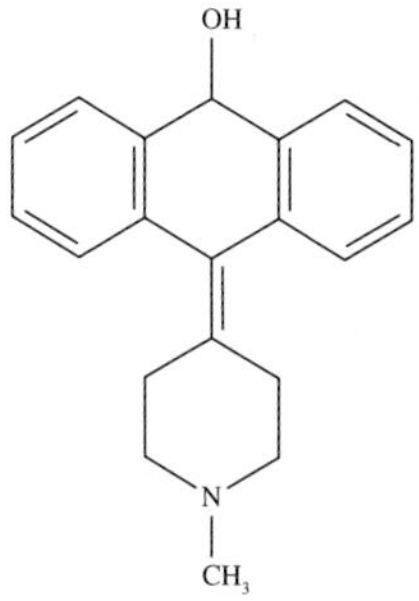

9,10-Dihydro-10-(1-methyl-4-piperidylidene)-9-anthrol
= 9,10-Dihydro-10-(1-methyl-4-piperidinylidene)-9-anthracenol (●)
S Danitracen**, WA 335-BS
U Antidepressant, serotonin antagonist, sexual stimulant

9379 (7615)

$C_{20}H_{21}NO$
968-63-8

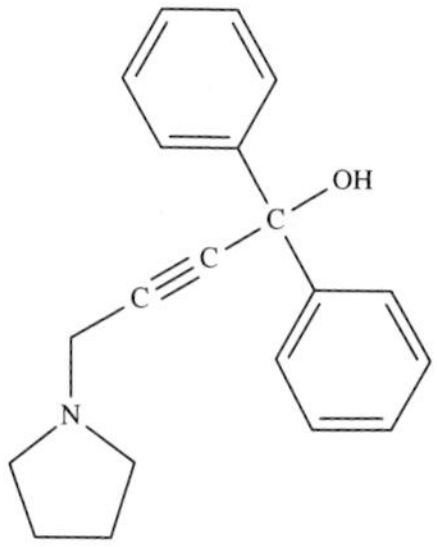

,1-Diphenyl-4-(1-pyrrolidinyl)-2-butyn-1-ol = α-Phenyl-α-[3-(1-pyrrolidinyl)-1-propynyl]benzenemethanol
)

Phosphate (1:1) (54118-66-0)
Azulon comp. (spasmolytic component), *Butinoline phosphate**, Homburg 811, Spasmo-Nervogastrol (spasmolytic component)
Anticholinergic, spasmolytic

9380 (7616)

$C_{20}H_{21}NO$
34662-67-4

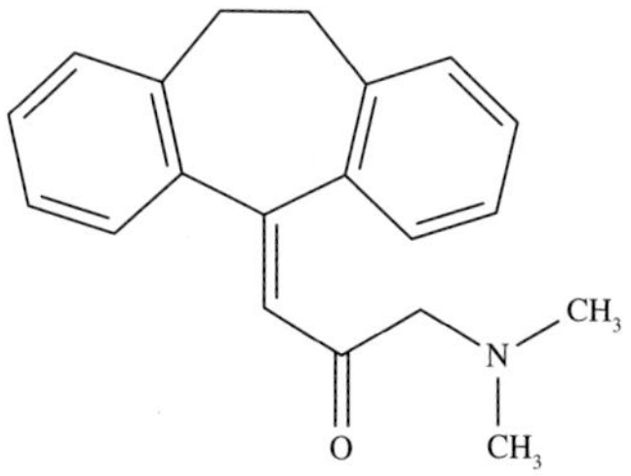

1-(10,11-Dihydro-5H-dibenzo[a,d]cyclohepten-5-ylidene)-3-(dimethylamino)-2-propanone (●)
S *Cotriptyline**, SD 2203-01
U Antidepressant

9381 (7617)

$C_{20}H_{21}NOS$
50838-36-3

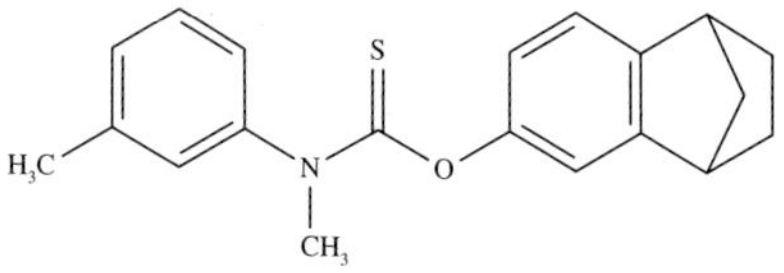

O-(1,2,3,4-Tetrahydro-1,4-methanonaphthalen-6-yl) m,N-dimethylthiocarbanilate = Methyl(3-methylphenyl)carbamothioic acid O-(1,2,3,4-tetrahydro-1,4-methanonaphthalen-6-yl) ester (●)
S FR 3143, Fungifos, K 9147, KC 9147, Kilmicen, *Tolciclate**, Tolmicen, Tolmicil, Tolmicol "Montedison", Toskil
U Antifungal

9382 (7618)

$C_{20}H_{21}NOS_2$
66788-41-8

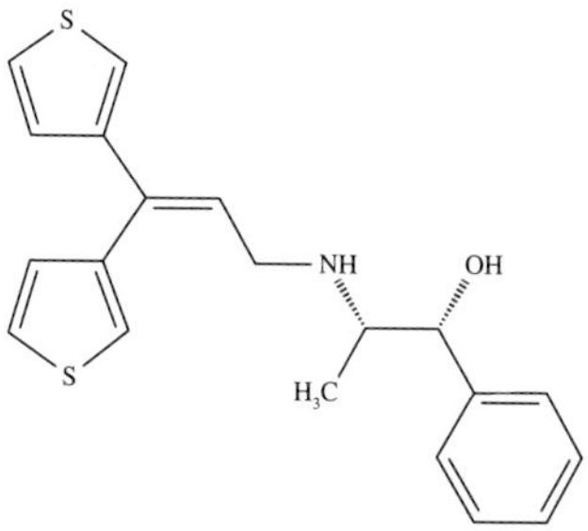

(+)-(R)-α-[(S)-1-[(3,3-Di-3-thienylallyl)amino]ethyl]benzyl alcohol = 2-[3,3-Di(3-thienyl)allylamino]-1-phenyl-1-propanol = [R-(R*,S*)]-α-[1-[(3,3-Di-3-thienyl-2-propenyl)amino]ethyl]benzenemethanol (●)

R Hydrochloride (50776-39-1)
S D 8955, Novocebrin, *Tinofedrine hydrochloride***
U Cerebral vasodilator

9383

$C_{20}H_{21}NO_2$

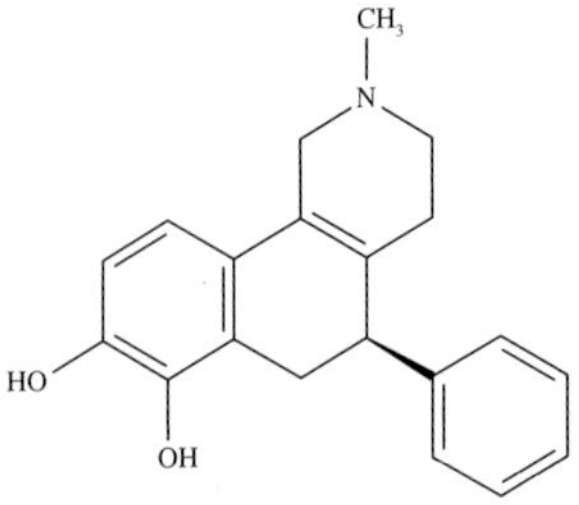

(*R*)-1,2,3,4,5,6-Hexahydro-2-methyl-5-phenyl-
benz[*h*]isoquinoline-7,8-diol
S A-85845
U Antiparkinsonian

9384 (7619)

$C_{20}H_{21}NO_2$
10539-19-2

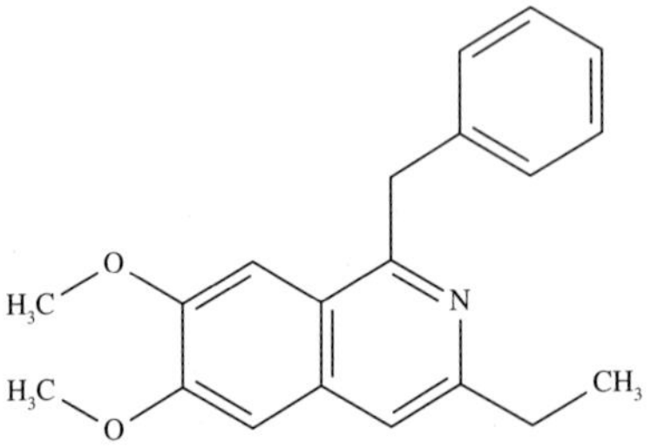

1-Benzyl-3-ethyl-6,7-dimethoxyisoquinoline = 3-Ethyl-
6,7-dimethoxy-1-(phenylmethyl)isoquinoline (●)
R Hydrochloride (1163-37-7)
S BEN, Certonal, Eupaverin(a), Kollateral, *Meteverine
hydrochloride*, *Moxaverine hydrochloride***, Paverin "Bracco"
U Spasmolytic

9385 (7620)

$C_{20}H_{21}NO_3$
1165-48-6

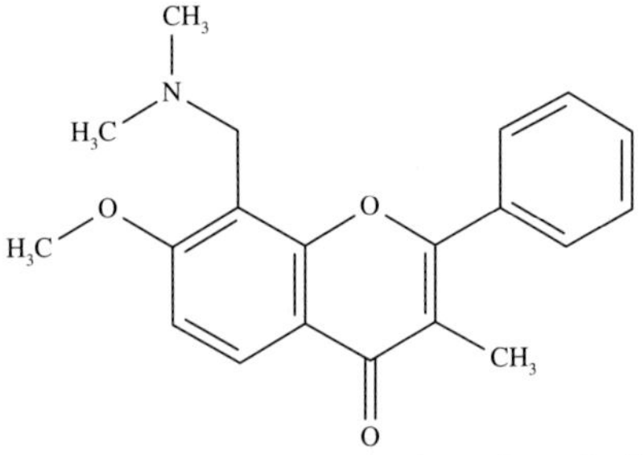

8-[(Dimethylamino)methyl]-7-methoxy-3-methylfla-
vone = 8-[(Dimethylamino)methyl]-7-methoxy-3-me-
thyl-2-phenylbenzo-γ-pyrone = 8-(Dimethylamino-me-
thyl)-7-methoxy-3-methyl-2-phenylchromen-4-one =
[(Dimethylamino)methyl]-7-methoxy-3-methyl-2-phe-
nyl-4*H*-1-benzopyran-4-one (●)
R Hydrochloride (2740-04-7)
S Demefline, *Dimefline hydrochloride***, DW-62,
NSC-114650, Reanimil, Rec 7/0267, Remeflin(e)
U Respiratory stimulant

9386 (7621)

$C_{20}H_{21}NO$
17854-59-

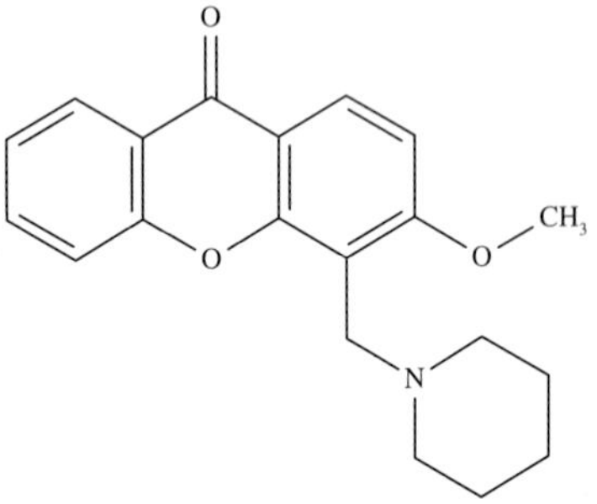

3-Methoxy-4-(piperidinomethyl)xanthone = 3-Methoxy-
4-(1-piperidinylmethyl)-9*H*-xanthen-9-one (●)
S *Mepixanox***, *Mepixantonum*, Pimexone
U Respiratory stimulant

9387 (7622)

$C_{20}H_{21}NO_4$
2565-01-7

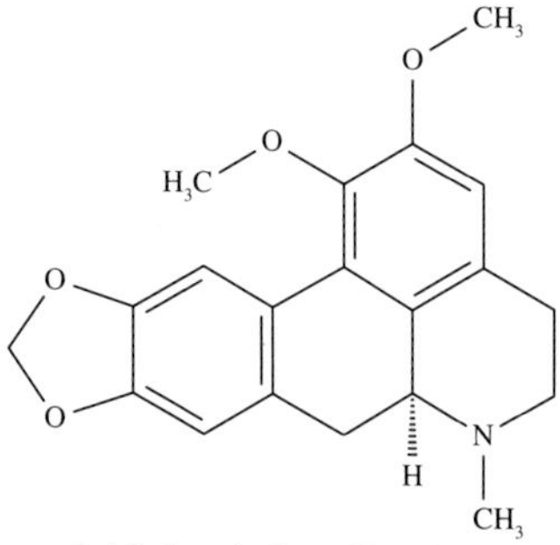

1,2-Dimethoxy-9,10-(methylenedioxy)aporphine = (S)-
5,6,6a,7-Tetrahydro-1,2-dimethoxy-6-methyl-4H-ben-
zo[de][1,3]benzodioxolo[5,6-g]quinoline (●)
S Domestine, Nantenine, O-Methyldomesticine
U Antitussive, antihypertensive (alkaloid from *Nandina
domestica*, *Corydalis* sp. and other plants)

9388

$C_{20}H_{21}NO_4$
25127-29-1

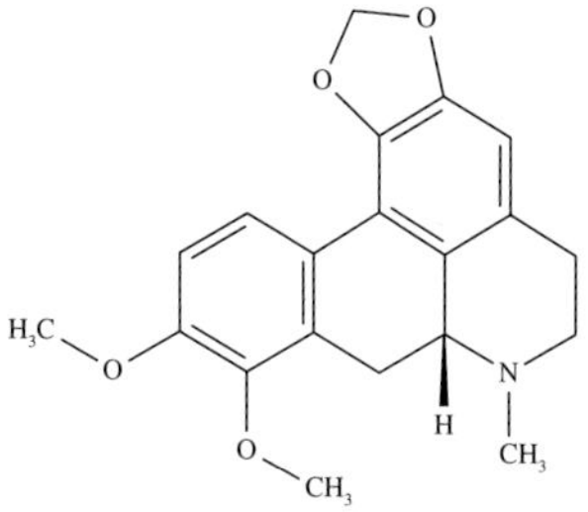

8,9-Dimethoxy-1,2-(methylenedioxy)aporphine = (7R)-
5,7,7a,8-Tetrahydro-9,10-dimethoxy-7-methyl-5H-ben-
zo[g]-1,3-benzodioxolo[6,5,4-de]quinoline (●)
S Crebanine
U Spasmolytic from *Stephania sasakii*

9389 (7623)

$C_{20}H_{21}NO_4$
517-66-8

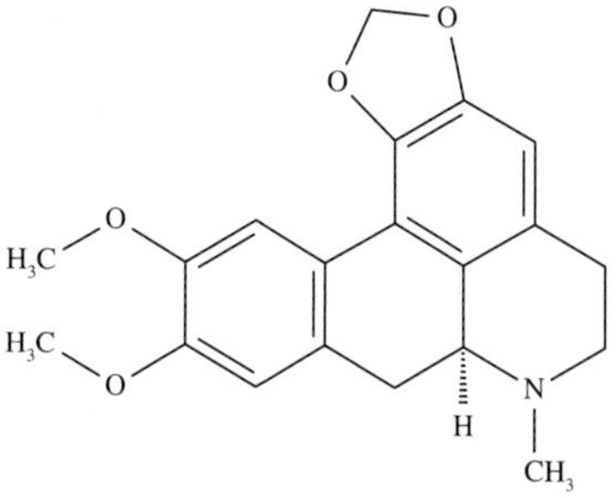

9,10-Dimethoxy-1,2-(methylenedioxy)-6aα-aporphine =
(S)-6,7,7a,8-Tetrahydro-10,11-dimethoxy-7-methyl-5H-
benzo[g]-1,3-benzodioxolo[6,5,4-de]quinoline (●)
S Dicentrine, Eximine
U α₁-Adrenoceptor antagonist

9390 (7624)

$C_{20}H_{21}NO_4$
522-97-4

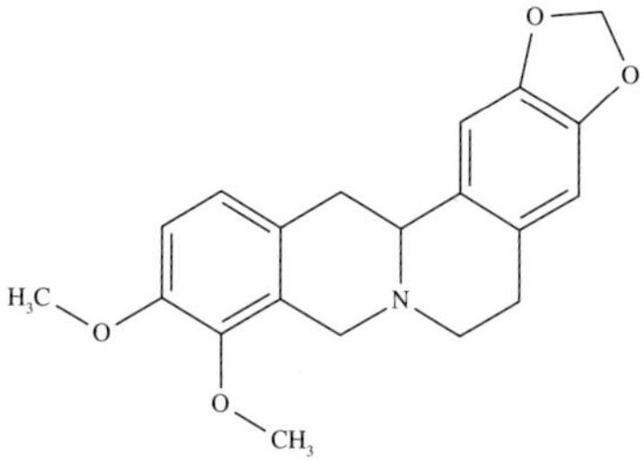

9,10-Dimethoxy-2,3-(methylenedioxy)berbine =
5,8,13,13a-Tetrahydro-9,10-dimethoxy-6H-benzo[g]-
1,3-benzodioxolo[5,6-a]quinolizine (●)
S Canadine, Tetrahydroberberine, Xanthopuccine
U Sedative, muscle relaxant (alkaloid from *Corydalis
cava*)

9391 (7625)

$C_{20}H_{21}NO_4$
95696-19-8

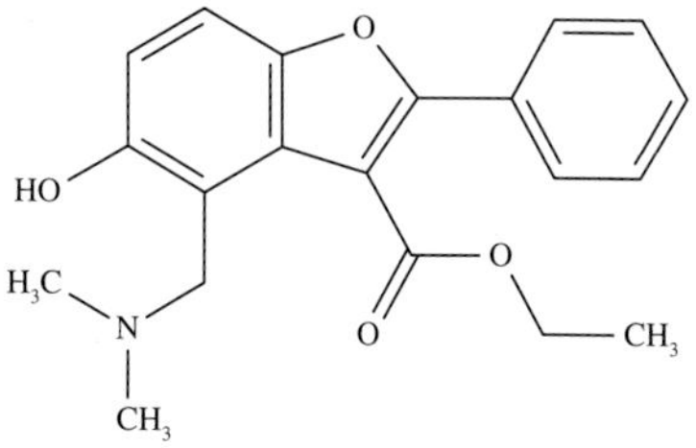

4-[(Dimethylamino)methyl]-5-hydroxy-2-phenyl-3-ben-
zofurancarboxylic acid ethyl ester (●)
R Hydrochloride (51771-50-7)
S Fenikaberan, Phenykaberan
U Coronary vasodilator, spasmolytic

9392 (7626) $C_{20}H_{21}NO_4$
58-74-2

1-(3,4-Dimethoxybenzyl)-6,7-dimethoxyisoquinoline =
6,7-Dimethoxy-1-veratrylisoquinoline = Papaveroline te-
tramethyl ether = 1-[(3,4-Dimethoxyphenyl)methyl]-6,7-
dimethoxyisoquinoline (●)
R also hydrochloride (61-25-6)
S Agmapav, Alapav, Amapol, Angioverin "Cor", Art-
egodan, Blupav, C "USV", Cardiospan, Cardoveri-
na, Cepaverin, Cerebid, Cerespan, Cervaspan, Cir-
bed, Delapav, Dilaspan, Dilaverin, Dilaves, Dipav,
Dispamil, Drapavel, Durapav, Dylate "Elder", Dyno-
vas, Eupavéryl, Forpavin, Genabid, Isquespan,
Kavrin, K-Pava, Lapav, Lemobid, Lempav, Myobid,
NSC-35443, Opdensit, Optenyl, Oxadilène, P-200,
Padoverin, Pameion, Panergon, Panpav TP, Papa-
chin N, Papacon, Papacontin, Papalease, Papaveril,
Papaverine, Papaverlumin fuerte, Papaversan, Papa-
véryl, Papavin, Papital T.R., Pap-Kaps, Pascopin,
P-A-V, Pava, PAVA-2, Pavabid, Pavacap(s), Pava-
cels, Pavacen, Pavaclar, Pavaclor, Pavaco, Pavacot,
Pavacron, Pavadel, Pavadur, Pavadyl, Pavagen, Pa-
vagrant, Pavakey, Pava-Lyn, Pava-Mead, Pava-Par,
Pavarine, Pava-RX, Pavased, Pavasil, Pavaspan, Pa-
vasule, Pavatest, Pavatime, Pavatine, Pavatran, Pa-
vatym, Pava-Wol, Paver, Paverine, Paver-Med, Pa-
verolan, Paveron, Pavex, Pavnell, Pavosal, Pavrin,
3 P-Pav, Promtpaverin, Qua-Bid, Robaxapap, Ro-Pa-
pav, S-M-R, Spasmolitico Prochena, Spasmone,
Spasmo-Nit, Sustaverine, Sy-Pav T.D., Tempav,
Therapav, Vasal, Vasocap, Vasoflo, Vaso-Pav,

Vas-O-Span, Vasospan "Ulmer", Vazosan, Veran-
trop
U Spasmolytic, vasodilator (opium alkaloid)

9392-01 (7626-01) 132-40-1
R Nitrite
S Panitrin
U Spasmolytic

9392-02 (7626-02) 1748-09-0
R Nicotinate
S Nicopavin, Nipavina, Paniverin, *Papaverini nicoti-
nas*
U Vasodilator

9392-03 (7626-03) 5949-36-0
R 5-Ethyl-5-phenylbarbiturate (no. 3255)
S Barberine, Pavemal
U Vasodilator

9392-04 (7626-04) 6591-59-9
R Phenylglycolate (mandelate) (no. 1133)
S Endoverine
U Spasmolytic

9392-05 (7626-05) 2053-26-1
R Sulfate (1:1)
S Maspaver inyectable, Synpaverine
U Spasmolytic

9392-06 (7626-06) 16960-31-1
R 3-(7-Theophyllinyl)propanesulfonate (no. 2308)
S Brunokal, Kaldil, LBC 319, *Papaverine teprosilate*
U Spasmolytic, coronary vasodilator

9392-07 (7626-07) 34317-37-
R Codecarboxylate (no. 1203)
S Albatran, PCDC
U Vasodilator

9392-08 (7626-08) 39024-96-
R Adenosine phosphate (no. 2310)
S CERM 3209, Dicertan, *Padefosum*, Sustein
U Vasodilator

9392-09 (7626-09) 19524-64-
R 6,7-Dihydroxycoumarin-4-methanesulfonate [6,7-di
hydroxy-2-oxo-2*H*-1-benzopyran-4-methanesulfo-
nate (●)]
S *Papaverine cromesilate*, Permavérine

U Vasoprotector

9393 $C_{20}H_{21}NO_4$
102585-03-5

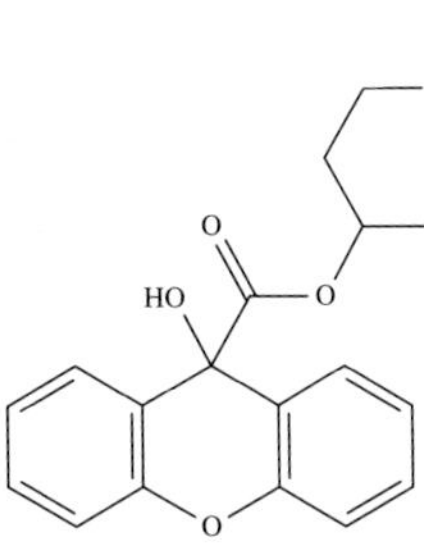

9-Hydroxy-9H-xanthene-9-carboxylic acid 1-methyl-4-piperidinyl ester (●)
S TCPN, *Tricyclopinate*
U Anticholinergic

9394 $C_{20}H_{21}NO_5$
216884-02-5

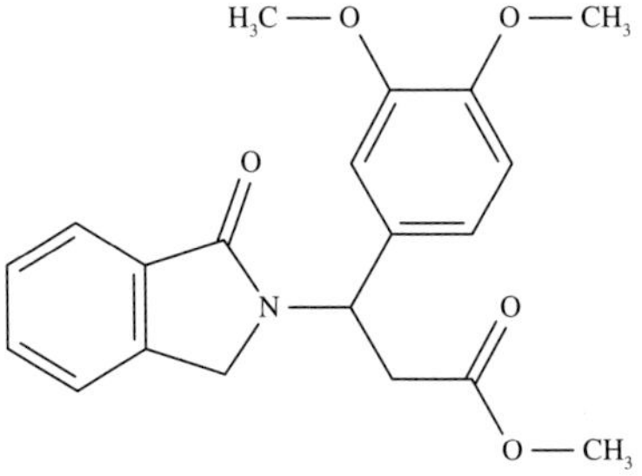

β-(3,4-Dimethoxyphenyl)-1,3-dihydro-1-oxo-2H-iso-indole-2-propanoic acid methyl ester (●)
S CC-3052
U Immunomodulator (TNF-α production inhibitor)

9395 (7627) $C_{20}H_{21}NO_5$
73080-51-0

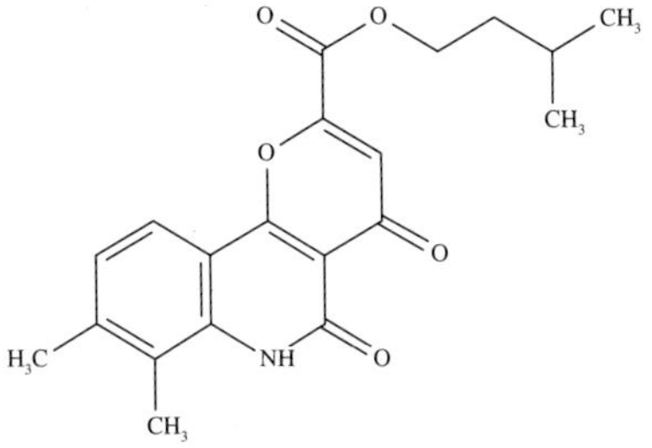

sopentyl 5,6-dihydro-7,8-dimethyl-4,5-dioxo-4H-pyra-no[3,2-c]quinoline-2-carboxylate = 5,6-Dihydro-7,8-di-

methyl-4,5-dioxo-4H-pyrano[3,2-c]quinoline-2-carb-oxylic acid 3-methylbutyl ester (●)
S Bay u 2372, MY-5116, *Repirinast***, Romet
U Anti-allergic, anti-asthmatic

9396 (7628) $C_{20}H_{21}NS$
1447-70-7

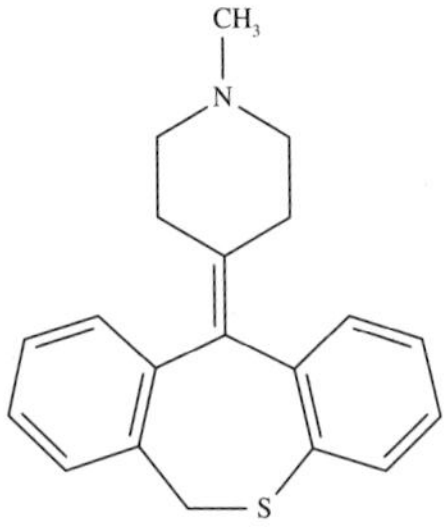

11-(1-Methyl-4-piperidylidene)-6,11-dihydrodiben-zo[b,e]thiepine = 4-Dibenzo[b,e]thiepin-11(6H)-ylidene-1-methylpiperidine (●)
R also hydrochloride (62126-73-2)
S Perithiaden (old form)
U Antidepressant

9397 (7629) $C_{20}H_{21}N_3$
137159-92-3

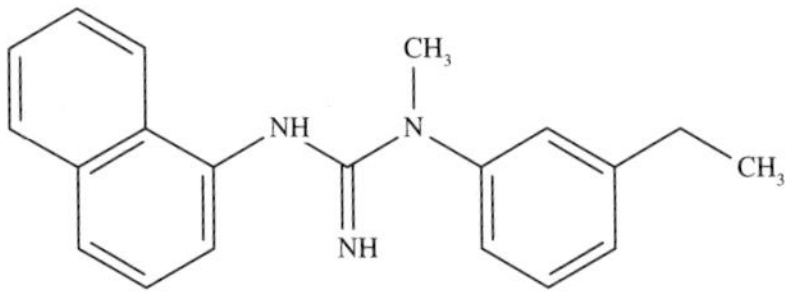

N-(3-Ethylphenyl)-N-methyl-N'-1-naphthalenylguani-dine (●)
R Monohydrochloride (137160-11-3)
S *Aptiganel hydrochloride***, Cerestat, CNS 1102
U NMDA (N-methyl-D-aspartate) antagonist

9398   (7630)

$C_{20}H_{21}N_3$
122456-37-5

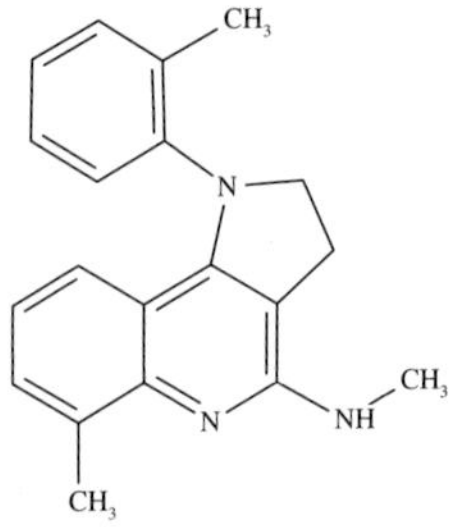

2,3-Dihydro-6-methyl-4-(methylamino)-1-(2-methylphe-
nyl)pyrrolo[3,2-c]quinoline = 2,3-Dihydro-*N*,6-dimethyl-
1-(2-methylphenyl)-1*H*-pyrrolo[3,2-c]quinolin-4-amine
(●)
S   MDPQ, SKF 96356
U   Anti-ulcer agent

9399   (7631)

$C_{20}H_{21}N_3O$
120635-74-7

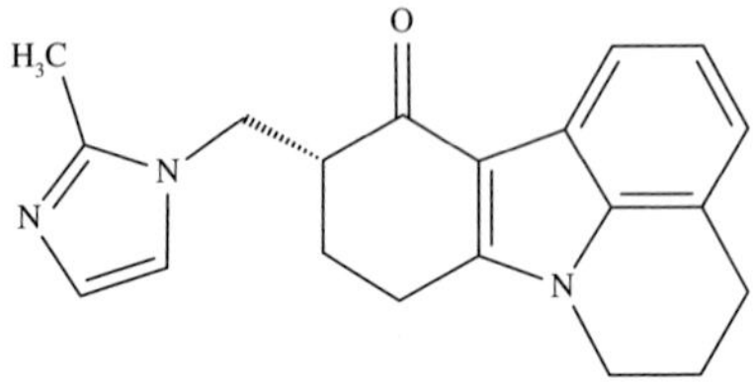

(*R*)-5,6,9,10-Tetrahydro-10-[(2-methyl-1*H*-imidazol-1-
yl)methyl]-4*H*-pyrido[3,2,1-jk]carbazol-11(8*H*)-one (●)
R   also monohydrochloride (120635-72-5)
S   *Cilansetron***, KC-9946
U   Anti-emetic (serotonin antagonist)

9400   (7632)

$C_{20}H_{21}N_3O_2$
52210-62-5

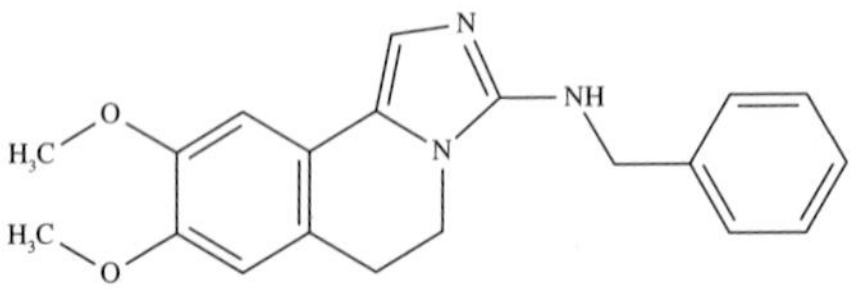

3-(Benzylamino)-5,6-dihydro-8,9-dimethoxyimida-
zo[5,1-*a*]isoquinoline = 5,6-Dihydro-8,9-dimethoxy-*N*-
(phenylmethyl)imidazo[5,1-*a*]isoquinolin-3-amine (●)
R   Monohydrochloride (52210-64-7)
S   BIIA, HE-36
U   Anti-arrhythmic

9401

$C_{20}H_{21}N_3O_2S_2$
174849-92-4

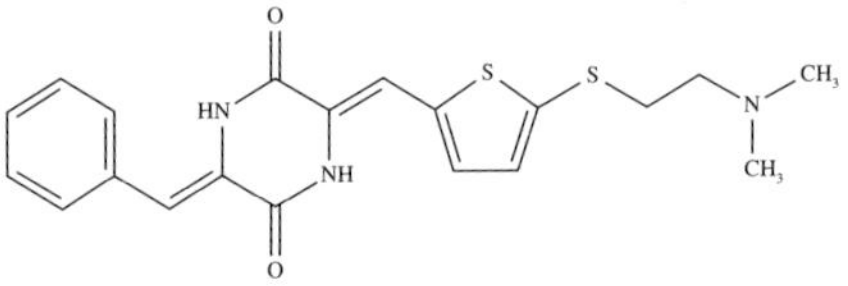

(Z,Z)-3-[[5-[[2-(Dimethylamino)ethyl]thio]-2-thie-
nyl]methylene]-6-(phenylmethylene)-2,5-piperazine-
dione (●)
R   Monohydrochloride (174766-49-5)
S   XR-5118
U   Plasminogen activator inhibitor

9402

$C_{20}H_{21}N_3O_2S_2$
140893-49-8

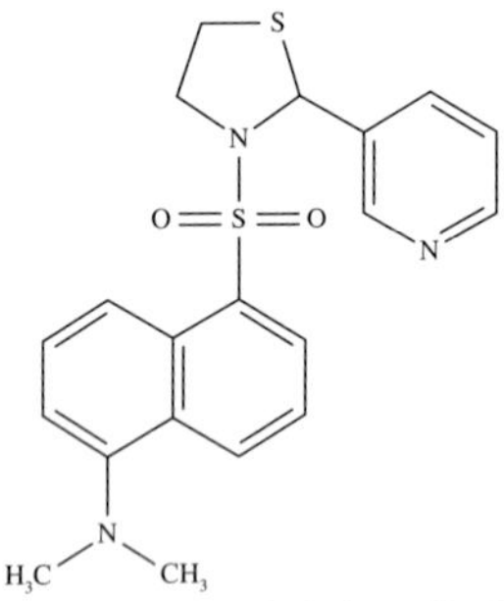

3-[[5-(Dimethylamino)-1-naphthalenyl]sulfonyl]-2-(3-
pyridinyl)thiazolidine (●)
S   YHI-1
U   Antiviral (HIV-1) from *Sardinops melanostricta*

9403   (7633)

$C_{20}H_{21}N_3O_3$
19395-58-5

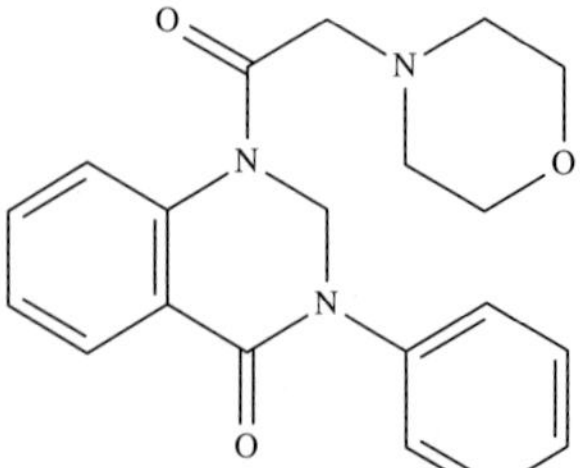

2,3-Dihydro-1-(4-morpholinylacetyl)-3-phenyl-4-(1*H*)-
quinazolinone (●)
R   also monohydrochloride (19395-78-9)
S   *Acemoquinazone, Moquizone***, Peristil, Rec 14/
0127, Recobilina

U Choleretic, antihyperlipidemic

9404 (7634) $C_{20}H_{21}N_3O_4$
                                 38964-88-4

5-(5,5-Diphenyl-3-hydantoinyl)norvaline = α-Amino-δ-(diphenylhydantoinyl)valeric acid = α-Amino-2,5-dioxo-4,4-diphenyl-1-imidazolidinepentanoic acid (●)

S Citrullamon V, *Fenivaline*, Neo-Citrullamon, *Phenytoin-3-norvaline*

U Anti-epileptic

9405 $C_{20}H_{21}N_3O_4$
                                 148396-36-5

(3S,5S)-5-[[(4'-Amidino-4-biphenylyl)oxy]methyl]-2-oxo-3-pyrrolidineacetic acid = (3S,5S)-5-[[[4'-(Aminoiminomethyl)[1,1'-biphenyl]-4-yl]oxy]methyl]-2-oxo-3-pyrrolidineacetic acid (●)

R see also no. 11778

S BIBU 52ZW, *Fradafiban***

U Fibrinogen receptor, ($α_{2B}/β_3$)-antagonist

9406 $C_{20}H_{21}N_3O_5$
                                 164210-17-7

Methylcarbonylmethyl 2(S)-[4-(4-guanidinobenzoyl)oxy)phenyl]propionate = (S)-4-[[4-[(Aminoiminomethyl)amino]benzoyl]oxy]-α-methylbenzeneacetic acid 2-oxopropyl ester (●)

R Monomethanesulfonate (164210-18-8)

S TT-S 24

U Proteinase inhibitor (in pancreatitis)

9407 (7635) $C_{20}H_{21}N_3O_5S$
                                 87027-09-6

2-Methyl-3-(2-pyridylcarbamoyl)-2H-1,2-benzothiazin-4-yl pivalate S,S-dioxide = 2,2-Dimethylpropanoic acid 2-methyl-3-[(2-pyridinylamino)carbonyl]-2H-1,2-benzothiazin-4-yl ester S,S-dioxide (●)

S CHF 10/21, Ciclafast, *Piroxicam pivalate*, Pivaloxicam, Unicam

U Anti-inflammatory

9408 (7636) $C_{20}H_{21}N_3O_6$
                                 137103-31-2

5-Methyl-3-(phenylmethyl)-1-β-D-ribofuranosylpyrido[2,3-d]pyrimidine-2,4(1H,3H)-dione (●)

S TI-79

U Antineoplastic, antileukemic

9409   C$_{20}$H$_{21}$N$_3$O$_6$
                                               173485-80-8

3-Hydroxy-L-valine (6-carboxy-4-methoxy-1-phenazi-
nyl)methyl ester (●)
S   LL 14I352alpha, *Pelagiomicin A*
U   Antibacterial, antineoplastic from marine bacterium
    Pelagiobacter variabilis

9410   (7637)   C$_{20}$H$_{21}$N$_3$O$_7$S
                                               99464-64-9

(±)-4-(1-Hydroxyethoxy)-2-methyl-*N*-2-pyridyl-2*H*-1,2-
benzothiazine-3-carboxamide ethyl carbonate (ester) 1,1-
dioxide = Carbonic acid ethyl 1-[[2-methyl-3-[(2-pyridi-
nyl)aminocarbonyl]-2*H*-1,2-benzothiazin-4-yl]oxy]ethyl
ester *S,S*-dioxide (●)
S   *Ampiroxicam***, CP-65703, Flucam, Nacyl, Nasil
U   Anti-inflammatory

9411   (7638)   C$_{20}$H$_{21}$N$_3$O$_8$S$_3$

1-Phenyl-3-[*p*-(2-pyridylsulfamoyl)anilino]propane-1,3-
disulfonic acid = 2-[*p*-(γ-Phenylpropylamino)benzenesul-
fonamido]pyridine-α,γ-disulfonic acid = 1-Phenyl-3-[[4-
[(2-pyridinylamino)sulfonyl]phenyl]amino]-1,3-propane-
disulfonic acid (●)

R   Disodium salt (53778-51-1)
S   *Solupyridine, Sulfapyridine Neutral Soluble*
U   Antibacterial

9412   (7639)   C$_{20}$H$_{21}$N$_3$O$_9$
                                               114580-45-9

1-Acetyl-7,9-bis(acetyloxy)-8-[[(aminocarbonyl)oxy]methyl]-1,1a,2,8,9,9a-hexahydro-3,9-epoxy-3*H*-
azirino[2,3-*c*][1]benzazocine-5-carboxaldehyde (●)
S   FK 973, FR 66973
U   Antineoplastic

9413   (7640)   C$_{20}$H$_{21}$N$_3$S
                                               24579-08-6

8-Cyano-10-(4-methyl-1-piperazinyl)-10,11-dihydrodi-
benzo[*b,f*]thiepine = 10,11-Dihydro-11-(4-methyl-1-pipe-
razinyl)dibenzo[*b,f*]thiepine-2-carbonitrile (●)
S   Cyanothepin
U   Neuroleptic

9414 (7641) $C_{20}H_{21}N_3S$
83747-76-6

3-[[2-[(1-Phenyl-2-pyrrolidinyliden)amino]ethyl]thio]-1H-indole = 2-(1H-Indol-3-ylthio)-N-(1-phenyl-2-pyrrolidinylidene)ethanamine (●)
S McN-4130
U Anti-arrhythmic, antifibrillatory

9415 $C_{20}H_{21}N_5O_3S$
137403-12-4

N-[4-[[5-[3-(2-Aminoethyl)-1H-indol-5-yl]-1,2,4-oxadiazol-3-yl]methyl]phenyl]methanesulfonamide (●)
S L 694247
U 5-HT$_{1A}$/5-HT$_{1B/D}$ receptor agonist

9416 $C_{20}H_{21}N_5O_6$
137281-23-3

N-[4-[2-(2-Amino-4,7-dihydro-4-oxo-1H-pyrrolo[2,3-d]pyrimidin-5-yl)ethyl]benzoyl]-L-glutamic acid (●)
R Disodium salt (150399-23-8)
S Alimta, LY 231514, *Pemetrexed disodium***, Rolazar, Tifolar
U Antineoplastic (thymidylate synthase inhibitor)

9417 $C_{20}H_{21}N_7O_5S_4$
189448-23-5

7-[2-(2-Amino-4-thiazolyl)-2-[(Z)-hydroxyimino]acetamido]-3-[[3-(2-ammonioethylthiomethyl)-4-pyridinyl]thio]ceph-3-em-4-carboxylate = (6R,7R)-3-[[3-[[(2-Aminoethyl)thio]methyl]-4-pyridinyl]thio]-7-[[(2Z)-(2-amino-4-thiazolyl)(hydroxyimino)acetyl]amino]-8-oxo-5-thia-1-azabicyclo[4.2.0]oct-2-ene-2-carboxylic acid (●)
S MC 02479
U Antibacterial (gram-positive)

9418 (7642) $C_{20}H_{21}N_7O_6$
2179-16-0

N-[p-[1-(2-Amino-4-hydroxy-6-pteridinyl)ethylamino]benzoyl]-L-glutamic acid = 9-Methylpteroyl-L-glutamic acid = N-[4-[[1-(2-Amino-1,4-dihydro-4-oxo-6-pteridinyl)ethyl]amino]benzoyl]-L-glutamic acid (●)
S Bremfol, Ninopterin
U Antineoplastic (antimetabolite)

9419 (7643) $C_{20}H_{21}N_7O_6$
2410-93-7

N-[p-[N-(2-Amino-4-hydroxy-6-pteridinylmethyl)methylamino]benzoyl]-L-glutamic acid = N^{10}-Methylpteroyl-L-glutamic acid = N-[4-[[(2-Amino-1,4-dihydro-4-oxo-6-pteridinyl)methyl]methylamino]benzoyl]-L-glutamic acid (●)
S Methopterine
U Antineoplastic (antimetabolite)

9420 (7644) $C_{20}H_{21}N_7O_6S_2$
60925-61-3

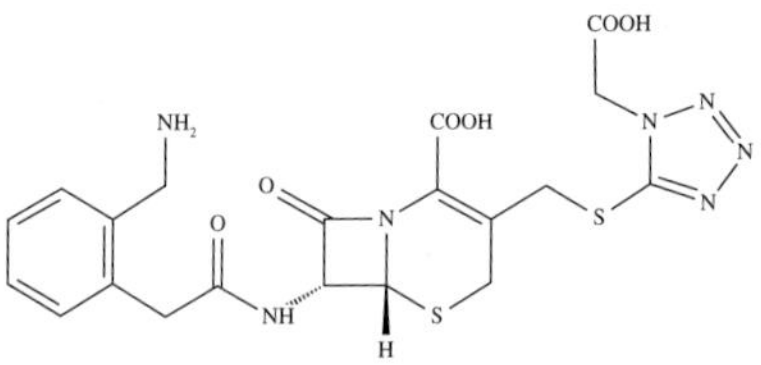

(6R,7R)-7-[2-(α-Amino-o-tolyl)acetamido]-3-[[[1-(carb-oxymethyl)-1H-tetrazol-5-yl]thio]methyl]-8-oxo-5-thia-1-azabicyclo[4.2.0]oct-2-ene-2-carboxylic acid = (7R)-7-[[2-(Aminomethyl)phenyl]acetamido]-3-[[[1-(carboxy-methyl)-1H-tetrazol-5-yl]thio]methyl]-3-cephem-4-carb-oxylic acid = (7R)-7-[2-(α-Amino-o-tolyl)acetamido]-3-(1-carboxymethyl-1H-tetrazol-5-ylthiomethyl)-3-ce-phem-4-carboxylic acid = (6R-trans)-7-[[[2-(Amino-methyl)phenyl]acetyl]amino]-3-[[[1-(carboxymethyl)-1H-tetrazol-5-yl]thio]methyl]-8-oxo-5-thia-1-azabicy-clo[4.2.0]oct-2-ene-2-carboxylic acid (●)
S BL-S 786, *Ceforanide***, Precef, Procef "Mead-John-son", Radacef
U Antibiotic

9421 $C_{20}H_{21}N_7O_7$
10538-99-5

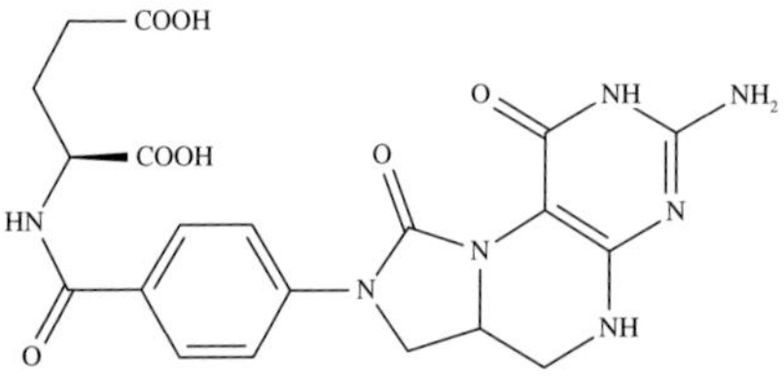

5,6,7,8-Tetrahydro-N^5,N^{10}-carbonylfolic acid = N-[4-(3-Amino-1,2,5,6,6a,7-hexahydro-1,9-dioxoimida-zo[1,5-f]pteridin-8(9H)-yl)benzoyl]-L-glutamic acid (●)
S LY 354899
U Antiproliferative

9422 (7645) $C_{20}H_{22}ClN$
91-82-7

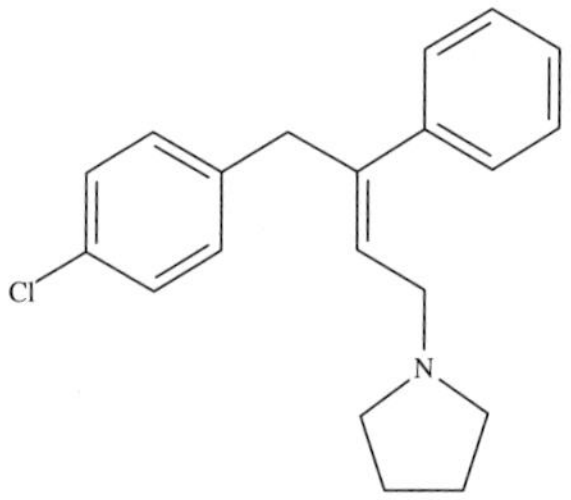

1-(p-Chlorophenyl)-2-phenyl-4-pyrrolidino-2-butene = 1-[4-(4-Chlorophenyl)-3-phenyl-2-butenyl]pyrrolidine (●)
R also phosphate (1:2) (135-31-9)
S *Pirrobutamina*, Proladyl, Pyronil, *Pyrrobutamine*
U Antihistaminic

9423 $C_{20}H_{22}ClNO$
192517-00-3

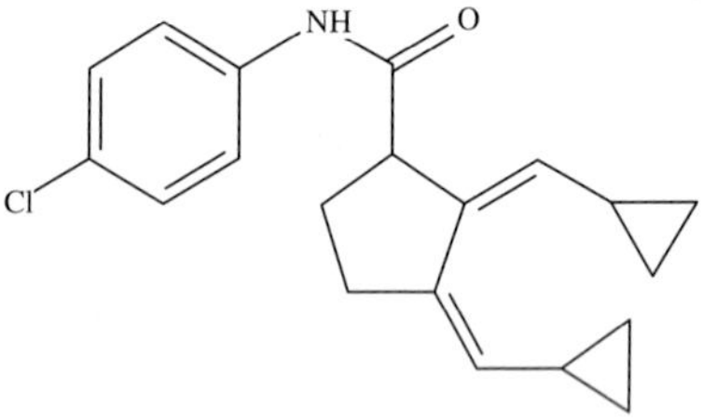

(Z,Z)-N-(4-Chlorophenyl)-2,3-bis(cyclopropylmethy-lene)cyclopentanecarboxamide (●)
S L 245976
U Anti-androgen

9424 $C_{20}H_{22}ClNO_6$
252721-95-2

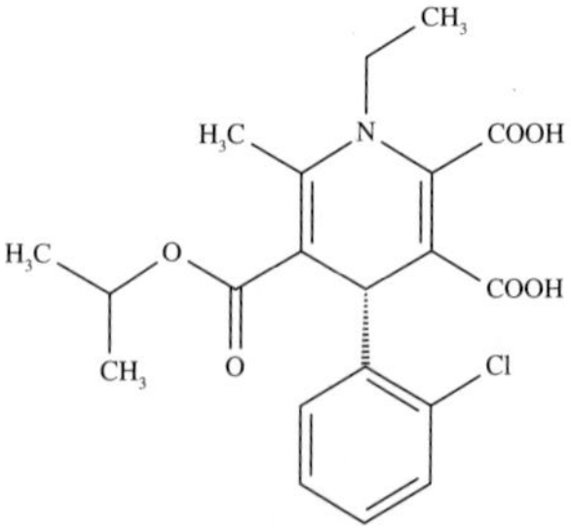

(4S)-4-(2-Chlorophenyl)-1-ethyl-1,4-dihydro-6-methyl-2,3,5-pyridinetricarboxylic acid 5-(1-methylethyl) ester (●)
S W 1807

U Antidiabetic

9425 (7646) $C_{20}H_{22}ClN_3$
14169-54-1

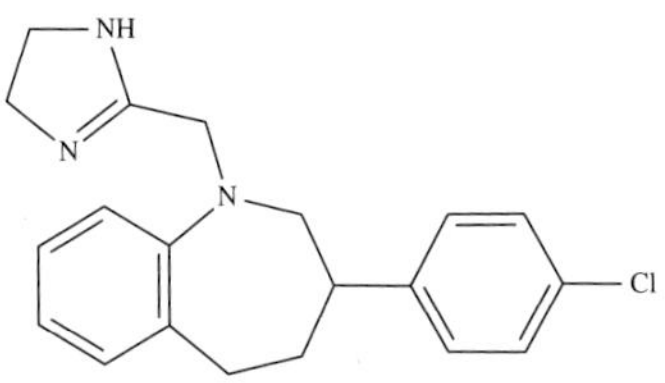

3-(p-Chlorophenyl)-1-(2-imidazolin-2-ylmethyl)-2,3,4,5-
tetrahydro-1H-1-benzazepine = 3-(4-Chlorophenyl)-1-
[(4,5-dihydro-1H-imidazol-2-yl)methyl]-2,3,4,5-tetrahy-
dro-1H-1-benzazepine (●)
R Monohydrochloride (12379-49-6)
S Su-13197
U Anti-arrhythmic

9426 (7647) $C_{20}H_{22}ClN_3$
21228-13-7

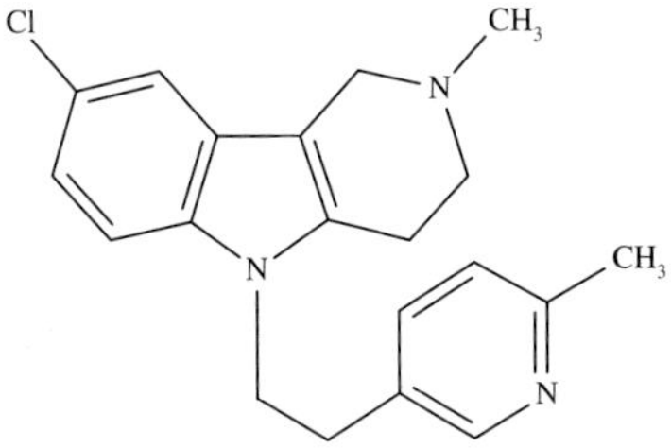

8-Chloro-2,3,4,5-tetrahydro-2-methyl-5-[2-(6-methyl-3-
pyridinyl)ethyl]-1H-pyrido[4,3-b]indole (●)
R Dihydrochloride (21228-28-4)
S *Dorastine hydrochloride**, Ro 5-9110/1
U Antihistaminic

9427 (7648) $C_{20}H_{22}ClN_3O$
86-42-0

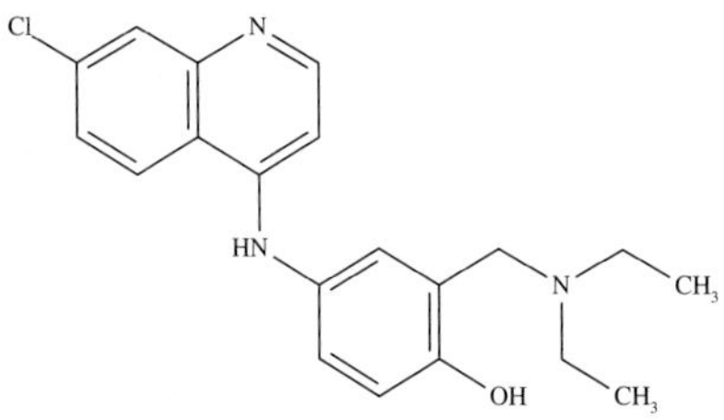

4-[(7-Chloro-4-quinolyl)amino]-α-(diethylamino)-o-cre-
sol = 7-Chloro-4-[3-(diethylaminomethyl)-4-hydroxyani-
lino]quinoline = 4-[(7-Chloro-4-quinolinyl)amino]-2-
[(diethylamino)methyl]phenol (●)
R Dihydrochloride dihydrate (6398-98-7)
S *Amodiachinhydrochlorid, Amodiaquine hydrochloride***, Amoquine "PDH", Basoquin, CAM-AQ 1, Ca-
mochin, Camoquinal, Camoquin(e), Dosaquine, Fla-
voquine, Miaquin, 4281 R.P., S.N. 10751, WR 2977
U Antimalarial

9428 (7649) $C_{20}H_{22}ClN_3O$
117796-62-0

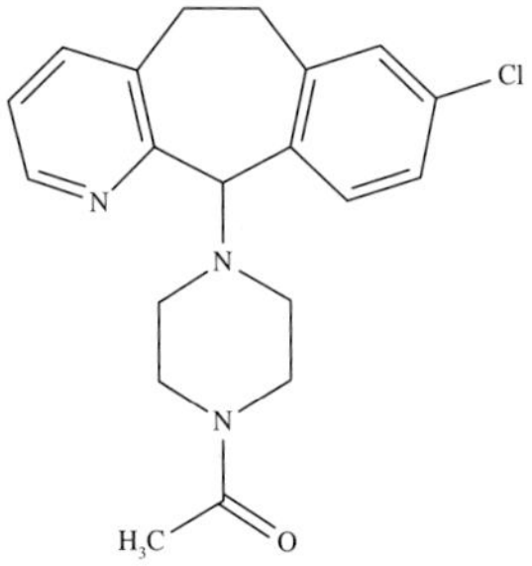

1-Acetyl-4-(8-chloro-6,11-dihydro-5H-benzo[5,6]cyclo-
hepta[1,2-b]pyridin-11-yl)piperazine (●)
S Sch 40338
U Anti-allergic

9429 $C_{20}H_{22}ClN_3O_3S_2$
209481-20-9

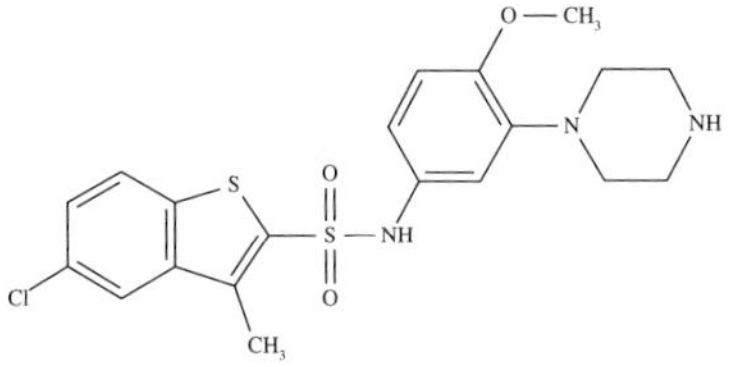

5-Chloro-N-[4-methoxy-3-(1-piperazinyl)phenyl]-3-me-
thylbenzo[b]thiophene-2-sulfonamide (●)
R Monohydrochloride (209481-24-3)
S SB 271046
U $5HT_6$-receptor antagonist

9430 (7650) $C_{20}H_{22}ClN_3O_4$
                                                        57083-89-3

1-(4-Amino-5-chloro-*o*-anisoyl)-4-piperonylpiperazine =
1-(4-Amino-5-chloro-2-methoxybenzoyl)-4-piperonylpi-
perazine = 1-(4-Amino-5-chloro-2-methoxybenzoyl)-4-
(1,3-benzodioxol-5-ylmethyl)piperazine (●)
S   *Peralopride***
U   Antidepressant

9431 (7651) $C_{20}H_{22}ClN_5O_2$
                                                        92990-90-4

4-Azido-5-chloro-*N*-(1-benzyl-4-piperidyl)-*o*-anisamide
= 4-Azido-5-chloro-2-methoxy-*N*-[1-(phenylmethyl)-4-
piperidinyl]benzamide (●)
S   *Azapride, Azidoclebopride*
U   Dopamine antagonist

9432 (7652) $C_{20}H_{22}Cl_2N_2O_4$
                                                        75616-02-3

2-[Bis(2-hydroxyethyl)amino]-4'-chloro-2'-(*o*-chloroben-
zoyl)-*N*-methylacetanilide = 2-[Bis(2-hydroxyethyl)ami-
no]-*N*-[4-chloro-2-(2-chlorobenzoyl)phenyl]-*N*-methyl-
acetamide (●)

S   *Dulozafone***, F 1933
U   Anxiolytic, anticonvulsant

9433 (7653) $C_{20}H_{22}Cl_2N_2O_6$
                                                        125729-29-5

(±)-3-Isopropyl 5-methyl 2-[(carbamoyloxy)methyl]-4-
(2,3-dichlorophenyl)-1,4-dihydro-6-methyl-3,5-pyridine-
dicarboxylate = (±)-2-[[(Aminocarbonyl)oxy]methyl]-4-
(2,3-dichlorophenyl)-1,4-dihydro-6-methyl-3,5-pyridine-
dicarboxylic acid 5-methyl 3-(1-methylethyl) ester (●)
R   also without (±)-definition (94739-29-4)
S   *Bamilodipine, Lemildipine***, NB 818, NPK-1886
U   Antihypertensive (calcium antagonist)

9434 (7654) $C_{20}H_{22}Cl_2N_4O_2$
                                                        32021-40-2

β-[1-Phenyl-5-[bis(β-chloroethyl)amino]-2-benzimida-
zolyl]-DL-alanine = (±)-α-Amino-5-[bis(2-chloro-
ethyl)amino]-1-phenyl-1*H*-benzimidazole-2-propanoic
acid (●)
S   CIMET 3164, ZIMET 3164
U   Immunosuppressive

9435　　　　　　　　　　　　　　$C_{20}H_{22}FNO_4$
171866-31-2

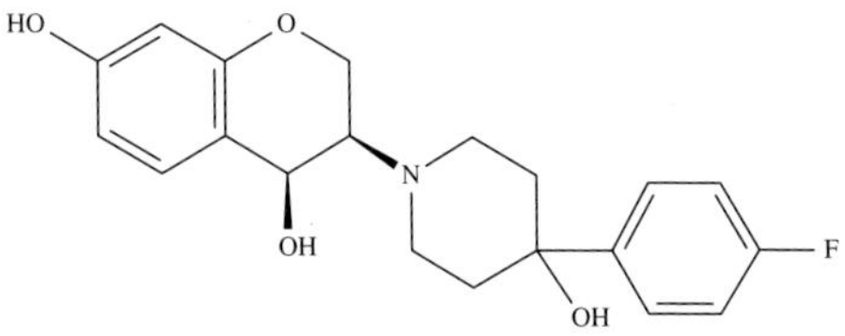

(3R,4S)-3-[4-(4-Fluorophenyl)-4-hydroxy-1-piperidi-
nyl]chroman-4,7-diol = (3R-cis)-3-[4-(4-Fluorophenyl)-
4-hydroxy-1-piperidinyl]-3,4-dihydro-2H-1-benzopyran-
4,7-diol (●)
S CP-283097
U NMDA receptor antagonist

9436　　　　　　　　　　　　　　$C_{20}H_{22}FN_3O_3$
108461-05-8

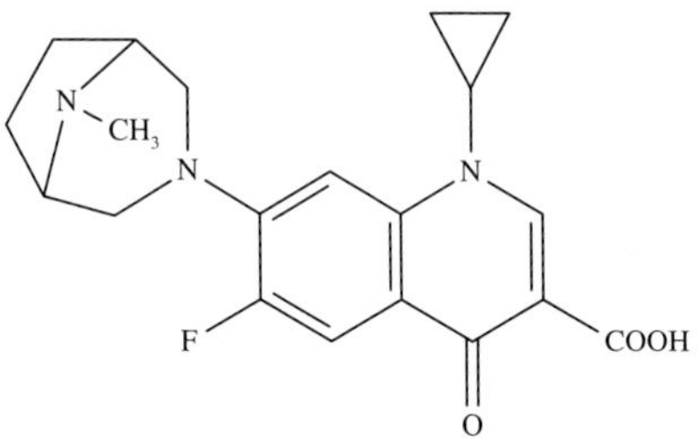

1-Cyclopropyl-6-fluoro-1,4-dihydro-7-(8-methyl-3,8-di-
azabicyclo[3.2.1]oct-3-yl)-4-oxo-3-quinolinecarboxylic
acid (●)
S CP-74667
U Antibacterial

9437　　　　　　　　　　　　　　$C_{20}H_{22}FN_3O_4$
151390-79-3

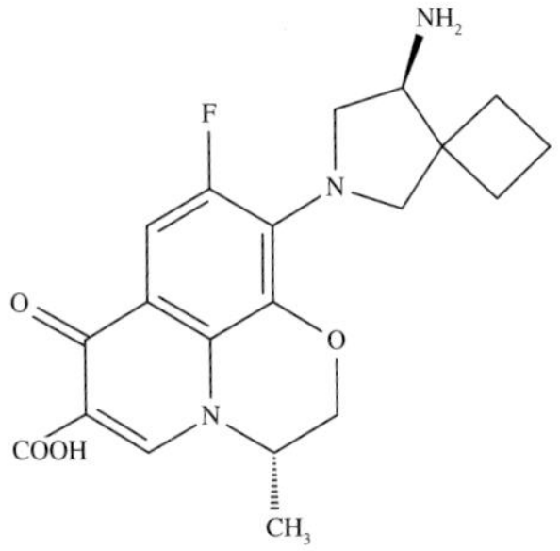

(3S)-10-[(8S)-8-Amino-6-azaspiro[3.4]oct-6-yl]-9-
fluoro-2,3-dihydro-3-methyl-7-oxo-7H-pyrido[1,2,3-de]-
1,4-benzoxazine-6-carboxylic acid (●)

R also hemihydrate
S DV 7751a
U Antibacterial

9438 (7655)　　　　　　　　　　　$C_{20}H_{22}FN_3O_6$
100587-52-8

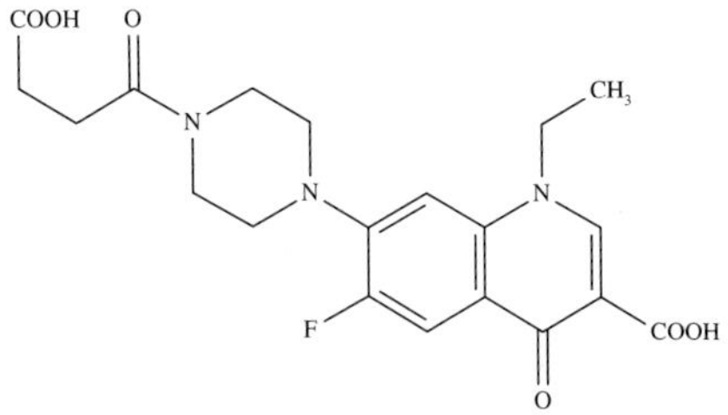

7-[4-(3-Carboxypropionyl)-1-piperazinyl]-1-ethyl-6-
fluoro-1,4-dihydro-4-oxo-3-quinolinecarboxylic acid = 7-
[4-(3-Carboxy-1-oxopropyl)-1-piperazinyl]-1-ethyl-6-
fluoro-1,4-dihydro-4-oxo-3-quinolinecarboxylic acid (●)
S Eminor, *Norfloxacin succinil***
U Antibacterial

9439 (7656)　　　　　　　　　　　$C_{20}H_{22}F_3NO_4$
23191-75-5

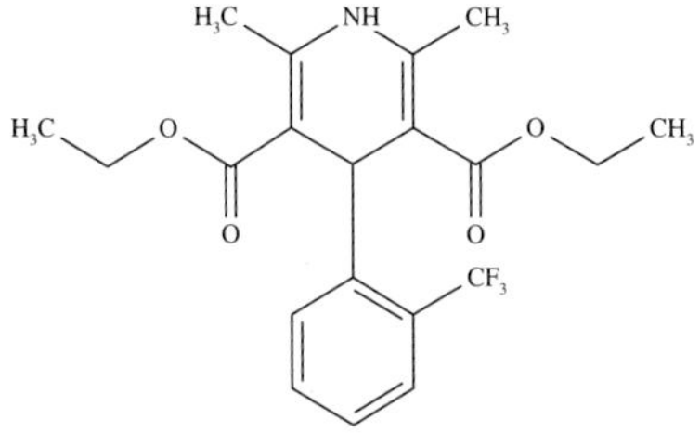

Diethyl 1,4-dihydro-2,6-dimethyl-4-[2-(trifluoromethyl)phenyl]-3,5-pyridinedicarboxylate = 1,4-Dihydro-
2,6-dimethyl-4-[2-(trifluoromethyl)phenyl]-3,5-pyridine-
dicarboxylic acid diethyl ester (●)
S Ftorin, SKF 24260
U Antihypertensive, vasodilator

9440   (7657) $C_{20}H_{22}NO_2$

6-Ethyl-1,2-(methylenedioxy)-6aβ-aporphinium = (R)-7-
Ethyl-6,7,7a,8-tetrahydro-7-methyl-5H-benzo[g]-1,3-
benzodioxolo[6,5,4-de]quinolinium (●)
R   Chloride (21178-62-1)
S   Remaxan
U   Muscle relaxant

9441   (7659) $C_{20}H_{22}N_2$
                                                         96645-87-3

1,2,3,4,5,10-Hexahydro-3,10-dimethylazepino[4,5-d]di-
benz[b,f]azepine (●)
S   CGP 15564, *Erizepine***
U   Neuroleptic

9442   (7658) $C_{20}H_{22}N_2$
                                                         3964-81-6

6,11-Dihydro-11-(1-methyl-4-piperidinylidene)-5H-ben-
zo[5,6]cyclohepta[1,2-b]pyridine (●)

R   Maleate (1:2) (3978-86-7)
S   Atiramin, *Azatadine maleate***, Bonamid, Idulami-
        na, Idulian, Idumed, Lergocil, Nalomet, Optimine
        "Schering-Plough; UK", Sch 10649, Trinalin, Ver-
        ben, Zadine
U   Antihistaminic, antipruritic

9443   (7660) $C_{20}H_{22}N_2O$
                                                         88941-45-1

5-(1-Imidazolyl)-2,2',4',6'-tetramethylbenzhydrol = α-[5-
(1H-Imidazol-1-yl)-2-methylphenyl]-2,4,6-trimethylben-
zenemethanol (●)
S   Y-19018
U   Thromboxane A synthetase inhibitor

9444 $C_{20}H_{22}N_2O$
                                                         120928-09-8

4-[2-[4-(1,1-Dimethylethyl)phenyl]ethoxy]quinazoline
(●)
S   *Fenazaquin*
U   Miticide

9445 $C_{20}H_{22}N_2O_2$
171723-79-8

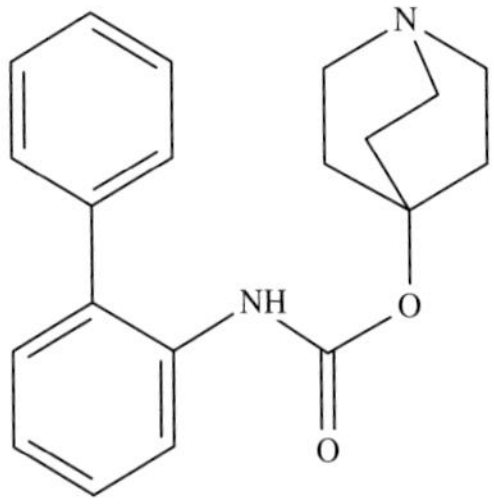

Quinuclidin-4-yl biphenyl-2-ylcarbamate = [1,1'-Biphenyl]-2-ylcarbamic acid 1-azabicyclo[2.2.2]oct-4-yl ester (●)
R Monohydrochloride (171722-81-9)
S YM-46303
U Muscarinic antagonist

9446 $C_{20}H_{22}N_2O_2$
157374-61-3

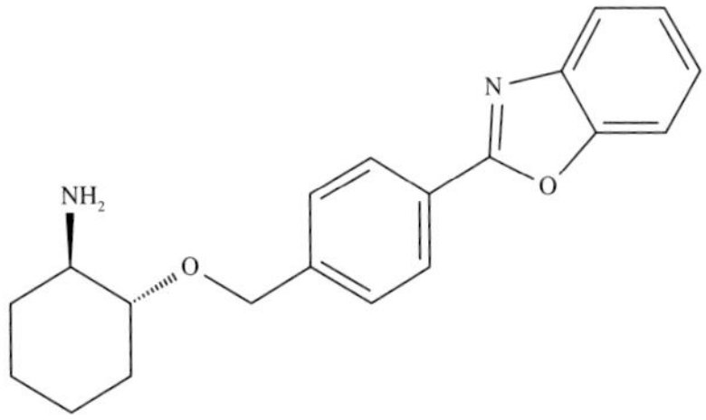

trans-2-[[4-(2-Benzoxazolyl)phenyl]methoxy]cyclohexanamine (●)
R Monohydrochloride (167504-53-2)
S RPR 101821
U Antihyperlipidemic

9447 (7661) $C_{20}H_{22}N_2O_2$
96972-93-9

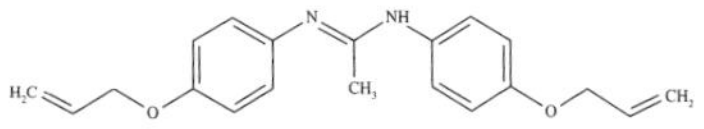

N,N'-Bis[*p*-(allyloxy)phenyl]acetamidine = *N,N'*-Bis[4-(2-propenyloxy)phenyl]ethanimidamide (●)
R Monohydrochloride (537-76-8)
S Diocain
U Local anesthetic

9448 (7662) $C_{20}H_{22}N_2O_2$
22013-23-6

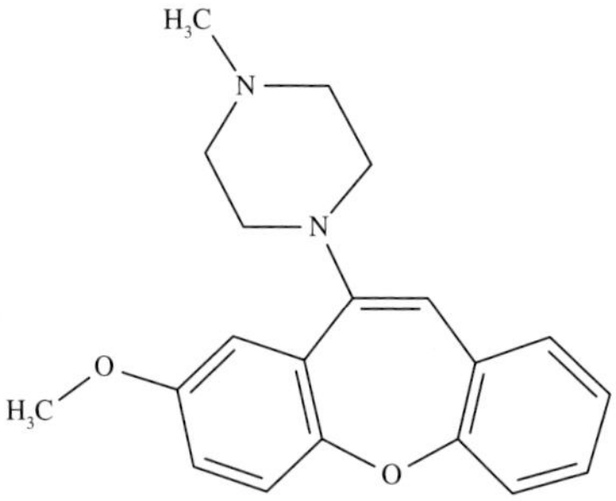

1-(8-Methoxydibenz[*b,f*]oxepin-10-yl)-4-methylpiperazine (●)
S *Metoxepin***
U Neuroleptic, antihistaminic, anti-emetic

9449 (7663) $C_{20}H_{22}N_2O_2$
509-15-9

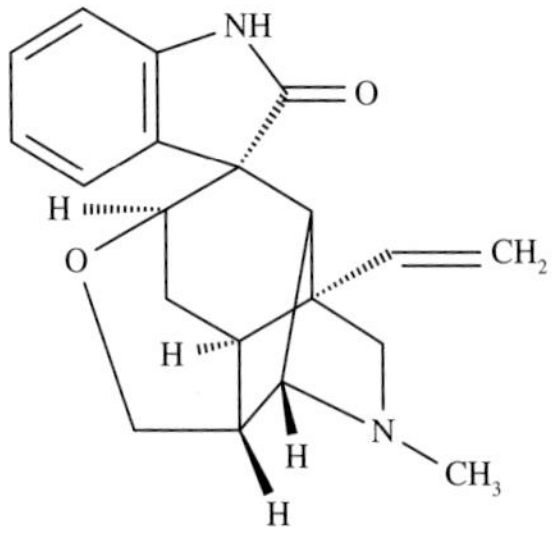

[3*R*-(3α,4aβ,5α,8α,8aβ,9*S**,10*S**)]-5-Ethenyl-3,4,4a,5,6,7,8,8a-octahydro-7-methylspiro[3,5,8-ethanylylidene-1*H*-pyrano[3,4-*c*]pyridine-10,3'-[3*H*]indol]-2'(1'*H*)-one (●)
S *Gelsemine*
U Central stimulant, antineuralgic (alkaloid from roots and rhizome of *Gelsemium sempervirens*)

9450 (7664) $C_{20}H_{22}N_2O_2S$
65322-75-0

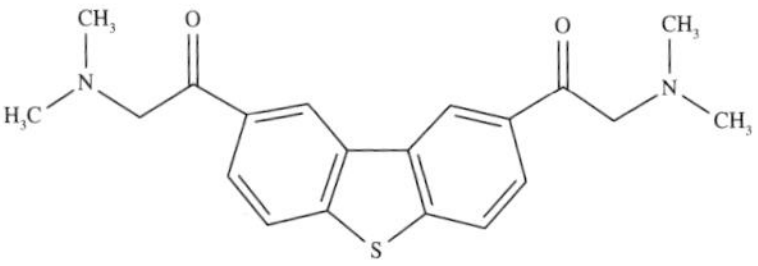

2,8-Bis[2-(dimethylamino)acetyl]dibenzothiophene = 1,1'-(2,8-Dibenzothiophenediyl)bis[2-(dimethylamino)ethanone] (●)

1915

R Dihydrochloride (35556-06-0)
S RMI 11877 DA
U Antiviral

9451 $C_{20}H_{22}N_2O_3$
201411-46-3

β-Hydroxy-2-(2-phenylethyl)-1*H*-benzimidazole-4-pro-
panoic acid ethyl ester (●)
S M 50367
U Anti-allergic (Th1/Th2 balance modulator)

9452   (7665) $C_{20}H_{22}N_2O_3$
1963-86-6

(*E*)-12-Ethylidene-1,2,3a,4,5,7-hexahydro-9-hydroxy-
3,5-ethano-3*H*-pyrrolo[2,3-*d*]carbazole-6-carboxylic acid
methyl ester = (19*E*)-2,16,19,20-Tetradehydro-12-hydr-
oxycuran-17-oic acid methyl ester (●)
S *Vinervine*
U Antineoplastic (alkaloid from *Vinca erecta*)

9453   (7666) $C_{20}H_{22}N_2O_4$
118635-52-2

N,N-Dimethylcarbamoylmethyl 2-(2-methyl-5*H*-[1]ben-
zopyrano[2,3-*b*]pyridin-7-yl)propionate = (±)-α,2-Di-
methyl-5*H*-[1]benzopyrano[1,2-*b*]pyridine-7-acetic acid
ester with *N,N*-dimethylglycolamide = α,2-Dimethyl-5*H*-
[1]benzopyrano[2,3-*b*]pyridine-7-acetic acid 2-(dimethylamino)-2-oxoethyl ester (●)

S *Mepranoprofen arbamel, Tilnoprofen arbamel***,
Y-23023
U Anti-inflammatory, antipyretic, analgesic

9454   (7667) $C_{20}H_{22}N_2O_4S$
106982-79-0

Methyl 4,7-dihydro-3-isobutyl-6-methyl-4-(*m*-nitrophe-
nyl)thieno[2,3-*b*]pyridine-5-carboxylate = 4,7-Dihydro-
6-methyl-3-(2-methylpropyl)-4-(3-nitrophenyl)thie-
no[2,3-*b*]pyridine-5-carboxylic acid methyl ester (●)
S S 312
U Vasodilator (calcium antagonist)

9455 $C_{20}H_{22}N_2O_5$
169322-61-6

5,10-Dihydro-*N*-[3-(dimethylamino)propyl]-1-methoxy-
5,10-dioxo-1*H*-benzo[*g*]isochromene-3-carboxamide =
N-[3-(Dimethylamino)propyl]-5,10-dihydro-1-methoxy-
5,10-dioxo-1*H*-naphtho[2,3-*c*]pyran-3-carboxamide (●)
R Monohydrochloride (169322-60-5)
S BCH 2051
U Cytotoxic

9456   (7668) $C_{20}H_{22}N_2O_6$
86781-07-9

Cyclopropylmethyl methyl 1,4-dihydro-2,6-dimethyl-4-(*m*-nitrophenyl)-3,5-pyridinedicarboxylate = 1,4-Dihydro-2,6-dimethyl-4-(3-nitrophenyl)-3,5-pyridinedicarboxylic acid cyclopropylmethyl methyl ester (●)
S　MPC-2101
U　Calcium antagonist

9457　(7669)　　　　　　　　　　　　　　$C_{20}H_{22}N_2O_7S$
　　　　　　　　　　　　　　　　　　　　127373-66-4

2-[*o*-[*p*-(Pivaloyloxy)benzenesulfonamido]benzamido]acetic acid = *o*-(*p*-Hydroxybenzenesulfonamido)hippuric acid pivalate (ester) = *N*-[2-[[[4-(2,2-Dimethyl-1-oxopropoxy)phenyl]sulfonyl]amino]benzoyl]glycine (●)
R　also monosodium salt (150374-95-1)
S　EI 546, Elaspol, ONO 5046, *Silevastat, Sirelastat, Sivelestat***
U　Human neutrophil elastase inhibitor

9458　(7670)　　　　　　　　　　　　　　$C_{20}H_{22}N_2S$
　　　　　　　　　　　　　　　　　　　　124939-85-1

11-[3-(Dimethylamino)propyl]-6,11-dihydrodibenzo[*b,e*]thiepine-2-carbonitrile (●)
R　Fumarate [(*E*)-2-butenedioate] (124939-86-2)
S　VÚFB-17084
U　Antidepressant

9459　(7671)　　　　　　　　　　　　　　$C_{20}H_{22}N_2S$
　　　　　　　　　　　　　　　　　　　　29216-28-2

10-(3-Quinuclidinylmethyl)phenotinazine = 3-(10-Phenothiazinylmethyl)quinuclidine = 10-(1-Azabicyclo[2.2.2]oct-3-ylmethyl)-10*H*-phenothiazine (●)
R　see also no. 10429
S　Aliman, Anistin, Butix-Tabl., Ceacnalon, Halemunin, Hisporan, Instotal, Istalar, Kitazemin, LM 209, Mekitazenon, Mequitan, Mequitazilan, *Mequitazine***, Metaplexan, Mircol, Narsulan, Neosurant, Nipolazin, Niporadin, Primalan, Primasone, Quitadrill, Release, Sattol, Vigigan, Virginan, Zesulan
U　Antihistaminic, sedative, antitussive, antidepressant

9460　　　　　　　　　　　　　　　　　$C_{20}H_{22}N_3OS_3$
　　　　　　　　　　　　　　　　　　　　199916-88-6

3-Ethyl-2-[[3-ethyl-5-(3-methyl-2(3*H*)-benzothiazolylidene)-4-oxo-2-thiazolidinylidene]methyl]-4-methylthiazolium (●)
R　Chloride (247129-77-7)
S　FJ-5002
U　Antineoplastic (telomerase inhibitor)

9461 (7672)

$C_{20}H_{22}N_4$
112228-65-6

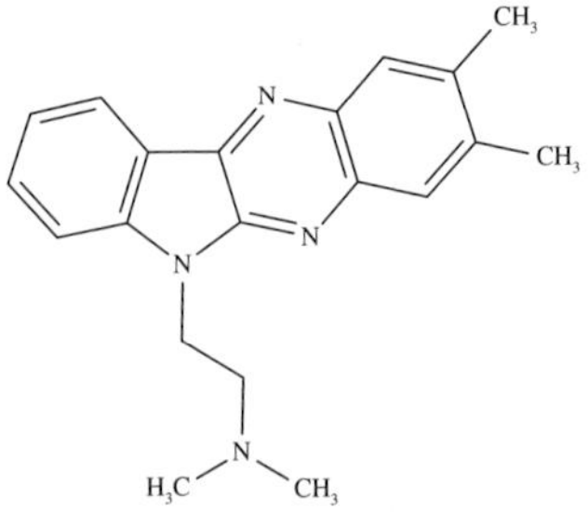

6-[2-(Dimethylamino)ethyl]-2,3-dimethyl-6H-indo-
lo[2,3-b]quinoxaline = N,N,2,3-Tetramethyl-6H-indo-
lo[2,3-b]quinoxaline-6-ethanamine (●)
S B-220
U Antiviral

9462 (7673)

$C_{20}H_{22}N_4O$
20170-20-1

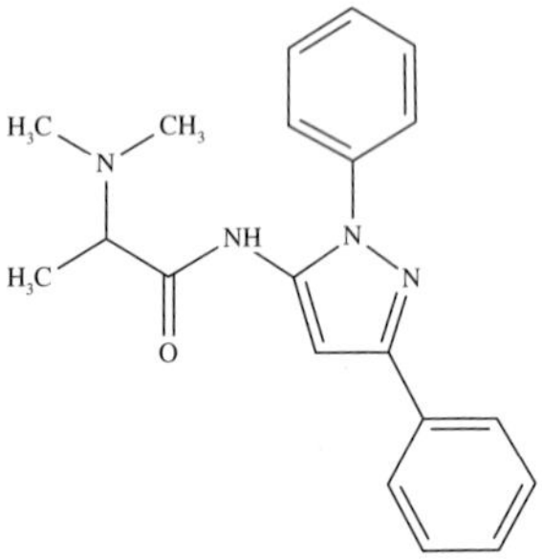

2-(Dimethylamino)-N-(1,3-diphenyl-5-pyrazolyl)pro-
pionamide = 1,3-Diphenyl-5-[2-(dimethylamino)propion-
amido]pyrazole = 2-(Dimethylamino)-N-(1,3-diphenyl-
1H-pyrazol-5-yl)propanamide (●)
S AP-14, *Difamizole, Difenamizole**, Diphenamizole*,
   Pasalin
U Analgesic, muscle relaxant, antagonist to narcotics

9463

$C_{20}H_{22}N_4O_2$
138154-39-9

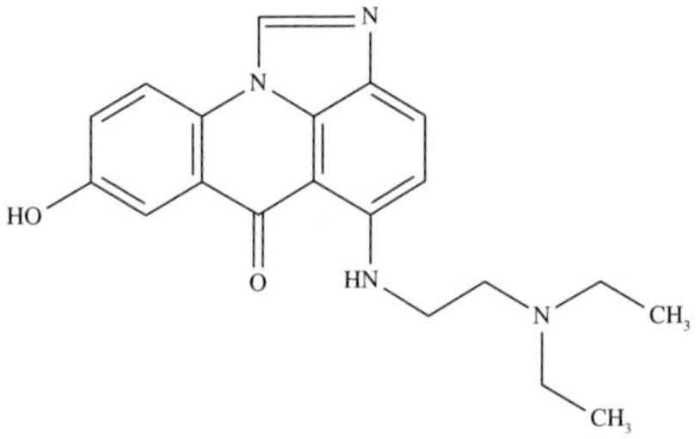

5-[[2-(Diethylamino)ethyl]amino]-8-hydroxy-6H-imida-
zo[4,5,1-de]acridin-6-one (●)
R Dihydrochloride (138154-55-9)
S C 1311, *Imidacrine hydrochloride*, NSC-645809
U Antineoplastic

9464 (7674)

$C_{20}H_{22}N_4O_2$
107052-56-2

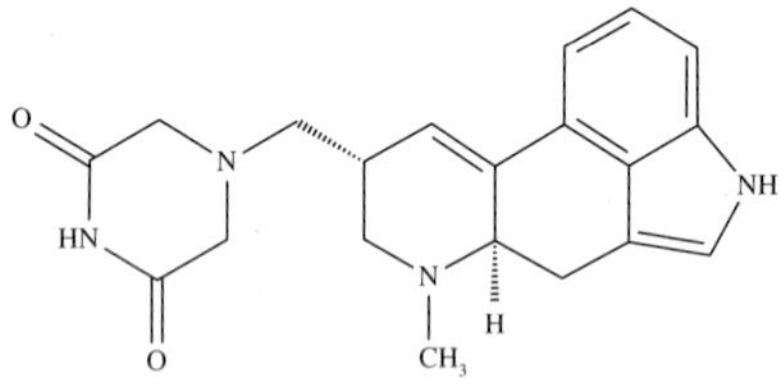

4-[(9,10-Didehydro-6-methylergolin-8β-yl)methyl]-2,6-
piperazinedione = 4-[[(8β)]-9,10-Didehydro-6-methyler-
golin-8-yl]methyl]-2,6-piperazinedione (●)
S FCE 23884, *Romergoline***
U Antiparkinsonian, antipsychotic (dopamine antago-
   nist/agonist)

9465

$C_{20}H_{22}N_4O_2$

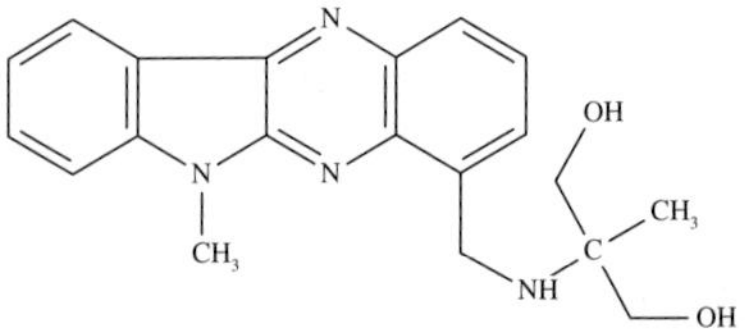

2-Methyl-2-[[(6-methyl-6H-indolo[2,3-b]quinoxalin-4-
yl)methyl]amino]-1,3-propanediol (●)
R Monohydrochloride (208446-44-0)
S NCA 0424
U Antineoplastic

9466 (7675) $C_{20}H_{22}N_4O_3$
104229-37-0

7-[3-[4-(2-Pyrimidinyl)-1-piperazinyl]propoxy]-2*H*-1-
benzopyran-2-one (●)
S PD 118717
U Antischizophrenic (dopamine agonist)

9467 (7676) $C_{20}H_{22}N_4O_3$
97581-70-9

2'-Hydroxy-3'-propyl-4'-[[4-(1*H*-tetrazol-5-ylme-
thyl)phenoxy]methyl]acetophenone = 1-[2-Hydroxy-3-
propyl-4-[[4-(1*H*-tetrazol-5-ylmethyl)phenoxy]me-
thyl]phenyl]ethanone (●)
R Monosodium salt (107223-78-9)
S LY 163443
U Peptide leukotriene receptor antagonist

9468 (7677) $C_{20}H_{22}N_4O_4$
106465-45-6

(*E*)-3-[4-(2,3,6,7-Tetrahydro-2,6-dioxo-1,3-dipropyl-1*H*-
purin-8-yl)phenyl]-2-propenoic acid (●)
S BW A 1433
U Adenosine A_2-receptor antagonist

9469 (7678) $C_{20}H_{22}N_4O_4S$
52231-20-6

(6*R*,7*R*)-3-Methyl-8-oxo-7-[2-[*p*-(1,4,5,6-tetrahydro-2-
pyrimidinyl)phenyl]acetamido]-5-thia-1-azabicy-
clo[4.2.0]oct-2-ene-2-carboxylic acid = (7*R*)-3-Methyl-7-
[2-[*p*-(1,4,5,6-tetrahydro-2-pyrimidinyl)phenyl]acetami-
do]-3-cephem-4-carboxylic acid = (6*R-trans*)-3-Methyl-
8-oxo-7-[[[4-(1,4,5,6-tetrahydro-2-pyrimidinyl)phe-
nyl]acetyl]amino]-5-thia-1-azabicyclo[4.2.0]oct-2-ene-2-
carboxylic acid (●)
S *Cefrotil***, HR 580
U Antibiotic

9470 (7679) $C_{20}H_{22}N_4O_5$
59721-28-7

p-Guanidinobenzoic acid ester with (*p*-hydroxyphe-
nyl)acetic acid ester with *N,N*-dimethylglycolamide = *p*-
Guanidinobenzoic acid ester with (dimethylcarbamoyl)methyl (*p*-hydroxyphenyl)acetate = (Dimethylcarbamoyl)methyl *p*-[(*p*-guanidinobenzoyl)oxy]phenyl-
acetate = 4-[[4-[(Aminoiminomethyl)amino]benzoyl]
oxy]benzeneacetic acid 2-(dimethylamino)-2-oxoethyl
ester (●)
R monomethanesulfonate (59721-29-8)
S Archment, *Armostat mesylate*, *Camostat mesilate**,
Foipan, FOY 305, Foypan, FOY S 980
U Enzyme inhibitor (proteinase)

9471   (7680) $C_{20}H_{22}N_4O_6S$
                                                   58306-30-2

Dimethyl [[2-(2-methoxyacetamido)-4-(phenylthio)phe-
nyl]imidocarbonyl]dicarbamate = 2'-[2,3-Bis(methoxycarbonyl)guanidino]-5'-(phenylthio)-2-methoxyacetani-
lide = [[2-[(Methoxyacetyl)amino]-4-(phenylthio)phe-
nyl]carbonimidoyl]bis[carbamic acid] dimethyl ester (●)
S   Amatron, Bayverm, Bay Vh 5757, Combotel, *Feban-
      tel**, Negabot Plus, Oratel, Rintal
U   Anthelmintic (veterinary)

9472   (7681) $C_{20}H_{22}N_4O_6S_2$
                                                   119-85-7

$N^4,N^{4'}$-Vanillylidenebis(sulfanilamide) = *N,N'*-Bis(4-sulf-
amoylphenyl)vanillylidenediamine = 4,4'-[[(4-Hydroxy-
3-methoxyphenyl)methylene]diimino]bis[benzenesulfon-
amide] (●)
S   Argolamide (active substance), *Vanyldisulfamide**,
      Vanyldisulfanilamide
U   Antibacterial

9473   (7682) $C_{20}H_{22}N_4O_{10}S$
                                                   64544-07-6

1-Acetoxyethyl (6*R*,7*R*)-7-[2-(2-furyl)glyoxylamido]-3-
(hydroxymethyl)-8-oxo-5-thia-1-azabicyclo[4.2.0]oct-2-
ene-2-carboxylate 7^2-(*Z*)-(*O*-methyloxime) carbamate
(ester) = 1-Acetoxyethyl (*Z*)-(7*R*)-3-[(carbamoyloxy)me-
thyl]-7-[2-(2-furyl)-2-(methoxyimino)acetamido]-3-ce-
phem-4-carboxylate = [6*R*-[6α,7β(*Z*)]]-3-[[(Aminocarbonyl)oxy]methyl]-7-[[2-furanyl(methoxyimino)ace-
tyl]amino]-8-oxo-5-thia-1-azabicyclo[4.2.0]oct-2-ene-2-
carboxylic acid 1-(acetyloxy)ethyl ester (●)
S   Axoril, Bioracef, CCI 15641, Cefaricida, Cefatin
      "Roche", Ceftin, Cefudura, Cefuhexal, Cefuracet,
      Cefurax, Cefurol, Cefuro-Puren, Cefuroxim "Allen",
      *Cefuroxime axetil**, Cefurox Oral, Cefurox-Wolff,
      Celocid, Cépazine, Cetoxil-tabs, Cupax, Curocef
      oral, CXM, CXM-AX, Elobact, Feacef, Foucacillin,
      Furaxetil, Genephoxal, Interbion, Kalcef caps., Ma-
      xitil, Mosalan-oral, Naroxit, Nivador, Novocef, Ora-
      cef "Nippon Glaxo", Oraceftin, Oraxim, Roxime, Se-
      dopan, Sefaktil, Sefuroks, Selan, SN 407, Tilexim,
      Xepos, Ximos, Zimaks, Zinadol, Zinat "Glaxo",
      Zinnat, Zinnox, Zipos, Zoref
U   Antibiotic

9474 $C_{20}H_{22}N_6O$
                                                   108307-65-9

8,9-Dihydro-2-(4-methyl-1-piperazinyl)-6-phenylimida-
zo[1,2-*a*]pyrido[3,2-*e*]pyrazine 5-oxide (●)
S   RB 90740
U   Antineoplastic (bioreductive agent)

9475   (7683) $C_{20}H_{22}N_6O_2$
                                                   118344-71-1

2,4-Diamino-5-[4-(benzylmethylamino)-3-nitrophenyl]-6-ethylpyrimidine = 6-Ethyl-5-[4-[methyl(phenylmethyl)amino]-3-nitrophenyl]-2,4-pyrimidinediamine (●)
S MBP, *Methobenzaprim, Methylbenzoprim*
U Antineoplastic

9476 $C_{20}H_{22}N_6O_6$
                                             156595-85-6

N-[4-[[(2,4-Diaminofuro[2,3-*d*]pyrimidin-5-yl)methyl]methylamino]benzoyl]-L-glutamic acid (●)
S MTXO
U Antineoplastic (folic acid antagonist)

9477   (7685) $C_{20}H_{22}N_8O_5$
                                            751-19-9

4-Deoxy-4-amino-9-methylfolic acid = 4-Amino-9-methylpteroyl-L-glutamic acid = *N*-[4-[[1-(2,4-Diamino-6-pteridinyl)ethyl]amino]benzoyl]-L-glutamic acid (●)
S A-ninopterin
U Antineoplastic, antileukemic, folic acid antagonist

9478   (7684) $C_{20}H_{22}N_8O_5$
                                            59-05-2

4-Deoxy-4-amino-N^{10}-methylfolic acid = 4-Amino-10-methylpteroyl-L-glutamic acid = *N*-[4-[[(2,4-Diamino-6-

pteridinyl)methyl]methylamino]benzoyl]-L-glutamic acid (●)
R also disodium salt (7413-34-5) or hydrate (133073-73-1)
S Abitrexate, A-methopterin, Antifolan, Arthritrex, Beltreks, Biotrexate, Brimexate, Bristoltexato, CL 14377, Emethexate, Emtexate, Emthexat(e), Ervemin, Farmitrexan, Farmitrexat, Farmotrex, Fauldexato, Folex "Adria", G-301, Intradose-MTX, Lantarel, Ledertrexate, Lumexon, Matrex, Maxtrex, Medipterina, Metex, Methoblastin, Methorheumat, *Methotrexate***, *Methylaminopterinum, Metotressato, Metotrexato*, Metotrexol, Metrexan, Mexate "Bristol; Mead Johnson", MPI 5004, MTX, Neotrexate, Novatrex, NSC-740, O-trexat, R 9985, Rheumatrex, Rhodamer, Texate, Tremetex, Trexan "Atabay; Läakefarmos; Orion", Trixilem, Trixylem, Unitrexate
U Antineoplastic, antileukemic, folic acid antagonist, anti-arthritic

---

9479   (7686) $C_{20}H_{22}O_2$
                                              6033-99-4

3,4-Bis(*p*-methoxyphenyl)-2,4-hexadiene = 1,1'-(1,2-Diethylidene-1,2-ethanediyl)bis[4-methoxybenzene] (●)
S *Dienestrol dimethyl ether, Dienöstroldimethyläther*
U Synthetic estrogen

9480   (7687) $C_{20}H_{22}O_2$
                                              848-21-5

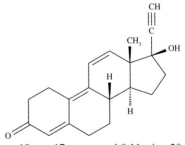

17-Hydroxy-19-nor-17α-pregna-4,9,11-trien-20-yn-3-one = 17α-Ethynyl-17β-hydroxy-4,9,11-estratrien-3-one

= (17α)-17-Hydroxy-19-norpregna-4,9,11-trien-20-yn-3-one (●)
S A 301, *Norgestrienone***, Ogyline, R 2010
U Progestin

9481 C$_{20}$H$_{22}$O$_3$
81840-57-5

(1*E*)-1-(4-Hydroxy-3-methoxyphenyl)-7-phenyl-1-hepten-3-one (●)
S *Yakuchinone B*
U Tyrosinase inhibitor from *Alpinia oxyphylla*

9482 (7688) C$_{20}$H$_{22}$O$_3$
70356-09-1

1-(*p-tert*-Butylphenyl)-3-(*p*-methoxyphenyl)-1,3-propanedione = 1-[4-(1,1-Dimethylethyl)phenyl]-3-(4-methoxyphenyl)-1,3-propanedione (●)
S *Avobenzone***, Clarins, CTFA 00348, Episol "Schering-Plough", Eusolex 920, Parsol 1789, Shade Uvagard
U Sunscreen agent

9483 (7689) C$_{20}$H$_{22}$O$_3$
3771-19-5

2-Methyl-2-[*p*-(1,2,3,4-tetrahydro-1-naphthyl)phenoxy]propionic acid = 2-Methyl-2-[4-(1,2,3,4-tetrahydro-1-naphthalenyl)phenoxy]propanoic acid (●)
S CH 13437, Melipan (new form), *Nafenoic acid*, *Nafenopin***, Su-13437, TPIA

U Antihyperlipidemic

9484 (7690) C$_{20}$H$_{22}$O$_6$
546-97-4

[1*R*-(1α,4β,4aα,6aβ,9β,10aβ,10bα)]-9-(3-Furanyl)decahydro-4-hydroxy-4a,10a-dimethyl-1,4-etheno-3*H*,7*H*-benzo[1,2-*c*:3,4-*c*']dipyran-3,7-dione (●)
S *Columbin*
U Bitter principle from roots of *Jatrorrhiza palmata*

9485 C$_{20}$H$_{22}$O$_6$
487-36-5

2,2'-Bis(4-hydroxy-3-methoxyphenyl)furo[3,4-*c*]furan = 4,4'-[[1*S*-(1α,3aα,4α,6aα)]-Tetrahydro-1*H*,3*H*-furo[3,4-*c*]furan-1,4-diyl]bis[2-methoxyphenol] (●)
S *Pinoresinol*
U Cytotoxic from *Streptomyces* sp. IT-44

9486 C$_{20}$H$_{22}$O$_7$
82470-74-4

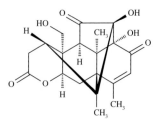

(2*R*,2a*R*,5a*S*,6a*R*,10*R*,10a*S*,10b*S*,10c*R*,11*S*)-2a,6,6a,9,10,10a,10b,10c-Octahydro-2,2a-dihydroxy-

10a-(hydroxymethyl)-5,10c,11-trimethyl-2,5a,10-metheno-5aH-acenaphtho[4,3-b]pyran-1,3,8(2H)-trione (●)
S *Shinjulactone C*
U Anti-AIDS agent

9487 $C_{20}H_{22}O_8$
                               27208-80-6

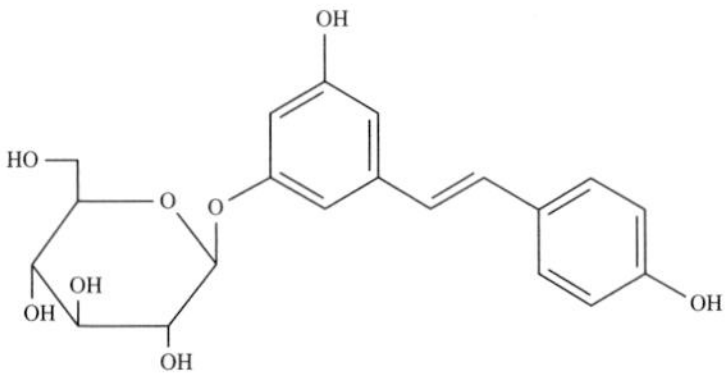

3-Hydroxy-5-(*p*-hydroxystyryl)phenyl β-D-glucopyranoside = Resveratrol 3-β-mono-D-glucoside = 3-Hydroxy-5-[(1E)-2-(4-hydroxyphenyl)ethenyl]phenyl β-D-glucopyranoside (●)
S *Piceid, Polydatin*
U Antithrombotic from *Polygonum cuspidatum* or *Iris hookeriana*

9488   (7691) $C_{20}H_{23}BrClN_3O_2$
                               71195-56-7

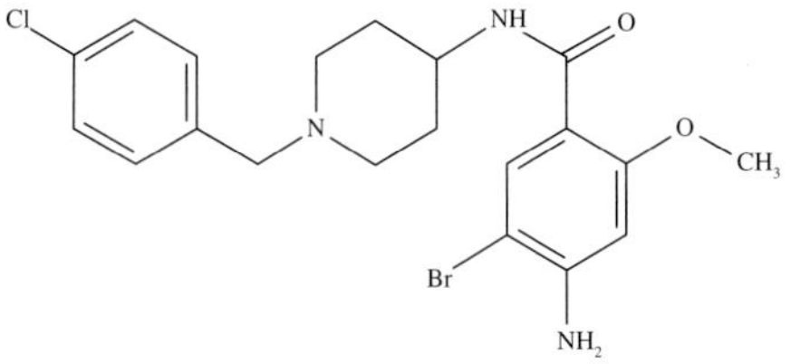

4-Amino-5-bromo-*N*-[1-(4-chlorobenzyl)-4-piperidyl]-*o*-anisamide = 4-Amino-5-bromo-*N*-[1-[(4-chlorophenyl)methyl]-4-piperidinyl]-2-methoxybenzamide (●)
S *Broclepride***
U Tranquilizer, anti-emetic

9489   (7692) $C_{20}H_{23}Br_2N_3O_2$
                               39669-49-3

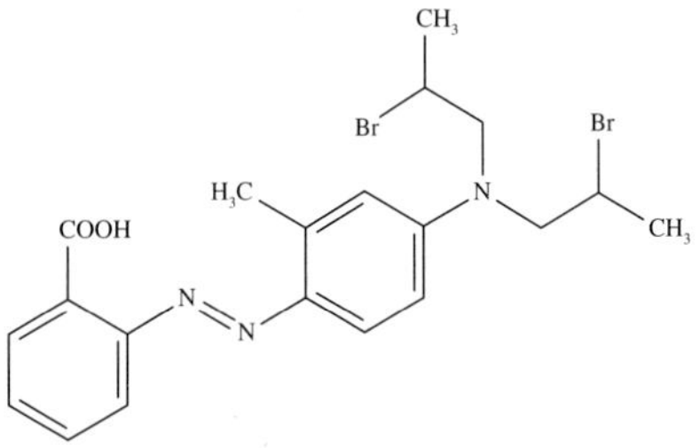

4-[Bis(2-bromopropyl)amino]-2'-carboxy-2-methylazobenzene = 2-[[4-[Bis(2-bromopropyl)amino]-2-methylphenyl]azo]benzoic acid (●)
S CB-10252
U Antineoplastic

9490   (7693) $C_{20}H_{23}ClFNO$
                               119431-25-3

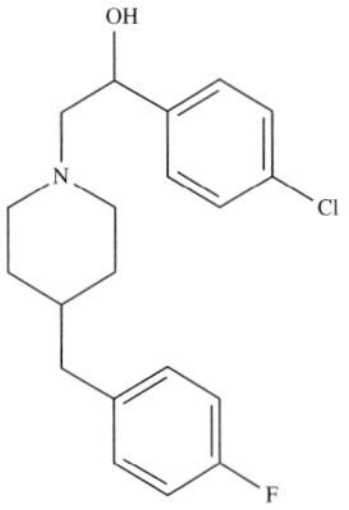

(±)-α-(*p*-Chlorophenyl)-4-(*p*-fluorobenzyl)-1-piperidineethanol = (±)-α-(4-Chlorophenyl)-4-[(4-fluorophenyl)methyl]-1-piperidineethanol (●)
R Hydrochloride (136634-88-3)
S *Eliprodil hydrochloride***, SL 820715, ST 820715
U Cerebral vasodilator (*N*-methyl-D-aspartate antagonist)

9491　(7694)　　　　　　　　　　　　$C_{20}H_{23}ClN_2O$

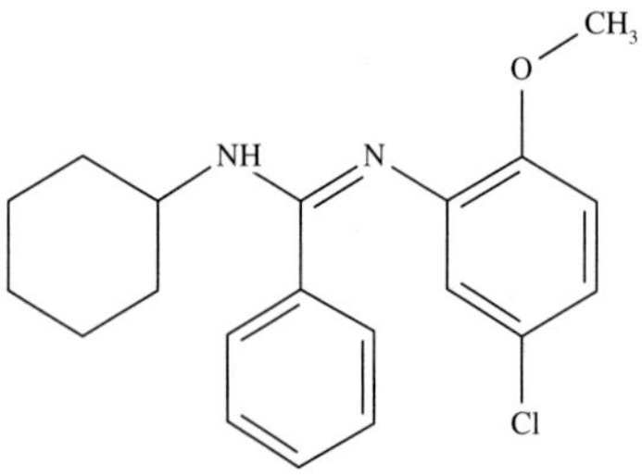

N-(5-Chloro-2-methoxyphenyl)-*N*'-cyclohexylbenzami-
dine = *N*-(5-Chloro-2-methoxyphenyl)-*N*'-cyclohexyl-
benzenecarboximidamide (●)
R　Monohydrochloride (55232-80-9)
S　HG-70
U　Anti-inflammatory, analgesic

9492　　　　　　　　　　　　　　　　$C_{20}H_{23}ClN_2O$
　　　　　　　　　　　　　　　　　　149811-12-1

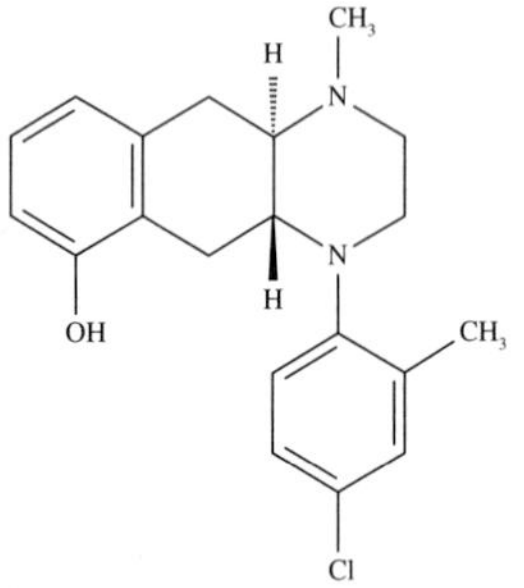

(4a*R*-*trans*)-4-(4-Chloro-2-methylphenyl)-
1,2,3,4,4a,5,10,10a-octahydro-1-methylbenzo[*g*]quin-
oxalin-6-ol (●)
S　SDZ-PSD 958
U　Dopamine D$_1$ receptor antagonist

9493　　　　　　　　　　　　　　　$C_{20}H_{23}ClN_2OS$
　　　　　　　　　　　　　　　　　136929-56-1

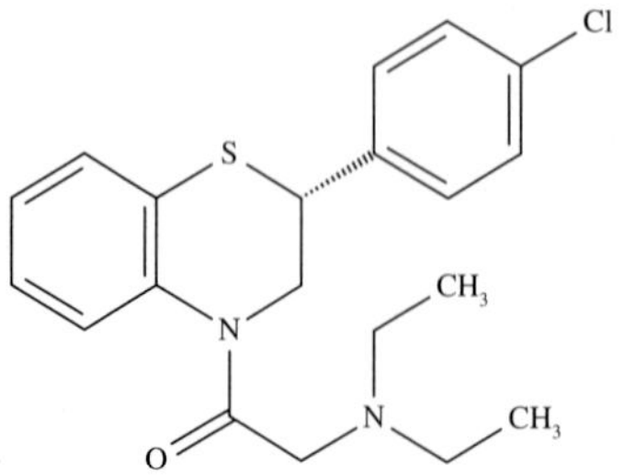

(*R*)-2-(4-Chlorophenyl)-4-[(diethylamino)acetyl]-3,4-di-
hydro-2*H*-1,4-benzothiazine (●)
S　T-477
U　Neuroprotective

9494　(7695)　　　　　　　　　　　$C_{20}H_{23}ClN_2OS$
　　　　　　　　　　　　　　　　　56934-18-0

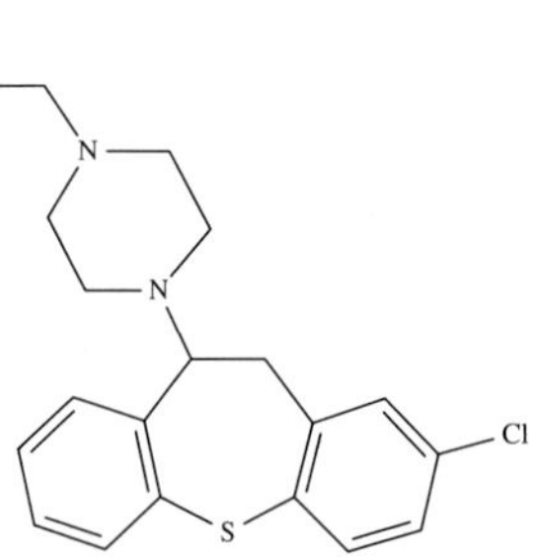

2-Chloro-10-[4-(2-hydroxyethyl)-1-piperazinyl]-10,11-
dihydrodibenzo[*b,f*]thiepin = 4-(2-Chloro-10,11-dihydro-
dibenzo[*b,f*]thiepin-10-yl)-1-piperazineethanol (●)
R　Succinate (1:1) (56934-24-8)
S　*Docloxythepin succinate*, VÚFB-10032
U　Neuroleptic

9495　(7696)　　　　　　　　　　　$C_{20}H_{23}ClN_2OS$
　　　　　　　　　　　　　　　　　34775-62-7

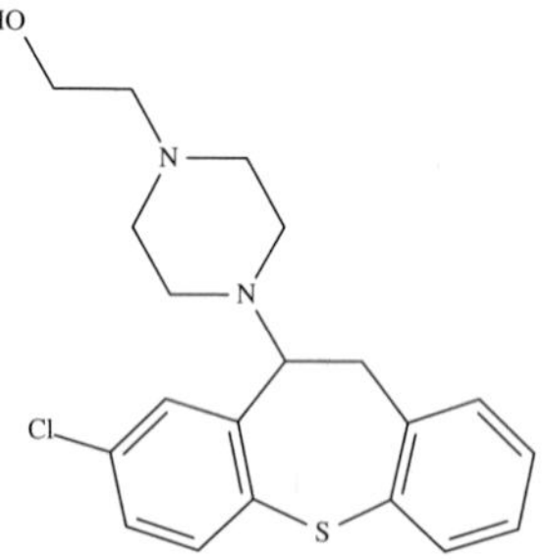

2-Chloro-11-[4-(2-hydroxyethyl)-1-piperazinyl]-10,11-
dihydrodibenzo[*b,f*]thiepin = 4-(8-Chloro-10,11-dihydro-
dibenzo[*b,f*]thiepin-10-yl)-1-piperazineethanol (●)
R　see also no. 14366
S　*Noroxyclothepin*
U　Neuroleptic

9496 (7697) $C_{20}H_{23}ClN_2O_2$
3761-70-4

6-Chloro-1-[[2-(diethylamino)ethyl]amino]-4-methyl-9-xanthone = 6-Chloro-1-[[2-(diethylamino)ethyl]amino]-4-methyl-9*H*-xanthen-9-one (●)
S Miracil B
U Antischistosomal

9497 (7698) $C_{20}H_{23}ClN_2O_3$
69907-17-1

1-[(3-Chloro-2-methyl-1*H*-indol-4-yl)oxy]-3-[(2-phen-oxyethyl)amino]-2-propanol (●)
S Indopanolol**, Infendolol, LQ 31-341
U α-and β-adrenergic blocker

9498 (7699) $C_{20}H_{23}ClN_4O_2$
3861-76-5

2-(p-Chlorobenzyl)-1-[2-(diethylamino)ethyl]-5-nitro-benzimidazole = 2-[2-(p-Chlorobenzyl)-5-nitro-1-benz-imidazolyl]ethyldiethylamine = 2-[(4-Chlorophenyl)me-thyl]-N,N-diethyl-5-nitro-1*H*-benzimidazole-1-ethan-amine (●)
C-193901, Ciba 19390, Clobedolum, Clonitazene**, Clonitazinum, NIH 7586

U Narcotic analgesic

9499 (7700) $C_{20}H_{23}ClO_2$
3124-93-4

21-Chloro-17-hydroxy-19-nor-17α-pregna-4,9-dien-20-yn-3-one = 17α-(2-Chloroethynyl)-17β-hydroxy-4,9-estradien-3-one = (17α)-21-Chloro-17-hydroxy-19-nor-pregna-4,9-dien-20-yn-3-one (●)
S Ethynerone**, MK-665
U Progestin

9500 (7701) $C_{20}H_{23}ClO_3$
55937-99-0

Ethyl (±)-2-[[α-(p-chlorophenyl)-p-tolyl]oxy]-2-methyl-butyrate = Ethyl 2-(4-p-chlorobenzylphenoxy)-2-methyl-butyrate = (±)-2-[4-[(4-Chlorophenyl)methyl]phenoxy]-2-methylbutanoic acid ethyl ester (●)
S Beclipur, Beclobrate**, Beclosclerin, Sgd 24774, Turec
U Antihyperlipidemic

9501 (7702) $C_{20}H_{23}Cl_2N_3O_2$
96164-19-1

(±)-(E)-p-Chlorobenzaldehyde O-[3-[4-(o-chlorophenyl)-1-piperazinyl]-2-hydroxypropyl]oxime = (±)-(E)-4-Chlo-robenzaldehyde O-[3-[4-(2-chlorophenyl)-1-piperazi-nyl]-2-hydroxypropyl]oxime (●)

S *Peraclopone***
U Antihyperlipidemic

9502   (7703)

$C_{20}H_{23}Cl_2N_3O_4$
1952-96-1

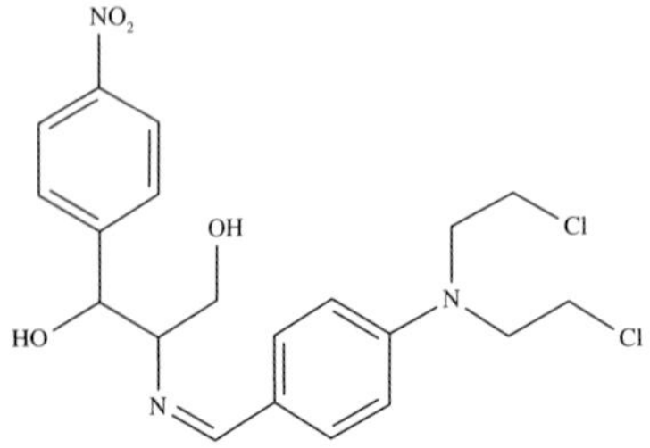

2-[[*p*-[Bis(2-chloroethyl)amino]benzylidene]amino]-1-(*p*-nitrophenyl)-1,3-propanediol = 2-[[[4-[Bis(2-chloroethyl)amino]phenyl]methylene]amino]-1-(4-nitrophenyl)-1,3-propanediol (●)
S AT-16
U Antineoplastic

9503   (7704)

$C_{20}H_{23}FN_2O$
2354-61-2

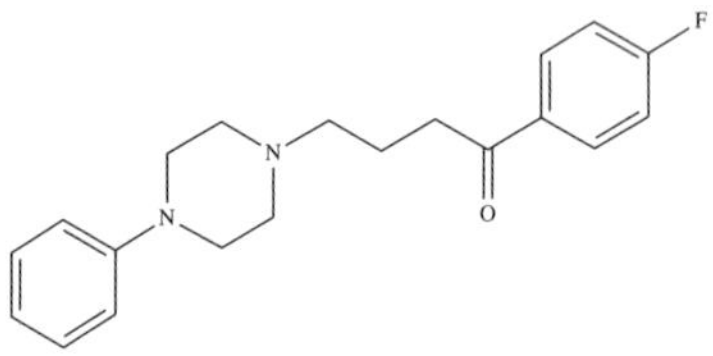

4'-Fluoro-4-(4-phenyl-1-piperazinyl)butyrophenone = 1-[3-(*p*-Fluorobenzoyl)propyl]-4-phenylpiperazine = 1-(4-Fluorophenyl)-4-(4-phenyl-1-piperazinyl)-1-butanone (●)
S *Butropipazone*, R 1892
U Neuroleptic

9504

$C_{20}H_{23}FN_4O_3$
167887-97-0

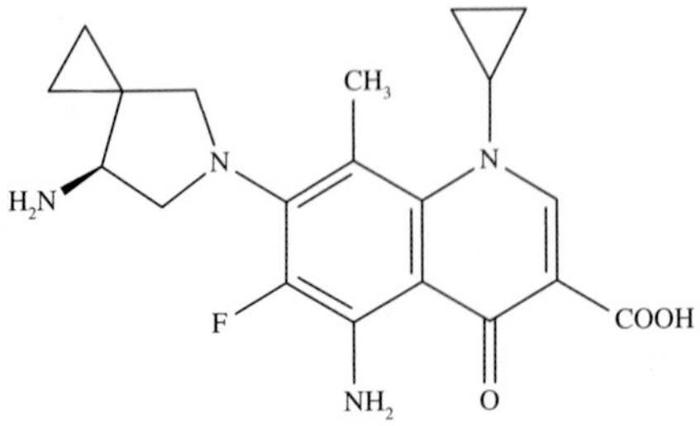

5-Amino-7-[(7*S*)-7-amino-5-azaspiro[2.4]hept-5-yl]-1-cyclopropyl-6-fluoro-1,4-dihydro-8-methyl-4-oxo-3-quinolinecarboxylic acid (●)
R Monomethanesulfonate (167888-07-5)
S HSR-903, *Olamufloxacin mesilate***, *Tolamufloxacin mesilate*
U Antibacterial

9505   (7705)

$C_{20}H_{23}F_2N_3O_3$
99734-97-1

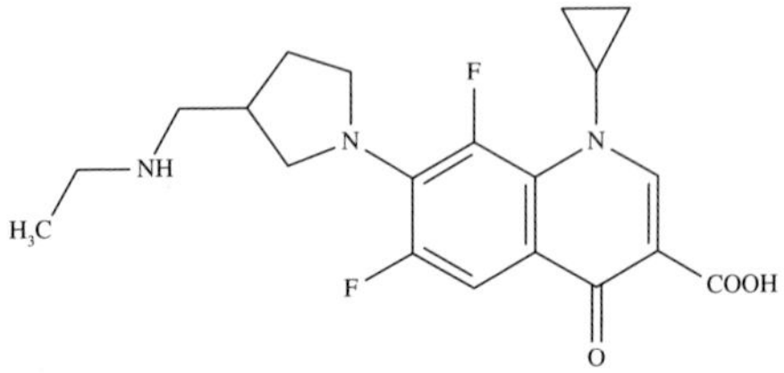

1-Cyclopropyl-7-[3-[(ethylamino)methyl]-1-pyrrolidinyl]-6,8-difluoro-1,4-dihydro-4-oxo-3-quinolinecarboxylic acid (●)
S PD 117558
U Antibacterial

9506

$C_{20}H_{23}F_3N_2O_2$
145742-28-5

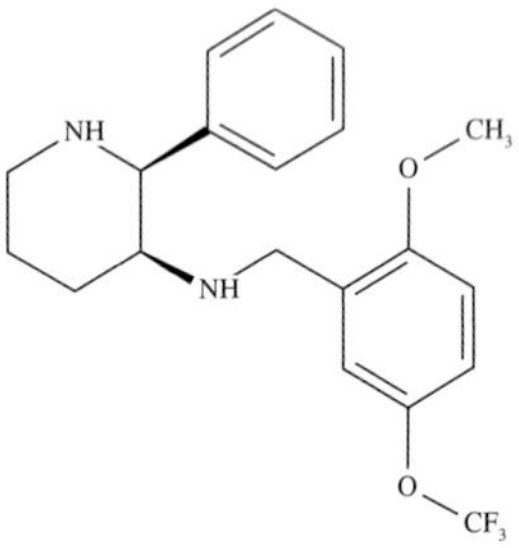

(+)-(2*S*,3*S*)-3-[[2-Methoxy-5-(trifluoromethoxy)benzyl]amino]-2-phenylpiperidine = (2*S*,3*S*)-*N*-[[2-Methoxy-5-(trifluoromethoxy)phenyl]methyl]-2-phenyl-3-piperidinamine (●)
S CP-122721
U Anti-inflammatory, anti-asthmatic (neurokinin NK$_4$ receptor antagonist)

9507   (7707) $C_{20}H_{23}N$
                                        10262-69-8

N-Methyl-9,10-ethanoanthracene-9(10*H*)-propylamine =
9-[3-(Methylamino)propyl]-9,10-dihydro-9,10-ethanoan-
thracene = 1-(3-Methylaminopropyl)dibenzo[*b,e*]bicy-
clo[2.2.2]octadiene = *N*-Methyl-9,10-ethanoanthracene-
9(10*H*)-propanamine (●)

R Hydrochloride (10347-81-6)
S Aneural, Ba-34276, Cronmolin, Delgian, Depressa-
    se, Deprilept, Kanopan, Klimastress, Ladiomil, Lu-
    diomil, Ludionil, Maludil, Mapril, Mapro-Gry,
    Maprolu, Mapromil, Maprostat, Mapro-Tablinen,
    Maprotibene, Maprotil, *Maprotiline hydrochlori-
    de***, Matrin, Melodil, Mirpan, Neuomil, Novo-Ma-
    protiline, Psymion, Retinyl
U Antidepressant

9507-01   (7707-01)   58902-67-3
R Methanesulfonate
S Ludiomil-Inj., Maprolit, Maprolu-Inj., *Maprotiline
    mesilate***
U Antidepressant

9508   (7706) $C_{20}H_{23}N$
                                        50-48-6

0,11-Dihydro-*N,N*-dimethyl-5*H*-dibenzo[*a,d*]cyclohep-
ene-$\Delta^{5,\gamma}$-propylamine = 3-(10,11-Dihydro-5*H*-diben-
o[*a,d*]cyclohepten-5-ylidene)propyldimethylamine = 5-
3-(Dimethylamino)propylidene]dibenz[*a,d*]-1,4-cyclo-
eptadiene = 3-(10,11-Dihydro-5*H*-dibenzo[*a,d*]cyclo-
epten-5-ylidene)-*N,N*-dimethyl-1-propanamine (●)

R Hydrochloride (549-18-8) [or pamoate (2:1)
    (17086-03-2)]
S Adepress, Adepril "Lepetit", ADT-Zimaia, Amavil,
    Ami-Anelun, Amilent, Amiline, Amilit, Amineurin,
    Amiprin, Amiptanol, Amitid, Amiton, Amitril "War-
    ner-Chilcott", Amitrip, *Amitriptilina (cloridrato di)*,
    Amitriptol, *Amitriptyline hydrochloride***, Amitrol,
    Amitryn, Amtrip, Amylin(e), Amytril, Amyzol,
    Anapsique, Ancholibre, Ancix, Angopasse, Annoly-
    tin, Apo-Amitriptyline, Apo-Triptyn, Arin, Atopthi-
    ne, Atriptal, Atryptal, Belpax, Damilen, Daprimen,
    Deprebio, Deprestat, Deprex "Beecham", Diatrase,
    Domical, Elatrol, Elatrolet, Elavil, Elegen-G, Elivel,
    Emitrip, Endep, Enovil, Euplit, KSR 117, Kyliran,
    Lantron, Laroxal, Laroxyl, Larozyl, Lentizol, Leva-
    te, Loraxyl, Mareline, Maxivale, Maxivalet, Mela-
    pramine A, Meravil, Miketorin, Mitaptyline,
    MK-230, N 750, Neuptanol, Neutrex, Nobritol, Nor-
    maln, Novoprotect, Novotriptyn, Oasil-M "Gap", Po-
    lysal, Polytanol, *Proheptadiene*, Prosaldon, Prota-
    nol, Psycolin, Psyrax, Quietal "Alembic", Redomex,
    Ro 4-1575, Saroten(a), Sarotex, Sch 7172, Schuvel,
    SK-Amitriptyline, Stelminal, Sylvemid, Syneudon,
    Teperin, Terpone, Thymontil, Torydanol, Trepi-
    lin(e), Tridep "Squibb", Tripta, Triptafen, Tripta-
    med, Triptanol, Triptilin, Triptizol, Triptopol, Tri-
    ptyl, Triptyline, Tryptacap, Tryptal(ette), Tryptanol,
    Tryptine, Tryptizol, Tryptomer, Trytil, TV 322,
    Uxen
U Antidepressant

9509   (7708) $C_{20}H_{23}N$
                                        5118-30-9

9,10-Dihydro-10,10-dimethyl-9-[3-(methylamino)propy-
lidene]anthracene = 9-[3-(Methylamino)propylidene]-
10,10-dimethyl-9,10-dihydroanthracene = 3-(9,10-Dihy-
dro-10,10-dimethyl-9-anthracenylidene)propylmethyl-
amine = 3-(10,10-Dimethyl-9(10*H*)-anthracenylidene)-*N*-
methyl-1-propanamine (●)

S *Litracen**, N 7049
U Antidepressant

9510 (7709)

$C_{20}H_{23}NO$
10447-39-9

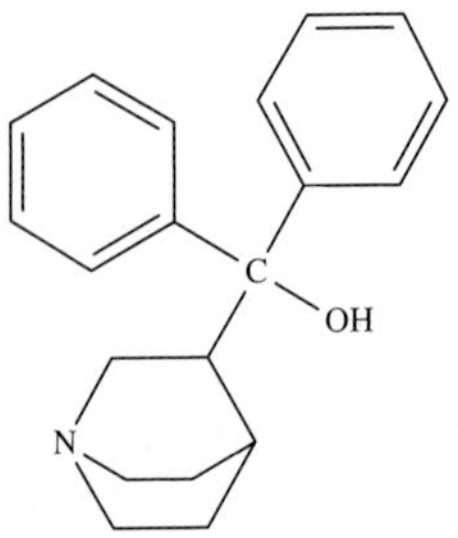

α,α-Diphenyl-3-quinuclidinemethanol = 3-Quinuclidi-nyldiphenylmethanol = α,α-Diphenyl-1-azabicy-clo[2.2.2]octane-3-methanol (●)
R Hydrochloride (10447-38-8)
S Fenkarol, Phencarol, *Quifenadine hydrochloride***
U Antihistaminic

9511 (7710)

$C_{20}H_{23}NO$
31828-74-7

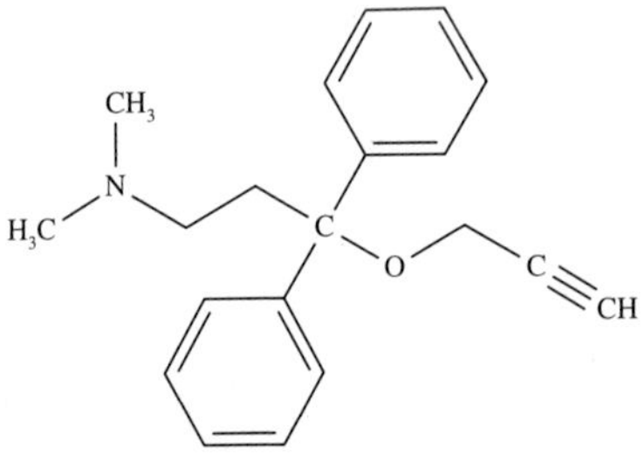

N,N-Dimethyl-3,3-diphenyl-3-(propargyloxy)propyl-amine = *N,N*-Dimethyl-γ-phenyl-γ-(2-propynyloxy)ben-zenepropanamine (●)
S Sklerosan, X 50
U Spasmolytic, analgesic

9512 (7711)

$C_{20}H_{23}NO$
56433-44-4

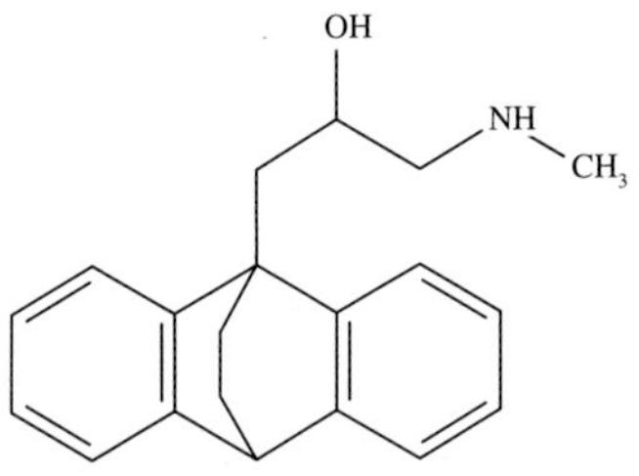

1-[2-Hydroxy-3-(methylamino)propyl]dibenzo[*b,e*]bicy-clo[2.2.2]octadiene = α-[(Methylamino)methyl]-9,10-ethanoanthracene-9(10*H*)-ethanol (●)
R Hydrochloride (39022-39-4)
S C 49802 B-Ba, *Oxaprotiline hydrochloride***, OXPN
U Antidepressant

9513 (7712)

$C_{20}H_{23}NO$
76496-68-9

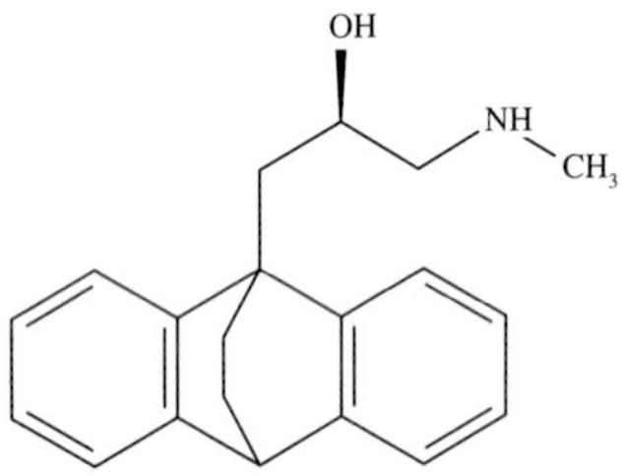

(*R*)-α-[(Methylamino)methyl]-9,10-ethanoanthracene-9(10*H*)-ethanol (●)
R also hydrochloride (76496-69-0)
S CGP 12103 A, *Levoprotiline***, *Levoxaprotiline*, (–)-*Oxaprotiline*
U Antidepressant

9514 (7713)

$C_{20}H_{23}NO$
92629-87-

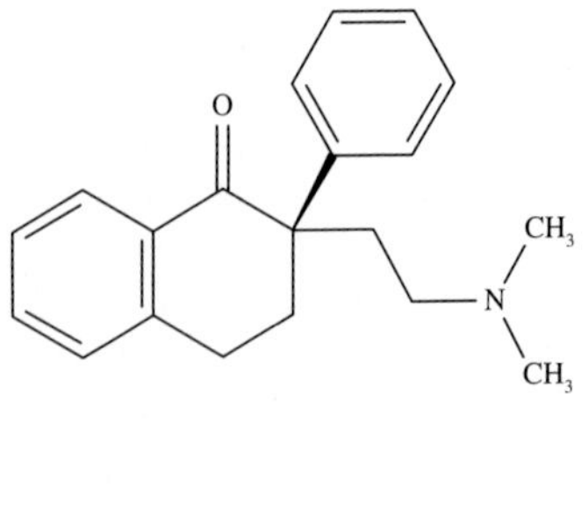

(+)-(S)-2-[2-(Dimethylamino)ethyl]-3,4-dihydro-2-phe-
nyl-1(2H)naphthalenone = (+)-2-[2-(Dimethylamino)ethyl]-3,4-dihydro-2-phenyl-1(2H)-naphthalenone (●)
S *Dexnafenodone**, LU 43706
U Antidepressant

9515 (7715)

$C_{20}H_{23}NO$
92615-20-8

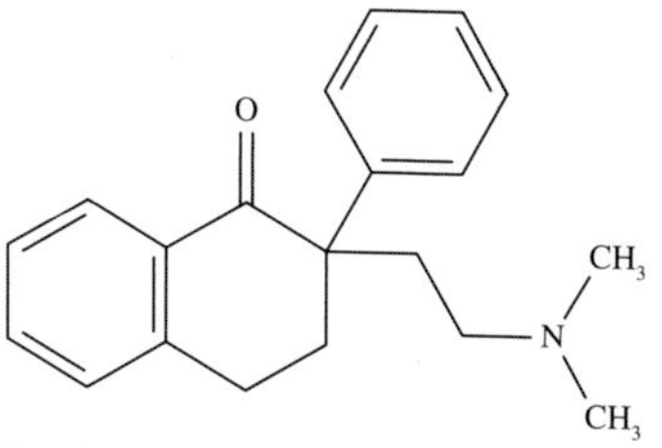

(±)-2-[2-(Dimethylamino)ethyl]-3,4-dihydro-2-phenyl-
1(2H)-naphthalenone (●)
S *Nafenodone**
U Antidepressant

9516 (7714)

$C_{20}H_{23}NO$
4317-14-0

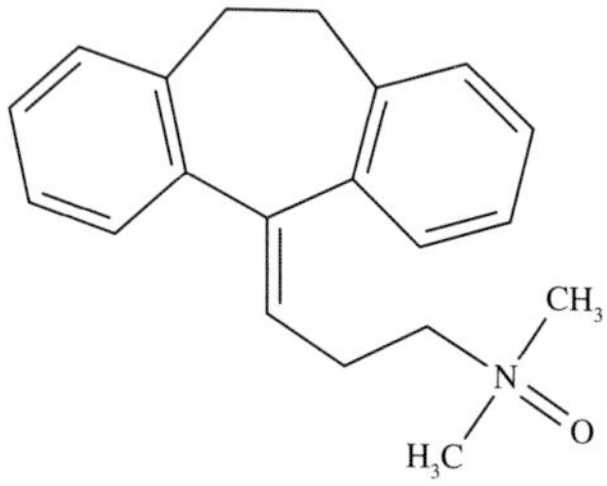

10,11-Dihydro-N,N-dimethyl-5H-dibenzo[a,d]cyclohep-
ene-Δ$^{5,\gamma}$-propylamine N-oxide = 3-(10,11-Dihydro-5H-
dibenzo[a,d]cyclohepten-5-ylidene)-N,N-dimethyl-1-
propanamine N-oxide (●)
R also dihydrate
S Ambivalon, Amioxid-neuraxpharm, *Amitriptylin-
oxide**, Dano, Equilibrin, Neurarmonil
U Antidepressant

9517 (7717)

$C_{20}H_{23}NO_2$
76-65-3

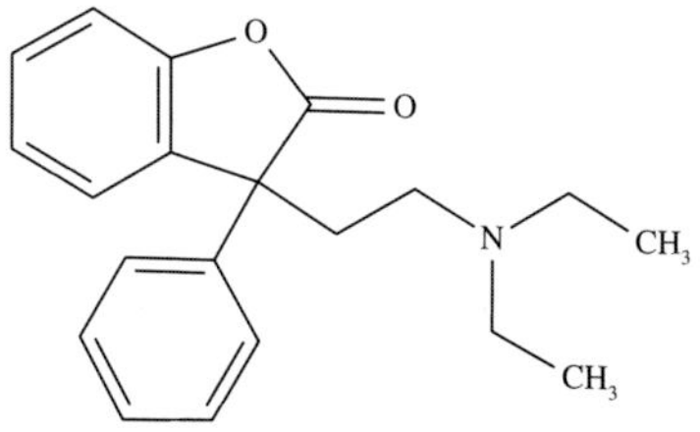

3-[2-(Diethylamino)ethyl]-3-phenyl-2(3H)-benzofura-
none (●)
R also hydrochloride (6009-67-2)
S Amethone, *Amocainum, Amolanone*, AP-43
U Local anesthetic

9518 (7716)

$C_{20}H_{23}NO_2$
82166-77-6

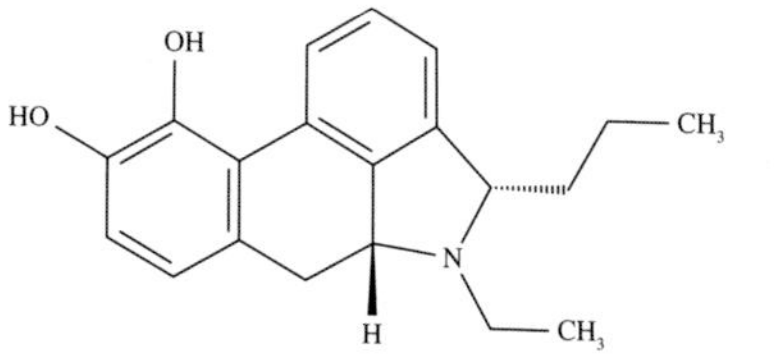

(4S-trans)-5-Ethyl-4,5,5a,6-tetrahydro-4-propyldi-
benz[cd,f]indole-9,10-diol (●)
R Hydrochloride (82188-33-8)
S CI 201-678
U Dopamine agonist

9519

$C_{20}H_{23}NO_2$

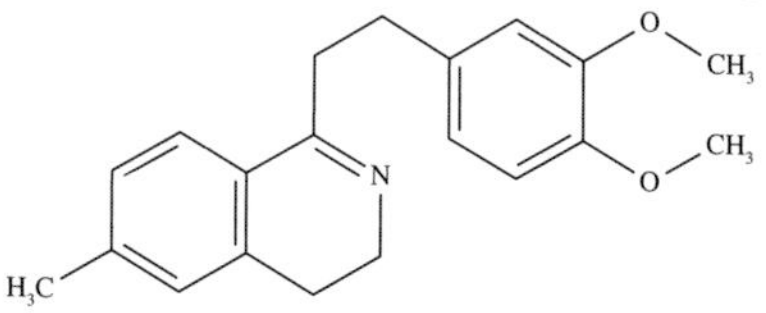

1-[2-(3,4-Dimethoxyphenyl)ethyl]-3,4-dihydro-6-me-
thylisoquinoline (●)
R Hydrochloride (132928-46-2)
S PF-10040
U Anti-allergic, gastric acid inhibitor, PAF antagonist

9520 (7722)

$C_{20}H_{23}NO_2$
4425-78-9

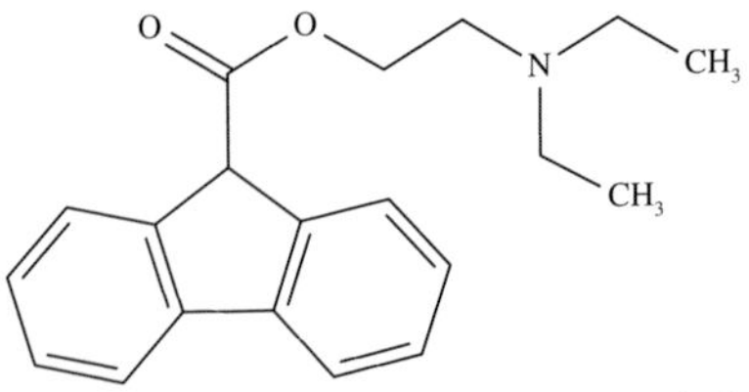

9-Fluorenecarboxylic acid β-(diethylamino)ethyl ester =
9H-Fluorene-9-carboxylic acid 2-(diethylamino)ethyl ester (●)
R Hydrochloride (548-65-2)
S *Aminocarbofluorene*, Pavatrin(e), Robitrin, Spasma-
drina
U Spasmolytic

9521 (7718)

$C_{20}H_{23}NO_2$
6495-46-1

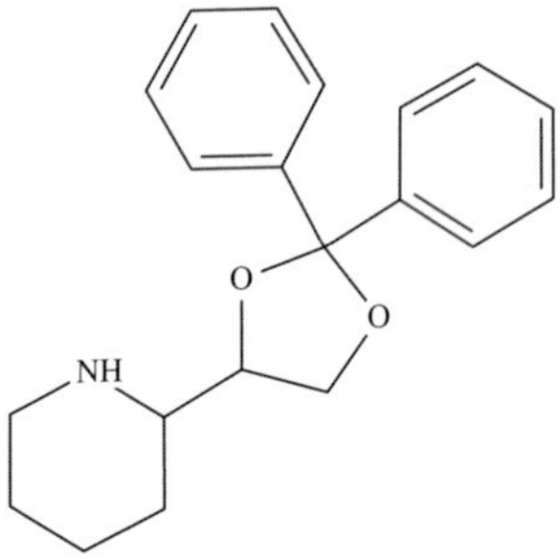

(±)-2-(2,2-Diphenyl-1,3-dioxolan-4-yl)piperidine (●)
R also hydrochloride (3666-69-1)
S CL-639 C, *Dioxadrol**, Oxadrolum*, Rydar
U Antidepressant

9522 (7720)

$C_{20}H_{23}NO_2$
4792-18-1

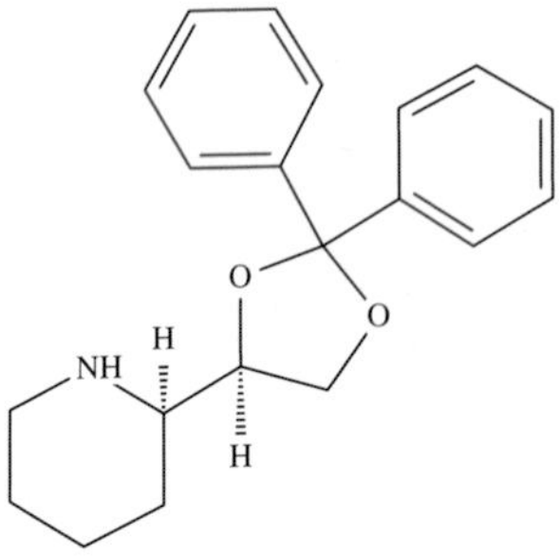

(±)-2-(2,2-Diphenyl-1,3-dioxolan-4-yl)piperi-
dine (●)
R also hydrochloride (23257-58-1)
S CL-912 C, *Levoxadrol**, Levoxan, NSC-526063,
U-22304 A
U Antidepressant, local anesthetic, smooth muscle re-
laxant

9523 (7719)

$C_{20}H_{23}NO_2$
4741-41-7

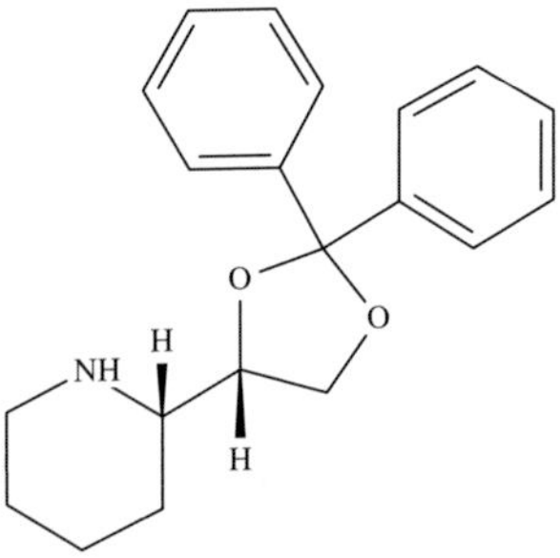

[S-(R*,R*)]-2-(2,2-Diphenyl-1,3-dioxolan-4-yl)piperi-
dine (●)
R also hydrochloride (631-06-1)
S CL-911 C, *Dexoxadrol**, NSC-526062, Relane,
U-22559 A
U Antidepressant, central stimulant, analgesic

9524 (7721)

$C_{20}H_{23}NO_2$
34753-46-3

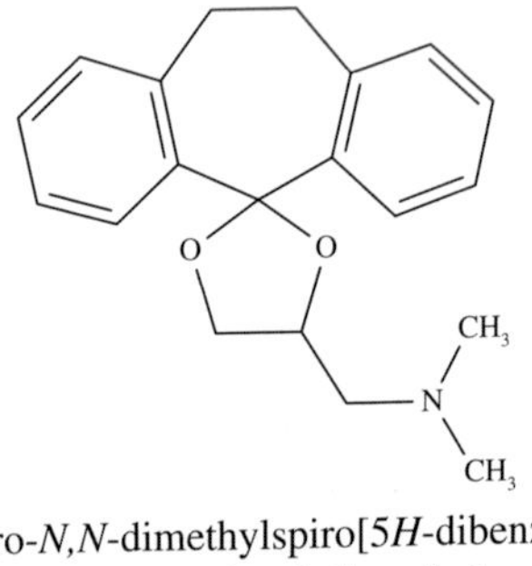

10,11-Dihydro-N,N-dimethylspiro[5H-dibenzo[a,d]cy-
cloheptene-5,2'-[1,3]dioxolane]-4'-methylamine = 10,11-
Dihydro-N,N-dimethylspiro[5H-dibenzo[a,d]cyclohep-
tene-5,2'-[1,3]dioxolane]-4'-methanamine (●)
S *Ciheptolane***
U Analgesic, antihypertensive

[R-(R*,R*)]-2-(2,2-Diphenyl-1,3-dioxolan-4-yl)piperi-
dine (●)

9525 (7723)   C$_{20}$H$_{23}$NO$_3$
7009-76-9

1-Methyl-3-pyrrolidinemethanol benzilate (ester) = Diphenylhydroxyacetic acid (1-methyl-3-pyrrolidinyl)methyl ester = α-Hydroxy-α-phenylbenzeneacetic acid (1-methyl-3-pyrrolidinyl)methyl ester (●)
S   *Triclazate**, Tricylatate*
U   Anticholinergic

9526 (7724)   C$_{20}$H$_{23}$NO$_3$
17720-33-1

2-[[[4-(Benzyloxy)butyl]amino]methyl]-1,4-benzodioxin = *N*-[4-(Phenylmethoxy)butyl]-1,4-benzodioxin-2-methanamine (●)
S   WB 4130
U   Convulsant

9527 (7725)   C$_{20}$H$_{23}$NO$_3$
119261-09-5

1-(2,5-Dimethoxyphenyl)-3-[4-(dimethylamino)phenyl]-2-methyl-2-propen-1-one (●)
S   DDMP, MDL 27048
U   Antineoplastic

9528 (7726)   C$_{20}$H$_{23}$NO$_4$

O-Acetylbenzilic acid β-(dimethylamino)ethyl ester = α-(Acetyloxy)-α-phenylbenzeneacetic acid 2-(dimethylamino)ethyl ester (●)
S   Acetyldiphemin
U   Neurotonic, antirheumatic

9529 (7727)   C$_{20}$H$_{23}$NO$_4$
479-39-0

(*S*)-2,3,12,12a-Tetrahydro-6,9,10-trimethoxy-1-methyl-1*H*-[1]benzoxepino[2,3,4-*ij*]isoquinoline (●)
S   *Cularine*
U   Smooth muscle relaxant

9530 (7728)   C$_{20}$H$_{23}$NO$_4$
6957-27-3

1-(3,4-Dimethoxybenzyl)-3,4-dihydro-6,7-dimethoxyisoquinoline = 3,4-Dihydro-6,7-dimethoxy-1-veratrylisoquinoline = 1-[(3,4-Dimethoxyphenyl)methyl]-3,4-dihydro-6,7-dimethoxyisoquinoline (●)
S   *Dihydropapaverine*, Paverin
U   Spasmolytic

9531 (7730) $C_{20}H_{23}NO_4$
                                              16590-41-3

17-(Cyclopropylmethyl)-4,5α-epoxy-3,14-dihydroxy-
morphinan-6-one = (5α)-17-(Cyclopropylmethyl)-4,5-ep-
oxy-3,14-dihydroxymorphinan-6-one (●)
R Hydrochloride (16676-29-2)
S Antaxone, Basinal, Celupan, Depade, EN-1639 A,
      Nalorex, Naltrel, *Naltrexone hydrochloride***, Nar-
      coral, Nemexin, Revia, Trexan "DuPont-Merck",
      Trexonil, UM 792
U Antagonist to narcotics, treatment of alcoholism

9532 (7729) $C_{20}H_{23}NO_4$
                                              466-90-0

4,5α-Epoxy-3-methoxy-17-methylmorphin-6-en-6-ol
acetate = 6-Acetoxy-4,5α-epoxy-3-methoxy-*N*-methyl-
morphin-6-en = Acetyldihydrocodeinone = (–)-(5*R*)-4,5-
Epoxy-3-methoxy-9a-methylmorphin-6-en-6-yl acetate =
6-*O*-Acetyl-7,8-dihydro-3-*O*-methyl-6,7-didehydromor-
phine = (5α)-6,7-Didehydro-4,5-epoxy-3-methoxy-17-
methylmorphinan-6-ol acetate (ester) (●)
R also hydrochloride (20236-82-2)
S Acedicon, *Acethydrocodone*, Cofadicon, Diacodon,
      Negadol, Novocodon, *Tebacone, Tebakon*, Thebace-
      tyl, Thebacodon, *Thebacon**
U Narcotic analgesic, antitussive

9533 $C_{20}H_{23}NO_4$
                                              131179-95-8

2-[4-[2-[(3,5-Dimethylphenyl)amino]-2-oxoethyl]phen-
oxy]-2-methylpropanoic acid (●)
S RSR 13
U Allosteric effector

9534 (7731) $C_{20}H_{23}NO_4S$
                                              2731-16-0

N-Deacetylthiocolchicine = (*S*)-7-Amino-6,7-dihydro-
1,2,3-trimethoxy-10-(methylthio)benzo[*a*]heptalen-
9(5*H*)-one (●)
R Hydrochloride (16665-61-5)
S Corps R. 261, *Deacetylthiocolchicine hydrochloride*,
      NSC-9170, Thio-Colciran, Tio-Colciran
U Antimitotic

9535 (7732) $C_{20}H_{23}NO_5$
                                              3476-50-4

N-Deacetylcolchicine = Trimethylcolchicinic acid methyl
ether = (*S*)-7-Amino-6,7-dihydro-1,2,3,10-tetramethoxy-
benzo[*a*]heptalen-9(5*H*)-one (●)
R Tartrate (27963-65-1)

S Citostal, TMCA
U Antineoplastic

9536 (7733) $C_{20}H_{23}NO_5$
86414-28-0

5,8,13,13a-Tetrahydro-9,10,11-trimethoxy-6*H*-dibenzo[*a,g*]quinolizine-2,3-diol (●)
R Hydrochloride (86414-27-9)
S Drohynol, IS-35
U Antiglaucoma agent

9537 $C_{20}H_{23}NO_5$
143526-87-8

S)-1-(3,5-Dimethoxyphenyl)-3,4-dihydro-6,7-dimethoxy-3-isoquinolinemethanol (●)
S SDZ-ISQ 844
U Phosphodiesterase isoenzyme type III/IV inhibitor

9538 (7734) $C_{20}H_{23}NO_5$
19485-08-6

thyl 6,7-bis(cyclopropylmethoxy)-4-hydroxy-3-quinolinecarboxylate = 6,7-Bis(cyclopropylmethoxy)-4-hydroxy-3-quinolinecarboxylic acid ethyl ester (●)

S *Ciproquinate***, Coxytrol, *Cyproquinate, Cyproxyquine*, Su-18137
U Coccidiostatic (for poultry)

9539 (7735) $C_{20}H_{23}NO_6$
81703-42-6

(α*R*,α'*S*,2*S*,2'*R*)-α,α'-(Iminodimethylene)bis[1,4-benzodioxan-2-methanol] = (1*R*,5*S*)-1-[(2*S*)-2,3-Dihydro-1,4-benzodioxin-2-yl]-5-[(2*R*)-2,3-dihydro-1,4-benzodioxin-2-yl]-3-aza-1,5-pentanediol = [2*R**[*S**[*R**(*S**)]]]-α,α'-[Iminobis(methylene)]bis[2,3-dihydro-1,4-benzodioxine-2-methanol] (●)
R Methanesulfonate (81737-62-4)
S *Bendacalol mesilate***, *Bendacolol mesilate*, CGS 10078 B
U Antihypertensive

9540 (7736) $C_{20}H_{23}NS$
4969-02-2

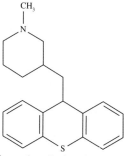

1-Methyl-3-(thioxanthen-9-ylmethyl)piperidine = 9-(1-Methyl-3-piperidylmethyl)thioxanthene = 1-Methyl-3-(9*H*-thioxanthen-9-ylmethyl)piperidine (●)
R Hydrochloride (1553-34-0) or hydrochloride monohydrate (7081-40-5)
S Alraad, Atosil "Teisan-Nagase", Cholinfall, Dalpan, Inoball, Mahatomin, Manamisen, Methixart, *Methixene hydrochloride*, Methyloxan, *Metixene hydrochloride***, NSC-78194, 6 O/SJ 1977, PT 432, Raunans, SJ 1977, Thioperkin, Tremaril, Tremarit, Tremilor, Tremonil, Tremoquil, Trest
U Smooth muscle relaxant, antiparkinsonian, spasmolytic

9541 (7737)

C$_{20}$H$_{23}$NS
13448-33-4

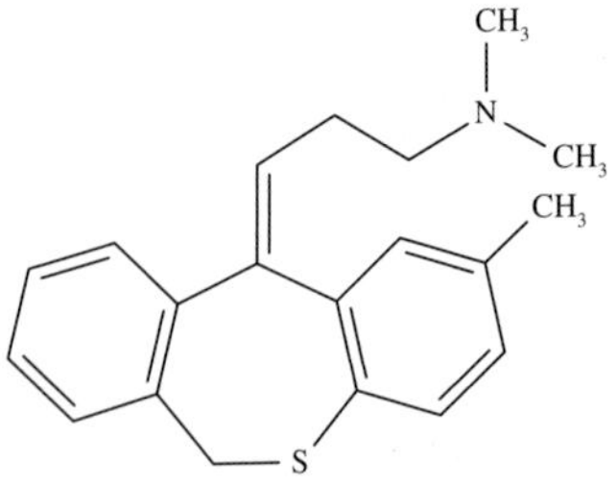

11-[3-(Dimethylamino)propylidene]-6,11-dihydro-2-me-
thyldibenzo[b,e]thiepin = N,N-Dimethyl-3-[2-methyldi-
benzo[b,e]thiepin-11(6H)-ylidene]-1-propanamine (●)
R Hydrochloride (53046-96-1)
S *Medosulepin hydrochloride*, Methiaden
U Antihistaminic

9542

C$_{20}$H$_{23}$NS
205120-96-3

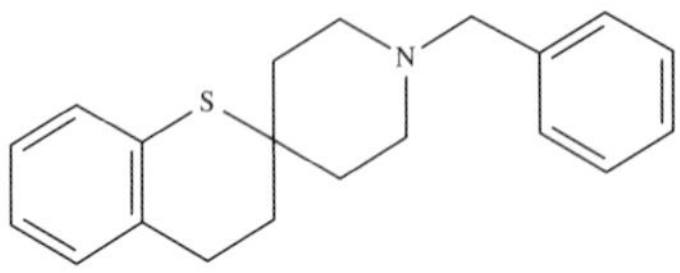

1'-Benzyl-3,4-dihydrospiro[2H-1-benzothiopyran-2,4'-
piperidine] = 3,4-Dihydro-1'-(phenylmethyl)spiro[2H-1-
benzothiopyran-2,4'-piperidine] (●)
R Ethanedioate (1:1) (205122-16-3)
S *Spipethiane*
U σ-Ligand

9543 (7738)

C$_{20}$H$_{23}$N$_2$O
6866-93-9

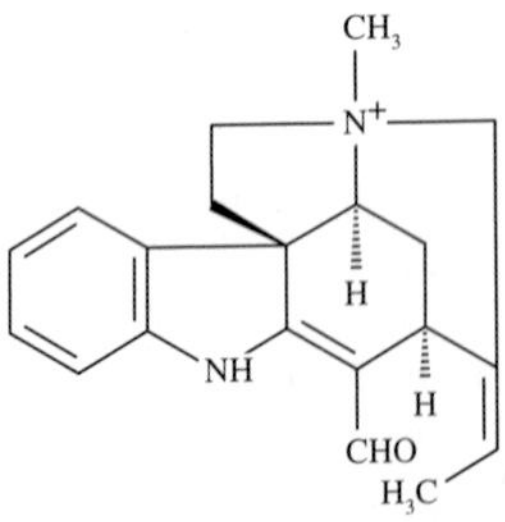

(19E)-2,16,19,20-Tetradehydro-4-methyl-17-oxocura-
nium (●)
R Chloride (22273-09-2)
S Metvin, *Vincanine methyl chloride*

U Ganglion blocking agent

9544 (7739)

C$_{20}$H$_{23}$N$_2$OS
103132-98-5

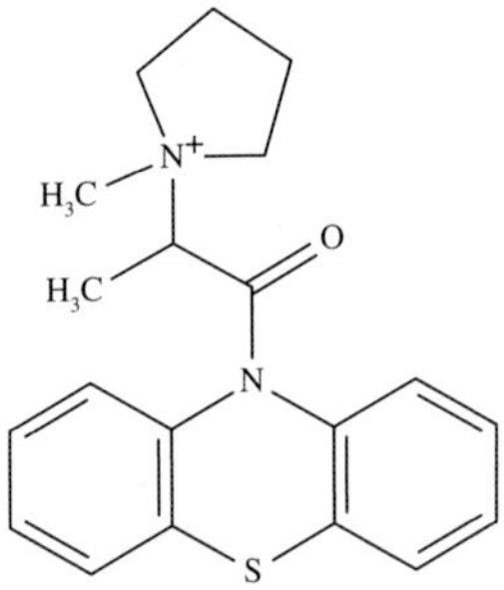

1-Methyl-1-[1-(10-phenothiazinylcarbonyl)ethyl]pyrroli-
dinium = 1-Methyl-1-[1-methyl-2-oxo-2-(10H-pheno-
thiazin-10-yl)ethyl]pyrrolidinium (●)
R Bromide (145-54-0)
S Diaspasmyl, LD 335, *Propyromazine bromide***,
SD 104-19
U Anticholinergic, spasmolytic

9545 (7740)

C$_{20}$H$_{23}$N$_2$
66834-24-0

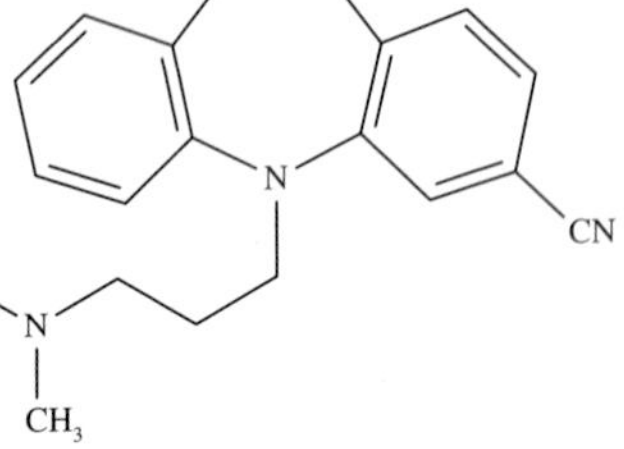

5-[3-(Dimethylamino)propyl]-10,11-dihydro-5H-di-
benz[b,f]azepine-3-carbonitrile (●)
S *Cianopramine***, 3-Cyanoimipramine, DAC,
Ro 11-2465
U Antidepressant

9546

$C_{20}H_{23}N_3$
160161-67-1

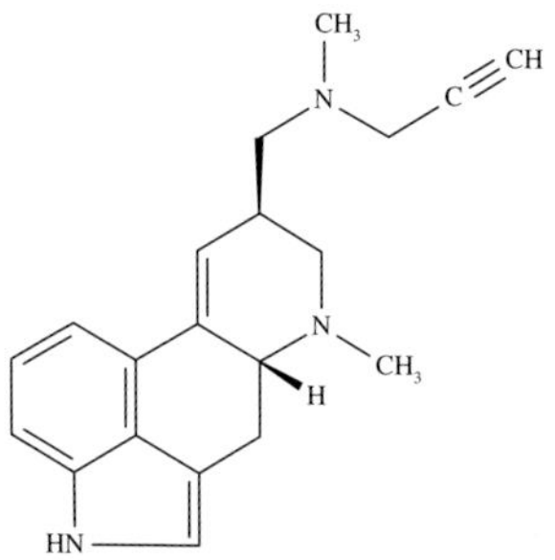

9,10-Didehydro-*N*-methyl-*N*-2-propynyl-6-methyl-8β-
(aminomethyl)ergoline = (8β)-9,10-Didehydro-*N*,6-dime-
thyl-*N*-2-propynylergoline-8-methanamine (●)
S LEK-8829
U Antipsychotic, antiparkinsonian

9547 (7741)

$C_{20}H_{23}N_3$
80410-36-2

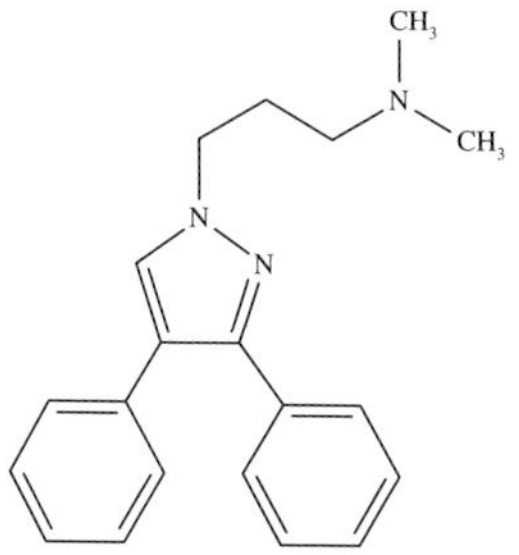

1-[3-(Dimethylamino)propyl]-3,4-diphenylpyrazole =
N,*N*-Dimethyl-3,4-diphenyl-1*H*-pyrazole-1-propan-
amine (●)
R Fumarate (1:1) (80410-37-3)
S *Fezolamine fumarate***, Win 41528-2
U Antidepressant

9548

$C_{20}H_{23}N_3O$
86797-91-3

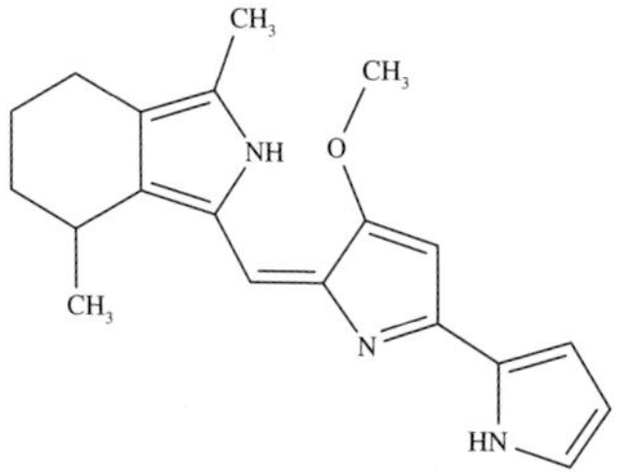

4,5,6,7-Tetrahydro-3-[[3-methoxy-5-(1*H*-pyrrol-2-yl)-
2*H*-pyrrol-2-ylidene]methyl]-1,4-dimethyl-2*H*-isoindole
(●)
S *Cycloprodigiosin*
U Immunosuppressive, antimalarial from *Pseudoaltero-
monas denitrificans*

9549 (7742)

$C_{20}H_{23}N_3OS$
23492-69-5

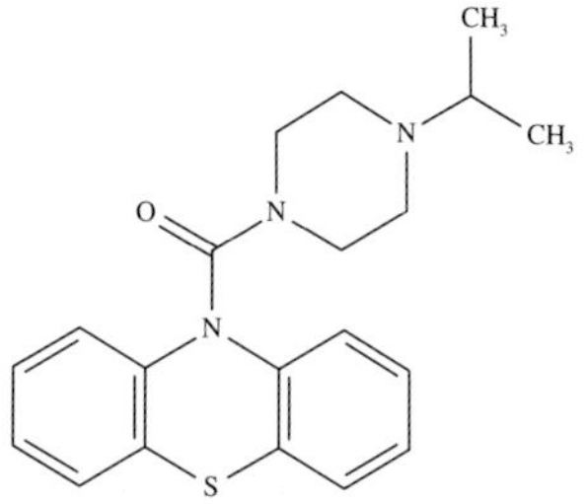

10-[(4-Isopropyl-1-piperazinyl)carbonyl]phenothiazine =
10-[[4-(1-Methylethyl)-1-piperazinyl]carbonyl]-10*H*-
phenothiazine (●)
S *Sopitazine***
U Anticholinergic

9550 (7743)

$C_{20}H_{23}N_3O_2$
5322-53-2

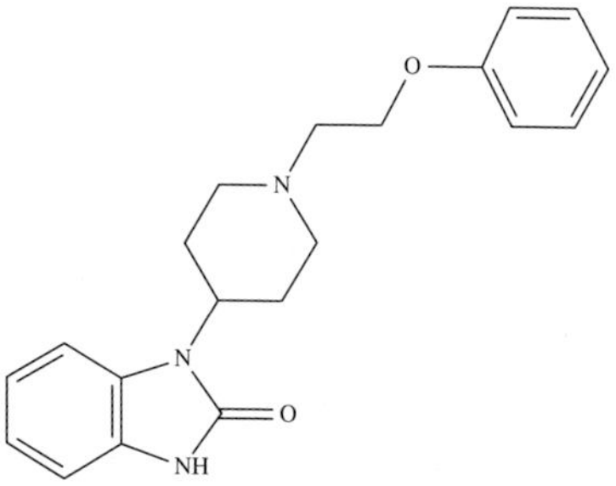

1-[1-(2-Phenoxyethyl)-4-piperidyl]-2-benzimidazoli-
none = 1,3-Dihydro-1-[1-(2-phenoxyethyl)-4-piperidi-
nyl]-2*H*-benzimidazol-2-one (●)
S *Oxiperomide***, *Peromide*, R 4714
U Neuroleptic, antipsychotic

9551 $C_{20}H_{23}N_3O_2$
101246-66-6

(3a*S*,8a*R*)-1,2,3,3a,8,8a-Hexahydro-1,3a,8-trimethylpyr-rolo[2,3-*b*]indol-5-ol phenylcarbamate (ester) (●)
S *Phenserine*
U Acetylcholinesterase inhibitor (cognition enhancer)

9552 (7744) $C_{20}H_{23}N_3O_4$
86880-51-5

(±)-*N*-[2-[[3-(*o*-Cyanophenoxy)-2-hydroxypropyl]ami-no]ethyl]-2-(*p*-hydroxyphenyl)acetamide = *N*-[2-[[3-(2-Cyanophenoxy)-2-hydroxypropyl]amino]ethyl]-4-hydr-oxybenzeneacetamide (●)
S *Epanolol***, ICI 141292, Visacor
U β-Adrenergic blocker (anti-arrhythmic)

9553 (7745) $C_{20}H_{23}N_3O_4S$
26785-16-0

(2*S*,5*R*,6*R*)-6-[α-(2-Butenylideneamino)phenylacetami-do]-3,3-dimethyl-7-oxo-4-thia-1-azabicyclo[3.2.0]hep-tane-2-carboxylic acid = (6*R*)-6-[α-(2-Butenylideneami-no)phenylacetemido]penicillanic acid = [2*S*-(2α,5α,6β)]-6-[[(2-Butenylideneamino)phenylacetyl]amino]-3,3-di-methyl-7-oxo-4-thia-1-azabicyclo[3.2.0]heptane-2-carb-oxylic acid (●)
S *Butampicillin*
U Antibiotic

9554 $C_{20}H_{23}N_3O_9$
142489-47-2

N-[2-(2-Phthalimidoethoxy)acetyl]-L-alanyl-D-glutamic acid = *N*-[[2-(1,3-Dihydro-1,3-dioxo-2*H*-isoindol-2-yl)ethoxy]acetyl]-L-alanyl-D-glutamic acid (●)
S LEK-423, LK-423
U Immunomodulator

9555 (7746) $C_{20}H_{23}N_3S$
105389-86-4

3-[[[5-Methyl-4-(1-piperidinyl)-2-pyridinyl]me-thyl]thio]-1*H*-indole (●)
S AU-1382
U Anti-ulcer agent, cytoprotective

9556 (7747) $C_{20}H_{23}N_5O$
68289-14-5

5,6-Bis[*p*-(dimethylamino)phenyl]-2-methyl-*as*-triazin-3(2*H*)-one = 5,6-Bis[4-(dimethylamino)phenyl]-2-me-thyl-1,2,4-triazin-3(2*H*)-one (●)
S *Metrazifone***, ST 729

U Analgesic

9557 (7748) $C_{20}H_{23}N_5O_2$
86365-92-6

N-(1-Benzyl-4-piperidyl)-6-methoxy-1*H*-benzotriazole-
5-carboxamide = *N*-(1-Benzyl-4-piperidyl)-4,5-azimido-
o-anisamide = 6-Methoxy-*N*-[1-(phenylmethyl)-4-piperi-
dinyl]-1*H*-benzotriazole-5-carboxamide (●)
S *Trazolopride***
U Anti-emetic, stomachic

9558 $C_{20}H_{23}N_5O_4$
142853-45-0

5-[(3-Aminopropyl)amino]-7,10-dihydroxy-2-[[(2-hydr-
oxyethyl)amino]methyl]-6*H*-pyrazolo[4,5,1-*de*]acridin-
6-one (●)
R Dihydrochloride (207862-44-0)
S KW 2170
U Antineoplastic

9559 $C_{20}H_{23}N_5O_6$
126016-79-3

N-[4-[2-(2-Amino-4,5,6,7-tetrahydro-4-oxo-1*H*-pyrro-
lo[2,3-*d*]pyrimidin-5-yl)ethyl]benzoyl]-L-glutamic acid
(●)
S LY 288601

U Antineoplastic

9560 (7749) $C_{20}H_{23}N_5O_6S$
37091-66-0

(2*S*,5*R*,6*R*)-3,3-Dimethyl-7-oxo-6-[(*R*)-2-(2-oxo-1-imi-
dazolidinecarboxamido)-2-phenylacetamido]-4-thia-1-
azabicyclo[3.2.0]heptane-2-carboxylic acid = (6*R*)-6-[*N*-
(2-Oxoimidazolidin-1-ylcarbonyl)-D-phenylglycylami-
no]penicillanic acid = [2*S*-[2α,5α,6β(*S**)]]-3,3-Dimethyl-
7-oxo-6-[[[[(2-oxo-1-imidazolidinyl)carbonyl]ami-
no]phenylacetyl]amino]-4-thia-1-azabicyclo[3.2.0]hep-
tane-2-carboxylic acid (●)
R Monosodium salt (37091-65-9)
S Abrodil, Alocin, Azlin, Azlocil, *Azlocillin sodium***,
Bay e 6905, Securopen
U Antibiotic

9561 (7750) $C_{20}H_{23}N_5S$
71079-19-1

1-Cyclohexyl-2-(2-methyl-4-quinolyl)-3-(2-thiazo-
lyl)guanidine = *N*-Cyclohexyl-*N'*-(2-methyl-4-quinoli-
nyl)-*N''*-2-thiazolylguanidine (●)
S SR 1368, *Timegadine***
U Anti-inflammatory

9562 $C_{20}H_{23}N_6O_5P$
                                                        210355-12-7

N-[[[(2*Z*)-[(6-Amino-9*H*-purin-9-yl)methylene]cyclopro-
pyl]methoxy]phenoxyphosphinyl]-L-alanine methyl ester
(●)
S   QYL-609
U   Antiviral (HIV-1 protease inhibitor)

9563   (7751) $C_{20}H_{23}N_7O_6$
                                                        86936-90-5

1-(6-Amino-9*H*-purin-9-yl)-1,3-dideoxy-3-(*O*-methyl-L-
tyrosylamino)-β-D-ribofuranuronic acid = (*S*)-3-[[2-Ami-
no-3-(4-methoxyphenyl)-1-oxopropyl]amino]-1-(6-ami-
no-9*H*-purin-9-yl)-1,3-dideoxy-β-D-ribofuranuronic acid
(●)
S   *Chryscandin*, FR 48736, WF 4629
U   Antifungal, gram-positive antibacterial (antibiotic
        from *Chrysosporium pannorum*)

9564   (7752) $C_{20}H_{23}N_7O_7$
                                                        58-05-9

N-[*p*-[[(2-Amino-5-formyl-5,6,7,8-tetrahydro-4-hydr-
oxy-6-pteridinyl)methyl]amino]benzoyl]-L-glutamic acid
= 5-Formyl-5,6,7,8-tetrahydropteroyl-L-glutamic acid =
N-[4-[[(2-Amino-5-formyl-1,4,5,6,7,8-hexahydro-4-oxo-
6-pteridinyl)methyl]amino]benzoyl]-L-glutamic acid (●)
S   *Acide folinique, Acidum folinicum*, CF, *Citrovorin,
        Citrovorum factor*, Cromatonbic Folinico, *Factor ci-
        trovorum, Folinic acid, Folinsäure, Formyltetrahy-
        drofolic acid*, FTHF, Leucal, *Leuconostoc-citro-
        vorum-Factor*
U   Growth factor, anti-anemic

9564-01   (7752-01)   1492-18-8
R   Calcium salt (1:1) [also calcium salt (1:1) pentahydra-
        te (6035-45-6)]
S   Antrex, Asovorin, Calcifolin, *Calcio folinato, Calci-
        um folinate**, Calcium Leucovorin, Calfolex, Calfo-
        nat, Calinat, Cehafolin, Chemifolin, Citofolin, Ci-
        trec, Cromaton, Dalisol, Degalin, Disintox, Divical,
        Ecofol, Emovis, Erbanfol, Ergix, Flynoken, Folaren,
        Folaxin, Foli-cell, Folidan, Folidar, Folinac, Folina-
        te-SF Calcium, Folinoral, Folinvit, Foliplus, Folix,
        Haemato-folin, Imofolin, Lederfolat, Lederfolin(e),
        Ledervorin-Calcium, Lesten, Leucocalcin, Leucosar,
        Leucovarin-Teva, Leucovorin(a), *Leucovorin calci-
        um*, Levorin, Medicofolin, Medifolin, Neofolin, No-
        vafoline, NSC-3590, Nyrin, O-folin, Osfolate, Osfo-
        lato, Perfolate, Perfolin, Refolinon, Rescufolin, Res-
        cuvolin, Resfolin, Ribofolin, Rontafor, Sanifolin,
        Sulton, Tonofolin, Wellcovorin
U   Antidote to folic acid antagonists, anti-anemic

9565   (7753) $C_{20}H_{23}N_7O_7$
                                                        68538-85-2

N-[*p*-[[[(6*S*)-2-Amino-5-formyl-1,4,5,6,7,8-hexahydro-4-
oxo-6-pteridinyl]methyl]amino]benzoyl]-L-glutamic acid
= (*S*)-*N*-[4-[[(2-Amino-5-formyl-1,4,5,6,7,8-hexahydro-
4-oxo-6-pteridinyl)methyl]amino]benzoyl]-L-glutamic
acid (●)
R   Calcium salt (1:1) (80433-71-2)

S *Calcium levofolinate***, CL 307782, Elfosin, Elvorine, Foliben, Isofolin, Isovorin, Levofolene, *Levoleucovorin calcium*, Levorin "Wyeth-Lederle", L-Leucovorin

U Antidote to folic acid antagonists (antineoplastic)

9566   (7754) $C_{20}H_{24}BrNO_4$
                                    58939-37-0

1-[(2-Bromo-4,5-dimethoxyphenyl)methyl]-1,2,3,4-tetrahydro-6-methoxy-2-methyl-7-isoquinolinol (●)
S A-69024
U Dopamine D$_1$-receptor antagonist

9567   (7755) $C_{20}H_{24}BrN_3O$
                                    478-84-2

2-Bromo-*N,N*-diethyl-D-lysergamide = (8β)-2-Bromo-9,10-didehydro-*N,N*-diethyl-6-methylergoline-8-carboxamide (●)
S BOL 148, 2-Brom-LSD
U Sympatholytic, serotonin inhibitor

9568   (7756) $C_{20}H_{24}ClN$

4-Chloro-α-ethyl-*N,N*-dimethyl-α-(3-phenylallyl)benzylamine = 4-Chloro-α-ethyl-*N,N*-dimethyl-α-(3-phenyl-2-propenyl)benzenemethanamine (●)

R Hydrochloride (129140-62-1)
S JO-1870
U Psychotropic (treatment of urinary incontinence)

9569   (7757) $C_{20}H_{24}ClNO$
                                    3703-76-2

1-[2-[(*p*-Chloro-α-phenylbenzyl)oxy]ethyl]piperidine = β-Piperidinoethyl *p*-chlorobenzhydryl ether = 1-[2-[(4-Chlorophenyl)phenylmethoxy]ethyl]piperidine (●)
R also hydrochloride (14984-68-0) or 2-[(6-hydroxy[1,1'-biphenyl]-3-yl)carbonyl]benzoate (fendizoate) (85187-37-7)
S Cloel, *Cloperastine***, Flutox, HT-11, Hustazol, Nitossil, Novotossil, Novotusil, Quik, Risoltuss, Seki, Sekin, Sekisan
U Antitussive

9570 $C_{20}H_{24}ClNO$
                                    132301-89-4

(*S*)-1-[2-[(4-Chlorophenyl)phenylmethoxy]ethyl]piperidine (●)
R 2-[(6-Hydroxy[1,1'-biphenyl]-3-yl)carbonyl]benzoate (fendizoate) (220329-19-1)
S Clofend, *Levocloperastine fendizoate*, Politosse
U Antitussive

9571   (7759) $C_{20}H_{24}ClNO_2$
6699-38-3

α-Chlorodiphenylacetic acid β-(diethylamino)ethyl ester
= α-Chloro-α-phenylbenzeneacetic acid 2-(diethylami-
no)ethyl ester (●)
R   Hydrochloride (902-83-0)
S   Diamifen, Diamiphene, Diaphen
U   Anticholinergic, antihistaminic, tranquilizer

9572 $C_{20}H_{24}ClNO_2$
210757-91-8

N-[2(R)-(3-Chlorophenyl)-2-hydroxyethyl]-N-[[7-meth-
oxy-1,2,3,4-tetrahydronaphthalen-2(R)-yl]methyl]amine
= (αR)-3-Chloro-α-[[[[(2R)-1,2,3,4-tetrahydro-7-meth-
oxy-2-naphthalenyl]methyl]amino]methyl]benzeneme-
thanol (●)
R   Hydrochloride (136758-99-1)
S   SR 59119A
U   β₃-Adrenoceptor agonist

9573   (7758) $C_{20}H_{24}ClNO_2$
2154-02-1

1-(p-Chlorophenethyl)-1,2,3,4-tetrahydro-6,7-dimeth-
oxy-2-methylisoquinoline = 1-[2-(4-Chlorophenyl)ethyl]-

1,2,3,4-tetrahydro-6,7-dimethoxy-2-methylisoquinoline
(●)
S   ARC 1-K-1, *Methopholine*, *Metofoline**, NIH 7672,
Ro 4-1778/1, Versidyne
U   Analgesic

9574   (7760) $C_{20}H_{24}ClNO_4$
86615-96-5

(R*,R*)-(±)-[4-[2-[[2-(3-Chlorophenyl)-2-hydroxy-
ethyl]amino]propyl]phenoxy]acetic acid methyl ester (●)
R   Hydrobromide (86615-41-0)
S   BRL 35135
U   Oral hypoglycemic, anti-obesity

9575   (7761) $C_{20}H_{24}ClN_3O_2$
55905-53-8

4-Amino-N-(1-benzyl-4-piperidyl)-5-chloro-o-anis-
amide = 4-Amino-5-chloro-2-methoxy-N-[1-(phenylmethyl)-4-piperidinyl]benzamide (●)
R   also monohydrochloride (57645-39-3) or malate (1:1)
(57645-91-7)
S   Amicos, Clabol, Clanzoflat, Clanzol, Clast, Clebodi-
an, Clebofex, Cleboflat, Clebon, *Clebopride**, Cle-
boril, Clebutec, Cleprid, Crust, Darpan, Flatoril (active substance), Gastridin "Microsules Argentina",
LAS 9273, Leboril, Madurase, Motilex "Guidotti",
Ortogastril, Valoprid "Labinca", Vuxolin
U   Anti-emetic, stomachic

9576 (7762) $C_{20}H_{24}ClN_3O_2$
73328-60-6

N-(1-Benzyl-3-pyrrolidinyl)-5-chloro-4-(methylamino)-o-anisamide = 5-Chloro-2-methoxy-4-(methylamino)-N-[1-(phenylmethyl)-3-pyrrolidinyl]benzamide (●)
S YM-08050
U Neuroleptic

9577 (7763) $C_{20}H_{24}ClN_3O_4$
110933-28-3

N'-[2,3-Bis(4-hydroxyphenyl)pentyl]-N-(2-chloroethyl)-N-nitrosourea (●)
S HEX-CNU
U Antineoplastic

9578 (7764) $C_{20}H_{24}ClN_3S$
58-38-8

2-Chloro-10-[3-(4-methyl-1-piperazinyl)propyl]phenothiazine = 1-[3-(2-Chloro-10-phenothiazinyl)propyl]-4-methylpiperazine = 2-Chloro-10-[3-(4-methyl-1-piperazinyl)propyl]-10H-phenothiazine (●)

R also salts [maleate (1:2) (84-02-6), dimethanesulfonate (51888-09-6), 1,2-ethanedisulfonate (1:1) (1257-78-9)]
S Anti-Naus, Apiran, Bayer A 173, Buccastem, Capazine, Chlorazine, *Chlormeprazine*, Chloropernazinum, Chlorpazine, Chlorperazin, Compazine, Depafen, Dhaperazine, Dicopal, Emelent, Emetiral, Emidoxyn, F.I. 5685, Gastrectil, Klometil, Kronocin, Mentil, Meterazin, Metherazin, Mitil, Nautisol, Nibromin A, Nipodal, Normalmin, Novamin "Shionogi", Nu-Prochlor, Pasotomin, Peratil, Phenozine, *Prochlorpémazine*, *Prochlorperazine***, Proclorperazina, Properan, Propizin, Prorazin "Technilab", Prozière, 6140 R.P., Sametil, Scripto-Metic, Sedovomin, SKF 4657, Stabil, Stella, Stemetil, Stemmetil, Steremal, Témentil, Temetil, Ultrazine, Vertigon "SmithKline Beecham"
U Anti-emetic, neuroleptic

9579 (7765) $C_{20}H_{24}Cl_2N_{10}$
5636-92-0

1,1'-[1,4-Piperazinediylbis(imidocarbonyl)]bis[3-(p-chlorophenyl)guanidine] = 1,4-Bis[3-(p-chlorophenyl)guanidinoformimidoyl]piperazine = 1,1'-Diethylenebis[5-(p-chlorophenyl)biguanide] = N,N''-Bis[[(4-chlorophenyl)amino]iminomethyl]-1,4-piperazinedicarboximidamide (●)
R also dihydrochloride (19803-62-4)
S Medibact, *Picloxydine***, Vitabact
U Antibacterial, antifungal

9580　(7766)　　　　　　　　　　　　$C_{20}H_{24}FN_3O_2$
73865-18-6

(±)-α-[[[3-(1-Benzimidazolyl)-1,1-dimethylpropyl]ami-
no]methyl]-2-fluoro-4-hydroxybenzyl alcohol = 2-[[3-(1-
Benzimidazolyl)-1,1-dimethylpropyl]amino]-1-(2-
fluoro-4-hydroxyphenyl)ethanol = α-[[[3-(1*H*-Benzimi-
dazol-1-yl)-1,1-dimethylpropyl]amino]methyl]-2-fluoro-
4-hydroxybenzenemethanol (●)
S *Nardeterol***, SOM 1122
U Bronchospasmolytic (β-adrenergic receptor agonist)

9581　(7767)　　　　　　　　　　　　$C_{20}H_{24}FN_3O_2$
21686-10-2

3-[4-(*p*-Fluorophenyl)-3,6-dihydro-1(2*H*)-pyridyl]-1-[1-
(2-hydroxyethyl)-5-methyl-4-pyrazolyl]-1-propanone =
3-[4-(4-Fluorophenyl)-3,6-dihydro-1(2*H*)-pyridinyl]-1-
[1-(2-hydroxyethyl)-5-methyl-1*H*-pyrazol-4-yl]-1-propa-
none (●)
S *Flupranone***, Go 1507
U Antihypertensive

9582　(7768)　　　　　　　　　　　　$C_{20}H_{24}FN_3O_4$
127294-70-6

1-Cyclopropyl-6-fluoro-1,4-dihydro-8-methoxy-7-[3-
(methylamino)-1-piperidinyl]-4-oxo-3-quinolinecarb-
oxylic acid (●)
R also dihydrate

S *Balfloxacin*, *Balofloxacin***, Baloxin, BLFX, Neu-
quineron, Neuroquinoron, Q-35
U Antibacterial

9583　(7769)　　　　　　　　　　　　$C_{20}H_{24}FN_3O_4S$
56488-61-0

N-[[1-(*p*-Fluorobenzyl)-2-pyrrolidinyl]methyl]-5-sulf-
amoyl-*o*-anisamide = 5-(Aminosulfonyl)-*N*-[[1-[(4-
fluorophenyl)methyl]-2-pyrrolidinyl]methyl]-2-meth-
oxybenzamide (●)
S *Flubepride***, SL 74205
U Psychotropic, anti-emetic

9584　　　　　　　　　　　　　　　　$C_{20}H_{24}FN_5O$
141071-67-2

5-Fluoro-3-[3-[4-(5-methoxy-4-pyrimidinyl)-1-piperazi-
nyl]propyl]-1*H*-indole (●)
R Dihydrochloride (146479-45-0)
S BMS 181101
U Antidepressant

9585　(7770)　　　　　　　　　　　　$C_{20}H_{24}F_3N_3OS_2$
10202-40-1

4-[3-(4-β-Hydroxyethyl-1-piperazinyl)propyl]-6-tri-
fluoromethyl-4*H*-thieno[2,3-*b*][1,4]benzothiazine = 4-[3-

[6-(Trifluoromethyl)-4H-thieno[2,3-b][1,4]benzothiazin-4-yl]propyl]-1-piperazineethanol (●)
S *Flutizenol***, NR 286
U Neuroleptic

9586 $C_{20}H_{24}F_4N_4O_3$
                                                                186386-21-0

3-[3-[4-[4-Fluoro-2-(2,2,2-trifluoroethoxy)phenyl]-1-piperazinyl]propyl]-5-methyl-2,4(1H,3H)-pyrimidinedione (●)
R Monohydrochloride (186384-62-3)
S Ro 70-0004, RS-100975
U α_1-Adrenoceptor subtype antagonist for treatment of benign prostatic hyperplasia

9587 (7771) $C_{20}H_{24}I_2O_2$
                                                                552-22-7

4,4'-Bis(iodooxy)-5,5'-diisopropyl-2,2'-dimethyl-1,1'-biphenyl = Diiododithymol = Hypoiodous acid 2,2'-dimethyl-5,5'-bis(1-methylethyl)[1,1'-biphenyl]-4,4'-diyl ester (●)
R Mixture with isomers
S Annidalin, Aristol, *Diiodothymol*, *Dithymoldijodid*, Intrasol, Iodistol, Iodohydromol, Iodosol, *Iodothymol*, Iosol, Iothymol, Thymiode, Thymiodol, Thymodin, *Thymol iodide*, Thymotol
U Wound disinfectant

9588 (7772) $C_{20}H_{24}N$

4-Benzhydrylidene-1,1-dimethylpiperidinium = 4-(Diphenylmethylene)-1,1-dimethylpiperidinium (●)
R Methylsulfate (62-97-5)
S Demotil, *Diphemanil metilsulfate***, Diphenatil, *Diphenmethanil methylsulfate*, Nivelon, Prantal, Prantil, Prentol, Talpran, *Vagophemanil methylsulfate*, Variton
U Anticholinergic

9589 (7773) $C_{20}H_{24}NO_3$

1,1-Dimethyl-3-hydroxypyrrolidinium benzilate = 3-(Benziloyloxy)-1,1-dimethylpyrrolidinium = 3-[(Hydroxydiphenylacetyl)oxy]-1,1-dimethylpyrrolidinium (●)
R Bromide (13696-15-6)
S *Benzopyrrolate bromide*, *Benzopyrronium bromide***
U Anticholinergic

9590   (7774) $C_{20}H_{24}N_2$
                                                          5636-83-9

2-[1-[2-[2-(Dimethylamino)ethyl]-3-indenyl]ethyl]pyri-
dine = 2-[β-(Dimethylamino)ethyl]-3-[1-(2-pyri-
dyl)ethyl]indene = Dimethyl[2-[3-[1-(2-pyridyl)ethyl]-
1*H*-inden-2-yl]ethyl]amine = *N*,*N*-Dimethyl-3-[1-(2-pyri-
dinyl)ethyl]-1*H*-indene-2-ethanamine (●)
R   Maleate (1:1) (3614-69-5)
S   *Dimethindene maleate, Dimethpyrindene maleate,*
        Dimethylpyrindene maleate, Dimetindene male-
        ate**, Fenestil, Fengel, Fenistil, Fenostil, Forhistal,
        Foristal, Neostil, NSC-107677, Pecofenil, Specisun,
        Su-6518, Triten, Z 822
U   Antihistaminic, anti-allergic, antipruritic

9591   (7776) $C_{20}H_{24}N_2$
                                                          4855-95-2

1-(10,11-Dihydro-5*H*-dibenzo[*a,d*]cyclohepten-10-yl)-4-
methylpiperazine (●)
S   *Peraptene*
U   Psychosedative

9592   (7775) $C_{20}H_{24}N_2$
                                                          70713-45-0

1-Allyl-4-(diphenylmethyl)piperazine = 1-(Diphenylme-
thyl)-4-(2-propenyl)piperazine (●)
R   Dihydrochloride (41332-10-9)
S   Aligeron, AS$_2$
U   Cerebral vasodilator

9593   (7777) $C_{20}H_{24}N_2$
                                                          47206-15-5

5-[3-(Dimethylamino)propyl]-5,6-dihydro-11-methy-
lene-11*H*-dibenz[*b,e*]azepine = 5-[3-(Dimethylamino)propyl]-5,6-dihydro-11-methylenemorphanthridine =
6,11-Dihydro-*N*,*N*-dimethyl-11-methylene-5*H*-di-
benz[*b,e*]azepine-5-propanamine (●)
S   *Enprazepine***
U   Antidepressant

9594 (7778) $C_{20}H_{24}N_2$
1232-85-5

11-[3-(Dimethylamino)propylidene]-5,6-dihydro-5-me-
thyl-11H-dibenz[b,e]azepine = 11-[3-(Dimethylamino)propylidene]-5,6-dihydro-5-methylmorphanthridine =
3-(5,6-Dihydro-5-methyl-11H-dibenz[b,e]azepin-11-yli-
dene)-N,N-dimethyl-1-propanamine (●)
R also dicyclamate (1443-91-0)
S Elantrine*, EX 10-029, Fantrine, RMI 80029
U Antiparkinsonian, anticholinergic

9595 (7781) $C_{20}H_{24}N_2O$
68654-64-8

N-[2-(1-Pyrrolidinyl)ethyl]diphenylacetamide = α-Phe-
nyl-N-[2-(1-pyrrolidinyl)ethyl]benzeneacetamide (●)
R Monohydrochloride (63207-78-3)
S F 1459
U Antitussive, respiratory stimulant

9596 (7779) $C_{20}H_{24}N_2O$
74517-78-5

9-[3-(Isopropylamino)propyl]fluorene-9-carboxamide =
9-[3-[(1-Methylethyl)amino]propyl]-9H-fluorene-9-carb-
oxamide (●)
R Monohydrochloride (73681-12-6)
S Decabid, Indecainide hydrochloride**, LY 135837,
Ricainide hydrochloride
U Anti-arrhythmic

9597 (7780) $C_{20}H_{24}N_2O$
26070-78-0

2-[2-(Diethylamino)ethyl]-3-phenylphthalimidine = 2-(2-
Diethylaminoethyl)-3-phenylisoindolin-1-one = 2-[2-(Di-
ethylamino)ethyl]-2,3-dihydro-3-phenyl-1H-isoindol-1-
one (●)
R Phosphate (26070-68-8)
S Bu-232, Gitan, Ubisindine phosphate**
U Anti-arrhythmic, antitussive

9598 (7782) $C_{20}H_{24}N_2OS$
4774-53-2

S-[2-(Dimethylamino)ethyl] 9,9-dimethyl-10-acridancar-
bothioate = 9,9-Dimethyl-10(9H)-acridinecarbothioic
acid S-[2-(dimethylamino)ethyl] ester (●)
S Botiacrine**
U Antiparkinsonian

9599   (7783) $C_{20}H_{24}N_2OS$
4809-80-7

4-(10,11-Dihydrodibenzo[*b,f*]thiepin-10-yl)-1-pipera-
zineethanol (●)
S   *Noroxythepin*
U   Neuroleptic

9600   (7787) $C_{20}H_{24}N_2OS$
362-29-8

1-[10-[2-(Dimethylamino)propyl]-2-phenothiazinyl]-1-
propanone = 10-[2-(Dimethylamino)propyl]-2-propionyl-
phenothiazine = 1-[10-[2-(Dimethylamino)propyl]-10*H*-
phenothiazin-2-yl]-1-propanone (●)
S   NSC-169450, *Propiomazine**, Propionylprometha-
zine*
U   Sedative (pre-anesthetic), hypnotic

9600-01   (7787-01)   1240-15-9
R   Monohydrochloride
S   Largon, Travet
U   Sedative (pre-anesthetic), hypnotic

9600-02   (7787-02)   3568-23-8
R   Maleate (1:1)
S   1678 C.B., Dorevan, Dorévane, Indorm, Phenoctyl,
Propavan, Propial, Serentin, Wy-1359
U   Sedative (pre-anesthetic), hypnotic

9601   (7786) $C_{20}H_{24}N_2OS$
3568-24-9

10-[3-(Dimethylamino)propyl]-2-propionylphenothia-
zine = 1-[10-[3-(Dimethylamino)propyl]-10*H*-phenothia-
zin-2-yl]-1-propanone (●)
S   *Dipropimazine, Propionylpromazine, Propioproma-
zine*
U   Neuroleptic

9601-01   (7786-01)   14796-43-1
R   Phosphate (1:1)
S   1497 C.B., Combelen, Combilen, Tranvet
U   Neuroleptic

9602   (7784) $C_{20}H_{24}N_2OS$
1166-34-3

2'-[[3-(Dimethylamino)propyl]thio]cinnamanilide = *N*-
[2-[[3-(Dimethylamino)propyl]thio]phenyl]-3-phenyl-2-
propenamide (●)
R   Monohydrochloride (54-84-2)
S   *Cinanserin hydrochloride**, NSC-125717,
SQ 10643
U   Serotonin inhibitor

9603   (7785) $C_{20}H_{24}N_2OS$
479-50-5

1-[[2-(Diethylamino)ethyl]amino]-4-methylthioxanthen-9-one = 1-[[2-(Diethylamino)ethyl]amino]-4-methyl-9*H*-thioxanthen-9-one (●)

R Monohydrochloride (548-57-2)

S B.W. 57-233, *Lucanthone hydrochloride***,
Miracil D, Miracol, Ms. 752, Nilodin, NSC-14574, 3735 R.P., Scapuren, 79 T 61, *Thioxanthone, Tixantone*

U Antischistosomal

9604 (7788)

$C_{20}H_{24}N_2O_2$
130-95-0

(*R*)-(6-Methoxy-4-quinolyl)[(2*S*,4*S*,5*R*)-5-vinyl-2-quinuclidinyl]methanol = 6-Methoxy-α-(5-vinyl-2-quinuclidinyl)-4-quinolinemethanol = (5-Vinyl-2-quinuclidinyl)(6-methoxy-4-quinolyl)methanol = 6-Methoxycinchonine = (8α,9*R*)-6'-Methoxycinchonan-9-ol (●)

R see also nos. 1186-08, 1979-01, 4131-04, 6247-03, 11892

S *Chinin, Kinin, Methylcupreine, Quinine*

U Antimalarial, antipyretic, muscle relaxant, tonic, bitter stomachic (main alkaloid of *Cinchona* bark)

9604-01 (7788-01)   7549-43-1

R Hydrochloride [especially dihydrochloride (60-93-5)]

S Aquanine, Chinimetten, Kuining, Paluquina, Quininga, Sagittaproct CH

U Influenza prophylactic, sclerosing agent

9604-02 (7788-02)   549-48-4

R Dihydriodide

S Reducto (old form)

U Influenza prophylactic

9604-03 (7788-03)   804-63-7

R Sulfate (2:1) [or sulfate (2:1) dihydrate (6119-70-6)]

S Adaquin, Aflukin "A.C.F.", Bi-Quin, Biquinate, *Chininum sulfuricum*, Circonyl N, Coco-Quinine, Dentojel, El Caribe, Endopalur, Impalud, Legatrin,

Limptar N, Myoquin, Night Leg Cramp Relief, Novoquinine, Perlete Goui's, Quin-260, Quinamin, Quinaminoph, Quinamm, Quinate "RPR", Quinbisan, Quinbisul, Quindan, Quine, *Quinine sulfate*, Quinoc-F, Quinoctal, Quinsan, Quinsul, Quiphile, Quirin, Q-vel, Strema

U Influenza prophylactic, antimalarial, in nocturnal leg cramps

---

9604-04 (7788-04)   8048-94-0

R Mixture with bismuth iodide (BiI_3)

S Angino-Bismutho, Bijochinol, Bijogadol, Bismiochinina, Bismos, Bismosalvan, Bismugalol, Bismuxel, Hepabismol, Iodobismuto "Pons", Jobichin, Quimutol, Quinby, Quinimutol, Quiniobine, Quiniobismuth, Quinostab, Recto-Klari, Rubyl, Suquinabis, Trepoquinol

U Antisyphilitic

9604-05 (7788-05)   130-93-8

R Acetylsalicylate (no. 1535) (1:1)

S Chinacetyl, Chinasprin, Xaxaquin

U Antipyretic, analgesic, internal antiseptic

9604-06 (7788-06)   146-40-7

R Ascorbate (1:2)

S Combatin, *Quinine ascorbate*

U Antipyretic, influenza prophylactic, smoking deterrent

9604-07 (7788-07)   117-72-6

R Salicylosalicylate (no. 4411) (1:2)

S Quinisal, Quinisan

U Analgesic, antipyretic

9604-08 (7788-08)   6151-67-3

R Compd. with diethylbarbituric acid (no. 1290) (1:1)

S Chineonal

U Antipyretic, uterotonic

9604-09 (7788-09)   130-90-5

R Formate (1:1)

S Quinaform, Quinoform(e)

U Antibacterial

9604-10 (7788-10)   60662-72-8

R Gentisate

S Gentochin

U Influenza prophylactic

9604-11 (7788-11) 146-39-4
R Glycerophosphate (2:1)
S Kineurine
U Tonic

9604-12 (7788-12) 60662-71-7
R Guaiacol salicylate (1:1)
S Ermerol
U Influenza prophylactic

9604-13 (7788-13) 60662-73-9
R 2-Hydroxyisophthalate (3-carboxysalicylate)
S Hivernine
U Antirheumatic, antipyretic, analgesic

9604-14 (7788-14)
R Phenylquinolinecarboxylate (no. 5881)
S Chinin "Byk; Lüdi; Weil", Chininett, Chiniphen, Diplochin, Fenilbichina
U Influenza prophylactic

9604-15 (7788-15)
R 3-Hydroxy-2-phenyl-4-quinolinecarboxylate (no. 5884)
S Chininum oxycinchophenas
U Analgesic, antipyretic

9604-16 (7788-16) 16846-31-4
R Compd. with thymol (1:1)
S Neotimolo, Quinothymol, Quinotimol, Tilakin, Timochin, Timolkina
U Bronchotherapeutic

9604-17 (7788-17)
R Compd. with gold thiosulfate
S Orosanil B
U Gold therapeutic

9605 (7789) $C_{20}H_{24}N_2O_2$
56-54-2

1948

(S)-(6-Methoxy-4-quinolyl)[(2R,4S,5R)-5-vinyl-2-quinuclidinyl]methanol = stereoisomer of 6-Methoxy-α-(5-vinyl-2-quinuclidinyl)-4-quinolinemethanol = stereoisomer of (5-Vinyl-2-quinuclidinyl)-(6-methoxy-4-quinolyl)methanol = (9S)-6'-Methoxycinchonan-9-ol (●)
R see also no. 3255-13
S Chinidin, β-Chinin, Chinotin, Cinchotin, Conchinin, Conquinine, Kinidin, Lipettes, Pitayine, Quinidine, β-Quinine, Quinipec
U Anti-arrhythmic (alkaloid of Cinchona bark)

9605-01 (7789-01) 50-54-4
R Sulfate (2:1) [also sulfate (2:1) dihydrate (6591-63-5) or sulfate (1:1) tetrahydrate]
S Apo-quinidine, Auriquin, Biquin Durules, Cardio-Quinol, Cardoquin, Chinidin-Duriles, Chinidinorm, Chinidin-retard-Isis, Chinidinum sulfuricum, Chinteina, Cin-Quin, Clinadol (old form), Cordichin, Kardiokin, Keyquin, Kiditard, Kinichron, Kinidin Duretter, Kinidin-Duriles, Kiniduron, Kinilentin, Kinilong, Kinipure, Kinitard, Kuiniding, Maso-Quin, Ni-Cram, Novoquinidin, Optochinidin retard, Proquin, Quincardine, Quinicardine, Quinidate, Quiniday, Quinidex, Quinidine sulfate, Quinidoxine, Quiniduran, Quinidurule, Quini-Durules, Quinilent, Quinitex, Quinora, Ritmo-quinidina, Sedotensil "Furmavel", Sincroquina, Solfachinid, Systodin, Vanquin "Vangard"
U Anti-arrhythmic

9605-02 (7789-02) 60662-74-0
R Camphorsulfonate (no. 2423) (1:1)
S Canfochinid, Rhythmidine-Amp.
U Cardiotonic

9605-03 (7789-03)
R Deoxyribonucleinate
S Nuclinid
U Anti-arrhythmic

9605-04 (7789-04) 7054-25-3
R Mono-D-gluconate (1:1)
S Depo-Kinidin, Durakinid, Duraquin, Dura-Tab, Gluquinate, Gluquine, QuinaDur, Quinaglute, Quinalan, Quinate "Rougier", Quinatime, Quinidine gluconate, Quin-Release SR
U Anti-arrhythmic

9605-05 (7789-05)   58829-32-6
R Polygalacturonate
S Cardioquin(e), Galactoquin, Galatturil-Chinidina,
     Naticardina, Neochinidin, Quinilac, Ritmocor "Male-
     sci", Sineflutter
U Cardiotonic

9605-06 (7789-06)   49558-34-1
R 7-Theophyllineacetate (no. 1619) (1:2)
S Chinteina (old form)
U Cardiotonic, anti-arrhythmic

9605-07 (7789-07)   54692-74-9
R Arabogalactan sulfate
S Longachin, Longacor, Longaquin, QAGS
U Anti-arrhythmic

9606 (7790) $C_{20}H_{24}N_2O_2$
                                                  89331-66-8

9-[[3-(*tert*-Butylamino)-2-hydroxypropyl]oximino]fluo-
rene = 9*H*-Fluoren-9-one *O*-[3-[(1,1-dimethylethyl)ami-
no]-2-hydroxypropyl]oxime (●)
R Monohydrochloride (60979-28-4)
S IPS 339
U β-Adrenergic blocker

9607 (7791) $C_{20}H_{24}N_2O_2$
                                                  7119-40-6

4,4'-(*p*-Phenylenedi-2-propynylene)dimorpholine = 4,4'-
1,4-Phenylenedi-2-propyne-3,1-diyl)bis[morpholine]
(●)
S Benzomopine
U Antihypertensive

9608 (7792) $C_{20}H_{24}N_2O_2$
                                                  84-55-9

(3*R*,4*R*)-1-(6-Methoxy-4-quinolyl)-3-(3-vinyl-4-piperi-
dyl)-1-propanone = 6-Methoxy-4-[β-(3-vinyl-4-piperi-
dyl)propionyl]quinoline = (3*R*-*cis*)-3-(3-Ethenyl-4-pipe-
ridinyl)-1-(6-methoxy-4-quinolinyl)-1-propanone (●)
R also monohydrochloride (52211-63-9)
S *Chinicine, Chinotoxin*, Desclidium, LM 192, *Mequi-
     verine*, Permiran, *Quinicine, Quinotoxine, Quinoto-
     xol*, Vasexeten, *Viquidil***, Xitadil
U Vasodilator

9609 (7793) $C_{20}H_{24}N_2O_2$
                                                  3569-84-4

1-[[2-(Diethylamino)ethyl]amino]-4-methyl-9-xanthone
= 1-[[2-(Diethylamino)ethyl]amino]-4-methyl-9*H*-xan-
then-9-one (●)
R also hydrochloride
S Miracil A
U Antischistosomal

9610 (7794) $C_{20}H_{24}N_2O_2S$
3105-97-3

1-[[2-(Diethylamino)ethyl]amino]-4-(hydroxyme-thyl)thioxanthen-9-one = 1-[[2-(Diethylamino)ethyl]ami-no]-4-(hydroxymethyl)-9H-thioxanthen-9-one (●)
R also monomethanesulfonate (23255-93-8)
S Etrenol, *Hycanthone**, Hydroxylucanthone, Lucanthone Metabolite*, NSC-134434, Win 24933
U Antischistosomal

9611 (7795) $C_{20}H_{24}N_2O_3$
80513-72-0

(6aR-cis)-6,6a,7,8,13,13a-Hexahydro-3,4-dimethoxy-10,11-dimethyl[1]benzopyrano[4,3-b][1,5]benzodiaze-pine (●)
S ZIMET 54/79
U Antineoplastic

9612 (7796) $C_{20}H_{24}N_2O_3$
50516-43-3

3-[2-Hydroxy-3-(isopropylamino)propoxy]-2-phenyl-phthalimidine = 3-[2-Hydroxy-3-(isopropylamino)prop-oxy]-2-phenylisoindol-1-one = 2,3-Dihydro-3-[2-hydr-oxy-3-[(1-methylethyl)amino]propoxy]-2-phenyl-1H-isoindol-1-one (●)
R Fumarate (1:1) (70096-14-9)
S *Nofecainide fumarate**, Nofedone fumarate*, 30356 R.P.
U Anti-arrhythmic

9613 (7797) $C_{20}H_{24}N_2O_3$
82900-57-0

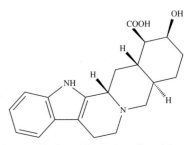

5-[3-(4-Phenyl-1-piperazinyl)propoxy]methylenedioxy-benzene = 1-[3-(1,3-Benzodioxol-5-yloxy)propyl]-4-phe-nylpiperazine (●)
S BP 554
U Hypothermic (selective 5HT$_{1a}$-receptor agonist)

9614 (7798) $C_{20}H_{24}N_2O_3$
522-87-2

17α-Hydroxyyohimban-16α-carboxylic acid = (16α,17α)-17-Hydroxyyohimban-16-carboxylic acid (●)
S *Acidum yohimbicum**, Yohimbic acid**, Yohimboa-säure*
U Sympatholytic

9615 (7799) $C_{20}H_{24}N_2O_3S$
90405-00-8

2-(Diethylamino)-N-(6,11-dihydrodibenzo[b,e]thiepin-11-yl)acetamide S,S-dioxide (●)
R also fumarate [(E)-2-butenedioate] (1:1) (90405-01-9)
S *Amidepin*, VÚFB-14524
U Anti-arrhythmic

9616 (7800) $C_{20}H_{24}N_2O_4S_2$
73845-37-1

2,2'-Dithiobis[N-(2-hydroxypropyl)benzamide] (●)
S KF 4939
U Platelet aggregation inhibitor

9617 (7801) $C_{20}H_{24}N_2O_5$
36141-82-9

β,β'-Oxybis[p-acetophenetidide] = 2,2'-Bis[p-(acetylamino)phenoxy]diethyl ether = N,N'-[Oxybis(2,1-ethanediyloxy-4,1-phenylene))]bis[acetamide] (●)
S Acemidophen, Atascol, B.W. 68-198, Compound 68-198, Coriban, Coryphamin, *Diamfe-*

netide**, *Diamphenethide*, *Oxybisphenacetin*
U Anthelmintic (fasciolicide), flukicide

9618 (7802) $C_{20}H_{24}N_2O_5$
7125-76-0

[[(4,5α-Epoxy-3-methoxy-17-methylmorphinan-6-ylidene)amino]oxy]acetic acid = Dihydrocodeinone O-(carboxymethyl)oxime = (5α)-[[(4,5-Epoxy-3-methoxy-17-methylmorphinan-6-ylidene)amino]oxy]acetic acid (●)
S *Codossima*, *Codoxime***
U Antitussive

9619 (7803) $C_{20}H_{24}N_2O_5$
56290-94-9

5-[1-Hydroxy-2-[[1-methyl-3-[3,4-(methylenedioxy)phenyl]propyl]amino]ethyl]salicylamide = 5-[2-[3-(1,3-Benzodioxol-5-yl)-1-methylpropylamino]-1-hydroxyethyl]salicylamide = 5-[2-[[3-(1,3-Benzodioxol-5-yl)-1-methylpropyl]amino]-1-hydroxyethyl]-2-hydroxybenzamide (●)
R Monohydrochloride (70161-10-3)
S CRMI 81968 A 8, *Medroxalol hydrochloride***, RMI 81968 A
U Antihypertensive (α- and β-blocker)

9620　(7805)　　　　　　　　　　　　　$C_{20}H_{24}N_2O_6$
　　　　　　　　　　　　　　　　　　　　35998-29-9

N,N'-Bis(2-hydroxybenzyl)ethylenediamine-*N,N'*-diacetic acid = *N,N'*-1,2-Ethanediylbis[*N*-[(2-hydroxyphenyl)methyl]glycine] (●)
S　Chel II, HBED
U　Chelating agent (iron)

9621　(7804)　　　　　　　　　　　　　$C_{20}H_{24}N_2O_6$
　　　　　　　　　　　　　　　　　　　　63675-72-9

Isobutyl methyl 1,4-dihydro-2,6-dimethyl-4-(*o*-nitrophenyl)-3,5-pyridinedicarboxylate = 1,4-Dihydro-2,6-dimethyl-4-(2-nitrophenyl)-3,5-pyridinedicarboxylic acid isobutyl methyl ester = 1,4-Dihydro-2,6-dimethyl-4-(2-nitrophenyl)-3,5-pyridinedicarboxylic acid methyl 2-methylpropyl ester (●)
S　Bayer 5552, Bay k 5552, Baymycard, Corasol "Sanitas", Cornel, Coronil, Lengil, Nisocor, Nisodipen, *Nisoldipine***, Nizoldin, Norvasc "Miles", Siscor, Sular, Syscor, Trendin, Zadipina
U　Coronary vasodilator

9622　(7806)　　　　　　　　　　　　　$C_{20}H_{24}N_2S$
　　　　　　　　　　　　　　　　　　　　32367-75-2

10-[2-(1-Methyl-2-piperidyl)ethyl]phenothiazine = 10-[2-(1-Methyl-2-piperidinyl)ethyl]-10*H*-phenothiazine (●)
S　*Ridazine*
U　Neuroleptic

9623　(7807)　　　　　　　　　　　　　$C_{20}H_{24}N_2S_2$
　　　　　　　　　　　　　　　　　　　　20229-30-5

8-(Methylthio)-10-(4-methyl-1-piperazinyl)-10,11-dihydrodibenzo[*b,f*]thiepin = 1-[10,11-Dihydro-8-(methylthio)dibenzo[*b,f*]thiepin-10-yl]-4-methylpiperazine (●)
R　also maleate (1:1) (19728-88-2)
S　*Methiothepine, Metitepine**, Ro 8-6837, VÚFB-6276
U　Neuroleptic

9624　(7808)　　　　　　　　　　　　　$C_{20}H_{24}N_4$
　　　　　　　　　　　　　　　　　　　　134326-62-8

(±)-1-(2-Pyridinyl)-4-[4-(2-pyridinyl)-3-cyclohexen-1-yl]piperazine (●)
S　PD 135222
U　Antipsychotic (dopamine autoreceptor agonist)

9625　　　　　　　　　　　　　　　　　$C_{20}H_{24}N_4O_4$
　　　　　　　　　　　　　　　　　　　　174858-27-6

5-Butyl-7-(3,4,5-trimethoxybenzamido)pyrazo-
lo[1,5-*a*]pyrimidine = *N*-(5-Butylpyrazolo[1,5-*a*]pyrimi-
din-7-yl)-3,4,5-trimethoxybenzamide (●)
S OT-7100
U Analgesic

9629

$C_{20}H_{24}N_6O$

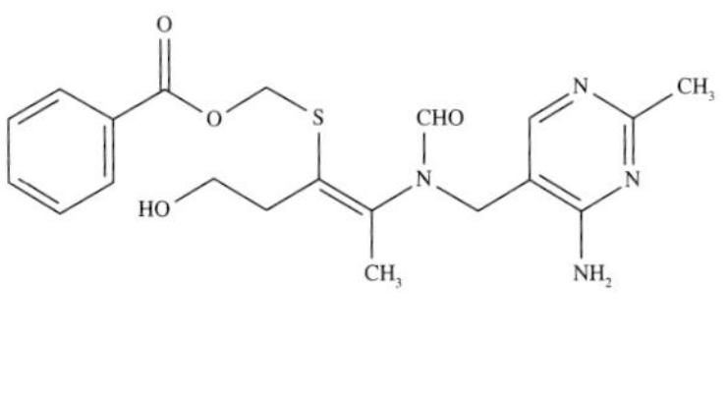

Benzoyloxymethylthiamine = *N*-[(4-Amino-2-methyl-5-
pyrimidinyl)methyl]-*N*-[2-[[(benzoyloxy)methyl]thio]-4-
hydroxy-1-methyl-1-butenyl]formamide (●)
S *Benmetiamina*, BT-851
U Antineuritic

9626

$C_{20}H_{24}N_4O_4$
182198-53-4

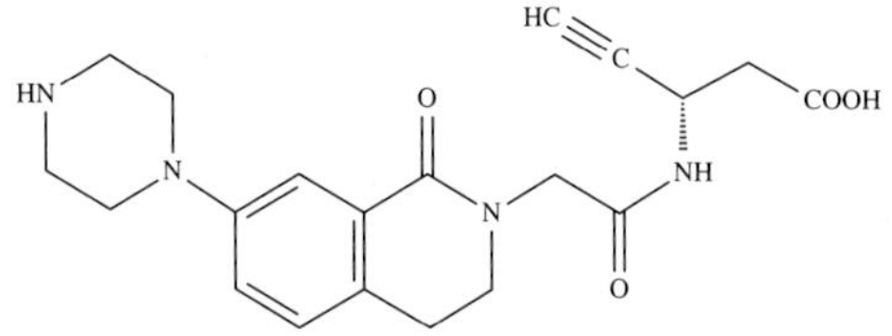

N-[[3,4-Dihydro-7-(1-piperazinyl)-1-oxo-2(1*H*)-isoqui-
nolinyl]acetyl]-3(*S*)-ethynyl-β-alanine = (3*S*)-3-[[[3,4-
Dihydro-1-oxo-7-(1-piperazinyl)-2(1*H*)-isoquinoli-
nyl]acetyl]amino]-4-pentynoic acid (●)
R Trifluoroacetate (182198-54-5)
S L 767679
U Fibrinogen receptor antagonist

9627

$C_{20}H_{24}N_4O_4$
155270-99-8

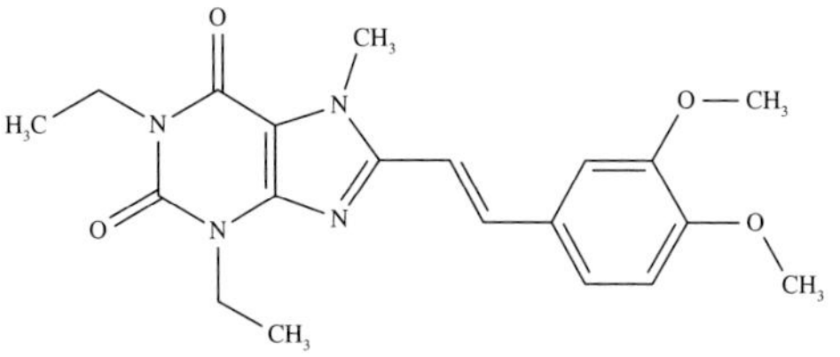

8-[(1*E*)-2-(3,4-Dimethoxyphenyl)ethenyl]-1,3-diethyl-
3,7-dihydro-7-methyl-1*H*-purine-2,6-dione (●)
S KW 6002
U Antiparkinsonian (adenosine A_{2A} receptor antagonist)

9628

$C_{20}H_{24}N_4O_4S$
18481-26-0

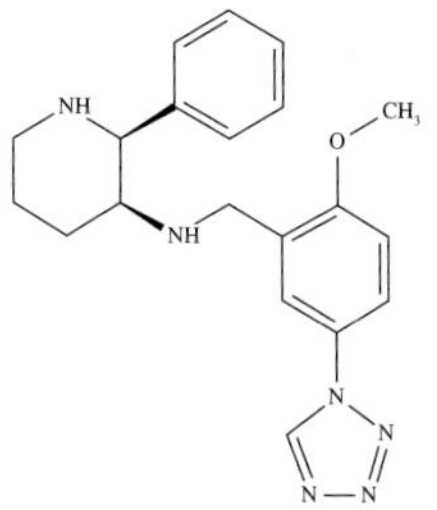

(2*S-cis*)-*N*-[[2-Methoxy-5-(1*H*-tetrazol-1-yl)phenyl]me-
thyl]-2-phenyl-3-piperidinamine (●)
R Dihydrochloride (168398-02-5)
S GR 203040
U Anti-emetic (tachykinin NK_1 receptor antagonist)

9630

$C_{20}H_{24}N_6O_2S$
196880-13-4

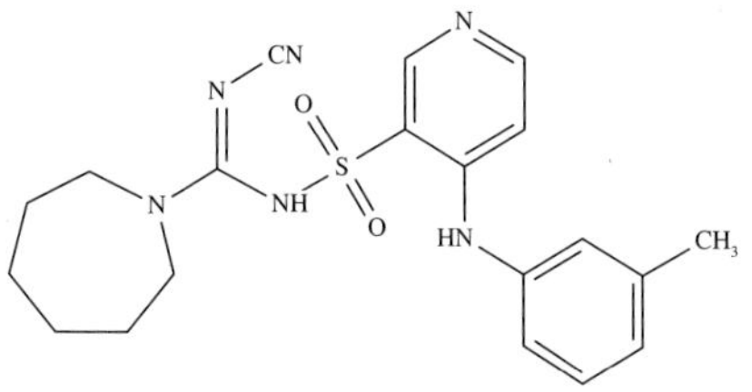

N-Cyano-*N'*-[[4-(3-methylphenylamino)-3-pyridyl]sulfo-
nyl]homopiperidine-1-amidine = *N*-Cyanohexahydro-*N'*-
[[4-[(3-methylphenyl)amino]-3-pyridinyl]sulfonyl]-1*H*-
azepine-1-carboximidamide (●)
S BM 144
U Thromboxane A_2 receptor antagonist

9631 $C_{20}H_{24}N_6O_3$
192705-79-6

1-[2-Amino-6-(3,5-dimethoxyphenyl)pyrido[2,3-*d*]pyri-
midin-7-yl]-3-*tert*-butylurea = *N*-[2-Amino-6-(3,5-di-
methoxyphenyl)pyrido[2,3-*d*]pyrimidin-7-yl]-*N*'-(1,1-di-
methylethyl)urea (●)
S PD 166866
U Tyrosine kinase inhibitor

9632 $C_{20}H_{24}N_6O_6S_2$
162081-63-2

7β-[(*Z*)-2-(2-Amino-4-thiazolyl)-2-(methoxyimino)ace-
tylamino]-3-[(*E*)-2-[(*S*)-2,2-dimethyl-5-isoxazolidi-
nio]ethenyl]-3-cephem-4-carboxylate = [6*R*-
[3[*E*(*S**)],6α,7β(*Z*)]]-5-[2-[7-[[(2-Amino-4-thiazo-
lyl)(methoxyimino)acetyl]amino]-2-carboxy-8-oxo-5-
thia-1-azabicyclo[4.2.0]oct-2-en-3-yl]ethenyl]-2,2-dime-
thylisoxazolidinium inner salt (●)
S YM-32825
U Antibiotic

9633 $C_{20}H_{24}N_6O_7S$
185536-58-7

3-[[[(5*R*)-3-[4-(Aminoiminomethyl)phenyl]-4,5-dihydro-
5-isoxazolyl]acetyl]amino]-*N*-[(3,5-dimethyl-4-isoxazo-
lyl)sulfonyl]-L-alanine (●)
R Monomethanesulfonate (185536-59-8)
S DMP 802
U Antithrombotic (platelet GPIIb/IIIa receptor antagonist)

9634 $C_{20}H_{24}N_7O_5P$
210355-14-9

N-[[[(2*Z*)-[(2,6-Diamino-9*H*-purin-9-yl)methylene]cy-
clopropyl]methoxy]phenoxyphosphinyl]-L-alanine methyl ester (●)
S QYL-685
U Antiviral (HIV-1 protease inhibitor)

9635 $C_{20}H_{24}N_9O_9P$
121135-53-3

3'-Azido-3'-deoxythymidylyl-(5'→5')-2',3'-dideoxyino-
sine (●)
S AZT-P-ddI, IVX-E-59, Scriptene
U Antiviral

9636   (7809)   C$_{20}$H$_{24}$O$_2$
                                                107868-30-4

6-Methyleneandrosta-1,4-diene-3,17-dione (●)
S   Aromasin, Exemastine, Exemestane**, FCE 24304,
    Nikidess, PNU-155971
U   Antineoplastic (aromatase inhibitor)

9637   (7810)   C$_{20}$H$_{24}$O$_2$
                                                130-79-0

rans-α,β-Diethyl-4,4'-dimethoxystilbene = Diethylstilbestrol dimethyl ether = (E)-1,1'-(1,2-Diethyl-1,2-ethene-
diyl)bis[4-methoxybenzene] (●)
S   Depoestron, Depot-Cyren, Depot-Östromenin, De-
    pot-Östromon, Diäthylstilböstroldimethyläther, Dia-
    nisylhexene, Diethylstilbestrol dimethyl ether, Dime-
    strol, Dimethoxydiethylstilbene, Dimethyl-Oestro-
    gen, Dimoestrol, Estromenin-Deposito, Östrastilben
    (D), Oestromenine-Depot, Stilböstroldimethyläther,
    Synthila
U   Synthetic estrogen

9638   (7811)   C$_{20}$H$_{24}$O$_2$
                                                102607-41-0

3-Isopropyl-7-methyl-8-(4-methyl-3-pentenyl)-1,2-naph-
thoquinone = 7-Methyl-3-(1-methylethyl)-8-(4-methyl-3-
pentenyl)-1,2-naphthalenedione (●)
S   Saprorthoquinone
U   Cytotoxic, antibacterial from Salvia prionitis

9639   C$_{20}$H$_{24}$O$_2$
                                                79491-58-0

7-Methyl-3-(1-methylethyl)-8-(4-methyl-4-pentenyl)-
1,2-naphthalenedione (●)
S   Aethiopinone
U   Analgesic, anti-inflammatory, antipyretic, hematolo-
    gical (diterpenoid from Salvia aethiopis)

9640   (7812)   C$_{20}$H$_{24}$O$_2$
                                                57-63-6

17-Ethynylestra-1,3,5(10)-triene-3,17β-diol = 17α-Eth-
ynyl-3,17β-estradiol = (17α)-19-Norpregna-1,3,5(10)-tri-
en-20-yne-3,17-diol (●)
R   see also nos. 12000, 12887
S   Acnormon, Aethinylöstradiol, BP 12, Chee-O-Gen,
    Dicromil e, Diogyn-E, Diolyn, Duramen, Dyloform,
    Edrol, Ertonyl, Esteed, Esten-E, Estigyn, Estinil,
    Estinyl, Eston-E, Estoral "Orion", Estradol, Estrasal,
    Estrineforte, Estroid, Estrolan-E, Estronex-Tabl.,
    Estoral "Pabyrn", Ethidol, Ethinestryl, Ethin-Oes-
    tryl, Ethinoral, Ethinylestradiol**, Ethinyloestradi-
    ol, Ethistradon, Éthy 11, Eticiclina, Eticyclin, Eti-

cyclol, Eticylol, Etifollin, Etinestrol, Etinestryl, *Etinilestradiolo*, Etinil-Ovigon, Etinoestryl, *Etinylestradiol*, Etistradiol, Etivex, Farmacyrol, Feminone, Fodinyl, Follicoral, Follikoral, Furiginone, Ginestrene, Ginormon-Compr., Gynan-Rub, Gynolett, Gynoral, Hewoestrol, Inestra, Kolpi-Gynaedron, Kolpolyn, Linoral, Lynestoral, Lynoral, Lyn-ratiopharm Sequenz, Menolyn, Metrociclina, Metroval, Microfollin, Mikrofollin, Nadestryl, Naphaestrone-Tabl., Nelova, Neo-Estrone, Norma-1, Norma-oestren, Norquentiel A, Novestrol, NSC-10973, Nylestrin, Oestralyn, Oestroperos, Oradiol, Orestralyn, Ostral, Ovahormon-Strong, Oviol e, Ovogyn, Pabendrol, Palonyl, Perovex, Primogyn C, Primogyn M, Progynon C, Progynon M, Rolidiol, Spanestrin, Tradinol, Ugenol, Ylestrol, Zetzurin-B

U Estrogen

9641 (7813) $C_{20}H_{24}O_3$
66332-77-2

2-(*p*-Isobutylphenyl)propionic acid guaiacol ester = *p*-Isobutylhydratropic acid *o*-methoxyphenyl ester = α-Methyl-4-(2-methylpropyl)benzeneacetic acid 2-methoxyphenyl ester (●)

S AF 2259, Belep, Benflogin "Angelini", Flogofen, Flubenil, *Ibuprofen guaiacol ester*, Medipul, *Metoxibutropate*

U Anti-inflammatory

9642 (7814) $C_{20}H_{24}O_3$
901-93-9

3-Hydroxyestra-1,3,5(10)-trien-17-one 3-acetate = 3-(Acetyloxy)-estra-1,3,5(10)-trien-17-one (●)

S *Estrone acetate*, Hovigal, *Östronacetat*

U Estrogen

9643 (7815) $C_{20}H_{24}O_3$
10161-34-9

17β-Hydroxyestra-4,9,11-trien-3-one 17-acetate = (17β)-17-(Acetyloxy)estra-4,9,11-trien-3-one (●)

S Finaject, Finaplix, RU 1697, *Trenbolone acetate***, *Trienbolone acetate*

U Anabolic

9644 (7816) $C_{20}H_{24}O_?$
78954-23-?

1-(4-Hydroxy-3-methoxyphenyl)-7-phenyl-3-heptanone (●)

S *Yakuchinone A*

U Tyrosinase inhibitor, cardiotonic from *Alpinia oxyphylla*

9645 (7817) $C_{20}H_{24}O?$
86790-29-?

trans-2-(4-*tert*-Butylcyclohexyl)-3-hydroxy-1,4-naphthoquinone = *trans*-2-[4-(1,1-Dimethylethyl)cyclohexyl]-3-hydroxy-1,4-naphthalenedione (●)

S B.W. 58 C

U   Antimalarial

9646 $C_{20}H_{24}O_3$
                                        65316-65-6

(2E,4E,6E,8E)-9-(4-Hydroxy-2,3,6-trimethylphenyl)-3,7-dimethyl-2,4,6,8-nonatetraenoic acid (●)
S   Ro 12-7310
U   Dermatic

9647 $C_{20}H_{24}O_3$
                                        74285-86-2

(3bR,9bS)-3b,4,5,9b,10,11-Hexahydro-6-hydroxy-9b-methyl-7-(1-methylethyl)phenanthro[1,2-c]furan-1(3H)-one (●)
S   Triptophenolide
U   Anti-inflammatory, immunomodulator from *Tripterygium wilfordii*

9648   (7818) $C_{20}H_{24}O_5$
                                        56488-59-6

4-[3-(p-tert-Butylphenoxy)-2-hydroxypropoxy]benzoic acid = 4-[3-[4-(1,1-Dimethylethyl)phenoxy]-2-hydroxypropoxy]benzoic acid (●)
  K 11267, *Terbufibrol***
  Antihyperlipidemic

9649 $C_{20}H_{24}O_5$
                                        154677-96-0

(2R,3S)-rel-1,4-Bis(4-hydroxy-3-methoxyphenyl)-2,3-dimethyl-1-butanone (●)
S   *Cinnamophilin*
U   Thromboxane H_2 receptor antagonist, calcium antagonist (from *Cinnamomum philippinense*)

9650   (7819) $C_{20}H_{24}O_6$
                                        38748-32-2

[3bR-(3bα,4aα,5aS*,6β,6aβ,7aβ,7bα,8aS*,8bβ)]-3b,4,4a,6,6a,7a,7b,8b,9,10-Decahydro-6-hydroxy-8b-methyl-6a-(1-methylethyl)trisoxireno[4b,5:6,7:8a,9]phenanthro[1,2-c]furan-1(3H)-one (●)
S   NSC-163062, PG 490, *Triptolide*
U   Antineoplastic, immunosuppressive from *Tripterygium wilfordii*

9651 $C_{20}H_{24}O_7S$
                                        251565-85-2

(S)-2-Ethoxy-3-[4-[2-[4-[(methylsulfonyl)oxy]phenyl]ethoxy]phenyl]propanoic acid = (αS)-α-Ethoxy-4-[2-[4-[(methylsulfonyl)oxy]phenyl]ethoxy]benzenepropanoic acid (●)

S   AR-H 039242XX
U   Insulin sensitizer (PPAR α and γ agonist)

9652   (7820) $C_{20}H_{24}O_{10}$
15291-77-7

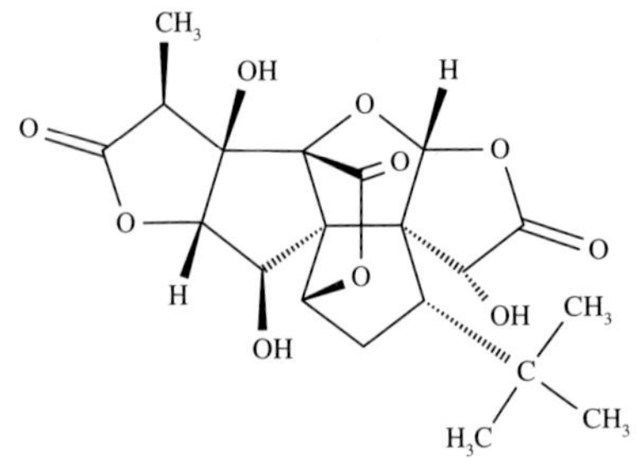

(1β)-1-Hydroxyginkgolide A (●)
S   BN 52021, *Ginkgolide B*
U   Anti-anaphylactic (PAF-acether receptor antagonist)

9653 $C_{20}H_{24}O_{11}$
129878-44-0

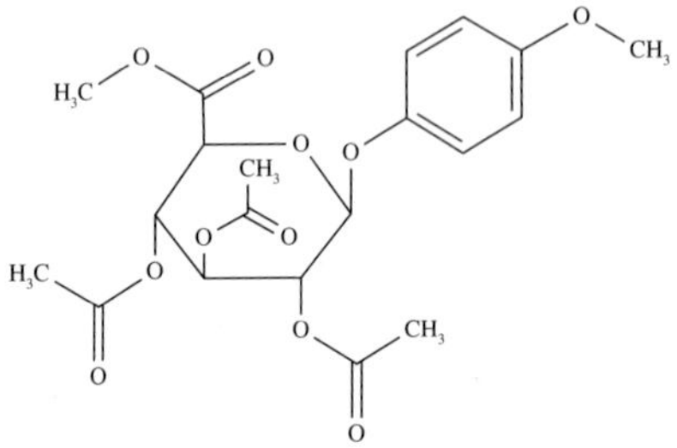

1-(*p*-Methoxyphenyl)-2,3,4-tris(*O*-acetyl)-β-D-glucuronic acid methyl ester = 4-Methoxyphenyl β-D-glucopyranosiduronic acid methyl ester triacetate (●)
S   Glycosole
U   Antineoplastic

9654 $C_{20}H_{24}O_{11}$
189289-76-7

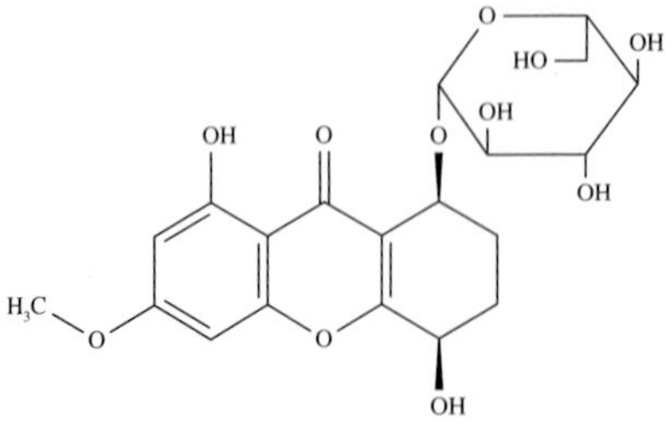

(1*S*,4*R*)-1-(β-D-Glucopyranosyloxy)-1,2,3,4-tetrahydro-4,8-dihydroxy-6-methoxy-9*H*-xanthen-9-one (●)
S   *Tetrahydroswertianolin*
U   Hepatoprotectant from *Swertia japonica*

9655   (7821) $C_{20}H_{25}BrN_4O$
83455-48-5

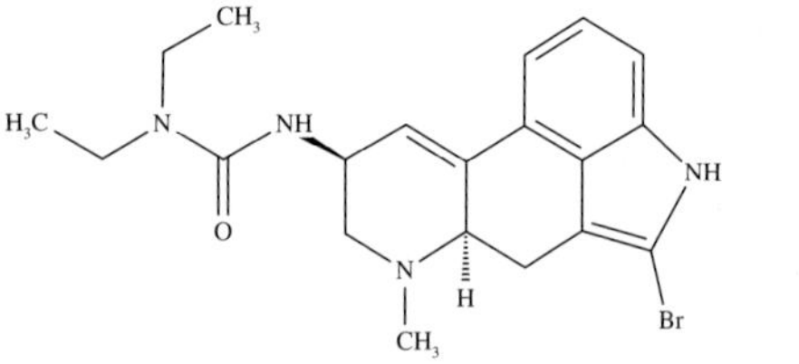

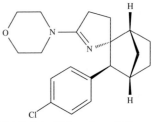

3-(2-Bromo-9,10-didehydro-6-methylergolin-8α-yl)-1,1-diethylurea = *N*'-[(8α)-2-Bromo-9,10-didehydro-6-methylergolin-8-yl]-*N*,*N*-diethylurea (●)
S   *Bromerguride***, 2-Bromolisuride, ZK 95451
U   Dopamine antagonist

9656   (7822) $C_{20}H_{25}ClN_2O$
90243-97-

(−)-(1*R*,2*R*,3*S*,4*S*)-3-(*p*-Chlorophenyl)-2'-morpholinospiro[norbornane-2,5'-[1]pyrroline] = (−)-(1*R*,2*R*,3*S*,4*S*)-3-(4-Chlorophenyl)bicyclo[2.2.1]heptane-2-spiro-2'-(5'-morpholino-5'-pyrroline) = 3-(4-Chlorophenyl)-3',4'-dihydro-5'-(4-morpholinyl)spiro[bicyclo[2.2.1]heptane-2,2'-[2*H*]pyrrole] (●)
S   *Spiclamine**
U   Antidepressant

9657　　　　　　　　　　　　　　$C_{20}H_{25}ClN_2O_2$
125363-06-6

2-Chloro-5a,6,7,8,9,9a-hexahydro-*N*-3-quinuclidinyl-4-
dibenzofurancarboxamide = *N*-1-Azabicyclo[2.2.2]oct-3-
yl-2-chloro-5a,6,7,8,9,9a-hexahydro-4-dibenzofuran-
carboxamide (●)
S　RG 12915
U　Anti-emetic (5-HT₃ receptor antagonist)

9658　(7823)　　　　　　　　　　$C_{20}H_{25}ClN_2O_2$
1243-83-0

1-*o*-Chlorophenyl-4-(3,4-dimethoxyphenethyl)pipera-
zine = 1-(2-Chlorophenyl)-4-[2-(3,4-dimethoxyphe-
nyl)ethyl]piperazine (●)
S　*Mefeclorazine***, *Méphéchlorazine*, SD 218-06
U　Neuroleptic

9659　　　　　　　　　　　　　　$C_{20}H_{25}ClN_2O_3$
199108-54-8

1-(4-Chlorophenyl)-4-[(3,4,5-trimethoxyphenyl)me-
thyl]piperazine (●)
S　PD 35680
U　Antipsychotic

9660　(7824)　　　　　　　　　　$C_{20}H_{25}ClN_2O_5$
88150-42-9

3-Ethyl 5-methyl (±)-2-[(2-aminoethoxy)methyl]-4-(*o*-
chlorophenyl)-1,4-dihydro-6-methyl-3,5-pyridinedicarb-
oxylate = 2-[(2-Aminoethoxy)methyl]-4-(2-chlorophe-
nyl)-1,4-dihydro-6-methyl-3,5-pyridinedicarboxylic acid
3-ethyl 5-methyl ester (●)
R　Maleate (1:1) (88150-47-4)
S　*Amlodipine maleate***, UK-48340-11
U　Anti-anginal, antihypertensive

9660-01　(7824-01)　　　　　　　　　111470-99-6
R　Monobenzenesulfonate
S　Amcard, Amloc, Amlodin(e), *Amlodipine besila-
te***, Amlodis, Amlogard, Amlokard, Amlopin, Am-
loprax, Amlor, Amlosyn, Amlovas, Amlozek, Anta-
cal, Arteriosan, Astudal, Biocard, Cardiorex, Cordar-
ex "Biosintetica", Cordipina "Farmasa", Corgarene,
Coroval, Dilotex "Dollder", Istin "Pfizer", Lipinox,
Lopidin, Monavas, Monopina, Monovas "Mustafa
Nevzat", Myodura, Nicord "Marjan", Nipidol, Norlo-
pin, Normodipine, Norvadin, Norvas, Norvasc "Pfiz-
er", Norvask, Pelmec, Pressat, Roxflan, Stamlo,
Symtin, Tensivask, Terloc, Tervalon "Lazar",
UK-48340-26, Unidoscor, Vaskyl, Vasocard "Abfar"
U　Anti-anginal, antihypertensive

9661　(7825)　　　　　　　　　　$C_{20}H_{25}ClN_4OS$
15311-77-0

2-[4-[3-(3-Chloro-10*H*-pyrido[3,2-*b*][1,4]benzothiazin-
10-yl)propyl]-1-piperazinyl]ethanol = 3-Chloro-10-[3-[4-
(2-hydroxyethyl)-1-piperazinyl]propyl]-1-azaphenothia-

zine = 4-[3-(3-Chloro-10*H*-pyrido[3,2-*b*][1,4]benzothia-
zin-10-yl)propyl]-1-piperazineethanol (●)
S *Cloxphendyl, Cloxypendyl**, D 1262
U Neuroleptic

9662 (7826) $C_{20}H_{25}ClO_3$
30781-27-2

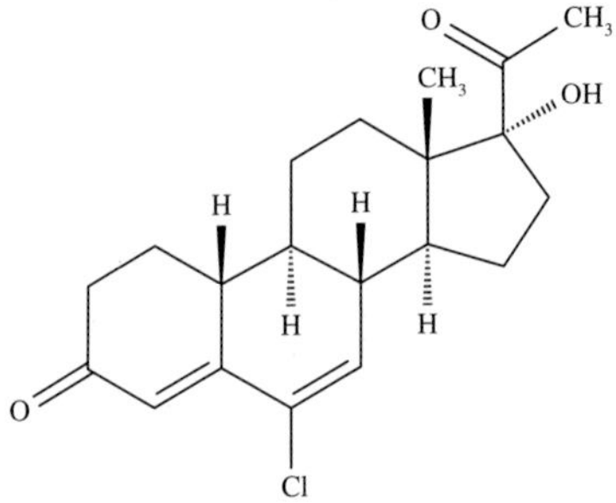

6-Chloro-17-hydroxy-19-norpregna-4,6-diene-3,20-
dione (●)
R see also no. 11210
S *Amadinone***, *19-Norchlormadinone*
U Progestin

9663 (7827) $C_{20}H_{25}ClO_4$
105149-04-0

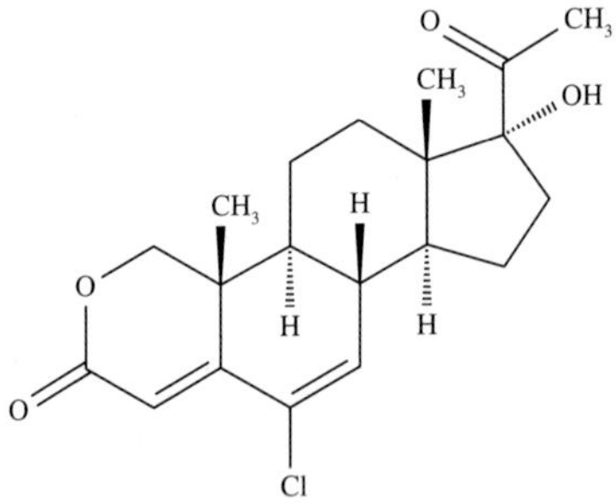

6-Chloro-17-hydroxy-2-oxapregna-4,6-diene-3,20-dione
(●)
R see also no. 11211
S *Gestoxarone, Osaterone***
U Anti-androgen, progestogen

9664 (7828) $C_{20}H_{25}Cl_3O_3$
54063-33-1

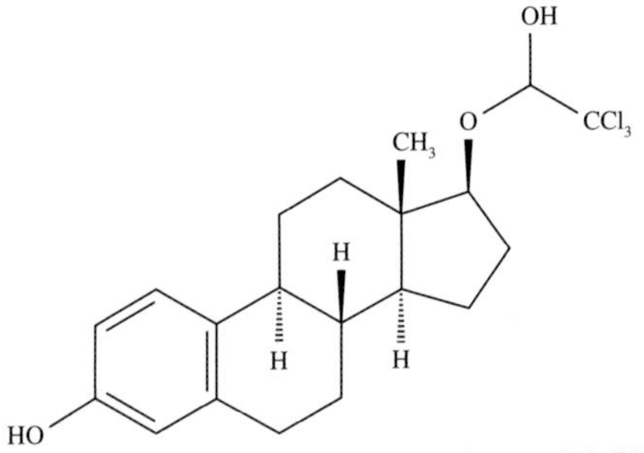

17β-(2,2,2-Trichloro-1-hydroxyethoxy)estra-1,3,5(10)-
trien-3-ol = (17β)-17-(2,2,2-Trichloro-1-hydroxyeth-
oxy)estra-1,3,5(10)-trien-3-ol (●)
R see also no. 12378
S *Cloxestradiol***
U Estrogen

9665 (7829) $C_{20}H_{25}FN_4C$
27367-90-

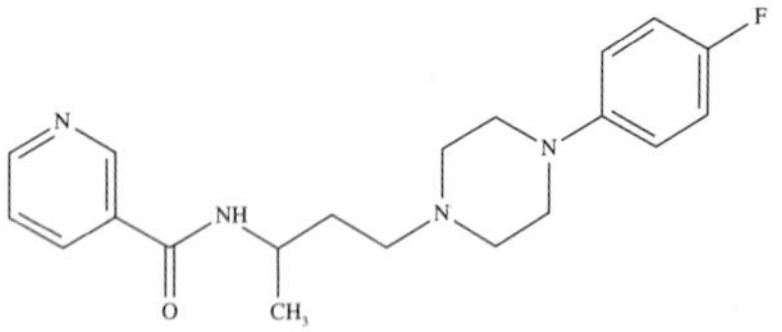

N-[3-[4-(*p*-Fluorophenyl)-1-piperazinyl]-1-methylpro-
pyl]nicotinamide = *N*-[3-[4-(4-Fluorophenyl)-1-piperazi-
nyl]-1-methylpropyl]-3-pyridinecarboxamide (●)
S CERM 1709, *Niaprazine***, Nopron
U Sedative, anti-allergic, bronchospasmolytic

9666 (7830) $C_{20}H_{25}FN_8O_6S$
116853-25-

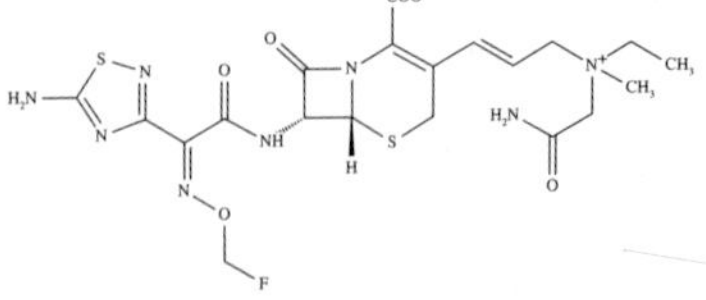

(−)-(*RS*)-[(*E*)-3-[(6*R*,7*R*)-7-[(*Z*)-2-(5-Amino-1,2,4-thiad
azol-3-yl)-2-[(fluoromethoxy)imino]acetamido]-2-carb-
oxy-8-oxo-5-thia-1-azabicyclo[4.2.0]oct-2-en-3-yl]-2-
propenyl](carbamoylmethyl)ethylmethylammonium
hydroxide inner salt = (−)-[(*E*)-3-[(6*R*,7*R*)-7-[2-(5-Ami-
no-1,2,4-thiadiazol-3-yl)glyoxylamido]-2-carboxy-8-

oxo-5-thia-1-azabicyclo[4.2.0]oct-2-en-3-yl]allyl](carbamoylmethyl)ethylmethylammonium hydroxide inner
salt, 7^2-(Z)-[O-(fluoromethyl)oxime] = [6R-
[3(E),6α,7β(Z)]]-N-(2-Amino-2-oxoethyl)-3-[7-[[(5-ami-
no-1,2,4-thiadiazol-3-yl)[(fluoromethoxy)imino]acetyl]
amino]-2-carboxy-8-oxo-5-thia-1-azabicyclo[4.2.0]oct-
2-en-3-yl]-N-ethyl-N-methyl-2-propen-1-aminium hydroxide inner salt (●)
S *Cefluprenam***, *Cefprenam*, CFLP, E-1077
U Antibiotic

9667 (7831) $C_{20}H_{25}IN_4O_2$
                                      53230-08-3

3-Iodo-4,4'-(hexamethylenedioxy)dibenzamidine = p,p'-
Diamidino-o-iodo-1,6-diphenoxyhexane = 4-[[6-[4-
(Aminoiminomethyl)phenoxy]hexyl]oxy]-3-iodobenzen-
ecarboximidamide (●)
S *Iodohexamidine*, *Jodhexamidin*, M. & B. 1314
U Antiprotozoal

9668 (7832) $C_{20}H_{25}N$
                                      3540-95-2

1,1-Diphenyl-3-piperidinopropane = 1-(3,3-Diphenylpro-
yl)piperidine (●)
R Hydrochloride (3329-14-4)
      Asipol (active substance), Aspasan (active sub-
      stance), Aspen (active substance), *Fenpiprane hydro-
      chloride***, Hoechst 10116
U Anti-asthmatic, anti-allergic, spasmolytic

9669 (7833) $C_{20}H_{25}N$
                                      31314-38-2

1-Isopropyl-4,4-diphenylpiperidine = 1-(1-Methylethyl)-
4,4-diphenylpiperidine (●)
S Anthen, *Prodipine***
U Antiparkinsonian

9670 (7834) $C_{20}H_{25}NO$
                                      126168-32-9

1-Heptyl-2-methyl-9H-carbazol-3-ol (●)
S *Carazostatin*, DC 118
U Anti-inflammatory from *Streptomyces chromofuscus*
      DC 48

9671 (7835) $C_{20}H_{25}NO$
                                      467-85-6

6-(Dimethylamino)-4,4-diphenyl-3-hexanone (●)
R also hydrochloride (847-84-7)

S *Desmethylmethadone*, Eucopon (active substance), Hoechst 10582, Mepidon, NIH 2820, *Noramidone*, Normedon, *Normetadone*, *Normethadone**, Phenyldimazone*, U-9558
U Narcotic analgesic, antitussive

9671-01 (7835-01)
R with addition of methylsynephrine (no. 2345)
S Cophylac, Dacartil, Ralopar (old form), Ticarda
U Antitussive

9671-02 (7835-02)
R 2, 6-Di-*tert*-butylnaphthalenedisulfonate (no. 8077)
S Extussin, Tinafon
U Antitussive

9672 (7836) $C_{20}H_{25}NO$
7182-51-6

N,3,3-Trimethyl-1-phenyl-1-phthalanpropylamine = 3,3-Dimethyl-1-[3-(methylamino)propyl]-1-phenylphthalan = 1,3-Dihydro-*N*,3,3-trimethyl-1-phenyl-1-isobenzofuranpropanamine (●)
R Hydrochloride (7013-41-4)
S AY-21554, Lu 3-010, *Phtalapromine*, *Talopram hydrochloride***
U Antidepressant (catecholamine potentiator), ophthalmologic

9673 $C_{20}H_{25}NO$
125067-40-5

5,6,7,8-Tetrahydro-5-[[methyl(2-phenylethyl)amino]methyl]-2-naphthalenol (●)

R Hydrochloride
S A-75169
U α_2-Adrenergic antagonist

9674 (7841) $C_{20}H_{25}NO$
4960-10-5

Benzhydryl β-piperidinoethyl ether = 1-[2-(Diphenylmethoxy)ethyl]piperidine (●)
R Hydrochloride (3626-66-2)
S Antihistamin Spofa, *Benzperidine hydrochloride*, *Perastine hydrochloride**, Spofa III/101
U Antihistaminic

9675 (7838) $C_{20}H_{25}NO$
13862-07-

2-(2-Benzhydrylpiperidino)ethanol = 1-(2-Hydroxyethyl)-2-(diphenylmethyl)piperidine = 2-(Diphenylmethyl)-1-piperidineethanol (●)
R Hydrochloride (20269-19-6)
S Ba-18189, Ciba 18189, Cléofil, *Difemetorex hydrochloride**, Difemetoxidine hydrochloride, Diphemethoxidine hydrochloride*
U Anorexic

9676 (7837)   C$_{20}$H$_{25}$NO
510-07-6

1,1-Diphenyl-2-piperidino-1-propanol = β-Methyl-α,α-diphenyl-1-piperidineethanol (●)
R also hydrochloride (6159-39-3) or salicylate (1059-28-5)
S *Difepipramolo*, *Diphépanol*, Hoechst 10682, Tus(s)ukal (active substance), Tusucal (active substance)
U Antitussive

9676-01 (7837-01)   55668-48-9
R Guaiacolsulfonate (1:1)
S Viatussin
U Antitussive

___

9677 (7839)   C$_{20}$H$_{25}$NO
107703-78-6

-Phenethyl-α-phenyl-4-piperidinemethanol = α-(1-Phenethyl-4-piperidyl)benzyl alcohol = α-Phenyl-1-(2-phenylethyl)-4-piperidinemethanol (●)
R also with (±)-definition (132553-86-7)
S *Glemanserin***, MDL 11939
U Anxiolytic, anti-arrhythmic

___

9678 (7840)   C$_{20}$H$_{25}$NO
511-45-5

1,1-Diphenyl-3-piperidino-1-propanol = α-(2-Piperidinoethyl)benzhydrol = α,α-Diphenyl-1-piperidinepropanol (●)
R Hydrochloride (968-58-1)
S 238 C, Parks, Par KS 12-Hommel, *Pridinol hydrochloride***, Ridinol
U Anticholinergic, antiparkinsonian

9678-01 (7840-01)   6856-31-1
R Methanesulfonate
S Cyroquitil, Flacid, HH 212, Hikicenon, Konlax "Nippon Shinyaku", Kuningan, Loxeen, Lyseen, Menopacin, Miolibran, Miopanol, Mitanoline, Myosen, Myoson, Nonplesin, Polmesilat, Poluraxin, *Pridinol mesilate***, Relax "Nichiiko", Rheumacaan, Rimentol, Tirashizin, Trilax "Ono", Wanlax, Zenmicone
U Anticholinergic, muscle relaxant, spasmolytic

___

9679 (7842)   C$_{20}$H$_{25}$NOS
82-99-5

S-[2-(Diethylamino)ethyl] diphenylthioacetate = Diphenylthiolacetic acid β-(diethylamino)ethyl ester = α-Phenylbenzeneethanethioic acid S-[2-(diethylamino)ethyl] ester (●)
R Hydrochloride (548-68-5)
S B-23, *Thiphenamil hydrochloride, Thiphen(um)*, Tifen, *Tifenamil hydrochloride***, Tiphen, Trocinate
U Anticholinergic, spasmolytic

9680 (7853) $C_{20}H_{25}NO_2$
                                                        5941-36-6

(±)-3-Methoxy-8-aza-19-nor-17α-pregna-1,3,5(10)-tri-en-20-yn-17-ol = (17α)-(±)-3-Methoxy-8-aza-19-nor-pregna-1,3,5(10)-trien-20-yn-17-ol (●)
R Hydrobromide (15179-97-2)
S *Estrazinol hydrobromide**, W 4454 A*
U Estrogen

9681 (7843) $C_{20}H_{25}NO_2$
                                                        64-95-9

2-(Diethylamino)ethyl diphenylacetate = Diphenylacetic acid β-(diethylamino)ethyl ester = α-Phenylbenzeneacetic acid 2-(diethylamino)ethylester (●)
R see also no. 9991;
       Hydrochloride (50-42-0)
S Adiphen, *Adiphenine hydrochloride**, Anfosentin-Grageas, Difacil, Diphacil, NSC-129224, Patrovina (active substance), Paxil "Frosst", Rophene, Sentiv, Solvamin AB, Spasmolytin, Trasentin, Vagospasmyl, Vegantil, Vegantin
U Anticholinergic, spasmolytic, smooth muscle relaxant

9682 (7845) $C_{20}H_{25}NO_2$
                                                        3562-48-9

β-[(Dimethylamino)methyl]-β-ethylphenethyl benzoate = 2-[(Dimethylamino)methyl]-2-phenylbutyl benzoate = N-[2-(Benzoyloxymethyl)-2-phenylbutyl]dimethylamine = β-[(Dimethylamino)methyl]-β-ethylbenzeneethanol benzoate (ester) (●)
S Benzobutamine
U Local anesthetic, antitussive

9683 (7844) $C_{20}H_{25}NO_2$
                                                        493-76-5

α-[2-(Diethylamino)ethyl]benzyl benzoate = 3-(Diethylamino)-1-phenylpropyl benzoate = Benzoic acid 3-(diethylamino)-1-phenylpropyl ester = α-[2-(Diethylamino)ethyl]benzenemethanol benzoate (ester) (●)
R also hydrochloride (1679-79-4)
S 467 D_3, Detraine, *Propanocaine**
U Local anesthetic

9684 (7846) $C_{20}H_{25}NO$
                                                        102699-10-

3-(Diethylamino)-1-propanol 2-phenylbenzoate = [1,1'-Biphenyl]-2-carboxylic acid 3-(diethylamino)propyl ester (●)

R Hydrochloride (102698-88-4)
S JAW-669
U Anticonvulsant

9685   (7847) $C_{20}H_{25}NO_2$
                                                 26989-37-7

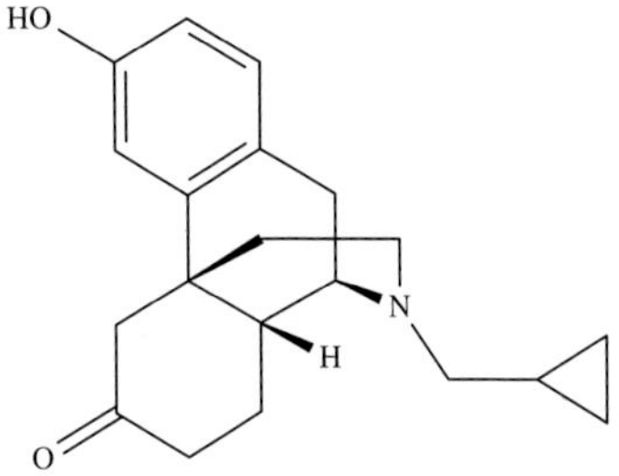

17-(Cyclopropylmethyl)-3-hydroxymorphinan-6-one (●)
S NIH 9466
U Analgesic, antagonist to narcotics

9686   (7848) $C_{20}H_{25}NO_2$
                                                 79798-39-3

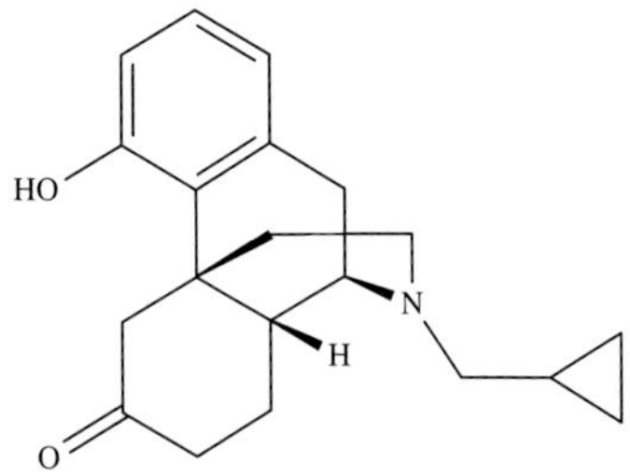

17-(Cyclopropylmethyl)-4-hydroxymorphinan-6-one (●)
S Ketorfanol**, SBW-22
U Analgesic

9687   (7849) $C_{20}H_{25}NO_2$
                                                 17692-39-6

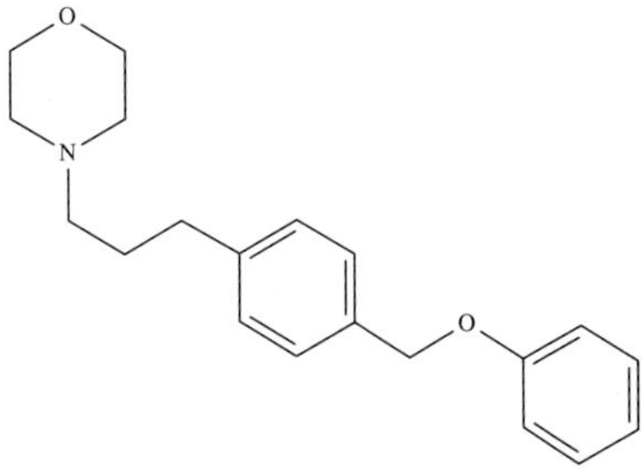

4-[3-(α-Phenoxy-*p*-tolyl)propyl]morpholine = 4-(3-Morpholinopropyl)benzyl phenyl ether = 4-[3-[4-(Phenoxymethyl)phenyl]propyl]morpholine (●)
R also hydrochloride (56583-43-8)
S Erbocain, *Fomocaine***, P-652, Panacain
U Local anesthetic

9688   (7850) $C_{20}H_{25}NO_2$
                                                 65928-58-7

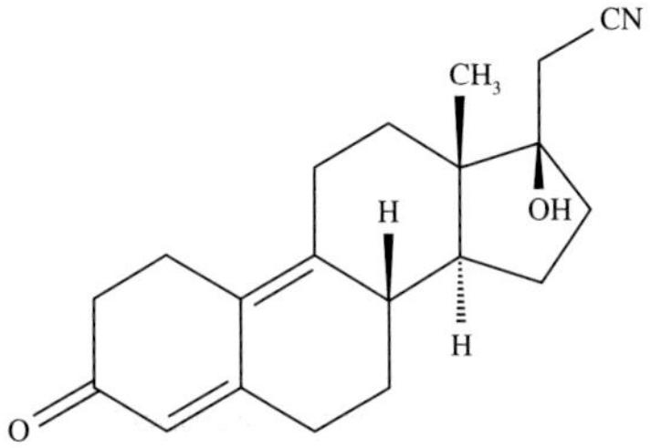

17-Hydroxy-3-oxo-19-nor-17α-pregna-4,9-diene-21-nitrile = 17α-(Cyanomethyl)-17-hydroxy-13β-methylgona-4,9-dien-3-one = (17α)-17-Hydroxy-3-oxo-19-norpregna-4,9-diene-21-nitrile (●)
S *Dienogest***, *Dienogestril*, Endometrion, Klimodien, STS 557, Valette
U Progestin

9689   (7851) $C_{20}H_{25}NO_2$
19732-39-9

N-[2-(Benzhydryloxy)ethyl]homomorpholine = 4-[2-(Diphenylmethoxy)ethyl]hexahydro-1,4-oxazepine (●)
S   Linadryl H
U   Antihistaminic, anticholinergic, spasmolytic

9690   (7852) $C_{20}H_{25}NO_2$
59859-58-4

(3*R*-*trans*)-3-[(4-Methoxyphenoxy)methyl]-1-methyl-4-phenylpiperidine (●)
R   Hydrochloride (56222-04-9)
S   *Femoxetine hydrochloride***, FG 4963, Malexil
U   Antidepressant

9691   (7854) $C_{20}H_{25}NO_2S_2$
115103-54-3

(−)-(*R*)-1-[4,4-Bis(3-methyl-2-thienyl)-3-butenyl]nipecotic acid = (*R*)-1-[4,4-Bis(3-methyl-2-thienyl)-3-butenyl]-3-piperidinecarboxylic acid (●)
R   also hydrochloride (145821-59-6)
S   Abbott-70569, ABT 569, Gabitril, NNC 05-0328, NNC 328, NO-05-0328, NO 328, Tiabex, *Tiagabine***
U   Anti-epileptic (GABA-uptake inhibitor)

9692 $C_{20}H_{25}NO_3$
179022-08-3

N-Methyl-*N*-[3-[4-(phenylmethyl)phenoxy]propyl]-β-alanine (●)
S   SC-57461
U   Leukotriene anhydrolase inhibitor

9693   (7855) $C_{20}H_{25}NO_3$
302-40-9

2-(Diethylamino)ethyl benzilate = Benzilic acid β-(diethylamino)ethyl ester = 2-Diethylaminoethyl diphenylglycolate = α-Hydroxy-α-phenylbenzeneacetic acid 2-(diethylamino)ethyl ester (●)
R   Hydrochloride (57-37-4)
S   Actozine, Amicil, Amikon, Amisyl, Amitacon, Amizil, Amizylum, Angustil, Arcadin, Aspamin, AY-5406, Beatilina, Benactina, Benactizan, Benactizin, Benactone, *Benactyzine hydrochloride***, Benetrank, Cafron, Cedad, Cevanol, Destendo, Diazil (CIS prep.), Ester 22, Fobex, Fortran, Ibiotyzil, Karmazine, Lucidex, Lucidil, Megasedan (old form), Mokapon-Inj., Morcain, Nervacton, Nervatil, Neurobenzile, Neuroleptone, NFN, Noicetin, Nutinal, Parasan "Medicinalco", Paratil (old form), Parpon, Pho-

bex, Procalm, Savitil, Sedansina, Stoikon, Suavitil, Svavitil, Tranquillactin, Tranquilline T.E.V.A., Valladan, Weisen, Win 5606

U Tranquilizer, anticholinergic

9694 (7856) $C_{20}H_{25}NO_3$
80387-96-8

2-(Dimethylamino)-1,1-dimethylethyl benzilate = α-Hydroxy-α-phenylbenzeneacetic acid 2-(dimethylamino)-1,1-dimethylethyl ester (●)

R old definition: 2-(dimethylamino)-2-methylpropyl benzilate (3477-97-2); hydrochloride (70280-88-5)

S Difemerine hydrochloride**, Luostyl, UP 57

U Anticholinergic, spasmolytic

9695 (7857) $C_{20}H_{25}NO_3$
509-78-4

2-(Dimethylamino)ethyl 1-ethoxy-1,1-diphenylacetate = Ethoxydiphenylacetic acid β-(dimethylamino)ethyl ester = α-Ethoxy-α-phenylbenzeneacetic acid 2-(dimethylamino)ethyl ester (●)

R Hydrochloride (2424-75-1)

S AD 67, Aestocin, Dimenossadolo cloridrato, Dimenoxadol hydrochloride**, Esthocin, Estocin, Lokarin, NIH 7577, Propalgyl

U Analgesic, spasmolytic

9696 $C_{20}H_{25}NO_3$
134234-12-1

(1S,2S)-1-(4-Hydroxyphenyl)-2-(4-hydroxy-4-phenylpiperidino)-1-propanol = (αS,βS)-4-Hydroxy-α-(4-hydroxyphenyl)-β-methyl-4-phenyl-1-piperidineethanol (●)

R Methanesulfonate (salt) trihydrate (189894-57-3)

S CP-101606, Traxoprodil mesilate*

U Antiparkinsonian, neuroprotective (NMDA receptor antagonist)

9697 (7858) $C_{20}H_{25}NO_4$
550-92-5

8-(3-Hydroxypropyl)-4-nortropidinecarboxylic acid ethyl ester benzoate (ester) = 8-(3-Benzoyloxypropyl)nortrop-3-ene-2β-carboxylic acid ethyl ester = N-[γ-(Benzoyloxy)propyl]noranhydroecgonine ethyl ester = 8-[3-(Benzoyloxy)propyl]-8-azabicyclo[3.2.1]oct-3-ene-2-carboxylic acid ethyl ester (●)

R Hydrochloride (5796-28-1)

S Eccain, Ekkain

U Local anesthetic

9698 $C_{20}H_{25}NO_4$
                                                                       153259-65-5

cis-4-Cyano-4-[3-(cyclopentyloxy)-4-methoxyphe-
nyl]cyclohexanecarboxylic acid (●)
S   Ariflo, *Cilomilast***, SB 207499
U   Anti-asthmatic (phosphodiesterase 4 inhibitor)

9699   (7859) $C_{20}H_{25}NO_4$
                                                                       3861-72-1

4,5α-Epoxy-3-methoxy-17-methylmorphinan-6α-yl acetate = 6-Acetoxy-3-methoxy-*N*-methyl-4,5-epoxymorphi-
nan = (5α,6α)-4,5-Epoxy-3-methoxy-17-methylmorphi-
nan-6-ol acetate (ester) (●)
R   also hydrochloride
S   Acetylcodon(e), *Acetyldihydrocodeine*, *Acidrocodei-
        ne*
U   Narcotic analgesic

9700   (7860) $C_{20}H_{25}NO_5$
                                                                       585-14-8

(1*S*,3*s*,5*R*,6*S*,7*R*)-6,7-Epoxytropan-3-yl (*S*)-*O*-propio-
nyltropate = 3-[2-Phenyl-2-[(propionyloxy)methyl]acet-
oxy]-6,7-epoxytropane = Propionylhyoscine = [7(*S*)-
(1a,2β,4β,5α,7β)]-α-[(1-Oxopropoxy)methyl]benzene-
acetic acid 9-methyl-3-oxa-9-azatricyclo[3.3.1.0^{2,4}]non-
7-yl ester (●)
R   Hydrobromide (2089-54-5)
S   *Poskine hydrobromide***, *Propionylscopolamine hy-
        drobromide*, Proscopine
U   Anticholinergic, central nervous system depressant

9701   (7861) $C_{20}H_{25}NS$
                                                                       40550-32-1

3-[2-(Dimethylamino)ethyl]-1,3-dihydro-1,1-dimethyl-3-
phenylbenzo[*c*]thiophene = 3,3-Dimethyl-1-(2-dimethyl-
aminoethyl)-1-phenylthiophthalan = 1,3-Dihydro-
N,N,3,3-tetramethyl-1-phenylbenzo[*c*]thiophene-1-ethan-
amine (●)
S   Lu 6-062
U   Analgesic (antinociceptive)

9702   (7862) $C_{20}H_{25}NS$
                                                                       21489-20-3

1,3-Dihydro-*N*,3,3-trimethyl-1-phenylbenzo[*c*]thio-
phene-1-propylamine = 1,3-Dihydro-1,1-dimethyl-3-[3-
(methylamino)propyl]-3-phenylbenzo[*c*]thiophene = 3,3-
Dimethyl-1-(3-methylaminopropyl)-1-phenylthiophtha-
lan = 1,3-Dihydro-*N*,3,3-trimethyl-1-phenylbenzo[*c*]thio-
phene-1-propanamine (●)

S Lu 5-003, *Talsupram***, "*Thiophtalane*"
U Antidepressant

U Antifungal antibiotic from *Chromobacterium prodigiosum*

9703   (7863) $C_{20}H_{25}N_3$
                                                   20069-03-8

[4a*R*-(4aα,5β,13bα,14β,14aα)]-
1,2,3,4,4a,5,7,8,13,13b,14,14a-Dodecahydro-5,14-etha-
noindolo[2',3':3,4]pyrido[1,2-*g*]-1,6-naphthyridine (●)
S *Nitrarine*
U Spasmolytic, antihypertensive (alkaloid from *Nitraria schoberi*)

9704 $C_{20}H_{25}N_3O$
                                                   156896-33-2

(*R*)-1,3,4,5-Tetrahydro-6-(5-oxazolyl)-*N*,*N*-dipropyl-
benz[*cd*]indol-4-amine (●)
S LY 301317
U 5-HT$_{1A}$ receptor agonist

9705   (7864) $C_{20}H_{25}N_3O$
                                                   82-89-3

4-Methoxy-5-[(5-methyl-4-pentyl-2*H*-pyrrol-2-yli-
dene)methyl]-2,2'-bi-1*H*-pyrrole (●)
S *Prodigiosin*

9706   (7865) $C_{20}H_{25}N_3O$
                                                   50-37-3

N,*N*-Diethyl-D-lysergamide = D-Lysergic acid diethylamide = D-7-Methyl-4,6,6a,7,8,9-hexahydroindo-
lo[4,3-*fg*]quinoline-9-carboxylic acid diethylamide =
(8β)-9,10-Didehydro-*N*,*N*-diethyl-6-methylergoline-8-
carboxamide (●)
R also D-tartrate (2:1) (17676-08-3)
S Delysid, *Lisergide*, LSD, LSD-25, Lysergamid "Spofa", *Lysergide***
U Sympatholytic, psychotomimetic

9707   (7866) $C_{20}H_{25}N_3O$
                                                   3147-75-9

2-(2*H*-Benzotriazol-2-yl)-4-(1,1,3,3-tetramethylbu-
tyl)phenol (●)
S Cyasorb 5411, Cyasorb UV 5411, *Octrizole***, Spectra-Sorb UV 5411
U Ultraviolet screen

9708 $C_{20}H_{25}N_3O$
                                                  162141-96-0

1-Cyclopentyl-3-ethyl-1,4,5,6-tetrahydro-6-(2-methyl-
phenyl)-7H-pyrazolo[3,4-c]pyridin-7-one (●)
S CP-220629
U Human eosinophil phosphodiesterase inhibitor

9709 (7868) $C_{20}H_{25}N_3O_2$
                                                  5793-04-4

9,10-Didehydro-N-[(S)-2-hydroxy-1-methylethyl]-1,6-
dimethylergoline-8β-carboxamide = D(+)-1-Methyllyser-
gic acid β-hydroxyisopropylamide = [8β(S)]-9,10-Dide-
hydro-N-(2-hydroxy-1-methylethyl)-1,6-dimethylergoli-
ne-8-carboxamide (●)
R Maleate (1:1) (76958-69-5)
S Ergalgin, 1-Methylergometrine maleate, Propisergi-
de maleate**
U Migraine therapeutic

9710 (7867) $C_{20}H_{25}N_3O_2$
                                                  113-42-8

(+)-N-[1-(Hydroxymethyl)propyl]-D-lysergamide = D-
Lysergic acid (+)-butanolamide-(2) = [8β(S)]-9,10-Dide-
hydro-N-[1-(hydroxymethyl)propyl]-6-methylergoline-8-
carboxamide (●)

S Methylergobasine, Methylergobrevine, Methylergo-
metrine**, Methylergonovine, Metilergometrina
U Oxytocic

9710-01 (7867-01)   57432-61-8
R Maleate (1:1) [or tartrate (2:1) (6209-37-6)]
S Basofortina, Demergin, Derganin, Dometrion, El-
amidon, Elpan S, Emifarol, Enovine, Erezin, Ergoba-
cin, Ergopartin, Ergotyl, Ergovit-Amp., Ingagen-M,
Levospan, Mergot, Metenarin, Methecrine, Mether-
gen, Methergin(e), Metiler, Mitrabagin-C, Mitrosy-
stal, Mitrotan, Myomergin, NSC-186067, Obstet,
Partergin, Ryegonovin, Santargot, Secotyl, Spame-
trin-M, Takimetrin-M, Telpalin, Unidergin, Utergi-
ne, Uterin, Uterjin
U Oxytocic

9711 (7869) $C_{20}H_{25}N_3O_2$
                                                  111466-41-2

(2S-trans)-1,3,4,5',6,6',7,12b-Octahydro-1',3'-dimethyl-
spiro[2H-benzofuro[2,3-a]quinolizine-2,4'-(1'H)-pyrimi-
din]-2'(3'H)-one (●)
S L 657743, MK-912
U Antidepressant (α₂-adrenoceptor antagonist)

9712 (7870) $C_{20}H_{25}N_3O_2S$

4,5-Dihydro-5-methyl-6-[6-[2-(1-piperidinyl)eth-
oxy]benzo[b]thien-2-yl]-3(2H)-pyridazinone (●)
R Monohydrochloride (129426-01-3)

S Org 20494
U Cardiotonic (positive inotropic)

9713 (7872) $C_{20}H_{25}N_3O_3$
88578-07-8

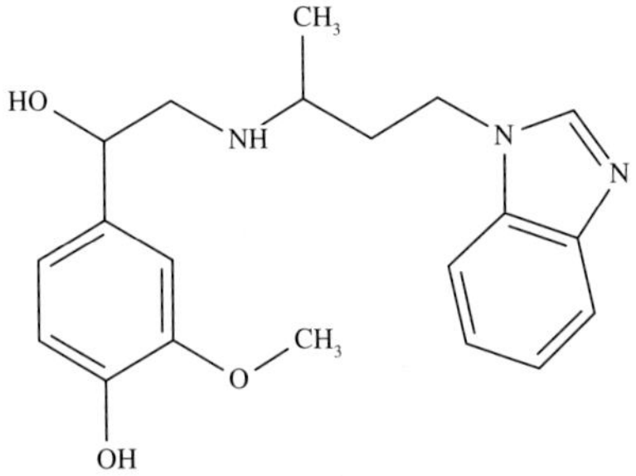

α-[[[3-(1-Benzimidazolyl)-1-methylpropyl]amino]me-
thyl]vanillyl alcohol = 2-[[3-(1-Benzimidazolyl)-1-me-
thylpropyl]amino]-1-(4-hydroxy-3-methoxyphenyl)etha-
nol = α-[[[3-(1H-Benzimidazol-1-yl)-1-methylpropyl]
amino]methyl]-4-hydroxy-3-methoxybenzenemethanol
(●)
S *Imoxiterol**, 1622 RB, RP 55802 B
U Bronchodilator

9714 (7871) $C_{20}H_{25}N_3O_3$
73674-86-9

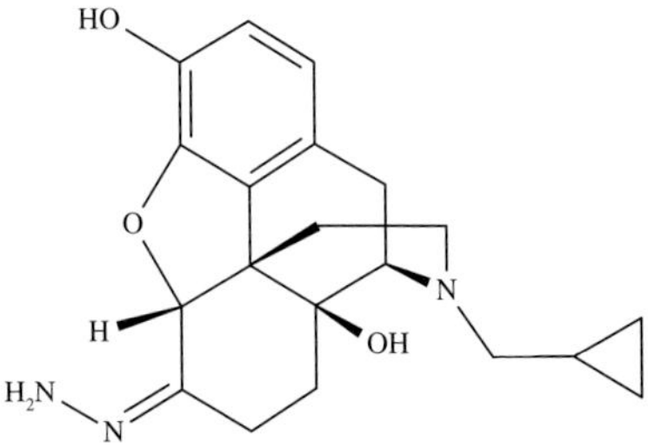

17-(Cyclopropylmethyl)-4,5α-epoxy-3,14-dihydroxy-
morphinan-6-one hydrazone = (5α)-17-(Cyclopropylmethyl)-4,5-epoxy-3,14-dihydroxymorphinan-6-one hydrazone (●)
S *Naltrexazone*
U Antagonist to narcotics

9715 $C_{20}H_{25}N_3O_3$
93799-37-2

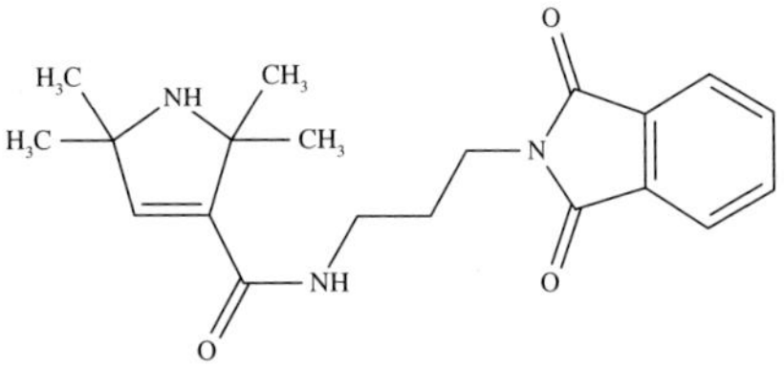

N-[3-(1,3-Dihydro-1,3-dioxo-2H-isoindol-2-yl)propyl]-
2,5-dihydro-2,2,5,5-tetramethyl-1H-pyrrole-3-carbox-
amide (●)
S A-2545
U Anti-arrhythmic

9716 (7873) $C_{20}H_{25}N_3O_3S_2$
94662-52-9

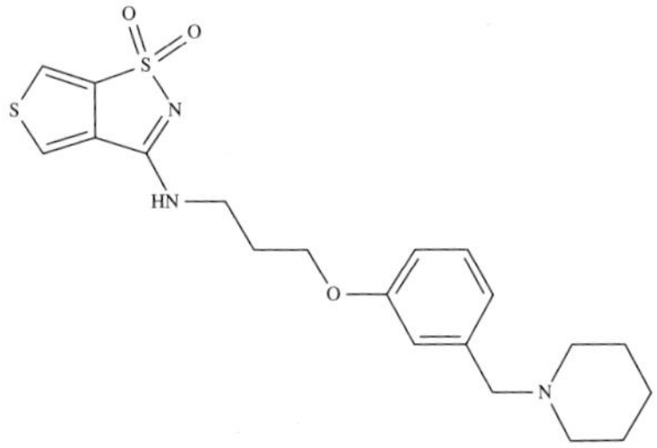

N-[3-[3-(1-Piperidinylmethyl)phenoxy]propyl]thie-
no[3,4-d]isothiazol-3-amine 1,1-dioxide (●)
S Wy-45662
U Gastric antisecretory (histamine H_2-receptor antagonist)

9717 (7874) $C_{20}H_{25}N_3O_4$
104775-36-2

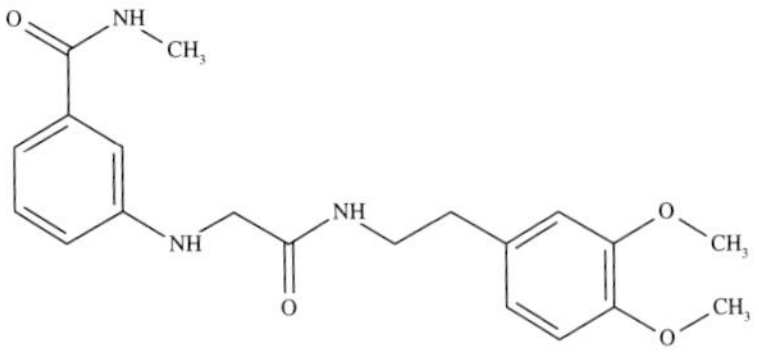

3-[[[[(3,4-Dimethoxyphenethyl)carbamoyl]methyl]ami-
no]-N-methylbenzamide = 3-[[2-[[2-(3,4-Dimethoxyphe-
nyl)ethyl]amino]-2-oxoethyl]amino]-N-methylbenz-
amide (●)
S DQ-2511, *Ecabamide*, *Ecabapide**, Muralis
U Anti-ulcer agent

9718 (7875) $C_{20}H_{25}N_3O_4$
88053-05-8

1-[(*E*)-3,4-(Methylenedioxy)cinnamoyl]-4-[(1-pyrrolidi-
nylcarbonyl)methyl]piperazine = (*E*)-1-[3-(1,3-Benzo-
dioxol-5-yl)-1-oxo-2-propenyl]-4-[2-oxo-2-(1-pyrrolidi-
nyl)ethyl]piperazine (●)
S   *Cinoxopazide***
U   Cerebrotonic

9719 $C_{20}H_{25}N_3O_5S_2$
173151-24-1

(*R*)-3,4,5,6,7,8,9,10-Octahydro-16-hydroxy-14-methoxy-
13-methyl-4-(3-methyl-1,2,4-oxadiazol-5-yl)-6-thioxo-
11,2,5-benzoxathiaazacyclotetradecin-12(1*H*)-one (●)
S   Ro 48-2865
U   Antibacterial (gram-positive)

9720 (7876) $C_{20}H_{25}N_3O_6S$
110646-15-6

3-Ethyl-5-methyl 2-[[(2-aminoethyl)thio]methyl]-1,4-di-
hydro-6-methyl-4-(*m*-nitrophenyl)-3,5-pyridinedicarb-
oxylate = 2-[[(2-Aminoethyl)thio]methyl]-1,4-dihydro-6-
methyl-4-(3-nitrophenyl)-3,5-pyridinedicarboxylic acid
3-ethyl 5-methyl ester (●)
S   *Tiamdipine*

U   Antihypertensive (calcium antagonist)

9721 (7877) $C_{20}H_{25}N_3S$
84-97-9

10-[3-(4-Methyl-1-piperazinyl)propyl]phenothiazine =
10-[3-(4-Methyl-1-piperazinyl)propyl]-10*H*-phenothia-
zine (●)
R   also malonate (1:2) (14777-25-4)
S   P 725, *Perazine*, Pernazinum, Psytomin, Taxilan
U   Neuroleptic

9722 $C_{20}H_{25}N_5$
160730-61-0

(8β)-6-Propyl-8-(1*H*-1,2,4-triazol-1-ylmethyl)ergoline
(●)
S   BAM-1120
U   Antiparkinsonian (dopamine agonist)

9723 $C_{20}H_{25}N_5O$
156007-21-5

1-[3-(1*H*-Benzotriazol-1-yl)propyl]-4-(2-methoxyphe-
nyl)piperazine = 1-[3-[4-(2-Methoxyphenyl)-1-piperazi-
nyl]propyl]-1*H*-benzotriazole (●)
S   MP-3022

U   5-HT$_{1A}$ receptor antagonist

9724   (7878) $C_{20}H_{25}N_5O_3S$
                              99248-32-5

5-[(1,2-Dihydro-2-oxo-4-pyridyl)methyl]-2-[[2-[[5-[(di-methylamino)methyl]furfuryl]thio]ethyl]amino]-4(1H)-pyrimidinone = 5-[(1,2-Dihydro-2-oxo-4-pyridinyl)me-thyl]-2-[[2-[[[5-[(dimethylamino)methyl]-2-furanyl]me-thyl]thio]ethyl]amino]-4(1H)-pyrimidinone (●)
S   *Donetidine**, SKF 93574
U   Histamine H$_2$-receptor antagonist (anti-ulcer agent, antihistaminic)

9725   (7879) $C_{20}H_{25}N_5O_4$
                              101616-07-3

Methyl 3-[(diethylamino)methyl]-5,8-dihydro-7-methyl-5-(m-nitrophenyl)imidazo[1,2-a]pyrimidine-6-carb-oxylate = 3-[(Diethylamino)methyl]-1,5-dihydro-7-me-thyl-5-(3-nitrophenyl)imidazo[1,2-a]pyrimidine-6-carb-oxylic acid methyl ester (●)
R   Monohydrochloride (101616-06-2)
S   Y-19638
U   Coronary vasodilator (calcium antagonist)

9726 $C_{20}H_{25}N_5O_4S$
                              180144-61-0

3-[N-[4-[4-(Aminoiminomethyl)phenyl]-2-thiazolyl]-N-1-(carboxymethyl)-4-piperidinyl]amino]propanoic acid = 4-[[4-[4-(Aminoiminomethyl)phenyl]-2-thiazolyl](2-carboxyethyl)amino]-1-piperidineacetic acid (●)
R   Trihydrochloride (190841-78-2)
S   SR 121566A
U   Antithrombotic (GP IIb/IIIa antagonist)

9727   (7880) $C_{20}H_{25}N_5O_7S_2$
                              65243-33-6

(Pivaloyloxy)methyl (6R,7R)-7-[2-(2-amino-4-thiazo-lyl)glyoxylamido]-3-methyl-8-oxo-5-thia-1-azabicy-clo[4.2.0]oct-2-ene-2-carboxylate, 7^2-(Z)-(O-methyl-oxime) = [6R-[6α,7β(Z)]]-7-[[(2-Amino-4-thiazo-lyl)(methoxyimino)acetyl]amino]-3-methyl-8-oxo-5-thia-1-azabicyclo[4.2.0]oct-2-ene-2-carboxylic acid (2,2-dimethyl-1-oxopropoxy)methyl ester (●)
R   also monohydrochloride (111696-23-2)
S   Cefec, *Cefetamet pivoxil**, Cefyl, CFMT-PI, Globo-cef, Globocep, Ro 15-8075, Tarcevis
U   Antibiotic

9728   (7881) $C_{20}H_{25}N_5O_{10}$
                              59456-70-1

[2S-(2R*,3R*,4R*)]-5-[[2-Amino-4-hydroxy-4-(5-hydr-oxy-2-pyridinyl)-3-methyl-1-oxobutyl]amino]-1,5-dide-oxy-1-(3,4-dihydro-2,4-dioxo-1(2H)-pyrimidinyl)-β-D-allofuranuronic acid (●)
S   *Neopolyoxin C, Nikkomycin Z*
U   Antifungal antibiotic from *Streptomyces tendae* and other sp.

9729 (7882) $C_{20}H_{25}N_7O_4$
120225-53-8

1-[6-Amino-2-[(2-phenylethyl)amino]-9*H*-purin-9-yl]-1-
deoxy-*N*-ethyl-β-D-ribofuranuronamide (●)
S CGS 21577
U Antihypertensive

9730 (7883) $C_{20}H_{25}N_7O_6$
134-35-0

N^5-Methyltetrahydrofolic acid = *N*-[4-[[(2-Amino-
1,4,5,6,7,8-hexahydro-5-methyl-4-oxo-6-pteridinyl)me-
thyl]amino]benzoyl]-L-glutamic acid (●)
R Calcium salt (1:1) (26560-38-3)
S Biofolic, *Calcii mefolinas*, Furoic, Prefolic, Saf
U Hematopoietic, hepatoprotectant, antineoplastic

9731 (7884) $C_{20}H_{26}BrNO_4S$
130466-54-5

(+)-5(*Z*)-6-[(1*R*,2*R*,3*R*,4*S*)-3-(*N*-4-Bromobenzenesulfo-
nylaminomethyl)bicyclo[2.2.1]hept-2-yl]-5-hexenoic
acid = [1*R*-[1α,2α(*Z*),3α,4α]]-6-[3-[[[(4-Bromophe-
nyl)sulfonyl]amino]methyl]bicyclo[2.2.1]hept-2-yl]-5-
hexenoic acid (●)

S NT 126, ONO-NT-126
U Thromboxane A$_2$-receptor antagonist

9732 (7885) $C_{20}H_{26}Br_2N_2O$
54785-02-3

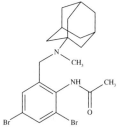

α-(1-Adamantylmethylamino)-4',6'-dibromo-*o*-acetoto-
luidide = 2'-[(1-Adamantylmethylamino)methyl]-4',6'-di-
bromoacetanilide = *N*-(2-Acetamido-3,5-dibromoben-
zyl)-*N*-methyl-1-adamantanamine = *N*-[2,4-Dibromo-6-
[(methyltricyclo[3.3.1.1^{3,7}]dec-1-ylamino)methyl]phe-
nyl]acetamide (●)
S *Adamexine***, Adamacol, Broncostyl
U Mucolytic

9733 (7886) $C_{20}H_{26}ClNO$
511-46-6

2-[(*p*-Chloro-α-methyl-α-phenylbenzyl)oxy]triethyl-
amine = *N*,*N*-Diethyl-2-(*p*-chloro-1,1-diphenyleth-
oxy)ethylamine = β-Diethylaminoethyl *p*-chloro-α-me-
thylbenzhydryl ether = 2-[1-(4-Chlorophenyl)-1-phenyl-
ethoxy]-*N*,*N*-diethylethanamine (●)
R Hydrochloride (2019-16-1)
S *Clofenetamine hydrochloride***, Diethylsystral, Kai-
thon, Keithon, KT 5, *Phénoxéthamine*
U Anticholinergic, antiparkinsonian

9734   (7887) $C_{20}H_{26}ClNO_3$
82168-26-1

2-(1-Adamantylamino)ethyl (*p*-chlorophenoxy)acetate =
2-(4-Chlorophenoxy)acetic acid 2-(tricy-
clo[3.3.1.1^{3,7}]dec-1-ylamino)ethyl ester (●)
S   *Adafenoxate**, WON 150
U   Nootropic, psychotonic

9735   (7888) $C_{20}H_{26}ClNO_5$
68206-94-0

8-Chloro-3-[β-(diethylamino)ethyl]-4-methyl-7-[(ethoxycarbonyl)methoxy]coumarin = [[8-Chloro-3-[2-(diethylamino)ethyl]-4-methyl-2-oxo-2*H*-1-benzopyran-7-
yl]oxy]acetic acid ethyl ester (●)
R   also hydrochloride (74697-28-2)
S   AD-6, Assogen, *Cloricromen**, Cromocap, Proen-
dotel
U   Coronary vasodilator, platelet aggregation inhibitor

9736   (7889) $C_{20}H_{26}ClN_3O$
101000-49-1

N-[(8α)-2-Chloro-6-methylergolin-8-yl]-2,2-dimethyl-
propanamide (●)
R   also monomethanesulfonate (137639-61-3)
S   SDZ 208-912, SDZ-HDC 912
U   Antischizophrenic (dopamine D_2 agonist)

9737 $C_{20}H_{26}ClN_3O_2$
151213-85-3

(−)-4-Amino-5-chloro-*N*-(*endo*-8-methyl-8-azabicy-
clo[3.2.1]oct-3α-yl)-2-[[1(*S*)-methyl-2-buty-
nyl]oxy]benzamide = 4-Amino-5-chloro-*N*-[(3-*endo*)-8-
methyl-8-azabicyclo[3.2.1]oct-3-yl]-2-[[(1*S*)-1-methyl-
2-butynyl]oxy]benzamide (●)
R   Monohydrochloride (151213-86-4)
S   E-3620
U   Anti-emetic (serotonin 5-HT$_3$ antagonist and 5-HT$_4$
agonist)

9738   (7890) $C_{20}H_{26}ClN_3O_2$
55837-13-3

2-[2-[4-(*p*-Chloro-α-2-pyridylbenzyl)-1-piperazinyl]eth-
oxy]ethanol = 2-[2-[4-[(4-Chlorophenyl)-2-pyridinylme-
thyl]-2-piperazinyl]ethoxy]ethanol (●)
S   *Piclopastine**
U   Antihistaminic

9739 (7891) $C_{20}H_{26}Cl_2N_2O$
33189-65-0

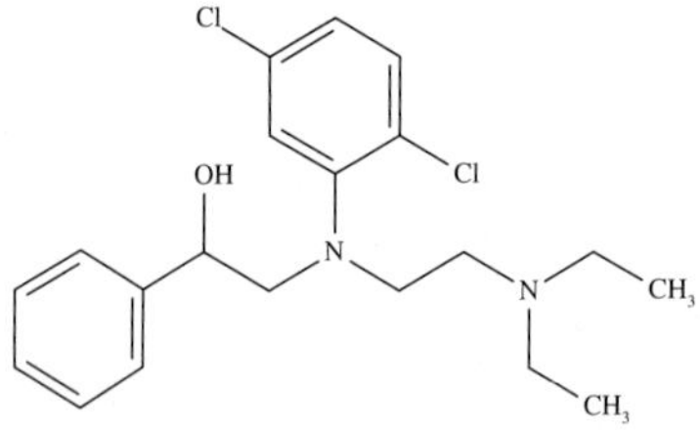

α-[[*N*-(2,5-Dichlorophenyl)-2-(diethylamino)ethylami-
no]methyl]benzyl alcohol = α-[[(2,5-Dichlorophenyl)[2-
(diethylamino)ethyl]amino]methyl]benzenemethanol (●)
S AN 162, RMI 6792
U Antithrombotic

9740 (7892) $C_{20}H_{26}I_3N_3O_{12}$
75751-89-2

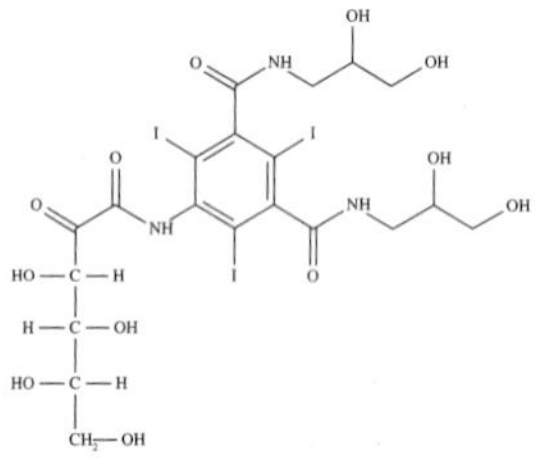

N,N'-Bis(2,3-dihydroxypropyl)-5-L-*xylo*-2-hexuloson-
amido-2,4,6-triiodoisophthalamide = *N,N'*-Bis(2,3-di-
hydroxypropyl)-5-[(L-*xylo*-2-hexulosonoyl)amino]-2,4,6-
triiodo-1,3-benzenedicarboxamide (●)
S *Iogulamide*, MP-10013
U Diagnostic aid (radiopaque medium)

9741 (7893) $C_{20}H_{26}NO_3$
15209-00-4

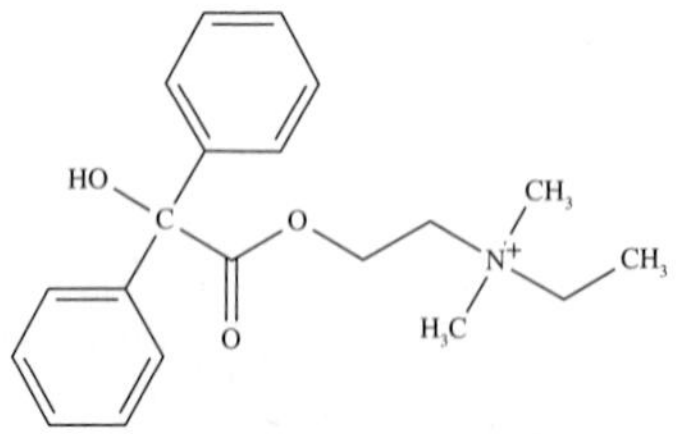

Ethyl(2-hydroxyethyl)dimethylammonium benzilate = *N*-
Ethyl-*N,N*-dimethyl-2-(benziloyloxy)ethylammonium =
N-Ethyl-2-[(hydroxydiphenylacetyl)oxy]-*N,N*-dimethyl-
ethanaminium (●)
R Chloride (1164-38-1)

S E₃, *Lachesine chloride*, Laxesin
U Mydriatic

9742 (7894) $C_{20}H_{26}NO_3$

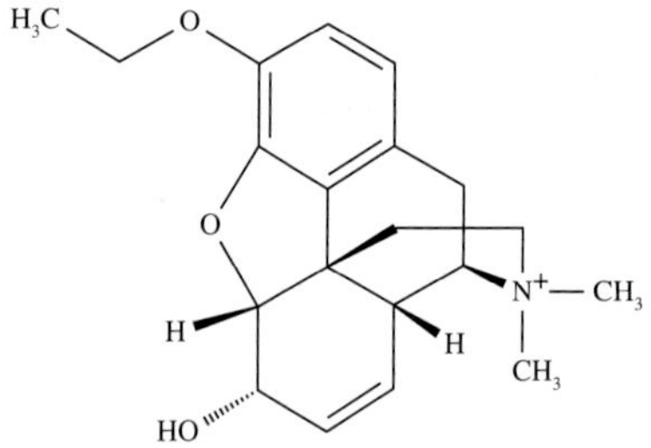

4,5α-Epoxy-3-ethoxy-6α-hydroxy-N,N-dimethylmor-
phin-7-enium = (5α,6α)-7,8-Didehydro-4,5-epoxy-3-eth-
oxy-6-hydroxy-17,17-dimethylmorphinanium (●)
R Iodide (6696-59-9)
S *Äthylmorphinmethyljodid*, *Codéthyline iodomethylate*, *Ethylmorphine methyl iodide*, Trachyl
U Narcotic analgesic, antitussive

9743 $C_{20}H_{26}NO_6P$
187608-26-0

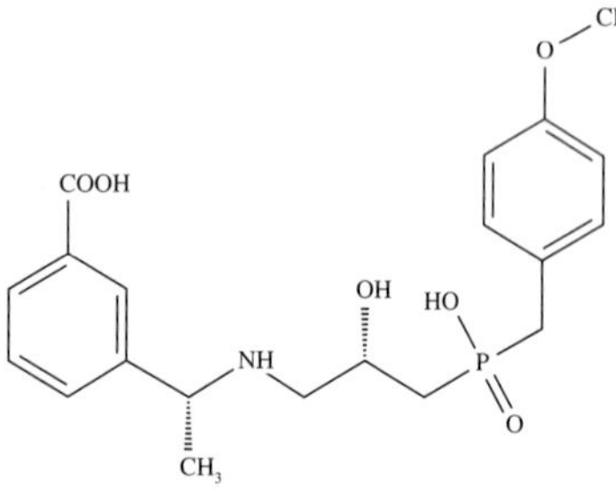

3-[(1*R*)-1-[[(2*S*)-2-Hydroxy-3-[hydroxy[(4-methoxyphe-
nyl)methyl]phosphinyl]propyl]amino]ethyl]benzoic acid
(●)
S CGP 62349
U GABA_B receptor antagonist

9744   (7897) $C_{20}H_{26}N_2$
                                                 4757-55-5

10-[3-(Dimethylamino)propyl]-9,9-dimethylacridan =
N,N,9,9-Tetramethyl-10(9*H*)-acridinepropanamine (●)
R   Tartrate (1:1) (3759-07-7)
S   *Acripramine, Dimetacrine tartrate**, Dimethacrine*
     tartrate, HK 5031, Isotonil, Istonil, Linostil, Miroi-
     stonil, SD 709
U   Antidepressant

9745   (7898) $C_{20}H_{26}N_2$
                                                   739-71-9

5-[3-(Dimethylamino)-2-methylpropyl]-10,11-dihydro-
5*H*-dibenz[*b,f*]azepine = *N*-(3-Dimethylamino-2-methyl-
propyl)iminodibenzyl = 10,11-Dihydro-*N,N*,β-trimethyl-
5*H*-dibenz[*b,f*]azepine-5-propanamine (●)
R   also monomethanesulfonate (25332-13-2), maleate
     (1:1) (521-78-8) or monohydrochloride (3589-21-7)
S   Apo-Trimip, F.I. 6120, Gerfonal, Herphonal,
     IL-6001, Novo-Tripramine, NSC-169472, Nu-Trimi-
     pramine, Pro-Trimip, Rhotrimine, 7162 R.P., Sa-
     pilent, Stangyl, Surmontil, *Triméprimine, Trimepro-*
     priminum, Trimipramine**, Tripress, Tydamine
U   Antidepressant

9746   (7896) $C_{20}H_{26}N_2$
                                                   3626-71-9

1-[2-(*N*-Benzylanilino)ethyl]piperidine = *N*-[β-(Benzyl-
phenylamino)ethyl]piperidine = *N*-Phenyl-*N*-(phenylme-
thyl)-1-piperidineethanamine (●)
R   Hydrochloride
S   Allantan, *Histapiperidine hydrochloride*
U   Antihistaminic

9747 $C_{20}H_{26}N_2$
                                                   170964-67-7

8-Cyclohexyl-2,3,3a,4,5,6-hexahydro-1*H*-pyrazi-
no[3,2,1-*jk*]carbazole (●)
R   also monohydrochloride (135897-69-7)
S   *Tetrindole*
U   Antidepressant, cognition enhancer

9748 (7895) C$_{20}$H$_{26}$N$_2$

1-[2-(Dibenzylamino)ethyl]pyrrolidine = *N,N*-Dibenzyl-
N-(β-pyrrolidinoethyl)amine = *N,N*-Bis(phenylmethyl)-
1-pyrrolidineethanamine (●)
R Monohydrochloride (60662-76-2)
S Myostimin
U Antihistaminic

9749 (7899) C$_{20}$H$_{26}$N$_2$
3277-59-6

1,2,3,4,4aβ,5,7,8,13,13bα,14,14aα-Dodecahydro-13-me-
thylbenz[*g*]indolo[2,3-*a*]quinolizine = 1-Methylyohim-
ban (●)
R Monohydrochloride (5560-73-6)
S *Mimbane hydrochloride***, W 2291 A
U Analgesic

9750 (7900) C$_{20}$H$_{26}$N$_2$O
53716-46-4

(−)-3-(*p*-Aminophenethyl)-2,3,4,5-tetrahydro-8-meth-
oxy-2-methyl-1*H*-3-benzazepine = (−)-4-[2-[1,2,4,5-Tet-
rahydro-8-methoxy-2-methyl-3*H*-3-benzazepin-3-
yl]ethyl]benzenamine (●)
R Dihydrochloride (53716-45-3)
S *Anilopam hydrochloride***, PR 786-723

U Analgesic

9751 (7901) C$_{20}$H$_{26}$N$_2$O
83-74-9

12-Methoxyibogamine (●)
S Endabuse, *Ibogaine*, NIH 10567
U Hallucinogen, treatment of addiction (alkaloid from
Tabernanthe iboga)

9752 (7902) C$_{20}$H$_{26}$N$_2$O
72714-74-0

6-Methoxy-4-[3-[(3*R*,4*R*)-3-vinyl-4-piperidyl]pro-
pyl]quinoline = (3*R-cis*)-4-[3-(3-Ethenyl-4-piperidi-
nyl)propyl]-6-methoxyquinoline (●)
S PK-5078, *Viqualine***
U Antidepressant

9753 (7903) C$_{20}$H$_{26}$N$_2$C
72714-75-1

6-Methoxy-4-[3-[(3*S*,4*R*)-3-vinyl-4-piperidyl]pro-
pyl]quinoline = (3*S-trans*)-4-[3-(3-Ethenyl-4-piperidi-
nyl)propyl]-6-methoxyquinoline (●)
S *Ivoqualine***, PK-7059
U Antipsychotic

9754 (7904) $C_{20}H_{26}N_2OS$
1165-22-6

N-[1-[2-(2-Thienyl)ethyl]-4 piperidyl]propionanililide =
N-Phenyl-N-[1-[2-(2-thienyl)ethyl]-4-piperidinyl]pro-
panamide (●)
S Fentatienil
U Analgesic

9755 (7912) $C_{20}H_{26}N_2O_2$
4360-12-7

3-Ethyl-1,2,3,4,6,7,12a,12b-octahydro-4,13-dihydroxy-
12-methyl-12H-7a,2,6-ethanylylideneindolo[2,3-a]quino-
lizine = (17R,21α)-Ajmalan-17,21-diol (●)
R also hydrochloride (4410-48-4) or phosphate
S Ajimaline, Ajma, Ajmaline, Ajmalulin, Aritmina,
   Arytmal, Cardia-Tab., Cardiorythmine, Cardio-Spar-
   tan, Cartagine, Coeur-Maline, Crossperin, Datrin N,
   Gilurytmal, Hanemaline, Ignazin, Merabitol, Nichi-
   maline, Pulsline, Rauforin, Rauverid, Rauwolfine,
   Rhythmaton, Ritmos, Rytmalin, Serenol, Tachmalin,
   Tajmalin, Takycor, Wolfina
U Anti-arrhythmic (alkaloid from Rauwolfia serpentina
   and other Rauwolfia sp.)

9755-01 (7912-01)
R Compd. with 2-aminoethanol dihydrogen phosphate
   (no. 91)
S Ajmalini olaminphosphas, Normorytmina
U Anti-arrhythmic

9755-02 (7912-02)   21290-16-4

R Compd. with phenobarbital (no. 3255) (1:1)
S Ajmalinum phenobarbitalum, Ritmosedina
U Anti-arrhythmic

9756 (7905) $C_{20}H_{26}N_2O_2$
111974-80-2

(R)-4-[3-[(2-Hydroxy-2-phenylethyl)amino]-3-methyl-
butyl]benzamide (●)
R Monohydrochloride (111112-18-6)
S LY 195448
U Antineoplastic, antihypertensive

9757 (7906) $C_{20}H_{26}N_2O_2$
79201-80-2

(±)-2-(p-Aminophenethyl)-1,2,3,4-tetrahydro-6,7-di-
methoxy-1-methylisoquinoline = (±)-4-[2-(3,4-Dihydro-
6,7-dimethoxy-1-methyl-2(1H)-isoquinolinyl)ethyl]ben-
zenamine (●)
R Dihydrochloride (76448-47-0)
S PR 0870-714 A, Veradoline hydrochloride**
U Analgesic

9758 (7907) $C_{20}H_{26}N_2O_2$
68318-20-7

3-(p-Aminophenethyl)-2,3,4,5-tetrahydro-7,8-dimeth-
oxy-1H-3-benzazepine = 4-[2-(1,2,4,5-Tetrahydro-7,8-di-
methoxy-3H-3-benzazepin-3-yl)ethyl]benzenamine (●)
R Dihydrochloride (67394-31-4)

1979

S PR 0818-156 A, *Verilopam hydrochloride***
U Analgesic

9759 (7908) $C_{20}H_{26}N_2O_2$
522-66-7

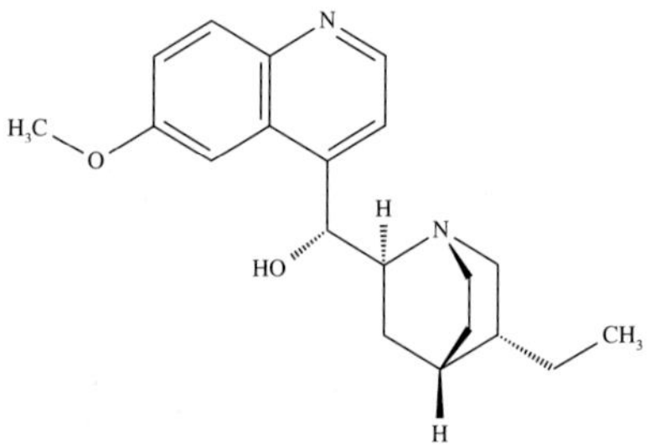

(*R*)-[(2*S*,4*S*,5*R*)-5-Ethyl-2-quinuclidinyl](6-methoxy-4-quinolyl)methanol = α-(5-Ethyl-2-quinuclidinyl)-6-methoxy-4-quinolinemethanol = (8α,9*R*)-10,11-Dihydro-6'-methoxycinchonan-9-ol (●)
S *Dihydrochinin, Dihydroquinine, Hydrochinin, Hydroquinine, Methylhydrocupreine*
U Antimalarial, depigmentor

9759-01 85153-19-1
R Monohydrobromide
S Inhibin "Asta Medica"
U Muscle relaxant

9760 (7909) $C_{20}H_{26}N_2O_2$
1435-55-8

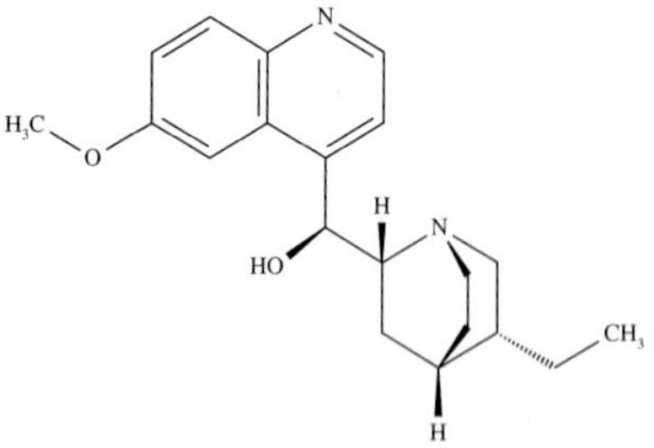

(*S*)-[(2*S*,4*S*,5*R*)-5-Ethyl-2-quinuclidinyl](6-methoxy-4-quinolyl)methanol = stereoisomer of α-(5-Ethyl-2-quinuclidinyl)-6-methoxy-4-quinolinemethanol = (9*S*)-10,11-Dihydro-6'-methoxycinchonan-9-ol (●)
R see also nos. 3042-02, 3255-14;
also mono-D-gluconate (21666-86-4) or monohydrochloride (1476-98-8)

S *Dihydrochinidin, Dihydroconchinin, Dihydroquinidine, Hydrochinidin, Hydroquine, Hydroquinidine, Idrochinidina*, Lentoquine, Sérécor, *Ydroquinidine*
U Anti-arrhythmic

9760-01
R Alginate
S Algiquin
U Anti-arrhythmic

9761 (7910) $C_{20}H_{26}N_2O_2$
98123-83-2

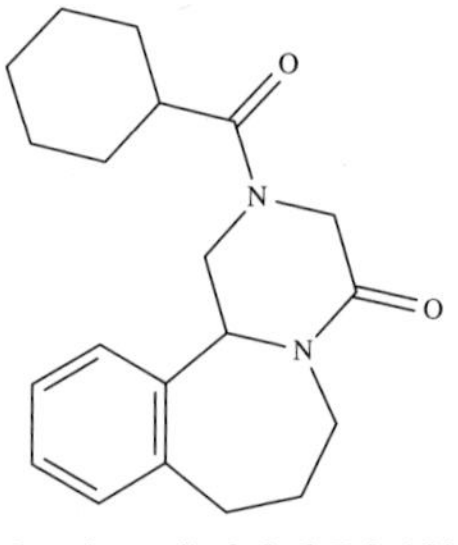

(±)-2-(Cyclohexylcarbonyl)-2,3,6,7,8,12b-hexahydropyrazino[2,1-*a*][2]benzazepin-4(1*H*)-one (●)
S BRL 38705, Cestex, *Epsiprantel***
U Anthelmintic

9762 (7911) $C_{20}H_{26}N_2O_2$
3569-85-5

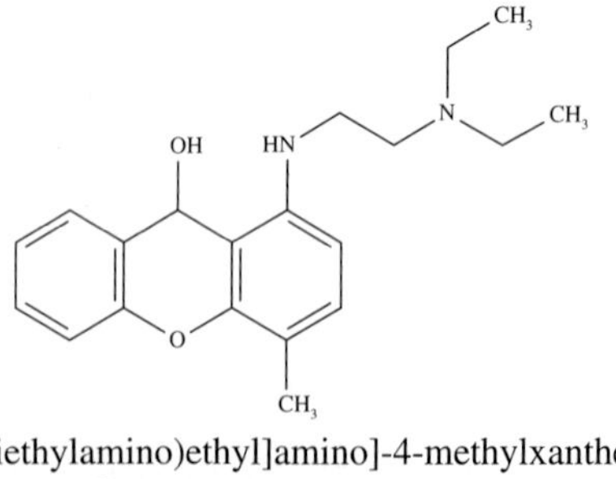

1-[[2-(Diethylamino)ethyl]amino]-4-methylxanthen-9-ol = 4-Methyl-1-(β-diethylaminoethylamino)xanthydrol = 1-[[2-(Diethylamino)ethyl]amino]-4-methyl-9*H*-xanthen-9-ol (●)
S Miracil C
U Antischistosomal

9763 (7913)

$C_{20}H_{26}N_2O_3$
68550-75-4

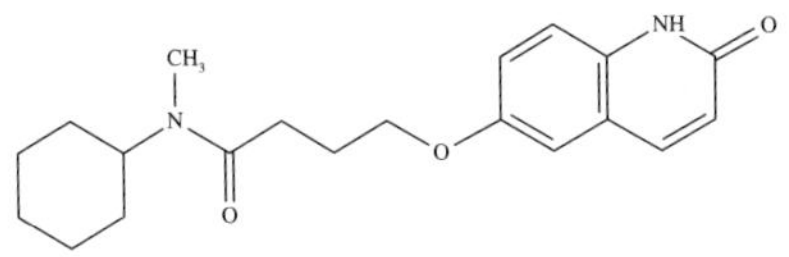

N-Cyclohexyl-4-[(1,2-dihydro-2-oxo-6-quinolyl)oxy]-*N*-methylbutyramide = *N*-Cyclohexyl-*N*-methyl-4-(6-carbostyryloxy)butyramide = *N*-Cyclohexyl-4-[(1,2-dihydro-2-oxo-6-quinolinyl)oxy]-*N*-methylbutanamide (●)
S *Cilostamide***, OPC-3689
U Antithrombotic

9764 (7914)

$C_{20}H_{26}N_2O_3$
85405-59-0

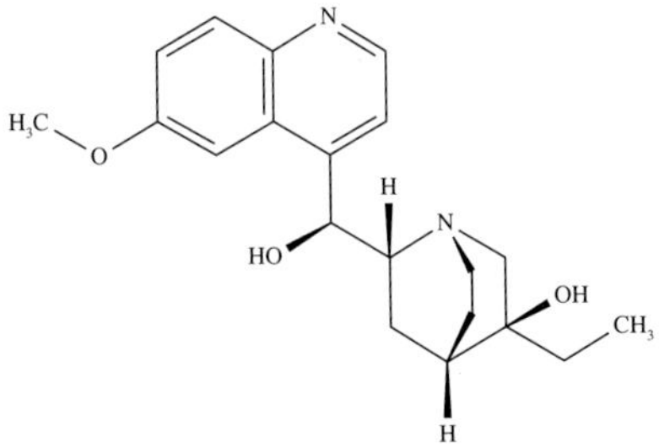

(3*S*)-3-Hydroxyhydroquinidine = [(2*R*,4*S*,5*S*)-5-Ethyl-5-hydroxy-2-quinuclidinyl](6-methoxy-4-quinolyl)-(*S*)-methanol = (9*S*)-10,11-Dihydro-6'-methoxycinchonan-3,9-diol (●)
R also sulfate (1:1) (130061-81-3)
S *Hydroquinidinol*, LNC-834, Pacicor
U Anti-arrhythmic

9765 (7915)

$C_{20}H_{26}N_2O_3$
67019-69-6

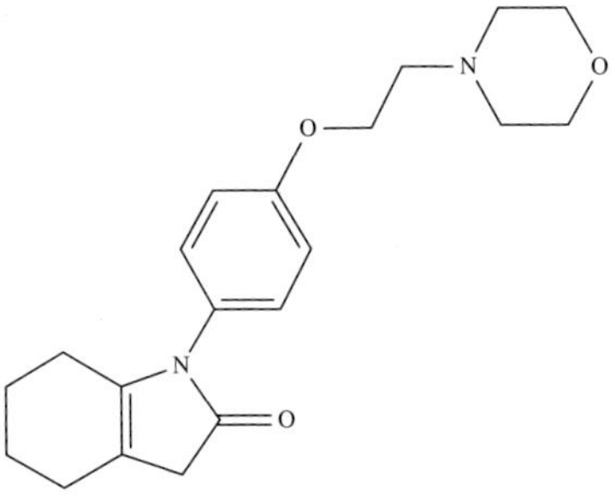

1-[*p*-(2-Morpholinoethoxy)phenyl]-4,5,6,7-tetrahydro-1*H*-indol-2(3*H*)-one = 1,3,4,5,6,7-Hexahydro-1-[4-[2-(4-morpholinyl)ethoxy]phenyl]-2*H*-indol-2-one (●)
S 1520 RB
U Platelet aggregation inhibitor

9766 (7916)

$C_{20}H_{26}N_2O_3$
87495-31-6

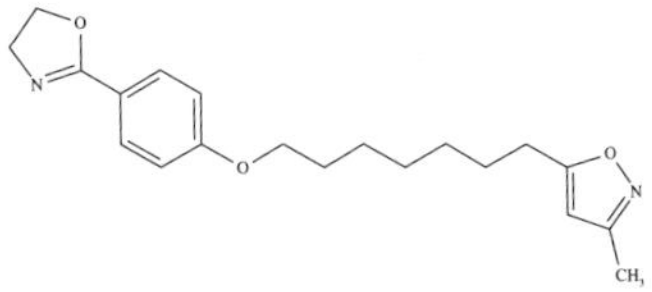

3-Methyl-5-[7-(*p*-2-oxazolin-2-ylphenoxy)heptyl]isoxazole = 5-[7-[4-(4,5-Dihydro-2-oxazolyl)phenoxy]heptyl]-3-methylisoxazole (●)
S *Disoxaril***, Win 51711
U Antiviral

9767 (7917)

$C_{20}H_{26}N_2O_4$
102908-59-8

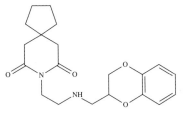

(±)-*N*-[2-[(1,4-Benzodioxan-2-ylmethyl)amino]ethyl]-1,1-cyclopentanediacetimide = (±)-8-[2-[[(2,3-Dihydro-1,4-benzodioxin-2-yl)methyl]amino]ethyl]-8-azaspiro[4.5]decane-7,9-dione (●)
R Monomethanesulfonate (124756-23-6)
S *Binospirone mesilate***, MDL 73005 EF
U Anxiolytic (5HT$_{1A}$-receptor antagonist)

9768   (7918) $C_{20}H_{26}N_2O_4$
122898-67-3

N-[p-[2-(Dimethylamino)ethoxy]benzyl]veratramide =
N-[[4-[2-(Dimethylamino)ethoxy]phenyl]methyl]-3,4-di-
methoxybenzamide (●)
R   Monohydrochloride (122892-31-3)
S   Ganaton, HSR-803, *Itopride hydrochloride***
U   Gastrointestinal prokinetic (dopamine D_2-receptor
antagonist)

9769   (7919) $C_{20}H_{26}N_2O_4$
90895-85-5

(±)-4'-[2-Hydroxy-3-(isopropylamino)propoxy]-p-
anisanilide = (±)-N-[4-[2-Hydroxy-3-[(1-methyl-
ethyl)amino]propoxy]phenyl]-4-methoxybenzamide (●)
S   Ronactolol**
U   β-Adrenergic blocker

9770   (7920) $C_{20}H_{26}N_2O_5S$
74258-86-9

N-[1-[(S)-3-Mercapto-2-methylpropionyl]-L-prolyl]-3-
phenyl-L-alanine acetate = 1-[(S)-3-(Acetylthio)-2-me-
thylpropanoyl]-L-prolyl-L-phenylalanine = (S)-N-[1-[3-

(Acetylthio)-2-methyl-1-oxopropyl]-L-prolyl]-L-phenyl-
alanine (●)
S   *Alacepril***, Cetapril, DU-1219
U   Antihypertensive (ACE inhibitor)

9771   (7921) $C_{20}H_{26}N_2O_5S_2$
83602-05-5

(8S)-7-[(S)-N-[(S)-1-Carboxy-3-phenylpropyl]alanyl]-
1,4-dithia-7-azaspiro[4.4]nonane-8-carboxylic acid =
[8S-[7[R*(R*)],8R*]]-7-[2-[(1-Carboxy-3-phenylpro-
pyl)amino]-1-oxopropyl]-1,4-dithia-7-azaspiro[4.4]no-
nane-8-carboxylic acid (●)
R   see also no. 11426
S   Sch 33861, *Spiraprilat***
U   Antihypertensive (ACE inhibitor)

9772 $C_{20}H_{26}N_2O_8S_2$
52988-50-8

(2R,3'R,5R,5'aR,8'S,8'aS,9'aR)-8'-(Acetyloxy)-
4,5,5'a,6',8'a,9'-hexahydro-8'a-hydroxy-3'-(hydroxymethyl)-4,4,5,11'-tetramethylspiro[furan-2(3H),7'(8'H)-
[3,9a](iminomethano)[9aH]cyclopenta[4,5]pyrro-
lo[2,1-c][1,2,4]dithiazine]-3,4',10'(3'H)-trione (●)
S   *Sirodesmin A*, TAN 1496B
U   Antiviral from *Microsphaeropsis* sp. FL-16144 or *Si
rodesmium diversum*

9773 (7922) $C_{20}H_{26}N_2S$
523-54-6

10-[3-(Dimethylamino)-2-methylpropyl]-2-ethylpheno-
thiazine = 2-Ethyl-*N,N*,β-trimethyl-10*H*-phenothiazine-
10-propanamine (●)

R also monohydrochloride (3737-33-5)

S *Äthylisobutrazin*, Diquel, *Ethotrimeprazine,
Ethylisobutrazine, Ethyltrimeprazine, Etymemazi-
ne**, Nuital, 6484 R.P., Sergétyl

U Neuroleptic, antihistaminic, hypnotic

9774 $C_{20}H_{26}N_2S_2$
231298-83-2

N-(4-Butylphenyl)-*N*'-[4-(propylthio)phenyl]thiourea (●)

S JDDD 46

U Antimycobacterial

9775 $C_{20}H_{26}N_4O$
155029-32-6

3-[4-(1*H*-Imidazol-1-yl)benzoyl]-7-isopropyl-3,7-diaza-
bicyclo[3.3.1]nonane = 3-[4-(1*H*-Imidazol-1-yl)benzoyl]-
7-(1-methylethyl)-3,7-diazabicyclo[3.3.1]nonane (●)

R Diperchlorate (155029-33-7)

S GLG-V-13

U Anti-arrhythmic

9776 (7923) $C_{20}H_{26}N_4O$
18016-80-3

3-(9,10-Didehydro-6-methylergolin-8α-yl)-1,1-diethyl-
urea = *N*-(D-6-Methyl-8-isoergolenyl)-*N',N'*-diethylurea
= *N'*-[(8α)-9,10-Didehydro-6-methylergolin-8-yl]-*N,N*-
diethylurea (●)

R also maleate (1:1) (19875-60-6)

S Apodel, Arolac, Cuvalit, Dipergon, Dopagon, Doper-
gin(e), Eunal, LHM, Lisenil, *Lisuride**, Lizenil, Ly-
senyl, *Lysurid(e)*, *Mesorgydin*, *Methylergolcarbami-
de*, MIP/2204, Prolacam, Revanil, SH 31072 B

U Serotenin inhibitor (anti-allergic, migraine therapeutic, antiparkinsonian), prolactin inhibitor

---

9777 (7924) $C_{20}H_{26}N_4OS$
5585-93-3

2-[4-[3-(1-Aza-10-phenothiazinyl)propyl]-1-piperazi-
nyl]ethanol = 10-[3-(1-β-Hydroxyethyl-4-piperazi-
nyl)propyl]thiophenylpyridylamine = 4-[3-(10*H*-Pyri-
do[3,2-*b*][1,4]benzothiazin-10-yl)propyl]-1-piperazine-
ethanol (●)

R Dihydrochloride (17297-82-4)

S D 706 E, *Oxipendyli chloridum, Oxypendyl hydrochloride**, Perthipendyl hydrochloride*, Pervetral

U Anti-emetic

9778   (7925) $C_{20}H_{26}N_4O_2$
                                       121324-51-4

3,3'-(Hexamethylenedioxy)dibenzamidine = 3,3'-[1,6-He-
xanediylbis(oxy)]bis[benzenecarboximidamide] (●)
S *Meta-hexamidine*
U Antiprotozoal

9779   (7926) $C_{20}H_{26}N_4O_2$
                                       3811-75-4

4,4'-(Hexamethylenedioxy)dibenzamidine = *p,p'*-Diami-
dino-1,6-diphenoxyhexane = 4,4'-[1,6-Hexanediyl-
bis(oxy)]bis[benzenecarboximidamide] (●)
S *Hexamidine***
U Antiprotozoal

9779-01   (7926-01)   659-40-5
R Isethionate (1:2)
S Desomedin, Désomédine, Esomedina, Hexabon, He-
       xacol "Specia", Hexacrem, Hexalon, Hexaseptine,
       Hexomedin(e), Hexomedin N, Jabodine "Specia",
       Laryngomedin N, Ophtamedine, Pulvo-Hexa,
       2535 R.P.
U Topical antiseptic

9780 $C_{20}H_{26}N_4O_2$
                                       154512-24-0

7-[4-(Cyclohexylamino)butoxy]-1,3-dihydro-2*H*-imida-
zo[4,5-*b*]quinolin-2-one (●)
S BMY-21190

U Antithrombotic

9781   (7927) $C_{20}H_{26}N_4O_2$
                                       86140-10-5

(±)-1-(1*H*-Indazol-4-yloxy)-3-[[2-(2,6-xylidi-
no)ethyl]amino]-2-propanol = 1-[[2-[(2,6-Dimethylphe-
nyl)amino]ethyl]amino]-3-(1*H*-indazol-4-yloxy)-2-pro-
panol (●)
S BM 14298, *Neraminol***
U β-Adrenergic blocker

9782   (7928) $C_{20}H_{26}N_4O_4S$
                                       114914-42-0

(2*R-trans*)-*N*-[2-(1,3,4,6,7,12b-Hexahydro-2'-oxospiro-
[2*H*-benzofuro[2,3-*a*]quinolizine-2,4'-imidazolidin]-3'-
yl)ethyl]methanesulfonamide (●)
S L 659066
U α_2-Adrenoceptor antagonist

9783   (7929) $C_{20}H_{26}N_4O_5S$
                                       84845-75-0

N-[2-[[5-[(Dimethylamino)methyl]furfuryl]thio]ethyl]-2-
nitro-*N'*-piperonyl-1,1-ethenediamine = *N*-(1,3-Benzo-
dioxol-5-ylmethyl)-*N'*-[2-[[[5-[(dimethylamino)methyl]-
2-furanyl]methyl]thio]ethyl]-2-nitro-1,1-ethenediamine
(●)
R Monohydrochloride

S   AP 880, Gafir, Nipergastrina, *Niperotidine hydro-chloride**, Perultid, *Piperonylranitidine hydrochloride*, Receptine, Rotil

U   Gastric antisecretory (histamine H_2-receptor antagonist)

9784   (7930) $C_{20}H_{26}N_4O_5S$
                                                                                24477-37-0

1-Cyclohexyl-3-[[*p*-[2-(5-methyl-3-isoxazolecarboxamido)ethyl]phenyl]sulfonyl]urea = *N*-[2-[4-[[[(Cyclohexylamino)carbonyl]amino]sulfonyl]phenyl]ethyl]-5-methyl-3-isoxazolecarboxamide (●)

S   Diabenor, *Glisolamide***

U   Oral hypoglycemic

9785   (7931) $C_{20}H_{26}N_4O_7$
                                                                                129108-50-5

[1a*S*-(1α,8β,8aα,8bα)]-8-[[(Aminocarbonyl)oxy]methyl]-6-[(1,3-dioxolan-2-ylmethyl)amino]-1,1a,2,8,8a,8b-hexahydro-8a-methoxy-1,5-dimethylaziridino[2',3':3,4]pyrrolo[1,2-*a*]indole-4,7-dione (●)

S   BMY-42355

U   Antineoplastic

9786   (7936) $C_{20}H_{26}O_2$
                                                                                96301-34-7

1-Methylandrosta-1,4-diene-3,17-dione (●)

S   ATA, *Atamestane***, Biomed 777, ZK 95639

U   Aromatase inhibitor

9787 $C_{20}H_{26}O_2$
                                                                                116229-13-1

14,17α-Ethano-1,3,5(10)-estratriene-3,17β-diol = 14,21-Cyclo-19-norpregna-1,3,5(10)-triene-3,17-diol (●)

S   ZK 115194

U   Estrogenic

9788   (7934) $C_{20}H_{26}O_2$
                                                                                68-22-4

17α-Ethynyl-17-hydroxyestr-4-en-3-one = 17α-Ethynyl-19-nortestesterone = Anhydrohydroxynorprogesterone = (17α)-17-Hydroxy-19-norpregn-4-en-20-yn-3-one (●)

R   see also nos. 11334, 11381, 13613

S   *Äthinylnortestosteron*, Anovule, Anzolan, Conceplan-micro, Conludag, Dianor, *Ethinylnortestosterone*, Evorel-Pak, Fortilut, Gesta-Plan, Gestest, Hormoluton "Vielfar", Kysteron, LG 202, Limitin, Locilan, Menzol, Micronett, Micronor, Micronovum, Microtab, Mini-Pe, Mini-Pill, Neo-Hemagol, Neo-Norinyl, *Noräthisteron*, Noralutin, Norcolut, *Norethindrone*, *Norethisterone**, Noretisterone*, Norfor, Norgestin, Noriday, Noridei, Norluten, Norlutin, Norluton, *Norpregneninolone*, Norprogen, Nor-Q.D., NSC 9564, Planotab, Primolutin, Primolut N, Progela, Proluteasi, Rimare, RS 759, Santalut, SC-4640, Steron, Utovlan, Vilcolut

U Progestin

9789 (7935) $C_{20}H_{26}O_2$
68-23-5

17α-Ethynyl-17-hydroxyestr-5(10)-en-3-one = 17-Hydroxy-19-nor-17α-pregn-5(10)-en-20-yn-3-one = (17α)-17-Hydroxy-19-norpregn-5(10)-en-20-yn-3-one (●)
S 8023 C.B., Conovid, Enavid (active substance), Enidrel (active substance), Enovid (active substance), *5(10)-Noräthisteron, Noräthynodrel, 5(10)-Norethisterone, Norethynodrel, Noretinodrel, Noretynodrel**, NSC-15432, Rodilen, SC-4642
U Progestin

9790 (7932) $C_{20}H_{26}O_2$
130-73-4

4,4'-(1,2-Diethylethylene)di-*o*-cresol = 3,4-Bis(4-hydroxy-3-methylphenyl)hexane = 4,4'-(1,2-Diethyl-1,2-ethanediyl)bis[2-methylphenol] (●)
R see also no. 13226
S *Dimethylhexöstrol*, Meprane, *Methestrol**, Meth-

oestrolum, Promethoestrol
U Synthetic estrogen

9791 (7933) $C_{20}H_{26}O_2$
85-95-0

4,4'-(1,2-Diethyl-3-methyltrimethylene)diphenol = 3-Ethyl-2,4-bis-(*p*-hydroxyphenyl)hexane = 4,4'-(1,2-Diethyl-3-methyl-1,3-propanediyl)bis[phenol] (●)
S *Benzestrol**, Benzöstrol*, Chemestrogen, Ocestrol, Octestrol, *Octoestrol(um)*, Octofollin
U Synthetic estrogen

9792 (7937) $C_{20}H_{26}O_3$
58769-17-8

17-Hydroxy-19-norpregna-4,6-diene-3,20-dione (●)
S *Gestadienol**
U Progestin

9793 (7938) $C_{20}H_{26}O_4$
5957-80-2

[4a*R*-(4aα,9α,10aβ)]-1,3,4,9,10,10a-Hexahydro-5,6-dihydroxy-1,1-dimethyl-7-(1-methylethyl)-2*H*-9,4a-(epoxymethano)phenanthren-12-one (●)

S *Carnosol*
U Antibiotic from *Lepechinia hastata*

9794 (7939) $C_{20}H_{26}O_6$
25644-08-0

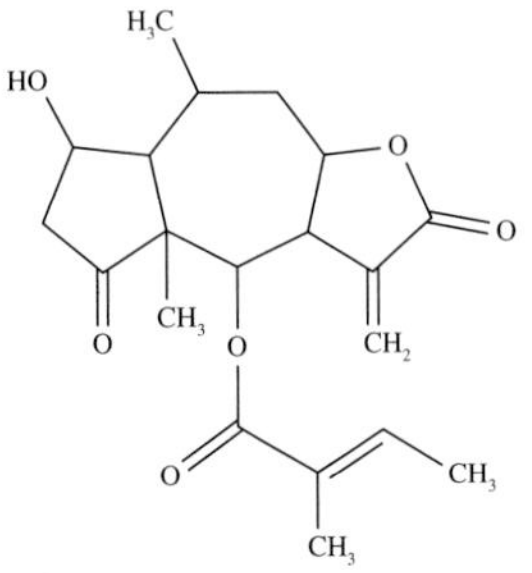

(E)-2,6α,8β-Trihydroxy-4-oxoambros-11(13)-en-12-oic
acid 12,8-lactone 6-(2-methylcrotonate) = 2-Methyl-2-
butenoic acid dodecahydro-7-hydroxy-4a,8-dimethyl-3-
methylene-2,5-dioxoazuleno[6,5-*b*]furan-4-yl ester (●)
S *Arnifolin*
U Hemostatic from *Arnica folliosa* and *Arnica montana*

9795 $C_{20}H_{26}O_6$
52617-37-5

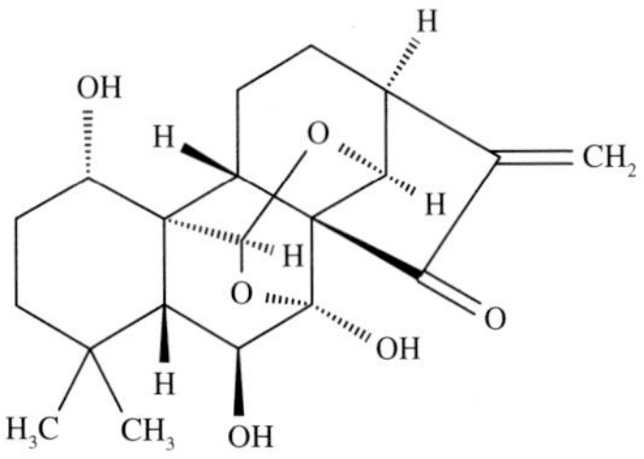

1α,6β,7α,14S,20S)-7,20:14,20-Diepoxy-1,6,7-trihydr-
oxykaur-16-en-15-one (●)
S *Ponicidin*
U Antineoplastic, antiseptic from *Rabdosia rubescens*

9796 (7940) $C_{20}H_{26}O_7$
24394-09-0

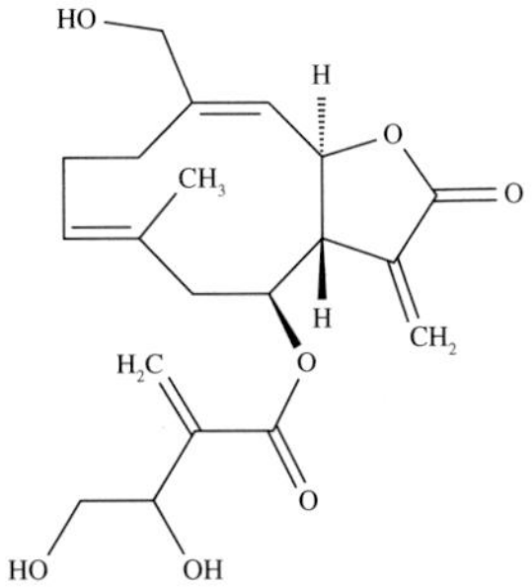

[3a*R*-(3aα,4α,6*E*,10*E*,11aβ)]-3,4-Dihydroxy-2-methy-
lenebutanoic acid 2,3,3a,4,5,8,9,11a-octahydro-10-(hydroxymethyl)-6-methyl-3-methylene-2-oxocyclodeca[*b*]fu-
ran-4-yl ester (●)
S *Centaurin, Cnicin, Cynisin*
U Bitter principle from *Cnicus benedictus*

9797 (7941) $C_{20}H_{26}O_7$
75680-27-2

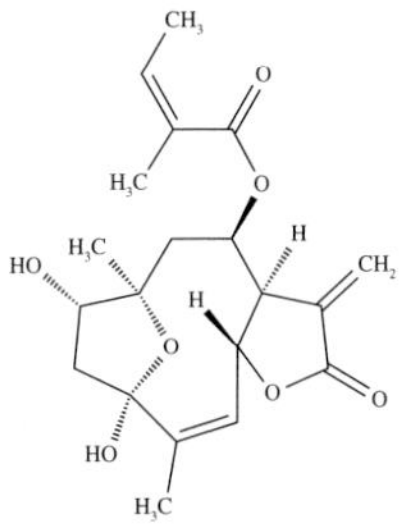

[3a*R*-[3a*R**,4*R**(*Z*),6*R**,7*S**,9*R**,10*Z*,11a*R**]]-2-Methyl-
2-butenoic acid 2,3,3a,4,5,6,7,8,9,11a-decahydro-7,9-di-
hydroxy-6,10-dimethyl-3-methylene-2-oxo-6,9-epoxy-
cyclodeca[*b*]furan-4-yl ester (●)
S *Annuithrin*
U Antibacterial and antineoplastic germacranolide from
Helianthus annuus

9798   (7942) $C_{20}H_{26}O_7$
137131-18-1

(12β)-12,13-Deepoxy-12,13-dihydroxytriptolide (●)
S   Triptriolide
U   Anti-inflammatory from *Tripterygium wilfordii*

9799   (7943) $C_{20}H_{27}ClO_2$
2446-23-3

4-Chloro-17β-hydroxy-17-methyl-1,4-androstadien-3-
one = 4-Chloro-1-dehydromethyltestosterone = (17β)-4-
Chloro-17-hydroxy-17-methylandrosta-1,4-dien-3-one
(●)
S   *Chlorodehydromethyltestosterone, Dehydrochloro-
methyltestosterone*, Oral-Turinabol
U   Anabolic

9800   (7944) $C_{20}H_{27}ClO_3$
2614-57-5

4-Chloro-11β,17β-dihydroxy-17-methyl-1,4-androsta-
dien-3-one = 4-Chloro-1-dehydro-11β-hydroxy-17-me-
thyltestesterone = (11β,17β)-4-Chloro-11,17-dihydroxy-
17-methylandrosta-1,4-dien-3-one (●)
S   *Chloroxydienone*
U   Anabolic

9801   (7945) $C_{20}H_{27}ClO_3$
1164-99-4

4-Chloro-17β-hydroxyestr-4-en-3-one acetate = 4-
Chloro-19-nortestesterone acetate = (17β)-17-(Acetyl-
oxy)-4-chloroestr-4-en-3-one (●)
S   Anabol 4-19, *Norclostebol acetate***
U   Anabolic

9802   (7946) $C_{20}H_{27}FN_2O_3$
54063-29-5

4'-Fluoro-4-(octahydro-4-hydroxy-1(2*H*)-quinolyl)buty-
rophenone carbamate (ester) = 1-[4-(*p*-Fluorophenyl)-4-
oxobutyl]perhydro-4-quinolyl carbamate = 4-[4-[(Aminocarbonyl)oxy]octahydro-1(2*H*)-quinolinyl]-1-(4-
fluorophenyl)-1-butanone (●)
R   Monohydrochloride (50656-93-4)
S   *Cicarperone hydrochloride***, L 7810
U   Antidiarrheal

9803　(7947)　　　　　　　　　　　　$C_{20}H_{27}HgN_5O_5$
117-20-4

N-[3-(1,2,3,6-Tetrahydro-1,3-dimethyl-2,6-dioxo-7-puri-nylmeruri)allyl]-α-camphoramic acid = 1,2,2-Trimethyl-3-[[3-(7-theophyllinylmercuri)-2-propenyl]carb-amoyl]cyclopentanecarboxylic acid = [3-(3-Carboxy-2,2,3-trimethylcyclopentanecarboxamido)propenylmer-curi]theophylline = [3-[[(3-Carboxy-2,2,3-trimethyl-cyclopentyl)carbonyl]amino]-1-propenyl](1,2,3,6-tetra-hydro-1,3-dimethyl-2,6-dioxo-7H-purin-7-yl)mercury (●)

R　Sodium salt (6416-03-1)
S　Tachidrolo, Tachydrol
U　Mercurial diuretic

9804　(7948)　　　　　　　　　　　　　　$C_{20}H_{27}N$
15793-40-5

N-tert-Butyl-1-methyl-3,3-diphenylpropylamine = N-(1,1-Dimethylethyl)-α-methyl-γ-phenylbenzenepro-panamine (●)

R　Hydrochloride (7082-21-5)
S　Bicor, Mictrol "Kabi Vitrum", Mictrol "Kissei; Re-cip", Micturin, Micturol, Terodiline hydrochlori-de**, Terolin
U　Coronary vasodilator

9805　(7949)　　　　　　　　　　　　　　　$C_{20}H_{27}N$
150-59-4

N-Ethyl-3,3'-diphenyldipropylamine = N,N-Bis-(3-phe-nylpropyl)ethylamine = N-Ethyl-N-(3-phenylpropyl)ben-zenepropanamine (●)
R　also citrate (1:1) (5560-59-8) or tartrate
S　Alvercol, Alverine citrate**, Antispasmin, Calma-bel, Dipropyline, Espastik-Supos., Fenpropamine, Fujipaverin, Gamatran, Gastrodog, Kalmadol, Phenopropyl, Phenpropaminum, Praeformin, Profe-nil, Profenine, Prophelan, Relaxyl "Seton", Sedo-gamma, Sestron, Spacolin, Spasmaverine, Spasmo-col, Spasmonal "Norgine", Spasmovisterina, Supa-van
U　Anticholinergic, spasmolytic

9806　(7950)　　　　　　　　　　　　　　$C_{20}H_{27}NO$
60662-77-3

2-[p-(Phenethyl)phenoxy]triethylamine = 4-(β-Diethyl-aminoethoxy)diphenylethylene = N,N-Diethyl-2-[4-(2-phenylethyl)phenoxy]ethanamine (●)
R　Hydrochloride
S　M.G. 346, Neo-Coronaril
U　Cardiotonic, spasmolytic

9807　(7951)　　　　　　　　　　　　　　$C_{20}H_{27}NO$
58313-74-9

2-[(α-Tricyclo[2.2.1.0^{2,6}]hept-3-ylidenebenzyl)oxy]tri-ethylamine = N,N-Diethyl-2-(phenyltri-cyclo[2.2.1.0^{2,6}]hept-3-ylidenemethoxy)ethanamine (●)
R Hydrochloride (58313-75-0)
S S 5521, *Treptilamine hydrochloride***
U Spasmolytic, anticholinergic

9808 (7952)

$C_{20}H_{27}NO$
4163-15-9

17-(Cyclopropylmethyl)morphinan-3-ol (●)
S *Cyclorphan*
U Analgesic, antagonist to narcotics

9809 (7954)

$C_{20}H_{27}NO_2$
3735-45-3

N,N-Dimethyl-α-(3-phenylpropyl)veratrylamine = N,N-Dimethyl-1-(3,4-dimethoxyphenyl)-4-phenylbutylamine = 1-(3,4-Dimethoxyphenyl)-1-(dimethylamino)-4-phenylbutane = α-(3,4-Dimethoxyphenyl)-N,N-dimethylben-zenebutanamine (●)
R Hydrochloride (5974-09-4)
S *Dimephebumine*, Monzal, Monzaldon, *Profenvera-mine hydrochloride*, *Revatrine hydrochloride*, Sp 281, *Vetrabutine hydrochloride***
U Uterine relaxant

9810 (7953)

$C_{20}H_{27}NO_2$

Bis(*o*-methoxy-β-methylphenethyl)amine = 2-Methoxy-N-[2-(2-methoxyphenyl)propyl]-β-methylbenzeneethan-amine (●)
R Lactate (24407-55-4)
S *Bimethoxycaine lactate*, Isocaine "Upjohn"
U Local anesthetic

9811 (7955)

$C_{20}H_{27}NO_2$
34616-39-2

α-Ethyl-*p*-[2-[(α-methylphenethyl)amino]ethoxy]benzyl alcohol = 1-[*p*-[2-(α-Methylphenethylamino)ethoxy]phe-nyl]-1-propanol = α-Ethyl-4-[2-[(1-methyl-2-phenyl-ethyl)amino]ethoxy]benzenemethanol (●)
R Hydrochloride (34535-83-6)
S Cordoxène, *Fenalcomine hydrochloride***, Medicor, Oxileina
U Coronary vasodilator

9812 (7958)

$C_{20}H_{27}NO$
3570-06-

2-Phenyl-2-norbornanecarboxylic acid β-pyrrolidinoethyl ester = 2-Phenylbicyclo[2.2.1]heptane-2-carboxylic acid 2-(1-pyrrolidinyl)ethyl ester (●)
S Bicyclophenamin
U Spasmolytic

9813 (7956) $C_{20}H_{27}NO_2$
57281-35-3

4-(Cyclohexylamino)-1-(1-naphthyloxy)-2-butanol = 4-(Cyclohexylamino)-1-(1-naphthalenyloxy)-2-butanol (●)
R also maleate (1:1) (86246-07-3)
S CH-103, Chinoin-103, TE-176
U β-Adrenergic blocker (anti-arrhythmic)

9814 (7959) $C_{20}H_{27}NO_2$
40796-85-8

1-Phenylcyclopentanecarboxylic acid 3α-tropanyl ester = endo-1-Phenylcyclopentanecarboxylic acid 8-methyl-8-azabicyclo[3.2.1]oct-3-yl ester (●)
R Hydrochloride (60662-78-4)
S Tropentan
U Anticholinergic, spasmolytic

9815 (7957) $C_{20}H_{27}NO_2$
42281-59-4

(−)-17-(Cyclopropylmethyl)morphinan-3,14-diol = 17-(Cyclopropylmethyl)morphinan-3,14-diol (●)
levo-BC-2605, *Oxilorphan***

U Antagonist to narcotics

9816 (7960) $C_{20}H_{27}NO_2S$
114538-73-7

3-(Ethylsulfonyl)-N,N,1-trimethyl-3,3-diphenylpropyl-amine = D-3-(Dimethylamino)-1,1-diphenylbutyl ethyl sulfone = γ-(Ethylsulfonyl)-N,N,α-trimethyl-γ-phenyl-benzenepropanamine (●)
R Hydrochloride (60662-79-5)
S I-C 26, *Methadone-S*, Win 1161-3
U Antitussive, narcotic

9817 (7961) $C_{20}H_{27}NO_2S$
115853-84-4

1-[3-[2-(Diethylamino)ethoxy]-4-methyl-2-thienyl]-3-phenyl-1-propanone (●)
R Hydrochloride (115853-79-7)
S LG 84-6-10
U Anti-arrhythmic

9818 (7963)

$C_{20}H_{27}NO_3$
13647-35-3

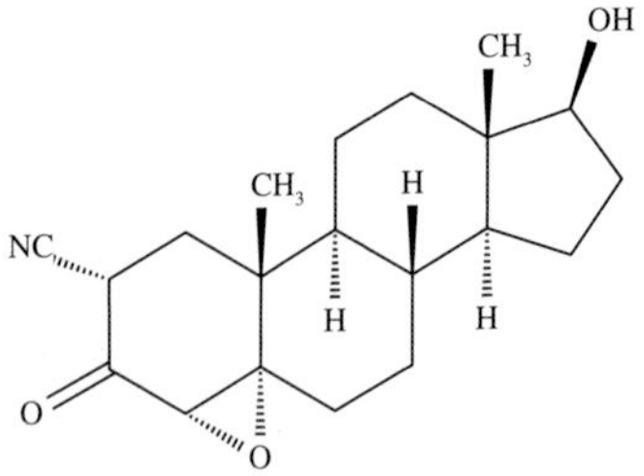

4α,5-Epoxy-17β-hydroxy-3-oxo-5α-androstane-2α-carbonitrile = (2α,4α,5α,17β)-4,5-Epoxy-17-hydroxy-3-oxoandrostane-2-carbonitrile (●)
S Desopan, Modrastane, Modrenal, MWD 1822, *Trilostane* *, Trilox, Win 24540, Winstan
U Adrenocortical suppressant (steroid biosynthesis inhibitor)

9819 (7962)

$C_{20}H_{27}NO_3$
72734-63-5

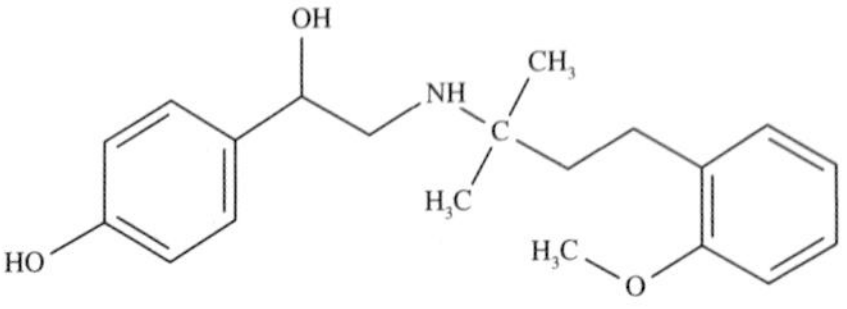

p-Hydroxy-α-[[[3-(o-methoxyphenyl)-1,1-dimethylpropyl]amino]methyl]benzyl alcohol = 1-(4-Hydroxyphenyl)-2-[[3-(2-methoxyphenyl)-1,1-dimethylpropyl]amino]ethanol = 4-Hydroxy-α-[[[3-(2-methoxyphenyl)-1,1-dimethylpropyl]amino]methyl]benzenemethanol (●)
R Hydrochloride (72734-58-8)
S D 2343-HCl
U α- and β-adrenergic blocker

9820

$C_{20}H_{27}NO_3$

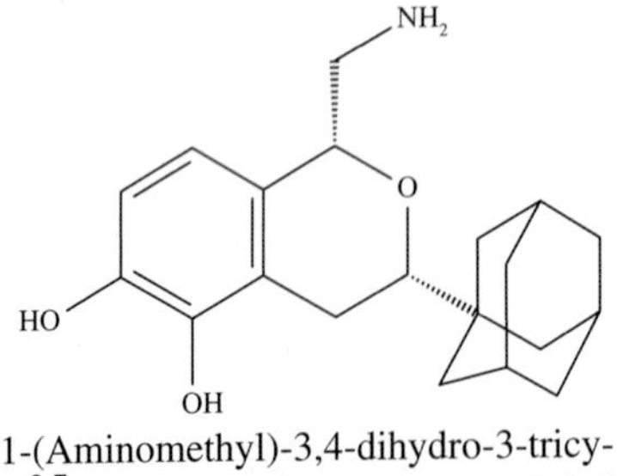

(1R-cis)-1-(Aminomethyl)-3,4-dihydro-3-tricyclo[3.3.1.1^{3,7}]dec-1-yl-1H-2-benzopyran-5,6-diol (●)
R Hydrochloride (145307-34-2)

S A-77636
U Antiparkinsonian (dopamine D_1 agonist)

9821

$C_{20}H_{27}NO_3Si_2$
125973-56-0

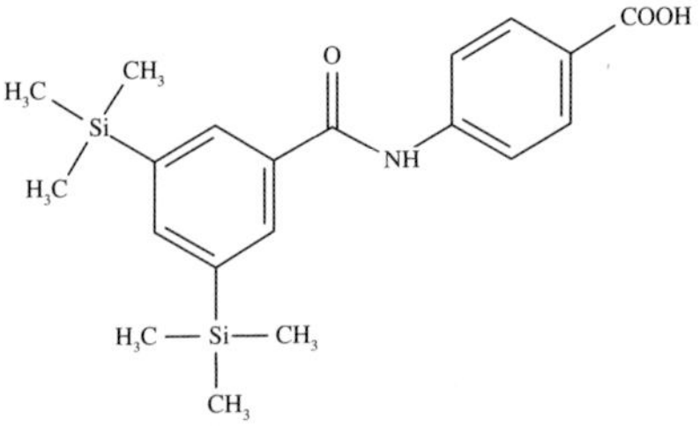

4-[[3,5-Bis(trimethylsilyl)benzoyl]amino]benzoic acid (●)
S TAC-101
U Antineoplastic

9822 (7964)

$C_{20}H_{27}NO_4$
7009-65-6

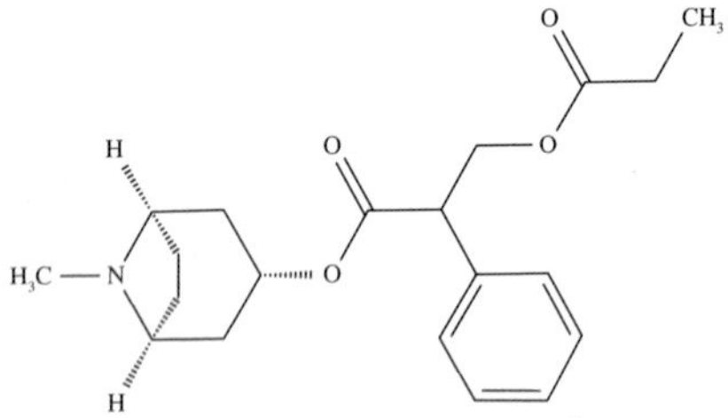

1αH,5αH-Tropan-3α-ol (±)-tropate propionate = (1R,3r,5S)-Tropan-3-yl (S)-O-propionyltropate = Atropine propionate ester = endo-α-[(1-Oxopropoxy)methyl]benzeneacetic acid 8-methyl-8-azabicyclo[3.2.1]oct-3-yl ester (●)
R see also no. 10699
S *Atropine propionate, Prampine* **
U Anticholinergic

9823 (7965)

$C_{20}H_{27}NO$
59170-23-

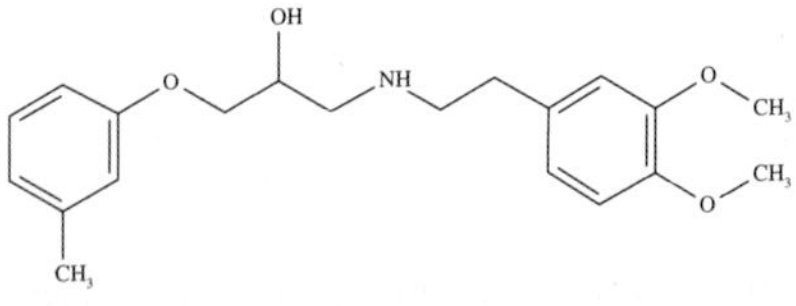

1-[(3,4-Dimethoxyphenethyl)amino]-3-(m-tolyloxy)-2-propanol = 1-[[2-(3,4-Dimethoxyphenyl)-ethyl]amino]-3-(3-methylphenoxy)-2-propanol (●)

R Hydrochloride (42864-78-8)

S *Bevantolol hydrochloride***, Calvan, CI-775,
NC 1400, Ranestol, Sentiloc, Vantol

U β-Adrenergic blocker, anti-arrhythmic

9824 (7966) $C_{20}H_{27}NO_4S$
66264-77-5

4-Hydroxy-α-[[[3-(*p*-methoxyphenyl)-1-methylpro-
pyl]amino]methyl]-3-(methylsulfinyl)benzyl alcohol = 1-
[4-Hydroxy-3-(methylsulfinyl)phenyl]-2-[[3-(4-meth-
oxyphenyl)-1-methylpropyl]amino]ethanol = 4-Hydroxy-
α-[[[3-(4-methoxyphenyl)-1-methylpropyl]amino]me-
thyl]-3-(methylsulfinyl)benzenemethanol (●)

R Hydrochloride (63251-39-8)

S Perifadil, *Sulfinalol hydrochloride**, Win 40808-7

U Antihypertensive, β-adrenergic blocker

9825 (7967) $C_{20}H_{27}NO_4S$
112966-96-8

(+)-(Z)-7-[3-*endo*-(Phenylsulfonylamino)bicy-
lo[2.2.1]hept-2-*exo*-yl]-5-heptenoic acid = (+)-(Z)-7-
[(1R,2S,3S,4S)-3-Benzenesulfonamido-2-norbornyl]-5-
heptenoic acid = (5Z)-7-[[(1R,2S,3S,4S)]-3-[(Phenylsulfonyl)amino]bicyclo[2.2.1]hept-2-yl]-5-heptenoic acid
(●)

Domitroban**, (+)-S-145

U Anti-allergic (TXA$_2$-receptor antagonist)

9825-01   132747-47-8

R Calcium salt (2:1) of the [1R-(1α,2α(Z),3β,4α)]-
compd.
Anboxan, *Domitroban calcium***, S-1452

U Anti-allergic (thromboxane A$_2$-receptor-antagonist)

9826 (7968) $C_{20}H_{27}NO_5$
804-10-4

3-(β-Diethylaminoethyl)-4-methyl-7-(carbethoxymeth-
oxy)coumarin = Ethyl [[3-[2-(diethylamino)ethyl]-4-me-
thyl-2-oxo-2H-1-benzopyran-7-yl]oxy]acetate = [[3-[2-
(Diethylamino)ethyl]-4-methyl-2-oxo-2H-1-benzopyran-
7-yl]oxy]acetic acid ethyl ester (●)

R also hydrochloride (655-35-6)

S A-27053, AG-3, Anangor, Antiangor, *Carbochro-
men*, *Carbocromen***, Carbocromin, Cardiocap,
Cassella 4489, *Chromonar*, Cromene, Intenkordin,
Intensacrom, Intensain, NSC-110430

U Coronary vasodilator

9827 (7969) $C_{20}H_{27}NO_5$
5486-03-3

Ethyl 4-hydroxy-6,7-diisobutoxy-3-quinolinecarboxylate
= 4-Hydroxy-6,7-diisobutoxyquinoline-3-carboxylic acid
ethyl ester = 4-Hydroxy-6,7-bis(2-methylpropoxy)-3-qui-
nolinecarboxylic acid ethyl ester (●)

S Antagonal, Bonaid, *Buchinolatum*, *Buquinolate***,
Butoril, EU-1093, U-1093, VÚFB-4824

U Coccidiostatic (for poultry)

9828 (7970) $C_{20}H_{27}NO_{11}$
29883-15-6

Mandelonitrile-β-gentiobioside = D-Mandelonitrile-β-D-glucosido-6-β-D-glucoside = (R)-α-[(6-O-β-D-Glucopy-ranosyl-D-β-glucopyranosyl)oxy]benzeneacetonitrile (●)
S *Amygdalin, Amygdaloside*, NSC-15780, *Vitamin B$_{17}$*
U Antineoplastic (from seeds of *Rosaceae*)

9829 (7971) $C_{20}H_{27}N_2O$
14007-49-9

(3-Carbamoyl-3,3-diphenylpropyl)ethyldimethylammonium = N-Ethyl-N,N-dimethyl-3-carbamoyl-3,3-diphenylpropylammonium = γ-(Aminocarbonyl)-N-ethyl-N,N-dimethyl-γ-phenylbenzenepropanaminium (●)
R Bromide (115-51-5)
S *Ambutonium bromide*, BL 700 B, R 100
U Anticholinergic, spasmolytic

9830 (7972) $C_{20}H_{27}N_2O_2S$

1-Methyl-1-[2-(N-methyl-α-2-thienylmandelamido)ethyl]pyrrolidinium = 1-[2-[(Hydroxyphenyl-2-thienylacetyl)methylamino]ethyl]-1-methylpyrrolidinium (●)
R Bromide (26058-50-4)

S *Dotefonium bromide***, H 4132
U Anticholinergic, spasmolytic

9831 $C_{20}H_{27}N_3O$
170856-57-2

N-(3-Ethoxy-2-pyridinyl)-N-methyl-1-(phenylmethyl)-4-piperidinamine = 3-Ethoxy-N-methyl-N-[1-(phenylmethyl)-4-piperidinyl]-2-pyridinamine (●)
R Maleate (1:1) (188989-96-0)
S U-99363E
U Antipsychotic (dopamine D$_4$ antagonist)

9832 (7973) $C_{20}H_{27}N_3O$
115749-98-9

5-[[2-(Diethylamino)ethyl]amino]-5,11-dihydro-7-methoxy[1]benzoxepino[3,4-p]pyridine = N'-(5,11-Dihydro-7-methoxy[1]benzoxepino[3,4-b]pyridin-5-yl)-N,N-diethyl-1,2-ethanediamine (●)
R Fumarate (2:3) (115750-37-3)
S KW 3407
U Anti-arrhythmic

9833 (7974) $C_{20}H_{27}N_3O$
58013-09-

1-[2-Ethoxy-2-(3-pyridinyl)ethyl]-4-(2-methoxyphenyl)piperazine (●)
S IP-66

U Antihypertensive (α-blocker)

9834 (7975) $C_{20}H_{27}N_3O_3$
                                                     136013-69-9

endo-2-(Cyclopropylmethoxy)-N-[[(8-methyl-8-azabicy-
clo[3.2.1]oct-3-yl)amino]carbonyl]benzamide (●)
R Maleate (1:1) (136013-70-2)
S WAY-100289
U Anxiolytic (5-HT$_3$ antagonist)

9835 (7976) $C_{20}H_{27}N_3O_3$
                                                     6874-80-2

13-Hydroxylupanine 2-pyrrolecarboxylic acid ester =
[2S-(2α,7β,7aβ,14β,14aα)]-1H-Pyrrole-2-carboxylic
acid dodecahydro-11-oxo-7,14-methano-2H,6H-dipyri-
do[1,2-a:1',2'-e][1,5]diazocin-2-yl ester (●)
S Calpurnine, Hoechst 933, Oroboidine
U Antihypertensive, anti-arrhythmic (alkaloid from Ca-
dia ellisiana)

9836 (7977) $C_{20}H_{27}N_3O_5$
                                                     90139-06-3

[1S,9S)-9-[[(S)-1-Carboxy-3-phenylpropyl]amino]octa-
hydro-10-oxo-6H-pyridazino[1,2-a][1,2]diazepine-1-
carboxylic acid = N-[(1S,9S)-1-Carboxy-10-oxoperhydro-
pyridazino[1,2-a][1,2]diazepin-9-yl]-4-phenyl-L-homo-
alanine = [1S-[1α,9α(R*)]]-9-[(1-Carboxy-3-phenylpro-
pyl)amino]octahydro-10-oxo-6H-pyridazino[1,2-a][1,2]
diazepine-1-carboxylic acid (●)
R see also no. 11480
S Cilazaprilat**, Ro 31-3113
U Antihypertensive (ACE inhibitor)

9837 (7978) $C_{20}H_{27}N_3O_5S$
                                                     109683-79-6

(S)-2-tert-Butyl-4-[(S)-N-[(S)-1-carboxy-3-phenylpro-
pyl]alanyl]-Δ^2-1,3,4-thiadiazoline-5-carboxylic acid =
(2S)-5-tert-Butyl-3-[N-[(S)-1-carboxy-3-phenylpropyl]-
L-alanyl]-2,3-dihydro-1,3,4-thiadiazole-2-carboxylic acid
= [2S-[2R*,3[R*(R*)]]]-3-[2-[(1-Carboxy-3-phenylpro-
pyl)amino]-1-oxopropyl]-5-(1,1-dimethylethyl)-2,3-di-
hydro-1,3,4-thiadiazole-2-carboxylic acid (●)
R see also no. 11480
S Utibaprilat**
U Antihypertensive (ACE inhibitor)

9838 (7979) $C_{20}H_{27}N_3O_6$
                                                     89371-37-9

(4S)-3-[(2)-N-[(1S)-1-Carboxy-3-phenylpropyl]alanyl]-
1-methyl-2-oxo-4-imidazolidinecarboxylic acid 3-ethyl
ester = (S)-3-[N-[(S)-1-(Ethoxycarbonyl)-3-phenylpro-
pyl]-L-alanyl]-1-methyl-2-oxoimidazolidine-4-carb-
oxylic acid = [4S-[3[R*(R*)],4R*]]-3-[2-[[1-(Ethoxycarbonyl)-3-phenylpropyl]amino]-1-oxopropyl]-1-methyl-
2-oxo-4-imidazolidinecarboxylic acid (●)
R Monohydrochloride (89396-94-1)

S   ACE/TA-6366, *Imidapril hydrochloride***, Nova-rok, SH 6366, TA-6366, Tanatril
U   Antihypertensive (ACE inhibitor)

9839   (7980) $C_{20}H_{27}N_3O_6$
13246-02-1

1-(3-Butoxy-2-hydroxypropyl)-5-ethyl-5-phenylbarbituric acid carbamate (ester) = 1-[3-Butoxy-2-(carbamoyloxy)propyl]-5-ethyl-5-phenylbarbituric acid = 1-[2-[(Aminocarbonyl)oxy]-3-butoxypropyl]-5-ethyl-5-phenyl-2,4,6(1*H*,3*H*,5*H*)-pyrimidinetrione (●)
R   see also no. 3255-15
S   BCPT, *Febarbamate***, Getril, Go-560, G-Tril, MS-543, *Phenobamate*, Solium "Lirca", Tymium
U   Sedative, tranquilizer, relaxant

9840   (7981) $C_{20}H_{27}N_3O_6$
23113-01-1

1-β-D-Arabinofuranosylcytosine 5'-(1-adamantanecarboxylate) = Tricyclo[3.3.1.1^{3,7}]decane-1-carboxylic acid 5'-ester with 4-amino-1-β-D-arabinofuranosyl-2(1*H*)-pyrimidinone (●)
S   *Adamantoylcytarabine*, Adam CA, adoAra C, AdO-CA, NSC-117614, U-26516
U   Immunosuppressive

9841   (7982) $C_{20}H_{27}N_5O_2$
73963-72-1

6-[4-(1-Cyclohexyl-5-tetrazolyl)butoxy]-3,4-dihydrocarbostyril = 6-[4-(1-Cyclohexyl-1*H*-tetrazol-5-yl)butoxy]-3,4-dihydro-2(1*H*)-quinolinone (●)
S   *Cilostazol**, OPC 21, OPC-13013, Pletaal, Pletal
U   Antithrombotic, cerebral vasodilator

9842   (7983) $C_{20}H_{27}N_5O_3$
2016-63-9

8-Benzyl-7-[2-[ethyl(2-hydroxyethyl)amino]ethyl]theophylline = 7-[2-[Ethyl(2-hydroxyethyl)amino]ethyl]-3,7-dihydro-1,3-dimethyl-8-(phenylmethyl)-1*H*-purine-2,6-dione (●)
R   Monohydrochloride (20684-06-4)
S   AC 3810, Bamifix, *Bamifylline hydrochloride***, Ba mi-med, *Bamiphylline hydrochloride*, BAX 2739 Z, *Benzétamophylline*, Briofil, 8102 C.B., Eupnal, Pulmac, Stofilin, Trentadil
U   Bronchodilator, coronary vasodilator

9843   (7984) $C_{20}H_{27}N_5O_4S$
67514-88-

1-Cyclohexyl-3-[[*p*-[2-(3-methyl-1-pyrazolecarboxamido)ethyl]phenyl]sulfonyl]urea = *N*-[2-[4-[[[(Cyclohexylamino)carbonyl]amino]sulfonyl]phenyl]ethyl]-3-methyl 1*H*-pyrazole-1-carboxamide (●)

S SPC-5002
U Oral hypoglycemic

9844 (7985)

$C_{20}H_{27}N_5O_5S$
25046-79-1

1-(Hexahydro-1H-azepin-1-yl)-3-[[p-[2-(5-methyl-3-isoxazolecarboxamido)ethyl]phenyl]sulfonyl]urea = 4-[4-[β-(5-Methylisoxazole-3-carboxamido)ethyl]phenylsulfonyl]-1,1-hexamethylenesemicarbazide = 5-Methyl-N-[4-(perhydroazepin-1-ylureidosulfonyl)phenethyl]-3-isoxazolecarboxamide = N-[2-[4-[[[[(Hexahydro-1H-azepin-1-yl)amino]carbonyl]amino]sulfonyl]phenyl]ethyl]-5-methyl-3-isoxazolecarboxamide (●)
S Bay b 4231, BS 4231, FB b 4231, Glisepin, *Glisoxepide***, Glucoben, Glysepin, Pro-Diaban, 22410 R.P.
U Oral hypoglycemic

9845

$C_{20}H_{27}N_5O_6$
170902-52-0

3-[[[(5R)-3-[4-(Aminoiminomethyl)phenyl]-4,5-dihydro-5-isoxazolyl]acetyl]amino]-N-(butoxycarbonyl)-L-alanine (●)
S XV 459
U Fibrinogen receptor antagonist

9846 (7986)

$C_{20}H_{27}N_5O_6$
114606-56-3

(R)-8-[[1-(3,4-Dimethoxyphenyl)-2-hydroxyethyl]amino]-3,7-dihydro-7-(2-methoxyethyl)-1,3-dimethyl-1H-purine-2,6-dione (●)
S MKS 492, SDZ-MKS 492
U Bronchospasmolytic

9847

$C_{20}H_{27}N_{11}O_4$
147056-66-4

N-[2-(Dimethylamino)ethyl]-1-methyl-4-[1-methyl-4-(4-formamido-1-methylimidazole-2-carboxamido)imidazole-2-carboxamido]imidazole-2-carboxamide = N-[2-[[[2-(Dimethylamino)ethyl]amino]carbonyl]-1-methyl-1H-imidazol-4-yl]-4-[[[4-(formylamino)-1-methyl-1H-imidazol-2-yl]carbonyl]amino]-1-methyl-1H-imidazole-2-carboxamide (●)
S AR-1-144
U DNA minor groove binder

9848 (7987)

$C_{20}H_{27}O_4P$
1241-94-7

2-Ethylhexyl diphenyl phosphate = Phosphoric acid 2-ethylhexyl diphenyl ester (●)
S *Octicizer*, Santicizer 141
U Plasticizer (in wound dressing spray)

9849   (7988) $C_{20}H_{28}Cl_4N_2O_4$
5560-78-1

N,N'-(*p*-Phenylenedimethylene)bis[2,2-dichloro-*N*-(2-ethoxyethyl)acetamide] = *N,N'*-[1,4-Phenylenebis(methylene)]bis[2,2-dichloro-*N*-(2-ethoxyethyl)acetamide] (●)
S   Falmonox, NSC-107433, *Teclozan***, *Teclozine*,
Win 13146
U   Anti-amebic

9850   (7989) $C_{20}H_{28}F_3N_3O_2$
121264-01-5

1-(*tert*-Butylamino)-3-[[α-methyl-α-(1-methyl-2-imidazolyl)-*m*-(trifluoromethyl)benzyl]oxy]-2-propanol = 1-[(1,1-Dimethylethyl)amino]-3-[1-(1-methyl-1*H*-imidazol-2-yl)-1-[3-(trifluoromethyl)phenyl]ethoxy]-2-propanol (●)
S   E-3753
U   Anti-arrhythmic, local anesthetic

9851   (7991) $C_{20}H_{28}I_3N_3O_9$
89797-00-2

N,N'-Bis(2,3-dihydroxypropyl)-5-[*N*-(2-hydroxy-3-methoxypropyl)acetamido]-2,4,6-triiodoisophthalamide = 5-[Acetyl(2-hydroxy-3-methoxypropyl)amino]-*N,N'*-bis(2,3-dihydroxypropyl)-2,4,6-triiodo-1,3-benzenedicarboxamide (●)
S   Cpd. 5411, Imagopaque, *Iopentol***, Ivepaque
U   Diagnostic aid (radiopaque medium)

9852   (7990) $C_{20}H_{28}I_3N_3O_9$
136949-58-1

N,N'-Bis(2,3-dihydroxypropyl)-5-[2-(hydroxymethyl)hydracrylamido]-2,4,6-triiodo-*N,N'*-dimethylisophthalamide = *N,N'*-Bis(2,3-dihydroxypropyl)-5-[[3-hydroxy-2-(hydroxymethyl)-1-oxopropyl]amino]-2,4,6-triiodo-*N,N'*-dimethyl-1,3-benzenedicarboxamide (●)
S   *Iobitridol***, Xenetix
U   Diagnostic aid (radiopaque medium)

9853   (7992) $C_{20}H_{28}I_3N_3O_{13}$
63941-74-2

N,N'-[2,4,6-Triiodo-5-(methylcarbamoyl)-*m*-phenylene]bis[D-gluconamide] = *N,N'*-[2,4,6-Triiodo-5-[(methylamino)carbonyl]-1,3-phenylene]bis[D-gluconamide] (●)
S   *Ioglucomide***, MP-8000
U   Diagnostic aid (radiopaque medium)

9854 (7993)

$C_{20}H_{28}N$
27892-33-7

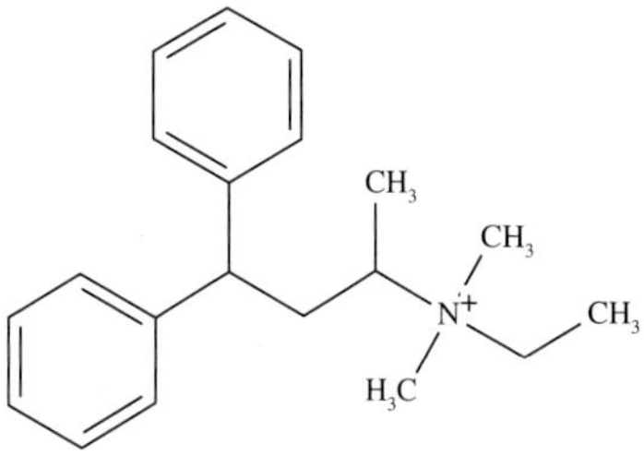

Ethyldimethyl(1-methyl-3,3-diphenylpropyl)ammonium = N-Ethyl-N,N,1-trimethyl-3,3-diphenylpropylammoni-um = Dimethylethyl(β-benzhydrylisopropyl)ammonium = N-Ethyl-N,N,α-trimethyl-γ-phenylbenzenepropanami-nium (●)
R Bromide (3614-30-0)
S Cetiprin, Cis relax, Detrulisin, *Emepronium bromi-de**, Hexanium "Fides", Inkopax, Restenacht, Ripi-rin, Uricrat, Urikrat, Urinox "Orion", Uro-Ripirin (old form)
U Anticholinergic, spasmolytic

9854-01 (7993-01)
R Carrageenan salt
S Cetiprin novum, *Emepronium carrageenate*, Urikrat novum, Uro-Ripirin (novum)
U Anticholinergic

9855

$C_{20}H_{28}NO$
52793-85-8

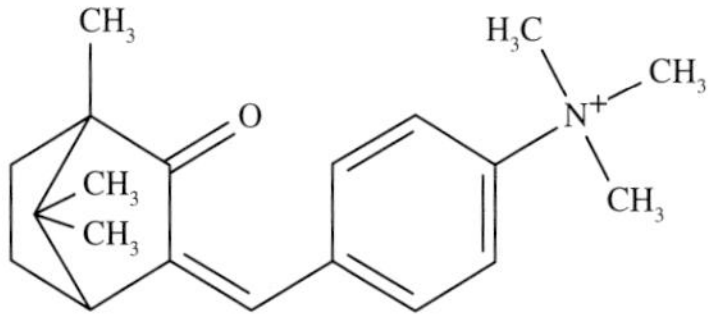

N,N,N-Trimethyl-4-[(4,7,7-trimethyl-3-oxobi-cyclo[2.2.1]hept-2-ylidene)methyl]benzenaminium (●)
R Methylsulfate (52793-97-2)
Mexoryl SK
U Ultraviolet screen

9856 (7994)

$C_{20}H_{28}NO$
19213-48-0

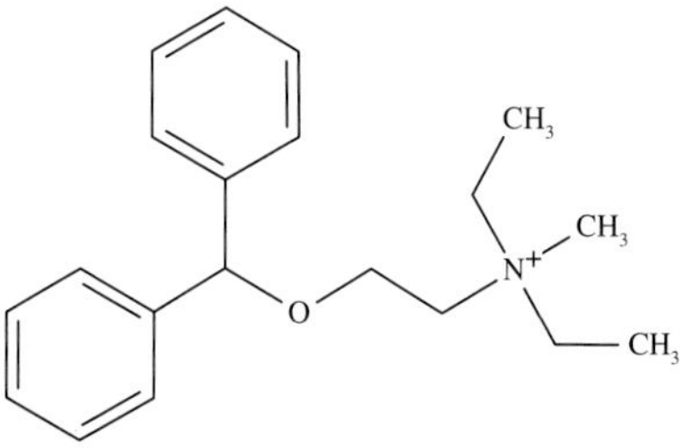

[2-(Diphenylmethoxy)ethyl]diethylmethylammonium = N,N-Diethyl-N-methyl-2-(diphenylmethoxy)ethylammo-nium = 2-(Diphenylmethoxy)-N,N-diethyl-N-methyl-ethanaminium (●)
R Bromide or iodide (5982-52-5)
S *Emetonium iodide*, Esyntin, Metropin
U Anticholinergic, spasmolytic

9857 (7995)

$C_{20}H_{28}NO_3$

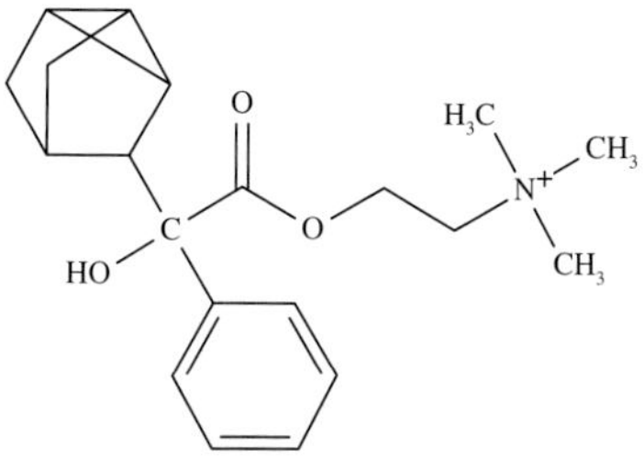

[2-[2-Hydroxy-2-phenyl-2-(tricyclo[2.2.1.0^{2,6}]hept-3-yl)acetoxy]ethyl]trimethylammonium = Tricyclo [2.2.1.0^{2,6}]hept-3-ylphenylhydroxyacetic acid β-(trime-thylammonio)ethyl ester = 2-[[Hydroxyphenyl(tricy-clo[2.2.1.0^{2,6}]hept-3-yl)acetyl]oxy]-N,N,N-trimethyl-ethanaminium (●)
R Bromide with 10 % of an isomeric compd. (dehydro-norbornyl group instead of tricycloheptyl group) (8058-76-2)
S *Endobenzyline bromide*, PC-1238, Ulcyn
U Anticholinergic

9858 (7996) $C_{20}H_{28}NO_4$

3α-Hydroxy-8-methyl-1αH,5αH-tropanium (±)-ethyl phenylmalonate = (±)-endo-3-(3-Ethoxy-1,3-dioxo-2-phenylpropoxy)-8,8-dimethyl-8-azoniabicyclo[3.2.1]octane (●)
R Methyl sulfate (salt) (113932-41-5)
S CDDD 3602, HGP-6, *Tematropium metilsulfate***
U Anticholinergic

9859 (7997) $C_{20}H_{28}N_2O$
15301-88-9

2-[α-[2-(Dimethylamino)ethoxy]-2,6-diethylbenzyl]pyridine = 2-[(2,6-Diethylphenyl)-2-pyridinylmethoxy]-N,N-dimethylethanamine (●)
R Monohydrochloride (641-35-0)
S B.S. 7161-D, *Pytamine hydrochloride**
U Diuretic

9860 (7998) $C_{20}H_{28}N_2O$
121502-05-4

(E)-2,6-Di-*tert*-butyl-4-[2-(5-methyl-3-pyrazolyl)vinyl]phenol = (E)-2,6-Bis(1,1-dimethylethyl)-4-[2-(5-methyl-1H-pyrazol-3-yl)ethenyl]phenol (●)
S PD 127443
U Anti-inflammatory

9861 $C_{20}H_{28}N_2O_2$
139095-05-9

(S)-2-[(Dimethylamino)methyl]-1-[(5,6,7,8-tetrahydro-5-oxo-2-naphthalenyl)acetyl]piperidine = (S)-N,N-Dimethyl-1-[(5,6,7,8-tetrahydro-5-oxo-2-naphthalenyl)acetyl]-2-piperidinemethanamine (●)
R Monohydrochloride (139094-88-5)
S BRL 53001
U κ-Opioid receptor agonist

9862 (7999) $C_{20}H_{28}N_2O$
125-53-

1,4,5,6-Tetrahydro-1-methyl-2-pyrimidinemethanol α-phenylcyclohexaneglycolate (ester) = 1,4,5,6-Tetrahydro-1-methyl-2-pyrimidinylmethyl α-cyclohexylmandelate = α-Cyclohexylphenylglycolic acid (1,4,5,6-tetrahydro-1-methyl-2-pyrimidinyl)methyl ester = α-Cyclohexyl-α-hydroxybenzeneacetic acid (1,4,5,6-tetrahydro-1-methyl-2-pyrimidinyl)methyl ester (●)
R Monohydrochloride (125-52-0)
S Acolergic, Antulcus, Caridan, Cycmin, Daricol, Dar-icon, Dominil, Enterex, Gastrix, Gastrovitória, Ino-

maru-S, Lictine, Madil, Manir, Murofren, Naridan,
Norman, Oximin, *Oxiphencyclimini chloridum, Oxy-
phencyclimine hydrochloride**,* Protropin "Ben-
zon-Greece", Quaternil, Ritarucon, Rotropin, Roxi-
fen "Sigurtà", S 1-1236, Sedomucol, Setrol, Spaza-
min, Syklifen, Ulcociclinina, Ulcomin, UM, Vagoga-
strin, Vio-Thene, W-T Anticholinergic, Zamanil
U   Anticholinergic, spasmolytic

9863 (8000) $C_{20}H_{28}N_2O_3$
                                103583-09-1

α-[(*tert*-Butylamino)methyl]-4-hydroxy-3-[(4-methoxy-
benzyl)amino]benzyl alcohol = α-[[(1,1-Dimethyl-
ethyl)amino]methyl]-4-hydroxy-3-[[(4-methoxyphe-
nyl)methyl]amino]benzenemethanol (●)
R   Hydrochloride (60853-38-5)
S   QH 25
U   Bronchodilator

9864 $C_{20}H_{28}N_2O_4$
                                186452-09-5

1-(3,3-Dimethyl-1,2-dioxopentyl)-L-proline 3-(3-pyridi-
nyl)propyl ester (●)
S   GPI 1046
U   Neuroprotectant (rotamase enzyme activity inhibitor)

9865 (8001) $C_{20}H_{28}N_2O_4$
                                18174-89-5

1,1'-(Ethylenediimino)bis(3-phenoxy-2-propanol) =
N,N'-Bis(2-hydroxy-3-phenoxypropyl)ethylenediamine =
1,1'-(1,2-Ethanediyldiimino)bis[3-phenoxy-2-propanol]
(●)

R   Dihydrochloride (35607-13-7)
S   IF ROM-203, ROM-203
U   β-Adrenergic blocker

9866 (8002) $C_{20}H_{28}N_2O_5$
                                132875-61-7

4-Carboxy-4-(*N*-phenylpropionamido)-1-piperidinepro-
pionic acid dimethyl ester = 4-(Methoxycarbonyl)-4-[(1-
oxopropyl)phenylamino]-1-piperidinepropanoic acid methyl ester (●)
R   Monohydrochloride (132539-07-2)
S   GI 87084 B, *Remifentanil hydrochloride**,* Ultiva
U   Anesthetic, analgesic (ultra-short acting)

9867 (8003) $C_{20}H_{28}N_2O_5$
                                75847-73-3

1-[*N*-[(*S*)-1-Carboxy-3-phenylpropyl]-L-alanyl]-L-pro-
line 1'-ethyl ester = (*S*)-1-[*N*-[1-(Ethoxycarbonyl)-3-phe-
nylpropyl]-L-alanyl]-L-proline (●)
R   Maleate (1:1) (76095-16-4)
S   Acehexal, Acepril "Lachema; Spirig", Acetensil,
Ace-tensina, Agioten, Alacor, Alapren, Alphrin, Am-
prace, Analept, Angiotec, Antiprex, Apo-Enalap,
Apo-Enalapril, Arnaril, Atens, Bajaten "Sanitas", Ba-
ripril, Benalapril, Berlipril, Besanil, Biocronil, Biten-
sil, Bonuten, BQL, Calpiren, Cardace "Zafe", Car-
dio-Pres, Cardiopril "Sanitas", Cardiovet, Cetampril,
Chipil, Clipto, Controlvas, Converten, Convertin,
Corodil "Gea", Corprilor, Corvo, Cosil, Crinoren
"Uriach", Dabonal, Defluin, Denapril, Diasistol, Di-

nid, Ditensor, Drugapril, Dynef, Ecadil, Ecaprilat, Ecaprinil, Ednyt, Enabeta, Enacard, En.Ace, Enadura, Ena-Hennig, Enahexal, Enal, Enaladil, Enalagamma, Enalap, *Enalapril maleate***, Enalasyn, Enaldun, Enaloc, Enalten, Enam, Enap, Enapren, Enapress, Enapril, Enaprin, Ena-Puren, Enarenal, Enaril, Enasifar, Enatec, Enical, Enpril, Envas, Epril, Erxetilan, Esalfon "Recalcine", Eupressin, Feliberal, Gadopril, Glioten, Gnostocardin, Grifopril, Herten, Hipoartel, Hipten, Hytrol, Iecatec, Innovace, Inoprilat, Insup, Invoril, Ircon, Kalipren, Kaparlon S, Kenopril, Kinfil, Kontic, Konveril, L-154739-01 D, Lapril, Leovinezal, Linatil, Lotrial, Mapryl, Megapres(s), Mepril, Minipril, Mirapril, MK-421, Myopril, Nacor, Naprilene, Neotensin, Noprilen, Norpril, Nuril "U.S. Vitamin", Octorax, Ofnifenil, Olivin, Palane, Pres, Pressitan "Bristol-Myers; Iquisona", Pressotec, Presuren "Kressfor", Prilace, Prilenap, Pritec, Protal, Pulsol, Rablas, Racen, Reca, Regomed, Reminal, Renacardon, Renidur, Renital, Renitec, Renitek, Reniten, Renivace, Reomin, Ristalen, Stadelant, Sulocten, Supotron, Tenace, Tensazol, Teracarden, Tesoren, Ulticadex, Unipril "Farma Colomb", Vapresan, Vasolapril, Vasopren, Vasopril "Biolab", Vasotec, Virfen, Vitobel, Xanef, Xanef Cor

U Antihypertensive (ACE inhibitor)

9868 (8004) $C_{20}H_{28}N_2O_5S$
106133-20-4

(*R*)-5-[2-[[2-(2-Ethoxyphenoxy)ethyl]amino]propyl]-2-methoxybenzenesulfonamide (●)
R Monohydrochloride (106463-17-6)
S Alna, Alphlo, *Amsulosin hydrochloride*, Expros, Flomax "Boehringer-Ingelheim; Glaxo-Wellcome; Yamanouchi", Harnal, Josir, LY 253351, Omic, Omix, Omnic, Pradif, Secotex, *Tamsulosin hydrochloride***, Urolisin, Urolosin, YM-617, YM-12617-1
U Antihypertensive, bradycardic (α_1-adrenoceptor antagonist)

9869 (8005) $C_{20}H_{28}N_2O_6$
23887-41-4

Ethyl 4-(3,4,5-trimethoxycinnamoyl)-1-piperazineacetate = 2-[4-(3,4,5-Trimethoxycinnamoyl)-1-piperazinyl]acetic acid ethyl ester = 4-[1-Oxo-3-(3,4,5-trimethoxyphenyl)-2-propenyl]-1-piperazineacetic acid ethyl ester (●)
R also maleate (1:1) (50679-07-7)
S Acoridil, *Äthylcinepazat, Cinepazet*, Ethyl cinepazate*, 6753 MD, Vascoril
U Coronary vasodilator

9870 $C_{20}H_{28}N_2O_7S$
157542-49-9

(Pivaloyloxy)methyl (1*R*,5*S*,6*S*)-6-[(*R*)-1-hydroxyethyl]-1-methyl-2-[[(*R*)-5-oxopyrrolidin-3-yl]thio]-1-carbapen-2-em-3-carboxylate = (4*R*,5*S*,6*S*)-6-[(1*R*)-1-Hydroxyethyl]-4-methyl-7-oxo-3-[[(3*R*)-5-oxo-3-pyrrolidinyl]thio]-1-azabicyclo[3.2.0]hept-2-ene-2-carboxylic acid (2,2-dimethyl-1-oxopropoxy)methyl ester (●)
S CS 834
U Antibiotic

9871 $C_{20}H_{28}N_4O$
122009-54-5

4-(Pentyloxy)-2-[4-(2-propenyl)-1-piperazinyl]quinazoline (●)

R (2E)-2-Butenedioate (1:1) (131916-69-3)
S KB 5666
U Anti-oxidant (anti-ischemic, antihemorrhagic)

9872 (8006) $C_{20}H_{28}N_4O$
37686-84-3

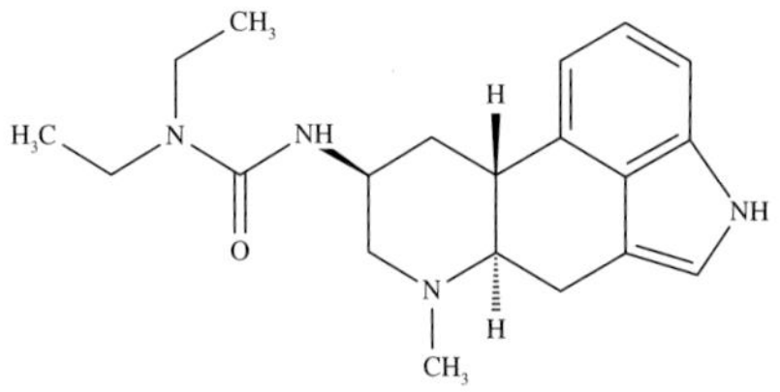

1,1-Diethyl-3-(6-methylergolin-8α-yl)urea = N,N-Di-ethyl-N'-[(8α)-6-methylergolin-8-yl]urea (●)
S SH 406, TDHL, *Terguride***, Terulon, *Transdihydrolisuride*, ZK 31224
U Antipsychotic

9872-01 (8006-01) 37686-85-4
R Maleate (1:1)
S Dironyl, Mysalfon, Teluron, VÚFB-6638
U Prolactin inhibitor (antiparkinsonian, immunomodulator)

9873 (8007) $C_{20}H_{28}N_4O_2$
7543-69-3

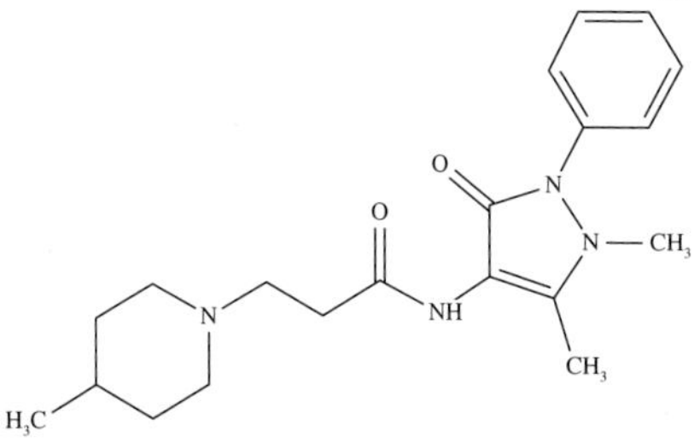

2,3-Dimethyl-4-[3-(4-methylpiperidino)propionylamino]-1-phenyl-5-pyrazolone = N-(2,3-Dihydro-1,5-dimethyl-3-oxo-2-phenyl-1H-pyrazol-4-yl)-4-methyl-1-piperidinepropanamide (●)
R Salicylate (1:1) (18429-62-4)
S HM 187, *Promepiazon*
U Anti-inflammatory, antirheumatic

9874 (8008) $C_{20}H_{28}N_4O_2$
136199-02-5

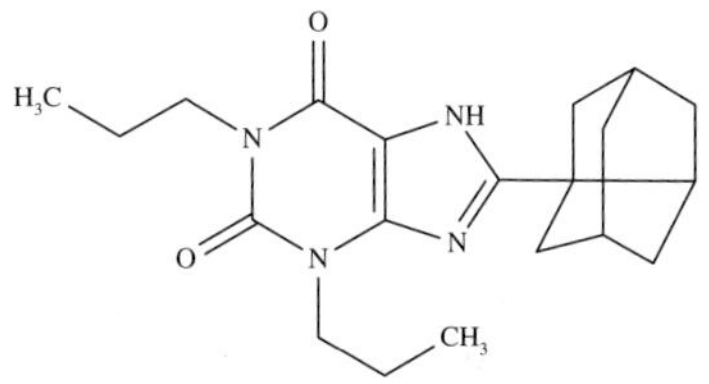

8-(Noradamant-3a-yl)-1,3-dipropylxanthine = 8-(Hexahydro-2,5-methanopentalen-3a(1H)-yl)-3,7-dihydro-1,3-dipropyl-1H-purine-2,6-dione (●)
S KW 3902
U Diuretic, renal protective (adenosine A$_1$ antagonist)

9875 $C_{20}H_{28}N_4O_4$
142880-36-2

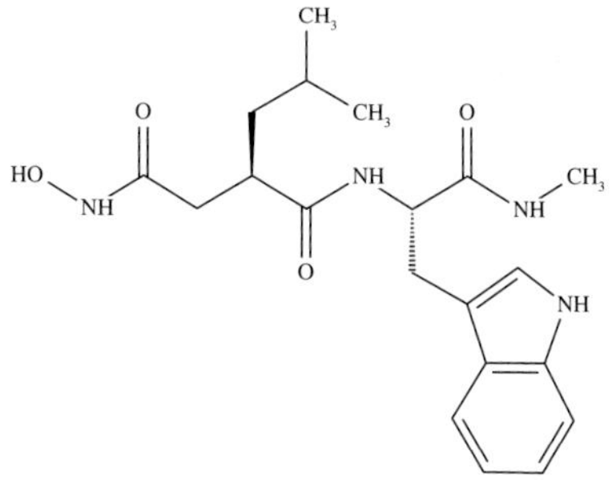

3-(N-Hydroxycarbamoyl)-2(R)-isobutylpropionyl-L-tryptophanmethylamide = (R)-N^1-Hydroxy-N-[(S)-2-indol-3-yl]-1-(methylcarbamoyl)ethyl-2-isobutylsuccinamide = (2R)-N^4-Hydroxy-N^1-[(1S)-1-(1H-indol-3-ylmethyl)-2-(methylamino)-2-oxoethyl]-2-(2-methylpropyl)butanediamide (●)
S CS-610, Galardin, GM 6001, *Ilomastat***
U Anti-inflammatory, protease inhibitor

9876 $C_{20}H_{28}N_4O_6$
172927-65-0

Ethyl (Z)-[[1-[N-[(p-Hydroxyamidino)benzoyl]-L-ala-nyl]-4-piperidyl]oxy]acetate = [[1-[(2S)-2-[[4-[(Z)-Ami-no(hydroxyimino)methyl]benzoyl]amino]-1-oxopropyl]-4-piperidinyl]oxy]acetic acid ethyl ester (●)
S   G-7333, Ro 48-3657, *Sibrafiban***, Xubix
U   Fibrinogen receptor antagonist

9877 $C_{20}H_{28}N_4O_7$
147865-49-4

N-[N-(4-Amidinophenoxy)butanoyl]-L-α-aspartyl-L-va-line = N-[4-[4-(Aminoiminomethyl)phenoxy]-1-oxobu-tyl]-L-α-aspartyl-L-valine (●)
S   FK 633, FR 144633
U   Fibrinogen inhibitor (GP IIB/IIIA antagonist)

9878   (8009) $C_{20}H_{28}N_6O_2$
28947-50-4

8-[[2-[Methyl(α-methylphenethyl)amino]ethyl]ami-no]caffeine = N-Methyl-N-(1-methyl-2-phenylethyl)-N'-(1,3,7-trimethyl-2,6-dioxo-8-purinyl)ethylenediamine =

3,7-Dihydro-1,3,7-trimethyl-8-[[2-[methyl(1-methyl-2-phenylethyl)amino]ethyl]amino]-1H-purine-2,6-dione (●)
R   Hydrochloride (33246-03-6)
S   Altimina, *Fencamina hydrochloride, Phencamine hy-drochloride*, Sicoclor, ST-374
U   Psychotonic, analeptic

9879 $C_{20}H_{28}N_6O_3$
199986-75-9

2-[Bis(2-hydroxyethyl)amino]-6-[(4-methoxyben-zyl)amino]-9-isopropylpurine = 2,2'-[[6-[[(4-Methoxy-phenyl)methyl]amino]-9-(1-methylethyl)-9H-purin-2-yl]imino]bis[ethanol] (●)
S   CVT-313
U   Cyclin-dependant kinase 2 inhibitor

9880   (8010) $C_{20}H_{28}O$
63014-96-0

1α-Methylandrost-4,16-dien-3-one = (1α)-1-Methyland-rost-4,16-dien-3-one (●)
S   *Delanterone***
U   Anti-androgen (anti-acne)

9881 $C_{20}H_{28}O$
116-31-

(*all-E*)-3,7-Dimethyl-9-(2,6,6-trimethyl-1-cyclohexen-1-yl)-2,4,6,8-nonatetraenal (●)
S *Axerophthal, Retinaldehyde, Retinals, Retinene, Vitamin A aldehyde*, Ystheal
U Dermatic

9882 (8012) $C_{20}H_{28}O$
                                                            52-76-6

19-Nor-17α-pregn-4-en-20-yn-17-ol = 17α-Ethynylestr-4-en-17-ol = (17α)-19-Norpregn-4-en-20-yn-17-ol (●)
S *Äthinylöstrenol*, Endometril, *Ethinylestrenol*, Exlutena, Exlution, Exluton(a), *Linestrenol(o), Lynestrenol**, Lynoestrenol*, Lynomin, Minette, NSC-37725, Orgametil, Orgametril, Orgametrol
U Progestin

9883 (8013) $C_{20}H_{28}O$
                                                            16915-71-2

19-Nor-17α-pregn-5-en-20-yn-17-ol = 17α-Ethynylestr-5-en-17-ol = (17α)-19-Norpregn-5-en-20-yn-17-ol (●)
S *Cingestol***
U Progestin

9884 (8014) $C_{20}H_{28}O$
                                                            896-71-9

19-Nor-17α-pregn-5(10)-en-20-yn-17-ol = 17α-Ethynylestr-5(10)-en-17-ol = (17α)-19-Norpregn-5(10)-en-20-yn-17-ol (●)
S *Tigestol***
U Progestin

9885 (8011) $C_{20}H_{28}O$
                                                            79-80-1

3,7-Dimethyl-9-(2,6,6-trimethyl-1,3-cyclohexadienyl)-2,4,6,8-nonatetraen-1-ol = 3,4-Didehydroretinol (●)
S *Dehydroretinol, Vitamin A_2*
U Antixerophthalmic vitamin

9886 (8022) $C_{20}H_{28}O_2$
                                                            72-63-9

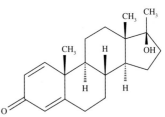

17β-Hydroxy-17-methylandrosta-1,4-dien-3-one = 1-Dehydro-17α-methyltestosterone = (17β)-17-Hydroxy-17-methylandrosta-1,4-dien-3-one (●)
S Abirol, Anabol "Piam", Anabolex "CIPLA", Anabolicum-Medivet, Anabolin "Gea; Medica", Anaboral, Andoredan, Bionabol, Ciba 17309-Ba, Crein, Danabol, *Dehydromethyltestosterone*, Dialone, Dianabol, Distranorm, Encephan, Geabol, Lanabolin, Metaboli-

na, Metanabol, *Metandienone**, *Metandrostenolone*,
Metastenol, *Methandienone*, *Methandrostenolone*,
Methastenon, Methbolin, *Methylandrostadienolone*,
Nabolin, Naposim, Neo-Analolene, Nerobol(ettes),
Novabol, NSC-42722, Perabol, Perbolin, Protobolin,
Sirabolin, Sirobolin, Stenolon, TMV-17, Tonobo-
lin-Tabl., Vanabol
U Anabolic, androgen

9887 (8015) $C_{20}H_{28}O_2$
                                                          112018-00-5

3',5'-Di-*tert*-butyl-4'-hydroxy-5-hexynophenone = 1-[3,5-
Bis(1,1-dimethylethyl)-4-hydroxyphenyl]-5-hexyn-1-one
(●)
S NE 11740, *Tebufelone***
U Analgesic, anti-inflammatory, antipyretic

9888 (8016) $C_{20}H_{28}O_2$
                                                          302-79-4

all-trans-3,7-Dimethyl-9-(2,6,6-trimethyl-1-cyclohexen-
1-yl)-2,4,6,8-nonatetraenoic acid = *all-trans*-Retinoic
acid = 15-Apo-β-caroten-15-oic acid = (*all-E*)-3,7-Di-
methyl-9-(2,6,6-trimethyl-1-cyclohexen-1-yl)-2,4,6,8-
nonatetraenoic acid (●)
R see also nos. 13210, 15630
S A-Acido, Abarel, Aberel(a), Acid-A-Vit, Acnavit,
      Acnelyse, Acnoten, Acretin, Acta "Jean-Marie", Ac-
      ticin, Acudyne, Adina, Airoderm, Airol, Akluvin,
      Aknebon, Aknemin "E. Merck", Aknoten, Aldo-
      quin-Anti-acne, Alfamatic, Alöten, Alquin-Gel, Al-
      ten, Altinac, AR-623, Arretin, ATRA, Atragen, Atre-
      derm, Aviderm, Avita "Bertec; Penederm", Avitcid,
      Avitoin, Betarretin, Cordes-VAS, Cravobene,
      Derm-A, Dermairol, Dermodan, Dermojuventus, De-
      rugin, Dumin, Effederm, Epi-Aberel, Esiderm, Eudy-

na, Hidrosam T, Kefran, Kerlocal, Kétrel, Kligacid,
Locacid "Fabre", Masc Retynowa, Niterey,
NSC-122758, Pekol, Rejuven-A, Relastef, Renova,
Retacnyl, Reticrem, Retiderm(a), Retigel, Retin A,
Retinei, Retino, *Retinoesäure*, *Retinoic acid*, Retino-
va, Retirides, Retisol, Retisol-A, Retitop, ReTrieve,
Ro 1-5488, Solbrin, Stieva-A (StieVAA), Tina-A,
Tretinax, Tretin M, *Tretinoin***, VAS, Vesanoid,
Vit-A-Acid, Vitacid, Vitacid A, *Vitamin A acid*,
Vitamin A-säure, Vitanol-A, Vitinoin
U Acne-therapeutic, keratolytic, antileukemic

9889 (8017) $C_{20}H_{28}O_2$
                                                          4759-48-2

(13*Z*)-15-Apo-β-caroten-15-oic acid = (2*Z*,4*E*,6*E*,8*E*)-
3,7-Dimethyl-9-(2,6,6-trimethyl-1-cyclohexen-1-yl)-
2,4,6,8-nonatetraenoic acid = 13-*cis*-Retinoic acid (●)
R see also no. 14127
S Accure, Accutane, *cis-Retinoic acid*, Dermoretin,
      Isotretinoin**, Isotrex, Oratane, Ro 4-3780, Roac-
      cutan(e), Roacnetan, Roacutan, Teriosal
U Acne-therapeutic, keratolytic

9890 (8018) $C_{20}H_{28}O_2$
                                                          33122-60-0

11α-Hydroxy-17,17-dimethyl-18-norandrosta-4,13-dien-
3-one = (11α)-11-Hydroxy-17,17-dimethyl-18-noran-
drosta-4,13-dien-3-one (●)
S *Nordinone***
U Anti-androgen

9891   (8020) $C_{20}H_{28}O_2$
6795-60-4

17-Hydroxy-17α-vinylestr-4-en-3-one = 17-Hydroxy-19-nor-17α-pregna-4,20-dien-3-one = 17α-Vinyl-19-nortestosterone = (17α)-17-Hydroxy-19-norpregna-4,20-dien-3-one (●)
S   Neoprogestin, Nor-progestelea, *Norvinisterone***, SC-4641, *Vinylnortestosterone*
U   Progestin

9892   (8019) $C_{20}H_{28}O_2$
13563-60-5

17-Hydroxy-17α-vinylestr-5(10)-en-3-one = 17α-Vinylestr-5(10)-en-17-ol-3-one = 17-Hydroxy-19-nor-17α-pregna-5(10),20-dien-3-one = (17α)-17-Hydroxy-19-nor-pregna-5(10),20-dien-3-one (●)
S   *Norgesterone***, *Norvinodrel*, *Vinylestrenolone*
U   Progestin

9893   (8021) $C_{20}H_{28}O_2$
1231-93-2

19-Nor-17α-pregn-4-en-20-yne-3β,17-diol = 17α-Ethynylestr-4-ene-3β,17-diol = (3β,17α)-19-Norpregn-4-en-20-yne-3,17-diol (●)
R   see also no. 12517
S   *Ethynodiol, Etynodiol**
U   Progestin

9894 $C_{20}H_{28}O_2$
5300-03-8

(2E,4E,6Z,8E)-3,7-Dimethyl-9-(2,6,6-trimethyl-1-cyclohexen-1-yl)-2,4,6,8-nonatetraenoic acid = 9-*cis*-Retinoic acid (●)
S   AGN 192013, *Alitretinoin***, ALRT 1057, LG 100057, LGD 1057, NSC-659772, Panretin, Panretyn, Panrexin
U   Antineoplastic, keratolytic

9895   (8023) $C_{20}H_{28}O_3$
2320-86-7

4,17β-Dihydroxy-17-methylandrosta-1,4-dien-3-one = (17β)-4,17-Dihydroxy-17-methylandrosta-1,4-dien-3-one (●)
S   *Enestebol***
U   Anabolic

9896   (8024) $C_{20}H_{28}O_3$
26577-85-5

3α-Hydroxyeremophila-6,9-dien-8-one, (Z)-2-methylcro-
tonate = [1R-[1α,2β(Z),7β,8aα]]-2-Methyl-2-butenoic
acid 1,2,3,4,6,7,8,8a-octahydro-1,8a-dimethyl-7-(1-me-
thylethenyl)-6-oxo-2-naphthalenyl ester (●)
S   Aranidolor, *Petasin*
U   Analgesic, spasmolytic, anti-inflammatory from *Peta-
sites officinalis*

9897   (8025) $C_{20}H_{28}O_3$
5108-94-1

3-Methoxy-16-methylestra-1,3,5(10)-triene-16β,17β-diol
= (16β,17β)-3-Methoxy-16-methylestra-1,3,5(10)-triene-
16,17-diol (●)
S   Anvene, Manvene, *Mytatrienediol*, SC-6924
U   Antihyperlipidemic, estrogen

9898   (8026) $C_{20}H_{28}O_3$
2137-18-0

17-Hydroxy-19-norpregn-4-ene-3,20-dione (●)
R   see also no. 13303
S   *Gestronol*
U   Progestin

9899 $C_{20}H_{28}O_3$
81499-22-1

(1S,5E,7E,11Z)-1,5-Dimethyl-8-(1-methylethyl)-15-oxa-
bicyclo[9.3.2]hexadeca-5,7,11-triene-2,16-dione (●)
S   *Sarcophytolide*
U   Neuroprotective from the soft coral *Sarcophyton
glaucum*

9900   (8027) $C_{20}H_{28}O_3$
67494-15-9

(4aR-*trans*)-1,3,4,9,10,10a-Hexahydro-6-hydroxy-1,1-di-
methyl-7-(1-methylethyl)-4a(2H)-phenanthrenecarb-
oxylic acid (●)
S   *Pisiferic acid*
U   Antimicrobial, antineoplastic from *Chamaecyparis
pisifera*

9901 (8028) $C_{20}H_{28}O_3S$
5560-69-0

Ethyl 3,6-di-*tert*-butyl-1-naphthalenesulfonate = 3,6-Di-*tert*-butyl-1-naphthalenesulfonic acid ethyl ester = 3,6-Bis(1,1-dimethylethyl)-1-naphthalenesulfonic acid ethyl ester (●)
S *Aethyli dibunas*, *Ethyl dibunate***, NDR 304, Neodyne, Tussets
U Antitussive

9902 (8029) $C_{20}H_{28}O_5$
71302-27-7

(3α,4β,5α)-3-Hydroxy-4,8-bis(hydroxymethyl)-4-methyl-18-nor-16-oxaandrosta-13(17),14-dien-2-one (●)
S *Spongiatriol*
U Antihypertensive from marine sponges

9903 (8030) $C_{20}H_{28}O_5S$
33159-27-2

13-Isopropyl-12-sulfopodocarpa-8,11,13-trien-15-oic acid = 12-Sulfodehydroabietic acid = [1R-(1α,4aβ,10aα)]-1,2,3,4,4a,9,10,10a-Octahydro-1,4a-dimethyl-7-(1-methylethyl)-6-sulfo-1-phenanthrenecarboxylic acid (●)
R Monosodium salt (86408-72-2)
S *Ecabet sodium***, *Ecarxate sodium*, Gastrom, TA-2711
U Anti-ulcer agent

9904 $C_{20}H_{28}O_6$
79722-03-5

3-Methyl-2-butenoic acid (3aS,6R,10S,11R,11aS)-dodecahydro-10-hydroxy-6,10-dimethyl-3-methylene-2,5-dioxocyclodeca[b]furan-11-yl ester (●)
S *Nepalolide A*
U Nitric oxide synthase inhibitor

9905 (8031) $C_{20}H_{28}O_6$
17673-25-5

[1aR-(1aα,1bβ,4aβ,7aα,7bα,8α,9β,9aα)]-1,1a,1b,4,4a,7a,7b,8,9,9a-Decahydro-4a,7b,9,9a-tetrahydroxy-3-(hydroxymethyl)-1,1,6,8-tetramethyl-5H-cyclopropa[3,4]benz[1,2-e]azulen-5-one (●)
S *Phorbol*
U Cathartic (from seeds of *Croton tiglium*)

9906 $C_{20}H_{28}O_6$
                                        28957-04-2

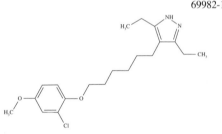

(1α,6β,7α,14R)-7,20-Epoxy-1,6,7,14-tetrahydroxykaur-
16-en-15-one (●)
S *Isodonol, Oridonin*
U Antineoplastic from *Rabdosia* sp.

9907 $C_{20}H_{28}O_6$
                                        100440-25-3

(2S)-*rel*-(–)-2-[(1R)-1-Hydroxy-2-[(1S,2S,3R,4aS,8aS)-
1,2,3,4,4a,7,8,8a-octahydro-3-hydroxy-1,2,4a,5-tetrame-
thyl-4-oxo-1-naphthalenyl]ethyl]-α-oxooxiraneacetalde-
hyde (●)
S *Terpentecin*
U Antineoplastic, antibiotic from *Streptomyces* sp.

9908 (8032) $C_{20}H_{28}O_6S$
                                        71116-82-0

(±)-(Z)-7-[(1R*,2R*,3R*,5S*)-3,5-Dihydroxy-2-[(E)-
(3R*S*)-3-hydroxy-4-(3-thienyloxy)-1-butenyl]cyclo-
pentyl]-5-heptenoic acid = 7-[3,5-Dihydroxy-2-[3-hydr-

oxy-4-(3-thienyloxy)-1-butenyl]cyclopentyl]-5-hepte-
noic acid (●)
R also salt with trometamol (no. 314) (1:1) (71116-83-1)
S HR 837, Iliren, *Tiaprost***
U Luteolytic (veterinary)

9909 $C_{20}H_{28}O_7$
                                        143200-52-6

Hydroxyterpentecin = 2-[1-Hydroxy-2-[1,2,3,4,4a,7,8,8a-
octahydro-3-hydroxy-5-(hydroxymethyl)-1,2,4a-trime-
thyl-4-oxo-1-naphthalenyl]ethyl]-α-oxooxiraneacetalde-
hyde (●)
S UCT 4B
U Antileukemic, antibacterial from *Streptomyces* sp.

9910 (8033) $C_{20}H_{29}ClN_2O_2$
                                        69982-17-8

4-[6-(2-Chloro-4-methoxyphenoxy)hexyl]-3,5-diethyl-
1H-pyrazole (●)
R Monomethanesulfonate (69982-18-9)
S Win 41258-3
U Antiviral

9911   (8034) $C_{20}H_{29}ClO_3$
10392-52-6

4-Chloro-11β,17β-dihydroxy-17-methylandrost-4-en-3-
one = 4-Chloro-11β-hydroxy-17-methyltestosterone =
(11β,17β)-4-Chloro-11,17-dihydroxy-17-methylandrost-
4-en-3-one (●)
S   *Chloroxymesterone*
U   Anabolic

9912   (8035) $C_{20}H_{29}ClO_4$
56219-57-9

4[6-(2-Chloro-4-methoxyphenoxy)hexyl]-3,5-heptane-
dione (●)
S   *Arildone***, Win 38020
U   Antiviral

9913 $C_{20}H_{29}FN_2O_3S$

+)-(8aR,12aS,13aS)-12-[(3-Fluoropropyl)sulfonyl]-
5,8,8a,9,10,11,12,12a,13,13a-decahydro-3-methoxy-6H-
soquino[2,1-g][1,6]naphthyridine
S   RS-15385-FP
U   Antihypertensive, antidepressant (α_2-adrenoceptor
        antagonist)

9914   (8036) $C_{20}H_{29}FN_6O_3$
101345-71-5

(±)-*cis-N*-[1-[2-(4-Ethyl-5-oxo-2-tetrazolin-1-yl)ethyl]-
3-methyl-4-piperidyl]-2'-fluoro-2-methoxyacetanilide =
cis-N-[1-[2-(4-Ethyl-4,5-dihydro-5-oxo-1H-tetrazol-1-
yl)ethyl]-3-methyl-4-piperidinyl]-*N*-(2-fluorophenyl)-2-
methoxyacetamide (●)
R   Monohydrochloride (117268-95-8)
S   A-3331, *Brifentanil hydrochloride***
U   Narcotic analgesic

9915   (8037) $C_{20}H_{29}FO_3$
76-43-7

9-Fluoro-11β,17β-dihydroxy-17-methylandrost-4-en-3-
one = 9-Fluoro-11β-hydroxy-17-methyltestosterone =
(11β,17β)-9-Fluoro-11,17-dihydroxy-17-methylandrost-
4-en-3-one (●)
S   Afluteston, Androfluorone, Android-F, Androstero-
lo, Ferona "Sidus", *Fluossimesterone*, Fluotestin,
*Fluoximesteronum, Fluoxymesterone***, Flusteron,
Flutestos, Halotestin, Hysterone, Neo-Ormonal,
NSC-12165, Oralsterone, Oratestin, Ora-Testryl,
Stenox, Testoral "Midy", U-6040, Ultandren
U   Androgen

9916 $C_{20}H_{29}FO_7P_2$
167269-90-1

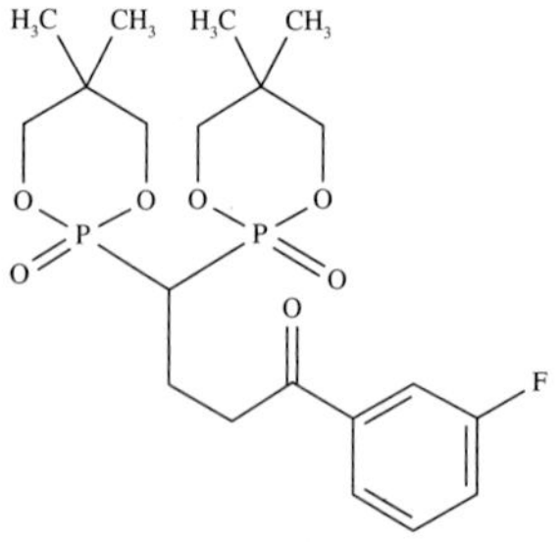

4,4-Bis(5,5-dimethyl-2-oxido-1,3,2-dioxaphosphorinan-
2-yl)-1-(3-fluorophenyl)-1-butanone (●)
S   PNU-91638
U   Anti-arthritic

9917   (8038) $C_{20}H_{29}NO$
54063-47-7

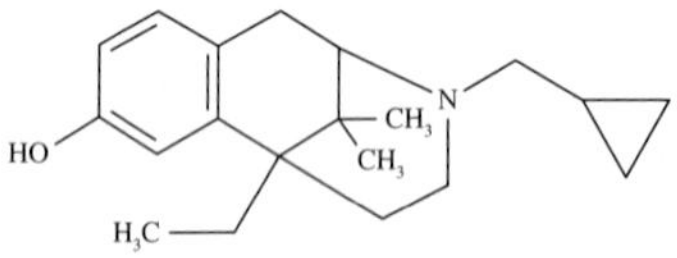

3-(Cyclopropylmethyl)-6-ethyl-1,2,3,4,5,6-hexahydro-
11,11-dimethyl-2,6-methano-3-benzazocin-8-ol (●)
S   Cyclogemine, *Gemazocine***, R 15497
U   Antagonist to narcotics

9918   (8039) $C_{20}H_{29}NO$
57653-28-8

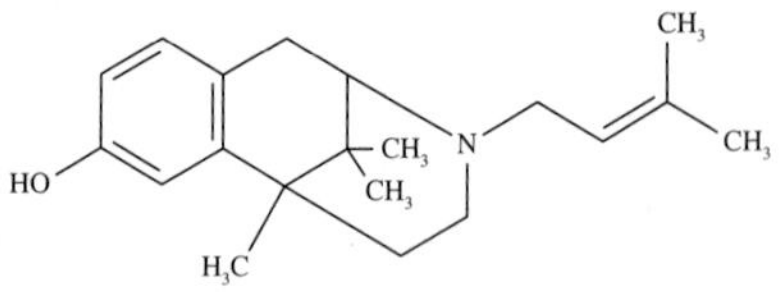

1,2,3,4,5,6-Hexahydro-6,11,11-trimethyl-3-(3-methyl-2-
butenyl)-2,6-methano-3-benzazocin-8-ol (●)
S   *Ibazocine**
U   Analgesic

9919 $C_{20}H_{29}NO$
143899-92-7

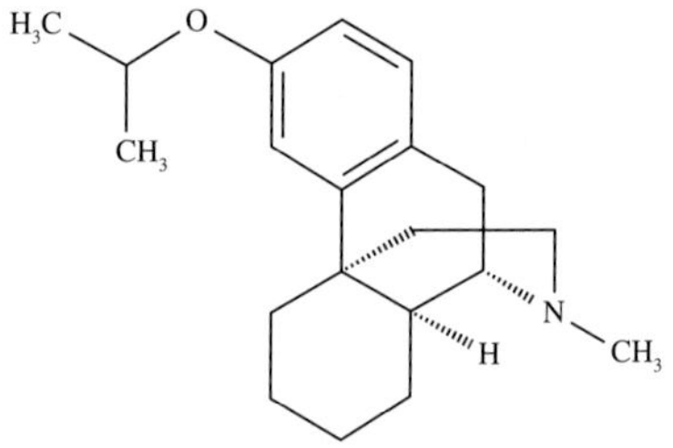

O-(1-Methylethyl)dextrorphan = (9α,13α,14α)-17-
Methyl-3-(1-methylethoxy)morphinan (●)
S   AHN 1-037
U   Anticonvulsant

9920 $C_{20}H_{29}NO$
162809-72-5

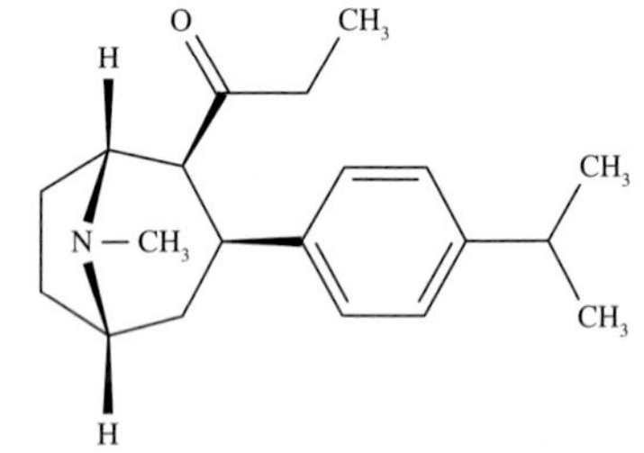

2β-Propanoyl-3β-(4-isopropylphenyl)tropane = 1-
[(1*R*,2*S*,3*S*,5*S*)-8-Methyl-3-[4-(1-methylethyl)phenyl]-8-
azabicyclo[3.2.1]oct-2-yl]-1-propanone (●)
S   WF-31
U   Antidepressant (selective serotonin uptake inhibitor)

9921 $C_{20}H_{29}NO_2$
87827-02-9

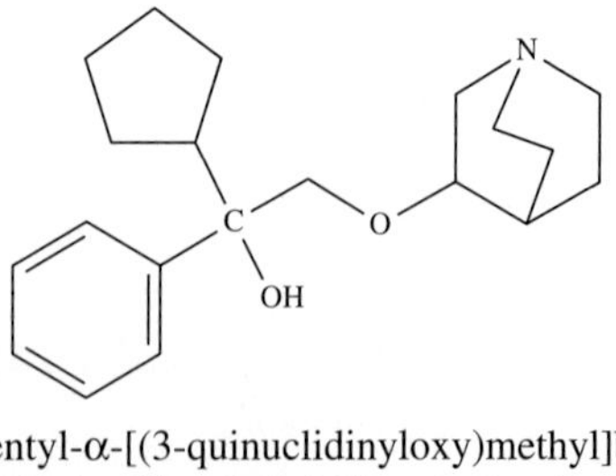

α-Cyclopentyl-α-[(3-quinuclidinyloxy)methyl]benzyl alcohol = α-[(1-Azabicyclo[2.2.2]oct-3-yloxy)methyl]-α-
cyclopentylbenzenemethanol (●)
R   Hydrochloride (151937-76-7)
S   *Penequine hydrochloride*

U   Anticholinergic

9922   (8040) $C_{20}H_{29}NO_2$
71990-00-6

6-Ethyl-1,2,3,4,5,6-hexahydro-3-[(1-hydroxycyclopro-pyl)methyl]-11,11-dimethyl-2,6-methano-3-benzazocin-8-ol (●)
S   *Bremazocine***, Ph 3753
U   Analgesic

9923   (8042) $C_{20}H_{29}NO_3$
4354-45-4

1-Methyl-3-piperidyl α-phenylcyclohexaneglycolate = α-Phenylcyclohexylglycolic acid 1-methyl-3-piperidyl ester = α-Cyclohexyl-α-hydroxybenzeneacetic acid 1-methyl-3-piperidinyl ester (●)
R   also hydrochloride (1420-03-7)
S   Delinal, JB-840, NDR 263, *Oxiclipinum, Oxyclipi-ne***, *Propenzolate*
U   Anticholinergic

9924   (8041) $C_{20}H_{29}NO_3$
94868-25-4

(1-Methyl-2-pyrrolidinyl)methyl α-phenylcyclohexylgly-colate = α-Phenylcyclohexylglycolic acid (1-methyl-2-pyrrolidinyl)methyl ester = α-Cyclohexyl-α-hydroxyben-zeneacetic acid (1-methyl-2-pyrrolidinyl)methyl ester (●)
R   also hydrochloride (5585-94-4)
S   Promandeline-263
U   Anticholinergic

9925   (8043) $C_{20}H_{29}NO_3$
7199-05-5

(1-Ethyl-2-pyrrolidinyl)methyl α-phenylcyclopentylgly-colate = α-Cyclopentyl-α-hydroxybenzeneacetic acid (1-ethyl-2-pyrrolidinyl)methyl ester (●)
R   Mixture of the hydrochloride with 30 % 1-ethyl-3-piperidyl α-phenylcyclopentylglycolate hydrochlori-de (8015-54-1)
S   Ditran, JB-329
U   Antidepressant

9926   (8044) $C_{20}H_{29}NO_3$
                                                        22150-28-3

9-Isopropylgranatoline (±)-tropate (ester) = *endo*-(±)-α-
(Hydroxymethyl)benzeneacetic acid 9-(1-methylethyl)-9-
azabicyclo[3.3.1]non-3-yl ester (●)
S   *Ipragratine***, Kor 12-CL
U   Anticholinergic, spasmolytic

9927   (8045) $C_{20}H_{29}NO_3$

7-[3-(3-Hydroxy-1-octenyl)-4-pyridinyl]-5-heptenoic
acid (●)
S   MPPA, OKY-1555
U   Tromboxane (TXB$_2$) synthetase inhibitor

9928   (8046) $C_{20}H_{29}NO_3$
                                                        127245-22-1

4-(3,5-Di-*tert*-butyl-4-hydroxybenzylidene)-5,6-dihydro-
2-methyl-2H-1,2-oxazin-3(4H)-one = 4-[[3,5-Bis(1,1-di-
methylethyl)-4-hydroxyphenyl]methylene]dihydro-2-
methyl-2H-1,2-oxazin-3(4H)-one (●)
S   BF-389, Biofor 389

U   Anti-inflammatory

9929   (8047) $C_{20}H_{29}NO_3$
                                                        107746-52-1

3-(3,5-Di-*tert*-butyl-4-hydroxybenzylidene)-1-methoxy-
2-pyrrolidinone = 3-[[3,5-Bis(1,1-dimethylethyl)-4-hydr-
oxyphenyl]methylene]-1-methoxy-2-pyrrolidinone (●)
S   E-5110
U   Anti-inflammatory

9930   (8048) $C_{20}H_{29}NO_4$
                                                        23271-74-1

1-Methyl-3-morpholinopropyl tetrahydro-4-phenyl-2H-
pyran-4-carboxylate = 4-Phenylperhydropyran-4-carb-
oxylic acid 1-methyl-3-morpholinopropyl ester = Tetra-
hydro-4-phenyl-2H-pyran-4-carboxylic acid 1-methyl-3-
(4-morpholinyl)propyl ester (●)
R   Maleate
S   Alevotos, Corbar S, Dykatuss S, *Fedrilate male-
       ate***, Gotas Binelli, *Phenhydropyxylate*, Sedatoss,
       Tussapax, Tussefan(e), U.C.B. 3928
U   Antitussive

9931   (8049) $C_{20}H_{29}NO_4S$
                                                        25442-88-

N-(β-Hydroxy-α-methylphenethyl)-N-methyl-2-oxo-
bornane-10-sulfonamide = Camphor-10-sulfonic acid N-

(2-hydroxy-2-phenyl-1-methylethyl)-*N*-methylamide =
N-(2-Hydroxy-1-methyl-2-phenylethyl)-*N*,7,7-trimethyl-
2-oxobicyclo[2.2.1]heptane-1-methanesulfonamide (●)
S *Camphamédrine*, Camphotone, Cardenyl
U Analeptic

9932 (8050) $C_{20}H_{29}NO_4S_2$
136511-43-8

N-[2(*S*)-(Acetylthiomethyl)-3-*o*-tolylpropionyl]-L-me-
thionine ethyl ester = *N*-[(2*S*)-2-[(Acetylthio)methyl]-3-
(2-methylphenyl)-1-oxopropyl]-L-methionine ethyl ester
(●)
S Sch 42495
U Antihypertensive (endopeptidase inhibitor)

9933 (8051) $C_{20}H_{29}NO_6S$
15130-91-3

3-(3-Sulfopropyl)atropinium hydroxide inner salt = 3α-
Hydroxy-8-(3-sulfopropyl)-1α*H*,5α*H*-tropanium hydroxide (−)-tropate inner salt = 3-[3-(3-Hydroxy-2-phenyl-
propionyloxy)-8-methyl-8-tropanio]propanesulfonate =
endo-(±)-3-(3-Hydroxy-1-oxo-2-phenylpropoxy)-8-me-
thyl-8-(3-sulfopropyl)-8-azoniabicyclo[3.2.1]octane
hydroxide inner salt (●)
A 118, Sultropan, *Sultroponium***, *Sultroponum*
J Anticholinergic, spasmolytic

9934 (8053) $C_{20}H_{29}N_3$
59033-43-1

N-Isopropyl-*N*-[2-[D-6-methyl-8β-ergolinyl]ethyl]amine
= (8β)-6-Methyl-*N*-(1-methylethyl)ergoline-8-ethan-
amine (●)
R Maleate (1:2) (59033-44-2)
S Iralin, VÚFB-10726
U Prolactin inhibitor

9935 (8052) $C_{20}H_{29}N_3$
89303-63-9

trans-1-Ethyl-1,2,3,4,4a,5,6,12b-octahydro-4-isopropyl-
12-methylpyrazino[2',3':3,4]pyrido[1,2-*a*]indole = *trans*-
1-Ethyl-1,2,3,4,4a,5,6,12b-octahydro-12-methyl-4-(1-
methylethyl)pyrazino[2',3':3,4]pyrido[1,2-*a*]indole (●)
R Maleate (1:1) (89303-64-0)
S *Atiprosin maleate**, AY-28228
U Antihypertensive

9936 (8054) $C_{20}H_{29}N_3O_2$
85-79-0

2-Butoxy-*N*-[2-(diethylamino)ethyl]cinchoninamide = 2-
Butoxy-4-quinolinecarboxylic acid β-(diethylamino)ethylamide = 2-Butoxy-*N*-[2-(diethylamino)ethyl]-4-
quinolinecarboxamide (●)

R Monohydrochloride (61-12-1)
S Afko-Dibucaine, Anesta-Caine, Benzolin, Butylcaine, Cincain, *Cincaini chloridum, Cinchocaine hydrochloride* **, *Cinchocainium chloride***, *Cinkain*, D-Caine, Dermacaine, *Dibucaine hydrochloride*, DoloPosterine N, Lar, NeoVitacain, Nupercain, Nupercainal, Nuperlone, Nuporals, Optokain, Percain(e), Percamin, Quinocaine, *Sovcain(um)*, *Zincho-kainhydrochlorid*
U Local anesthetic

9937 $C_{20}H_{29}N_3O_3$
              195966-93-9

(1*R*,3'*R*,5a*R*,8a*R*,9a*S*)-*rel*-(−)-Tetrahydro-1,1',8,8,11-pentamethylspiro[5*H*,6*H*-5a,9a-(iminomethano)-1*H*-cyclopent[*f*]indolizine-7(8*H*),3'-pyrrolidine]-2',5',10-trione (●)
S *Aspergillimide*, VM 55598
U Anthelmintic from *Aspergillus* str. IMI 337664

9938 (8055) $C_{20}H_{29}N_3O_4S$
              69387-87-7

5-[(1,1-Dimethyl-2-propynyl)sulfamoyl]-*N*-[(1-ethyl-2-pyrrolidinyl)methyl]-*o*-anisamide = 5-[[(1,1-Dimethyl-2-propynyl)amino]sulfonyl]-*N*-[(1-ethyl-2-pyrrolidinyl)methyl]-2-methoxybenzamide (●)
S *Tinisulpride***
U Psychotropic, anti-emetic

9939 $C_{20}H_{29}N_3O_5S$

1-[2-Amino-4-(methanesulfonamido)phenoxy]-2-[*N*-(3,4-dimethoxyphenethyl)-*N*-methylamino]ethane = *N*-[3-Amino-4-[2-[[2-(3,4-dimethoxyphenyl)ethyl]methylamino]ethoxy]phenyl]methanesulfonamide (●)
R Monohydrochloride (177596-55-3)
S KCB-328
U Anti-arrhythmic (class III)

9940 (8056) $C_{20}H_{29}N_5O_3$
              34661-75-1

6-[[3-[4-(*o*-Methoxyphenyl)-1-piperazinyl]propyl]amino]-1,3-dimethyluracil = 6-[[3-[4-(2-Methoxyphenyl)-1-piperazinyl]propyl]amino]-1,3-dimethyl-2,4(1*H*,3*H*)-pyrimidinedione (●)
R also monohydrochloride (64887-14-5) or fumarate (102411-11-0)
S Alpha-Depressan, B 66256, BY 256, α-Depressan, Ebrantil, Elgadil, Eupressyl, Médiatensyl, *Urapidil***, Uraprene
U Antihypertensive

9940-01 (8056-01)   64057-49-
R Compd. with furosemide (no. 3192) (1:1)
S *Urapidil furosemide*
U Antihypertensive

9941 (8057) $C_{20}H_{29}N_5O_3$
59184-78-0

1-Butyl-3-[1-(6,7-dimethoxy-4-quinazolinyl)-4-piperi-
dyl]urea = N-Butyl-N'-[1-(6,7-dimethoxy-4-quinazoli-
nyl)-4-piperidinyl]urea (●)
S BDPU, *Buquineran***, UK-14275
U Coronary vasodilator

9942 (8058) $C_{20}H_{29}N_5O_4$
76953-65-6

Acetone (±)-[6-[3-[(3,4-dimethoxyphenethyl)amino]-2-
hydroxypropoxy]-3-pyridazinyl]hydrazone = 6-[3-[[2-
3,4-Dimethoxyphenyl)ethyl]amino]-2-hydroxyprop-
oxy]-3(2H)-pyridazinone(1-methylethylidene)hydrazone
(●)
S *Dramedilol***
U β-Adrenergic blocker (antihypertensive, vasodilator)

9943 (8059) $C_{20}H_{29}N_5O_6$
35795-16-5

-Hydroxy-2-methylpropyl 4-(4-amino-6,7,8-trimeth-
xy-2-quinazolinyl)-1-piperazinecarboxylate = 4-(4-
mino-6,7,8-trimethoxy-2-quinazolinyl)-1-piperazine-
carboxylic acid 2-hydroxy-2-methylpropyl ester (●)
R Monohydrochloride monohydrate (53746-46-6)

S Cardovar, CP-19106-1, Supres "Pfizer", *Trimazosin
hydrochloride***
U Antihypertensive, cardiac stimulant

9944 $C_{20}H_{30}ClN_2O_4$

(1-Butyl-1-methyl-4-piperidinyl)methyl 8-amino-7-chlo-
ro-1,4-benzodioxane-5-carboxylate = 4-[[[(8-Amino-7-
chloro-2,3-dihydro-1,4-benzodioxin-5-yl)carbo-
nyl]oxy]methyl]-1-butyl-1-methylpiperidinium (●)
R Iodide (148688-00-0)
S SB 205008
U 5-HT_4 receptor antagonist

9945 (8060) $C_{20}H_{30}ClN_3O_4$
106707-51-1

4-Amino-2-butoxy-5-chloro-N-[1-(1,3-dioxolan-2-ylme-
thyl)-4-piperidinyl]benzamide (●)
S *Dobupride***
U Gastrokinetic

9946 $C_{20}H_{30}Cl_2N_8$
181367-74-8

N,N'-Bis(2-chloroethyl)-4,8-di-1-piperidinylpyrimi-
do[5,4-*d*]pyrimidine-2,6-diamine (●)
S DIP-C 1
U Antineoplastic

9947 (8061) $C_{20}H_{30}NO_2$

1-Ethyl-3-hydroxy-1-methylpyrrolidinium α-cyclopenty-
lphenylacetate = 1-Ethyl-1-methyl-3-(α-cyclopentylphe-
nylacetoxy)pyrrolidinium = 3-[(Cyclopentylphenylace-
tyl)oxy]-1-ethyl-1-methylpyrrolidinium (●)
R Bromide (15599-22-1)
S *Cyclopyrronium bromide***
U Anticholinergic

9948 $C_{20}H_{30}NO_3$

3-[2-(Hydroxymethyl)-1-oxo-2-phenylbutoxy]-8,8-dime-
thyl-8-azoniabicyclo[3.2.1]octane (●)
R Iodide (129109-88-2)
S Troventol, Truvent
U Bronchodilator, anticholinergic

9949 (8062) $C_{20}H_{30}NO_3$
                                                         60205-81-4

(8*r*)-3α-Hydroxy-8-isopropyl-1α*H*,5α*H*-tropanium (±)-
tropate = 8-Isopropyl-3α-DL-tropoyloxy-1α*H*,5α*H*-tropa-
nium = (*endo,syn*)-(±)-3-(3-Hydroxy-1-oxo-2-phenyl-
propoxy)-8-methyl-8-(1-methylethyl)-8-azoniabicy-
clo[3.2.1]octane (●)
R Bromide (22254-24-6) or bromide monohydrate
(66985-17-9)
S Apo-Ipravent, Arutropid, Atem, Atronase, Atrovent,
Biovit-A, Bitrop, Bronquibiom, Disne Asmol,
Iprabon, Ipranasal, Ipratin, Ipratrin, *Ipratropium bromide***, Ipvent, ITBR, Itrop, Narilet, Normosecre-
tol, Norsecrolo, Novo-Ipramide, Respontin, Rhino-
trop, Rhinovent, Rinatec, Rinoberen, Rinovagos,
Sch 1000, Steri-Neb Ipratropium, Tropiovent, Vagos
U Anticholinergic, bronchodilator, spasmolytic, anti-ar-
rhythmic

9950 (8063) $C_{20}H_{30}NO$
                                                         14461-98-

1,1-Dimethyl-3-hydroxypyrrolidinium α-phenylcyclohexaneglycolate (ester) = 3-(α-Cyclohexylphenylglycoloy-
loxy)-1,1-dimethylpyrrolidinium = 3-[(Cyclohexylhydroxyphenylacetyl)oxy]-1,1-dimethylpyrrolidinium (●)
R Bromide (3734-12-1)
S AHR-483, *Hexopyrrolate, Hexopyrronium bromide***
U Anticholinergic

9951 (8064) $C_{20}H_{30}N_2O$
30099-00-4

1-Cyclohexyl-2',4',6'-trimethyl-2-pyrrolidinecarboxanilide = 1-Cyclohexyl-N-(2,4,6-trimethylphenyl)-2-pyrrolidinecarboxamide (●)
R Monohydrochloride (19089-27-1)
S Cyclomecain, Tsiklomekain
U Local anesthetic

9952 $C_{20}H_{30}N_2O_2$
139062-78-5

Isobutyl N-[1-(2-indanyl)-4-piperidyl]-N-methylcarbamate = [1-(2,3-Dihydro-1H-inden-2-yl)-4-piperidinyl]methylcarbamic acid 2-methylpropyl ester (●)
R Fumarate (1:1) (139062-79-6)
S S 14905
U Antipsychotic (sigma ligand)

9953 (8066) $C_{20}H_{30}N_2O_2$
1239-29-8

7-Methyl-5α-androstano[2,3-c]furazan-17β-ol = (5α,17β)-17-Methylandrostano[2,3-c][1,2,5]oxadiazol-17-ol (●)
Androfurazanol, DH 245, Frazalon, Frazobol, *Furazabol**, Miotolon, Pirzolon

U Anabolic, antilipidemic

9954 (8065) $C_{20}H_{30}N_2O_2$
117-30-6

1-Piperidineethanol α-phenyl-1-piperidineacetate (ester) = 2-Piperidinoethyl α-phenyl-α-piperidinoacetate = α-Phenyl-α-piperidinoacetic acid β-piperidinoethyl ester = α-Phenyl-1-piperidineacetic acid 2-(1-piperidinyl)ethyl ester (●)
R also dihydrochloride (2404-18-4)
S *Dipiproverine***, L.D. 935, Lévospasme, Levospasmol, Spasmo-Kalipsina, Spasmonal "AWD"
U Anticholinergic, spasmolytic

9955 (8067) $C_{20}H_{30}N_2O_2S$
132392-39-3

5-[[3,5-Bis(1,1-dimethylethyl)-4-hydroxyphenyl]methylene]-3-(dimethylamino)-4-thiazolidinone (●)
S LY 221068
U Anti-inflammatory, anti-oxidant

9956   (8068) $C_{20}H_{30}N_2O_3$
469-81-8

1-(2-Morpholinoethyl)-4-phenylpiperidine-4-carboxylic
acid ethyl ester = Ethyl 1-(2-morpholinoethyl)-4-phenyl-
piperidine-4-carboxylate = 1-[2-(4-Morpholinyl)ethyl]-4-
phenyl-4-piperidinecarboxylic acid ethyl ester (●)
R   also dihydrochloride
S   *Morferidina, Morpheridine***, *Morpholinoethylnor-
pethidine*, NIH 7289, TA 1
U   Narcotic analgesic

9957 $C_{20}H_{30}N_3O_8P$
168680-87-3

Glycerylphosphonyl-leucinyl-tryptophan = *N*-[*N*-[(2,3-
Dihydroxypropoxy)hydroxyphosphinyl]-L-leucyl]-L-
tryptophan (●)
R   Disodium salt (168680-81-7)
S   S 17162
U   Antihypertensive (endothelin receptor antagonist)

9958 $C_{20}H_{30}N_4O_4$
153074-56-7

(3*S*)-7-[[4-(Aminoiminomethyl)benzoyl]amino]-3-[(3-
methyl-1-oxobutyl)amino]heptanoic acid (●)
S   GPI 562, SDZ-GPI 562
U   Glycoprtein IIb/IIIa receptor inhibitor

9959 $C_{20}H_{30}N_4O_5$
169499-56-3

N-[4-[(3-Aminopropyl)amino]butyl]-*N²*-(*trans*-epoxy-
succinyl)phenylalaninamide = [2*S*-[2α,3β(*R**)]]-3-[[[2-
[[4-[(3-Aminopropyl)amino]butyl]amino]-2-oxo-1-(phenylmethyl)ethyl]amino]carbonyl]oxiranecarboxylic acid
(●)
R   Sulfate (2:1) (169499-57-4)
S   TMC 52D
U   Enzyme inhibitor

9960 $C_{20}H_{30}N_4O_6$
176777-37-0

3-[*N*-[1-[*N*-(4-Aminobutyl)-*N*-(3-aminopropyl)carb-
amoyl]-2-(4-hydroxyphenyl)ethyl]carbamoyl]oxirane-2-
carboxylic acid = 3-[[[2-[(4-Aminobutyl)(3-aminopro-
pyl)amino]-1-[(4-hydroxyphenyl)methyl]-2-oxo-
ethyl]amino]carbonyl]oxiranecarboxylic acid (●)
S   WF 14861
U   Cathepsin B and L inhibitor

9961 (8069) $C_{20}H_{30}N_6O$
95992-79-3

3-Amino-1-methyl-5-[[3-[(1-piperidino-4-inda-
nyl)oxy]propyl]amino]-1*H*-1,2,4-triazole = N^5-[3-[[2,3-
Dihydro-1-(1-piperidinyl)-1*H*-inden-4-yl]oxy]propyl]-1-
methyl-1*H*-1,2,4-triazole-3,5-diamine (●)
S RGW-2568, WHR-2568
U Histamine H_2-receptor antagonist (anti-ulcer agent)

9962 (8070) $C_{20}H_{30}N_6O_3$
60503-05-1

S)-D-Phenylalanyl-*N*-[4-[(aminoiminomethyl)amino]-1-
ormylbutyl]-L-prolinamide (●)
R also sulfate (1:1) (83997-16-4)
S DPPA, GYKI 14166, RGH-2958
U Anticoagulant, thrombin inhibitor

963 (8071) $C_{20}H_{30}O$
68-26-8

5-Apo-β-caroten-15-ol = (*all-E*)-3,7-Dimethyl-9-(2,6,6-
imethyl-1-cyclohexen-1-yl)-2,4,6,8-nonatetraen-1-ol
●)
R see also nos. 11503, 12111, 15062
"A 313", A-Caps, Achepans, A-Cidan, Acon
"Endo", Acrisina, Actifral A, Adatone, Afebo,
AFI-A, Afilina, Agiolan, Agoncal, A-Kairon, Akso-
derm, Alcovit A, Alfadelta A, Alfaergin, Alfamin,

Alfa-Monovit, Alfa-Sir, Alfasole, Alfasterolo, Alfa-
tar, Alfavena, Alfavitina, Alfene, Alphalin, Alphaste-
rol, Alphavit, Amravit, A-Mulsal, Amulvit, Amuni-
ne, Anatola, Anavit, *Antixerophthalmic Vitamin*,
A-Om, Aoral, Apexol, A-Pharma, Aquacaps, Aqua-
sol A, Arivon F, Asol, Asteril, Atamin, Atav, Atera-
pion, A-Tiber, Atin, Ativina, Audax "Faran", Auxi-
na A, Avax, Avibon, Avigen, Avilon 500, Avimin,
A-Vi-Pel, A-Visol, A-Vit, Avita, Avitabiol, A-Vita-
dit, Avital, Avitam, *A-Vitamin*, Avitaminum-Kolin,
Avitan, Avitana, Avitil, Avitina, Avitol, A-Viton,
Avitpal, Avogina, Avoleum, Axerodina, *Axeroftolo*,
Axerol, *Axerophthol*, *Axerophthylium*, Axerovit,
Bagovit A, Bentavit A, Betacept, Bioanabee, Bio-
barc-A, Biominol "A", Biosterol, Bio-Tan, Capi A,
Carofral, Catavin A, Chocola-A, Crystal, Cuti-
fress B, Cytobiase, Davitamon A, Daxeron,
Del-VI-A, Derm-A-Caps, Dermis Albula, Dermo-
san, Dermosavit, Dermovit A, Dextamina A, Difvit-
amin A, Dohyfral-A, Dolce-A, Dri-A, Duphafral A,
Egavit A, Endo A, Epiteliol, Essogen, Euvit A, Evi-
tex A, Farmobion A, Fiviton A, Fletase, *Gadol*, Gor-
do-Vite A, Haliviten, Helgovit A, Hi-A-Vita, Hid-
ro-A, IDO-A, Idratene, Idrurto A, Ingavit A, Inovi-
tan A, Lagarmin-A, Larmavita A, Ledovit A, Massi-
ve-A, Meditalfa, Miluyvit A, Mono-tabs, Mulsal A,
Neo-Axil, Neominas A, Oleovit A, *Oleovitamin A*,
Ophthalamin, Pedi-Vit A, Perlaminas A, Plivit A,
Prepalin, Preparato A Glaxo, Purvit A, Radiosa A,
Recalferol A, Resistovites, Retiblan, *Retinol**,*
Rexamin A, Rinocusi, Sahne, Seamvita A, Skin-
derm A, Solaneed, Solu-A, Solvisyn-A, Solvitan A,
Somavit A, Sterogyl A, Super A, Supervit A, Tanvi-
mil A, Testavol, Ucemine A, Ursovit A, Vaconex,
Vaflol, Veroftal, Viacaps, Viadenin, Vi-alpha, Via-
mit, Vicelpas A, Virgin, Vita-A, Vitabiol A, Vit-
aderm, Vit-A-Drops, Vitaendil A, Vitalen A, Vital-
fa, Vitama, *Vitamin A*, *Vitamin A₁*, Vitaplex A,
Vit-A-Plos, Vitapur A, Vitaquimiol A, Vit-Asal-A,
Vitasan A, Vitavel A, Vitavex, Vitemade A, Vitpex,
Vitwas A, Vizo A, Vogan, Wandervit A, Xeroftal,
Xerophthol
U Antixerophthalmic vitamin

9964 (8072) $C_{20}H_{30}O$
                                                514-62-5

13-Isopropylpodocarpa-8,11,13-trien-12-ol = (4b*S*-
trans)-4b,5,6,7,8,8a,9,10-Octahydro-4b,8,8-trimethyl-2-
(1-methylethyl)-3-phenanthrenol (●)
S   *Ferruginol*
U   Cytotoxic

9965 (8074) $C_{20}H_{30}O_2$
                                                153-00-4

17β-Hydroxy-1-methyl-5α-androst-1-en-3-one =
(5α,17β)-17-Hydroxy-1-methylandrost-1-en-3-one (●)
R   see also nos. 11504, 13652
S   *Metenolone***, *Methenolone*, *Methylandrostenolone*
U   Anabolic

9966 (8075) $C_{20}H_{30}O_2$
                                                5197-58-0

17β-Hydroxy-2-methyl-5α-androst-1-en-3-one =
(5α,17β)-17-Hydroxy-2-methylandrost-1-en-3-one (●)
R   see also no. 11505

S   *Estenbolona, Stenbolone***
U   Anabolic

9967 (8076) $C_{20}H_{30}O_2$
                                                58-18-4

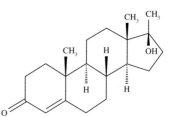

17β-Hydroxy-17-methylandrost-4-en-3-one = 17α-
Methyl-4-androsten-17β-ol-3-one = (17β)-17-Hydroxy-
17-methylandrost-4-en-3-one (●)
R   see also no. 13332
S   Afro, Agovirin-Dragees, Andrest-Tabl., Andrhormo-
    ne-Tabl., Andrifar-Compr., Android-5 (-10, -25),
    Andrometh, Androna, Andronex-Tabl., Androral,
    Androsan-Tabl., Androsten, Androteston-Tabl., An-
    droxil-oral, Anertan perlingual, Arcosterone, Biol-
    mon-Cap., Enarmon-Tabl., Entestil-M, Gastastero-
    ne, Glosso-Stérandryl, Gynosterone, Homan, Ho-
    mandren-Tabl., Hormale, Hormobin, Hormondri-
    ne-Tabl., Hormoneta, Malestrone-Tabl., Malogen,
    Masenone, Mastestona, Mesteron, Metandren, Mete-
    stine, Metestone, Metexterona, Methylets, *Methylte-
    stosterone***, *Metiltestosterone*, Metrone "Barre",
    Nadosterone, Neo-Hombreol-M, Neo-Restor, Neovi-
    ron-Tabl., NSC-9701, Nu-Man, Opotestan-perlingu-
    al, Orandrone, Oraviron, Orchisterone-M, Orchiste-
    ron-Lingualtabl., Oreton-M, Oreton Methyl, Pante-
    stin-oral, Perandren-Linguetten, Perandrone-Linguet-
    tes, Primoteston-Tabl., Seksfort, Steronyl, Sublings,
    Suprasteron-Tabl., Synandrotabs, Syndren-Tabl., Te-
    lipex-Tabl., Testaform-Tabl., Testahomen-Tabl., Te-
    steplex-Tabl., Testhormona, Testin-Tabl., Testipron,
    Testobase, Testogenina, Testomet, Teston "Remek",
    Testonorpon-Tabl., Teston-Tabl., Testora, Testoral
    "Leo", Testorex, Testormon-Tabl., Testosid-Tabl.,
    Testostelets, Testosterona Dexter Compr., Testoster-
    one-Vifor, Testosteron Grossmann-Tabl., Testoste-
    ron-Lingualtabl., Testosteron lingvalete, Testosteron
    resoribl. "Orion", Testosteron-Tabl. "Jenapharm",
    Testotonic B, Testoviron-Tabl., Testovis Compr.,
    Testoxyl perlinguale, Testral, Testred, Tostrina-M,

Tylandril, Vi-Andro, Virex-oral, Virilon, Virormo-lo-compr., Virormone-oral

U Androgen

9968 $C_{20}H_{30}O_2$
25378-27-2

Eicosapentaenoic acid (●)
R L-Lysine compd. (1:1) (146116-84-9)
S LSL 90202
U Prevention of cyclosporin nephrotoxicity

9969 (8077) $C_{20}H_{30}O_2$
10417-94-4

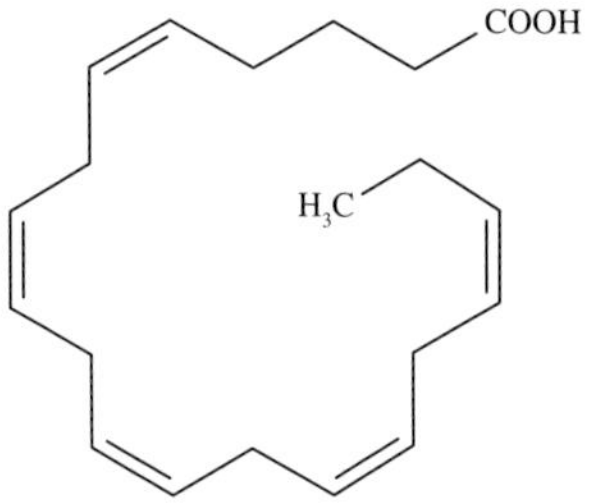

(all-Z)-5,8,11,14,17-Icosapentaenoic acid = (all-Z)-5,8,11,14,17-Eicosapentaenoic acid (●)
R see also no. 11544
S Epa-E Nissui, *Ethyl Icosapentate*, *Icosapent***, Proepa, *Timnodonic acid*
U Platelet aggregation inhibitor

9970 (8078) $C_{20}H_{30}O_2$
33765-68-3

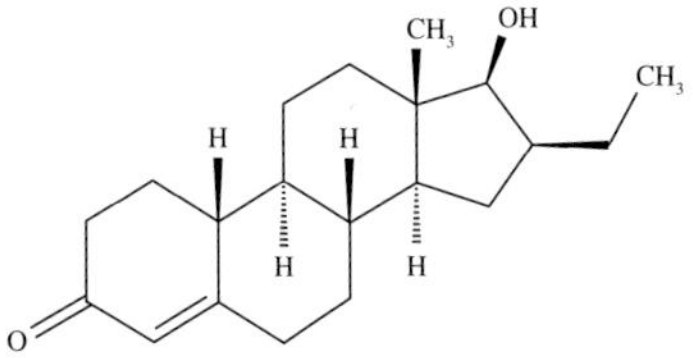

6β-Ethyl-17β-hydroxyestr-4-en-3-one = (16β,17β)-16-Ethyl-17-hydroxyestr-4-en-3-one (●)
S Adiafor "Abbott", Horosteon, *Oxendolone***, Prostetin, *Roxenone*, TSAA-291
U Anti-androgen (for treatment of benign prostatic hypertrophy)

9971 (8079) $C_{20}H_{30}O_2$
3704-09-4

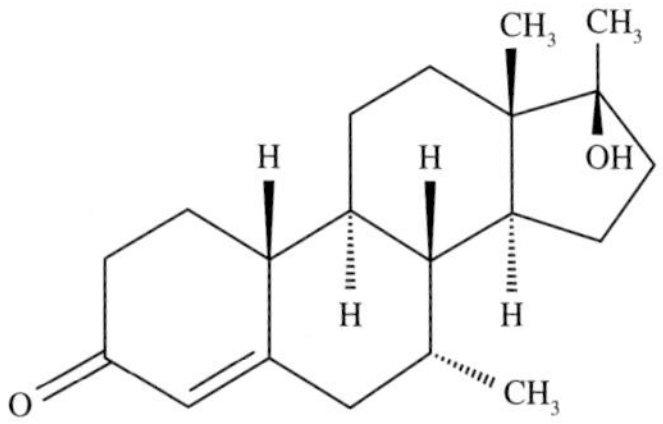

17β-Hydroxy-7α,17-dimethylestr-4-en-3-one = (7α,17β)-17-Hydroxy-7,17-dimethylestr-4-en-3-one (●)
S CDB 904, Cheque, Matenon, *Mibolerone***, NSC-72260, U-10997
U Anabolic, androgen

9972 (8080) $C_{20}H_{30}O_2$
52279-58-0

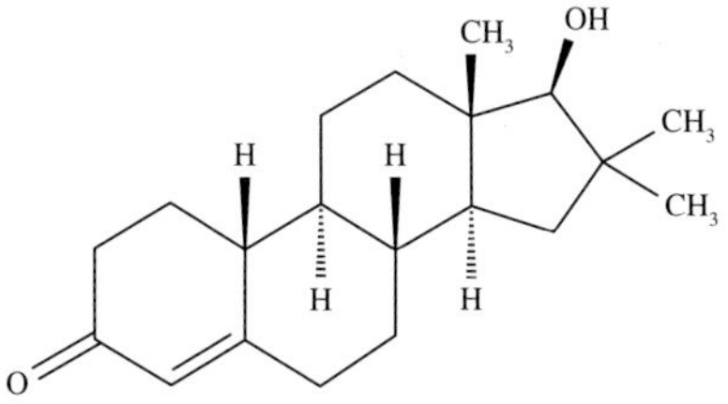

17β-Hydroxy-16,16-dimethylestr-4-en-3-one = (17β)-17-Hydroxy-16,16-dimethylestr-4-en-3-one (●)
S *Metogest***, SC-14207
U Acne-therapeutic

9973 (8081) $C_{20}H_{30}O_2$
81485-25-8

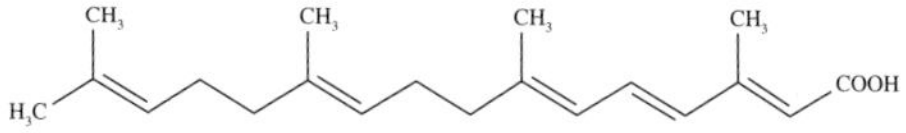

(all-E)-3,7,11,15-Tetramethyl-2,4,6,10,14-hexadecapentaenoic acid (●)
S E-5166, *Polypreic acid*, *Polyprenoic acid*
U Antineoplastic

9974 (8082)

$C_{20}H_{30}O_2$
52-78-8

17α-Ethyl-17-hydroxyestr-4-en-3-one = 17-Hydroxy-19-nor-17α-pregn-4-en-3-one = 17α-Ethyl-17-hydroxy-19-norandrost-4-en-3-one = 17α-Ethyl-19-nortestesterone = (17α)-17-Hydroxy-19-norpregn-4-en-3-one (●)

S *Äthylnortestosteron*, *Äthyloestrenolon*, 8022 C.B., 17-ENT, *Ethylnortestosterone*, Nilevar, *Noräthandrolon*, *Noretandrolone*, *Norethandrolone***, Pronabol

U Anabolic, androgen

9975 (8083)

$C_{20}H_{30}O_2$
3643-00-3

20β-Hydroxy-19-norpregn-4-en-3-one = (20R)-20-Hydroxy-19-norpregn-4-en-3-one (●)

R see also no. 14156

S *Oxogestone***

U Progestin

9976 (8073)

$C_{20}H_{30}O_2$
514-10-3

1,2,3,4,4a,4b,5,6,10,10a-Decahydro-7-isopropyl-1,4a-dimethyl-1-phenanthrenecarboxylic acid = 13-Isopropylpodocarpa-7,13-dien-15-oic acid = Abietic acid = [1R-

(1α,4aβ,4bα,10aα)]-1,2,3,4,4a,4b,5,6,10,10a-Decahydro-1,4a-dimethyl-7-(1-methylethyl)-1-phenanthrenecarboxylic acid (●)

R Manganese salt (54675-76-2)

S Abietal, Manganol

U Anti-allergic

9977 (8084)

$C_{20}H_{30}O_2$
24035-36-7

13-Isopropylpodocarpa-8,11,13-triene-12,17-diol = (4aR-trans)-1,3,4,9,10,10a-Hexahydro-6-hydroxy-1,1-dimethyl-7-(1-methylethyl)-4a(2H)-phenanthrenemethanol (●)

S *Pisiferol*

U Cytotoxic

9978 (8085)

$C_{20}H_{30}O$
145-12-0

4,17β-Dihydroxy-17-methylandrost-4-en-3-one = 4-Hydroxy-17α-methyltestesterone = (17β)-4,17-Dihydroxy-17-methylandrost-4-en-3-one (●)

S Anamidol, Balnimax, *Hydroxymethyltestosterone*, *Methandrostenediolone*, Oranabol, *Ossimesterone*, *Oximesteronum*, Oxymesterone**, Oxymestrone, Sanaboral, Théranabol, Tubil

U Anabolic, androgen

9979 $C_{20}H_{30}O_3$
71641-23-1

(E)-8,13(17)-Epoxylabd-12-ene-15,16-dial = (2E)-[2-
[(1R,2S,4aS,8aS)-Octahydro-5,5,8a-trimethylspiro[naph-
thalene-2(1H),2'-oxiran]-1-yl]ethylidene]butanedial (●)
S *Aframodial*
U Antileukemic diterpene from *Aframomum daniellii*

9980   (8086) $C_{20}H_{30}O_3$
132922-55-5

(9α,13R)-15,16-Diepoxylabdan-14-en-17-one = [1"R-
1"α(R*),2"α,4"aα,8"aβ]]-3',4',4"a,5",6",7",8",8"a-Octa-
hydro-2",5",5",8"a-tetramethyldispiro[furan-
3(2H),2'(5'H)-furan-5',1"(2"H)-naphthalen]-3"(4"H)-one
(●)
S LC 5504, *Prehispanolone*
U PAF antagonist from *Leonurus heterophyllus*

9981   (8087) $C_{20}H_{30}O_4$
65669-72-9

[1R-(1R*,4S*,6S*,9E,13S*,14R*)]-13-Hydroxy-4,9,13-
trimethyl-17-methylene-5,15-dioxatricy-
clo[12.3.1.0^{4,6}]octadec-9-en-16-one (●)
S *Cembranoid lactone*, *Flexibilide*, *Sinularin*
U Anti-inflammatory, anti-arthritic (from the soft coral
Sinularia flexibilis)

9982   (8088) $C_{20}H_{30}O_4$
112514-46-2

(7α,14R,15β)-7,14,15-Trihydroxykaur-16-en-3-one (●)
S *Glaucocalyxin C*
U Antineoplastic from *Rabdosia japonica* var. *glauco-
calyx*

9983   (8089) $C_{20}H_{30}O_5$
5508-58-7

3α,14,15,18-Tetrahydroxy-5β,9βH,10α-labda-8(20),12-
dien-16-oic acid γ-lactone = [1R-
[1α[E(S*)],4aβ,5α,6α,8aα]]-3-[2-[Decahydro-5-hydr-
oxy-5-(hydroxymethyl)-5,8a-dimethyl-2-methylene-1-
naphthalenyl]ethylidene]dihydro-4-hydroxy-2(3H)-fura-
none (●)
R see also no. 9985
S *Andrographolide*
U Antimicrobial from *Andrographis paniculata*

9984 (8090)

$C_{20}H_{30}O_6$
99401-76-0

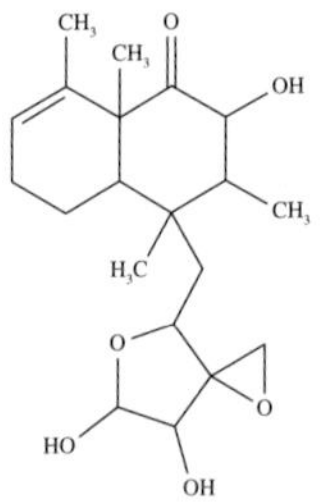

4-[(6,7-Dihydroxy-1,5-dioxaspiro[2.4]hept-4-yl)methyl]-
3,4,4a,5,6,8a-hexahydro-2-hydroxy-3,4,8,8a-tetramethyl-
1(2H)-naphthalenone (●)
S *Spirocardin A*
U Antibacterial, antimycoplasmal from *Nocardia* sp.
    SANK 64282

9985 (8091)

$C_{20}H_{30}O_7S$

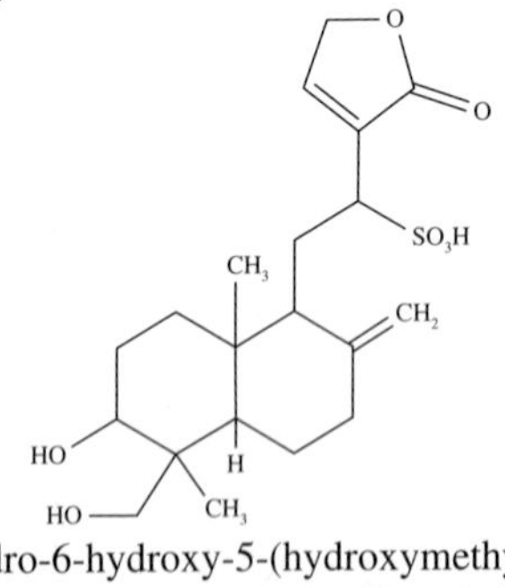

α-[[Decahydro-6-hydroxy-5-(hydroxymethyl)-5,8a-di-
methyl-2-methylene-1-naphthalenyl]methyl]-2,5-dihy-
dro-2-oxo-3-furanmethanesulfonic acid (●)
R Monosodium salt (71202-97-6)
S *Andrographolide sodium bisulfite*
U Antimicrobial

9986

$C_{20}H_{31}ClO_5$
120962-76-7

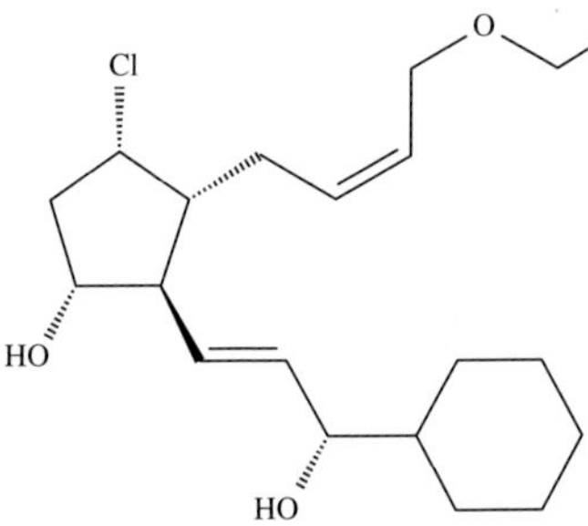

[[(2Z)-4-[(1R,2R,3R,5S)-5-Chloro-2-[(1E,3S)-3-cyclohe-
xyl-3-hydroxy-1-propenyl]-3-hydroxycyclopentyl]-2-bu-
tenyl]oxy]acetic acid (●)
S ZK 118182
U Anti-ischemic (PGD$_2$ mimetic)

9987 (8092)

$C_{20}H_{31}Cl_2N_3O_2$
31661-12-8

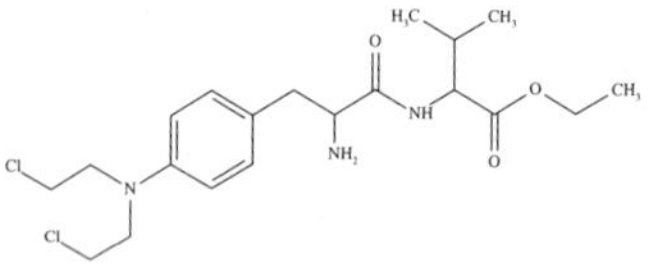

Sarcolysyl-α-valine ethyl ester = N-[4-[Bis(2-chloro-
ethyl)amino]-DL-phenylalanyl]-DL-valine ethyl ester (●)
R Hydrochloride (30632-01-0)
S Salin, Salyne
U Cytotoxic

9988 (8093)

$C_{20}H_{31}FO_4$
69900-71-

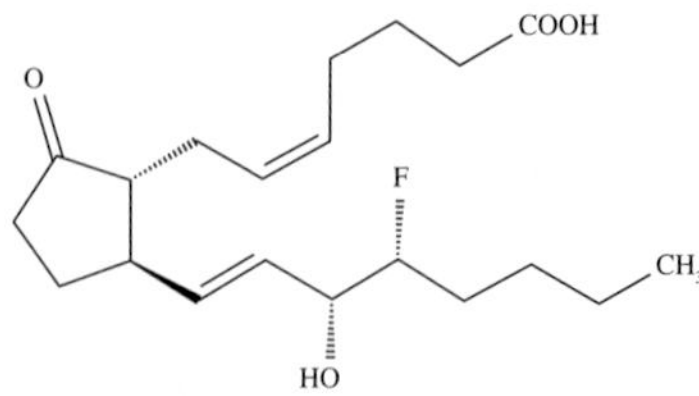

(5Z,13E,15R,16R)-16-Fluoro-15-hydroxy-9-oxoprosta-
5,13-dien-1-oic acid (●)
S Ro 22-1327
U Mucosal protectant

9989 (8094) $C_{20}H_{31}NO$
120444-71-5

N,N-Dimethyl-2-[[(1R,2S,4R)-2-phenyl-2-bor-nyl]oxy]ethylamine = 2-Phenyl-2-[2-(dimethylami-no)ethoxy]-1,7,7-trimethylbicyclo[2.2.1]heptane = (1R-exo)-N,N-Dimethyl-2-[(1,7,7-trimethyl-2-phenylbicy-clo[2.2.1]hept-2-yl)oxy]ethanamine (●)
R Fumarate (1:1) (120444-72-6)
S *Deramciclane fumarate***, Egis-3886, Egyt-3886
U Anxiolytic, anticonvulsant

9990 (8095) $C_{20}H_{31}NO$
144-11-6

-Cyclohexyl-1-phenyl-3-piperidino-1-propanol = α-Cy-lohexyl-α-phenyl-1-piperidinepropanol (●)
R Hydrochloride (52-49-3)
S Anti-Spas, Antitrem, Aparkan(e), Aphen "Major", Apo-Hexidyl, Apo-Trihex, Artane, Artilan, Bentex, *Benzhexol hydrochloride*, Broflex, Ciklodol, *Cyclo-dol(um)*, Hexifen, Hexinal, Hexyphen, Hipokinon, Novohexidyl, Pacitane, Paralest, Pargitan, Parkan, Parkidyl, Parkinane LP, Parkines, Parkinsan, Par-kisan, Parkisonal, Parkopan, Partane, Partigan, Pera-git, Pipanol, Placidyl "La-Medica", PMS-Trihexy-phenidyl, Pozhexol, Pyramistin, Rodenal, Rompar-kin, Sedrena, Stobrun, Tonaril "Lab. Chile", Top-cron, Tremin, Trems, *Triesifenidile*, Triexidil, Trifen "Sun", Trihexane, Trihexidyl, Trihexin, Trihexy, *Tri-hexyfenidil hydrochloride*, *Trihexyphenidyl hydro-*

*chloride***, Trinidyl, Triphedinon, Triphenidyl, Trixyl, Win 511
U Anticholinergic, spasmolytic, antiparkinsonian

9991 (8096) $C_{20}H_{31}NO_2$
1679-76-1

2-(Diethylamino)ethyl α-phenylcyclohexaneacetate = α-Phenylcyclohexylacetic acid β-(diethylamino)ethyl ester = α-Cyclohexylbenzeneacetic acid 2-(diethylamino)ethyl ester (●)
R Hydrochloride (548-66-3)
S Anfosentin-Amp., *Drofenine hydrochloride***, *He-xahydroadiphenine hydrochloride*, Trasentine-A, Trasentin H
U Spasmolytic

9992 (8097) $C_{20}H_{31}NO_2$
561-79-5

2-(Diethylamino)ethyl 1-(3,4-xylyl)cyclopentanecarb-oxylate = 1-(3,4-Dimethylphenyl)cyclopentanecarboxylic acid 2-(diethylamino)ethyl ester (●)
R also hydrochloride (1950-31-8)
S *Dimethylcaramiphen*, 3012 G, *Metacaraphen*, *Met-caraphen*, Netrin, Parnetil, Vanetril(a), Vanetrilla
U Anticholinergic, spasmolytic

9993 (8098) $C_{20}H_{31}NO_2S$
14176-10-4

2-(Hexahydro-1*H*-azepin-1-yl)ethyl α-cyclohexyl-3-thio-
pheneacetate = α-Cyclohexyl-3-thiopheneacetic acid 2-
(hexahydro-1*H*-azepin-1-yl)ethyl ester (●)
R Citrate (1:1) (16286-69-4)
S Celsis, *Cetiedil citrate***, Fusten, Gremedolin, Hu-
berdilat, INO 502, Stratene, Vasocet
U Peripheral vasodilator

9994 (8099) $C_{20}H_{31}NO_3$
77-23-6

2-[2-(Diethylamino)ethoxy]ethyl 1-phenylcyclopentane-
carboxylate = 1-Phenylcyclopentanecarboxylic acid 2-[2-
(diethylamino)ethoxy]ethyl ester (●)
R Citrate (1:1) (23142-01-0) [some preparations also as
hydrochloride (1045-21-2)]
S Antees, Antis, Aslos, Asthma "Nichiiko", Astoma-
top, Atomin S, Atussil, Balsoclase, Bestfull, Bosil,
Caldan, Calnathal, *Carbapentane citrate*, Carbe-
tan(e), *Carbetapentane Citrate*, Carbeten, *Carbeto-
pentane citrate*, Carbex "Etna", Carbin, Carcarol,
Cartpenta, Citobetan, Cofbetan, Cofden, Cossym,
Coughcode, Culten, Elmocar, Fuscardin, Fustpenta-
ne, Fuszemin CP, Gailess, Germapect, Haltos, Hu-
stopentane, Kaibohl, Kaseet, Kibol, Ledax, Libegan,
Loucarbate, Merol, Mezocar, Milysted, Miroclase,
Mityperden, Nyal Dry Cough, Patcon, Pectosan,
Pencal, Pencarbil, *Pentoxiverini citras*, *Pentoxyveri-
ne citrate***, Pertix-L (-T, -Z), Sedotussin, Solutus-
syl, Takabetane, Tesilex, Toclase, Tosnone, Tosse-
din, Tossex, Tuclase, Tusolven, Tussa-Tablinen,
U.C.B. 2543, Vicks Cough Syrup
U Antitussive

9995 $C_{20}H_{31}NO_3S$
158089-95-3

2,6-Bis(1,1-dimethylethyl)-4-[(*E*)-(2-ethyl-1,1-dioxido-
5-isothiazolidinylidene)methyl]phenol (●)
S S-2474
U Cyclooxygenase-2 inhibitor

9996 (8100) $C_{20}H_{31}NO_5$
53034-85-8

5-[2-(*tert*-Butylamino)-1-hydroxyethyl]-*m*-phenylene di-
isobutyrate = 2-Methylpropanic acid 5-[2-[(1,1-dimethyl
ethyl)amino]-1-hydroxyethyl]-1,3-phenylene ester (●)
R also hydrochloride (61435-51-6)
S *Ibuterol***, KWD 2058, Spiranyl, *Terbutaline diiso-
butyrate*
U Bronchodilator

9997 $C_{20}H_{31}N_2O_6P$
150151-86-

(2α,5β,6β,7β,8β,10aβ)-Octahydro-5-[(4-methoxyphe-
nyl)methyl]-2-(methylamino)-1*H*-7,10a-methanopyrro-
lo[1,2-*a*]azocine-6,8-diol 8-(dihydrogen phosphate) (●)
R Monohydrochloride
S FR 901483
U Immunosuppressive from *Cladobotryum* sp. 11231

9998

$C_{20}H_{31}N_3O_4$
158798-83-5

N-[(Phenylmethoxy)carbonyl]-L-leucyl-*N*-ethyl-L-2-ami-
nobutanamide (●)
S AK 275
U Calpain inhibitor

9999 (8101)

$C_{20}H_{31}N_3O_6S$
50846-45-2

2*S*,5*R*,6*R*)-6-[[(Hexahydro-1*H*-azepin-1-yl)methy-
ene]amino]-3,3-dimethyl-7-oxo-4-thia-1-azabicy-
lo[3.2.0]heptane-2-carboxylic acid ester with ethyl 1-
ydroxyethyl carbonate = 1-(Ethoxycarbonyloxy)ethyl
6*R*)-6-(perhydroazepin-1-ylmethyleneamino)penicilla-
ate = [2*S*-(2α,5α,6β)]-6-[[(Hexahydro-1*H*-azepin-1-
l)methylene]amino]-3,3-dimethyl-7-oxo-4-thia-1-azabi-
yclo[3.2.0]heptane-2-carboxylic acid 1-[(ethoxycarbo-
yl)oxy]ethyl ester (●)
• *Bacmecillinam**, KW 1100
J Antibiotic

10000 (8102)

$C_{20}H_{31}N_5O_3S$
80343-63-1

1-[*m*-[3-[[1-Methyl-3-[(methylsulfonyl)methyl]-1*H*-
1,2,4-triazol-5-yl]amino]propoxy]benzyl]piperidine = 1-
Methyl-3-(mesylmethyl)-1*H*-1,2,4-triazol-5-yl-[3-(α-
piperidino-*m*-tolyloxy)propyl]amine = 1-Methyl-3-[(methylsulfonyl)methyl]-*N*-[3-[3-(1-piperidinylmethyl)phen-
oxy]propyl]-1*H*-1,2,4-triazol-5-amine (●)
S AH 25352, *Sufotidine***
U Histamine H_2-receptor antagonist

10001 (8103)

$C_{20}H_{32}NO$
60-48-0

1-(3-Cyclohexyl-3-hydroxy-3-phenylpropyl)-1-methyl-
pyrrolidinium (●)
R Chloride (3818-88-0)
S Aconil, Anticholinergic 14045, Elorine, Lergine,
Lilly 14045, *Procyclidine methyl chloride*, Tricolo-
id, *Tricyclamol chloride***, Tricyvagol, Vagosin,
Verbindung 14045
U Anticholinergic, spasmolytic

10002 (8104) $C_{20}H_{32}NO_3$

2-(3,4-Dihydro-6-hydroxy-2,5,7,8-tetramethyl-2H-1-
benzopyran-2-yl)ethyltrimethylammonium acetate (ester)
= 6-(Acetyloxy)-3,4-dihydro-N,N,N,2,5,7,8-heptamethyl-
2H-1-benzopyran-2-ethanaminium (●)
R p-Toluenesulfonate (4-methylbenzenesulfonate) (1:1)
   (128008-97-9)
S MDL 74270
U Anti-ischemic

10003 (8105) $C_{20}H_{32}N_2O_2$
117009-82-2

1-(3,5-Di-tert-butyl-4-hydroxybenzoyl)-1,4-diazepane =
1-[3,5-Bis(1,1-dimethylethyl)-4-hydroxybenzoyl]hexa-
hydro-1H-1,4-diazepine (●)
S A 854777
U Anti-inflammatory, immunosuppressive

10004 (8106) $C_{20}H_{32}N_2O_3$
81185-85-5

Ethyl 4-(o-methoxyphenyl)-1-piperazineenanthate = Ethyl 7-[4-(2-methoxyphenyl)-1-piperazinyl]heptanoate = 4-

(2-Methoxyphenyl)-1-piperazineheptanoic acid ethyl es-
ter (●)
R Dihydrochloride (81186-05-2)
S SGB-483
U Antihypertensive

10005 $C_{20}H_{32}N_2O_4$
142996-66-5

4,4'-[1,2-Ethanediylbis(iminomethylidyne)]bis[dihydro-
2,2,5,5-tetramethyl-3(2H)-furanone] (●)
S Furomine*, MP-1549
U Ligand in Technetium Tc furifosmin (no. 15509)

10006 $C_{20}H_{32}N_4O_5$

(2S,3R)-3-[[(1S)-2,2-Dimethyl-1-(2-pyridylcarbamoyl)-
propyl]carbamoyl]-2-methoxy 5-methylhexanohydro-
xamic acid
S BB-3644, Solimastat**
U Matrix metalloproteinase inhibitor

10007 $C_{20}H_{32}N_4O_6P_2$
                                                          146777-91-5

Tetraethyl 3-cyano-2,5-dimethylpyrazolo[1,5-*a*]pyrimi-
din-7-ylpropylbisphosphonate = [3-(3-Cyano-2,5-dime-
thylpyrazolo[1,5-*a*]pyrimidin-7-yl)propylidene]bis[phos-
phonic acid] tetraethyl ester (●)
S U-89395
U Anti-inflammatory

10008 $C_{20}H_{32}N_5O_8P$
                                                          142340-99-6

[2-(6-Amino-9*H*-purin-9-yl)ethoxy]methyl]phosphonic
·cid diester with hydroxymethyl pivalate = 9-[2-[[Bis[(pi-
·aloyloxy)methoxy]phosphinyl]methoxy]ethyl]adenine =
·,2-Dimethylpropanoic acid [[[2-(6-amino-9*H*-purin-9-
·l)ethoxy]methyl]phosphinylidene]bis(oxymethylene)
·ster (●)
· *Adefovir dipivoxil*, Bis(POM)PMEA, GS-840,
    Piv2PMEA, Preveon
· Antiviral

10009 (8107) $C_{20}H_{32}N_6O_6$
                                                          74550-97-3

4,4'-Ethylenebis[1-(morpholinomethyl)-2,6-piperazine-
dione] = 4,4'-(1,2-Ethanediyl)bis[1-(4-morpholinylme-
thyl)-2,6-piperazinedione] (●)
S AT-1727, Bimolane, NSC-351358
U Antineoplastic

10010 (8108) $C_{20}H_{32}N_6O_{12}S_2$
                                                          27025-41-8

N,N'-[Dithiobis[(*R*)-1-[(carboxymethyl)carbamoyl]ethy-
lene]]di-L-glutamine = L-γ-Glutamyl-L-cysteinylglycine
(2→2')-disulfide (●)
S BSS Plus, *Glutathione disulfide*, GSSG, *Oxidized
       glutathione, Oxiglutatione***
U Ophthalmic

10011 $C_{20}H_{32}O$
                                                          72629-69-7

(1*S*,2*Z*,4*E*,8*E*,12*E*)-5,9,13-Trimethyl-2-(1-methylethyl)-
2,4,8,12-cyclotetradecatetraen-1-ol (●)
S *Sarcophytol A*

U Antineoplastic (ornithine decarboxylase inhibitor), topical anti-inflammatory

---

10012 (8109) $C_{20}H_{32}O$
965-90-2

19-Nor-17α-pregn-4-en-17-ol = 17α-Ethylestr-4-en-17-ol = (17α)-19-Norpregn-4-en-17-ol (●)
S *Äthyloestrenol*, Durabolin-O, Durabolin-Oral, Duraboral, *Ethylestrenol***, *Ethylnandrol*, *Ethyloestrenol(um)*, *Etilestrenolo*, Maxibolin, Nandoral, Orabolin, Org-483, Orgabolin, Orgaboral, Virastine
U Anabolic

10013 (8110) $C_{20}H_{32}O$
16915-78-9

19-Nor-17α-pregn-5-en-17-ol = 17α-Ethylestr-5-en-17-ol = (17α)-19-Norpregn-5-en-17-ol (●)
S *Bolenol**
U Anabolic

10014 (8112) $C_{20}H_{32}O_2$
1424-00-6

17β-Hydroxy-1α-methyl-5α-androstan-3-one = (1α,5α,17β)-17-Hydroxy-1-methylandrostan-3-one (●)
S Androviron, Dapoder, Gavrol(in), Mesteranum, *Mesterolone***, Mestoranum, 1α-*Methylandrostanolone*, NSC-75054, Proviron (new form), Provironum, SH 723, SH 60723, Sten-or, Testiwop, Vistimon
U Androgen

10015 (8113) $C_{20}H_{32}O_2$
58-19-5

17β-Hydroxy-2α-methyl-5α-androstan-3-one = 2α-Methyldihydrotestosterone = (2α,5α,17β)-17-Hydroxy-2-methylandrostan-3-one (●)
R see also no. 12140
S *Dromostanolone, Drostanolone***, Methalone "Syntex"
U Anabolic, anticholesteremic

10016 (8114) $C_{20}H_{32}O_2$
521-11-9

17β-Hydroxy-17-methyl-5α-androstan-3-one = 17α-Methyl-5α-androstan-17β-ol-3-one = (5α,17β)-17-Hydroxy-17-methylandrostan-3-one (●)
R see also no. 13224
S Anavormol, Andoron, Andron(e), Androstalone, Antalon "Kobayashi K.", Assimil, Duramin, Ermalone, Etnabolate, Hermalone-Glosset, Macrobin (Tabl. + Syrup), Mechiaron, Mesanolon, Mestalone, *Mestanolone***, Methyantalon, Methybol, 17α-*Methylandrostanolone*, Preroide, Prohormo, Protenolon, Proterigine, Proteron, Restore, Tantarone, Yonchlon-Syr.

U Androgen, anabolic

10017 (8111) $C_{20}H_{32}O_2$
521-10-8

17-Methylandrost-5-ene-3β,17β-diol = (3β,17β)-17-Me-thylandrost-5-ene-3,17-diol (●)

R some synonyms also the 3,17-dipropionate (no. 13323);
see also nos. 12141, 15196

S Anabolin "Sig", Anadiol, Andris, Androdiol, Androgonyl, Androteston "M", Anormon, Cenabolic, Diandrin, Diolandrone, Diolostene, Drostene, Esjaydiol, Gynediolo, Hybolin, Isormon, MAD, Madiol, Masdiol, Megabion "Teikoku Zoki", *Mestenediol*, Metandiol, Metasteron, Metendiol, Methanabol, Methandiol, *Methandriol***, Methandrol, Methostan, *Methylandrostenediol*, "Methyl Diol", Methyltestediol, Metidione, *Metilandrostendiolo*, Metilbisexovis-Compr., Metildiolo, Metocryst, Nabadial, Neostene, Neosteron, Neusteron, Neutrormone, Neutrosteron, Neviron, Notandron, Novandrol, Noviril, Protandren, Protendiol, Sinesex, Spenbolic, Stenediol, Stenesium, Stenibell, Steniform, Stenosterone, Testodiol, Tonormon, Troformone

U Anabolic

10018 (8115) $C_{20}H_{32}O_2$
506-32-1

all-Z)-5,8,11,14-Icosatetraenoic acid = (*all-Z*)-5,8,11,14-Eicosatetraenoic acid (●)

S *Acidum arachidonicum*, ARA, *Arachidonic acid*, *Arachidonsäure*, Matiga

U Nutrient, anti-eczematic (topical)

10019 (8116) $C_{20}H_{32}O_3S$
53602-61-2

2-[(3,5-Di-*tert*-butyl-4-hydroxyphenyl)thio]caproic acid = 2-[[3,5-Bis(1,1-dimethylethyl)-4-hydroxyphenyl]thio]hexanoic acid (●)

S DH 990, DL 990, MDL 29350

U Antihyperlipidemic

10020 (8117) $C_{20}H_{32}O_4S$
64868-63-9

(Z)-(3aR,4R,5R,6aS)-Hexahydro-5-hydroxy-4-[(E)-(3S)-3-hydroxy-1-octenyl]-2H-cyclopenta[b]thiophene-$\Delta^{2,\delta}$-valeric acid = (5Z)-9-Deoxy-6,9a-epithio-Δ^5-prostaglandin $F_{1\alpha}$ = (5Z,11α,13E,15S)-6,9-Epithio-11,15-dihydroxyprosta-5,13-dien-1-oic acid (●)

S PGI$_2$-S, *6.9-Thiaprostacyclin*, TPGI$_2$

U Antihypertensive

10021   (8119) $C_{20}H_{32}O_5$
363-24-6

(E,Z)-(1R,2R,3R)-7-[3-Hydroxy-2-[(3S)-(3-hydroxy-1-
octenyl)]-5-oxocyclopentyl]-5-heptenoic acid =
(5Z,11α,13E,15S)-11,15-Dihydroxy-9-oxoprosta-5,13-
dien-1-oic acid (●)
S   Cervidil "Forest", Cerviprime, Dinopron-EM, *Dinoprostone***, Enzaprost-E, Femidyn, *Medullin*, Minprostin, Minprostin E_2, PGE$_2$, Prandin E2, Prepidil,
Primiprost, Prolisina E2, Propess, Propess-RS, Pros-
tagland E2, *Prostaglandin E_2*, Prostarmon E, Proste-
non, Prostin "Orifarm", Prostin(e) E_2, Prostin VR,
U-12062
U   Oxytocic, abortifacient, vasodilator, smooth muscle
stimulant

10021-01   (8119-01)
R   Dinoprostone incorporated in a three-dimensional
cross-linked starch derivative
S   Cerviprost, *Polydextrin dinoprostone*
U   Oxytocic, abortifacient

10022   (8118) $C_{20}H_{32}O_5$
35121-78-9

(Z)-(3aR,4R,5R,6aS)-Hexahydro-5-hydroxy-4-[(E)-(3S)-
3-hydroxy-1-octenyl]-2H-cyclopenta[b]furan-Δ^{2,δ}-vale-
ric acid = (Z)-5-[(3aR,4R,5R,6aS)-5-Hydroxy-4-[(E)-
(3S)-3-hydroxy-1-octenyl]perhydrocyclopenta[b]furan-2-
ylidene]valeric acid = (5Z)-9-Deoxy-6,9α-epoxy-Δ^5-
prostaglandin F_1 = (5Z,9α,11α,13E,15S)-6,9-Epoxy-
11,15-dihydroxyprosta-5,13-dien-1-oic acid (●)
R   Sodium salt (61849-14-7)

S   Cyclo-Prostin, *Epoprostenol sodium***, Flolan,
PGI$_2$, PGX, *Prostacyclin sodium*, Prostaglandin I_2,
Prostaglandin X, U-53217 A
U   Platelet aggregation inhibitor, antimetastatic

10023   (8120) $C_{20}H_{32}O_6$
78690-80-9

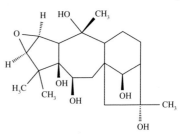

S   *Rhododendrotoxin, Rhomotoxin*
U   Antihypertensive from *Rhododendron molle*

10024   (8121) $C_{20}H_{33}NO_3$
468-61-1

2-[2-(Diethylamino)ethoxy]ethyl 2-ethyl-2-phenylbuty-
rate = α,α-Diethylphenylacetic acid 2-[2-(diethylamino)ethoxy]ethyl ester = α,α-Diethylbenzeneacetic acid 2-
[2-(diethylamino)ethoxy]ethyl ester (●)
R   Citrate (1:1) (52432-72-1)
S   Antivix, Antusel, Aplacol, Bronchell-retard, Calmo-
quintil, Dorex retard, Dresan antitusivo, Elitos, Erga-
tosse, Ethochlon, Frenotos "Disprovent", Hihustan,
Hustopan-Ox, Hydon OX, Marukofon-A, Neoas-
drin, Neobex, Neusedan, Notox, *Oxeladin citrate***,
Oxethamol, Paxeladine, Pectamol, Pectamon, Pec-
tussil, Plardox, Silopentol, stas-Hustenstiller, To-
xedin, Tussedina-Smit, Tussilisin, Tussimol, Tusso-
caps, Tusuprex
U   Antitussive

10024-01   (8121-01)
R   Tannate
S   Tarmina
U   Antitussive

10025 (8122)　　　　　　　　　　$C_{20}H_{33}NO_3$
　　　　　　　　　　　　　　　　　　299-61-6

3-(Diethylamino)-1,2-dimethylpropyl *p*-isobutoxybenzoate = *p*-Isobutoxybenzoic acid γ-diethylamino-α,β-dimethylpropyl ester = 4-(2-Methylpropoxy)benzoic acid 3-(diethylamino)-1,2-dimethylpropyl ester (●)
R Hydrochloride (1510-29-8)
S *Ganglefene hydrochloride***, Ganglerone
U Ganglion blocking agent, coronary vasodilator

10026 (8123)　　　　　　　　　　$C_{20}H_{33}NO_3$
　　　　　　　　　　　　　　　　　　36691-30-2

1-(Isopropylamino)-3-[(2,2,5,7,8-pentamethylchroman-6-yl)oxy]-2-propanol = 2,2,5,7,8-Pentamethyl-6-(3-isopropylamino-2-hydroxypropoxy)chroman = 1-[(3,4-Dihydro-2,2,5,7,8-pentamethyl-2*H*-1-benzopyran-6-yl)oxy]-3-[(1-methylethyl)amino]-2-propanol (●)
R Tartrate (2:1) (36691-33-5)
S *Cromipranol*, VÚFB-6493
U β-Adrenergic blocker

10027 (8124)　　　　　　　　　　$C_{20}H_{33}NO_4S$
　　　　　　　　　　　　　　　　　　80225-28-1

Methyl 4-[[(3a*R**,4*R**,5*R**,6a*S**)-3,3a,4,5,6,6a-hexahydro-5-hydroxy-4-[(*E*)-(3*S**)-3-hydroxy-1-octenyl]cyclopenta[*b*]pyrrol-2-yl]thio]butyrate =
3aα,4α(1*E*,3*S**),5β,6aα]-(±)-4-[[3,3a,4,5,6,6a-Hexahydro-5-hydroxy-4-(3-hydroxy-1-octenyl)cyclopenta[*b*]pyrrol-2-yl]thio]butanoic acid methyl ester (●)
S Hoechst 892, S 792892 A, *Tilsuprost***
U Antihypertensive, cardiotonic, anticoagulant

10028 (8125)　　　　　　　　　　$C_{20}H_{33}NO_7$
　　　　　　　　　　　　　　　　　　123122-54-3

(α*S*)-1-[(*cis*-4-Carboxycyclohexyl)carbamoyl]-α-[(2-methoxyethoxy)methyl]cyclopentanepropionic acid = [4(*S*)-*cis*]-4-[[[1-[2-Carboxy-3-(2-methoxyethoxy)propyl]cyclopentyl]carbonyl]amino]cyclohexanecarboxylic acid (●)
R see also no. 14187
S *Candoxatrilat***, UK-73967
U Antihypertensive (endopeptidase inhibitor)

10029 (8126)　　　　　　　　　　$C_{20}H_{33}N_2O$
　　　　　　　　　　　　　　　　　　6004-98-4

4-(β-Cyclohexyl-β-hydroxyphenethyl)-1,1-dimethylpiperazinium = 4-(2-Cyclohexyl-2-hydroxy-2-phenylethyl)-1,1-dimethylpiperazinium (●)
R Methylsulfate (115-63-9)
S AB 803, Duopax, *Hexocyclium metilsulfate***, Mosidal (new form), Plegan, Sanulcer, Tral, Tralin(e), Ulcazina, Ulcolin
U Anticholinergic, anti-ulcer agent

10030　(8127)　　　　　　　　　　$C_{20}H_{33}N_3O_3$
57460-41-0

(±)-1-[p-[3-(tert-Butylamino)-2-hydroxypropoxy]phe-
nyl]-3-cyclohexylurea = (±)-N-Cyclohexyl-N'-[4-[3-
[(1,1-dimethylethyl)amino]-2-hydroxypropoxy]phe-
nyl]urea (●)
S　02-115, Cordanum, Kordanum, *Talinolol***
U　β-Adrenergic blocker

10031　(8128)　　　　　　　　　　$C_{20}H_{33}N_3O_3S$
87056-78-8

(±)-N,N-Diethyl-N'-[(3R*,4aR*,10aS*)-
1,2,3,4,4a,5,10,10a-octahydro-6-hydroxy-1-propylben-
zo[g]quinolin-3-yl]sulfamide = (3S,4aS,10aR)-3-
(Diethylaminosulfonylamino)-1,2,3,4,4a,5,10,10a-octa-
hydro-1-propylbenzo[g]quinolin-6-ol = (3α,4aα,10aβ)-
(±)-N,N-Diethyl-N'-(1,2,3,4,4a,5,10,10a-octahydro-6-
hydroxy-1-propylbenzo[g]quinolin-3-yl)sulfamide (●)
R　also monohydrochloride (94424-50-7)
S　CV 205-502, Norprolac, *Quinagolide***,
　　SDZ 205-502, SDZ-CV 205-502
U　Prolactin inhibitor (D_2-dopamine receptor agonist)

10032　(8129)　　　　　　　　　　$C_{20}H_{33}N_3O_4$
56980-93-9

3-[3-Acetyl-4-[3-(tert-butylamino)-2-hydroxyprop-
oxy]phenyl]-1,1-diethylurea = N'-[3-Acetyl-4-[3-[(1,1-di-
methylethyl)amino]-2-hydroxypropoxy]phenyl]-N,N-di-
ethylurea (●)
R　also monohydrochloride (57470-78-7)

S　Cardem "Rorer", Celectol, Celipro-Lich, *Celipro-
lol***, Celipro-Sanorania, Cordiax, Corliprol, Dila-
norm, Jofurol, Moderator, NBP 582, REV 5320 A,
RHC 5320 A, Selecor, Selectol, Selecturon,
ST 1396, UL/1677
U　β-Adrenergic blocker

10033　(8130)　　　　　　　　　　$C_{20}H_{33}N_3O_4S$
139147-26-5

4-[[1-[(2-Cyclohexyl-2-hydroxyethylidene)amino]-3-
ethyloctahydro-2-oxo-5-cyclopentimidazolyl]thio]buta-
noic acid (●)
S　192 C 86
U　Prostaglandin DP-receptor partial agonist (platelet aggregation inhibitor, antiglaucoma)

10034　(8131)　　　　　　　　　　$C_{20}H_{33}N_5O_9$
120081-14-3

1-[N^2-[N-(N-Acetyl-L-seryl)-L-α-aspartyl]-L-lysyl]-L-
proline (●)
S　AcSDKP, *Goralatide***, Nac-SDKP, *Serapimod*, Se-
raspenide
U　Immunomodulator

10035　(8132)　　　　　　　　　　$C_{20}H_{33}N_5O_{11}$
76490-22-?

D-Lactyl-L-alanyl-γ-L-glutamyl-L-*meso*-α,ε-diaminopi-
melylglycine = (*R*)-*N*-(2-Hydroxy-1-oxopropyl)-L-ala-
nyl-D-γ-glutamyl-*meso*-α,ε-diaminopimelylglycine =
(*R*)-*N*-[(*R*)-6-Carboxy-N^2-[*N*-[*N*-(2-hydroxy-1-oxopro-
pyl)-L-alanyl]-D-γ-glutamyl]-L-lysyl]glycine (●)
S　FK 156, *Gludapcin*
U　Immunostimulant from *Streptomyces olivaceo-gri-
seus* and *Str. violaceus*

10036　　　　　　　　　　　　　　　　$C_{20}H_{33}N_5O_{11}$

3-[[5-(Aminomethyl)tetrahydro-3,4-dihydroxy-2-fura-
nyl]oxy]-2-[(3-aminopropyl)methylamino]-3-[5-(1,2,3,4-
tetrahydro-2,4-dioxo-1-pyrimidinyl)tetrahydro-3,4-di-
hydroxy-2-furanyl]propionic acid = 5-*O*-(5-Amino-5-de-
oxypentofuranosyl)-6-[(3-aminopropyl)methylamino]-
1,6-dideoxy-1-(3,4-dihydro-2,4-dioxo-1(2*H*)pyrimidi-
nyl)heptofuranuronic acid (●)
S　FR 900493
U　Antibacterial from *Bacillus cereus*

10037　(8133)　　　　　　　　　　　　$C_{20}H_{33}N_7O_3$
　　　　　　　　　　　　　　　　　　　　98833-92-2

8-[3-[4-[(Diethylamino)carbonyl]-1-piperazinyl]pro-
pyl]caffeine = *N,N*-Diethyl-4-[3-(2,3,6,7-tetrahydro-
1,3,7-trimethyl-2,6-dioxo-1*H*-purin-8-yl)propyl]-1-pipe-
razinecarboxamide (●)
S　*Estacofilina***, S 9977, *Stacofylline***
U　Migraine therapeutic, nootropic

10038　(8134)　　　　　　　　　　　　$C_{20}H_{34}AuO_9PS$
　　　　　　　　　　　　　　　　　　　　34031-32-8

(1-Thio-β-D-glucopyranosate)(triethylphosphine)gold
2,3,4,6-tetraacetate = *S*-(Triethylphosphoranediyllaurio)-
1-thio-β-D-glucopyranose 2,3,4,6-tetraacetate = (2,3,4,6-
Tetra-*O*-acetyl-1-thio-β-D-glucopyranosato-*S*)(triethylphosphine)gold (●)
S　AF, Aktil, *Auranofin***, Auranoid, Aurilfar, Auro-
pan "Krka", Crisinor, Crisofin, Ridaura, Ridauran,
SKF 39162, SKF D-39162
U　Anti-arthritic, antirheumatic

10039　(8135)　　　　　　　　　　　　$C_{20}H_{34}GdN_5O_{10}$
　　　　　　　　　　　　　　　　　　　　131069-91-5

[*N,N*-Bis[2-[[(carboxymethyl)[(2-methoxyethyl)carb-
amoyl]methyl]amino]ethyl]glycinato(3–)]gadolinium =
[8,11-Bis(carboxymethyl)-14-[2-[(2-methoxyethyl)ami-
no]-2-oxoethyl]-6-oxo-2-oxa-5,8,11,14-tetraazahexade-
can-16-oato(3–)]gadolinium (●)
S　*Gadoversetamide***, MP-1177, Optimark
U　Diagnostic aid

10040　(8136)　　　　　　　　　　　　$C_{20}H_{34}N_2O_4$
　　　　　　　　　　　　　　　　　　　　119940-80-6

(±)-(5*R**,6*S**,7*R**)-7-Hexyl-2,4-dioxo-1,3-diazaspi-
ro[4.4]nonane-6-heptanoic acid = (*RS*)-*cis*-6-(6-Carboxy-

hexyl)-*trans*-7-hexyl-1,3-diazaspiro[4.4]nonane-2,4-
dione = (5α,6α,7β)-(±)-7-Hexyl-2,4-dioxo-1,3-diazaspi-
ro[4.4]nonane-6-heptanoic acid (●)
R Sodium salt
S IBI-P 01028, *Spiriprostil sodium***
U Anti-ulcer agent

10041 (8137) $C_{20}H_{34}N_2O_5$
                                                108391-88-4

(2S,3aS,7aS)-1-[(S)-N-[(S)-1-Carboxypentyl]alanyl]hexa-
hydro-2-indolinecarboxylic acid 1-ethyl ester = [2S-[1-
[R*(R*)],2α,3aβ,7aβ]]-1-[2-[[1-(Ethoxycarbonyl)pen-
tyl]amino]-1-oxopropyl]octahydro-1*H*-indole-2-carb-
oxylic acid (●)
S Armedil, *Orbutopril***
U Antihypertensive (ACE inhibitor)

10042 (8138) $C_{20}H_{34}N_2O_6$
                                                126509-46-4

(4S)-1,2-Epoxy-2-(hydroxymethyl)-4-(N-isooctanoyl-L-
serinamido)-6-methylhept-6-en-3-one = [2R-
[2R*[S*(S*)]]]-N-[1-(Hydroxymethyl)-2-[[2-(hydroxymethyl)oxiranyl]carbonyl]-3-methyl-3-butenyl]ami-
no]-2-oxoethyl]-6-methylheptanamide (●)
S *Eponemycin*
U Antineoplastic antibiotic from *Streptomyces hygro-
        scopicus* no. P 247-71

10043 (8139) $C_{20}H_{34}N_4O_4$
                                                83654-05-1

Cyclohexanone O,O'-[1,6-hexanediylbis(iminocarbo-
nyl)]dioxime (●)
S RHC 80267, U-57908
U Enzyme inhibitor

10044 (8140) $C_{20}H_{34}N_4O_{11}$
                                                60355-78-4

N^2-[N-(N-Acetylmuramoyl)-L-alanyl]-D-α-glutamine
methyl ester (●)
S *Murametide*
U Immunostimulant

10045 (8141) $C_{20}H_{34}N_4O_{12}$
                                                66112-59-2

2-Acetamido-3-O-[[(1R)-1-[(1S,2R)-1-[[(1R)-1-carb-
amoyl-3-carboxypropyl]carbamoyl]-2-hydroxypro-
pyl]carbamoyl]ethyl]-2-deoxy-D-glucopyranose = N-
[(R)-2-[(3R,4R,5S,6R)-3-Acetamidotetrahydro-2,5-di-
hydroxy-6-hydroxymethyl-2*H*-pyran-4-yloxy]propio-
nyl]-L-threonyl-D-isoglutamine = N^2-[N-(N-Acetylmuramoyl)-L-threonyl]-D-α-glutamine (●)
S RS-37449, SDZ 280-636, *Temurtide***
U Vaccine adjuvant, immunomodulator

10046   (8142)

$C_{20}H_{34}O_2$
77769-22-3

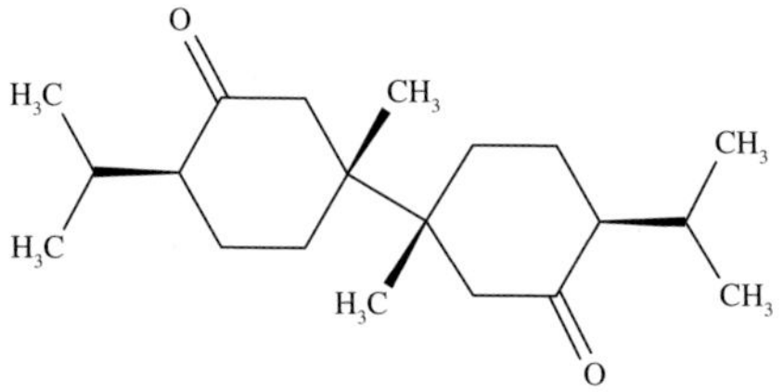

1,1'-Bisisomenthone = [1α(1'S*,4'R*),4β]-1,1'-Dimethyl-
4,4'-bis(1-methylethyl)[1,1'-bicyclohexyl]-3,3'-dione (●)
S   Antiasthmone
U   Anti-asthmatic

10047   (8143)

$C_{20}H_{34}O_2$
64218-02-6

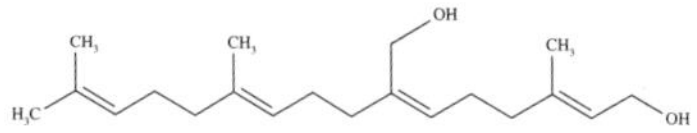

(2Z,6E)-2-[(3E)-4,8-Dimethyl-3,7-nonadienyl]-6-me-
thyl-2,6-octadiene-1,8-diol = (2E,6Z,10E)-7-(Hydroxymethyl)-3,11,15-trimethyl-2,6,10,14-hexadecatetraen-1-
ol = (Z,E,E)-2-(4,8-Dimethyl-3,7-nonadienyl)-6-methyl-
2,6-octadiene-1,8-diol (●)
S   CS 684, Kelnac, Kelnal, *Plaugenol, Plaunotol***
U   Gastrointestinal agent from *Croton sublyratus*

10048   (8144)

$C_{20}H_{34}O_2$
14760-52-2

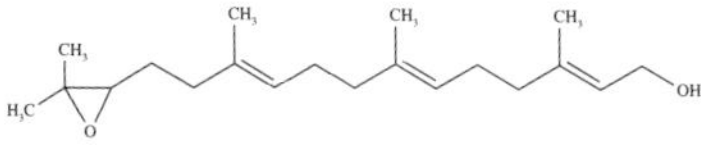

all-trans-(−)-14,15-Epoxygeranylgeranol = (−)-(E,E,E)-
13-(3,3-Dimethyloxiranyl)-3,7,11-trimethyl-2,6,10-tride-
catrien-1-ol (●)
S   *Pterodon*
U   for schistosomiasis prophylaxis (terpene from *Pterodon pubescens*)

10049   (8145)

$C_{20}H_{34}O_3$
111025-83-3

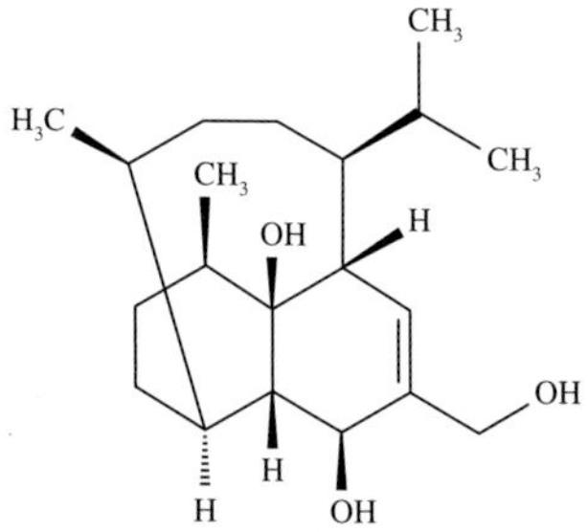

[1R-(1α,4β,4aβ,5α,8β,8aβ,9S*,12S*)]-4,4a,5,6,7,8-He-
xahydro-3-(hydroxymethyl)-8,9-dimethyl-12-(1-methyl-
ethyl)-1,5-butanonaphthalene-4,8a(1H)-diol (●)
S   *Vinigrol*
U   Antihypertensive, platelet aggregation inhibitor from
Virgaria nigra

10050   (8146)

$C_{20}H_{34}O_3$
54857-96-4

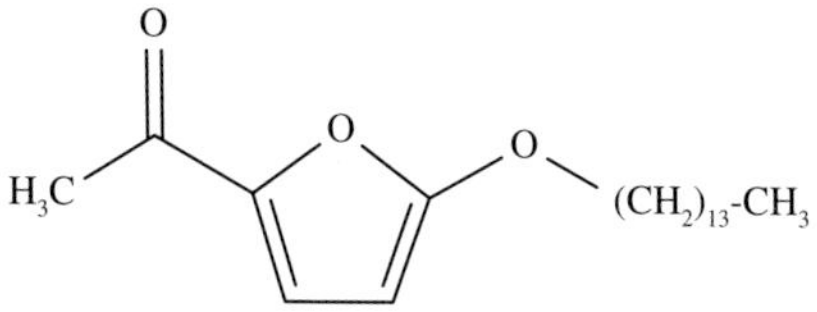

2-Acetyl-5-(tetradecyloxy)furan = 1-[5-(Tetradecyloxy)-
2-furanyl]ethanone (●)
S   RMI 15731
U   Antirhinoviral

10051

$C_{20}H_{34}O_3$
62279-93-0

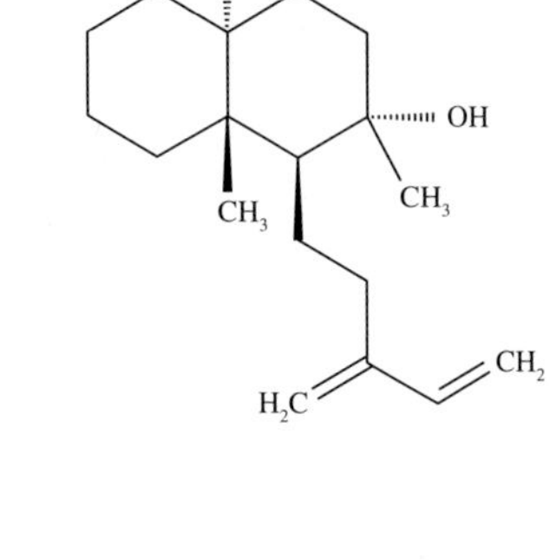

(1R,3S,4S,4aR,8S,8aS)-Decahydro-8-(hydroxymethyl)-3,4a,8-trimethyl-4-(3-methylene-4-pentenyl)-1,3-naphthalenediol (●)
S *Andalusol*
U Anti-inflammatory from *Sideritis foetens*

10052 (8147) $C_{20}H_{34}O_3S$
110902-61-9

[1R-[1α,2α(Z),3α,4α]]-7-[3-[(Hexylthio)methyl]-7-oxabicyclo[2.2.1]hept-2-yl]-5-heptenoic acid (●)
S SQ 28913
U Platelet aggregation inhibitor, bronchodilator

10053 (8148) $C_{20}H_{34}O_4$
38966-21-1

[3R-(3α,4α,4aα,6aβ,8β,9β,11aβ,11bβ)]-Tetradecahydro-3,9-dihydroxy-4,11b-dimethyl-8,11a-methano-11aH-cyclohepta[a]naphthalene-4,9-dimethanol (●)
S *Aphidicolin*, ICI 69653
U Antiviral, antineoplastic

10054 $C_{20}H_{34}O_4$
64695-05-2

(5Z,9α,11α,13E)-9,11-Dihydroxyprosta-5,13-dien-1-oic acid (●)
R Monosodium salt (138282-73-2)
S S-1033
U Antiglaucoma agent

10055 (8150) $C_{20}H_{34}O_5$
551-11-1

(E,Z)-(1R,2R,3R,5S)-7-[3,5-Dihydroxy-2-[(3S)-(3-hydroxy-1-octenyl)]cyclopentyl]-5-heptenoic acid =
(5Z,9α,11α,13E,15S)-9,11,15-Trihydroxyprosta-5,13-dien-1-oic acid (●)
S Bovigland, Dinogyn F, *Dinoprost***, Dinoripe Gel, Enzaprost-F, Equigland, Ginoprost, Glandal, Glandin N, Glandinon, Gravidex, Hormo P 2 alpha, Horsafertil (old form), Panacelan F, PGF$_2$, PG Tab "Dr. Reddy", Proglaxinum, Prosmon, Prostadiel F, Prostaglan, *Prostaglandin F$_{2\alpha}$*, Prostamodin-F, Prostarmon F, Prostin(e) F$_2$ Alpha, Protamodin F, Sincro, Suigland, U-14583
U Oxytocic, abortifacient, vasodilator, smooth muscle stimulant

10055-01 (8150-01) 38562-01-5
R Compd. with trometamol (no. 314) (1:1)
S Amoglandin, Dinolytic, *Dinoprost Trometanol*, *Dinoprost Tromethamine*, Lutalyse, Minprostin F$_{2\alpha}$, PGF$_2$-THAM, Pronalgon F, Prostin(e), Prostin(e) F$_2$, Prostin(e) F$_2$ Alpha Inj., U-14583 E, Zinoprost
U Oxytocic, abortifacient, vasodilator, smooth muscle stimulant

10056 (8149) $C_{20}H_{34}O_5$
745-65-

(1R,2R,3R)-3-Hydroxy-2-[(E)-(3S)-3-hydroxy-1-octe-nyl]-5-oxocyclopentaneheptanoic acid = (11α,13E,15S)-11,15-Dihydroxy-9-oxoprost-13-en-1-oic acid (●)

S Alista, *Alprostadil***, Alprostan, Alprostapint, Alprox-TD, Befar, Bondil, Caverject, Femprox, Karon, Liprostin, Minprog, Musa "Astra", Muse, PGE$_1$, Prosilina VR, *Prostaglandin E$_1$*, Prostin E$_1$, Prostin(e) VR, Prostivas, Protensit, Topiglan, U-10136, Vasaprostan, Vasopint

U Oxytocic, abortifacient, vasodilator, smooth muscle stimulant

10056-01   (8149-01)   93591-00-5
R α-Cyclodextrin clathrate
S *Alprostadil alfadex***, Alprostar, Altesil, Apistandin, Edex, Ipceryl, Mediprost, *PGE$_1$-alfadex*, Prostandin, Prostavasin, Prostavisin, Rigidur, Sugiran, Vazaprostan, Viridal
U Peripheral vasodilator

10056-02   (8149-02)
R Incorporated in microspheres (soybean lipid emulsion)
S Eglandin, Liple, *Lipo-Alprostadil, Lipo-PGE$_1$*, Palux "Taisho", Riprodil, TLC C-53
U Peripheral vasodilator

10057   (8151) $C_{20}H_{34}O_6S$
                                                      55837-16-6

2-[2-[2-[4-(1,1,3,3-Tetramethylbutyl)phenoxy]eth-oxy]ethoxy]ethanesulfonic acid (●)
R Sodium salt (2917-94-4)
S *Entsufon-Natrium***, Entsufon Sodium***
U Detergent

10058   (8152) $C_{20}H_{34}O_8$
                                                      77-90-7

Tributyl *O*-acetylcitrate = 2-(Acetyloxy)-1,2,3-propanetricarboxylic acid tributyl ester (●)
S Blo-trol
U Meteorism therapeutic (veterinary)

10059   (8153) $C_{20}H_{35}NOS$
                                                      54767-75-8

erythro-p-(Isopropylthio)-α-[1-(octylamino)ethyl]benzyl alcohol = *erythro*-1-[p-(Isopropylthio)phenyl]-2-(octyl-amino)-1-propanol = (αS)-4-[(1-Methylethyl)thio]-α-[(1R)-1-(octylamino)ethyl]benzenemethanol (●)
S Bemperil, Cerebro, Circleton, CP 556 S, Daufan, Dulasi, Duloctil, Euesdon "Farmavel", Euvasal, Farectil, Flindixan, Fluversin, Fluvisco, Hapturon, Hemoantin, Iangene, Ibisul, Lebanoxan, Loctidene, Loctidon, Locton, Metactiv, MJF-12637, MY 103, Octamet, Palinium, Periman, Polivasal, Ravenil "Genepharm", Sat, Sofectidil, Sudil, Sulc, Sulcalin, Suloctal, *Suloctidil***, Sulocton, Sulodene, Suloktil, Sutidil, Tamid, Tioloct, Vascudil, Vasedil, Zomotur
U Peripheral vasodilator, platelet aggregation inhibitor

10060   (8154) $C_{20}H_{35}NO_2$
                                                      561-77-3

1-Cyclohexylcyclohexanecarboxylic acid β-piperidinoethyl ester = 2-Piperidinoethyl ester of bicyclohexyl-1-
carboxylic acid = [1,1'-Bicyclohexyl]-1-carboxylic acid
2-(1-piperidinyl)ethyl ester (●)
R also hydrochloride (5588-25-0)
S *Dihexiverine, Dihexyverine***, Dispas, Diverine,
J.L. 1078, Metaspas, Neospasmina, Olimplex, Sec-
lin, Spasmalex, Spasmodex, Spasmolevel
U Anticholinergic, spasmolytic

10061 (8155)

$C_{20}H_{35}NO_2$
79700-61-1

1-[1-(Isobutoxymethyl)-2-[[1-(1-propynyl)cyclohe-
xyl]oxy]ethyl]pyrrolidine = 1-[1-[(2-Methylpropoxy)me-
thyl]-2-[[1-(1-propynyl)cyclohexyl]oxy]ethyl]pyrroli-
dine (●)
R Hydrochloride (103915-56-6)
S CERM 4205, *Dopropidil hydrochloride***,
Org 30701
U Anti-anginal, anti-ischemic, anti-arrhythmic

10062 (8156)

$C_{20}H_{35}NO_2S$
70895-45-3

1-[p-(Isopropylthio)phenoxy]-3-(octylamino)-2-propa-
nol = 1-[4-[(1-Methylethyl)thio]phenoxy]-3-(octylami-
no)-2-propanol (●)
R Hydrochloride (70895-39-5)
S MJ 12880-1, *Tipropidil hydrochloride***
U Vasodilator

10063

$C_{20}H_{35}N_3O_6$
142489-44-9

N-(1,7-Dioxododecyl)-L-alanyl-D-α-glutamine (●)
S LK-404
U Bone marrow protector

10064 (8157)

$C_{20}H_{36}BrN_3O_2$
6042-36-0

4-(2-Bromo-4,5-dimethoxyphenyl)-1,1,7,7-tetraethyl-
diethylenetriamine = N-(2-Bromo-4,5-dimethoxyphenyl)-
N-[2-(diethylamino)ethyl]-N',N'-diethyl-1,2-ethanedi-
amine (●)
S RC-12, WR 27653
U Antimalarial

10065 (8158)

$C_{20}H_{36}N$
7631-49-4

1-Tetradecyl-4-picolinium = N-Myristyl-γ-picolinium =
4-Methyl-1-tetradecylpyridinium (●)
R Chloride (2748-88-1)
S *Miripirium chloride***, Quatrasan, Quatresin
U Antiseptic

10066   (8159) $C_{20}H_{36}N_2O_3S$
                                                  122647-31-8

(±)-4'-[4-(Ethylheptylamino)-1-hydroxybutyl]methane-
sulfonanilide = (±)-N-[4-[4-(Ethylheptylamino)-1-hydr-
oxybutyl]phenyl]methanesulfonamide (●)
R   Fumarate (2:1) (122647-32-9)
S   Corvert, Covert, *Ibutilide fumarate***, U-70226 E
U   Anti-arrhythmic

10067   (8160) $C_{20}H_{36}N_6O$
                                                  135-43-3

1,1'-[4-(Dodecyloxy)-*m*-phenylene]diguanidine = 2,4-Di-
guanidinophenyl dodecyl ether = N,N'''-[4-(Dodecyloxy)-
1,3-phenylene]bis[guanidine] (●)
R   Dihydrochloride (135-42-2)
S   Farmidril, *Lauroguadine hydrochloride***, P₇
U   Topical antiseptic

10068 $C_{20}H_{36}N_{10}O_2$
                                                  219325-24-3

O,O'-Bis(4,5-diamino-1,2-dihydro-2,2-tetramethylene-*s*-
triazin-1-yl)-1,6-hexanediol = 10,10'-[1,6-Hexanediyl-
bis(oxy)]bis[6,8,10-triazaspiro[4.5]deca-6,8-diene-7,9-
diamine] (●)
R   Dihydrochloride (172280-69-2)
S   SIPI 1029, T-46, *Trybizine hydrochloride*
U   Antitrypanosomal

10069   (8162) $C_{20}H_{36}O_2$
                                                  623-32-5

Chaulmoogric acid ethyl ester = 2-Cyclopentene-1-tride-
canoic acid ethyl ester (●)
R   Mixture with hydnocarpic acid ethyl ester (no. 8270)
        (8054-56-6)
S   Antileprol, Chaulmestrol, Chaulmoogrol, Hyrganol,
        Moogrol
U   Tuberculostatic, leprostatic

10070   (8161) $C_{20}H_{36}O_2$
                                                  544-35-4

Ethyl linoleate = Linoleic acid ethyl ester = (Z,Z)-9,12-
Octadecadienoic acid ethyl ester (●)
S   Cholestan F, *Ethyl linoleate*, Floria "Vitamin F"-Sal-
        be, Floriabene, Lipotate, Mandenol
U   Anti-atherosclerotic, dermatic

10071   (8163) $C_{20}H_{36}O_4$
                                                  52591-03-4

[3R-[3α(R*),4aβ,5β,6aα,10aβ,10bα]]-1-(Dodecahydro-
5-hydroxy-3,4a,7,7,10a-pentamethyl-1H-naph-
tho[2,1-*b*]pyran-3-yl)-1,2-ethanediol (●)
S   *Borjatriol*
U   Anti-inflammatory (diterpenoid from *Sideritis mugro-
        nensis*)

10072 (8164) $C_{20}H_{36}O_5$
745-62-0

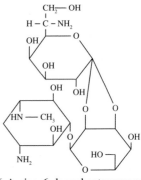

(1*R*,2*R*,3*R*,5*S*)-3,5-Dihydroxy-2-[(*E*)-(3*S*)-3-hydroxy-1-octenyl]cyclopentaneheptanoic acid = (9α,11α,13*E*,15*S*)-9,11,15-Trihydroxyprost-13-en-1-oic acid (●)
S PGF$_1$, *Prostaglandin F*$_{1α}$
U Oxytocic, abortifacient, vasodilator, smooth muscle stimulant

10073 (8165) $C_{20}H_{37}I_3O_2$
60996-84-1

Ethyl 10,13,16-triiodostearate = 10,13,16-Triiodostearic acid ethyl ester = 10,13,16-Triiodooctadecanoic acid ethyl ester (●)
S Hepatoselectan
U Diagnostic aid (radiopaque medium)

10074 (8166) $C_{20}H_{37}NO_3$
53716-44-2

2-(Diethylamino)-1-methylethyl *cis*-1-hydroxy[bicyclohexyl]-2-carboxylate = *cis*-2-Hydroxy-2-cyclohexylcyclohexanecarboxylic acid 2-(diethylamino)-1-methylethyl ester = 1-Hydroxy[1,1'-bicyclohexyl]-2-carboxylic acid 2-(diethylamino)-1-methylethyl ester (●)
S LG-30158, Rilaten, *Rociverine* **
U Anticholinergic, spasmolytic

10075 (8167) $C_{20}H_{37}N_3O_2$
16791-42-7

N,*N*-Bis(2-diethylaminoethyl)-3,4-dimethoxyaniline = 1,2-Dimethoxy-4-[bis(diethylaminoethyl)amino]benzene = *N*-[2-(Diethylamino)ethyl]-*N*-(3,4-dimethoxyphenyl)-*N'*,*N'*-diethyl-1,2-ethanediamine (●)
S Dimeplasmin
U Antimalarial

10076 (8168) $C_{20}H_{37}N_3O_{13}$
14918-35-5

5-*O*-[2,3-*O*-(6-Amino-6-deoxyheptopyranosylidene)-β-D-talopyranosyl]-2-deoxy-*N*^{3}-methyl-D-streptamine = *O*-6-Amino-6-deoxy-L-*glycero*-D-*galacto*-heptopyranosylidene-(1→2-3)-*O*-β-D-talopyranosyl(1→5)-2-deoxy-*N*^{1}-methyl-D-streptamine (●)
S Anthelmin "TAD", *Destomycin A*, Destonate 20
U Veterinary anthelmintic (antibiotic from *Streptomyces rimofaciens*)

10077   (8169)   C$_{20}$H$_{37}$N$_3$O$_{13}$
                                              31282-04-9

O-6-Amino-6-deoxy-L-*glycero*-D-*galacto*-heptopyrano-
sylidene-(1→2-3)-*O*-β-D-talopyranosyl-(1→5)-2-deoxy-
N^3-methyl-D-streptamine (●)
S Hyanthelmix, *Hygromycin B*, Hygrovetine
U Veterinary anthelmintic (antibiotic from *Streptomyces hygroscopicus*)

10078   (8170)   C$_{20}$H$_{37}$N$_5$O$_{10}$
                                              129009-83-2

N,N-Bis[2-[[(carboxymethyl)][(2-methoxyethyl)carb-
amoyl]methyl]amino]ethyl]glycine = 8,11-Bis(carboxymethyl)-14-[2-[(2-methoxyethyl)amino]-2-oxoethyl]-6-
oxo-2-oxa-5,8,11,14-tetraazahexadecan-16-oic acid =
N,N-Bis[2-[(carboxymethyl)][2-[(2-methoxyethyl)ami-
no]-2-oxoethyl]amino]ethyl]glycine (●)
S MP-1196, *Versetamide***
U Stabilizer, carrier agent

10079   C$_{20}$H$_{37}$O$_6$P
                                              144279-58-3

2-Hexylcyclopropaneoctanoic acid (2-hydroxy-2-oxido-
1,3,2-dioxaphospholan-4-yl)methyl ester (●)
R Sodium salt (152141-70-3)
S PHILPA
U Antiproliferative from *Physarum polycephalum*

10080   (8171)   C$_{20}$H$_{38}$I$_2$O$_2$
                                              7008-02-8

Ethyl 9,10-diiodooctadecanoate = Ethyl 9,10-diiodo-
stearate = 9,10-Diiodostearic acid ethyl ester = 9,10-Diio-
dooctadecanoic acid ethyl ester (●)
S Angiopac (old form), *Iodetryl***, *Jodetrylum*, Vaso-
       selectan
U Diagnostic aid (radiopaque medium-vasographic)

10081   (8172)   C$_{20}$H$_{38}$NO$_2$
                                              15518-72-6

Triethyl(2-hydroxyethyl)ammonium dicyclopentylace-
tate = 2-(Dicyclopentylacetoxy)tetraethylammonium =
Dicyclopentylacetic acid β-triethylammoniooethyl ester =
2-[(Dicyclopentylacetyl)oxy]-*N,N,N*-triethylethanamini-
um (●)
R Bromide (2001-81-2)
S DCDE, *Dipenine bromide*, *Diponium bromide***, Es-
       paston, HL 267, Sa 267, Spaston, Unospaston
U Spasmolytic

10082 (8173) $C_{20}H_{38}N_2O_2$

[p-Phenylenebis(oxytrimethylene)]bis[ethyldimethylam-
monium] = N,N'-[3,3'-(p-Phenylenedioxy)dipro-
pyl]bis[ethyldimethylammonium] = 3,3'-[1,4-Phenylene-
bis(oxy)]bis[N-ethyl-N,N-dimethyl-1-propanaminium]
(●)
R Diiodide (5807-94-3)
S *Dipropamine*
U Ganglion blocking agent

10083 (8174) $C_{20}H_{38}N_2O_4$

1-(2-Hydroxyethyl)-1-methylpiperidinium succinate =
N,N'-[2,2'-(Succinyldioxy)diethyl]bis[N-methylpiperidi-
nium] = 1,1'-[(1,4-Dioxo-1,4-butanediyl)bis(oxy-2,1-
ethanediyl)]bis[1-methylpiperidinium] (●)
R Dibromide (60996-86-3)
S Cyclotylin
U Cholinesterase inhibitor

10084 (8175) $C_{20}H_{38}N_2O_4$
60381-08-0

N,N'-[2,2'-Ethylenebis(1,3-dioxolan-4-ylmethyle-
ne)]bis[N-methylpyrrolidinium] = 1,1'-[1,2-Ethanediyl-
bis(1,3-dioxolane-2,4-diylmethylene)]-bis[1-methylpyr-
rolidinium] (●)
R Diiodide (37069-07-1)
S Dioxonium

U Muscle relaxant

10085 (8176) $C_{20}H_{38}N_4O_4$
119-48-2

4,4'-(N,N'-Dibutyl-N,N'-ethylenedicarbamoyl)dimorpho-
line = N,N'-Dibutyl-N,N'-bis(morpholinocarbonyl)ethy-
lenediamine = N,N'-1,2-Ethanediylbis[N-butyl-4-mor-
pholinecarboxamide] (●)
S Amipan T, Atmurin, Cardiac, *Dimorpholaminum*,
Puls, Respiron, 1064 TH, Théraleptique, Therapline,
Theraptique
U Analeptic

10086 $C_{20}H_{38}N_4O_8$
205380-20-7

1-[6-(D-Alanyl-L-glutaminylamino)hexyl]-L-fucopyra-
nose = D-Alanyl-N^1-[6-[(6-deoxy-β-L-galactopyrano-
syl)oxy]hexyl]-L-glutamamide (●)
S BCH 2537
U NK cell enhancer

10087 $C_{20}H_{38}N_8O$
188674-15-

N-[8-[1-(Aminocarbonyl)-2-imino-4-imidazolidinyl]-2,3-diamino-6,7-dihydroxyoctanoyl]-L-alanyl-L-valine = *N*-[2,3-Diamino-8-[2-amino-1-(aminocarbonyl)-4,5-di-hydro-1*H*-imidazol-5-yl]-2,3,4,5,8-pentadeoxyoctonoyl]-L-alanyl-L-valine (●)
S NA 22598A$_1$
U Cytotoxic from *Streptomyces* sp. NA22598

10088 (8177) $C_{20}H_{38}N_8S_2$
                                                        13456-08-1

2,3-Butanedione bis[4-(2-piperidinoethyl)thiosemicarb-azone] = 2,2'-(1,2-Dimethyl-1,2-ethanediylidene)bis[*N*-[2-(1-piperidinyl)ethyl]hydrazinecarbothioamide] (●)
S *Bitipazone***, Hoechst 153
U Coccidiostatic

10089 (8178) $C_{20}H_{38}O_2$
                                                        54460-46-7

Hexadecyl cyclopropanecarboxylate = Cyclopropane-carboxylic acid hexadecyl ester (●)
S *Cycloprate*, Zardex, ZR-856
U Miticide

10090 (8179) $C_{20}H_{38}O_4$
                                                        87272-20-6

3,3,14,14-Tetramethylhexadecanedioic acid (●)
S MEDICA 16
U Hypolipidemic

10091 (8180) $C_{20}H_{38}O_7S$
                                                        10041-19-7

1,4-Bis(2-ethylhexyl) sulfosuccinate = Sulfosuccinic acid bis(2-ethylhexyl) ester = Bis(2-ethylhexyl) succinate-2-sulfonic acid = Sulfobutanedioic acid 1,4-bis(2-ethylhe-xyl) ester (●)
R Sodium salt (577-11-5)
S Abilax, Aerosol OT, Afko-Lube, Alphasol OT, Ano-naid T, Aqua-Lax, Audinorm, Bantex, Bu-Lax, Cel-lubril, Cerumen-Löser Thilo, Cetimel, Clyss-Go, Co-lace, Colax-S, Coloctyl, Coloxyl, Comfolax, Com-plemix, Constab-100, Constiban, Consto-Caps., Con-stonate, Coprol(a), Correctol extra gentle, Da-ma-Lax, Decerosol OT, Defilin, Definate, DGSS, Dialose "J & J-Merck", Dicole, Dilax, Dioctin (old form), Dioctlyn, Diocto, Dioctosofteze, Dioctyl "Everest; Schwarz", Dioctylal, *Dioctyl Disodium Sul-fosuccinate*, Dioctyl-Medo, Dioctyl-Polfa, *Dioctyl Sodium Sulphosuccinate*, Diocyl "Amco", Dioeze, Diofectyl, Diolaxil, Diomedicone, Dionex, Diosate, Diosuccin, Dio-Sul, Diosux, Diotilan, Diovac, Diox, Dipolaxan, Disonate, Disoplex, Di-Sosul, Disulans, Doctate, Doctyl, Doctynol, *Docusate Sodium***, *Do-cusatum natricum***, Docusoft S, Docusol, DOK, Dollax, DONS, D.O.S., Doss (300), Doxate, Doxa-te-S, Doxilato, Doxinate "Hoechst-Roussel; Lloyd", Doxol "Purdue Frederick", Doxolan, DSS, Dulsivac, Duosol "Kirkman", Dynoctol, Easy-Lax, Ediclone, Emtix, Emulax simplex, Eosan, Evactil, Evac-U-8, Exalcol, Exonic OT, Factol, Fecalate, Fisiorallaxan-te, Fletcher-Enemette, Fluilax, Gardalax, Genasoft, Hisof, Humectol, Humevac, Hydrolax Pediatric, Il-ozoft, Jamylène, Klyx, Konlax "Cooper", Konsto, Kosate, Laksadif, Lambanol, Laxagel, Laxal "Mc-Gregor", Laxapol, Laxativum 59, Lax-Gel, Laxicon, Laxinate, Laxol "Herbapol", Laxopol, Laxyl "Ca-dril", Liqui-Doss, Manoxol OT, Merdex, Mervami-ne, Milkinol, Min-Evac, Modane Soft, Molatoc,

Molcer, Mollax, Molofac, *Natrii dioctylsulfosucci-nas*, *Natriumdioctylsulfosuccinat*, Neo-Vadrin D-S-S, Neu-Vadrin, Nevax, Nigalax, Norgalax, Nor-golax, Norm-Evac, Norval (old form), Obston, Oc-tyl-Softener, Otitex, Otosol, Parlax "Robinson", Peri Sofcap, Physiolax, PMS-Docusate Sodium, Prontoclisma, Pro-Sof, Provilax, Purgeron, Redcel, Redux, Regal "Andromaco", Regul-aid, Regulax SS, Regulex, Regutol, Rektolax, Revac, Rodox, Rytona-te, Samofax, Selax, Silace, Sintolax, Siponol-O-100, *Sodium dioctylsulfosuccinate*, Sofcaps, Soffecine, Soflax, Sof-ner, Soft-B-M, Softene, Softeze, Softil, Softol, Softon, Soliwax, Solusol, Stool Softener "Amlab; Weeks & Leo", Stulex, Sulfimel DOS, The-revac-SB, Tirolaxo, Tropagel, Ummetus, Vatsol OT, Velmol, Viforlax, Wasser-Lax, Waxnate, Waxsol, Wayds
U Laxative (stool softener), surfactant, cerumenolytic

10091-01   (8180-01)   128-49-4
R Calcium salt (2:1)
S Albert Docusate, Calax, Calcium Docuphen, Co-lax-C, DC 240, DC Softgels, Dioctocal, Dioctodol, *Dioctyl Calcium Sulfosuccinate*, *Docusate Calcium*, Doslax, Doxate-C, Doxica, PMS-Docusate Calcium, Pro-Cal-Sof, Sulfalax Calcium, Surfak
U Laxative (stool softener)

10091-02   (8180-02)   7491-09-0
R Potassium salt
S Dialose "Stuart", Diocto-K, Dioctolose, Docusate-K, *Docusate Potassium*, DSMC plus, Kasof, Rectalad Enema, Sigmol, Therevac
U Laxative (stool softener)

10092   (8181) $C_{20}H_{39}IO_2$

Ethyl monoiodostearate = Iodostearic acid ethyl ester = 9-Iodooctadecanoic acid ethyl ester (●)
R Mixture with the 10-iodo-isomer (29611-66-3)
S Duroliopaque, *Ethyl iodostearate*, Etiodorate
U Diagnostic aid (radiopaque medium)

10093   (8182) $C_{20}H_{39}N_2$
                                                         95050-40-1

1-[2-(Dicyclohexylamino)ethyl]-1-methylpiperidinium (●)
R Chloride (60996-85-2)
S Calacidol, I.U. 7
U Spasmolytic

10094   (8183) $C_{20}H_{39}N_5O_7$
                                                         70639-48-4

O-3-Deoxy-3-(ethylamino)-4-*C*-methyl-β-L-arabinopyra-nosyl-(1→4)-*O*-[2,6-diamino-2,3,4,6-tetradeoxy-α-D-*glycero*-hex-4-enopyranosyl-(1→6)]-2-deoxy-L-strept-amine = 4-*O*-[(2*S*,3*R*)-*cis*-3-Amino-6-(aminomethyl)-3,4-dihydro-2*H*-pyran-2-yl]-2-deoxy-6-*O*-[3-deoxy-3-(ethylamino)-4-*C*-methyl-β-L-arabinopyranosyl]strept-amine = *O*-3-Deoxy-3-(ethylamino)-4-*C*-methyl-β-L-ara-binopyranosyl-(1→6)-*O*-[2,6-diamino-2,3,4,6-tetra-deoxy-α-D-*glycero*-hex-4-enopyranosyl-(1→4)]-2-de-oxy-D-streptamine (●)
S Bay V1 4718, *Etisomicin**
U Antibiotic

10095　(8184)　　　　　　　　　　$C_{20}H_{40}N_2$
　　　　　　　　　　　　　　　　　　123018-34-8

N,N-Dimethyl-8,8-dipropyl-2-azaspiro[4.5]decane-2-
propanamine (●)
R　Dihydrochloride (124796-27-6)
S　SKF 105685
U　Immunosuppressive

10096　(8185)　　　　　　　　　　$C_{20}H_{40}N_2O_8$
　　　　　　　　　　　　　　　　　　13149-69-4

ɒ-Gluconic acid 6-bis(diisopropylamino)acetate = D-Glu-
conic acid 6-[bis[bis(1-methylethyl)amino]acetate] (●)
R　also sodium salt
S　*Acidum pangamicum*, Acipangan, Beforce$_{15}$,
　　Bio-15, bio-pangamina, Cruvi B 15, Dura B 15, Hid-
　　rovitamina B$_{15}$, Miluyvit B$_{15}$, Oyo, Painex-Sauer-
　　stoffaufnahme-Dr., Pangacitin, *Pangamic acid*, Pan-
　　gamil, Pangamine, Pangamox, *Pangamsäure*, Plenti-
　　ne, Sopangamine, Tamavit, *Vitamin B$_{15}$*
U　Diuretic, antisclerotic

0096-01　(8185-01)　　　　　　　　11041-98-8
R　Calcium salt
S　Beta 15, *Calcii pangamas*, Calgam, *Ca-Pangamat*,
　　Desfatigan "Pentafarma", Egostar, Pangametin, Pul-
　　sor
U　Cardiac stimulant

0096-02　(8185-02)
　　Arginine salt
　　Cardiopangamina

U　Cardiac stimulant

10097　(8186)　　　　　　　　　　$C_{20}H_{40}N_4O_{10}$
　　　　　　　　　　　　　　　　　　49863-47-0

O-2-Amino-2,7-dideoxy-D-*glycero*-α-D-glucoheptopyra-
nosyl-(1→4)-*O*-[3-deoxy-4-*C*-methyl-3-(methylamino)-
β-L-arabinopyranosyl-(1→6)]-2-deoxy-D-streptamine (●)
R　also sulfate (salt) (57793-87-0)
S　G 418, *Geneticin*, NSC-606702
U　Antineoplastic antibiotic from *Micromonospora rho-
　　dorangea*

10098　(8187)　　　　　　　　　　$C_{20}H_{41}NO_3$
　　　　　　　　　　　　　　　　　　35301-24-7

N-[(1*S*,2*S*)-2-Hydroxy-1-(hydroxymethyl)heptade-
cyl]acetamide = [*S*-(*R**,*R**)]-*N*-[2-Hydroxy-1-(hydroxymethyl)heptadecyl]acetamide (●)
S　*Cedefingol***, SPC-101210
U　Antipsoriatic, chemopotentiator

10099　(8188)　　　　　　　　　　$C_{20}H_{41}N_5O_7$
　　　　　　　　　　　　　　　　　　52093-21-7

N-Methyl-gentamycin C$_{1A}$ = *O*-2-Amino-2,3,4,6-tetra-
deoxy-6-(methylamino)-α-D-*erythro*-hexopyranosyl-

(1→4)-*O*-[3-deoxy-4-*C*-methyl-3-(methylamino)-β-L-arabinopyranosyl-(1→6)]-2-deoxy-D-streptamine (●)
R also sulfate (66803-19-8)
S Antibiotic KW 1062, Cromicin, *Gentamicin C$_{2B}$*, KW 1062, Luxomicina, Marocin, *Micronomicin***, Microphta, Miksin, Sagamicin(a), Santemycin, Senacin, XK-62-2
U Antibiotic from *Micromonospora sagamiensis* var. *nonreducans* nov. sp. MM 62

10100 (8189)　　　　　　　　　　　　$C_{20}H_{43}N$
　　　　　　　　　　　　　　　　　　124-28-7

N,N-Dimethyloctadecylamine = *N,N*-Dimethyl-1-octadecanamine (●)
R Hydrochloride (1613-17-8)
S *Dimantine hydrochloride***, Doda, *Dymanthine hydrochloride*, GS-1339, NSC-5547, Thelmesan
U Anthelmintic (veterinary)

10101 (8190)　　　　　　　　　　　　$C_{20}H_{44}N$
　　　　　　　　　　　　　　　　　　10328-33-3

Ethylhexadecyldimethylammonium = Cetylethyldimethylammonium = *N*-Ethyl-*N,N*-dimethyl-1-hexadecanaminium (●)
R Bromide chloride mixture
S "Cetylamin", Radiol
U Antiseptic

10101-01 (8190-01)　　　　　　　　　3006-10-8
R Ethyl sulfate
S *Mecetronium etilsulfate***, Querton 16 ES, Sterillium (main active substance)
U Antiseptic

10102 (8191)　　　　　　　　　　　　$C_{20}H_{45}N_3$
　　　　　　　　　　　　　　　　　　3687-16-9

N^1,N^3-Bis(2-ethylhexyl)-1,2,3-triamino-2-methylpropane = N^1,N^3-Bis(2-ethylhexyl)-2-methyl-1,2,3-propanetriamine (●)
S *Propoctamine*
U Antiseptic

10103　　　　　　　　　　　　　　　$C_{20}H_{47}N_5$
　　　　　　　　　　　　　　　　　　147510-59-6

1,19-Bis(ethylamino)-5,10,15-triazanonadecane = *N*-[4-(Ethylamino)butyl]-*N'*-[4-[[4-(ethylamino)butyl]amino]butyl]-1,4-butanediamine (●)
S BE 4-4-4-4
U Antineoplastic

10104 (8192)　　　　　　　　　　　　$C_{21}H_{13}F_3N_2O_4$
　　　　　　　　　　　　　　　　　　66898-62-2

Phthalidyl 2-(α,α,α-trifluoro-*m*-toluidino)nicotinate = 2-[3-(Trifluoromethyl)anilino]-3-pyridinecarboxylic acid 3-phthalidyl ester = 2-[[3-(Trifluoromethyl)phenyl]amino]-2-pyridinecarboxylic acid 1,3-dihydro-3-oxo-1-isobenzofuranyl ester (●)
S BA 7602-06, Somalgen, *Talniflumate***
U Anti-inflammatory, analgesic

10105　　　　　　　　　　$C_{21}H_{14}ClNO_3$
142326-59-8

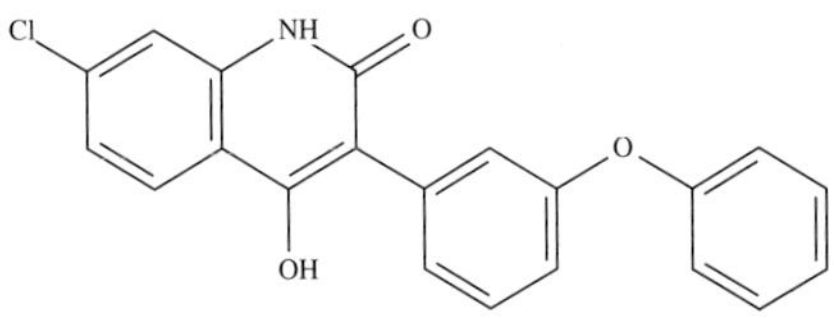

7-Chloro-4-hydroxy-3-(3-phenoxyphenyl)-2(1*H*)-quino-
linone (●)
S　L 701324
U　Antipsychotic (glycine/NMDA receptor antagonist)

10106　　　　　　　　　　$C_{21}H_{14}ClN_5O_2$
183721-15-5

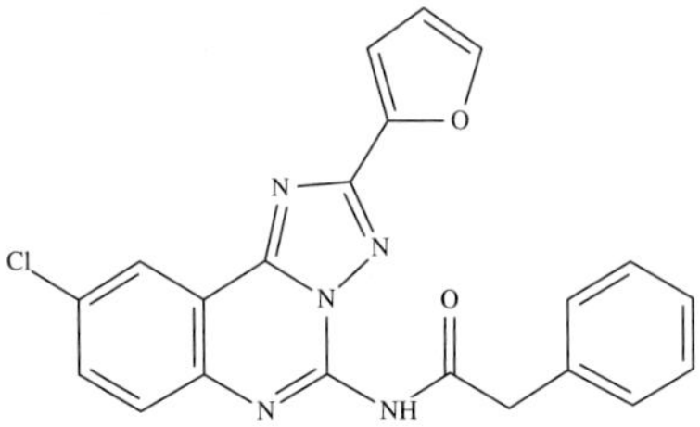

9-Chloro-2-(2-furyl)-5,6-dihydro-5-[(phenylacetyl)imi-
no]-[1,2,4]triazolo[2,3-*c*]quinazoline = *N*-[9-Chloro-2-(2-
furanyl)[1,2,4]triazolo[1,5-*c*]quinazolin-5-yl]benzene-
acetamide (●)
S　MRS 1220
U　Adenosine A$_3$ receptor antagonist

10107　　　　　　　　　　$C_{21}H_{15}N_3O$
112575-48-1

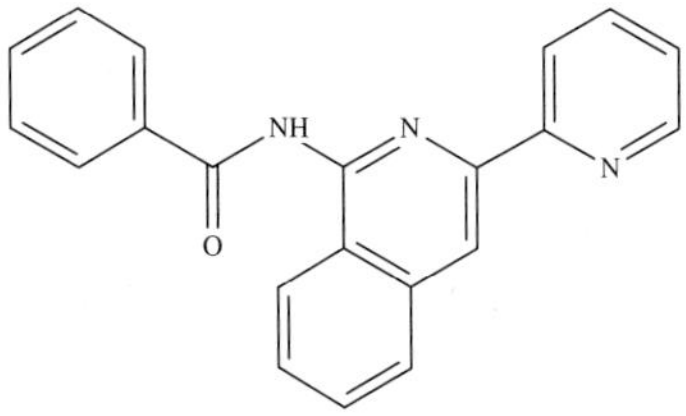

N-[3-(2-Pyridinyl)-1-isoquinolinyl]benzamide (●)
VUF 8507
U　Adenosine A$_3$ receptor ligand

10108　　　　　　　　　　$C_{21}H_{15}N_5O$
139482-55-6

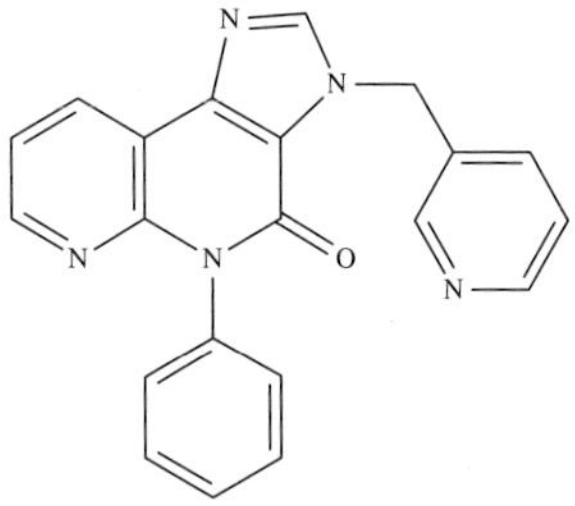

3,5-Dihydro-5-phenyl-3-(3-pyridinylmethyl)-4*H*-imida-
zo[4,5-*c*][1,8]naphthyridin-4-one (●)
S　KF 19514
U　Bronchoprotective (phosphodiesterase I/IV inhibitor)

10109　(8193)　　　　　　　　$C_{21}H_{16}Cl_2N_2O_4$
105687-68-1

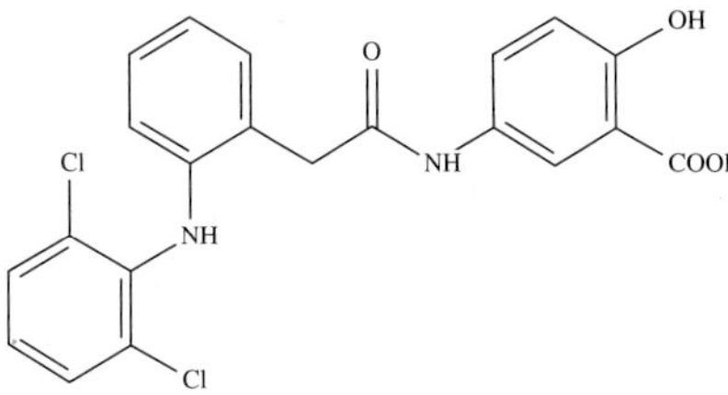

5-[2-[2-(2,6-Dichloroanilino)phenyl]acetamido]salicylic
acid = 5-[[[2-[(2,6-Dichlorophenyl)amino]phenyl]ace-
tyl]amino]-2-hydroxybenzoic acid (●)
S　NB 431
U　Anti-inflammatory, analgesic

10110　　　　　　　　　　$C_{21}H_{16}Cl_2N_4OS$
260415-63-2

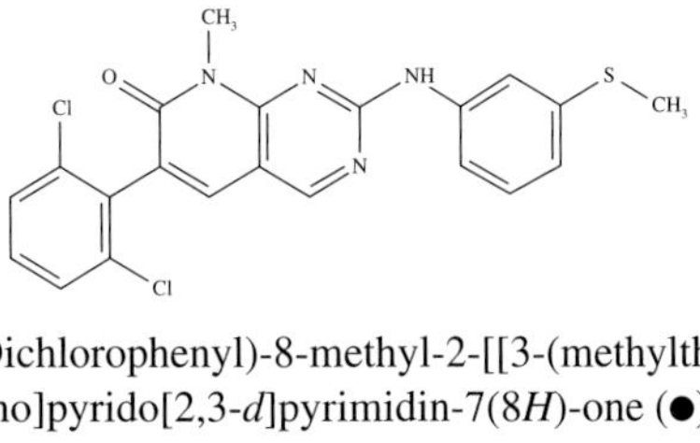

6-(2,6-Dichlorophenyl)-8-methyl-2-[[3-(methylthio)phe-
nyl]amino]pyrido[2,3-*d*]pyrimidin-7(8*H*)-one (●)
S　PD 173955
U　Tyrosine kinase inhibitor

10111 (8194) $C_{21}H_{16}Cl_4N_2S$
28614-19-9

N-[2,3-Bis(3,4-dichlorophenyl)propyl]thioisonicotin-
amide = N-[2,3-Bis(3,4-dichlorophenyl)propyl]-4-pyridi-
necarbothioamide (●)
S Hoechst 30036
U Antitrypanosomal

10112 $C_{21}H_{16}FNO_3S$
161600-01-7

5-[[6-[(o-Fluorobenzyl)oxy]-2-naphthyl]methyl]-2,4-
thiazolidinedione = 5-[[6-[(2-Fluorophenyl)methoxy]-2-
naphthalenyl]methyl]-2,4-thiazolidinedione (●)
S MCC-555
U Antidiabetic (insulin sensitizer)

10113 $C_{21}H_{16}FN_3OS$
152121-47-6

4-[5-(4-Fluorophenyl)-2-[4-(methylsulfinyl)phenyl]-1H-
imidazol-4-yl]pyridine (●)
S SB 203580
U Anti-inflammatory (protein kinase inhibitor)

10114 $C_{21}H_{16}F_2N_2O_4$
220997-97-7

(5R)-5-Ethyl-9,10-difluoro-1,4,5,13-tetrahydro-5-hydr-
oxy-3H,15H-oxepino[3',4':6,7]indolizino[1,2-b]quino-
line-3,15-dione (●)
S Diflomotecan*, Flomotecan, Moflotecan
U Antineoplastic (topoisomerase 1 inhibitor)

10115 (8195) $C_{21}H_{16}N_2$
1729-61-9

α-Fluoren-9-ylidene-p-toluamidine = 9-(p-Amidinoben-
zylidene)fluorene = 4-(9H-Fluoren-9-ylidenemethyl)ben-
zenecarboximidamide (●)
R also monohydrochloride (5585-60-4)
S Mer-27, Paranyline, Renytoline**
U Anti-inflammatory, anti-arthritic

10116 (8196) $C_{21}H_{16}N_2O_4S$
21413-28-

5,5-Diphenyl-1-(phenylsulfonyl)hydantoin = 5,5-Di-
phenyl-1-(phenylsulfonyl)-2,4-imidazolidinedione (●)
S PS-796

U   Anti-inflammatory

10117 $C_{21}H_{16}N_2O_5S$
                                                 149556-49-0

5-[[*p*-[(3-Methyl-2-pyridyl)sulfamoyl]phenyl]ethenyl]sa-
licylic acid = 2-Hydroxy-5-[[4-[[(3-methyl-2-pyridi-
nyl)amino]sulfonyl]phenyl]ethynyl]benzoic acid (●)
S   FA 31A, *Susalimod***
U   Immunomodulator

10118   (8197) $C_{21}H_{16}N_4O_8S_2$
                                                 41906-86-9

(6R,7R)-3-[(*E*)-2,4-Dinitrostyryl]-8-oxo-7-(2-thienyl-
acetamido)-5-thia-1-azabicyclo[4.2.0]oct-2-ene-2-carb-
oxylic acid = (7R)-3-[(*E*)-2,4-Dinitrostyryl]-7-(2-thienyl-
acetamido)-3-cephem-4-carboxylic acid = [6R-
[3(*E*),6α,7β]]-3-[2-(2,4-Dinitrophenyl)ethenyl]-8-oxo-7-
[(2-thienylacetyl)amino]-5-thia-1-azabicyclo[4.2.0]oct-
2-ene-2-carboxylic acid (●)
R   also sodium salt (61618-29-9)
Cephalosporin 87-312, GL 87/312, *Nitrocefin*
U   Antibiotic (diagnostic aid)

10119 $C_{21}H_{16}O_5$
                                                 189697-45-8

11-Hydroxy-1,10-dimethoxy-8-methyl-5,12-naphtha-
cenedione (●)
S   XR-651
U   Immunosuppressive

10120   (8198) $C_{21}H_{16}O_6S_2$
                                                 100477-09-6

Methylenebis(2-naphthyl-3-sulfonic acid) = 3,3'-Methy-
lenebis[2-naphthalenesulfonic acid] (●)
R   Phenylmercuric compd. (1:2) (14235-86-0)
S   Conotrane, Hydraphen, *Hydrargaphen***, Mer
    "Takeda", Neo-Penotran strong, Peneton, Penotrane,
    Versotrane
U   Topical antiseptic, antifungal

10120-01   (8198-01)   53370-43-7
R   Disilver salt
S   Methargen, Viacutan
U   Topical antiseptic

10121   (8199) $C_{21}H_{16}O_7$
                                                 4366-18-1

3,3'-(2-Methoxyethylidene)bis(4-hydroxycoumarin) =
4,4'-Dihydroxy-3,3'-(2-methoxyethylidene)dicoumarin =
2,2-Bis(4-hydroxy-3-coumarinyl)ethyl methyl ether =
3,3'-(2-Methoxyethylidene)bis[4-hydroxy-2*H*-1-benzo-
pyran-2-one] (●)
S   *Coumetarol*, *Cumetarolum, Cumetharol, Cume-
    thoxaethanum, Cumethoxan, Dicoumoxyl, Dicu-
    moxan(e), Ph. 137*

U Anticoagulant

10122 (8200) $C_{21}H_{16}O_{11}$
1400-58-4

Anhydromethylenecitric acid disalicylic acid ester =
O,O'-Anhydromethylenecitroyldisalicylic acid = 5-Oxo-
1,3-dioxolane-4,4-diacetic acid bis(2-carboxyphenyl) es-
ter (●)
S Citrodisalyl, Novaspirin, Salicitrin
U Antipyretic, analgesic, antineuralgic

10123 (8201) $C_{21}H_{17}AsN_2O_5S_2$
91-71-4

o,o'-(*p*-Ureidophenylarsylenedithio)dibenzoic acid = 4-
Ureidophenyl bis(2-carboxyphenylthio)arsenite = 2,2'-
[[4-[(Aminocarbonyl)amino]phenyl]arsinidine-
bis(thio)]bis[benzoic acid] (●)
S Thio-Carbamisin, Thiocarbamizine
U Anti-amebic

10124 $C_{21}H_{17}ClN_2$
130186-25-3

cis-2-(4-Chlorophenyl)-4,5-dihydro-4,5-diphenyl-1*H*-
imidazole (●)
R Monohydrochloride (130186-26-4)
S TA-383
U Antirheumatic

10125 (8202) $C_{21}H_{17}ClO$
7469-01-

3-Chloro-2,3-diphenylpropiophenone = 3-Chloro-1,2,3-
triphenyl-1-propanone (●)
S E 383
U Contraceptive

10126 $C_{21}H_{17}Cl_3O$
146949-21-

O-(2,6-Dichlorophenyl)-*O'*-(2-chloro-4-methoxyphenyl)
1,4-benzenedimethanol = 1-[(2-Chloro-4-methoxyphen-
oxy)methyl]-4-[(2,6-dichlorophenoxy)methyl]benzene
(●)
S Sch 48973

U Antiviral

10127 $C_{21}H_{17}NO_4$
210245-19-5

6-[(Tetrahydro-4-methylene-5-oxo-2-phenyl-2-fura-
nyl)methoxy]-2(1H)-quinolinone (●)
S CCT 62
U Antiplatelet phosphodiesterase inhibitor

10128 (8203) $C_{21}H_{17}NO_5$
37646-31-4

1-Cinnamoyl-2-methyl-5,6-(methylenedioxy)-3-indole-
acetic acid = 6-Methyl-5-(1-oxo-3-phenyl-2-propenyl)-
5H-1,3-dioxolo[4,5-f]indole-7-acetic acid (●)
S ID-1619
U Anti-inflammatory

10129 (8204) $C_{21}H_{17}NO_9S_2$
54935-03-4

3,3-Bis(p-hydroxyphenyl)-7-methyl-2-indolinone bis(hydrogen sulfate) (ester) = 1,3-Dihydro-7-methyl-3,3-bis[4-
sulfooxy)phenyl]-2H-indol-2-one (●)

R Disodium salt (54935-04-5)
S DAN-603, Gutanorme, Laxitex, *Sulisatin Sodium***
U Laxative

10130 (8205) $C_{21}H_{17}N_2O_3$
104114-27-4

3,6-Diamino-9-[2-(methoxycarbonyl)phenyl]xanthylium
(●)
R Chloride (62669-70-9)
S RH-123, Rhodamine 123
U Antineoplastic, dye

10131 (8206) $C_{21}H_{18}ClNO_6$
53164-05-9

1-(p-Chlorobenzoyl)-5-methoxy-2-methylindole-3-ace-
tic acid ester with glycolic acid = O-(1-p-Chlorobenzoyl-
5-methoxy-2-methylindol-3-ylacetyl)glycolic acid = 1-
(4-Chlorobenzoyl)-5-methoxy-2-methyl-1H-indole-3-
acetic acid carboxymethyl ester (●)
S *Acemetacin***, Acemetadoc, Acemix, Acenol "HC
Charm", Acephlogont, Altren, Analgel, Azeat, Bay
f 4975, Emflex, Espledol, Flamarion, Gamespir, Gy-
nalgia, K-708, Mostanol "Boehringer; De Angeli",
Oldan "Castejon; Europharma", Peran, Rantudal,
Rantudil, Rheumibis, Rheutrop, Solart, Sportix, Tar-
bis, Tilur, TVX 1322
U Anti-inflammatory, analgesic

10132 (8207) $C_{21}H_{18}ClN_3O_7S$
36920-48-6

(6R,7R)-7-[3-(o-Chlorophenyl)-5-methyl-4-isoxazole-
carboxamido]-3-(hydroxymethyl)-8-oxo-5-thia-1-azabi-
cyclo[4.2.0]oct-2-ene-2-carboxylic acid acetate (ester) =
(6R,7R)-3-(Acetoxymethyl)-7-[3-(o-chlorphenyl)-5-me-
thyl-4-isoxazolecarboxamido]-8-oxo-5-thia-1-azabicy-
clo[4.2.0]oct-2-ene-2-carboxylic acid = (7R)-7-(3-o-
Chlorophenyl-5-methylisoxazole-4-carboxamido)cepha-
losporanic acid = (6R-trans)-3-[(Acetyloxy)methyl]-7-
[[[3-(2-chlorophenyl)-5-methyl-4-isoxazolyl]carbo-
nyl]amino]-8-oxo-5-thia-1-azabicyclo[4.2.0]oct-2-ene-2-
carboxylic acid (●)
S   Cefoxazole**, Cephoxazole, Glaxo 291/1
U   Antibiotic

10133 $C_{21}H_{18}F_3N_3O_3$
121357-19-5

(2S-trans)-7-(4-Amino-2-methyl-1-pyrrolidinyl)-1-(2,4-
difluorophenyl)-6-fluoro-1,4-dihydro-4-oxo-3-quinoline-
carboxylic acid (●)
R   Monohydrochloride (114676-82-3)
S   A-65326 HCl, A-80556
U   Antibacterial

10134 (8208) $C_{21}H_{18}F_3N_3O_3$
108319-06-8

1-(2,4-Difluorophenyl)-6-fluoro-1,4-dihydro-7-(3-me-
thyl-1-piperazinyl)-4-oxo-3-quinolinecarboxylic acid (●)
R   Monohydrochloride (105784-61-0)
S   A-62254, Abbott-62254, Omniflox, Signum, Teflox,
Temabiotic, Temac, Temadie, Temaflox, Temafloxa-
cin hydrochloride**, TFLX
U   Antibacterial

10135 $C_{21}H_{18}NO_4$
34316-15-9

1,2-Dimethoxy-12-methyl-[1,3]benzodioxolo[5,6-c]phen-
anthridinium (●)
S   Chelerythrine, Heleritrine, Toddaline
U   Protein kinase C inhibitor from Chelidonium majus

10136 (8209) $C_{21}H_{18}NO$
6872-57-

2,3-Dimethoxy-12-methyl[1,3]benzodioxolo[5,6-c]phen-
anthridinium (●)
R   Chloride (13063-04-2)
S   Nitidine chloride, NSC-146397

U Antineoplastic

10137

$C_{21}H_{18}N_2O_4$
50906-88-2

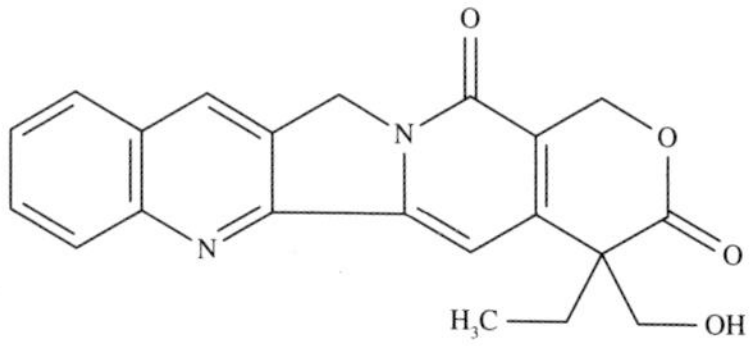

4-Ethyl-4-(hydroxymethyl)-1H-pyrano[3',4':6,7]indolizino[1,2-b]quinoline-3,14(4H,12H)-dione (●)
S BN 80927, hCPT, *Homocamptothecin*
U Antineoplastic

10138

$C_{21}H_{18}N_2O_4S$
203923-75-5

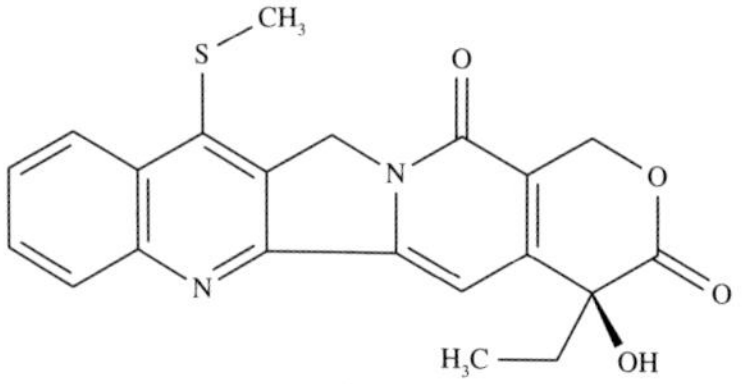

7-(Methylthio)camptothecin = (4S)-4-Ethyl-4-hydroxy-1-(methylthio)-1H-pyrano[3',4':6,7]indolizino[1,2-b]quinoline-3,14(4H,12H)-dione (●)
S BNP 1100
U Topoisomerase I inhibitor

10139

$C_{21}H_{18}N_2O_4S_2$
141716-96-3

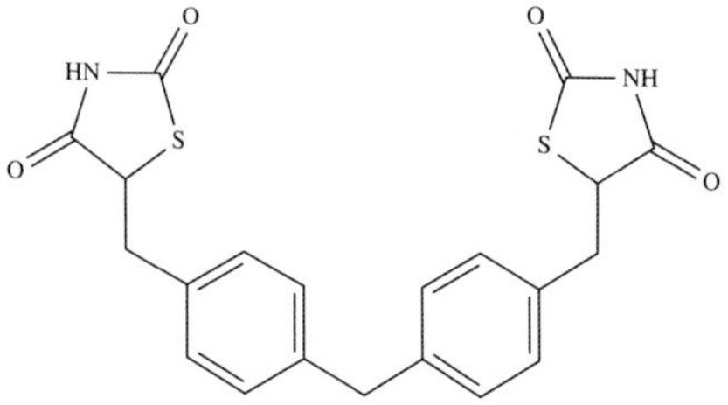

Bis[4-[(2,4-dioxo-5-thiazolidinyl)methyl]phenyl]methane = 5,5'-[Methylenebis(4,1-phenylenemethylene)]bis[2,4-thiazolidinedione] (●)
S YM-268
U Antidiabetic

10140

$C_{21}H_{18}N_4O_3$
131185-37-0

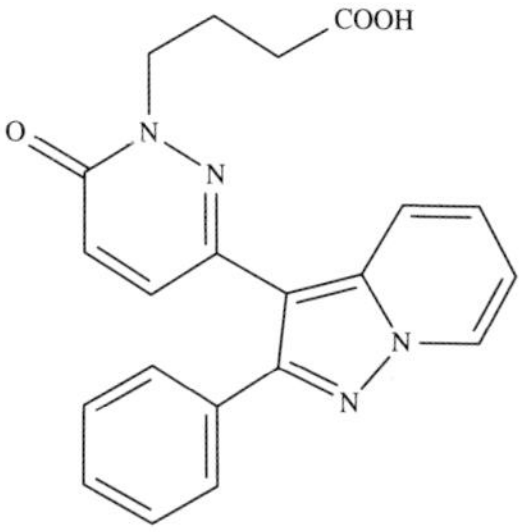

6-Oxo-3-(2-phenylpyrazolo[1,5-a]pyridin-3-yl)-1(6H)-pyridazinebutanoic acid (●)
S FK 838
U Adenosine A_1 receptor antagonist

10141 (8210)

$C_{21}H_{18}N_8O_8S_4$
114875-57-9

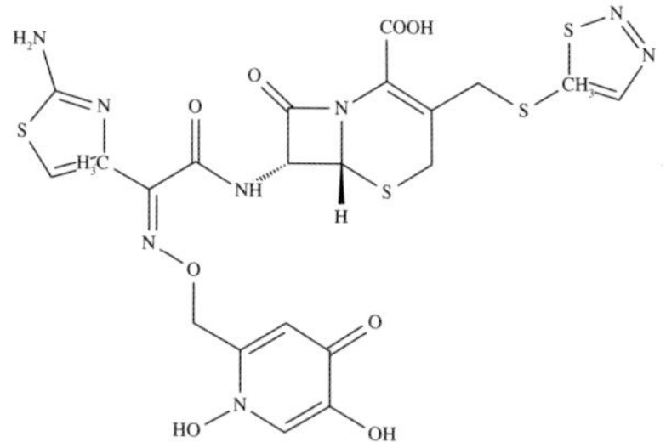

(6R,7R)-7-[2-(2-Amino-4-thiazolyl)glyoxylamido]-8-oxo-3-[(1,2,3-thiadiazol-5-ylthio)methyl]-5-thia-1-azabicyclo[4.2.0]oct-2-ene-2-carboxylic acid, 7^2-(Z)-[O-[(1,4-dihydro-1,5-dihydroxy-4-oxo-2-pyridyl)methyl]oxime] = (Z)-(7R)-7-[2-(2-Amino-4-thiazolyl)-2-[[(1,4-dihydro-1,5-dihydroxy-4-oxo-2-pyridinyl)methoxy]imino]acetamido]-3-[(1,2,3-thiadiazol-5-ylthio)methyl]-3-cephem-4-carboxylic acid = [6R-[6α,7β(Z)]]-7-[[(2-Amino-4-thiazolyl)[[(1,4-dihydro-1,5-dihydroxy-4-oxo-2-pyridinyl)methoxy]imino]acetyl]amino]-8-oxo-3-[(1,2,3-thiadiazol-5-ylthio)methyl]-5-thia-1-azabicyclo[4.2.0]oct-2-ene-2-carboxylic acid (●)
R Disodium salt (133686-28-9)
S KP 736
U Antibiotic

2057

10142   (8211) $C_{21}H_{18}O_6$
                                                87838-97-9

(+)-9-[[4-(2,5-Dihydro-4-methyl-5-oxo-2-furanyl)-3-me-
thyl-2-butenyl]oxy]-7H-furo[3,2-g][1]benzopyran-7-one
(●)
S   *Clausenacoumarin*
U   Smooth muscle relaxant, hypoglycemic from *Clausena* sp.

10143 $C_{21}H_{18}O_6S$
                                                156731-78-1

[4-[4-Hydroxy-2-oxo-3-[(2-phenylethyl)thio]-2H-pyran-
6-yl]phenoxy]acetic acid (●)
S   PD 153103
U   Antiviral (HIV-1 protease inhibitor)

10144 $C_{21}H_{18}O_7$
                                                149415-60-1

8-(3,4-Dihydroxyphenyl)-5-hydroxy-7-methoxy-2,2-di-
methyl-2H,6H-benzo[1,2-b:5,4-b']dipyran-6-one (●)
S   *Sarothranol*
U   Antibacterial from *Hypericum perforatum*

10145   (8212) $C_{21}H_{18}O_{11}$
                                                21967-41-9

5,6,7-Trihydroxyflavone 7-β-D-glucopyranuronoside =
5,6-Dihydroxy-4-oxo-2-phenyl-4H-1-benzopyran-7-yl-
β-D-glucopyranosiduronic acid (●)
S   *Baicalin*
U   Anti-allergic, anti-asthmatic

10146   (8213) $C_{21}H_{19}ClFNO_2$
                                                113243-00-8

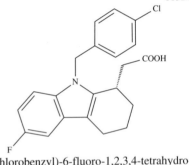

(+)-9-(p-Chlorobenzyl)-6-fluoro-1,2,3,4-tetrahydro-1-
carbazoleacetic acid = (S)-9-[(4-Chlorophenyl)methyl]-6-
fluoro-2,3,4,9-tetrahydro-1H-carbazole-1-acetic acid (●)
S   L 657926
U   Anticoagulant

10147 $C_{21}H_{19}ClN_4O$
                                                181632-25-

6-Chloro-5-methyl-1-[[2-[(2-methyl-3-pyridyl)oxy]-5-
pyridyl]carbamoyl]indoline = 6-Chloro-2,3-dihydro-5-

methyl-*N*-[6-[(2-methyl-3-pyridinyl)oxy]-3-pyridinyl]-1*H*-indole-1-carboxamide (●)
S SB 242084
U 5-HT$_{2C}$ receptor antagonist

10148

$C_{21}H_{19}ClN_6O$
114798-35-5

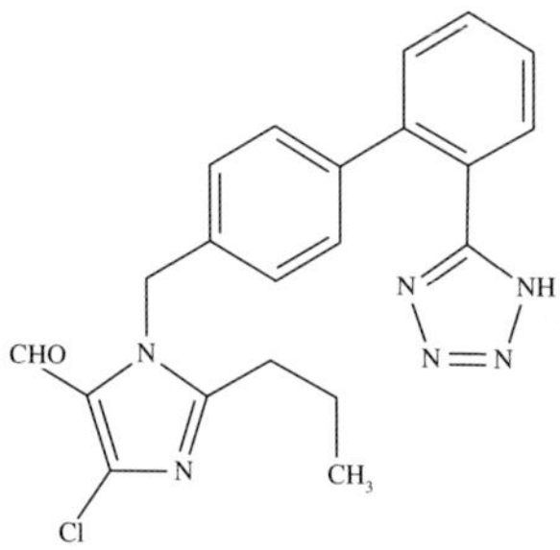

4-Chloro-2-propyl-1-[[2'-(1*H*-tetrazol-5-yl)[1,1'-biphenyl]-4-yl]methyl]-1*H*-imidazole-5-carboxaldehyde (●)
S EXP 3312
U Antihypertensive (angiotensin II antagonist)

10149

$C_{21}H_{19}Cl_2IN_4O_2$
202463-68-1

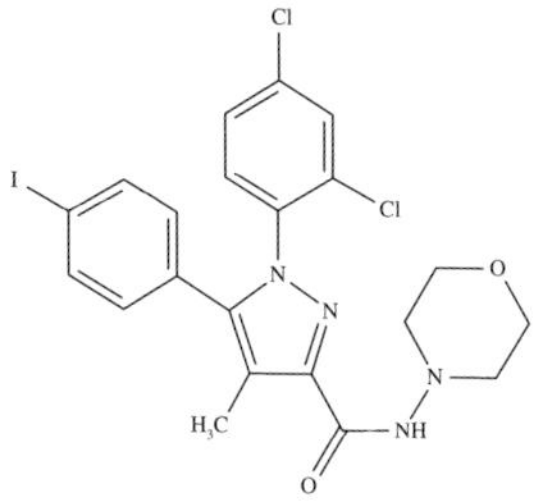

-(2,4-Dichlorophenyl)-5-(4-iodophenyl)-4-methyl-*N*-4-morpholinyl-1*H*-pyrazole-3-carboxamide (●)
AM-281
J Cannabinoid CB$_1$ receptor antagonist

10150

$C_{21}H_{19}Cl_4N_6Ru$
124875-20-3

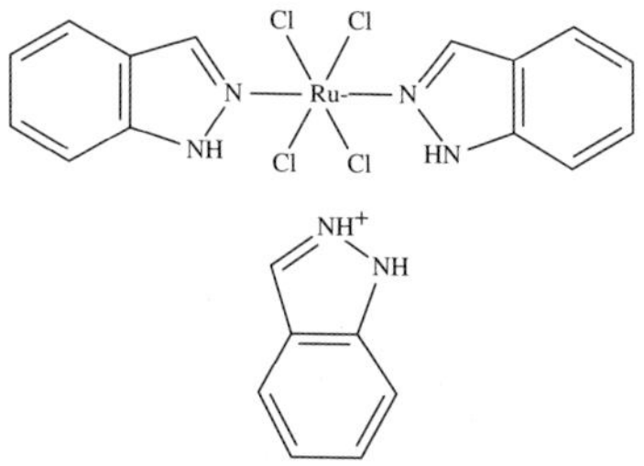

trans-Indazolium [tetrachlorobisindazoleruthenate(III)] = Hydrogen (*OC*-6-11)-tetrachlorobis(1*H*-indazole-κ*N*^{2})-ruthenate(1−) compd. with 1*H*-indazole (1:1) (●)
S IndCR, KP 1019, NSC-666158
U Cytotoxic

10151 (8214)

$C_{21}H_{19}F_2N_3O_3$
98106-17-3

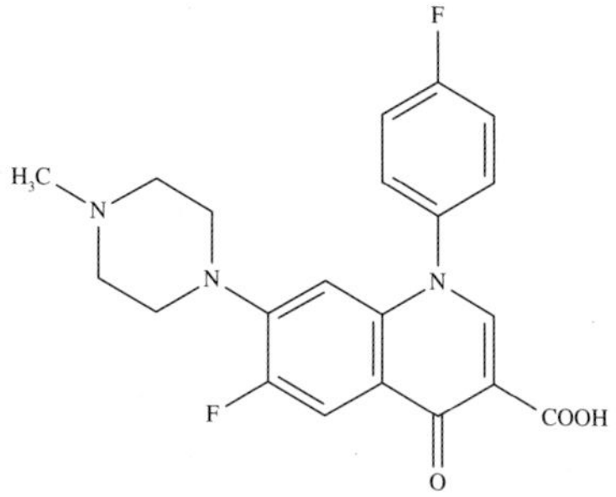

6-Fluoro-1-(4-fluorophenyl)-1,4-dihydro-7-(4-methyl-1-piperazinyl)-4-oxo-3-quinolinecarboxylic acid (●)
R Monohydrochloride (91296-86-5)
S A-56619, Abbott-56619, Dicural, *Difloxacin hydrochloride***
U Anti-infective (DNA gyrase inhibitor)

10152 (8215)

$C_{21}H_{19}N$
27466-27-9

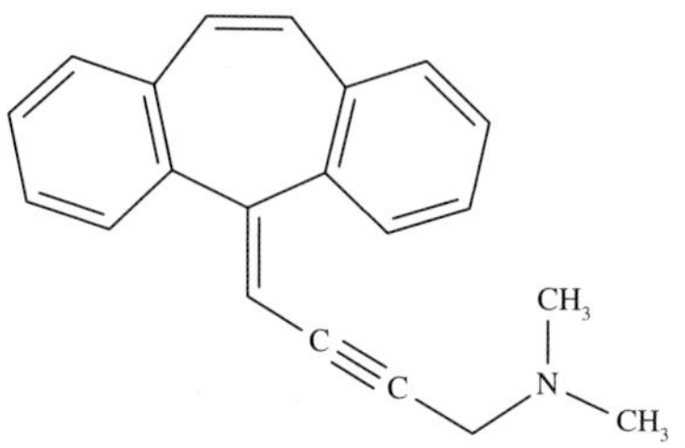

4-(5*H*-Dibenzo[*a,d*]cyclohepten-5-ylidene)-*N,N*-dimethyl-2-butynylamine = 4-(5*H*-Dibenzo[*a,d*]cyclohepten-5-ylidene)-*N,N*-dimethyl-2-butyn-1-amine (●)
R Hydrochloride (27466-29-1)
S AY-22124, *Intriptyline hydrochloride***
U Antidepressant

10153 (8216) $C_{21}H_{19}NO$
7433-09-2

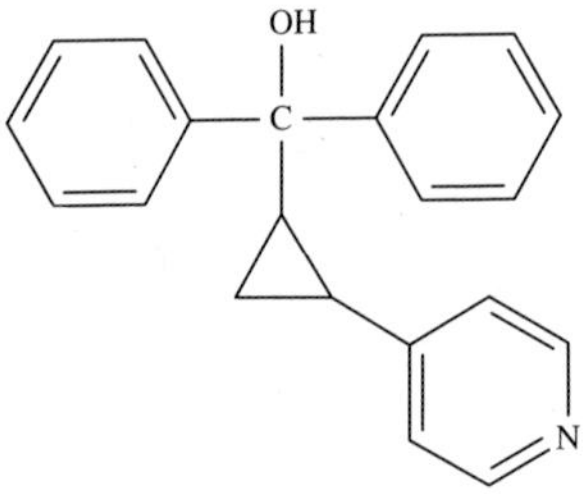

Diphenyl[2-(4-pyridyl)cyclopropyl]methanol = 2-(4-Pyridyl)-α,α-diphenylcyclopropanemethanol = α-[2-(4-Pyridyl)cyclopropyl]benzhydrol = α-Phenyl-α-[2-(4-pyridinyl)cyclopropyl]benzenemethanol (●)
R Hydrochloride (2364-72-9) [also *trans*-compd.
(4904-00-1); hydrochloride (4904-01-2)]
S *Cyprolidol hydrochloride***, IN 1060, NSC-84973
U Antidepressant

10154 (8217) $C_{21}H_{19}NO$
7387-60-2

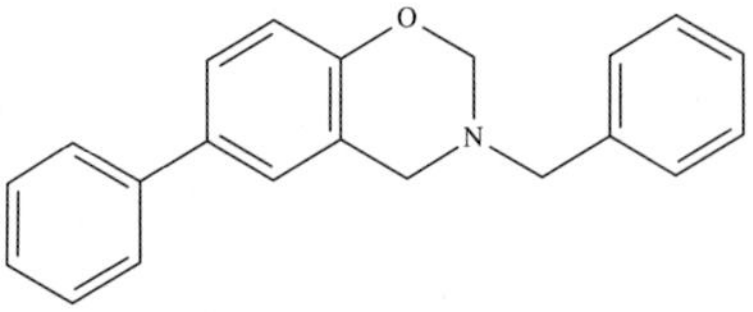

3-Benzyl-3,4-dihydro-6-phenyl-2*H*-1,3-benzoxazine = 3,4-Dihydro-6-phenyl-3-(phenylmethyl)-2*H*-1,3-benzoxazine (●)
S T 615
U Antibacterial (tuberculostatic)

10155 (8218) $C_{21}H_{19}NO_2$
78250-23-4

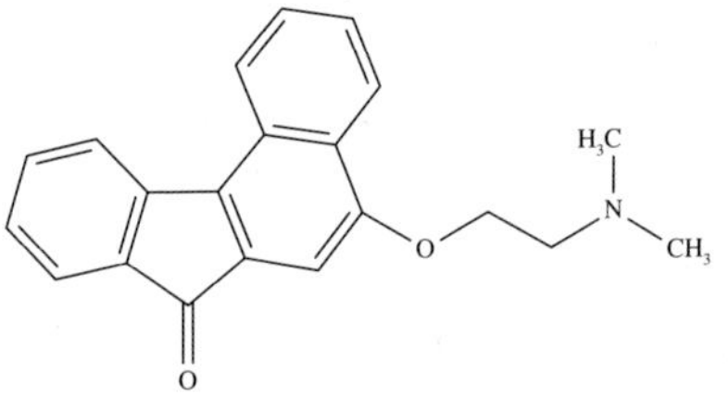

5-[2-(Dimethylamino)ethoxy]-7*H*-benzo[*c*]fluoren-7-one (●)
R Hydrochloride (80427-58-3)
S Benfluron, VÚFB-13468
U Antineoplastic, antileukemic

10156 (8219) $C_{21}H_{19}NO_3S$
7527-94-8

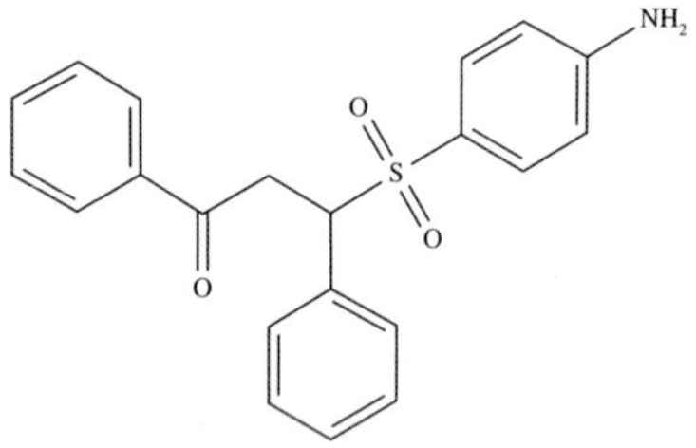

3-Phenyl-3-sulfanilylpropiophenone = 3-[(4-Aminophenyl)sulfonyl]-1,3-diphenyl-1-propanone (●)
S Alfone, *Alkafanone, Alkofanone,* Clafanone, Nu-404, Ro 2-0404
U Antidiarrheal

10157 (8220) $C_{21}H_{19}NO_4$
35578-20-

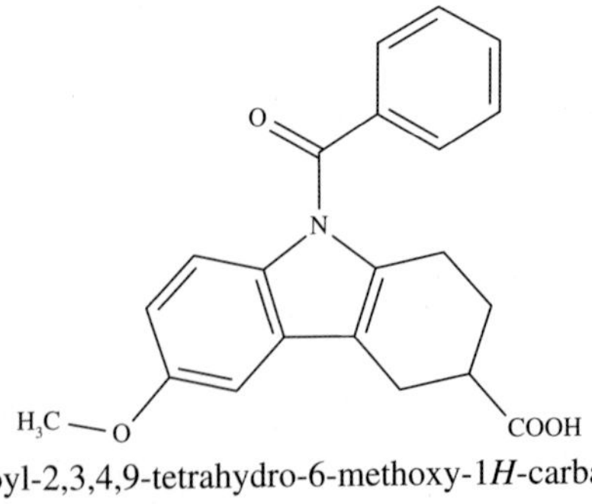

9-Benzoyl-2,3,4,9-tetrahydro-6-methoxy-1*H*-carbazole-3-carboxylic acid (●)
S *Oxarbazole***, Win 34284

U   Anti-asthmatic

10158   (8221) $C_{21}H_{19}NO_4$
                                                      20168-99-4

1-Cinnamoyl-5-methoxy-2-methylindole-3-acetic acid =
5-Methoxy-2-methyl-1-(1-oxo-3-phenyl-2-propenyl)-
1*H*-indole-3-acetic acid (●)
S   Cetanovo, Cindomet, *Cinmetacin***, Indolacin, Indo-
        latsin, S 1290, *Tsinmetatsin*, TVX 1764
U   Anti-inflammatory

10159 $C_{21}H_{19}NO_5$
                                                      152815-51-5

(3*E*,4*E*)-3-Benzylidene-4-(3,4,5-trimethoxybenzyli-
ene)-2,5-pyrrolidinedione = (*E,E*)-3-(Phenylmethylene)-
-[(3,4,5-trimethoxyphenyl)methylene]-2,5-pyrrolidine-
ione (●)
        T-686
        Antithrombotic

10160   (8222) $C_{21}H_{19}NO_6$
                                                      125697-92-9

5-[(2,5-Dihydroxybenzyl)(2-hydroxybenzyl)amino]sali-
cylic acid = 5-[[(2,5-Dihydroxyphenyl)methyl][(2-hydr-
oxyphenyl)methyl]amino]-2-hydroxybenzoic acid (●)
S   *Lavendustin A*
U   Tyrosine kinase inhibitor from *Streptomyces griseola-
        vendus*

10161   (8223) $C_{21}H_{19}N_3O_3S$
                                                      51264-14-3

4'-(9-Acridinylamino)methanesulfon-*m*-anisidide = *N*-[4-
(9-Acridinylamino)-3-methoxyphenyl]methanesulfon-
amide (●)
R   also monolactate (80277-11-8)
S   Amekrin, *Amsacrine**, Amsakrin, Amsa-PD, Amsidi-
        le, Amsidine, Amsidyl, Amsine, Amsydil, CI-880,
        Lamasine, *m*-AMSA, NSC-249992, S.N. 11841
U   Antineoplastic, antiviral (DNA-polymerase inhibitor)

10162 $C_{21}H_{19}N_3O_6$
                                                      129794-24-7

3,9-Bis[(dimethylcarbamoyl)oxy]benzofuro[3,2-*c*]quino-lin-6(5*H*)-one = Dimethylcarbamic acid 5,6-dihydro-6-oxobenzofuro[3,2-*c*]quinoline-3,9-diyl ester (●)
S KCA-098
U Bone calcium regulator

10163 $C_{21}H_{19}N_3O_6S_2$
                                              56369-20-1

7-[(α-Hydroxy-α-phenylacetyl)amino]-3-[[(1-oxido-2-pyridinyl)thio]methyl]ceph-3-em-4-carboxylic acid = (6*R*,7*R*)-7-[[(2*R*)-Hydroxyphenylacetyl]amino]-3-[[(1-oxido-2-pyridinyl)thio]methyl]-8-oxo-5-thia-1-azabicyclo[4.2.0]oct-2-ene-2-carboxylic acid (●)
S MCO
U Antibacterial

10164 (8224) $C_{21}H_{19}N_5O_2$
                                              103926-64-3

6-Amidino-2-naphthyl *p*-(2-imidazolin-2-ylamino)ben-zoate = 4-[(4,5-Dihydro-1*H*-imidazol-2-yl)amino]ben-zoic acid 6-(aminoiminomethyl)-2-naphthalenyl ester (●)
R Dimethanesulfonate (103926-82-5)
S FUT 187, *Sepimostat mesilate***, TO-187
U Protease inhibitor

10165 (8225) $C_{21}H_{19}N_5O_6S$
                                              146404-38-8

2-Acetoxy-5-[[*p*-[(4,6-dimethyl-2-pyrimidinyl)sulf-amoyl]phenyl]azo]benzoic acid = 2-(Acetyloxy)-5-[[4-[[(4,6-dimethyl-2-pyrimidinyl)amino]sulfonyl]phe-nyl]azo]benzoic acid (●)
S *Disalazine*, VÚFB-17259
U Anti-arthritic, immunomodulator

10166 $C_{21}H_{20}ClNO_5$
                                              146426-40-6

rel-(−)-2-(2-Chlorophenyl)-5,7-dihydroxy-8-[(3*R*,4*S*)-3-hydroxy-1-methyl-4-piperidinyl]-4*H*-1-benzopyran-4-one (●)
R also hydrochloride (131740-09-5)
S *Avodenib*, *Flavopiridol*, HL 275, HMR 1275, L 868275, MDL 107826 A, NSC-649890
U Antineoplastic (cyclin-dependent kinase inhibitor)

10167 (8226) $C_{21}H_{20}ClNS$
                                              36471-39-

3-[(2-Chlorothioxanthen-9-ylidene)methyl]quinuclidine = 3-[(2-Chloro-9*H*-thioxanthen-9-ylidene)methyl]-1-aza-bicyclo[2.2.2]octane (●)
S LM 21005, *Nuclotixene***
U Antidepressant

10168 (8227) $C_{21}H_{20}Cl_2FN_3O_2$
72444-63-4

5-[2-[4-(3,5-Dichlorophenyl)-1-piperazinyl]ethyl]-4-(*p*-fluorophenyl)-4-oxazolin-2-one = 5-[2-[4-(3,5-Dichlorophenyl)-1-piperazinyl]ethyl]-4-(4-fluorophenyl)-2(3*H*)-oxazolone (●)
S *Lodiperone***
U Neuroleptic

10169 (8228) $C_{21}H_{20}Cl_2N_6O_3$
99593-25-6

-[(2-Aminoacetamido)methyl]-1-[4-chloro-2-(*o*-chloro-benzoyl)phenyl]-*N,N*-dimethyl-1*H*-1,2,4-triazole-3-carboxamide = 5-[[(Aminoacetyl)amino]methyl]-1-[4-chloro--(2-chlorobenzoyl)phenyl]-*N,N*-dimethyl-1*H*-1,2,4-triazole-3-carboxamide (●)
R also monohydrochloride (85815-37-8)
Rhythmy, *Rilmazafone***, S 191, 450191-S, *Trimazafone*
J Hypnotic

0170 (8229) $C_{21}H_{20}Cl_2O_3$
52645-53-1

3-Phenoxybenzyl (1*RS*)-*cis-trans*-3-(2,2-dichlorovinyl)-2,2-dimethylcyclopropanecarboxylate = 3-(2,2-Dichloroethenyl)-2,2-dimethylcyclopropanecarboxylic acid (3-phenoxyphenyl)methyl ester (●)
S Acticin "Penederm", Assy Espuma, Auriplak, Bichol "Medipharm", Canitex, Canovel, Catovel, Defencat, Defender, Defendog, Delixi, Dertil "Drag Pharma", Duncankil, Ectomethrin, Ekoped, Elimite, Eolia, EX-spot, FMC-33297, Foractil, Frento, Hairclin, Heidi, Infectopedicul, Kilnits, Kinderval, Lotrix, Loxazol, Lumat, Lyclear, Lyderm "PSM", Lysum, Neo Kill Antiparasit, Niksen, Nittifor, Nittyfor, Nix, Nok, Novo-Herklin 2000, NRDC 143, OMS 1821, Oroclean, Percapyl, *Permethrin***, Permit, Pertrin E, Puce-Stop Direct, Pulvex, Pyreflor, Pyrifoam, Quellada, Repemas, Residex P55, Rid, Ridect, Rycovet Ryposect, Sarcop, Stomoxin, Tectonik, Vet-Kem, Wellcare, WL-43479, Zalvor, Zebric-Teva, Zehu Ze
U Insecticide, ectoparasiticide

10171 (8230) $C_{21}H_{20}FN_3O_6S$
123447-62-1

(±)-7-[4-[(Z)-2,3-Dihydroxy-2-butenyl]-1-piperazinyl]-6-fluoro-1-methyl-4-oxo-1*H*,4*H*-[1,3]thiazeto[3,2-*a*]quinoline-3-carboxylic acid cyclic carbonate = 6-Fluoro-1-methyl-7-[4-[(5-methyl-2-oxo-1,3-dioxol-4-yl)methyl]-1-piperazinyl]-4-oxo-1*H*,4*H*-[1,3]thiazeto[3,2-*a*]quinoline-3-carboxylic acid (●)
S NAD-441A, NM 441, *Prulifloxacin***, *Pulfloxacin dioxolil*, Quisnon, Sword, *Tulifloxacin*
U Antibacterial

10172　(8231)

$C_{21}H_{20}NO_4$
52259-65-1

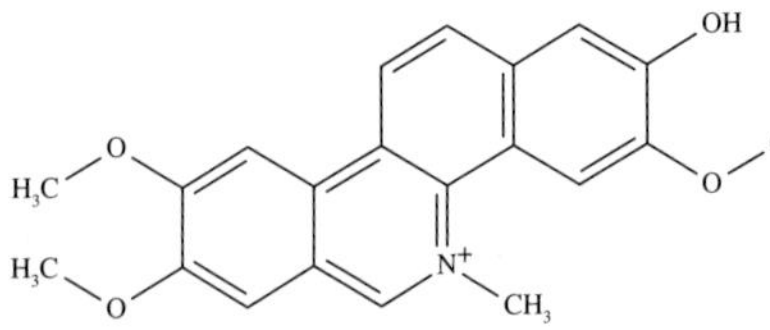

2-Hydroxy-3,8,9-trimethoxy-5-methylbenzo[c]phenanthridinium (●)
R Chloride (52259-64-0)
S *Fagaronine (chloride)*, FGR, NSC-157995
U Antineoplastic alkaloid from *Fagara xanthoxyloides*

10173　(8232)

$C_{21}H_{20}N_2O$
76145-76-1

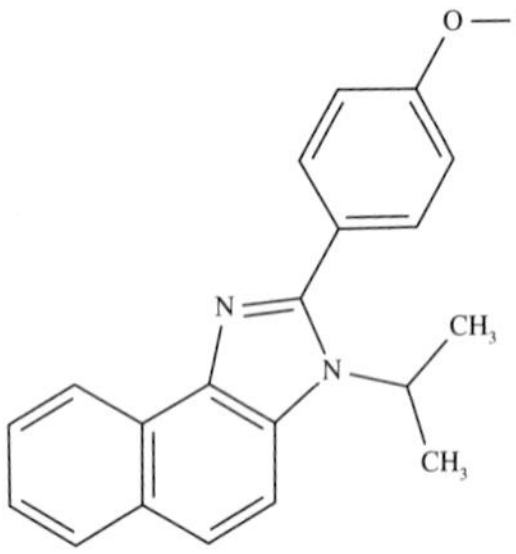

3-Isopropyl-2-(p-methoxyphenyl)-3H-naphth[1,2-d]imidazole = 2-(4-Methoxyphenyl)-3-(1-methylethyl)-3H-naphth[1,2-d]imidazole (●)
S MDL 035, *Tomoxiprole***
U Analgesic, anti-inflammatory

10174

$C_{21}H_{20}N_2O_4S$
193807-60-2

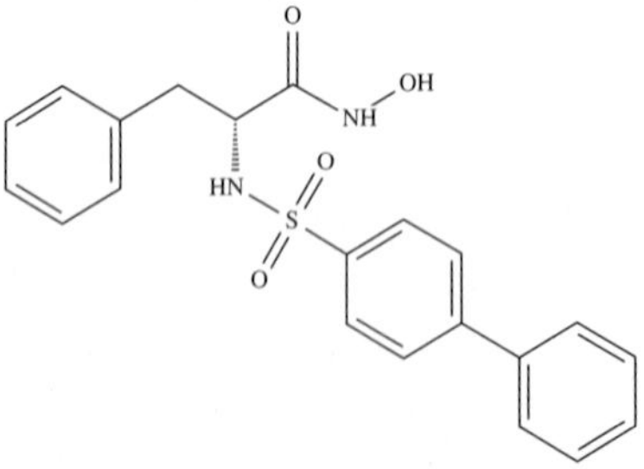

N-(4-Biphenylylsulfonyl)phenylalanine hydroxamic acid = (αR)-α-[([1,1'-Biphenyl]-4-ylsulfonyl)amino]-N-hydroxybenzenepropanamide (●)
S BPHA

U Matrix metalloproteinase inhibitor

10175　(8233)

$C_{21}H_{20}N_2O_4S$
304-43-8

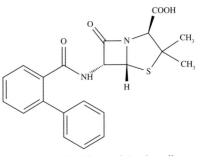

(2S,5R,6R)-3,3-Dimethyl-7-oxo-6-(o-phenylbenzamido)-4-thia-1-azabicyclo[3.2.0]heptane-2-carboxylic acid = (6R)-6-(o-Phenylbenzamido)penicillanic acid = o-Biphenylpenicillin = [2S-(2α,5α,6β)]-6-[([1,1'-Biphenyl]-2-ylcarbonyl)amino]-3,3-dimethyl-7-oxo-4-thia-1-azabicyclo[3.2.0]heptane-2-carboxylic acid (●)
R Sodium salt (2289-50-1)
S Ancillin, Bifenicillin, Biphenicillin, BL-P 413, *Diphenicillin-Natrium*, Penicillin 12141-Z, SKF 12141, *Sodium diphenicillin*
U Antibiotic

10176

$C_{21}H_{20}N_2O$
172732-68-

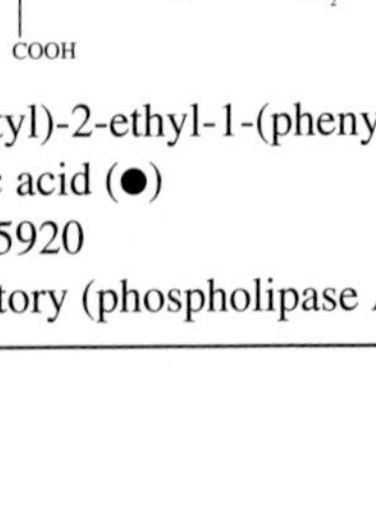

[[3-(Aminooxoacetyl)-2-ethyl-1-(phenylmethyl)-1H-indol-4-yl]oxy]acetic acid (●)
S LY 315920, S 5920
U Anti-inflammatory (phospholipase A_2 inhibitor)

10177   (8234) $C_{21}H_{20}N_2O_5$
                                                                  105785-61-3

N-[*m*-(2-Quinolylmethoxy)phenyl]succinohydroxamic
acid methyl ester = 4-[Hydroxy[3-(2-quinolinylmeth-
oxy)phenyl]amino]-4-oxobutanoic acid methyl ester (●)
S   Wy-45911
U   Leukotriene D$_4$ antagonist

10178   (8235) $C_{21}H_{20}N_2O_5$
                                                                  129564-92-7

(5*R-cis*)-5,6,11,11a-Tetrahydro-5-(4-hydroxy-3,5-di-
methoxyphenyl)-1*H*,3*H*-oxazolo[3',4':1,6]pyri-
do[3,4-*b*]indol-3-one (●)
S   *Azatoxin*, NSC-640737 M
U   Topoisomerase II inhibitor

10179   (8236) $C_{21}H_{20}N_3$
                                                                  3546-21-2

3,8-Diamino-5-ethyl-6-phenylphenanthridinium (●)
R   Bromide (1239-45-8)
S   *Äthidiumbromid*, Babidium bromide, *Ethidium bro-
mide*, *Homidium bromide***, Novidium bromide,
RD 1572

U   Antitrypanosomal

10180   (8237) $C_{21}H_{20}N_4O$
                                                                  14910-31-7

6,6'-Ureylenebis(1-methylquinolinium) = 6,6'-(Carbonyl-
diimino)bis[1-methylquinolinium] (●)
R   Bis(methylsulfate) (135-14-8)
S   Acaprin, Acapron, Atral, Babesan, Baburan, *Chin-
uridi methylsulfas*, Diveronal, Kinurid-Avlon, Lu-
dobal, Pirevan, Pirocyl, Pirozoo, Pyroplasmin,
Quinuronium sulfate, S.N. 5870, Zothelone
U   Antiprotozoal (babesicide)

10181 $C_{21}H_{20}N_4OS$
                                                                211311-08-9

9-[2-(Dimethylamino)ethyl]-8,13-dihydropyri-
do[4,3,2-*mn*][1,4]thiazino[3,2-*b*]acridin-12(11*H*)-one (●)
S   *Shermilamine D*
U   Cytotoxic alkaloid from *Cystodites violatinctus*

10182   (8238) $C_{21}H_{20}N_4O_3$
                                                                  32828-81-2

N,N'-Bis(3-picolyl)-4-methoxyisophthalamide = 4-Methoxy-N,N'-bis(3-pyridinylmethyl)-1,3-benzenedicarboxamide (●)
R also monohydrate (80530-63-8) or tartrate (86247-87-2)
S G-137, *Picotamide*, Plactidil
U Anticoagulant, fibrinolytic

10183 (8239) $C_{21}H_{20}N_6O$
3811-56-1

1,3-Bis(4-amino-2-methyl-6-quinolyl)urea = N,N'-Bis(4-amino-2-methyl-6-quinolinyl)urea (●)
R also dihydrochloride (5424-37-3)
S *Aminochincarbamid, Aminochinuridum, Aminoquinuride**, Gelastypt S, Revasa(n), Surfen
U Antiseptic

10184 (8240) $C_{21}H_{20}N_6O_4$
108427-72-1

9-[(4-Acetyl-3-hydroxy-2-propylphenoxy)methyl]-3-(1H-tetrazol-5-yl)-4H-pyrido[1,2-a]pyrimidin-4-one (●)
S AS-35
U Anti-allergic

10185 (8241) $C_{21}H_{20}N_6O_{10}S_4$
84728-38-1

(6R,7R)-7-[2-(2-Amino-4-thiazolyl)glyoxylamido]-3-[[(4-carboxy-3-hydroxy-5-isothiazolyl)thio]methyl]-8-oxo-5-thia-1-azabicyclo[4.2.0]oct-2-ene-2-carboxylic acid, 7^2-(Z)-[O-(1-carboxy-1-methylethyl)oxime] = (Z)-(7R)-7-[2-(2-Amino-4-thiazolyl)-2-[(1-carboxy-1-methylethoxy)imino]acetamido]-3-[[(4-carboxy-3-hydroxy-5-isothiazolyl)thio]methyl]-3-cephem-4-carboxylic acid = [6R-[6α,7β(Z)]]-7-[[(2-Amino-4-thiazolyl)[(1-carboxy-1-methylethoxy)imino]acetyl]amino]-3-[[(4-carboxy-3-hydroxy-5-isothiazolyl)thio]methyl]-8-oxo-5-thia-1-azabicyclo[4.2.0]oct-2-ene-2-carboxylic acid (●)
R Trisodium salt (92071-93-7)
S YM-13115
U Antibiotic

10186 $C_{21}H_{20}O$
186835-06-3

[1,1':2',1''-Terphenyl]-4'-propanol (●)
S F 050
U Antithrombotic

10187 (8242)

$C_{21}H_{20}O_6$
458-37-7

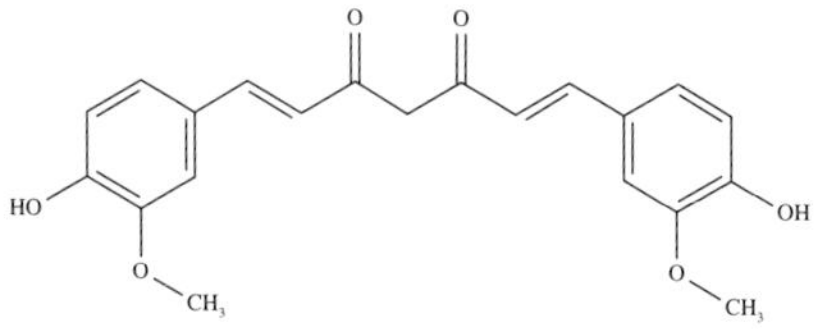

Diferuloylmethane = 1,7-Bis(4-hydroxy-3-methoxyphenyl)-1,6-heptadiene-3,5-dione (●)
S *Curcumin*
U Choleretic, antineoplastic, immunosuppressant (coloring matter from *Curcuma longa*)

10188 (8243)

$C_{21}H_{20}O_9$
3681-99-0

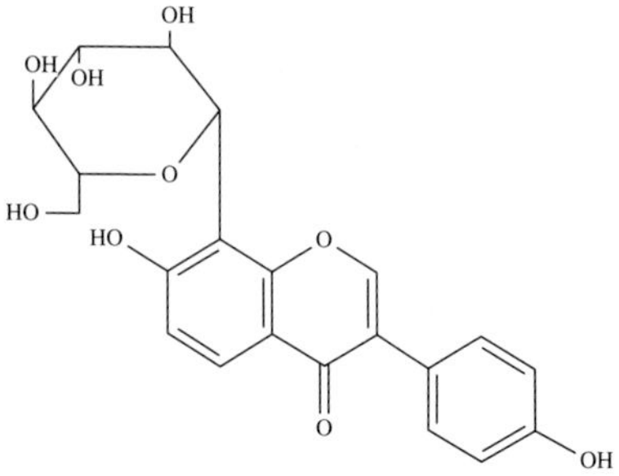

8-D-Glucopyranosyl-4',7-dihydroxyisoflavone = 8-D-Glucopyranosyldaidzein = 8-β-D-Glucopyranosyl-7-hydroxy-3-(4-hydroxyphenyl)-4H-1-benzopyran-4-one
(●)
S *Puerarin*
U Coronary vasodilator from *Pueraria lobata*

10189 (8244)

$C_{21}H_{21}ClFNO_2$
103253-15-2

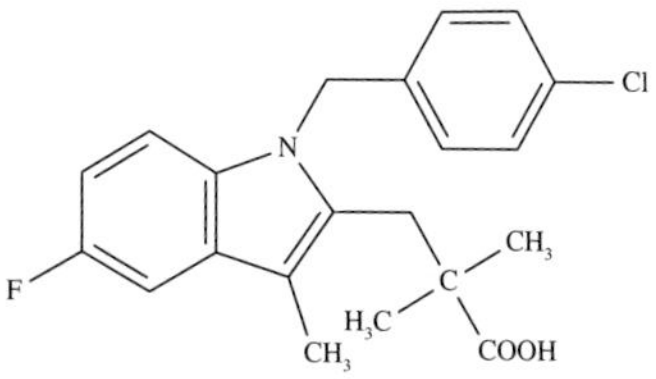

-[1-(p-Chlorobenzyl)-5-fluoro-3-methyl-2-indolyl]-2,2-imethylpropionic acid = 1-[(4-Chlorophenyl)methyl]-5-uoro-α,α,3-trimethyl-1H-indole-2-propanoic acid (●)
L 655240
Bronchodilator, platelet aggregation inhibitor

10190 (8245)

$C_{21}H_{21}ClFN_3O_2$
65899-72-1

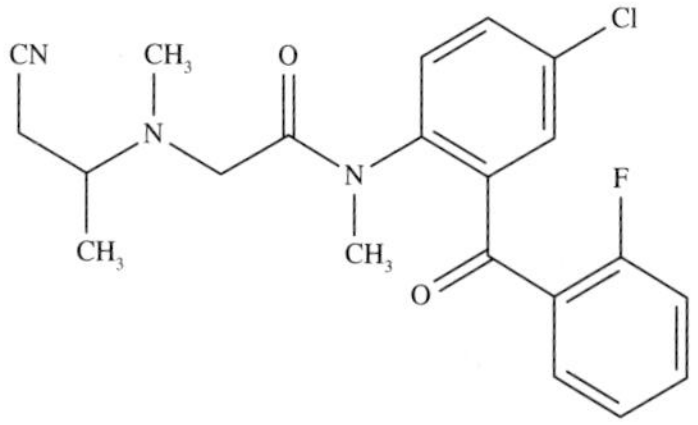

4'-Chloro-2-[(2-cyano-1-methylethyl)methylamino]-2'-(o-fluorobenzoyl)-N-methylacetanilide = N-[4-Chloro-2-(2-fluorobenzoyl)phenyl]-2-[(2-cyano-1-methylethyl)methylamino]-N-methylacetamide (●)
S *Alozafone***, CAS 108
U Tranquilizer, hypnotic

10191 (8246)

$C_{21}H_{21}ClN_2O$
85532-75-8

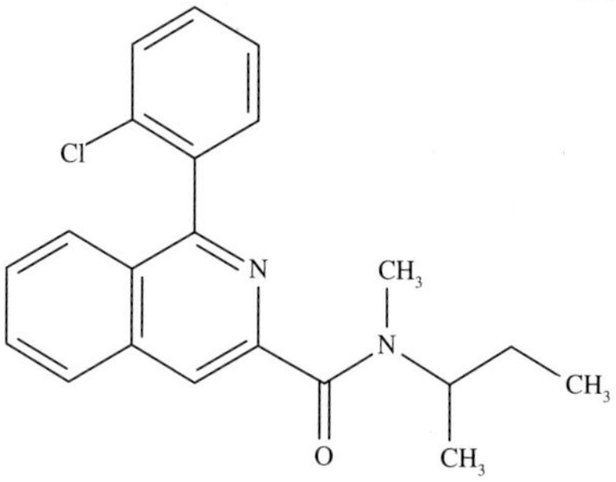

N-sec-Butyl-1-(o-chlorophenyl)-N-methyl-3-isoquinolinecarboxamide = 1-(2-Chlorophenyl)-N-methyl-N-(1-methylpropyl)-3-isoquinolinecarboxamide (●)
S PK-11195, RP 52028
U Analgesic, immunomodulator (benzodiazepine receptor antagonist)

10192 (8247)

$C_{21}H_{21}ClN_2O$
117796-52-8

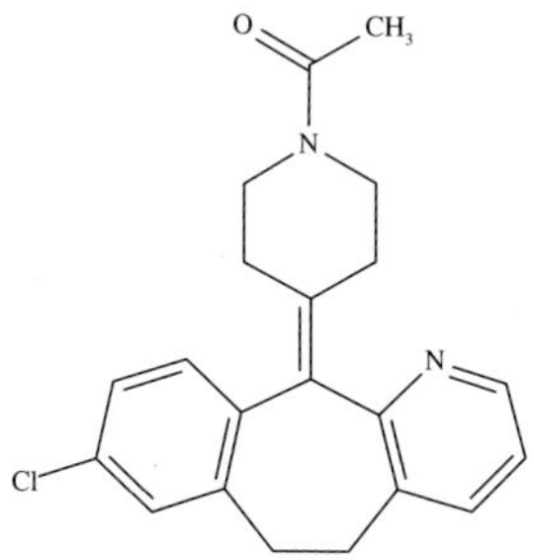

1-Acetyl-4-(8-chloro-5,6-dihydro-11H-benzo[5,6]cyclo-hepta[1,2-b]pyridin-11-ylidene)piperidine (●)
S   *Respatadine*, Sch 37370
U   Anti-allergic, PAF antagonist

10193 $C_{21}H_{21}ClN_2O_2$
                                         211363-59-6

2-(Benzoylamino)-3-chloro-3-phenyl-2-propenoylpiperi-dide = N-[(1E)-2-Chloro-2-phenyl-1-(1-piperidinylcarbo-nyl)ethenyl]benzamide (●)
S   AT 61
U   Antiviral (hepatitis B)

10194   (8248) $C_{21}H_{21}ClN_2O_2$
                                         57916-70-8

7-Chloro-1-[2-(cyclopropylmethoxy)ethyl]-1,3-dihydro-5-phenyl-2H-1,4-benzodiazepin-2-one (●)
S   *Clazepam, Iclazepam***
U   Tranquilizer

10195   (8249) $C_{21}H_{21}ClN_2O_3$
                                         71923-34-7

4-(4-Chlorophenyl)-5-[2-(4-phenyl-1-piperazinyl)ethyl]-1,3-dioxol-2-one (●)
S   *Clodoxopone***, LR 19731
U   Antihyperlipidemic

10196   (8250) $C_{21}H_{21}ClN_2O_8$
                                         127-33-3

7-Chloro-4-(dimethylamino)-1,4,4a,5,5a,6,11,12a-octa-hydro-3,6,10,12,12a-pentahydroxy-1,11-dioxo-2-naph-thacenecarboxamide = 7-Chloro-6-demethyltetracycline = [4S-(4α,4aα,5aα,6β,12aα)]-7-Chloro-4-(dimethylami-no)-1,4,4a,5,5a,6,11,12a-octahydro-3,6,10,12,12a-penta-hydroxy-1,11-dioxo-2-naphthacenecarboxamide (●)
R   also monohydrochloride (64-73-3)
S   Achromycin A VIII, Actaciclina, Ameciclina, Amsaciclina D.M., Benaciclin, Biormicin, Biotercic-lin, Biotil, Cloracetine, Clortetrin, Complecyclin, Dalmicina, Dankermicina, Deciclina, Declobiocina, Declomicina, Declomycin, Declor, Deganol, Deme-cidin, Demeclocillin, *Demeclocycline***, Demeclor, Demeplus, Deme-Proter, Demetetra, *Demethylchlor-tetracycline*, Demetilciclina, Demetilcina, *Deme-tilclortetraciclina*, Demetilina "Archifar", Demetra-ciclina, Demetraclin, Demexin, Demisin, De-Tetra-longa, Detracin, Detravis, Detricin, Dimeral, Diucic-lin "Benvegna", DMCT, Doriciclina, D-siklin, Dura-mycin "Ilsan", Elkamicina, Famibon, Fidocin, Glu-comycin, Isodemetil, Iticiclina, Latomicina, Leder-micina, Ledermycin, Leomicina, Leuciclina, Ma-

gis-Ciclina, Meciclin, Megaciclina, Metilciclina, Mexocine, Mirciclina, Neocromaciclin, Novociclina, Novotriclina, Ozark, Perciclina, Provimicina, Riclor, Ricomicina, Royalciclina, 10192 R.P., Rynabron, Sadociclina, Sumaclina, Superciclina, Tassocidin, Temet, Tetra-cipan, Tetradek, Tollerclin, Unicycline, Veraciclina, Wolnerciclina
U　Antibiotic

10196-01　(8250-01)
R　Magnesium salt
S　Diuciclin "Benvegna" caps.
U　Antibiotic

10197　(8251)　　　　　　　　　$C_{21}H_{21}ClN_4OS$
　　　　　　　　　　　　　　　　　146939-27-7

5-[2-[4-(1,2-Benzisothiazol-3-yl)-1-piperazinyl]ethyl]-6-chloro-2-indolinone = 5-[2-[4-(1,2-Benzisothiazol-3-yl)-1-piperazinyl]ethyl]-6-chloro-1,3-dihydro-2H-indol-2-one (●)
R　Monohydrochloride monohydrate (138982-67-9)
S　CP-88059-1, Zeldox, *Ziprasidone hydrochloride***
U　Antipsychotic

10198　(8252)　　　　　　　　　$C_{21}H_{21}ClN_4O_3$
　　　　　　　　　　　　　　　　　54533-85-6

2'-Chloro-2-[2-[(diethylamino)methyl]-1-imidazolyl]-5-nitrobenzophenone = 1-[2-(2-Chlorobenzoyl)-4-nitrophenyl]-2-[(diethylamino)methyl]imidazole = (2-Chlorophenyl)[2-[2-[(diethylamino)methyl]-1H-imidazol-1-yl]-5-nitrophenyl]methanone (●)
R　Fumarate (93929-95-4)

S　Ekonal, *Midafenone fumarate*, *Nizofenone fumarate***, Y-9179
U　Anti-anoxic, cerebroprotective

10199　(8253)　　　　　　　　　$C_{21}H_{21}ClO_3$
　　　　　　　　　　　　　　　　　35838-63-2

3-[p-(2-Chloroethyl)-α-propylbenzyl]-4-hydroxycoumarin = 3-[1-[4-(2-Chloroethyl)phenyl]butyl]-4-hydroxy-2H-1-benzopyran-2-one (●)
S　*Clocoumarol***, DB 112
U　Anticoagulant

10200　(8254)　　　　　　　　　$C_{21}H_{21}FN_2O_3$
　　　　　　　　　　　　　　　　　71923-29-0

4-(4-Fluorophenyl)-5-[2-(4-phenyl-1-piperazinyl)ethyl]-1,3-dioxol-2-one (●)
S　*Fludoxopone***, LR 19635
U　Antihyperlipidemic

10201　　　　　　　　　　　　　$C_{21}H_{21}FN_2O_3$
　　　　　　　　　　　　　　　　　161522-25-4

4-(8-Fluoro[1]benzoxepino[4,3-*b*]pyridin-11(5*H*)-ylidene)-1-piperidinepropanoic acid (●)
R Dihydrate (188199-97-5)
S HSR-609
U Anti-allergic (histamine H_1 receptor antagonist)

10202 (8255) $C_{21}H_{21}FN_2O_4S$
                                                  116649-85-5

(+)-9-(2-Carboxyethyl)-3(*R*)-(4-fluorophenylsulfonami-do)-1,2,3,4-tetrahydrocarbazole = (+)-(3*R*)-3-(*p*-Fluoro-benzenesulfonamido)-1,2,3,4-tetrahydrocarbazole-9-pro-pionic acid = (*R*)-3-[[(4-Fluorophenyl)sulfonyl]amino]-1,2,3,4-tetrahydro-9*H*-carbazole-9-propanoic acid (●)
S Baynas, Bay u 3405, EN-137774, *Ramatroban***
U Thromboxane antagonist, anti-allergic

10203 $C_{21}H_{21}FN_4O_3$
                                                  195532-12-8

1-Cyclopropyl-7-[(*S,S*)-2,8-diazabicyclo[4.3.0]non-8-yl]-6-fluoro-8-cyano-1,4-dihydro-4-oxo-3-quinolinecarb-oxylic acid = (4a*S*-*cis*)-8-Cyano-1-cyclopropyl-6-fluoro-1,4-dihydro-7-(octahydro-6*H*-pyrrolo[3,4-*b*]pyridin-6-yl)-4-oxo-3-quinolinecarboxylic acid (●)
S Bay 14-1877, *Pradofloxacin**
U Antibacterial

10204 (8256) $C_{21}H_{21}F_3N_2O_2$
                                                  67987-40-0

N-[[Ethyl[α-methyl-3-(trifluoromethyl)phenethyl]ami-no]methyl]phthalimide = 2-[[Ethyl[1-methyl-2-[3-(tri-fluoromethyl)phenyl]ethyl]amino]methyl]-1*H*-isoindole-1,3(2*H*)-dione (●)
S ITA-213
U Anorexic

10205 (8257) $C_{21}H_{21}N$
                                                  65472-88-0

(*E*)-*N*-Cinnamyl-*N*-methyl-1-naphthalenemethylamine = (*E*)-*N*-Methyl-*N*-(1-naphthylmethyl)-3-phenyl-2-propen-1-amine = (*E*)-Cinnamyl(methyl)(1-naphthylme-thyl)amine = (*E*)-*N*-Methyl-*N*-(3-phenyl-2-propenyl)-1-naphthalenemethanamine (●)
R also hydrochloride (65473-14-5)
S AW 105-843, Benecut, Exoderil, Exoteril, Fetimin, Micosona, *Naftifine***, *Naftifungin*, Naftin, SN 105-843, Suadian
U Topical antifungal

10206 (8258) $C_{21}H_{21}N$
                                                  129-03-3

4-(5-Dibenzo[*a,d*]cycloheptatrienylidene)-1-methylpipe-
ridine = 5-(1-Methyl-4-piperidinylidene)-5*H*-diben-
zo[*a,d*]cycloheptene = 4-(5*H*-Dibenzo[*a,d*]cyclohepten-
5-ylidene)-1-methylpiperidine (●)
R Hydrochloride (969-33-5) [or hydrochloride sesqui-
    hydrate (41354-29-4)]
S Adekin, Agotex, Alergil, Allerginol-Syr., Alphahist,
    Anarexol, Antegan, Apetexil, Apetigen, Aplexol,
    Astonin, Belindox, Betoliman, Ceptadin, Ciabel, Ci-
    cloventin, Ciplactin, Cipractin, Cipramin, Cipro "Be-
    ta", Ciprodal, Ciprogan, *Ciproheptadina*, Ciproral,
    Ciptadine, Cobaglobal, Contrallerg, Cypro, Cyproa-
    tin, Cyprodin, *Cyproeptadina*, Cypro-Gal, Cyprogin,
    *Cyproheptadine hydrochloride***, Cypromin "Sa-
    wai", Cyprotol, Cyptazin, Ditur, Ennamax, Eptadi-
    na, F.I. 5967, Grisetin "Andromaco", Heptasan,
    HSp 1229, Ifrasarl, Istabin, Istam-Far, Klarivitina,
    Klarivitine, Kulinet, Kyliver, Lexahist, Menactin,
    Mirsol "North Med.", MK-141, Nebor, Nuran, Oper-
    ma, Oractine, Oviedan, Peptoral, Periactin(e), Periac-
    tinol, Periatin, Pericap, Peritol, PMS-Cyproheptadi-
    ne, Practin "Merind", Prakten, Prohessen, Prolyn,
    Pronicy, Protadina, Pyrohep, Refex, Rosolio
    "Schweitzer", Sarohist, Sialotin, Sigloton, Sipraktin,
    Siprodin, Supersan, Syptajin, Triactin, Vieldrin, Vi-
    micon, Vinorex
U Antihistaminic, anti-allergic, appetite stimulant

10206-01 (8258-01)   24440-32-2
R Cyclamate (no. 717)
S Stolimina
U Appetite stimulant

10206-02   58131-49-0
R 2-Oxopentanedioate (no. 365) (1:1)
S *Cyproheptadine ketoglutarate, Glutodina*
U Appetite stimulant

10206-03 (8258-02)
R Salt with pyridoxal phosphate (no. 1203)
S Axoprol, *Dihexazin*, Viternum, Viternun
U Appetite stimulant

10206-04 (8258-03)   56595-83-6
R 7-Theophyllineacetate (no. 1619)
S *Acefyllin Cyproheptadine*, Metopina, UR-185
U Appetite stimulant

10207 (8259) $C_{21}H_{21}NO$
                    60996-87-4

N-(Diphenylmethyl)-*p*-methoxybenzylamine = *N*-(*p*-Me-
thoxybenzyl)benzhydrylamine = *N*-(Diphenylmethyl)-4-
methoxybenzenemethanamine (●)
R Hydrochloride
S Calmavine
U Spasmolytic

10208 (8260) $C_{21}H_{21}NO_2$
                    26020-55-3

N,N-Dimethylbenzofuro[3,2-*c*][1]benzoxepin-$\Delta^{6(12H),\gamma}$-
propylamine = 6-[3-(Dimethylamino)propylidene]-6,12-
dihydrobenzofuro[3,2-*c*][1]benzoxepin = 3-Benzofu-
ro[3,2-*c*][1]benzoxepin-6(12*H*)-ylidene-*N,N*-dimethyl-1-
propanamine (●)
R Fumarate (1:1) (34522-46-8)
S L 6257, Nocertone, Oxedix, *Oxetorone fumarate***
U Migraine therapeutic

10209 (8261) $C_{21}H_{21}NO_2$
                    96404-52-3

2-[(3-Fluoranthenylmethyl)amino]-2-methyl-1,3-pro-
panediol (●)
R Hydrochloride (96404-51-2)

S 773 U 82
U Antineoplastic

10210 $C_{21}H_{21}NO_2S$
                                                138970-92-0

6,11-Dihydro-11-(1-methyl-4-piperidinylidene)diben-
zo[*b,e*]thiepin-2-carboxylic acid (●)
R Hydrochloride (142783-62-8)
S VÚFB-17689
U Antihistaminic

10211 $C_{21}H_{21}NO_2S$
                                                118292-40-3

Ethyl 6-[(4,4-dimethylthiochroman-6-yl)ethynyl]nicoti-
nate = 6-[(3,4-Dihydro-4,4-dimethyl-2*H*-1-benzothiopy-
ran-6-yl)ethynyl]-3-pyridinecarboxylic acid ethyl ester
(●)
S AGN 190168, Suretin, *Tazarotene***, Tazorac, Zorac
U Keratolytic

10212 (8262) $C_{21}H_{21}NO_4$
                                                86111-26-4

1-Ethyl-2-(*p*-hydroxyphenyl)-3-methylindol-5-ol diacetate (ester) = 2-[4-(Acetyloxy)phenyl]-1-ethyl-3-methyl-
1*H*-indol-5-ol acetate (ester) (●)
S D-16726, NSC-341952, *Zindoxifene***

U Anti-estrogen (mammary tumor inhibitor)

10213 $C_{21}H_{21}NO_4S$
                                                206444-72-6

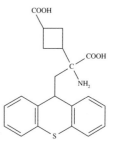

α-Amino-α-(3-carboxycyclobutyl)-9*H*-thioxanthene-9-
propanoic acid (●)
S LY 393053
U Metabotropic glutamate receptor antagonist

10214 (8263) $C_{21}H_{21}NO_6$
                                                6877-25-4

7-(Acetylamino)-6,7-dihydro-1,10-dimethoxy-2,3-(methylenedioxy)benzo[*a*]heptalen-9(5*H*)-one = (*S*)-*N*-
(4,6,7,8-Tetrahydro-3,13-dimethoxy-4-oxoheptale-
no[1,2-*f*][1,3]benzodioxol-6-yl)acetamide (●)
S *Cornigerine*
U Antimitotic alkaloid from *Colchicum cornigerum* and
other plants

10215   (8264) $C_{21}H_{21}NO_6$
118-08-1

1-(6,7-Dimethoxy-3-phthalidyl)-2-methyl-6,7-methy-
lenedioxy-1,2,3,4-tetrahydroisoquinoline = [S-(R^*,S^*)]-
6,7-Dimethoxy-3-(5,6,7,8-tetrahydro-6-methyl-1,3-di-
oxolo[4,5-g]isoquinolin-5-yl)-1-(3H)-isobenzofuranone
(●)
S *Hydrastine, Idrastina*
U Hemostatic (alkaloid from *Hydrastis canadensis*)

10216   (8265) $C_{21}H_{21}N_2S_2$
18403-49-1

3-Ethyl-2-[3-(3-ethyl-2-benzothiazolinylidene)prope-
nyl]benzothiazolium = 3-Ethyl-2-[3-(3-ethyl-2(3H)-ben-
zothiazolylidene)-1-propenyl]benzothiazolium (●)
R Mixture of one molecule of the 2,4,5-trichloropheno-
late with two molecules of 2,4,5-trichlorophenol
(5779-59-9)
S *Alazanine triclofenate***
U Anthelmintic

10217   (8266) $C_{21}H_{21}N_3O_2$
90274-22-9

1-[(E)-(5,6-Dihydro-6-oxo-11-morphanthridinyli-
dene)acetyl]-4-methylpiperazine = (E)-1-[(5,6-Dihydro-
6-oxo-11H-dibenz[b,e]azepin-11-ylidene)acetyl]-4-me-
thylpiperazine (●)
S *Darenzepine***
U Gastric antisecretory, anti-ulcer agent

10218 $C_{21}H_{21}N_3O_3$
211169-95-8

(3S)-L-Tryptophyl-1,2,3,4-tetrahydroisoquinoline-3-
carboxylic acid = (3S)-2-[(2S)-2-Amino-3-(1H-indol-3-
yl)-1-oxopropyl]-1,2,3,4-tetrahydro-3-isoquinolinecarb-
oxylic acid (●)
S TSL-225
U Anti-arthritic (dipeptidyl peptidase IV inhibitor)

10219 $C_{21}H_{21}N_3O_3$
182234-26-0

(2α,3β,5aα)-(−)-5a,6,7,8-Tetrahydro-3-(2-methyl-1-pro-
penyl)spiro[5H,10H-dipyrrolo[1,2-a:1',2'-d]pyrazine-
2(3H),3'-[3H]indole]-2',5,10(1'H)-trione (●)
S *Spirotryprostatin B*
U Cytotoxic from *Aspergillus fumigatus*

10220 (8267)

$C_{21}H_{21}N_3O_3$
92257-40-4

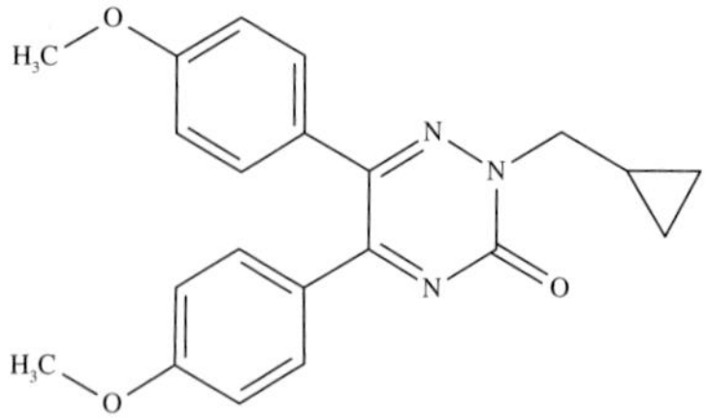

2-(Cyclopropylmethyl)-5,6-bis(*p*-methoxyphenyl)-*as*-triazin-3(2*H*)-one = 2-(Cyclopropylmethyl)-5,6-bis(4-methoxyphenyl)-1,2,4-triazin-3(2*H*)-one (●)
S *Dizatrifone**, ST 1118
U Peripheral analgesic

10221 (8268)

$C_{21}H_{21}N_3O_5S$
26785-17-1

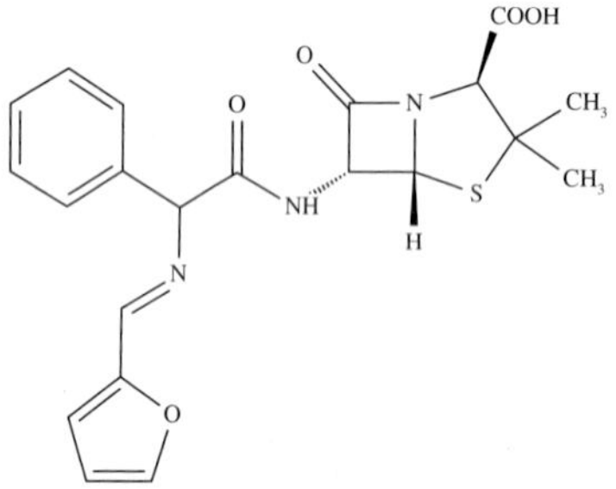

(2*S*,5*R*,6*R*)-6-[α-(Furfurylideneamino)phenylacetamido]-3,3-dimethyl-7-oxo-4-thia-1-azabicyclo[3.2.0]heptane-2-carboxylic acid = (6*R*)-6-[α-(Furfurylideneamino)phenylacetamido]penicillanic acid = [2*S*-(2α,5α,6β)]-6-[[[(2-Furanylmethylene)amino]phenylacetyl]amino]-3,3-dimethyl-7-oxo-4-thia-1-azabicyclo[3.2.0]heptane-2-carboxylic acid (●)
R Monosodium salt (50838-04-5)
S *Furampicillin*
U Antibiotic

10222 (8269)

$C_{21}H_{21}N_3O_6S$
78186-33-1

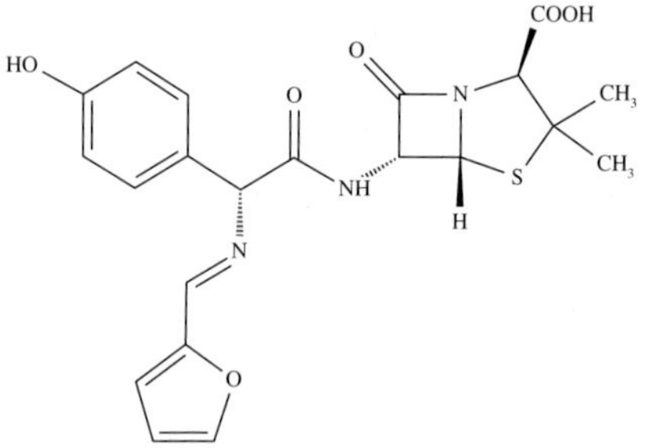

(2*S*,5*R*,6*R*)-6-[(*R*)-2-(Furfurylideneamino)-2-(*p*-hydroxyphenyl)acetamido]-3,3-dimethyl-7-oxo-4-thia-1-azabicyclo[3.2.0]heptane-2-carboxylic acid = (6*R*)-6-[D-2-(Furfurylideneamino)-2-(*p*-hydroxyphenyl)acetamido]penicillanic acid = D(–)-α-Furfurylideneamino-*p*-hydroxybenzylpenicillin = [2*S*-[2α,5α,6β(*S**)]]-6-[[[(2-Furanylmethylene)amino](4-hydroxyphenyl)acetyl]amino]-3,3-dimethyl-7-oxo-4-thia-1-azabicyclo[3.2.0]heptane-2-carboxylic acid (●)
R Monosodium salt (82631-53-6)
S FU-02, *Fumoxicillin sodium**, Furfurylideneamoxicillin sodium, Furoxicillin sodium*
U Antibiotic

10223

$C_{21}H_{21}N_3O_8$
132224-71-6

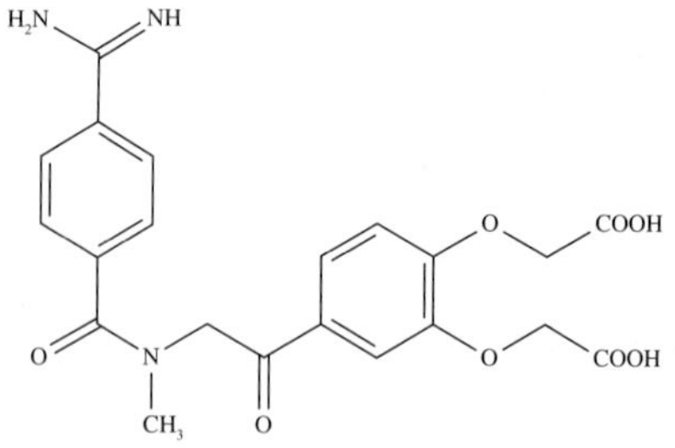

2,2'-[[4-[[[4-(Aminoiminomethyl)benzoyl]methylamino]acetyl]-1,2-phenylene]bis(oxy)]bis[acetic acid] (●)
S Ro 43-8857
U (GP)IIb/IIIa receptor antagonist

10224 (8270) $C_{21}H_{21}N_3O_9$
                                                           5585-59-1

4-(Dimethylamino)-1,4,4a,5,5a,6,11,12a-octahydro-3,10,12,12a-tetrahydroxy-7-nitro-1,11-dioxo-2-naphtha-cenecarboxamide = [4S-(4α,4aα,5aα,12aα)]-4-(Dimeth-ylamino)-1,4,4a,5,5a,6,11,12a-octahydro-3,10,12,12a-te-trahydroxy-7-nitro-1,11-dioxo-2-naphthacenecarbox-amide (●)
S *Nitrocycline***
U Antibiotic

10225 (8271) $C_{21}H_{21}N_3S$
                                                           57935-49-6

9,10-Didehydro-6-methyl-8β-[(2-pyridylthio)methyl]er-goline = (8β)-9,10-Didehydro-6-methyl-8-[(2-pyridinyl-thio)methyl]ergoline (●)
S CF 25-397, *Tiomergine***
U Antidepressant (dopamine receptor agonist)

10226 $C_{21}H_{21}N_5O_4S_2$
                                                           205054-36-0

4-Hydroxy-3-[[(3S)-2-oxo-3-[[[5-(3-pyridinyl)-2-thie-nyl]sulfonyl]amino]-1-pyrrolidinyl]methyl]benzenecarb-oximidamide (●)

S RPR 130737
U Factor X_A inhibitor

10227 $C_{21}H_{21}N_7O_5$
                                                           176857-41-3

L-γ-Methylene-10-deazaaminopterin = N-[4-[2-(2,4-Di-amino-6-pteridinyl)ethyl]benzoyl]-4-methylene-L-gluta-mic acid (●)
S L-MDAM, Mobiletrex
U Antineoplastic

10228 (8272) $C_{21}H_{22}CIFN_4O_2$
                                                           59831-65-1

N-[2-[4-(5-Chloro-2-oxo-1-benzimidazolinyl)piperidi-no]ethyl]-p-fluorobenzamide = N-[2-[4-(5-Chloro-2,3-di-hydro-2-oxo-1H-benzimidazol-1-yl)-1-piperidinyl]ethyl]-4-fluorobenzamide (●)
S *Halopemide***, R 34301
U Antipsychotic, neuroleptic

10229 (8273) $C_{21}H_{22}CINOS$
                                                           119475-97-7

3-[(2-Chloro-6,11-dihydrodibenzo[b,e]thiepin-11-yl)oxy]quinuclidine = (R*,R*)-(±)-3-[(2-Chloro-6,11-di-

hydrodibenzo[*b*,*e*]thiepin-11-yl)oxy]-1-azabicy-clo[2.2.2]octane (●)
R　Maleate [(*Z*)-2-butenedioate] (1:1) (119476-02-7)
S　VÚFB-17089
U　Antidepressant

10230　　　　　　　　　　　　　　　　$C_{21}H_{22}ClNO_2$
　　　　　　　　　　　　　　　　　　　146145-17-7

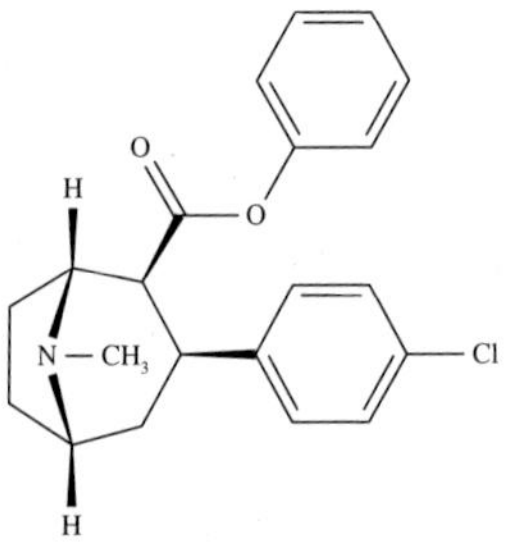

3β-(4-Chlorophenyl)tropane-2β-carboxylic acid phenyl ester = (1*R*,2*S*,3*S*,5*S*)-3-(4-Chlorophenyl)-8-methyl-8-azabicyclo[3.2.1]octane-2-carboxylic acid phenyl ester (●)
R　Hydrochloride (141807-57-0)
S　RTI 113
U　Dopamine transporter

10231　　　　　　　　　　　　　　　　　$C_{21}H_{22}ClN_3O$

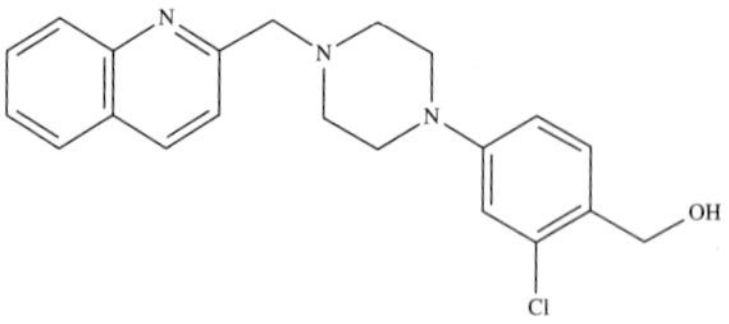

1-[3-Chloro-4-(hydroxymethyl)phenyl]-4-(2-quinolyl-methyl)piperazine = 2-Chloro-4-[4-(2-quinolinylmethyl)-1-piperazinyl]benzenemethanol (●)
S　PD 89232
U　Antipsychotic (dopamine D_4 receptor antagonist)

10232　(8274)　　　　　　　　　　　　$C_{21}H_{22}ClN_3O_2$
　　　　　　　　　　　　　　　　　　　87646-93-3

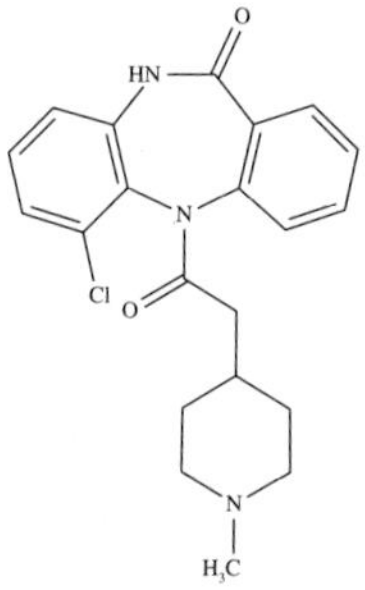

6-Chloro-5,10-dihydro-5-[(1-methyl-4-piperidinyl)ace-tyl]-11*H*-dibenzo[*b*,*e*][1,4]diazepin-11-one (●)
R　Monohydrochloride (120382-14-1)
S　UH-AH 37
U　Antihypertensive

10233　(8275)　　　　　　　　　　　　$C_{21}H_{22}ClN_3O_2$
　　　　　　　　　　　　　　　　　　　65352-97-8

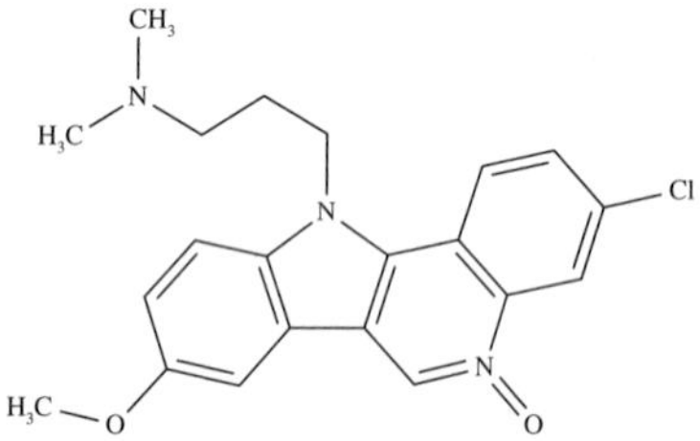

3-Chloro-8-methoxy-*N*,*N*-dimethyl-11*H*-indo-lo[3,2-*c*]quinoline-11-propanamine 5-oxide (●)
S　CM 6606
U　Antimalarial

10234　(8276)　　　　　　　　　　　　$C_{21}H_{22}ClN_3O_3$
　　　　　　　　　　　　　　　　　　　85076-06-8

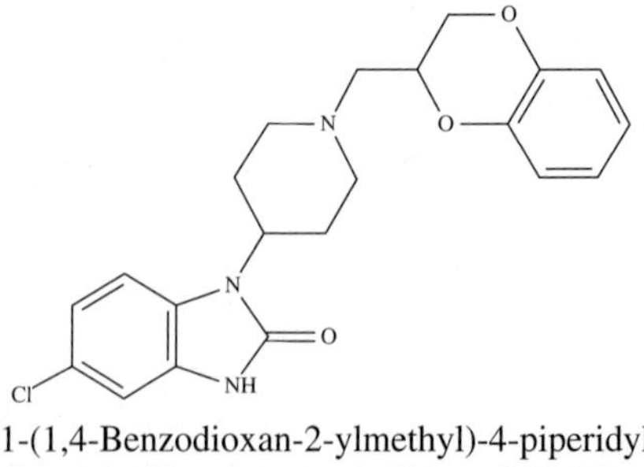

(±)-1-[1-(1,4-Benzodioxan-2-ylmethyl)-4-piperidyl]-5-chloro-2-benzimidazolinone = 5-Chloro-1-[1-[(2,3-dihy-

dro-1,4-benzodioxin-2-yl)methyl]-4-piperidinyl]-1,3-di-
hydro-2H-benzimidazol-2-one (●)
S *Axamozide***
U Neuroleptic

10235

$C_{21}H_{22}ClN_5O$
149847-76-7

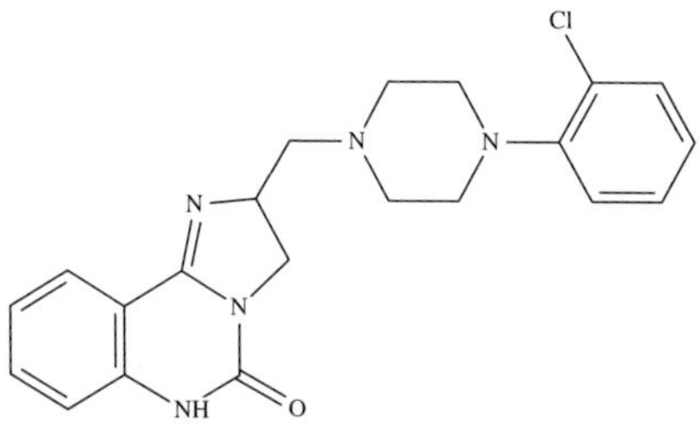

2-[[4-(2-Chlorophenyl)-1-piperazinyl]methyl]-2,6-dihy-
droimidazo[1,2-c]quinazolin-5(3H)-one (●)
S CK 53
U Antihypertensive, platelet aggregation inhibitor

10236 (8277)

$C_{21}H_{22}Cl_2N_3$
14668-07-6

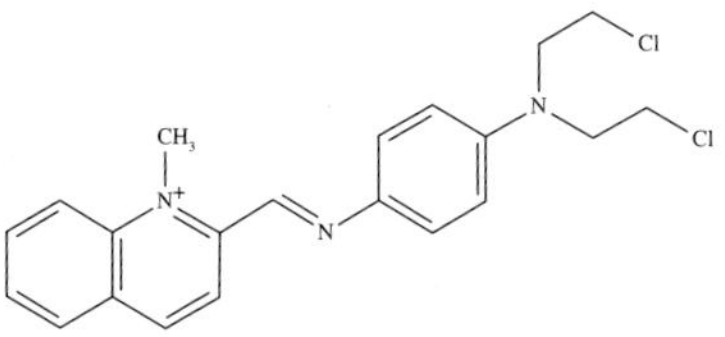

2-[[[4-[Bis(2-chloroethyl)amino]phenyl]imino]methyl]-
1-me thylquinolinium (●)
R Chloride (25843-64-5)
S IMET 3106
U Antineoplastic

10237

$C_{21}H_{22}FN_3O$
186544-26-3

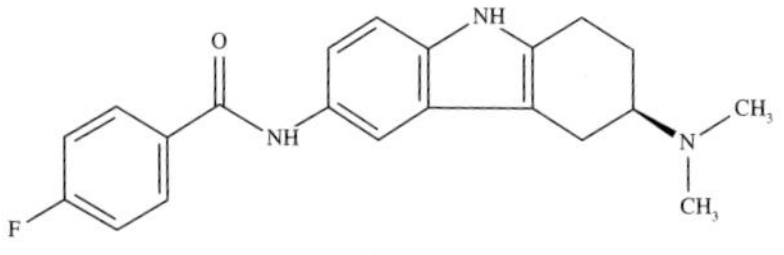

N-[(3R)-3-(Dimethylamino)-2,3,4,9-tetrahydro-1H-car-
bazol-6-yl]-4-fluorobenzamide (●)
S LY 344864
U 5-HT$_{1B/1D}$ receptor agonist

10238

$C_{21}H_{22}FN_3O$
182563-08-2

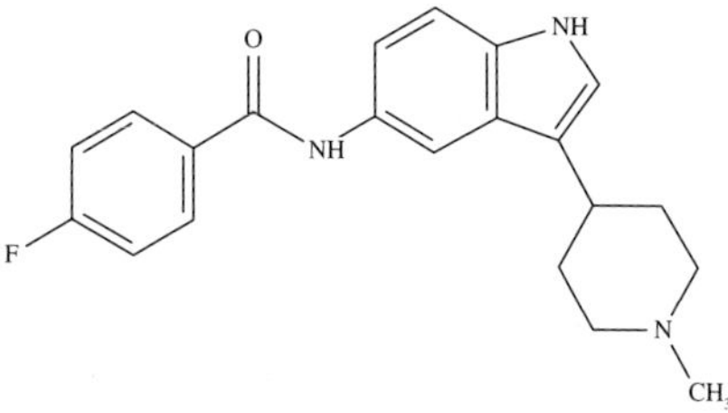

4-Fluoro-N-[3-(1-methyl-4-piperidinyl)-1H-indol-5-
yl]benzamide (●)
R Fumarate (1:1) (182563-09-3)
S LY 334370
U 5-HT$_{1F}$ receptor agonist (migraine therapeutic)

10239 (8278)

$C_{21}H_{22}F_3N_3OS$
33414-30-1

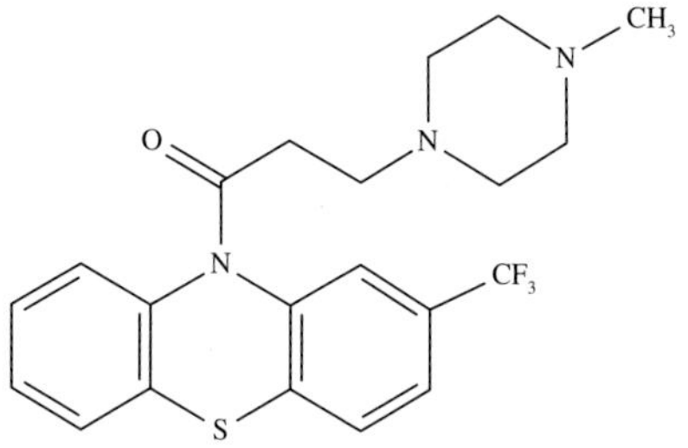

10-[3-(4-Methyl-1-piperazinyl)propionyl]-2-(trifluoromethyl)phenothiazine = 10-[3-(4-Methyl-1-piperazinyl)-
1-oxopropyl]-2-(trifluoromethyl)-10H-phenothiazine (●)
S *Ftormetazine***
U Antidepressant

10240 (8279)

$C_{21}H_{22}NO_4$
3486-67-7

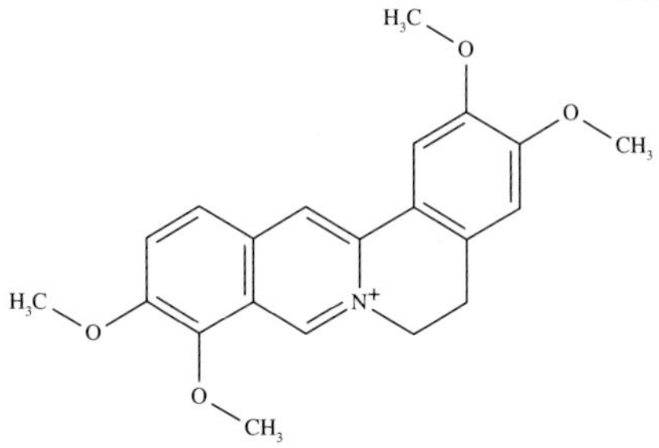

5,6-Dihydro-2,3,9,10-tetramethoxydibenzo[a,g]quinoli-
zinium (●)
R most iodide (4880-79-9)

S *Calystigine, Palmatine*
U Antibacterial (*Menispermaceae* alkaloid)

10241 (8280) $C_{21}H_{22}N_2O_2$
                                                    115607-61-9

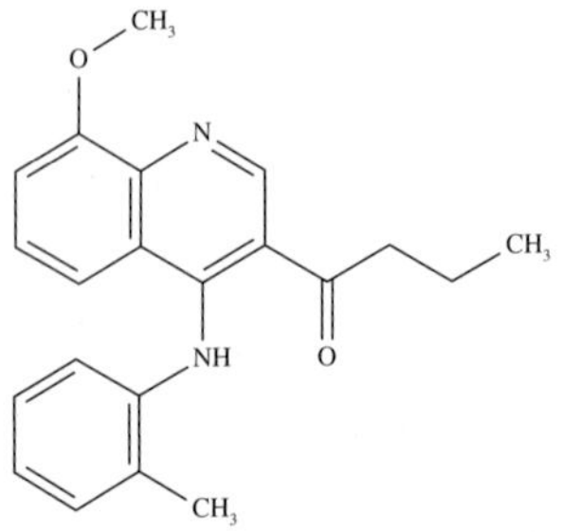

1-[8-Methoxy-4-[(2-methylphenyl)amino]-3-quinolinyl]-1-butanone (●)
S SKF 96067
U Gastric antisecretory

10242 (8281) $C_{21}H_{22}N_2O_2$
                                                    50270-32-1

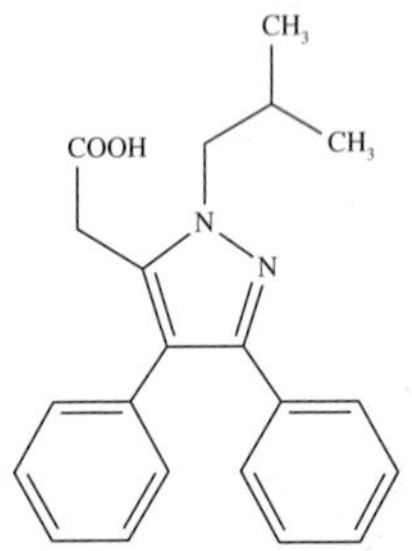

1-Isobutyl-3,4-diphenyl-5-pyrazoleacetic acid = 1-(2-Methylpropyl)-3,4-diphenyl-1*H*-pyrazole-5-acetic acid (●)
S *Bufezolac***, LM 22070
U Anti-inflammatory

10243 (8282) $C_{21}H_{22}N_2O_2$
                                                    57-24-9

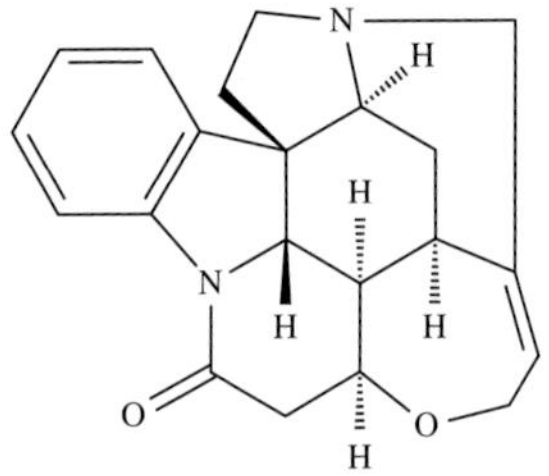

Strychnidin-10-one (●)
S *Estricnina, Estrychnina, Stricnina, Striknin, Strychnine*
U CNS stimulant, convulsant, cholinesterase inhibitor (alkaloid from *Strychnos nux-vomica* and other *Strychnos* sp.), tonic, stomachic

10243-01 (8282-01)   66-32-0
R Mononitrate
S Stchinin, Strychintran, Strychnovet
U Analeptic, tonic

10243-02 (8282-02)   60-41-3
R Sulfate (2:1)
S Strynervene
U CNS stimulant

10244 (8283) $C_{21}H_{22}N_2O_3$
                                                    159094-94-7

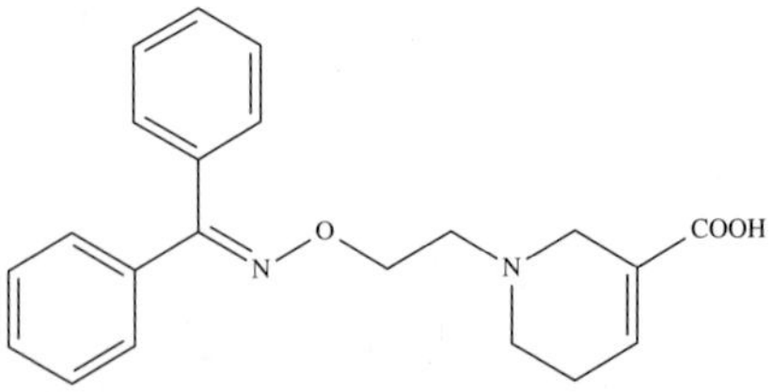

1,2,5,6-Tetrahydro-1-[2-[[(diphenylmethylene)amino]oxy]ethyl]-3-pyridinecarboxylic acid (●)
R also monohydrochloride (145645-62-1)
S NNC-711, NO-711
U GABA uptake inhibitor

10245   (8284) $C_{21}H_{22}N_2O_3$
7248-28-4

Strychnine *N*-oxide = Strychnidin-10-one 19-oxide (●)
R   also hydrochloride or benzoate
S   Genostrychnin, Movellan-Tabl., Stricnogen, *Strych-naminoxydum*, Z 203
U   Depot strychnine preparation

10246   (8285) $C_{21}H_{22}N_2O_4$
127408-30-4

rans-4-[Acetyl(phenylmethoxy)amino]-3,4-dihydro-3-hydroxy-2,2-dimethyl-2*H*-1-benzopyran-6-carbonitrile = *3S-trans*)-N-(6-Cyano-3,4-dihydro-3-hydroxy-2,2-dime-hyl-2*H*-1-benzopyran-4-yl)-*N*-(phenylmethoxy)acet-mide (●)
S   Y-27152
U   Antihypertensive (potassium channel activator)

0247 $C_{21}H_{22}N_2O_4S$
160135-56-8

(3S)-2,3,4,5-Tetrahydro-3-[[(2S)-2-mercapto-1-oxo-3-phenylpropyl]amino]-2-oxo-1*H*-1-benzazepine-1-acetic acid (●)
S   BMS 182657
U   Neutral endopeptidase and ACE inhibitor

10248   (8286) $C_{21}H_{22}N_2O_5S$
147-52-4

(2S,5R,6R)-6-(2-Ethoxy-1-naphthamido)-3,3-dimethyl-7-oxo-4-thia-1-azabicyclo[3.2.0]heptane-2-carboxylic acid = (6R)-6-(2-Ethoxy-1-naphthamido)penicillanic acid = 2-Ethoxy-1-naphthylpenicillin = [2S-(2α,5α,6β)]-6-[[(2-Ethoxy-1-naphthalenyl)carbonyl]amino]-3,3-dimethyl-7-oxo-4-thia-1-azabicyclo[3.2.0]heptane-2-carboxylic acid (●)
R   Monosodium salt (985-16-0) [or monosodium salt monohydrate (7177-50-6)]
S   Amplifen, Nafcil, *Nafcillin sodium***, Naftopen, Nallpen, *Naphcillin sodium*, *Sodium nafcillin***, Unipen "Wyeth-Ayerst", Vigopen, Wy-3277
U   Antibiotic

10249   (8287) $C_{21}H_{22}N_2O_7$
808-26-4

4-(Dimethylamino)-1,4,4a,5,5a,6,11,12a-octahydro-3,10,12,12a-tetrahydroxy-1,11-dioxo-2-naphthacene-carboxamide = 6-Demethyl-6-deoxytetracycline = [4S-(4α,4aα,5aα,12aα)]-4-(Dimethylamino)-1,4,4a,5,5a,6,11,12a-octahydro-3,10,12,12a-tetrahydr-oxy-1,11-dioxo-2-naphthacenecarboxamide (●)
S   Bonomycin, GS 2147, *Norcycline*, NSC-51812, *Sancycline***
U   Antibiotic

10250 (8288) $C_{21}H_{22}N_2O_8$
987-02-0

4-(Dimethylamino)-1,4,4a,5,5a,6,11,12a-octahydro-
3,6,10,12,12a-pentahydroxy-1,11-dioxo-2-naphthacene-
carboxamide = [4S-(4α,4aα,5aα,6β,12aα)]-4-(Dimethylamino)-1,4,4a,5,5a,6,11,12a-octahydro-3,6,10,12,12a-
pentahydroxy-1,11-dioxo-2-naphthacenecarboxamide (●)
S A IX, CL 22415, *Demecycline***, *Demethyltetracyc-
line*, Floricina (active substance), 6798 R.P.
U Antibiotic

10251 $C_{21}H_{22}N_3OS_2$
142306-96-5

1-Ethyl-2-[[3-ethyl-5-(3-methyl-2(3H)-benzothiazolyli-
dene)-4-oxo-2-thiazolidinylidene]methyl]pyridinium (●)
R Chloride (147366-41-4)
S FJ 776, MKT-077, SDZ-MKT 077
U Antineoplastic (rhodacyanine dye)

10252 $C_{21}H_{22}N_4OS$
129639-79-8

Hexahydro-2-[[4-[o-(2,4-xylyloxy)phenyl]-2-thiazo-
lyl]imino]pyrimidine = N-[4-[2-(2,4-Dimethylphen-
oxy)phenyl]-2-thiazolyl]-1,4,5,6-tetrahydro-2-pyrimidin-
amine (●)
S *Abafungin**, BAY w 6341

U Antibacterial, antifungal

10253 (8289) $C_{21}H_{22}N_4O_4$
93781-35-2

(±)-2-[3-[[2-(2,4-Dioxo-3-phenyl-1-imidazolidinyl)-
ethyl]amino]-2-hydroxypropoxy]benzonitrile (●)
R Monohydrochloride (115043-86-2)
S P-0160
U Anti-arrhythmic (α-adrenoceptor antagonist)

10254 (8290) $C_{21}H_{22}N_4O_6$
70343-57-6

7-N-(p-Hydroxyphenyl)mitomycin C = [1aS-
(1aα,8β,8aα,8bα)]-8-[[(Aminocarbonyl)oxy]methyl]-
1,1a,2,8,8a,8b-hexahydro-6-[(4-hydroxyphenyl)amino]-
8a-methoxy-5-methylazirino[2',3':3,4]pyrrolo[1,2-a]in-
dole-4,7-dione (●)
S KW 2083, M 83, NSC-278891
U Antineoplastic

10255 (8291) $C_{21}H_{22}N_4O$
103417-69-

3-Ethyl-5-methyl 1,4-dihydro-2-(1-imidazolylmethyl)-6
methyl-4-(m-nitrophenyl)-3,5-pyridinedicarboxylate =
1,4-Dihydro-2-(1H-imidazol-1-ylmethyl)-6-methyl-4-(3

nitrophenyl)-3,5-pyridinedicarboxylic acid 3-ethyl 5-methyl ester (●)
S Wy-27569
U Calcium antagonist, thromboxane synthetase inhibitor

10256 (8292) $C_{21}H_{22}N_4O_6S$
112887-68-0

N-[5-[[(3,4-Dihydro-2-methyl-4-oxo-6-quinazolinyl)methyl]methylamino]-2-thenoyl]-L-glutamic acid = N-[[5-[[(1,4-Dihydro-2-methyl-4-oxo-6-quinazolinyl)methyl]methylamino]-2-thienyl]carbonyl]-L-glutamic acid
(●)
S D 1694, ICI D 1694, NSC-639186, *Potidistat*, *Raltitrexed***, Tomudex, ZD 1694, ZN-D 1694
U Antineoplastic (thymidylate synthase inhibitor)

10257 (8293) $C_{21}H_{22}N_6O_4S_3$
40158-24-5

7R)-7-[(Phenylacetimidoylamino)acetamido]-3-(5-methyl-1,3,4-thiadiazol-2-ylthiomethyl)ceph-3-em-4-carboxylic acid = (6R-*trans*)-7-[[[(1-Imino-2-phenylethyl)amino]acetyl]amino]-3-[[(5-methyl-1,3,4-thiadiazol-2-yl)thio]methyl]-8-oxo-5-thia-1-azabicyclo[4.2.0]oct-2-ene-2-carboxylic acid (●)
S BL-S 339
U Antibiotic

10258 (8294) $C_{21}H_{22}N_6O_5$
18921-70-5

N-[4-[[(2,4-Diamino-5-methyl-6-quinazolinyl)methyl]amino]benzoyl]-L-aspartic acid (●)

R also disodium salt (18921-69-2)
S *Methasquin*, NSC-122870, SK 29836
U Antineoplastic

10259 $C_{21}H_{22}N_6O_5$
155036-68-3

N-Ethyl-1'-deoxy-1'-[6-amino-2-(3-hydroxy-3-phenyl-1-propynyl)-9H-purin-9-yl]-β-D-ribofuran-5'-uronamide = 1-[6-Amino-2-(3-hydroxy-3-phenyl-1-propynyl)-9H-purin-9-yl]-1-deoxy-N-ethyl-β-D-ribofuranuronamide (●)
S Sch 59761
U Adenosine receptor agonist

10260 (8295) $C_{21}H_{22}N_{10}O_{11}S_2$
123444-35-9

[S-(Z)]-2-[[[1-(2-Amino-4-thiazolyl)-2-[[1-[[[[3-(1,4-dihydro-5-hydroxy-4-oxo-2-pyridinyl)-4,5-dihydro-4-methyl-5-oxo-1H-1,2,4-triazol-1-yl]sulfonyl]amino]carbonyl]-2-oxo-3-azetidinyl]amino]-2-oxoethylidene]amino]oxy]-2-methylpropanoic acid (●)
S U-78608
U Antibiotic (iron complexing)

10261 (8296) $C_{21}H_{22}O_4$
52995-37-6

5-(1-Phenylcyclopentyl)-1,3-benzodioxole-2-carboxylic acid ethyl ester (●)
S RMI 14654

U Antihyperlipidemic

10262 $C_{21}H_{22}O_4$
                                                         7380-40-7

4-[[(2E)-3,7-Dimethyl-2,6-octadienyl]oxy]-7H-fu-
ro[3,2-g][1]benzopyran-7-one (●)
S *Bergamottin, Bergaptin*
U Anti-anginal, anti-arrhythmic, antineoplastic from
      bergamot oil

10263 $C_{21}H_{22}O_4$
                                                         58749-22-7

(2E)-3-[5-(1,1-Dimethyl-2-propenyl)-4-hydroxy-2-meth-
oxyphenyl]-1-(4-hydroxyphenyl)-2-propen-1-one (●)
S *Licochalcone A*
U Antileishmanial, antiparasitic, antiviral from *Glycyr-
      rhiza inflata* roots

10264   (8297) $C_{21}H_{22}O_7$
                                                         73069-27-9

[9S-[9α(Z),10α]]-2-Methyl-2-butenoic acid 10-(acetyloxy)-9,10-dihydro-8,8-dimethyl-2-oxo-2H,8H-ben-
zo[1,2-b:3,4-b']dipyran-9-yl ester (●)

S *Praeruptorin A*
U Cardiotonic from *Peucedanum praeruptorum*

10265 $C_{21}H_{22}O_8$
                                                         478-01-3

3',4',5,6,7,8-Hexamethoxyflavone = 2-(3,4-Dimethoxy-
phenyl)-5,6,7,8-tetramethoxy-4H-1-benzopyran-4-one
(●)
S *Nobiletin*
U Anti-inflammatory (*Citrus* flavonoid)

10266   (8298) $C_{21}H_{22}O_9$
                                                         1415-73-2

10-Glucopyranosyl-1,8-dihydroxy-3-(hydroxymethyl)-9-
anthrone = 1,8-Dihydroxy-3-(hydroxymethyl)-9-anthrone
10-C-glucoside = 10-(1,5-Anhydroglucosyl)aloe emodin-
9-anthrone = (S)-10-β-D-Glucopyranosyl-1,8-dihydroxy-
3-(hydroxymethyl)-9(10H)-anthracenone (●)
R also commercial mixture with Aloine B and small
      quantities of other compounds named Aloin
      (8015-61-0)
S *Aloin A, Barbaloin*
U Laxative (from various sp. of aloe)

10267   (8299) $C_{21}H_{23}BrFNO_2$
10457-90-6

4-[4-(*p*-Bromophenyl)-4-hydroxypiperidino]-4'-fluoro-
butyrophenone = 4-(*p*-Bromophenyl)-1-[3-(*p*-fluoroben-
zoyl)propyl]-4-piperidinol = 4-[4-(4-Bromophenyl)-4-
hydroxy-1-piperidinyl]-1-(4-fluorophenyl)-1-butanone
(●)

R   see also no. 14524;
also lactate

S   Agostine, Agotrix, Azurene, Bridol, Brodel, Brom-
idol, Bromodol, *Bromoperidol*, *Bromperidol***, Bu-
rom, C-C 2489, Consilium, Erodium, Fobinil, Impro-
men, R 11333, Tesoprel

U   Neuroleptic

10268   (8300) $C_{21}H_{23}ClFNO_2$
52-86-8

4-[4-(*p*-Chlorophenyl)-4-hydroxypiperidino]-4'-fluoro-
butyrophenone = 4-(*p*-Chlorophenyl)-1-[3-(*p*-fluoroben-
zoyl)propyl]-4-piperidinol = 4-[4-(4-Chlorophenyl)-4-
hydroxy-1-piperidinyl]-1-(4-fluorophenyl)-1-butanone
(●)

R   see also no. 14525

S   Alased, Aloperidin, *Aloperidolo*, Apo-Haloperidol,
Apo-Peridol, Bioperidolo, Brotopon, Buteridol, Ce-
reen, Cosminal, CT-Halop, Dozic, Duraperidol,
Einalon S, Elaubat, Esextin, Eukystol, Euteberol,
Fortunan, Haldol, Halidol, Halipia, Halojust, Halom-
idol, Haloneural, Halopal, *Haloperidol***, Haloperil,
Haloperin, Haloper von ct, Halophen, Halopidol, Ha-
losten, Halozen, Halperon, Halree, Hapidol, Hapo-
dol, Haridol, Keselan, Lemonamin, Leptol, Limerix,
Linton, McN-JR 1625, Medilorin, Mixidol, Nerve-
maine, Norodol, Novoperidol, NSC-170973, Ovocte-
rol, Pacedol, Parosmin, Peluces, Peridol "Technilab", Peridor, Pernox, Phlogis, PMS-Haloperidol, Po-
lyhaldol, R 1625, Seatop, Sedaperidol, Sedocat, Sele-

zyme, Senorm, Seponsque, Serenace, Serenase, Se-
renelfi, Sevium, Sigaperidol, Suirolin, Sylador
"Dumex", Tamide, Tensidol, Trakimin, Trancodol,
Xyduril, Youperidol, Zafrionil, Zetoridal

U   Neuroleptic

10269   (8301) $C_{21}H_{23}ClFN_3O$
17617-23-1

7-Chloro-1-[2-(diethylamino)ethyl]-5-(2-fluorophenyl)-
1,3-dihydro-2*H*-1,4-benzodiazepin-2-one (●)

R   most hydrochloride [monohydrochloride
(36105-20-1); dihydrochloride (1172-18-5)]

S   Apo-Flurazepam, Beconerv Neu, Benozil, Dalma-
dor, Dalmadorm, Dalmane, Dalmate, Dormigen "Sanitas", Dormodor, Durapam, Enoctan, Felison, Fel-
mane, Fludane, Fluleep, Flunox "Boehringer-Mannheim", Fluralema, *Flurazepam hydrochloride***, Flu-
zepam, Fordrim, ID-480, Insonium, Insumin, Irdal,
Linzac, Lunipax, Midorm, Morfex, Natam, Nergart,
Nindral, Noctosom, Noctyn-N.F., Nomadon, No-
voflupam, NSC-78559, Paxane, Remdue,
Ro 5-6901, Somlan, Somnol "Horner", Som-Pam,
Sonoluso, SR 759, Staurodorm (-Neu), Valdorm, Ze-
mon

U   Hypnotic

10270   (8302) $C_{21}H_{23}ClFN_3OS$
104821-36-5

10-[4-(2-Aminocarbonylethyl)-1-piperazinyl]-2-chloro-7-fluoro-10,11-dihydrodibenzo[b,f]thiepine = 4-(2-Chloro-7-fluoro-10,11-dihydrodibenzo[b,f]thiepin-10-yl)-1-piperazinepropanamide (●)
R Monomethanesulfonate (104821-37-6)
S *Chloflumid, Cloflumide*, VÚFB-15496
U Neuroleptic

6-(4-Chlorobenzyl)-1,4-dimethyl-5-oxo-1,4,5,6-tetrahydro-pyridazino[4,5-c]pyridazine-3,4 dicarboxylic acid diethyl ether = 6-[(4-Chlorophenyl)methyl]-1,4,5,6-tetrahydro-1,4-dimethyl-5-oxopyridazino[4,5-c]pyridazine-3,4-dicarboxylic acid diethyl ester (●)
S CK 119
U Cytotoxic

10271 $C_{21}H_{23}ClN_2O_4$
                             179819-89-7

[[3-[2-[[2-(3-Chlorophenyl)-2-hydroxyethyl]amino]propyl]-1H-indol-7-yl]oxy]acetic acid (●)
S AD 9677
U β_3-Adrenergic agonist

10272 (8303) $C_{21}H_{23}ClN_4O_2$
                             4052-13-5

3-[3-[4-(3-Chlorophenyl)-1-piperazinyl]propyl]-2,4(1H,3H)-quinazolinedione (●)
R Monohydrochloride (525-26-8)
S *Cloperidone hydrochloride***, MA 1337
U Sedative, tranquilizer

10273 $C_{21}H_{23}ClN_4O_5$
                             197917-10-5

10274 $C_{21}H_{23}Cl_2NO_6$
                             167221-71-8

(Butyryloxy)methyl methyl 4-(2,3-dichlorophenyl)-1,4-dihydro-2,6-dimethyl-3,5-pyridinedicarboxylate = (±)-Hydroxymethyl methyl 4-(2,3-dichlorophenyl)-1,4-dihydro-2,6-dimethyl-3,5-pyridinedicarboxylate butyrate (ester) = 4-(2,3-Dichlorophenyl)-1,4-dihydro-2,6-dimethyl-3,5-pyridinedicarboxylic acid methyl (1-oxobutoxy)methyl ester (●)
S *Clevidipine***, H 324/38
U Antihypertensive (calcium antagonist)

10275 (8304) $C_{21}H_{23}Cl_2N_3C$
                             82626-01-5

6-Chloro-2-(4-chlorophenyl)-N,N-dipropylimidazo[1,2-a]pyridine-3-acetamide (●)
S *Alpidem***, Ananxyl, SL 800342
U Anxiolytic

10276 $C_{21}H_{23}FN_2O_3$
195988-65-9

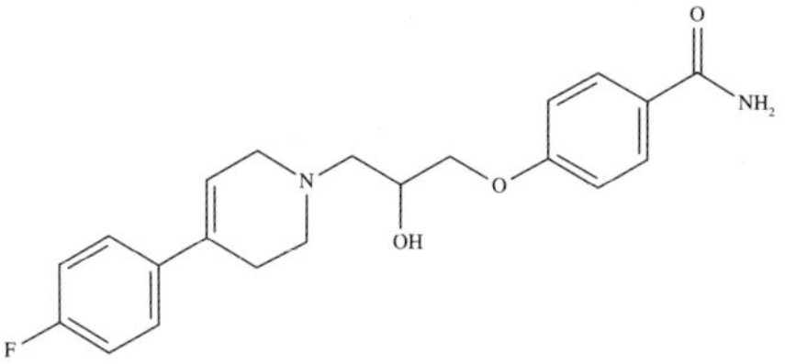

4-[3-[4-(4-Fluorophenyl)-3,6-dihydro-1(2H)-pyridinyl]-
2-hydroxypropoxy]benzamide (●)
S Ro 8-4304
U NMDA receptor antagonist

10277 $C_{21}H_{23}FN_2O_4S$
136861-95-5

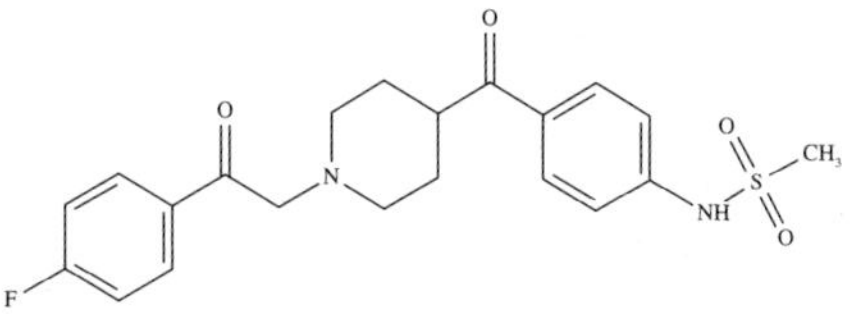

1-(4-Fluorophenyl)-2-[4-[[4-(methanesulfonamido)phe-
nyl]carbonyl]-1-piperidinyl]ethanone = N-[4-[[1-[2-(4-
fluorophenyl)-2-oxoethyl]-4-piperidinyl]carbonyl]phe-
nyl]methanesulfonamide (●)
Monohydrochloride (136861-96-6)
MDL 28133A
Antithrombotic (5-HT$_2$ receptor antagonist)

10278 $C_{21}H_{23}F_3N_6O$
168266-90-8

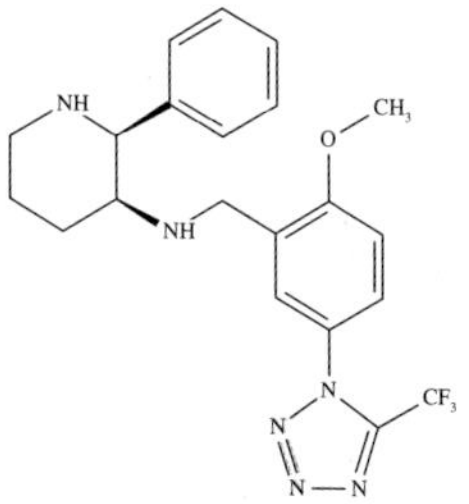

(2S,3S)-3-[[2-Methoxy-5-[5-(trifluoromethyl)-1H-tetra-
zol-1-yl]benzyl]amino]-2-phenylpiperidine = (2S,3S)-N-
[[2-Methoxy-5-[5-(trifluoromethyl)-1H-tetrazol-1-
yl]phenyl]methyl]-2-phenyl-3-piperidinamine (●)
Dihydrochloride (168266-51-1)
GR 205171A, Vofopitant hydrochloride**

U Anti-emetic (tachykinin NK$_1$ receptor antagonist)

10279   (8305) $C_{21}H_{23}N$
34061-33-1

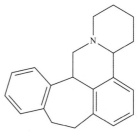

2,3,4,4a,8,9,13b,14-Octahydro-1H-benzo[6,7]cyclohep-
ta[1,2,3-de]pyrido[2,1-a]isoquinoline (●)
R Hydrochloride (34061-34-2)
S AY-22214, Taclamine hydrochloride**
U Tranquilizer

10280   (8306) $C_{21}H_{23}NO$
119356-77-3

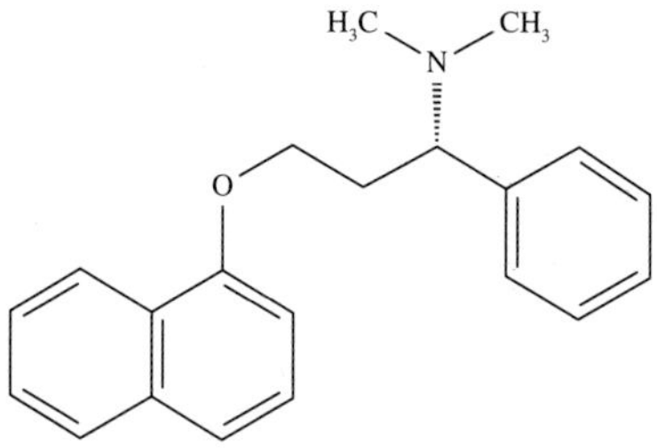

(+)-(S)-N,N-Dimethyl-α-[2-(1-naphthyloxy)ethyl]ben-
zylamine = (S)-N,N-Dimethyl-α-[2-(1-naphthalenyl-
oxy)ethyl]benzenemethanamine (●)
R Hydrochloride (129938-20-1)
S Dapoxetine hydrochloride**, LY 210448 HCl
U Antidepressant

10281 (8307)

$C_{21}H_{23}NO$
972-04-3

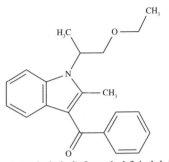

1,1-Diphenyl-4-piperidino-2-butyn-1-ol = α-Phenyl-α-[3-(1-piperidinyl)-1-propynyl]benzenemethanol (●)
S Diferidin, Dipheridin
U Anticholinergic, antihypertensive

10282 (8308)

$C_{21}H_{23}NO$
5061-32-5

4-(1,2,3,6-Tetrahydro-4-phenyl-1-pyridyl)butyrophe-none = 4-(3,6-Dihydro-4-phenyl-1(2H)-pyridinyl)-1-phe-nyl-1-butanone (●)
R also hydrochloride (14777-22-1)
S *Phenoperidone*, R 1516, Trabuton
U Neuroleptic

10283 (8309)

$C_{21}H_{23}NO_2$
15686-60-9

6-[(Diethylamino)methyl]-3-methylflavone = 6-[(Di-ethylamino)methyl]-3-methyl-2-phenyl-4H-chromen-4-one = 6-[(Diethylamino)methyl]-3-methyl-2-phenyl-4H-1-benzopyran-4-one (●)
R Hydrochloride (16146-79-5)

S *Flavamine hydrochloride***, Rec 7/0052
U Spasmolytic

10284 (8310)

$C_{21}H_{23}NO_2$
77992-61-

1-(2-Ethoxy-1-methylethyl)-2-methyl-3-indolyl phenyl ketone = 3-Benzoyl-1-(β-ethoxyisopropyl)-2-methylin-dole = [1-(2-Ethoxy-1-methylethyl)-2-methyl-1H-indol-3-yl]phenylmethanone (●)
S TEI-4120
U Antithrombotic, platelet aggregation inhibitor

10285 (8311)

$C_{21}H_{23}NO$
13479-13

2-(Dimethylamino)ethyl diphenyl(2-propynyloxy)ace-tate = 2-(Dimethylamino)ethyl O-(2-propynyl)benzilate α,α-Diphenyl-α-(propargyloxy)acetic acid β-(dimethyl amino)ethyl ester = α-Phenyl-α-(2-propynyloxy)ben-zeneacetic acid 2-(dimethylamino)ethyl ester (●)
R Hydrochloride (2765-97-1)
S BE 50, Bipasmin, Doledon, Elidol-A, Gastrinerval, *Pargeverine hydrochloride***, Pasmosedan, *Pro-pinox hydrochloride*, R 164, Sertal, Terisal, Vago-pax "Jaba", Viadil "Pharma Investi-Domin. Rep."
U Spasmolytic, analgesic

10285-01 (8311-01)
R Hydrochloride in combination with aluminum glyc nate (no. 75-02)

S Ipecsal, Ulcadol, Ventrinerval
U Stomachic

10286 (8312) $C_{21}H_{23}NO_3$
113806-05-6

Z)-11-[3-(Dimethylamino)propylidene]-6,11-dihydrodi-
benz[b,e]oxepin-2-acetic acid (●)
R also hydrochloride (140462-76-6)
S AL-4943 A, Allelock, ALO 4943 A, *Doxepadine*,
KW 4679, *Olopatadine***, Patanol
U Anti-allergic

0287 (8313) $C_{21}H_{23}NO_3$
33743-96-3

-(1,4-Benzodioxan-6-yl)-3-(3-phenyl-1-pyrrolidinyl)-1-
ropanone = 6-[3-(3-Phenyl-1-pyrrolidinyl)propio-
yl]benzodioxane = 1-(2,3-Dihydro-1,4-benzodioxin-6-
)-3-(3-phenyl-1-pyrrolidinyl)-1-propanone (●)
Hydrochloride (33025-33-1)
AY-24269, Piroxan "Ayerst", Pirroksan, *Proroxan
hydrochloride**, Pyrroxane
Antihypertensive (α-receptor blocker)

288 (8314) $C_{21}H_{23}NO_4$
79619-31-1

(±)-7-[2-Hydroxy-3-(propylamino)propoxy]flavone =
(±)-7-[2-Hydroxy-3-(propylamino)propoxy]-2-phenyl-
4*H*-1-benzopyran-4-one (●)
R Maleate (1:1) (79619-32-2)
S *Flavodilol maleate***, PR 877-530 L
U Antihypertensive

10289 $C_{21}H_{23}NO_4$
35661-60-0

N-[(9*H*-Fluoren-9-ylmethoxy)carbonyl]-L-leucine (●)
S NPC 15199
U Anti-inflammatory

10290 $C_{21}H_{23}NO_4S$
151774-56-0

(±)-3-[2-Benzyl-3-(propionylthio)propionamido]-5-me-
thylbenzoic acid = 3-Methyl-5-[[1-oxo-2-[[(1-oxopro-
pyl)thio]methyl]-3-phenylpropyl]amino]benzoic acid (●)
S BL-2401
U Enkephalinase inhibitor

10291 (8315) $C_{21}H_{23}NO_4S$
81110-73-8

(±)-*N*-[α-(Mercaptomethyl)hydrocinnamoyl]glycine benzyl ester acetate = (±)-*N*-[α-(Mercaptomethyl)hydrocin-

namoyl]glycine benzyl ester acetate (ester) = (±)-[2-
[(Acetylthio)methyl]benzenepropanoyl]glycine phenylmethyl ester = (±)-Benzyl 2-[3-(acetylthio)-2-benzylpro-
pionamido]acetate = (±)-N-[2-[(Acetylthio)methyl]-1-
oxo-3-phenylpropyl]glycine phenylmethyl ester (●)
S   Acetorphan, BP 0.52-S0.48, Ecatorfate, Hidrasec,
    Racecadotril**, Tiorfan
U   Antidiarrheal (enkephalinase inhibitor)

10292   (8317)

$C_{21}H_{23}NO_4S$
112573-72-5

(+)-N-[(R)-α-(Mercaptomethyl)hydrocinnamoyl]glycine
benzyl ester acetate (ester) = (R)-N-[2-[(Acetylthio)me-
thyl]-1-oxo-3-phenylpropyl]glycine phenylmethyl ester
(●)
S   BP 1.01-S 0.81, Dexecadotril**, Retorphan
U   Antidiarrheal (enkephalinase inhibitor)

10293   (8316)

$C_{21}H_{23}NO_4S$
112573-73-6

N-[(S)-α-(Mercaptomethyl)hydrocinnamoyl]glycine benzyl ester acetate = (S)-[2-[(Acetylthio)methyl]benzene-
propanoyl]glycine phenylmethyl ester = (S)-Benzyl 2-[3-
(acetylthio)-2-benzylpropionamido]acetate = (S)-N-[2-
[(Acetylthio)methyl]-1-oxo-3-phenylpropyl]glycine phenylmethyl ester (●)
S   5.049, Bay y 7432, BP 1.02, Ecadotril**, Levacetor-
    phan, Levecatorfate, (S)-Acetorphan, Sinorphan
U   Cardiotonic (endopeptidase inhibitor)

10294   (8319)

$C_{21}H_{23}NO_5$
3246-21-7

3,9,10-Trimethoxy-1,2-(methylenedioxy)-6aα-apor-
phine = (S)-6,7,7a,8-Tetrahydro-4,10,11-trimethoxy-7-
methyl-5H-benzo[g]-1,3-benzodioxolo[6,5,4-de]quino-
line (●)
S   Ocoteine, Thalicmine
U   Antitussive (alkaloid from Thalictrum minus, Nec-
    tandra saligna and Phoebe porfiria)

10295   (8320)

$C_{21}H_{23}NO$
485-91-

5,7,8,15-Tetrahydro-3,4-dimethoxy-6-methylben-
zo[e][1,3]dioxolo[4,5-k][3]benzazecin-14(6H)-one (●)
Σ   α-Allocryptopine, α-Fagarine
U   Anti-arrhythmic (alkaloid from Fagara coca and
    other Papaveraceae)

10296   (8318)

$C_{21}H_{23}NO$
561-27-

4,5α-Epoxy-17-methylmorphin-7-ene-3,6α-diyl diacetate = 3,6α-Diacetoxy-4,5α-epoxy-*N*-methylmorphin-7-
ene = (5α,6α)-7,8-Didehydro-4,5-epoxy-17-methylmor-
phinan-3,6-diol diacetate (ester) (●)
R also hydrochloride (1502-95-0)
S Acetomorfin, Acetomorphin, Aspron, Diacephin,
Diacetylmorphine, Diagesil, *Diamorphine*, Diaphi-
ne, Diaphorm, Eclorion, Eroin, Heroin, Herolan, Ie-
roin, Iroini, Morphacetin, *Morphine diacetate*, Preza
U Narcotic analgesic

10297　(8321)　　　　　　　　　　　　$C_{21}H_{23}NO_5S$
　　　　　　　　　　　　　　　　　　　132014-21-2

(+)-1-[(3*S*,4*R*)-3-Hydroxy-2,2-dimethyl-6-(phenylsulfo-
nyl)-4-chromanyl]-2-pyrrolidinone = (3*S-trans*)-1-[3,4-
Dihydro-3-hydroxy-2,2-dimethyl-6-(phenylsulfonyl)-
2H-1-benzopyran-4-yl]-2-pyrrolidinone (●)
　Hoechst 234, *Rilmakalim***
U Antihypertensive (potassium channel activator)

10298　(8322)　　　　　　　　　　　　$C_{21}H_{23}NO_5S$
　　　　　　　　　　　　　　　　　　　58761-87-8

(7-Carboxy-4-hexyl-9-oxoxanthen-2-yl)-*S*-methylsulf-
imine = 5-Hexyl-7-(*S*-methylsulfonimidoyl)-9-oxo-
9H-xanthene-2-carboxylic acid (●)
　*Sudexanox***
　Anti-allergic

10298-01　(8322-01)　　　　　　　　　66934-53-0
　Salt with trometamol (no. 314) (1:1)
　RU 31156

U Anti-allergic

10299　(8323)　　　　　　　　　　　　$C_{21}H_{23}N_3$
　　　　　　　　　　　　　　　　　　　42011-54-1

1-[(*tert*-Butylimino)methyl]-2-(3-indolyl)indoline = *N*-
[[2,3-Dihydro-2-(1*H*-indol-3-yl)-1*H*-indol-1-yl]methy-
lene]-2-methyl-2-propanamine (●)
R Monohydrochloride (36815-43-7)
S MJ 8592-1
U Diuretic

10300　(8324)　　　　　　　　　　　　$C_{21}H_{23}N_3$
　　　　　　　　　　　　　　　　　　　72320-59-3

2-(4-Ethyl-1-piperazinyl)-4-phenylquinoline (●)
R Dihydrochloride (72320-60-6)
S AD 1308
U Antidepressant

10300-01　(8324-01)　　　　　　　　　94788-64-4
R Maleate (1:2)
S AD 2646, AS 2646
U Anti-ulcer agent

10301　(8325)　　　　　　　　　　$C_{21}H_{23}N_3O$
118989-65-4

5-[(4-Methyl-1-piperazinyl)acetyl]-5H-dibenz[b,f]aze-
pine (●)
R　Dihydrochloride (118989-88-1)
S　VÚFB-17113
U　Anti-ulcer agent

10302　(8326)　　　　　　　　　　$C_{21}H_{23}N_3OS$
2622-26-6

10-[3-(4-Hydroxypiperidino)propyl]phenothiazine-2-car-
bonitrile = 2-Cyano-10-[3-(4-hydroxypiperidino)pro-
pyl]phenothiazine = 10-[3-(4-Hydroxy-1-piperidinyl)pro-
pyl]-10H-phenothiazine-2-carbonitrile (●)
S　Amplan, Aolept, Apamin, Bayer 1409, IC 6002, Iri-
yakin, Nemactil, Neulactil, Neuleptil, Neuperil, *Peri-
ciazine**, Pericyazine, Propériciazine*, Propetyl, Psy-
cholept, 8909 R.P., SKF 20716, WH 7508
U　Neuroleptic

10303　(8327)　　　　　　　　　　$C_{21}H_{23}N_3O_3$
105277-61-0

7-[3-[4-(2-Pyridinyl)-1-piperazinyl]propoxy]-4H-1-ben-
zopyran-4-one (●)
R　Dihydrochloride (105277-43-8)
S　PD 119819
U　Dopamine autoreceptor agonist

10304　(8328)　　　　　　　　　　$C_{21}H_{23}N_3O$
19395-74-

1-Morpholinoacetyl-2-methyl-3-phenyl-4-oxo-1,2,3,4-te-
trahydroquinazoline = 2,3-Dihydro-2-methyl-1-(4-mor-
pholinylacetyl)-3-phenyl-4(1H)-quinazolinone (●)
R　Monohydrochloride (20866-13-1)
S　HQ-275
U　Choleretic

10305　(8329)　　　　　　　　　　$C_{21}H_{23}N_3O$
5874-95

9-Amino-4-(dimethylamino)-1,4,4a,5,5a,6,11,12a-octa-
hydro-3,10,12,12a-tetrahydroxy-1,11-dioxo-2-naphtha-
cenecarboxamide = [4S-(4α,4aα,5aα,12aα)]-9-Amino-
(dimethylamino)-1,4,4a,5,5a,6,11,12a-octahydro-

3,10,12,12a-tetrahydroxy-1,11-dioxo-2-naphthacene-carboxamide (●)
S *Amicycline***
U Antibiotic

10306 (8330) $C_{21}H_{23}N_3O_7S$
86273-18-9

2,3-Dihydroxy-2-butenyl (2S,5R,6R)-6-[(R)-2-amino-2-phenylacetamido]-3,3-dimethyl-7-oxo-4-thia-1-azabicyclo[3.2.0]heptane-2-carboxylate cyclic carbonate = 2S,5R,6R)-6-[(R)-2-Amino-2-phenylacetamido]-3,3-dimethyl-7-oxo-4-thia-1-azabicyclo[3.2.0]heptane-2-carboxylic acid (5-methyl-2-oxo-1,3-dioxol-4-yl)methyl ester = Ampicillin (5-methyl-2-oxo-1,3-dioxol-4-yl) ester = [2S-[2α,5α,6β(S*)]]-6-[(Aminophenylacetyl)amino]-3,3-dimethyl-7-oxo-4-thia-1-azabicyclo[3.2.0]heptane-2-carboxylic acid (5-methyl-2-oxo-1,3-dioxol-4-yl)methyl ester (●)
R Monohydrochloride (80734-02-7)
KB 1585, KBT-1585, LAPC, *Lenampicillin hydrochloride***, Takacillin, Varacillin
U Antibiotic

10307 (8331) $C_{21}H_{23}N_3O_{11}$
130548-53-7

-[R*,R*-(Z)]]-N,N'-[5-[[1-(5-Acetyl-4-hydroxy-2,6-dioxo-2H-pyran-3(6H)-ylidene)ethyl]amino]-1,3-phenylene]bis[2,3-dihydroxypropanamide] (●)
AGN 1-190144, SKF 84210

U Topical ocular anti-allergic

10308 $C_{21}H_{23}N_5O$
139047-55-5

2,3-Dihydro-2-[(4-phenyl-1-piperazinyl)methyl]imidazo[1,2-c]quinazolin-5(6H)-one (●)
S AT-112
U Antihypertensive

10309 (8332) $C_{21}H_{23}N_5O_2$
39632-88-7

1,4-Dimorpholino-7-phenylpyrido[3,4-d]pyridazine = 1,4-Di-4-morpholinyl-7-phenylpyrido[3,4-d]pyridazine (●)
S DS-511
U Diuretic

10310 (8333) $C_{21}H_{23}N_7O_6$
22006-84-4

N-[p-[[1-(2-Amino-4-hydroxy-6-pteridinyl)ethyl]methylamino]benzoyl]-L-glutamic acid = 9,N^{10}-Dimethylfolic acid = N-[4-[[1-(2-Amino-1,4-dihydro-4-oxo-6-pteridinyl)ethyl]methylamino]benzoyl]-L-glutamic acid (●)
S *Denopterin*, Dimetfol

U Antineoplastic, antileukemic

10311 (8334)

$C_{21}H_{23}N_7O_8S_4$
101004-07-3

1-[(Ethoxycarbonyl)oxy]ethyl (7*R*)-7-[2-(2-amino-4-thiazolyl)-2-(methoxyimino)acetamido]-3-[(1,2,3-thiadiazol-5-ylthio)methyl]-3-cephem-4-carboxylate = [6*R*-[6α,7β(*Z*)]]-7-[[(2-Amino-4-thiazolyl)(methoxyimino)acetyl]amino]-8-oxo-3-[(1,2,3-thiadiazol-5-ylthio)methyl]-5-thia-1-azabicyclo[4.2.0]oct-2-ene-2-carboxylic acid 1-[(ethoxycarbonyl)oxy]ethyl ester (●)
S *Baccefuzonam*, CL 118673
U Antibiotic

10312 (8335)

$C_{21}H_{24}BrFN_2O_3$
107188-82-9

(*R*)-5-Bromo-*N*-[[1-(*p*-fluorobenzyl)-2-pyrrolidinyl]methyl]-2,3-dimethoxybenzamide = (*R*)-5-Bromo-*N*-[[1-[(4-fluorophenyl)methyl]-2-pyrrolidinyl]methyl]-2,3-dimethoxybenzamide (●)
R Monohydrochloride monohydrate (142227-10-9)
S NCQ 115
U Antipsychotic (dopamine D$_2$-antagonist)

10313 (8336)

$C_{21}H_{24}BrN_5O$
86181-42-2

2-[[4-(5-Bromo-3-methyl-2-pyridinyl)butyl]amino]-5-[(6-methyl-3-pyridinyl)methyl]-4(1*H*)-pyrimidinone (●)
S SKF 93944, *Temelastine***
U Antihistaminic (H$_1$-receptor antagonist)

10314 (8337)

$C_{21}H_{24}ClNO$
5627-46-3

3-[(*p*-Chloro-α-phenylbenzyl)oxy]tropane = 3-(*p*-Chlorodiphenylmethoxy)tropane = Tropine 4-chlorobenzhydryl ether = 3-[(4-Chlorophenyl)-phenylmethoxy]-8-methyl-8-azabicyclo[3.2.1]octane (●)
R Hydrochloride (14008-79-8)
S *Chlorobenztropine hydrochloride, Clobenztropine hydrochloride***, FC 1, SL 6057, Teprin
U Antihistaminic

10315 (8338)

$C_{21}H_{24}ClNO$
1168-02-

trans-2-(4-Chlorophenyl)-1,3,4,6,7,11b-hexahydro-9,10-dimethoxy-2*H*-benzo[*a*]quinolizine (●)
R also without *trans*-definition (15301-89-0)
S B.S. 7284, *Chillifolinum, Killifolin, Quillifoline***
U Analgesic

10316

$C_{21}H_{24}ClNO$
141335-10-

4-[[1-[[[(4-Chlorophenyl)sulfonyl]amino]methyl]cyclo-pentyl]methyl]benzeneacetic acid (●)
R Sodium salt (141335-11-7)
S LCB 2853
U Thromboxane A_2 receptor antagonist

10317 (8339) $C_{21}H_{24}ClNO_5$
31848-01-8

4'-Chloro-3,5-dimethoxy-4-(2-morpholinoethoxy)benzo-phenone = (4-Chlorophenyl)[3,5-dimethoxy-4-[2-(4-mor-pholinyl)ethoxy]phenyl]methanone (●)
R Hydrochloride (31848-02-9)
S Dimeclofenone, K 3712, Medicil, *Morclofone hydro-chloride***, Nitux, Novotossil (old form), Plausital, Plausitin, Plauten
U Antitussive

10318 (8340) $C_{21}H_{24}ClN_3OS$
84-04-8

1-[3-(2-Chloro-10-phenothiazinyl)propyl]-4-piperidine-carboxamide = 10-[3-(4-Carbamoylpiperidino)propyl]-2-chlorophenothiazine = 1-[3-(2-Chloro-10-phenothiazi-nyl)propyl]isonipecotamide = 1-[3-(2-Chloro-10*H*-phe-nothiazin-10-yl)propyl]-4-piperidinecarboxamide (●)
S Mornidine, Nausidol, Nometine, NSC-169475, *Pipa-mazine***, 9153 R.P., SC-9387
U Anti-emetic

10319 (8341) $C_{21}H_{24}ClN_3O_3$
18053-31-1

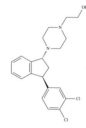

3'-Chloro-α-[methyl[(morpholinocarbonyl)methyl]ami-no]-*o*-benzotoluidide = 3'-Chloro-2'-[[methyl[(morpholi-nocarbonyl)methyl]amino]methyl]benzanilide = *N*-[3-Chloro-2-[[methyl[2-(4-morpholinyl)-2-oxoethyl]ami-no]methyl]phenyl]benzamide (●)
R Monohydrochloride (24600-36-0)
S Broncomenal, Deronyl, Finaten "Finadiet", Fomi-bron, *Fominoben hydrochloride***, Noleptan, Olep-tan, PB 89, Tasadox, Terion, Tosifar, Tussirama
U Respiratory stimulant, antitussive

10320 (8342) $C_{21}H_{24}Cl_2N_2O$
85663-55-4

(±)-*trans*-4-[3-(3,4-Dichlorophenyl)-1-indanyl]-1-pipera-zineethanol = *trans*-4-[3-(3,4-Dichlorophenyl)-2,3-dihy-dro-1*H*-inden-1-yl]-1-piperazineethanol (●)
S Lu 17-133
U Antidepressant

10321 (8344) $C_{21}H_{24}Cl_2N_2O_3$
5683-88-5

4-[[[4-[Bis(2-chloroethyl)amino]phenyl]acetyl]ami-no]benzoic acid ethyl ester (●)
S Fenastezin, Phenastezin

U Antineoplastic

10322 (8345) $C_{21}H_{24}Cl_2N_2O_3$
32656-65-8

4-[[[4-[Bis(2-chloropropyl)amino]phenyl]acetyl]ami-
no]benzoic acid (●)
S Promicil
U Antineoplastic, antileukemic

10323 (8343) $C_{21}H_{24}Cl_2N_2O_3$
10047-08-2

N-[[4-[Bis(2-chloroethyl)amino]phenyl]acetyl]-DL-phe-
nylalanine (●)
S Lofenal, Lophenal
U Antineoplastic

10324 $C_{21}H_{24}Cl_2N_2O_3$
65427-88-5

N-[[4-[Bis(2-chloroethyl)amino]phenyl]acetyl]-L-phe-
nylalanine (●)
S Fenalol, L-Lofenal, Phenalol
U Antineoplastic

10325 $C_{21}H_{24}Cl_2N_6O_3S$
169190-51-6

2,5'-Dichloro-5'-deoxy-N-[4-(phenylthio)-1-piperidi-
nyl]adenosine (●)
S NNC 21-0147
U Adenosine A_1 receptor agonist

10326 (8346) $C_{21}H_{24}Cl_2O_?$
78548-88-?

Ethyl all-trans-9-(2,6-dichloro-4-methoxy-3-methylphe-
nyl)-3,7-dimethyl-2,4,6,8-nonatetraenoate = (all-E)-9-
(2,6-Dichloro-4-methoxy-3-methylphenyl)-3,7-dimethyl-
2,4,6,8-nonatetraenoic acid ethyl ester (●)
S DCMMP, Ro 12-7554
U Antineoplastic

10327 (8347) $C_{21}H_{24}FNO_?$
3109-12-?

4'-Fluoro-4-(4-hydroxy-4-phenylpiperidino)butyrophe-
none = 1-(4-Fluorophenyl)-4-(4-hydroxy-4-phenyl-1-pi-
peridinyl)-1-butanone (●)
R also hydrochloride (4021-57-2)
S Peridol, R 1589

U Neuroleptic

10328 (8348) $C_{21}H_{24}FNO_3$
                                                    21492-67-1

4'-Fluoro-4-[3-(o-methoxyphenoxy)-1-pyrrolidinyl]buty-
rophenone = 1-(4-Fluorophenyl)-4-[3-(2-methoxyphen-
oxy)-1-pyrrolidinyl]-1-butanone (●)
S AHR-1900
U Neuroleptic

10329 (8349) $C_{21}H_{24}FN_3O_2$
                                                    88793-12-8

6-Fluoro-3-[3-[4-(2-methoxyphenyl)-1-piperazinyl]pro-
pyl]-1,2-benzisoxazole (●)
S HRP 392
U Antipsychotic

10330 (8350) $C_{21}H_{24}FN_3O_2S$
                                                    86487-64-1

6-[2-[4-(p-Fluorobenzoyl)piperidino]ethyl]-2,3-dihydro-
7-methyl-5H-thiazolo[3,2-a]pyrimidin-5-one = 6-[2-[4-
[4-Fluorobenzoyl)-1-piperidinyl]ethyl]-2,3-dihydro-7-
methyl-5H-thiazolo[3,2-a]pyrimidin-5-one (●)
R 52245, *Setoperone***
U Antipsychotic

10331 $C_{21}H_{24}FN_3O_3$
                                                    186293-38-9

(4aS-cis)-1-Cyclopropyl-7-fluoro-9-methyl-8-(octahy-
dro-6H-pyrrolo[3,4-b]pyridin-6-yl)-4-oxo-4H-quinoli-
zine-3-carboxylic acid (●)
R Monohydrochloride (181141-52-6)
S ABT 255
U Antibacterial

10332 $C_{21}H_{24}FN_3O_4$
                                                    151096-09-2

1-Cyclopropyl-6-fluoro-1,4-dihydro-8-methoxy-7-
[(4aS,7aS)-octahydro-6H-pyrrolo[3,4-b]pyridin-6-yl]-4-
oxo-3-quinolinecarboxylic acid (●)
S Bay y 6957, *Moxifloxacin***
U Antibacterial

10332-01   186826-86-8
R Monohydrochloride
S Actira, Avalox, Avelox, Bay 12-8039, *Moxifloxacin
hydrochloride***, Octegra, Proflox
U Antibacterial

10333 (8351) $C_{21}H_{24}FN_5O_3$
                                                    82190-91-8

7-[2-[4-(*p*-Fluorobenzoyl)piperidino]ethyl]theophylline
= 7-[2-[4-(4-Fluorobenzoyl)-1-piperidinyl]ethyl]-3,7-di-
hydro-1,3-dimethyl-1*H*-purine-2,6-dione (●)
S *Flufylline***, Sgd 19578
U Analeptic

10334　(8352)　　　　　　　　　　　　　$C_{21}H_{24}F_2N_2O_3$
　　　　　　　　　　　　　　　　　　150756-35-7

[2-[4-[Bis(4-fluorophenyl)methyl]-1-piperazinyl]eth-
oxy]acetic acid (●)
R Dihydrochloride (225367-66-8)
S *Efletirizine hydrochloride***, U.C.B. 28754
U Anti-allergic

10335　　　　　　　　　　　　　　　　　$C_{21}H_{24}F_2N_6$
　　　　　　　　　　　　　　　　　　114667-79-7

N-[3-(3,5-Difluorophenyl)-3-(2-pyridinyl)propyl]-*N*'-[3-
(1*H*-imidazol-4-yl)propyl]guanidine (●)
R also trihydrochloride (121598-32-1)
S BU-E-76, He 90481
U Antihypertensive (histamine H$_2$-agonist)

10336　(8353)　　　　　　　　　　　　　$C_{21}H_{24}F_3N$
　　　　　　　　　　　　　　　　　　35764-73-9

(±)-*cis*-9,10-Dihydro-*N,N*,10-trimethyl-2-(trifluoromethyl)-9-anthracenepropylamine = (*9R,10S*)-*rel*-9,10-Di-
hydro-*N,N*,10-trimethyl-2-(trifluoromethyl)-9-anthra-
cenepropanamine (●)
R Hydrochloride (35764-29-5)
S *Fluotracen hydrochloride***, SKF 28175
U Antidepressant, antipsychotic

10337　(8354)　　　　　　　　　　　　　$C_{21}H_{24}F_3N_3O_2$
　　　　　　　　　　　　　　　　　　73356-67-9

4-Methyl-*N*-[3-(1-piperidinyl)propyl]-6-(trifluorome-
thyl)-4*H*-furo[3,2-*b*]indole-2-carboxamide (●)
S FI-302
U Anti-inflammatory

10338　(8355)　　　　　　　　　　　　　$C_{21}H_{24}F_3N_3$
　　　　　　　　　　　　　　　　　　117-89-

10-[3-(4-Methyl-1-piperazinyl)propyl]-2-(trifluoromethyl)phenothiazine = 10-[3-(4-Methyl-1-piperazinyl)pro-
pyl]-2-(trifluoromethyl)-10*H*-phenothiazine (●)
R also dihydrochloride (440-17-5)

S   Anchostam (old form), Apo-Trifluoperazine, Aquil "RGK", Asterfenazine, Athimol, Bitafurazine, Calmazine, Cariogrin-S, Chemflurazine, Clinazine, Diethriton, Diprazine, Discimer, Dymoperazine, Ecuril, Equazine, Equizine, Escazine, Eskazine, Eskazinyl, Fluazine, Fluoren, Flupazine, Fluperin, Flurazine, Ifizine, Iquil, Iremo-Pierol, Jatroneural, Jetirison, Leptane, Leptazine, Lufacintyl, Modalina, Nerolet, Normaln P, Novoflurazine, NSC-17474, Nylipton, Oxyperazine, Pentazine "Pentagone", Pierol, Promax, Riboring, Robablesix, Sedizine, Sedofren, Semazine, SKF 5019, Solazine, Sporalon, Stelan, Stelazine, Stelium, Sterazin, Stilizan, Suprazine, Telazin, Terflurazine, Terfluzine, T.F.P., Thioquil, Trankozine, Tranquis, Trazin, Triazin(e), *Trifluoperazine**, Trifluoper-Ez-Ets, *Trifluoroperazine, Trifluperazine*, Triflurin, Triftazin, Trinaphazine, Triozine, Tripazine, Triperazin(e), Triphtazine, *Triphthazinum*, Trivan, Tryptazin, Vespezine, Yatroneural

U   Neuroleptic (Calmodulin inhibitor)

10338-01   (8355-01)   605-75-4

R   Maleate (1:2)

S   Asterfenazina

U   Neuroleptic

10339   (8356) $C_{21}H_{24}I_3NO_4$
                                         15301-96-9

2-(Diethylamino)ethyl [3,5-diiodo-4-(3-iodo-4-methoxyphenoxy)phenyl]acetate = 3,5-Diiodo-4-(3-iodo-4-methoxyphenoxy)phenylacetic acid 2-(diethylamino)ethyl ester = 3,5-Diiodo-4-(3-iodo-4-methoxyphenoxy)benzeneacetic acid 2-(diethylamino)ethyl ester (●)

R   Hydrochloride (57-65-8)

S   SKF 13364-A, *Thyromedan hydrochloride, Tyromedan hydrochloride***

U   Thyromimetic

10340   (8357) $C_{21}H_{24}NO_4$
                                         6882-14-0

5,6,7-Trimethoxy-1-(*p*-methoxybenzyl)-2-methylisoquinolinium = 5,6,7-Trimethoxy-1-[(4-methoxyphenyl)methyl]-2-methylisoquinolinium (●)

R   Iodide (5083-11-4)

S   *Takatonium iodide*

U   Spasmolytic

10341   (8358) $C_{21}H_{24}N_2$
                                         31721-17-2

10,11-Dihydro-5-(3-quinuclidinyl)-5*H*-dibenz[*b,f*]azepine = 5-(1-Azabicyclo[2.2.2]oct-3-yl)-10,11-dihydro-5*H*-dibenz[*b,f*]azepine (●)

S   Adeprim, *Chinupramina*, Kevopril, Kinupril, LM 208, *Quinupramine**, Quinuprine

U   Antidepressant

10342   (8359) $C_{21}H_{24}N_2$
26070-23-5

1-(9,10-Dihydro-9,10-ethano-9-anthryl)-4-methylpipera-
zine = 1-(Dibenzo[b,e]bicyclo[2.2.2]octadien-1-yl)-4-
methylpiperazine = 1-(9,10-Ethanoanthracen-9(10H)-yl)-
4-methylpiperazine (●)
S   SD 1223-01, Trazitiline**
U   Antihistaminic

10343 $C_{21}H_{24}N_2$
185855-91-8

1-Benzyl-4-(pentylimino)-1,4-dihydroquinoline = N-[1-
(Phenylmethyl)-4(1H)-quinolinylidene]-1-pentanamine
(●)
S   CP-339818
U   Antineoplastic

10344   (8360) $C_{21}H_{24}N_2O$
35452-73-4

N-(cis-2,trans-3-Diphenylcyclopropyl)-1-pyrrolidine-
acetamide = N-(Pyrrolidinoacetyl)-2,3-cis,trans-diphe-
nylcyclopropylamine = (1α,2α,3β)-N-(2,3-Diphenyl-
cyclopropyl)-1-pyrrolidineacetamide (●)
S   Ciprafamide**, Z 839
U   Anti-arrhythmic

10345   (8361) $C_{21}H_{24}N_2O_2$
4880-92-6

Methyl (3α,16α)-eburnamenine-14-carboxylate = Methyl
(13aS,13bS)-13a-ethyl-2,3,5,6,13a,13b-hexahydro-1H-
indolo[3,2,1-de]pyrido[3,2,1-ij][1,5]naphthyridine-12-
carboxylate = Methyl apovincaminate = (3α,16α)-Ebur-
namenine-14-carboxylic acid methyl ester (●)
S   Apovincamine**, ApV
U   Vasodilator

10345-01   (8361-01)   65826-02-(
R   Compd. with α-D-glucopyranose 1-(dihydrogen phos-
     phate) (1:1)
S   Apovincamine gluphosphate
U   Cerebrotonic, metabolic

10346 $C_{21}H_{24}N_2O$
146998-34-

1-(2,3-Dihydro-1,4-benzodioxin-5-yl)-4-(2,3-dihydro-
1H-inden-2-yl)piperazine (●)
S   Cinalzotan, S 15535
U   Pre- and postsynaptic 5-HT$_{1A}$ receptor agonist and
     antagonist

10347 (8362) $C_{21}H_{24}N_2O_3$
7008-14-2

3-(2-Aminoethyl)-1-(p-methoxybenzyl)-2-methylindol-5-ol acetate (ester) = 5-Acetoxy-3-(2-aminoethyl)-1-(p-methoxybenzyl)-2-methylindole = 3-(2-Aminoethyl)-1-[(4-methoxyphenyl)methyl]-2-methyl-1H-indol-5-ol acetate (ester) (●)
S Hydroxindasate**, Oxindasoli acetas
U Serotonin antagonist, diuretic

10348 (8363) $C_{21}H_{24}N_2O_3$
483-04-5

Tetrahydroserpentine = (19α)-16,17-Didehydro-19-methyloxayohimban-16-carboxylic acid methyl ester (●)
S Ajmalicine, Alkaloid F, Circolene, Dilasin, Hydrosarpan, Isoar, Isoarteril, Lamuran, Loparol, Melanex "Boehringer-Mannheim", Perife, Ranitol, Raubaserp, Raubasil, Raubasine, Raumalina, Raunatine, Rauvasan, Rauvasil, Sarpan, Tensyl, δ-Yohimbine
U Peripheral and cerebral vasodilator (Rauwolfia alkaloid)

10349 (8364) $C_{21}H_{24}N_2O_3$
642-17-1

(3β,19α,20α)-16,17-Didehydro-19-methyloxayohimban-16-carboxylic acid methyl ester (●)
S Akuammigine
U Rauwolfia alkaloid

10350 (8365) $C_{21}H_{24}N_2O_3$
509-52-4

Strychnic acid = 9,10-Secostrychnidin-10-oic acid (●)
R Monosodium salt (6033-06-3)
S Movellan-Amp., Natrium strychninicum, Perton
U Strychnine depot prep.

10350-01 (8365-01)
R Ethylbetaine sulfate
S Betaino-Strychnine, Strychnal
U Strychnine depot prep.

10351 (8366) $C_{21}H_{24}N_2O_4$
5779-54-4

1,1-Cyclopentanedimethanol dicarbanilate = Cyclopentylidenedimethyl bis(phenylcarbamate) = 1,1-Cyclopentylidenedimethyl dicarbanilate = 1,1-Cyclopentanedimethanol bis(phenylcarbamate) (●)

S B.S.M. 906 M, C 1428, Calmalone, Camalon, Casmalon, *Cyclarbamate***, Cyclocarbamate, *Cyclopentaphène*

U Spasmolytic, tranquilizer, muscle relaxant

10352 (8367) $C_{21}H_{24}N_2O_4$
147568-66-9

8-Hydroxy-5-[(1R)-1-hydroxy-2-[[(1R)-2-(4-methoxyphenyl)-1-methylethyl]amino]ethyl]carbostyril = [R-(R*,R*)]-8-Hydroxy-5-[1-hydroxy-2-[[2-(4-methoxyphenyl)-1-methylethyl]amino]ethyl]-2(1H)-quinolinone (●)

R Monohydrochloride (137888-11-0)

S TA-2005

U Bronchodilator

10353 $C_{21}H_{24}N_2O_4S$
144980-29-0

(−)-2-[4-[[(R)-2-Chromanylmethyl]amino]butyl]-1,2-benzisothiazolin-3-one 1,1-dioxide = 2-[4-[[[(2R)-3,4-Dihydro-2H-1-benzopyran-2-yl]methyl]amino]butyl]-1,2-benzisothiazol-3(2H)-one 1,1-dioxide (●)

R Monohydrochloride (144980-77-8)

S BAY x 3702, *Repinotan hydrochloride***

U Neuroprotective (5-HT$_{1A}$ receptor agonist)

10354 $C_{21}H_{24}N_2O_5$
83861-02-3

N-[(1S)-1-Carboxy-2-phenylethyl]-L-phenylalanyl-β-alanine (●)

S Sch 32615

U Enkephalinase inhibitor

10355 (8368) $C_{21}H_{24}N_2O_5$
143343-83-3

(±)-6-[2-Hydroxy-3-(veratrylamino)propoxy]carbostyril = 6-[3-[[(3,4-Dimethoxyphenyl)methyl]amino]-2-hydroxypropoxy]-2(1H)-quinolinone (●)

S *Loborinone*, OPC-18790, *Toborinone***, Toboron, *Toprorinone*

U Cardiotonic (positive inotropic)

10356 $C_{21}H_{24}N_2O_5S_2$
110221-53-1

(2S,6R)-6-[[(1S)-1-Carboxy-3-phenylpropyl]amino]tetrahydro-5-oxo-2-(2-thienyl)-1,4-thiazepine-4(5H)-acetic acid (●)

R see also no. 11894

S RNH-5139, RS-5139, *Temocaprilat***, *Temocapril diacid*

U Antihypertensive (ACE inhibitor)

10357 (8369) $C_{21}H_{24}N_2O_7$
                                            138661-03-7

(±)-Methyl tetrahydrofurfuryl 1,4-dihydro-2,6-dimethyl-
4-(*o*-nitrophenyl)-3,5-pyridinedicarboxylate = 1,4-Dihy-
dro-2,6-dimethyl-4-(2-nitrophenyl)-3,5-pyridinedicarb-
oxylic acid methyl (tetrahydro-2-furanyl)methyl ester (●)
R Mixture of enantiomers
S CRE-319, *Furnidipine***
U Antihypertensive, anti-ischemic (calcium antagonist)

10358 (8370) $C_{21}H_{24}N_2O_7S$
                                            113658-85-8

1,4-Dihydro-2,4,6-trimethyl-3,5-pyridinedicarboxylic
acid ethyl 2-(3-oxo-1,2-benzisothiazol-2(3*H*)-yl)ethyl ester *S,S*-dioxide (●)
S PCA-4230, *Trombodipine*
U Antithrombotic

10359 (8371) $C_{21}H_{24}N_2S$
                                            85275-48-5

2-[[2-(Dimethylamino)-2-methylpropyl]thio]-3-phenyl-
quinoline = *N,N*,2-Trimethyl-1-[(3-phenyl-2-quinoli-
nyl)thio]-2-propanamine (●)
R Monohydrochloride (85275-49-6)

S ICI 170809
U Antihypertensive

10360 (8372) $C_{21}H_{24}N_4O$
                                            116870-68-9

3-(2-Amino-5-cyanophenyl)-2-[2-(diethylamino)ethyl]isoindol-1-one = 4-Amino-3-[2-[2-(diethylamino)ethyl]-2,3-dihydro-3-oxo-1*H*-isoindol-1-yl]benzo-
nitrile (●)
S KW 3299
U Anti-arrhythmic

10361 (8373) $C_{21}H_{24}N_4O$
                                            125974-72-3

11-[[3-(Dimethylamino)propyl]amino]-8-methyl-7*H*-
benzo[*e*]pyrido[4,3-*b*]indol-3-ol (●)
R Dimethanesulfonate (133711-99-6)
S *Intoplicine mesilate***, NSC-645008, RP 60475
U Antineoplastic

10362 (8374) $C_{21}H_{24}N_4O_2$
                                            2208-51-7

3-[3-(4-Phenyl-1-piperazinyl)propyl]-2,4(1*H*,3*H*)-quinazolinedione (●)

R Monohydrochloride (42877-18-9)

S *Pelanserin hydrochloride***, TR-2515

U Antihypertensive, vasodilator

10363 (8375)   C21H24N4O3
                                  88068-67-1

3-[2-[4-(2-Methoxyphenyl)-1-piperazinyl]ethyl]-2,4-(1*H*,3*H*)quinazolinedione (●)

R Monohydrochloride (88068-72-8)

S SGB 1534

U Antihypertensive (α_1-adrenoceptor blocker)

10364 (8376)   C21H24N4O3S
                                  26242-33-1

N-[(4-Amino-2-methyl-5-pyrimidinyl)methyl]-*N*-[2-[(2-benzoylvinyl)thio]-4-hydroxy-1-methyl-1-butenyl]formamide = *S*-Benzoylvinylthiamine = *N*-[(4-Amino-2-methyl-5-pyrimidinyl)methyl]-*N*-[4-hydroxy-1-methyl-2-[(3-oxo-3-phenyl-1-propenyl)thio]-1-butenyl]formamide (●)

S *Vinthiamol, Vintiamol***

U Analgesic (Vitamin B$_1$ source)

10365 (8377)   C21H24N4O4
                                  92406-14-9

Methyl 3-cyclopentyl-4,7-dihydro-1,6-dimethyl-4-(*m*-nitrophenyl)-1*H*-pyrazolo[3,4-*b*]pyridine-5-carboxylate = 3-Cyclopentyl-4,7-dihydro-1,6-dimethyl-4-(3-nitrophenyl)-1*H*-pyrazolo[3,4-*b*]pyridine-5-carboxylic acid methyl ester (●)

S 8363-S

U Coronary vasodilator, antihypertensive (calcium antagonist)

10366 (8378)   C21H24N4O7S
                                  10072-48-7

S-Ester of thio-2-furoic acid with *N*-[(4-amino-2-methyl-5-pyrimidinyl)methyl]-*N*-(4-hydroxy-2-mercapto-1-methyl-1-butenyl)formamide *O*-glycolate acetate = *N*-[4-(2-Acetoxyacetoxy)-2-(2-furoylthio)-1-methyl-1-butenyl]-*N*-[(4-amino-2-methyl-5-pyrimidinyl)methyl]formamide = (Acetyloxy)acetic acid 4-[[(4-amino-2-methyl-5-pyrimidinyl)methyl]formylamino]-3-[(2-furanylcarbonyl)thio]-3-pentenyl ester (●)

S *Acefurtiamine***, AFT, Biotamin S, *Glycofurthiamine*

U Analgesic (Vitamin B$_1$ source)

10367 (8379) $C_{21}H_{24}N_4O_8S_3$

1-[4-[(2,4-Dimethyl-6-pyrimidinyl)sulfamoyl]anilino]-3-
phenyl-1,3-propanedisulfonic acid = 6-[N-(γ-Phenylpro-
pyl)sulfanilamido]-2,4-dimethylpyrimidine-α,γ-disul-
fonic acid = 1-[[4-[[(2,6-Dimethyl-4-pyrimidinyl)ami-
no]sulfonyl]phenyl]amino]-3-phenyl-1,3-propanedisul-
fonic acid (●)
R Disodium salt (60662-80-8)
S Aristoplomb
U Antibacterial

10368 (8380) $C_{21}H_{24}N_4O_{12}$
                                                     88594-08-5

2-(Nitrooxy)propyl 3-(nitrooxy)propyl 1,4-dihydro-2,6-
dimethyl-4-(m-nitrophenyl)-3,5-pyridinedicarboxylate =
1,4-Dihydro-2,6-dimethyl-4-(3-nitrophenyl)-3,5-pyri-
dinedicarboxylic acid 2-(nitrooxy)propyl 3-(nitrooxy)propyl ester (●)
S CD-349
U Calcium antagonist

10369 $C_{21}H_{24}N_6O_5$
                                                     125991-51-7

N-[4-[3-(2,4-Diamino-1H-pyrrolo[2,3-d]pyrimidin-5-
yl)propyl]benzoyl]-L-glutamic acid (●)
S TNP-351
U Antineoplastic (folic acid antagonist)

10370 (8381) $C_{21}H_{24}N_8O_5$
                                                     751-44-0

4-Amino-9,N^{10}-dimethylfolic acid = N-[4-[[1-(2,4-Di-
amino-6-pteridinyl)ethyl]methylamino]benzoyl]-L-gluta-
mic acid (●)
S A-denopterin
U Antineoplastic, antileukemic

10371 (8382) $C_{21}H_{24}N_8O_5$
                                                     60688-54-2

N^2-[p-[[1-(2-Amino-4-hydroxy-6-pteridinyl)ethyl]me-
thylamino]benzoyl]-L-glutamine = 9,N^{10}-Dimethylfolic
acid amide = N^2-[4-[[1-(2-Amino-1,4-dihydro-4-oxo-6-
pteridinyl)ethyl]methylamino]benzoyl]-L-glutamine (●)
S Dimetfolamide
U Antineoplastic, antileukemic

10372 (8383) $C_{21}H_{24}O_2$
                                                     16320-04-0

13-Ethyl-17-hydroxy-18,19-dinor-17α-pregna-4,9,11-tri-
en-20-yn-3-one = 13-Ethyl-17α-ethynyl-17-hydroxygo-
na-4,9,11-trien-3-one = (17α)-13-Ethyl-17-hydroxy-
18,19-dinorpregna-4,9,11-trien-20-yn-3-one (●)
S A-46745, *Äthylnorgestrienon*, Dimetriose, Dimetro-
se, *Ethylnorgestrienone*, *Gestrinone***, Nemestran,
R 2323, RU 2323, Tridomose

U　Progestin (antiprogesterone)

10373　　　　　　　　　　　　　　　$C_{21}H_{24}O_2$
　　　　　　　　　　　　　　　　　110072-15-6

(17α)-17-Hydroxy-11-methylene-19-norpregna-4,15-
dien-20-yn-3-one (●)
S　*Exogestin, Letogestin*, Org 30659*
U　Progestagen

10374　(8384)　　　　　　　　　　$C_{21}H_{24}O_5$
　　　　　　　　　　　　　　　　　95851-37-9

[2S-(2α,3β,3aα)]-2-(3,4-Dimethoxyphenyl)-3,3a-dihy-
dro-3a-methoxy-3-methyl-5-(2-propenyl)-6(2*H*)-benzo-
furanone (●)
S　*Kadsurenone*
U　PAF factor receptor antagonist from Haifengteng
　　(*Piper futokadsura*)

10375　　　　　　　　　　　　　　$C_{21}H_{24}O_6$
　　　　　　　　　　　　　　　　　7770-78-7

(3R,4R)-4-[(3,4-Dimethoxyphenyl)methyl]dihydro-3-
[(4-hydroxy-3-methoxyphenyl)methyl]-2(3*H*)-furanone
(●)
S　*(–)-Arctigenin*

U　Immunomodulator (lignan component)

10376　(8385)　　　　　　　　　　$C_{21}H_{24}O_7$
　　　　　　　　　　　　　　　　　477-32-7

3,4,5-Trihydroxy-2,2-dimethyl-6-chromanacrylic acid δ-
lactone 4-acetate 3-(2-methylbutyrate) = 10-Acetoxy-
9,10-dihydro-8,8-dimethyl-2-oxo-2*H*,8*H*-pyra-
no[2,3-*f*]chromen-9-yl 2-methylbutyrate = 9,10-Dihydro-
9,10-dihydroxy-8,8-dimethyl-2*H*,8*H*-benzo[1,2-*b*:3,4-
b']dipyran-2-one 10-acetate 9-(2-methylbutyrate) = 3'-α-
Methylbutyryloxy-4'-acetoxy-3',4'-dihydroseseline =
(2R)-2-Methylbutanoic acid (9R,10R)-10-(acetyloxy)-
9,10-dihydro-8,8-dimethyl-2-oxo-2*H*,8*H*-benzo[1,2-
b:3,4-*b*']dipyran-9-yl ester (●)
S　Argiodin, Cardine, Carduben, Isonergine, Provismi-
　　ne (A), Selva, Vibeline, Visgan, Visnacorin, *Visnadi-
　　ne***, *Visnagan*, Visnamine
U　Coronary vasodilator (from *Ammi visnaga*)

10377　(8386)　　　　　　　　　　$C_{21}H_{24}O_9$
　　　　　　　　　　　　　　　　　155-58-8

4'-Methoxy-3,3',5-stilbenetriol 3-glucoside = 3-Hydroxy-
5-[2-(3-hydroxy-4-methoxyphenyl)ethenyl]phenyl β-D-
glucopyranoside (●)
S　*Ponticin, Rhaponticin, Rhapontin*
U　Estrogen (glucoside from *Rheum* sp.)

10378　(8387)　　　　　　　　　　$C_{21}H_{24}O_{1}$
　　　　　　　　　　　　　　　　　101404-86-

2,3,3',4,4',5,7-Heptahydroxyflavan glucoside
S　Eurhyton

U Vasodilator, cardiotonic

10379 (8388)

$C_{21}H_{25}BrF_2O_5$
57781-15-4

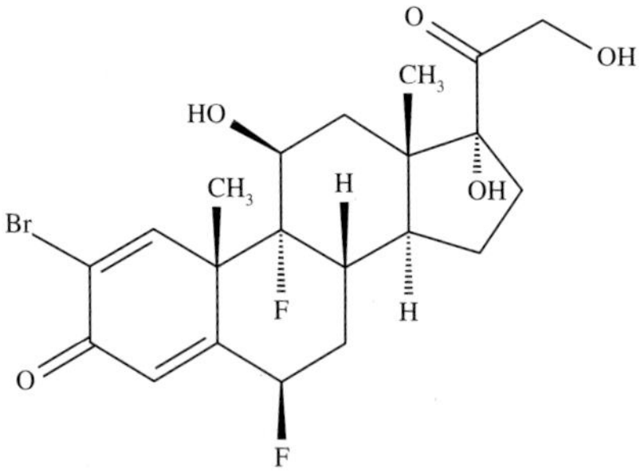

2-Bromo-6β,9-difluoro-11β,17α,21-trihydroxypregna-1,4-diene-3,20-dione = (6β,11β)-2-Bromo-6,9-difluoro-11,17,21-trihydroxypregna-1,4-diene-3,20-dione (●)
R see also no. 12790
S *Halopredone***
U Glucocorticoid (anti-inflammatory, anti-allergic)

10380 (8389)

$C_{21}H_{25}BrN_2O_3$
57475-17-9

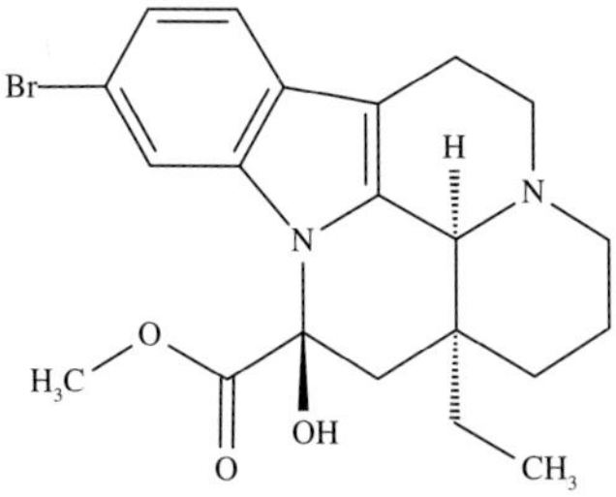

11-Bromovincamine = (3α,14β,16α)-11-Bromo-14,15-dihydro-14-hydroxyeburnamenine-14-carboxylic acid methyl ester (●)
R also fumarate (1:1) (84964-12-5)
S *Brovincamine***, BV 26-723, Sabromin, Zabromin
U Cerebrotonic

10381 (8390)

$C_{21}H_{25}ClFN_3O_3$
112885-41-3

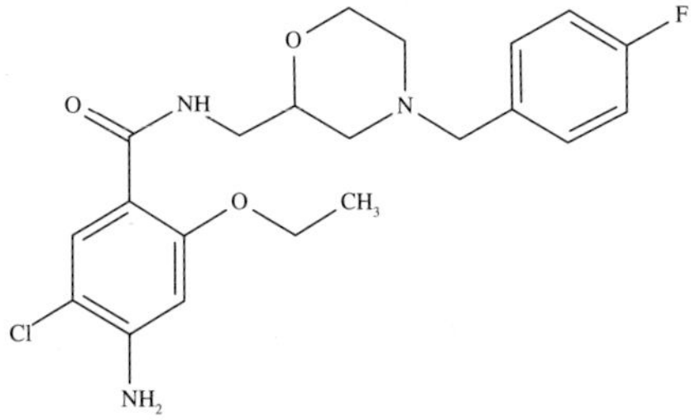

4-Amino-5-chloro-2-ethoxy-*N*-[[4-(*p*-fluorobenzyl)-2-morpholinyl]methyl]benzamide = 4-Amino-5-chloro-2-ethoxy-*N*-[[4-[(4-fluorophenyl)methyl]-2-morpholinyl]methyl]benzamide (●)
R Citrate (1:1) (112885-42-4)
S AS 4370, Gasmotin, *Mosapride citrate***, *Rimopride citrate*
U Gastrokinetic

10382 (8391)

$C_{21}H_{25}ClN_2OS$
30297-71-3

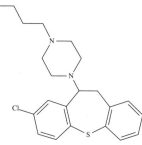

3-[4-(8-Chloro-10,11-dihydrodibenzo[*b,f*]thiepin-10-yl)-1-piperazinyl]-1-propanol = 2-Chloro-11-[4-(3-hydroxypropyl)-1-piperazinyl]-10,11-dihydrodibenzo[*b,f*]thiepin = 8-(8-Chloro-10,11-dihydrodibenzo[*b,f*]thiepin-10-yl)-1-piperazinepropanol (●)
S *Oxyclothepin*
U Neuroleptic

10383 (8392) $C_{21}H_{25}ClN_2O_3$
83881-51-0

(±)-[2-[4-(*p*-Chloro-α-phenylbenzyl)-1-piperazinyl]eth-oxy]acetic acid = [2-[4-[(4-Chlorophenyl)phenylmethyl]-1-piperazinyl]ethoxy]acetic acid (●)
R Dihydrochloride (83881-52-1)
S Agelmin, Alerid "Asche; UCB", Alerlisin, Allerset, *Cetirizine hydrochloride***, Cetriler, Cetrin "Reddy", Cetrizet, Cetrizin "Sintofarma", Cetryn, Cetzine, Cezin, Cirrus, Formistin, Hitrizin, P 071, Reactine, Ressital, Rigix, Ryzen, Salvalerg, Setir, Setiral, Stopaler, Triz, Virdos, Virlix, Voltric, Zetir, Ziptek, Zirtec, Zirtek, Zirtin, Zyrlex, Zyrtec
U Anti-allergic (histamine H_1-receptor antagonist)

10384 $C_{21}H_{25}ClN_2O_3$
130018-77-8

[2-[4-[(*R*)-*p*-Chloro-α-phenylbenzyl]-1-piperazinyl]eth-oxy]acetic acid = [2-[4-[(*R*)-(4-Chlorophenyl)phenylme-thyl]-1-piperazinyl]ethoxy]acetic acid (●)
R Dihydrochloride (130018-87-0)
S *Levocetirizine hydrochloride***, U.C.B. 28556, Xu-sal, Xyzal
U Anti-allergic (histamine H_1 receptor antagonist)

10385 $C_{21}H_{25}ClN_2O_3$
113240-02-1

1-[2-(4-Chlorophenyl)ethyl]-4-(3,5-dimethoxyben-zoyl)piperazine (●)
R Monohydrochloride (113240-27-0)
S CGP 29030A
U Analgesic (antinociceptive)

10386 $C_{21}H_{25}ClN_2O_3$
190786-43-7

(+)-4-[[(*S*)-*p*-Chloro-α-2-pyridylbenzyl]oxy]-1-piperi-dinebutyric acid = 4-[(*S*)-(4-Chlorophenyl)-2-pyridinyl-methoxy]-1-piperidinebutanoic acid (●)
R Monobenzenesulfonate (190786-44-8)
S *Bepotastine besilate***, *Betotastine besilate*, Talion, TAU-284
U Anti-allergic

10387 (8393) $C_{21}H_{25}ClN_2O_4S$
66981-73-5

7-[(3-Chloro-6,11-dihydro-6-methyldiben-zo[*c*,*f*][1,2]thiazepin-11-yl)amino]heptanoic acid *S*,*S*-di-oxide (●)
R Monosodium salt (30123-17-2)
S Coaxil, S 1574, Stablon, *Tianeptine sodium***
U Antidepressant

10388 (8394) $C_{21}H_{25}ClN_6O_2$
                                              48223-06-9

3-[[2-Chloro-4-(4,6-diamino-2,2-dimethyl-1,3,5-triazin-1(2H)-yl)phenoxy]methyl]-N,N-dimethylbenzamide (●)
R Ethanesulfonate (1:1) (41191-04-2)
S BAF, Baker's antifol, NSC-139105, Triazinate
U Antineoplastic

10389 (8395) $C_{21}H_{25}ClO_2$
                                              5192-84-7

6-Chloro-9β,10α-pregna-1,4,6-triene-3,20-dione = 6-Chloro-1,6-bisdehydroretroprogesterone = (9β,10α)-6-Chloropregna-1,4,6-triene-3,20-dione (●)
S Reteroid, Retroid, Retrone "Roche", Ro 4-8347, *Trengestone***
U Progestin

10390 $C_{21}H_{25}ClO_3$
                                              172998-54-8

4-Chloro-α-[4-(1,1-dimethylethyl)phenoxy]benzenepentanoic acid (●)
S BM 131180

U Hypoglycemic

10391 (8396) $C_{21}H_{25}ClO_3$
                                              62516-91-0

Ethyl 2-chloro-3-[p-(2-methyl-2-phenylpropoxy)phenyl]propionate = α-Chloro-4-(2-methyl-2-phenylpropoxy)benzenepropanoic acid ethyl ester (●)
S AL 294
U Antihyperlipidemic

10392 (8397) $C_{21}H_{25}ClO_3$
                                              15262-77-8

6-Chloro-17-hydroxypregna-1,4,6-triene-3,20-dione (●)
R see also no. 11822
S Δ^1-*Chlormadinone*, *Delmadinone***
U Progestin, anti-androgen, anti-estrogen

10393 (8398) $C_{21}H_{25}ClO_5$
                                              52080-57-6

6α-Chloro-17,21-dihydroxypregna-1,4-diene-3,11,20-trione = (6α)-6-Chloro-17,21-dihydroxypregna-1,4-diene-3,11,20-trione (●)

R see also no. 11823
S *Chloroprednisone***, *Chlorprednisonum*
U Glucocorticoid (anti-inflammatory, anti-allergic)

10394 (8399) $C_{21}H_{25}ClO_5$
                                                       5251-34-3

6-Chloro-11β,17α,21-trihydroxypregna-1,4,6-triene-3,20-dione = (11β)-6-Chloro-11,17,21-trihydroxypregna-1,4,6-triene-3,20-dione (●)

S *Cloprednol***, Cloradryn, Novacort, RS-4691, Synclopred, Syntestan
U Glucocorticoid (anti-inflammatory, anti-allergic)

10395 $C_{21}H_{25}Cl_2N_3O$
                                                       13015-79-7

6-Chloro-9-[[3-[(2-chloroethyl)ethylamino]propyl]amino]-2-methoxyacridine = N-(2-Chloroethyl)-N'-(6-chloro-2-methoxy-9-acridinyl)-N-ethyl-1,3-propanediamine (●)

R Dihydrochloride (146-59-8)
S ICR-170
U Antineoplastic

10396 (8400) $C_{21}H_{25}Cl_2N_3O_3$
                                                       5380-30-3

3-[p-[Bis(2-chloroethyl)amino]phenyl]-2-(nicotinoylamino)propionic acid ethyl ester = 4-[Bis(2-chloroethyl)amino]-N-(3-pyridinylcarbonyl)phenylalanine ethyl ester (●)

S Nicosin, Nikozin
U Antineoplastic

10397 (8401) $C_{21}H_{25}FN_2O_2$
                                                       1480-19-9

4'-Fluoro-4-[4-(o-methoxyphenyl)-1-piperazinyl]butyrophenone = 1-[3-(p-Fluorobenzoyl)propyl]-4-(o-methoxyphenyl)piperazine = 1-(4-Fluorophenyl)-4-[4-(2-methoxyphenyl)-1-piperazinyl]-1-butanone (●)

S Anti-Pica, *Fluanisone***, *Haloanisone*, 2028 MD, Methorin, Metoran, R 2167, Sedalande, Solusediv
U Neuroleptic

10398 (8402) $C_{21}H_{25}FN_4O_3$
                                                       133364-61-1

3-[2-[4-(4-Fluorobenzoyl)-1-piperidinyl]ethyl]-6,7,8,9-tetrahydro-2H-pyrido[1,2-a]-1,3,5-triazine-2,4(3H)dione (●)

R Maleate (1:1) (133364-63-3)
S DV 7028

U Antihypertensive, platelet aggregation inhibitor (serotonin antagonist)

10399 (8403)　　　　　　　　　　$C_{21}H_{25}FN_6$
106669-71-0

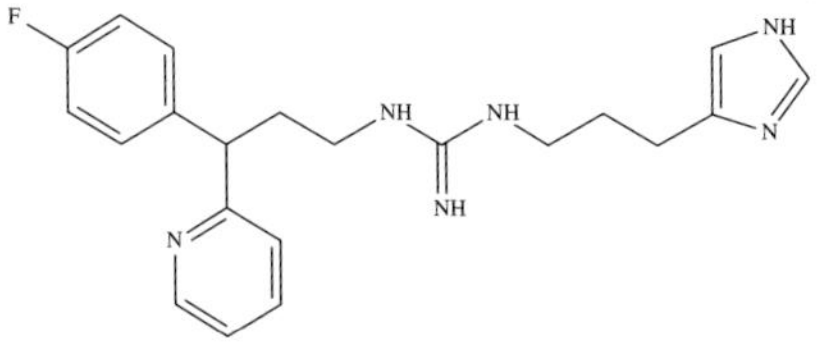

(±)-1-[3-(*p*-Fluorophenyl)-3-(2-pyridyl)propyl]-3-(3-imidazol-4-ylpropyl)guanidine = *N*-[3-(4-Fluorophenyl)-3-(2-pyridinyl)propyl]-*N'*-[3-(1*H*-imidazol-4-yl)propyl]guanidine (●)
S *Arpromidine***, BU-E-50, He 90371
U Histamine H$_2$-receptor agonist (cardiotonic)

10400　　　　　　　　　　　　　　$C_{21}H_{25}FN_6O_2$
172407-17-9

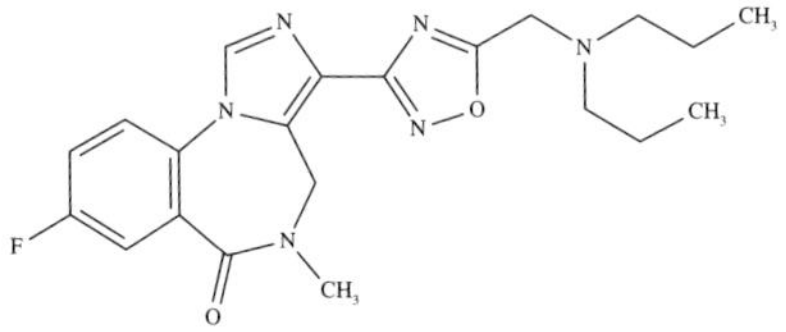

3-[5-[(Dipropylamino)methyl]-1,2,4-oxadiazol-3-yl]-8-fluoro-4,5-dihydro-5-methyl-6*H*-imidazo[1,5-*a*][1,4]benzodiazepin-6-one (●)
S Ro 48-6791
U Benzodiazepine receptor agonist (anesthetic)

10401　　　　　　　　　　　　　　$C_{21}H_{25}F_2N_3O$
86725-40-8

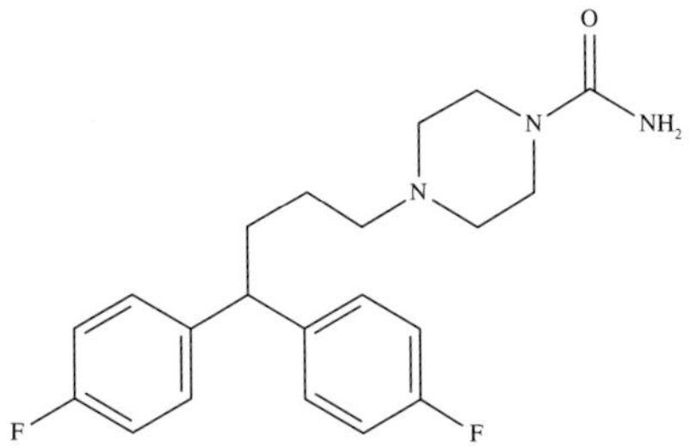

4-[4,4-Bis(4-fluorophenyl)butyl]-1-piperazinecarboxamide (●)
R Hydrochloride (75529-80-5)

S FG 5620
U Antipsychotic

10402　　　　　　　　　　　　　　$C_{21}H_{25}N$
138951-54-9

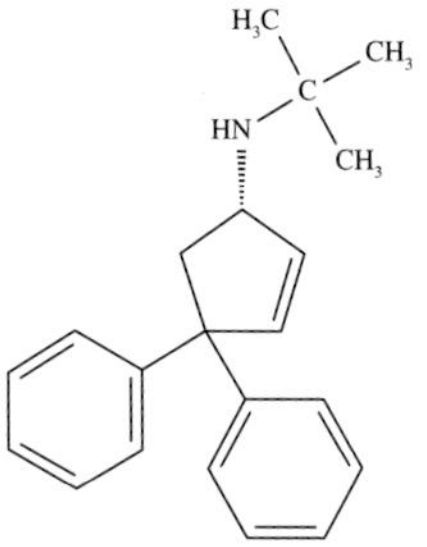

(*S*)-*N*-(1,1-Dimethylethyl)-4,4-diphenyl-2-cyclopenten-1-amine (●)
R Hydrochloride (138951-61-8)
S FK 584
U Agent for treatment of overactive detrusor

10403 (8408)　　　　　　　　　　$C_{21}H_{25}N$
91161-71-6

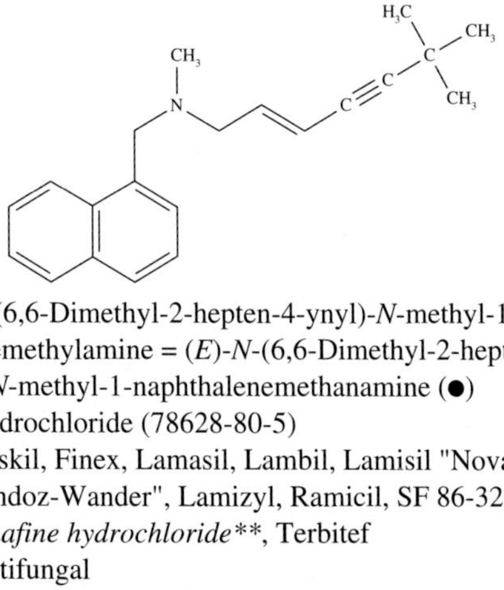

(*E*)-*N*-(6,6-Dimethyl-2-hepten-4-ynyl)-*N*-methyl-1-naphthalenemethylamine = (*E*)-*N*-(6,6-Dimethyl-2-hepten-4-ynyl)-*N*-methyl-1-naphthalenemethanamine (●)
R Hydrochloride (78628-80-5)
S Daskil, Finex, Lamasil, Lambil, Lamisil "Novartis; Sandoz-Wander", Lamizyl, Ramicil, SF 86-327, *Terbinafine hydrochloride***, Terbitef
U Antifungal

10404   (8405) $C_{21}H_{25}N$
849-06-9

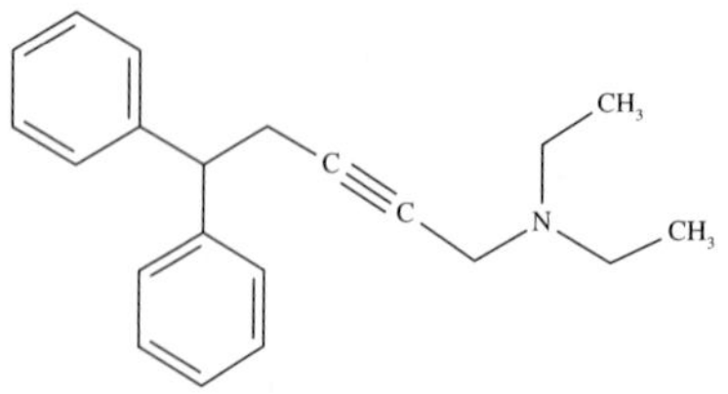

N,N-Diethyl-5,5-diphenyl-2-pentynylamine = *N,N*-Di-
ethyl-5,5-diphenyl-2-pentyn-1-amine (●)
R   Hydrochloride (3146-15-4)
S   Pediphen
U   Spasmolytic

10405   (8406) $C_{21}H_{25}N$
5118-29-6

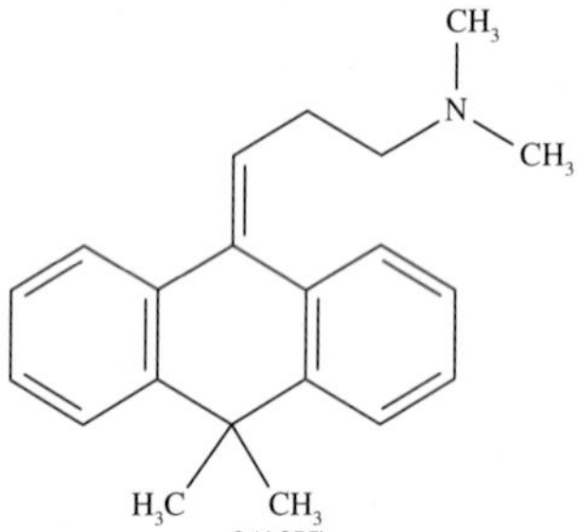

N,N,10,10-Tetramethyl-$\Delta^{9(10H),\gamma}$-anthracenepropyl-
amine = 9-[3-(Dimethylamino)propylidene]-10,10-dime-
thyl-9,10-dihydroanthracene = *N,N*-Dimethyl-3-(9,10-di-
hydro-10,10-dimethylanthracen-9-ylidene)propylamine =
3-(10,10-Dimethyl-9-(10*H*)-anthracenylidene)-*N,N*-di-
methyl-1-propanamine (●)
R   also hydrochloride (10563-70-9) or methanesulfonate
(31149-47-0)
S   Adaptol, Adepril, Dixeran, *Melitracen**, Meli-
xeran, Metraxil, N 7001, TAN 15, Thymeol, Trausa-
bun, U-24973
U   Antidepressant

10406   (8407) $C_{21}H_{25}N$
7009-69-0

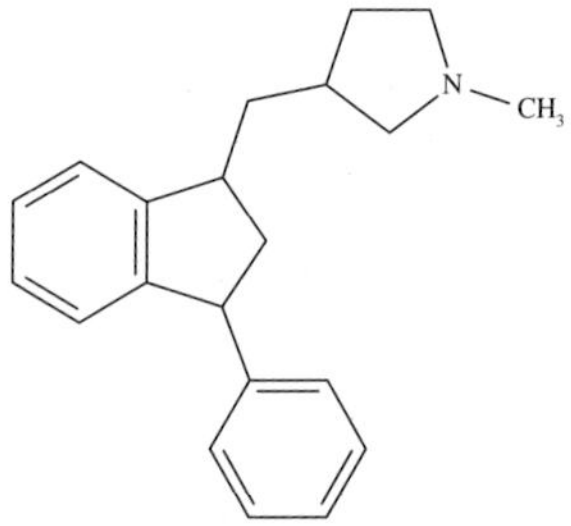

1-Methyl-3-[(3-phenyl-1-indanyl)methyl]pyrrolidine = 1-
(1-Methyl-3-pyrrolidinylmethyl)-3-phenylindan = 3-
[(2,3-Dihydro-3-phenyl-1*H*-inden-1-yl)methyl]-1-me-
thylpyrrolidine (●)
S   *Pyrophendane***, *Pyrophenindane*
U   Spasmolytic

10407 $C_{21}H_{25}N$
95417-67-7

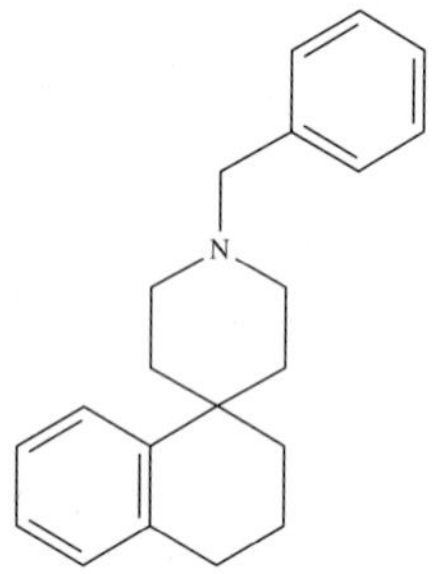

3,4-Dihydro-1'-(phenylmethyl)spiro[naphthalene-
1(2*H*),4'-piperidine] (●)
S   L 687384
U   Psychotropic (σ opiate receptor antagonist)

10408   (8409) $C_{21}H_{25}NO$
86-13-5

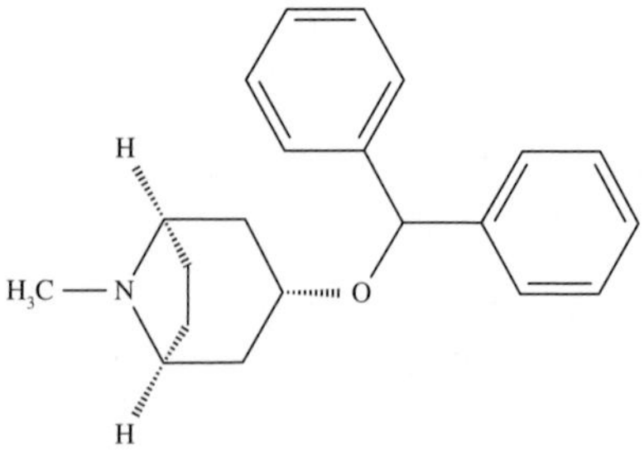

3α-(Diphenylmethoxy)-1αH,5αH-tropane = (1R,3r,5S)-
3-(Benzhydryloxy)tropane = Tropine benzhydryl ether =
endo-3-(Diphenylmethoxy)-8-methyl-8-azabicy-
clo[3.2.1]octane (●)
R Methanesulfonate (132-17-2)
S Akitan, Apo-Benzotropine, Bensylate, *Benzatropine
      mesilate**, Benzotropine, Benztropine methanesulfo-
      nate*, Cobrentin, C.O.G., Cogentin, Cogentinol, Co-
      tolate, MK-02, PMS-Benzotropine
U Anticholinergic, antiparkinsonian

10409   (8410) $C_{21}H_{25}NO$
                                                     1096-72-6

4-[(10,11-Dihydro-5*H*-dibenzo[*a,d*]cyclohepten-5-
yl)oxy]-1-methylpiperidine (●)
R Maleate (1:1) (1057-81-4)
S B.S. 7051, *Hepzidine maleate**
U Antidepressant

10410   (8412) $C_{21}H_{25}NO_2$
                                                     82-98-4

1-Ethyl-3-piperidyl diphenylacetate = Diphenylacetic
acid 1-ethyl-3-piperidyl ester = α-Phenylbenzeneacetic
acid 1-ethyl-3-piperidinyl ester (●)
R Hydrochloride (129-77-1)
S Crapinon, Dactil, Dactilake, Dactilate, Dactiran,
      Dactylate, Dactylet, Edelel, JB-305, Lactil, *Piperido-
      late hydrochloride**
U Anticholinergic, spasmolytic, analgesic

10411   (8411) $C_{21}H_{25}NO_2$
                                                     3626-07-1

2-Piperidinoethyl diphenylacetate = Diphenylacetic acid
β-piperidinoethyl ester = α-Phenylbenzeneacetic acid 2-
(1-piperidinyl)ethyl ester (●)
S Anicain
U Local anesthetic

10412   (8413) $C_{21}H_{25}NO_2$
                                                     1679-75-0

2-(Diethylamino)ethyl 2,3-diphenylacrylate = 2-(Diethylamino)ethyl 2-phenylcinnamate = 2-Phenylcinnamic acid
β-(diethylamino)ethyl ester = α-(Phenylmethylene)ben-
zeneacetic acid 2-(diethylamino)ethyl ester (●)
S 2136 C.B., *Cinnamaverine**
U Spasmolytic

10413 $C_{21}H_{25}NO_2$
                                                     183883-01-4

1-[3-(2-Phenylethyl)-2-benzofuranyl]-2-(propylamino)ethanol = 3-(2-Phenylethyl)-α-[(propylamino)me-
thyl]-2-benzofuranmethanol (●)
R Hydrochloride (157493-95-3)

S GE 68
U Anti-arrhythmic

10414   (8414) $C_{21}H_{25}NO_2$
                                                      4575-34-2

2-[3-(*m*-Hydroxyphenyl)-2,3-dimethylpiperidino]aceto-
phenone = 2-[3-(3-Hydroxyphenyl)-2,3-dimethyl-1-pipe-
ridinyl]-1-phenylethanone (●)
S *Myfadol***, NIH 8173, TA 306
U Analgesic, antitussive

10415   (8417) $C_{21}H_{25}NO_3$
                                                      4546-39-8

1-Piperidineethanol benzilate = 2-Piperidinoethyl benzi-
late = Benzilic acid β-piperidinoethyl ester = α-Hydroxy-
α-phenylbenzeneacetic acid 2-(1-piperidinyl)ethyl ester
(●)
R Hydrochloride (4544-15-4)
S Contrem (old form), Cylcain, Daipisate, DPX-8,
      Epithanate, Horipinen, Iminon, Inomaru, Inormal,
      Norticon, Offterror, Parma, Pensanate, Pethanail,
      Phorbic, Pinate, Pinnes, Pipenate, *Piperilate hydro-
      chloride*, Pipesan(e), *Pipethanate hydrochloride***,
      Pipetharon, Sycotrol
U Anticholinergic,tranquilizer

10416   (8415) $C_{21}H_{25}NO_3$
                                                      55096-26-9

17-(Cyclopropylmethyl)-4,5α-epoxy-6-methylenemor-
phinan-3,14-diol = (5α)-17-(Cyclopropylmethyl)-4,5-ep-
oxy-6-methylenemorphinan-3,14-diol (●)
R also hydrochloride (58895-64-0)
S Arthene, Cervene, Incystene, JF-1, *Nalmefene***,
      Nalmetrene, ORF 11676, Revex
U Antagonist to narcotics

10417   (8416) $C_{21}H_{25}NO_3$
                                                      3626-55-9

3-Methyl-4-morpholino-2,2-diphenylbutyric acid = β-
Methyl-α,α-diphenyl-4-morpholinebutanoic acid (●)
S *Moramide intermediate, Moramid-Zwischenproduct,
      Pre-Moramide*
U Narcotic analgesic

10418   (8418) $C_{21}H_{25}NO_3S$
                                                      95588-08-2

(±)-6,7-Dihydro-5-[[[(1*R**,2*R**,3*R**)-2-hydroxy-3-phen-
oxycyclopentyl]amino]methyl]-2-methylbenzo[*b*]thio-

phen-4(5*H*)-one = 6,7-Dihydro-5-[[[(*1R,2R,3R*)-2-hydr-
oxy-3-phenoxycyclopentyl]amino]methyl]-2-methylben-
zo[*b*]thiophen-4(5*H*)-one (●)
R Hydrochloride (95671-26-4)
S MDL 19744, *Tipentosin hydrochloride***
U Antihypertensive (ACE inhibitor)

10419 $C_{21}H_{25}NO_4$
                                               174393-13-6

(2*S*,4*S*)-2-Amino-4-(4,4-diphenylbutyl)pentane-1,5-dioic
acid = (4*S*)-4-(4,4-Diphenylbutyl)-L-glutamic acid (●)
S LY 307452
U Glutamate type II receptor antagonist

10420   (8422) $C_{21}H_{25}NO_4$
                                               475-81-0

1,2,9,10-Tetramethoxy-6aα-aporphine = (*S*)-5,6,6a,7-
Tetrahydro-1,2,9,10-tetramethoxy-6-methyl-4*H*-diben-
zo[*de,g*]quinoline (●)
R Hydrobromide (5996-06-5)
S *Glaucine hydrobromide*, Glauvent, Tussiglaucin
U Antitussive (alkaloid from *Papaveraceae* and *Fuma-
riaceae*)

10420-01   (8422-01)   73239-87-9
R Phosphate (2:3) of the (±)-compd.
S DL-832, Tusidil
U Antitussive

10421   (8420) $C_{21}H_{25}NO_4$
                                               483-14-7

2,3,9,10-Tetramethoxy-13aα-berbine = (*S*)-5,8,13,13a-
Tetrahydro-2,3,9,10-tetramethoxy-6*H*-dibenzo[*a,g*]qui-
nolizine (●)
S *Caseanine, Gindarin, Gyndarin, Hyndarin,*
        Jin Bu Huan, Rotundine, Tetrahydropalmatine
U Sedative, antihypertensive (alkaloid from *Stephania*
        glabra)

10422   (8421) $C_{21}H_{25}NO_4$
                                               523-02-4

(*S*)-5,8,13,13a-Tetrahydro-2,3,10,11-tetramethoxy-6*H*-
dibenzo[*a,g*]quinolizine (●)
S *Norcoralydine, Xylopinine*
U Protoberberine alkaloid

10423   (8419) $C_{21}H_{25}NO_4$

1-(3,4-Dimethoxyphenyl)-6,7-diethoxy-3,4-dihydroiso-
quinoline (●)
R Hydrochloride (60662-81-9)

S Tedryl
U Spasmolytic

10424 (8423) $C_{21}H_{25}NO_4$
16676-33-8

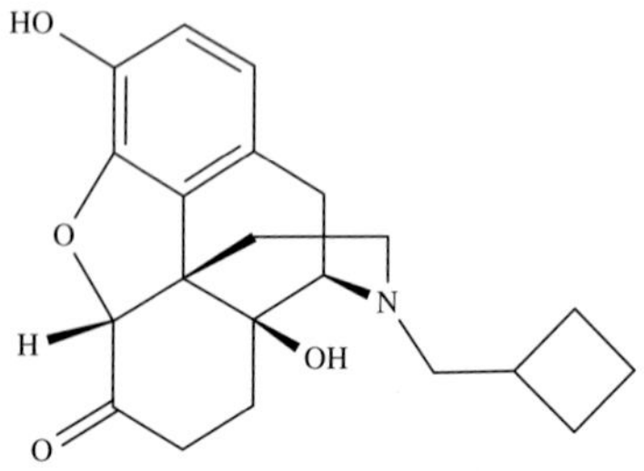

(–)-17-(Cyclobutylmethyl)-4,5α-epoxy-3,14-dihydroxy-morphinan-6-one = (5α)-17-(Cyclobutylmethyl)-4,5-epoxy-3,14-dihydroxymorphinan-6-one (●)
S *Nalbuphone*
U Antagonist to narcotics

10425 (8424) $C_{21}H_{25}NO_4$
16676-26-9

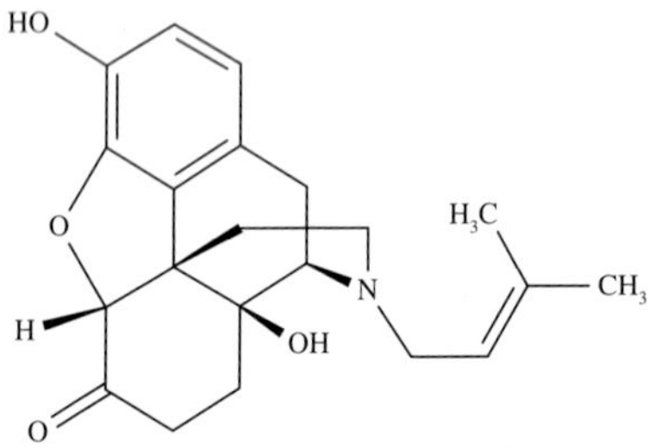

4,5α-Epoxy-3,14-dihydroxy-17-(3-methyl-2-butenyl)morphinan-6-one = 7,8-Dihydro-14-hydroxy-N-(3-methyl-2-butenyl)normorphinone = (5α)-4,5-Epoxy-3,14-dihydroxy-17-(3-methyl-2-butenyl)morphinan-6-one (●)
R Hydrochloride (16676-27-0)
S EN-1620 A, *Nalmexone hydrochloride***, UM 592
U Analgesic, antagonist to narcotics

10426 (8425) $C_{21}H_{25}NO_4$
129026-48-8

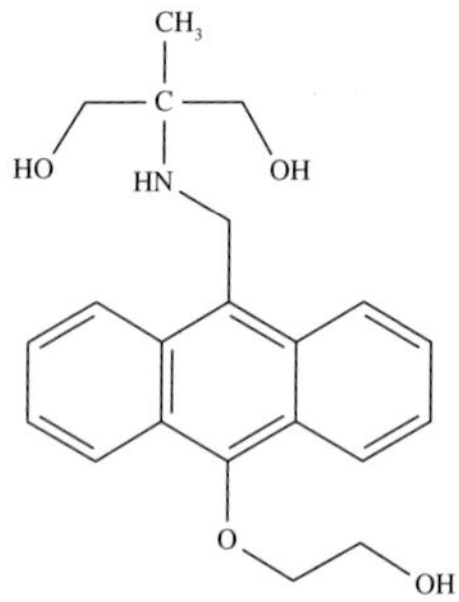

2-[[[10-(2-Hydroxyethoxy)-9-anthracenyl]methyl]amino]-2-methyl-1,3-propanediol (●)
R Hydrochloride (119499-33-1)
S 502 U 83
U Antineoplastic

10427 (8426) $C_{21}H_{25}NO_5$
477-30-5

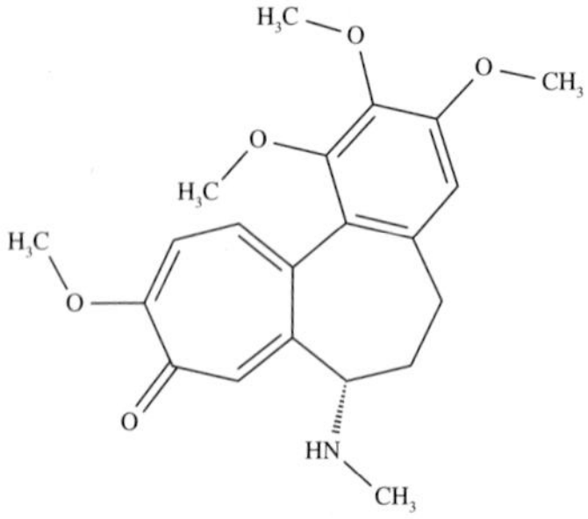

N-Deacetyl-N-methylcolchicine = (S)-6,7-Dihydro-1,2,3,10-tetramethoxy-7-(methylamino)benzo[a]heptalen-9(5H)-one (●)
S *Alkaloid F*, Ciba 12669 A, Colcemid, *Colchamine*, *Demecolcine***, Kolkamin, NSC-3096, Omain, *Santavy's Substance F, Substance F, Verbindung F aus Herbstzeitlose*
U Antineoplastic, antimitotic (alkaloid from *Colchicum autumnale*)

10428 (8427)   C$_{21}$H$_{25}$NO$_5$
                                                509-71-7

4,5α-Epoxy-17-methylmorphinan-3,6α-diyl diacetate =
3,6-Diacetoxy-*N*-methyl-4,5-epoxymorphinan = (5α,6α)-
4,5-Epoxy-17-methylmorphinan-3,6-diol diacetate (ester)
(●)
R Hydrochloride (5893-87-8)
S *Diacetyldihydromorphine*, *Dihydrodiamorphine*, Di-
      hydroheroin, *Dihydromorphine diacetate*, Paralaudin
U Narcotic

10429 (8428)   C$_{21}$H$_{25}$N$_2$S
                                                101396-46-7

(±)-1-Methyl-3-(10-phenothiazinylmethyl)quinuclidini-
um = 1-Methyl-3-(10*H*-phenothiazin-10-ylmethyl)-1-
azoniabicyclo[2.2.2]octane (●)
R Iodide (101396-42-3)
S LG-30435, *Mequitamium iodide***, *Mequitazine me-
      thyliodide*, *Mequitazium iodide*
U Bronchodilator, anti-allergic, anti-asthmatic

10430   C$_{21}$H$_{25}$N$_3$
                                                222545-63-3

N$^\alpha$-Methyl-3-(3,3-diphenylpropyl)histamine = 2-(3,3-
Diphenylpropyl)-*N*-methyl-1*H*-imidazole-4-ethanamine
(●)
S *Methylhistaprodifen*
U Histamine H$_1$ receptor agonist

10431 (8429)   C$_{21}$H$_{25}$N$_3$
                                                3613-73-8

2,3,4,5-Tetrahydro-2,8-dimethyl-5-[2-(6-methyl-3-pyri-
dinyl)ethyl]-1*H*-pyrido[4,3-*b*]indole (●)
S Dimebol, Dimebolin, Dimebone
U Antihistaminic

10432   C$_{21}$H$_{25}$N$_3$O
                                                147432-77-7

2-[[(*S*)-2-Cyclohexyl-1-(2-pyridyl)ethyl]amino]-5-me-
thylbenzoxazole = *N*-[(1*S*)-2-Cyclohexyl-1-(2-pyridi-
nyl)ethyl]-5-methyl-2-benzoxazolamine (●)
S BIRM-270, *Ontazolast***, *Oxazolast*

U Anti-asthmatic (leukotriene biosynthesis inhibitor)

10433 (8430) $C_{21}H_{25}N_3O$
83991-25-7

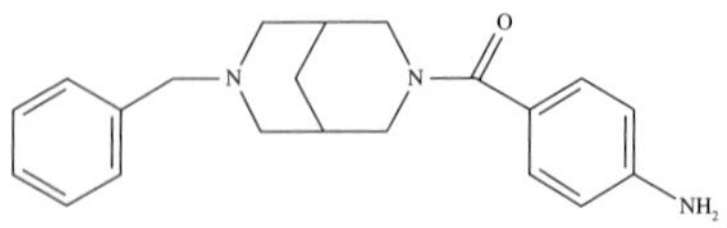

3-(p-Aminobenzoyl)-7-benzyl-3,7-diazabicy-clo[3.3.1]nonane = 3-(4-Aminobenzoyl)-7-(phenylme-thyl)-3,7-diazabicyclo[3.3.1]nonane (●)
S *Ambasilide***, LU 47110
U Anti-arrhythmic

10434 (8431) $C_{21}H_{25}N_3O$
3861-88-9

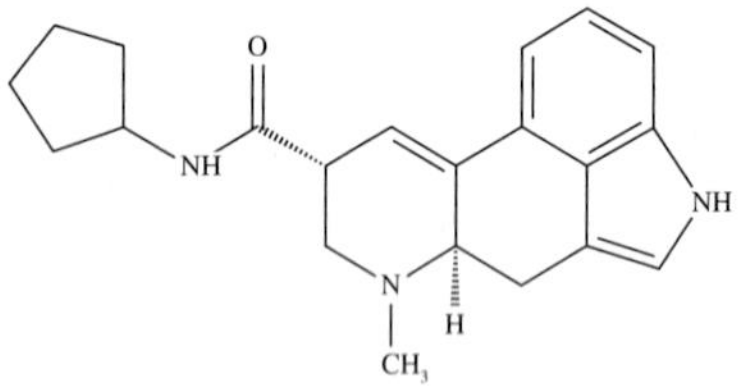

N-Cyclopentyllysergamide = Lysergic acid cyclopentyl-amide = (8β)-N-Cyclopentyl-9,10-didehydro-6-methyl-ergoline-8-carboxamide (●)
S C$_5$Al, Cepentil, Cepentyl, Tsepentil
U Uterotonic

10435 (8432) $C_{21}H_{25}N_3OS$
72293-38-0

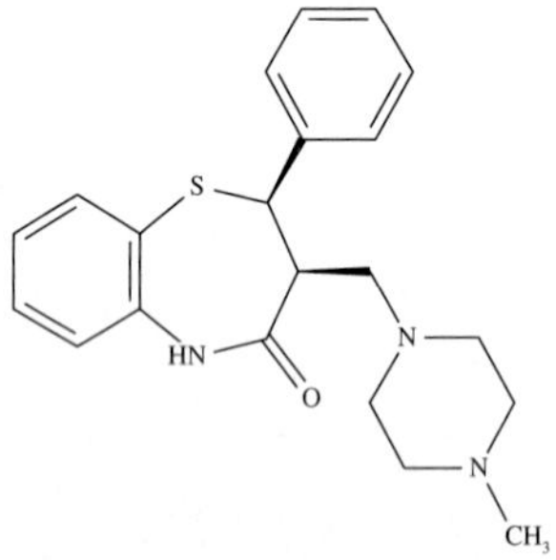

cis-(–)-2,3-Dihydro-3-[(4-methyl-1-piperazinyl)methyl]-2-phenyl-1,5-benzothiazepin-4(5H)-one (●)
R Dihydrochloride (72293-40-4)
S BTM-1042
U Antispasmodic

10436 (8433) $C_{21}H_{25}N_3O_2S$
111974-69-7

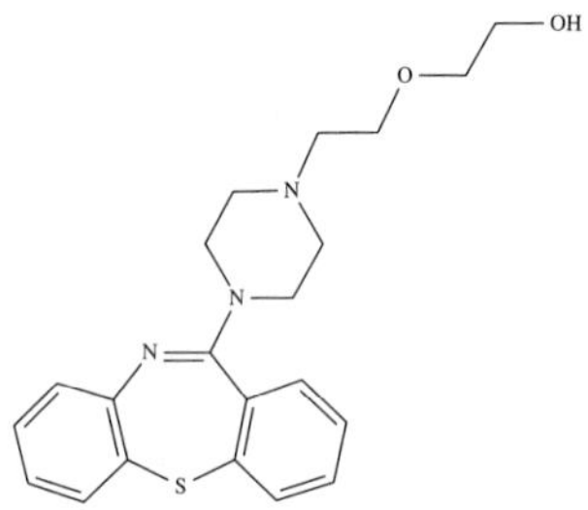

2-[2-(4-Dibenzo[b,f][1,4]thiazepin-11-yl-1-piperazi-nyl)ethoxy]ethanol (●)
R Fumarate (2:1) (111974-72-2)
S ICI 204636, *Quetiapine fumarate***, Seroquel, ZD 5077, ZM-204636
U Antipsychotic

10437 (8434) $C_{21}H_{25}N_3O_2S_2$
64099-44-1

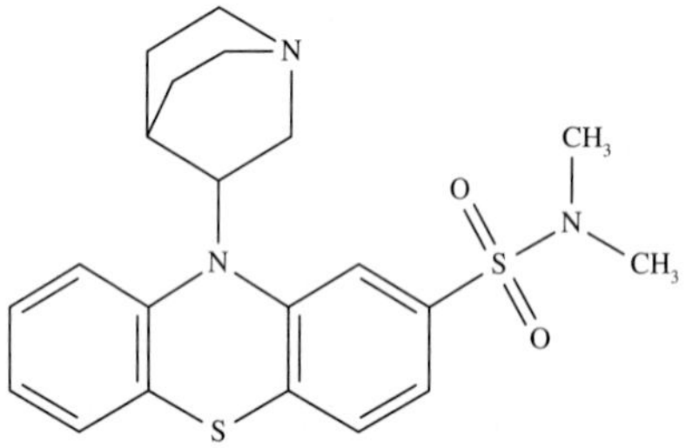

N,N-Dimethyl-10-(3-quinuclidinyl)-2-phenothiazinesul-fonamide = 10-(1-Azabicyclo[2.2.2]oct-3-yl)-N,N-dime-thyl-10H-phenothiazine-2-sulfonamide (●)
S LM 24056, *Quisultazine***, *Quisultidine*
U Gastric antisecretory, anti-ulcer agent

10438 (8435) $C_{21}H_{25}N_3O_3$
83275-56-3

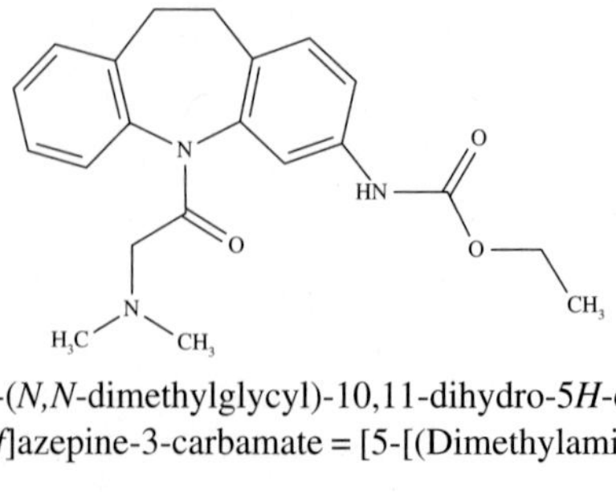

Ethyl 5-(N,N-dimethylglycyl)-10,11-dihydro-5H-di-benz[b,f]azepine-3-carbamate = [5-[(Dimethylamino)ace-

tyl]-10,11-dihydro-5*H*-dibenz[*b*,*f*]azepin-3-yl]carbamic acid ethyl ester (●)
R Monohydrochloride (78816-67-8)
S AWD 19-166, Bonecor, Bonnecor, GS 015, *Tiracizine hydrochloride***
U Anti-arrhythmic

10439 (8436)　　　　　　　　　　　　　$C_{21}H_{25}N_3O_3S$
　　　　　　　　　　　　　　　　　　　　　　2167-85-3

2-(2-Piperidinoethoxy)ethyl 10*H*-pyrido[3,2-*b*][1,4]benzothiazine-10-carboxylate = 1-Azaphenothiazine-10-carboxylic acid 2-(2-piperidinoethoxy)ethyl ester = Thiophenylpyridylamine-10-carboxylic acid 2-(2-piperidinoethoxy)ethyl ester = 10*H*-Pyrido[3,2-*b*][1,4]benzothiazine-10-carboxylic acid 2-[2-(1-piperidinyl)ethoxy]ethyl ester (●)
R Monohydrochloride (6056-11-7)
S D 254, Dipect, Lenopect, LG 254, *Pipazetate hydrochloride***, *Pipazethate hydrochloride*, Selvignon, Selvigon, Selvjgon, SKF 70230-A, SQ 15874, Theratuss, Toraxan, Transpulmin Hustensaft N
U Antitussive

10440 (8437)　　　　　　　　　　　　　$C_{21}H_{25}N_3O_4S$
　　　　　　　　　　　　　　　　　　　　　　136727-01-0

-[2-Hydroxy-3-[methyl(2-quinolylmethyl)amino]propoxy]methanesulfonanilide = *N*-[4-[2-Hydroxy-3-[methyl(2-quinolinylmethyl)amino]propoxy]phenyl]methanesulfonamide (●)
　WAY-123223
　Anti-arrhythmic

10441 (8560)　　　　　　　　　　　　　$C_{21}H_{25}N_3O_5$
　　　　　　　　　　　　　　　　　　　　　　54913-26-7

[4b*S*-(4bα,6aα,7α,8aβ,8bβ,9aα,9aα,12bβ)]-4b,5,6a,7,8,8a,8b,9,9a,11,12,12b-Dodecahydro-3-methoxy-2,8-dimethyl-7,9-methano-6,10-dioxa-8,8c,12a-triazaindeno[6,5,4-*fg*]acenathrylene-1,4-dione (●)
S *Naphthyridinomycin A*
U Antineoplastic antibiotic from *Streptomyces lusitanus* NRRL 8034

10442 (8438)　　　　　　　　　　　　　$C_{21}H_{25}N_3O_7S$
　　　　　　　　　　　　　　　　　　　　　　118587-22-7

3-Ethyl 5-methyl 2-[[[2-(formylamino)ethyl]thio]methyl]-1,4-dihydro-6-methyl-4-(*m*-nitrophenyl)-3,5-pyridinedicarboxylate = 2-[[[2-(Formylamino)ethyl]thio]methyl]-1,4-dihydro-6-methyl-4-(3-nitrophenyl)-3,5-pyridinedicarboxylic acid 3-ethyl 5-methyl ester (●)
S BBR 2160
U Antihypertensive (calcium antagonist)

10443   (8439) $C_{21}H_{25}N_3O_8S_2$
                                            94961-79-2

1,2-Di-(*p*-sulfophenyl)-4-(2-diethylaminoethyl)-3,5-py-
razolidinedione = 4,4'-[4-[2-(Diethylamino)ethyl]-3,5-di-
oxo-1,2-pyrazolidinediyl]bis[benzenesulfonic acid] (●)
R   Disodium salt (53039-87-5)
S   Sulfodethamedion
U   Anti-inflammatory

10444 $C_{21}H_{25}N_5$
                                            157012-18-5

7-Methyl-6-phenyl-2,4-di-1-pyrrolidinyl-7*H*-pyrro-
lo[2,3-*d*]pyrimidine (●)
S   PNU-87663, U 87663
U   Lipophilic anti-oxidant

10445   (8440) $C_{21}H_{25}N_5O$
                                            136816-76-7

1-(1*H*-Indol-2-ylcarbonyl)-4-[3-[(1-methylethyl)amino]-
2-pyridinyl]piperazine (●)
S   U-88204

U   Antiviral

10446   (8441) $C_{21}H_{25}N_5O_2$
                                            136816-75-6

1-[3-(Ethylamino)-2-pyridyl]-4-[(5-methoxy-2-indo-
lyl)carbonyl]piperazine = 1-[3-(Ethylamino)-2-pyridi-
nyl]-4-[(5-methoxy-1*H*-indol-2-yl)carbonyl]piperazine
(●)
R   Monomethanesulfonate (138540-32-6)
S   *Atevirdine mesilate***, U-87201 E
U   Antiviral

10447   (8442) $C_{21}H_{25}N_5O$
                                            15387-10-

N-[[(2,3-Dimethyl-5-oxo-1-phenyl-3-pyrazolin-4-yl)iso-
propylamino]methyl]nicotinamide = 4-[*N*-Isopropyl-*N*-
(nicotinamidomethyl)amino]-2,3-dimethyl-1-phenyl-3-
pyrazolin-5-one = *N*-[[(2,3-Dihydro-1,5-dimethyl-3-oxo
2-phenyl-1*H*-pyrazol-4-yl)(1-methylethyl)amino]me-
thyl]-3-pyridinecarboxamide (●)
S   *"Niapirina"*, *"Niapyrinum"*, *Niprofazone***, Ra 101,
         Ravalgene
U   Anti-inflammatory, antihistaminic

10448   (8443) $C_{21}H_{25}N_5C$
                                            71351-79-

2-[[4-(3-Methoxy-2-pyridyl)butyl]amino]-5-(6-methyl-3-picolyl)-4(1H)-pyrimidinone = 2-[[4-(3-Methoxy-2-pyridinyl)butyl]amino]-5-[(6-methyl-3-pyridinyl)]methyl-4(1H)-pyrimidinone (●)
R Trihydrochloride (71351-65-0)
S *Icotidine hydrochloride*, SKF 93319
U Histamine H_1-and H_2-receptor antagonist

10449 (8444) $C_{21}H_{25}N_5O_4$
91441-23-5

5-[(3-Aminopropyl)amino]-7,10-dihydroxy-2-[2-[(2-hydroxyethyl)amino]ethyl]anthra[1,9-cd]pyrazol-6(2H)-one (●)
R Dihydrochloride (105118-12-5)
S CI-942, DuP 942, NSC-349174, *Oxantrazole hydrochloride*, PD 111815, *Piroxantrone hydrochloride***
U Antineoplastic, antileukemic

10450 (8445) $C_{21}H_{25}N_5O_4$
91441-48-4

,10-Dihydroxy-2-[2-[(2-hydroxyethyl)amino]ethyl]-5-[2-(methylamino)ethyl]amino]anthra[1,9-cd]pyrazol-(2H)-one (●)
Dihydrochloride monohydrate (132937-88-3)
CI-937, DuP 937, *Moxantrazole hydrochloride*, NSC-355644, PD 113309, *Teloxantrone hydrochloride***

U Antineoplastic

10451 (8446) $C_{21}H_{25}N_5O_6$
106400-81-1

N-[p-[2-[(R)-2-Amino-3,4,5,6,7,8-hexahydro-4-oxopyrido[2,3-d]pyrimidin-6-yl]ethyl]benzoyl]-L-glutamic acid = (6β)-5,10-Dideaza-5,6,7,8-tetrahydrofolic acid = (R)-N-[4-[2-(2-Amino-1,4,5,6,7,8-hexahydro-4-oxopyrido[2,3-d]pyrimidin-6-yl)ethyl]benzoyl]-L-glutamic acid (●)
R Disodium salt (120408-07-3)
S DDATHF, *Lometrexole sodium***, LY 264618 disodium, T-64
U Antineoplastic (GARFT inhibitor)

10452 (8447) $C_{21}H_{25}N_5O_8S_2$
51481-65-3

(2S,5R,6R)-3,3-Dimethyl-6-[(R)-2-[3-(methylsulfonyl)-2-oxo-1-imidazolidinecarboxamido]-2-phenylacetamido]-7-oxo-4-thia-1-azabicyclo[3.2.0]heptane-2-carboxylic acid = (6R)-6-[N-(3-Methylsulfonyl-2-oxoimidazolidin-1-ylcarbonyl)-D-phenylglycylamino]penicillanic acid = D-α-[(2-Oxo-3-mesyl-1-imidazolidinyl)carbonylamino]benzylpenicillin = [2S-[2α,5α,6β(S*)]]-3,3-Dimethyl-6-[[[[[3-(methylsulfonyl)-2-oxo-1-imidazolidinyl]carbonyl]amino]phenylacetyl]amino]-7-oxo-4-thia-1-azabicyclo[3.2.0]heptane-2-carboxylic acid (●)
R Monosodium salt (42057-22-7) [or monosodium salt monohydrate (80495-46-1)]
S Baycipen, Bay f 1353, Baypen, Melocin, Mezlin, *Mezlocillin sodium***, Multocillin, MZPC
U Antibiotic

10453 (8448) $C_{21}H_{25}N_7OS$
77749-49-6

8-(Benzylthio)-4-morpholino-2-(1-piperazinyl)pyrimido[5,4-d]pyrimidine = 4-(4-Morpholinyl)-8-[(phenylmethyl)thio]-2-(1-piperazinyl)pyrimido[5,4-d]pyrimidine (●)

S RX-RA 69
U Phosphodiesterase inhibitor

10454 $C_{21}H_{26}BrNO_3S$
179410-97-0

7-Bromo-3-butyl-3-ethyl-2,3,4,5-tetrahydro-5-phenyl-1,5-benzothiazepin-8-ol 1,1-dioxide (●)
S GW 577
U Antihyperlipidemic

10455 (8449) $C_{21}H_{26}ClNO$
15686-51-8

(+)-(2R)-2-[2-[[(R)-p-Chloro-α-methyl-α-phenylbenzyl]oxy]ethyl]-1-methylpyrrolidine = (+)-2-[2-(p-Chloro-α-methylbenzhydryloxy)ethyl]-1-methylpyrrolidine = [R-(R*,R*)]-2-[2-[1-(4-Chlorophenyl)-1-phenylethoxy]ethyl]-1-methylpyrrolidine (●)
R Fumarate (1:1) (14976-57-9)
S Agasten, Alagyl, Aller.eze, Aloginan, Alphamin "SS", Alusas, Anhistan, Antihist-I, Antriptin, Arrest, Arumejil, Batom, Benanzyl, Chlonaryl, Clemamallet, Clemanil, *Clemastine fumarate**, Climerol, Dayhist-1, Fuluminol, Fumalestine, Fumaresutin, Fumartin, Fuyosyosin, Hinews, Hishimeel, Histamedine, Histaverin "Hokuriku", HS-592, Inbestan, Kinotomin, Lacretin, Lecasol, Lemagyl, Lulu A, Magotin, Maikohis, Mallermin-F, Marsthine, Martine, Masletine, *Meclastine fumarate*, *Mecloprodine fumarate*, Mejurin, Natarilon, Piloral, Raseltin, Reconin, Romien, Tavegil, Tavegyl, Tavist, Telgin-G, Tonotilin, Tonotirin, Trabest, Vegestigman, Xolamin
U Antihistaminic, anti-allergic

10456 (8450) $C_{21}H_{26}ClN_2C$

9-[(4-Chlorophenyl)phenylmethyl]-3-oxa-9-aza-6-azoniaspiro[5.5]undecane (●)
R Chloride hydrochloride (55981-23-2)
S CRC 7001
U Antineoplastic, anti-inflammatory

10457 (8451) $C_{21}H_{26}ClN_3O$
58-39-

2-[4-[3-(2-Chloro-10-phenothiazinyl)propyl]-1-piperazi-nyl]ethanol = 2-Chloro-10-[3-(4-β-hydroxyethyl-1-pipe-razinyl)propyl]phenothiazine = 4-[3-(2-Chloro-10H-phe-nothiazin-10-yl)propyl]-1-piperazineethanol (●)

R see also nos. 11869, 12895-02, 13900, 14481, 14540

S Aethaperazinum, Apo-Perphenazine, Biofenazina, Calmazina, *Chlorpiprazine*, *Chlorpiprozine*, Decen-tan, Emesinal, Etaperazin, Ethaperazin, Fentazin, F-Mon, Leptopsique, Metid, Neika, Neuropax, Pam-folick-R, Penamin, Penazine, Peratsin, *Perfena-zin(a)*, Perfenil, Perfesin, Pernamed, Perphenan, *Per-phenazine**, Phenazine "ICN", PZC, Scanpazine, Sch 3940, Sedofaro, Sedozina, Thilatazin, Tran-quisan "N.D.F.", Trifon, Trilafan, Trilafon, Trilifan, Trimin, Trinazin, Triomin, Triphenot

U Neuroleptic, anti-emetic

10458 (8452) $C_{21}H_{26}ClN_3O_2$
                                     75272-39-8

(±)-*cis*-N-(1-Benzyl-2-methyl-3-pyrrolidinyl)-5-chloro-4-(methylamino)-*o*-anisamide = *rel*-5-Chloro-2-methoxy-4-(methylamino)-N-[(2R,3R)-2-methyl-1-(phenylme-thyl)-3-pyrrolidinyl]benzamide (●)

S *Beminapride*, Emilace, Emirace, *Emonapride*, *Nemo-napride**, YM-09151-2

U Neuroleptic (D_1-dopamine receptor antagonist)

10459 (8453) $C_{21}H_{26}ClN_3O_2$
                                     522-20-3

6-Chloro-9-[[3-(diethylamino)-2-hydroxypropyl]amino]-2-methoxyacridine = 1-[(6-Chloro-2-methoxy-9-acridi-nyl)amino]-3-(diethylamino)-2-propanol (●)

R Dihydrochloride (1684-42-0)

S Acranil, SKF 16214-A2, S.N. 186, Sostol

U Antiprotozoal, antimutagen

10460 $C_{21}H_{26}Cl_2N_2O$
                             135330-85-7

1-[2-(2,4-Dichlorophenyl)-2-[[(2E)-3,7-dimethyl-2,6-oc-tadienyl]oxy]ethyl]-1H-imidazole (●)

S AFK 108

U Antifungal

10461 (8454) $C_{21}H_{26}Cl_2N_2O_4$
                                     137795-35-8

(R)-γ-(3,4-Dichlorobenzamido)-δ-oxo-8-azaspiro[4.5]de-cane-8-valeric acid = (R)-γ-[(3,5-Dichlorobenzoyl)ami-no]-δ-oxo-8-azaspiro[4.5]decane-8-pentanoic acid (●)

S CR 2194, *Spiroglumide**

U Anti-ulcer agent, anxiolytic

10462   (8455)

$C_{21}H_{26}Cl_2O$
37693-01-9

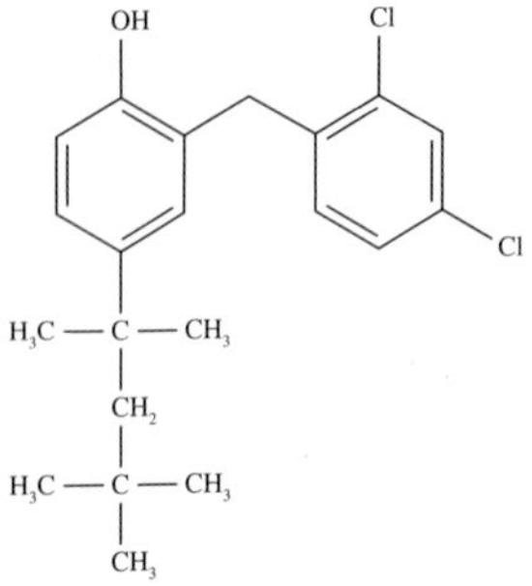

α-(2,4-Dichlorophenyl)-4-(1,1,3,3-tetramethylbutyl)-o-cresol = o-(2,4-Dichlorobenzyl)-p-(1,1,3,3-tetramethyl-butyl)phenol = 2-[(2,4-Dichlorophenyl)methyl]-4-(1,1,3,3-tetramethylbutyl)phenol (●)
S   Clofoctol**, Gramplus, Octofène, Rectavis
U   Antibacterial for the respiratory tract

10463   (8456)

$C_{21}H_{26}Cl_2O_2$
15686-33-6

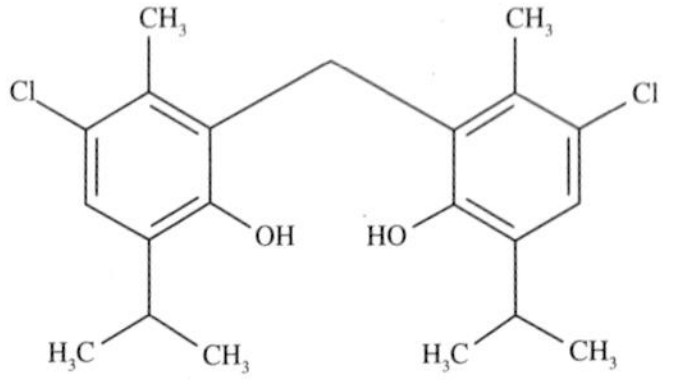

2,2'-Methylenebis(6-chlorothymol) = 2,2'-Methylene-bis[4-chloro-3-methyl-6-(1-methylethyl)phenol] (●)
S   Biclotymol**, Hexadraps, Hexadreps, Hexapock, Hexaspray
U   Antiseptic

10464   (8457)

$C_{21}H_{26}Cl_2O_4$
7008-26-6

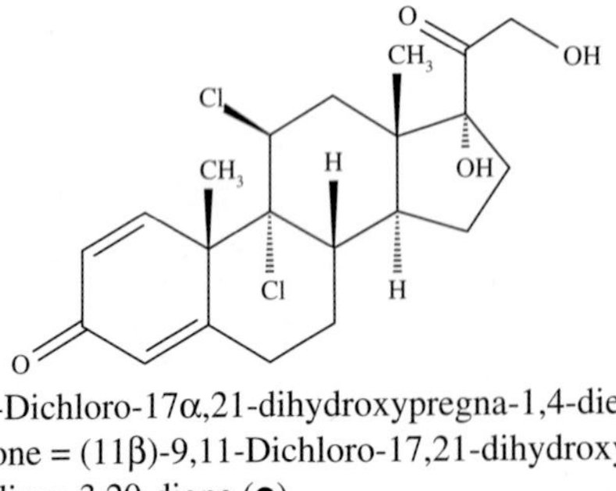

9α,11β-Dichloro-17α,21-dihydroxypregna-1,4-diene-3,20-dione = (11β)-9,11-Dichloro-17,21-dihydroxypreg-na-1,4-diene-3,20-dione (●)

R   see also no. 11874
S   Dichlorisone**, Sch 5350
U   Glucocorticoid (anti-inflammatory, anti-allergic)

10465

$C_{21}H_{26}FNO_3$
189192-18-5

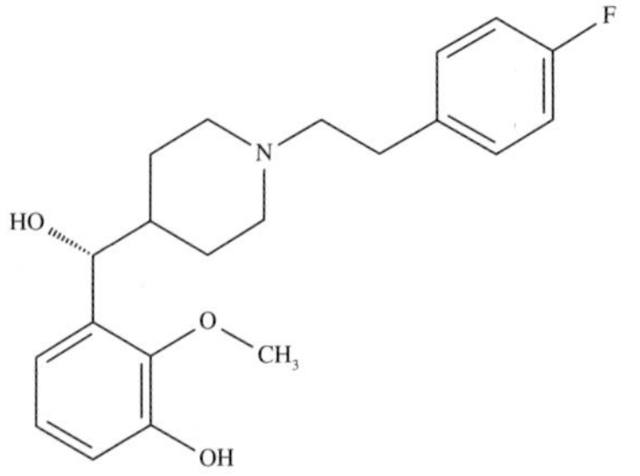

(αR)-1-[2-(4-Fluorophenyl)ethyl]-α-(3-hydroxy-2-meth-oxyphenyl)-4-piperidinemethanol (●)
S   MDL 105725
U   Antipsychotic (5-HT$_2$ receptor antagonist)

10466   (8458)

$C_{21}H_{26}FN_3O$
133091-61-9

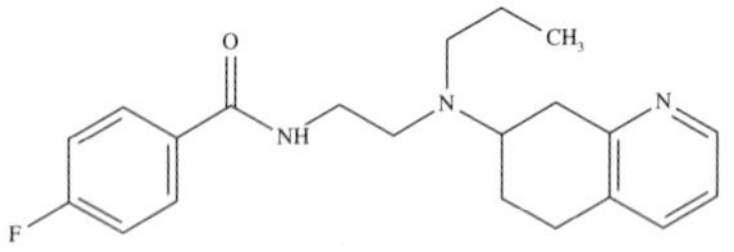

4-Fluoro-N-[2-[propyl(5,6,7,8-tetrahydro-7-quinoli-nyl)amino]ethyl]benzamide (●)
S   WAY-100012
U   5-HT$_{1A}$-receptor agonist

10467

$C_{21}H_{26}FN_3O$
143383-65-

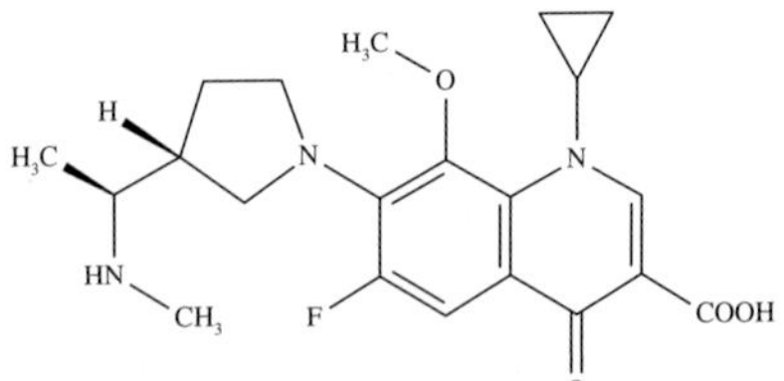

1-Cyclopropyl-6-fluoro-1,4-dihydro-8-methoxy-7-[(3R)3-[(1S)-1-(methylamino)ethyl]-1-pyrrolidinyl]-4-oxo-3-quinolinecarboxylic acid (●)
S   PD 140288, Premafloxacin**, U-95376
U   Antibacterial (veterinary)

10468   (8459) $C_{21}H_{26}F_3N_5$
108785-69-9

(±)-*cis*-5,5a,6,7,8,8a-Hexahydro-3-[2-[4-(α,α,α-tri-
fluoro-*m*-tolyl)-1-piperazinyl]ethyl]cyclopenta[3,4]pyr-
rolo[2,1-*c*]-*s*-triazole = 5,5a,6,7,8,8a-Hexahydro-3-[2-[4-
[3-(trifluoromethyl)phenyl]-1-piperazinyl]ethyl]cyclo-
penta[3,4]pyrrolo[2,1-*c*]-1,2,4-triazole (●)
S   *Lorpiprazole***, Normarex
U   Anxiolytic

10469   (8460) $C_{21}H_{26}NO$

1-Methyl-1-[2-(9*H*-xanthen-9-yl)ethyl]piperidinium
R   Bromide (60662-82-0)
S   Espalisal, Espalisyl, Spalisal
U   Anticholinergic, spasmolytic

10470 $C_{21}H_{26}NO_2$
168887-12-5

2,12-Diethoxy-6,11-dihydro-13,13-dimethyl-6,11-
ethanobenzo[*b*]quinolizinium (●)
R   Chloride (166885-43-4)
Win 63480

U   Neuroprotective, anti-ischemic

10471   (8461) $C_{21}H_{26}NO_2$
113587-15-8

1-[2-[(Cyclohexylphenylacetyl)oxy]ethyl]pyridinium (●)
R   Bromide
S   *Hexadifenium bromide, Hexadiphenium bromide*
U   Spasmolytic

10472   (8466) $C_{21}H_{26}NO_3$
5818-17-7

Diethyl(2-hydroxyethyl)methylammonium xanthene-9-
carboxylate = Diethylmethyl-2-(xanthen-9-ylcarbonyl-
oxy)ethylammonium = *N,N*-Diethyl-*N*-methyl-2-[(9*H*-
xanthen-9-ylcarbonyl)oxy]ethanaminium (●)
R   Bromide (53-46-3)
S   Asabine, Asatylon, Avagal, Banthin(e), Bantosal,
Bromantina, Bronerg, Dexabine, *Dixamone bromide*, Doladene, Emtebe-51, Evogal, Frenogastrico,
Gastron, Gastrophyl, Gastrosedan, Klinantine, Man-
theline, *Metantelina*, Metantyl, Metaxan, Meth-
anide, *Methantheline bromide**, Methanthine Bromide, Methelina, MTB 51, Pantamin, Pepulsan, Pro-
bantim, Resobantin (active substance), SC-2910, Ul-
cantina, Ulcine, Ulcudexter, Ulcuwas, Uldumont, Ul-
kophob, Vagamin, Vagantin, Vagominal, Vaxante-
ne, Ventrisan, Vilcurin, Xanteline, Xantenol
U   Anticholinergic, anti-ulcer agent

10473   (8464) $C_{21}H_{26}NO_3$
25990-43-6

3-Hydroxy-1,1-dimethylpiperidinium benzilate (ester) = 3-(Benziloyloxy)-1,1-dimethylpiperidinium = 3-[(Hydroxydiphenylacetyl)oxy]-1,1-dimethylpiperidinium (●)
R   Bromide (76-90-4)
S   Atenecolin, Cantil, Cantilake, Cantilaque, Cantilon, Cantilyn, Cantril, Colibantil, Colopiril, Colum, Delevil, Eftoron, Gastropidil, Giacol, *Glycophenylate bromide*, JB-340, Mepenzolan, *Mepenzolate bromide***, *Mepenzolone bromide*, Sachicoron, Tendalin, Tralanta, Trancolon, Trokonil
U   Anticholinergic

10474   (8465) $C_{21}H_{26}NO_3$
15394-61-3

4-Hydroxy-1,1-dimethylpiperidinium benzilate (ester) = 4-(Benziloyloxy)-1,1-dimethylpiperidinium = 4-[(Hydroxydiphenylacetyl)oxy]-1,1-dimethylpiperidinium (●)
R   Bromide (5634-41-3)
S   Dorela, Neupran, *Parapenzolate bromide***, *Parapenzolone bromide*, Relanol, Relenol, Sch 3444, Spacine, Vagopax
U   Anticholinergic

10475   (8462) $C_{21}H_{26}NO_3$
54505-25-8

1-(2-Hydroxyethyl)-1-methylpyrrolidinium benzilate (ester) = 1-[2-(Benziloyloxy)ethyl]-1-methylpyrrolidinium = 1-[2-[(Hydroxydiphenylacetyl)oxy]ethyl]-1-methylpyrrolidinium (●)
R   Iodide (3478-15-7)
S   *Etipirium iodide***, *Etipyrium iodide*, I.M.P.E.
U   Antihemorrhagic, anticholinergic

10476   (8463) $C_{21}H_{26}NO_3$
596-50-9

2-(Hydroxymethyl)-1,1-dimethylpyrrolidinium benzilate (ester) = 2-[(Benziloyloxy)methyl]-1,1-dimethylpyrrolidinium = 2-[[(Hydroxydiphenylacetyl)oxy]methyl]-1,1-dimethylpyrrolidinium (●)
R   Methyl sulfate (545-80-2)
S   Alfaland, CLB 499, I.S. 499, McN-R-726-47, Nactate, Nacton, *Poldine metilsulfate**, *Poldoni methylsulfas*
U   Anticholinergic, anti-ulcer agent

10477 $C_{21}H_{26}NO_4$
83387-25-1

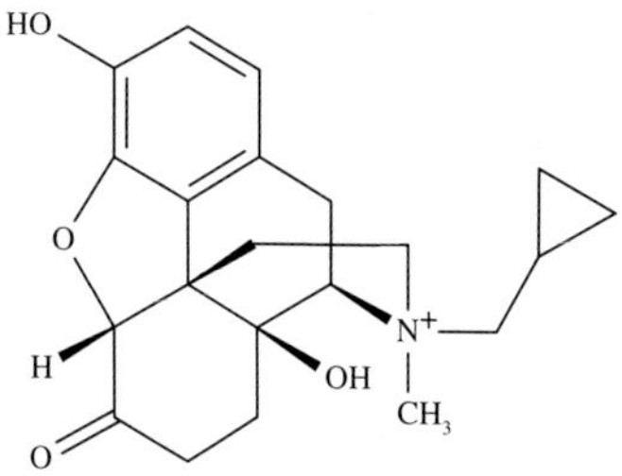

N-Methylnaltrexone = (5α)-17-(Cyclopropylmethyl)-4,5-epoxy-3,14-dihydroxy-17-methyl-6-oxomorphinanium (●)
R Bromide (73232-52-7)
S *Methylnaltrexonium*
U Therapeutic in opioid-induced constipation

10478 (8468) $C_{21}H_{26}N_2O$
20537-22-8

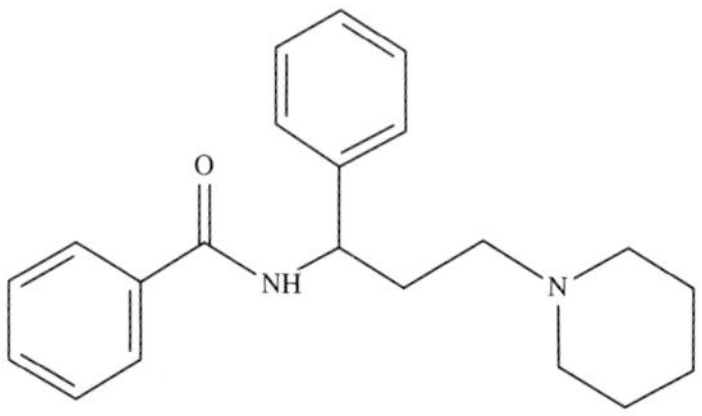

N-(1-Phenyl-3-piperidinopropyl)benzamide = 1-Benzamido-1-phenyl-3-piperidinopropane = *N*-[1-Phenyl-3-1-piperidinyl)propyl]benzamide (●)
S Digammacain
U Local anesthetic

10479 (8469) $C_{21}H_{26}N_2O$
70312-00-4

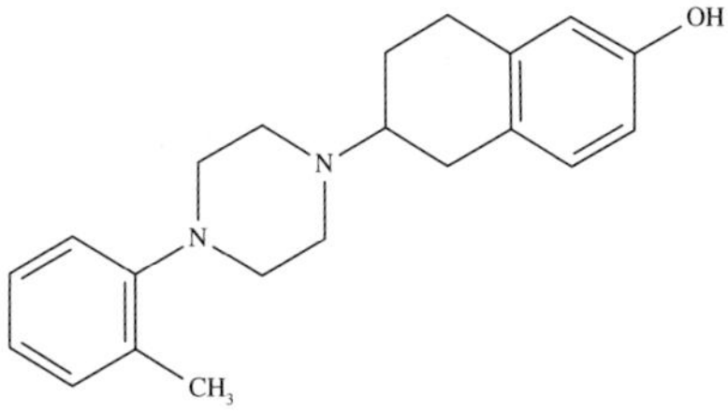

,6,7,8-Tetrahydro-6-(4-*o*-tolyl-1-piperazinyl)-2-naphol = 5,6,7,8-Tetrahydro-6-[4-(2-methylphenyl)-1-piperazinyl]-2-naphthalenol (●)
BK 34-530, *Tolnapersine***

U Antihypertensive

10480 (8470) $C_{21}H_{26}N_2O$
48203-83-4

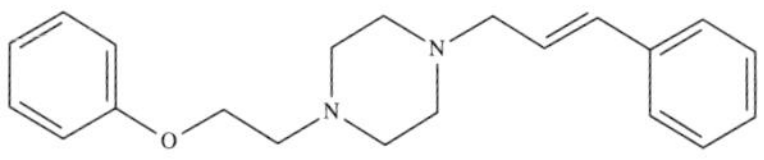

1-Cinnamyl-4-(2-phenoxyethyl)piperazine = 1-(2-Phenoxyethyl)-4-(3-phenyl-2-propenyl)piperazine (●)
R Dihydrochloride (41331-77-5)
S AS 34
U Coronary vasodilator, spasmolytic

10481 (8467) $C_{21}H_{26}N_2O$
77-01-0

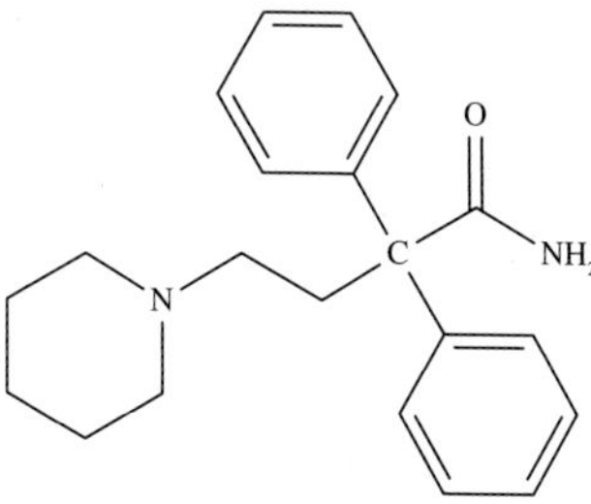

α,α-Diphenyl-1-piperidinebutyramide = 2,2-Diphenyl-4-piperidinobutyramide = α,α-Diphenyl-1-piperidinebutanamide (●)
R Monohydrochloride (14007-53-5)
S *Fenpipramide hydrochloride***, Hoechst 9980, R 14, U-0229
U Spasmolytic

10482 (8471) $C_{21}H_{26}N_2OS$
20769-36-2

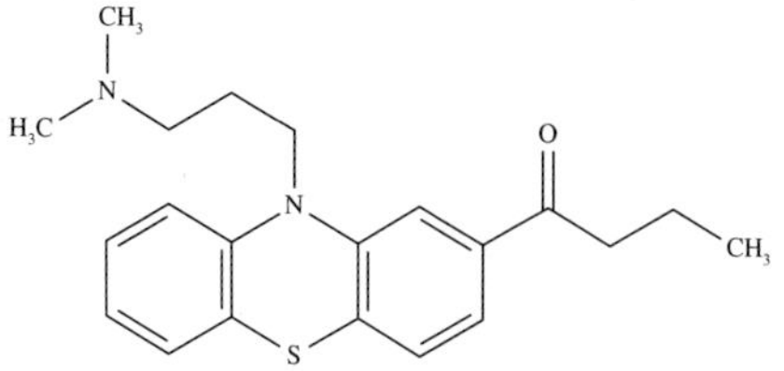

2-Butyryl-10-[3-(dimethylamino)propyl]phenothiazine = 1-[10-[3-(Dimethylamino)propyl]-10*H*-phenothiazin-2-yl]-1-butanone (●)
S *Butyrylpromazine*, 1613 C.B.
U Neuroleptic

10483 (8473) $C_{21}H_{26}N_2OS$
                                             16926-48-0

10-[4-(3-Hydroxypropyl)piperazino]-10,11-dihydrodi-
benzo[b,f]thiepine = 4-(10,11-Dihydrodiben-
zo[b,f]thiepin-10-yl)-1-piperazinepropanol (●)
S Oxythepin
U Neuroleptic

10484 (8472) $C_{21}H_{26}N_2OS$
                                             14759-04-7

2-Methoxy-10-[2-(1-methyl-2-piperidyl)ethyl]phenothia-
zine = 2-Methoxy-10-[2-(1-methyl-2-piperidinyl)ethyl]-
10H-phenothiazine (●)
S KS 33, *Oxiridazinum, Oxyridazine***
U Psychosedative

10485 (8474) $C_{21}H_{26}N_2OS_2$
                                             5588-33-0

10-[2-(1-Methyl-2-piperidyl)ethyl]-2-(methylsulfinyl)phenothiazine = 10-[2-(1-Methyl-2-piperidi-
nyl)ethyl]-2-(methylsulfinyl)-10H-phenothiazine (●)
R Benzenesulfonate (1:1) (32672-69-8)
S Calodal, Lidanar, Lidanil, Lidanor, *Mesoridazine besilate***, NC 123, Serentil, *Thioridazine sulfoxide besilate*, TPS-23
U Psychosedative

10486 (8475) $C_{21}H_{26}N_2O_2$
                                             66304-03-8

N-[(1-Ethyl-2-pyrrolidinyl)methyl]benzilamide = N-[(1-
Ethyl-2-pyrrolidinyl)methyl]-α-hydroxy-α,α-diphenyl-
acetamide = N-[(1-Ethyl-2-pyrrolidinyl)methyl]-α-hydr-
oxy-α-phenylbenzeneacetamide (●)
S *Epicainide***, F 1427
U Anti-arrhythmic

10487 (8476) $C_{21}H_{26}N_2O_2S$
                                             116773-55-8

2,3,4,5-Tetrahydro-5-[3-[[2-(m-tolyloxy)ethyl]ami-
no]propionyl]-1,5-benzothiazepine = 2,3,4,5-Tetrahydro-
5-[3-[[2-(3-methylphenoxy)ethyl]amino]-1-oxopropyl]-
1,5-benzothiazepine (●)
R Fumarate (1:1) (134296-41-6)
S KT 2-230
U Vasodilator

10488 $C_{21}H_{26}N_2O_2S$

(2R-cis)-3-[3-(Dimethylamino)propyl]-2,3-dihydro-2-(4-methoxyphenyl)-1,5-benzothiazepin-4(5H)-one (●)
R Oxalate (167710-86-3)
S MR-14134
U Antihypertensive

10489   (8477) $C_{21}H_{26}N_2O_2S_2$
                                                 14759-06-9

10-[2-(1-Methyl-2-piperidyl)ethyl]-2-(methylsulfonyl)phenothiazine = 2-Mesyl-10-[2-(1-methyl-2-piperi-
dyl)ethyl]phenothiazine = 10-[2-(1-Methyl-2-piperidi-
nyl)ethyl]-2-(methylsulfonyl)-10H-phenothiazine (●)
S Imagotan, Inofal, Inosulon, Psychoson, *Solforidazi-
ne*, *Sulforidazine***, TPN 12
U Neuroleptic

10490   (8478) $C_{21}H_{26}N_2O_3$
                                                 25552-59-4

N-[(4-Hydroxy-1-phenethyl-4-piperidyl)methyl]salicyl-
amide = 2-Hydroxy-N-[[4-hydroxy-1-(2-phenylethyl)-4-
piperidinyl]methyl]benzamide (●)
R Monohydrochloride (25552-58-3)
S S 1592
U Antitussive

10491   (8481) $C_{21}H_{26}N_2O_3$
                                                 1617-90-9

Vincamic acid methyl ester = (3α,14β,16α)-14,15-Dihy-
dro-14-hydroxyeburnamenine-14-carboxylic acid methyl
ester (●)
R also hydrochloride (10592-03-7) or tartrate (1:1)
      (64034-84-0)
S Adelvinor, Aethroma, Alfavinca, Anasclerol, Angio-
      lipid, Angiopac, Artensen, Arteriovinca, Asnai, Ater-
      vit, Ausomina, Branex, Camphoginol, Cebralart, Ce-
      malyt, Centractiva, Cerebramina, Cerebroxine, Cere-
      dia, Ceredilan, Cervinca, Cetal, Cincuental, Circuli-
      ne "Sanitas", Davinova, Defardole, Devincan, Dilar
      "Teoforma", Dilarterial, Domeni, Donemi, Egefani-
      um, Encevin, Equipur, Esberidin, Etuamol, Flora-
      min, Gerialan, Gibivi, Horusvin, Istogenol, Jusitel,
      Livenza, Minorin, Mnemofer, Neuro-Kranit "Kle-
      va", Nooxine, Novicet, Nuclesil, Ocu-Vinc, Ophdil-
      vas N, Optonium, Oxencephal, Oxicebral, Oxivinca,
      Oxybral, Oxygeron, Perphal, Perval, Pervin, Pervin-
      camin(e), Pervone, Plenocor, Retarvic, Ribex "Volpi-
      no", Rotrimin, Sostenil, Sozinalin, Tefavinca, Toni-
      for, Tripervan, Tserovin, Tussimatol, Vadicate, Va-
      relin, Vasculogène, Vasonett, Vencano, Venoxigen,
      Verolinex, Vigamma, Vinca 10 (20, 30), Vincabio-
      mar, Vincabrain, Vincacen, Vincacervol, Vinca-
      chron, Vincadar, Vinca-Dil, Vinca-Ecobi, Vinca-
      farm, Vincafolina, Vincafor, Vincagalup, Vincagen,
      Vincagil, Vinca-Hexal, Vincalen, Vincalex, Vin-
      calvar, Vincamas, Vincamed, Vincamer, Vincame-
      trin, Vincamidol, *Vincamine***, Vincaminol, Vinca-

minor, Vincamiso, Vincamyl, Vincane, Vincanor, Vincanox, Vincapan, Vincapront, Vinca-Ri, Vincasal, Vincasaunier, Vinca-Tablinen, Vinca-Treis, Vincavix, Vincimax, Vincophal, Vinkhum, Vinodrel, Vinsal, Viribleton, Visal, Vitren, VNC 10, Vraap, Zinka

U Vasodilator, cerebrotonic, antihypertensive (alkaloid from *Vinca minor*)

10491-01 (8481-01)   54341-01-4
R 2-Oxoglutarate [2-oxopentanedioate (●)] (1:1)
S Activen, Cetovinca, EL-485, *Oxovinca*, Roiten
U Cerebrotonic

10491-02 (8481-02)   19548-24-4
R 6,7-Dihydroxy-2-oxo-2*H*-1-benzopyran-4-methanesulfonate (1:1)
S *Vincamine cromesilate*, Vincaryl
U Cerebrotonic

10491-03 (8481-03)   51179-28-3
R 3-(7-Theophyllinyl)propanesulfonate (no. 2308) (1:1)
S Teproside, *Vincamine teprosilate*
U Cerebrotonic

10492 (8479) $C_{21}H_{26}N_2O_3$
                                        33257-13-5

[4*S*-(4α,4aα,12aα)]-4-Ethyl-2,3,4,4a,5,6,7,12-octahydro-2-methyl-6-oxopyrido[3',4':4,5]cyclohept[1,2-*b*]indole-12a(1*H*)-carboxylic acid methyl ester (●)
S *Ervatamine*
U Anesthetic

10493 (8480) $C_{21}H_{26}N_2O_3$
                                        146-48-5

Yohimbic acid methyl ester = (+)-2α-Hydroxyyohimban-1α-carboxylic acid methyl ester = (16α,17α)-17-Hydroxyyohimban-16-carboxylic acid methyl ester (●)
S Corymbin, Corynine, *Iohimbina, Johimbin*, NMI-879, *Quebrachine, Yohimbine*, Yohimvetol
U Sympatholytic, aphrodisiac (Yohimbe alkaloid)

10493-01 (8480-01)   65-19-0
R Monohydrochloride
S Aphrodyne, Dayto Himbin, Gynimbine, Menolysin, Parkimbine, Pluriviron mono, Prowess Plain, Tosanpin, Yobine, Yobinol, Yocon, Yohimex, Yohydrol
U Sympatholytic, aphrodisiac

10493-02 (8480-02)
R Valerate
S Valymbin
U Sedative, analgesic

10494 (8482) $C_{21}H_{26}N_2O_4$
                                        80109-27-9

(−)-(*S*)-2-[4-(β-Hydroxy-3,4-dimethoxyphenethyl)-1-piperazinyl]-2,4,6-cycloheptatrien-1-one = (*S*)-2-[4-[2-(3,4-Dimethoxyphenyl)-2-hydroxyethyl]-1-piperazinyl]-2,4,6-cycloheptatrien-1-one (●)
R Monohydrochloride (83529-09-3)
S AY-27110, *Ciladopa hydrochloride***, Tremerase

U Dopaminergic (antiparkinsonian)

10495 $C_{21}H_{26}N_2O_6$
179463-81-1

(αS)-α-[[[[(2S)-2-Carboxy-2-hydroxyethyl](2-methyl-propyl)amino]carbonyl]amino]-2-naphthalenepropanoic acid (●)

S SA 6817

U Neutral endopeptidase, ACE and endothelin-converting enzyme inhibitor

---

10496 $C_{21}H_{26}N_2O_6$
168902-10-1

1,4-Dihydro-2,6-dimethyl-4-(3-nitrophenyl)-3,5-pyri-dinedicarboxylic acid methyl pentyl ester (●)

S MN-9202

U Coronary vasodilator (calcium antagonist)

10497 (8483) $C_{21}H_{26}N_2O_7$
66085-59-4

isopropyl 2-methoxyethyl 1,4-dihydro-2,6-dimethyl-4-*m*-nitrophenyl)-3,5-pyridinedicarboxylate = 1,4-Dihy-dro-2,6-dimethyl-4-(3-nitrophenyl)-3,5-pyridinedicarb-oxylic acid isopropyl 2-methoxyethyl ester = 1,4-Dihy-dro-2,6-dimethyl-4-(3-nitrophenyl)-3,5-pyridinedicarb-oxylic acid 2-methoxyethyl 1-methylethyl ester (●)

S Acival, Admon, Arfine, Bay e 9736, Befimat, Blo-quel, Brainal, Brainox, Calnit, Cebrofort, Cletonol, Curban "Rafarm", Eugerial, Figozant, Galmodipin, Genovox, Grifonimod, Kenesil, Macobal, Modina, Modus "Berenguer", Myodipine "Help", Naborel, Nelbinex, Nemotan, Nimodil, Nimodilat, *Nimodipi-ne***, Nimotide, Nimotop, Nimovas, Nivas, NMDP, Noodipina, Nortolan, Norton "Farmasa", Oxigen, Pe-riplum, Regental, Remontal, Rosital, Sobrepina "Far-moz", Stigmicarpin, Thrionipen, Trinalion, Tropocer "Leti", Vactripine, Vasoactin, Vasopspan N, Vaso-top "Cipla", Ziremex

U Cerebral vasodilator

10498 (8485) $C_{21}H_{26}N_2S_2$
50-52-2

10-[2-(1-Methyl-2-piperidyl)ethyl]-2-(methylthio)pheno-thiazine = 10-[2-(1-Methyl-2-piperidinyl)ethyl]-2-(me-thylthio)-10*H*-phenothiazine (●)

R also monohydrochloride (130-61-0) or tartrate (1257-76-7)

S Aldazine, Apo-Thioridazine, Calmaril, Detril, Elpe-ril, Flaracantyl, Mallorol, Malloryl, Meleril, Melir, Mellaril, Mellerets, Mellerette(n), Melleril, Melzine, Mepiozin, Milazine, Millazine, Nervosan, Novorida-zine, NSC-186060, Orsanil, Raomin, Ridazin(e), Ri-deril, Sonapax, Stalleril, Thiodazine, *Thioridazi-ne***, Thioril, Thiozine, Tinsenol, Tioriander, *Tiori-dazina*, Tioridil, TP 21, Trixifen

U Neuroleptic

10499   (8484) $C_{21}H_{26}N_2S_2$
70891-43-9

1-[2-(2-Naphthalenyl)-2,2-bis(propylthio)ethyl]-1*H*-imidazole (●)

R   Monohydrochloride (70891-44-0)
S   RS-84184
U   Secretagogue, anti-ulcer agent

10500   (8486) $C_{21}H_{26}N_4O_3$
14680-51-4

1-[2-(Diethylamino)ethyl]-2-(*p*-methoxybenzyl)-5-nitro-benzimidazole = *N,N*-Diethyl-2-[(4-methoxyphenyl)methyl]-5-nitro-1*H*-benzimidazole-1-ethanamine (●)
S   Ba-20227
U   Analgesic

10501   (8487) $C_{21}H_{26}N_4O_5S$
125228-82-2

(±)-4'-[2-Hydroxy-3-[[2-(*p*-1-imidazolylphen-oxy)ethyl]amino]propoxy]methanesulfonanilide = *N*-[4-

[2-Hydroxy-3-[[2-[4-(1*H*-imidazol-1-yl)phen-oxy]ethyl]amino]propoxy]phenyl]methanesulfonamide
(●)
S   CK 3579, HE 93
U   Anti-arrhythmic

10502 $C_{21}H_{26}N_4O_5S$
125279-79-0

4'-[(2*S*)-2-Hydroxy-3-[[2-(*p*-1-imidazolylphen-oxy)ethyl]amino]propoxy]methanesulfonanilide = *N*-[4-[(2*S*)-2-Hydroxy-3-[[2-[4-(1*H*-imidazol-1-yl)phen-oxy]ethyl]amino]propoxy]phenyl]methanesulfonamide
(●)
S   CK-4000, *Ersentilide***
U   Anti-arrhythmic

10503   (8488) $C_{21}H_{26}N_8O_6S_2$
105239-91-6

(+)-1-[[[(6*R*,7*R*)-7-[2-(5-Amino-1,2,4-thiadiazol-3-yl)gly-oxylamido]-2-carboxy-8-oxo-5-thia-1-azabicy-clo[4.2.0]oct-2-en-3-yl]methyl]-4-carbamoylquinuclidi-nium hydroxide inner salt 7^2-(*Z*)-(*O*-methyloxime) = (7*R*)-3-[[4-(Aminocarbonyl)-1-quinuclidinio]methyl]-7-[2-(5-amino-1,2,4-thiadiazol-3-yl)-(*Z*)-2-(methoxyimi-no)acetamido]-3-cephem-4-carboxylate = [6*R*-[6α,7β(*Z*)]]-4-(Aminocarbonyl)-1-[[7-[[(5-amino-1,2,4-thiadiazol-3-yl)(methoxyimino)acetyl]amino]-2-carb-oxy-8-oxo-5-thia-1-azabicyclo[4.2.0]oct-2-en-3-yl]me-thyl]-1-azoniabicyclo[2.2.2]octane hydroxide inner salt
(●)
S   *Cefaclidine, Cefclidin***, E-1040
U   Antibiotic

10504 $C_{21}H_{26}N_{10}O_4$
                                        210237-78-8

(2R,3R,4S,5R)-2-[6-Amino-2-(1-benzyl-2-hydroxyethyl-
amino)purin-9-yl]-5-(2-ethyl-2H-tetrazol-5-yl)tetrahy-
drofuran-3,4-diol = (2R,3R,4S,5R)-2-[6-Amino-2-[[(1S)-
1-(hydroxymethyl)-2-phenylethyl]amino]-9H-purin-9-
yl]-5-(2-ethyl-2H-tetrazol-5-yl)tetrahydro-3,4-furandiol
(●)
S   GW 328267C, *Miradenose*
U   Adenosine A$_{2A}$ receptor agonist, anti-inflammatory

10505 $C_{21}H_{26}O_2$
                                        197631-44-0

(5R)-3-[(1E,3E,5E,7E,9E)-2,10-Dimethyl-11-oxo-
1,3,5,7,9-dodecapentaenyl]-2,5-dimethyl-2-cyclopenten-
1-one (●)
S   *Falconensone A*
U   Antileukemic from *Emericella falconensis* or E. *fruti-
culosa*

10506   (8489) $C_{21}H_{26}O_2$
                                        521-35-7

6,6,9-Trimethyl-3-pentylbenzo[c]chromen-1-ol = 6,6,9-
Trimethyl-3-pentyl-6H-dibenzo[b,d]pyran-1-ol (●)
S   *Cannabinol***
U   Active principle from *Cannabis sativa*

10507   (8490) $C_{21}H_{26}O_2$
                                        60282-87-3

13-Ethyl-17-hydroxy-18,19-dinor-17α-pregna-4,15-
dien-20-yn-3-one = 17α-Ethynyl-17-hydroxy-18-methyl-
estra-4,15-dien-3-one = 17-Hydroxy-18-methyl-19-nor-
17α-pregna-4,15-dien-20-yn-3-one = (17α)-13-Ethyl-17-
hydroxy-18,19-dinorpregna-4,15-dien-20-yn-3-one (●)
S   Avaden, Convaden, Femodene, *Gestodene***, Gyno-
        vin, Minulet, Mirelle, Monogestin, SH 546,
        SHB 331, SHG 415 G
U   Progestin

10508   (8491) $C_{21}H_{26}O_2$
                                        850-52-2

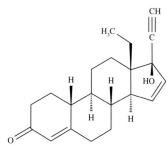

17α-Allyl-17-hydroxyestra-4,9,11-trien-3-one = 17β-Hy-droxy-19,21,24-trinorchola-4,9,11,22-tetraen-3-one = 17α-Allyl-17-hydroxy-19-norandrosta-4,9,11-trien-3-one = (17β)-17-Hydroxy-17-(2-propenyl)estra-4,9,11-tri-en-3-one (●)
S A 35957, 17α-*Allyltrenbolone*, *Altrenogest***,
DRC 6246, Regumate, RH 2267, RU 2267
U Anabolic, progestogen (veterinary)

10509 (8492) $C_{21}H_{26}O_2$
77016-85-4

10-Propargylestr-4-ene-3,17-dione = 10-(2-Propy-nyl)estr-4-ene-3,17-dione (●)
S MDL 18962, *Plomestane***
U Antineoplastic (aromatase inhibitor)

10510 (8493) $C_{21}H_{26}O_2$
72-33-3

3-Methoxy-19-nor-17α-pregna-1,3,5(10)-trien-20-yn-17-ol = 17α-Ethynyl-3-methoxyestra-1,3,5(10)-trien-17-ol = 17α-Ethynyl-3,17β-estradiol 3-methyl ether = (17α)-3-Methoxy-19-norpregna-1,3,5(10)-trien-20-yn-17-ol (●)
S 8027 C.B., Devocin, EE₃ME, L 33355, Menophase,
*Mestranol***, Ovastol, RS-1044, Tranel
U Estrogen

10511 (8498) $C_{21}H_{26}O_3$
80809-81-0

2-(12-Hydroxy-5,10-dodecadiynyl)-3,5,6-trimethyl-*p*-benzoquinone = 2-(12-Hydroxy-5,10-dodecadiynyl)-3,5,6-trimethyl-2,5-cyclohexadiene-1,4-dione (●)
S A-61589, AA-861, *Docebenone***, HCD
U Anti-asthmatic, anti-inflammatory (5-lipoxygenase inhibitor)

10512 (8494) $C_{21}H_{26}O_3$
1843-05-6

2-Hydroxy-4-(octyloxy)benzophenone = [2-Hydroxy-4-(octyloxy)phenyl]phenylmethanone (●)
S *Benzophenone-12*, Cyasorb UV 531, *Octabenzo-ne***, Spectra-Sorb UV 531
U Sunscreen agent

10513 (8499) $C_{21}H_{26}O_3$
88426-33-9

2-[(4-*tert*-Butylcyclohexyl)methyl]-3-hydroxy-1,4-naph-thoquinone = 2-[[4-(1,1-Dimethylethyl)cyclohexyl]me-thyl]-3-hydroxy-1,4-naphthalenedione (●)
R also *trans*-compd. (86790-15-0)
S *Buparvaquone***, Butalex, B.W. 720 C
U Antiprotozoal (veterinary), theilericide

10514 (8500) $C_{21}H_{26}O_3$
55079-83-9

3-Methoxy-15-apo-φ-caroten-15-oic acid = (all-E)-9-(4-Methoxy-2,3,6-trimethylphenyl)-3,7-dimethyl-2,4,6,8-nonatetraenoic acid (●)
R see also no. 11996
S *Acitretin**, Etretin*, Neotigason, Ro 10-1670, Soriatane, TMMP-RA
U Antipsoriatic, antineoplastic

10515 (8495) $C_{21}H_{26}O_3$
69427-46-9

(Z,E,E,E)-9-(4-Methoxy-2,3,6-trimethylphenyl)-3,7-dimethyl-2,4,6,8-nonatetraenoic acid (●)
S *Isoetretin*, Ro 13-7652
U Antipsoriatic

10516 (8496) $C_{21}H_{26}O_3$
61665-15-4

11α-Methoxy-19-nor-17α-pregna-1,3,5(10)-trien-20-yne-3,17-diol = 17α-Ethynyl-11α-methoxyestra-1,3,5-(10)-triene-3,17-diol = (11α,17α)-11-Methoxy-19-nor-pregna-1,3,5(10)-trien-20-yne-3,17-diol (●)
S *Mexestrol*, RU 16117
U Anti-estrogen

10517 (8497) $C_{21}H_{26}O_3$
34816-55-2

11β-Methoxy-19-nor-17α-pregna-1,3,5(10)-trien-20-yne-3,17-diol = 17α-Ethynyl-11β-methoxyestra-1,3,5(10)-triene-3,17-diol = (11β,17α)-11-Methoxy-19-norpregna-1,3,5(10)-trien-20-yne-3,17-diol (●)
S *Methoxyethinylestradiol, Moxestrol**,* NSC-118191, R 2858, RU 2858, Surestryl (new form), Urestryl
U Estrogen

10518 (8501) $C_{21}H_{26}O_4$
96609-16-4

(±)-p-[4-(p-tert-Butylphenyl)-2-hydroxybutoxy]benzoic acid = 4-[4-[4-(1,1-Dimethylethyl)phenyl]-2-hydroxybutoxy]benzoic acid (●)
S K 12148, *Lifibrol**,* U-83860
U Hypolipidemic

10519 (8502) $C_{21}H_{26}O_5$
53-03-2

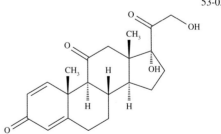

1,4-Pregnadiene-17α,21-diol-3,11,20-trione = 1,2-Dehydrocortisone = 17,21-Dihydroxypregna-1,4-diene-3,11,20-trione (●)

R   see also nos. 12837, 15123;
also 21-acetate (125-10-0)

S   Adasone, Alti-Prednisone, Alto-Pred, Ancortone,
Antison, Apo-Prednisone, Benison, Bicortone,
Bi-Delta, Buffacort, Chrocort, Clonisone, Co-Deltra,
Colisone, Corisone, Cortan, Cortancyl, Cortialer,
Corticor, Cortidelt, Cortilona, Cortinter, Cortiol,
Cortisid "Boehringer", Δ^1-*Cortisone*, Cutason, Dabroson (old form), Dacortin (Merck-Spain), De-
cortancyl, Decortin "E. Merck", DeCortisyl, Decorton "Salfa", *Dehydrocortisone*, *Deidrocortisone*,
Delco-Cortex, Delcort, Delcortin, Delta-Corlin, Del-
ta-Cortelan, Delta-cortene, *Delta-Cortisone*, Delta-
cortone, Delta-Dome, delta E, Delta-Prenovis, Del-
ta-Scheroson, Deltasone, Deltasson, Deltastendiolo,
Deltatrione, Deltison, Deltra, Di-Adreson, Disperso-
na, Ejizon, Encorton, Erftopred, Fernisone-Tabl., Fi-
asone, Hicorton, Hormozol, Hostacortin, Idrosone,
Inocortyl, Insone, Intensol "Philips-Roxane", Juva-
son, Keteocort, Keysone, Kolpisone, Leocortine-D,
Lepicortin, Liquid-Pred, Lisacort, Marnisonal, Mar-
sone, Marvidiene, Marvisona, Maso-Pred, Médiaso-
ne, Me-Korti, Meprison, *Metacortandracin*, Meta-
cortin, Metasone, Meticortem, Meticorten(e), Meti-
sone "Luso", Neoaltesona, Neo-Cortisona, Nisocort,
Nisonavibus, Nisone, Nizon, Novoprednisone,
NSC-10023, Nurison, Orasone, Ostepred, Panafcort,
Panasol, Pancortox, Pansone, Paracort, Parmenison,
Precortal, Precortex, Pred-5, Predeltin, Predicorten,
Predna, Predni-Artrit, Prednicen-M, Prednico, Prednicorm, Prednicort "Cont. Ph.", Prednidib, Pred-
nifor-Tabl., Prednilong, Prednilonga, Predniment,
Prednimut, Predniseguer, Prednisol "Smallwood",
Prednisone**, Predni-Tablinen, Prednital, Pred-
ni-Wolner, *Prednizon*, Prednovister, Predorgasona,
Predsol "Morgan", Predsone, Prelone "Muro", Presone "Langley", Pronison, Propred, Rectodelt, Retro-
cortin, Ropred, Servisone, Solisone, Sone, Stera-
pred, Supercortil, Supernisona, Supopred, Tarocor-
ten, Trolic, Ultracorten, Umecortil, Urtilone, Vita-
zon, Wescopred, Winpred, Wojtab, Xynisone, Zena-
drid, Zenidrid

U   Glucocorticoid (anti-inflammatory, anti-allergic)

10520 $C_{21}H_{26}O_6$
                                                          123014-06-2

4-[[5-(2,3-Dihydroxyphenyl)pentyl]oxy]-2-hydroxy-3-
propylbenzoic acid (●)
S   Ro 24-0553
U   Anti-inflammatory

10521   (8503) $C_{21}H_{27}BrO_3$
                                                          2527-11-9

9α-Bromo-11-oxoprogesterone = 9-Bromopregn-4-ene-
3,11,20-trione (●)
S   BOP, Braxarone, Brolon, *Bromoxoprogesterone*,
Broxoron
U   Formerly as palliative in breast cancer

10522   (8504) $C_{21}H_{27}ClN_2O_2$
                                                          68-88-2

2-[2-[4-(*p*-Chloro-α-phenylbenzyl)-1-piperazinyl]eth-
oxy]ethanol = 2-[2-[4-(*p*-Chlorodiphenylmethyl)-1-pipe-
razinyl]ethoxy]ethanol = 1-(*p*-Chlorobenzhydryl)-4-[2-
(2-hydroxyethoxy)ethyl]piperazine = 2-[2-[4-[(4-Chloro-
phenyl)phenylmethyl]-1-piperazinyl]ethoxy]ethanol (●)
R   see also nos. 14491, 15083;
most dihydrochloride (2192-20-3)

S Adzine, AH 3 N, Alamon, Alperon, Amidoctan, Anxanil, Apacil "Halgam", Apo-Hydroxyzine, Arcanax, Atara, Atarax, Ataraxona, Atazina, Aterax, Atozine, Bestalin, Cedar, Clorixin, Coraphene "Kevel-Portugal", Deinait, Disron, Durrax, Elroquil N, E-Vista, Forticalman, Hiderax, Hidroxina, *Hidroxizin*, Histan "Siam", Hizin, Hydroxacen, *Hydroxizinum*, *Hydroxysinum*, *Hydroxyzine hydrochloride***, Hyzine, *Idrossizina*, *Idroxizina*, Iremofar, Iremoxin, Iterax, Judolor "Velka", Marax "Pfizer", Multipax, Navicalm "UCB-Netherlands", Neocalma, Neucalm, Neurolax "Pharmachim", Neurozina, Nevrolaks, Novohydroxyzin, NP 212, Orgatrax, Otarex, PAS-Depress, Placidol, PMS-Hydroxyzine, Prurancit, Prurizin, Quiess, QYS, Rezine, Tran-Q, Tyndal, U.C.B. 4492, Ucerax, Ultramax, Validol "Sano-Hakay", Vison, Vistacon, Vistaject, Vistaquel, Vistarex, Vistaril i.m., Vistazine

U Tranquilizer, antihistaminic

10522-01 (8504-01) 10246-75-0
R Embonate (pamoate) (1:1)
S Atarax P, Bobsule, Disron P, Domarax "Pfizer", Equipose, *Hydroxyzine embonoate***, Hy-Pam, Masmoran, Pamozine, Paxistil, Primedoron, Tranquijust, Tranquizine, Vamate, Vistaril, Warazix
U Tranquilizer, antihistaminic

10523 (8505) $C_{21}H_{27}ClN_2O_5S$
134162-66-6

3-[[(4-Chlorophenyl)sulfonyl]amino]-4-[2-(3-pyridinyl-oxy)ethyl]octanoic acid (●)
S CGS 23305
U Anti-allergic, cardiovascular agent, thromboxane receptor antagonist

10524 (8506) $C_{21}H_{27}ClO_3$
54063-31-9

6α-Chloro-17-hydroxypregna-1,4-diene-3,20-dione = (6α)-6-Chloro-17-hydroxypregna-1,4-diene-3,20-dione (●)
R see also no. 11918
S *Cismadinone***
U Progestin

10525 (8507) $C_{21}H_{27}ClO_3$
1961-77-9

6-Chloro-17-hydroxypregna-4,6-diene-3,20-dione (●)
R see also no. 11919
S *Chlormadinone***
U Progestin

10526 (8508) $C_{21}H_{27}ClO_5$
129260-79-3

Chloromethyl 11β,17-dihydroxy-3-oxoandrosta-1,4-diene-17β-carboxylate = (11β,17α)-11,17-Dihydroxy-3-

oxoandrosta-1,4-diene-17-carboxylic acid chloromethyl ester (●)
R see also no. 12449
S *Loteprednol***
U Glucocorticoid (topical anti-inflammatory)

10527 (8509) $C_{21}H_{27}FN_2O_2$
442-03-5

(±)-1-(*p*-Fluorophenyl)-4-[4-(*o*-methoxyphenyl)-1-piperazinyl]-1-butanol = (±)-α-(4-Fluorophenyl)-4-(2-methoxyphenyl)-1-piperazinebutanol (●)
S *Anisopirol***, Haloisol, R 2159
U Neuroleptic

10528 (8510) $C_{21}H_{27}FO_5$
53-34-9

6α-Fluoro-11β,17,21-trihydroxypregna-1,4-diene-3,20-dione = 6α-Fluoroprednisolone = (6α,11β)-6-Fluoro-11,17,21-trihydroxypregna-1,4-diene-3,20-dione (●)
R see also nos. 11924, 12845, 13236
S Alphadrol, B 673, Celtolan, Etadrol, F.I. 6150, Flucort "Ikapharm", *6-Fluorprednisolone, Fluprednisolone***, Isopredon, NSC-47439, Selectren, U-7800, Vladicort
U Glucocorticoid (anti-inflammatory, anti-allergic)

10529 (8511) $C_{21}H_{27}FO_5$
595-52-8

9-Fluoro-11β,16α,17-trihydroxypregna-1,4-diene-3,20-dione = (11β,16α)-9-Fluoro-11,16,17-trihydroxypregna-1,4-diene-3,20-dione (●)
S *Descinolone***
U Glucocorticoid (anti-inflammatory, anti-allergic)

10530 (8512) $C_{21}H_{27}FO_5$
338-95-4

9-Fluoro-11β,17,21-trihydroxypregna-1,4-diene-3,20-dione = 9-Fluoroprednisolone = (11β)-9-Fluoro-11,17,21-trihydroxypregna-1,4-diene-3,20-dione (●)
R see also no. 11925
S Abicorten, *Deltafludrocortisone, 9-Fluorprednisolone, Isoflupredone***
U Glucocorticoid (anti-inflammatory, anti-allergic)

10531 (8513) $C_{21}H_{27}FO_6$
124-94-?

9-Fluoro-11β,16α,17,21-tetrahydroxypregna-1,4-diene-3,20-dione = 9-Fluoro-16α-hydroxyprednisolone = (11β,16α)-9-Fluoro-11,16,17,21-tetrahydroxypregna-1,4-diene-3,20-dione (●)

R see also nos. 12384, 12461, 12495, 12844, 13235, 13864, 13865, 14082, 14367, 14735, 14946, 15326, 15865

S Adcortyl-Tabl., Albacort, Alotone, Aristocort, Aristo-Pak, Atolone, Berlicort, Bucalsone, Celeste, Cinolone, CL 19823, Cortinovus, Delfacort, Delphicort, Delsolone, Derma-S, Ditrizin, Elacort, Extracort (4 + 8 mg), Flamicort, Flogicort, Fluosterolone, *Fluoxiprednisolonum*, *Fluoxyprednisolone*, Glicortene, Ipercortis, Kenacort, Kenalog-E, Ledercort, Medicort "Medici", Omcilon, Omnicort-V, Orion "Cyanamid", Oticortrix, Polcortolon, Reza-Pak, Sadocort, Sedozolona, Senciderm, Sterocort "Taro", Supercort, Tal 8, T-Orapaste, Triacort (in Russia), Triacortyl, Trialona, Triam 4, Triamcet, *Triamcinolone***, Triamciterap, Triamcort "Farmila; Helvepharm", Triam-Denk, Triaminoral, Triam-oral 4, Triamsicort, *Triamsinolone*, Triam-Tablinen, Tricilone Tablet, Tricortale, Tricort-Tabl., Trigon, Trilon, Trisucort, Volon, Voncort-Compr.

U Glucocorticoid (anti-inflammatory, anti-allergic)

10532 (8514) $C_{21}H_{27}FeN_3O_{18}P_3$
138708-31-3

Tris[(4,5-dihydroxy-6-methyl-3-pyridinemethanol 3-phosphato)(3–)-O^3,O^3,O^5]ferrate(6–) = Tris[4-hydroxy-2-methyl-5-[(phosphonooxy)methyl]pyridine-3-olato]iron(3+) = Nonahydrogen tris[mono[(4,5-dihydroxy-6-methyl-3-pyridinyl)methyl]phosphato-(4–)]ferrate(9–) (●)

R Hexasodium salt (138708-32-4)
S *Ferpifosate sodium***
U Magnetic resonance contrast agent

10533 (8516) $C_{21}H_{27}N$
3426-08-2

1-(3,3-Diphenylpropyl)cyclohexamethylenimine = 1-(3,3-Diphenylpropyl)perhydroazepine = 1,1-Diphenyl-3-hexamethyleniminopropane = 1-(3,3-Diphenylpropyl)hexahydro-1H-azepine (●)

S *Hexadiphane*, *Prozapine***, R 714
U Choleretic, spasmolytic

10533-01 (8516-01)   13657-24-4
R Hydrochloride [also mixture with sorbitol (8057-40-7)]
S Norbiline
U Choleretic, spasmolytic

10534 (8515) $C_{21}H_{27}N$
14334-40-8

N-Isopropyl-4,4-diphenylcyclohexylamine = *N*-(1-Methylethyl)-4,4-diphenylcyclohexanamine (●)

R also hydrochloride (14334-41-9)
S Cintaverin, EMD 9806, HSp 2986, Monoverin, *Pramiverine***, *Propaminodiphene*, Raptalgin, Sintaverina, Sistalcin, Sistalgin, Syntaverin, Uralgim
U Spasmolytic

10535 (8517) $C_{21}H_{27}N$
35941-65-2

(±)-10,11-Dihydro-*N*,*N*,β-trimethyl-5*H*-dibenzo[*a*,*d*]cy-
cloheptene-5-propylamine = (±)-3-(10,11-Dihydro-5*H*-
dibenzo[*a*,*d*]cyclohepten-5-yl)-2-methylpropyldimethyl-
amine = DL-*N*,*N*,2-Trimethyl-3-(dibenzo[*a*,*d*]-1,4-cyclo-
heptadien-5-yl)propylamine = (±)-10,11-Dihydro-*N*,*N*,β-
trimethyl-5*H*-dibenzo[*a*,*d*]cycloheptene-5-propanamine
(●)
R Hydrochloride (5585-73-9)
S AY-62014, *Butriptyline hydrochloride***, Centroly-
se, Evadene, Evadyne, Evasidol
U Antidepressant

10536 (8518) $C_{21}H_{27}N$
21489-22-5

3,3-Dimethyl-1-[3-(methylamino)propyl]-1-phenylindan
= 2,3-Dihydro-*N*,3,3-trimethyl-1-phenyl-1*H*-indene-1-
propanamine (●)
S *Prindamine*
U Antidepressant, gastric antisecretory

10537 (8519) $C_{21}H_{27}N$
57982-78-2

1-*tert*-Butyl-4,4-diphenylpiperidine = 1-(1,1-Dimethyl-
ethyl)-4,4-diphenylpiperidine (●)
R also hydrochloride (63661-61-0)
S *Budipine***, BY 701, Parkinsan "Byk; Promonta"
U Antiparkinsonian, antidepressant

10538 $C_{21}H_{27}N$
136534-70-8

4-Phenyl-1-(4-phenylbutyl)piperidine (●)
S PPBP
U Neuroprotective (σ receptor ligand)

10539 (8520) $C_{21}H_{27}NO$
76-99-3

6-(Dimethylamino)-4,4-diphenyl-3-heptanone (●)
R most hydrochloride (1095-90-5)
S A 4624, Adanon, Adolan "Abic", Algidon, Algiton,
Algolysin, Algovetan "Warthausen", Algoxale, Alt-

hose, *Amidone*, Amidosan, Amilone, AN 148, Bétha-
done, Butalgin "Gea", B.W. 47-173, Cloro-Nona,
Deamin, Depridol, Deptadol, Diadone, Diaminon,
Dianone, Disefonin, Disipan, Disket, Dolafin, Dola-
mid, Dolamina, Dolesona, Dolmed, Dolofin, Dolo-
heptan, Dolophin(e), Dolorex, Dorexol, Eptadone,
Fenadone, Fiseptona, Fysepton, Heptadol, Hepta-
don, "Heptanal", "Heptanon", H.E.S., Hoechst
10820, Ketalgin "Amino", Mecodin, Mekodin, Me-
pecton, Mephenon, *Metadone*, Metasedin, Metatone
"Parke-Davis", Methaddict, *Methadone**, Methado-
se, Methaforte, Methajade, Methex, Methidon, Me-
thodex, Metylan, Miadon(e), Midadone, Moheptan,
Optalgin, Panalgen, Petalgin, *Phenadon(um)*, Phy-
septon(e), Polamidon, Polamivet, Porfolan, Quotidi-
ne, Quotidon (active substance), Sedamidone, Sedo
"Rapide", Sedo-Rapide, Septa-om, Sin-Algin, Sintal-
gon, Symoron, Synthanal, Syrco, Turanone, Tussal,
Veronyl, Westadone, Zefalgin
U Narcotic analgesic

10540 (8521)   C$_{21}$H$_{27}$NO
                                                    125-58-6

(*R*)-6-(Dimethylamino)-4,4-diphenyl-3-heptanone (●)
R Hydrochloride (5967-73-7)
S *Levomethadone hydrochloride**, Levothyl, L-Pol-
amidon
U Narcotic analgesic

10540-01 (8521-01)   20233-35-6
R Tartrate (1:1)
S Levadone, Win 1766
U Narcotic analgesic

10541 (8522)   C$_{21}$H$_{27}$NO
                                                    466-40-0

(±)-6-(Dimethylamino)-5-methyl-4,4-diphenyl-3-hexa-
none (●)
R also L-form (561-10-4)
S B.W. 47-442, 442-C-47, Isoadanon, *Isoamidone, Iso-
metadone, Isomethadone**, Isopolamidon, Liden,
Win 1783
U Narcotic analgesic

10542 (8523)   C$_{21}$H$_{27}$NO
                                                    87857-27-0

5,6,7,8-Tetrahydro-6-(phenethylpropylamino)-1-naph-
thol = 5,6,7,8-Tetrahydro-6-[(2-phenylethyl)propylami-
no]-1-naphthalenol (●)
S N-0434, PPHT
U Dopamine D$_2$-agonist (ocular antihypertensive)

10543 (8524)   C$_{21}$H$_{27}$NO
                                                    116376-62-6

(E)-2,6-Di-*tert*-butyl-4-[2-(3-pyridyl)vinyl]phenol = (E)-2,6-Bis(1,1-dimethylethyl)-4-[2-(3-pyridinyl)ethenyl]phenol (●)
S   BI-L-93 BS
U   Anti-inflammatory

10544   (8527) $C_{21}H_{27}NO$
                                              2156-27-6

1-[2-(2-Benzylphenoxy)-1-methylethyl]piperidine = 1-(*o*-Benzylphenoxy)-2-piperidinopropane = 1-[1-Methyl-2-[2-(phenylmethyl)phenoxy]ethyl]piperidine (●)
R   also phosphate (1:1) (19428-14-9) or embonate (pamoate) (2:1) (64238-92-2)
S   ASA 158/5, *Benproperine***, Blascorid, Cofrel, Flaveric, Papanin, Pectipront, Pirexil, Pirexyl, Pyrexil, Tusafel, Tussafug
U   Antitussive

10545   (8525) $C_{21}H_{27}NO$
                                              972-02-1

1,1-Diphenyl-4-piperidino-1-butanol = α,α-Diphenyl-1-piperidinebutanol (●)
R   also hydrochloride (3254-89-5) or embonate (pamoate) (2:1) (26363-46-2)
S   Ansmin, Antiul, Avomol, Cefadol, Celmidol, Cephadol, Cerachidol, Cerrosa, Deanosarl, Dedidol, Degidole, Difanil, *Difenidol***, Difenidolin, *Diphenidol*, Gipsydol, Intease, Isedoll, Isodalol, Jefron "Kyorin", Keidole, Lyric, Maniol, Mecalmin, Meniedlin, Mera-

nom, Midnighton, Nescodol, Nometic, Normavom, Pineroro, Promodor, Remean, Satanolon, Shulander, SKF 478, Sobulaline, Sofalead, Solnomin, Tarophadole, Tatimil, Tenesdol, Tosperal, Verterge, Vontril, Vontrol, Wansar, Yesdol, Yophadol
U   Anti-emetic, antihistaminic

10546   (8526) $C_{21}H_{27}NO$
                                              3735-08-8

1,1-Diphenyl-3-piperidino-1-butanol = γ-Methyl-α,α-diphenyl-1-piperidinepropanol (●)
R   Hydrochloride (21100-36-7)
S   Aspaminol
U   Spasmolytic

10547   (8528) $C_{21}H_{27}NOS$
                                              63996-84-9

(±)-*erythro*-2,3-Dihydro-α-[1-[(4-phenylbutyl)amino]ethyl]benzo[*b*]thiophene-5-methanol = 1-(2,3-Dihydro-5-benzo[*b*]thienyl)-2-[(4-phenylbutyl)amino]-1-propanol = (R*,S*)-(±)-2,3-Dihydro-α-[1-[(4-phenylbutyl)amino]ethyl]benzo[*b*]thiophene-5-methanol (●)
S   CP 804 S, *Tibalosin***
U   Anxiolytic, antihypertensive

10548   (8531) $C_{21}H_{27}NO_2$
                                              3563-01-2

2-(Diethylamino)ethyl 2,2-diphenylpropionate = α,α-Diphenylpropionic acid β-(diethylamino)ethyl ester = α-Methyl-α-phenylbenzeneacetic acid 2-(diethylamino)ethyl ester (●)
R   also hydrochloride (2589-00-6)
S   *Aprofene***, *Aprophene*, Diphen (CIS prep.)
U   Spasmolytic

10549   (8530) $C_{21}H_{27}NO_2$
                                       64-94-8

2-(Diethylamino)-1-methylethyl diphenylacetate = Diphenylacetic acid 2-(diethylamino)-1-methylethyl ester = α-Phenylbenzeneacetic acid 2-(diethylamino)-1-methylethyl ester (●)
R   Hydrochloride (3213-44-3)
S   IEM-265, Methyldiphacil
U   Anticholinergic

10550   (8529) $C_{21}H_{27}NO_2$
                                       3578-28-7

3-(Diethylamino)propyl diphenylacetate = Diphenylacetic acid 3-(diethylamino)propyl ester = α-Phenylbenzeneacetic acid (3-diethylamino)propyl ester (●)
R   Hydrochloride (3098-65-5)
S   Arpenal
U   Anticholinergic, anti-asthmatic

10551 $C_{21}H_{27}NO_2$
                                       13062-02-7

2,2-Diphenylvaleric acid 2-(ethylamino)ethyl ester = α-Phenyl-α-propylbenzeneacetic acid 2-(ethylamino)ethyl ester (●)
R   Hydrochloride (5938-35-2)
S   SKF 8742A
U   Drug potentiator (inhibitor of drug metabolism)

10552   (8532) $C_{21}H_{27}NO_2$
                                       528-52-9

2-(Diethylamino)ethyl 2,3-diphenylpropionate = α-Benzylphenylacetic acid β-(diethylamino)ethyl ester = α-Phenylbenzenepropanoic acid 2-(diethylamino)ethyl ester (●)
R   Hydrochloride (5807-85-2)
S   Spasmadryl
U   Spasmolytic

10553   (8537) $C_{21}H_{27}NO_2$
                                       147241-81-4

2,2-Dimethyl-7-[2-(dimethylamino)ethoxy]-4-phenylchroman = 2-[(3,4-Dihydro-2,2-dimethyl-4-phenyl-2*H*-1-benzopyran-7-yl)oxy]-*N*,*N*-dimethylethanamine (●)
R   Hydrochloride (59257-33-9)

S BRL 14831
U Antidepressant

10554

$C_{21}H_{27}NO_2$
56649-76-4

(2R,6R,11R)-6,11-Diethyl-3-(3-furanylmethyl)-
1,2,3,4,5,6-hexahydro-2,6-methano-3-benzazocin-8-ol
(●)
S MR 2266
U κ-Opioid antagonist

10555 (8533)

$C_{21}H_{27}NO_2$
104822-02-8

3,5-Di-*tert*-butyl-4-hydroxybenzophenone oxime = [3,5-
Bis(1,1-dimethylethyl)-4-hydroxyphenyl]phenylmetha-
none oxime (●)
S TZI 41078
U Anti-inflammatory

10556 (8534)

$C_{21}H_{27}NO_2$
23210-56-2

4-Benzyl-α-(*p*-hydroxyphenyl)-β-methyl-1-piperidine-
ethanol = 2-(4-Benzylpiperidino)-1-(4-hydroxyphenyl)-
1-propanol = α-[1-(4-Benzylpiperidino)ethyl]-*p*-hydr-
oxybenzyl alcohol = α-(4-Hydroxyphenyl)-β-methyl-4-
(phenylmethyl)-1-piperidineethanol (●)

R also with (R*,S*)-(±)-definition (66414-06-0), tartrate
(1:1) (66414-07-1)
S Angiotrofin "Montpellier", Aponol (new form), Ce-
ratolebon, Cerocral, Ceruado, Dilvax, Enceron, Eru-
rose, Fenprol, Furezanil, FX-505, Iburonol, Ifenbel,
*Ifenprodil tartrate***, Iprodil, Kasias, Linbulane, Me-
cretin, Meloups, RC 61-91, Resenol, Serimic, Sero-
cral, Techlarm, Technis, Tophene, Vadilex, Vasculo-
dil, Youagil
U Vasodilator

10557

$C_{21}H_{27}NO_2$

(2S)-1-(2-Ethylphenoxy)-3-[[(1S)-1,2,3,4-tetrahydro-1-
naphthalenyl]amino]-2-propanol (●)
R Oxalate (1:1) (174689-39-5)
S SR 59230A
U β3-Adrenergic receptor antagonist

10558 (8535)

$C_{21}H_{27}NO_2$
90-54-0

2'-[2-(Diethylamino)ethoxy]-3-phenylpropiophenone =
2-[2-(Diethylamino)ethoxy]phenylphenethyl ketone = 1-
[2-[2-(Diethylamino)ethoxy]phenyl]-3-phenyl-1-propa-
none (●)
R Hydrochloride (2192-21-4)
S Asamedel, Baxacor, Cardilicor, Chneef, Corodilan,
Corofenon, Coronabason, Dialicor, Esanthin-S, Eta-
fenarin, *Etafenone hydrochloride***, Etamfen, Exfe-
none, Folicrophen, Helzcor, Hypochit, Infacardin,
KCA, Korofenon, LG-11457, Pagano-Cor, Perucor

Relicor "Azevados", Rolin, SA-1, Sofaricol, Youachol
U   Coronary vasodilator

10559   (8536) $C_{21}H_{27}NO_2$
3686-78-0

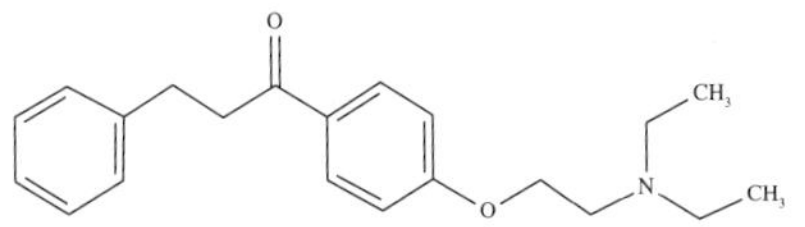

4'-[2-(Diethylamino)ethoxy]-3-phenylpropiophenone = 4-[2-(Diethylamino)ethoxy]phenyl phenethyl ketone = 1-[4-[2-(Diethylamino)ethoxy]phenyl]-3-phenyl-1-propanone (●)
R   Citrate (1:1) (58919-63-4)
S   *Dietifen citrate***, Epilin
U   Epilatory, antifungal

10560 $C_{21}H_{27}NO_2$
189372-48-3

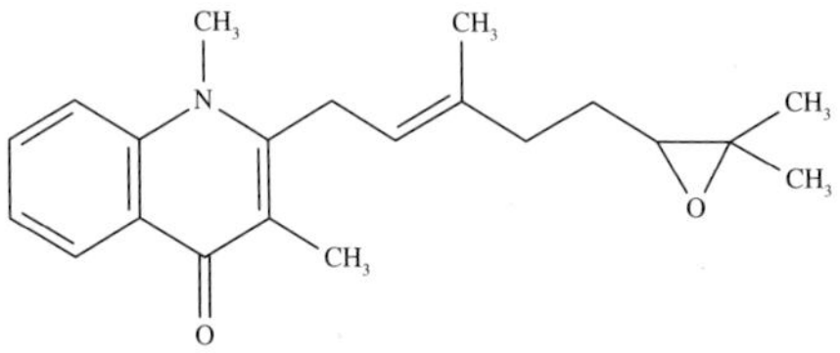

E)-2-(6,7-Epoxy-3,7-dimethyloct-2-enyl)-1,3-dimethyl-4(1*H*)-quinolone = 2-[(2*E*)-5-(3,3-Dimethyloxiranyl)-3-methyl-2-pentenyl]-1,3-dimethyl-4(1*H*)-quinolinone (●)
S   CJ 13564
U   Antibacterial

10561 $C_{21}H_{27}NO_2S$
152802-07-8

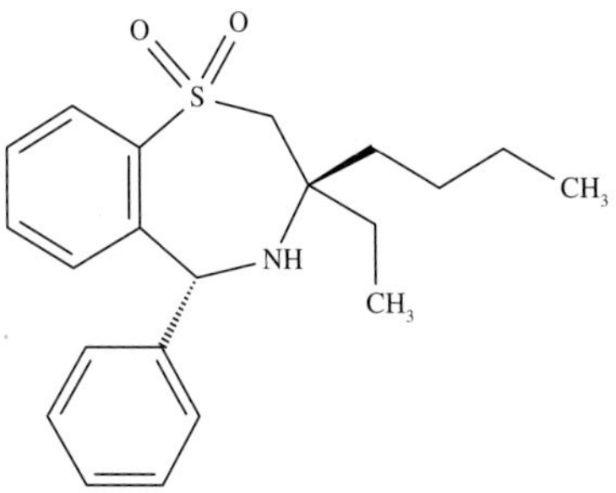

3*R*,5*R*)-3-Butyl-3-ethyl-2,3,4,5-tetrahydro-5-phenyl-,4-benzothiazepine 1,1-dioxide (●)

S   2164U90
U   Hypocholesterolemic

10562   (8538) $C_{21}H_{27}NO_3$
57-36-3

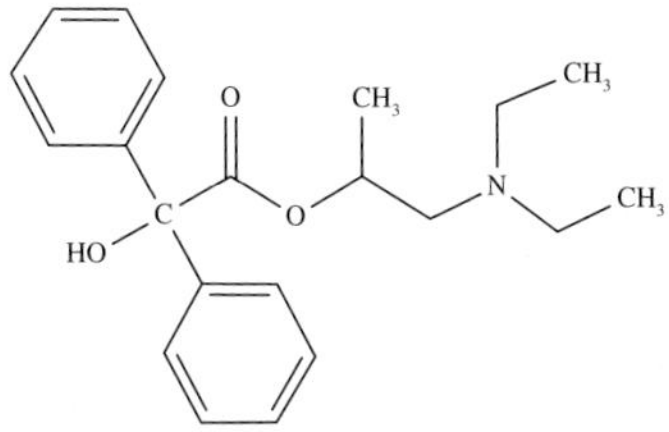

Benzilic acid 2-(diethylamino)-1-methylethyl ester = α-Hydroxy-α-phenylbenzeneacetic acid 2-(diethylamino)-1-methylethyl ester (●)
R   Hydrochloride (10503-18-1)
S   IEM-275, Metamisyl, Methamicil, *Methylbenactyzine hydrochloride*, Methyldiazil
U   Anticholinergic

10563   (8539) $C_{21}H_{27}NO_3$
22487-42-9

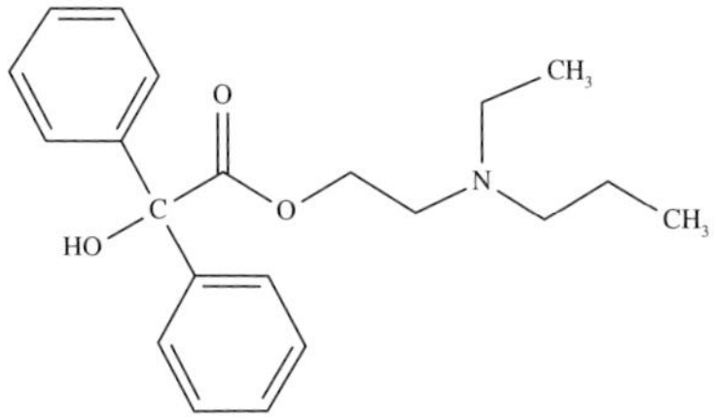

2-(Ethylpropylamino)ethyl benzilate = Benzilic acid 2-(ethylpropylamino)ethyl ester = α-Hydroxy-α-phenylbenzeneacetic acid 2-(ethylpropylamino)ethyl ester (●)
R   Hydrochloride (3202-55-9)
S   AP 1288, *Benaprizine hydrochloride***, *Benapryzine hydrochloride*, *Beneprizine hydrochloride*, Brizin, BRL 1288
U   Anticholinergic, antiparkinsonian

10564 (8540) $C_{21}H_{27}NO_3$
                                       71205-89-5

3-Hydroxy-3,3-diphenylpropionic acid 2-(diethylami-
no)ethyl ester = β-Hydroxy-β-phenylbenzenepropanoic
acid 2-(diethylamino)ethyl ester (●)
R Hydrochloride (53421-38-8)
S *Chinsedal*, Chin Z-F
U Analgesic

10565 $C_{21}H_{27}NO_3$
                                       193359-26-1

4-Hydroxy-1-[2-(4-hydroxyphenoxy)ethyl]-4-(4-methyl-
benzyl)piperidine = 1-[2-(4-Hydroxyphenoxy)ethyl]-4-
[(4-methylphenyl)methyl]-4-piperidinol (●)
S Co 101244, PD 174494
U NR1/2B NMDA receptor antagonist

10566 (8541) $C_{21}H_{27}NO_3$
                                       54063-53-5

2'-[2-Hydroxy-3-(propylamino)propoxy]-3-phenylpro-
piophenone = 1-[2-[2-Hydroxy-3-(propylamino)prop-
oxy]phenyl]-3-phenyl-1-propanone (●)
R Hydrochloride (34183-22-7)
S Arythmol, Asonacor, Baxarytmon, Corecel, Cu-
   xafenon, *Fenopraine hydrochloride*, Jutanorm, Me-

tronom, Nistaken, Norfenon, Normoritmin, Nor-
morytmin, Polfenon, Profenan, Prolekofen, Pronon,
Propafen, Propafen-BASF, *Propafenone hydrochloride***, Propamerck, Propanorm, Propa-Oramon,
Propa-Sanorania, Propastad, Prorynorm, Pulonon,
Rhythmocor, Ritmocor "Recalcine", Ritmonorm, Ry-
thmol "Biosedra; Knoll", Rythmonorm, Rytmo-
genat, Rytmonorm(a), Rytmo-Puren, SA 79, Tachy-
fenon, WZ-884, WZ 884-642, YM-13400
U Coronary vasodilator, anti-arrhythmic

10567 (8542) $C_{21}H_{27}NO_3S_2$
                                       42024-98-6

6,6,9-Trimethyl-9-azabicyclo[3.3.1]non-3β-yl di-2-thie-
nylglycolate = *exo*-α-Hydroxy-α-2-thienyl-2-thiophene-
acetic acid 6,9,9-trimethyl-9-azabicyclo[3.3.1]non-3-yl
ester (●)
R Hydrochloride (32891-29-5)
S KAO-264, *Mazaticol hydrochloride***, Pentona,
   PG-501
U Antiparkinsonian

10568 (8543) $C_{21}H_{27}NO_4$
                                       47467-79-8

O-(β-Methoxyethyl)benzilic acid β-(dimethylamino)ethyl ester = α-(2-Methoxyethoxy)-α-phenylbenzene-
acetic acid (2-dimethylamino)ethyl ester (●)
S Despasmin
U Anticholinergic, spasmolytic

10569   (8544) $C_{21}H_{27}NO_4$
                                                                    73694-50-5

2-(*p*-Isobutylphenyl)propionic acid [5-hydroxy-4-(hydroxymethyl)-6-methyl-3-pyridyl]methyl ester = [5-Hy-
droxy-4-(hydroxymethyl)-6-methyl-3-pyridyl]methyl *p*-
isobutylhydratropate = α-Methyl-4-(2-methylpropyl)ben-
zeneacetic acid [5-hydroxy-4-(hydroxymethyl)-6-methyl-
3-pyridinyl]methyl ester (●)
S   *Ibuprofen pyridoxine ester*
U   Anti-inflammatory, analgesic, antipyretic

10570   (8547) $C_{21}H_{27}NO_4$
                                                                    549-28-0

5,7,8,9-Tetrahydro-2,3,10,12-tetramethoxy-7-methyl-
5H-dibenz[*d,f*]azonine (●)
S   *Protostephanine*
U   Antihypertensive

10571   (8545) $C_{21}H_{27}NO_4$
                                                                    20594-83-6

17-(Cyclobutylmethyl)-4,5α-epoxymorphinan-3,6α,14-
triol = (−)-(5*R*,6*S*,14*S*)-9a-(Cyclobutylmethyl)-4,5-ep-
oxymorphinan-3,6,14-triol = (5α,6α)-17-(Cyclobutylmethyl)-4,5-epoxymorphinan-3,6,14-triol (●)

R   Hydrochloride (23277-43-2)
S   Bufigen, EN-2234 A, Hodepine, *Nalbuphine hydro-
        chloride***, Nubain(a), Nubak, Ruphine
U   Analgesic, antagonist to narcotics

10572   (8546) $C_{21}H_{27}NO_4$
                                                                    86384-10-3

5'-Hydroxy-2'-[2-hydroxy-3-(propylamino)propoxy]-3-
Phenylpropiophenone = 5-Hydroxypropafenone = 1-[5-
Hydroxy-2-[2-hydroxy-3-(propylamino)propoxy]phe-
nyl]-3-phenyl-1-propanone (●)
S   5 HP, *Hydroxyfenone*
U   Anti-arrhythmic

10573 $C_{21}H_{27}NO_4$
                                                                    153950-29-9

1-Methyl-6-[[[3-(tetrahydro-4-methoxy-2-methyl-2*H*-
pyran-4-yl)-2-propenyl]oxy]methyl]-2(1*H*)-quinolinone
(●)
S   A-121798
U   Anti-allergic (5-lipoxygenase inhibitor)

10574   (8548) $C_{21}H_{27}N_3$
                                                                    75859-04-0

9-[3-(*cis*-3,5-Dimethyl-1-piperazinyl)propyl]carbazole = *cis*-9-[3-(3,5-Dimethyl-1-piperazinyl)propyl]-9*H*-carba-zole (●)
R Dihydrochloride (75859-03-9)
S B.W. 234 U, *Rimcazole hydrochloride***
U Antipsychotic

10575 (8549) $C_{21}H_{27}N_3O$
132197-26-3

2-(4-Methyl-1-piperazinyl)-4'-phenethylacetophenone (*Z*)-oxime = (*Z*)-2-(4-Methyl-1-piperazinyl)-1-[4-(2-phe-nylethyl)phenyl]ethanone oxime (●)
R Monohydrochloride (122378-96-5)
S MCI-727
U Anti-ulcer agent

10576 $C_{21}H_{27}N_3O_2$
142773-98-6

2-Amino-*N*-[(1*S*)-2-[(3*S*)-3-hydroxy-1-pyrrolidinyl]-1-phenylethyl]-*N*-methylbenzeneacetamide (●)
S EMD 60400
U Analgesic, anti-inflammatory (κ-opioid receptor ago-nist)

10577 (8550) $C_{21}H_{27}N_3O_2$
361-37-5

N-[α-(Hydroxymethyl)propyl]-1-methyl-D-lysergamide = 1-Methyl-D-lysergic acid (+)-butanolamide-(2) = [8β(*S*)]-9,10-Didehydro-*N*-[1-(hydroxymethyl)propyl]-1,6-dimethylergoline-8-carboxamide (●)
R also maleate (1:1) (129-49-7)
S Deseril, Deserila, Désernil, Deserril, Deseryl, *Dime-thylergometrine*, Methylmethergin, *Methysergide***, *Metisergide*, MLD 41, NSC-186061, Sansert, UML 491, Vultax
U Serotonin antagonist, migraine prophylactic

10578 (8551) $C_{21}H_{27}N_3O_2$
27258-23-7

2-(Diethylamino)ethylmalonanilide = 2-[2-(Diethylami-no)ethyl]-*N*,*N*'-diphenylpropanediamide (●)
R Monohydrochloride (52507-55-8)
S Detamid, Dethamid
U Anti-inflammatory

10579 (8552) $C_{21}H_{27}N_3O_2S$
16185-21-0

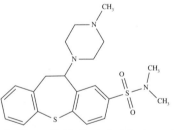

8-(Dimethylsulfamoyl)-10-(4-methyl-1-piperazinyl)-10,11-dihydrodibenzo[*b*,*f*]thiepin = 10,11-Dihydro-*N*,*N*-dimethyl-11-(4-methyl-1-piperazinyl)diben-zo[*b*,*f*]thiepin-2-sulfonamide (●)
R Monomethanesulfonate monohydrate (39841-98-0)
S *Sulfamothepin*, VÚFB-9056
U Neuroleptic

10580   (8553) $C_{21}H_{27}N_3O_3$
                                                                                        124436-59-5

Ethyl *p*-[2-[1-(6-methyl-3-pyridazinyl)-4-piperidyl]eth-
oxy]benzoate = 4-[2-[1-(6-Methyl-3-pyridazinyl)-4-pipe-
ridinyl]ethoxy]benzoic acid ethyl ester (●)
S   *Pirodavir***, R 77975
U   Antiviral (in common cold)

10581   (8554) $C_{21}H_{27}N_3O_3$
                                                                                        79781-00-3

N^1-(3-Acetyl-4,5-dihydro-2-furanyl)-N^4-(6-methoxy-8-
quinolinyl)-1,4-pentanediamine = 1-[4,5-Dihydro-2-[[4-
[(6-methoxy-8-quinolinyl)amino]pentyl]amino]-3-fura-
nyl]ethanone (●)
S   CDRI 80-53, Compound 80-53
U   Antimalarial

10582 $C_{21}H_{27}N_3O_3$
                                                                                        223661-25-4

Dihydro-3-[1-[[4-[(6-methoxy-8-quinolinyl)amino]pen-
tyl]amino]ethylidene]-2(3*H*)-furanone (●)
S   Ablaquin, *Bulaquine**
U   Antimalarial

10583   (8555) $C_{21}H_{27}N_3O_3$
                                                                                        34297-34-2

3-(Diethylamino)propiophenone *O*-[(*p*-methoxyphe-
nyl)carbamoyl]oxime = 3-(Diethylamino)-1-phenyl-1-
propanone *O*-[[(4-methoxyphenyl)amino]carbo-
nyl]oxime (●)
S   *Anidoxime***, *Bamoxine*, BRL 11870, E-142,
        USVP-E 142
U   Analgesic

10584   (8556) $C_{21}H_{27}N_3O_3$
                                                                                        76252-06-7

(±)-1,2,3,4-Tetrahydro-8-[2-hydroxy-3-(isopropylamino)propoxy]-1-nicotinoylquinoline = (±)-1,2, 3,4-Tetra-
hydro-8-[2-hydroxy-3-[(1-methylethyl)amino]propoxy]-
1-(3-pyridinylcarbonyl)quinoline (●)
S   CAS 924, *Nicainoprol***, RU 42924
U   Anti-arrhythmic

10585 $C_{21}H_{27}N_3O_3S$
                                                                                        170858-33-0

(−)-4-[4-[2-[(*S*)-1-Isochromanyl]ethyl]-1-piperazi-
nyl]benzenesulfonamide = 4-[4-[2-[(1*S*)-3,4-Dihydro-
1*H*-2-benzopyran-1-yl]ethyl]-1-piperazinyl]benzenesul-
fonamide (●)

R Monomethanesulfonate (170858-34-1)
S PNU-101387G, *Sonepiprazole mesilate***, U-101387
U Antipsychotic (dopamine D_4 antagonist)

10586 $C_{21}H_{27}N_3O_3S$

(R)-1-(4-Chromanylethyl)-4-[4-(aminosulfonyl)phe-
nyl]piperazine = (R)-4-[4-[2-(3,4-Dihydro-2H-1-benzo-
pyran-4-yl)ethyl]-1-piperazinyl]benzenesulfonamide (●)
S PNU-106675
U Antipsychotic (dopamine D_4 antagonist)

10587 (8557) $C_{21}H_{27}N_3O_3S$
113558-89-7

p-[[1-[2-(6-Methyl-2-pyridyl)ethyl]-4-piperidyl]carbo-
nyl]methanesulfonanilide = N-[4-[[1-[2-(6-Methyl-2-py-
ridinyl)ethyl]-4-piperidinyl]carbonyl]phenyl]methane-
sulfonamide (●)
R Dihydrochloride (113559-13-0)
S E-4031
U Anti-arrhythmic

10588 $C_{21}H_{27}N_3O_3S_2$
146537-07-7

1-[[2-(Diethylamino)ethyl]amino]-4-[[(methylsulfo-
nyl)amino]methyl]-9H-thioxanthen-9-one = N-[[1-[[2-
(Diethylamino)ethyl]amino]-9-oxo-9H-thioxanthen-4-
yl]methyl]methanesulfonamide (●)
S SR 233377, SW 33377, Win 33377
U Antineoplastic

10589 (8558) $C_{21}H_{27}N_3O_5$
35843-07-3

4-Methyl-7-(4-morpholinecarboxamido)-3-(2-morpholi-
noethyl)coumarin = N-[4-Methyl-3-[2-(4-morpholi-
nyl)ethyl]-2-oxo-2H-1-benzopyran-7-yl]-4-morpholine-
carboxamide (●)
R Monohydrochloride (35843-09-5)
S *Morocromen hydrochloride***, TVX 647
U Coronary vasodilator

10590 (8559) $C_{21}H_{27}N_3O_5S$
40966-79-8

Methoxymethyl (2S,5R,6R)-6-(2,2-dimethyl-5-oxo-4-
phenyl-1-imidazolidinyl)-3,3-dimethyl-7-oxo-4-thia-1-
azabicyclo[3.2.0]heptane-2-carboxylate = Methoxymethyl (6R)-6-(2,2-dimethyl-5-oxo-4-phenyl-1-imidazo-
lidinyl)penicillanate = [2S-(2α,5α,6β)]-6-(2,2-Dimethyl-
5-oxo-4-phenyl-1-imidazolidinyl)-3,3-dimethyl-7-oxo-4-
thia-1-azabicyclo[3.2.0]heptane-2-carboxylic acid methoxymethyl ester (●)
R also [2S-[2α,5α,6β(S*)]]-compd. (60252-40-6)
S BL-P 1761, *Hetacillin methoxymethyl ester*, *Sarpicil-
lin**
U Antibiotic

10591 (8561) $C_{21}H_{27}N_3O_6S$
67337-44-4

Methoxymethyl (2S,5R,6R)-6-[4-(p-hydroxyphenyl)-2,2-dimethyl-5-oxo-1-imidazolidinyl]-3,3-dimethyl-7-oxo-4-thia-1-azabicyclo[3.2.0]heptane-2-carboxylate = Methoxymethyl (6R)-6-[4-(4-hydroxyphenyl)-2,2-dimethyl-5-oxo-1-imidazolidinyl]penicillanate = [2S-(2α,5α,6β)]-6-[4-(4-Hydroxyphenyl)-2,2-dimethyl-5-oxo-1-imidazolidinyl]-3,3-dimethyl-7-oxo-4-thia-1-azabicyclo[3.2.0]heptane-2-carboxylic acid methoxymethyl ester (●)

S BL-P 1780, *Oxetacillin methoxymethyl ester*, *Sarmoxicillin***

U Antibiotic

10592 (8562) $C_{21}H_{27}N_3O_7S$
50972-17-3

(2S,5R,6R)-6-[(R)-(2-Amino-2-phenylacetamido)]-3,3-dimethyl-7-oxo-4-thia-1-azabicyclo[3.2.0]heptane-2-carboxylic acid ester with ethyl 1-hydroxyethyl carbonate = 1-(Ethoxycarbonyloxy)ethyl (6R)-6-(α-D-phenylglycylamino)penicillanate = [2S-[2α,5α,6β(S*)]]-6-[(Amino-phenylacetyl)amino]-3,3-dimethyl-7-oxo-4-thia-1-azabicyclo[3.2.0]heptane-2-carboxylic acid 1-[(ethoxycarbonyl)oxy]ethyl ester (●)

also monohydrochloride (37661-08-8)
Albaxin, Ambacamp, Ambaxin, Ambaxino, Amplibac, Bacacil, Bacamcillin, *Bacampicillin***, Bacampicin(e), Bacampil, Bacamsilin, Bacocil, Bakamsilin, Bamaxin, Benpel, Bestpolin, *Carampicillin*, Daxid "Pfizer-Chile", Doktacillin (new form), EPC 272, Flexipen "Armstrong-Argentina", Lekobacyn, Penbac, Pen-Bak, Penglobal, Penglobe, Spectrobid, Velbacil, Y-11838
Antibiotic

10592-01 (8562-01) 77442-47-8

R 6β-Bromopenicillanate (no. 1189) (1:1)
S VD 2081
U Antibiotic

10593 (8563) $C_{21}H_{27}N_5OS$
113457-05-9

5-[(2-Aminoethyl)amino]-2-[2-(diethylamino)ethyl]-2H-[1]benzothiopyrano[4,3,2-cd]indazol-8-ol (●)

R Trihydrochloride (119221-49-7)
S CI-958, *Ledoxantrone hydrochloride***, *Sedoxantrone trihydrochloride*
U Antineoplastic

10594 (8564) $C_{21}H_{27}N_5O_2S$
83903-06-4

2-[[2-[[[5-[(Dimethylamino)methyl]-2-furanyl]methyl]thio]ethyl]amino]-5-[(6-methyl-3-pyridinyl)methyl]-4(1H)-pyrimidinone (●)

R Trihydrochloride (72716-75-7)
S *Lupitidine hydrochloride***, SKF 93479
U Anti-ulcer agent (histamine H_2-receptor antagonist)

10595 (8565) $C_{21}H_{27}N_5O_2S$
85581-64-2

2-[4-[2-(2,4-Dioxo-1-thia-3-azaspiro[4.4]non-3-yl)butyl]-1-piperazinyl]-3-pyridinecarbonitrile (●)

R Monohydrochloride (85581-65-3)
S MJ 13980-1
U Antipsychotic

10596 (8566)

$C_{21}H_{27}N_5O_4S$
29094-61-9

1-Cyclohexyl-3-[[p-[2-(5-methylpyrazinecarboxami-
do)ethyl]phenyl]sulfonyl]urea = N-[2-[4-[[[(Cyclohexylamino)carbonyl]amino]sulfonyl]phenyl]ethyl]-5-methyl-
pyrazinecarboxamide (●)
S Antidiab, Apamid, CP-28720, FK 1320, Glibenese,
Glibinese, Glide, Glipid, *Glipizide**, Glucolip, Glu-
colite, Glucotrol, Glupitel, *Glydiazinamide*, Glynase
"U. S. Vitamin", K 4024, Melizide, Menodiab, Min-
diab, Minibetic, Minidiab, Minodiab, Ozidia, Sucra-
zine
U Oral hypoglycemic

10597 (8567)

$C_{21}H_{27}N_5O_7S$
63358-49-6

(2S,5R,6R)-6-[(2R)-2-[(2R)-2-Amino-3-(methylcarbamoyl)propionamido]-2-(p-hydroxyphenyl)acetamido]-
3,3-dimethyl-7-oxo-4-thia-1-azabicyclo[3.2.0]heptane-2-
carboxylic acid = [2S-[2α,5α,6β[S*(S*)]]]-6-[[[[2-Ami-
no-4-(methylamino)-1,4-dioxobutyl]amino](4-hydroxy-
phenyl)acetyl]amino]-3,3-dimethyl-7-oxo-4-thia-1-aza-
bicyclo[3.2.0]heptane-2-carboxylic acid = N-(N-Methyl-
D-asparaginyl)amoxicillin = [2S-(2α,5α,6β)]-N-Methyl-
D-asparaginyl-N-(2-carboxy-3,3-dimethyl-7-oxo-4-thia-
1-azabicyclo[3.2.0]hept-6-yl)-D-2-(4-hydroxyphe-
nyl)glycinamide (●)
S Apoxin, *Aspaxicillin*, ASPC, *Aspoxicillin**, Doyle,
TA-058
U Antibiotic

10598 (8568)

$C_{21}H_{27}N_5O_8S_2$
125181-57-9

1-Hydroxyethyl (+)-(6R,7R)-7-[2-(2-amino-4-thiazo-
lyl)glyoxylamido]-3-(methoxymethyl)-8-oxo-5-thia-1-
azabicyclo[4.2.0]oct-2-ene-2-carboxylate, 7^2-(Z)-oxime
pivalate (ester) = [6R-[6α,7β(Z)]]-7-[[(2-Amino-4-thia-
zolyl)(hydroxyimino)acetyl]amino]-3-(methoxymethyl)-
8-oxo-5-thia-1-azabicyclo[4.2.0]oct-2-ene-2-carboxylic
acid 1-(2,2-dimethyl-1-oxopropoxy)ethyl ester (●)
R Mono(4-methylbenzenesulfonate) (salt)
(135467-21-9)
S *Cefdaloxine pentexil tosilate**, HRT-916 K
U Antibiotic

10599 (8569)

$C_{21}H_{27}N_5O_9S$
87239-81-

(±)-1-Hydroxyethyl (+)-(6R,7R)-7-[2-(2-amino-4-thiazo-
lyl)glyoxylamido]-3-(methoxymethyl)-8-oxo-5-thia-1-
azabicyclo[4.2.0]oct-2-ene-2-carboxylate 7^2-(Z)-(O-methyloxime) isopropyl carbonate (ester) = 1-[[(1-Methyle-
thoxy)carbonyl]oxy]ethyl (7R)-7-[2-(2-amino-4-thiazo-
lyl)-2-(methoxyimino)acetamido]-3-(methoxymethyl)-3-
cephem-4-carboxylate = [6R-[6α,7β(Z)]]-7-[[(2-Amino-
4-thiazolyl)(methoxyimino)acetyl]amino]-3-(methoxymethyl)-8-oxo-5-thia-1-azabicyclo[4.2.0]oct-2-ene-2-
carboxylic acid 1-[[(1-methylethyl)carbonyl]oxy]ethy-
ester (●)
S Alevix, Banan, Biforan, Biocef, *Cefmeproxime*, Ce-
fodox, Cefpodoxim "Sankyo", *Cefpodoxime proxe-
til**, CPD, CPDX-PR, CS 807, Doxef, Kelbium "F

es", Obrelan, Ofraxid, Orelox, Orexol, Otreon, Po-
domexef, RU 51807, U-76252, Vantin
U Antibiotic

10600　(8570)

$C_{21}H_{27}N_7O_6$
52196-22-2

5-Methyltetrahydrohomofolic acid = N-[4-[[2-(2-Amino-
1,4,5,6,7,8-hexahydro-5-methyl-4-oxo-6-pteridi-
nyl)ethyl]amino]benzoyl]-L-glutamic acid (●)
R Disodium salt (52386-42-2)
S *Ketotrexate sodium***, Me-H4F, 5 MTHHF,
NSC-139490
U Antineoplastic

10601　(8571)

$C_{21}H_{27}N_7O_{14}P_2$
53-84-9

3-Carbamoyl-1-β-D-ribofuranosylpyridinium hydroxide
5'-ester with adenosine 5'-pyrophosphate inner salt = 1-(3-
Carbamoylpyridinio)-β-D-ribofuranoside 5-(adenosine-
5'-pyrophosphate) = Adenosine 5'-(trihydrogen diphosphate) 5'→5'-ester with 3-(aminocarbonyl)-1-β-D-ribo-
furanosylpyridinium hydroxide inner salt (●)
R see also no. 12067
Codehydrase I, Codehydrogenase I, Coenzyme I,
CO-I, *Cozymase, Diphosphopyridine nucleotide,*
DPN, Enzopride, NAD, *Nadide***, Nad-Medical,
641 ND, Nicodrasi, *Nicotinamide-adenine dinucleo-
tide*, NSC-20272
Apozymase coenzyme (antagonist to alcohol and narcotic analgesics)

10602　(8572)

$C_{21}H_{28}BrFO_2$
3538-57-6

17-Bromo-6α-fluoropregn-4-ene-3,20-dione = 17-Bro-
mo-6α-fluoroprogesterone = (6α)-17-Bromo-6-fluoro-
pregn-4-ene-3,20-dione (●)
S *Haloprogesterone***, Prohalone
U Progestin

10603　(8573)

$C_{21}H_{28}ClN_3O_4$
99599-78-7

6-Chloro-1'-[3-(4-hydroxypiperidino)propyl]-2,2-dime-
thylspiro[chroman-4,4'-imidazolidine]-2',5'-dione = 6-
Chloro-2,3-dihydro-1'-[3-(4-hydroxy-1-piperidinyl)pro-
pyl]-2,2-dimethylspiro[4H-1-benzopyran-4,4'-imidazoli-
dine]-2',5'-dione (●)
S E-0747
U Anti-arrhythmic

10604   (8574) $C_{21}H_{28}Cl_2N_2O_3$
71360-45-7

6β-[Bis(2-chloroethyl)amino]-17-methyl-4,5α-epoxy-
morphinan-3,14-diol = (5α,6β)-6-[Bis(2-chloro-
ethyl)amino]-17-methyl-4,5-epoxymorphinan-3,14-diol
(●)
S   *Chloroxymorphamine*, COA
U   Antagonist to narcotics, antineoplastic

10605 $C_{21}H_{28}FN_3O$
259525-01-4

4-(4-Fluorophenyl)-2-methyl-6-[[5-(1-piperidinyl)pen-
tyl]oxy]pyrimidine (●)
R   Monohydrochloride (178429-67-9)
S   NS-7
U   Neuroprotective

10606 $C_{21}H_{28}IN_3O$
194038-94-3

N-[1-[(4-Iodophenyl)methyl]-4-piperidinyl]-N-methyl-3-
(1-methylethoxy)-2-pyridinamine (●)
S   RBI-257

U   Dopamine D_4 receptor antagonist

10607   (8575) $C_{21}H_{28}NO_2$
58875-43-7

Diethyl(2-hydroxyethyl)methylammonium diphenylace-
tate = 2-(Diphenylacetoxy)triethylmethylammonium = 2-
[(Diphenylacetyl)oxy]-*N,N*-diethyl-*N*-methylethanami-
nium (●)
R   Bromide (6113-04-8)
S   Lunal, *Metadiphenii bromidum*
U   Ganglion blocking agent, spasmolytic

10608   (8576) $C_{21}H_{28}NO$
13473-61-

Diethyl(2-hydroxyethyl)methylammonium benzilate = 2
(Benziloyloxy)triethylmethylammonium = *N,N*-Diethyl-
2-[(hydroxydiphenylacetyl)oxy]-*N*-methylethanaminium
(●)
R   Bromide (3166-62-9)
S   Alsain, Apulcin, *Benactyzine methobromide*, Filci-
lin, Final, Finalin, Forinarin, Ganglioplegina, Gastri
made, Igsain, Magemin (old form), *Methylbenactyzi
um bromide***, Neo-Aspamin, Neo-Nichigreen,
Neunalin, Noinarin, Notilas, Paragone, PMB, Sanri-
ne, Semulgin, Spatomac, Venaburo
U   Anticholinergic, anti-ulcer agent

10609 (8577)

$C_{21}H_{28}NO_4$

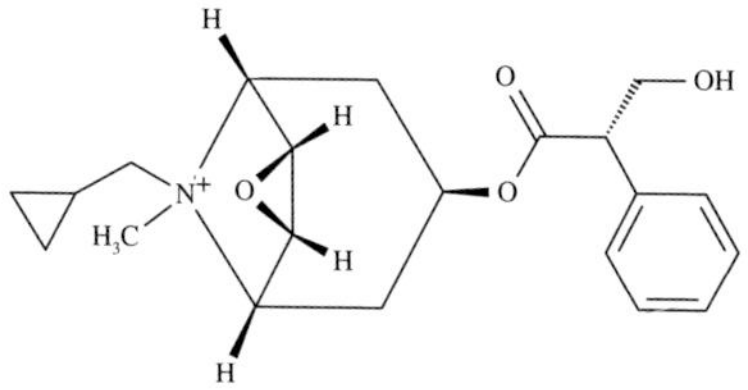

8-(Cyclopropylmethyl)-6β,7β-epoxy-3α-hydroxy-1αH,5αH-tropanium (–)-(S)-tropate = N-(Cyclopropyl-methyl)-O-tropoylscopinium = [7(S)-(1α,2β,4β,5α,7β)]-9-(Cyclopropylmethyl)-7-(3-hydroxy-1-oxo-2-phenyl-propoxy)-9-methyl-3-oxa-9-azoniatricyclo[3.3.1.0²,⁴]no-nane (●)

R Bromide (51598-60-8)

S Alginor, Algiron, *Cimetropium bromide**,* DA 3177, Lisador

U Anticholinergic, spasmolytic

10610 (8578)

$C_{21}H_{28}N_2O$
70976-76-0

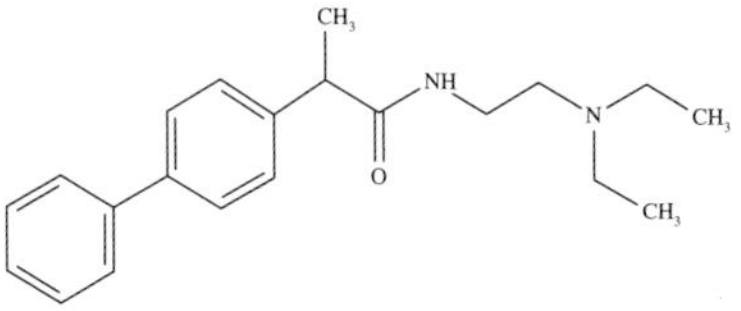

(±)-N-[2-(Diethylamino)ethyl]-α-methyl-4-biphenylacet-amide = (±)-2-(4-Biphenylyl)-N-[2-(diethylami-no)ethyl]propionamide = N-[2-(Diethylamino)ethyl]-α-methyl[1,1'-biphenyl]-4-acetamide (●)

S *Bifepramide**, Biprofenide*

U Spasmolytic

10611 (8579)

$C_{21}H_{28}N_2O$
552-25-0

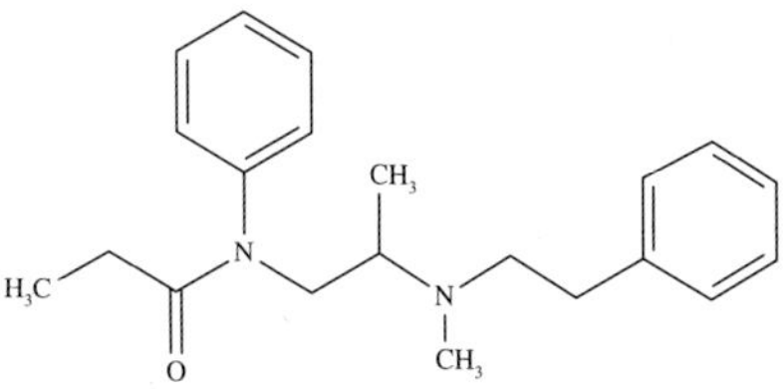

-[2-(Methylphenethylamino)propyl]propionanilide = N-2-[Methyl(2-phenylethyl)amino]propyl]-N-phenylpro-anamide (●)

S CL 22119, *Diampromide**,* NIH 7603

U Analgesic

10612 (8580)

$C_{21}H_{28}N_2OS$
111728-01-9

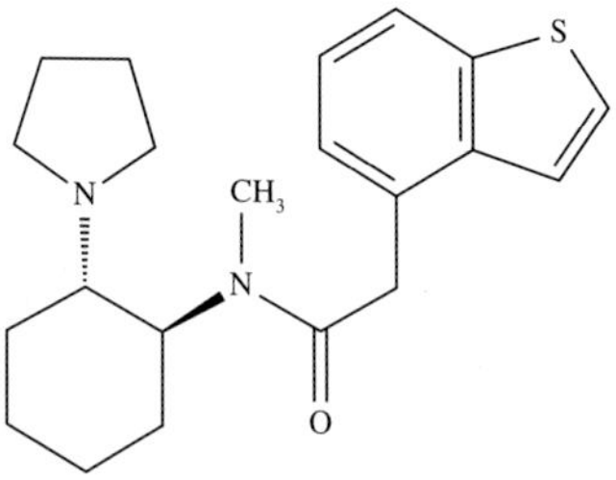

trans-(±)-N-Methyl-N-[2-(1-pyrrolidinyl)cyclohexyl] benzo[b]thiophene-4-acetamide (●)

S PD 117302

U Analgesic (kappa-opioid agonist)

10613 (8582)

$C_{21}H_{28}N_2O_2$
522-60-1

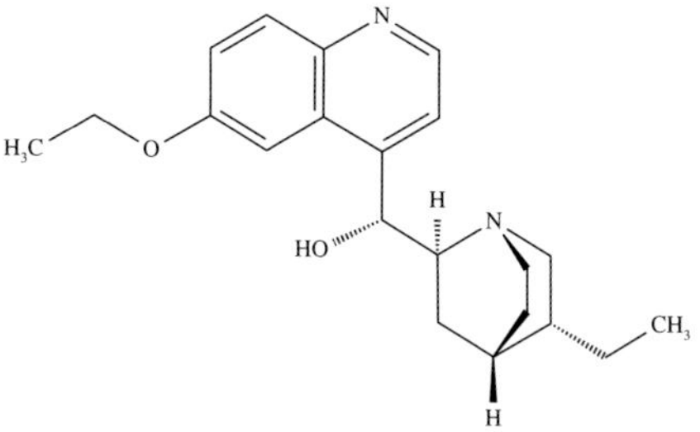

(6-Ethoxy-4-quinolyl)(5-ethyl-2-quinuclidinyl)methanol = (8α,9R)-6'-Ethoxy-10,11-dihydrocinchonan-9-ol (●)

R also monohydrochloride (3413-58-9)

S *Äthylhydrocuprein, Ethylhydrocupreine,* Neumolisi-na, Neumolysin, Numoquin, Optochin (old form)

U Antibacterial (ophthalmic)

10614　(8581)　　$C_{21}H_{28}N_2O_2$
3733-63-9

1-[2-(2-Hydroxyethoxy)ethyl]-4-[α-phenylbenzyl]pipe-
razine = 2-[2-[4-(Diphenylmethyl)-1-piperazinyl]eth-
oxy]ethanol (●)
R　Dihydrochloride (13073-96-6)
S　Autokar, *Decloxizine dihydrochloride***, *Hydroxydi-
　　äthylphenamin, Rescupal, U.C.B. 1402
U　Respiratory stimulant, bronchodilator, anti-emetic

10615　(8583)　　$C_{21}H_{28}N_2O_3$
76805-48-6

4-[2-Hydroxy-3-[(1-methyl-3-phenylpropyl)amino]prop-
oxy]benzeneacetamide (●)
S　KF 4317
U　α- and β-adrenergic blocker

10616　(8584)　　$C_{21}H_{28}N_2O_4$
5936-70-9

3,4-Diethoxy-*N*-[*p*-[2-(methylamino)ethoxy]ben-
zyl]benzamide = 3,4-Diethoxy-*N*-[[4-[2-(methylami-
no)ethoxy]phenyl]methyl]benzamide (●)
R　Monohydrochloride (19457-00-2)
S　Ancobon (old form), Ro 5-6574
U　Anti-emetic

10617　(8585)　　$C_{21}H_{28}N_2O$

o-[3-(*tert*-Butylamino)-2-hydroxypropoxy]-*p*-methoxy-
benzanilide = 2-[3-[(1,1-Dimethylethyl)amino]-2-hydr-
oxypropoxy]-4-methoxy-*N*-phenylbenzamide (●)
R　Monohydrochloride (137550-80-2)
S　TR-2732
U　β-Adrenergic blocker

10618　(8586)　　$C_{21}H_{28}N_2O$
90103-92-

(*S*)-2-[(*S*)-*N*-[(*S*)-1-Carboxy-3-phenylpropyl]alanyl]-2-
azabicyclo[2.2.2]octane-3-carboxylic acid = [3*S*-
[2[*R**(*R**)],3*R**]]-2-[2-[(1-Carboxy-3-phenylpropyl)am-

no]-1-oxopropyl]-2-azabicyclo[2.2.2]octane-3-carb-
oxylic acid (●)

R see also no. 12055

S *Zabiciprilat***

U Antihypertensive (ACE inhibitor)

10619 $C_{21}H_{28}N_2O_5$

8-[2-[[(2,3-Dihydro-8-methoxy-1,4-benzodioxin-2-
yl)methyl]amino]ethyl]-8-azaspiro[4.5]decane-7,9-dione
(●)

R Monohydrochloride (159650-30-3)

S MDL 73975

U Cardiotonic

10620 (8587) $C_{21}H_{28}N_2O_5$
138-56-7

N-[*p*-[2-(Dimethylamino)ethoxy]benzyl]-3,4,5-trimeth-
oxybenzamide = 4-(2-Dimethylaminoethoxy)-*N*-(3,4,5-
trimethoxybenzoyl)benzylamine = *N*-[[4-[2-(Dimethylamino)ethoxy]phenyl]methyl]-3,4,5-trimethoxybenz-
amide (●)

R Monohydrochloride (554-92-7)

S Ametik, Anaus "Molteni", Anti-Vomit, Arrestin,
Contrauto, Denausin, Elen (old form), Emamin,
Emedur, Emetan, Hymetic, Ibikin, Kantem, Meto-
benzomit, Nausan, Nauseton, Navogan, Novom, Po-
ligerim, Ro 2-9578, Stemetic, Supremesil, Tebami-
de, T-Gen, Ticon, Tigan, Tiject, Triban "Great South-
ern", Trimazide, *Trimethobenzamide hydrochlori-
de***, Trimetoxan, Violamil, Vomet, Vomitin "Akde-
niz", Vonon, Xametina

U Anti-emetic

10621 $C_{21}H_{28}N_2O_5$
252044-45-4

6-Cyano-3,4-*trans*-3,4-dihydro-2,2-dimethyl-3-hydroxy-
4-[2-oxo-5(*S*)-(1-ethoxyethoxymethyl)-1-pyrrolidinyl]-
2*H*-1-benzopyran = (3*R*,4*S*)-*rel*-4-[(2*R*)-2-[(1-Ethoxye-
thoxy)methyl]-5-oxo-1-pyrrolidinyl]-3,4-dihydro-3-
hydroxy-2,2-dimethyl-2*H*-1-benzopyran-6-carbonitrile
(●)

S MJ-355

U Potassium channel opener

10622 (8588) $C_{21}H_{28}N_2O_5$
87269-97-4

(2*S*,3a*S*,6a*S*)-1-[(*S*)-*N*-[(*S*)-1-Carboxy-3-phenylpro-
pyl]alanyl]octahydrocyclopenta[*b*]pyrrole-2-carboxylic
acid = 2-[*N*-[(*S*)-1-Carboxy-3-phenylpropyl]-L-alanyl]-
(1*S*,3*S*,5*S*)-2-azabicyclo[3.3.0]octane-3-carboxylic acid =
[2*S*-[1[*R**(*R**)],2α,3aβ,6aβ]]-1-[2-[(1-Carboxy-3-phe-
nylpropyl)amino]-1-oxopropyl]octahydrocyclopen-
ta[*b*]pyrrole-2-carboxylic acid (●)

R see also no. 12056

S M 1, *Ramiprilat***

U Antihypertensive (ACE inhibitor)

10623 (8589) $C_{21}H_{28}N_2O_5S$
90055-97-3

(±)-Ethyl 2-[3-(*tert*-butylamino)-2-hydroxypropoxy]-5-(2-thiophenecarboxamido)benzoate = 2-[3-[(1,1-Dimethylethyl)amino]-2-hydroxypropoxy]-5-[(2-thienylcarbonyl)amino]benzoic acid ethyl ester (●)
R Monohydrochloride (97067-66-8)
S *Tienoxolol hydrochloride***, UP 788-42
U β-Adrenergic blocker, diuretic

10624 (8590) $C_{21}H_{28}N_2O_7$
76812-98-1

(±)-5-[2-[[2-Hydroxy-3-[*p*-(2-methoxyethoxy)phenoxy]propyl]amino]ethoxy]salicylamide = (*RS*)-1-[[2-(3-Carbamoyl-4-hydroxyphenyl)ethyl]amino]-3-[4-(2-methoxyethoxy)phenoxy]-2-propanol = 2-Hydroxy-5-[2-[[2-hydroxy-3-[4-(2-methoxyethoxy)phenoxy]propyl]amino]ethoxy]benzamide (●)
R Monomethanesulfonate (salt) (108289-44-7)
S CGP 17/582 B, *Trigevolol mesilate***
U Antihypertensive (β-adrenergic blocker)

10625 $C_{21}H_{28}N_4O$
202579-74-6

N-(1-Ethylpropyl)-3-(4-methoxy-2-methylphenyl)-2,5-dimethylpyrazolo[1,5-*a*]pyrimidin-7-amine (●)
S DMP 904
U hCRF$_1$ receptor antagonist

10626 (8591) $C_{21}H_{28}N_4O$
2697-80-5

N-(D-1,6-Dimethyl-8-isoergolenyl)-*N'*,*N'*-diethylurea = *N'*-[(8α)-9,10-Didehydro-1,6-dimethylergolin-8-yl]-*N*,*N*-diethylurea (●)
S Mesenyl
U Antiserotonin

10627 (8592) $C_{21}H_{28}N_4O_3$
94192-59-3

N-Cyclohexyl-*N*-methyl-4-[(1,2,3,5-tetrahydro-2-oxoimidazo[2,1-*b*]quinazolin-7-yl)oxy]butyramide = *N*-Cyclohexyl-*N*-methyl-4-[(1,2,3,5-tetrahydro-2-oxoimidazo[2,1-*b*]quinazolin-7-yl)oxy]butanamide (●)
R Sulfate (1:1) monohydrate (101626-67-9)
S *Lixazinone sulfate***, RS-82856
U Cardiotonic (positive inotropic)

10628 (8593) $C_{21}H_{28}N_4O_3$

1-(4-Amino-6,7-dimethoxy-2-quinazolinyl)-4-piperidyl cyclopentyl ketone = 2-[4-(Cyclopentylcarbonyl)-1-piperidinyl]-4-amino-6,7-dimethoxyquinazoline = [1-(4-Amino-6,7-dimethoxy-2-quinazolinyl)-4-piperidinyl]cyclopentylmethanone (●)
R Monohydrochloride (86543-38-6)
S OR N 0676
U Antihypertensive

10629　(8594)

$C_{21}H_{28}N_4O_3$
131727-13-4

Methyl 4-(N-propionylanilino)-1-[2-(1-pyrazo-lyl)ethyl]isonipecotate = 4-[(1-Oxopropyl)phenylamino]-1-[2-(1H-pyrazol-1-yl)ethyl]-4-piperidinecarboxylic acid methyl ester (●)

S　A-3622

U　Analgesic (short acting)

10630

$C_{21}H_{28}N_4O_5$
104691-81-8

Des-Tyr-D-Phe-β-casomorphin = L-Prolyl-D-phenylala-nyl-L-prolylglycine (●)

　　BCH 325

U　Antidepressant

10631

$C_{21}H_{28}N_6O_4S$
155581-72-9

-[[3-(6,9-Dihydro-6-oxo-9-propyl-1H-purin-2-yl)-4-thoxyphenyl]sulfonyl]-4-methylpiperazine (●)

　　UK-93928

U　Anti-anginal

10632　(8595)

$C_{21}H_{28}N_7O_{17}P_3$
53-59-8

3-Carbamoyl-1-β-D-ribofuranosylpyridinium hydroxide 5'→5'-ester with adenosine 2'-(dihydrogen phosphate) 5'-(trihydrogen pyrophosphate) inner salt = Adenosine 5'-(trihydrogen diphosphate) 2'-(dihydrogen phosphate) 5'→5'-ester with 3-(aminocarbonyl)-1-β-D-ribofuranosyl-pyridinium hydroxide inner salt (●)

S　Codehydrase II, Codehydrogenase II, Coenzyme II, Co-II, Nadide phosphate, Nadifosfate, NADP, Phos-phocozymase, TPN

U　Coenzyme, vitamin

10633　(8601)

$C_{21}H_{28}O_2$
797-63-7

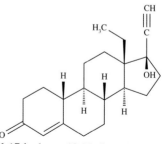

(−)-13-Ethyl-17-hydroxy-18,19-dinor-17α-pregn-4-en-20-yn-3-one = D-17α-Ethynyl-17-hydroxy-18-methyl-estr-4-en-3-one = D-13-Ethyl-17α-ethynyl-17-hydroxy-gon-4-en-3-one = (17α)-13-Ethyl-17-hydroxy-18,19-dinorpregn-4-en-20-yn-3-one (●)

R　see also no. 12028

S　Capronor, Dexnorgestrel, Duofem, Follistrel, Gravis-tat, Jadelle, Levonelle, Levonorgestrel**, Levonova, LNG, Micro-30, Microlut, Microluton, Microval, Mikro-30, Minera, 28 Mini, Mirena, Neogest, Norge-ston, D-Norgestrel, Norgestrel Jenapharm, Norlevo, Norplant, Nortrel, Plan B, Postinor, Vikela, WL 17, Wy-5104

U　Progestin

10634 (8600)

$C_{21}H_{28}O_2$
6533-00-2

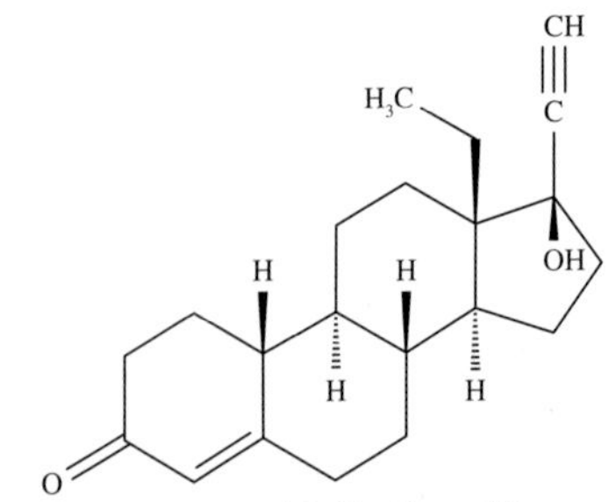

(±)-13-Ethyl-17-hydroxy-18,19-dinor-17α-pregn-4-en-20-yn-3-one = DL-17α-Ethynyl-17-hydroxy-18-methylestr-4-en-3-one = DL-13-Ethyl-17α-ethynyl-17-hydroxygon-4-en-3-one = (17α)-(±)-13-Ethyl-17-hydroxy-18,19-dinorpregn-4-en-20-yn-3-one (●)

S   Duolton, Monovar, Norgeal, *Norgestrel***, Ovrette, Plamover, SH 850, Wy-3707

U   Progestin

10635 (8596)

$C_{21}H_{28}O_2$
23983-19-9

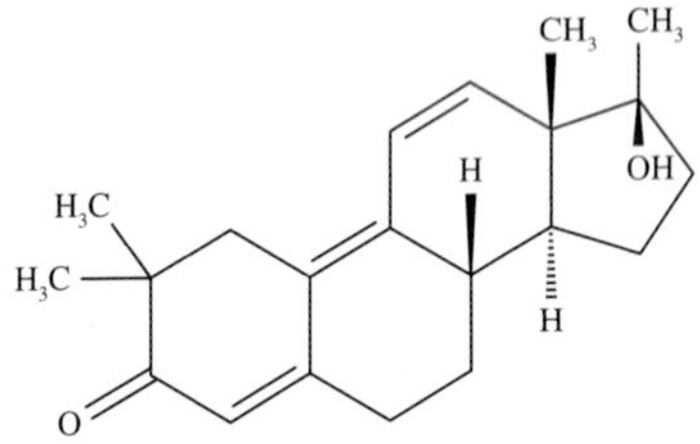

17β-Hydroxy-2,2,17-trimethylestra-4,9,11-trien-3-one = (17β)-17-Hydroxy-2,2,17-trimethylestra-4,9,11-trien-3-one (●)

S   R 2956

U   Anti-androgen

10636 (8598)

$C_{21}H_{28}O_2$
10116-22-0

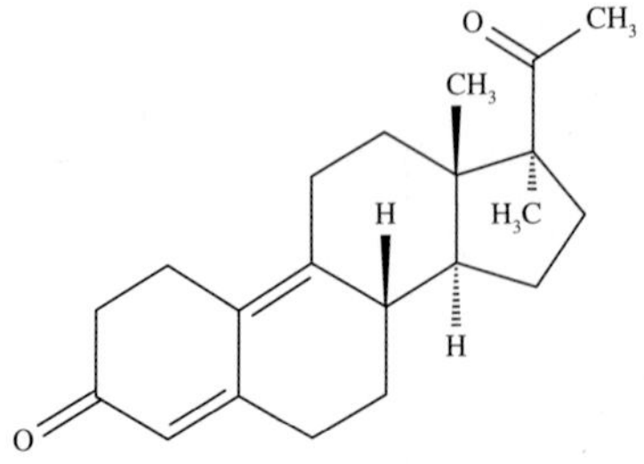

17-Methyl-19-norpregna-4,9-diene-3,20-dione (●)

S   *Demegestone***, Lutionex, R 2453

U   Progestin

10637 (8599)

$C_{21}H_{28}O_2$
5630-53-5

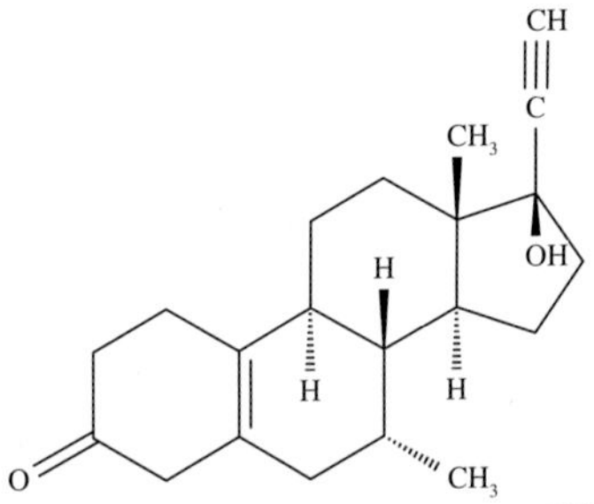

17-Hydroxy-7α-methyl-19-nor-17α-pregn-5(10)-en-20-yn-3-one = 17α-Ethynyl-17-hydroxy-7α-methylestr-5(10)-en-3-one = (7α,17α)-17-Hydroxy-7-methyl-19-norpregn-5(10)-en-20-yn-3-one (●)

S   Boltin, Livial, Liviel, Liviella, Livifem, 7α-*Methylnoretynodrel*, Org OD 14, Tibofem, *Tibolone***, Xyvion

U   Menopausal symptoms suppressant

10638

$C_{21}H_{28}O_2$
55041-29-?

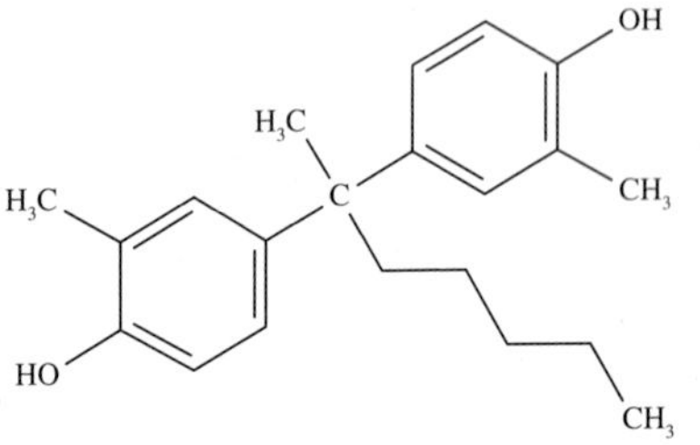

2,2-Bis(4-hydroxy-3-methylphenyl)heptane = 4,4'-(1-Methylhexylidene)bis[2-methylphenol] (●)

S   PD 009094

U   Antibacterial

10639 (8602) $C_{21}H_{28}O_2$
152-62-5

9β,10α-Pregna-4,6-diene-3,20-dione = 6-Dehydro-*retro*-progesterone = (9β,10α)-Pregna-4,6-diene-3,20-dione (●)

S Biphaston, Dabroston, Dufaston, Duphaston, Duvaron, *Dydrogesterone**, Gestatron, Gynorest, *Isopregnenone*, Medichrol, NSC-92336, Prodel, Retrone "Merrell", Terolut

U Progestin

10640 $C_{21}H_{28}O_2$
95975-55-6

Pregna-4,17(20)-diene-3,16-dione (●)
Guggulsterone, Gugulipid
Antihyperlipidemic

10641 (8597) $C_{21}H_{28}O_2$
434-03-7

17-Hydroxy-17α-pregn-4-en-20-yn-3-one = 17α-Ethinyl-17-hydroxyandrost-4-en-3-one = Anhydrohydroxy-progesterone = (17α)-17-Hydroxypregn-4-en-20-yn-3-one (●)

S *Aethinyltestosteron, Aethisteron, Anhydrohydroxyprogesterone*, Cistone, Colutoid, Diketolut-Tabl., Etherone, Ethinone, *Ethinyltestosterone, Ethisterone***, Etisteron(a)*, Gestone-Oral, Gestoral, Lucorteum-Oral, Lupronex-Tablets, Luteosterone-Compresse, Lutidon-Oral, Lutociclina-Linguetas, Lutocyclin-Linguetten, Lutocylol, Lutogyl-Tabl., Lutoral "Assia", Lutormon (Compr. + Suppos.), Lutral, Nalutoral, NSC-9565, Nugestoral, Oophormin-Luteum-Tabl., Ora-Lutin, Oraluton, *Praegninum*, Pranone, *Pregneninolone, Pregnin*, Pregnoral, Primolut C, Prodroxan, Produxan, Progestab, Progesterona-Serral-Tabl., Progesteron lingvalete, Progestin P, Progestolets, Progestoral, Progestrol, Prolidon-Comprimidos, Prolutol, Proluton C, Prone, Syngestrotabs, Syntolutin-Tabl., Trosinone

U Progestin

10642 (8603) $C_{21}H_{28}O_2$
83860-24-6

11-*cis*-12-(Hydroxymethyl)retinoic acid δ-lactone (●)
S BMY-30047
U Acne-therapeutic, keratolytic

10643 (8606) $C_{21}H_{28}O_3$
3758-34-7

Estra-1,3,5(10)-triene-3,17β-diol 17-propionate = 3,17β-Estradiol 17-monopropionate = (17β)-Estra-1,3,5(10)-triene-3,17-diol 17-propanoate (●)
S Acrofollin, Akrofollin, *Estradiol monopropionate*, Follhormon "Saper", *Östradiolmonopropionat, Oestrolum propionicum*

U Estrogen

10644 (8604) $C_{21}H_{28}O_3$
                                                            58691-88-6

17-Hydroxy-6-methyl-19-norpregna-4,6-diene-3,20-dione (●)
R see also no. 11998
S *Nomegestrol**, 19-Normegestrol*
U Progestin

10645 (8605) $C_{21}H_{28}O_3$
                                                            516-15-4

11-Oxoprogesterone = Pregn-4-ene-3,11,20-trione (●)
S Bio. 66, Ketogestin, *11-Ketoprogesterone*, U-1258
U Formerly in bovine acetonemia, anesthetic

10646 (8607) $C_{21}H_{28}O_4$
                                                            2454-11-7

11α,17β-Dihydroxy-17-methyl-3-oxoandrosta-1,4-diene-2-carboxaldehyde = 2-Formyl-11α,17β-dihydroxy-17-methylandrosta-1,4-diene-3-one = (11α,17β)-11,17-

Dihydroxy-17-methyl-3-oxoandrosta-1,4-diene-2-carboxaldehyde (●)
S Esiclene, *Formebolone**, Formyldienolone*, Hubernol, Metanor
U Anabolic

10647 (8608) $C_{21}H_{28}O_4$
                                                            20423-99-8

11β,17-Dihydroxypregna-1,4-diene-3,20-dione = (11β)-11,17-Dihydroxypregna-1,4-diene-3,20-dione (●)
R see also no. 12519
S *21-Deoxyprednisolone, Deprodone**, Desolone*
U Glucocorticoid (anti-inflammatory, anti-allergic)

10648 $C_{21}H_{28}O$
                                                            10184-70-

17,21-Dihydroxypregna-4,9(11)-diene-3,20-dione (●)
R see also no. 12002
S Anecortave
U Angiostatic steroid

10649 (8609) $C_{21}H_{28}O_4$
72-23-1

11-Dehydrocorticosterone = 21-Hydroxypregn-4-ene-3,11,20-trione (●)

S *Dehydrocorticosterone, Kendall's compound A,* NSC-9702, *Verbindung A von Kendall*

U Glucocorticoid

10650 (8610) $C_{21}H_{28}O_5$
60023-92-9

11β,17β-Dihydroxy-17-methyl-3-oxoandrosta-1,4-diene-2-carboxylic acid = 2-Carboxy-11β,17β-dihydroxy-17-methylandrosta-1,4-dien-3-one = (11β,17β)-11,17-Dihydroxy-17-methyl-3-oxoandrosta-1,4-diene-2-carboxylic acid (●)

R see also nos. 14564, 14807

S BR 906, *Roxibolone***

U Anabolic

10651 (8612) $C_{21}H_{28}O_5$
50-24-8

11β,17,21-Trihydroxypregna-1,4-diene-3,20-dione = Pregna-1,4-diene-11β,17α,21-triol-3,20-dione = 1,2-Dehydrohydrocortisone = (11β)-11,17,21-Trihydroxypregna-1,4-diene-3,20-dione (●)

R see also nos. 10692, 12006, 12895, 13277, 13278, 13589, 13618, 13623, 13830, 13922, 13926, 14128, 14959, 15040, 15126, 15344, 15386; also sesquihydrate (52438-85-4)

S Abilene "Farmagen", Adelcort, Adnisolone, Aersolin, Afisolone, Alferm, Alpicort N, Antisolon, Aprednislon, Arcosolone, Benisolon, Berlisolon, Biosolone, Bisuo-Cream, Caberdelta, Capsoid, Clemisolone, Clonisolone, Codelcortone, Codelton, Co-Hydeltra, Cordelta, Cordex, Cordonin, Cordrol, Corisilone, Cortadeltona, Cortalone, Cortelinter, Corticar, Cortilona-B, Cortisolone, Cotelone, Cotogesic, Cotolone, Croycort B, Dacortin(a), Dacortin-H, Decaprednil, Decortasmyl, Decortin H, Decortril, *Dehydrocortisol, Dehydro-hydrocortisone*, Delcort-E, Delcortol, Delhydro-Cortex, Delta-Cortef, Deltacortenolo, Deltacortril, Deltaderm, Delta-Ef-Cortelan, delta F, Delta-Genacort, Deltaglycortril, Delta-Hycortol, *Deltahydrocortisone*, Delta-Larma, Deltalone, Delta-ophticor, Delta-Phoricol, Delta-Prenin, Deltasolone, Deltastab, Deltidrosol, Deltisilone, Deltisolon, Deltolasson, Deltosona, Dermipred, Dermosolon, Derpo PD, Dhasolone, Di-Adreson-F, Dicortol, Domucortone, Donisolone, Dontisolon D-Salbe, Duraprednisolon, Ejizolon, Encortolon, Exacort, Facort, Farnisol "Lauzier", Fernisolone, Filocorten, Flamasone, Glucortin-Tabl., H 358, Hefasolon, Hesse-Pred, Hicortin, Hostacortin H, Hubersona, Hydeltra, Hydeltrone, Hydrocortancyl, Hydrocortidelt, Δ^1-*Hydrocortisone*, Hydrodeltalone, Hydrodeltisone, Hydroretrocortin, In-solone, Juvasolon, Keteocort H, Klismacort, Kühlprednon-Salbe, Kylicorton, Lenisolone, Leocortol, Lepicortinolo, Linola-H N, Livopredin, Longiprednil, Marsolone, Maxiderm, Médiasolone, Mégasolone, Meprisolon, Metacortalon, *Metacortandralone*, Meticortelone, Meti-Derm, Microsolone, Morlone, Neocorlin, Neocorten, Neo-Cortisolona, Neodelta, Neohidroaltesona, Niscolone, Nisolone, Normonsona, Novoprednisolone, NSC-9120, Nurisolon, Optocort, Oropred, Panafcortelone, Panisolone, Paracortol, Phaken, Phlogex, Prebital, Precin, Precortalon-Tabl., Precortancyl, Precortilon(e), Pre-Cortisyl, Predaler, Predartrine, Predeltilone, Predenema, Preditabs, Predne-Dome, Pred-

nelan, Prednicare, Prednicen, Predni-Coelin, Pred-
nicor, Prednicorm H, Prednicort, Prednicortelone,
Predni-Helvacort, Predni-H-Tablinen, Predniretard,
Prednis, Prednisol "Farmila", Prednisolo, *Prednisolone**, Prednistab, Prednitex, Prednivet, *Pred-
nizolon*, Prednol, Prednorsolon, Predonine, Predorga-
solona, Predsim, Predsolan "Glaxovet; Pitman-Moo-
re", Predzon, Prehokon, Prelon(e), Prenolone, Pren-
sal, Pre-Ped, Preskort, Presolon, Presone "Smith Stai-
nistreet", Prezolon, Ranbisolone, Reumazine, Ro-
predlone, Scansolone, Scheriproct N, Scherisolon,
Serilone, Sherisolon, Shikiprezonin, Solona, Solone
"F.& M.", Solucort "Therapeutic", Solupren, Sopa-
cortelone, Spiricort, Spolotan, S.P. Ster-Tabl., Ster,
Sterane, Sterasolone, Stereon, Stermin, Sterolone,
Supercortisol, Tarocortelone, Ulacort, Ultracorten H, Vinopred, Vitacort "Vitarine", Wysolone
U Glucocorticoid (anti-inflammatory, anti-allergic)

10652 (8613) $C_{21}H_{28}O_5$
52-39-1

11β,21-Dihydroxy-3,20-dioxopregn-4-en-18-al 11,18-
Hemiacetal of 11β,21-dihydroxy-3,20-dioxopregn-4-en-
18-al (11,18-Epoxy-18,21-dihydroxypregn-4-ene-3,20-
dione) = (11β)-11,21-Dihydroxy-3,20-dioxopregn-4-en-
18-al (●)
S Aldocorten, Aldocortin, *Aldosterone**, Elektrocor-
tin, Oxocorticosterone, Reichstein's substance X,
Verbindung X von Reichstein*
U Mineralocorticoid (in Addisonian disease)

10653 (8611) $C_{21}H_{28}O_5$
53-06-5

17α,21-Dihydroxypregn-4-ene-3,11,20-trione = Pregn-4-
ene-17α,21-diol-3,11,20-trione = 17α-Hydroxy-11-dehy-
drocorticosterone = 17,21-Dihydroxypregn-4-ene-
3,11,20-trione (●)
R see also no. 12007
S Adrenalex, *Compound E von Kendall, Compound F
von Wintersteiner*, Corlodrin, Cortelin, Cortisartro-
ne, *Cortisone**, Cortistan, Cortivite, Cortizon,
Cortogen, Eschatin, KE, *Kendall's compound E,
Kortison*, NSC-9703, Rafa-Cortin, *Reichstein's
substance Fa*, Rincorten, *Verbindung E von Kendall,
Verbindung Fa von Reichstein, Wintersteiner's
compound F*
U Glucocorticoid (anti-inflammatory, anti-allergic)

10654 (8614) $C_{21}H_{28}O_6$
18118-80-4

11β,17,21-Trihydroxy-*B*-homo-*A*-norpregn-1-ene-
3,6,20-trione = (11β)-11,17,21-Trihydroxy-*B*-homo-*A*-
norpregn-1-ene-3,6,20-trione (●)
S *Oxisopred***
U Corticosteroid

10655   (8617) $C_{21}H_{29}ClO_3$
                                                        855-19-6

4-Chloro-17β-hydroxyandrost-4-en-3-one acetate = 4-
Chlorotestosterone acetate = (17β)-17-(Acetyloxy)-4-
chloroandrost-4-en-3-one (●)
R   see also nos. 8959, 11456, 12961
S   Alfa-Trofodermin, Anabolit, *Chlortestosterone ace-
        tate*, *Clostebol acetate***, Clostene, Esteranabol,
        Esterbol-Depo, Macrobin, Megagrisevit mono, Ster-
        abol, Steranabol, Test-Anabol, Testomed, Turinabol
        (old form)
U   Anabolic (topical)

10656   (8615) $C_{21}H_{29}ClO_3$
                                                        20047-75-0

6-Chloro-3β,17-dihydroxypregna-4,6-dien-20-one =
(3β)-6-Chloro-3,17-dihydroxypregna-4,6-dien-20-one
(●)
R   see also no. 12900
S   *Clogestone***
U   Progestin

10657   (8616) $C_{21}H_{29}ClO_3$
                                                        16469-74-2

6α-Chloro-17-hydroxypregn-4-ene-3,20-dione = 6α-
Chloro-17-hydroxyprogesterone = (6α)-6-Chloro-17-
hydroxypregn-4-ene-3,20-dione (●)
R   see also no. 12013
S   *Hydromadinone***
U   Progestin

10658   (8618) $C_{21}H_{29}ClO_6S$
                                                        67110-79-6

(±)-(Z)-7-[(1R*,2S*,3S*,5R*)-2-[[(R*)-3-(m-Chlorophenoxy)-2-hydroxypropyl]thio]-3,5-dihydroxycyclo-
pentyl]-5-heptenoic acid = (±)-(5Z)-(9S,11R,15S)-16-(3-
Chlorophenoxy)-9,11,15-trihydroxy-ω-tetranor-13-thia-
prost-5-enoic acid = [1α(Z),2β(R*),3α,5α]-(±)-7-[2-[[3-
(3-Chlorophenoxy)-2-hydroxypropyl]thio]-3,5-dihydro-
xycyclopentyl]-5-heptenoic acid (●)
S   EMD 34946, Equestrolin, *Luprostiol***, Pronilen,
        Prosolvin, *Prostianol*, Reprodin
U   Luteolytic (veterinary)

10659   (8619) $C_{21}H_{29}Cl_2N_3O_2$
                                                        130641-36-0

(−)-(R)-4-Amino-3,5-dichloro-α-[[[6-[2-(2-pyridyl)eth-
oxy]hexyl]amino]methyl]benzyl alcohol = (R)-1-(4-Ami-

no-3,4-dichlorophenyl)-2-[[6-[2-(2-pyridyl)-ethoxy]he-
xyl]amino]ethanol = (R)-4-Amino-3,5-dichloro-α-[[[6-
[2-(2-pyridinyl)ethoxy]hexyl]amino]methyl]benzeneme-
thanol (●)
R Fumarate (2:1) (130641-37-1)
S GR 63411 B, GR 114297 A, *Picumeterol fumarate***
U Bronchodilator (β₂-adrenoceptor agonist)

10660 (8620) $C_{21}H_{29}Cl_3O_3$
53608-96-1

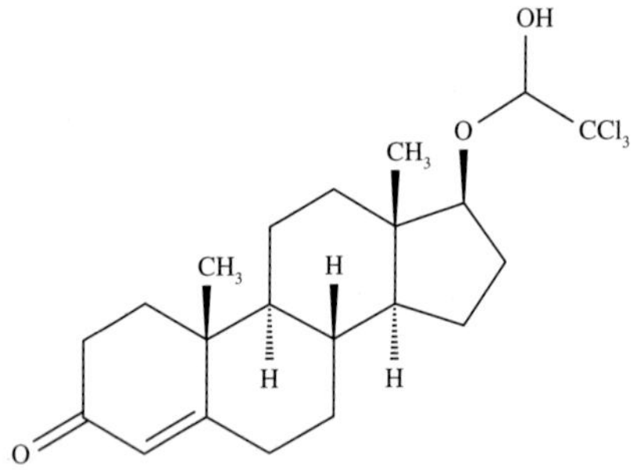

17β-(2,2,2-Trichloro-1-hydroxyethoxy)androst-4-en-3-
one = (17β)-17-(2,2,2-Trichloro-1-hydroxyethoxy)an-
drost-4-en-3-one (●)
R see also no. 12015
S *Cloxotestosterone**, Testosterone 17-chloral hemi-
      acetal*
U Androgen

10661 (8621) $C_{21}H_{29}FN_2O_2$
54340-64-6

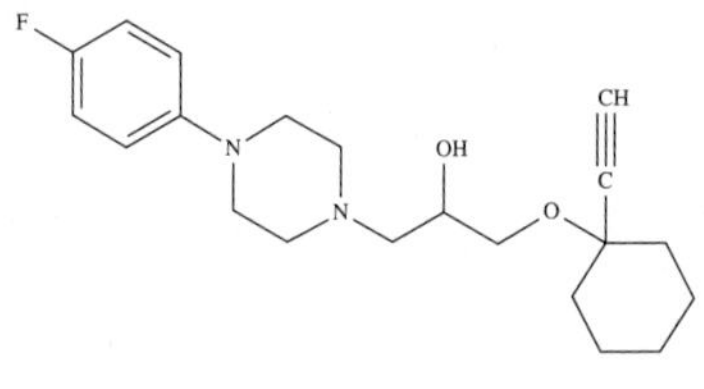

1-[(1-Ethynylcyclohexyl)oxy]-4-[3-(p-fluorophenyl)-1-
piperazinyl]-2-propanol = α-[[(1-Ethynylcyclohe-
xyl)oxy]methyl]-4-(4-fluorophenyl)-1-piperazineethanol
(●)
S *Fluciprazine***
U Neuroleptic

10662 (8622) $C_{21}H_{29}FN_2O_3$
54063-38-6

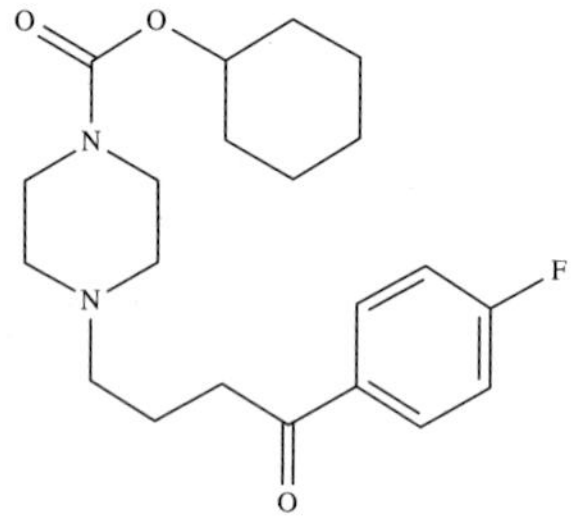

4-[3-(p-Fluorobenzoyl)propyl]-1-piperazinecarboxylic
acid cyclohexyl ester = 4-[4-(4-Fluorophenyl)-4-oxobu-
tyl]-1-piperazinecarboxylic acid cyclohexyl ester (●)
S *Fenaperone**, MD 67332
U Neuroleptic

10663 (8623) $C_{21}H_{29}FO_4$
337-03-1

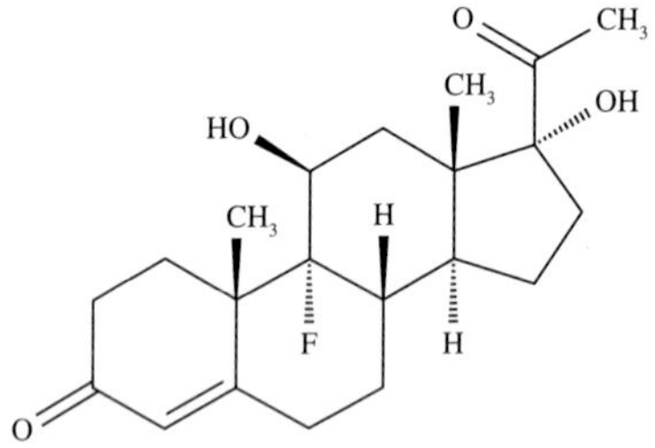

9-Fluoro-11β,17-dihydroxypregn-4-ene-3,20-dione =
(11β)-9-Fluoro-11,17-dihydroxypregn-4-ene-3,20-dione
(●)
R see also no. 12017
S *Flugestone***
U Progestin

10664 (8624) $C_{21}H_{29}FO_5$
127-31-1

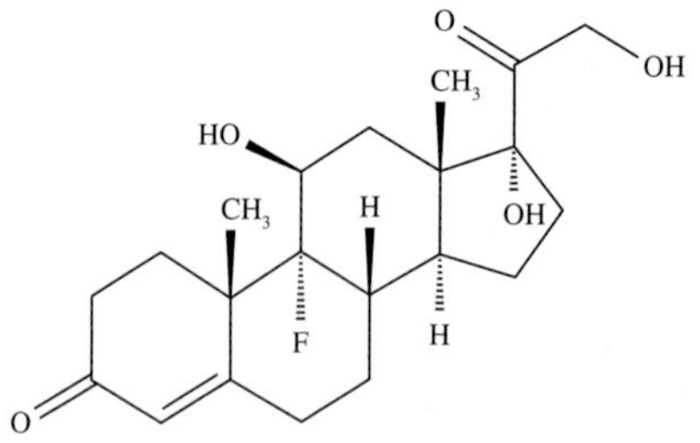

9-Fluoro-11β,17,21-trihydroxypregn-4-ene-3,20-dione = 9-Fluorohydrocortisone = (11β)-9-Fluoro-11,17,21-trihydroxypregn-4-ene-3,20-dione (●)

R see also no. 12018

S 9-AF, Alflorone, Astonin (-H), Corti-9, *Fludrocortisone***, Fludrocortone, Fludrone, Fluohydrisone, *Fluorcortisol*, *Fluorhydrocortisonum*, Lonikan, StC 1400

U Corticoid (mainly mineralocorticoid), hypertensive

10665 (8625) $C_{21}H_{29}I_3N_4O_9$
97702-82-4

3,5-Diacetamido-2,4,6-triiodo-*N*-methyl-*N*-[[methyl(D-*gluco*-2,3,4,5,6-pentahydroxyhexyl)carbamoyl]methyl]benzamide = 1-[[[[3,5-Bis(acetylamino)-2,4,6-triiodobenzoyl]methylamino]acetyl]methylamino]-1-deoxy-D-glucitol (●)

S *Iosarcol**, Melitrast

U Diagnostic aid (radiopaque medium)

10666 (8626) $C_{21}H_{29}N$
5966-41-6

N,N-Diisopropyl-3,3-diphenylpropylamine = *N,N*-Bis(1-methylethyl)-γ-phenylbenzenepropanamine (●)

R Hydrochloride (24358-65-4) (also mixture with sorbitol)

S Agofel(l), Belagol, Belivron, Bilagol, Biliflux, Cholagol, *Diisopromine hydrochloride***, *Disoprominii chloridum*, Do-Bil, Galbil, Hepabyl, Hepacol, Méga-

byl, Norbilin "Phoenix", Perebrin, Polagol, Propabyl, Prylaphan, R 253, Vesicalma

U Choleretic, spasmolytic

10667 (8627) $C_{21}H_{29}NO$
545-90-4

6-(Dimethylamino)-4,4-diphenyl-3-heptanol = β-[2-(Dimethylamino)propyl]-α-ethyl-β-phenylbenzeneethanol (●)

S *Amidol*, *Bimethadolum*, *Dimefeptanolo*, *Dimepheptanol***, *Methadol*, NIH 2933, Pangerin, *Racemethadol*

U Narcotic analgesic

10668 (8628) $C_{21}H_{29}NO$
17199-54-1

(3R,6R)-6-(Dimethylamino)-4,4-diphenyl-3-heptanol = [R-(R*,R*)]-β-[2-(Dimethylamino)propyl]-α-ethyl-β-phenylbenzeneethanol (●)

S *Alfametadolo*, *Alphamethadol***, α-*Methadol*

U Narcotic analgesic

10669 (8629) $C_{21}H_{29}NO$
                                                                  17199-55-2

(3S,6R)-6-(Dimethylamino)-4,4-diphenyl-3-heptanol =
[S-(R^*,S^*)]-β-[2-(Dimethylamino)propyl]-α-ethyl-β-
phenylbenzeneethanol (●)
S *Betametadolo*, *Betamethadol***, β-*Methadol*
U Narcotic analgesic

10670 (8632) $C_{21}H_{29}NO$
                                                                  604-74-0

2-[(*o-tert*-Butyl-α-phenylbenzyl)oxy]-N,N-dimethyl-
ethylamine = 2-(*o-tert*-Butyldiphenylmethoxy)-N,N-di-
methylethylamine = 2-[[2-(1,1-Dimethylethyl)phe-
nyl]phenylmethoxy]-N,N-dimethylethanamine (●)
R also hydrochloride (5697-60-9)
S B.S. 6534, *Bufenadinum*, *Bufenadrine***
U Anti-emetic, antihistaminic, antiparkinsonian

10671 (8630) $C_{21}H_{29}NO$
                                                                  514-65-8

1-(Bicyclo[2.2.1]hept-5-en-2-yl)-1-phenyl-3-piperidino-
1-propanol = α-5-Norbornen-2-yl-α-phenyl-1-piperidine-
propanol = α-Bicyclo[2.2.1]hept-5-en-2-yl-α-phenyl-1-
piperidinepropanol (●)
R also hydrochloride (1235-82-1) or lactate (7085-45-2)
S Akineton, Akinophyl, Berofin, Bicamol, *Biperi-
       den***, Dekinet, Desiperiden, Ipsatol, KL 373,
       Norakin N, Paraden, Tasmolin
U Anticholinergic, antiparkinsonian

10672 (8631) $C_{21}H_{29}NO$
                                                                  14617-18-6

1-(Tricyclo[2.2.1.0^{2,6}]hept-2-yl)-1-phenyl-3-piperidino-
1-propanol = α-Phenyl-α-tricyclo[2.2.1.0^{2,6}]hept-2-yl-1-
piperidinepropanol (●)
R Hydrochloride of the mixture with the isomeric tri-
       cyclohept-3-yl-compd. (14617-19-7)
S Norakin, P 259, *Triperiden hydrochloride*
U Anticholinergic, antiparkinsonian

10673 $C_{21}H_{29}NO$
                                                                  116376-79-5

2,6-Bis(1,1-dimethylethyl)-4-[2-(3-pyridinyl)ethyl]phe-
nol (●)
S BI-L-93BS
U Anti-inflammatory

10674 (8633) $C_{21}H_{29}NO_2$
3572-62-1

Tropine-α-cyclopentylphenylacetate = α-Cyclopentyl-phenylacetic acid tropine ester = *endo*-α-Cyclopentylbenzeneacetic acid 8-methyl-8-azabicyclo[3.2.1]oct-3-yl ester (●)

S Fenatrop, IT-79, Phenatrop
U Anticholinergic

10675 (8634) $C_{21}H_{29}NO_2$
42408-82-2

–)-17-(Cyclobutylmethyl)morphinan-3,14-diol = 17-Cyclobutylmethyl)morphinan-3,14-diol (●)
R also D-(–)-tartrate (1:1) (58786-99-5)
S Beforal, Butomidor, *Butorphanol***, Dolorex "Intervet", levo-BC-2627, Moradol, Morphasol, Stadol(e), TNB, Torate, Torbugesic, Torbutrol, Verstadol
J Analgesic, antitussive

10676 $C_{21}H_{29}NO_2$
53016-31-2

(17α)-13-Ethyl-17-hydroxy-18,19-dinorpregn-4-en-20-yn-3-one oxime (●)
S *17-Deacetylnorgestimate, Levonorgestrel 3-oxime, Norelgestromin*, Norgestin*, Norplant 3-oxime, RWJ 100553
U Female contraceptive

10677 (8635) $C_{21}H_{29}NO_2S$
47450-00-0

3-(Diethylamino)propyl α-allylbenzo[*b*]thiophene-3-propanoate = 2-(Benzo[*b*]thien-3-ylmethyl)-4-pentenoic acid 3-(diethylamino)propyl ester = α-2-Propenylbenzo[*b*]thiophene-3-propanoic acid 3-(diethylamino)propyl ester (●)
R Citrate (1:1) (35057-88-6)
S Assedil, INO 1646 AT
U Cerebral vasodilator

10678 (8636) $C_{21}H_{29}NO_3$
67634-12-2

Methyl N-[[4-(4-hydroxy-4-methylpentyl)-3-cyclohexenyl]methylene]anthranilate = 2-[[[4-(4-Hydroxy-4-methylpentyl)-3-cyclohexenyl]methylene]amino]benzoic acid methyl ester (●)
R Mixture of three isomers
S Lyrame
U Insect repellent

10679 (8637) $C_{21}H_{29}NO_3$
                                         4539-95-1

2-Oxo-1-(β-piperidinoethyl)cyclohexanecarboxylic acid benzyl ester = 2-(β-Piperidinoethyl)cyclohexanone-2-carboxylic acid benzyl ester = 2-Oxo-1-[2-(1-piperidinyl)ethyl]cyclohexanecarboxylic acid phenylmethyl ester (●)
S Cesedon, Cetran
U Analgesic, sedative, spasmolytic

10680 (8638) $C_{21}H_{29}NO_4$
                                         33978-72-2

N,N-Bis(3,4-dimethoxyphenethyl)-N-methylamine = N-[2-(3,4-Dimethoxyphenyl)ethyl]-3,4-dimethoxy-N-methylbenzeneethanamine (●)
R Hydrochloride (89805-39-0)
S MR 817, YS-035
U Calcium antagonist

10681 (8639) $C_{21}H_{29}NS_2$
                                         486-17-9

2-[[p-(Butylthio)-α-phenylbenzyl]thio]-N,N-dimethylethylamine = 2-[[4-(Butylthio)benzhydryl]thio]ethyldimethylamine = 2-[[p-(Butylthio)diphenylmethyl]thio]ethyldimethylamine = p-(Butylthio)benzhydryl β-(dimethylamino)ethyl sulfide = 2-[[[4-(Butylthio)phenyl]phenylmethyl]thio]-N,N-dimethylethanamine (●)
R also hydrochloride (904-04-1)
S AY-55074, *Captodiame**, Captodiamine, Captodramine hydrochloride*, Covatin(e), Covatix, N-68, Suvren
U Tranquilizer

10682 (8640) $C_{21}H_{29}N_2O$
                                         47324-98-1

Benzyldiethyl[(2,6-xylylcarbamoyl)methyl]ammonium = N-[2-[(2,6-Dimethylphenyl)amino]-2-oxoethyl]-N,N-diethylbenzenemethanaminium (●)
R Benzoate (3734-33-6) or benzoate monohydrate (86398-53-0)
S Bitrex, Denaton, *Denatonium benzoate***, Indigestin, *Lidocaine benzylbenzoate*, Nailicure, NSC-157658, Stop-Zit, THS-839, Win 16568
U Alcohol denaturant, nail biting deterrent, veterinary = stomachic for ruminants

10682-01 (8640-01)   90823-38-
R Salt (1:1) with 1,2-benzisothiazol-3(2H)-one 1,1-dioxide (no. 796)
S *Denatonium saccharide*, Vilest
U Animal repellent

10683 $C_{21}H_{29}N_3O$
120478-64-0

N-[(8α)-2,6-Dimethylergolin-8-yl]-2,2-dimethylpropan-
amide (●)
S SDZ 208-911
U Dopamine agonist

10684 $C_{21}H_{29}N_3O$
179556-82-2

N-Methyl-3-(1-methylethoxy)-*N*-[1-(phenylmethyl)-4-
piperidinyl]-2-pyridinamine (●)
S PNU-101958, U-101958
U Dopamine D$_1$ receptor antagonist

10685 (8641) $C_{21}H_{29}N_3O$
3737-09-5

α-[2-(Diisopropylamino)ethyl]-α-phenyl-2-pyridineacet-
amide = 4-(Diisopropylamino)-2-phenyl-2-(2-pyridyl)bu-
tyramide = α-[2-[Bis(1-methylethyl)amino]ethyl]-α-phe-
nyl-2-pyridineacetamide (●)
R also phosphate (1:1) (22059-60-5)

S Chiyoban, Corapace, Dicorantil, Dicorynan, Dimo-
dan, Dirythmin SA, Dirytmin, Disaloc, Disocor "Polfa", Diso-Duriles, Disomet, Disonorm, *Disopyrami-
de***, Disopyran, Disopyr von ct, Durbis, Fanmil,
H 3292, Isomide, Isorythm, Kafier, Korapace, Kora-
peis, Lispine, Napamide "Major", Normoritmo, Nor-
pace, Norpaso, Palpitin, Postormin, Prorritmina, Ri-
modan, Ritmilen, Ritmodan, Ritmoforine, Rizora-
mid, RU 18850, Rythmical, Rythmodan, Rythmo-
dul, Rytmidan, Rytmilen, SC-7031, SC-13957,
Searle 703, Sopyrat, Stedicor, Tailinder, Zarcero
U Anti-arrhythmic

10686 (8642) $C_{21}H_{29}N_3O_4$
98672-91-4

[1*S*-[1α,2α(*Z*),3α,4α]]-7-[3-[[2-[(Phenylamino)carbo-
nyl]hydrazino]methyl]-7-oxabicyclo[2.2.1]hept-2-yl]-5-
heptenoic acid (●)
S SQ 29548
U Thromboxane receptor antagonist

10687 $C_{21}H_{29}N_3O_7S$

4-[[(6"*R*,7"*S*)-7"-Methoxy-3"-methyl-5",5"-dioxido-8"-
oxodispiro[cyclopentane-1,1'-cyclopropane-2',4"-
[5]thia[1]azabicyclo[4.2.0]oct[2]en]-2"-yl]carbonyl]-1-
piperazineacetic acid (●)
R Sodium salt (217199-33-2)
S SYN 1390
U Neutrophil elastase inhibitor

10688　　　　　　　　　　　　　　　　　　　$C_{21}H_{29}N_5O$
　　　　　　　　　　　　　　　　　　　　　143394-68-7

(8R)-8-Ethyl-2-(hexahydro-2,5-methanopentalen-3a(1H)-yl)-1,4,7,8-tetrahydro-4-propyl-5H-imida-zo[2,1-i]purin-5-one (●)
S KF 20274
U Adenosine A_1 receptor antagonist

10689　(8643)　　　　　　　　　　　　　　$C_{21}H_{29}N_5O_2$
　　　　　　　　　　　　　　　　　　　　　87760-53-0

(1R*,2S*,3R*,4S*)-N-[4-[4-(2-Pyrimidinyl)-1-piperazi-nyl]butyl]-2,3-norbornanedicarboximide = N-[4-[4-(2-Pyrimidinyl)-1-piperazinyl]butyl]bicyclo[2.2.1]heptane-2,3-di-exo-carboximide = (3aα,4β,7β,7aα)-Hexahydro-2-[4-[4-(2-pyrimidinyl)-1-piperazinyl]butyl]-4,7-methano-1H-isoindole-1,3(2H)-dione (●)
R Citrate (1:1) (112457-95-1)
S *Metanopirone citrate*, Sediel, SM 3997, *Tandospiro-ne citrate***
U Anxiolytic, anticonflict

10690　　　　　　　　　　　　　　　　　　$C_{21}H_{29}N_5O_6$
　　　　　　　　　　　　　　　　　　　　　170902-47-3

(2S)-3-[2-[(5R)-3-(p-Amidinophenyl)-2-isoxazolin-5-yl]acetamido]-2-(carboxyamino)propionic acid 2-butyl methyl ester = 3-[[[(5R)-3-[4-(Aminoiminomethyl)phe-nyl]-4,5-dihydro-5-isoxazolyl]acetyl]amino]-N-(butoxy-carbonyl)-L-alanine methyl ester (●)
R Monoacetate (176022-59-6)
S DMP 754, Lumaxis, *Roxifiban acetate***
U Antithrombotic (fibrinogen receptor antagonist)

10691　　　　　　　　　　　　　　　　　　$C_{21}H_{29}N_5O_6$
　　　　　　　　　　　　　　　　　　　　　183547-57-1

Ethyl 2-[[4-[(5R)-3-[4-[[(methoxycarbonyl)amino]imino-methyl]phenyl]-2-oxo-5-oxazolidinyl]methyl]-1-pipera-zinyl]acetate = 4-[[(5R)-3-[4-[Imino[(methoxycarbo-nyl)amino]methyl]phenyl]-2-oxo-5-oxazolidinyl]me-thyl]-1-piperazineacetic acid ethyl ester (●)
R Citrate (1:1) (215956-37-9)
S EMD 122347, *Gantofiban citrate***, YM-028
U Antithrombotic (fibrinogen receptor antagonist)

10692　(8644)　　　　　　　　　　　　　　　　$C_{21}H_{29}O_8P$
　　　　　　　　　　　　　　　　　　　　　302-25-0

11β,17,21-Trihydroxypregna-1,4-diene-3,20-dione 21-(dihydrogen phosphate) = Prednisolone 21-phosphate = (11β)-11,17-Dihydroxy-21-(phosphonooxy)pregna-1,4-diene-3,20-dione (●)
R see also no. 15386;
　　also disodium salt (125-02-0)
S AK-Pred, Alto-Pred soluble, Caberdelta-endovena, Cenalone "Century", Codelsol, Cortisate-10, Del-ta-Corti-Ofteno, Deltelan, Dojilon, Fisopred, Hefaso-lon i.v., Hydeltrasol, Hydrosol, Inflamase "Cooper;

Iolab; Ciba Vision", I-Pred, Key-Pred-SP, Lite Pred, Metreton, Minims-Prednisolone, Nor-Pred S., Norsol "Bilim", Nova-Pred, Ocu-Pred, Optival, Orapred, Parisilon, Pediapred, Phortisolone, Pred "Grin.", Predair forte, Pred-Clysma, Predmix, Prednabene, Prednesol, Predniment (rectal), Predni Monem, Prednis-A-Vet, *Prednisolondinatriumphosphat*, *Prednisolone Sodium Phosphate*, Predsol "Glaxo", Predsolan, Prozorin-inj., PSP-IV, Rectopred, R.O.-Predphate, Savacort-S, Sol-Pred, Solucort "Chibret", Solucort Ophta, Solu-Pred "Kenyon; Myers-Carter"

U Glucocorticoid (anti-inflammatory, anti-allergic)

10692-01 **(8644-01)**   6693-90-9
R Compd. (1:1) with fenoxazoline (no. 4169)
S Aturgyl-Delta, Déturgylone, L.D. 4003, Oto-Rinil, *Prednazoline***
U Nasal decongestant, anti-inflammatory, anti-allergic

10693 $C_{21}H_{30}ClN$
                                                            139593-05-8

cis-1-(3-chloro-4-Cyclohexylphenyl)-3-(hexahydro-1*H*-azepin-1-yl)prop-1-ene = (*Z*)-1-[3-(3-Chloro-4-cyclohexylphenyl)-2-propenyl]hexahydro-1*H*-azepine (●)
S SR 31742A
U σ-Opioid antagonist, antipsychotic

10694 **(8645)** $C_{21}H_{30}Cl_2N_2O_5$
                                                            107097-80-3

(±)-4-(3,4-Dichlorobenzamido)-*N*-(3-methoxypropyl)-*N*-pentylglutaramic acid = 4-[(3,4-Dichlorobenzoyl)amino]-

5-[(3-methoxypropyl)pentylamino]-5-oxopentanoic acid (●)
S CR 1505, *Loxiglumide***, *Oxiglumide*
U Cholecystokinin antagonist

10695 **(8646)** $C_{21}H_{30}Cl_2N_2O_5$
                                                            119817-90-2

(*R*)-4-(3,4-Dichlorobenzamido)-*N*-(3-methoxypropyl)-*N*-pentylglutaramic acid = (*R*)-4-[(3,4-Dichlorobenzoyl)amino]-5-[(3-methoxypropyl)pentylamino]-5-oxopentanoic acid (●)
S CR-2017, *Dexloxiglumide***
U Cholecystokinin antagonist

10696 **(8647)** $C_{21}H_{30}FN_3O_2$
                                                            1893-33-0

1'-[3-(*p*-Fluorobenzoyl)propyl][1,4'-bipiperidine]-4'-carboxamide = 1-[4-(*p*-Fluorophenyl)-4-oxobutyl]-4-piperidinopiperidine-4-carboxamide = *p*-Fluoro-γ-(4-carbamoyl-4-piperidinopiperidino)butyrophenone = 1'-[4-(4-Fluorophenyl)-4-oxobutyl][1,4'-bipiperidine]-4'-carboxamide (●)
R also dihydrochloride (2448-68-2)
S *Carpiperone*, Dipiperal, Dipiperon, Ellesdine, *Floropipamide*, *Fluoropipamide*, McN-JR 3345, *Pipamperone***, Piperonil, Piperonyl, Propitan, R 3345
U Neuroleptic, antipsychotic

10697 (8648) $C_{21}H_{30}FN_3O_2$

N-Cyclohexyl-4-[3-(p-fluorobenzoyl)propyl]-1-pipera-
zinecarboxamide = N-Cyclohexyl-4-[4-(4-fluorophenyl)-
4-oxobutyl]-1-piperazinecarboxamide (●)
R   Monohydrochloride (97507-92-1)
S   FG 5803
U   Antipsychotic

10698 (8649) $C_{21}H_{30}I_3N_3O_9$
                                     79211-10-2

2,4,6-Triiodo-1,3,5-benzenetricarboxylic acid tris[bis(2-
hydroxyethyl)amide] = N,N,N',N',N'',N''-Hexakis(2-hydr-
oxyethyl)-2,4,6-triiodo-1,3,5-benzenetricarboxamide (●)
S   *Iosimide**, Univist, ZK 36699
U   Diagnostic aid (radiopaque medium)

10699 (8650) $C_{21}H_{30}NO_4$
                                     47473-94-9

N-Methyl-3α-[(O-propionyltropoyl)oxy]-1αH,5αH-tro-
panium = endo-(±)-8,8-Dimethyl-3-[1-oxo-3-(1-oxoprop-
oxy)-2-phenylpropoxy]-8-azoniabicyclo[3.2.1]octane (●)
R   Nitrate (14319-87-0)
S   *Methylprampini nitras*, PAMN, *Prampine methyl ni-
       trate**, Protropin "Benzon"
U   Anticholinergic

10700 (8651) $C_{21}H_{30}NO_4$
                                     7182-53-8

8-Butyl-6β,7β-epoxy-3α-hydroxy-1αH,5αH-tropanium
(−)-tropate = N-Butyl-O-tropoylscopinium = [7(S)-
(1α,2β,4β,5α,7β)]-9-Butyl-7-(3-hydroxy-1-oxo-2-phe-
nylpropoxy)-9-methyl-3-oxa-9-azoniatricy-
clo[3.3.1.0²,⁴]nonane (●)
R   Bromide (149-64-4)
S   Ad 1, Alcopan, Amisepan, Anespas, Antispasmin
       "Green Cross", Bacotan, Barukomin, Brobutil, BS-ra-
       tiopharm, Bubusco-S, Bulamin, Bulospan, Bupain,
       Buscapin(a), Buscol, Buscolamin, Buscolt, Buscoly-
       sin, Buscom, Buscomeric, Buscopan, Buscopin,
       Buscoridin, Buscote, Buscotek, Buscovital, Busfoli-
       ron, Busforion, Buskas, Buskolizin, Buskolysin, Bu-
       spin, Buspon "Toyo Pharmar", Butibol, Butilescopo-
       lamina Duncan, Butipolan "Tobishi", Butopan, Buto-
       scapin, Butybolon, Butylmaido, Butylmin, Butyl-
       pan, *Butylscopolamin(e) (bromide)*, *Butylscopolami-
       ni bromidum*, *Butylscopolammonium bromide*, Bu-
       tylspan, Butylspasmin, Butymide, Butysco, Cifespas-
       mo, Colobolina, Contrispasman, Cryopina, Dhaco-
       pan, Diaste-M, Dividol Remedica, Donopon-B, Es-
       capin, Escopantil, Escopina, Espa-butyl, Espasmo-
       dor, Espasmotab, Estalan, Fujisco A, Gosper F,
       Hiofenil simple, Hioscian, Hioscina "Zimaia", Hiyo-
       sin, Hostosyl, Hybropan, *Hyoscine butyl bromide*,
       Hyoscomin, Hyospan, Hyospasmol, Hyscopan, Hys-
       nal, Hysopan, Inosco, Iscopan, Molit, Monospan,
       Moryspan, Nevrorestyl, Noapin A, Rejicopan S, Re-
       ladan, Rupe-N, Salfalgin, Scobro(n), Scobut, Scobu-
       til, Scobutiramin, Scopas, Scopbutan, Scopex, Scopi-

nal, *Scopolamine butyl bromide, Scopolaminium bu-tylbromatum*, Scopolan, Scordin B, Scorpan, Sedal-gim simplex, Selpiran-S, SKF 1637, Spanil, Spari-con, Spasmalexin, Spasmamina, Spasman scop, Spasmopan, Spasmowern, Spazmotek, Sporamin, Stibron, Tirantil, Toscopan, Tyaspan, Unispan, Vagotrope S, Verbindung 1637, Viviv

U Anticholinergic, spasmolytic

10701 (8652)

$C_{21}H_{30}N_2O$
5697-57-4

17-Methylandrost-4-eno[3,2-*c*]pyrazol-17β-ol = 17β-Hy-droxy-17-methylandrost-4-eno[3,2-*c*]pyrazole = (17β)-17-Methyl-2'*H*-androsta-2,4-dieno[3,2-*c*]pyrazol-17-ol (●)

S *Hydroxystenozole**

U Androgen, anabolic

10702 (8653)

$C_{21}H_{30}N_2O$
32421-46-8

N-Butyl-*N*-[2-(diethylamino)ethyl]-1-naphthamide = *N*-Butyl-*N*-[2-(diethylamino)ethyl]-1-naphthalenecarbox-amide (●)

R also citrate (80095-31-4) or monohydrochloride (58779-43-4)

S *Bunaftine**, Bunamide, *Bunaphtide*, Daredan, EU-16738, Meregon

U Anti-arrhythmic

10703 (8654)

$C_{21}H_{30}N_2O$
86024-64-8

(±)-2-*tert*-Butyl-α-[2-(4-piperidyl)ethyl]-4-quinolineme-thanol = (*RS*)-1-(2-*tert*-Butyl-4-quinolyl)-3-(4-piperidyl)-1-propanol = (±)-2-(1,1-Dimethylethyl)-α-[2-(4-piperidi-nyl)ethyl]-4-quinolinemethanol (●)

R also without (±)-definition (86073-85-0)

S PK-10139, *Quinacainol**

U Anti-arrhythmic

10704

$C_{21}H_{30}N_2O_4$
76470-87-6

N-Cyclohexyl-*N*-(2-hydroxyethyl)-4-[(1,2,3,4-tetrahy-dro-2-oxo-6-quinolinyl)oxy]butanamide (●)

S OPC-3911

U CAMP phosphodiesterase inhibitor

10705

$C_{21}H_{30}N_2O_5$
200954-39-8

Caproyl-L-prolyl-L-tyrosine methyl ester = 1-(1-Oxohe-xyl)-L-prolyl-L-tyrosine methyl ester (●)

S GZR-123

U Antipsychotic

10706　(8655)　　　　　　　　$C_{21}H_{30}N_2O_8S$
74639-40-0

(−)-(S)-2-Acetamido-N-(3,4-dihydroxyphenethyl)-4-(me-thylthio)butyramide bis(ethyl carbonate) (ester) = N-(N-Acetyl-L-methionyl)-O,O'-bis(ethoxycarbonyl)dopamine = (S)-Carbonic acid 4-[2-[[2-(acetylamino)-4-(methyl-thio)-1-oxobutyl]amino]ethyl]-1,2-phenylene diethyl es-ter (●)
S *Docarpamine***, TA-870, TA-8704, Tanadopa
U Cardiotonic, vasopressor (dopamine prodrug)

10707　　　　　　　　　　　$C_{21}H_{30}N_4O$
136816-67-6

N-Ethyl-2-[4-[(4-methoxy-3,5-dimethylphenyl)methyl]-1-piperazinyl]-3-pyridinamine (●)
S U-80493E
U Antiviral

10708　(8656)　　　　　　　　$C_{21}H_{30}N_4O_3S$
25859-76-1

N-[p-[[3-(3-Cyclohexen-1-yl)-2-imino-1-imidazolidi-nyl]sulfonyl]phenethyl]butyramide = N-[2-[4-[[3-(3-Cy-clohexen-1-yl)-2-imino-1-imidazolidinyl]sulfonyl]phe-nyl]ethyl]butanamide (●)
S *Glibutimine***, Glucidol, GP 51084
U Oral hypoglycemic

10709　(8657)　　　　　　　　$C_{21}H_{30}N_4O_3S$
85390-06-3

N-[p-[(3-Cyclohexyl-2-imino-1-imidazolidinyl)sulfo-nyl]phenethyl]crotonamide = (E)-N-[2-[4-[(3-Cyclo-hexyl-2-imino-1-imidazolidinyl)sulfonyl]phenyl]ethyl]-2-butenamide (●)
S CGP 11112
U oral hypoglycemic

10710　(8658)　　　　　　　　$C_{21}H_{30}N_4O_4$
66564-14-5

4-Amino-N-[1-(3-cyclohexen-1-ylmethyl)-4-piperidi-nyl]-2-ethoxy-5-nitrobenzamide (●)
R also tartrate (96623-56-2)
S Blaston, Cidine "Almirall", *Cinitapride***,
　　LAS 17177
U Neuroleptic, anti-emetic, stomachic

10711　(8659)　　　　　　　　$C_{21}H_{30}N_8O_3$
122009-61-4

2'-Hydroxy-3'-propyl-4'-[4-[2-[4-(1H-tetrazol-5-yl)bu-tyl]-2H-tetrazol-5-yl]butoxy]acetophenone = 1-[2-Hy-droxy-3-propyl-4-[4-[2-[4-(1H-tetrazol-5-yl)butyl]-2H-tetrazol-5-yl]butoxy]phenyl]ethanone (●)
S LY 203647
U Antihypertensive (leukotriene D_4-antagonist)

10712   (8660)

$C_{21}H_{30}O_2$
55303-98-5

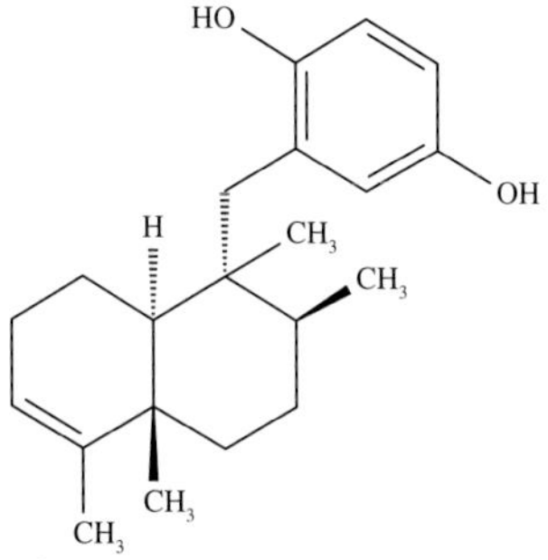

[1R-(1α,2β,4aβ,8aα)]-2-[((1,2,3,4,4a,7,8,8a-Octahydro-1,2,4a,5-tetramethyl-1-naphthalenyl)methyl]-1,4-ben-zenediol (●)
S   Avarol
U   Antibacterial, antifungal, antiviral

10713

$C_{21}H_{30}O_2$
177365-18-3

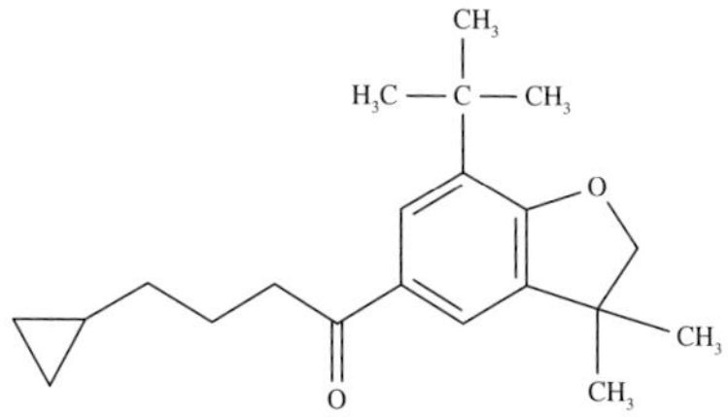

1-(7-tert-Butyl-2,3-dihydro-3,3-dimethylbenzo[b]furan-5-yl)-4-cyclopropyl-1-butanone = 4-Cyclopropyl-1-[7-(1,1-dimethylethyl)-2,3-dihydro-3,3-dimethyl-5-benzo-furanyl]-1-butanone (●)
S   PGV-20229
U   Anti-inflammatory

10714   (8661)

$C_{21}H_{30}O_2$
1972-08-3

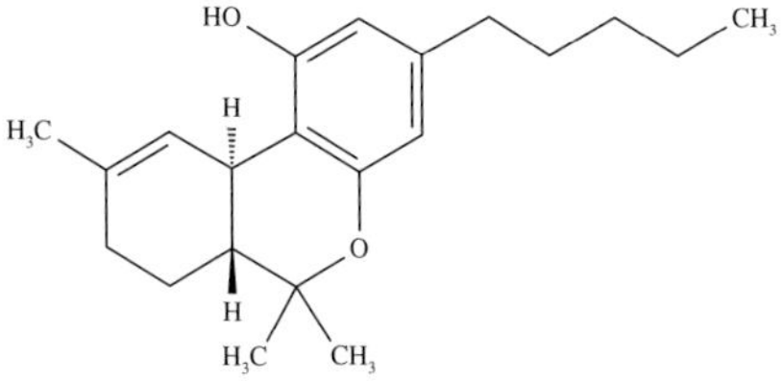

6aR,10aR)-(−)-6a,7,8,10a-Tetrahydro-6,6,9-trimethyl-3-pentylbenzo[c]chromen-1-ol = (6aR-trans)-6a,7,8,10a-

Tetrahydro-6,6,9-trimethyl-3-pentyl-6H-dibenzo[b,d]pyr-an-1-ol (●)
S   Cannabic, Compassia, delta-9-THC, Deltanyne, Dronabinol**, Elevat, Marinol, NSC-134454, QCD 84924, Sisacan, $Δ^9$-Tetrahydrocannabinol, $Δ^9$-THC
U   Euphoric, hallucinogen, anti-emetic adjunct in cancer chemotherapy, cognition enhancer

10715   (8662)

$C_{21}H_{30}O_2$
7663-50-5

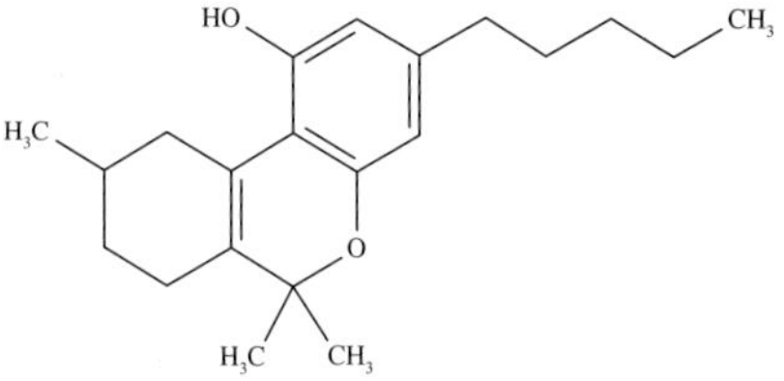

7,8,9,10-Tetrahydro-6,6,9-trimethyl-3-pentylben-zo[c]chromen-1-ol = 7,8,9,10-Tetrahydro-6,6,9-trime-thyl-3-pentyl-6H-dibenzo[b,d]pyran-1-ol (●)
S   Tetrahydrocannabinol
U   Euphoric, hallucinogen [active principle of the resin of Cannabis sativa (hashish, marihuana)]

10716   (8663)

$C_{21}H_{30}O_2$
23163-42-0

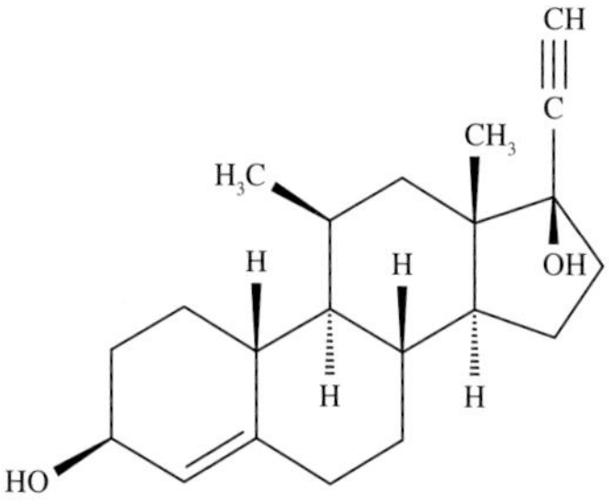

11β-Methyl-19-nor-17α-pregn-4-en-20-yne-3β,17-diol = (3β,11β,17α)-11-Methyl-19-norpregn-4-en-20-yne-3,17-diol (●)
R   see also no. 12927
S   Methynodiol, Metynodiol**
U   Progestin

10717 $C_{21}H_{30}O_2$
                                                  164266-48-2

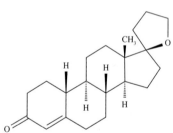

(*all-E*)-8-[3-Ethyl-2-(1-methylethyl)-2-cyclohexen-1-yli-
dene]-3,7-dimethyl-2,4,6-octatrienoic acid (●)
S   UAB-8
U   Dermatic

10718   (8664) $C_{21}H_{30}O_2$
                                                  57-83-0

Pregn-4-ene-3,20-dione (●)
R   see also nos. 10641, 12080, 13298, 13636, 13957
S   Agolutin, Akrolutin, Alprolut, Bio-Luton, Biostero-
ne "Biochem", BP 14, COL-1620, Colprosterone,
Corlutina, Corlutive, Corlutone, Corpolutin, Corpo-
mone, Corporin, Corpuluton, *Corpus luteum hormone*, Cosex PG 50, Crinone, Cutifitol, Cyclogest, Di-
ketolut-Inyect., Eazi-Breed, Elutin "Medicinalco",
Endolutina, Esolut, Femotrone, Flavolutan, Folo-
genon, *Gelbkörperhormon*, Gestacne, Gester-Gel,
Gesterol, Gestiron, Gestone, Gestormone, Ge-
stronaq, Glanducorpin, Glanestin, Gonadyl, Gynlu-
tin, Gynolutin, Gynolutone, Hormeva, Hormoflavei-
ne, Hormoluton, Lilutin, Lingusorbs, Lipo-Lutin, Lu-
corteum-Sol, Lugesteron, Lupronex-Amp., Luteine,
Luteinica, Luteinol, Lutemil, Luteocrinin, Luteocrin
normale, Luteocryst, Luteocyclin, Luteodyn, Luteo-
gan, Luteogen, *Luteohormone*, Luteoici, Luteo-Inex,
Luteol, Luteolin, Luteolipex, Luteomenin, Luteo-
mensin, Luteopur, Luteormon, Luteosan, Luteosid,
Luteostab, Luteosterone-fiale, Luteo-total, Luteovis,
Luteovister, Luteroid, Lutex, Lutidon(a), Lutin, Lu-
tociclina, Lutocor, Lutocyclin (M), Lutocylin (M),
Lutodin, Lutoform, Lutogil, Lutogyl, Lutogynon,

Lutoidral, Lutolin, Luto-Quibi, Lutorm, Lutormon,
Lutren, Lutromone, Lutron, Lutyrexin, Macrogestin,
Mafel, Maslutein, Mastogel, Membrettes, Me-
phalutren, Migestrone, Nalutone, Nalutron, Neolu-
teone, Neolutin "Orma", NSC-9704, Oophormin-Lu-
teum, Ormoluteina, Ornisteril, Pan-Gest, Pregnedio-
ne, Premastrone, Primolut, Prodion, Profac-O, Pro-
gatin, Progeffik, Proge-Hormon, Progekan, Proge-
lan, Progelun, Progelut, Progenar, Progenin, Proge-
rin, Progest, Progestaject-50, Progestamate, Proge-
stan, Progestaq, Progestar, Progestasert, Progestero-
id 50, Progesterol, *Progesterone***, Progestilin, Progestin, Progestogel, Progestol(o), Progestona, Proge-
stosol, Progestro-Med, Progestronal, Progestronaq,
Progeubril, Progman, Progokab, Progonaq, Progo-
nol, Progynol forte, Projenil, Projestonil, Prolets,
Prolidon, Prolusteron, Proluton, Prometrium,
Promulse, Pronasone, Pronogene, Prontogest, Proro-
ne, Protormone, Sigmaprogestin, Syngesterone, Syn-
gestrets, Syntolutan, Syntolutin-Amp., Teralutil, Ti-
tho-Gel, Utrogest, Utrogestan, Venolutin
U   Progestin (corpus luteum hormone)

10719   (8665) $C_{21}H_{30}O_2$
                                                  1235-13-8

Estr-4-en-3-onespiro-17α,2'-(tetrahydrofuran) = (17β)-
4',5'-Dihydrospiro[estr-4-ene-17,2'(3'*H*)-furan]-3-one (●)
S   *Furanestrenone, 19-Norspiroxenone*
U   Anti-estrogen

10720 (8669) $C_{21}H_{30}O_3$
                                                1045-69-8

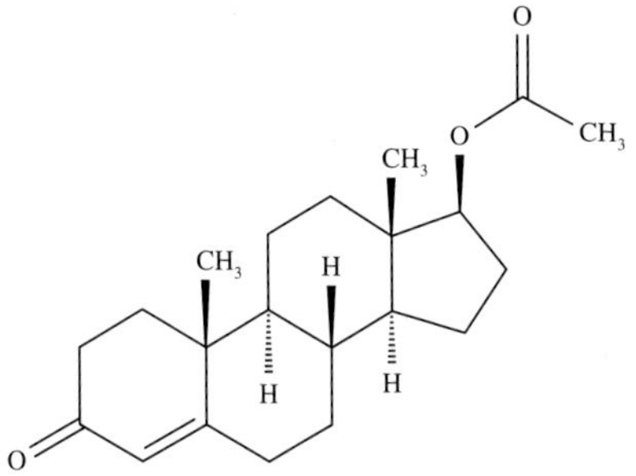

17β-Hydroxyandrost-4-en-3-one acetate = 3-Oxoandrost-4-en-17β-yl acetate = (17β)-17-(Acetyloxy)androst-4-en-3-one (●)

S Acéto-Stérandryl, Aceto-Testoviron, Amolisin, Androtest A, Deposteron, Farmatest, Perandrone A, *Testosterone acetate*

U Androgen

10721 (8670) $C_{21}H_{30}O_3$
                                                6157-87-5

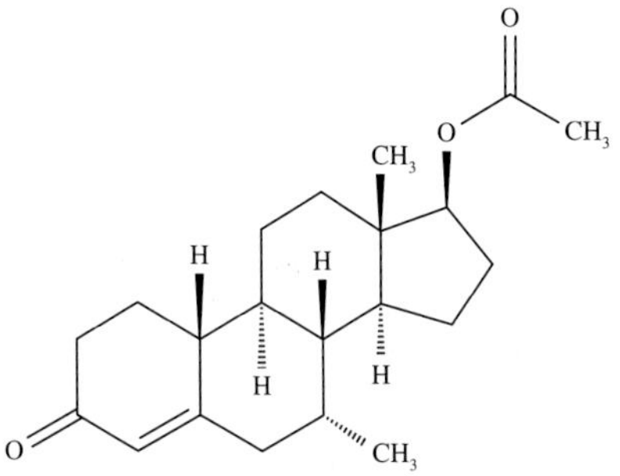

17β-Hydroxy-7α-methylestr-4-en-3-one acetate = 7α,17β)-17-(Acetyloxy)-7-methylestr-4-en-3-one (●)

S CDB 903, NSC-69948, *Trestolone acetate**, U-15614

U Androgen, antineoplastic

10722 (8671) $C_{21}H_{30}O_3$
                                                7207-92-3

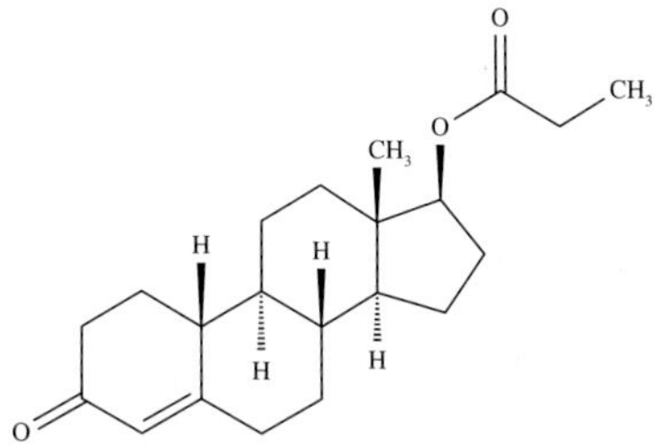

17β-Hydroxyestr-4-en-3-one propionate = 17β-Hydroxy-19-norandrost-4-en-3-one propionate = (17β)-17-(1-Oxo-propoxy)estr-4-en-3-one (●)

S Anabolicus, *Nandrolone propionate*, Nor-Anabol, Nortesto, *Nortestosterone propionate*, Norybol-19, Pondus, Testobolin

U Anabolic

10723 $C_{21}H_{30}O_3$
                                                149639-79-2

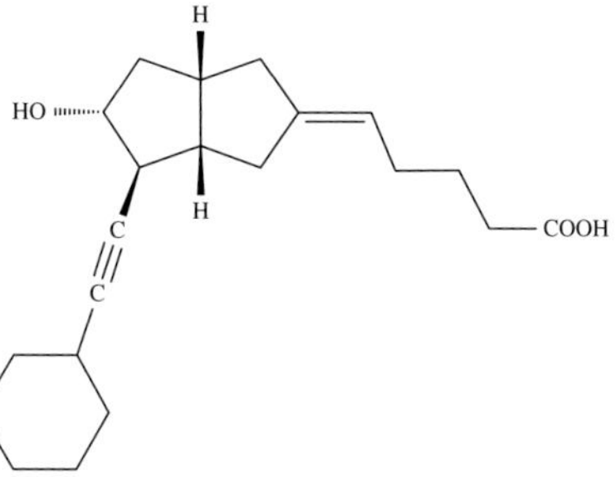

5-[(3aS,4S,5R,6aS)-4-(Cyclohexylethynyl)hexahydro-5-hydroxy-2(1H)-pentalenylidene]pentanoic acid (●)

S MM 706

U Platelet aggregation inhibitor

10724 (8666) $C_{21}H_{30}O_3$
                                                80-75-1

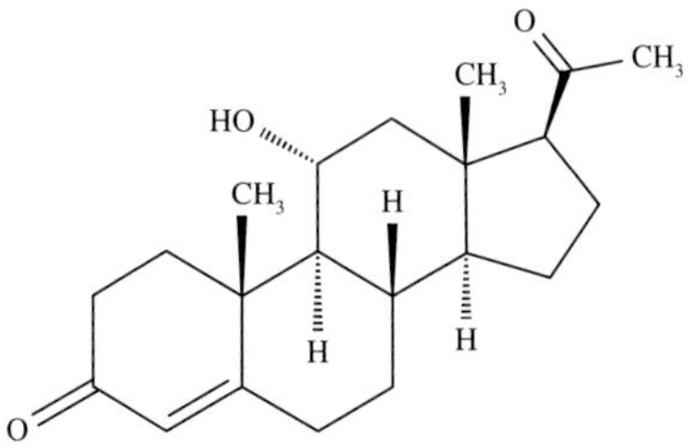

11α-Hydroxypregn-4-ene-3,20-dione = (11α)-11-Hydroxypregn-4-ene-3,20-dione (●)

S   11α-*Hydroxyprogesterone, Idrogesterone,*
11αOHP, Setaderm, 11-α-Testaktiv, U-0384

U   Topical anti-androgen

10725   (8667) $C_{21}H_{30}O_3$
                                                   68-96-2

17-Hydroxypregn-4-ene-3,20-dione (●)

R   see also nos. 12080, 13636, 13957

S   *Hydroxyprogesterone**

U   Progestin

10726   (8668) $C_{21}H_{30}O_3$
                                                   64-85-7

4-Pregnen-21-ol-3,20-dione = 21-Hydroxypregn-4-ene-3,20-dione (●)

R   see also nos. 12081, 13304, 13642, 13958, 14197

S   *Compound A,* Cortexon, Corthormon, Corticormon,
DCXA, *Decortone, Deoxycortone, Desossicorticosterone, Desoxicortonum, 11-Desoxycorticosterone,
Desoxycorticosteron(um), Desoxycortone**,* DOC,
21-Hydroxyprogesterone, Kendall's desoxy compound B, Kendalls Desoxyverbindung B,
NSC-11319, Ocriten, Ormosurrenol, *Reichstein's
substance Q, Verbindung Q von Reichstein*

U   Mineralocorticoid (salt-regulating)

10727   (8677) $C_{21}H_{30}O_4$
                                                   32696-18-7

Androst-5-ene-3β,17β-diol diformate = (3β,17β)-Androst-5-ene-3,17-diol diformate (●)

S   Bisexovis-orale

U   Anabolic, androgen

10728   (8672) $C_{21}H_{30}O_4$
                                                   105284-21-7

[1S-[1α,2α(R*),3β,6α,7Z]]-4-[2-(3-Cyclohexyl-3-hydroxy-1-propynyl)-3-hydroxybicyclo[4.2.0]oct-7-ylidene]butanoic acid (●)

S   RS-93427-007

U   Antihypertensive, platelet aggregation inhibitor, antiatherosclerotic

---

10729 $C_{21}H_{30}O$
                                                   163106-91-0

(1S*,3S*)-1-Hydroxy-3-[(3R*,E)-3-hydroxy-7-phenyl-1-heptenyl]-1-cyclohexaneacetic acid = [1α,3β(1E,3S*)]-1-Hydroxy-3-(3-hydroxy-7-phenyl-1-heptenyl)cyclohexaneacetic acid (●)

R   Monosodium salt (163251-41-0)

S PH-163
U Leukotriene B$_4$ receptor antagonist

10730 (8673) $C_{21}H_{30}O_4$
                                                          13445-75-5

α-Butylbenzyl hydrogen camphorate = Camphoric acid
mono(1-phenylpentyl) ester = *cis*-1,2,2-Trimethyl-1,3-cy-
clopentanedicarboxylic acid 1-(1-phenylpentyl) ester (●)
R Sodium salt (4201-91-6)
S *Camphamyl sodium*, Colerex "Beta", *Fenipentol sodium camphorate*, Flubigen, Flubilar
U Choleretic

10731 $C_{21}H_{30}O_4$
                                                          31685-80-0

15-Hydroxylabda-8(20),13-diene-16,19-dioic acid γ-lac-
one methyl ester = (1S,4aR,5S,8aR)-5-[2-(2,5-Dihydro-2-
oxo-3-furanyl)ethyl]decahydro-1,4a-dimethyl-6-methy-
ene-1-naphthalenecarboxylic acid methyl ester (●)
S *Pinusolide*
U PAF receptor antagonist from *Biota orientalis*

10732 (8674) $C_{21}H_{30}O_4$
                                                          50-22-6

11β,21-Dihydroxypregn-4-ene-3,20-dione = 4-Pregnene-
11β,21-diol-3,20-dione = (11β)-11,21-Dihydroxypregn-
4-ene-3,20-dione (●)
S *Compound B, Corticosterone*, Hydroxycortexon,
Kendall's compound B, NSC-9705, *Reichstein's
substance H, Verbindung B von Kendall,
Verbindung H von Reichstein*
U Adrenocortical steroid

10733 (8675) $C_{21}H_{30}O_4$
                                                          595-77-7

16α,17-Dihydroxypregn-4-ene-3,30-dione = 16α,17-Di-
hydroxyprogesterone = (16α)-16,17-Dihydroxypregn-4-
ene-3,20-dione (●)
R see also nos. 12574, 14126
S *Alfasone, Algestone**, Alphasone*
U Glucocorticoid (anti-inflammatory)

10734 (8676) $C_{21}H_{30}O_4$
                                                          152-58-9

17,21-Dihydroxypregn-4-ene-3,20-dione (●)
S *Cortexolone*, Cortifen, *Cortodoxone**, NSC-18317, *Reichstein's substance S, Reichstein's Substanz S,* SKF 3050, *Verbindung S von Reichstein*
U Glucocorticoid (anti-inflammatory)

10735 (8678) $C_{21}H_{30}O_4S$
61951-99-3

11β,17-Dihydroxy-21-mercaptopregn-4-ene-3,20-dione = (11β)-11,17-Dihydroxy-21-mercaptopregn-4-ene-3,20-dione (●)
R see also nos. 12957, 13308, 13943
S *Tiocortisol, Tixocortol***
U Glucocorticoid (anti-inflammatory, anti-allergic)

10736 (8679) $C_{21}H_{30}O_5$
566-35-8

11α,17,21-Trihydroxypregn-4-ene-3,20-dione = (11α)-11,17,21-Trihydroxypregn-4-ene-3,20-dione (●)
S 11α-Cortisol, *Epicortisol*, Epi-F, *Epihydrocortisone*, *11-Isocortisol*, U-1676
U Angiostatic

10737 (8680) $C_{21}H_{30}O_5$
50-23-7

11β,17,21-Trihydroxypregn-4-ene-3,20-dione = 4-Pregnene-11β,17α,21-triol-3,20-dione = (11β)-11,17,21-Trihydroxypregn-4-ene-3,20-dione (●)
R see also nos. 10756, 11452, 12084, 12933, 12959, 13310, 13640, 13643, 13945, 14199, 15094, 15392
S Acortine, Acticort, Acusone, Aeroseb-HC, Ala-Cort, Ala-Scalp, Algicortis, Alocort, Alphacortison, Alphaderm, Alumate-HC, Amberin, Anflam, *Anti-inflammatory hormone*, Aquacort, Aquanil HC, Bacid "Farma-Lepori", Bactine Hydrocortisone, Balneol-HC, Barriere-HC, Barseb HC, Basan-Corti, Beta-HC, Bio-cort, Bio-Cortex, Caldecort-Spray, Calmoderm, Cetacort, Clocort, Cobadex, *Compound F von Kendall*, CortaGel, Cortaid maximum strength, Cortaid-Spray, Cortanal, Cortate "Schering Corp.", Cort-Dome, Cortef, Cortemen, Cortenema, Cortesal, Corticare, Corticorinol, Cortifair, Cortifan, Cortinal, *Cortisol*, Cortispray, Cortizone-S, Cort-Nib, Cortoderm "K-Line", Cortolotion, Cortopin, Cortoxide, Cortril, Cort-top, Cotacort, Covocort, Crema-Transcutanea Astier, Cremesone, Cremicort H, Cutaderm, Cutisol, Cycort, Delacort, Delcort "Hauck", Dermacalm, Dermacort "Merckle; Rowell; Solvay", Dermaflex HC, Derm-aid, Dermallerg-ratiopharm, Dermapan, Dermarest Dricort, Dermaspray, Dermaspray démangeaison, Derm Cort, Dermicort, Dermil "Nettopharma", Dermocortal, Dermolate "Schering Corp.", Dermolen, Dermol HC, Dermo-posterisan, Dermtex HC, Diacort "Nadeau", Dioderm, Dome-cort, Domolene-HC, DP HC, Durel-Cort, Ecosone, Eczacort, Efcorlin, Efcortelan, Efcorticon, Egocort, Eldecort, Emo-cort, Endofuson, Epicort "Blue line", Episone, Evacort, Eyecort, Ficortril, Filocort, Foille Insetti, Genacort, Gly-cort, Grecort, Gyno-Cortisone, Hautosone, HC, HC-cream, HC-Derma-Pax, H-Cort, Heb-Cort, Hi-Cor, Hi-Cort, Hidroaltesona, Hidro-Colisona, *Hidrocortisona, Hidrocortizon*, Hid

rotisona, Hycor, Hycort, Hycortin, Hycortol(e), Hydracort, Hydrasson, Hydro-Adreson-Salbe, Hydrocort "DHA; Ferring; Pharmagalen", Hydrocortal, Hydrocortemel, Hydro-Cortex, Hydrocortifor, Hydro-Cortilean, *Hydrocortisone**,* Hydrocortistab (oint and tabs), Hydrocortisyl, Hydrocortodrin, Hydrocortone-Tabl., Hydrocutan-Salbe, Hydroderm, Hydrofoam, Hydrogalen, *Hydrokortison,* Hydrotex, Hydrotisona, Hydrotoin, Hydrovate, *17-Hydroxycorticosterone,* Hymac, Hynax, Hyproderm HC, Hysone, Hytisone, Hytone, *Idrocortisone,* Ivocort, Jungle Formula Sting Relief Cream, *Kendall's compound F,* Kericort, Komed HC, Kort "Srip", Kyypakkaus, Lacticare-HC, Lactisona, Lexocort, Lexoderm, Liocort "Ronava", Lubricort, Maintasone, Manticor, Maso-Cort, Maximum Strength Cortaid "Upjohn", Maximum Strength Kericort-10 "B-M Sq", Medicort "Almay; Care", Mensicort, Microcort, Microsona, Mildison, Milliderm, Monocort "Nortech", Munitren H, My-Cort, Neutrogena T/Scalp, No More Itchies, Noviltisona, NSC-10483, Nutracort "Alcon; Galderma; Owen", Optef, Optisone, Otosone F, *17-Oxycorticosterone,* Panhydrosone, Penecort, Pharmacort, Phiacort, Polcort H, Polysorb-HC, Prepcort, Prevex HC, Procort, Proctocort "Boehringer-Ingelheim; Solvay", Proctocream-HC, Proctofoam, Procutan, ratioAllergy Hydrocortison, Recort, Rectoid, *Reichstein's substance M,* Remederm HC, Rocort, Ruhlicort, Sagittacortin-Creme, Sanatison mono, Sarna HC, Sarnol-HC, Scalpicin Capilar, Schericur (new form), Scheroson F-Tabl., Sensacort, Sigmacort, Signef, Silian, Skincalm, S-T Cort, Sterocort "Omega", Stie-Cort, Stiefcortil, Synacort, Systral Hydrocort, Tega-Cort, Tegrin HC, Texacort, Timocort, Topcort, Topicort (old form), Topisone "Syntex", Transderma H, Traumaide, T/Scalp, U-1851, Ulcort (Cream + Lotion), Ultraderm "Berlimed", Unicort "A. & H.; Glaxo", Uniderm, *Verbindung F von Kendall, Verbindung M von Reichstein,* Waspeze, Zenoxone
U Glucocorticoid (anti-inflammatory, anti-allergic)

10738 (8681)

$C_{21}H_{30}O_6$
136164-66-4

(*E*)-2-[(4,5-Dimethoxy-2-methyl-3,6-dioxo-1,4-cyclohexadien-1-yl)methylene]undecanoic acid (●)
S E-3330
U Hepatoprotectant

10739 (8682)

$C_{21}H_{30}O_{14}$
60197-59-3

2'-[(6-*O*-α-D-Glucopyranosyl-β-D-glucopyranosyl)oxy]-6'-hydroxy-4'-methoxyacetophenone = 1-[2-[(6-*O*-α-D-Glucopyranosyl-β-D-glucopyranosyl)oxy]-6-hydroxy-4-methoxyphenyl]ethanone (●)
S *Hyrcanoside*
U Cardiotonic from *Dorema hyrcanum*

10740

$C_{21}H_{31}BrClN_7O_2$
119646-52-5

3,5-Diamino-*N*-[2-[[2-[[[3-bromo-5-(1,1-dimethylethyl)-2-hydroxyphenyl]methyl]amino]ethyl]methylamino]ethyl]-6-chloropyrazinecarboxamide (●)
S ICI 206970
U Eucalemic diuretic

10741 (8683) $C_{21}H_{31}ClN_2O$
21363-18-8

1-(*o*-Chlorobenzyl)-α-[(di-*sec*-butylamino)methyl]pyr-role-2-methanol = 1-[1-(*o*-Chlorobenzyl)-2-pyrrolyl]-2-(di-*sec*-butylamino)ethanol = α-[[Bis(1-methylpro-pyl)amino]methyl]-1-[(2-chlorophenyl)methyl]-1*H*-pyr-role-2-methanol (●)
R also *p*-hydroxybenzoate (1:1) (23784-10-3)
S Dividol, *Diviminol*, Lenigesial "Inpharzam", Rich-dor, *Viminol***, Z. 424
U Central analgesic, antitussive

10742 (8684) $C_{21}H_{31}NO$
57653-29-9

3-(Cyclobutylmethyl)-6-ethyl-1,2,3,4,5,6-hexahydro-11,11-dimethyl-2,6-methano-3-benzazocin-8-ol (●)
S *Cogazocine***
U Analgesic

10743 (8685) $C_{21}H_{31}NO_2$
360-66-7

17-Methyl-5α-androstano[3,2-*c*]isoxazol-17β-ol = 17β-Hydroxy-17α-methyl-5α-androstano[3,2-*c*]isoxazole = (5α,17β)-17-Methylandrostano[3,2-*c*]isoxazol-17-ol (●)
S AIZ, *Androisoxazole*, Androxan, Neo-Ponden, neo-Pondus

U Anabolic

10744 (8686) $C_{21}H_{31}NO_2$
20448-86-6

3-(Diethylamino)propyl 2-phenyl-2-norbornanecarboxy-late = 2-Phenyl-2-norbornanecarboxylic acid 3-(diethyl-amino)propyl ester = 2-Phenylbicyclo[2.2.1]heptane-2-carboxylic acid 3-(diethylamino)propyl ester (●)
R Hydrochloride (26908-91-8)
S *Bornaprine hydrochloride***, Kr 339, Sormodren
U Antiparkinsonian

10745 (8687) $C_{21}H_{31}NO_?$
99876-41-?

(17β)-17-[(3-Hydroxypropyl)amino]estra-1,3,5(10)-tri-en-3-ol (●)
S *Prolame*
U Anticoagulant, antineoplastic

10746 (8688) $C_{21}H_{31}NO?$
79700-63-?

1-[1-(Isobutoxymethyl)-2-[(1-methyl-1-phenyl-2-propy-nyl)oxy]ethyl]pyrrolidine = 1-[1-[(1-Methyl-1-phenyl-2-propynyl)oxy]methyl]-2-[(2-methylpropoxy)ethyl]pyrro-lidine (●)
S *Fronedipil, Fronepidil***

U   Anti-arrhythmic, anti-ischemic

10747   (8689) $C_{21}H_{31}NO_3$
65429-87-0

(±)-4'-[3-(*tert*-Butylamino)-2-hydroxypropoxy]spiro[cyclohexane-1,2'-indan]-1'-one = (±)-4'-[3-[(1,1-Dimethylethyl)amino]-2-hydroxypropoxy]spiro[cyclohexane-1,2'-[2H]inden]-1'(3'H)-one (●)
R   also without (±)-definition (81840-58-6)
S   LI-32468, S-32-468, *Spirendolol***
U   β-Adrenergic blocker

10748   (8690) $C_{21}H_{31}NO_4$
2385-81-1

-Phenyl-1-[2-[(tetrahydrofurfuryl)oxy]ethyl]-4-piperidinecarboxylic acid ethyl ester = Ethyl 4-phenyl-1-[2-[(tetrahydrofurfuryl)oxy]ethyl]-4-piperidinecarboxylate = 4-Phenyl-1-[2-[(tetrahydrofurfuryl)oxy]ethyl]isonipecotic acid ethyl ester = 4-Phenyl-1-[2-[(tetrahydro-2-furayl)methoxy]ethyl]-4-piperidinecarboxylic acid ethyl ester (●)
   *Furaethidin, Furethidine***, Furetidina*, TA 48
U   Narcotic analgesic

10749 $C_{21}H_{31}N_3O_4$
147783-67-3

$(2R,3S)$-N^4-Hydroxy-N^1-[(1S)-2-(methylamino)-2-oxo-1-(phenylmethyl)ethyl]-2-(2-methylpropyl)-3-(2-propenyl)butanediamide (●)
S   BB-1101
U   Metalloproteinase and tumor necrosis factor inibitor

10750 $C_{21}H_{31}N_3O_5$
191406-88-9

$(6S,7R,10S)$-N^6-Hydroxy-N^{10}-methyl-7-(2-methylpropyl)-8-oxo-2-oxa-9-azabicyclo[10.2.2]hexadeca-12,14,15-triene-6,10-dicarboxamide (●)
S   SE-205
U   Matrix metalloproteinase and tumor necrosis factor inhibitor

---

10751   (8691) $C_{21}H_{31}N_3O_5$
23887-47-0

N-Isopropyl-4-(3,4,5-trimethoxycinnamoyl)-1-piperazineacetamide = N-(1-Methylethyl)-4-[1-oxo-3-(3,4,5-trimethoxyphenyl)-2-propenyl]-1-piperazineacetamide (●)
R   Maleate (1:1) (26328-00-7)
S   *Cinpropazide maleate***, MD 68111

U   Coronary vasodilator

10752   (8692) $C_{21}H_{31}N_3O_5$
76547-98-3

1-[N^2-[(S)-1-Carboxy-3-phenylpropyl]-L-lysyl]-L-pro-
line = (S)-1-[N^2-(1-Carboxy-3-phenylpropyl)-L-lysyl]-L-
proline (●)
R   also dihydrate (83915-83-7)
S   Acemin "Astra-Zeneca; ICI", Acerbon,
Acerbon Cor, Acerdil, Aceril, Acerilin, Acetan
"Kwizda; Merck Sharp & Dohme", Adicanil, Ala-
pril, Carace, Cipril, Cipril "Cipla", Coric, Coric
card, Dapril, Doneka, Doxapril, Doxapril "Bago",
Ecalisin, Gnostoval, Hidropresse, Hypomed, Icoran,
Inhibril, Irumed, L-154826, Landolaxin, Leruze,
Linopril, Linvas, Lipreren, Lipril, Lisibeta, Lisigam-
ma, Lisihexal, Lisilet, Lisi-Lich, Lisinal, *Lisino-
pril**, Lisipril, Lisi-Puren, Lisir, Lisodura, Liso-
press, Lisoril, Listril, Longes, Loril, MK-521, Nafo-
dryl, Nisinol, Nopril, Novatec, Perenal, Presokin
"Sanitas", Press-12, Prinil, Prinivil, Prinizil, Prinzi-
de, Privinil, Rantex, Rilace, Secubar, Sedotensil, Sedotensil "Ramon", Sinopril, Sinopryl, Skopril, Tan-
stop, Tensikey, Tensiopril, Tensopril "Syncro; Teva", Tensyn, Tersif, Thriusedon, Tonoten, Tonoten-
sil, Unopril, Veroxil "Anfarm", Vivatec, Z-bec, Zen-
tril, Zestril
U   Antihypertensive (ACE inhibitor)

10753 $C_{21}H_{31}N_3O_{12}S$
202656-49-3

2-[[(2S)-2-(Acetylamino)-3-methyl-3-(nitrosothio)-1-
oxobutyl]amino]-2-deoxy-β-D-glucopyranose 1,3,4,6-te-
traacetate (●)
S   RIG 200
U   Vasodilator (nitric oxide donor)

10754   (8693) $C_{21}H_{31}N_5O$
127266-56-2

N-[2-[4-(2-Pyrimidinyl)-1-piperazinyl]ethyl]-1-adaman-
tanecarboxamide = *N*-[2-[4-(2-Pyrimidinyl)-1-piperazi-
nyl]ethyl]tricyclo[3.3.1.1^{3,7}]decane-1-carboxamide (●)
R   Monohydrochloride (144966-96-1)
S   *Adatanserin hydrochloride***, Wy-50324
U   Anxiolytic, antidepressant

10755   (8694) $C_{21}H_{31}N_5O_2$
36505-84-

N-[4-[4-(2-Pyrimidinyl)-1-piperazinyl]butyl]-1,1-cyclo-
pentanediacetimide = 1-[4-[4-(2-Pyrimidinyl)-1-piperazi-
nyl]butyl]piperidine-4-spirocyclopentane-2,6-dione = 8-

[4-[4-(2-Pyrimidinyl)-1-piperazinyl]butyl]-8-azaspi-
ro[4.5]decane-7,9-dione (●)
R Monohydrochloride (33386-08-2)
S Abivax, Anchocalm, Ansial, Ansiced, Ansienon, An-
sitec, Ansiten, Anxiolan, Anxiron, Apo-Buspirone,
Axoren, Barpil, Bergamol, Bespar, Biron, Boronex,
Brispar, Bulantil, Buparon, Busansil, Busansilium,
Buscalm(a), Busp, Buspan, Buspanil, Buspar, Buspi-
men, Buspinol, *Buspirone hydrochloride**, Buspi-
sal, Buspium, Buspon, Buspon "Deva", Censpar,
Dalpas, Effiplen, Establix, Hiremon, Hobatstress, Id-
roxin, Itagil, Italgil, Komasin, Lebilon, Ledion, Los-
iren, Loxapin "Norma Hellas", Lucelan, Mabuson,
MJ 9022-1, Nadrifor, Narol, Natrabus, Nerbet, Ner-
vilon "Fidelis", Nervostal, Neurorestol, Neurosine
"Armstrong", Nevrorestol, Nodeprex, Nopiron, Nor-
bal, Normaton, Nosedar, Paceum "Sanus", Pasrin,
Pax "Codilab", Paxil "Olsen", Paxon "Eurolab", Pax-
tran, Pendium, Piranil, Piroxona, Psibeter, Psicofar,
Psicofar "Medinfar", Sbrol, Senspar, Simpless, Spa-
milan, Stressigal, Supiron, Svitalark, Tendan, Ten-
sispes, Trafuril "Faran", Tranbus, Transiol, Travin,
Travin "Lek", Tutran, Umolit, Umotil, Veparon,
Zyntabac
U Anxiolytic

10756   (8695) $C_{21}H_{31}O_8P$
3863-59-0

1β,17,21-Trihydroxypregn-4-ene-3,20-dione 21-(dihydrogen phosphate) = Hydrocortisone 21-phosphate =
11β)-11,17-Dihydroxy-21-(phosphonooxy)pregn-4-ene-
,20-dione (●)
R see also no. 15392;
disodium salt (6000-74-4)
S Actocortin, Actocortina, Antax, Cleiton, Corhydron,
Corphos, Corticool, Cortiphate, Efcortesol inj., Enty-
corten, Flebocortid P 1000, *Hydrocortisone disodium
phosphate, Hydrocortisone Sodium Phosphate*, Hydrocortone Phosphate, Hydrocortone (water soluble),
Idracemi, Medecort, Physiocortison

U Glucocorticoid (anti-inflammatory, anti-allergic)

10757   (8696) $C_{21}H_{32}N$
27112-40-9

1-[3-(1-Cyclohexen-1-yl)-3-phenylpropyl]-1-methylpi-
peridinium (●)
R Methyl sulfate (30817-43-7)
S *Fenclexonium metilsulfate**, Hoechst 019
U Anticholinergic, spasmolytic

10758   (8697) $C_{21}H_{32}N$
16826-19-0

N-Laurylisoquinolinium = 2-Dodecylisoquinolinium (●)
R Bromide (93-23-2)
S Isothan (Q15), *Laurylisoquinolinium bromide*
U Antifungal

10759   (8698) $C_{21}H_{32}NO_3$
116533-64-3

2-(2-Hydroxymethyl)-1,1-dimethylpyrrolidinium α-phenylcyclohexaneglycolate = 2-[[(α-Cyclohexylphenylglycoloyl)oxy]methyl]-1,1-dimethylpyrrolidinium = 2-

[[(Cyclohexylhydroxyphenylacetyl)oxy]methyl]-1,1-di-
methylpyrrolidinium (●)
R   Bromide (561-43-3)
S   BM 3055, BRL 556, Immetro, Immétropan,
    L.D. 3055, *Oxipyrroni bromidum*, *Oxypyrronium*
    bromide**
U   Anticholinergic, spasmolytic

10760   (8699) $C_{21}H_{32}N_2O$
                                                              10418-03-8

17-Methyl-5α-androstano[3,2-*c*]pyrazol-17β-ol = 17β-
Hydroxy-17α-methyl-5α-androstano[3,2-*c*]pyrazole =
(5α,17β)-17-Methyl-2'*H*-androst-2-eno[3,2-*c*]pyrazol-
17-ol (●)
S   Anabol, Anasyth, Anazole, *Androstanazol*, *Estano-*
    zolol, Estazol, Estazolina, Menabol, *Methylstanazo-*
    lum, Neurabol Caps., NSC-43193, *Stanazolol*, *Stano-*
    zolol**, Stromba, Strombaject, Tevabolin,
    Win 14833, Winstroid, Winstrol
U   Anabolic, androgen

10761   (8700) $C_{21}H_{32}N_2O$
                                                              87189-13-7

17β-[(3-Aminopropyl)amino]estra-1,3,5,(10)-trien-3-ol =
N-[3-Hydroxyestra-1,3,5(10)-trien-17β-yl]-1,3-propy-
lenediamine = (17β)-17-[(3-Aminopropyl)amino]estra-
1,3,5(10)-trien-3-ol (●)
S   *Prodiame*
U   Antithrombotic

10762   (8701) $C_{21}H_{32}N_2O_2$
                                                              118516-27-1

1-(3-Cyclohexylpropionyl)-4-(2-ethoxyphenyl)pipera-
zine = 1-(3-Cyclohexyl-1-oxopropyl)-4-(2-ethoxyphe-
nyl)piperazine (●)
R   Monohydrochloride (96222-01-4)
S   D-16120
U   Analgesic

10763 $C_{21}H_{32}N_4$
                                                              136577-30-5

N-[2-(Diethylamino)-2-methylpropyl]-6-phenyl-5-pro-
pyl-3-pyridazinamine = N^2,N^2-Diethyl-2-methyl-N^1-(6-
phenyl-5-propyl-3-pyridazinyl)-1,2-propanediamine (●)
R   Fumarate (2:3) (137733-33-6)
S   SR 46559A
U   Muscarinic agonist (in Alzheimers disease)

10764 $\qquad$ $C_{21}H_{32}N_4O_2S$
93822-42-5

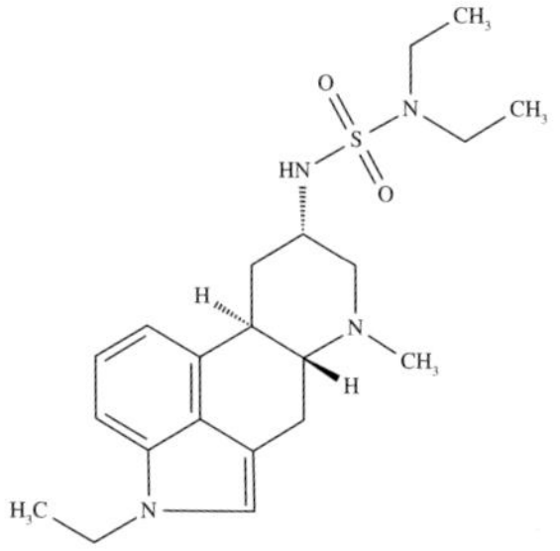

N,N-Diethyl-*N'*-[(8α)-1-ethyl-6-methylergolin-8-yl]sulf-
amide (●)

S CQA 206-291

U Dopaminergic ergot agonist

10765 $\qquad$ $C_{21}H_{32}N_4O_3$
164178-34-1

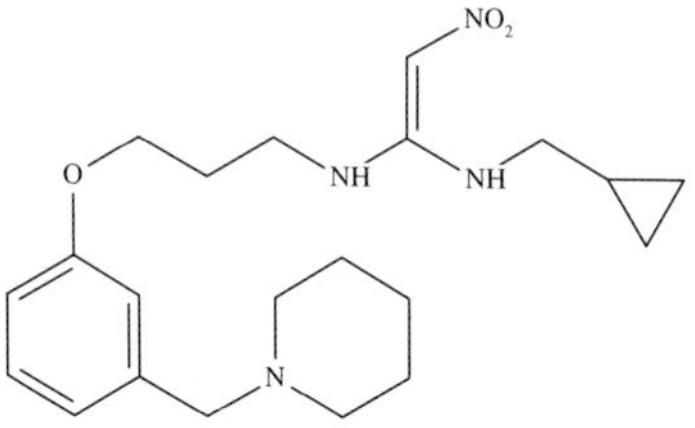

N-(Cyclopropylmethyl)-2-nitro-*N'*-[3-[3-(1-piperidinyl-
methyl)phenoxy]propyl]-1,1-ethenediamine (●)

S JB-9322

U Histamine H_2 receptor antagonist (anti-ulcer)

10766 $\qquad$ $C_{21}H_{32}N_4O_3$
146269-39-8

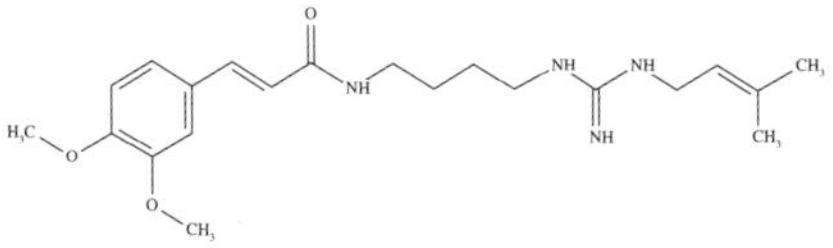

E)-1-[(3,4-Dimethoxycinnamoyl)amino]-4-[(3-methyl-
-butenyl)guanidino]butane = (*E*)-3-(3,4-Dimethoxyphe-
yl)-*N*-[4-[[imino[(3-methyl-2-butenyl)amino]me-
hyl]amino]butyl]-2-propenamide (●)

R also mixture with the (*Z*)-form (146269-40-1)

S Caracasanamide

U Antihypertensive from *Verbesina caracasana*

10767 (8702) $\qquad$ $C_{21}H_{32}N_4O_5$
69479-26-1

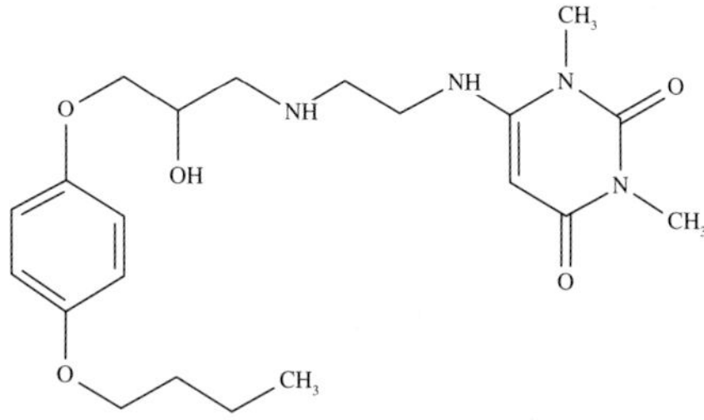

(±)-6-[[2-[[3-(*p*-Butoxyphenoxy)-2-hydroxypropyl]ami-
no]ethyl]amino]-1,3-dimethyluracil = *N*-[3-(4-Butoxy-
phenoxy)-2-hydroxypropyl]-*N'*-[1,2,3,4-tetrahydro-1,3-
dimethyl-2,4-dioxo-6-pyrimidinyl]ethylenediamine = 6-
[[2-[[3-(4-Butoxyphenoxy)-2-hydroxypropyl]ami-
no]ethyl]amino]-1,3-dimethyl-2,4(1*H*,3*H*)-pyrimidine-
dione (●)

S *Pirepolol***

U β-Adrenergic blocker

10768 (8703) $\qquad$ $C_{21}H_{32}N_6O_3$
105806-65-3

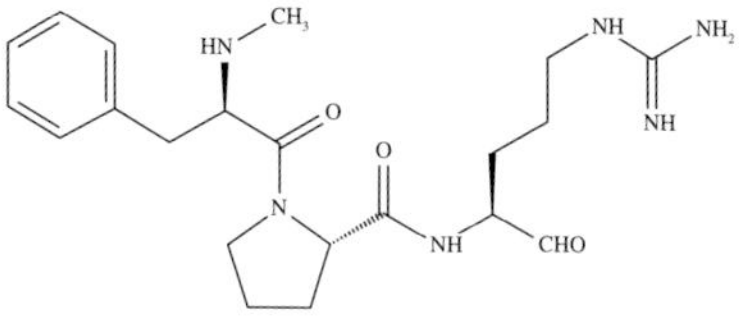

N-Methyl-D-phenylalanyl-*N*-[(1*S*)-1-formyl-4-guani-
dinobutyl]-L-prolinamide = *N*-Methyl-D-phenylalanyl-L-
prolyl-L-argininal = (*S*)-*N*-Methyl-D-phenylalanyl-*N*-[4-
[(aminoiminomethyl)amino]-1-formylbutyl]-L-prolin-
amide (●)

R Sulfate (1:1) (126721-07-1)

S *Efegatran sulfate***, GYKI 14766, LY 294468

U Anticoagulant

10769　(8704)　　　　　　　　　　　　　　　$C_{21}H_{32}N_6O_3$
71195-58-9

N-[1-[2-(4-Ethyl-5-oxo-2-tetrazolin-1-yl)ethyl]-4-(meth-oxymethyl)-4-piperidyl]propionanilide = *N*-[1-[2-(4-Ethyl-4,5-dihydro-5-oxo-1*H*-tetrazol-1-yl)ethyl]-4-(meth-oxymethyl)-4-piperidinyl]-*N*-phenylpropanamide (●)
R　Monohydrochloride monohydrate (70879-28-6)
S　Alfenta, *Alfentanil hydrochloride**, Brevafen, Fana-xal, Fentalim, Limifen, R 39209, Rapifen
U　Narcotic analgesic

10770　(8705)　　　　　　　　　　　　　　　　　$C_{21}H_{32}O$
432-60-0

17-Allylestr-4-en-17β-ol = 17α-Allyl-17-hydroxy-19-norandrost-4-ene = (17β)-17-(2-Propenyl)estr-4-en-17-ol
(●)
S　Aerofilm, *Alilestrenol, Allylestrenol**, Allyloestre-nol*, Gestanin, Gestanol, Gestanon, Gestanyn, Ge-stormone "Zorka", Maintane Tab., Orageston, Perse-lin, Premaston, Profar, SC-6393, Turinal
U　Progestin, anti-androgen

10771　(8708)　　　　　　　　　　　　　　　　　$C_{21}H_{32}O_2$
1605-89-6

17β-Hydroxy-7α,17-dimethylandrost-4-en-3-one =
7α,17-Dimethyltestosterone = (7α,17β)-17-Hydroxy-7,17-dimethylandrost-4-en-3-one (●)
S　*Bolasterone***, Dimethyltestosterone*, Myagen,
NSC-66233, U-19763
U　Anabolic

10772　(8709)　　　　　　　　　　　　　　　　　$C_{21}H_{32}O_2$
17021-26-C

17β-Hydroxy-7β,17-dimethylandrost-4-en-3-one =
7β,17-Dimethyltestosterone = (7β,17β)-17-Hydroxy-7,17-dimethylandrost-4-en-3-one (●)
S　*Calusterone***, Methosarb, NSC-88536, Riedemil,
U-22550
U　Antineoplastic

10773　(8707)　　　　　　　　　　　　　　　　　$C_{21}H_{32}O$
465-53-

6β-Hydroxy-3,5-cyclopregnan-20-one = (3α,5β,6β)-6-Hydroxy-3,5-cyclopregnan-20-one (●)

S *Cyclopregnol***, Neurosterone
U Psychotropic

10774 (8710)

$C_{21}H_{32}O_2$
797-58-0

13-Ethyl-17-hydroxy-18,19-dinor-17α-pregn-4-en-3-one
= 13,17α-Diethyl-17-hydroxygon-4-en-3-one = (17α)-
13-Ethyl-17-hydroxy-18,19-dinorpregn-4-en-3-one (●)
R also with (±)-definition (1235-15-0)
S Genabol, *Norbolethone, Norboletone***, Wy-3475
U Anabolic

10775 (8706)

$C_{21}H_{32}O_2$
145-13-1

β-Hydroxypregn-5-en-20-one = 5-Pregnen-3β-ol-20-
ne = (3β)-3-Hydroxypregn-5-en-20-one (●)
R see also nos. 12114, 12116, 12956
S Bina-Skin, Enelone, Natolone, Preghnenolone, *Pregnenolone***, Pregnetan, Pregneton, Pregnolon, Pre-
nolon, Regnosone, Sharmone, Skinostelon
U Glucocorticoid (anti-inflammatory, anti-allergic)

10776 (8713)

$C_{21}H_{32}O_3$
434-07-1

17β-Hydroxy-2-(hydroxymethylene)-17-methyl-5α-an-
drostan-3-one = 2-(Hydroxymethylene)-17-methyldihy-
drotestosterone = (5α,17β)-17-Hydroxy-2-(hydroxymethylene)-17-methylandrostan-3-one (●)
S Adroidin, Adroyd, Adroyed, Anadrol, Anadroyd,
Anapolon, Anasteron, Anasteronal, Anebox, Beco-
rel, CI-406, Dynasten, Hemogenin, HMD, *Hydroxy-
metholone*, Metalar, Methabol, Nasténon,
NSC-26198, *Ossimetolone, Oximetholonum, Oxime-
tolona*, Oxitosona-50, *Oxymetholone***, Pardroyd,
Pavisoid, Plenastril, Protanabol, Roboral, RS-992,
Synasteron, Zenalosyn
U Anabolic, androgen

10777 (8711)

$C_{21}H_{32}O_3$
23930-19-0

3α-Hydroxy-5α-pregnane-11,20-dione = (3α,5α)-3-Hy-
droxypregnane-11,20-dione (●)
S *Alfaxalone***, *Alphaxalone*, GR 2/234
U Anesthetic

10777-01 (8711-01)

8067-82-1

R Mixture (9:3) with alfadolone acetate (no. 12120)
S Alfadion, Alfatecin, Alfatesin, Alfatésine, Alfathe-
sin, Alfatone, Alphadione, Althesin, Aurantex,
CT 1341, Saffan
U Anesthetic

10778 (8712) $C_{21}H_{32}O_3$
565-99-1

3α-Hydroxy-5β-pregnane-11,20-dione = (3α,5β)-3-Hy-droxypregnane-11,20-dione (●)
S   *Renanolone**
U   Progestin, anesthetic

10779 (8714) $C_{21}H_{32}O_4$
83997-19-7

(+)-(2E,3aS,4R,5R,6aS)-4-[(1E,3S)-3-Cyclopentyl-3-hydroxypropenyl]-3,3a,4,5,6,6a-hexahydro-5-hydroxy-$\Delta^{2(1H),\delta}$-pentalenevaleric acid = (+)-(E)-5-[(1S,5S,6S,7R)-7-Hydroxy-6-[(1E,3S)-3-cyclopentyl-3-hydroxy-1-propenyl]bicyclo[3.3.0]octane]-$\Delta^{3,\delta}$-penta-noic acid = [3aS-[2E,3aα,4α(1E,3R*),5β,6aα]]-5-[4-(3-Cyclopentyl-3-hydroxy-1-propenyl)hexahydro-5-hydr-oxy-2(1H)-pentalenylidene]pentanoic acid (●)
S   *Ataprost***
U   Platelet aggregation inhibitor, peripheral vasodilator

10779-01 (8714-01)   108866-14-4
P   α-Cyclodextrin compd. (1:1)
S   *Ataprost alfadex***, *Cepaprost alfadex*, ONO 41483, OP 41483
U   Platelet aggregation inhibitor, peripheral vasodilator

10780 (8715) $C_{21}H_{32}O_4$
129845-81-4

(1R,2S,4aR,4bR,7S,8aR,10aR)-1,2,3,4,4a,4b,5,6,7,8,8a,10a-Dodecahydro-4a,7-dime-thyl-1,8a-epoxy-1-hydroxy-2-(1-methylethyl)-10-phen-anthrenecarboxylic acid methyl ester = [1R-(1α,2α,4aβ,4bα,7α,8aα,10aβ)]-1,3,4,4a,4b,5,6,7,8,10a-Decahydro-1-hydroxy-4a,7-dimethyl-2-(1-methylethyl)-2H-1,8a-epoxyphenanthrene-10-carboxylic acid methyl ester (●)
S   *Chatancin*
U   Platelet aggregation inhibitor from coral *Sarcophyton*

10781 (8716) $C_{21}H_{32}O_{4}$
14107-37-(

3α,21-Dihydroxy-5α-pregnane-11,20-dione = (3α,5α)-3,21-Dihydroxypregnane-11,20-dione (●)
R   see also nos. 10777-01, 12120
S   *Alfadolone***, *Alphadolone*, GR 2/1574
U   Anesthetic

10782 $C_{21}H_{32}O$
92711-55-

(7E,13E,15S)-15-Hydroxy-9-oxoprosta-7,10,13-trien-1-oic acid methyl ester (●)
S *LipoΔ^7PGA*, TEI-9038
U Antineoplastic

10783 (8717)

$C_{21}H_{32}O_5$
61557-12-8

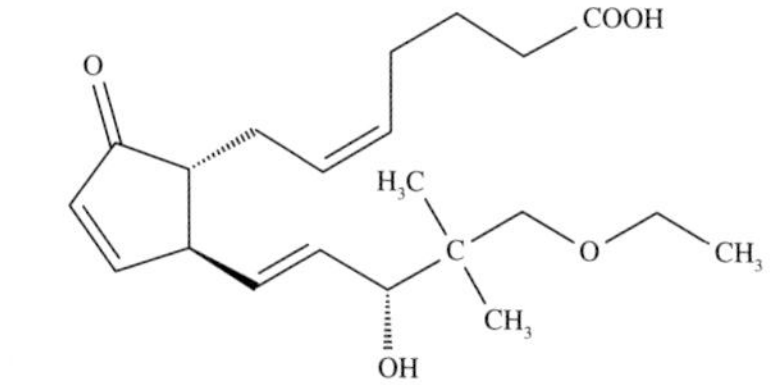

(±)-(Z)-7-[(1R*,2R*)-2-[(E)-(3R*)-5-Ethoxy-3-hydroxy-4,4-dimethyl-1-pentenyl]-5-oxo-3-cyclopenten-1-yl]-5-heptenoic acid = [1α(Z),2β(1E,3R*)]-7-[2-(5-Ethoxy-3-hydroxy-4,4-dimethyl-1-pentenyl)-5-oxo-3-cyclopenten-1-yl]-5-heptenoic acid (●)
S *Penprostene***
U Antihypertensive

10784 (8718)

$C_{21}H_{33}BN_6O_5$
130982-43-3

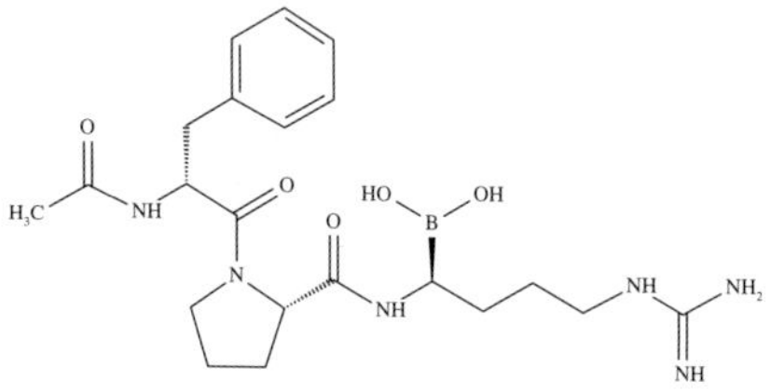

S)-N-Acetyl-D-phenylalanyl-N-[4-[(aminoiminome-hyl)amino]-1-boronobutyl]-L-prolinamide (●)
R Monohydrochloride (131062-98-1)
S DuP 714
U Antithrombotic

10785 (8719)

$C_{21}H_{33}FNO_3$

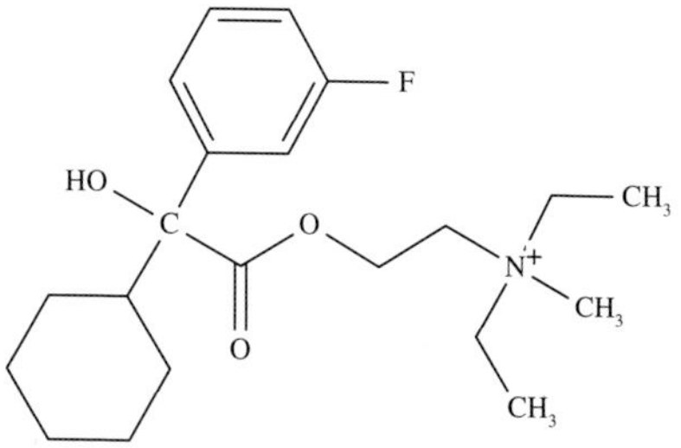

Diethyl(2-hydroxyethyl)methylammonium α-(m-fluoro-phenyl)cyclohexaneglycolate = [2-[(α-Cyclohexyl-m-fluoromandeloyl)oxy]ethyl]diethylmethylammonium = 2-[[Cyclohexyl(3-fluorophenyl)hydroxyacetyl]oxy]-N,N-diethyl-N-methylethanaminium (●)
R Bromide (2681-10-9)
S *Fluoxyphenonium bromide*
U Anticholinergic, spasmolytic

10786

$C_{21}H_{33}{}^{123}IO_2$
74855-17-7

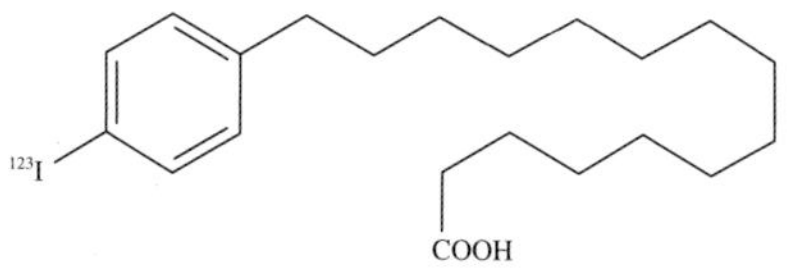

15-(p-^{123}I-Iodophenyl)pentadecanoic acid = 4-(Iodo-^{123}I)-benzenepentadecanoic acid (●)
S *Acidum iocanlidicum* (^{123}I)**, ^{123}I-IPPA, *Iocanlidic acid* (^{123}I)**, Via Scint
U Imaging agent

10787 (8720)

$C_{21}H_{33}NO$
96743-96-3

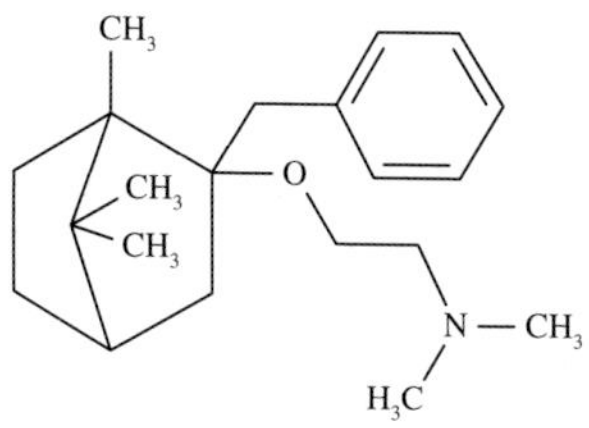

2-[(2-Benzyl-2-bornyl)oxy]-N,N-dimethylethylamine = 2-Benzyl-2-[2-(dimethylamino)ethoxy]-1,7,7-trimethyl-bicyclo[2.2.1]heptane = N,N-Dimethyl-2-[[1,7,7-trime-thyl-2-(phenylmethyl)bicyclo[2.2.1]hept-2-yl]oxy]ethan-amine (●)

S *Ramciclane**
U Tranquilizer

10788 $C_{21}H_{33}NO_2$
                                              138255-51-3

(*R**,*S**)-2-Amino-1-[3-(1-dodecynyl)phenyl]-1,3-pro-
panediol (●)
S P-10050
U Anti-inflammatory (protein kinase C inhibitor)

10789 $C_{21}H_{33}NO_3$
                                              303-75-3

2-Ethyl-1,3,4,6,7,11b-hexahydro-9,10-dimethoxy-3-(2-
methylpropyl)-2*H*-benzo[*a*]quinolizin-2-ol (●)
R also hydrochloride (1976-93-8)
S Ro 4-1284
U Dopamine depletion agent

10790 (8721) $C_{21}H_{33}N_2O_6P$
                                              111223-26-8

1-[(2*S*)-6-Amino-2-hydroxyhexanoyl]-L-proline hydrogen (4-phenylbutyl)phosphonate (ester) = (*S*)-1-[6-Ami-
no-2-[[hydroxy(4-phenylbutyl)phosphinyl]oxy]-1-oxo-
hexyl]-L-proline (●)
S *Ceranapril*, *Ceronapril***, Novopril, SQ 29852

U Antihypertensive (ACE inhibitor)

10791 (8722) $C_{21}H_{33}N_3O$
                                              57695-04-2

8-[[6-(Diethylamino)hexyl]amino]-6-methoxy-4-methyl-
quinoline = 8-[6-(Diethylamino)hexylamino]-6-methoxy-
lepidine = *N*,*N*-Diethyl-*N'*-(6-methoxy-4-methyl-8-quino-
linyl)-1,6-hexanediamine (●)
R Dihydrochloride (5330-29-0)
S Q 45, *Sitamaquine hydrochloride**, WR 6026
U Antileishmanial

10792 (8723) $C_{21}H_{33}N_3O_2$
                                              101246-68-8

(3a*S*,8a*R*)-1,2,3,3a,8,8a-Hexahydro-1,3a,8-trimethylpyr-
rolo[2,3-*b*]indol-5-ol heptylcarbamate = *N*-Demethyl-*N*-
heptylphysostigmine = (3a*S*-*cis*)-Heptylcarbamic acid
1,2,3,3a,8,8a-hexahydro-1,3a,8-trimethylpyrrolo[2,3-*b*]in-
dol-5-yl ester (●)
R Tartrate (1:1) (121652-76-4)
S *Eptastigmine tartrate**, *Heptylphysostigmine tartrate*, *Heptylstigmine tartrate*, L 693487, MF-201,
Synapton
U Acetylcholinesterase inhibitor

10793 $C_{21}H_{33}N_3O$
                                              147650-57-

(4a*S*-*cis*)-Heptylcarbamic acid 2,3,4,4a,9,9a-hexahydro-2,4a,9-trimethyl-1,2-oxazino[6,5-*b*]indol-6-yl ester (●)
S   CHF-2060, *Geneseroline heptylcarbamate*, *Terestigmine**, *Teserstigmine*
U   Acetylcholinesterase inhibitor, nootropic

10794   (8724) $C_{21}H_{33}N_3O_5S$
                                                                32886-97-8

Hydroxymethyl (2*S*,5*R*,6*R*)-6-[[(hexahydro-1*H*-azepin-1-yl)methylene]amino]-3,3-dimethyl-7-oxo-4-thia-1-azabicyclo[3.2.0]heptane-2-carboxylate pivalate (ester) = (2*S*,5*R*,6*R*)-6-[[(Hexahydro-1*H*-azepin-1-yl)methylene]amino]-3,3-dimethyl-7-oxo-4-thia-1-azabicyclo[3.2.0]heptane-2-carboxylic acid (pivaloyloxy)methyl ester = (Pivaloyloxy)methyl (6*R*)-6-(perhydroazepin-1-ylmethyleneamino)penicillanate = [2*S*-(2α,5α,6β)]-6-[[(Hexahydro-1*H*-azepin-1-yl)methylene]amino]-3,3-dimethyl-7-oxo-4-thia-1-azabicyclo[3.2.0]heptane-2-carboxylic acid (2,2-dimethyl-1-oxopropoxy)methyl ester (●)
R   also monohydrochloride (32887-03-9)
S   *Amdinocillin Pivoxil*, Beltomecin, Celfuron, Coactabs, FL 1039, Kanorpsin, Maxibol "Berenguer-Beneyto", Mecilim, Mecillin 200, Melicin, Melysin, Negaxid, Peniamicin, Pivexid, *Pivmecillinam***, Pivuran, Ro 10-9071, Selecid, Selecidin, Selexid, Selexidin, Tindacilin, Uromelynam
U   Antibiotic

10795 $C_{21}H_{33}N_3S$
                                                                60563-40-8

4-Phenyl-5-tridecyl-4*H*-1,2,4-triazole-3-thiol = 2,4-Dihydro-4-phenyl-5-tridecyl-3*H*-1,2,4-triazole-3-thione (●)
S   PD 140195

U   CETP inhibitor

10796   (8725) $C_{21}H_{33}N_5O$
                                                                124824-02-8

3-(13-Amino-2,3β-dihydro-2,6-dimethyl-8α-ergolinyl)-1,1-diethylurea = *N'*-[(3β,8α)-13-Amino-2,3-dihydro-2,6-dimethylergolin-8-yl]-*N*,*N*-diethylurea (●)
S   ZK 112566
U   Vasodilator (dopamine DA$_2$-receptor agonist)

10797   (8726) $C_{21}H_{34}NO_3$
                                                                14214-84-7

Diethyl(2-hydroxyethyl)methylammonium α-phenylcyclohexaneglycolate = 2-[(α-Cyclohexylphenylglycoloyl)oxy]triethylmethylammonium = 2-[(α-Cyclohexylmandeloyl)oxy]ethyldiethylmethylammonium = 2-[(Cyclohexylhydroxyphenylacetyl)oxy]-*N*,*N*-diethyl-*N*-methylethanaminium (●)
R   Bromide (50-10-2)
S   Antispasmin "Pharmachim-Sofia", Antrenil, Antrenyl, Ba-5473, C 5473, Helkamon, Oxifenon, *Oxiphenoni bromidum*, Oxyfenon, Oxyphenon, *Oxyphenonium bromide***, Spasmophen, Spastrex, Subranyl
U   Anticholinergic, spasmolytic

10798   (8727) $C_{21}H_{34}N_2O$
143257-97-0

N-Ethyl-1-hexyl-N-methyl-4-phenylisonipecotamide =
N-Ethyl-1-hexyl-N-methyl-4-phenyl-4-piperidinecarbox-
amide (●)
S   *Sameridine***
U   Analgesic, local anesthetic

10799   (8728) $C_{21}H_{34}N_2O_3$
76629-85-1

2-(1-Piperidyl)ethyl o-(heptyloxy)carbanilate = o-(Hepty-
loxy)carbanilic acid 2-piperidinoethyl ester = [2-(Hepty-
loxy)phenyl]carbamic acid 2-(1-piperidinyl)ethyl ester
(●)
R   Monohydrochloride (55792-21-7)
S   HCP, Heptacaine
U   Anti-arrhythmic

10800 $C_{21}H_{34}N_2O_3$
147951-44-8

(±)-trans-2-(Diethylamino)cyclopentyl m-(pentyloxy)carbanilate = trans-[3-(Pentyloxy)phenyl]carbamic
acid 2-(diethylamino)cyclopentyl ester (●)
S   K-1905
U   Anti-ulcer agent

10801   (8729) $C_{21}H_{34}N_6O_6$
108093-90-9

1,2-Bis[4-(morpholinomethyl)-3,5-dioxo-1-piperazi-
nyl]propane = 4,4'-(1-Methyl-1,2-ethanediyl)bis[1-(4-
morpholinylmethyl)-2,6-piperazinedione] (●)
R   also with (±)-definition (95604-83-4)
S   AT-2153, MM 159, MST-02, *Probimane*
U   Antineoplastic

10802   (8731) $C_{21}H_{34}O_2$
2881-21-2

17β-Hydroxy-1α,17-dimethyl-5α-androstan-3-one =
1α,17-Dimethyldihydrotestosterone = (1α,5α,17β)-17-
Hydroxy-1,17-dimethylandrostan-3-one (●)
S   Demalon, *Dimethylandrostanolone*, DMA, Super-
        orabolon
U   Anabolic

10803 (8732) $C_{21}H_{34}O_2$
70434-82-1

cis-3-[2-Hydroxy-4-(1,1-dimethylheptyl)phenyl]cyclohe-
xanol = *rel*-5-(1,1-Dimethylheptyl)-2-[(*1R,3S*)-3-hydr-
oxycyclohexyl]phenol (●)
S   CP-47497
U   Cannabimimetic, analgesic, muscle relaxant

10804 $C_{21}H_{34}O_2$
516-54-1

3α,5α-Tetrahydroprogesterone = (3α,5α)-3-Hydroxypre-
gnan-20-one (●)
S   ALLO, *(3α)-Allopregnanolone*
U   Neurosteroid (progesterone metabolite)

10805 (8730) $C_{21}H_{34}O_2$
128-20-1

3α-Hydroxy-5β-pregnan-20-one = (3α,5β)-3-Hydro-
xypregnan-20-one (●)
S   *Eltanolone**, Pregnanolone,* ZK 4915
U   Anesthetic, hypnotic

10806 (8733) $C_{21}H_{34}O_2Si$
6717-72-2

17β-(Trimethylsiloxy)estr-4-en-3-one = (17β)-17-[(Tri-
methylsilyl)oxy]estr-4-en-3-one (●)
S   Silabolin
U   Anabolic

10807 $C_{21}H_{34}O_3S$
89617-02-7

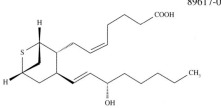

(5Z)-7-[(1S,2R,3R,5S)-3-[(1E,3S)-3-Hydroxy-1-octenyl]-
6-thiabicyclo[3.1.1]hept-2-yl]-5-heptenoic acid (●)
S   ONO 11113, STA$_2$
U   Thromboxane A$_2$ receptor agonist

10808 $C_{21}H_{34}O_4$
56985-32-1

(5Z)-7-[(1S,4R,5S,6R)-5-[(1E,3S)-3-Hydroxy-1-octenyl]-
2-oxabicyclo[2.2.1]hept-6-yl]-5-heptenoic acid (●)
S   U-44069
U   Thromboxane receptor agonist

10809 (8734) $C_{21}H_{34}O_4$
56985-40-1

(5Z,9α,11α,13E,15S)-11,9-(Epoxymethano)-15-hydr-oxyprosta-5,13-dien-1-oic acid = [1R-[1α,4α,5β(Z),6α-(1E,3S*)]]-7-[6-(3-Hydroxy-1-octenyl)-2-oxabicy-clo[2.2.1]hept-5-yl]-5-heptenoic acid (●)
S U-46619
U Thromboxane receptor agonist

10810 (8735) $C_{21}H_{34}O_4$
88911-35-7

(+)-9-(O)-Methano-Δ$^{6(9α)}$-prostaglandin = [3aS-[3aα,5β,6α(1E,3R*),6aα]]-1,3a,4,5,6,6a-Hexahydro-5-hydroxy-6-(3-hydroxy-1-octenyl)-2-pentalenepentanoic acid (●)
S Isocarbacyclin, TEI-7165
U Antithrombotic

10810-01 (8735-01)
R Incorporated in microspheres
S Lipo-Isocarbacyclin, TTC-109
U Antithrombotic

10811 (8736) $C_{21}H_{34}O_4$
69552-46-1

(5E)-6a-Carbaprostacyclin = [3aS-[2E,3aα,4α(1E,3R*),5β,6aα]]-5-[Hexahydro-5-hydroxy-4-(3-hydroxy-1-octenyl)-2(1H)-pentalenylidene]penta-noic acid (●)
S 6a-Carbaprostaglandin I$_2$, Carbacyclin, Carbo-prostacyclin, 9(O)-Methanoprostacyclin
U Platelet aggregation inhibitor

10812 (8737) $C_{21}H_{34}O_5$
53-02-1

3α,11β,17,21-Tetrahydroxy-5β-pregnan-20-one = (3α,5β,11β)-3,11,17,21-Tetrahydroxypregnan-20-one (●)
S Tetrahydrocortisol, Urocortisol
U Glucocorticoid (anti-inflammatory)

10813 $C_{21}H_{34}O_5$
159359-95-2

(5Z,9α,11α,13E,15S)-Prosta-5,13-diene-1,9,11,15-tetrol cyclic 9,11-carbonate (●)
S AGN 192093

U Thromboxane A_2 mimetic

10814 (8738)

$C_{21}H_{34}O_5$
55028-70-1

(*E*,*Z*)-1-(1*R*,2*R*,3*R*)-7-[3-Hydroxy-2-[(3*R*)-(3-hydroxy-3-methyl-1-octenyl)]-5-oxocyclopentyl]-5-heptenoic acid = (15*R*)-15-Methylprostaglandin E_2 = (5*Z*,11α,13*E*,15*R*)-11,15-Dihydroxy-15-methyl-9-oxoprosta-5,13-dien-1-oic acid (●)
S Arbacet, *Arbaprostil***, Arbastyl, CU-83, U-42842
U Gastric antisecretory, anti-ulcer agent

10815 (8739)

$C_{21}H_{34}O_6$
90243-98-4

(*Z*)-7-[(1*RS*,2*RS*,3*RS*)-2-[(*E*)-(3*R*)-5-Ethoxy-3-hydroxy-4,4-dimethyl-1-pentenyl]-3-hydroxy-5-oxocyclopentyl]-5-heptenoic acid = 16,16-Dimethyl-18-oxaprostaglandin E_2 = 7-[2-(5-Ethoxy-3-hydroxy-4,4-dimethyl-1-pentenyl)-3-hydroxy-5-oxocyclopentyl]-5-heptenoic acid (●)
R also with [1*R*-[1α(*Z*),2β(1*E*,3*R**),3α]]-definition (58687-40-4) (see structural formula)
S *Dimoxaprost***, Hoechst 260, HR 260
U Antisecretory, cytoprotective

10816

$C_{21}H_{34}O_6$
63266-93-3

Methyl (*Z*)-7-[(1*R*,2*R*,3*R*)-2-[(1*E*,3*S*,7*R*)-3,7-dihydroxy-1-octenyl]-3-hydroxycyclopentyl]-5-heptenoate = (5*Z*,11α,13*E*,15*S*,19*R*)-11,15,19-Trihydroxy-9-oxoprosta-5,13-dien-1-oic acid methyl ester (●)
S *Eganoprost**, JB 971
U Peripheral vasodilator

10817 (8740)

$C_{21}H_{35}NO$
78613-35-1

(±)-*cis*-2,6-Dimethyl-4-[2-methyl-3-(*p-tert*-pentylphenyl)propyl]morpholine = *cis*-(±)-4-[3-[4-(1,1-Dimethylpropyl)phenyl]-2-methylpropyl]-2,6-dimethylmorpholine (●)
R also hydrochloride (78613-38-4)
S *Amorolfine***, Bekiron, Loceryl, Locetar, MT 861, Odenil, Pekiron, Ro 14-4767/000
U Antimycotic

10818 (8741)

$C_{21}H_{35}NO$
119463-16-0

2-Benzyl-1-methyl-5-nonyl-3-pyrrolidinol = 1-Methyl-5-nonyl-2-(phenylmethyl)-3-pyrrolidinol (●)
R also (2*S*,3*S*,5*R*)-compd. [= [2*S*-(2α,3α,5α)]-compd.] (125356-66-3)

S L 657398, *Preussin*
U Antifungal from *Aspergillus ochraceus* and *Preussia* sp.

10819　(8742)　　　　　　　　　　　$C_{21}H_{35}NO_2$
18668-43-4

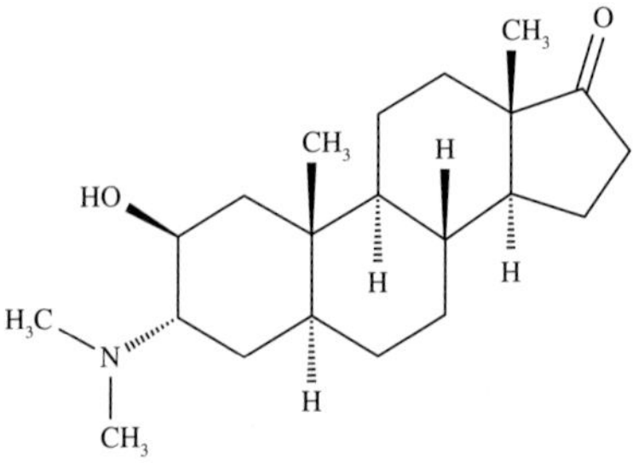

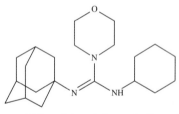

3α-(Dimethylamino)-2β-hydroxy-5α-androstan-17-one = (2β,3α,5α)-3-(Dimethylamino)-2-hydroxyandrostan-17-one (●)
S Org NA 13
U Anesthetic, anti-arrhythmic

10820　　　　　　　　　　　　　　$C_{21}H_{35}NO_2$
121808-24-0

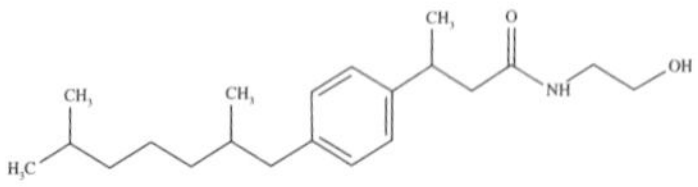

4-(2,6-Dimethylheptyl)-*N*-(2-hydroxyethyl)-β-methyl-benzenepropanamide (●)
S E-5050
U Uricosuric

10821　(8743)　　　　　　　　　　$C_{21}H_{35}N_2O_3$
109260-82-4

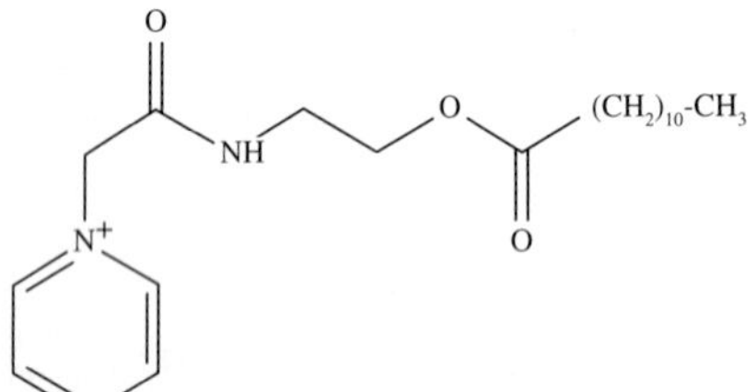

1-[[(2-Hydroxyethyl)carbamoyl]methyl]pyridinium laurate (ester) = 1-[[[2-(Lauroyloxy)ethyl]carbamoyl]methyl]pyridinium = 1-[2-Oxo-2-[[2-[(1-oxododecyl)oxy]ethyl]amino]ethyl]pyridinium (●)
R Chloride (6272-74-8)

S DG-6, Emcol E-607, *Lapirium chloride**,
Lapyrium Chloride, NSC-33659
U Detergent

10822　(8744)　　　　　　　　　　　$C_{21}H_{35}N_3$
51481-62-0

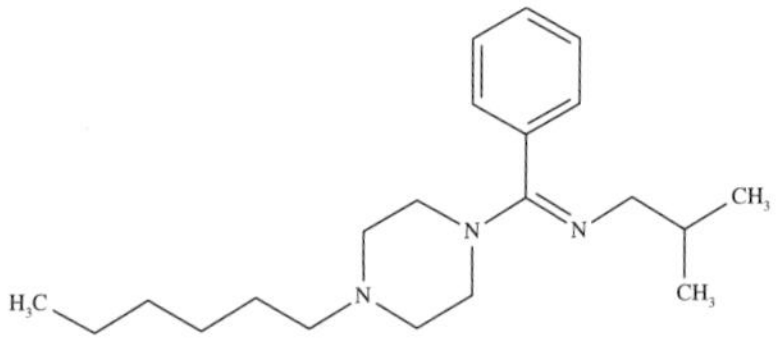

1-Hexyl-4-(*N*-isobutylbenzimidoyl)piperazine = 1-Hexyl-4-[[(2-methylpropyl)imino]phenylmethyl]piperazine = *N*-[(4-Hexyl-1-piperazinyl)phenylmethylene]-2-methyl-1-propanamine (●)
R Maleate (1:2) (51481-63-1)
S *Bucainide maleate**, G 233, RHC-G 233, USVP-G 233
U Anti-arrhythmic

10823　　　　　　　　　　　　　　$C_{21}H_{35}N_3O$
122252-09-9

N-1-Adamantyl-*N*'-cyclohexyl-4-morpholinecarboxamidine = *N*-Cyclohexyl-*N*'-tricyclo[3.3.1.1^{3,7}]dec-1-yl-4-morpholinecarboximidamide (●)
R Monohydrochloride (57568-80-6)
S PNU-37883A, U-37883A
U ATP-sensitive potassium channel blocker

10824　(8745)　　　　　　　　　　$C_{21}H_{36}Cl_2N$
58390-78-6

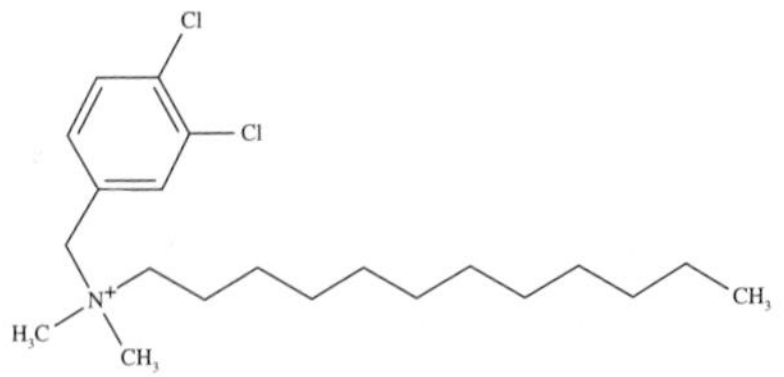

(3,4-Dichlorobenzyl)dodecyldimethylammonium = 3,4-Dichloro-*N*-dodecyl-*N,N*-dimethyl-benzenemethanaminium (●)

R Chloride (102-30-7)
S Dermofongin "B", Dermofungine "B", Dichloran, *Dichlorbenzalkonium chloride, Dichlorobenzododecinium chloride*, Riseptin, Tetrosan
U Topical antifungal, antiseptic, deodorant

10825 (8746) $C_{21}H_{36}NO$
60-49-1

(3-Cyclohexyl-3-hydroxy-3-phenylpropyl)triethylammonium = γ-Cyclohexyl-*N,N,N*-triethyl-γ-hydroxybenzenepropanaminium (●)

R Chloride (4310-35-4) or iodide (125-99-5)
S C 921, Claviton, Compound 921 C, Duoesetil, Pathilon, Patilon, *Propethoni jodidum, Tridiesetile ioduro, Tridihexäthylchlorid, Tridihexethide, Tridihexethyl chloride**, Tridihexide
U Anticholinergic, spasmolytic

10826 (8747) $C_{21}H_{36}N_2O_3$
76629-87-3

2-(Diethylamino)-1-methylethyl *o*-(heptyloxy)carbanilate = *o*-(Heptyloxy)carbanilic acid 2-(diethylamino)-1-methylethyl ester = [2-(Heptyloxy)phenyl]carbamic acid 2-(diethylamino)-1-methylethyl ester (●)

R Monohydrochloride (68931-03-3)
S BK-95, Carbisocaine, Carbizocaine
U Local anesthetic, anti-arrhythmic

10827 (8748) $C_{21}H_{36}N_7O_{16}P_3S$
85-61-0

Adenosine 5'-(trihydrogen diphosphate) 3'-(dihydrogen phosphate) 5'-mono[(*R*)-3-hydroxy-4-[[3-[(2-mercaptoethyl)amino]-3-oxopropyl]amino]-2,2-dimethyl-4-oxobutyl] ester (●)

S Aluzime, Co. A, Coalip, Coenzimade, *Coenzyme A*, Lucina, *Pantadefosfatum*
U Metabolic (co-factor in enzymatic acetyl transfer reactions)

10828 (8749) $C_{21}H_{36}O_2$
80-92-2

5β-Pregnane-3α,20α-diol = (3α,5β,20S)-Pregnane-3,20-diol (●)

S Diol, *Pregnanediol*
U Progesterone metabolite (in acne vulgaris and menstrual syndrome)

10829 $C_{21}H_{36}O_4$
76808-15-6

(3S,4S)-3-Ethyl-4-[(1S,3E,5R,7S,8R,9R)-8-hydroxy-1,3,5,7,9-pentamethyl-6-oxo-3-undecenyl]-2-oxetanone (●)

S *Ebelactone B*

U Lipase inhibitor from *Streptomyces aburaviensis*

10830 (8750) $C_{21}H_{36}O_4$
51953-95-8

(1R*,2R*)-2-[(E)-3-Hydroxy-3-methyl-1-octenyl]-5-oxocyclopentaneheptanoic acid = (13E,15S)-(±)-15-Hydroxy-15-methyl-9-oxoprost-13-en-1-oic acid (●)
S   AY 24559, *Doxaprost***
U   Bronchodilator

10831 (8751) $C_{21}H_{36}O_5$
35700-23-3

(E,Z)-(1R,2R,3R,5S)-7-[3,5-Dihydroxy-2-[(3S)-(3-hydroxy-3-methyl-1-octenyl)]cyclopentyl]-5-heptenoic acid = (5Z,9α,11α,13E,15S)-9,11,15-Trihydroxy-15-methylprosta-5,13-dien-1-oic acid (●)
R   see also no. 11591
S   *Carboprost***, Deviprost, *15-Methyl-PGF_{2α}*, 15M-PG, *15(S)-15-Methylprostaglandin F_{2α}*, U-32921
U   Oxytocic

10831-01 (8751-01) 58551-69-2
R   Compd. (1:1) with trometamol (no. 314)
S   *Carboprost Trometamol,*
Carboprost Tromethamine, Hemabate, Prostinfenem, Prostin 15M, Prostodin, U-32921 E
U   Oxytocic

10832 (8752) $C_{21}H_{37}ClN$
68379-02-2

[4-(p-Chlorophenyl)butyl]diethylheptylammonium = 4-Chloro-N,N-diethyl-N-heptylbenzenebutanaminium (●)
R   Phosphate (1:1) (68379-03-3)
S   *Clofilium phosphate***, LY 150378
U   Anti-arrhythmic

10833 $C_{21}H_{37}FN_2O_3S$
180918-68-7

(−)-4'-[(S)-4-[Ethyl(6-fluoro-6-methylheptyl)amino]-1-hydroxybutyl]methanesulfonanilide = (S)-N-[4-[4-[Ethyl(6-fluoro-6-methylheptyl)amino]-1-hydroxybutyl]phenyl]methanesulfonamide (●)
R   Fumarate (2:1) (191349-60-7)
S   *Trecetilide fumarate***, U-108342E
U   Anti-arrhythmic (class III)

10834 (8753) C_{21}H_{37}NO
474-44-2

3α-Amino-5α-pregnan-20α-ol = (3α,5α,20S)-3-Aminopregnan-20-ol (●)
S   *Funtumidine*

U Tranquilizer

10835 (8754) $C_{21}H_{37}N_5O_{14}$
12706-94-4

4-Amino-1-[4-amino-6-O-(3-amino-3-deoxy-β-D-gluco-pyranosyl)-4-deoxy-D-*glycero*-D-*galacto*-β-D-*gluco*-un-decopyranosyl]-2(1*H*)-pyrimidinone (●)
S *Antelmycin***, *Anthelmycin*, *Hikizimycin*, L 33876
U Anthelmintic (antibiotic from *Streptomyces longissimus* or *Str.* A-5)

10836 (8755) $C_{21}H_{38}N$
10328-35-5

Benzyldodecyldimethylammonium = *N*-Dodecyl-*N,N*-di-methylbenzenemethanaminium (●)
R Chloride (139-07-1) or bromide (7281-04-1)
S Ajatin, Benzo-Davur, *Benzododecinium***, Chi-bro-Benzo, DDBAC, Effimyl, Kemerhinose, Prorhi-nel, Rhinédrine, Rhinibar, Rodalon, Rubaseptyl, Ste-rinol, TAGG
U Disinfectant, antiseptic

10837 (8756) $C_{21}H_{38}N$
7773-52-6

N-Cetylpyridium = 1-Hexadecylpyridinium (●)
R Chloride (123-03-5) [also chloride monohydrate (6004-24-6)]
S Algol, Alsol "Also" (new form), Angifonil, Anticari-ol, Anti-Plaque "Foulding", Antussan, Aseptol "Ak-deniz", Astring-O-Mint, Bactalin, Batzeta, Benalet, Benylin Lozenges, Biosept, Borcolin, Borocaina Gola, Bronchenolo Gola, Bucacem, Bucosem, Catio-nil, Ceepryn, Cepacol, Cepamint, Ceprim, Cerigel, Cetafilm, Cetazol, Cetilsan, *Cetylpyridinium chlori-de***, Cetylyre, CPK, Curisol, Dareets, Dobendan, Emereze Plus, Faringola, Fluprim Gola, Forma-mint N, Fresh Mouth, Frubizin, Germidin, Geyderm sepsi, Golacetin, Halset, Halstabl.-ratiopharm, Her-bagola, Lemsip Lozenges, Medi-Keel A, Menoril, Merocet, Merothol, Micrin plus, Neocepacol, Neo-Coricidin Gola, Neoformidol, Neoformitrol, Neogola Sel, Neo-Mentoformio, No-Alcool Sella, Novoptine, Oracain "Noco", Oracol, Pectochups, Pe-nipastil, Pristacin, Pronto G, Pyrets, Pyrisept, Quifa-sol, Ragaden, Rikospray, Search, Sterinet, Suprol "Iwaki", Throat Lozenges "Novopharm", Tirocetil, Tsetazol, Tyrocane junior, Universal Throat Lollies, Zepacole
U Disinfectant, antiseptic

10837-01 (8756-01) 140-72-7
R Bromide
S Bromocet, Cetapharm, Cetazolin, Fixanol C, Neder-min, Sterogenol
U Disinfectant, antiseptic

10837-02 (8756-02) 8065-02-9
R Ethyl sulfate
S Sterillium (old form)
U Antiseptic

10837-03 (8756-03)
R Pantothenate (no. 1869)

S Pantodinium
U Antiseptic

10837-04 (8756-04)
R o-Thymothinate (no. 2844)
S Contrafungin, Pedyol
U Antifungal

10838 (8757) $C_{21}H_{38}N_2$
103923-27-9

1,4-Dihydro-1-octyl-4-(octylimino)pyridine = N-[1-Oc-tyl-4(1H)-pyridinylidene]-1-octanamine (●)
R Monohydrochloride (100227-05-2)
S Pirtenidine hydrochloride**, Win 52172-2
U Antibacterial (antigingivitis)

10839 $C_{21}H_{38}N_4O_5$
137096-61-8

[6S-[1(S*),6R*(R*)]]-6-[[(1-Acetyl-3-methylbutyl)ami-no]carbonyl]tetrahydro-N-hydroxy-γ-oxo-β-pentyl-1(2H)-pyridazinebutanamide (●)
S YL-01869P, YM-24074
U Antibacterial (gram-positive) from *Streptomyces* sp. YL-01869P

10840 $C_{21}H_{38}N_4O_8$
67655-94-1

(2S,3R)-3-Amino-2-hydroxy-5-methylhexanoyl-L-valyl-L-valyl-L-aspartic acid (●)
S Amastatin
U Peptidase inhibitor

10841 $C_{21}H_{38}N_6O_4$
155415-08-0

N-[(1R)-2-Cyclohexyl-1-[[(2S)-2-[(3-guanidinopro-pyl)carbamoyl]piperidino]carbonyl]ethyl]glycine = N-[(1R)-2-[(2S)-2-[[[3-[(Aminoiminomethyl)amino]pro-pyl]amino]carbonyl]-1-piperidinyl]-1-(cyclohexylme-thyl)-2-oxoethyl]glycine (●)
S Inogatran**, MW 439 Da
U Thrombin inhibitor

10842 (8758) $C_{21}H_{38}N_8O_5$
76152-06-2

L-Threonyl-L-lysyl-L-prolyl-L-arginine cyclic (4→2)-peptide (●)
S Cyclotuftsin, Tsiklotaftsin
U Antihypertensive

10843 (8759) $C_{21}H_{38}O_4$
                                                 33813-84-2

(1*R*,2*S*)-2-(3-Hydroxy-3-methyloctyl)-5-oxocyclopen-
taneheptanoic acid = 15-Hydroxy-15-methyl-9-oxo-
prostan-1-oic acid (●)
S AY 22469, *Deprostil***
U Anti-ulcer agent, gastric antisecretory

10844 (8760) $C_{21}H_{38}O_4$
                                                 77287-05-9

(2*R*,3*R*,4*R*)-4-Hydroxy-2-(7-hydroxyheptyl)-3-[(*E*)-
(4*RS*)-4-hydroxy-4-methyl-1-octenyl]cyclopentanone =
(11α,13*E*)-1,11,16-Trihydroxy-16-methylprost-13-en-9-
one (●)
S Bay o 6893, ORF 15927, *Rioprostil***, Rostec,
      RWJ 15927, TR-4698
U Gastric antisecretory, cytoprotective

10845 $C_{21}H_{39}NO_6$
                                                 35891-70-4

[2*S*-(2*R**,3*S**,4*S**,6*E*)]-2-Amino-3,4-dihydroxy-2-(hydr-
oxymethyl)-14-oxo-6-eicosenoic acid = (2*S*,3*R*,4*R*,6*E*)-2-
Amino-3,4-dihydroxy-2-(hydroxymethyl)-14-oxo-6-ei-
cosenoic acid (●)
S ISP-I, *Myriocin, Thermozymocidin*
U Immunosuppressive from *Mycelia sterilia*

10846 (8761) $C_{21}H_{39}N_5O_{14}$
                                                 11011-72-6

O-2-Deoxy-2-(methylamino)-α-L-glucopyranosyl-
(1→2)-*O*-5-deoxy-3-*C*-(hydroxymethyl)-α-L-lyxofura-
nosyl-(1→2)-1-[(aminoiminomethyl)amino]-1-deoxy-D-
scyllo-inositol 5-carbamate (●)
S Bluensin, *Bluensomycin***, Glebomycin, S 438,
      U-12898
U Antibiotic from *Streptomyces bluensis*

10847 (8762) $C_{21}H_{39}N_7O_{12}$
                                                 57-92-1

all-trans-2,4-Diguanidino-3,5,6-trihydroxycyclohexyl-5-
deoxy-2-*O*-(2-deoxy-2-methylamino-α-L-glucopyrano-
syl)-3-*C*-formyl-β-L-lyxopentofuranoside = [1,3-Digua-
nidino-2,4,5,6-tetrahydroxycyclohexane]-α-2'-(*N*-me-
thyl-2-glucosaminosido)-4-streptoside = *O*-2-Deoxy-2-
(methylamino)-α-L-glucopyranosyl-(1→2)-*O*-5-deoxy-
3-*C*-formyl-α-L-lyxofuranosyl-(1→4)-*N*,*N*'-bis(aminoiminomethyl)-D-streptamine (●)
S *A*-Streptomycin, *Estreptomicina*, Merstrep,
      NSC-14083, *Streptomicin(a), Streptomisin, Strepto-
      mycin***, *Streptomycin A*
U Antibiotic from *Streptomyces griseus* and other *Strep-
      tomyces* sp.

10847-01 (8762-01)   3810-74-0
R Sulfate (2:3)
S Ambistryn (-S), Ampistrep, Bonastrept, Cidan-Est,
      Cocabiotico-Estrepto, Comycin-S, Darostrep, Devo-
      mycin, Dif-Estrepto-E, Efastrep, Estrepto-E, Estrep-

to-Level, Estreptomade, Estreptorgan, Estrepto-siner-
ge, Estrepto-Wolner, Gamafin, Metastrep, Neodie-
streptopab, Novostrep, Orastrep, Orostrept, Servi-
strep, Solumycine, Solustrep, Solvostrept "S",
Strep-Deva, Strepolin, Strepsulfat, Streptancil, Strep-
taquaine, Streptevan, Streptobrettin, Streptocin,
Streptocol, Streptofar, Strepto-Fatol, Strepto-Hefa,
Streptoliquin, Streptomagma, Streptopanat, Streptor-
ex, Streptorit, Streptosol, Streptowerfft, Strycin,
Strysolin, Sul-Mycin II, Tebejekt, Unik-Strepto(sol),
Vaccastrep, Vetstrep
U　Antibiotic

10847-02　(8762-02)
R　Pantothenate sulfate mixture
S　Estrepto "P", Estreptopanto, Neodualestrepto,
　　Pantostracina E, Pantostrep, Strepancil, Strepantine,
　　Streptothenat
U　Tuberculostatic

10847-03　(8762-03)　　　　　　　　　4938-25-4
R　Salt with opiniazide (no. 6094)
S　Streptosaluzid
U　Tuberculostatic

10847-04　(8762-04)
R　Compd. with sodium polymethacrylate
S　Streptolymphin
U　Antibiotic

10848　(8763)　　　　　　　　　　　　$C_{21}H_{39}N_7O_{13}$
　　　　　　　　　　　　　　　　　　　　　　6835-00-3

all-trans-2,4-Diguanidino-3,5,6-trihydroxycyclohexyl-2-
O-(2-deoxy-2-methylamino-α-L-glucopyranosyl)-3-C-
formyl-β-L-lyxopentofuranoside = O-2-Deoxy-2-(methylamino)-α-L-glucopyranosyl-(1→2)-O-3-C-formyl-α-
L-lyxofuranosyl-(1→4)-N,N-bis(aminoiminomethyl)-D-
streptamine (●)
S　Hydroxystreptomycin, Oxystreptomycin, Reticulin

U　Antibiotic from Streptomyces reticuli H 365

10849　(8764)　　　　　　　　　　　　$C_{21}H_{40}N_8O_6$
　　　　　　　　　　　　　　　　　　　　　　9063-57-4

N^2-[1-(N^2-L-Threonyl-L-lysyl)-L-prolyl]-L-arginine (●)
R　also diacetate (72103-53-8)
S　Taftsin, Tuftsin
U　Physiological activator of phagocytic cells, CNS sti-
　　mulant

10850　(8765)　　　　　　　　　　　　$C_{21}H_{40}N_8O_7$
　　　　　　　　　　　　　　　　　　　　　　85466-18-8

N-[N-(N^2-L-Arginyl-L-lysyl)-L-α-aspartyl]-L-valine (●)
S　RGH-0206, Thymocartin**, Thymopoietin 32-35,
　　TP-4
U　Immunomodulator

10851　(8766)　　　　　　　　　　　　$C_{21}H_{41}N_5O_7$
　　　　　　　　　　　　　　　　　　　　　　56391-56-1

O-3-Deoxy-4-C-methyl-3-(methylamino)-β-L-arabinopy-
ranosyl-(1→4)-O-[2,6-diamino-2,3,4,6-tetradeoxy-α-D-
glycero-hex-4-enopyranosyl-(1→6)]-2-deoxy-N^3-ethyl-
L-streptamine = 4-O-[(2R,3R)-cis-3-Amino-6-aminome-
thyl-3,4-dihydro-2H-pyran-2-yl]-2-deoxy-6-O-(3-deoxy-

4-C-methyl-3-methylamino-β-L-arabinopyranosyl)-1-N-ethylstreptamine = N'-Ethylsisomicin = O-3-Deoxy-4-C-methyl-3-(methylamino)-β-L-arabinopyranosyl-(1→6)-O-[2,6-diamino-2,3,4,6-tetradeoxy-α-D-glycero-hex-4-enopyranosyl-(1→4)]-2-deoxy-N^1-ethyl-D-streptamine (●)

R Sulfate (2:5) (56391-57-2)

S Acosin, Certomycin, Dalinar, *Ethylsisomicin sulfate*, Guardocin "Krka", Netillin, *Netilmicin sulfate***, Netilyn, Netrocin, Netromicina, Nétromicine, Netromycin(e), Nettacin, Netylyn, Rizaldon, Sch 20569, Vectacin, Vivicil, Zaby, Zetamicin

U Antibiotic

10852 (8767) $C_{21}H_{41}N_5O_{11}$
37321-09-8

4-O-[3α-Amino-6α-[(4-amino-4-deoxy-α-D-glucopyranosyl)oxy]-2,3,4,4aβ,6,7,8,8aα-octahydro-8β-hydroxy-7β-(methylamino)pyrano[3,2-b]pyran-2α-yl]-2-deoxy-streptamine = 4,O-[(2R,3R,4aS,6R,7S,8R,8aR)-3-Amino-6-(4-amino-4-deoxy-α-D-glucopyranosyloxy)-8-hydroxy-7-(methylamino)perhydropyrano[3,2-b]pyran-2-yl]-2-deoxystreptamine = O-4-Amino-4-deoxy-α-D-glucopyranosyl-(1→8)-O-(8R)-2-amino-2,3,7-trideoxy-7-(methylamino)-D-glycero-α-D-allo-octodialdo-1,5:8,4-dipyranosyl-(1→4)-2-deoxy-D-streptamine (●)

R also sulfate (65710-07-8)

S Ambylan, Apralame, Apralan(e), *Apramicina, Apramycin***, EL-857/820, Lilly 47657, Magnimix, Nebramycin Factor 2, Santamix Apra

U Antibiotic from *Streptomyces tenebrarius*

10853 (8768) $C_{21}H_{41}N_5O_{12}$
12772-35-9

A-Form   B-Form

O-2,6-Diamino-2,6-dideoxy-α-D-glucopyranosyl-(1→4)-O-[β-D-xylofuranosyl-(1→5)]-N^1-(4-amino-L-2-hydroxybutyryl)-2-deoxy-D-streptamine (A form) mixture with O-2,6-diamino-2,6-dideoxy-α-D-glucopyranosyl-(1→4)-O-[β-D-ribofuranosyl-(1→5)]-N^1-(4-amino-L-2-hydroxybutyryl)-2-deoxy-D-streptamine (B form) = (S)-O-2,6-Diamino-2,6-dideoxy-α-D-glucopyranosyl-(1→4)-O-[β-D-xylofuranosyl-(1→5)]-N^1-(4-amino-2-hydroxy-1-oxobutyl)-2-deoxy-D-streptamine mixture with (S)-O-2,6-diamino-2,6-dideoxy-α-D-glucopyranosyl-(1→4)-O-[β-D-ribofuranosyl-(1→5)]-N^1-(4-amino-2-hydroxy-1-oxobutyl)-2-deoxy-D-streptamine (●)

R also sulfate (1:2) dihydrate (51022-98-1)

S *Ambutyrosin, Butirosin***, CI-642

U Antibiotic from *Bacillus circulans*

10854 (8769) $C_{21}H_{41}N_7O_{11}$
26086-49-7

O-2-Deoxy-2-(methylamino)-α-L-glucopyranosyl-(1→2)-O-3,5-dideoxy-3-(hydroxymethyl)-α-L-arabinofuranosyl-(1→4)-N,N'-bis(aminoiminomethyl)-D-streptamine (●)

R Sulfate (34520-86-0)

S Desoximicina, *Desoxydihydrostreptomycin*, Desoxymycin, *Dihydrodesoxystreptomycin*, Exostrept, Mervastrept

U Antibiotic

10855   (8770) $C_{21}H_{41}N_7O_{12}$
                                        128-46-1

all-trans-2,4-Diguanidino-3,5,6-trihydroxyclohexyl-5-
deoxy-2-O-(2-deoxy-2-methylamino-α-L-glucopyrano-
syl)-3-C-hydroxymethyl-β-L-lyxopentofuranoside = O-2-
Deoxy-2-(methylamino)-α-L-glucopyranosyl-(1→2)-O-
5-deoxy-3-C-(hydroxymethyl)-α-L-lyxofuranosyl-
(1→4)-N,N'-bis(aminoiminomethyl)-D-streptamine (●)
R   also sulfate (2:3) (5490-27-7)
S   Abiocine, Acuostrepto, Bactemycin "Leo", Ben-
    distrep, Bioestrepto, Bucomicina, Citrocil, Cocabio-
    tico-Dihidro, Coryfax, CX 60, DHS, DHSM,
    Dicromycine, Didromicin, Didromycin(e), Diestrep-
    topab, Dif-Estrepto Dh, Dihidro-Cidan, *Dihidroe-
    streptomicina*, Dihydrosol, Dihydrostrep, Dihy-
    dro-Streptofor, Dihydro-Streptoliquin, *Dihydrostrep-
    tomycin**, Diidro-Estreptomicina, Diidrostreptomi-
    cina*, Diostrept, Distrepcin, Doramicin, DST, Emici-
    na, Enterastrept, Estreptoluy, Estreptomaipe, Hidra-
    step, Hydromycin, Hydrosin, Hydrostrep, Hystrecin,
    Izostreptomicina, Kabistrep, Mycin, Novomycin S,
    Sanestrepto, Sol-Mycin, Solvostrept, Stremicina,
    Streptomicina Morgan, Streptoral, Veticare, Vibrio-
    mycin, Zidestrepto
U   Antibiotic

10855-01   (8770-01)   3144-30-7
R   *p*-Aminosalicylate (1:3)
S   Pasomycin, Streptopas
U   Tuberculostatic

10855-02   (8770-02)   3563-84-6
R   Pantothenate (no. 1869)
S   Pantomycin
U   Tuberculostatic

10855-03   (8770-03)
R   Pantothenate sulfate mixture

S   Didromycine Pantothénique, Didropantine, Didrothe-
    nat, Diidro-Pantostrept, Dipantosol, Estrepto-Plas-
    min, Panstreptina, Pantostracina D, Streptomycine
    Pantothénique Specia
U   Tuberculostatic

10855-04   (8770-04)   1401-47-4
R   Salt with pyruvic acid isonicotinoylhydrazone (no.
    1574)
S   Dihidroestreptomicina-Alter-Hidrazida, Streptoti-
    bine, Tridies
U   Tuberculostatic

───────────────

10856 $C_{21}H_{42}N_4O_3S_3$
                                        201487-53-8

N-[[(4R)-3-[(2S,3S)-2-[[(2R)-2-Amino-3-mercaptopro-
pyl]amino]-3-methylpentyl]-5,5-dimethyl-4-thiazolidi-
nyl]carbonyl]-L-methionine methyl ester (●)
S   BIM-46068
U   Antineoplastic (farnesyltransferase inhibitor)

───────────────

10857   (8771) $C_{21}H_{43}N_5O_7$
                                        25876-10-2

O-2-Amino-2,3,4,6,7-pentadeoxy-6-(methylamino)-α-D-
ribo-heptopyranosyl-(1→4)-O-[3-deoxy-4-C-methyl-3-
(methylamino)-β-L-arabinopyranosyl-(1→6)]-2-deoxy-
D-streptamine (●)
S   Gentamicin C_1
U   Antibiotic from *Micromonospora purpurea*

10857-01   (8771-01)   1403-66-3
R Mixture with Gentamicin C_2 (R = CH_3, R' = H) and Gentamicin C_{1A} (R = R' = H)
S *Gentamicin**, *Gentamicin C*, *Gentamycin*, *Gentamysin*, Septacin "Abbott"
U Antibiotic from *Micromonospora purpurea*

10857-02   (8771-02)   1405-41-0
R Sulfate of the three Gentamicins (Gentamicin sulfate)
S Aagent, Alcomicin(a), Altamicina, Amgent, Aminogen, Ampligen "Norte-Mexico", Amplomicina, Andregen, Anfomilan, Aploxim, Apogen, Barmicil, B-Gentam, Biogaracin, Biogén-Cusi, Biogenta, Biomargen, Birocin, Bramcyn, Bristagen, BT 603, Categor, Catogen, Celermicin, Centaurin "Kleva", Chrispacin, Cidomicina, Cidomycin, Coliriocilina Gentam, Cortolexan, Cydomycin, Dabroson "Norma Hellas", Datamicine, Derfarlicin, Dermabiotik, Diacarmon, Diakarmon "Help", Dianfarma, Diogent, Dispagent, Duracoll, Duragentam, Duragentamicin, Duramycin "Durachemie", Eltacin, Enclitin, Enterolyt, Epabe, Espectrocina, Eyesfor, Fermentmycin "Intramed", Forticine, Frieso-Gent, Fripeintil, Fustermicina, G 4, Garacin, Garacol, Garalen, Garalone, Garamicina, Garamycin(a), Garasol, Garatec, Garaver, Garbilocin, Geminimycin, Gemonyl, Genalidene, Genbexil, Gencal-vet, Gencin, Genemicin, Gen-Gard, Genkova, Genoptic, Genrex, Gensif, Gensumycin, Genta, Gentabac, Gentabel, Gentabilles, Gentabiotic, Gentabran, Gentacat, Gentacicol, Gentacidin, Gentacidon, Gentacilina, Gentacin, Gentaclin, Gentacoll, Genta-Cortem, Gentacycole, Gentacytin, Gentadavur, Gentaderm, Gentadog, Gentaerba, Gentafair, Gentaflam, Gentaflex, Gentafromm, Genta-Gen, Gentagil, Genta-Gobens, Genta-Grin, Gentagut, Gentaject, Gent-Ak, Gentakel, Gental, Gentalan, Gentalius, Gentallenas, Gentalline, Gentallorens, Gentalodina, Gentalol, Gentalyn, Gentamax, Gentamedical, Gentamedin, Gentamen, Gentamil, Gentamilan, Gentamin(a), Gentamisin, Gentamival, Gentamix, Gentamorgens, Gentamytrex, Gentane, Gentanex, Gentanil, GentaNit, Gentanol, Gentapen, Gentaplen, Gentaplus, Gentapro, Gentaracin, Gentarad, Genta-Rande, Gentarim, Gentaroger, Gentaseptin, Gentasil, Gentasilin, Gentasillin, Gentasin, Gentasol "EN; Ocusoft", Gentasol Damla, Gentasona, Gentasporin, Gentasul, Genta-Sulfat, Gentasum, Gentatrim, Gentavan, Gentavet, Gentavis, Gentavit,
Genta von ct, Gentax, Gentaxil, Gentazaf, Gentazett, Gentex, Genthaver, Gentiber, Gentibioptal, Genticin(a), Genticol, Genticyn, Gentile, Gentina, Gentisin, Gentisul, Gentisum, Gentobic, Gentobryan, Gentocil, Gentocin, Gentodiar, Gentofarma, Gentogram, Gentokulin, Gentoler, Gentollorens, Gentoma, Gentomil, Gentophar, Gent-Ophtal, Gentoptine, Gentosept, Gentovet, Gentralay, Gentrasul, Gentreks, Genty, Geomycin(e), Geracin, Geraphan, Getamisin, Getasin, Gevramycin, G.I. "Provit", Glevomicina, G-Mital, G-Mycin "BPL; Orion", G-Myticin, Gramilina, Gramomycin, Hexamicin, Hexamycin, Horsafertil, Hosbogen, I-Gent, Ikagen, Ikatin, Isotic Timact, Jenamicin, Kantogent, Kylimor, Lacromycin, Lantogent, Lidogen, Lisibac, Logenta, Lugacin, Luinesin, Lyramycin, Macmicina, Magnicine, Marcogen, Martigenta, Medigenta, Medigentacin, Megental, Metrorrigen, Migent, Miramycin, Mycefal, NA-Gentam, Nichogencin, Novoxyn, Nozolon, NSC-82261, Nuclogen, Octoret, Ocugram, Ocu-Mycin, Ogrigenta, Ophtagram, Optagen, Optigen, Optimycin "R + N", Otomygen, Pancutan, Pangram, Pargenta, Penagent, Plenomicina "Argentia", Plurisemina, Progara, Provisual, Pyogenta, Quilagen, Quintamicina, Refobacin, Rexgenta, Ribomicin, Riftamycin, Rigaminol, R.O. Gentycin, Rolixasin, Romsagen, Rovixida, Rupegen, Sabax Gentamix, Sanicalv, Sch 9724, Sedanazin, Septigen, Septopal, Septoral, Septospes, Servigenta, Sintepul, Solgenta, Soligental, Sulgemicin, Sulmycin, Supragenta, Tamadit-80, Tefoval, Theragent, Timact-80, TLC G-65, Tondex, Totamicine, Transgram, Traviocina, Trifagen, U-Gencin, Ultradermis, Utérogen, Vactiriostam, Vactrim, Velminol, Vepha-Gent, Vetagent, Vetrigent, Vijomicin, Violyzen, Viro Genta M, Vistagen, Warigent, Yectamicina, Yedoc, Yordivina
U Antibiotic from *Micromonospora purpurea*

10857-03
R an implantable collagen fleece loaded with gentamicin
S EMD 46217, EMD 53155, *Gentamicin crobefat*
U Antibiotic

10858 (8772) $C_{21}H_{43}N_5O_{12}$
66887-96-5

O-3-Amino-3-deoxy-α-D-glucopyranosyl-(1→4)-O-[2,6-diamino-2,6-dideoxy-α-D-glucopyranosyl-(1→6)]-2-deoxy-N^3-[2-hydroxy-1-(hydroxymethyl)ethyl]-L-streptamine = O-3-Amino-3-deoxy-α-D-glucopyranosyl-(1→6)-O-[2,6-diamino-2,6-dideoxy-α-D-glucopyranosyl-(1→4)]-2-deoxy-N^1-[2-hydroxy-1-(hydroxymethyl)ethyl]-D-streptamine (●)
S   Propikacin*, UK-31214
U   Antibiotic

10859 (8773) $C_{21}H_{44}NO_2$
14565-92-5

[1-(Ethoxycarbonyl)pentadecyl]trimethylammonium = α-Trimethylammoniohexadecanoic acid ethyl ester = 1-(Ethoxycarbonyl)-N,N,N-trimethyl-1-pentadecanaminium = 1-Ethoxy-N,N,N-trimethyl-1-oxo-2-hexadecanaminium (●)
R   Bromide (10567-02-9)
S   Alkonium bromide, Carbaethopendecinium bromatum, Carbethopendecinii bromidum, Mukoseptonex, Ophthalmo-Septonex, Septonex
U   Disinfectant

10860 $C_{21}H_{44}NO_6P$
157244-53-6

(R)-3-Carboxy-2-[[hydroxy(tetradecyloxy)phosphinyl]oxy]-N,N,N-trimethyl-1-propanaminium inner salt (●)
S   SDZ-CPI 975
U   Hypoglycemic (fatty acid oxidation inhibitor)

10861 (8774) $C_{21}H_{44}O_3$
544-62-7

3-(Octadecyloxy)-1,2-propanediol (●)
S   Batilol*, Batylalkohol, Batylol
U   Radioprotective

10862 (8775) $C_{21}H_{45}N_3$
141-94-6

5-Amino-1,3-bis(2-ethylhexyl)hexahydro-5-methylpyrimidine = 1,3-Bis(2-ethylhexyl)perhydro-5-methylpyrimidin-5-ylamine = 1,3-Bis(2-ethylhexyl)hexahydro-5-methyl-5-pyrimidinamine (●)
S   Amphicutan, Bactidol, Bactoderm, Belosept, Bucosept, Collu-Hextril, Doreperol N, Drossadin, Duranil, Elsix, Exetidina, Glypesin, Heksoral, Hekzoton, Hexalen "Warner", Hexalyse "Gerolymatos", Hexatin "Agepha", Hexederm, Hexelab, Hexetidine**, Hexifrice, Hexigel, Hexocil, Hexopyrimidine, Hexoral, Hextril, Irin, Isozid H "Gebro", Kleenosept,

Mallebrin-Fertiglösung, Mund- und Rachenantisepti-
cum "Azupharma", Muramyl, Neo-Angin Gurgel-
lsg., Oraldene, Oraldine, Oralspray, Oraseptic,
P-252, Paradenyl-Lösung, Stas-Gurgellösung, Steril-
ate, Sterisil, Steri/Sol, Stomatidin "Bosnalijek", Stop-
angin, Tilaksil, Triocil, Udder (-Lotion, -Sol), Va-
gi-Hex
U Antibacterial, antifungal

10863 (8776)

$C_{21}H_{46}NO_4P$
58066-85-6

Choline hydroxide hexadecyl hydrogen phosphate inner
salt = 2-[[(Hexadecyloxy)hydroxyphosphinyl]oxy]-
N,N,N-trimethylethanaminium inner salt (●)
S D-18506, HDPC, *Hexadecylphosphocholine*, *Milte-
fosine***, Miltex
U Antineoplastic, antileishmanial

10864 (8777)

$C_{22}H_{13}F_2NO_3$
1764-42-7

4-[(5,7-Difluoro-8-quinolyl)oxy]benzoic acid phenyl ester = 4-[(5,7-Difluoro-8-quinolinyl)oxy]benzoic acid phenyl ester (●)
S *Fluobenzoquinum*, Ondroly-A, Ondro-Spray-Zusatz
U Antiseptic

10865 (8778)

$C_{22}H_{14}Cl_2I_2N_2O_2$
57808-65-8

5'-Chloro-$\alpha^{4'}$-(*p*-chlorophenyl)-$\alpha^{4'}$-cyano-3,5-diiodo-
2',4'-salicyloxylidide = 5'-Chloro-4'-(*p*-chloro-α-cyano-
benzyl)-3,5-diiodo-2'-methylsalicylanilide = *N*-[5-
Chloro-4-[(4-chlorophenyl)cyanomethyl]-2-methylphe-
nyl]-2-hydroxy-3,5-diiodobenzamide (●)
R also monosodium salt (61438-64-0)
S *Closantel***, Flukiver, R 31520, Seponver
U Anthelmintic

10866 (8779)

$C_{22}H_{14}F_3NO_2S_2$
107008-29-7

3-Hydroxy-*N*-[2-phenyl-2-(2-thienyl)ethenyl]-5-(tri-
fluoromethyl)benzo[*b*]thiophene-2-carboxamide (●)
S L 652343
U Anti-inflammatory, analgesic

10867

$C_{22}H_{14}N_2O_5$
151271-53-3

2-[2-(2-Hydroxyphenyl)4-benzoxazolyl]-4-benzoxazole-
carboxylic acid methyl ester = 2'-(2-Hydroxyphenyl)-
[2,4'-bibenzoxazole]-4-carboxylic acid methyl ester (●)

S UK-1
U Cytotoxic from *Streptomyces* sp. 517-02

10868 (8780)

$C_{22}H_{14}N_4O_4$
81645-09-2

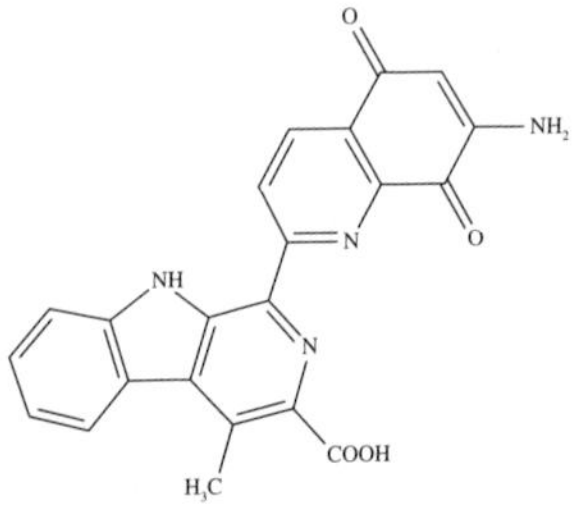

1-(7-Amino-5,8-dihydro-5,8-dioxo-2-quinolinyl)-4-me-
thyl-9*H*-pyrido[3,4-*b*]indole-3-carboxylic acid (●)
S *Antibiotic K 82 A, Lavendamycin*
U Antineoplastic antibiotic from *Streptomyces lavendulae*

10869 (8781)

$C_{22}H_{14}O_9$
4431-00-9

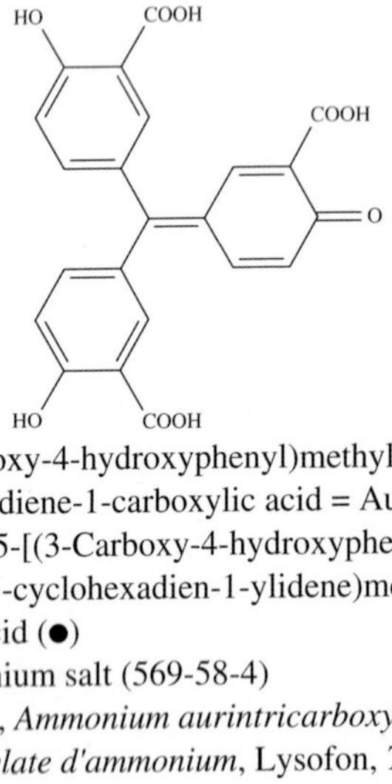

3-[Bis(3-carboxy-4-hydroxyphenyl)methylene]-6-oxo-
1,4-cyclohexadiene-1-carboxylic acid = Aurintricarb-
oxylic acid = 5-[(3-Carboxy-4-hydroxyphenyl)(3-carb-
oxy-4-oxo-2,5-cyclohexadien-1-ylidene)methyl]-2-hydr-
oxybenzoic acid (●)
R Triammonium salt (569-58-4)
S Aluminon, *Ammonium aurintricarboxylate, Aurine
tricarboxylate d'ammonium*, Lysofon, *Tricarbaurini-
um*
U Hyaluronidase inhibitor, anti-inflammatory, analgesic, reagent

---

10870 (8782)

$C_{22}H_{15}FN_2O_2$
131745-25-0

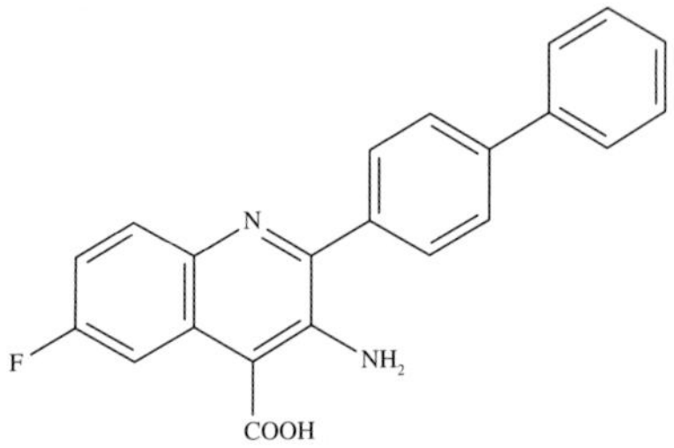

3-Amino-2-[1,1'-biphenyl]-4-yl-6-fluoro-4-quinoline-
carboxylic acid (●)
S CL 306293
U Anti-inflammatory, anti-arthritic

10871 (8783)

$C_{22}H_{16}Cl_2O_4S$
22619-35-8

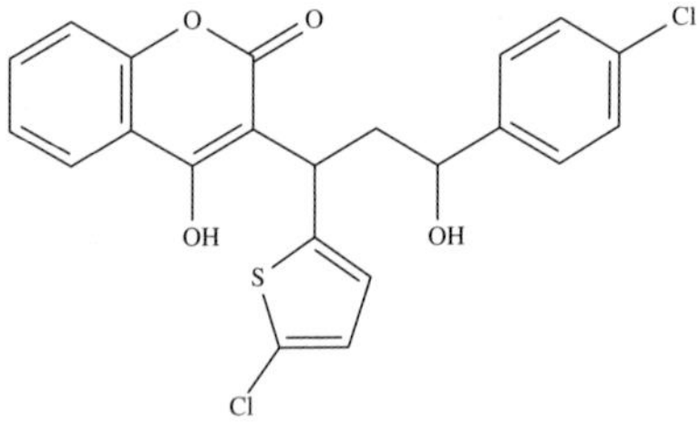

3-[5-Chloro-α-(*p*-chloro-β-hydroxyphenethyl)-2-thenyl]-
4-hydroxycoumarin = 3-[3-(4-Chlorophenyl)-1-(5-chlo-
ro-2-thienyl)-3-hydroxypropyl]-4-hydroxycoumarin = 3-
[3-(4-Chlorophenyl)-1-(5-chloro-2-thienyl)-3-hydroxy-
propyl]-4-hydroxy-2*H*-1-benzopyran-2-one (●)
S Apegmone, LM 550, *Tioclomarol***
U Anticoagulant

10872 (8784)

$C_{22}H_{16}F_2N$
119006-77-8

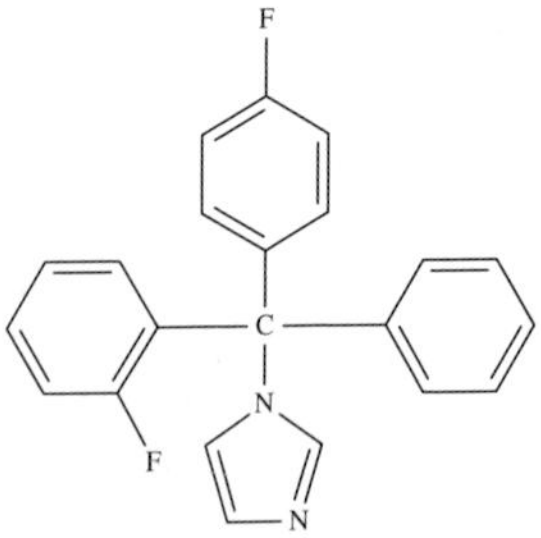

1-[o-Fluoro-α-(p-fluorophenyl)-α-phenylbenzyl]imidazole = 1-(2,4'-Difluorotrityl)imidazole = 1-[(2-Fluorophenyl)(4-fluorophenyl)phenylmethyl]-1H-imidazole (●)
S Cicer, Cutimian, Flusporan, *Flutrimazole***, Funcenal, Micetal, UR-4056
U Antifungal

10873 (8785) $C_{22}H_{16}F_2N_2O_2$
                        131587-93-4

3-[2-[(2',4'-Difluoro-4-biphenylyl)oxy]ethyl]-4-(3H)-quinazolinone = 3-[2-[(2',4'-Difluoro[1,1'-biphenyl]-4-yl)oxy]ethyl]-4(3H)-quinazolinone (●)
S *Flubichin*, VÚFB-15950
U Analgesic (nonnarcotic)

10874 (8786) $C_{22}H_{16}F_3N_3$
                        31251-03-3

-[α,α-Diphenyl-m-(trifluoromethyl)benzyl]-1,2,4-trizole = Diphenyl[3-(trifluoromethyl)phenyl](1,2,4-trizol-1-yl)methane = 1-[Diphenyl[3-(trifluoromethyl)pheyl]methyl]-1H-1,2,4-triazole (●)
 BUE 0620, *Fluotrimazole*
J Antifungal

10875 $C_{22}H_{16}N_2O_3S$
                        118384-10-4

5-[[2-(2-Naphthalenylmethyl)-5-benzoxazolyl]methyl]-2,4-thiazolidinedione (●)
S LY 282449, T-174
U Antidiabetic

10876 (8787) $C_{22}H_{16}N_4O$
                        139339-11-0

3,5-Dihydro-5-phenyl-3-(phenylmethyl)-4H-imidazo[4,5-c][1,8]naphthyridin-4-one (●)
S KF 18280
U Anti-inflammatory, anti-allergic, bronchodilator

10877 (8788) $C_{22}H_{16}O_6$
                        20004-62-0

3,5,7,10-Tetrahydroxy-1,1,9-trimethyl-2H-benzo[cd]pyrene-2,6(1H)-dione (●)
S *Croceomycin*, Geliomycin, Heliomycin "Austin-USA", Itamycin, *Resistomycin*, X 340
U Antibiotic from *Streptomyces resistomycificus* and other *Streptomyces* sp.

10878   (8789) $C_{22}H_{16}O_8$
548-00-5

Bis(4-hydroxy-2-oxo-2H-1-benzopyran-3-yl)acetic acid
ethyl ester = Bis(4-hydroxy-3-coumarinyl)acetic acid
ethyl ester = Ethyl bis(4-hydroxy-2-oxo-2H-chromen-3-
yl)acetate = 4-Hydroxy-α-(4-hydroxy-2-oxo-2H-1-ben-
zopyran-3-yl)-2-oxo-2H-1-benzopyran-3-acetic acid
ethyl ester (●)
S   *Aethyldicumacinum, Aethyldicumarolum, Aethyli bis-
cumacetas, Aethylis biscoumacetas, Aethylium dihy-
droxycumarinylaceticum, Biscoumacétate d'éthyle,*
Biscouron, B.O.E.A., *Carbäthoxydicoumarol,*
D.E.A., Dicumacyl, Dicumaryl, Difarsan, *Ethyl bis-
coumacetate***, *Ethyldicoumarol, Ethylis biscoumacetas***, Etocumarol, G 11705, *Neodicumarinum,*
Pelentan, Pelentanettae, Stabilène, Tremexane,
Trombarin, Trombex, Trombil, Trombolysan,
Tromexan(o)
U   Anticoagulant

10879   (8790) $[C_{22}H_{16}O_{12}]_x$
52486-80-3

Polyhexahydroxytricarboxytriphenylmethane = 3,3',3''-
Methylidynetris[2,5-dihydroxybenzoic acid] homopoly-
mer (●)
S   21 P, Rehibin (old form), *Trigentisic acid, Trigenti-
sinsäure*
U   Hyaluronidase inhibitor, antirheumatic

10880   (8791) $C_{22}H_{17}ClN_2$
60628-98-0

(±)-1-(α-4-Biphenyl-*o*-chlorobenzyl)imidazole = (±)-1-
[[1,1'-Biphenyl]-4-yl(2-chlorophenyl)methyl]-1H-imida-
zole (●)
S   Bay h 6020, *Lombazole***
U   Antimicrobial preservative

10881   (8792) $C_{22}H_{17}ClN_2$
23593-75-1

1-(*o*-Chloro-α,α-diphenylbenzyl)imidazole = 1-(α-2-
Chlorotrityl)imidazole = Diphenyl(2-chlorophenyl)-1-
imidazolylmethane = 1-[(2-Chlorophenyl)diphenylme-
thyl]-1H-imidazole (●)
S   Agisten, Aknecolor, Antifungol, Antimicotico Same,
Antimyk Neu, Apocanda, ARU Spray C, Athletes
Food "Wallis", Atlesan, Axasol, Azutrimazol, Bay
b 5097, Bayer b 5097, Benzoderm myco, Calcrem,
Canalba, Canazol, Candaspor, Candazol, Candibene,
Candid, Candimon "Andromaco", Candimon(a),
Candizole, Canesten, Canex, Canifug, *Chlortritilmi-
dazole*, Cinabel "Columbia", Clazol, Clocim, Clo-
derm "Dermapharm", Clogen, Clomaderm, Cloma-
zen, Clomazol, Clomizol, Clonea, Clortrilen, Clos-
trin, Clot-basan, Clotri, Clotri AbZ, Clotriferm,
Clotrifug, Clotrigalen, Clotrimaderm, *Clotrim-
azole***, Clotrimin, Clotrimix, Clotrimox, Clotrimyl,
Clotri-OPT, Clotrizol, Clovagil, Clozol(e), Cobart,
Contrafungin "Pharma Galen", Cortimazole Soluti-

on, Cotrisan, Covospor, Cremasten, Cutamycon, Cutanil, Cutistad, Damosten, Dermobene antimicotico, Desamix antimicotico, Dignotrimazol, Dinin, Diomicete, Dolexalan, Durafungol, Dynaspor, Elcid, Empecid, Enschent, Eparol, Epicort "Best", Eurosan "Mepha", Eximius, Factodin, FB b 5097, FemCare, Femizol-7, Flagesten, Fragin, Fugolin, Fungederm, Fungiderm "Konimax; Terra-Bio", Fungidermo, Fungiframan, Fungi-med, Fungisid-ratiopharm, Fungisten, Fungistin, Fungizid-ratiopharm, Fungoid topical solut., Fungosten, Fungotox "Mepha", Gilt, Gine-Clotrimox, Gine-Lotramina, Gino-Canesten, Gino-Lotremine, Gromazol, Gyne-Lotremin, Gyne-Lotrimin, Gynezol, Gyno-Canesten, Gyno-Empecid, Gynofil, Hakuserin, Haltrionin, Hexycol CT, Hiderm, Holfungin, Ictan, Imazol, Imidil, Ipalat "Cofasa", Jenamazol, KadeFungin, Kanesol, Kanesten, Kansen, Keimeizuo, Klofan, Klotricid, *Klotrimazol*, Kotozole, Lecibis "Andromaco", Locasten, Logomed-Hautpilz-Salbe, Lokalicid, Lotramina, Lotremin(e), Lotrimin, Masnoderm, Medaspor, Medisten, Micomazol, Micomisan, Micosan "Bioplix; Incobra", Micoter, Micotrizol, Miderm "Infasa", Mono-Baycuten, Mycelex, Mycil Gold, Myclo, Myclo-Derm, Myclo-Gyne, Mycoban, Mycocid "Chemo-Pharma", Mycofug, Myco-Hermal, Mycolind gyno, Mycoril, Mycosporin (old form), Mycotrim, MykoCordes, Mykofungin, Mykohaug C, Nalbix, Neo-Stadamycon, Neo-Zol, Normospor, Onymyken, Ovis neu, Ovutrin, Pan-Fungex, Panmicol, PCPIM, Peckle, Pedicurol, Pedikurol, Pedisafe, Perfungol, Plimycol, Powzol, Prescription Strength Desenex Cream, Radikal bei Fußpilz "Maurer", Scholl Fußpilz-Puder (-Spray), Scorpio, SD-Hermal, Sinium, Solutrim, Stiemazol, Suprazol, Surfaz, Taon, Tapit, Tibatin "DAD", Tinaderm Extra, Topimazol, Tri-Clor, Tricosten, Trimatsol, Trimysten, Uromykol, Vagibene, Vagiclox, Veltrim "Bayer Animal Health", Vicaderm, Xeraspor

J　Antifungal

10882　　　　　　　　　　　　　　　　　　　　$C_{22}H_{17}FN_2OS$
　　　　　　　　　　　　　　　　　　　　　　188352-45-6

2-(4-Fluorophenyl)-3-(4-pyridinyl)-5-[4-(methylsulfinyl)phenyl]-1*H*-pyrrole = 4-[2-(4-Fluorophenyl)-5-[4-(methylsulfinyl)phenyl]-1*H*-pyrrol-3-yl]pyridine (●)
S　L 167307
U　Cytokine inhibitor

10883　　　　　　　　　　　　　　　　　　　$C_{22}H_{17}F_2N_5OS$
　　　　　　　　　　　　　　　　　　　　　　182760-06-1

p-[2-[(αR,βR)-2,4-Difluoro-β-hydroxy-α-methyl-β-(1*H*-1,2,4-triazol-1-ylmethyl)phenethyl]-4-thiazolyl]benzonitrile = 4-[2-[(1R,2R)-2-(2,4-Difluorophenyl)-2-hydroxy-1-methyl-3-(1*H*-1,2,4-triazol-1-yl)propyl]-4-thiazolyl]benzonitrile (●)
S　BMS 207147, ER-30346, *Ravuconazole*
U　Antifungal

10884　(8793)　　　　　　　　　　　　　　　$C_{22}H_{17}NOS$
　　　　　　　　　　　　　　　　　　　　　　1050-10-8

N-Methyl-*N*-(2-naphthyl)thiocarbamic acid *O*-(2-naphthyl) ester = *N*-Methyl-*N,O*-di(2-naphthyl)thiocarbamate

= Methyl(2-naphthalenyl)carbamothioic acid O-(2-naph-thalenyl) ester (●)
S Naphthiomate N
U Antifungal

10885 (8794) $C_{22}H_{17}NO_2$
1897-89-8

18831-34-0

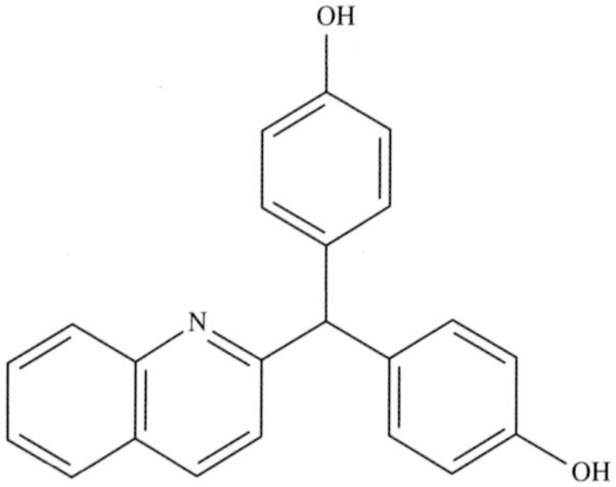

4,4'-(2-Quinolylmethylene)diphenol = 4,4'-(2-Quinoli-nylmethylene)bis[phenol] (●)
R Hydrochloride (19035-45-1)
S Normolaxol
U Laxative

10886 $C_{22}H_{17}N_3$
89972-77-0

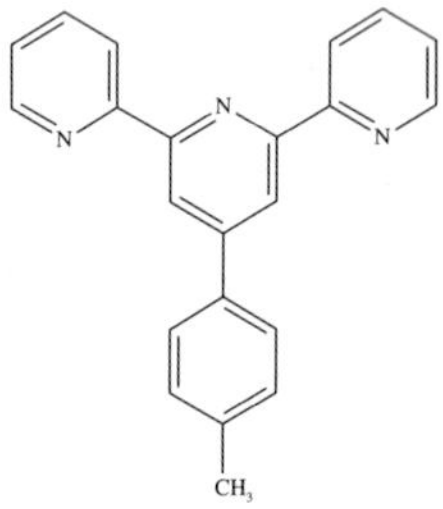

4'-(4-Methylphenyl)-2,2':6',2"-terpyridine (●)
R Trihydrochloride (158014-66-5)
S SS 701
U Neuroprotective

10887 (8795) $C_{22}H_{17}N_3O$
1897-89-8

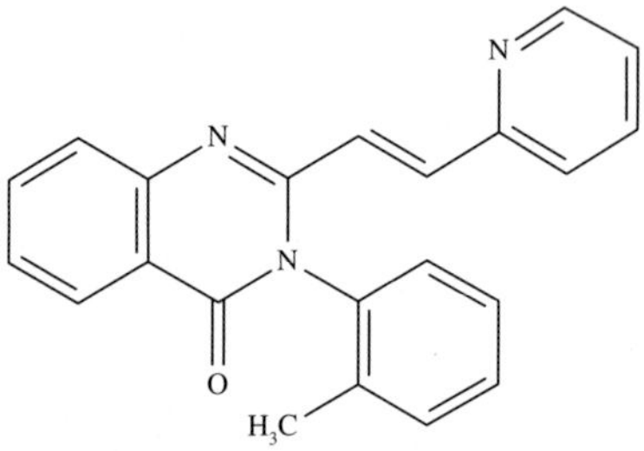

2-[2-(2-Pyridyl)vinyl]-3-o-tolyl-4(3H)-quinazolinone = 3-(2-Methylphenyl)-2-[2-(2-pyridinyl)ethenyl]-4(3H)-quinazolinone (●)
S *Piriqualone**, SRC 909
U Anticonvulsant, muscle relaxant

10888 (8796) $C_{22}H_{17}N_3O_4$
114918-24-0

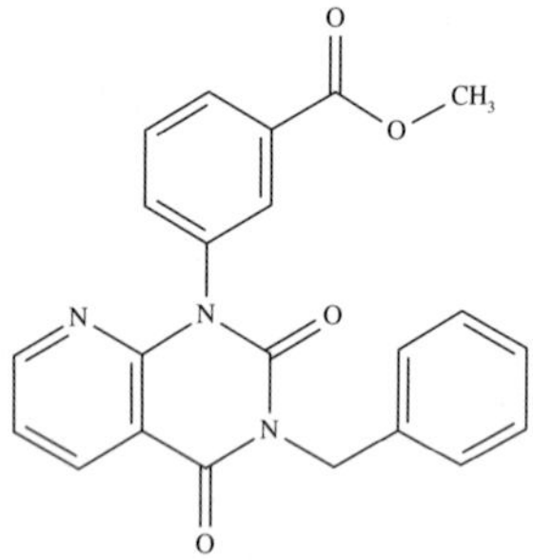

Methyl 3-(3-benzyl-1,2,3,4-tetrahydro-2,4-dioxopyri-do[2,3-d]pyrimidin-1-yl)benzoate = 3-[3,4-Dihydro-2,4-dioxo-3-(phenylmethyl)pyrido[2,3-d]pyrimidin-1-(2H)-yl]benzoic acid methyl ester (●)
S CP-77059
U Antidepressant, anti-inflammatory

10889 $C_{22}H_{17}N_4O_2S$

4-Phenyl-5-(4-nitrostyryl)-1,3,4-thiadiazolium-2-phenyl-
amine = 2-[2-(4-Nitrophenyl)ethenyl]-3-phenyl-5-(phe-
nylamino)-1,3,4-thiadiazolium (●)
R Chloride (178376-63-1)
S MI-D
U Mesoionic compd.

10890 $C_{22}H_{18}BrN_3O_2$
                                         140688-17-1

5-[[(4-Bromophenyl)amino]carbonyl]-4,5-diphenyl-3-
pyrazolidinone = N-(4-Bromophenyl)-3-oxo-4,5-diphe-
nyl-1-pyrazolidinecarboxamide (●)
S LY 262691
U Cholecystokinin B receptor antagonist, antipsychotic

10891 $C_{22}H_{18}BrN_3O_2$
                                         147523-65-7

(4S,5R)-N-(4-Bromophenyl)-3-oxo-4,5-diphenyl-1-pyra-
zolidinecarboxamide (●)
S LY 288513
U Cholecystokinin B receptor antagonist

10892   (8797) $C_{22}H_{18}Cl_2FNO_3$
                                         68359-37-5

α-Cyano-4-fluoro-3-phenoxybenzyl 3-(2,2-dichlorovi-
nyl)-2,2-dimethylcyclopropanecarboxylate = 3-(2,2-Di-
chloroethenyl)-2,2-dimethylcyclopropanecarboxylic acid
cyano(4-fluoro-3-phenoxyphenyl)methyl ester (●)
S Baythroid, Bay V1 1704, *Cyfluthrin*, *Cyfoxylate*,
     Eulan SP, FCR 1272, Solfac EW
U Insecticide (veterinary)

10893   (8798) $C_{22}H_{18}FN_3O$
                                         137103-53-8

α-(4-Fluorophenyl)-α-[4-(4-pyridinyl)phenyl]-1H-imi-
dazole-1-ethanol (●)
S DuP 983
U Anti-inflammatory (topical)

10894   (8799) $C_{22}H_{18}I_6N_2O_9$
                                         51022-74-3

3,3'-[Oxybis(ethyleneoxymethylenecarbonylimino)]
bis[2,4,6-triiodobenzoic acid] = 3,3'(3,6,9-Trioxaundeca-

nedioyldiimino)bis(2,4,6-triiodobenzoic acid) = 3,3'-
[Oxybis[2,1-ethanediyloxy(1-oxo-2,1-ethanediyl)imi-
no]]bis[2,4,6-triiodobenzoic acid] (●)
R also meglumine salt (1:2) (68890-05-1)
S *Acide iotroxique***, *Acidum iotroxicum***, Bilisco-
pin(e), Bilisegrol, Chologram, *Iotroxamide*, *Iotroxic
acid***, *Iotroxinsäure***, *Meglumine iotroxinate*,
SH 213 AB, SHH 273, ZK 4
U Diagnostic aid (radiopaque medium)

U Hypoglycemic

10897 $C_{22}H_{18}N_4$
                                                118112-07-5

3-Methyl-*N*-[3-(2-pyridinyl)-1-isoquinolinyl]benzene-
carboximidamide (●)
S VUF 8514
U Iron chelator

10895   (8800) $C_{22}H_{18}N_2$
                                                60628-96-8

1-(*p*,α-Diphenylbenzyl)imidazole = 1-(α-Biphenyl-4-yl-
benzyl)imidazole = 1-([1,1'-Biphenyl]-4-ylphenylme-
thyl)-1*H*-imidazole (●)
S Agispor, Amycor, Arel, Azolmen, Baritona, Bay-
clear, Bay h 4502, Bedriol, Bicoper, Bicutrin, Bifa-
zol, Bifokey, Bifomyk, Bifon, Bifonal, *Bifonazo-
le***, Bite, Canesten Extra, Compaser, Dermokey,
Fonazole, Fungiderm "Nycomed", Funginon, Fungo-
sel, Kavaderm, Micofun, Micomicen "Labomed",
Micotopic, Mifozome, Mikomax, Moldina, Mono-
stop, Multifung, Myco-flusemidon, Mycospor, My-
cosporan, Mycosporin "Bayer-Austria", Neltolon,
Poulmycin, Quilmicor, Topiderm "Sanitas"
U Topical antifungal

10898   (8801) $C_{22}H_{18}O_3$
                                                5129-14-6

2-(*o*-Methoxyphenyl)-3,3-diphenylacrylic acid = 2-(2-
Methoxyphenyl)-3,3-diphenyl-2-propenoic acid = α-(Di-
phenylmethylene)-2-methoxybenzeneacetic acid (●)
S *Anisacril***, SKF 16046
U Anorexic, antihyperlipidemic

10896 $C_{22}H_{18}N_2O_4S_3$
                                                213411-84-8

5-[[4-[2-[5-Methyl-2-(2-thienyl)-4-oxazolyl]ethoxy]ben-
zo[*b*]thien-7-yl]methyl]-2,4-thiazolidinedione (●)
S BM 152054

10899   (8802) $C_{22}H_{18}O_4$
                                                17667-23-

6,6'-Dimethyl-2,2'-binaphthyl-1,1',8,8'-tetrol = 6,6'-Di-
methyl[2,2'-binaphthalene]-1,1',8,8'-tetrol (●)
S Diospyrin, *Diospyrol*

U Anthelmintic from the fruits of *Diospyros mollis*

10900 $C_{22}H_{18}O_6$
182232-96-8

(–)-11-Hydroxy-5-(hydroxymethyl)-2-(1-methylpropyl)-4*H*-anthra[1,2-*b*]pyran-4,7,12-trione (●)
S *Espicufolin*
U Neuroprotective from *Streptomyces* sp. cu 39

10901 $C_{22}H_{18}O_6$
199795-31-8

11-Hydroxy-2-(2-hydroxy-1-methylpropyl)-5-methyl-4*H*-anthra[1,2-*b*]pyran-4,7,12-trione (●)
S AH 1763IIa
U Antiherpetic

10902 (8803) $C_{22}H_{18}O_6S$
3624-51-9

3,3'-[3-(Methylthio)propylidene]bis(4-hydroxycoumarin) = 3,3'-[3-(Methylthio)propylidene]bis[4-hydroxy-2*H*-benzopyran-2-one] (●)

S Propentan, Thiocoumar, Thioporan, *Thioxycoumarine, Tioporanum*
U Anticoagulant

10903 (8804) $C_{22}H_{18}O_7$
93513-59-8

1,5-Dihydroxy-3-methyl-8-[(2,6,7,7a-tetrahydro-4*H*-furo[3,2-*c*]pyran-4-yl)oxy]-9,10-anthracenedione (●)
S MT 81
U Antiprotozoal

10904 (8805) $C_{22}H_{18}O_8$
11006-83-0

S *Thermorubin*
U Antibiotic from *Thermoactinomyces antibioticus*

10905 (8806) $C_{22}H_{18}O_{11}$
989-51-5

3,4-Dihydro-2α-(3,4,5-trihydroxyphenyl)-2*H*-1-benzo-pyran-3α,5,7-triol 3-(3,4,5-trihydroxybenzoate) = 3,4,5-Trihydroxybenzoic acid (2*R-cis*)-3,4-dihydro-5,7-dihydr-oxy-2-(3,4,5-trihydroxyphenyl)-2*H*-1-benzopyran-3-yl ester (●)
S EGCG, *Epigallocatechin gallate*

U Antineoplastic (main constituant of green tea infusion)

10906 (8807) $C_{22}H_{19}Br$
479-68-5

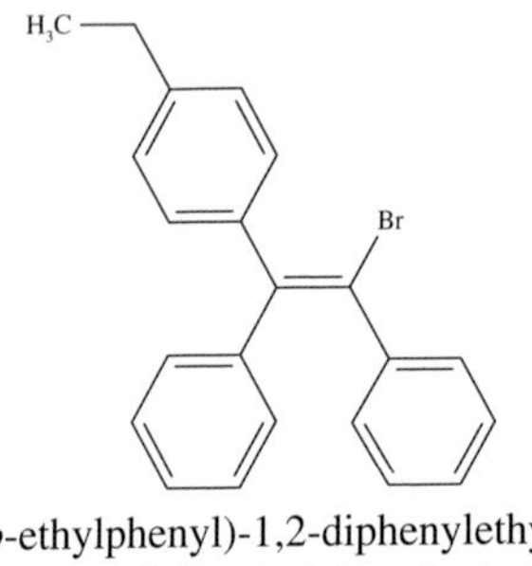

1-Bromo-2-(*p*-ethylphenyl)-1,2-diphenylethylene = α-Bromo-α,β-diphenyl-β-(*p*-ethylphenyl)ethylene = (*p*-Ethylphenyl)stilbene bromide = 1-(2-Bromo-1,2-diphenylethenyl)-4-ethylbenzene (●)
S Acnestrol "Scharper", Aknestrol, B.D.P.E., *Broparestrol***, *Broparoestrolum*, Keitol, L.N. 107, Longestrol
U Synthetic estrogen

10907 (8808) $C_{22}H_{19}BrO_2$
41038-34-0

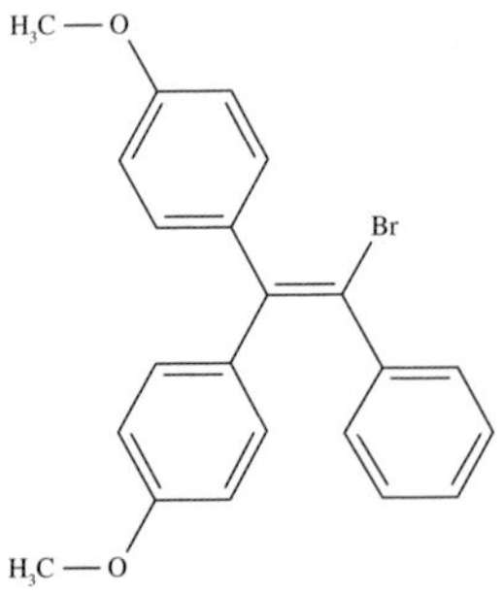

1-Bromo-2,2-bis(*p*-methoxyphenyl)-1-phenylethylene = α,α-Di(*p*-methoxyphenyl)-β-bromo-β-phenylethylene = 1,1'-(Bromophenylethenylidene)bis[4-methoxybenzene] (●)
S Oeplexyl
U Synthetic estrogen

10908 (8809) $C_{22}H_{19}Br_2NO_3$
52918-63-5

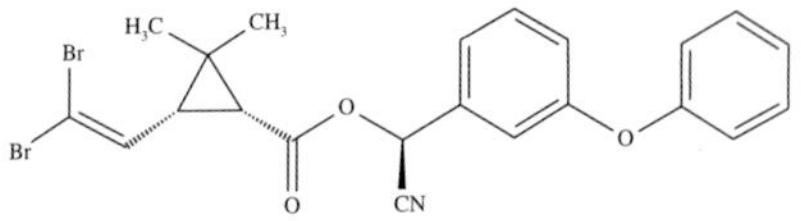

(S)-α-Cyano-3-phenoxybenzyl *cis*-(1R,3R)-2,2-dimethyl-3-(2,2-dibromovinyl)cyclopropanecarboxylate = [1R-[1α(S*),3α]]-3-(2,2-Dibromoethenyl)-2,2-dimethyl-cyclopropanecarboxylic acid cyano(3-phenoxyphenyl)methyl ester (●)
S Butox, Coopersect Vet, *Decamethrin*, Decis, Deltacid, *Deltamethrin*, Difexon "Armstrong", Esbecythrin, FMC 45498, Hexafen "Ems", Kothrine, Nopucid, NRDC 161, OMS 1998, Pediculosane, Pediderm "Biosintetica", RU 22974, Scalibor, Versatrine
U Insecticide

10909 $C_{22}H_{19}ClN_4O_3$
146033-03-6

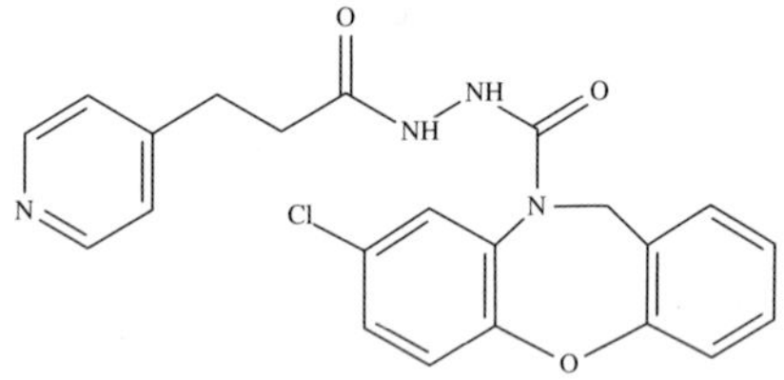

N'-[3-(4-Pyridyl)propionyl]-8-chlorodibenz[*b,f*][1,4]oxazepine-10(11*H*)-carbohydrazide = 8-Chlorodibenz[*b,f*][1,4]oxazepine-10(11*H*)-carboxylic acid 2-[1-oxo-3-(4-pyridinyl)propyl]hydrazide (●)
S SC-51089
U Analgesic (PGE$_2$ antagonist)

10910 (8810) $C_{22}H_{19}ClO_3$
95233-18-6

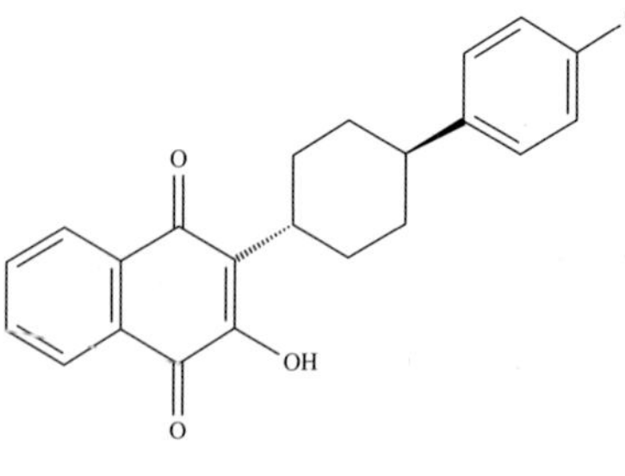

trans-2-[4-(p-Chlorophenyl)cyclohexyl]-3-hydroxy-1,4-naphthoquinone = trans-2-[4-(4-Chlorophenyl)cyclohexyl]-3-hydroxy-1,4-naphthalenedione (●)
R also without trans-definition (94015-53-9)
S Acuvel, Atovaquone**, BW A 566 C, 566 C 80, Mepron "Glaxo Wellcome", Metrone, Wellvone
U Antimalarial, antiparasitic

10910-01 156879-70-8
R Compd. with proguanil (no. 2930) (1:1)
S Malarone
U Antimalarial

10911 (8811) $C_{22}H_{19}Cl_2NO_3$
52315-07-8

(SR)-α-Cyano-3-phenoxybenzyl (RS)-cis-trans-3-(2,2-dichlorovinyl)-2,2-dimethylcyclopropanecarboxylate = 3-(2,2-Dichloroethenyl)-2,2-dimethylcyclopropanecarboxylic acid cyano(3-phenoxyphenyl)methyl ester (●)
R also stereoisomer compounds
S Alpha-Cypermethrin, Ardap-Konzentrat, Barricade, CCN-52, Cipermethrin, Contratic, Crovect, Cypermethrin, Cypertic, Cypor, Deosect, Dy-Sect, Dysect pour-on, Ectotrine, Ektomin, Equivite, Feutratic, Flectron, FMC-30980, Hygienex Piretro, INS 15, NRDC 149, OMS 2002, Parasol, Renegade, Ripcord, Robust, Rycovet Ovipor, TAD Anti-Insekt, Talmark, Topclip, Topclip Parasol, WL-43467
U Insecticide (veterinary)

10912 $C_{22}H_{19}F_5N_4O_3$
170422-36-3

2R,3R,6S)-2-(2,4-Difluorophenyl)-6-hydroxy-3-methyl--(1,2,4-1H-triazol-1-ylmethyl)-4-[4-(trifluoromethyl)benzoyl]morpholine = [2S-(2α,5α,6β)]-6-(2,4-Di-

fluorophenyl)-5-methyl-6-(1H-1,2,4-triazol-1-ylmethyl)-4-[4-(trifluoromethyl)benzoyl]-2-morpholinol (●)
S UR-9746
U Antifungal

10913 (8812) $C_{22}H_{19}NO_2$
5034-76-4

2,3-Bis(p-methoxyphenyl)indole = 2,3-Bis(4-methoxyphenyl)-1H-indole (●)
S Indoxole*, U-22020
U Antipyretic, anti-inflammatory

10914 (8813) $C_{22}H_{19}NO_4$
603-50-9

4,4'-(2-Pyridylmethylene)diphenol diacetate = 4,4'-(2-Pyridylmethylene)di(phenyl acetate) = 4,4'-Diacetoxydiphenyl-2-pyridinylmethane = 4,4'-(2-Pyridinylmethylene)bis[phenol] diacetate (ester) (●)
S Abführ-Dragees N "Wepa", Abilaxin, Acetphenolpicoline, Agaroletten, Ainsoft, Alaxa, Alsylax, Amado Lax, Analux "Bjarner", Anan, Anatas, Anumbral, Apo-Bisacodyl, Arcalax, Babynormo, Becolax forte N, Bekunis B, Bekunis-Bisacodyl, Benelax, Bicol "Wampole", Bicolax, Bideks, Bisa 5, Bisac-Evac, Bisacodyl**, Bisacolax, Bisacon, Bisacosit, Bisadil, Bisadyl, Bisalax, Bisaphar, Biscolax, Bisco-Zitron, Bisolax, Bizalaks, Brocalax, Capolax, Carter's Little Pills, Cenalax, Clysodrast novum, Codylax, Colac, Colax, Confetto Falqui C.M., Contac-

tolax, Contalax, Correctol "Schering-Plough", Corr-
ex, Critex, Dacodyl, Darmol-Bisacodyl, Darmolet-
ten, Deficol, Dekalax, Delco-Lax, Demolaxin, Dis-
lax, Dokusogan mini, Dralinsa-Dragees, Drenacol,
Drix N, Ducolax, Dulcagen, Dulcolan, Dulcolax,
Dulco-Laxo, Dulcotek, Durolax, Endokolat,
Enteralax, Ercolax, Eulaxan, Evac-Q-Tabs,
Evae-Q-Kwik, Extralax, Feen-A-Mint, Fenilaxan,
Fenolax "Polfa", Fleet-Bisacodyl, Fleet-Laxative,
Florisan N, Godalax, Hillcolax, Horton, Ivilax, Ka-
lax, La 96a, Laco, Lacsacodil, Laksodil, Laksotek,
Laxacodil, Laxacol, Laxadane-Sup., Laxadin, La-
xafucon N Drag., Laxagetten, Laxamed, Laxanin N,
Laxans "Thomae", Laxans ratiopharm, Laxatin, Lax-
bene, Laxematic, Laxibon, Laxierperlen Waaning,
Laxin "Pharco", Laxine "Maney", Laxit, Lax Makol
mono, Laxoberal Bisa, Laxodyl, Laxoluy, Laxorex,
Laxothyrin, Laxysat, Logomed-Abführdragees,
Manaron, Mandrolax Bisa, Marienbader Pillen N,
Medesup, Medibudget-Abführdr., Mediolax, Med-Laxan-Suppos., Megalax, Metalax "Star", Minilax, Mo-
dane, Moderlax, Mucinum "Pharmethic", Multilax,
Muxol "Vifor", Nedalax, Neodrast, Neolax, Neo-Sal-
vilax, Nigalax, Normalene, Nourilax N, Novolax
"Krka", Obstilax forte, Oralax, Organolax, Oterolax
Urbanuspillen S, Panlax, Paxolax, Pentalax, Perilax
"Lennon; Nordisk Droge", Phenilax, PMS-Bisaco-
dyl, Prepacol, Primalax, Prontolax, Purgo-Pil, Py-
rilax, Rhabarex B, Rheingold xpress, Rytmil, Rytmi-
lax, Sanvacual, Satolax-10, Schlank-Schlank VDH,
Sedolax, Sekolaks, Sensiblex, Sensnan, Serax "Ha-
meln", Sopalax, Spirolax, Stadalax, Superlaks, Su-
ratt, Tavolax Neue Formel, Telemin suppos., Tem-
po, Tempolax forte, Theralax, Tirgon N, Toilax, Toi-
lex, Tüzi Q, Ulcolax, Uzu Benpi, Vactrol, Verecole-
ne C.M., Videx "Chinoin", Vinco-Abführperlen,
Vinco forte, Yodel-Soft, Zetrax, Zwitsalax-N
U Laxative

10914-01 (8813-01)   1336-29-4
R Complex with tannic acid
S *Bisacodyl tannex*, Clysodrast
U Laxative

10915 $C_{22}H_{19}N_3O_4$
                                                              171596-29-5

(6R,12R)-2,3,6,7,12,12a-Hexahydro-2-methyl-6-[3,4-
(methylenedioxy)phenyl]pyrazino[1',2':1,6]pyri-
do[3,4-b]indole-1,4-dione = (6R,12aR)-6-(1,3-Benzodio-
xol-5-yl)-2,3,6,7,12,12a-hexahydro-2-methylpyrazi-
no[1',2':1,6]pyrido[3,4-b]indole-1,4-dione (●)
S *Adalufil*, Cialis, IC 351
U Vasodilator (PDE5 inhibitor)

10916 $C_{22}H_{19}N_5O$
                                                              65426-89-3

6-Amidino-2-[4-(4-amidinophenoxy)phenyl]indole = 2-
[4-[4-(Aminoiminomethyl)phenoxy]phenyl]-1H-indole-
6-carboximidamide (●)
S RWJ 61907
U Antifungal, antibacterial

10917 (8814) $C_{22}H_{19}N_7O_6S_2$
                                                              102253-70-

1-[[(6R,7R)-7-[2-(2-Amino-4-thiazolyl)glyoxylamido]-2
carboxy-8-oxo-5-thia-1-azabicyclo[4.2.0]oct-2-en-3-
yl]methyl]-4-(5-oxazolyl)pyridinium hydroxide inner salt
7^2-(Z)-(O-methyloxime) = (Z)-(7R)-7-[2-(2-Amino-4-

thiazolyl)-2-(methoxyimino)acetamido]-3-[[4-(5-oxazo-lyl)-1-pyridinio]methyl]-3-cephem-4-carboxylate = [6R-[6α,7β(Z)]]-1-[[7-[[(2-Amino-4-thiazolyl)(methoxyimi-no)acetyl]amino]-2-carboxy-8-oxo-5-thia-1-azabicy-clo[4.2.0]oct-2-en-3-yl]methyl]-4-(5-oxazolyl)pyridi-nium hydroxide inner salt (●)
S DQ-2556
U Antibiotic

10918 (8815)

$C_{22}H_{20}ClN_3$
79781-95-6

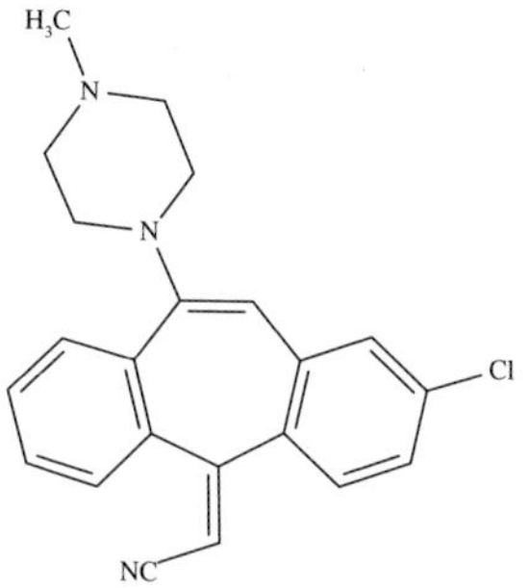

(Z)-2-Chloro-10-(4-methyl-1-piperazinyl)-5H-diben-zo[a,d]cycloheptene-Δ^{5,α}-acetonitrile = (Z)-[2-Chloro-10-(4-methyl-1-piperazinyl)-5H-dibenzo[a,d]cyclohep-en-5-ylidene]acetonitrile (●)
S Rilapine**
U Neuroleptic

10919 (8816)

$C_{22}H_{20}FN_3OS$
81656-30-6

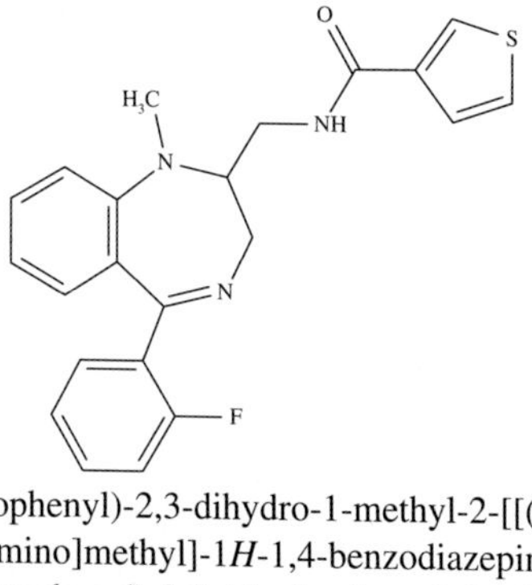

5-(o-Fluorophenyl)-2,3-dihydro-1-methyl-2-[[(3-thienyl-carbonyl)amino]methyl]-1H-1,4-benzodiazepine = (±)-N-[5-(2-Fluorophenyl)-2,3-dihydro-1-methyl-1H-1,4-ben-zodiazepin-2-yl]methyl]-3-thiophenecarboxamide (●)
S KC-5103, Tifluadom**
U Analgesic (opioid benzodiazepine)

10920 (8817)

$C_{22}H_{20}FN_3O_2$
85118-42-9

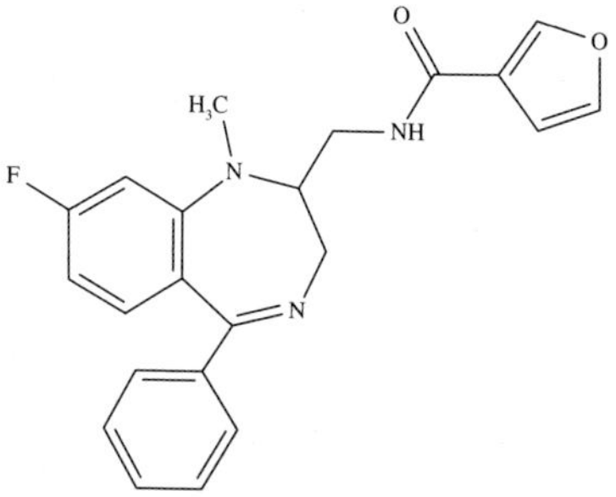

8-Fluoro-2-[[(3-furanylcarbonyl)amino]methyl]-2,3-di-hydro-1-methyl-5-phenyl-1H-1,4-benzodiazepine = (±)-N-[(8-Fluoro-2,3-dihydro-1-methyl-5-phenyl-1H-1,4-benzodiazepin-2-yl)methyl]-3-furancarboxamide (●)
S Lufuradom**
U Analgesic (opioid benzodiazepine)

10921

$C_{22}H_{20}F_2N_4O_4$
195048-23-8

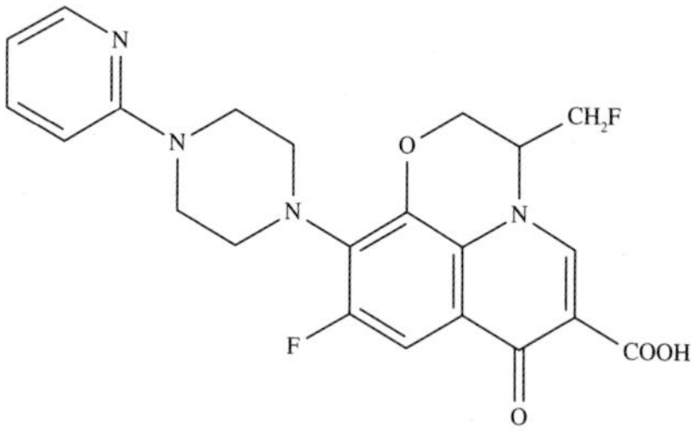

9-Fluoro-3-(fluoromethyl)-2,3-dihydro-7-oxo-10-[4-(2-pyridinyl)-1-piperazinyl]-7H-pyrido[1,2,3-de]-1,4-benz-oxazine-6-carboxylic acid (●)
S R 71762
U Antiviral (HIV-1)

10922

$C_{22}H_{20}F_2O_3$
159911-27-0

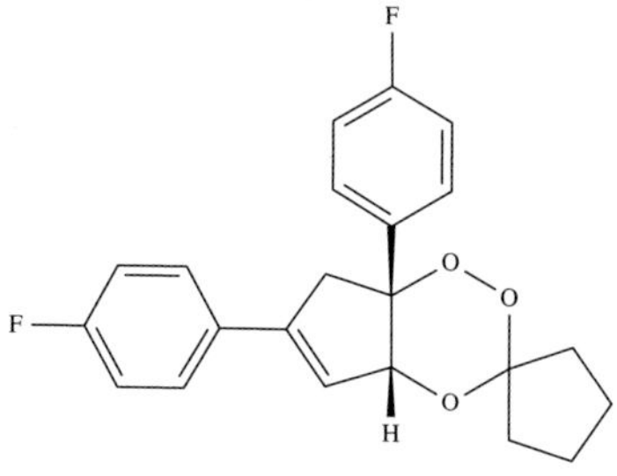

(4'aR,7'aR)-*rel*-6',7'a-Bis(4-fluorophenyl)-4'a,7'a-dihy-drospiro[cyclopentane-1,3'-[7H]cyclopenta[1,2,4]tri-oxin] (●)
S   *Fenozan B 07, Fenozan 50F*
U   Antiparasitic, antimalarial

10923

$C_{22}H_{20}F_6N_4O_3$
159706-39-5

2(S)-[[3,5-Bis(trifluoromethyl)benzyl]oxy]-3(S)-phenyl-4-[(1,2-dihydro-3-oxo-1,2,4-triazol-5-yl)methyl]morpho-line = 5-[[(2S,3S)-2-[[3,5-Bis(trifluoromethyl)phe-nyl]methoxy]-3-phenyl-4-morpholinyl]methyl]-1,2-dihy-dro-3H-1,2,4-triazol-3-one (●)
S   L 742694
U   Analgesic (neurokinin-1 receptor antagonist)

10924

$C_{22}H_{20}NO_5$
55950-32-8

7,8,10-Trimethoxy-2,3-methylenedioxy-N-methylben-zo[c]phenanthridinium = 1,2,4-Trimethoxy-12-methyl-[1,3]benzodioxolo[5,6-c]phenanthridinium (●)
R   also chloride (30044-83-8)
S   *Chelilutine, Methoxychelerythrine*
U   Antineoplastic, antimicrobial, antiparasitic

10925   (8818)

$C_{22}H_{20}N_2$
16571-59-8

1-Benzyl-3-[2-(4-pyridyl)ethyl]indole = 1-(Phenylme-thyl)-3-[2-(4-pyridinyl)ethyl]-1H-indole (●)
R   Monohydrochloride (5585-71-7)
S   *Benzindopyrine hydrochloride**,* IN 461, NSC-17789, *Pyrbenzindole*
U   Antipsychotic

10926   (8819)

$C_{22}H_{20}N_2O_2$
60576-13-8

m-Benzoyl-N-(4-methyl-2-pyridyl)hydratropamide = 2-(m-Benzoylphenyl)-N-(4-methyl-2-pyridyl)propion-amide = 3-Benzoyl-α-methyl-N-(4-methyl-2-pyridi-nyl)benzeneacetamide (●)
R   also monohydrochloride (59512-37-7)
S   Calmatel, Flexidol "FCE", Picalm, *Piketoprofen**,* Triparsean
U   Anti-inflammatory, analgesic

10927   (8820)

$C_{22}H_{20}N_2O_4$
78287-27-

(S)-4,11-Diethyl-4-hydroxy-1H-pyrano[3',4':6,7]indoliz-no[1,2-b]quinoline-3,14(4H,12H)-dione (●)

S ECPT, *7-Ethylcamptothecin*, SN-22
U Antineoplastic

10928 $C_{22}H_{20}N_2O_4S$
103787-97-9

5-[[4-[2-(5-Methyl-2-phenyl-4-oxazolyl)ethoxy]phenyl]methyl]-2,4-thiazolidinedione (●)
S BM 131246
U Hypoglycemic

10929 $C_{22}H_{20}N_2O_5$
170861-63-9

4-[[4-[2-(5-Methyl-2-phenyl-4-oxazolyl)ethoxy]phenyl]methyl]-3,5-isoxazolidinedione (●)
S JTT-501, PNU-182716, *Reglitazar**, *Reglixane*
U Insulin sensitizer (PPAR gamma agonist)

10930 $C_{22}H_{20}N_2O_5$
86639-52-3

(4S)-4,11-Diethyl-4,9-dihydroxy-1H-pyrano[3',4':6,7]indolizino[1,2-b]quinoline-3,14(4H,12H)-dione (●)
7-Ethyl-10-hydroxycamptothecin, SN-38
U Antineoplastic

10931 $C_{22}H_{20}N_2O_5S$
103788-05-2

5-[[4-[2-Hydroxy-2-(5-methyl-2-phenyl-4-oxazolyl)ethoxy]phenyl]methyl]-2,4-thiazolidinedione (●)
S AD 5075
U Antidiabetic (insulin sensitizer)

10932 (8821) $C_{22}H_{20}N_4O_5$
66093-35-4

N-[4-Amino-5-(3,4,5-trimethoxybenzyl)-2-pyrimidinyl]phthalimide = 4-Amino-2-phthalimido-5-(3,4,5-trimethoxybenzyl)pyrimidine = 2-[4-Amino-5-[(3,4,5-trimethoxyphenyl)methyl]-2-pyrimidinyl]-1H-isoindole-1,3(2H)-dione (●)
S DAN-122, *Phthaloyl-Trimetoprim*, *Talmetoprim***
U Antibacterial (intestinal)

10933 $C_{22}H_{20}N_4O_8$
163300-58-1

(Z)-1,4-Bis[4-[(3,5-dioxo-1,2,4-oxadiazolidin-2-yl)methyl]phenoxy]-2-butene = 2,2'-[(2Z)-2-Butene-1,4-diylbis(oxy-4,1-phenylenemethylene)]bis[1,2,4-oxadiazolidine-3,5-dione] (●)
S YM-440
U Antidiabetic

10934   (8822) $C_{22}H_{20}N_4O_8S_2$
62587-73-9

4-Carbamoyl-1-[[(6R,7R)-2-carboxy-8-oxo-7-[(2R)-2-phenyl-2-sulfoacetamido]-5-thia-1-azabicyclo[4.2.0]oct-2-en-3-yl]methyl]pyridinium hydroxide inner salt = (7R)-3-[(4-Carbamoylpyridinio)methyl]-7-[(2R)-2-phenyl-2-sulfoacetamido]-3-cephem-4-carboxylate = [6R-[6α,7β(R*)]]-4-(Aminocarbonyl)-1-[[2-carboxy-8-oxo-7-[(phenylsulfoacetyl)amino]-5-thia-1-azabicyclo[4.2.0]oct-2-en-3-yl]methyl]pyridinium hydroxide inner salt (●)
R   Monosodium salt (52152-93-9)
S   Abbott-46811, Cefomonil, *Cefsulodin sodium**,* CGP 7174/E, Monaspor, Pseudocef, Pseudomonil, Pyocefal, SCE 129, *Sulcephalosporin*, Takesef, Takesulin, Tilmapor, Ulfaret(e)
U   Antibiotic

10935   (8823) $C_{22}H_{20}N_6O_6$
17692-30-7

7-(2,3-Dihydroxypropyl)theophylline bis(nicotinate ester) = 7-(2,3-Dinicotinoyloxypropyl)-1,3-dimethylxanthine = 7-[2,3-Bis[(3-pyridinylcarbonyl)oxy]propyl]-3,7-dihydro-1,3-dimethyl-1H-purine-2,6-dione = 3-Pyridinecarboxylic acid 1-[(1,2,3,6-tetrahydro-1,3-dimethyl-2,6-dioxo-7H-purin-7-yl)methyl]-1,2-ethanediyl ester (●)
S   A 17 HF, Corverum (active substance), *Diniprofylline**, Diteonicon*
U   Broncholytic, vasodilator

10936   (8824) $C_{22}H_{20}N_6O_{10}S_3$
34941-71-4

N,N'-[Sulfonylbis(p-phenylene)]bis[2,6-dihydroxy-4-methyl-5-pyrimidinesulfonamide] = N,N'-(Sulfonyldi-4,1-phenylene)bis[1,2,3,4-tetrahydro-6-methyl-2,4-dioxo-5-pyrimidinesulfonamide] (●)
S   Diuciphone, Diutsifon
U   Leprostatic, agent for treating rheumatoid-type collagenosis, immunostimulant

10937   (8825) $C_{22}H_{20}O_6$
5664-34-6

4,4'-Dihydroxy-3,3',3''-trimethoxyfuchsone = Bis(4-hydroxy-3-methoxyphenyl)-3-methoxyquinomethane = 4-[Bis(4-hydroxy-3-methoxyphenyl)methylene]-2-methoxy-2,5-cyclohexadien-1-one (●)
S   Rubrocol, Rubrophen, *Trimethoxyaurin*
U   Tuberculostatic, dye

10938 $C_{22}H_{20}O_{…}$
190848-69-…

(7R)-2,3-Dideoxy-7-C-phenyl-D-xylo-hept-2-enonic acid δ-lactone 7-[(2E)-3-phenyl-2-propenoate] (●)
S   GHM-10, *Howiinol A*

U Antineoplastic from *Goniothalamus howii*

10939 (8826) $C_{22}H_{20}O_9$
                                        41206-96-6

4',5-Diacetyloxy-3,3',7-trimethoxyflavone = 5-(Acetyl-oxy)-2-[4-(acetyloxy)-3-methoxyphenyl]-3,7-dimethoxy-4*H*-1-benzopyran-4-one (●)
S Ro 9-0298
U Antiviral

10940 (8827) $C_{22}H_{20}O_{10}$
                                        19879-06-2

3a*S*-(3aα,5α,8α,9α,11β,13bα,15*S**)]-3,3a,5,8,11,13b-hexahydro-7,8,12,15-tetrahydroxy-5,9-dimethyl-8,11-ethanofuro[2,3-*e*]naphtho[2,3-*c*:6,7-*c'*]dipyran-2,6,13(9*H*)-trione (●)
S *Antibiotic WR 141*, *Granaticin*, *Granaticin A*
U Antibiotic from *Streptomyces olivaceus*

0941 $C_{22}H_{21}BrO_7$
                                        183241-67-0

-)-8-Bromo-2,3-dihydro-2,5,10-trihydroxy-3'-methoxy-methyl-2-(1-methylethyl)spiro[anthracene-(4*H*),2'(5'*H*)-furan]-4,5'-dione (●)

U Antibacterial (gram-positive)

10942 $C_{22}H_{21}ClN_2O_2$
                                        156635-05-1

1-(4-Chlorobenzoyl)-4-[4-(2-oxazolin-2-yl)benzyli-dene]piperidine = 1-(4-Chlorobenzoyl)-4-[[4-(4,5-dihy-dro-2-oxazolyl)phenyl]methylene]piperidine (●)
S BIBB 515
U 2,3-Oxidosqualene cyclase inhibitor

10943 (8828) $C_{22}H_{21}ClN_2O_6$
                                        57645-05-3

N-[[1-(*p*-Chlorobenzoyl)-5-methoxy-2-methyl-3-indo-lyl]acetyl]-L-serine = *N*-[[1-(4-Chlorobenzoyl)-5-meth-oxy-2-methyl-1*H*-indol-3-yl]acetyl]-L-serine (●)
S *Sermetacin***, SHG 318 AB, TVX 3158
U Anti-inflammatory

10944 (8829) $C_{22}H_{21}ClN_2O_8$
                                        2013-58-3

7-Chloro-4-(dimethylamino)-1,4,4a,5,5a,6,11,12a-octa-hydro-3,5,10,12,12a-pentahydroxy-6-methylene-1,11-di-oxo-2-naphthacenecarboxamide = 7-Chloro-6-demethyl-6-deoxy-5β-hydroxy-6-methylenetetracycline = [4*S*-(4α,4aα,5α,5aα,12aα)]-7-Chloro-4-(dimethylamino)-1,4,4a,5,5a,6,11,12a-octahydro-3,5,10,12,12a-pentahydr-

oxy-6-methylene-1,11-dioxo-2-naphthacenecarbox-
amide (●)
R also sulfosalicylate (1:1) (73816-42-9)
S GS-2989, Meclan "Ortho", *Meclociclina*, Meclocil,
Meclocycline**, Mecloderm, Meclosorb, Meclutin
semplice, Medan "Ortho", Novacnyl, NSC-78502,
Quoderm, Traumatociclina
U Antibacterial

10945 $C_{22}H_{21}ClN_4O_4$
                                                    150452-18-9

1-[6-Chloro-4-[[3,4-(methylenedioxy)benzyl]amino]-2-
quinazolinyl]-4-piperidinecarboxylic acid = 1-[4-[(1,3-
Benzodioxol-5-ylmethyl)amino]-6-chloro-2-quinazoli-
nyl]-4-piperidinecarboxylic acid (●)
R Monosodium salt (150452-19-0)
S E-4021, ER 21355
U Anti-ischemic

10946 $C_{22}H_{21}ClN_4O_6$
                                                    198559-42-1

(*S*)-9-Chloro-5-[[[*p*-(aminomethyl)-*o*-(carboxymethoxy)phenyl]carbamoyl]methyl]-6,7-dihydro-1*H*,5*H*-pyri-
do[1,2,3-*de*]quinoxaline-2,3-dione = 5-(Aminomethyl)-
2-[[[(5*S*)-9-chloro-2,3,6,7-tetrahydro-2,3-dioxo-1*H*,5*H*-
pyrido[1,2,3-*de*]quinoxalin-5-yl]acetyl]amino]phen-
oxy]acetic acid (●)
R Monohydrochloride (161292-39-3)
S SM 18400
U NMDA receptor antagonist (glycine site binding)

10947 $C_{22}H_{21}ClN_6O_2$
                                                    124750-92-1

2-Butyl-4-chloro-1-[[2'-(1*H*-tetrazol-5-yl)[1,1'-biphe-
nyl]-4-yl]methyl]-1*H*-imidazole-5-carboxylic acid (●)
S E-3174, EXP-3174
U Antihypertensive (angiotensin II receptor antagonist)

10948 $C_{22}H_{21}Cl_2IN_4O$
                                                    183232-66-8

1-(2,4-Dichlorophenyl)-5-(4-iodophenyl)-4-methyl-*N*-1-
piperidinyl-1*H*-pyrazole-3-carboxamide (●)
S AM-251
U Cannabinoid receptor antagonist

10949 $C_{22}H_{21}Cl_3N_4O$
                                         168273-06-1

5-(4-Chlorophenyl)-1-(2,4-dichlorophenyl)-4-methyl-N-
1-piperidinyl-1H-pyrazole-3-carboxamide (●)
R Monohydrochloride (158681-13-1)
S *Rimocaban hydrochloride, Rimonabant hydrochloride**, SR 141716A
U Cannabinoid receptor antagonist

10950 $C_{22}H_{21}FN_2O$

(−)-3-[[[(R)-2-Chromanylmethyl]amino]methyl]-5-(p-
fluorophenyl)pyridine = (R)-N-[(3,4-Dihydro-2H-1-ben-
zopyran-2-yl)methyl]-5-(4-fluorophenyl)-3-pyridineme-
thanamine (●)
R Dihydrochloride (177976-12-4)
S EMD 128130, *Sarizotan hydrochloride***
U Antipsychotic (D₂-antagonist and 5-HT₁ₐ-agonist)

0951   (8830) $C_{22}H_{21}F_2NO_4S$
                                         121083-05-4

(−)-6,8-Difluoro-2,3,4,9-tetrahydro-9-[[4-(methylsulfo-
yl)phenyl]methyl]-1H-carbazole-1-acetic acid (●)

S L 670596
U Thromboxane receptor antagonist

10952   (8831) $C_{22}H_{21}NO_2S$
                                         2799-07-7

2-Tritylthio-L-alanine = 3-[(Triphenylmethyl)thio]-L-ala-
nine (●)
S NSC-83265
U Antineoplastic

10953   (8832) $C_{22}H_{21}NO_3$
                                         120934-96-5

Methyl 4-(o-benzylbenzoyl)-2,5-dimethyl-3-pyrrolecarb-
oxylate = 2,5-Dimethyl-4-[2-(phenylmethyl)benzoyl]-
1H-pyrrole-3-carboxylic acid methyl ester (●)
S FPL 64176
U Cardiac inotropic (calcium channel activator)

10954   (8833) $C_{22}H_{21}NO_7$
                                         53228-00-5

(4α,4aβ,12aβ)-2-Acetyl-4-amino-4a,12a-dihydro-3,10,11,12a-tetrahydroxy-6,9-dimethyl-1,12(4H,5H)-naphthacenedione (●)

R Hydrochloride (53274-41-2); also [4R-(4α,4aβ,12aβ)]-compd. base: (29144-42-1); hydrochloride: (56433-46-6)

S Abbott 40728, *Cetocycline hydrochloride***, *Cetotetrine hydrochloride*, *Chelocardin*

U Antibiotic from *Nocardia sulphurea* sp.

10955

$C_{22}H_{21}N_3O_2$
214548-46-6

3-[2-(3,4-Dihydrobenzofuro[3,2-c]pyridin-2(1H)-yl)ethyl]-2-methyl-4H-pyrido[1,2-a]pyrimidin-4-one (●)

S *Lusaperidone***

U Antidepressant

10956

$C_{22}H_{21}N_3O_6$
151328-90-4

5-[[5-[[(6-Amidino-2-naphthyl)oxy]carbonyl]-2-furfuryl]amino]-5-oxovaleric acid = 5-[[(4-Carboxy-1-oxobutyl)amino]methyl]-2-furancarboxylic acid 2-[6-(amino-iminomethyl)-2-naphthalenyl] ester (●)

R Monohydrochloride (151328-91-5)

S TO-195

U Protease inhibitor

10957 (8834)

$C_{22}H_{21}N_3O_7S_2$
56453-01-1

(2S,5R,6R)-6-[(R)-(−)-2-(p-Hydroxyphenyl)-2-[(4-oxo-4H-thiopyran-3-yl)carboxamido]acetamido]-3,3-dimethyl-7-oxo-4-thia-1-azabicyclo[3.2.0]heptane-2-carboxylic acid = (6R)-6-[D-2-(4-Hydroxyphenyl)-2-[(4-oxo-4H-thiopyran-3-yl)-carboxamido]acetamido]penicillanic acid = [2S-[2α,5α,6β(S*)]]-6-[[(4-Hydroxyphenyl)[[(4-oxo-4H-thiopyran-3-yl)carbonyl]amino]acetyl]amino]-3,3-dimethyl-7-oxo-4-thia-1-azabicyclo[3.2.0]heptane-2-carboxylic acid (●)

R Monosodium salt (63076-45-9)

S *Timoxicillin sodium*

U Antibiotic

10958

$C_{22}H_{21}N_5$
160906-61-0

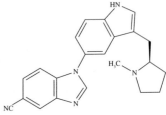

(R)-1-[3-[(1-Methyl-2-pyrrolidinyl)methyl]-1H-indol-5-yl]-1H-benzimidazole-5-carbonitrile (●)

S CP-161242

U Antipsychotic, migraine therapeutic

10959

$C_{22}H_{21}N_5O_2$
139958-16-0

3-Methoxy-2,6-dimethyl-4-[[2'-(1H-tetrazol-5-yl)[1,1'-biphenyl]-4-yl]methoxy]pyridine (●)

S ME-3221

U Antihypertensive (angiotensin II receptor antagonist)

10960 (8835)

$C_{22}H_{21}N_5O_3$
41689-64-9

N'-[1-Methyl-3-(3,5-dioxo-1,2-diphenylpyrazolidin-4-yl)propylidene]cyanoacetohydrazide = 1,2-Diphenyl-4-(3-oxobutyl)pyrazolidine-3,5-dione cyanoacetohydrazone = Cyanoacetic acid [3-(3,5-dioxo-1,2-diphenyl-4-pyrazolidinyl)-1-methylpropylidene]hydrazide (●)

S Kyazon

U Anti-inflammatory

10961 (8836)

$C_{22}H_{21}N_7O_3S_2$
77527-71-0

(6R,7R)-1-Benzyl-4-[[2-carboxy-3-[[(1-methyl-1H-tetrazol-5-yl)thio]methyl]-8-oxo-5-thia-1-azabicyclo[4.2.0]oct-2-en-7-yl]amino]pyridinium hydroxide inner salt = (7R)-7-[(1-Benzyl-4-pyridinium)amino]-3-[[(1-methyl-1H-tetrazol-5-yl)thio]methyl]-3-cephem-4-carboxylic acid hydroxide inner salt = (6R-trans)-4-[[2-Carboxy-3-[[(1-methyl-1H-tetrazol-5-yl)thio]methyl]-8-oxo-5-thia-1-azabicyclo[4.2.0]oct-2-en-7-yl]amino]-1-(phenylmethyl)pyridinium hydroxide inner salt (●)

S L-640876

U Antibiotic

10962 (8837)

$C_{22}H_{21}N_7O_6S_2$
103238-57-9

1-[[(6R,7R)-7-[2-(2-Amino-5-thiazolyl)glyoxylamido]-2-carboxy-8-oxo-5-thia-1-azabicyclo[4.2.0]oct-2-en-3-yl]methyl]pyridinium hydroxide inner salt 7^2-(E)-[O-(2-oxo-3-pyrrolidinyl)oxime] = (E)-7-[2-(2-Amino-5-thiazolyl)-2-(2-oxopyrrolidin-3-yloxyimino)acetamido]-3-pyridiniomethyl-3-cephem-4-carboxylate = [6R-[6α,7β(E)]]-1-[[7-[[(2-Amino-5-thiazolyl)[[(2-oxo-3-pyrrolidinyl)oxy]imino]acetyl]amino]-2-carboxy-8-oxo-5-thia-1-azabicyclo[4.2.0]oct-2-en-3-yl]methyl]pyridinium hydroxide inner salt (●)

S *Cefempidone***, GR 50692, TA-5901

U Antibiotic

10963 (8838)

$C_{22}H_{22}ClF_4NO_2$
10457-91-7

4-[4-(4-Chloro-α,α,α-trifluoro-m-tolyl)-4-hydroxypiperidino]-4'-fluorobutyrophenone = 4-[4-[4-Chloro-3-(trifluoromethyl)phenyl]-4-hydroxypiperidino]-4'-fluorobutyrophenone = 4-[4-[4-Chloro-3-(trifluoromethyl)phenyl]-1-[3-(4-fluorobenzoyl)propyl]-4-piperidinol = 4-[4-[4-Chloro-3-(trifluoromethyl)phenyl]-4-hydroxy-1-piperidinyl]-1-(4-fluorophenyl)-1-butanone (●)

R also hydrochloride (17230-87-4)

S *Clofluperol***, PJ-929, R 9298, *Seperidol*

U Neuroleptic

10964   (8839)   C$_{22}$H$_{22}$ClN
                                  69319-52-4

(1S-cis)-3'-Chloro-N,N-dimethylspiro[2-cyclohexene-1,5'-[5H]dibenzo[a,d]cyclohepten]-4-amine (●)
S   A-31472, SIR 155
U   Neuroleptic

10965   C$_{22}$H$_{22}$ClNO$_2$
                                  153973-24-1

2-Chloro-3-[(4-hexylphenyl)amino]-1,4-naphthoquinone = 2-Chloro-3-[(4-hexylphenyl)amino]-1,4-naphthalene-dione (●)
S   NQ 304
U   Antithrombotic, platelet aggregation inhibitor

10966   (8840)   C$_{22}$H$_{22}$ClN$_3$O$_5$
                                  82163-55-1

3-(2-Chlorophenyl)-3,4-dihydro-5,7-dimethyl-2-[[[(methylamino)carbonyl]oxy]methyl]-4-oxo-6-quinazoline-carboxylic acid ethyl ester (●)
S   EG 1088
U   Antihypertensive

10967   C$_{22}$H$_{22}$ClN$_5$O$_2$
                                  178308-65-1

4-[(3-Chloro-4-methoxybenzyl)amino]-1-(4-hydroxypi-peridino)-6-phthalazinecarbonitrile = 4-[[(3-Chloro-4-methoxyphenyl)methyl]amino]-1-(4-hydroxy-1-piperidi-nyl)-6-phthalazinecarbonitrile (●)
R   Monohydrochloride (178308-66-2)
S   E-4010
U   Phosphodiesterase 5 inhibitor

10968   (8841)   C$_{22}$H$_{22}$ClN$_5$O$_2$S
                                  105219-56-5

4-[3-[4-(o-Chlorophenyl)-9-methyl-6H-thieno[3,2-f]-s-triazolo[4,3-a][1,4]diazepin-2-yl]propionyl]morpholine = 3-[4-(2-Chlorophenyl)-9-methyl-6H-thie-no[3,2-f][1,2,4]triazolo[4,3-a][1,4]diazepin-2-yl]-1-(4-morpholinyl)-1-propanone = 4-[3-[4-(2-Chlorophenyl)-9-methyl-6H-thieno[3,2-f][1,2,4]triazolo[4,3-a][1,4]diaze-pin-2-yl]-1-oxopropyl]morpholine (●)
S   Apafant**, WEB 2086 BS
U   PAF antagonist

10969 (8842) $C_{22}H_{22}ClN_5O_4S_2$
77590-92-2

4-Propionyl-1-piperazinecarboxylic acid ester with (±)-6-(7-chloro-1,8-naphthyridin-2-yl)-2,3,6,7-tetrahydro-7-hydroxy-5*H*-p-dithiino[2,3-*c*]pyrrol-5-one = 4-(1-Oxo-propyl)-1-piperazinecarboxylic acid (±)-6-(7-chloro-1,8-naphthyridin-2-yl)-2,3,6,7-tetrahydro-7-oxo-5*H*-1,4-dithiino[2,3-*c*]pyrrol-5-yl ester (●)
S 37162 R.P., *Suproclone***
U Tranquilizer, hypnotic

10970 (8843) $C_{22}H_{22}FN_3O_2$
548-73-2

1-[1-[3-(*p*-Fluorobenzoyl)propyl]-1,2,3,6-tetrahydro-4-pyridyl]-2-benzimidazolinone = 1-[1-[4-(*p*-Fluorophenyl)-4-oxobutyl]-1,2,3,6-tetrahydro-4-pyridyl]-2-benzimidazolinone = *p*-Fluoro-γ-[4-(2-oxo-1-benzimidazolinyl)-1,2,3,6-tetrahydro-1-pyridyl]butyrophenone = 1-[1-[4-(4-Fluorophenyl)-4-oxobutyl]-1,2,3,6-tetrahydro-4-pyridinyl]-1,3-dihydro-2*H*-benzimidazol-2-one (●)
S Dehidrobenzperidol, Dehydrobenzperidol, Deidrobenzperidolo, Diaperidol, Dridol, Droleptan, *Droperidol***, Halkân, Inapsin(e), Inopsin, McN-JR 4749, Neurolidol, NSC-169874, Paxical, R 4749, Sintodian, Sintodril
U Neuroleptic, anti-emetic

10971 $C_{22}H_{22}FN_3O_2$
208661-17-0

3-[2-[(1*S*,5*R*,6*S*)-6-(4-Fluorophenyl)-3-azabicyclo[3.2.0]hept-3-yl]ethyl]-2,4(1*H*,3*H*)-quinazolinedione (●)
S *Balaperidone, Belaperidone***, LU 111995
U Antipsychotic (D_4/5-HT_{2A} receptor antagonist)

10972 (8844) $C_{22}H_{22}FN_3O_2S$
76330-71-7

3-[2-[4-(*p*-Fluorobenzoyl)piperidino]ethyl]-2-thio-2,4-(1*H*,3*H*)-quinazolinedione = 3-[2-[4-(4-Fluorobenzoyl)-1-piperidinyl]ethyl]-2,3-dihydro-2-thioxo-4(1*H*)-quinazolinone (●)
R Tartrate (1:1) (79449-96-0)
S *Altanserin tartrate***, R 53200
U Serotonin antagonist

10973 (8845) $C_{22}H_{22}FN_3O_3$
74050-98-9

3-[2-[4-(*p*-Fluorobenzoyl)piperidino]ethyl]-2,4(1*H*,3*H*)-quinazolinedione = 3-[2-[4-(4-Fluorobenzoyl)-1-piperidinyl]ethyl]-2,4(1*H*,3*H*)-quinazolinedione (●)
S *Cetanserina, Ketanserin***, R 41468
U Serotonin antagonist, antihypertensive

10973-01 (8845-01)   83846-83-7
R Tartrate (1:1)

S Aseranox, Ket, *Ketanserin tartrate***, Ketensin, KJK 945, Perketan, Prevasmin, R 49945, Serefrex, Serepress, Sufrex, Sufrexal, Taseron, Vulketan
U Serotonin antagonist, antihypertensive

10974 $C_{22}H_{22}F_3N_3O_2$
220831-12-9

1-(2-Methylphenyl)-4-[(3-hydroxypropyl)amino]-6-(trifluoromethoxy)-2,3-dihydropyrrolo[3,2-*c*]quinoline = 3-[[2,3-Dihydro-1-(2-methylphenyl)-6-(trifluoromethoxy)-1*H*-pyrrolo[3,2-*c*]quinolin-4-yl]amino]-1-propanol (●)
S AU-006
U Anti-ulcer agent

10975   (8846) $C_{22}H_{22}NO_4$
6872-73-7

5,6,7,8,13,13a-Hexahydro-2,3,10,11-tetramethoxy-8-methylberbinium = 2,3,10,11-Tetramethoxy-8-methyldibenzo[*a,g*]quinolizinium (●)
R Chloride (38989-38-7)
S *Coralyne chloride*, NSC-96349
U Antineoplastic, antileukemic

10976 $C_{22}H_{22}NO_5P$
142235-90-3

(*S*)-α-Amino-5-(phosphonomethyl)-[1,1':4',1''-terphenyl]-3-propanoic acid (●)
S SDZ 215-439
U NMDA receptor antagonist

10977 $C_{22}H_{22}N_2O$
197576-78-6

3-Hydroxy-3-[4-(6-quinolinyl)phenyl]quinuclidine = 3-[4-(6-Quinolinyl)phenyl]-1-azabicyclo[2.2.2]octan-3-ol (●)
R Dihydrochloride (190841-57-7)
S RPR 107393
U Antihyperlipidemic (squalene synthase inhibitor)

10978   (8847) $C_{22}H_{22}N_2O$
60996-89-6

2-Phenyl-4-quinolinecarboxylic acid monocyclohexylamide = *N*-Cyclohexyl-2-phenyl-4-quinolinecarboxamide (●)
S *Cycinchophène*, Lithopan
U Anti-arthritic, antirheumatic

10979   (8848) $C_{22}H_{22}N_2O_2$
                                            2056-56-6

1,2-(Pentylmalonyl)-1,2-dihydro-4-phenylcinnoline = 2-Pentyl-6-phenyl-1*H*-pyrazolo[1,2-*a*]cinnoline-1,3(2*H*)-dione (●)
S   AHR-3015, *Cinnopentazone***, *Cintazone*, NSC-102825, Scha 306
U   Anti-inflammatory, analgesic

10980 $C_{22}H_{22}N_2O_3S_2$
                                            193420-10-9

N-[(2*R*)-2-(Mercaptomethyl)-1-oxo-3-phenylpropyl]-4-(2-thiazolyl)-L-phenylalanine (●)
S   Z-13752A
U   Neutral endopeptidase and ACE inhibitor

10981 $C_{22}H_{22}N_2O_4$
                                            177036-94-1

(+)-(*S*)-2-[(4,6-Dimethyl-2-pyrimidinyl)oxy]-3-methoxy-3,3-diphenylpropionic acid = (*S*)-α-[(4,6-Dimethyl-2-pyrimidinyl)oxy]-β-methoxy-β-phenylbenzenepropanoic acid (●)

S   *Ambrasentan*, BSF 208075
U   Endothelin A receptor antagonist

10982   (8849) $C_{22}H_{22}N_2O_4S_3$
                                            107859-85-8

p-[[[5-[(4-Acetyl-3-hydroxy-2-propylbenzyl)thio]-1,3,4-thiadiazol-2-yl]thio]methyl]benzoic acid = 4-[[[5-[[(4-Acetyl-3-hydroxy-2-propylphenyl)methyl]thio]-1,3,4-thiadiazol-2-yl]thio]methyl]benzoic acid (●)
S   YM-17551
U   Leukotriene antagonist

10983   (8850) $C_{22}H_{22}N_2O_5S$
                                            1926-48-3

(2*S*,5*R*,6*R*)-3,3-Dimethyl-7-oxo-6-(2-phenoxy-2-phenyl-acetamido)-4-thia-1-azabicyclo[3.2.0]heptane-2-carb-oxylic acid = (6*R*)-6-(α-Phenoxyphenylacetamido)peni-cillanic acid = α-Phenoxybenzylpenicillin = [2*S*-(2α,5α,6β)]-3,3-Dimethyl-7-oxo-6-[(phenoxyphenylace-tyl)amino]-4-thia-1-azabicyclo[3.2.0]heptane-2-carb-oxylic acid (●)
R   Monopotassium salt (1177-30-6)
S   D.C.(B).L. 306, *Fenbenicillin Potassium***, Pen-spek, *Phenbenicillin Potassium*
U   Antibiotic

10984 $C_{22}H_{22}N_2O_6$
                                          171714-84-4

(+)-(S)-2-[(4,6-Dimethoxy-2-pyrimidinyl)oxy]-3-meth-
oxy-3,3-diphenylpropionic acid = (αS)-α-[(4,6-Dime-
thoxy-2-pyrimidinyl)oxy]-β-methoxy-β-phenylbenzene-
propanoic acid (●)
S   *Darusentan***, HMR 4005, LU 135252
U   Endothelin A receptor antagonist

10985   (8851) $C_{22}H_{22}N_2O_6S_2$
                                          59070-07-4

p-Tolyl (R)-*N*-[(2S,5R,6R)-(2-carboxy-3,3-dimethyl-7-
oxo-4-thia-1-azabicyclo[3.2.0]hept-6-yl)]-3-thiophene-
malonamate = [2S-[2α,5α,6β(S*)]]-3,3-Dimethyl-6-[[3-
(4-methylphenoxy)-1,3-dioxo-2-(3-thienyl)propyl]ami-
no]-7-oxo-4-thia-1-azabicyclo[3.2.0]heptane-2-carb-
oxylic acid (●)
R   Monosodium salt (59070-06-3)
S   BRL 12594, *Ticarcillin cresyl Sodium*
U   Antibiotic

10986   (8852) $C_{22}H_{22}N_2O_8$
                                          914-00-1

4-(Dimethylamino)-1,4,4a,5,5a,6,11,12a-octahydro-
3,5,10,12,12a-pentahydroxy-6-methylene-1,11-dioxo-2-

naphthacenecarboxamide = 6-Deoxy-6-demethyl-6-me-
thylene-5-hydroxytetracycline = [4S-
(4α,4aα,5α,5aα,12aα)]-4-(Dimethylamino)-
1,4,4a,5,5a,6,11,12a-octahydro-3,5,10,12,12a-pentahydr-
oxy-6-methylene-1,11-dioxo-2-naphthacenecarbox-
amide (●)
R   also monohydrochloride (3963-95-9)
S   Adramycin, Adriamicina "Farmochimica", Apricli-
    na, Benciclina, Bialatan, Bivimicina, Boscillina, Bre-
    vicillina, Ciclobiotic, Ciclum, Dobracina, Duecap,
    Duplaciclina, Duramicina, Dynamicin, Esarondil,
    Esquilin, Fitociclina, Franciclina, Francomicina,
    Gammaciclina, Germicina, Globociclina, Gramcicli-
    na, Grandemicina, GS-2876, Idrossimicina, Isometa,
    Jucin, Largomicina, Lentomicina-Zimaia, Londomy-
    cin, Lysocline, Medomycin "Medosan", Megamyci-
    ne, Metabiotic, Metabiotico(n), Metac, *Metaciclin*,
    Metacil "Ibirn", Metaclin, Metaclor, *Metacycline**,
    Metadomus, Metagram, Metamicina, Metatetril, *Me-
    thacycline*, *Méthylènecycline*, Metilenbiotic, Micoci-
    clina, Minibiotic, Minimicina, Mit-Ciclina, Molcicli-
    na, MOTC, Optimycin "Biochemie", Ossirondil,
    Oxilenta, Paveciclina, Physiomycine, Piziacina,
    Pluramycine, Plurigram, Prontomicina, Quickmici-
    na, Radiomicina, Rindex, Rondomicina, Rondomy-
    cin(e), Rotilen, Sernamicina, Simethacin, Sititetral,
    Stafilon, Tachiciclina, Tetrabios, Tetralen "Vitória",
    Tetranova, Tiberciclina, Ticomicina, Treis-ciclina,
    Valcin, Vitabiotic, Wassermicina, Yatrociclina, Zer-
    micina
U   Antibiotic

10987 $C_{22}H_{22}N_4$
                                          72549-77-0

2-Methyl-*N*-(1-methyl-2-phenylethyl)-6-phenyl-5*H*-pyr-
rolo[3,2-*d*]pyrimidin-4-amine (●)
S   *Proferodine*
U   Antineoplastic

10988 (8853) $C_{22}H_{22}N_4O_2$
517-83-9

4,4'-Bis(2,3-dimethyl-1-phenyl-3-pyrazolin-5-one) = 1,1',2,2'-Tetrahydro-1,1',5,5'-tetramethyl-2,2'-diphe-nyl[4,4'-bi-3H-pyrazole]-3,3'-dione (●)
S   Forbisen, S.N. 475
U   Antibacterial (formerly in bovine anaplasmosis)

10989 (8854) $C_{22}H_{22}N_4O_7S$
54661-82-4

(2S,5R,6R)-6-[(R)-(–)-2-(Furoylureido)-2-phenylacet-amido]-3,3-dimethyl-7-oxo-4-thia-1-azabicy-clo[3.2.0]heptane-2-carboxylic acid = (6R)-6-[D-2-(Fu-roylureido)-2-phenylacetamido]penicillanic acid = [2S-[2α,5α,6β(S*)]]-6-[[[[[(2-Furanylcarbonyl)amino]carbo-nyl]amino]phenylacetyl]amino]-3,3-dimethyl-7-oxo-4-thia-1-azabicyclo[3.2.0]heptane-2-carboxylic acid (●)
R   Monopotassium salt (25673-17-0)
S   BL-P 1597, *Furbenicillin*
U   Antibiotic

10990 (8855) $C_{22}H_{22}N_6O_5S_2$
84957-29-9

1-[[(6R,7R)-7-[2-(2-Amino-4-thiazolyl)glyoxylamido]-2-carboxy-8-oxo-5-thia-1-azabicyclo[4.2.0]oct-2-en-3-yl]methyl]-6,7-dihydro-5H-1-pyrindinium hydroxide inner salt 7^2-(Z)-(O-methyloxime) = (Z)-(7R)-7-[2-(2-

Amino-4-thiazolyl)-2-(methoxyimino)acetamido]-3-[(6,7-dihydro-5H-1-pyrindinio)methyl]-3-cephem-4-carboxylate = [6R-[6α,7β(Z)]]-1-[[7-[[(2-Amino-4-thia-zolyl)(methoxyimino)acetyl]amino]-2-carboxy-8-oxo-5-thia-1-azabicyclo[4.2.0]oct-2-en-3-yl]methyl]-6,7-dihy-dro-5H-1-pyrindinium hydroxide inner salt (●)
R   also sulfate (1:1) (98753-19-6)
S   Broact, Cedixen, *Cefpirome**, Cefrom, CPR, HR 810, Keiten, Metran
U   Antibiotic

10991 (8856) $C_{22}H_{22}N_6O_7S_2$
72558-82-8

1-[[(6R,7R)-7-[2-(2-Amino-4-thiazolyl)glyoxylamido]-2-carboxy-8-oxo-5-thia-1-azabicyclo[4.2.0]oct-2-en-3-yl]methyl]pyridinium hydroxide inner salt 7^2-(Z)-[O-(1-carboxy-1-methylethyl)oxime] = (Z)-(7R)-7-[2-(2-Ami-no-4-thiazolyl)-2-[(1-carboxy-1-methylethoxy)imi-no]acetamido]-3-(1-pyridiniomethyl)-3-cephem-4-carb-oxylate = [6R-[6α,7β(Z)]]-1-[[7-[[(2-Amino-4-thiazo-lyl)[(1-carboxy-1-methylethyl)imino]acetyl]amino]-2-carboxy-8-oxo-5-thia-1-azabicyclo[4.2.0]oct-2-en-3-yl]methyl]pyridinium hydroxide inner salt (●)
R   also pentahydrate (78439-06-2) or monosodium salt (73547-61-2)
S   Amjeceft, Biofort, Biotum, CAZ, CEF, Cefortam, *Ceftazidime**, Ceftazim, Ceftidin, Ceftim, Ceftum "Dexa Medica", Fortam, Fortax "Glaxo-Wellcome", Fortaz, Fortum, Fortumset, Ftazidime, Glazidim, GR 20263, Izadima, Kefadim, Kefamin, Kefazim, Kefzim, LY 139381, Magnacef, Mirocef, Modacin, Novocral, Panzid, Pluseptic, Potendal, Seftaz, SN 401, Solvetan, Spectrum "Sigma-Tau", Starcef "Firma", Tagal, Tasicef, Taxifur, Tazedem, Tazicef, Tazidime, Thidim, Tortam, Zefidime, Zibac
U   Antibiotic

10991-01 (8856-01)
R   Compd. with L-arginine (1:1)
S   Ceptaz, Pentacef

U Antibiotic

10992 (8857)

$C_{22}H_{22}N_6O_7S_3$
106061-25-0

4-[[[(6*R*,7*R*)-7-[2-(2-Amino-4-thiazolyl)glyoxylamido]-2-carboxy-8-oxo-5-thia-1-azabicyclo[4.2.0]oct-2-en-3-yl]methyl]thio]-1-(carboxymethyl)pyridinium hydroxide inner salt 7^2-(Z)-(*O*-ethyloxime) = (Z)-(7*R*)-7-[2-(2-Amino-4-thiazolyl)-2-(ethoxyimino)acetamido]-3-[[[1-(carboxymethyl)pyridinio-4-yl]thio]methyl]-3-cephem-4-carboxylate = [6*R*-[6α,7β(*Z*)]]-4-[[[7-[[(2-Amino-4-thiazolyl)(ethoxyimino)acetyl]amino]-2-carboxy-8-oxo-5-thia-1-azabicyclo[4.2.0]oct-2-en-3-yl]methyl]thio]-1-(carboxymethyl)pyridinium hydroxide inner salt (●)
S CP 107, ME-1220
U Antibiotic

10993 (8858)

$C_{22}H_{22}N_8$
78186-34-2

9,10-Anthracenedicarboxaldehyde bis(2-imidazolin-2-yl-hydrazone) = 9,10-Anthracenedicarboxaldehyde bis(4,5-dihydro-1*H*-imidazol-2-ylhydrazone) (●)
R Dihydrochloride (71439-68-4)
S ADAH, *Bisantrene hydrochloride***, CL 216942, NSC-337766, Orange crush, Zantrène
U Antineoplastic

10994

$C_{22}H_{22}N_8$
151406-07-4

2,7-Diethyl-5-[[2'-(1*H*-tetrazol-5-yl)[1,1'-biphenyl]-4-yl]methyl]-5*H*-pyrazolo[1,5-*b*][1,2,4]triazole (●)
R Potassium salt (151406-38-1) or potassium salt monohydrate (161800-08-4)
S YM-358
U Antihypertensive (angiotensin AT$_1$ receptor antagonist)

10995 (8859)

$C_{22}H_{22}O_4$
84-19-5

4,4'-(Diethylideneethylene)diphenol diacetate = 3,4-Bis(*p*-acetoxyphenyl)-2,4-hexadiene = 4,4'-(1,2-Diethylidene-1,2-ethanediyl)bis[phenol] diacetate (ester) (●)
R see also no. 7694
S *Dienestrol diacetate*, *Dienöstroldiacetat*, Faragynol, Farmacyrol (old form), Folacapon, Folöstrol T, Foragynol, Foralactol, Fortostilbene, Gynocyrol, Klianyl, Oestrasid, Oestroral-Tabl., Orestrol, Ovumin, Retalon-Oral-Tabl.
U Synthetic estrogen

10996 (8860)

$C_{22}H_{22}O_4$
24643-94-5

3-Ethyl-2-(*p*-hydroxyphenyl)-1-methylinden-6-ol diacetate = 1-Methyl-2-(4-acetoxyphenyl)-3-ethyl-6-acetoxyindene = 2-[4-(Acetyloxy)phenyl]-3-ethyl-1-methyl-1*H*-inden-6-ol acetate (●)
S *Indenestrol diacetate*, *Indenoestroldiacetat*

U Synthetic estrogen

10997   (8861) $C_{22}H_{22}O_5$
                                                                        579-23-7

2,6-Divanillylidenecyclohexanone = 2,6-Bis[(4-hydroxy-3-methoxyphenyl)methylene]cyclohexanone (●)
S   Beveno, Curcumoid, *Cyclovalone***, Cycvalon, Cyqualon, Divanil, Divanon, DVC, Flavugal, Sincolin semplice, Vanidène, Vanilon(e)
U   Choleretic

10998   (8862) $C_{22}H_{22}O_8$
                                                                        134044-97-6

8-(2,6-Dideoxy-β-D-*ribo*-hexopyranosyl)-5-hydroxy-2-(4-hydroxyphenyl)-7-methoxy-4*H*-1-benzopyran-4-one (●)
S   *Aciculatin*
U   Cytotoxic glycoside from *Chrysopogon aciculatis*

10999   (8863) $C_{22}H_{22}O_8$
                                                                        518-28-5

(5R,5aR,8aR,9R)-5,5a,6,8,8a,9-Hexahydro-9-hydroxy-5-(3,4,5-trimethoxyphenyl)furo[3',4':6,7]naphtho[2,3-*d*]-1,3-dioxol-6-one = Podophyllinic acid lactone = [5R-(5α,5aβ,8aα,9α)]-5,8,8a,9-Tetrahydro-9-hydroxy-5-

(3,4,5-trimethoxyphenyl)furo[3',4':6,7]naphtho[2,3-*d*]-1,3-dioxol-6(5a*H*)-one (●)
S   Condelone, Condyline, Condylone, Condylox, Pod-Ben-25, *Podofilox*, *Podophyllotoxin*, Podoxin, Psorex "Conpharm", PTOX, Reumason, Warix, Wartec, Warticon
U   Antimitotic, treatment of warts, anti-arthritic, antipsoriatic

11000   (8864) $C_{22}H_{22}O_{10}$
                                                                        6991-10-2

6-β-D-Glucopyranosyl-5-hydroxy-2-(4-hydroxyphenyl)-7-methoxy-4*H*-1-benzopyran-4-one (●)
S   *Swertisin*
U   Sedative from *Ziziphus vulgaris* var. *spinosus*

11001   (8865) $C_{22}H_{23}BrN_2O_8$
                                                                        4572-56-9

7-Bromo-4-(dimethylamino)-1,4,4a,5,5a,6,11,12a-octahydro-3,6,10,12,12a-pentahydroxy-6-methyl-1,11-dioxo-2-naphthacenecarboxamide = [4S-(4α,4aα,5aα,6β,12aα)]-7-Bromo-4-(dimethylamino)-1,4,4a,5,5a,6,11,12a-octahydro-3,6,10,12,12a-pentahydroxy-6-methyl-1,11-dioxo-2-naphthacenecarboxamide (●)
S   *Bromtetracycline*
U   Antibiotic

11002   (8866) $C_{22}H_{23}ClFNO_2$
                                                      61764-61-2

4-[4-(*p*-Chlorobenzoyl)piperidino]-4'-fluorobutyrophe-
none = 4-[4-(4-Chlorobenzoyl)-1-piperidinyl]-1-(4-
fluorophenyl)-1-butanone (●)
R   Hydrochloride (55695-56-2)
S   AHR-6134, *Cloroperone hydrochloride***
U   Antipsychotic, neuroleptic

11003   (8867) $C_{22}H_{23}ClFN_3O_2$
                                                      59831-64-0

5-Chloro-1-[3-[4-(*p*-Fluorobenzoyl)piperidino]propyl]-
2-benzimidazolinone = 5-Chloro-1-[3-[4-(4-fluoroben-
zoyl)-1-piperidinyl]propyl]-1,3-dihydro-2*H*-benzimida-
zol-2-one (●)
S   *Milenperone***, R 34009
U   Antipsychotic, neuroleptic

11004   (8868) $C_{22}H_{23}ClFN_5O_2$
                                                      75444-64-3

5-Chloro-1-[1-[3-(5-fluoro-2-oxo-1-benzimidazoli-
nyl)propyl]-4-piperidyl]-3-benzimidazolinone = 1-[3-[4-
(5-Chloro-2,3-dihydro-2-oxo-1*H*-benzimidazol-1-yl)-1-
piperidinyl]propyl]-5-fluoro-1,3-dihydro-2*H*-benzimida-
zol-2-one (●)
S   *Flumeridone***, *Fluorodomperidone*, R 45486

U   Anti-emetic

11005   (8869) $C_{22}H_{23}ClN_2O_2$
                                                      79794-75-5

Ethyl 4-(8-chloro-5,6-dihydro-11*H*-benzo[5,6]cyclohep-
ta[1,2-*b*]pyridin-11-ylidene)-1-piperidinecarboxylate = 4-
(8-Chloro-5,6-dihydro-11*H*-benzo[5,6]cyclohep-
ta[1,2-*b*]pyridin-11-ylidene)-1-piperidinecarboxylic acid
ethyl ester (●)
S   Aerotina, Alarin, Alergan, Alerpriv, Alertrin, Alle-
        dryl "Prater", Allerfre, Allertidin, Antor, Bedix, Bili-
        ranin, Biloina, Civeran, Cladine, Clantin, Claratyne,
        Claritin(e), Clarityn(e), Difmedol "Faran", Exul, Flo-
        nidan, Frenaler, Fristamin, Helporigin, Histadin
        "Nobel", Histaloran, Hobatadine, Horestyl, Larmax,
        Latoren, Lergy, Lertamine, Lesidas, Lexafen, Lisi-
        no, Loisan, Lontadex, Lorabet, Loracert, Loractin,
        Loradif, Loradin(e), Loramine, Loranil "Libbs", Lo-
        rantis, Lorastine, Lorastyne, Loratab, *Loratadine***,
        Loratyn, Loraver, Lordinex, Loretina, Lorex, Lor-
        fast, Lorine, Loritin(e), Novacloxab, Noxin "Bus-
        sie", Nularef, Optimin(e), Polaratyne, Ponderal "Bio-
        gen", Pylor, Ralinet, Ritin, Sanelor, Sch 29851, Sen-
        sibit, Tadine, Utel, Velodán, Versal "Nycomed",
        Viatine, Zelmar
U   Antihistaminic, anti-allergic

11006   (8870) $C_{22}H_{23}ClN_2O_8$
                                                      57-62-5

7-Chloro-4-(dimethylamino)-1,4,4a,5,5a,6,11,12a-octa-
hydro-3,6,10,12,12a-pentahydroxy-6-methyl-1,11-dioxo-
2-naphthacenecarboxamide = 7-Chlorotetracycline = [4S-
(4α,4aα,5aα,6β,12aα)]-7-Chloro-4-(dimethylamino)-
1,4,4a,5,5a,6,11,12a-octahydro-3,6,10,12,12a-pentahydr-
oxy-6-methyl-1,11-dioxo-2-naphthacenecarboxamide (●)
R also monohydrochloride (64-72-2)
S A-377, Acronize(d), Aquachlor, Arocin "Knoll-Au-
stralia", Aurecil, Aureobiotil, Auréocarmyl, Aureoci-
clin(a), Aureocina, Aureodermil, Aureo-Intes, Au-
reomicina, Aureomycin(e), Aureomykoin, Aureo-
sup, Aureum-Farmigea, Aurociclina, Aurofac, Auro-
far, Auromix, B-Aureo 100, Biomitsin, Biomycin,
Biovit-40, C 10 "Chevita", Cardopal, Centrauréo,
Cetece vet., Chevicet-Pulver, Chloratet 50, Chlorcyc-
lin, Chlorocycline, Chlorocyklina, Chlortet, Chlorte-
tra, *Chlortetracycline***, Chlortetraseam, Chlor-
tralim, Chrysomycin, Chrysomykine, Clor 20, Clor-
ciclina, Clorene 100, Clorfarm, Cloricina, Clormedi-
sol, Clorocipan, Cloroscel, Clortetra, *Clortetracicli-
na*, CLTC, Concentrat VO 43, CTC (-Soluble), Duo-
mycin, Eurochlor, Fermycin Soluble, Guardocin,
Hiramicin (old form), Intermycin, Isospen, Isphamy-
cin, Ladermycin, Lucomycin, Megaclor "Icar Leo",
Miba-Micina, Neobitetra D, NSC-13252, Oblémyci-
ne, Orociclina, Percrison, Pulsauréo, Rectobion, Rec-
tociclina, R.S.G, Septamycin "Medichem Pharm.",
Tetra 5, Tibicyklin, Topmycin, Velocillin, Vet-
cyklin, Vi-Mycin, Viro Aureo M, Vi-Tetra 40, Xan-
thomycin "Egyt"
U Antibiotic from *Streptomyces aureofaciens*

1006-01 (8870-01) 1111-27-9
R Compd. with *N,N*'-dibenzylethylenediamine (no.
6346) (2:1)
S Dibiomycin
U Antibiotic

1006-02 (8870-02) 5892-31-9
R Calcium salt
S Chlorachel-50, Clortetra
U Antibiotic

1006-03 (8870-03) 64173-03-1
R Borate
S Aureoftalmina
U Antibiotic

11007 (8871) $C_{22}H_{23}ClN_6O$
114798-26-4

2-Butyl-4-chloro-1-[*p*-(*o*-1*H*-tetrazol-5-ylphenyl)ben-
zyl]imidazole-5-methanol = 2-Butyl-4-chloro-1-[[2'-(1*H*-
tetrazol-5-yl)[1,1'-biphenyl]-4-yl]methyl]-1*H*-imidazole-
5-methanol (●)
R Monopotassium salt (124750-99-8)
S Avastar, Corus, Cosaar, Cozaar, Cozaarex, DuP 753,
E-3340, Loortan, Lortaan, Lorzaar, Losacor, Losa-
prex, *Losartan potassium***, Lotim, Lozaprex,
MK-954, Neo-Lotan, Newlotan, Niten, Nulotan, Paxon "Gador", Pinzaar, Redupress "Aché", Resilo, Ta-
cardia, Tenopres, X 7711
U Antihypertensive, natriuretic (angiotensin II receptor
antagonist)

11008 (8872) $C_{22}H_{23}ClO_5$
117621-64-4

[2α,4α,5α(*Z*)]-6-[2-(2-Chlorophenyl)-4-(2-hydroxyphe-
nyl)-1,3-dioxan-5-yl]-4-hexenoic acid (●)
S ICI 192605
U Thromboxane receptor antagonist

11009 $C_{22}H_{23}Cl_2N_3OS$
155512-49-5

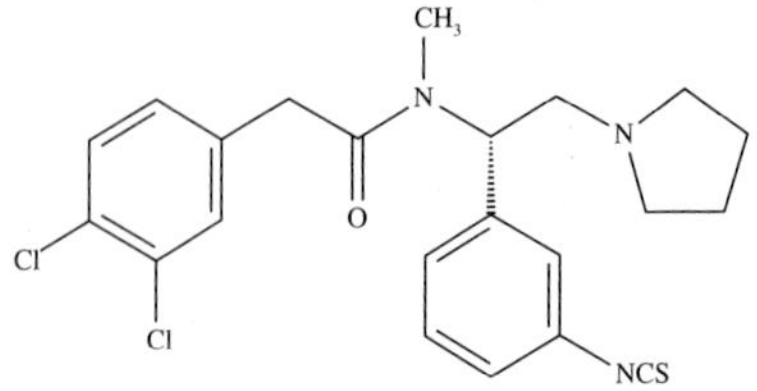

(*S*)-2-(3,4-Dichlorophenyl)-*N*-methyl-*N*-[1-(3-isothio-
cyanatophenyl)-2-(1-pyrrolidinyl)ethyl]acetamide = 3,4-
Dichloro-*N*-[(1*S*)-1-(3-isothiocyanatophenyl)-2-(1-pyrro-
lidinyl)ethyl]-*N*-methylbenzeneacetamide (●)
S DIPPA
U κ-Opioid receptor agonist

11010 $C_{22}H_{23}DN_4O_2$

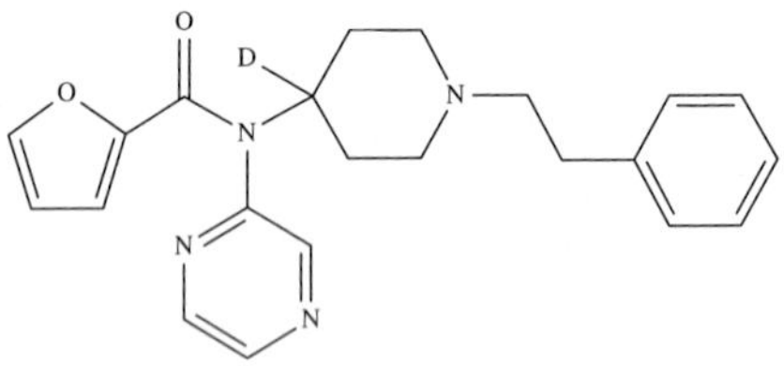

N-[4-Deuterio-1-(2-phenylethyl)-4-piperidinyl]-*N*-pyra-
zinyl-2-furancarboxamide (●)
S OHM 10579
U Analgesic

11011 $C_{22}H_{23}FN_2O_3$
178165-43-0

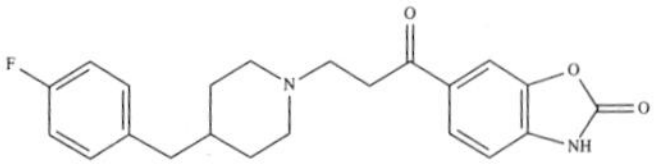

6-[3-[4-[(4-Fluorophenyl)methyl]-1-piperidinyl]-1-oxo-
propyl]-2(3*H*)-benzoxazolone (●)
S EMD 95885
U NMDA receptor antagonist

11012 $C_{22}H_{23}FN_4$
199463-33-7

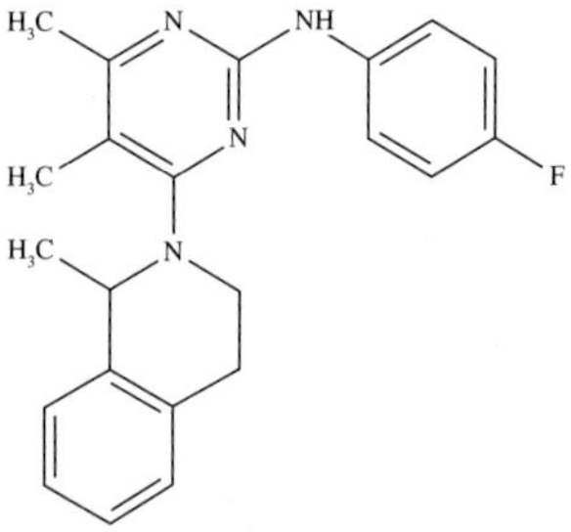

5,6-Dimethyl-2-(4-fluorophenylamino)-4-(1-methyl-
1,2,3,4-tetrahydroisoquinolin-2-yl)pyrimidine = 4-(3,4-
Dihydro-1-methyl-2(1*H*)-isoquinolinyl)-*N*-(4-fluorophe-
nyl)-5,6-dimethyl-2-pyrimidinamine (●)
R Monohydrochloride (178307-42-1)
S YH-1238, YH-1885
U Proton pump inhibitor

11013   (8873) $C_{22}H_{23}FN_4O_2$
86636-93-3

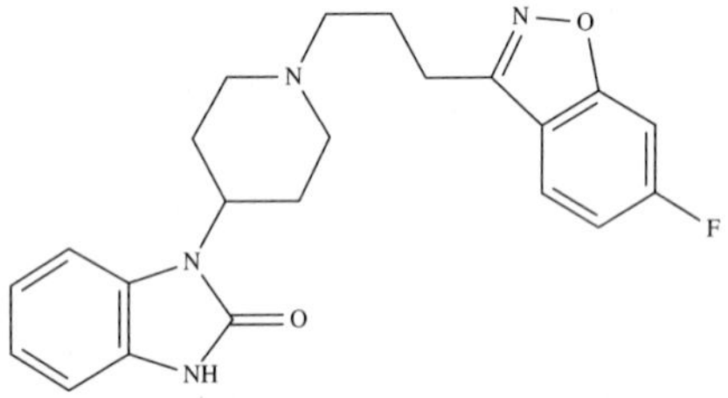

1-[1-[3-(6-Fluoro-1,2-benzisoxazol-3-yl)propyl]-4-pipe-
ridyl]-2-benzimidazolinone = 1-[1-[3-(6-Fluoro-1,2-ben-
zisoxazol-3-yl)propyl]-4-piperidinyl]-1,3-dihydro-2*H*-
benzimidazol-2-one (●)
R Monohydrochloride (86015-38-5)
S HRP 913, *Neflumozide hydrochloride***, P 793913
U Dopamine antagonist, antipsychotic

11014 (8874) $C_{22}H_{23}FO_3$
85493-98-7

[4R-[4α,6β(E)]]-6-[2-(4'-Fluoro-3,3',5-trimethyl[1,1'-bi-phenyl]-2-yl)ethenyl]tetrahydro-4-hydroxy-2H-pyran-2-one (●)
S L 645164
U Antihyperlipidemic (HMGCoA reductase inhibitor)

11015 (8875) $C_{22}H_{23}F_2NO_2$
24678-13-5

4'-Fluoro-4-[4-(p-fluorobenzoyl)piperidino]butyrophe-none = 4-[4-(4-Fluorobenzoyl)-1-piperidinyl]-1-(4-fluorophenyl)-1-butanone (●)
R Hydrochloride (24677-86-9)
S AHR-2277, Elanone-V, Lenperol, *Lenperone hydro-chloride**, RMI 11270
U Neuroleptic

11016 $C_{22}H_{23}F_3N_2O_3$
115621-25-5

(+)-1,4-Dihydro-5-isopropoxy-2-methyl-4-[2-(trifluoro-methyl)phenyl]-1,6-naphthyridine-3-carboxylic acid

ethyl ester = (4R)-1,4-Dihydro-2-methyl-5-(1-methyleth-oxy)-4-[2-(trifluoromethyl)phenyl]-1,6-naphthyridine-3-carboxylic acid ethyl ester (●)
S (+)-CI-951, NC 1500
U Calcium antagonist

11017 (8876) $C_{22}H_{23}F_3N_2O_3$
103183-79-5

(±)-Ethyl 1,4-dihydro-5-isopropoxy-2-methyl-4-(α,α,α-trifluoro-o-tolyl)-1,6-naphthyridine-3-carboxylate = (±)-1,4-Dihydro-2-methyl-5-(1-methylethoxy)-4-[2-(tri-fluoromethyl)phenyl]-1,6-naphthyridine-3-carboxylic acid ethyl ester (●)
S CI-951, Goe 5438
U Calcium antagonist

11018 (8877) $C_{22}H_{23}F_4NO_2$
749-13-3

4'-Fluoro-4-[4-hydroxy-4-(α,α,α-trifluoro-m-tolyl)pipe-ridino]butyrophenone = 4'-Fluoro-4-[4-hydroxy-4-[3-(tri-fluoromethyl)phenyl]piperidino]butyrophenone = 4-[3-(Trifluoromethyl)phenyl]-1-[3-(4-fluorobenzoyl)propyl]-4-piperidinol = 1-(4-Fluorophenyl)-4-[4-hydroxy-4-[3-(trifluoromethyl)phenyl]-1-piperidinyl]-1-butanone (●)
R also hydrochloride (2062-77-3)
S *Flumoperonum*, McN-JR-2498, NSC-169875, Psi-coperidol, Psychoperidol, R 2498, Ramidosan, Trifludol, *Trifluperidol**, Triperidol, Trisedil, Trise-dyl
U Neuroleptic

11019   (8878) $C_{22}H_{23}NO_2$
10087-89-5

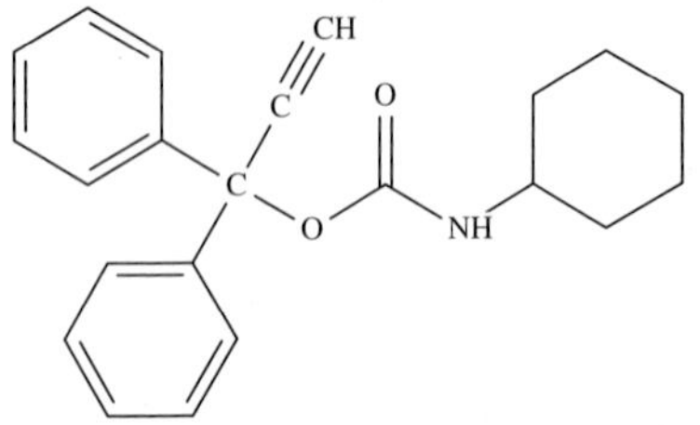

1,1-Diphenyl-2-propynyl cyclohexanecarbamate = Cy-
clohexylcarbamic acid α-ethynylbenzhydryl ester = Cy-
clohexylcarbamic acid 1,1-diphenyl-2-propynyl ester (●)
S   *Enpromate***, Lilly 59159, NSC-112682
U   Antineoplastic

11020   (8879) $C_{22}H_{23}NO_2$
63141-67-3

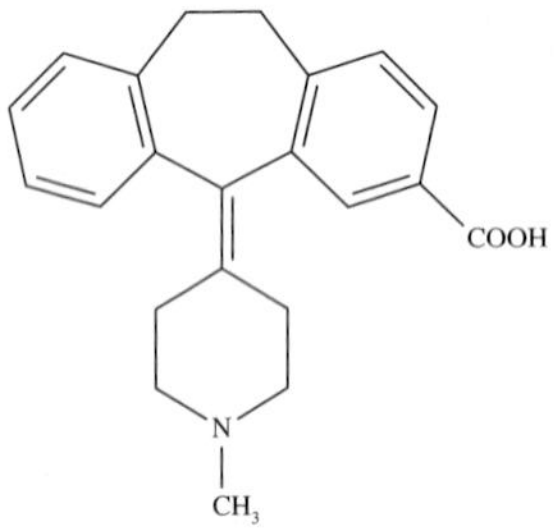

10,11-Dihydro-5-(1-methyl-4-piperidinylidene)-5*H*-di-
benzo[*a,d*]cycloheptene-3-carboxylic acid (●)
S   MK-711
U   Orexigenic

11021 $C_{22}H_{23}NO_2$
85375-88-8

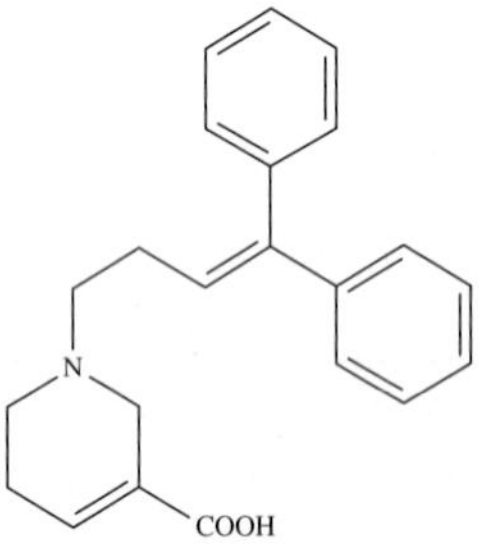

1-(4,4-Diphenyl-3-butenyl)-1,2,5,6-tetrahydro-3-pyri-
dinecarboxylic acid (●)
S   SKF-100330A

U   GABA reuptake inhibitor, anticonvulsant

11022   (8880) $C_{22}H_{23}NO_3$
114716-16-4

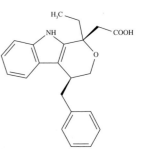

(±)-*cis*-4-Benzyl-1-ethyl-1,3,4,9-tetrahydropyra-
no[3,4-*b*]indole-1-acetic acid = *cis*-(±)-1-Ethyl-1,3,4,9-
tetrahydro-4-(phenylmethyl)pyrano[3,4-*b*]indole-1-ace-
tic acid (●)
R   formerly without *cis*-(±)-definition (103024-44-8)
S   AY-30715, *Pemedolac***
U   Analgesic

11023   (8881) $C_{22}H_{23}NO_3$
114030-44-3

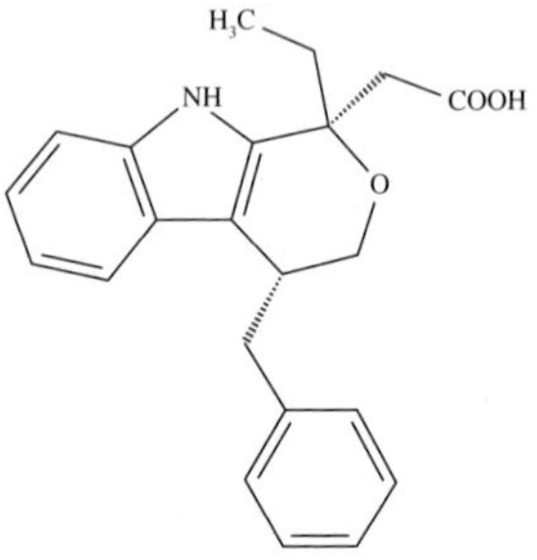

(1*S*,4*R*)-4-Benzyl-1-ethyl-1,3,4,9-tetrahydropyra-
no[3,4-*b*]indole-1-acetic acid = (1*S*-*cis*)-1-Ethyl-1,3,4,9-
tetrahydro-4-(phenylmethyl)pyrano[3,4-*b*]indole-1-ace-
tic acid (●)
S   *Dexpemedolac***, WAY-PEM-420
U   Analgesic

11024   (8882) $C_{22}H_{23}NO_4$
13997-19-8

Methyl 7-(benzyloxy)-6-butyl-1,4-dihydro-4-oxo-3-qui-
nolinecarboxylate = 7-(Benzyloxy)-6-butyl-1,4-dihydro-
4-oxo-3-quinolinecarboxylic acid methyl ester = 6-Butyl-
1,4-dihydro-4-oxo-7-(phenylmethoxy)-3-quinolinecarb-
oxylic acid methyl ester (●)
S   AY-20385, *Benzoquate*, ICI 55052,
Methyl Benzoquate, Neoquate, *Nequinate**, Stato-
quate, Statyl, Tranil
U   Coccidiostatic

11025   (8883) $C_{22}H_{23}NO_4S_2$
81872-10-8

4S)-N-[(S)-3-Mercapto-2-methylpropionyl]-4-(phenyl-
hio)-L-proline benzoate (ester) = [1(R*),2α,4α]-1-[3-
Benzoylthio)-2-methyl-1-oxopropyl]-4-(phenylthio)-L-
roline (●)
R   Calcium salt (2:1) (81938-43-4)
S   Bifril, SQ 26991, Zofenil, *Zofenopril calcium***, Zo-
prace
U   Antihypertensive (ACE inhibitor)

11026   (8884) $C_{22}H_{23}NO_7$
128-62-1

(5R)-5[(3S)-6,7-Dimethoxyphthalidyl]-5,6,7,8-tetrahy-
dro-4-methoxy-6-methyl-1,3-dioxolo[4,5-g]isoquinoline
= (1R)-1-[(3S)-6,7-Dimethoxy-3-phthalidyl]-1,2,3,4-te-
trahydro-8-methoxy-2-methyl-6,7-(methylenedioxy)iso-
quinoline = (3S)-6,7-Dimethoxy-3-[(5R)-5,6,7,8-tetrahy-
dro-4-methoxy-6-methyl-1,3-dioxolo[4,5-g]isoquinolin-
5-yl]-1(3H)-isobenzofuranone (●)
R   also hydrochloride (912-60-7)
S   Akintène, Anarkotin, Capval-Tropfen, Codipect, Co-
fanil, Colme "Laquifa", Consolets, Coscopin, Cosco-
tabs, Detusso, Dru-Tosse, Finipect, Gnoscopin,
Hederix, Hoeswin, Key-Tusscapine, Kofsin,
Lyobex, Manartin, Meenk, Mentoral, Mercotin (Drop
+ Powd.), *Methoxyhydrastine*, Narcompren, Narcosi-
ne, *Narcotine*, Narcotos, Narcotussin, *Narkotin*, Nar-
kotyl, Nectadon, Nicolane, Nipaxon, Norkotine,
Noscadon, Noscafides, Noscal, Noscalin, Noscamin,
Noscapal, Noscapalin, Noscapect, *Noscapine***,
Noscatuss, Noscorex, *Noskapin*, NSC-5366, Opian,
Opianine, Phymocil, Retentós, Sedatrix, Sinecod
"Alpine", Sirucap, Solcatuss, Stilco, Terial, Toseina,
Tucotine, Tusan, Tusilao, Tussanil N, Tusscapine,
Vadebex, Vichosan
U   Antitussive (opium alkaloid)

11026-01   (8884-01)   25333-79-3
R   Camphorsulfonate (no. 2423)
S   *Camphoscapine*, J.L. 374, Noscaflex-Tabl., *Nosca-
pine camsilate*, Tulisan, Tuli-Tos, Tuscalman, Tussi-
cure
U   Antitussive

11026-02   (8884-02)
R   Resin complex
S   Capval (Dragees + Saft), Ip-Toss, Longactin, Longa-
mex, Longatin, Lyobex retard, Nitepax, Sintusan,
Terbenol, Tos-Perdur

U Antitussive

11027 (8885) $\qquad$ $C_{22}H_{23}NS$
27574-24-9

3-Dibenzo[b,e]thiepin-11(6H)-ylidene-1αH,5αH-tropane = 3-Dibenzo[b,e]thiepin-11(6H)-ylidene-8-methyl-8-azabicyclo[3.2.1]octane (●)
R Hydrochloride (27574-25-0)
S Lepticur, SD 1248-17, *Tropatepine hydrochloride***
U Antiparkinsonian

11028 (8886) $\qquad$ $C_{22}H_{23}N_2O_5$

9-Hydroxy-2-(α-L-arabinopyranosyl)ellipticinium = 2-α-L-Arabinopyranosyl-9-hydroxy-5,11-dimethyl-6H-pyrido[4,3-b]carbazolium (●)
R Bromide (103461-20-7)
S *Ellipravin*, SUN 4599
U Antineoplastic

11029 (8887) $\qquad$ $C_{22}H_{23}N_3OS$
92-97-7

4-Isobutoxy-4'-(2-pyridyl)thiocarbanilide = N-(p-Isobutoxyphenyl)-N'-(p-2-pyridylphenyl)thiourea = N-[4-(2-

Methylpropoxy)phenyl]-N'-[4-(2-pyridinyl)phenyl]thiourea (●)
S THC, Thioban, *Thiocarbanidin*, Y 9525
U Tuberculostatic

11030 (8888) $\qquad$ $C_{22}H_{23}N_3O_2$
115313-22-9

Methyl (±)-1,3,4,16b-tetrahydro-2-methyl-2H,10H-indolo[2,1-c]pyrazino[1,2-a][1,4]benzodiazepine-16-carboxylate = (±)-1,3,4,16b-Tetrahydro-2-methyl-2H,10H-indolo[2,1-c]pyrazino[1,2-a][1,4]benzodiazepine-16-carboxylic acid methyl ester (●)
R Monohydrochloride (117581-05-2)
S CGS 15040 A, *Serazapine hydrochloride***
U Anxiolytic

11031 (8889) $\qquad$ $C_{22}H_{23}N_3O_2$
55902-02-8

(–)-N-Methyl-N-(1-methyl-2-phenylethyl)-6-oxo-3-phenyl-1(6H)-pyridazineacetamide (●)
S *Isamfazone***, Pir-353
U Anti-inflammatory

11032

$C_{22}H_{23}N_3O_2S$
167984-50-1

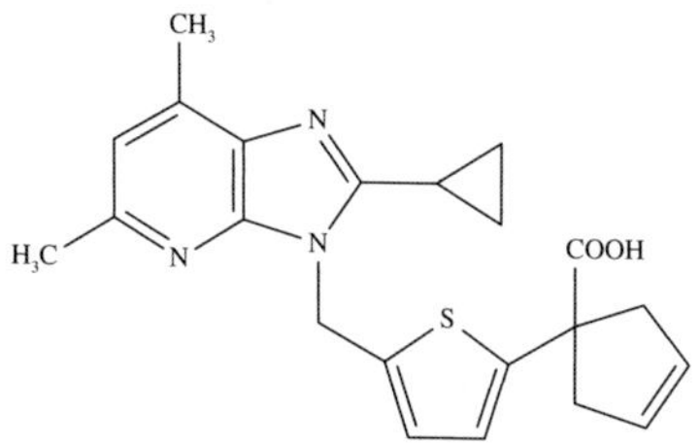

1-[5-[(2-Cyclopropyl-5,7-dimethyl-3*H*-imidazo[4,5-*b*]py-ridin-3-yl)methyl]-2-thienyl]-3-cyclopentene-1-carb-oxylic acid (●)
S CP-191166
U Angiotensin II receptor antagonist

11033

$C_{22}H_{23}N_3O_4$
183321-74-6

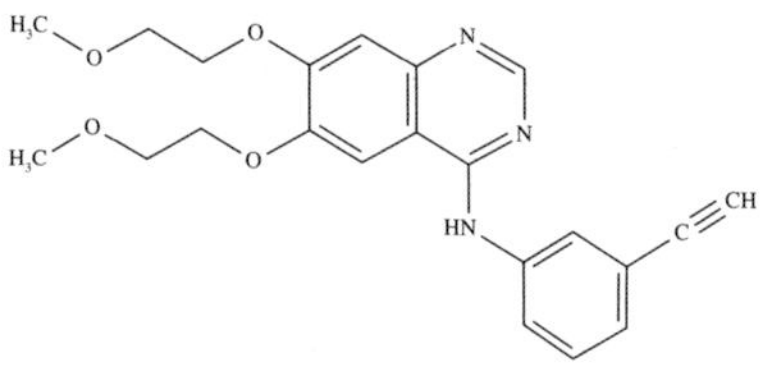

6,7-Bis(2-methoxyethoxy)-4-quinazolinyl](3-ethynyl-phenyl)amine = *N*-(3-Ethynylphenyl)-6,7-bis(2-methoxy-ethoxy)-4-quinazolinamine (●)
R Monohydrochloride (183319-69-9)
S CP-358774-01, *Erlotinib hydrochloride*, OSI-774
U Antineoplastic (tyrosine kinase inhibitor)

11034

$C_{22}H_{23}N_3O_4S$
181821-99-8

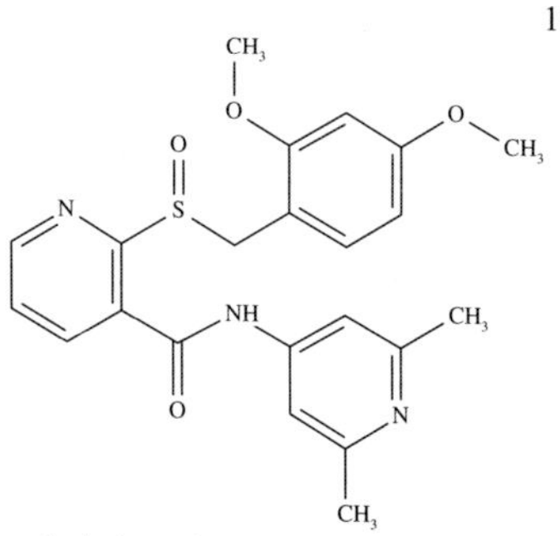

N-(2,6-Dimethyl-4-pyridyl)-2-[(2,4-dimethoxyben-yl)sulfinyl]nicotinamide = 2-[[(2,4-Dimethoxyphe-nyl)methyl]sulfinyl]-*N*-(2,6-dimethyl-4-pyridinyl)-3-py-ridinecarboxamide (●)
S AD 8717
U Anti-ulcer agent (proton pump inhibitor)

11035 (8890)

$C_{22}H_{23}N_3O_4S$
120223-04-3

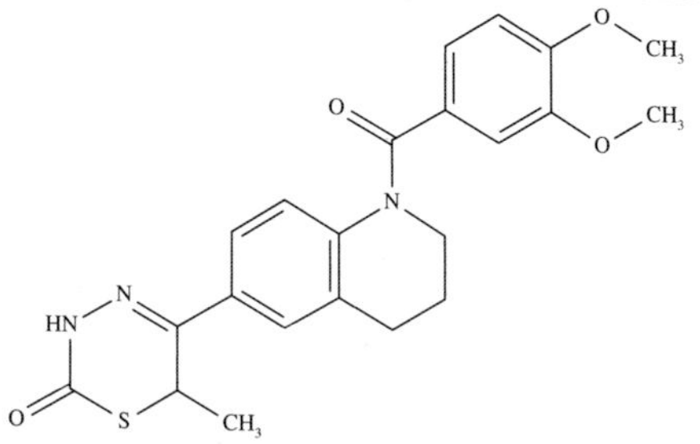

6-(3,6-Dihydro-6-methyl-2-oxo-2*H*-1,3,4-thiadiazin-5-yl)-1-(3,4-dimethoxybenzoyl)-1,2,3,4-tetrahydroquino-line (●)
S EMD 53998
U Cardiotonic (calcium sensitizer)

11036

$C_{22}H_{23}N_3O_4S$
147527-31-9

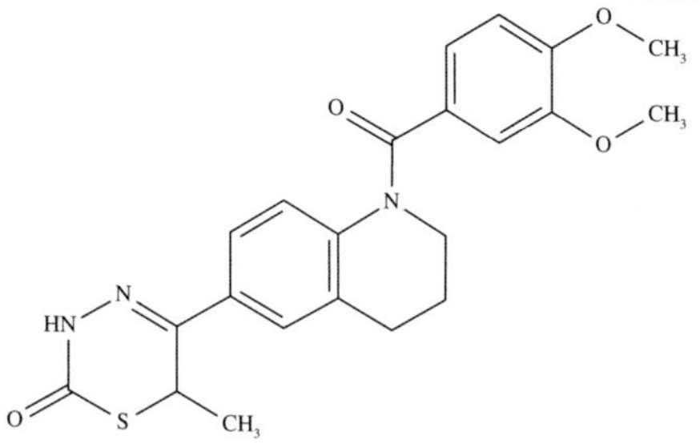

(+)-6-(3,6-Dihydro-6-methyl-2-oxo-2*H*-1,3,4-thiadiazin-5-yl)-1-(3,4-dimethoxybenzoyl)-1,2,3,4-tetrahydroquino-line (●)
S EMD 57033
U Cardiotonic (positive inotropic)

11037

$C_{22}H_{23}N_7O_9S_2$
146382-99-2

(7R)-7-[2-(2-Amino-4-thiazolyl)-2-[[(1,4-dihydro-1,5-di-hydroxy-4-oxo-2-pyridyl)methoxy]imino]acetamido]-3-(2-methyl-5-isoxazolidinyl)-3-cephem-4-carboxylic acid = [6R-[3(S*),6α,7β(Z)]]-7-[[(2-Amino-4-thiazolyl)[[(1,4-dihydro-1,5-dihydroxy-4-oxo-2-pyridinyl)methoxy]imi-no]acetyl]amino]-3-(2-methyl-5-isoxazolidinyl)-8-oxo-5-thia-1-azabicyclo[4.2.0]oct-2-ene-2-carboxylic acid (●)
S SPD 391
U Antibiotic

11038

$C_{22}H_{23}N_9O_5S_5$
200872-41-9

(7R)-7-[(Z)-2-(2-Amino-4-thiazolyl)(hydroxyimino)acet-amido]-3-[2,4-bis(isothioureidomethyl)phenylthio]-3-ce-phem-4-carboxylic acid = (6R,7R)-7-[[(2Z)-(2-Amino-4-thiazolyl)(hydroxyimino)acetyl]amino]-3-[[2,4-bis[[(aminoiminomethyl)thio]methyl]phenyl]thio]-8-oxo-5-thia-1-azabicyclo[4.2.0]oct-2-ene-2-carboxylic acid (●)
R Diacetate (salt) (214541-88-5)
S MC 02331
U Antibiotic (broad spectrum)

11039 (8891)

$C_{22}H_{24}BrClN_4O_4$
87646-83-1

1-[1,1-Bis(hydroxymethyl)ethyl]-3-[(S)-6-bromo-5-(o-chlorophenyl)-2,3-dihydro-1,3-dimethyl-2-oxo-1H-1,4-benzodiazepin-7-yl]urea = (S)-N-[6-Bromo-5-(2-chloro-phenyl)-2,3-dihydro-1,3-dimethyl-2-oxo-1H-1,4-benzo-diazepin-7-yl]-N'-[2-hydroxy-1-(hydroxymethyl)-1-me-thylethyl]urea (●)
S *Lodazecar***, Ro 16-0521
U Antihyperlipidemic

11040

$C_{22}H_{24}BrNO_4$
176181-84-3

(βR)-1-[(4-Bromophenyl)methyl]-5-methoxy-β,2-dime-thyl-1H-indole-3-butanoic acid (●)
S L 761066
U Cyclooxygenase 1 and 2 inhibitor

11041

$C_{22}H_{24}Br_2N_4O_6S_2$
110659-91-

N,N'-(Dithiodi-2,1-ethanediyl)bis[3-bromo-4-hydroxy-α-(hydroxyimino)benzenepropanamide] (●)
S *Bisprasin, Psammaplin A*

U Calcium antagonist with caffeine-like properties from the marine sponge *Dysidea* sp., topoisomerase II inhibitor

---

11042

$C_{22}H_{24}ClFN_4O_3$
184475-35-2

N-(3-Chloro-4-fluorophenyl)-7-methoxy-6-[3-(4-morpholinyl)propoxy]-4-quinazolinamine (●)

S *Gefitinib*, Iressa, ZD 1839

U Antineoplastic

11043 (8892)

$C_{22}H_{24}ClN_3O$
58581-89-8

4-(*p*-Chlorobenzyl)-2-(hexahydro-1-methyl-1*H*-azepin-4-yl)-1(2*H*)-phthalazinone = (*RS*)-4-(4-Chlorobenzyl)-2-1-methyl-4-azepanyl)-1(2*H*)-phthalazinone = 4-[(4-Chlorophenyl)methyl]-2-(hexahydro-1-methyl-1*H*-azepin-4-yl)-1(2*H*)-phthalazinone (●)

R Monohydrochloride (79307-93-0)

S A-5610, Afluon(a), Alferos, Allergodil, Asmelor, Asta A 5610, Astelin, Az, Azel, *Azelastine hydrochloride**, Azep, Azeptin, E-0659, Lasticom, Loxin, Nolen, Oculastin, Optilast, Optivar, Preventop, Radethazin, Rhinolast, Rino-Lastin, Vividrin akut, W 2979 M

U Antihistaminic, anti-asthmatic

11044 (8893)

$C_{22}H_{24}ClN_3OS$
49864-70-2

2-Chloro-10-[3-(hexahydropyrrolo[1,2-*a*]pyrazin-2(1*H*)-yl)propionyl]phenothiazine = 2-Chloro-10-[3-(hexahydropyrrolo[1,2-*a*]pyrazin-2(1*H*)-yl)-1-oxopropyl]-10*H*-phenothiazine (●)

R Dihydrochloride (49780-10-1)

S AY-25329, *Azaclorzine dihydrochloride**, Nonachlazin

U Coronary vasodilator

11045 (8894)

$C_{22}H_{24}ClN_3OS_2$
24527-27-3

8-[3-(2-Chloro-10-phenothiazinyl)propyl]-1-thia-4,8-diazaspiro[4.5]decan-3-one = 8-[3-(2-Chloro-10*H*-phenothiazin-10-yl)propyl]-1-thia-4,8-diazaspiro[4.5]decan-3-one (●)

R Monohydrochloride (27007-85-8)

S APY-606, *Clospirazine hydrochloride*, Diceplon, Disepron, *Spiclomazine hydrochloride**

U Psychotropic

11046 $C_{22}H_{24}ClN_3O_2S$
                        143393-27-5

1-[[(2S,4S)-4-[[(p-Aminophenyl)thio]methyl]-2-(p-chlorophenethyl)-1,3-dioxolan-2-yl]methyl]imidazole = 4-
[[[(2S,4S)-2-[2-(4-Chlorophenyl)ethyl]-2-(1H-imidazol-
1-ylmethyl)-1,3-dioxolan-4-yl]methyl]thio]benzenamine
(●)

R   Dihydrochloride (143484-82-6)
S   *Azalanstat hydrochloride**, RS-21607-197
U   Antihyperlipidemic (L-14αDM inhibitor)

11047   (8895) $C_{22}H_{24}ClN_5O_2$
                        57808-66-9

5-Chloro-1-[1-[3-(2-oxo-1-benzimidazolinyl)propyl]-4-
piperidyl]-2-benzimidazolinone = 5-Chloro-1-[1-[3-(2,3-
dihydro-2-oxo-1H-benzimidazol-1-yl)-propyl]-4-piperi-
dinyl]-1,3-dihydro-2H-benzimidazol-2-one (●)

R   also maleate (99497-03-7)
S   Borium, Cilroton, Costi, Dany, Digestivo Giuliani,
        Distonal, Domdon, Domerdon, Domon, Dompenyl,
        *Domperidone**, Domperon, Dompesin, Domstal,
        Dopadon, Ecuamon, Emefren, Euciton, Evoxin, Fo-
        bidon, Gasperin, Gastrocure, Gastronorm, Harmeto-
        ne, Kanthidin, KW 5338, Lomotil "Han Kook", Ma-
        ridone, Mexilon, Mod, Mogasinte, Montes, Moperi-
        dona, Morinol, Motilium, Motirol, Muvidon, Nause-
        line, Nauzelin, Netaf "Sintyal", Nordonil, Pelton,
        Peptomet, Peridal, Peridium, Peridon, Péridys, Pes-

don, Pleiadon, R 33812, Remotil, Seronex, Sibrinal,
Stoperi, Tameridone, Tametil, Tarma, Tavo-Domp,
Tilium, Trimodon, Unidone "Jin Ro", Vometa,
Voyou
U   Anti-emetic

11048 $C_{22}H_{24}Cl_2N_4$
                        149366-39-2

6,7-Dichloro-2-[[3-(dimethylamino)propyl]methylami-
no]-4-styrylquinoxaline = N-[6,7-Dichloro-3-(2-phenyl-
ethenyl)-2-quinoxalinyl]-N,N',N'-trimethyl-1,3-propane-
diamine (●)

R   Monohydrochloride (149839-55-4)
S   CP-99711
U   Glucagon receptor antagonist

11049 $C_{22}H_{24}FNO_4$

1-(Benzodioxan-5-yl)-3-[3-(4-fluorophenacyl)-1-pyrroli-
dinyl]-1-oxapropane = 2-[1-[2-[(2,3-Dihydro-1,4-benzo-
dioxin-5-yl)oxy]ethyl]-3-pyrrolidinyl]-1-(4-fluorophe-
nyl)ethanone (●)

R   L-Tartrate
S   *Daraperidone tartrate*, S 16924
U   Antipsychotic (Dopamine D_4 and 5-HT$_{2A}$ receptor
        antagonist)

11050   (8896) $C_{22}H_{24}FN_3OS$
                                         57648-21-2

4'-Fluoro-4-[4-(2-thioxo-1-benzimidazolinyl)piperidi-no]butyrophenone = 4-[4-(2,3-Dihydro-2-thioxo-1*H*-benzimidazol-1-yl)-1-piperidinyl]-1-(4-fluorophenyl)-1-butanone (●)
S   DD-3480, TIM, *Timiperone***, Tolopelon
U   Neuroleptic

11051   (8897) $C_{22}H_{24}FN_3O_2$
                                         63388-37-4

1-[3-[4-(*p*-Fluorobenzoyl)piperidino]propyl]-2-benzimi-dazolinone = 1-[3-[4-(4-Fluorobenzoyl)-1-piperidi-nyl]propyl]-1,3-dihydro-2*H*-benzimidazol-2-one (●)
S   *Declenperone***, R 33204
U   Neuroleptic (veterinary)

11052   (8898) $C_{22}H_{24}FN_3O_2$
                                         2062-84-2

1-[1-[3-(*p*-Fluorobenzoyl)propyl]-4-piperidyl]-2-benz-imidazolinone = 1-[1-[4-(*p*-Fluorophenyl)-4-oxobutyl]-4-

piperidyl]-2-benzimidazolinone = 1-[3-(4-Fluoroben-zoyl)propyl]-4-(2-oxo-1-benzimidazolinyl)piperidine = 1-[1-[4-(4-Fluorophenyl)-4-oxobutyl]-4-piperidinyl]-1,3-dihydro-2*H*-benzimidazol-2-one (●)
S   Anquil "Janssen", *Benperidol***, Benzperidol, 8089 C.B., Concilium, Frénactil, Frenactyl, Gliani-mon, McN-JR 4584, PB 806, Psicoben, R 4584
U   Neuroleptic

11053   (8899) $C_{22}H_{24}FN_3O_3$
                                         52867-74-0

4-(*p*-Fluorophenyl)-5-[2-[4-(*o*-methoxyphenyl)-1-pipera-zinyl]ethyl]-4-oxazolin-2-one = 4-(4-Fluorophenyl)-5-[2-[4-(2-methoxyphenyl)-1-piperazinyl]ethyl]-2(3*H*)-oxa-zolone (●)
S   LR 511, *Zoloperone***
U   Neuroleptic

11054   (8900) $C_{22}H_{24}F_3N_3O_2S$
                                         33414-36-7

10-[3-[4-(2-Hydroxyethyl)-1-piperazinyl]propionyl]-2-(trifluoromethyl)phenothiazine = 4-[3-Oxo-3-[2-(tri-fluoromethyl)-10-phenothiazinyl]propyl]-1-piperazine-ethanol = 10-[3-[4-(2-Hydroxyethyl)-1-piperazinyl]-1-oxopropyl]-2-(trifluoromethyl)-10*H*-phenothiazine (●)
S   *Ftorpropazine***
U   Antidepressant

11055 (8901)

$C_{22}H_{24}F_4N_2O$
80680-06-4

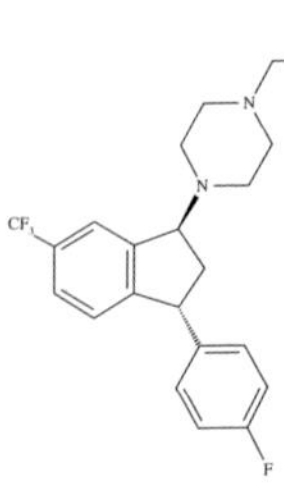

trans-4-[3-(*p*-Fluorophenyl)-6-(trifluoromethyl)-1-indanyl]-1-piperazineethanol = *trans*-4-[3-(4-Fluorophenyl)-2,3-dihydro-6-(trifluoromethyl)-1*H*-inden-1-yl]-1-piperazineethanol (●)
S Lu 18-012, *Tefludazine***
U Antipsychotic, neuroleptic

11056 (8902)

$C_{22}H_{24}F_4N_2O_2$
42048-72-6

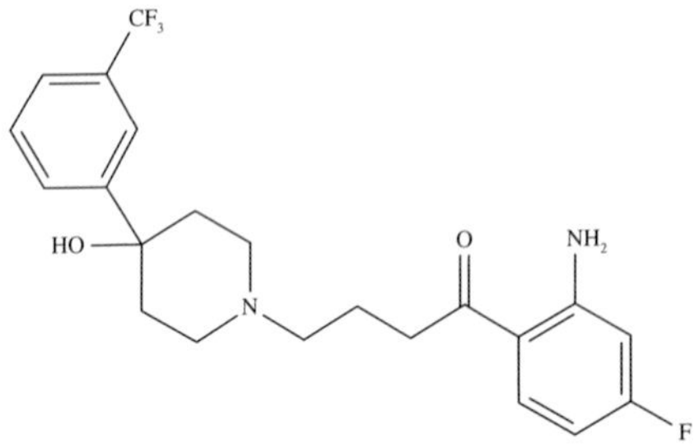

2'-Amino-4'-fluoro-4-[4-hydroxy-4-[3-(trifluoromethyl)phenyl]piperidino]butyrophenone = 1-[3-(2-Amino-4-fluorobenzoyl)propyl]-4-hydroxy-4-(3-trifluoromethylphenyl)piperidine = 1-(2-Amino-4-fluorophenyl)-4-[4-hydroxy-4-[3-(trifluoromethyl)phenyl]-1-piperidinyl]-1-butanone (●)
S ID-4708
U Neuroleptic

11057 (8903)

$C_{22}H_{24}IN_3O$
80930-91-2

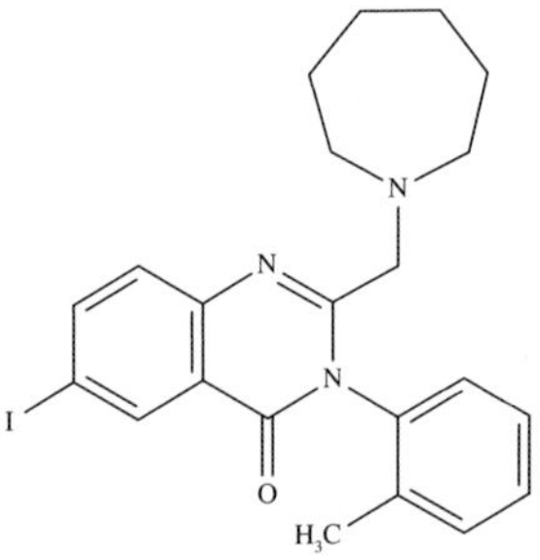

2-(Homopiperidinomethyl)-6-iodo-3-*o*-tolyl-4(3*H*)-quinazolinone = 2-[(Hexahydro-1*H*-azepin-1-yl)methyl]-6-iodo-3-(2-methylphenyl)-4(3*H*)-quinazolinone (●)
S QZ-16
U Anti-inflammatory

11058 (8904)

$C_{22}H_{24}NO_5$

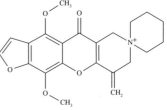

1*H*,3*H*-2-Aza-4-methylene-5,8-dimethoxyfuro[2'',3'':6,7]xanthone-2-spiro-1'-piperidinium = 8,9-Dihydro-4,11-dimethoxy-9-methylene-5-oxospiro[5*H*-furo[3',2':6,7][1]benzopyrano[3,2-*c*]pyridine-7(6*H*),1'-piperidinium] (●)
R Chloride (34959-30-3)
S AXPCl, *Azaspirium chloride**
U Bronchospasmolytic

11059

$C_{22}H_{24}N_2$

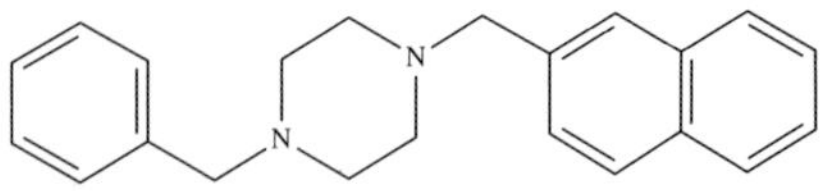

2-[(4-Benzyl-1-piperazinyl)methyl]naphthalene = 1-(2-Naphthalenylmethyl)-4-(phenylmethyl)piperazine (●)
R Dihydrochloride
S S 21378
U σ_1 receptor ligand

11060 (8905) $C_{22}H_{24}N_2$
                                        77472-98-1

2-Phenyl-4-[2-(4-piperidinyl)ethyl]quinoline (●)
S *Pipequaline***, PK-8165, RP 45319
U Anxiolytic

11061 (8906) $C_{22}H_{24}N_2O$
                                        19410-02-7

3α-(5H-Benzo[4,5]cyclohepta[1,2-b]pyrid-5-yloxy)-
1αH,5αH-tropane = 5-(1αH,5αH-Tropan-3α-yloxy)-5H-
benzo[4,5]cyclohepta[1,2-b]pyridine = *endo*-5-[(8-
Methyl-8-azabicyclo[3.2.1]oct-3-yl)oxy]-5H-ben-
zo[4,5]cyclohepta[1,2-b]pyridine (●)
R Maleate (1:1) (17929-04-3)
S B.S. 7723, *Tropirine maleate***
U Anticholinergic, respiratory stimulant

11062 (8907) $C_{22}H_{24}N_2O$
                                        122228-60-8

5-[(2,4,4-Trimethyl-1-cyclohexen-1-yl)methyl]-1-(5H)-
phenazinone (●)
S *Lavanducyanin*, WS-9659 A, YP 0298 L-C
U Antineoplastic from *Streptomyces* sp. CL 190

11063 (8908) $C_{22}H_{24}N_2O_2$
                                        87848-99-5

(E)-6-[(E)-3-(1-Pyrrolidinyl)-1-p-tolylpropenyl]-2-pyri-
dineacrylic acid = (E,E)-3-[6-[1-(4-Methylphenyl)-3-(1-
pyrrolidinyl)-1-propenyl]-2-pyridinyl]-2-propenoic acid
(●)
S *Acrivastine***, Benadryl "Warner-Lambert, UK",
       Benvent, B.W. 825 C, Semprair, Semprex
U Antihistaminic (histamine H_1-receptor antagonist)

11064 $C_{22}H_{24}N_2O_3$
                                        125500-29-0

3-Butyryl-4-(o-methylanilino)-8-(2-hydroxyethoxy)qui-
noline = 1-[8-(2-Hydroxyethoxy)-4-[(2-methylphe-
nyl)amino]-3-quinolinyl]-1-butanone (●)
S SKF 97574
U Gastric secretion inhibitor (ATPase inhibitor)

11065 (8909) $C_{22}H_{24}N_2O_3$
13221-27-7

1,2-Diphenyl-4-(2-pivaloylethyl)-3,5-pyrazolidinedione
= 4-(4,4-Dimethyl-3-oxopentyl)-1,2-diphenyl-3,5-pyra-
zolidinedione (●)
S Benetazon, TMZ, *Tribuzone**, *Trimethazone*
U Anti-inflammatory, antirheumatic

11066 (8910) $C_{22}H_{24}N_2O_4$
137578-35-9

Methyl 2,3-dihydro-2-[[(3-phenylpropyl)amino]methyl]-
7*H*-1,4-dioxino[2,3-*e*]indole-8-carboxylate = 2,3-Dihy-
dro-2-[[(3-phenylpropyl)amino]methyl]-7*H*-1,4-dioxi-
no[2,3-*e*]indole-8-carboxylic acid methyl ester (●)
R Monohydrochloride (137578-34-8)
S U-86192 A
U Antihypertensive

11067 $C_{22}H_{24}N_2O_5$
179000-76-1

4-(Carboxymethoxy)-3-methoxy-3'-(1-piperazinyl)chal-
cone = (*E*)-[2-Methoxy-4-[3-oxo-3-[3-(1-piperazi-
nyl)phenyl]-1-propenyl]phenoxy]acetic acid (●)
R Mono(trifluoroacetate) (179002-59-6)

S XT-111
U Anticoagulant

11068 (8911) $C_{22}H_{24}N_2O_5$
86541-78-8

(3*S*)-3-[[(1*S*)-1-Carboxy-3-phenylpropyl]amino]-2,3,4,5-
tetrahydro-2-oxo-1*H*-1-benzazepine-1-acetic acid = [*S*-
(*R**,*R**)]-3-[(1-Carboxy-3-phenylpropyl)amino]-2,3,4,5-
tetrahydro-2-oxo-1*H*-1-benzazepine-1-acetic acid (●)
R see also no. 12344
S *Benazeprilat***, CGS 14831
U Antihypertensive, cardioprotective (ACE inhibitor)

11069 (8913) $C_{22}H_{24}N_2O_8$
564-25-0

4-(Dimethylamino)-1,4,4a,5,5a,6,11,12a-octahydro-
3,5,10,12,12a-pentahydroxy-6-methyl-1,11-dioxo-2-
naphthacenecarboxamide = 6-Deoxy-5-hydroxytetracy-
cline = [4*S*-(4α,4aα,5α,5aα,6α,12aα)]-4-(Dimethylamino)-1,4,4a,5,5a,6,11,12a-octahydro-3,5,10,12,12a-penta-
hydroxy-6-methyl-1,11-dioxo-2-naphthacenecarbox-
amide (●)
R also monohydrochloride (10592-13-9) or monohydrate (17086-28-1); also hyclate (monohydrochloride
compd. with ethyl alcohol (2:1) monohydrate)
(24390-14-5)
S Abadox, Acne-Cy-Clean, AK-Ramycin, AK-Ratabs,
Alti-Doxycycline, Amplidox, Antodox, Apo-doxin,
Apo-Doxy, Atridox, Azudoxat, Bactidox, Baldoxin,
Basedillin, Bassado, Bibracin, Biociclina, Biocina,
Biocyclin, Biodoxi, Biostar, Biostar sospens.,
Bio-Tab "Intern. Ethical", Bromycin, Cadox, Capa-
dox, Cendox, Cirenyl, Clinofug D, Clisemina, Clo-
ran "Han All", Combaforte, CT-Doxy, Curasol

Akne, Cyclidox, Cyclodox, D 714, Demix, Deoxy-mykoin, Diksasil, Dinamisin, Diocimex, Dix, Dixicol, Docostyl, Doctacin, Dokcrin, Doksasil, *Doksiciklin*, Doksin, Domoxin novum, Dophar, Dorix, Doryx, Dosil, *Dossiciclina*, Dossil, Dosyklin, Dotur, Dovicin, Doxa "Sawai", Doxacin, Doxaclen, Doxaclicin, Doxakne, Doxal, Doxatet, Doxi, Doxibene, Doxibial, Doxibiotic, Doxicap, Doxicento, *Doxicicline*, Doxiclat, Doxiclin, Doxiclival, Doxi-Crisol, Doxidem, Doxidima, Doxifilm, Doxifin, Doxigalumicina, Doxigram, Doxilen, Doxileo, Doxilets, Doxilina, Doxilizon, Doximed, Doximicina, Doximycin, Doximycin-Tabl., Doxin(a), Doxinate "Torlan", Doxine, Doxipan, Doxi-Sergo, Doxitard, Doxiten, Doxithal, Doxitrecina, Doxivet, Doxivis, Dox-Life, Doxoral, Doxsig, Doxt, Doxy, Doxy-1, Doxy 100 LUT, Doxy-100 SMB, Doxy 200, Doxy AbZ, Doxy-acis, Doxy-basan, Doxy-BASF, Doxybene, Doxybiocin, Doxy-C, Doxy-Caps, Doxychel, Doxycholate, Doxycin, Doxyclin, *Doxycycline***, *Doxycyklin*, Doxycyl, Doxyderm, Doxyderma, Doxy-Diolan, Doxy-Disp, Doxydoc, Doxy Eu Rho, Doxyferm, Doxyfim, Doxyforte, Doxy-HP, Doxy-II, Doxy-Komb, Doxylag, Doxylan, Doxylar, Doxy-Lemmon, Doxylets, Doxylin, DoxyLindoxyl, Doxymerck, Doxymono, Doxy M-ratiopharm, Doxy M von ct, Doxymycin(e), Doxynor, Doxy-N-Tablinen, Doxypal, Doxypharm, Doxy-Pohl, Doxy P-ratiopharm, Doxy-Puren, Doxy-Recip, Doxyremed, Doxy-S, Doxysept, Doxyseptin, Doxy S + K, Doxysol, Doxystad, Doxytab, Doxy-Tablinen, DoxyTabs "Hexal", Doxy-Tabs "Rachelle", Doxytec, Doxytem, Doxytrex "Bioty; Luso-Farmaco", Doxyval, Doxy von ct, Doxy-Wolff, Dumoxin, Duraciclina, Duracyclin, Duradoxal, Dyclasin, Ecodox, Ekaciclina, Elantracin, Emidox, Esaciclina, Esadoxi, Esdoxin, Ethidoxin, Etidoks, Etidoxine, Extraciclina, Falorciclina, Farmacina, Farmadoxi, Farmodoxi, Fortacilina, Furdox, Geobiotico Depot, Germiciclin, Gewacyclin, Ghimadox, Gibidox, Godadox, Gram-Val, Granudoxy, Grodoxin, GS-3065, Helvedoclyn, Hiclamicina, Hiramicin, Huma-Doxylin, Hydramycin, Ichthraletten DN, Icidox, Iclados, Idocyklin, Impalamycin, Investin, Isodox, Ivamycin, Jenacyclin, Keimycin, Korcin, Lampodox, Lapribacter, Lenteclin, Lentomyk, Leu-

xolin, Lexcycline, Liomycin, Liviatin, Logamycil, Logamycyl, Lydox, Magdrin, Maxibiotic "Spencer", Medomicin, Medomycin, Mespafin, Microciclina, Microdox, Micromicin, Midoxin, Minidox, Mini-Tetra, Miraclin, Monocline, Monocyline, Monodoks, Monodox, Monodoxin, Monomycin "Lääke", Morsetil, Mynacap, Neociclina "Washington", Neocyclin "Neolab", Neodox, Neo-Vibrin, Nivocilin, Nordox, Novaciclin, Novelciclina, Novimax, Novo-Doxyline, NSC-56228, Nu-Doxycycline, Nymixcyclin N, Odicyclin, Oneacin, Oradoclin, Oralciclina, Oramycin, Otosal, Paldomycin, Paldoxina, Panfarmin, Paraseptum, Parkedox, Parvidoxil, Pefaciclin, Peraseptum, Periostat, Pharmodox, Pharmon Doxy, Philcociclina, Piperamycin, Pocaciclina, Policycline, Posacillin, Posineg, Primadox, Profidox, PT 122 M, Pulmodox, Radox, Ramsysis, Rapidocin "Shin Poong", Rasenamycin, Reci-Dox, Relociclina, Relyomycin, Remicyclin D, Remycin, Retens, Rho-Doxycycline, Ribociclina, Rodomicina, Ronaxan, Roximycin, Roxycaps, Roxyne, Rudocyclin, Samecin, Saramicina, Semelciclina, Semelin, Serefan, Servidoxyne, Severciclina, Sferamicina, Siadocin, Siclidon, Sigadoxin, Sincromicyn, Sithruin, SK-Doxycycline, Smilitene, Solupen "Aristegui", Sotomaten, Spanor, Spettrodox, Stamicina, Stevicin, Summicina, Supracyclin, Taxivin, Tecacin, Tecnomicina, Tenutan, Tetradox, Tetrasan, Tolexine, Tricylline-N, Tsurupioxin, Tuazil, Unacil, Uniciclina "Medal", Unidox, Unidoxi, Unidoxy, Viacin "Garec", Vibazine "Lasor", Vibrabiotic, Vibracina, Vibradox, Vibradoxil, Vibrafesa, Vibralex, Vibralon, Vibramicina, Vibramycin(e), Vibramycine N, Vibra-S, Vibratab, Vibra-Tabs, Vibraveineuse, Vibraven, Vibravenös, Vibravenosa, Vibravenous, Vibravet, Vibraxin, Viclamin, Victor-ciclina, Vip-ciclina, Viradoxyl-N, Visubiotic, Vivocycline, Vivox "Squibb", Vizam, Wanmycin, Ximicina, Yamamycin, Zadorin

U Antibiotic

11069-01 (8913-02)   39055-69-1
R Acetylcysteinate (no. 403) (1:1)
S Basecidina, Eficacina
U Antibiotic

11069-02 (8913-03)   65034-48-2
R Phosphate (1:1)
S Plenomicin, Pluricin

U　Antibiotic

11069-03　(8913-04)　　　　　　　　83038-87-3
R　Compd. with metaphosphoric acid ($H_4P_4O_{12}$) (3:1) monosodium salt
S　AB 08, Dagracycline, Dagramycine, DMSC, Doxycline, *Doxycycline fosfatex*, Doxy Dagra, Doxyphos, Mundicyclin, Mundicyl, Neo-Dagracycline, Pluridoxina, Rojazol, Sigacyclat, Thedox, Theodox
U　Antibiotic

11069-04　(8913-05)　　　　　　　　94088-85-4
R　Calcium salt
S　Samocycline, Vibramycin Calcium
U　Antibiotic

11069-05　(8913-06)
R　Hydroxymethoxybenzenesulfonate (guaiacolsulfonate)
S　DAS-Robert
U　Antibiotic

11069-06
R　Carrageenate
S　Vibracare, Vibramycin Neo
U　Antibiotic

11070　(8914)　　　　　　　　　　$C_{22}H_{24}N_2O_8$
　　　　　　　　　　　　　　　　　　　　79-85-6

4-Epimer of (dimethylamino)-1,4,4a,5,5a,6,11,12a-octahydro-3,6,10,12,12a-pentahydroxy-6-methyl-1,11-dioxo-2-naphthacenecarboxamide = [4R-(4α,4aβ,5aβ,6α,12aβ)]-4-(Dimethylamino)-1,4,4a,5,5a,6,11,12a-octahydro-3,6,10,12,12a-pentahydroxy-6-methyl-1,11-dioxo-2-naphthacenecarboxamide (●)
S　*4-Epitetracycline, Quatrimycin*
U　Antibiotic

11071　(8912)　　　　　　　　　　$C_{22}H_{24}N_2O_8$
　　　　　　　　　　　　　　　　　　　　60-54-8

4-(Dimethylamino)-1,4,4a,5,5a,6,11,12a-octahydro-3,6,10,12,12a-pentahydroxy-6-methyl-1,11-dioxo-2-naphthacenecarboxamide = [4S-(4α,4aα,5aα,6β,12aα)]-4-(Dimethylamino)-1,4,4a,5,5a,6,11,12a-octahydro-3,6,10,12,12a-pentahydroxy-6-methyl-1,11-dioxo-2-naphthacenecarboxamide (●)
R　see also no. 5378-02;
　also monohydrochloride (64-75-5) or trihydrate (6416-04-2)
S　Abiosan, Achro 500, Achromycin, Achropac, Acicycline, Acrocil, Acrocyclin, Acrodermil, Acromicina, Acromycin V, Acroreivil, Actisite, Afdicyclin, Agmacycline, Agrimicina-tetra, Agromicina, Akne-Pyodron Tabl., Ala-Tet, Alcyclin, Altracycline, Ambotetra, Ambramicetina, Ambramicina, Ambramycin, Amer-Tet, Amilin "Acila", Ampliociclina, Amracin, Amsaciclina, Amtet, Amycin, Apocyclin, Apo-Tetra, Aquacycline "Geneva", Arcanacycline, Archiciclina, Archicyclin, Aremcyclin, Arka-T-Mycin, Artomycin, Atcocyclin, Austramycin, Bactercycline, Bekacycline, Bicycline, Binicap-Ovuli, Biocin "Sam", Biocycline, Biotecline, Bio-Tetra, Biotricina, Bipicyclin, Bi-Steclin, Bitacycline, Bristaciclina, Bristacyclin, Bristazoo, Bristocycline, Brocycline, Brodspec, Cadicycline, Calociclina, Cancycline, Cap-O-Mycin, Carecyclin, Cebin, Cefracycline, Centaclin, Centacycline, Centet, Centracycline, Chemcycline, Chrocyclin, Chromocyclin, Cicli bion, Ciclicina, Ciclindif, Ciclobiotic(o), Cilcycline-500, Clemycine, Clinitetrin, Clormicrol-AM, Cofarcilina, Combicline, Copaltetra, Copharlan, Cordiblets, Cortemciclina, Crestamycin, Criseociclina, Croycyclin, Curocyclin, Cyclabid, Cyclibel, Cycline-250, Cyclisun, Cyclomycine, Cyclopar, Cyclutrin, Cyltrecin, Cytome, Dabbon, Decabiotic, Decyclin(e), Deltrina, Dema, Democracin, Dentacycline, Dericycline, Desamycin, Deschlorbiomycin, Deschloroaureomycin, Devacyclin, Diacycline, Diba-

terr, Dicicyclin, Diocyclin, Dipacycline, Dispatetrin, Dosamycin, Dotricin, Dumicyclin, Dumocyclin(a), Duratet, Ecocyclin, Economycin, Eficycline, Enzicycline (active substance), Ergocyclin, Erifor, Estycycline, Excelmycin, Fabacyclin, Facyclin, Fairmycin, Farciciclina, Farmatetra, Fed-Mycin, Félibiotic, Fercacyclin, Fermentmycin "Santos", Fermycine, Fidemycin, Filotetra, Finciclina, Flavacyn, Flavomicina, Floramicina, Florocycline (active substance), Forbesycline, Fort-Cyclin, Friciclin, Friesomycin, Fumisin, Gammatet, Gargacycline, Geciclin, Gene-Cycline, Geocyclin "Otsuka", G-Mycin "Coast", Goniciclina, Granitil, Grocyclin, Grunicyclin, GT "Horner", Halitrex, Harticiclin, Heksasiklin, Helvecyclin, Hexacycline, Hosta 500, Hostaciclina, Hostacyclin, Ibicyn, Idilin, Idrotetra, Ificyclin, Imex, Indo-Cycline, Infex "Elofar", Ingacycline, Injecur, Intercycline, Italcyclina, Junmycin, Juvacyclin, Kabacyclin, Kay-Cycline, Kesso-Tetra, Kristacyclin, Kylicycline, Lacycline-500, La-Tetra, Latycin, Lemotetra, Lemtrex, Lexacycline, Licifen, Life-Tetra, Lificycline, Limecycline, Linemett, Liquamycin, Lupicyclin, Lusotetra, Lykaclin, M 801, Marviciclina, Maso-Cycline, Maytrex, Medamycin, Mediacycline, Medicyclin(e), Mediletten, Medocycline, Menciclina, Mephacyclin, Mericycline, Metacycline "Bombay Drug House", Metzcycline, Micipan, Microcyclin, Moncycline, M-Tetra, Muracine, Mysteclin "Bristol-Myers Squibb-Australia", Naphacycline, Natomycin, NC "Meiji", Nectacycline-Caps., Neociclina "Sanitas", Neocyclin(e), Neo-Tetrine, Niacycline, Nicycline, Nococycline, Nor-Tet, Noviltetra, Novotetra, Nu-Tetra, Ofticlin, Omegamycin, Omnaze Richet, Onemisin, Opicycline, Oppacyn, Oricyclin, Ostotet, Otracyn, Paltet, Panbio-C (active substance), Panciclina, Pancycline (active substance), Panmycin, Panter, Partrex, Pédiatétracycline, Pericycline, Pervasol "Poen", Pexobiotic, Picycline, Piracaps, Pirocyclina, Plus-Tetra, Polarcyclin, Polfamycine, Polibac, Policiclina, Polycycline, Precyclin, Prelmycin, Prociclina, Progtet, Pro-Tet, 3 P-Tet, PTS, Purocyclina, Qidtet, Quadcin, Quadracycline, Quadran-500, Quatrax, Quimocyclin, Quimpe-Antibiotico, Quirvetin, Rancycline, R-Cycline, Remicyclin, Resomicina, Resteclin, Retet, Retifon, Rexamycin, Ricycline, Rimacin "Enza", Riostatin, Robicyclan, Robitet, Rocional-Tetra, Ro-cycline, Rophacycline, Roracyn, Roviciclina, Sagittacin N,

Sanbiotetra, Sanclomycine, Sarocycline, Sautrex, Scantetrin, Scotrex, Seamcycline, Servitet, Sodéycline, Solclin, Soltetra, Sopacycline, Spaciclina, Statinclyne, Steclin, Stilciclina, Subamycin, Sumycin hydrochloride, Supramycin, Supravet 21, Sustamycin, Sustetra, Sutran, Svedocyklin, T-125, T-250, Tancilina, T-Caps, T-Ciclina, T-Cycline, TE BR, Teciclina, Teclin, Tefilin, Teline, Telio, Telotrex, Teracin, Teramin, Tesyklin, Tet-Cy, Tetr(a), Tetra 200 solubile, Tetra "M.N."-Kaps., Tetra-Atlantis, Tetrabakat, Tetrabal, Tetrabid-Organon, Tetrabin, Tetrabion, Tetrabioptal, Tetrabior, Tetrabiotic, Tetrablet, Tetrabon, Tetra-C, Tetracap "Circle", Tetracapin, Tetracaps, Tétracat, Tetrachel, Tetraciclene, *Tetraciclina*, *Tetraciklin*, Tetracinil, Tetracitro S, Tetracline, Tetraclon, Tetraclor, Tetra-Co, Tetracompren, Tetracorm, Tetracrine, Tetracycl, *Tetracycline***, Tetracyclinol, *Tetracyklin*, Tetracyn(a), Tetra-D, Tetradecin, Tetradexter, Tetradif, Tetraerba, Tetrafac, Tetrafesa, Tetrafil, Tetragen, Tetragyn, Tetra Hubber grageas, Tetrakap, Tetrakar, Tetral, Tetra-Lam, Tetralan, Tetralar, Tetralean, Tetralen "C.E.P.A.", Tetraletas, Tetraline, Tetra-Liser, Tetralonga, Tetralution, Tetram "Dunhall", Tetramac, Tetramavan, Tetramax, Tetramed, Tetramicina, Tétramig, Tetramil "Milano-Mexico", Tetramin "Adeka; Collado", Tetra M.P., Tetramycin, Tetramykoin, Tetran "Vitarine", Tetranix, Tetranovin, Tetraos, Tetrapan, Tetraplon, Tetraplus, Tetrapon, Tetra-Proter, Tetrarco, Tetrascel, Tetraseptin, *Tetrasiklin*, Tetrasol 200, Tetrasone, Tetrasuiss, TetraSURE, Tetra-Tike, Tétraval, Tetravard, Tetra-Wedel, Tetra-Wolf, Tetrazine, Tetrerba, Tetrivo, Tetrosar, Tetrosol, Tetrun, Tetrus, Tetsol, Tevacycline, Theracycline, Thiocycline, Threocycline, Thuricyclin, T-Liquid, Topitetrina, Totacyclin V, Totomycin, Trexin, Triacycline, Triclina, Tricyclin, Triphacyclin, Trycin, Tsiklomitsin, T-Tabs, TV, T-Van, Ubamycin, Ultramycin, Ultratet, Umetracil, Unicin, Unicyclin, Unimycin "Un.Am.Ph.", Unitetra, U-Tet 250, Veracin, Vetaciclina, Vetacyclinum, Vetquamycin-324, Vimycin, Viosiklin, Vioterracin, Virociclina, Vistamycin "Vista", Vitacyclin, Wescocycline, Wintracin, Wintrex "Winthrop-USA", Yellamycin, Zemycin, Zyler

U　Antibiotic from *Streptomyces* sp.

11071-01　(8912-01)　　　　　　　　　　　　　1336-20-5
R　Phosphate complex

S Achromycin P-Kapseln, Acromaxiclina fosfato, Alfaciclina, Ankerciclina, Austrastaph V, Binicap-Caps., Biocheclina, Biotra, Brisai-TX, Bristaciclina A, Bristaciclina-Tetrex, Capcycline, Ciclinamen, Comycin, Conciclina, Devacyclin-Kaps., Devasiklin, Difociclina, Edusan-Tetramin (Suppos), Esabiotic, Fisioral-tetraciclina, Fosfaciclina, Fosfociclina, Fosfosiklin, Fostet, Hostacyclin "P", Hydracycline, Isiciclin, Junmycin V, Klinociclina-M, Kyclo-V, Litrex, Miciclin, Panmycin P, Panmycin V, Phosmycine, Phosphocyclin, Resomicina P, Riocyclin, Sanbiotetra-Susp., Sumycin, Super Tetra, Teclinazets, Teociclina, Teraksilin, Tetrabon V, *Tetracycline Phosphate Complex*, Tetracyclin P "Ackermann", Tetracyn Plus, Tetracyn V, Tetradecin novum, Tetradecin P, Tetrafosammina, Tetralet, Tetramin "EFEYN", Tetran F, Tetra-Pradel, Tetrascel fosfato, Tetrazetas-Retard, Tetrex, Ultraciclina, Upcyclin, Urozem, Vekarciclina, Veran-Ciclina, Zentraferin
U Antibiotic

11071-02 (8912-02)   23843-90-5
R Citrate (1:1)
S Achromycin P-Tropfen, Achromycin V, Acromicina P or V, Neo-Acromicina, Reociclina, Tetra-Citro
U Antibiotic

11071-03 (8912-03)
R Cyclohexylsulfamate (cyclamate) (no. 717)
S E 126 A, Sifacycline
U Antibiotic

11071-04 (8912-04)   8017-79-6
R Mixture of the monohydrochloride with glucosamine hydrochloride
S Abbomycin-V, Alteciclina, Altociclina, Aminociclina, Betaciclina, Cicloglucina, Ciclomax, Cosa-Tetracyn, Cosatetril, Forticiclina, Glucociclin, Glucoterracin, Glutetrex, Glutrex, Itaglucina, Lambdaciclina, Latycin G, Lexclin, Plus-Ciclina, Tassociclina, Tetraglamin, Tetraglucina, Tetra-Gluco
U Antibiotic

11071-05 (8912-05)   20228-26-6
R Guaiacolsulfonate (1:1)
S Guayaciclina, Guayacycline, Guayanovag
U Antibiotic

11071-06 (8912-06)
R Mixture of the monohydrochloride with *Lactobacillus casei* germs
S Lactotetracycline, Servicin
U Antibiotic

11071-07 (8912-07)   5821-53-4
R Dodecyl hydrogen sulfate (lauryl sulfate) (1:1)
S Ivax (old form), Laurabiotico, Lauraciclina, Lauracyclinegélules, LH-90, Myriamycine-Caps, Rolciclina, Teloril, Tetra-Lauril
U Antibiotic

11071-08 (8912-08)
R Phenoxymethylpenicillinate (no. 6249)
S Acrovigor, Fenciclina, Fenotetra, Largocyn, *Penicyclinum*, Penvicina, Tetracyllin
U Antibiotic

11071-09 (8912-09)
R Tartrate complex
S Mervacycline, Tartrociclina, Tetra-dispers
U Antibiotic

11071-10 (8912-10)   33049-22-8
R Mono(3,4,5-trimethoxybenzoate)
S Kinciclina, TCTMB
U Antibiotic

11071-11 (8912-11)   16063-83-5
R Dodecyl sulfamate (1:1)
S Dodeclin, *Dosuciclina*, *Dosucyclin*, Fenoseptil, Lauracycline-sirop, Myriamycine-Sir.
U Antibiotic

11071-12 (8912-12)   40910-00-7
R Double sulfate with aminochlorthenoxazine (no. 2086)
S *Aminochlorthenoxycycline*, *Aminoclortenoxiciclina*, Singletin
U Antibiotic (for infections of the respiratory system)

11071-13 (8912-13)   18303-44-3
R Guaiacolglycolate (no. 1629) (1:1)
S Promesaciclin Balsamico oral
U Antibiotic

11071-14 (8912-14)   18303-45-1
R Thymolsulfonate (1:1)
S Pectocilina

U Antibiotic

11071-15 (8912-15)
R Sulfide
S Istociclina, *Tiorclorcicline*
U Antibiotic

11072 (8915) $C_{22}H_{24}N_2O_9$
                                                  79-57-2

4-(Dimethylamino)-1,4,4a,5,5a,6,11,12a-octahydro-
3,5,6,10,12,12a-hexahydroxy-6-methyl-1,11-dioxo-2-
naphthacenecarboxamide = [4S-
(4α,4aα,5α,5aα,6β,12aα)]-4-(Dimethylamino)-
1,4,4a,5,5a,6,11,12a-octahydro-3,5,6,10,12,12a-hexahy-
droxy-6-methyl-1,11-dioxo-2-naphthacenecarboxamide
(●)

R see also no. 15932;
also monohydrochloride (2058-46-0) or dihydrate
(6153-64-6)
S Abbocin, Acti-Tetra B, Acu-Oxytet, Adamycin,
Agrimicina-oxi, Aknin-Winthrop-Kapseln, Alamy-
cin, Alcacycline, Alcyclin-O, Allamycin, Anprocicli-
na, Anprox, Aquachel, Aquacycline"Syntex-veterina-
ry", Aquaoxy, Aquatet, Arco-Spectron, Armici-
na-100, Avi-Tetra, Bactricycline, Benaossitetra,
Be-Oxytet, Berkamycen, Berkmycen, Betacycline,
Biaject, Biamycine, Biociclina "Klinos", Biocycline
"Upjohn", Bio-Mycin, Biostat, Biotet, Bioticlin,
Bioxi, Bivatop, Bovomicina, Bramcycline, Brontol
"Diétan", BTH-S, BTH-S Broncho-Tetra-Holz,
Castanicina, Catrexin O, Centroxycline, Chrysocin,
Clinimycin "Glaxo", Clinmycin, Cloncycline, Cofa-
mix OTC 326, Compomix V Terasol, Concentrat
VO 31, Coopertet, Copaloxi, Copharoxy, Corcine,
Corivit, Corocycline, Cortigrin, Cotet, Crisamicin,
Cyclinox, Cyclival, Cyclosol 200 LA, Dabicycline,
Dalimycin, Devacyclin-Amp., Dietancycline,
Dio-Tet, Ditropan "Am. Europharma", Duphacycli-
ne, Duracykline, DuraTetracyclin, Dynoxytet, Ego-
cin, Elaciclina, Elicycline, Elinton, Embacycline,
Embryostat, Engemycin(a), E.P. Mycin, Ergomycin,

Eurotetra, Euroxitetra, Eutetra, Fanterrin, Farbiotico
"Farber", Feamicycline, Fercaspectron, Galenomy-
cin, Garcycline, Geobiotico, Geo-Cipan, Geocycli-
ne, Geomicina, Geomycin "Pliva", Geoxin, Gyna-
mousse, Hesamycin, Huberbiotic, Humusmycin, *Hy-
droxytetracycline*, Ia-Ioxine, Idrossicina, Ificylin,
Ikamycin, Imperacin, Innolyre, Inoxtet, Intaloxin, In-
tracycline, Intramicina "Bruluart", Ivax, Izomicina,
Jenamycin, Lam-cycline, Lenocycline, Lificycli-
ne-Inj., Liquachel, Liquaciclina, Liquamycin
LA 200, Liqui-oxi, Liquitetra, Longacycline, Longi-
cine, Lusoxi, Lykaclin-O, Macocyn, Maxicyklin,
Maximycin, Maxitet, Meamycin, Medamycin "Medi-
co; Tech America", Mediciclina, Medicycline, Medi-
cyklin, Medivet Oxy-Tetra, Mepatar, Metaoxycycli-
ne, Micromycin, Mix R Mycin, Mycen, Nectacycli-
ne-Inj., Neo Bitetra Vitelli, Neocol "Günsa", Neocyclin O, Neo Tetrabion Oxi Suini, Neozine, Nimy-
cin, Novoxytetra, NSC-9169, O-4-Cycline, Occryce-
tin, Ocy-Latycin, Ocytetra, Oksimicin, Oksisyklin,
Oksitetraciklin, *Oksytetracyklin*, Opicycline-Inj.,
Orimycin, Ornimed, OS-100 (-200), Os-Cyclin, Os-
micina, Ossi 200, Ossibiotic, Ossicalf, Ossiciclina,
Ossicor, Ossifac, Ossimicetina, Ossimicina, Ossi-
mix, Ossiscel, Ossitetra, *Ossitetraciclina*, Ossitetra-
cin, Ossitetrasol, Ossivet, OTC, Otesolut, O-Tet,
Otetryn, Otracin, Oxacycle, Oxacycline, Oxamycen,
Oxater WS, Oxatets, Oxatracyl, Oxcy, Oxetan,
Oxi 200 I.F.V., Oxibac, Oxibiotic, Oxiclina, Oxi-
farm, Oxifesa, Oxigel, Oxiline, Oxim "Biopharm",
Oxi-Med, OxiMedicyclin, Oximedisol, Oximicetina,
Oxi-Micorten-F, Oximicrol-AM, Oxipalciclina, Oxi-
ritard, Oxital, Oxiter, Oxitetra, *Oxitetraciclina*, *Oxi-
tetracyclinum*, Oxitraklin, Oxlopar, OXTC, Oxtra,
Oxy 20 "Dox-Al", Oxy 20 La Noé, Oxybiocycline,
Oxybiotic, Oxycap, Oxycare, Oxy-Cline, Oxyclyne,
Oxycyclin, Oxy-DE 250, Oxydon, Oxydumocyclin,
Oxyfacyclin, Oxy-Ingacycline, Oxyject, Oxy-Kes-
so-Tetra, Oxylag, Oxylim-V, Oxylon, Oxy-M-50,
Oxymast, Oxymycin, Oxymykoin, Oxyn(e), Oxy-
pan, Oxy-Rivo, Oxy-Selz, Oxysentin, Oxysteclin,
Oxy-T, Oxytab, Oxyter, Oxyterracyna, Oxytet, Oxy-
tetra, Oxytetrabiotic, Oxy-Tetrachel, Oxytetrachlor,
Oxytetracid, *Oxytetracycline***, Oxytetral, Oxytetra-
mix, Oxy-tetran, Oxytetrarco, Oxytetraseptin, Oxyte-
travet, *Oxytetrazyklin*, Oxytetril, Oxytetrin, Oxytra-
cin, Oxytracyl, Oxytral, Oxy-Ultramycin, Oxyvet,
Oxy-WS, Panterramicina, Pan-Terramycin, Parlon,

Pharmaoxyline, P.M. 17, Posicycline, Préquinix N 5, Prokalen, Proteroxyna, Pulmozan L.A., Pulmozone-LA, Quimocyclar R, Quirvetin-intramuscular, Rémacycline, Rigocycline, Riomitsin, Rocap, Rontec, Rotet, Roxy, Ryomycin, Scanoxcin, Seamoxy, Sekamisin, Sekamycin, Servimycin, Sferomycin, Solciclina "Un. Comm. Lombarda", Solkaciclina, Spectratet, Stecsolin, Stevacin, Superciclin "Andrade", Supravet 21/0, Tarchocin(e), Tarocyn, Tenaline, Teravit, Terine, Teroftal, Terrabiot, Terrados, Terraflavine, Terraftalmina, Terrafungine, Terrahel, Terraject, Terralon, Terramedic, Terramicina, Terramycin(e), Terramyfar, Terraos, Terrasol, Terraveineuse, Terraven, Terravenös, Terricil, Tetcin, Tetra "M.N."-Amp., Tetrabion oxi, Tetracem, Tetracycletten N, Tetrafarm, Tetrafen, Tetralabor, Tetra-Medivet, Tetramel, Tetramine "Tutag", Tetran "Chinoin", Tétranase, Tetraphar, Tetraplex, Tetrasona, Tetra-Tablinen, Tetrazoo, Tetrinfan, Tetroxolane, Tetroxy, Tetroxyn, Ticycline, Tija, Toxinal, Tracin, TRMC, Ultramicina "Dispert", Unacillina, Uniciclina "Un. Comm. Lombarda", Unimycin "Unigreg", Uri-Tet, Ursocyclin, Uvomycin, Varimycin, Vendarcin, Vernacycline, Vetimycin, Vexin, Vioprol, Wolicyclin, Zimaiacil

U Antibiotic from *Streptomyces rimosus*

11072-01   (8915-01)   57122-98-2
R Calcium salt (2:1)
S *Oxytetracycline calcium***, Terrabon, Terramycin (Sirup; Kindertropfen (Pediatric Drops))
U Antibiotic

11073   (8916) $C_{22}H_{24}N_2O_9$
                                       95105-77-4

(+)-1,4-Dihydro-2,6-dimethyl-4-(*o*-nitrophenyl)-3,5-pyridinedicarboxylic acid methyl ester 5-ester with 1,4:3,6-dianhydro-D-glucitol = (*R*)-1,4:3,6-Dianhydro-D-glucitol

5-[methyl 1,4-dihydro-2,6-dimethyl-4-(2-nitrophenyl)-3,5-pyridinedicarboxylate] (●)
R designated also as (*S*)-compound (95105-78-5)
S *Sornidipine***
U Calcium antagonist

11074 $C_{22}H_{24}N_2O_{10}$
                             85233-19-8

1,2-Bis(*o*-aminophenoxy)ethane-*N,N,N',N'*-tetraacetic acid = *N,N'*-[1,2-Ethanediylbis(oxy-2,1-phenylene)]bis[*N*-(carboxymethyl)glycine] (●)
S BAPTA
U Calcium chelator

11075 $C_{22}H_{24}N_4O$
                             193958-35-9

5-Amidino-2-[4-(phenylmethyl)-1-piperidinylcarbonyl]-1*H*-indole = 2-[[4-(Phenylmethyl)-1-piperidinyl]carbonyl]-1*H*-indole-5-carboximidamide = 1-[[5-(Aminoiminomethyl)-1*H*-indol-2-yl]carbonyl]-4-(phenylmethyl)piperidine (●)
R also monohydrochloride
S LY 178550
U Anticoagulant (human α-thrombin inhibitor)

11076 (8917) $C_{22}H_{24}N_4O$
87186-60-5

1-(5,6-Bis-*p*-tolyl-1,2,4-triazin-3-yl)-4-piperidinol = 1-
[5,6-Bis(4-methylphenyl)-1,2,4-triazin-3-yl]-4-piperidi-
nol (●)
S LY 81067
U Anxiolytic

11077 (8918) $C_{22}H_{24}N_4O_2$
72807-01-3

1-Benzoyl-3-[1-(3-indolylmethyl)-4-piperidyl]urea = *N*-
[[[1-(1*H*-Indol-3-ylmethyl)-4-piperidinyl]amino]carbo-
nyl]benzamide (●)
S Wy-25093
U Antidepressant (5HT-inhibitor)

11078 (8919) $C_{22}H_{24}N_4O_2$
47588-84-1

[2-Amino-6-[(1,2-diphenylethyl)amino]-3-pyridi-
nyl]carbamic acid ethyl ester (●)
R Monohydrochloride (33400-47-4)
S D-10242
U Anti-inflammatory

11079 (8920) $C_{22}H_{24}N_4O_2$
117523-47-4

N-(1-Phenethyl-4-piperidyl)-*N*-pyrazinyl-2-furamide =
N-[1-(2-Phenylethyl)-4-piperidinyl]-*N*-pyrazinyl-2-
furancarboxamide (●)
R Monohydrochloride (119413-53-5)
S A-3508, *Mirfentanil hydrochloride***
U Narcotic analgesic

11080 (8921) $C_{22}H_{24}N_4O_2$
76283-03-9

6-[4-[4-(Phenylacetyl)-1-piperazinyl]phenyl]-4,5-dihy-
dro-3(2*H*)-pyridazinone (●)
S CCI 17810
U Platelet aggregation inhibitor

11081 $C_{22}H_{24}N_4O_2$
156055-46-8

N-[2-(Dimethylamino)ethyl]-9-hydroxy-5,6-dimethyl-
6*H*-pyrido[4,3-*b*]carbazole-1-carboxamide (●)
R Dihydrochloride (178169-99-8)
S *Anelliptole hydrochloride**, NSC-659687, S 16020-2
U Antineoplastic (topoisomerase II inhibitor)

11082 $C_{22}H_{24}N_4O_4S_2$
186549-27-9

4-[[(3S)-3-[[(7-Methoxy-2-naphthalenyl)sulfonyl]me-
thylamino]-2-oxo-1-pyrrolidinyl]methyl]-2-thiophene-
carboximidamide (●)
R   Monohydrochloride (223139-45-5)
S   RPR 120844
U   Antithrombotic

11083 $C_{22}H_{24}N_4O_5S$
186497-38-1

4-[3-[[(3-Methoxy-5-methylpyrazinyl)amino]sulfonyl]-
2-pyridinyl]-α,α-dimethylbenzenepropanoic acid (●)
S   ZD 1611
U   Endothelin receptor antagonist

11084 $C_{22}H_{24}N_4O_5S$
136075-61-1

3,4-Dihydro-1'-[2-(5-benzofurazanyl)ethyl]-6-(methanesulfonamido)spiro[(2H)-1-benzopyran-2,4'-piperidin]-4-
one = N-[1'-[2-(2,1,3-Benzoxadiazol-5-yl)ethyl]-3,4-di-
hydro-4-oxospiro[2H-1-benzopyran-2,4'-piperidin]-6-
yl]methanesulfonamide (●)
R   Monohydrochloride (136075-60-0)
S   L 691121
U   Anti-arrhythmic

11085 $C_{22}H_{24}N_8O_5S$
156579-02-1

N-[[4-[(2,4-Diamino-6-pteridinyl)methyl]-3,4-dihydro-
2H-1,4-benzothiazin-7-yl]carbonyl]-L-homoglutamic
acid = (2S)-2-[[[4-[(2,4-Diamino-6-pteridinyl)methyl]-
3,4-dihydro-2H-1,4-benzothiazin-7-yl]carbonyl]ami-
no]hexanedioic acid (●)
S   ML-68, MX-68
U   Immunosuppressive (antirheumatic, anti-arthritic)

11086   (8922) $C_{22}H_{24}N_{10}O_{12}S_2$
108319-07-9

2-[[[(2-Amino-4-thiazolyl)[[1-[[[3-(1,4-dihydro-5-hydr-
oxy-4-oxopicolinamido)-2-oxo-1-imidazolidinyl]sulfo-
nyl]carbamoyl]-2-oxo-3-azetidinyl]carbamoyl]methy-
lene]amino]oxy]-2-methylpropionic acid = 2-[[[1-(2-
Amino-4-thiazolyl)-2-[[[[3-[[(1,4-dihydro-5-hydr-
oxy-4-oxo-2-pyridinyl)carbonyl]amino]-2-oxo-1-imida-
zolidinyl]sulfonyl]amino]carbonyl]-2-oxo-3-azetidi-
nyl]amino]-2-oxoethylidene]amino]oxy]-2-methylpropa-
noic acid (●)
R   Disodium salt (104393-00-2)
S   *Pirazmonam sodium***, SQ 83360
U   Antibiotic

11087 $C_{22}H_{24}O_3S_2$
156910-58-6

4-[2-(3,4-Dihydro-4,4-dimethyl-2*H*-1-benzopyran-6-yl)-1,3-dithian-2-yl]benzoic acid (●)
S   SR 11238
U   Antineoplastic

11088   (8924) $C_{22}H_{24}O_4$
2021-14-9

Bimolecular cyclic ester of the 3-hydroxy-*p*-cymene-2-carboxylic acid = 1,7-Dimethyl-4,10-bis(1-methylethyl)-6*H*,12*H*-dibenzo[*b*,*f*][1,5]dioxocin-6,12-dione (●)
S   *Dicarvocrotonide*
U   Anti-inflammatory

11089   (8925) $C_{22}H_{24}O_4$
120360-17-0

Methyl (3*R**,5*S**)-(*E*)-3,5-dihydroxy-9,9-diphenyl-6,8-nonadienoate = [*R**,*S**-(*E*)]-3,5-Dihydroxy-9,9-diphenyl-6,8-nonadienoic acid methyl ester (●)
S   CP-83101

U   HMGCoA reductase inhibitor (hepatoselective)

11090   (8923) $C_{22}H_{24}O_4$
5965-06-0

4,4'-(1,2-Diethylvinylene)diphenol diacetate = *p*,*p*'-(1,2-Diethylvinylene)bis(phenyl acetate) = *p*,*p*'-Diacetoxy-α,β-diethylstilbene = Diethylstilbestrol diacetate = 4,4'-(1,2-Diethyl-1,2-ethenediyl)bis[phenol] diacetate (●)
S   Hormostilboral-stark
U   Synthetic estrogen

11091 $C_{22}H_{24}O_8$
178367-17-4

7-(2,3-Dihydroxypropoxy)-5-hydroxy-2-(3-hydroxy-4-methoxy-2-propylphenyl)-4*H*-1-benzopyran-4-one (●)
S   S 15673
U   Anti-edemic

11092 $C_{22}H_{24}O_8$
207225-51-2

rel-(+)-10-Hydroxy-8-[(2*R*,4*S*,5*R*)-5-hydroxy-4-(hydroxymethyl)-1,3-dioxan-2-yl]-3-methyl-1-propyl-1*H*-naphtho[2,3-*c*]pyran-6,9-dione (●)
S   *Alnumycin*
U   Cytotoxic, antibacterial

11093   (8926) $C_{22}H_{25}BrN_2O_3S$
                                      131707-25-0

6-Bromo-4-[(dimethylamino)methyl]-5-hydroxy-1-me-
thyl-2-[(phenylthio)methyl]-1*H*-indole-3-carboxylic acid
ethyl ester (●)
R   Monohydrochloride (131707-23-8)
S   Arbidol
U   Antiviral, immunostimulant, interferon inducer

11094   (8927) $C_{22}H_{25}ClF_3N_3S$
                                      83016-35-7

1-(2-Chloroethyl)-4-[3-[2-(trifluoromethyl)-10-pheno-
thiazinyl]propyl]piperazine = 10-[3-[4-(2-Chloroethyl)-1-
piperazinyl]propyl]-2-(trifluoromethyl)-10*H*-phenothia-
zine (●)
S   Flu-M, *Fluphenazine mustard*
U   Antineoplastic

11095   (8928) $C_{22}H_{25}ClN_2OS$
                                      982-24-1

2-[4-[3-(2-Chloro-9-thioxanthenylidene)propyl]-1-pipe-
razinyl]ethanol = 2-Chloro-9-[3-(*N*'-β-hydroxyethylpipe-
razino)propylidene]thioxanthene = 4-[3-(2-Chloro-9*H*-
thioxanthen-9-ylidene)propyl]-1-piperazineethanol (●)

R   see also nos. 12300, 14663;
       dihydrochloride (633-59-0); used also as *cis*-isomer
       [base: (53772-83-1); dihydrochloride: (58045-23-1)]
S   AY-62021, *Chlorpenthixol dihydrochloride, Chlor-
       perphenthixene dihydrochloride*, Ciatyl, Ciatyl-Z,
       Cisordinol, *Clopenthixol dihydrochloride***, Clopi-
       xol, Lu 0-108, N 746, Neo-Ciatyl, NSC-64087, Se-
       danxol, Sordenac, Sordinol, Thiapax, *Zuclopenthixol
       hydrochloride***
U   Neuroleptic, antipsychotic

11096   (8929) $C_{22}H_{25}ClN_2O_2$
                                      86621-92-3

p-[2-[4-(*p*-Chlorocinnamyl)-1-piperazinyl]ethyl]benzoic
acid = 4-[2-[4-[3-(4-Chlorophenyl)-2-propenyl]-1-pipera-
zinyl]ethyl]benzoic acid (●)
S   BM 15766
U   Antihyperlipidemic

11097   (8930) $C_{22}H_{25}ClN_2O_3$
                                      108210-73-7

(±)-2'-Chloro-α-methyl-4-biphenylacetic acid ester with
1-glycoloyl-4-methylpiperazine = 2-(4-Methyl-1-pipera-
zinyl)-2-oxoethyl 2-(2'-chloro-4-biphenylyl)propionate =
2'-Chloro-α-methyl[1,1'-biphenyl]-4-acetic acid 2-(4-me-
thyl-1-piperazinyl)-2-oxoethyl ester (●)
S   *Bifeprofen***
U   Anti-inflammatory, analgesic, antipyretic, anti-aggre-
       gant

11098 (8931) $C_{22}H_{25}ClN_2O_4S$
96125-53-0

(+)-*cis*-8-Chloro-5-[2-(dimethylamino)ethyl]-2,3-dihy-
dro-3-hydroxy-2-(*p*-methoxyphenyl)-1,5-benzothiaze-
pin-4(5*H*)-one acetate (ester) = *cis*-(+)-3-(Acetyloxy)-8-
chloro-5-[2-(dimethylamino)ethyl]-2,3-dihydro-2-(4-
methoxyphenyl)-1,5-benzothiazepin-4(5*H*)-one (●)
R Maleate (1:1) (96128-92-6)
S *Chlorodiltiazem maleate, Clentiazem maleate**, Lo-
gna, TA-3090
U Coronary vasodilator, antihypertensive (calcium antagonist)

11099 (8932) $C_{22}H_{25}Cl_2N_3O_4$
31847-13-9

3-[*p*-[[*p*-[Bis(2-chloroethyl)amino]-2,5-dimethoxybenzy-
lidene]amino]phenyl]-2-oxazolidinone = 3-[4-[[[4-
[Bis(2-chloroethyl)amino]-2,5-dimethoxyphenyl]methy-
lene]amino]phenyl]-2-oxazolidinone (●)
S Bay a 5850, GEA 29
U Antineoplastic

11100 (8933) $C_{22}H_{25}FN_2O$
26049-76-3

4'-Fluoro-4-(1,2,4,4a,5,6-hexahydro-3*H*-pyrazi-
no[1,2-*a*]quinolin-3-yl)butyrophenone = 3-[γ-(*p*-Fluoro-
benzoyl)propyl]-2,3,4,4a,5,6-hexahydro-1*H*-pyrazi-
no[1,2-*a*]quinoline = 1-(4-Fluorophenyl)-4-(1,2,4,4a,5,6-
hexahydro-3*H*-pyrazino[1,2-*a*]quinolin-3-yl)-1-butanone
(●)
R Monohydrochloride (51732-60-6)
S *Centpyraquin*, Compound 69-183
U Antihypertensive, neuroleptic

11101 (8934) $C_{22}H_{25}F_2NO_2$
38077-12-2

4'-Fluoro-4-[4-(*p*-fluoro-α-hydroxybenzyl)piperidi-
no]butyrophenone = 1-(4-Fluorophenyl)-4-[4-[(4-fluoro-
phenyl)hydroxymethyl]-1-piperidinyl]-1-butanone (●)
S *Dihydrolenperone*, NSC-343513, RMI 11974
U Antineoplastic, neuroleptic

11102 (8935) $C_{22}H_{25}F_2NO_4$
118457-14-0

α,α'-(Iminodimethylene)bis[6-fluoro-2-chromanmetha-
nol] = (α*S*,α'*S*,2*R*,2'*S*)-α,α'-[Iminobis(methylene)]bis[6-
fluoro-3,4-dihydro-2*H*-1-benzopyran-2-methanol] (●)
R also hydrochloride (169293-50-9)
S Hypoloc, Lobivon, *Narbivolol*, Nebilet, Nebilox, *Ne-
bivolol**, R 65824, R 67555

U Antihypertensive (β-adrenergic blocker)

11103 $C_{22}H_{25}F_2N_3O_2S$
144665-07-6

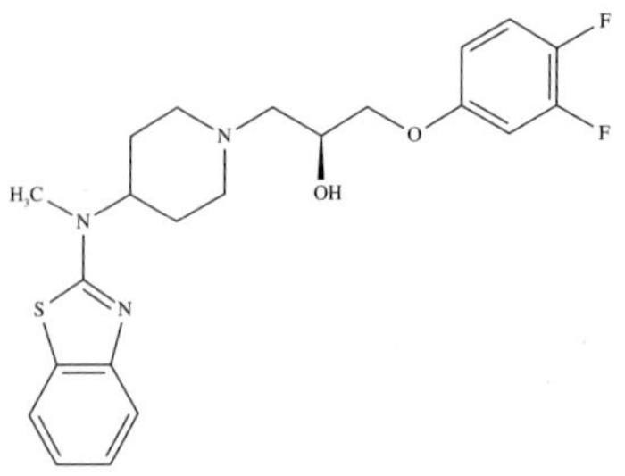

(αS)-4-(2-Benzothiazolylmethylamino)-α-[(3,4-difluoro-phenoxy)methyl]-1-piperidineethanol (●)
S JK-8792, *Lubeluzole***, Prosynap, R 87926
U Anti-ischemic, neuroprotective

11104 (8936) $C_{22}H_{25}N$
16378-21-5

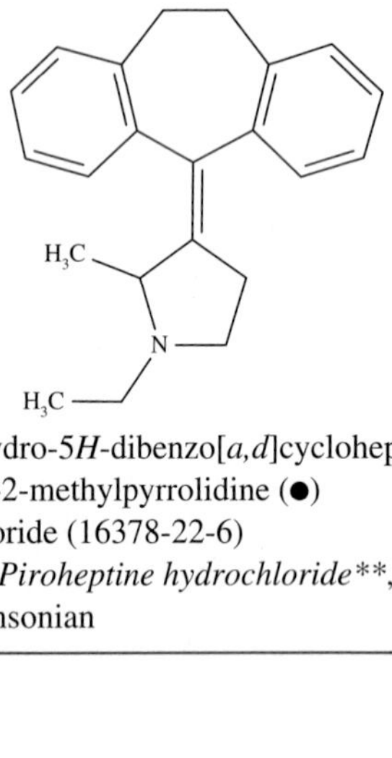

3-(10,11-Dihydro-5*H*-dibenzo[*a,d*]cyclohepten-5-ylidene)-1-ethyl-2-methylpyrrolidine (●)
R Hydrochloride (16378-22-6)
S FK 1190, *Piroheptine hydrochloride***, Trimol
U Antiparkinsonian

11105 (8937) $C_{22}H_{25}N$
41695-46-9

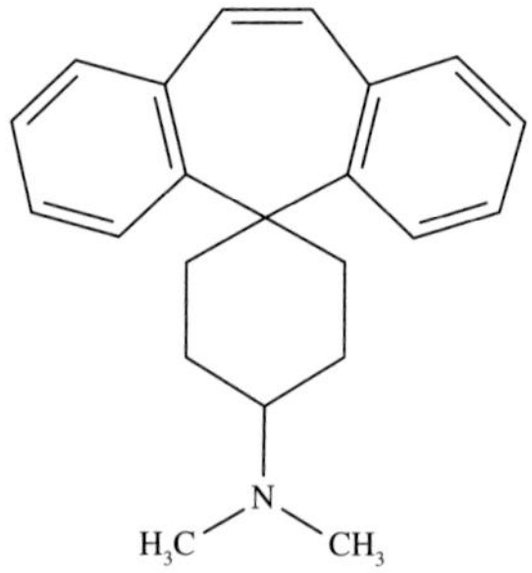

N,N-Dimethylspiro[cyclohexane-1,5'-[5*H*]dibenzo[*a,d*]cyclohepten]-4-amine (●)
S A 1866, CPG 147
U 5-HT and norepinephrine inhibitor

11106 (8938) $C_{22}H_{25}NOS$
119502-13-5

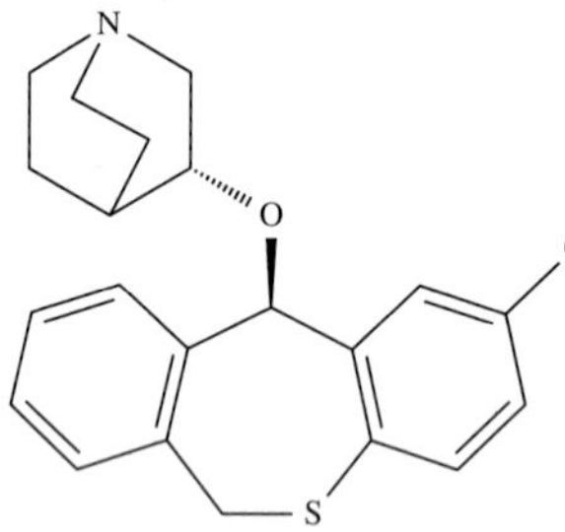

3-[(6,11-Dihydro-2-methyldibenzo[*b,e*]thiepin-11-yl)oxy]quinuclidine = (*R*,R**)-(±)-3-[(6,11-Dihydro-2-methyldibenzo[*b,e*]thiepin-11-yl)oxy]-1-azabicyclo[2.2.2]octane (●)
R Maleate [(2-*Z*)-2-butenedioate] (1:1) (119502-15-7)
S VÚFB-17088
U Antihistaminic

11107 (8940) $C_{22}H_{25}NO_2$
6581-06-2

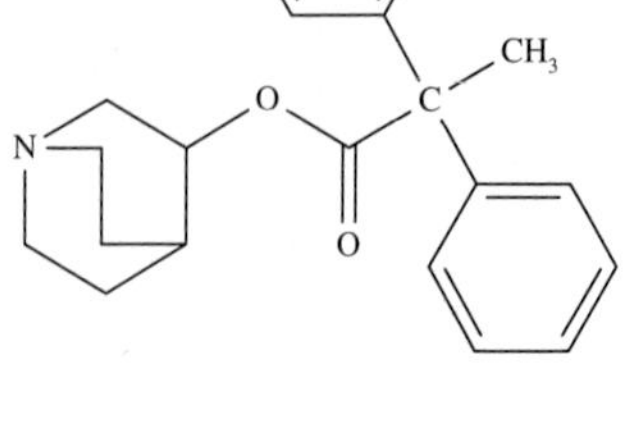

3-Quinuclidinyl α,α-diphenylpropionate = 2,2-Diphenyl-propionic acid 3-quinuclidinyl ester = α-Methyl-α-phe-nylbenzeneacetic acid 1-azabicyclo[2.2.2]oct-3-yl ester (●)
R Hydrochloride (3818-79-9)
S Aprolidine
U Anticholinergic

11108 (8941) $C_{22}H_{25}NO_2$
6878-98-4

1αH,5αH-Tropan-3α-ol diphenylacetate = Tropine di-phenylacetate = Diphenylacetic acid tropine ester = endo-α-Phenylbenzeneacetic acid 8-methyl-8-azabicy-clo[3.2.1]oct-3-yl ester (●)
R Hydrochloride (548-64-1)
S Diphenyltropine hydrochloride, Tropacin, Tropazi-ne, Win 5690
U Anticholinergic, spasmolytic

11109 (8942) $C_{22}H_{25}NO_2$
101910-24-1

1-[m-(2-Quinolylmethoxy)phenyl]hexanol = α-Pentyl-3-(2-quinolinylmethoxy)benzenemethanol (●)
S PF 5901, REV 5901, RG 5901, RHC 5901
U 5-Lipoxygenase inhibitor, peptidoleukotriene antago-nist

11110 (8939) $C_{22}H_{25}NO_2$
3565-47-7

4'-Methoxy-4-(1,2,3,6-tetrahydro-4-phenyl-1-pyridyl)bu-tyrophenone = 1-(3-p-Methoxybenzoylpropyl)-4-phenyl-Δ³-tetrahydropyridine = 1-(4-Methoxyphenyl)-4-(1,2,3,6-tetrahydro-4-phenyl-1-pyridinyl)-1-butanone (●)
R Hydrochloride (14777-23-2)
S Anisoperidone, R 1647
U Neuroleptic

11111 $C_{22}H_{25}NO_2$
85375-85-5

1-(4,4-Diphenyl-3-butenyl)-3-piperidinecarboxylic acid (●)
R Hydrochloride (85375-15-1)
S SKF 89976-A
U Anticonvulsant

11112 $C_{22}H_{25}NO_2$
148152-63-0

(3R)-rel-3-Phenyl-1-[[(6R)-6,7,8,9-tetrahydronaph-tho[1,2-d]-1,3-dioxol-6-yl]methyl]pyrrolidine (●)
R Methanesulfonate (149189-73-1)

S A-75200 mesylate, ABT 200, *Napitane mesilate***
U Antidepressant (α-adrenergic blocker and norepinephrine uptake antagonist)

11113   (8943) $C_{22}H_{25}NO_3$
                                          3736-36-5

1α*H*,5α*H*-Tropan-3α-ol benzilate = Tropine benzilate =
Benzilic acid tropine ester = *endo*-α-Hydroxy-α-phenyl-
benzeneacetic acid 8-methyl-8-azabicyclo[3.2.1]oct-3-yl
ester (●)
R Hydrochloride (1674-94-8)
S *Benztropeine hydrochloride*, BETE, BTE, Diphen-
trop, Glipin, Glypin, IT-78, *Tropine benzilate hydrochloride*
U Anticholinergic, spasmolytic, anti-asthmatic

11114   (8944) $C_{22}H_{25}NO_3$
                                          74051-39-1

1α*H*,5α*H*-Tropan-3β-ol benzilate = Benzilic acid pseudo-
tropine ester = *exo*-α-Hydroxy-α-phenylbenzeneacetic
acid 8-methyl-8-azabicyclo[3.2.1]oct-3-yl ester (●)
R Hydrochloride (36173-66-7)
S BPTE
U Spasmolytic

11115   (8945) $C_{22}H_{25}NO_3$
                                          94497-51-5

N-(5,6,7,8-Tetrahydro-5,5,8,8-tetramethyl-2-naph-
thyl)terephthalamic acid = 4-[[(5,6,7,8-Tetrahydro-
5,5,8,8-tetramethyl-2-naphthalenyl)amino]carbonyl]ben-
zoic acid (●)
S Am 80, *Tamibarotene***
U Antineoplastic, keratogenesis inhibitor, antipsoriatic

11116 $C_{22}H_{25}NO_3$
                                          102121-60-8

4-[[(5,6,7,8-Tetrahydro-5,5,8,8-tetramethyl-2-naphthale-
nyl)carbonyl]amino]benzoic acid (●)
S AM-580, Ro 40-6055
U Anti-angiogenic

11117   (8946) $C_{22}H_{25}NO_3$

6-[4-(3-Phenyl-1-pyrrolidinyl)butyryl]benzodioxane = 1-
(2,3-Dihydro-1,4-benzodioxin-6-yl)-4-(3-phenyl-1-pyr-
rolidinyl)-1-butanone (●)
R Hydrochloride (59939-57-0)
S Butiroksan, Butiroxan, Butyroxane
U α-Adrenergic blocker

11118   (8947) $C_{22}H_{25}NO_3$
                                                                       23744-24-3

5,5-Diphenyl-2-(2-piperidinoethyl)-1,3-dioxolan-4-one =
5,5-Diphenyl-2-[2-(1-piperidinyl)ethyl]-1,3-dioxolan-4-
one (●)
R   also hydrochloride (18174-58-8)
S   BR 18, Paraespas, *Pipossolano*, *Pipoxolan**, Roco-
        fin, Rowapraxin, Rowell
U   Spasmolytic, muscle relaxant

11119   (8948) $C_{22}H_{25}NO_4$
                                                                       54063-52-4

Methyl o-[p-(2-piperidinoethoxy)benzoyl]benzoate = *o-*
[*p*-(2-Piperidinoethoxy)benzoyl]benzoic acid methyl ester = 2-[4-[2-(1-Piperidinyl)ethoxy]benzoyl]benzoic acid
methyl ester (●)
R   Hydrochloride (1248-42-6)
S   *Bulgar ketone*, Hoechst 12771, *Pitofenone hydro-
        chloride***
U   Anticholinergic, spasmolytic

11120   (8949) $C_{22}H_{25}NO_4$
                                                                       117096-90-9

7-[2-Hydroxy-3-(isopropylamino)propoxy]-4'-methylfla-
vone = 7-[2-Hydroxy-3-[(1-methylethyl)amino]prop-
oxy]-2-(4-methylphenyl)-4*H*-1-benzopyran-4-one (●)
R   Hydrochloride (106287-81-4)
S   *Methylflavonolamine hydrochloride*, SIPI 549
U   Anti-arrhythmic

11121   (8950) $C_{22}H_{25}NO_4$
                                                                       147-27-3

1-(4-Ethoxy-3-methoxybenzyl)-6,7-dimethoxy-3-me-
thylisoquinoline = 1-[(4-Ethoxy-3-methoxyphenyl)me-
thyl]-6,7-dimethoxy-3-methylisoquinoline (●)
R   Phosphate (1:1) (5667-46-9)
S   *Dimoxyline phosphate***, *Dioxyline phosphate*,
        L 08146, Livera, Paveril phosphate, Paverona, Pave-
        rone, Pavrona
U   Spasmolytic, vasodilator

11122 $C_{22}H_{25}NO_4S$
                                                                       171752-56-0

4,5,5aα,6,7,11bβ-Hexahydro-9,10-diacetoxy-2-pro-
pylthieno[2,3-*c*]benzo[*f*]quinoline = (5a*R*,11b*S*)-
4,5,5a,6,7,11b-Hexahydro-2-propylbenzo[*f*]thie-
no[2,3-*c*]quinoline-9,10-diol diacetate (ester) (●)
R   Hydrochloride (166591-11-3)
S   A-93431.1, ABT 431, *Adrogolide hydrochloride***
U   Antiparkinsonian (dopamine D_1 receptor agonist)

11123 (8951) $C_{22}H_{25}NO_6$
64-86-8

Colchiceine methyl ether = (*S*)-*N*-(5,6,7,9-Tetrahydro-1,2,3,10-tetramethoxy-9-oxobenzo[*a*]heptalen-7-yl)acet-amide (●)
S Aqua-Colchin, Artrichine, *Colchicine*, Colchinéos, Colchiquim, Colchisol, Colcin, Colgout, Colmediten, *Colquicina*, Colsalide (new form), Colsaloid, Coluric, Condylon (active substance), Goutnil, *Kolkicin*, Kolsin, Novocolchine, NSC-757, Xaoliuwan
U Anti-arthritic, antineoplastic (alkaloid from *Colchicum autumnale*)

11123-01 (8951-01) 8013-62-5
R Salicylate
S Colchalon, Viraxatabs
U Anti-arthritic, antirheumatic

11124 (8952) $C_{22}H_{25}NO_6$
641-28-1

(*S*)-5-(Acetylamino)-6,7-dihydro-9,10,11-trimethoxy-5*H*-dibenzo[*a,c*]cycloheptene-3-carboxylic acid methyl ester (●)
S *Allocolchicine*, NSC-406042
U Cytotoxic, antimitotic

11125 (8953) $C_{22}H_{25}N_2OS$
7187-66-8

1,3-Dibenzyldecahydro-2-oxoimidazo[4,5-*c*]thieno[1,2-*a*]thiolium = (+)-1,3-Dibenzylperhydro-2-oxo-thieno[1',2':1,2]thieno[3,4-*d*]imidazol-5-ium = (+)-4,6-Dibenzyl-5-oxo-1-thionia-4,6-diazatricyclo[6.3.0.0^{3,7}]undecane = D-3,4-(1,3-Dibenzyl-2-oxoimidazolidino)-1,2-trimethylenethiophanium = Decahydro-2-oxo-1,3-bis(phenylmethyl)thieno[1',2':1,2]thieno[3,4-*d*]imidazol-5-ium (●)
R D-Camphorsulfonate (no. 2423) (1:1) (68-91-7)
S Arfonad, *Méthioplégium*, Nu 2222, Ro 2-2222, *Thimethaphan d-Camphorsulfonate*, Thiophanium, *Trimetaphan camsilate***, *Trimethaphan camsylate*
U Antihypertensive, ganglion blocking agent

11126 (8954) $C_{22}H_{25}N_3$
57961-90-7

2-[2-(4-*m*-Tolyl-1-piperazinyl)ethyl]quinoline = 2-[2-[4-(3-Methylphenyl)-1-piperazinyl]ethyl]quinoline (●)
S *Centaquin*, *Centhaquine*, Compound 7173
U Antihypertensive

11127 (8955) $C_{22}H_{25}N_3O$
26844-12-2

N-[1-[2-(3-Indolyl)ethyl]-4-piperidyl]benzamide = 3-[2-(4-Benzamidopiperidino)ethyl]indole = *N*-[1-[2-(1*H*-Indol-3-yl)ethyl]-4-piperidinyl]benzamide (●)
R Monohydrochloride (38821-52-2)
S Baratol "Monmouth; Roberts; Wyeth", Doralese, *Indoramin hydrochloride***, Indorene, Orfidora, Radibar, Vidora, Wy-21901, Wydora, Wypres, Wypresin, Zaidorim
U Antihypertensive, migraine prophylactic

11128 (8956) $C_{22}H_{25}N_3O$
                                                  53-89-4

4-Benzyl-1-(1-methyl-4-piperidyl)-3-phenyl-3-pyrazolin-5-one = 1,2-Dihydro-2-(1-methyl-4-piperidinyl)-5-phenyl-4-(phenylmethyl)-3*H*-pyrazol-3-one (●)
S Benzometan, *Benzopiperilone, Benzpiperylone***, Humedil, KB 95, Reubenil (old form), Telon, Tolcasil
U Antirheumatic

11129 (8957) $C_{22}H_{25}N_3O_2$
                                                  71119-11-4

o-[2-Hydroxy-3-[(2-indol-3-yl-1,1-dimethylethyl)amino]propoxy]benzonitrile = 3-[2-Hydroxy-3-[[2-(1*H*-indol-3-yl)-1,1-dimethylethyl]amino]propoxy]benzonitrile (●)
R Monohydrochloride (70369-47-0)
S Bextra, *Bucindolol hydrochloride***, MJ 13105-1
U Antihypertensive (α-and β-adrenergic blocker)

11130 $C_{22}H_{25}N_3O_2$
                                                  187654-40-6

1-[8-Ethoxy-4-[[(1*R*)-1-phenylethyl]amino]-1,7-naphthyridin-3-yl]-1-butanone (●)
S YJA 20379-8
U Proton-pump inhibitor

11131 (8958) $C_{22}H_{25}N_3O_2S$
                                                  104383-19-9

(−)-2-[[[1-(1,4-Benzodioxan-2-ylmethyl)-4-piperidyl]methyl]amino]benzothiazole = (*S*)-*N*-[[1-[(2,3-Dihydro-1,4-benzodioxin-2-yl)methyl]-4-piperidinyl]methyl]-2-benzothiazolamine (●)
S R 47243
U α$_2$-Adrenoceptor antagonist

11132 (8959) $C_{22}H_{25}N_3O_2S$
                                                  135722-27-9

1-(7-Methoxy-1-naphthyl)-4-[2-(thenoylamino)ethyl]piperazine = *N*-[2-[4-(7-Methoxy-1-naphthalenyl)-1-piperazinyl]ethyl]-2-thiophenecarboxamide (●)
S S 14671
U 5-HT$_{1A}$-agonist

11133 (8960) $C_{22}H_{25}N_3O_3$
130610-93-4

N-Methyl-2-(m-nitrophenyl)-N-[(1S,2S)-2-(1-pyrrolidi-
nyl)-1-indanyl]acetamide = (1S-trans)-N-[2,3-Dihydro-2-
(1-pyrrolidinyl)-1H-inden-1-yl]-N-methyl-3-nitroben-
zeneacetamide (●)
R Monohydrochloride (130610-94-5)
S *Niravoline hydrochloride***, RU 51599
U Diuretic (κ-opioid receptor agonist)

11134 (8962) $C_{22}H_{25}N_3O_3$
4448-96-8

7-[2-[4-(2-Methoxyphenyl)-1-piperazinyl]ethyl]-5H-1,3-
dioxolo[4.5-f]indole (●)
R also tartrate (1:1) (5591-43-5)
S *Solypertine***, Win 18413
U Anti-adrenergic

11135 $C_{22}H_{25}N_3O_3$
118974-02-0

(5aS,12S,14aS)-1,2,3,5a,6,11,12,14a-Octahydro-9-meth-
oxy-12-(2-methyl-1-propenyl)-5H,14H-pyrro-

lo[1'',2'':4',5']pyrazino[1',2':1,6]pyrido[3,4-b]indole-5,14-
dione (●)
S *Fumitremorgin C*
U Chemosensitizer (for antineoplastics)

11136 (8961) $C_{22}H_{25}N_3O_3$
1054-88-2

8-(1,4-Benzodioxan-2-ylmethyl)-1-phenyl-1,3,8-triaza-
spiro[4.5]decan-4-one = 8-[(2,3-Dihydro-1,4-benzodi-
oxin-2-yl)methyl]-1-phenyl-1,3,8-triazaspiro[4.5]decan-
4-one (●)
S R 5188, *Spiroxamide*, *Spiroxatrine***
U Analgesic, neuroleptic

11137 (8963) $C_{22}H_{25}N_3O_3S$
101387-98-8

1,2,3,5-Tetrahydro-6-[[(4-methoxy-3-methyl-2-pyridi-
nyl)methyl]sulfinyl]-1,1,3,3-tetramethylimidazo[4,5-f]in-
den-2-one = 5,7-Dihydro-2-[[(4-methoxy-3-methyl-2-py-
ridinyl)methyl]sulfinyl]-5,5,7,7-tetramethylinde-
no[5,6-d]imidazol-6(1H)-one (●)
S Ro 18-5364
U Gastric antisecretory

11138 (8964) $C_{22}H_{25}N_3O_3S$
101706-33-6

[2S-[2α[R*(R*)],3Z,5α]]-1,3-Dihydro-1-methyl-3-[1-[5-
[(methylthio)methyl]-6-oxo-3-(2-oxo-3-cyclopenten-1-
ylidene)-2-piperazinyl]ethyl]-2H-indol-2-one (●)
S FR 900452
U PAF receptor antagonist

11139 (8966) $C_{22}H_{25}N_3O_4$
81840-15-5

1-(1,2,3,4-Tetrahydro-2-oxo-6-quinolyl)-4-veratroylpi-
perazine = 3,4-Dihydro-6-[4-(3,4-dimethoxybenzoyl)-1-
piperazinyl]-2(1H)-quinolinone = 1-(3,4-Dimethoxyben-
zoyl)-4-[1,2,3,4-tetrahydro-2-oxo-6-quinolinyl]pipera-
zine (●)
S Arkin-Z, OPC-8212, Piperanometozine, Vesnarino-
ne**
U Coronary vasodilator (positive inotropic)

11140 (8965) $C_{22}H_{25}N_3O_4$
90402-40-7

4-Amino-2-(3,4-dihydro-6,7-dimethoxy-2(1H)-isoquino-
linyl)-6,7-dimethoxyquinoline = 2-(3,4-Dihydro-6,7-di-
methoxy-2(1H)-isoquinolinyl)-6,7-dimethoxy-4-quino-
linamine (●)
R Monomethanesulfonate (118931-00-3)

S Abanoquil mesilate**, UK-52046-27
U Anti-arrhythmic, coronary vasodilator ($α_1$-adreno-
ceptor antagonist)

11141 (8967) $C_{22}H_{25}N_3O_4S$
31883-05-3

Ethyl 10-(3-morpholinopropionyl)phenothiazine-2-carb-
amate = N-[10-(3-Morpholinopropionyl)-2-phenothiazi-
nyl]carbamic acid ethyl ester = [10-[3-(4-Morpholinyl)-1-
oxopropyl]-10H-phenothiazin-2-yl]carbamic acid ethyl
ester (●)
R Monohydrochloride (29560-58-5)
S Äthmozin, EN 313, ETHM, Ethmozine, Etmozin,
G 214, Moracizine hydrochloride**, Moricizine hydrochloride, Procizin
U Anti-arrhythmic

11142 $C_{22}H_{25}N_3O_5$
128759-76-2

(3α,16α)-Eburnamenine-14-carboxylic acid 2-(nitro-
oxy)ethyl ester (●)
S VA-045
U Cerebroprotectant

11143 $C_{22}H_{25}N_3O_7S$
153832-46-3

(1R,5S,6S)-2-[2(S)-[(N)-(3-Carboxyphenyl)aminocarbo-nyl]-4(S)-pyrrolidinylthio]-6-[1(R)-hydroxyethyl]-1-me-thylcarbapen-2-em-3-carboxylic acid = (4R,5S,6S)-3-[[(3S,5S)-5-[[(3-Carboxyphenyl)amino]carbonyl]-3-pyr-rolidinyl]thio]-6-[(1R)-1-hydroxyethyl]-4-methyl-7-oxo-1-azabicyclo[3.2.0]hept-2-ene-2-carboxylic acid (●)
R   Monosodium salt (153832-38-3)
S   *Ertapenem sodium**, Invanz, L 749345, MK-826, ZD 4433
U   Antibacterial

11144 $C_{22}H_{25}N_5O_2$
145938-37-0

2,6-Dihydro-2-[[4-(2-methoxyphenyl)-1-piperazinyl]me-thyl]imidazo[1,2-c]quinazolin-5(3H)-one (●)
S   DL-028
U   Antihypertensive (α_1-adrenoceptor antagonist)

11145 $C_{22}H_{25}N_5O_2$
149847-87-0

2,6-Dihydro-3-[[4-(2-methoxyphenyl)-1-piperazinyl]me-thyl]imidazo[1,2-c]quinazolin-5(3H)-one (●)
S   DC-015
U   Antihypertensive (α_1-adrenoceptor antagonist)

11146 $C_{22}H_{25}N_5O_4S$
187162-39-6

N-[(6-Amino-2-methyl-3-pyridinyl)methyl]-6-methyl-2-oxo-3-[[(phenylmethyl)sulfonyl]amino]-1(2H)-pyridine-acetamide (●)
S   L 374087
U   Anticoagulant

11147   (8968) $C_{22}H_{25}N_5O_5$
99682-33-4

N-[4-[3-(4-Acetyl-3-hydroxy-2-propylphenoxy)prop-oxy]phenyl]-1H-tetrazole-5-carboxamide (●)
S   LY 170198
U   Proteinkinase C inhibitor

11148 $C_{22}H_{25}N_5O_6S$
171028-74-3

3-[[[(5R)-3-[4-(Aminoiminomethyl)phenyl]-4,5-dihydro-5-isoxazolyl]acetyl]amino]-N-[(3-methylphenyl)sulfo-nyl]-L-alanine (●)
S   XV 454
U   Platelet GIIb/IIIa receptor antagonist

11149 (8969) $C_{22}H_{25}N_7O_5$
                                            80576-83-6

10-Ethyl-10-deazaaminopterin = N-[4-[1-[(2,4-Diamino-
6-pteridinyl)methyl]propyl]benzoyl]-L-glutamic acid (●)
S CGP 30694, 10-EDAM, *Edatrexate***, NSC-626715
U Antineoplastic

11150 (8970) $C_{22}H_{26}As_2N_2O_8S$
                                            120-76-3

2-[Bis(2,3-dihydroxypropyl)amino]phenol-[4-arseno-5′]-
β-(2-benzoxazolylthio)propionic acid = 3-[[5-[[3-
[Bis(2,3-dihydroxypropyl)amino]-4-hydroxyphenyl]diar-
senyl]-2-benzoxazolyl]thio]propanoic acid (●)
R Sodium salt (2921-50-8)
S Hoechst 10577, Spirotrypan
U Antisyphilitic, antitrypanosomal

11151 (8971) $C_{22}H_{26}ClFO_4$
                                            54063-32-0

21-Chloro-9-fluoro-17-hydroxy-16β-methylpregna-1,4-
diene-3,11,20-trione = (16β)-12-Chloro-9-fluoro-17-
hydroxy-16-methylpregna-1,4-diene-3,11,20-trione (●)
R see also no. 13183
S Clobetasone**
U Glucocorticoid (anti-inflammatory, anti-allergic)

11152 $C_{22}H_{26}ClNO_4$

(2R)-[3-[3-[2-(3-Chlorophenyl)-2-hydroxyethylami-
no]cyclohexyl]phenoxy]acetic acid = [3-[(1S,3S)-3-
[[(2R)-2-(3-Chlorophenyl)-2-hydroxyethyl]amino]cyclo-
hexyl]phenoxy]acetic acid (●)
R Monosodium salt (190063-31-1)
S GS-332
U Adrenergic β₃ receptor agonist

11153 $C_{22}H_{26}ClNO_4$
                                            121524-08-1

N-(2S)-[7-[(Carbethoxy)methoxy]-1,2,3,4-tetrahydro-2-
naphthyl]-(2R)-2-hydroxy-2-(3-chlorophenyl)ethan-
amine = [R-(R*,S*)]-[[7-[[2-(3-Chlorophenyl)-2-hydr-
oxyethyl]amino]-5,6,7,8-tetrahydro-2-naphthale-
nyl]oxy]acetic acid ethyl ester (●)
R Hydrochloride (121524-09-2)
S SR 58611A
U Thermogenic β-adrenoceptor agonist

11154 $C_{22}H_{26}ClN_3O_4S$
                                            130047-27-7

6-[4(R)-(4-Chlorophenylsulfonamido)-1-(3-pyridylme-
thyl)pyrrolidin-2(S)-yl]-5(Z)-hexenoic acid = [2S-
[2α(Z),4β]]-6-[4-[[(4-Chlorophenyl)sulfonyl]amino]-1-
(3-pyridinylmethyl)-2-pyrrolidinyl]-5-hexenoic acid (●)

R Monohydrochloride (130047-30-2)
S FK 070, FR 121070, KDI 792
U Thromboxane A_2 antagonist

11155   (8972) $C_{22}H_{26}Cl_2N_6O_5$
                                                       93118-77-5

3-Ethyl 5-methyl 2-[[2-[(5-amino-1*H*-1,2,4-triazol-3-
yl)amino]ethoxy]methyl]-4-(2,3-dichlorophenyl)-1,4-di-
hydro-6-methyl-3,5-pyridinedicarboxylate = 2-[[2-[(5-
Amino-1*H*-1,2,4-triazol-3-yl)amino]ethoxy]methyl]-4-
(2,3-dichlorophenyl)-1,4-dihydro-6-methyl-3,5-pyridine-
dicarboxylic acid 3-ethyl 5-methyl ester (●)
S UK-52831
U Antihypertensive, anti-anginal (calcium antagonist)

11156   (8973) $C_{22}H_{26}Cl_2O_3$
                                                       33145-88-9

4,6-Dichloro-17-hydroxy-16-methylenepregna-4,6-di-
ene-3,20-dione (●)
S MDAP
U Progestin

11157   (8974) $C_{22}H_{26}Cl_2O_6$
                                                       238401-69-9

6-*O*-[4-Chloro-2-[3-(4-chlorophenoxy)propyl]-6-methyl-
phenyl]-2,4-dideoxy-D-*erythro*-hexonic acid (●)
R Monosodium salt (136006-50-3)
S S 853758A
U Antihyperlipidemic (HMGCoA reductase inhibitor)

11158   (8975) $C_{22}H_{26}FNO$
                                                       86495-14-9

(±)-5,9α-Dimethyl-2-[2-(4-fluorophenyl)ethyl]-2'-hydr-
oxy-6,7-benzomorphan = (2α,6α,11*R**)-[2-(4-Fluorophe-
nyl)ethyl]-1,2,3,4,5,6-hexahydro-6,11-dimethyl-2,6-me-
thano-3-benzazocin-8-ol (●)
S *Fluorophen*
U δ-opioid receptor agonist

11159   (8976) $C_{22}H_{26}FNO_2$
                                                       1050-79-9

4'-Fluoro-4-(4-hydroxy-4-*p*-tolylpiperidino)butyrophe-
none = 4-*p*-Tolyl-1-(3'-*p*-fluorobenzoylpropyl)-4-piperi-
dinol = 1-(4-Fluorophenyl)-4-[4-hydroxy-4-(4-methyl-
phenyl)-1-piperidinyl]-1-butanone (●)
R Hydrochloride (3871-82-7)
S Kodamaron, Lubatren, Luvatren(a), Meldol, *Methyl-
        peridol, Moperone hydrochloride***, *Mopipérone hy-
        drochloride*, R 1658
U Neuroleptic

11160 (8977)　　　　　　　　$C_{22}H_{26}FN_3O_2$
510-74-7

8-[3-(4-Fluorophenoxy)propyl]-1-phenyl-1,3,8-triazaspiro[4.5]decan-4-one (●)
R also hydrochloride (53658-90-5)
S *Fluroxyspiramine*, R 5808, *Spiramide***
U Neuroleptic

11161 (8978)　　　　　　　$C_{22}H_{26}FN_3O_2S$
104383-17-7

(±)-1-[4-(2-Benzothiazolylmethylamino)piperidino]-3-(4-fluorophenoxy)-2-propanol = (±)-4-(2-Benzothiazolyl-methylamino)-α-[(4-fluorophenoxy)methyl]-1-piperidineethanol (●)
S R 58735, Reminyl "Janssen", *Sabeluzole***
U Anticonvulsant, antihypoxic, cognition enhancer

11162 (8979)　　　　　　　$C_{22}H_{26}FN_3O_4$
98206-10-1

(+)-p-Fluoro-N-[2-[4-[2-(hydroxymethyl)-1,4-benzodioxan-5-yl]-1-piperazinyl]ethyl]benzamide = (+)-N-[2-[4-(2,3-Dihydro-2-(hydroxymethyl)-1,4-benzodioxin-5-yl)-1-piperazinyl]ethyl]-4-fluorobenzamide (●)
R Monohydrochloride (98205-89-1)
S DU 29373, *Flesinoxan hydrochloride***

U Antihypertensive

11163　　　　　　　　　　　$C_{22}H_{26}FN_5O_2$
179634-04-9

3-[5-[(Dipropylamino)methyl]-2-oxazolyl]-8-fluoro-4,5-dihydro-5-methyl-6H-imidazo[1,5-a][1,4]benzodiazepin-6-one (●)
S Ro 48-8684
U Benzodiazepine receptor agonist (anesthetic)

11164 (8980)　　　　　　　$C_{22}H_{26}FN_5O_3$
85118-43-0

7-[3-[4-(p-Fluorobenzoyl)piperidino]propyl]theophylline = 7-[3-[4-(4-Fluorobenzoyl)-1-piperidinyl]propyl]-3,7-dihydro-1,3-dimethyl-1H-purine-2,6-dione (●)
S *Fluprofylline***, Sgd 14480
U Analeptic

11165 (8981)　　　　　　　$C_{22}H_{26}F_3NO_2$
65512-99-4

7-[2-(Dimethylamino)ethoxy]-2,2-dimethyl-4-[3-(trifluoromethyl)phenyl]chroman = 2-[[3,4-Dihydro-2,2-dimethyl-4-[3-(trifluoromethyl)phenyl]-2H-1-benzopyran-7-yl]oxy]-N,N-dimethylethanamine (●)
R Hydrochloride (59257-18-0)
S BRL 16644

U Antidepressant

11166　(8982)　　　　　　　　　　　$C_{22}H_{26}F_3N_3OS$
　　　　　　　　　　　　　　　　　　　　　　　　69-23-8

4-[3-[2-(Trifluoromethyl)-10-phenothiazinyl]propyl]-1-piperazineethanol = 2-[4-[3-[2-(Trifluoromethyl)-10-phenothiazinyl]propyl]-1-piperazinyl]ethanol = 10-[3-[4-(2-Hydroxyethyl)-1-piperazinyl]propyl]-2-(trifluoromethyl)phenothiazine = 4-[3-[2-(Trifluoromethyl)-10H-phenothiazin-10-yl]propyl]-1-piperazineethanol (●)

R see also nos. 11094, 14146, 14669;
Dihydrochloride (146-56-5)

S Anatensil, Anatensol, Antasol, Apo-Fluphenazine, Calmansial, Cardilac, Cenilene, Cerevin, Dapotum (acutum), Eldoral "Gäbert", Elinol, Esprivex, Flenaken, Flufenan, *Flufenazine dihydrochloride*, Flumazine, Flumezin, Flunazine, *Fluphenazine hydrochloride* **, Ftorfenazin, Horizone, Lyogen, Lyorodin, Mirenil, Misoril, Modecate acutum, Moditen, NSC-62323, Omca, Pacinol, Pacinone (old form), Permitil, Phthorphenazin, PMS-Fluphenazine, Prolinat, Prolixin, S 94, Sch 6894, Sediten, Selecten, Sevinal, Sevinol, Sevinon "Schering-USA", Siqualine, Siqualone, Tensofin, Teviral, Trancin, Valamina "Schering-USA", Vespazin

U Neuroleptic

11167　(8983)　　　　　　　　　　　　$C_{22}H_{26}NO$

N-(4-Phenylphenacyl)heliotridanium = 4-(2-[1,1'-Biphenyl]-4-yl-2-oxoethyl)hexahydro-1-methyl-1H-pyrrolizinium (●)

R Bromide (72586-42-6)
S C_7
U Anticholinergic, spasmolytic

11168　(8984)　　　　　　　　　　　$C_{22}H_{26}NO_3$
　　　　　　　　　　　　　　　　　　　　　　　　7020-55-5

3-Hydroxy-1-methylquinuclidinium benzilate = 3-(Benziloyloxy)-1-methylquinuclidinium = 3-[(Hydroxydiphenylacetyl)oxy]-1-methyl-1-azoniabicyclo[2.2.2]octane (●)

R Bromide (3485-62-9)
S *Clidinium bromide* **, Quarzan(bromide), Ro 2-3773
U Anticholinergic, anti-ulcer agent

11169　(8985)　　　　　　　　　　　$C_{22}H_{26}N_2O_2$
　　　　　　　　　　　　　　　　　　　　　　　　42971-09-5

Ethyl apovincamin-22-oate = Apovincamic acid ethyl ester = (3α,16α)-Eburnamenine-14-carboxylic acid ethyl ester (●)

S Airpacil, Amadolin, Apostron, AY-27255, Balptin, Calan "Takeda", Calanarol, Calansetin, Cammex, Cavinton, Cepoallen, Ceractin, Ceranovin, Davinion, D.c.vin, Finacilen, Gelutolan, Hematonyl, Inex, Karantomin, Laidal, Mandaris, Medichill, Mocomon, Naikinol, Nedelin, Neuro-race, Neusaura, Piltin, Pinoca, Ranen, Recervin, Remedial, RGH-4405, Rinobulen, Sarsetin, TCV-3B, Tepsa, Ultra-Vinca, Vinatin, Vinbren, Vincalin, Vincorex, *Vinpocetine* **, Vinpoton, Vipocem, Vipociton

U Cerebrotonic

11170 (8986)

$C_{22}H_{26}N_2O_2$
31386-25-1

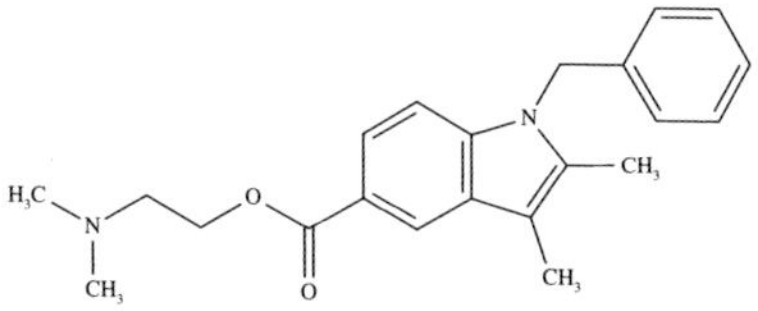

2-(Dimethylamino)ethyl 1-benzyl-2,3-dimethylindole-5-carboxylate = 1-Benzyl-2,3-dimethyl-5-indolecarboxylic acid 2-(dimethylamino)ethyl ester = 2,3-Dimethyl-1-(phenylmethyl)-1*H*-indole-5-carboxylic acid 2-(dimethylamino)ethyl ester (●)
S Indocarb, *Indocate***, Indokarb, Inmecarb, K-281
U Antiserotonin, anti-alcoholic

11171 (8987)

$C_{22}H_{26}N_2O_2$
87627-28-9

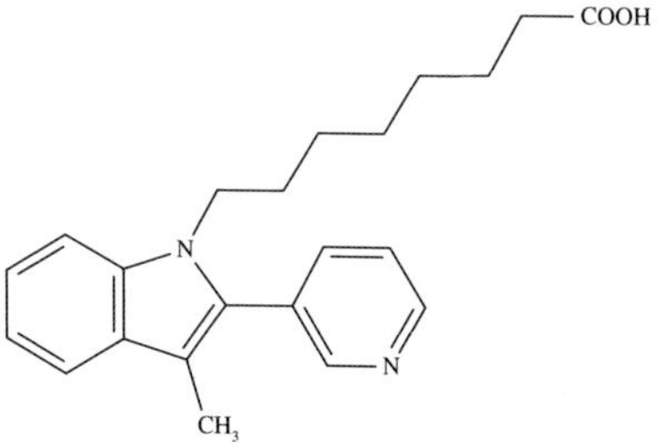

3-Methyl-2-(3-pyridyl)indole-1-octanoic acid = 3-Methyl-2-(3-pyridinyl)-1*H*-indole-1-octanoic acid (●)
S CGS 12970, *Lindogrel*
U Antithrombotic

11172

$C_{22}H_{26}N_2O_2$
153607-44-4

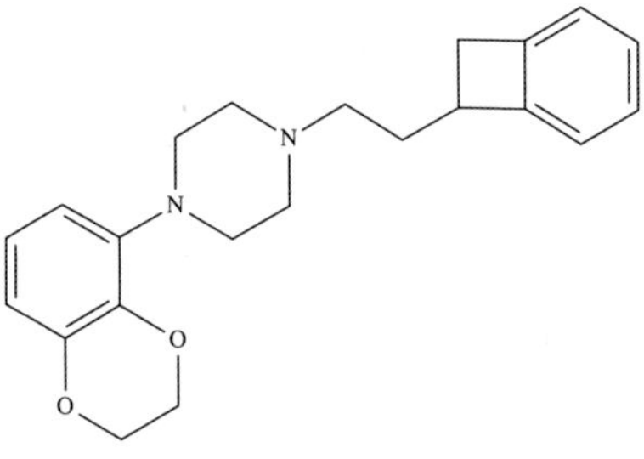

1-[2-(1-Benzocyclobutyl)ethyl]-4-(1,4-benzodioxan-5-yl)piperazine = 1-(2-Bicyclo[4.2.0]octa-1,3,5-trien-7-yl-ethyl)-4-(2,3-dihydro-1,4-benzodioxin-5-yl)piperazine (●)
S S 14489
U Anxiolytic, antidepressant, antipsychotic

11173

$C_{22}H_{26}N_2O_2$
185515-24-6

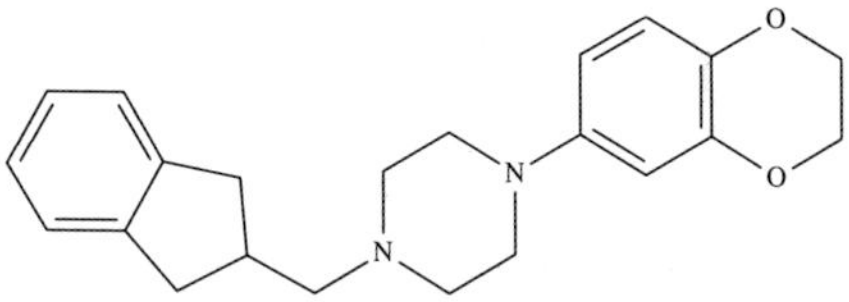

2-[[4-(2,3-Dihydrobenzo[1,4]dioxin-6-yl)-1-piperazinyl]methyl]indan = 1-(2,3-Dihydro-1,4-benzodioxin-6-yl)-4-[(2,3-dihydro-1*H*-inden-2-yl)methyl]piperazine (●)
S S 18126
U Antipsychotic (dopamine D_4 receptor antagonist)

11174 (8988)

$C_{22}H_{26}N_2O_2S$
130746-90-6

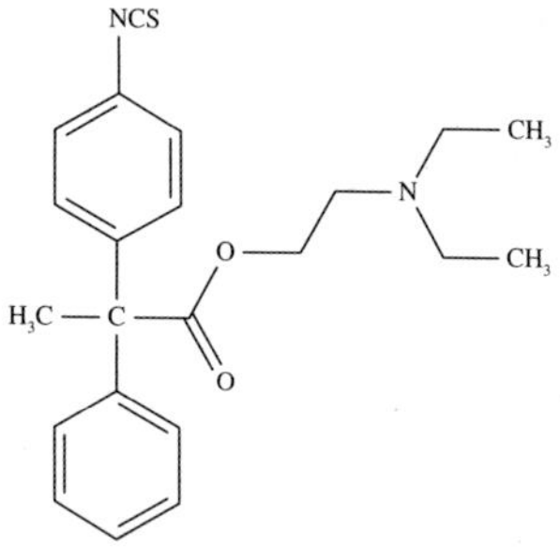

(±)-4-Isothiocyanato-α-methyl-α-phenylbenzeneacetic acid 2-(diethylamino)ethyl ester (●)
R Oxalate (ethanedioate) (1:1) (130746-91-7)
S *Aprophit*
U Muscarinic antagonist

11175

$C_{22}H_{26}N_2O_2S$
143322-58-1

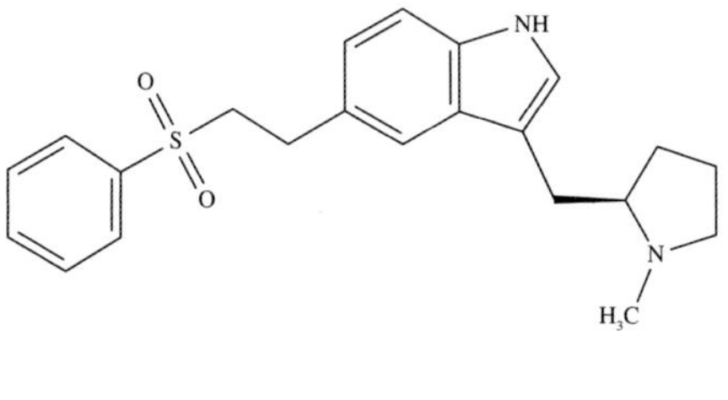

(*R*)-2-[3-[(1-Methyl-2-pyrrolidinyl)methyl]-1*H*-indol-5-yl]ethyl phenyl sulfone = 3-[[(2*R*)-1-Methyl-2-pyrrolidinyl]methyl]-5-[2-(phenylsulfonyl)ethyl]-1*H*-indole (●)
R Monohydrobromide (177834-92-3)
S *Eletriptan hydrobromide***, Relpax, UK-116044-04
U Serotonin agonist (migraine therapeutic)

11176 (8989) $C_{22}H_{26}N_2O_3$
162849-90-3

1-[2-(4-Methoxyphenyl)-2-[3-(4-methoxyphenyl)propoxy]ethyl]-1*H*-imidazole (●)
R Monohydrochloride (130495-35-1)
S SKF 96365
U Platelet aggregation inhibitor, calcium antagonist

11177 (8990) $C_{22}H_{26}N_2O_3S$
28810-23-3

6,11-Dihydro-6-methyl-11-(1α*H*,5α*H*-tropan-3α-yloxy)dibenzo[*c,f*][1,2]thiazepine 5,5-dioxide = *endo*-3-[(6,11-Dihydro-6-methyldibenzo[*c,f*][1,2]thiazepin-11-yl)oxy]-8-methyl-8-azabicyclo[3.2.1]octane *S,S*-dioxide = *endo*-6,11-Dihydro-6-methyl-11-[(8-methyl-6-azabicyclo[3.2.1]oct-3-yl)oxy]dibenzo[*c,f*][1,2]thiazepine 5,5-dioxide (●)
S *Zepastine***
U Antihistaminic, anticholinergic

11178 (8991) $C_{22}H_{26}N_2O_4$
22345-47-7

1-(3,4-Dimethoxyphenyl)-5-ethyl-7,8-dimethoxy-4-methyl-5*H*-2,3-benzodiazepine (●)
S Egyt-341, Felicyl, Grandaxin, SD 19050, Sériel, Tavor "Ramon", Tofisan, *Tofisopam***
U Tranquilizer

11179 (8995) $C_{22}H_{26}N_2O_4$
3512-87-6

[2*S*-(2α,3*E*,7aα,12aα,12bβ,15*S**)]-3-Ethylidene-1,3,4,6,7,12b-hexahydro-9-hydroxy-12-methyl-2*H*,12*H*-12a,2,7a-(epoxyethanylylidene)indolo[2,3-*a*]quinolizine-15-carboxylic acid methyl ester (●)
S *Akuammine, Vincamajoridine*
U Antihypertensive (*Apocynaceae* alkaloid)

11180 (8992) $C_{22}H_{26}N_2O_4$
84750-41-4

1-[2,3-Bis[(*p*-methoxybenzyl)oxy]propyl]imidazole = 1-[2,3-Bis[(4-methoxyphenyl)methoxy]propyl]-1*H*-imidazole (●)
S SC-38249
U Platelet aggregation inhibitor

11181 (8993)

$C_{22}H_{26}N_2O_4$
29726-99-6

1-(3,4-Dimethoxyphenyl)-4-ethyl-6,7-dimethoxy-3-methylisoquinoline 2-imide = 2-Amino-1-(3,4-dimethoxyphenyl)-4-ethyl-6,7-dimethoxy-3-methylisoquinolinium hydroxide inner salt (●)
S *Tofisoline*
U Tranquilizer

11182 (8994)

$C_{22}H_{26}N_2O_4$
7762-32-5

1,4-Bis(1,4-benzodioxan-2-ylmethyl)piperazine = 1,4-Bis[(2,3-dihydro-1,4-benzodioxin-2-yl)methyl]piperazine (●)
S Dibozan, McN-181
U Antihypertensive

11183 (8996)

$C_{22}H_{26}N_2O_4S$
42399-41-7

(+)-5-[2-(Dimethylamino)ethyl]-*cis*-2,3-dihydro-3-hydroxy-2-(*p*-methoxyphenyl)-1,5-benzothiazepin-4(5*H*)-one acetate (ester) = D-*cis*-3-Acetoxy-2,3-dihydro-5-[2-(dimethylamino)ethyl]-2-(*p*-methoxyphenyl)-1,5-benzothiazepin-4(5*H*)-one = (2*S*,3*S*)-5-[2-(Dimethylamino)ethyl]-2,3,4,5-tetrahydro-2-(4-methoxyphenyl)-4-oxo-1,5-benzothiazepin-3-yl acetate = (2*S*,3*S*)-3-(Acetyloxy)-5-[2-(dimethylamino)ethyl]-2,3-dihydro-2-(4-methoxyphenyl)-1,5-benzothiazepin-4-(5*H*)-one (●)
R Monohydrochloride (33286-22-5)
S Acalix, Acasmul, Adizem, Aldizem, Alfener, Altiazem, Alti-Diltiazem, Altizem, Alzidem, Angiact, Angicontin, Angidil "Benedetti", Angikard, Anginyl, Angiodrox, Angiolong, Angiotrofin "Armstrong", Angiozem, Angipress "Crinos", Angisant, Angitil, Angizem, Angoral, Anzem, Apo-Diltiaz, Auscard, Balcor, Bi-Tildiem, Blocalcin, Britiazim, Bruzem, Cal-Antagon, Calazem, Calcard, Calcicard "Norton", Calnurs, Calzem, Cardcal, Cardem "Anglo-French", Cardiacton, Cardiazem "Farma", Cardiben, Cardil, Cardionox "Berlimed", Cardiosta, Cardiser, Cardizem, Carex, Carreldon, Cartia XT, Carzem, Cirilen, Citizem, Clandilon "Fidelis", Clarute, Clauden "Alfarma", Clobendian, Convectal, Coramil, Coras, Corazem, Corazet, Cordiazem, Coridil "Ecosol", Cormax "RPR", Coroherser, Corolater, Coronexil, Corsalus, Corsenile, Cortiazem, CRD 401, Cronodine, Delay Tiasim, DelayTiazem, Delcoril, Deltazen, Denazox, Desbon, Diacor, Diacordin, Diazem "Medochemie", Dilaclan, Dilacor "RPR; Watson", Dilacor XR, Diladel, Dilapress, Dilatam(e), Dilatem, Dilatren, Dilauran, Dilazesefu, Dilcard, Dilcardia, Dilcontin, Dilcor "Durascan", Dilem, Dilfar, Dilgina, Diliter, Dilmen, Dilmin, Diloc, Dilpral, Dilren, Dilrène, Dilsal, Dil-Sanorania, Dilta, Dilta 1A Pharma, Dilta AbZ, Diltabeta, DiltaHexal,

Diltalpha, Diltam, Diltan, Diltapham, Diltaretard, Diltazid, Diltelan, Diltiagamma, Diltiamax, Diltiamerck, Diltiangina, Diltiares, Diltiasyn, Diltia XT, *Diltiazem hydrochloride***, Dilti-BASF, Diltical, Dilticard, Diltiem, Dilti-Essex, Diltikard, Diltikor, Diltilite, Diltiuc, Dilti von ct, Diltiwas, Dilti-Wolff, Diltizem, Dilzanol, Dilzem, Dilzene, Dilzereal, Dilzicardin, Dinisor "Parke-Davis; Substancia", Dipen, Doclis, Dodexen, DTM, DTZ, DTZ-SR, Elvesil, Entrydil, Ergoclavin, Etizem, Etyzem, Eukardil, Farmabes, Frotty, Gadoserin, Gen-Diltiazem, Gerbeser, Gewazem, Grifodilzem, Hart, Helsibon, Herben, Herbesser, Incoril, Iposal, Kaltiazem, Kardil, Lacerol, *Latiazem hydrochloride*, Levozem, Litizem, Longazem, Lytelsen, Masdil, Mavitalon, Metazem, Miocardy, Mono-Tildiem, Multipor, Myonil, Natasadol, Novo-Diltiazem, Nu-Diltiaz, Oxycardil, Paretnamin, Pazeadin, Pectazem, Pentilzeno, Poltiazem, Presokin "Sintyal", Presoquin, Pressoquin, Retalzem, RG 83606, Rubiten, Saubasin, Seresnatt, Silden, Slozem, Surazem, Syn-Diltiazem, Tazildil, Terner, Tiadil, Tiakem, Tiaves, Tiazac, Tiazem, Tiazen, Tilazam, Tilazem(e), Tildiem, Tilker, Tilzem, Trumsal, Ubicor "Siegfried", Unimasdil, Usno "Biomedica", Vasocardol, Viacor, Viazem, Viazem SR, WL-Diltiazem, Youtiazem, Zemtard, Zemycard, Zildem, Zilden, Zipertensin, Ziruvate

U Coronary vasodilator (calcium antagonist)

11183-01   (8996-01)   144604-00-2
R Malate (1:1)
S *Diltiazem malate***, MK-793, Tiamate
U Antihypertensive

11184 $C_{22}H_{26}N_2O_5$
                                           251353-30-7

Butanedioic acid mono[4-oxo-2,4-bis[(phenylmethyl)amino]butyl] ester (●)
S TZ-50-2

U Cardiotonic

11185   (8997) $C_{22}H_{26}N_2O_5$
                                           93750-14-2

4-[2-Hydroxy-3-[[2-(2-methoxyphenoxy)-1-methylethyl]amino]propoxy]isocarbostyril = 4-[2-Hydroxy-3-[[2-(2-methoxyphenoxy)-1-methylethyl]amino]propoxy]-1(2*H*)-isoquinolinone (●)
S N-1518
U α- and β-adrenergic blocker

11186 $C_{22}H_{26}N_2O_5$
                                           19542-51-9

N-[(Phenylmethoxy)carbonyl]-L-valyl-L-phenylalanine (●)
S MDL 28170
U Calpain inhibitor

11187 $C_{22}H_{26}N_2O_6$
                                           146461-98-5

3-[[[6-[(4-Acetyl-2-ethyl-5-hydroxyhenoxy)methyl]-2-pyridinyl]carbonyl]ethylamino]propanoic acid = *N*-[[6-[(4-Acetyl-2-ethyl-5-hydroxyphenoxy)methyl]-2-pyridinyl]carbonyl]-*N*-ethyl-β-alanine (●)
S SM 15178
U Leukotriene B$_4$ antagonist

11188 (8998) $C_{22}H_{26}N_2O_6$
115406-23-0

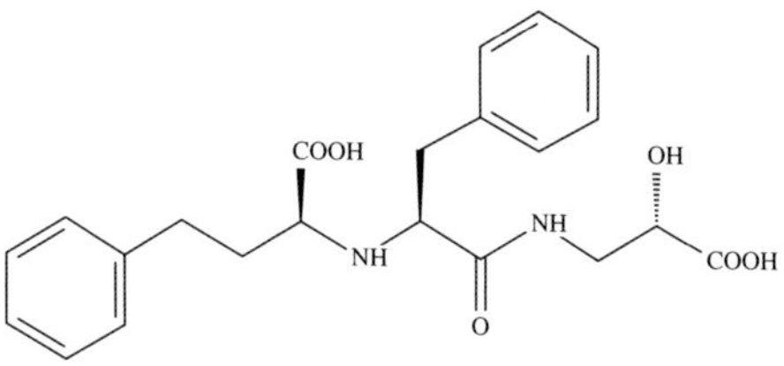

[S-(R*,R*)]-N-[N-(1-Carboxy-3-phenylpropyl)-L-phenyl-
alanyl]-2-hydroxy-β-alanine (●)
S Sch 39370
U Endopeptidase inhibitor

11189 (8999) $C_{22}H_{26}N_2O_6S_3$
120500-15-4

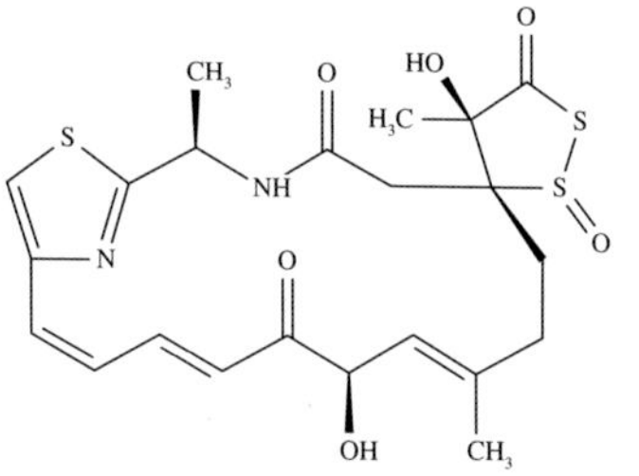

[2'R-[2'R*,6'R*(R*),9'E,11'R*,13'E,15'Z]]-4,11'-Dihy-
droxy-2',4,9'-trimethylspiro[1,2-thiolane-3,6'-
[19]thia[3,20]diazabicyclo[15.2.1]eico-
sa[1(20),9,13,15,17]pentaene]-4',5,12'-trione 2-oxide (●)
S Antibiotic DC 107, DC-107, Leinamycin, Renamycin
U Antineoplastic antibiotic from a Streptomyces strain

11190 (9000) $C_{22}H_{26}N_4O_5$
82423-05-0

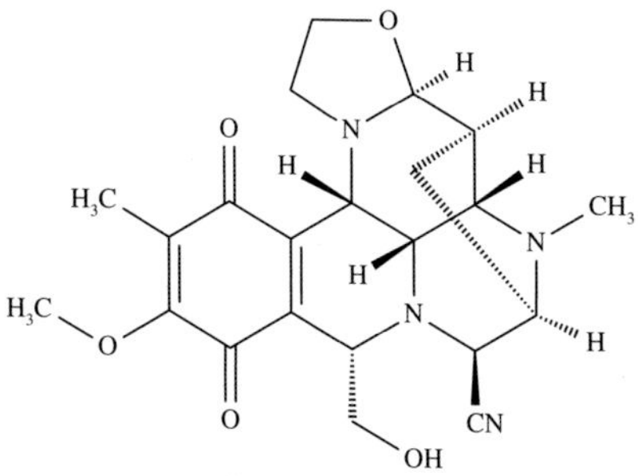

[3aS-(3aα,4α,4aβ,6α,7β,9α,13bβ,13cβ)]-
1,2,3a,4,4a,6,7,9,10,13,13b,13c-Dodecahydro-9-(hydroxymethyl)-11-methoxy-5,12-dimethyl-10,13-dioxo-4,6-

---

methano-5H-benz[h]oxazolo[3,2-a]pyrazi-
no[3,2,1-de][1,5]naphthyridine-7-carbonitrile (●)
S Cyanocycline A, Cyanonaphthyridinomycin
U Antineoplastic (antibiotic from Streptomyces flavo-
griseus)

11191 (9001) $C_{22}H_{26}N_6O_2$
102144-78-5

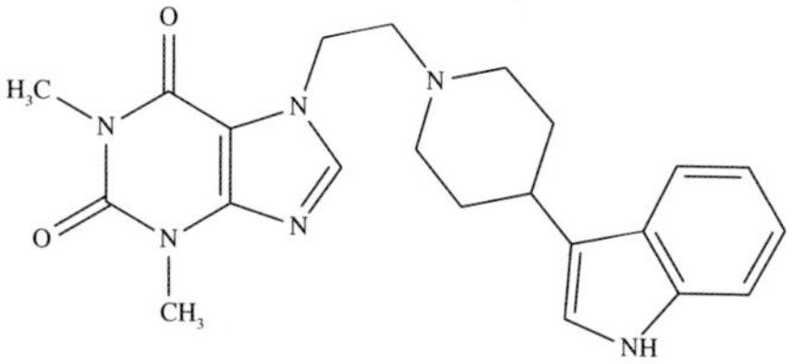

7-[2-(4-Indol-3-ylpiperidino)ethyl]theophylline = 3,7-Di-
hydro-7-[2-[4-(1H-indol-3-yl)-1-piperidinyl]ethyl]-1,3-
dimethyl-1H-purine-2,6-dione (●)
S R 51163, Tameridone**
U Sedative (veterinary)

11192 $C_{22}H_{26}O_3$
35764-59-1

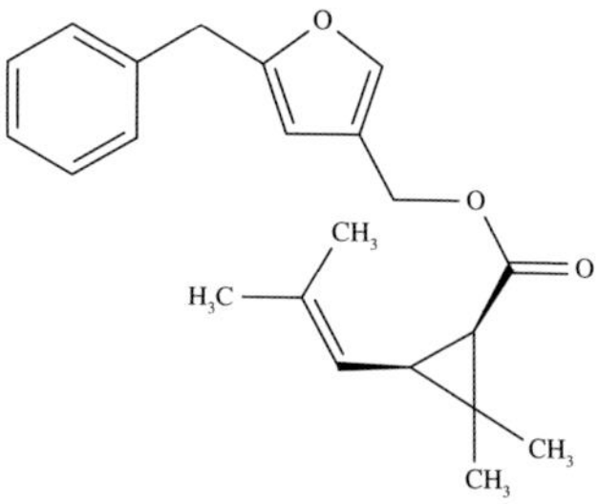

(5-Benzyl-3-furyl)methyl (1R,3S)-2,2-dimethyl-3-(2-me-
thyl-1-propenyl)cyclopropanecarboxylate = (1R,3S)-2,2-
Dimethyl-3-(2-methyl-1-propenyl)cyclopropanecarboxy-
lic acid [5-(phenylmethyl)-3-furanyl]methyl ester (●)
S AI 3-27987, Cismethrin, (+)-cis-Resmethrin,
FMC 26021, NIA 26021, NRDC 119, RU 12063
U Insecticide

11193   (9002) $C_{22}H_{26}O_3$
                                                  28434-01-7

(5-Benzyl-3-furyl)methyl (±)-*trans*-2,2-dimethyl-3-(2-methylpropenyl)cyclopropanecarboxylate = (1*R*,3*R*)-2,2-Dimethyl-3-(2-methyl-1-propenyl)cyclopropanecarboxylic acid [5-(phenylmethyl)-3-furanyl]methyl ester (●)
S   Al 3-27662, *Bioresmethrin***, Isathrine, NIA 18739, NRDC 107, RU 11484, Savenna, SBP 1390, (+)-*trans-Resmethrin*
U   Insecticide

11194   (9003) $C_{22}H_{26}O_4$
                                                  112665-43-7

7-(3,5,6-Trimethyl-1,4-benzoquinon-2-yl)-7-phenylheptanoic acid = (±)-2,4,5-Trimethyl-3,6-dioxo-ζ-phenyl-1,4-cyclohexadiene-1-heptanoic acid = ζ-(2,4,5-Trimethyl-3,6-dioxo-1,4-cyclohexadien-1-yl)benzeneheptanoic acid (●)
S   A-73001, AA-2414, Abbott-73001, Bronica, *Serabenast, Seratrodast***
U   Bronchodilator, anti-asthmatic

11195   (9004) $C_{22}H_{26}O_4$
                                                  4547-76-6

4,4'-(1,2-Diethylethylene)diphenol diacetate = *p,p*'-(1,2-Diethylethylene)bis(phenyl acetate) = 3,4-Bis(*p*-acetoxy-phenyl)hexane = 4,4'-(1,2-Diethyl-1,2-ethanediyl)bis[phenol] diacetate (●)
R   see also nos. 7942, 12435
S   *Hexestrol diacetate, Hexöstroldiacetat*, Retalon-Lingual-Tabl., Robal-Tabl., Sintestrol "Abello", Sintofolin
U   Synthetic estrogen

11196 $C_{22}H_{26}O_5$
                                                  142632-32-4

(10*R*,11*S*,12*S*)-11,12-Dihydro-12-hydroxy-6,6,10,11-tetramethyl-4-propyl-2*H*,6*H*,10*H*-benzo[1,2-*b*:3,4-*b*':5,6-*b*"]tripyran-2-one (●)
S   Cal A, *Calanolide A*, NSC-675451
U   Antiviral from *Calophyllum lanigerum*

11197 $C_{22}H_{26}O_5$
                                                  909-14-8

(10*S*,11*R*,12*S*)-11,12-Dihydro-12-hydroxy-6,6,10,11-tetramethyl-4-propyl-2*H*,6*H*,10*H*-benzo[1,2-*b*:3,4-*b*':5,6-*b*"]tripyran-2-one (●)
S   (−)-*Calanolide B, Costatolide*, NSC-661122
U   Antiviral

11198 (9005) $C_{22}H_{26}O_5$
1247-71-8

3,16α-Dihydroxyestra-1,3,5(10)-trien-17-one diacetate =
16α-Hydroxyestrone diacetate = (16α)-3,16-Bis(acetyloxy)estra-1,3,5(10)-trien-17-one (●)
S Colpoginon, Colpogynon, Colpormon, Hormobion,
Hormocervix Compr., *Hydroxyestrone diacetate, Hy-
droxyöstrondiacetat*, RD 310
U Estrogen

11199 (9006) $C_{22}H_{26}O_6$
23284-23-3

3-[3,4-(Methylenedioxy)benzyl]-4-(3,4,5-trimethoxyben-
zyl)tetrahydrofurane = *trans*-(±)-5-[[Tetrahydro-4-
[(3,4,5-trimethoxyphenyl)methyl]-3-furanyl]methyl]-1,3-
benzodioxole (●)
S *Burseran*, NSC-123428
U Antineoplastic from *Bursera microphylla*

11200 (9007) $C_{22}H_{26}O_8$
10215-89-1

[3aR-[3aα,4β(Z),6α,6aα,7β,9aα,9bβ]]-7-(Acetyloxy)-
3,3a,4,5,6a,7,9a,9b-octahydro-9a-hydroxy-9-methyl-3-
methylene-4-[(2-methyl-1-oxo-2-butenyl)oxy]spiro[azu-
leno[4,5-*b*]furan-6(2*H*)-2'-oxiran]-2-one (●)

S *Euparotin acetate*, NSC-104943
U Antineoplastic from *Eupatorium rotundifolium*

11201 (9008) $C_{22}H_{27}ClF_2O_3$
24320-27-2

9-Chloro-6α,11β-difluoro-21-hydroxy-16α-methylpreg-
na-1,4-diene-3,20-dione = (6α,11β,16α)-9-Chloro-6,11-
difluoro-21-hydroxy-16-methylpregna-1,4-diene-3,20-
dione (●)
S *Halocortolone***
U Glucocorticoid (anti-inflammatory, anti-allergic)

11202 (9009) $C_{22}H_{27}ClF_2O_4$
98651-66-2

21-Chloro-6α,9-difluoro-11β,17-dihydroxy-16β-methyl-
pregna-1,4-diene-3,20-dione = (6α,11β,16β)-21-Chloro-
6,9-difluoro-11,17-dihydroxy-16-methylpregna-1,4-di-
ene-3,20-dione (●)
R see also no. 12839
S *Halobetasol, Ulobetasol***
U Glucocorticoid (anti-inflammatory, anti-allergic)

11203   (9010) $C_{22}H_{27}ClF_2O_4S$
                                                             87556-66-9

S-(Chloromethyl) 6α,9-difluoro-11β,17-dihydroxy-16α-methyl-3-oxoandrosta-1,4-diene-17β-carbothioate = (6α,11β,16α,17α)-6,9-Difluoro-11,17-dihydroxy-16-methyl-3-oxoandrosta-1,4-diene-17-carbothioic acid *S*-(chloromethyl) ester (●)
R   see also no. 12840
S   *Cloticasone***
U   Glucocorticoid (anti-inflammatory, anti-allergic)

11204   (9011) $C_{22}H_{27}ClF_2O_5$
                                                             50629-82-8

2-Chloro-6α,9-difluoro-11β,17,21-trihydroxy-16α-methylpregna-1,4-diene-3,20-dione = (6α,11β,16α)-2-Chloro-6,9-difluoro-11,17,21-trihydroxy-16-methylpregna-1,4-diene-3,20-dione (●)
R   also monohydrate
S   C 48401, *Halometasone***, Sicorten(e)
U   Glucocorticoid (anti-inflammatory, anti-allergic)

11205   (9012) $C_{22}H_{27}ClN_2O$
                                                             59729-31-6

4'-Chloro-*N*-(1-isopropyl-4-piperidyl)-2-phenylacetanilide = *N*-(4-Chlorophenyl)-*N*-[1-(1-methylethyl)-4-piperidinyl]benzeneacetamide (●)
R   Monohydrochloride (58934-46-6)
S   Lopantrol, *Lorcainide hydrochloride***, Lorivox, Polantral "Janssen-Portugal", R 15889, Remivox, Ro 13-1042/001
U   Anti-arrhythmic, local anesthetic

11206   (9013) $C_{22}H_{27}ClN_2O_2S$
                                                             121433-92-9

4-[2-[(2-Chloro-10,11-dihydrodibenzo[*b,f*]thiepin-10-yl)oxy]ethyl]-1-piperazineethanol (●)
R   also butanedioate (1:1) (121434-01-3)
S   *Chlopithepin, Clopithepin*, VÚFB-1776
U   Neuroleptic

11207   (9014) $C_{22}H_{27}ClN_2O_3$
                                                             47562-08-3

Ajmaline 17-chloroacetate = 17-(Monochloroacetyl)ajmaline = (17*R*,21α)-Ajmalan-17,21-diol 17-(chloroacetate) (●)

R Monohydrochloride (40819-93-0)

S *Chloracetylajmaline hydrochloride*, *Lorajmine hydrochloride***, MCAA (-HCl), Nevergor, Ritmosel, Ritmos Elle, Viaductor, Win 11831

U Anti-arrhythmic

11208 (9015)

$C_{22}H_{27}ClN_4O_3$
65617-86-9

2'-Benzoyl-4'-chloro-2-[(*S*)-2,6-diaminohexanamido]-*N*-methylacetanilide = L-Lysyl-2'-benzoyl-4'-chloro-*N*-methylglycinanilide = L-Lysyl-*N*-(2-benzoyl-4-chlorophenyl)-*N*-methylglycinamide (●)

R Dihydrochloride (60067-16-5)

S *Avizafone hydrochloride***, Ro 3-7355/002

U Anxiolytic, anticonvulsant

11209 (9016)

$C_{22}H_{27}ClO_3$
2098-66-0

6-Chloro-1β,2β-dihydro-17-hydroxy-3'*H*-cyclopropa[1,2]pregna-1,4,6-triene-3,20-dione = (1β,2β)-6-Chloro-1,2-dihydro-17-hydroxy-3'*H*-cyclopropa[1,2]pregna-1,4,6-triene-3,20-dione (●)

R see also no. 12375

S *Cyproterone***, SH 881, SH 80881

U Anti-androgen

11210 (9017)

$C_{22}H_{27}ClO_4$
22304-34-3

6-Chloro-17-hydroxy-19-norpregna-4,6-diene-3,20-dione acetate = 17-(Acetyloxy)-6-chloro-19-norpregna-4,6-diene-3,20-dione (●)

S *Amadinone acetate***, RS-2208

U Progestin

11211 (9018)

$C_{22}H_{27}ClO_5$
105149-00-6

6-Chloro-17-hydroxy-2-oxapregna-4,6-diene-3,20-dione acetate = 17-(Acetyloxy)-6-chloro-2-oxapregna-4,6-diene-3,20-dione (●)

S *Gestoxarone acetate*, Hipros, *Osaterone acetate***, TZP 4238

U Anti-androgen, progestogen

11212 (9019)

$C_{22}H_{27}ClO_8$
20501-52-4

[3a*R*-[3aα,4β(*Z*),6α,6aα,7β,9aα,9bβ]]-2-Methyl-2-butenoic acid 7-(acetyloxy)-6-(chloromethyl)-

2,3,3a,4,5,6,6a,7,9a,9b-decahydro-6,9a-dihydroxy-9-me-
thyl-3-methylene-2-oxoazuleno[4,5-*b*]furan-4-yl ester
(●)
S *Eupochlorin acetate*
U Antineoplastic from *Eupatorium rotundifolium*

11213 (9020) $C_{22}H_{27}Cl_2N_3O_4$
65569-29-1

5-Chloro-4-[2-(*p*-chlorophenoxy)acetamido]-*N*-[2-(diethylamino)ethyl]-*o*-anisamide = 5-Chloro-4-[[(4-chloro-
phenoxy)acetyl]amino]-*N*-[2-(diethylamino)ethyl]-2-
methoxybenzamide (●)
S *Cloxacepride***
U Anti-inflammatory, anti-allergic, anti-asthmatic

11214 (9021) $C_{22}H_{27}Cl_3O_2$
567-41-9

3,3'-(2,2,2-Trichloroethylidene)dicarvacrol = Bis(2-hydr-
oxy-6-isopropyl-3-methylphenyl)(trichloromethyl)me-
thane = 2,2'-(2,2,2-Trichloroethylidene)bis[6-methyl-3-
(1-methylethyl)phenol] (●)
S Dithyral
U Antibacterial

11215 (9022) $C_{22}H_{27}FN_2O$
82859-89-0

5-[3-[4-(*p*-Fluorophenyl)-1-piperazinyl]propoxy]indan =
1-[3-[(2,3-Dihydro-1*H*-inden-5-yl)oxy]propyl]-4-(4-
fluorophenyl)piperazine (●)
S BP 528
U Anxiolytic

11216 (9023) $C_{22}H_{27}FN_2O_2$
101343-69-5

2'-Fluoro-2-methoxy-*N*-(1-phenethyl-4-piperidyl)acet-
anilide = *N*-(2-Fluorophenyl)-2-methoxy-*N*-[1-(2-phenyl-
ethyl)-4-piperidinyl]acetamide (●)
R Monohydrochloride (112964-97-3)
S A-3217, *Ocfentanil hydrochloride***
U Narcotic analgesic

11217 (9024) $C_{22}H_{27}FN_2O_3$
61897-04-9

3-[4-[2-[(*p*-Fluorobenzhydryl)oxy]ethyl]-1-piperazi-
nyl]propionic acid = 1-(2-Carboxyethyl)-4-(2-*p*-fluoro-
benzhydryloxyethyl)piperazine = 4-[2-[(4-Fluorophe-
nyl)phenylmethoxy]ethyl]-1-piperazinepropanoic acid
(●)
S S 350

U   Psycho-analeptic

11218   (9025) $C_{22}H_{27}FN_2O_4$
                                                       132758-89-5

[R-(R*,S*)]-1-[3-(3,4-Dihydroxyphenyl)-2-[[3-(4-fluoro-
phenyl)propyl]amino]-3-hydroxy-1-oxopropyl]pyrroli-
dine (●)
R   Monohydrobromide (142784-65-4)
S   SM 11044
U   β-Adrenoceptor agonist

11219   (9026) $C_{22}H_{27}FO_5$
                                                       2193-87-5

9-Fluoro-11β,17,21-trihydroxy-16-methylenepregna-1,4-
diene-3,20-dione = 9-Fluoro-16-methyleneprednisolone
= (11β)-9-Fluoro-11,17,21-trihydroxy-16-methylene-
pregna-1,4-diene-3,20-dione (●)
R   see also no. 12383
S   *Fluprednidene**, Fluprednylidene*
U   Glucocorticoid (anti-inflammatory, anti-allergic)

11220   (9027) $C_{22}H_{27}FO_5$
                                                       33124-50-4

6α-Fluoro-11β-hydroxy-16α-methyl-3,20-dioxopregna-
1,4-dien-21-oic acid = (6α,11β,16α)-6-Fluoro-11-hydr-
oxy-16-methyl-3,20-dioxopregna-1,4-dien-21-oic acid
(●)
R   see also no. 13234
S   *Fluocortin**, Fluocortolon-21-carbonsäure, Fluo-
cortolone-21-carboxylic acid*
U   Glucocorticoid (anti-inflammatory, anti-allergic)

11221   (9028) $C_{22}H_{27}F_3O_4S$
                                                       90566-53-3

S-(Fluoromethyl) 6α,9-difluoro-11β,17-dihydroxy-16α-
methyl-3-oxoandrosta-1,4-diene-17β-carbothioate =
(6α,11β,16α,17α)-6,9-Difluoro-11,17-dihydroxy-16-me-
thyl-3-oxoandrosta-1,4-diene-17-carbothioic acid S-
(fluoromethyl) ester (●)
R   see also no. 12846
S   *Fluticasone***
U   Glucocorticoid (anti-inflammatory, anti-allergic)

11222   (9029) $C_{22}H_{27}F_3O_5$
35135-68-3

6,6,9-Trifluoro-11β,17,21-trihydroxy-16α-methylpreg-
na-1,4-diene-3,20-dione = (11β,16α)-6,6,9-Trifluoro-
11,17,21-trihydroxy-16-methylpregna-1,4-diene-3,20-
dione (●)
R   see also no. 12386
S   *Cormetasone**
U   Glucocorticoid (anti-inflammatory, anti-allergic)

11223   (9032) $C_{22}H_{27}NO$
524-83-4

3α-(Diphenylmethoxy)-8-ethyl-1α*H*,5α*H*-nortropane =
(1*R*,3*r*,5*S*)-3-(Benzhydryloxy)-8-ethylnortropane = *N*-
Ethylnortropine benzhydryl ether = *endo*-3-(Diphenylmethoxy)-8-ethyl-8-azabicyclo[3.2.1]octane (●)
R   Hydrobromide (24815-25-6) [also hydrochloride
(26598-44-7)]
S   *Ethybenzatropine hydrobromide, Ethybenztropine
hydrobromide, Etybenzatropine hydrobromide**,
Etybenztropini hydrobromidum, Methylbenztropine
hydrobromide*, Panalide, Panolid, Ponalide, *Tropet-
hydrylin*, UK-738
U   Anticholinergic, antiparkinsonian

11224   (9031) $C_{22}H_{27}NO$
57734-69-7

α,α-Di-*o*-tolyl-3-quinuclidinemethanol = α,α-Bis(2-me-
thylphenyl)-1-azabicyclo[2.2.2]octane-3-methanol (●)
R   Hydrochloride (57734-70-0)
S   Bicarphene, Bikarfen, *Sequifenadine hydrochlori-
de***
U   Antihistaminic, antiserotonin

11225   (9030) $C_{22}H_{27}NO$
127-35-5

1,2,3,4,5,6-Hexahydro-6,11-dimethyl-3-phenethyl-2,6-
methano-3-benzazocin-8-ol = 2'-Hydroxy-5,9-dimethyl-
2-phenethyl-6,7-benzomorphan = 1,2,3,4,5,6-Hexahydro-
6,11-dimethyl-3-(2-phenylethyl)-2,6-methano-3-benz-
azocin-8-ol (●)
R   also hydrobromide (1239-04-9)
S   *Fenatsokin*, Narcidine, Narfen, Narphen, Narzocina,
NIH 7519, *Phenazocine***, *Phenethylazocinum*,
Phenobenzorphan, Prinadol, SKF 6574, Xenagol
U   Narcotic analgesic

11226   (9033)   C22H27NO2
                                                         14089-84-0

3-[(Dimethylamino)methyl]-1,2-diphenyl-3-buten-2-ol
propionate (ester) = 1-Benzyl-2-[(dimethylamino)me-
thyl]-1-phenylallyl propionate = α-[1-[(Dimethylamino)methyl]ethenyl]-α-phenylbenzeneethanol propanoate
(ester) (●)
S   Ba-40088, Ciba 40088, *Proxibutene***
U   Analgesic

11227   (9034)   C22H27NO2
                                                         47419-52-3

(+)-3-[(Dimethylamino)methyl]-1,2-diphenyl-3-buten-2-
ol propionate (ester) = (+)-1-Benzyl-2-[(dimethylamino)methyl]-1-phenylallyl propionate = (+)-α-[1-[(Dimethylamino)methyl]ethenyl]-α-phenylbenzeneethanol
propanoate (ester) (●)
S   44328, *Dexproxibutene***
U   Analgesic

11228   (9036)   C22H27NO2
                                                         90-69-7

2-[6-(β-Hydroxyphenethyl)-1-methyl-2-piperidyl]aceto-
phenone = 1-Methyl-2-(β-hydroxy-β-phenylethyl)-6-

phenacylpiperidine = [2S-[2α,6α(R*)]]-2-[6-(2-Hydroxy-
2-phenylethyl)-1-methyl-2-piperidinyl]-1-phenyletha-
none (●)
R   also sulfate (2:1) (134-64-5) or hydrochloride
    (134-63-4)
S   Antisol, Atmulatin, Bacco-Resist, Banico, Bantron,
    Cig-Ridettes, Citotal (active substance), Desista, Fu-
    marret, Habit-X, Inflatine, Lobatox, Lobelcon, *Lo-
    beline***, Lobesil, Lobeton, Lobidan, Lobidram, Lo-
    bishan, Lobron, NicErasc-IA (-SL), Nicotiless, Niko-
    ban, Nofum(o), No-kotin, PA-90, Smokeless, Stops-
    moke, Toban "USV", Unilobin, VG Smoke stop,
    Zoolobelin
U   Respiratory stimulant, smoking deterrent (alkaloid
    from *Lobelia inflata*)

11229   (9037)   C22H27NO2
                                                         57574-09-1

7-[(10,11-Dihydro-5H-dibenzo[a,d]cyclohepten-5-
yl)amino]heptanoic acid (●)
R   Hydrochloride (30272-08-3)
S   *Amineptine hydrochloride***, Directim, EU-1694,
    Maneon, Nealior, Neolior, Provector, S 1694, Sur-
    vector, Viaspera
U   Antidepressant

11230   (9039)   C22H27NO2
                                                         469-80-7

1-Phenethyl-4-phenyl-4-piperidinecarboxylic acid ethyl
ester = 1-(2-Phenylethyl)-4-carbethoxy-4-phenylpiperi-
dine = 4-Phenyl-1-(2-phenylethyl)-4-piperidinecarb-
oxylic acid ethyl ester (●)

S *Pheneridine**
U Narcotic analgesic

11231   (9038) $C_{22}H_{27}NO_2$
                                        15686-87-0

Ethyl α,α-diphenyl-2-piperidinepropionate = 2,2-Di-
phenyl-3-(2-piperidyl)propionic acid ethyl ester = α,α-
Diphenyl-2-piperidinepropanoic acid ethyl ester (●)
R Hydrochloride (60996-90-9)
S AGN 197, *Pifenate hydrochloride***
U Analgesic

11232   (9035) $C_{22}H_{27}NO_2$
                                        17230-88-5

17α-Pregna-2,4-dien-20-yno[2,3-*d*]isoxazol-17-ol =
(17α)-Pregna-2,4-dien-20-yno[2,3-*d*]isoxazol-17-ol (●)
S Anargil, Anason, Azol "Alphapharm", Baxal, Betri-
nat, Bonzol, Chronogyn, Cyclomen, Damodiol,
Danan, Danasin, Danatrol, Danazant, *Danazol**,
Danazolen, Danocrine, Danodiol, Danogar, Dano-
gen, Danokrin, Danol, Danoval, Dogalact, D-Zol,
Ectopal, Esdelart, Gonablok, Hosebon, Kinemon,
Ladazol, Ladogal "Winthrop", Ladogar, Lodogal,
Mastodanatrol, Norciden, Osylon, Sofunarin, Them,
Vabon, Win 17757, Winobanin
U Anterior pituitary suppressant

11233 $C_{22}H_{27}NO_2$

3-(2-Allylphenoxy)-1-[[(1*S*)-1,2,3,4-tetrahydro-1-naph-
thyl]amino]-2(*S*)-propanol = [*S*-(*R**,*R**)]-1-[2-(2-Prope-
nyl)phenoxy]-3-[(1,2,3,4-tetrahydro-1-naphthalenyl)ami-
no]-2-propanol (●)
R Hydrochloride (132017-03-9)
S SR 58894A
U β₃-Adrenergic receptor antagonist

11234   (9043) $C_{22}H_{27}NO_3$
                                        60996-91-0

2-(Diethylamino)ethyl *p*-methoxy-α-phenylcinnamate =
p-Methoxy-α-phenylcinnamic acid β-(diethylamino)ethyl ester = α-[(4-Methoxyphenyl)methylene]ben-
zeneacetic acid 2-(diethylamino)ethyl ester (●)
S *Anisocinnamol, Anisylcinnamat*, Valbine (active substance)
U Spasmolytic

11235   (9040) $C_{22}H_{27}NO_3$
                                        68876-74-4

(*E*)-3-[2-[2-(Diethylamino)ethoxy]phenoxy]-4-phenyl-3-
buten-2-one (●)

S *Zocainone**
U Anti-arrhythmic

11236 (9042) $C_{22}H_{27}NO_3$
467-86-7

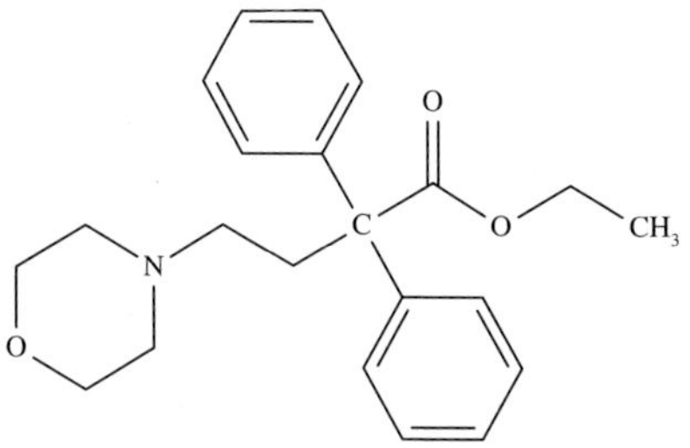

Ethyl 4-morpholino-2,2-diphenylbutyrate = 4-Morpholi-no-2,2-diphenylbutyric acid ethyl ester = α,α-Diphenyl-4-morpholinebutanoic acid ethyl ester (●)
S Amidalgon, *Diossafetile butirrato, Dioxaphetyl buty-rate**,* Spasmoxal
U Narcotic analgesic, spasmolytic

11237 $C_{22}H_{27}NO_3$
169191-56-4

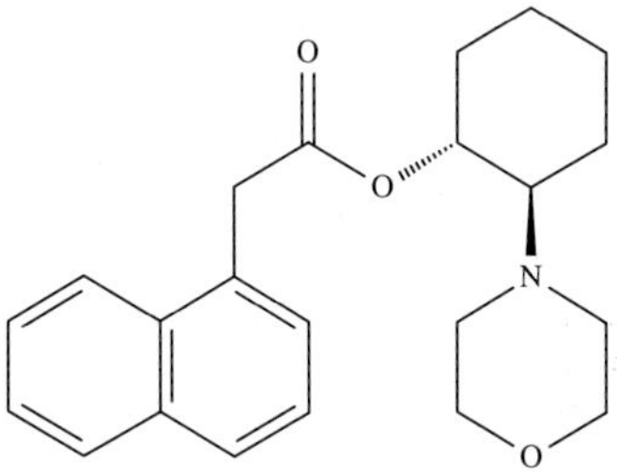

trans-[2-(4-Morpholinyl)cyclohexyl] naphthalene-1-ace-tate = 1-Naphthaleneacetic acid *rel*-(1*R*,2*R*)-2-(4-mor-pholinyl)cyclohexyl ester (●)
R Hydrochloride (169191-27-9)
S RSD 1000
U Anti-arrhythmic

11238 (9041) $C_{22}H_{27}NO_3$
546-32-7

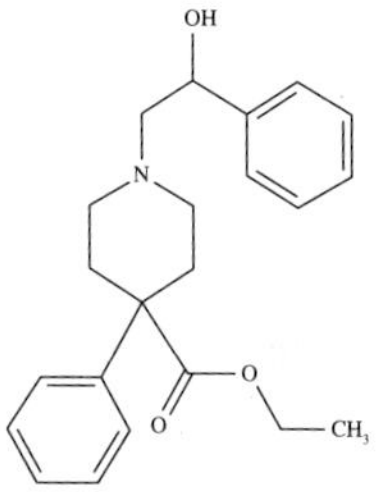

1-(β-Hydroxyphenethyl)-4-phenyl-4-piperidinecarb-oxylic acid ethyl ester = 1-(2-Hydroxy-2-phenylethyl)-4-carbethoxy-4-phenylpiperidine = Ethyl 1-(β-hydroxy-phenethyl)-4-phenylisonipecotate = 1-(2-Hydroxy-2-phe-nylethyl)-4-phenyl-4-piperidinecarboxylic acid ethyl es-ter (●)
S *Oxpheneridine**
U Narcotic analgesic

11239 (9044) $C_{22}H_{27}NO_4$
4129-86-6

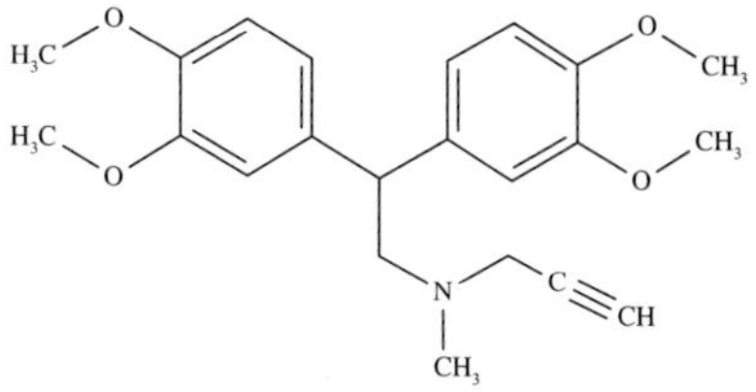

N-[2,2-Bis(3,4-dimethoxyphenyl)ethyl]-*N*-methyl-2-pro-pynylamine = 2,2-Bis(3,4-dimethoxyphenyl)-*N*-methyl-*N*-2-propynylethylamine = β-(3,4-Dimethoxyphenyl)-3,4-dimethoxy-*N*-methyl-*N*-2-propynylbenzeneethan-amine (●)
S BAEA
U Ganglion blocking agent, central depressant

11240 (9045) $C_{22}H_{27}NO_4$
119306-51-3

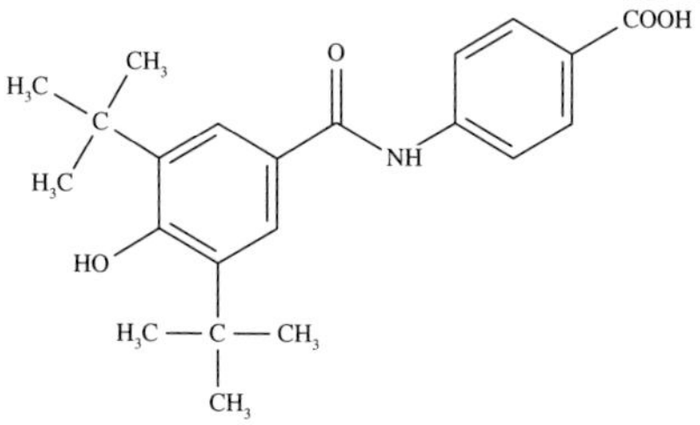

4-(3,5-Di-*tert*-butyl-4-hydroxybenzamido)benzoic acid = 4-[(3,5-Bis(1,1-dimethylethyl)-4-hydroxybenzoyl)amino]benzoic acid (●)
S   R 8605
U   Anti-allergic

11241 $C_{22}H_{27}NO_4$
                                     184877-64-3

4-Butyl-2,5-dimethyl-2-[(2,3,3a,8a-tetrahydro-3a-hydroxy-8*H*-furo[2,3-*b*]indol-8-yl)methyl]-4-cyclopentene-1,3-dione (●)
S   *Madindoline A*
U   Cytotoxic (inhibitor of IL-6 activity) from *Streptomyces* sp. K 93-0711

11242 $C_{22}H_{27}NO_5$
                                       221079-55-6

[4-[(1*Z*)-1-[2-[[(2*S*)-2-Hydroxy-3-phenoxypropyl]amino]ethyl]-1-propenyl]phenoxy]acetic acid (●)
R   Ethanedioate (1:1) (salt) (221079-56-7)
S   SWR-0342 SA
U $β_3$-Adrenoceptor agonist

11243   (9046) $C_{22}H_{27}NO_5$
                                       96609-38-0

1-(4-*tert*-Butylphenyl)-3-[4-(carboxymethylaminocarbonyl)phenoxy]-2-propanol = *N*-[4-[3-[4-(1,1-Dimethylethyl)phenyl]-2-hydroxypropoxy]benzoyl]glycine (●)
S   K 13004
U   Antihyperlipidemic

11244   (9047) $C_{22}H_{27}NO_6$
                                       60734-87-4

(±)-α-[(*tert*-Butylamino)methyl]-3,4-dihydroxybenzyl alcohol 3-acetate 4-*p*-anisate = Colterol 3-acetate 4-*p*-anisate = (±)-4-Methoxybenzoic acid 2-(acetyloxy)-4-[2-[(1,1-dimethylethyl)amino]-1-hydroxyethyl]phenyl ester (●)
R   Methanesulfonate (60734-88-5)
S   *Nisbuterol mesilate***, Win 34886
U   Bronchodilator

11245 $C_{22}H_{27}N_2O_5P$
                                       164083-84-5

[3-[[3-(2-Amino-2-oxoethyl)-2-ethyl-1-(phenylmethyl)-1*H*-indol-5-yl]oxy]propyl]phosphonic acid (●)

S LY 311727
U Phospholipase A₂ inhibitor

11246 (9048)   C₂₂H₂₇N₃
66842-87-3

8-Ethyl-2,3,4,5-tetrahydro-2-methyl-5-[2-(6-methyl-3-pyridinyl)ethyl]-1*H*-pyrido[4,3-*b*]indole (●)
S Emelin
U Antihistaminic

11247 (9049)   C₂₂H₂₇N₃OS
104076-38-2

2-[[3-[3-(Piperidinomethyl)phenoxy]propyl]amino]benzothiazole = *N*-[3-[3-(1-Piperidinylmethyl)phenoxy]propyl]-2-benzothiazolamine (●)
S SKF 95282, *Zolantidine*
U Histamine H₂-receptor antagonist

11248 (9050)   C₂₂H₂₇N₃O₂
21590-91-0

2-Methyl-3-(β-piperidino-*p*-phenetidino)phthalimidine = 2-Methyl-3-[*p*-(2-piperidinoethoxy)anilino]isoindolin-2-one = 2,3-Dihydro-2-methyl-3-[[4-[2-(1-piperidinyl)ethoxy]phenyl]amino]-1*H*-isoindol-1-one (●)
S Amiotren, K 2777, *Omidoline***, *Tamidoline*
U Antiparkinsonian

11249   C₂₂H₂₇N₃O₂
149649-22-9

N-[(1-Butyl-2-pyrrolidinyl)methyl]-4-cyano-1-methoxy-2-naphthalenecarboxamide (●)
R also (–)-compd. (173429-65-7)
S *Nafadotride*
U Antipsychotic (dopamine D₃ receptor antagonist)

11250 (9051)   C₂₂H₂₇N₃O₂
23465-76-1

1-[2-(Diethylamino)ethyl]-3-(*p*-methoxybenzyl)-2(1*H*)-quinoxalinone = 1-[2-(Diethylamino)ethyl]-3-[(4-methoxyphenyl)methyl]-2(1*H*)-quinoxalinone (●)
R also monohydrochloride (55750-05-5)
S Calmaverin, Calmavérine, *Caroverine**, Delirex, Espasmofibra, P 201-1, Spadon, Spasmium "Donau-Pharmazie", Tinnitin
U Spasmolytic

11251 (9052)   C₂₂H₂₇N₃O₂S

1-(2-Benzothiazolyl)-*N*-[2-(2-methoxyphenoxy)ethyl]-4-piperidinemethanamine (●)
R Dihydrochloride (137289-83-9)
S S 14063
U 5-HT₁ₐ-receptor antagonist

11252 $C_{22}H_{27}N_3O_3$

(2-Ethylphenyl)carbamic acid (4aS,9aS)-2,3,4,4a,9,9a-he-
xahydro-2,4a,9-trimethyl-1,2-oxazino[6,5-b]indol-6-yl
ester (●)
R Monohydrochloride (223585-99-7)
S CHF-2819, *Ganstigmine hydrochloride***
U Cognition enhancer (cholinesterase inhibitor)

11253 $C_{22}H_{27}N_3O_3$
                                                               171864-80-5

(3S,8aS)-Hexahydro-3-[[6-methoxy-2-(3-methyl-2-bute-
nyl)-1H-indol-3-yl]methyl]pyrrolo[1,2-a]pyrazine-1,4-
dione (●)
S TPS-A, *Tryprostatin A*
U Cytotoxic from *Aspergillus fumigatus*

11254 $C_{22}H_{27}N_3O_3S$
                                                               181935-23-9

3-[3-[N-[2-(3,4-Dimethoxyphenyl)ethyl]-N-methylami-
no]propoxy]-5-phenyl-1,2,4-thiadiazole = 3,4-Dimeth-
oxy-N-methyl-N-[3-[(5-phenyl-1,2,4-thiadiazol-3-
yl)oxy]propyl]benzeneethanamine (●)
R Monohydrochloride (181936-98-1)
S KC-12291
U Cardioprotective

11255   (9053) $C_{22}H_{27}N_3O_3S$
                                                               33414-33-4

Ethyl 10-[3-(diethylamino)propionyl]phenothiazine-2-
carbamate = 10-(N,N-Diethyl-β-alanyl)phenothiazine-2-
carbamic acid ethyl ester = [10-[3-(Diethylamino)-1-oxo-
propyl]-10H-phenothiazin-2-yl]carbamic acid ethyl ester
(●)
R also monohydrochloride (57530-40-2)
S *Aethacizin*, DAAE, *Etacicin, Etacizin*, ETHA, *Etha-
       cizine, Ethacyzin*, Ethmozine DAA, EZ 55, NIK 244
U Anti-arrhythmic

11255-01   211984-98-4
R Monohydrochloride of the mixture with Moracizine
       (no. 11141)
S *Methacizine*
U Anti-arrhythmic

11256 $C_{22}H_{27}N_3O_3S$
                                                               155990-20-8

N-[[1-[[2-(Diethylamino)ethyl]amino]-7-methoxy-9-
oxo-9H-thioxanthen-4-yl]methyl]formamide (●)
S BCN 326862, SR 271425, SW 71425, Win 71425
U Antineoplastic

11257　(9054)　　　　　　　　　　　　$C_{22}H_{27}N_3O_3S_2$
14008-44-7

1-[3-[2-(Methylsulfonyl)-10-phenothiazinyl]propyl]iso-
nipecotamide = 10-[3-(4-Carbamoylpiperidino)propyl]-2-
mesylphenothiazine = 1-[3-[2-(Methylsulfonyl)-10H-
phenothiazin-10-yl]propyl]-4-piperidinecarboxamide (●)
S　EXP-999, Metopimazine**, Nortrip "Rhodia", Pima-
　　zin, 9965 R.P., S 2300, 2167 TH, Vogalen, Vogalène
U　Anti-emetic

11258　(9055)　　　　　　　　　　　　$C_{22}H_{27}N_3O_4$
101-08-6

3-Piperidino-1,2-propanediol dicarbanilate (ester) = 3-
Piperidinopropylene bis(phenylcarbamate) = 3-(1-Piper-
idinyl)-1,2-propanediol bis(phenylcarbamate) (ester) (●)
R　Monohydrochloride (537-12-2) [or monohydrate
　　(51552-99-9)]
S　Diothan(e), Diothoid, Diperocainum, Diperodon hy-
　　drochloride**, Proctodon
U　Local anesthetic

11259　(9056)　　　　　　　　　　　　$C_{22}H_{27}N_3O_4S$
42110-58-7

2-(4-Methylpiperidino)ethyl 6-ethyl-2,3,6,9-tetrahydro-
3-methyl-2,9-dioxothiazolo[5,4-f]quinoline-8-carb-
oxylate = 6-Ethyl-2,3,6,9-tetrahydro-2,9-dioxothiazo-
lo[5,4-f]quinoline-8-carboxylic acid 2-(4-methyl-1-pipe-
ridinyl)ethyl ester (●)
R　Monohydrochloride (55236-13-0)
S　DB 2563, Metioxate hydrochloride**
U　Antibacterial

11260　(9057)　　　　　　　　　　　　$C_{22}H_{27}N_3O_5$
30751-23-6

N-[1-(4-Aminobenzoyl)-4-piperidinyl]-1,4,5-trimethoxy-
benzamide (●)
S　KU-55
U　Anti-ulcer agent

11261　(9058)　　　　　　　　　　　　$C_{22}H_{27}N_3O_5S$
32797-92-5

1-Cyclopentyl-3-[[p-[2-(o-anisamido)ethyl]phenyl]sulfo-
nyl]urea = N-[2-[4-[[[(Cyclopentylamino)carbonyl]ami-
no]sulfonyl]phenyl]ethyl]-2-methoxybenzamide (●)
S　Glipentide, Glisentide**, Glypentide, Staticum,
　　UR-661
U　Oral hypoglycemic

11262　(9059)　　　　　　　　　　　　$C_{22}H_{27}N_3O_6S$
63836-75-9

Hydroxymethyl (6R,7R)-7-[(R)-2-amino-2-phenylacet-amido]-3-methyl-8-oxo-5-thia-1-azabicyclo[4.2.0]oct-2-ene-2-carboxylate pivalate (ester) = (Pivaloyloxy)methyl (7R)-3-methyl-7-(α-D-phenylglycylamino)-3-cephem-4-carboxylate = [6R-[6α, 7β(R*)]]-7-[(Aminophenylace-tyl)amino]-3-methyl-8-oxo-5-thia-1-azabicy-clo[4.2.0]oct-2-ene-2-carboxylic acid (2,2-dimethyl-1-oxopropoxy)methyl ester (●)

R Monohydrochloride (27726-31-4)
S Ausocef, Bencef, Cefalen "Lenza", Cefalex "von Boch", *Cefalexin pivoxil hydrochloride*, Pivacef, Pivalex, Pivalexin, *Pivcefalexin hydrochloride*, Sigmacef, ST 21
U Antibiotic

11263 (9060) $C_{22}H_{27}N_5$
65222-35-7

10-[[3-(Diethylamino)propyl]amino]-6-methyl-5H-pyri-do[3',4':4,5]pyrrolo[2,3-g]isoquinoline = N,N-Diethyl-N'-(6-methyl-5H-pyrido[3',4':4,5]pyrrolo[2,3-g]isoquinolin-10-yl)-1,3-propanediamine (●)

S BD 40, NSC-327471 D, *Pazelliptine**, SR-95255 A
U Antineoplastic

11264 (9061) $C_{22}H_{27}N_5O$
151526-82-8

(R)-N-[Hexahydro-1-methyl-4-[(3-methylphenyl)me-thyl]-1H-1,4-diazepin-6-yl]-1H-indazole-3-carboxamide (●)

R Dihydrochloride (141034-42-6)
S DAT-582
U Serotonin S_3-receptor antagonist

11265 $C_{22}H_{27}N_5O$
162641-16-9

4-(Diphenylmethoxy)-1-[3-(1H-tetrazol-5-yl)propyl]pi-peridine (●)

S HQL 79
U Anti-allergic

11266 (9062) $C_{22}H_{27}N_5O$
7125-67-9

3,5-Dimethyl-N-(4,6,6a,7,8,9,10,10a-octahydro-4,7-di-methylindolo[4,3-fg]quinolin-9-yl)-1H-pyrazole-1-carb-oxamide = 4,7-Dimethyl-9-(3,5-dimethylpyrazole-1-carboxamido)-4,6,6a,7,8,9,10,10a-octahydroindo-lo[4,3-fg]quinoline = N-[(8β)-1,6-Dimethylergolin-8-yl]-3,5-dimethyl-1H-pyrazole-1-carboxamide (●)

S Lilly 42406, *Metoquizine**
U Anticholinergic, anti-ulcer agent

11267 $C_{22}H_{27}N_5OS$
155289-31-9

2-Amino-N-[4-[4-(1,2-benzisothiazol-3-yl)-1-piperazi-nyl]butyl]benzamide (●)

S 1192U90
U Antipsychotic

11268 (9063)

$C_{22}H_{27}N_5O_4$
88303-60-0

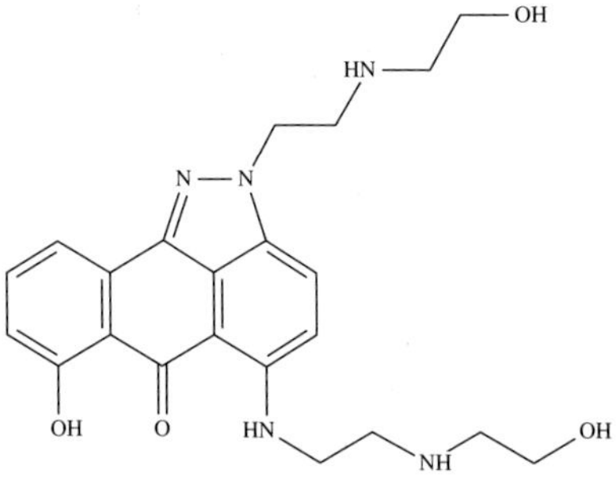

7-Hydroxy-2-[2-[(2-hydroxyethyl)amino]ethyl]-5-[[2-[(2-hydroxyethyl)amino]ethyl]amino]dibenzo[cd,g]indazol-6(2H)-one = 7-Hydroxy-2-[2-[(2-hydroxyethyl)amino]ethyl]-5-[[2-[(2-hydroxyethyl)amino]ethyl]amino]anthra[1,9-cd]pyrazol-6(2H)-one (●)

R Dihydrochloride (88303-61-1) or dihydrochloride monohydrate (132937-89-4)

S "Anthrapyrazolone", Biantrazole hydrochloride, CI-941, DuP 941, Losoxantrone hydrochloride**, NSC-357885, PD 113785

U Antineoplastic

11269

$C_{22}H_{27}N_5O_5$
198150-51-5

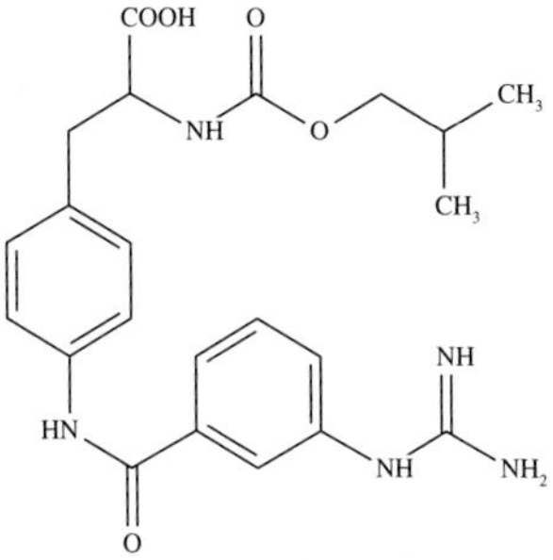

4-[[3-[(Aminoiminomethyl)amino]benzoyl]amino]-N-[(2-methylpropoxy)carbonyl]phenylalanine (●)

S SC-69000

U Anti-atherosclerotic (αV β3 antagonist)

11270

$C_{22}H_{27}N_5O_6$

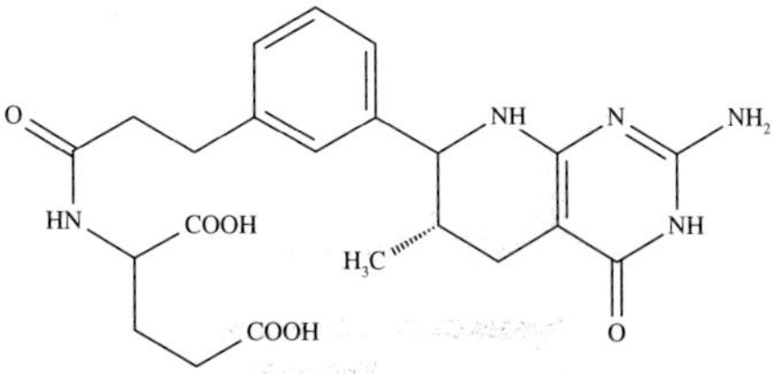

N-[3-[3-[2-Amino-3,4,5,6,7,8-hexahydro-6(S)-methyl-4-oxopyrido[2,3-d]pyrimidin-7-yl]phenyl]propionyl]-DL-glutamic acid

S AG-2024

U Antineoplastic (GARFT inhibitor)

11271

$C_{22}H_{27}N_5O_6S$
169456-85-3

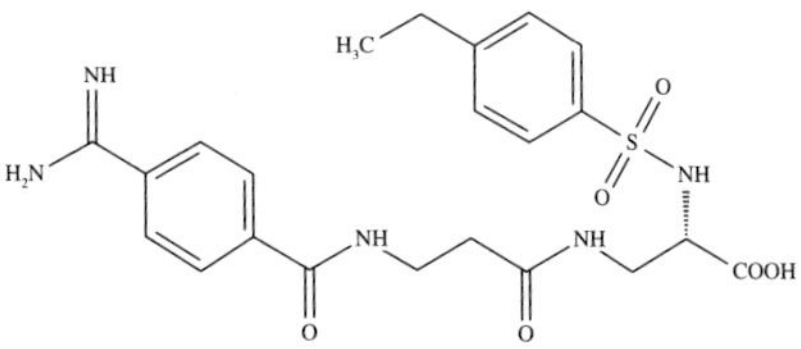

(2S)-N-[4-(Aminoiminomethyl)benzoyl]-β-alanyl-2-[[(4-ethylphenyl)sulfonyl]amino]-β-alanine (●)

S SM 20302

U GPIIb/IIIa receptor antagonist

11272 (9064)

$C_{22}H_{27}N_5S$
92928-47-7

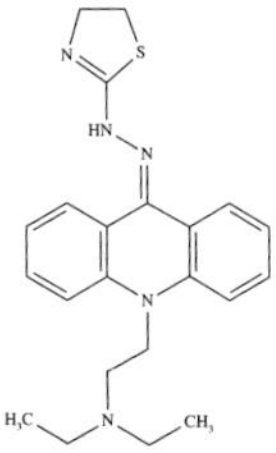

10-[2-(Diethylamino)ethyl]-9(10H)-acridinone (4,5-dihydro-2-thiazolyl)hydrazone (●)

S Ro 15-5458

U Antischistosomal

11273 (9065) $C_{22}H_{27}N_9O_4$
636-47-5

N''-(2-Amidinoethyl)-4-formamido-1,1',1''-trimethyl-
N,4':*N*',4''-ter[pyrrole-2-carboxamide] = β-[1-Methyl-4-
[1-methyl-4-[1-methyl-4-(formylamino)pyrrole-2-carb-
oxamido]pyrrole-2-carboxamido]pyrrole-2-carboxami-
do]propionamidine = *N*-[5-[[(3-Amino-3-iminopro-
pyl)amino]carbonyl]-1-methyl-1*H*-pyrrol-3-yl]-4-[[[4-
(formylamino)-1-methyl-1*H*-pyrrol-2-yl]carbonyl]ami-
no]-1-methyl-1*H*-pyrrole-2-carboxamide (●)
R also monohydrochloride (6576-51-8)
S *Distamycin A*, Dst-3, DST-A, F.I. 6426, Herperal,
*Stallimycin**
U Topical antiviral antibiotic from *Streptomyces distal-
licus*

11274 (9066) $C_{22}H_{27}N_9O_7S_2$
82547-81-7

(Pivaloyloxy)methyl (+)-(6*R*,7*R*)-7-[2-(2-Amino-4-thia-
zolyl)glyoxylamido]-3-[(5-methyl-2*H*-tetrazol-2-yl)me-
thyl]-8-oxo-5-thia-1-azabicyclo[4.2.0]oct-2-ene-2-carb-
oxylate 7^2-(*Z*)-(*O*-methyloxime) = [6*R*-[6α,7β(*Z*)]]-7-
[[(2-Amino-4-thiazolyl)(methoxyimino)acetyl]amino]-3-
[(5-methyl-2*H*-tetrazol-2-yl)methyl]-8-oxo-5-thia-1-aza-
bicyclo[4.2.0]oct-2-ene-2-carboxylic acid (2,2-dimethyl-
1-oxopropoxy)methyl ester (●)
S *Cefpivtetram*, *Cefteram pivoxil*, CFTM-PI,
Ro 19-5248, T-2588, Tomiron
U Antibiotic

11275 (9067) $C_{22}H_{28}ClFO_4$
4828-27-7

9-Chloro-6α-fluoro-11β,21-dihydroxy-16α-methylpreg-
na-1,4-diene-3,20-dione = (6α,11β,16α)-9-Chloro-6-
fluoro-11,21-dihydroxy-16-methylpregna-1,4-diene-
3,20-dione (●)
R see also nos. 12412, 13568, 13899
S *Clocortolone***
U Glucocorticoid (anti-inflammatory, anti-allergic)

11276 (9068) $C_{22}H_{28}ClFO_4$
25122-41-2

21-Chloro-9-fluoro-11β,17-dihydroxy-16β-methylpreg-
na-1,4-diene 3,20-dione = (11β,16β)-21-Chloro-9-fluoro-
11,17-dihydroxy-16-methylpregna-1,4-diene-3,20-dione
(●)
R see also no. 12862
S *Clobetasol***
U Glucocorticoid (anti-inflammatory, anti-allergic)

11277 (9069) $C_{22}H_{28}ClNO$
64294-95-7

1-[2-[(*p*-Chloro-α-methyl-α-phenylbenzyl)oxy]ethyl]he-
xahydro-1*H*-azepine = 1-[2-[1-(4-Chlorophenyl)-1-phe-
nylethoxy]ethyl]hexahydro-1*H*-azepine (●)
R Hydrochloride (59767-13-4)
S Egis-2062, Egyt-2062, Loderix, Loridex, *Setastine*
*hydrochloride***
U Antihistaminic

11278 (9070) $C_{22}H_{28}ClNO_3$
5649-10-5

2-(4-Chlorobutoxy)-2,2-diphenylacetic acid 2-(dimethyl-
amino)ethyl ester = α-(4-Chlorobutoxy)-α-phenylben-
zeneacetic acid 2-(dimethylamino)ethyl ester (●)
R Hydrochloride (1446-44-2)
S Satal
U Antihyperlipidemic, spasmolytic

11279 $C_{22}H_{28}ClNO_3$
139485-39-5

(*S*)-6-Chloro-1-[(2,5-dimethoxy-4-propylphenyl)me-
thyl]-1,2,3,4-tetrahydro-2-methyl-7-isoquinolinol (●)

S BW 737C, BW 737C89
U Antipsychotic (dopamine D_1 receptor antagonist)

11280 $C_{22}H_{28}ClN_5O_3S$
213453-89-5

(1*S*,2*R*,3*S*,4*R*)-4-[4-[[(1*R*)-1-(3-Chloro-2-thienyl)me-
thyl]propyl]amino]-7*H*-pyrrolo[2,3-*d*]pyrimidin-7-yl]-*N*-
ethyl-2,3-dihydroxycyclopentanecarboxamide (●)
S AMP 579, RPR 100579
U Cardioprotectant (adenosine A_1/A_2 receptor agonist)

11281 $C_{22}H_{28}Cl_2N_2$
1170-60-1

trans-1,4-Bis(2-chlorobenzylaminomethyl)cyclohexane
= *trans*-*N*,*N*'-Bis[(2-chlorophenyl)methyl]-1,4-cyclohe-
xanedimethanamine (●)
R Dihydrochloride (366-93-8)
S AY-9944
U Sterol synthesis inhibitor, antiviral

11282 $C_{22}H_{28}Cl_2N_2O_6$
132194-66-2

(4*S*)-2-[[2-(2-Aminoethoxy)ethoxy]methyl]-4-(2,3-di-
chlorophenyl)-1,4-dihydro-6-methyl-3,5-pyridinedicarb-
oxylic acid 3-ethyl 5-methyl ester (●)
S S 12968

2299

U   Calcium antagonist

11283   (9071)

$C_{22}H_{28}Cl_2N_2O_6$
115972-78-6

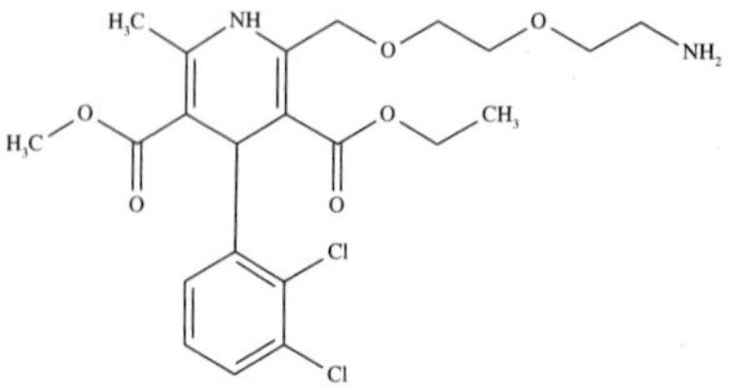

3-Ethyl 5-methyl (±)-2-[[2-(2-aminoethoxy)ethoxy]me-
thyl]-4-(2,3-dichlorophenyl)-1,4-dihydro-6-methyl-3,5-
pyridinedicarboxylate = (±)-2-[[2-(2-Aminoethoxy)eth-
oxy]methyl]-4-(2,3-dichlorophenyl)-1,4-dihydro-6-me-
thyl-3,5-pyridinedicarboxylic acid 3-ethyl 5-methyl ester
(●)
R   Monohydrochloride (132636-01-2)
S   *Olradipine hydrochloride**, S 11568
U   Smooth muscle relaxant, vasodilator (calcium antagonist)

11284   (9072)

$C_{22}H_{28}Cl_2O_4$
4732-48-3

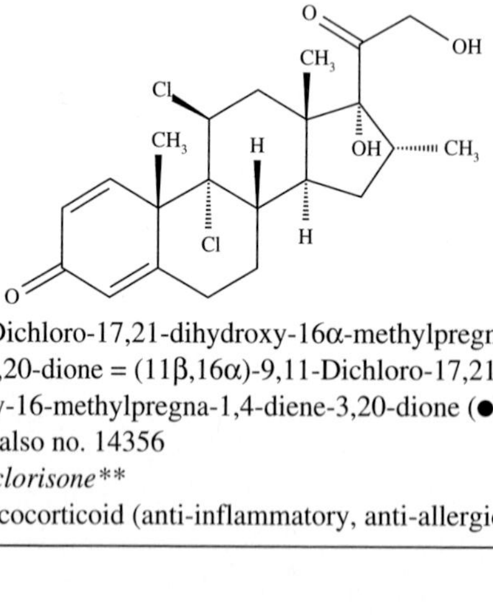

9,11β-Dichloro-17,21-dihydroxy-16α-methylpregna-1,4-
diene-3,20-dione = (11β,16α)-9,11-Dichloro-17,21-di-
hydroxy-16-methylpregna-1,4-diene-3,20-dione (●)
R   see also no. 14356
S   *Meclorisone***
U   Glucocorticoid (anti-inflammatory, anti-allergic)

11285   (9073)

$C_{22}H_{28}Cl_2O_4$
105102-22-5

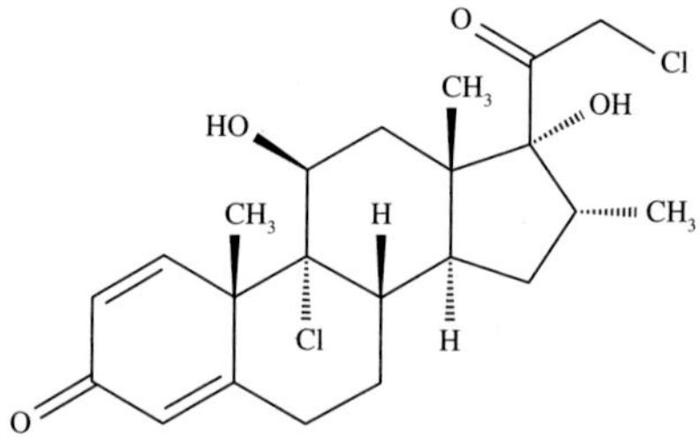

9,21-Dichloro-11β,17-dihydroxy-16α-methylpregna-1,4-
diene-3,20-dione = (11β,16α)-9,21-Dichloro-11,17-di-
hydroxy-16-methylpregna-1,4-diene-3,20-dione (●)
R   see also no. 13467
S   *Mometasone***
U   Glucocorticoid (topical anti-inflammatory, anti-allergic)

11286

$C_{22}H_{28}FNO_3$
139290-65-6

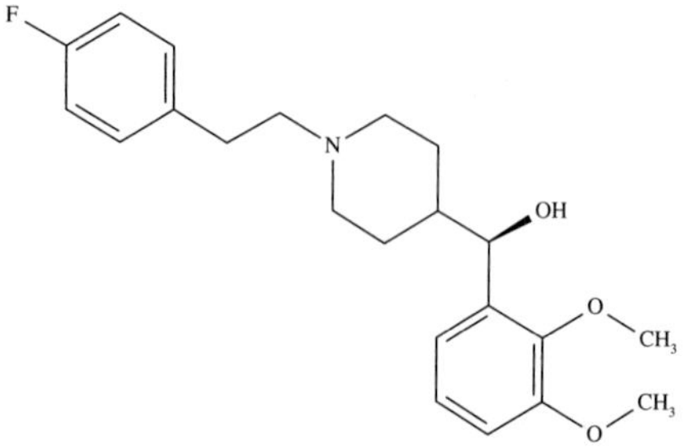

(αR)-α-(2,3-Dimethoxyphenyl)-1-[2-(4-fluorophe-
nyl)ethyl]-4-piperidinemethanol (●)
S   M 100907, MDL 100907
U   Antipsychotic (5-HT$_2$ receptor antagonist)

11287   (9074)

$C_{22}H_{28}FN_3O_3$
80428-29-1

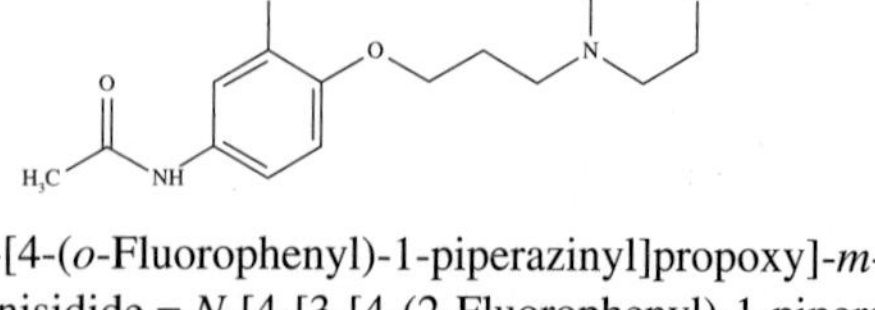

4'-[3-[4-(*o*-Fluorophenyl)-1-piperazinyl]propoxy]-*m*-
acetanisidide = *N*-[4-[3-[4-(2-Fluorophenyl)-1-piperazi-
nyl]propoxy]-3-methoxyphenyl]acetamide (●)

R also monomethanesulfonate (80428-31-5)
S IK-640, *Mafoprazine**, Mafroperone*
U Neuroleptic (for pigs)

11288 $C_{22}H_{28}FN_3O_6S$
287714-41-4

(+)-7-[4-(4-Fluorophenyl)-6-(1-methylethyl)-2-[me-
thyl(methylsulfonyl)amino]-5-pyrimidinyl]-3(R),5(S)-di-
hydroxy-6(E)-heptenoic acid = (3R,5S,6E)-7-[4-(p-
Fluorophenyl)-6-isopropyl-2-(N-methylmethanesulfon-
amido)-5-pyrimidinyl]-3,5-dihydroxy-6-heptenoic acid =
[S-[R*,S*-(E)]]-7-[4-(4-Fluorophenyl)-6-(1-methyl-
ethyl)-2-[methyl(methylsulfonyl)amino]-5-pyrimidinyl]-
3,5-dihydroxy-6-heptenoic acid (●)
R Calcium salt (2:1) (147098-20-2)
S Crestor, *Rosuvastatin calcium**, S 4522, ZD 4522
U Antihyperlipidemic (HMGCoA reductase inhibitor)

11289 (9075) $C_{22}H_{28}F_2O_4$
2607-06-9

6α,9-Difluoro-11β,21-dihydroxy-16α-methylpregna-1,4-
diene-3,20-dione = (6α,11β,16α)-6,9-Difluoro-11,21-di-
hydroxy-16-methylpregna-1,4-diene-3,20-dione (●)
R see also nos. 13569, 13570
S *Diflucortolone***
U Glucocorticoid (anti-inflammatory, anti-allergic)

11290 (9076) $C_{22}H_{28}F_2O_4S$
74131-77-4

S-Methyl 6α,9-difluoro-11β,17-dihydroxy-16α-methyl-
3-oxoandrosta-1,4-diene-17β-carbothioate =
(6α,11β,16α,17α)-6,9-Difluoro-11,17-dihydroxy-16-me-
thyl-3-oxoandrosta-1,4-diene-17-carbothioic acid S-methyl ester (●)
R see also no. 12867
S *Ticabesone***
U Glucocorticoid (anti-inflammatory, anti-allergic)

11291 (9077) $C_{22}H_{28}F_2O_5$
2135-17-3

6α,9-Difluoro-11β,17,21-trihydroxy-16α-methylpregna-
1,4-diene-3,20-dione = (6α,11β,16α)-6,9-Difluoro-
11,17,21-trihydroxy-16-methylpregna-1,4-diene-3,20-
dione (●)
R see also nos. 12415, 13571
S Acutol "Byk", Anaprime, Aniprime, Cortexilar, Cor-
tival, Flucort "Syntex", Flucorticin, Flucortin, *Flu-
metasone**, Flumethasone*, Flumilar, *Fluordexame-
thasone*, Fluvet "Gellini; Syntex", Mathagon, Metha-
gon, NSC-54702, RS-2177, U-10974
U Glucocorticoid (anti-inflammatory, anti-allergic)

11292 (9078) $C_{22}H_{28}F_2O_5$
2557-49-5

6α,9-Difluoro-11β,17,21-trihydroxy-16β-methylpregna-1,4-diene-3,20-dione = (6α,11β,16β)-6,9-Difluoro-11,17,21-trihydroxy-16-methylpregna-1,4-diene-3,20-dione (●)
R see also no. 13185
S Diflorasone**
U Glucocorticoid (anti-inflammatory, anti-allergic)

11293 (9079) $C_{22}H_{28}GdN_3O_{11}$
113662-23-0

Dihydrogen (±)-[4-carboxy-5,8,11-tris(carboxymethyl)-1-phenyl-2-oxa-5,8,11-triazatridecan-13-oato(5–)]gadolinate(2–) = Dihydrogen [4-carboxy-5,8,11-tris(carboxymethyl)-1-phenyl-2-oxa-5,8,11-triazatridecan-13-oato-(5–)-$N^5,N^8,N^{11},O^4,O^5,O^8,O^{11},O^{13}$]gadolinate(2–) (●)
S Acidum gadobenicum**, B 19036, Gadobenic acid**, GD-BOPTA
U Diagnostic aid (paramagnetic contrast medium)

11293-01 (9079-01) 127000-20-8
R Dimeglumine salt
S B 19036/7, Dimeglumine gadobenate, E-7155, Gadobenate dimeglumine, Gd-BOPTA/Dimeg, MultiHance
U Diagnostic aid (paramagnetic contrast medium)

11294 (9080) $C_{22}H_{28}N$
10236-81-4

3-Benzhydrylidene-1,1-diethyl-2-methylpyrrolidinium = 3-(Diphenylmethylene)-1,1-diethyl-2-methylpyrrolidinium (●)
R Bromide (4630-95-9)
S Anespas F, Dornal, Kosmium, No-Spasm, Padrin, PDB, Prifinial, Prifini bromidum, Prifinium bromide**, Pyrodifenium bromide, Riabal, Spanomid
U Anticholinergic, spasmolytic

11295 (9081) $[C_{22}H_{28}N]_x$

Polystyrenetrimethylammonium
R Chloride as structural type for a styrene-divinylbenzene copolymer (about 2% divinylbenzene) containing quaternary ammonium groups (11041-12-6); see also no. 4425
S Cholemin "Berk", Cholespor, Cholestyramine (Resin), Chol-less, Cholybar, Colemen, Colesthexal, Colestran, Colestrol, Colestyramine**, Colestyr von ct, Cuemid, Divistyramine, Dowex 1-X2-Cl, Duolite AP-143, Efensol, Estramin, Filicol, Holestan, Ipocol, Kolestran, Lipocol-Merz, Lismol, Lochobest, MK-325, Novo-Cholamine, PMS-Cholestyramine, Prevalite, Quantalan, Questran, Resin-colestiramina, Sevit, Syn-Cholestyramine, Vasosan P/-S

U Antihyperlipidemic, antipruritic (in case of partial biliary obstruction)

11296 (9082)

C$_{22}$H$_{28}$NO$_2$
172343-54-3

1-[(2,2-Diphenyl-1,3-dioxolan-4-yl)methyl]-1-methylpiperidinium (●)
R Iodide (21216-78-4)
S Anacolin
U Anticholinergic

11297 (9084)

C$_{22}$H$_{28}$NO$_3$
13473-38-6

1-Ethyl-3-hydroxy-1-methylpiperidinium benzilate = 3-Benziloyloxy)-1-ethyl-1-methylpiperidinium = 1-Ethyl-3-[(hydroxydiphenylacetyl)oxy]-1-methylpiperidinium (●)
R Bromide (125-51-9)
S Apren-Pediatr., Bionzolato, Brontil-Compr., Bropen-Tabl., Bropenzol, Bropidol, Contral, Dropenzil, Endolit, Espazylen, Fincol, Ila-Med M, JB-323, Nagin, Penzofen, Pipenfolin, *Pipensolon*, Pipenzol, *Pipenzolate bromide**, Pipenzolon bromide*, Piper (Anticolinergico), Piptal, Piptalake, Piptalin, Piptem, Profartyl, Sertal "North Medic.", Tisafren
U Anticholinergic, spasmolytic

11298 (9085)

C$_{22}$H$_{28}$NO$_3$
33371-53-8

2-(Hydroxymethyl)-1,1-dimethylpiperidinium benzilate = 2-[(Benziloyloxy)methyl]-1,1-dimethylpiperidinium = 2-[[(Hydroxydiphenylacetyl)oxy]methyl]-1,1-dimethylpiperidinium (●)
R Methyl sulfate (5205-82-3)
S Acabel, *Bevonium metilsulfate***, CG 201, Confielle, Dalys, *Piribenzyl methylsulfate*, Spalgo
U Anticholinergic, spasmolytic

11299 (9083)

C$_{22}$H$_{28}$NO$_3$
16175-92-1

1,1-Diethyl-3-hydroxypyrrolidinium benzilate = 3-(Benziloyloxy)-1,1-diethylpyrrolidinium = 1,1-Diethyl-3-[(hydroxydiphenylacetyl)oxy]pyrrolidinium (●)
R Bromide (1050-48-2)
S *Benzilonium bromide***, CI-379, CN-20172-3, Minelcin, Minelco, Minelsin, Ortyn, Partyn, Pirbenina, Portyn, PU 239, Pyrbenine, Ulcoban
U Anticholinergic

11300   (9086) $C_{22}H_{28}N_2$
15686-38-1

14-(Cyclopropylmethyl)-1,2,3,4,4a,5,6,11-octahydro-5,11b-(iminoethano)-11b*H*-benzo[*a*]carbazole (●)
S   *Carbazocine***, W 4500
U   Analgesic

11301   (9088) $C_{22}H_{28}N_2O$
3691-21-2

4-(1-Perhydroazepinyl)-2,2-diphenylbutyramide = 2,2-Diphenyl-4-(hexamethyleneimino)butyramide = 1-(3-Carbamoyl-3,3-diphenylpropyl)hexahydroazepine = Hexahydro-α,α-diphenyl-1*H*-azepine-1-butanamide (●)
S   Acemydrite, *Buzepidonum*, *Buzepidum*, Calmacid, *Hexamethamide*, Mydriamide, Pervioral, R 658
U   Anticholinergic, mydriatic

11302 $C_{22}H_{28}N_2O$
159059-34-4

(3*R*,6*R*,7*R*)-*rel*-3-(2-Methoxyphenyl)-7-phenyl-1,8-diazaspiro[5.5]undecane (●)
S   CP-210053
U   Substance P antagonist

11303   (9087) $C_{22}H_{28}N_2O$
437-38-7

N-(1-Phenethyl-4-piperidyl)propionanilide = 1-Phenethyl-4-(*N*-propionylanilino)piperidine = *N*-Phenyl-*N*-[1-(2-phenylethyl)-4-piperidinyl]propanamide (●)
R   most citrate (1:1) (990-73-8)
S   AB-Fentanyl, Actiq, Beatryl, Duragesic, Durogesic, Fentamed, Fentanest, Fentanil, *Fentanil citrato*, Fentanilo Fabra, *Fentanyl citrate***, Fentanylix, Haldid, Hefanil, Leptanal, Leskin, McN-JR 4263, Oralet, *Phentanyl citrate*, R 4263, R 5240, Sentonyl, Sintenyl, Stobium, Sublimaze, Tanyl, Trofentyl
U   Narcotic analgesic

11304   (9089) $C_{22}H_{28}N_2OS_2$
29604-16-8

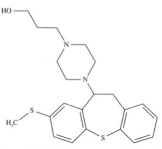

2-(Methylthio)-11-[4-(3-hydroxypropyl)-1-piperazinyl]-10,11-dihydrodibenzo[*b,f*]thiepin = 4-[10,11-Dihydro-8-(methylthio)dibenzo[*b,f*]thiepin-10-yl]-1-piperazinepropanol (●)
R   see also no. 14678;
Monomethanesulfonate (34775-83-2)
S   Meclopin (oral), *Oxyprothepin*, VÚFB-8334
U   Neuroleptic

11305   (9090) $C_{22}H_{28}N_2O_2$
37612-13-8

(±)-2'-[2-(1-Methyl-2-piperidyl)ethyl]-*p*-anisanilide =
(±)-4-Methoxy-2'-[2-(1-methyl-2-piperidyl)ethyl]benz-
anilide = (±)-4-Methoxy-*N*-[2-[2-(1-methyl-2-piperidi-
nyl)ethyl]phenyl]benzamide (●)
R   also without (±)-definition; monohydrochloride
    (115904-25-1)
S   Encade, Encaid, *Encainide hydrochloride**, Enka-
    de, Enkaid, MJ 9067-1
U   Anti-arrhythmic

11306 $C_{22}H_{28}N_2O_2$
170858-41-0

3,4-Dihydro-1-[2-[4-(4-methoxyphenyl)-1-piperazi-
nyl]ethyl]-2-benzopyran = 1-[2-(3,4-Dihydro-1*H*-2-ben-
zopyran-1-yl)ethyl]-4-(4-methoxyphenyl)piperazine (●)
S   PNU-96970
U   Neuroleptic (dopamine D_4 receptor antagonist)

11307   (9091) $C_{22}H_{28}N_2O_2$
144-14-9

Ethyl 1-(*p*-aminophenethyl)-4-phenylisonipecotate = 1-
p-Aminophenethyl)-4-phenyl-4-piperidinecarboxylic

acid ethyl ester = 1-[2-(4-Aminophenyl)ethyl]-4-phenyl-
4-piperidinecarboxylic acid ethyl ester (●)
R   also dihydrochloride (126-12-5) or phosphate (1:1)
    (4268-37-5)
S   Alidine, *Anileridine**, Apodol, Lerinol, Leritine,
    MK-89, Nipecotan
U   Narcotic analgesic

11308   (9092) $C_{22}H_{28}N_2O_2$
104051-20-9

(±)-6-[2-[(1,1-Dimethyl-3-phenylpropyl)amino]-1-hydr-
oxyethyl]-3,4-dihydrocarbostyril = 6-[2-[(1,1-Dimethyl-
3-phenylpropyl)amino]-1-hydroxyethyl]-3,4-dihydro-
2(1*H*)-quinolinone (●)
S   BDF 8186, *Brefonalol**
U α_1- and β_1-adrenergic blocker

11309   (9093) $C_{22}H_{28}N_2O_2S$
13093-88-4

1-[3-(2-Methoxy-10-phenothiazinyl)-2-methylpropyl]-4-
piperidinol = 10-[3-(4-Hydroxypiperidino)-2-methylpro-
pyl]-2-methoxyphenothiazine = 1-[3-(2-Methoxy-10*H*-
phenothiazin-10-yl)-2-methylpropyl]-4-piperidinol (●)
S   1317 A.N., Leptryl, *Perimetazine**, 9159 R.P.
U   Neuroleptic, anesthetic (premedication)

11310 (9094) $C_{22}H_{28}N_2O_2S$
15351-04-9

1-[[2-[Ethyl(2-hydroxy-2-methylpropyl)ami-
no]ethyl]amino]-4-methylthioxanthen-9-one = 1-[[2-
[Ethyl(2-hydroxy-2-methylpropyl)amino]ethyl]amino]-
4-methyl-9*H*-thioxanthen-9-one (●)
R also monohydrochloride (5591-22-0)
S *Becanthone*, *Becantone**, Loranil "Winthrop",
NSC-15796, Win 13820
U Antischistosomal

11311 (9095) $C_{22}H_{28}N_2O_2S_2$
2620-88-4

2,2'-Dicarboxydiphenyldisulfide bis-*n*-butylamide = 2,2'-
Dithiobis[*N*-butylbenzamide] (●)
S Azobicina, *Disulfbumide*, O.D. 507
U Topical antiseptic

11312 (9096) $C_{22}H_{28}N_2O_3$
81329-71-7

(±)-2'-[2-(1-Methyl-2-piperidyl)ethyl]vanillanilide = (±)-
4-Hydroxy-3-methoxy-*N*-[2-[2-(1-methyl-2-piperidi-
nyl)ethyl]phenyl]benzamide (●)

R also without (±)-definition (82522-70-1)
S BMY-40327, MJ 14030, MODE, *Modecainide***
U Anti-arrhythmic

11313 (9097) $C_{22}H_{28}N_2O_3$
68616-83-1

7,8-Didehydro-4,5α-epoxy-3-hydroxy-17-methyl-14-
(pentylamino)morphinan-6-one = (5α)-7,8-Didehydro-
4,5-epoxy-3-hydroxy-17-methyl-14-(pentylamino)mor-
phinan-6-one (●)
S A-4492, *Pentamorphone***, RX 77989
U Narcotic analgesic

11314 (9098) $C_{22}H_{28}N_2O_3$
34675-77-9

4'-[2-Hydroxy-3-(4-phenyl-1-piperazinyl)propoxy]pro-
piophenone = 1-(*p*-Propionylphenoxy)-3-(4-phenyl-1-pi-
perazinyl)-2-propanol = 1-[4-[2-Hydroxy-3-(4-phenyl-1-
piperazinyl)propoxy]phenyl]-1-propanone (●)
S *Centpropazine*, Compound 67/14
U Antidepressant

11315 $C_{22}H_{28}N_2O_3$
151951-24-5

endo-(8-Methyl-8-azabicyclo[3.2.1]oct-3-yl) 1,2-dihy-
dro-1-isobutyl-2-oxo-4-quinolinecarboxylate = *endo*-1,2-
Dihydro-1-(2-methylpropyl)-2-oxo-4-quinolinecarb-
oxylic acid 8-methyl-8-azabicyclo[3.2.1]oct-3-yl ester (●)
R Monohydrochloride (139094-48-7)
S KF 18259
U Anti-emetic (5-HT$_3$ receptor antagonist)

11316 (9099) $C_{22}H_{28}N_2O_3S$
93392-97-3

5-[*N*-(3,4-Dimethoxyphenethyl)-β-alanyl]-2,3,4,5-tetra-
hydro-1,5-benzothiazepine = 5-[3-[[2-(3,4-Dimethoxy-
phenyl)ethyl]amino]-1-oxopropyl]-2,3,4,5-tetrahydro-
1,5-benzothiazepine (●)
R Fumarate (1:1) (105394-80-7)
S KT-362
U Vasodilator, calcium antagonist

11317 (9100) $C_{22}H_{28}N_2O_3S$
140646-80-6

E)-2,2-Diethyl-4-[[3-[2-[4-(1-methylethyl)-2-thiazo-
yl]ethenyl]phenyl]amino]-4-oxobutanoic acid (●)
S MCI-826
U Anti-asthmatic

11318 (9101) $C_{22}H_{28}N_2O_3S$
60771-38-2

4-[(1-Oxopropyl)phenylamino]-1-[2-(2-thienyl)ethyl]-4-
piperidinecarboxylic acid methyl ester (●)
R Monohydrochloride (125144-20-9)
S R 31826, *Thiofentanil hydrochloride*
U Narcotic analgesic

11319 $C_{22}H_{28}N_2O_5S_2$
154189-40-9

4-Hydroxy-7-[2-[[2-[[3-(2-phenylethoxy)propyl]sulfo-
nyl]ethyl]amino]ethyl]-2(3*H*)-benzothiazolone (●)
R Monohydrochloride (154189-24-9)
S AR-C 68397AA, FPL 68397, *Sibenadet hydrochlori-
de**, Viozan
U Dual D$_2$ receptor and β$_2$-adrenoceptor agonist

11320 (9102) $C_{22}H_{28}N_4OS$
118196-10-4

1-(3-Phenylpropyl)-4-[[2-(3-pyridinyl)-4-thiazolidi-
nyl]carbonyl]piperazine (●)
R Fumarate (1:1) (118196-11-5)

S YM-461
U Anti-allergic (PAF antagonist)

11321 (9103) $C_{22}H_{28}N_4O_3$
911-65-9

1-[2-(Diethylamino)ethyl]-2-(*p*-ethoxybenzyl)-5-nitro-
benzimidazole = 2-[2-(4-Ethoxybenzyl)-5-nitro-1-benz-
imidazolyl]ethyldiethylamine = 2-[(4-Ethoxyphenyl)me-
thyl]-*N,N*-diethyl-5-nitro-1*H*-benzimidazole-1-ethan-
amine (●)
R also monohydrochloride (2053-25-0)
S *Aethonitazen*, Ba-20684, C. 20410, Ciba 20684, *Eto-
bedolum*, *Etonitazene***, *Etonitazinum*, NIH 7607
U Narcotic analgesic

11322 (9104) $C_{22}H_{28}N_4O_4$
64862-96-0

1,4-Bis[[2-[(2-hydroxyethyl)amino]ethyl]amino]anthra-
quinone = 1,4-Bis[[2-[(2-hydroxyethyl)amino]ethyl]ami-
no]-9,10-anthracenedione (●)
R Diacetate (70711-40-9)
S *Ametantrone acetate***, CI-881, HAQ, NSC-287513
U Antineoplastic

11323 $C_{22}H_{28}N_4O_4$
183745-58-6

(*S*)-3-[[[3,4-Dihydro-1-oxo-7-(1-piperazinyl)-2(1*H*)-iso-
quinolinyl]acetyl]amino]-4-pentynoic acid ethyl ester (●)
S L 767685
U Glycoprotein IIb/IIIa antagonist

11324 $C_{22}H_{28}N_4O_4$
141807-96-7

(*E*)-8-(3,4-Dimethoxystyryl)-7-methyl-1,3-dipropylxan-
thine = 8-[(1*E*)-2-(3,4-Dimethoxyphenyl)ethenyl]-3,7-di-
hydro-7-methyl-1,3-dipropyl-1*H*-purine-2,6-dione (●)
S KF 17837
U Adenosine A_2 receptor antagonist

11325 (9105) $C_{22}H_{28}N_4O_5$
6035-39-8

1-(Diethylamino)-3-[(2,3-dimethoxy-6-nitro-9-acridi-
nyl)amino]-2-propanol (●)
R also dihydrochloride (73972-50-6)
S L 280401, Nitroacridine 3582, Nitroakridin 3582,
W 1889
U Antibacterial, antiseptic

11326 (9106)

$C_{22}H_{28}N_4O_5S_2$
139047-48-6

N-[2-[[Methyl[3-[4-[(methylsulfonyl)amino]phenoxy]propyl]amino]methyl]-6-quinolinyl]methanesulfonamide (●)

S WAY-125971
U Anti-arrhythmic

11327 (9107)

$C_{22}H_{28}N_4O_6$
65271-80-9

1,4-Dihydroxy-5,8-bis[[2-[(2-hydroxyethyl)amino]ethyl]amino]anthraquinone = 1,4-Dihydroxy-5,8-bis[[2-[(2-hydroxyethyl)amino]ethyl]amino]-9,10-anthracenedione (●)

R Dihydrochloride (70476-82-3)
S CL 232315, DAD, DHAQ, Genefadrone, Metrisone "Beta", Misostol, *Mitoxantrone hydrochloride**,* *Mitozantrone hydrochloride,* Mitroxone, Neopom, Norexan, Novantron(e), Novatrone, NSC-301739 D, Oncotron, Onkotrone, Pralifan, Refador
U Antineoplastic

11328 (9108)

$C_{22}H_{28}N_6O_3S$
136817-59-9

1-[3-(Isopropylamino)-2-pyridyl]-4-[[5-(methanesulfonamido)-2-indolyl]carbonyl]piperazine = 1-[3-[(1-Methyl-ethyl)amino]-2-pyridinyl]-4-[[5-[(methylsulfonyl)amino]-1*H*-indol-2-yl]carbonyl]piperazine (●)

R Monomethanesulfonate (147221-93-0)
S *Delavirdine mesilate**,* Rescriptor, U-90152 S
U Antiviral

11329

$C_{22}H_{28}N_6O_4$
158921-85-8

N^2-[(1-Hydroxy-2-naphthalenyl)carbonyl]-L-arginyl-L-prolinamide (●)

R Monohydrochloride (178925-65-0)
S APC 366, Bay 17-1998
U Tryptase inhibitor (anti-asthmatic)

11330

$C_{22}H_{28}N_6O_8S_2$
135821-54-4

(+)-(Pivaloyloxy)methyl (6*R*,7*R*)-7-[2-[2-(L-alanylamino)thiazol-4-yl]glyoxylamido]-8-oxo-5-thia-1-azabicyclo[4.2.0]oct-2-ene-2-carboxylate 7^2-[(*Z*)-*O*-methyloxime] = [6*R*-[6α,7β[*Z*(*S**)]]]-7-[[[2-[(2-Amino-1-oxopropyl)amino]-4-thiazolyl](methoxyimino)acetyl]amino]-8-oxo-5-thia-1-azabicyclo[4.2.0]oct-2-ene-2-carboxylic acid (2,2-dimethyl-1-oxopropoxy)methyl ester (●)

R also monohydrochloride (135767-36-1)
S AS-924, *Ceftizoxime alapivoxil***
U Antibacterial

11331 $C_{22}H_{28}N_6O_{10}S_2$
142966-42-5

(Z)-(7R)-7-[2-(2-Amino-4-thiazolyl)-2-(hydroxyimi-no)acetamido]-3-[(dimethylcarbamoyloxy)methyl]-3-ce-phem-4-carboxylic acid 1-[[(isopropoxy)carbo-nyl]oxy]ethyl ester = [6R-[6α,7β(Z)]]-7-[[(2-Amino-4-thiazolyl)(hydroxyimino)acetyl]amino]-3-[[[(dimethyl-amino)carbonyl]oxy]methyl]-8-oxo-5-thia-1-azabicy-clo[4.2.0]oct-2-ene-2-carboxylic acid 1-[[(1-methyleth-oxy)carbonyl]oxy]ethyl ester (●)
R   Monohydrochloride (142494-87-9)
S   E-1101
U   Antibiotic

11332   (9109) $C_{22}H_{28}O_2$
54048-10-1

13-Ethyl-17-hydroxy-11-methylene-18,19-dinor-17α-pregn-4-en-20-yn-3-one = 17β-Hydroxy-11-methylene-18-homo-19-nor-17α-pregn-4-en-20-yn-3-one = (17α)-13-Ethyl-17-hydroxy-11-methylene-18,19-dinorpregn-4-en-20-yn-3-one (●)
S   ENG, *Etonogestrel**, Implanon, 3-KDOG, *3-Ketode-sogestrel*, 3-Keto-DSG, Org 3236
U   Progestin

11333   (9110) $C_{22}H_{28}O_3$
6184-18-5

Salicylic acid 2-methyl-5-(2-methyl-3-methylene-2-nor-bornyl)-2-pentenyl ester = Santalol salicylate = 2-Hydr-oxybenzoic acid 2-methyl-5-(2-methyl-3-methylenebi-cyclo[2.2.1]hept-2-yl)-2-pentenyl ester (●)
R   see also no. 5798
S   Santacyl, Santyl
U   Antibacterial (urinary)

11334   (9111) $C_{22}H_{28}O_3$
51-98-9

17-Hydroxy-19-nor-17α-pregn-4-en-20-yn-3-one ace-tate = 17α-Ethynyl-17-hydroxyestr-4-en-3-one acetate = 17α-Ethynyl-19-nortestesterone acetate = (17α)-17-(Acetyloxy)-19-norpregn-4-en-20-yn-3-one (●)
S   Aygestin, ENTA, Gestakadin, Milligynon, Mono-gest, Net-Ac, *Noraethisteronum aceticum*, *Nore-thindrone acetate*, Norethisteron "Jenapharm", *Nore-thisterone acetate*, Norlutate, Norluten A, Norlu-tin-A, Norlutin acetate, Orlutate, Primolut-Nor, SH 420, Sovel "Novartis", Styptin
U   Progestin

11335   (9112) $C_{22}H_{28}O_2$
976-71-0

17-Hydroxy-3-oxo-17α-pregna-4,6-diene-21-carboxylic acid γ-lactone = 17α-(2-Carboxyethyl)-17-hydroxyan-drosta-4,6-diene-3-one lactone = 3-(3-Oxo-17-hydroxy-4,6-androstadien-17α-yl)propionic acid lactone = 3-

Oxoandrosta-4,6-diene-17-spiro-2'-tetrahydrofuran-5'-one = (17α)-17-Hydroxy-3-oxopregna-4,6-diene-21-carboxylic acid γ-lactone (●)
S *Canrenone**, Contaren, Luvion "GiEnne; Simes", Phanurane, R 5830, 11614 R.P., SC-9376, Spiroktan
U Aldosterone antagonist, diuretic

11336　　　　　　　　　　　　　　　　　$C_{22}H_{28}O_5$
　　　　　　　　　　　　　　　　　　　909-13-7

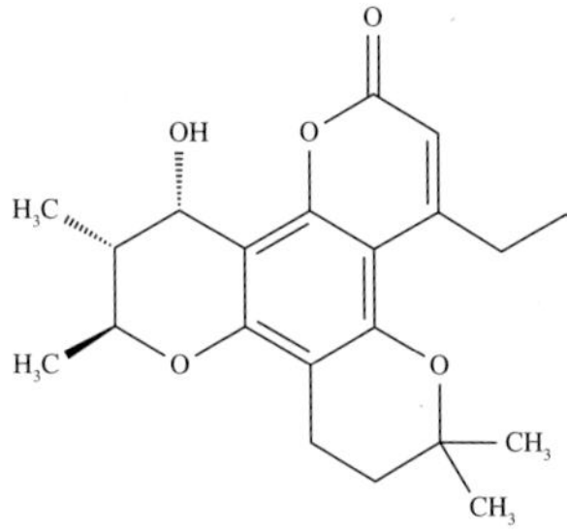

[10S-(10α,11β,12β)]-7,8,11,12-Tetrahydro-12-hydroxy-6,6,10,11-tetramethyl-4-propyl-2H,6H,10H-benzo[1,2-b:3,4-b':5,6-b'']tripyran-2-one (●)
S *7,8-Dihydrocostatolide*, NSC-661123
U Anti-HIV agent

11337　(9113)　　　　　　　　　　　　$C_{22}H_{28}O_5$
　　　　　　　　　　　　　　　　　　　7698-93-3

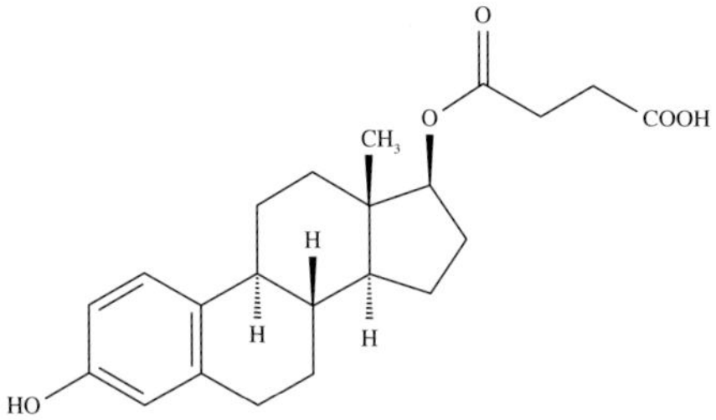

Estra-1,3,5(10)-triene-3,17β-diol 17-(hydrogen succinate) = (17β)-Estra-1,3,5(10)-triene-3,17-diol 17-(hydrogen butanedioate) (●)
S *Estradiol hemisuccinate*, Eutocol, Östradiolhemisuccinat
U Estrogen

11338　(9115)　　　　　　　　　　　　$C_{22}H_{28}O_5$
　　　　　　　　　　　　　　　　　　　599-33-7

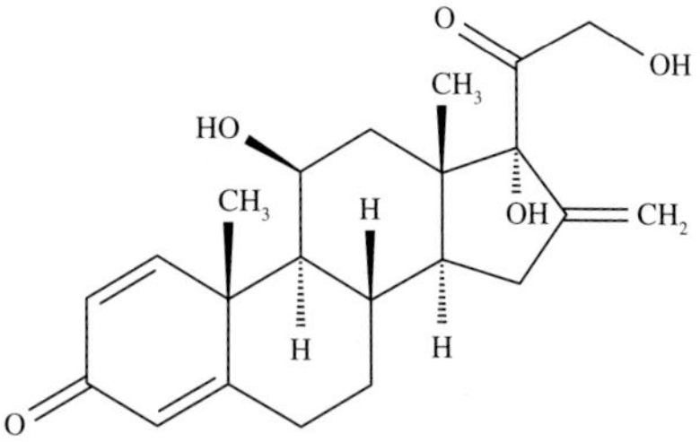

11β,17,21-Trihydroxy-16-methylenepregna-1,4-diene-3,20-dione = 16-Methyleneprednisolone = (11β)-11,17,21-Trihydroxy-16-methylenepregna-1,4-diene-3,20-dione (●)
R see also no. 13927
S Dacortilen, Dacortin (Merck-Brazil), Decortilen, *Methylprednisolone, Prednylidene***, St. 105, Sterocort "Draco"
U Glucocorticoid (anti-inflammatory, anti-allergic)

11339　(9116)　　　　　　　　　　　　$C_{22}H_{28}O_5$
　　　　　　　　　　　　　　　　　　　17332-61-5

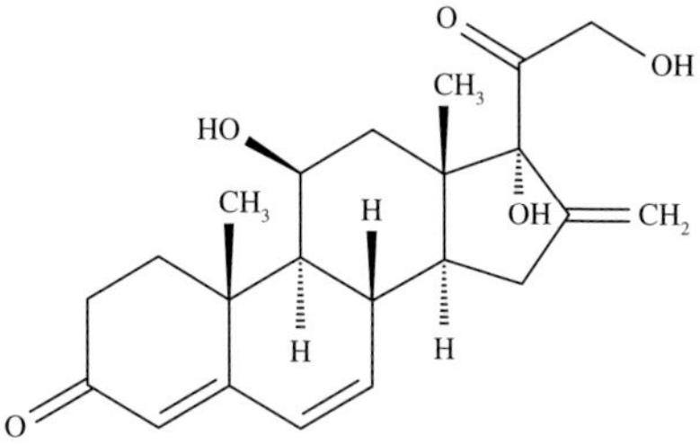

11β,17,21-Trihydroxy-16-methylenepregna-4,6-diene-3,20-dione = 6-Dehydro-16-methylenecortisol = (11β)-11,17,21-Trihydroxy-16-methylenepregna-4,6-diene-3,20-dione (●)
S *Isoprednidene**, StC 407
U ACTH inhibitor

11340   (9114) $C_{22}H_{28}O_5$
                                            1247-42-3

17,21-Dihydroxy-16β-methylpregna-1,4-diene-3,11,20-
trione = 16β-Methylprednisone = (16β)-17,21-Dihy-
droxy-16-methylpregna-1,4-diene-3,11,20-trione (●)
R   see also no. 13196
S   Betalona, Bétalone "Lepetit", Betanison(a), Betapar,
        Betapred "Schering-USA", Corti-bi, Cortipyren B,
        Deltacortene-Beta, Deltisona B, Lepicortin-Beta,
        *Meprednisone**, Methylprednisone*, NSC-527579,
        Sch 4358
U   Glucocorticoid (anti-inflammatory, anti-allergic)

11341   (9117) $C_{22}H_{28}O_7$
                                            16964-56-0

[1R-(1α,4aβ,6aβ,9β,11α,11aα,11bα)]-11b-[(Acetyloxy)methyl]dodecahydro-11-hydroxy-2,2-dimethyl-8-
methylene-6,7-dioxo-6H-6a,9-methanobenzo[b]cyclo-
hepta[d]pyran-1-carboxaldehyde = (5α)-8-(Acetyloxy)-
8,10-deepoxy-10,13-dideoxy-5-hydroxy-10-oxoenmein
(●)
S   *Isodonal*
U   Platelet aggregation inhibitor from *Isodon japonicus*

11342   (9118) $C_{22}H_{28}O_8$
                                            84294-78-0

[1R-(1α,4aβ,6aβ,9β,11α,11aα,11bα)]-11b-[(Acetyloxy)methyl]dodecahydro-11-hydroxy-2,2-dimethyl-8-
methylene-6,7-dioxo-6H-6a,9-methanobenzo[b]cyclo-
hepta[d]pyran-1-carboxylic acid (●)
S   *Isodonoic acid*
U   Antineoplastic from *Rabdosia ternifolia*

11343   (9119) $C_{22}H_{28}O_9$
                                            84062-60-2

(1β,11β,12α,15β)-15-(Acetyloxy)-11,20-epoxy-1,11,12-
trihydroxypicras-3-ene-2,16-dione (●)
S   *Holacanthone*, NSC-126765
U   Antineoplastic

11344 $C_{22}H_{29}ClNO_6$

[R-(Z)]-4-[2-(6-Chloro-3-oxo-1(3H)-isobenzofuranyli-
dene)ethoxy]-N,N,N-trimethyl-2-(3-methyl-1-oxobut-
oxy)-4-oxo-1-butanaminium (●)
R   Bromide (143484-41-7)
S   ST-899
U   PAF antagonist (treatment of endotoxic shock)

11345   (9120) $C_{22}H_{29}ClN_2O_4S$
                                                                         134162-24-6

8-(4-Chlorophenylsulfonamido)-4-[3-(3-pyridyl)pro-
pyl]octanoic acid = γ-[4-[[(4-Chlorophenyl)sulfonyl]ami-
no]butyl]-3-pyridineheptanoic acid (●)
S   CGS 22652
U   Antithrombotic

11346   (9121) $C_{22}H_{29}ClN_4O$
                                                                         22906-83-8

6-Chloro-9-[[4-(diethylamino)-1-methylbutyl]amino]-2-
methoxy-1-azaacridine = 7-Chloro-10-[[4-(diethylamino)-1-methylbutyl]amino]-2-methoxybenzo[b]-1,5-naph-
thyridine = N^4-(7-Chloro-2-methoxybenzo[b]-1,5-naph-
thyridin-10-yl)-N^1,N^1-diethyl-1,4-pentanediamine (●)
R   Dihydrochloride (22907-02-4)
S   Azaatabrine, Azacrin, *Azamepacrine*
U   Antimalarial

11347   (9122) $C_{22}H_{29}ClO_3$
                                                                         5367-84-0

6-Chloro-17-hydroxy-16α-methylpregna-4,6-diene-3,20-
dione = (16α)-6-Chloro-17-hydroxy-16-methylpregna-
4,6-diene-3,20-dione (●)

R   see also no. 12448
S   *Clomegestone***
U   Progestin

11348   (9123) $C_{22}H_{29}ClO_5$
                                                                         67452-97-5

7α-Chloro-11β,17,21-trihydroxy-16α-methylpregna-1,4-
diene-3,20-dione = (7α,11β,16α)-7-Chloro-11,17,21-tri-
hydroxy-16-methylpregna-1,4-diene-3,20-dione (●)
R   see also no. 13886
S   *Alclometasone***
U   Glucocorticoid (topical anti-inflammatory, anti-allergic)

11349   (9124) $C_{22}H_{29}ClO_5$
                                                                         4647-20-5

9-Chloro-11β,17,21-trihydroxy-16α-methylpregna-1,4-
diene-3,20-dione = (11β,16α)-9-Chloro-11,17,21-tri-
hydroxy-16-methylpregna-1,4-diene-3,20-dione (●)
R   see also no. 13885
S   *Icometasone*
U   Glucocorticoid (anti-inflammatory, anti-allergic)

11350   (9125) $C_{22}H_{29}ClO_5$
                                                  4419-39-0

9-Chloro-11β,17,21-trihydroxy-16β-methylpregna-1,4-diene-3,20-dione = 9-Chloro-16β-methylprednisolone = (11β,16β)-9-Chloro-11,17,21-trihydroxy-16-methylpregna-1,4-diene-3,20-dione (●)
R   see also nos. 13887, 14078, 14160
S   *Beclometasone***, *Beclomethasone*
U   Glucocorticoid (anti-inflammatory, anti-allergic)

11350-01   182176-70-1
R   Enemate
S   *Beclometasone enemas*, CHF 1514
U   Glucocorticoid (anti-inflammatory, anti-allergic)

11351   (9126) $C_{22}H_{29}ClO_6$
                                                  40665-92-7

(±)-(Z)-7-[(1R*,2R*,3R*,5S*)-2-[(E)-(3R*)-4-(m-Chlorophenoxy)-3-hydroxy-1-butenyl]-3,5-dihydroxycyclopentyl]-5-heptenoic acid = (±)-(5Z,13E)-(9S,11R,15R)-16-(3-Chlorophenoxy)-9,11,15-trihydroxy-ω-tetranorprosta-5,13-dienoic acid = [1α(Z),2β(1E,3R*),3α,5α]-(±)-7-[2-[4-(3-Chlorophenoxy)-3-hydroxy-1-butenyl]-3,5-dihydroxycyclopentyl]-5-heptenoic acid (●)
R   see also no. 12450;
        Sodium salt (55028-72-3)
S   *Cloprostenol Sodium***, Estrumat(e), Heifex,
        ICI 80996, Oestrophan, Planate, Suimate, Uniandine
U   Luteolytic (synthetic prostaglandin)

11352 $C_{22}H_{29}ClO_6$
                                                  54276-21-0

(7Z)-7-[(1R,2R,3R,5S)-2-[(1E,3R)-4-(3-Chlorophenoxy)-3-hydroxy-1-butenyl]-3,5-dihydroxycyclopentyl]-5-heptenoic acid (●)
R   Monosodium salt (62561-03-9)
S   (+)-*Cloprostenol sodium*, Dalmazin, *Dexcloprostenol sodium*, Genestran, Remophan, Veteglan
U   Luteolytic

11353   (9127) $C_{22}H_{29}Cl_2N_3O_4$
                                                  104672-11-9

1-(4-Amino-3,5-dichlorophenyl)-2-[4-(2,3,4-trimethoxybenzyl)-1-piperazinyl]ethanol = α-(4-Amino-3,5-dichlorophenyl)-4-[(2,3,4-trimethoxyphenyl)methyl]-1-piperazineethanol (●)
S   XC-1, Xinchuanling
U   Cardiotonic

11354   (9128) $C_{22}H_{29}Cl_2N_3O_4S$
                                                  52993-98-3

(2S,5R,6R)-6-[4-[p-[Bis(2-chloroethyl)amino]phenyl]butyramido]-3,3-dimethyl-7-oxo-4-thia-1-azabicyclo[3.2.0]heptane-2-carboxylic acid = (6R)-6-[4-[p-[Bis(β-chloroethyl)amino]phenyl]butyramido]penicillanic acid = [2S-(2α,5α,6β)]-6-[[4-[4-[Bis(2-chloroethyl)amino]phenyl]-1-oxobutyl]amino]-3,3-dimethyl-7-oxo-4-thia-1-azabicyclo[3.2.0]heptane-2-carboxylic acid (●)

R also sodium salt (25390-24-3)
S *Chlorbutinpenicillin*
U Antineoplastic

11355 (9130) $C_{22}H_{29}FO_4$
152-97-6

6α-Fluoro-11β,21-dihydroxy-16α-methylpregna-1,4-diene-3,20-dione = 6α-Fluoro-16α-methyl-1-dehydrocorticosterone = (6α,11β,16α)-6-Fluoro-11,21-dihydroxy-16-methylpregna-1,4-diene-3,20-dione (●)
R see also nos. 11220, 13590, 13924;
also monohydrate
S Colsipan, *Fluocortolone**, Fluormethyldehydrocorticosterone*, SH 742, SR 2238, Syracort, Ultracur oral, Ultracur S, Ultralan oral, Ultralanum oral
U Glucocorticoid (anti-inflammatory, anti-allergic)

11356 (9129) $C_{22}H_{29}FO_4$
426-13-1

9-Fluoro-11β,17-dihydroxy-6α-methylpregna-1,4-diene-3,20-dione = 21-Deoxy-6α-methyl-9-fluoroprednisolone = (6α,11β)-9-Fluoro-11,17-dihydroxy-6-methylpregna-1,4-diene-3,20-dione (●)
R see also no. 12453
S Cortilet, Cortisdin, Delmeron, Delmeson-Corticoid, Efflumidex, Eflumidex, Facocinerin, Fluaton, Flu-Base, Flucon, Fluforte, Flulon, Flumerol, Flumetholon,

Flumetol-Ofteno, Flumetol semplice, Flumetol Simplex, Flumex "Allergan", Flumexflor, Fluocon, Fluoderm plain, Fluolon, *Fluormetholonum, Fluorometholone**, Fluor-OP, Fluor-Op-Ophth, Fluoropos, Flurolon, Flurop, Fluxinam, FML, Grifonolona, Helpometil, Isopto-Flucon, Lerma, Mesoflon R, Methasite, NB 123, Neo-Oxylon, NSC-33001, Odomel, Okilon, Oxylone, Pitos, Regresin-sine, Talirax, Toscacort, Trilcin, U-8614, Ursnon, Zitapal
U Glucocorticoid (anti-inflammatory, anti-allergic)

11357 (9131) $C_{22}H_{29}FO_4$
1879-77-2

9-Fluoro-11β,17-dihydroxy-16β-methylpregna-1,4-diene-3,20-dione = (11β,16β)-9-Fluoro-11,17-dihydroxy-16-methylpregna-1,4-diene-3,20-dione (●)
R see also no. 12901
S *21-Deoxybetamethasone, Doxibetasol**, Doxybetasol*
U Glucocorticoid (anti-inflammatory, anti-allergic)

11358 (9132) $C_{22}H_{29}FO_4$
382-67-2

9-Fluoro-11β,21-dihydroxy-16α-methylpregna-1,4-diene-3,20-dione = (11β,16α)-9-Fluoro-11,21-dihydroxy-16-methylpregna-1,4-diene-3,20-dione (●)
S A 41-304, Actiderm, Dermo-Hidrol, *Desoximetasone**, Desoxymethasone*, Esperson, Flubason, Hoechst 304, Ibaril, Indexon, Inerson, R 2113, Stiedex, Topicort(e), Topiderm "Roussel", Topisolon

U Glucocorticoid (anti-inflammatory, anti-allergic)

11359 $C_{22}H_{29}FO_4$
378-61-0

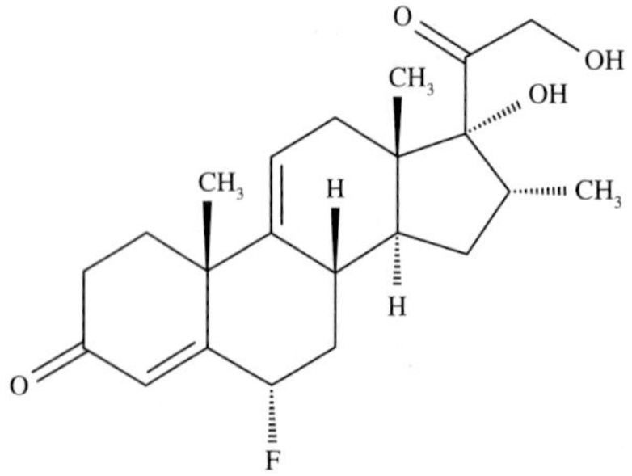

(6α,16α)-6-Fluoro-17,21-dihydroxy-16-methylpregna-4,9(11)-diene-3,20-dione (●)
S U-24067
U Angiostatic

11360 (9133) $C_{22}H_{29}FO_4S$
87116-72-1

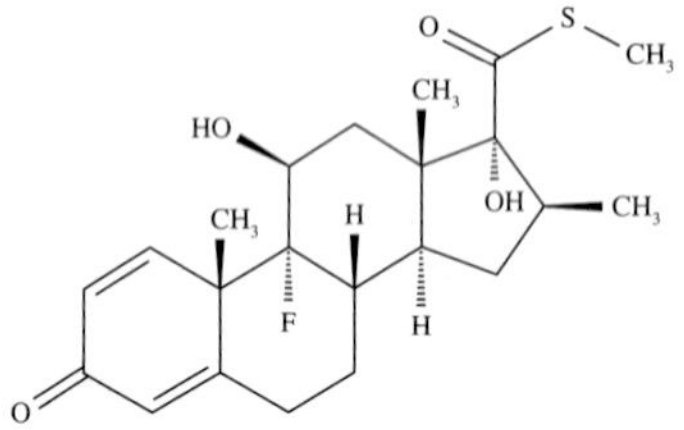

S-Methyl 9-fluoro-11β,17-dihydroxy-16β-methyl-3-oxo-androsta-1,4-diene-17β-carbothioate = (11β,16β,17α)-9-Fluoro-11,17-dihydroxy-16-methyl-3-oxoandrosta-1,4-diene-17-carbothioic acid S-methyl ester (●)
R see also no. 12455
S *Timobesone***
U Glucocorticoid (topical anti-inflammatory, anti-allergic)

11361 (9134) $C_{22}H_{29}FO_5$
3841-11-0

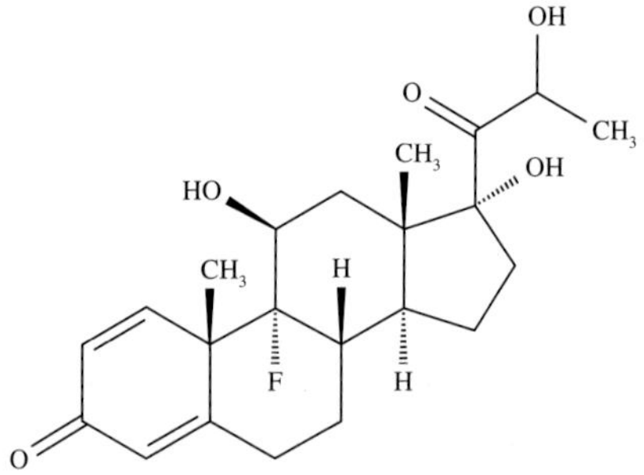

9-Fluoro-11β,17α-dihydroxy-17-lactoylandrosta-1,4-dien-3-one = (11β,17α)-9-Fluoro-11,17-dihydroxy-17-(2-hydroxy-1-oxopropyl)androsta-1,4-dien-3-one (●)
R see also no. 12456
S *Fluperolone***
U Glucocorticoid (anti-inflammatory, anti-allergic)

11362 (9138) $C_{22}H_{29}FO_5$
86348-98-3

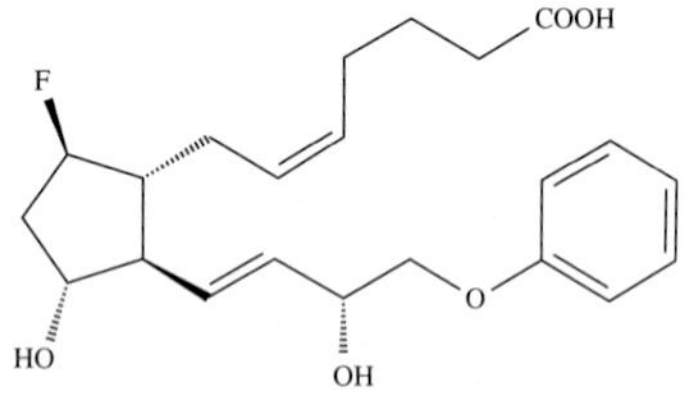

(Z)-7-[(1R,2R,3R,5R)-5-Fluoro-3-hydroxy-2-[(E)-(3R)-3-hydroxy-4-phenoxy-1-butenyl]cyclopentyl]-5-heptenoic acid = (5Z,13E)-(9R,11R,15R)-9-Fluoro-11,15-dihydroxy-16-phenoxy-17,18,19,20-tetranor-5,13-prostadienoic acid = [1R-[1α(Z),2β(1E,3R*),3α,5β]]-7-[5-Fluoro-3-hydroxy-2-(3-hydroxy-4-phenoxy-1-butenyl)cyclopentyl]-5-heptenoic acid (●)
S *Flunoprost***, ZK 95377
U Prostaglandin

11363 (9135) $C_{22}H_{29}FO_5$
53-33-8

6α-Fluoro-11β,17,21-trihydroxy-16α-methylpregna-1,4-diene-3,20-dione = (6α,11β,16α)-6-Fluoro-11,17,21-tri-hydroxy-16-methylpregna-1,4-diene-3,20-dione (●)

R see also nos. 11405, 12457

S *Paramethasone***

U Glucocorticoid (anti-inflammatory, anti-allergic)

11364 (9136) $C_{22}H_{29}FO_5$
50-02-2

9-Fluoro-11β,17,21-trihydroxy-16α-methylpregna-1,4-diene-3,20-dione = 9-Fluoro-16α-methylprednisolone = (11β,16α)-9-Fluoro-11,17,21-trihydroxy-16-methylpregna-1,4-diene-3,20-dione (●)

R see also nos. 11366, 11406, 12458, 13203, 13591, 13593, 13817, 13890, 13925, 13933, 14085, 14383, 15292

S Abifluorene, Adexone solution, Aeroseb-Dex, Aflucoson, Afpred forte-Dexa-Tabs, Alfalyl, Alin-Tabl. "Chinoin-Mexico", Alpermell, Ametaz-D, Amradexone, Amumetazon, Anaflogistico "Devi", Anemul mono, Aphtasolon, Aplerdex, Arcodexan, Artrosone, Aspicorte, Auxil, Azium, Baycadron, Biodexone, Bisuo-DS, Butiol, Calonat, Carulon, Corson, Cortadex, Cortaméthasone, Corti-Attritin, Cortidex, Cortidexason, Cortidrona, Cortisumman, Cosdex, Cresophene, Dab, Dabu M, Dacortina fuerte, Daidrone, Dalalone L.A., Dalaron L.A., Daxin, Deason, De-

bicort, Deca, Decacort, Decacortin, Decadeltasona, Decaderm "MSD", Decadin, Decadran, Decadrol L.A., Decadron, Decaesadril, Decagel, Decalix, Decalone "Palenzona", Decameth Tabl., Decasone, Decaspray, Decdan, Decilone, Decobel, Decofluor, Dectan, Dekacort, Dekasone, Dekort, Deksalon, Deksamet, Deltafluoren(e), Demaminor, Dermax "Pharmasint", Dermazon, Deronil, Dersone, Desacort, Desacortone, Desadrene, Desalark, *Desametasone*, Detasona, Dethamedin, DEX, Dexa A, Dexa-Aldon, Dexachel, Dexa-Chosei, Dexacon, Dexacorm, Dexacort "Adelco; Ikapharm" (Tabl.), Dexacortal, Dexacorten, Dexacortin, Dexa-Dabrosan, Dexadermil, Dexadrol, Dexafar "Lafar", Dexafarma, Dexaflam "Hebron; Lichtenstein", Dexagalen crinale, Dexagrin, Dexa-Helvacort, Dexa-Korti, Dexa-Life, Dexalocal, Dexalona, Dexaltin, Dexa-Mamallet, Dexamas, Dexame, Dexamecortin, Dexamed "Medice", Dexameson, Dexametaluy, Dexametan, *Dexametasone*, Dexameth, Dexamethadrone, *Dexamethasone***, *Dexamethazon*, Dexametriol, Dexamin "Optifarma", Dexaminor, Dexamiso, Dexamonozon N, Dexa-Morgens, Dexan(e), Dexanil, Dexanteric, Dexantil, Dexaplast, Dexaplon, Dexapolcort, Dexaport, Dexapot, Dexapred, Dexa-Rhinosan, Dexasan, Dexa-Scheroson, Dexasine, Dexasol, Dexason(e), Dexathiorozon, Dexa-Wolner, Dexazone, Dexinolon-Salbe, Dexinoral, Dexmethsone, Dexo-L.A., Dexolan, Dexona, Dexone, Dexoral, Dextelan, Dezone, Dinormon, D-Medrol, Donray, Edacelan, Elixir-Methasone, Esacortene, Eurason D, Exadion, Fadametasona, Falban, Fatrocortin, Fercasone, Fexadron, Firmalone, *Flumeprednisolone*, *Fluormethylprednisolone*, Fluormetilone, Fluormone, Fluorocort, Fluorodelta, Fortecortin, Friburona, Gammacorten, Garsone, Gedetax, Glucocort, Grathazon, Grosodexon, *Hexadecadrol*, Hexadrol, Hightisone, Idizone, Ificort, Indarzona, Inhalocort, Isnacort, Isodexam, Isopto-Dex, Isopto-Maxidex, Juvadexon, Lepifluorene, Lokalison, Loverine, Luxazone, Maradex, Marvidione, Marvisona D, Maxidex, Médiaméthasone, Megacortin, Mephameson, Meridexasone, Metacort, Metalexina, Metasolon, Metason(a), Methazon-Ion, Metisone "Lafi", Metona, Mifardex, Millicorten, Minicort, Minidex, Miral "Geneva", Mitasone, MK-125, Moco, Moderix, Mymethasone, Nisometasona, Novocort "Pharmax", Novodon, Novomethasone, NSC-34521, Nucort, Nuridexon, Ocu-Dexa, Oftadex, Oftan-Dexa, Onadron, Opticorten, Oradexon, Orgadrone, Osatron, Peridex "Sang A", Pet Derm III, Polidex, Predni-F-Tablinen, Prednisolon F, Prodexona, Rally-A, Rancortin, Resticort, Rophamethazone, Rupedex, Santeson-Tabl., Sawasone, Sch 4651, Selectasona, Sisotek, Sokaral, Soldex, Solone "Liade", Soludex, Solupen S Mono, Solutio Cordes Dexa N, Sondex, Spansona, Sploven, Stamcort, Steralol, Sterasone, Sterocon, Superprednol, Suprason, Thilodexine, Triamcimetil, Tuttozem N, Urecortin, Vasodex, Visumetazone, Wymesone-Tabl., Zonometh

U Glucocorticoid (anti-inflammatory, anti-allergic)

11365   (9137) $C_{22}H_{29}FO_5$
378-44-9

9-Fluoro-11β,17,21-trihydroxy-16β-methylpregna-1,4-diene-3,20-dione = 9-Fluoro-16β-methylprednisolone = (11β,16β)-9-Fluoro-11,17,21-trihydroxy-16-methylpregna-1,4-diene-3,20-dione (●)

R see also nos. 11407, 12459, 13204, 13594, 13889, 13891, 14081, 14083, 14162, 14163, 14674, 14778

S Becort, Betacorlan, Betacort, Betacortal, Betacortex, Betacortril, Betadex, *Betadexamethasone*, Betagel, Betagen, Betalone "Firma", Betamamallet, *Betametasone*, Betametha, *Betamethasone***, Betapredol, Betapredsol, Betasolon, Betasona, Betilon "Tecnifar", Betnelan, Betnesol "Glaxo-Germany", Betsolan, Bivicortene, Butasona, Celestamine N, Celestan, Célestène, Celestona, Celeston(e), Celestovet, Cidoten, Citoden, Cuantin "ICN", Cuantona, Dabbeta, Dacam "Chile", Desacort-Beta, Detarmon, Exabet, *Flubenisolonum*, Galinocort, Helpoderm, Inflacor, β-*Methasone*, Methazon, Movithiol, Norbet, No-Reumar, Novovate, NSC-39470, PD 863, Pertene, Probasona, Retagel, RG 833, Rinderon, Rinesteron, Sch 4831, Sclane-Compr., Skizone, Steralol B, Steromien, Supercortene, Unicort "Unipharm", Vabetadermil, Visubeta, Walacort

U Glucocorticoid (anti-inflammatory, anti-allergic)

11366   (9139) $C_{22}H_{29}FO_8S$
7793-27-3

9-Fluoro-11β,17,21-trihydroxy-16α-methylpregna-1,4-diene-3,20-dione 21-(hydrogen sulfate) = Dexamethasone 21-hemisulfate = (11β,16α)-9-Fluoro-11,17-dihydroxy-16-methyl-21-(sulfooxy)pregna-1,4-diene-3,20-dione (●)

R Sodium salt (466-11-5)

S *Dexamethasone sodium sulfate, Dexamethasonnatriumsulfat*, Dexa-Scheroson-Amp.

U Glucocorticoid (anti-inflammatory, anti-allergic)

11367   (9140) $C_{22}H_{29}F_3N_2O_2$
56693-15-3

(±)-α-[[(1-Ethynylcyclohexyl)oxy]methyl]-4-(α,α,α-trifluoro-*m*-tolyl)-1-piperazineethanol = 1-[(1-Ethynylcyclohexyl)oxy]-3-[4-[3-(trifluoromethyl)phenyl]-1-piperazinyl]-2-propanol = 1-[3-(Trifluoromethyl)phenyl]-4-[3-(1-ethynylcyclohexyloxy)-2-hydroxypropyl]piperazine = α-[[(1-Ethynylcyclohexyl)oxy]methyl]-4-[3-(trifluoromethyl)phenyl]-1-piperazineethanol (●)

S CERM 3519, *Terciprazine***

U Neuroleptic

11368   (9141) $C_{22}H_{29}F_3O_3$
15687-21-5

17-Hydroxy-6α-(trifluoromethyl)pregn-4-ene-3,20-dione
= (6α)-17-Hydroxy-6-(trifluoromethyl)pregn-4-ene-3,20-
dione (●)
R   see also no. 12462
S   *Flumedroxone**
U   Migraine therapeutic

11369   (9142) $C_{22}H_{29}NOS$
5835-72-3

S-[2-(Dipropylamino)ethyl] diphenylthioacetate = Diphe-
nylthioacetic acid *S*-(β-dipropylaminoethyl) ester = α-
Phenylbenzeneethanethioic acid *S*-[2-(dipropylami-
no)ethyl] ester (●)
R   Hydrochloride (2424-69-3)
S   *Diprofene hydrochloride**, Diprophene
U   Spasmolytic, vasodilator

11370   (9143) $C_{22}H_{29}NOS$
7009-79-2

S-[2-(Diethylamino)ethyl] 2-(4-biphenylyl)thiobutyrate =
2-(4-Biphenylyl)thiobutyric acid *S*-(β-diethylaminoethyl)

ester = α-Ethyl-[1,1'-biphenyl]-4-ethanethioic acid *S*-[2-
(diethylamino)ethyl] ester (●)
S   *Xenthiorate**
U   Anticholesteremic

11371   (9145) $C_{22}H_{29}NO_2$
13426-07-8

Diphenylacetic acid 1-ethyl-2-(diethylamino)ethyl ester =
α-Phenylbenzeneacetic acid 1-[(diethylamino)me-
thyl]propyl ester (●)
S   Eterofen, Etherophen, Ethyldiphacil, IEM-506
U   Anticholinergic

11372   (9144) $C_{22}H_{29}NO_2$
23227-52-3

Ethyl 4-(diethylamino)-2,2-diphenylbutyrate = 2-[β-(Diethylamino)ethyl]-2,2-diphenylacetic acid ethyl ester = α-
[2-(Diethylamino)ethyl]-α-phenylbenzeneacetic acid
ethyl ester (●)
S   *Benzetile*
U   Spasmolytic

11373   (9146) $C_{22}H_{29}NO_2$
77-50-9

(±)-4-(Dimethylamino)-1,2-diphenyl-3-methyl-2-buta-
nol propionate (ester) = (±)-1-Benzyl-3-(dimethylamino)-
2-methyl-1-phenylpropyl propionate = (±)-4-(Dimethylamino)-3-methyl-2-(propionyloxy)-1,2-diphenylbutane =
α-[2-(Dimethylamino)-1-methylethyl]-α-phenylbenzene-
ethanol propanoate (ester) (●)
R   Hydrochloride (6509-36-0)
S   *Dimeprotane hydrochloride, Propoxipheni chloridum, (±)-Propoxyphene hydrochloride*
U   Analgesic

11374   (9148) $C_{22}H_{29}NO_2$
2338-37-6

(−)-4-(Dimethylamino)-1,2-diphenyl-3-methyl-2-butanol
propionate (ester) = (−)-1-Benzyl-3-(dimethylamino)-2-
methyl-1-phenylpropyl propionate = (−)-4-(Dimethylamino)-3-methyl-2-(propionyloxy)-1,2-diphenylbutane = [R-
(R*,S*)]-α-[2-(Dimethylamino)-1-methylethyl]-α-phe-
nylbenzeneethanol propanoate (ester) (●)
R   2-Naphthalenesulfonate monohydrate (55557-30-7)
S   Contratuss, Letusin, *Levopropoxipheni napsylas, Le-
vopropoxyphene napsilate**, Lilly 29866, Novrad,
Rava, Regretos
U   Antitussive

11374-01   (9148-01)   31852-19-4

R   2,6-Di-*tert*-butyl-1,5-naphthalenedisulfonate (2:1)
S   *Levopropoxiphene dibudinate**, Probunafon*, Sotor-
ni
U   Antitussive

11375   (9147) $C_{22}H_{29}NO_2$
469-62-5

(2S,3R)-(+)-4-(Dimethylamino)-1,2-diphenyl-3-methyl-
2-butanol propionate (ester) = (+)-1-Benzyl-3-(dimethylamino)-2-methyl-1-phenylpropyl propionate = (+)-4-(Di-
methylmino)-3-methyl-2-(propionyloxy)-1,2-diphenyl-
butane = [S-(R*,S*)]-α-[2-(Dimethylamino)-1-methyl-
ethyl]-α-phenylbenzeneethanol propanoate (ester) (●)
R   see also no. 8541-04;
Hydrochloride (1639-60-7)
S   Abalgin, Algafan, Algaphan, Algiphene "Ferraz",
Algodex, Algodin, Antalvic, Daloxen, Darval, Dar-
von, Deksofen, Deprancol, Depromic, Depronal, De-
velin retard, Dextrogesic, Dextropropoxifen "DAK",
*Dextropropoxipheni chloridum, Dextropropoxyphe-
ne hydrochloride**, Dolan, Dolene, Dolocap, Do-
loksen, Dolorphen, Dolotard, Doraphen, Doxaphe-
ne, Dymopoxyphene, Erantin (old form), Femadol,
Gafanal, Harmar, ICN-65, Kesso-Gesic, L 16298,
Lentadol, Levadol, Levitan, Liberen, Mardon, Mar-
gesic improved, Nefertal, Neo-Mal, NIH 5821, No-
vopropoxyn, Paljin, Parvon, Piril, Praia, Preparten
(old form), Pro 65, Procaps-65, Progesic-65, Pronal-
gic, Prophene 65, Propox, Propoxychel, Propoxyn,
D-*Propoxyphene, Propoxyphene hydrochloride*, Pro-
xagesic, Proxene, Proxyphene, RC 600, Regredol,
Romidon, Ropoxy, Scrip-Dyne, SK 65, S-Pain-65,
Sudhinol, 642 Tablets, Tawasan, Tilene (old form),
Troliber, Unigesic, Vandar 65, Zideron
U   Analgesic

11375-01   (9147-01)   26570-10-5
R   2-Naphthalenesulfonate monohydrate (napsylate)

S Darvon-N, Dexofen, *Dextropropoxyphene napsilate**, Doloxene, Doloxene N, Napsalgesic, *Propoxyphene napsylate*, S 9700
U Analgesic

11375-02 (9147-02)   52387-20-9
R 1-Theobromineacetate (1:1)
S *Dextropropoxyphenum carbocoffeinum*, Lenigesial, Z 867
U Analgesic, antipyretic

11376 (9149) $C_{22}H_{29}NO_2$
                                                 1477-39-0

(±)-1-Ethyl-4-(methylamino)-2,2-diphenylpentyl acetate = 6-(Methylamino)-4,4-diphenyl-3-heptanol acetate (ester) = (±)-α-3-Acetoxy-6-(methylamino)-4,4-diphenyl-heptane = α-4-Acetoxy-*N*,1-dimethyl-3,3-diphenylhexyl-amine = α-Ethyl-β-[2-(methylamino)propyl]-β-phenyl-benzeneethanol acetate (ester) (●)
R also hydrochloride (5633-25-0)
S Lilly 30109, NIH 7667, *Noracimetadolo, Noracyme-thadol***
U Analgesic

11377 $C_{22}H_{29}NO_2$
                                         169274-78-6

(αR,βS)-α-(4-Hydroxyphenyl)-β-methyl-4-(phenylme-thyl)-1-piperidinepropanol (●)
S Ro 25-6981

U Neuroprotective (NMDA receptor antagonist)

11378 (9151) $C_{22}H_{29}NO_3$
                                                 2911-15-1

Ethoxydiphenylacetic acid 2-(diethylamino)ethyl ester = α-Ethoxy-α-phenylbenzeneacetic acid 2-(diethylamino)ethyl ester (●)
R Hydrochloride (2075-04-9)
S Gorutin
U Spasmolytic

11379 (9152) $C_{22}H_{29}NO_3$
                                                 60883-72-9

2-(Diethylamino)ethyl *p*-methoxy-α-phenylhydrocinna-mate = 3-(*p*-Methoxyphenyl)-2-phenylpropionic acid β-(diethylamino)ethyl ester = 4-Methoxy-α-phenylben-zenepropanoic acid 2-(diethylamino)ethyl ester (●)
R also hydrochloride (3820-14-2) or oleate
S *Anisidrocinnamol, Anisohydrocinnamol*, Eubispas-me (active substance)
U Spasmolytic

11380 (9153) $C_{22}H_{29}NO_3$
                                                 118111-54-9

17-(Cyclopropylmethyl)-4,14-dimethoxymorphinan-6-one (●)
S *Cyprodime*
U Selective μ-opioid antagonist

11381 (9150) $C_{22}H_{29}NO_3$
20799-24-0

17α-Ethynyl-17-hydroxyestr-4-en-3-one 17-acetate 3-oxime = (17α)-17-(Acetyloxy)-19-norpregn-4-en-20-yn-3-one 3-oxime (●)
S *Norethisterone acetate oxime*
U Progestin (postcoitally contraceptive)

11382 (9154) $C_{22}H_{29}NO_5$
39133-31-8

β-(Dimethylamino)-β-ethylphenethyl alcohol 3,4,5-tri-methoxybenzoate (ester) = 2-(Dimethylamino)-2-phenyl-butyl 3,4,5-trimethoxybenzoate = 3,4,5-Trimethoxyben-zoic acid 2-(dimethylamino)-2-phenylbutyl ester (●)
R Maleate (1:1) (34140-59-5)
S Behm, Bumetin, Butikinon, Butina "Bifan", Celvel, Cerekinon, Coolbutin, Debridat, Digerent, Dromostat, Drugbutin, Foldox "Sidus", Garapepsin, Gastlock, Gastrovase, Hamoniton, Ibutin, Istinon, Jintroi, Kalius, Lamotris, Libertrim, Liement, Malbuthin, Malebutine, Manpotin, Mebucolon, Mebutin, Mebutit, Meptart, Miopropan, Modulase, Modulite (active substance), Modulon, Movarol, Mustrick, Naruhi-

ton, Nepten, Nervotens, Nichimalon, Pelkysil, Pilemain, Polibutin, Prescol, Recutin, Rematin, Sakion, Salnatin, Spabucol, Supeslone, Tefmetin, TM 906, Transacalm, Transicalm, Trasben, Trikinosin, *Trimebutine maleate***, Trimedat, Trimetine, Trishi
U Spasmolytic

11383 (9155) $C_{22}H_{29}N_2O$

1-(3-Carbamoyl-3,3-diphenylpropyl)-1-methylpiperidinium = 1-(4-Amino-4-oxo-3,3-diphenylbutyl)-1-methyl-piperidinium (●)
R Bromide (125-60-0)
S *Bulgar amide*, Fenpipramide methylbromide, *Fenpiverinium bromide***, Hoechst 12494, Resantin
U Spasmolytic

11384 $C_{22}H_{29}N_2O_4$
6871-44-9

(3β,16R)-3,17-Dihydroxy-16-(methoxycarbonyl)-4-methyl-2,4(1H)-cyclo-3,4-secoakuammilanium = (1S,3S,4E,8aS,13aR,14R)-4-Ethylidene-2,3,4,5,7,8-hexa-hydro-1-hydroxy-14-(hydroxymethyl)-14-(methoxycar-bonyl)-6-methyl-13H-3,8a-methano-1H-azepi-no[1',2':1,2]pyrrolo[2,3-b]indolium (●)
R Chloride (6878-36-0)
S *Echitamine chloride*
U Cytotoxic from *Alstonia scholaris*

11385 $C_{22}H_{29}N_3O_3$
                        157832-55-8

N-[5-(Diethylamino)-1-phenylpentyl]-4-nitrobenzamide (●)
R Monohydrochloride (157832-56-9)
S HE-11, Nibentan
U Anti-arrhythmic

11386 (9156) $C_{22}H_{29}N_3O_3$
                        124436-97-1

Ethyl 4-[3-[1-(6-methyl-3-pyridazinyl)-4-piperidyl]prop-oxy]benzoate = 4-[3-[1-(6-Methyl-3-pyridazinyl)-4-pipe-ridinyl]propoxy]benzoic acid ethyl ester (●)
S R 78206
U Antiviral

11387 $C_{22}H_{29}N_3O_3S_2$
                        170006-73-2

N-[[5-[[(2*R*)-2-Amino-3-mercaptopropyl]amino][1,1'-bi-phenyl]-2-yl]carbonyl]-L-methionine methyl ester (●)
S FTI 277
U Farnesyltransferase inhibitor

11388 (9157) $C_{22}H_{29}N_3O_4$
                        67254-81-3

m-Anisaldehyde *O*-[2-hydroxy-3-[4-(*o*-methoxyphenyl)-1-piperazinyl]propyl]oxime = 3-Methoxybenzaldehyde *O*-[2-hydroxy-3-[4-(2-methoxyphenyl)-1-piperazi-nyl]propyl]oxime (●)
R Dihydrochloride (67254-80-2)
S HWA 923, *Peradoxime hydrochloride***
U Antihypertensive

11389 (9158) $C_{22}H_{29}N_3O_4S$
                        118288-08-7

(±)-2-(Furfurylsulfinyl)-*N*-[(*Z*)-4-[[4-(piperidinomethyl)-2-pyridyl]oxy]-2-butenyl]acetamide = (*Z*)-2-[(2-Furanyl-methyl)sulfinyl]-*N*-[4-[[4-(1-piperidinylmethyl)-2-pyri-dinyl]oxy]-2-butenyl]acetamide (●)
S FRG-8813, *Lafutidine***, *Loctidine*, Protecadin, Sto-ga(r)
U Histamine H_2-receptor antagonist (anti-ulcer)

11390 $C_{22}H_{29}N_3O_6S$
                        184633-88-3

Pivaloyloxymethyl (1*R*,5*S*,6*S*)-2-[6,7-dihydro-5*H*-pyrro-lo[1,2-*a*]imidazol-6(*S*)-ylthio]-6-[1(*R*)-hydroxyethyl]-1-methyl-1-carba-2-penem-3-carboxylate = (4*R*,5*S*,6*S*)-3-[[(6*S*)-6,7-Dihydro-5*H*-pyrrolo[1,2-*a*]imidazol-6-

yl]thio]-6-[(1R)-1-hydroxyethyl]-4-methyl-7-oxo-1-aza-
bicyclo[3.2.0]hept-2-ene-2-carboxylic acid (2,2-dime-
thyl-1-oxopropoxy)methyl ester (●)
S   DZ 2640
U   Antibiotic

11391   (9159) $C_{22}H_{29}N_3O_6S$
                                                            33817-20-8

(2S,5R,6R)-6-[(R)-2-Amino-2-phenylacetamido]-3,3-di-
methyl-7-oxo-4-thia-1-azabicyclo[3.2.0]heptane-2-carb-
oxylic acid hydroxymethyl ester pivalate (ester) = Pivaloyloxymethyl (6R)-6-(D-α-phenylglycylamino)penicil-
lanate = D(–)-(α-Aminobenzyl)penicillin hydroxymethyl
ester pivalate (ester) = [2S-[2α,5α,6β(S*)]]-6-[(Amino-
phenylacetyl)amino]-3,3-dimethyl-7-oxo-4-thia-1-azabi-
cyclo[3.2.0]heptane-2-carboxylic acid (2,2-dimethyl-1-
oxopropoxy)methyl ester (●)
R   most monohydrochloride (26309-95-5)
S   Acerum, Alfacilina, Alphacilina, Alphacillin
"MSD", Bencilin "Medicamenta", Bensamin, Beroci-
lina, Berocillin, Brotacilina, Centurina, Co-Pivam,
Crisbiotic, CS 390, Dancilin, Devonian, Diancina,
Inacilin, Isvitrol, Kesmicina, Kratistocin, Lanzabio-
tic, Lervipan, Maxifen "MSD", MK-191, Novopi-
vam, Oxidina, Penimenal, Pibena, Pivabiot, Pivacid,
Pivacilin(a), Pivacostyl, Pivadilon, Piva-Efesal, Piva-
mag, Pivamboi, Pivaminol, Pivamiser, Pivamkey, *Pi-
vampicillin hydrochloride**, Pivampil, Pivampin,
Pivapen, Pivascel, Pivastol, Pivatil, Pivavives, Pivio-
tic, Pondocil, Pondocillina, Pondocillin(e), Pro-Am-
pi "Parke-Davis; Stafford-Miller", Roxil "Robins",
Sanguicillin, Tam-Cilin, Tryco, Vampi-Framan,
VD 923, Vedecillin
U   Antibiotic

11391-01   (9159-01)   42190-91-0
R   Compd. with probenecid (no. 4222) (1:1)
S   MK-356, *Pivampicillin probenate*
U   Antibiotic

11391-02   (9159-02)   39030-72-3
R   Pamoate (embonate) (2:1)
S   *Pivampicillin pamoate*
U   Antibiotic

11391-03   (9159-03)   77442-45-6
R   6β-Bromopenicillanate (no. 1189) (1:1)
S   VD 2085
U   Antibiotic

11392   (9160) $C_{22}H_{29}N_3S_2$
                                                            1420-55-9

2-(Ethylthio)-10-[3-(4-methyl-1-piperazinyl)propyl]phe-
nothiazine = 2-(Ethylthio) 10-[3-(4-methyl-1-piperazi-
nyl)propyl]-10H-phenothiazine (●)
R   Maleate (1:2) (1179-69-7) [also malate (1:2)
(52239-63-1)]
S   GS 95, Norzine, NSC-130044, *Thiethylperazine
maleate***, Thietylperazine maleate, *Tietilperazina
dimaleato*, Torecan, Toresten, Tresten
U   Anti-emetic, central depressant

11393   (9161) $C_{22}H_{29}N_7O_5$
                                                            53-79-2

3'-(L-α-Amino-p-methoxyhydrocinnamamido)-3'-deoxy-
N,N-dimethyladenosine = 6-(Dimethylamino)-9-[3-de-
oxy-3-(p-methoxy-L-β-phenylalanylamino)-β-D-ribofu-
ranosyl]purine = (S)-3'-[[2-Amino-3-(4-methoxyphenyl)-

1-oxopropyl]amino]-3'-deoxy-*N,N*-dimethyladenosine (●)

R also dihydrochloride (58-58-2)

S Achromycin (old form), CL 16536, 3123 L, NSC-3055, P-638, *Puromicina*, *Puromycin**, Stillomycin, Stylomycin

U Antineoplastic, antitrypanosomal (antibiotic from *Streptomyces alboniger*)

11394   (9162) $C_{22}H_{29}N_9O_3$
126856-36-8

4,5-Dihydro-6-[4-[4-[[[3-(1*H*-imidazol-4-yl)propyl]amino]iminomethyl]-1-piperazinyl]-3-nitrophenyl]-5-methyl-3(2*H*)-pyridazinone = *N*-[3-(1*H*-Imidazol-4-yl)propyl]-4-[2-nitro-4-(1,4,5,6-tetrahydro-4-methyl-6-oxo-3-pyridazinyl)phenyl]-1-piperazinecarboximidamide (●)

S He 30582

U Cardiotonic

11395   (9163) $C_{22}H_{29}N_9O_9S_2$
76610-84-9

(6*R*,7*R*)-7-[(2*R*,3*S*)-2-(4-Ethyl-2,3-dioxo-1-piperazinecarboxamido)-3-hydroxybutyramido]-7-methoxy-3-[[(1-methyl-1*H*-tetrazol-5-yl)thio]-methyl]-8-oxo-5-thia-1-azabicyclo[4.2.0]oct-2-ene-2-carboxylic acid = (7*R*)-7-[(2*R*,3*S*)-2-(4-Ethyl-2,3-dioxo-1-piperazinecarboxamido)-3-hydroxybutyramido]-7-methoxy-3-[[(1-methyl-5-tetrazolyl)thio]methyl]-3-cephem-4-carboxylic acid = [6*R*-[6α,7α,7(2*R**,3*S**)]]-7-[[2-[[(4-Ethyl-2,3-dioxo-1-piperazinyl)carbonyl]amino]-3-hydroxy-1-oxobutyl]amino]-7-methoxy-3-[[(1-methyl-1H-tetrazol-5-yl)thio]methyl]-8-oxo-5-thia-1-azabicyclo[4.2.0]oct-2-ene-2-carboxylic acid (●)

R Monosodium salt (76648-01-6)

S BMY-25182, CBPZ-Na, *Cefbuperazone sodium***, Keiperazon, T-1982, Tomiporan

---

U Antibiotic

11396   (9164) $C_{22}H_{30}BrClNO$

4-Bromobenzyl[3-(4-chloro-2-isopropyl-5-methylphenoxy)propyl]dimethylammonium = 4-Bromo-*N*-[3-[4-chloro-5-methyl-2-(1-methylethyl)phenoxy]propyl]-*N,N*-dimethylbenzenemethanaminium (●)

R Chloride (7008-13-1)

S *Halopenium chloride***, *Haloponi chloridum*

U Antibacterial, antifungal

11397 $C_{22}H_{30}BrN_6O_{10}P$
213982-93-5

N-[3'-Azido-5-bromo-3'-deoxy-5,6-dihydro-6-methoxy-*P*-(4-methoxyphenyl)-5'-thymidylyl]-L-alanine methyl ester (●)

S WHI-05

U Antiviral (HIV), spermicide

11398   (9165) $C_{22}H_{30}ClN_3O_2$
66564-15-6

2-(Allyloxy)-4-amino-5-chloro-*N*-[1-(3-cyclohexen-1-yl-methyl)-4-piperidyl]benzamide = 4-Amino-5-chloro-*N*-[1-(3-cyclohexen-1-ylmethyl)-4-piperidinyl]-2-(2-propenyloxy)benzamide (●)

S *Alepride***

U   Anti-emetic, stomachic

11399 $C_{22}H_{30}ClN_3O_3$
252240-56-5

8-[2-[4-(5-Chloro-2-methoxyphenyl)-1-piperazi-
nyl]ethyl]-8-azaspiro[4.5]decane-7,9-dione (●)
S   Rec 15/3039
U   Adrenergic inverse agonist

11400   (9166) $C_{22}H_{30}ClN_3O_3$
110675-51-9

6-Chloro-2-cyclohexyl-1'-[3-(dimethylamino)propyl]spi-
ro[chroman-4,4'-imidazolidine]-2',5'-dione = 6-Chloro-2-
cyclohexyl-1'-[3-(dimethylamino)propyl]-2,3-dihydro-
spiro[4H-1-benzopyran-4,4'-imidazolidine]-2',5'-dione
(●)
S   M 79193
U   Detoxicant, hepatoprotectant

11401   (9167) $C_{22}H_{30}Cl_2N_2O_2$
87151-85-7

(±)-2-(3,4-Dichlorophenyl)-N-methyl-N-[(5R*,7S*,8S*)-
7-(1-pyrrolidinyl)-1-oxaspiro[4.5]dec-8-yl]acetamide =
rel-3,4-Dichloro-N-methyl-N-[(5R,7S,8S)-7-(1-pyrrolidi-
nyl)-1-oxaspiro[4.5]dec-8-yl]benzeneacetamide (●)

R   Monomethanesulfonate (87173-97-5)
S   *Spiradoline mesilate***, U-62066 E
U   Analgesic

11402   (9168) $C_{22}H_{30}Cl_2N_{10}$
55-56-1

1,1'-Hexamethylenebis[5-(p-chlorophenyl)biguanide] =
1,6-Bis[5-(p-chlorophenyl)diguanido]hexane = N,N''''-
1,6-Hexanediylbis[N'-(4-chlorophenyl)imidodicarbon-
imidediamide] = N,N''-Bis(4-chlorophenyl)-3,12-diimi-
no-2,4,11,13-tetraazatetradecanediimidamide (●)
R   also salts [diacetate (56-95-1), dihydrochloride
(3697-42-5), di-D-gluconate (18472-51-0)]
S   Abacil, Acriflex "Seton", Adro-derm, Allcut, An-
gisan, anti-Plaque, Antisept, Aqua Emoplac Antipla-
ca, Arnosept, AY-5312, Bacterisol, Bacticlens, Bac-
tigras, Bactisidal, Bactoshield, Baxedin, Beach-Gur-
gellösung, Benclosid, Betasept "Purdue Frederick",
Bidex, Bio-Facial, Biopatch, Bioscrub, Biotensid,
Blend-A-med Periochip, Brian Care, Broxodin, Bu-
casept, Bucoral, Bush Formula, Cathejell S, Cefa-
sept, Cepton, Cervitec, Cetal "Orapharm",
CHG Scrub, Chlohexin, Chlorasept, Chlorhexamed,
Chlorhex-a-myl, *Chlorhexidine***, Chlorhexitulle,
Chlorhexplak, Chlorhexseptic, Chlorohex, Chlor-
zoin, CHX Dental Gel, Cida-Stat, Cidegol C, Cle-
gen, Clinitol, Cloraldin, Clorexan, Clorexdermol,
Clorexident, *Clorexidina*, *Clorexidina Gluconato*,
Clorhexidin, Clorhexitulle, Cloridin, Clorosan "La-
chifarma", Clorxil, Collunovar atomiseur, Colun-
ovex, Compound 10040, Contigel, Corsodyl, Cristal-
crom, Cristalmina, Curafil, Cuvefilm, CX powder,
Dentisept, Dentohexin, Dentosmin, Deratin, Deratin
Orto, Derma Plast, DermaSept, Dermexin, Dermo-
diatic, Descutan, Desmanol (main active substance),
Dispray 1 (or 2), Disteryl, Dosiseptine, D-Seb, Dy-
na-Hex, Effetre, Eksplorasjonskrem, Ekuba, Elgydi-
um, Eludril (active substance), Elugel, Endosgel,
Exidine, Exoseptoplix, Femfresh, Fermajin, Fimeil,
Finoderm, Freshmel, Frubilurgyl, Garonsept, Gin-

gisan, Golasan, Golaseptine, Golasol, Grim, Gurgel-lösung Chauvin, Gurgol, Hansamed-Spray, Heksasol, Hexadol, Hexamedal, Hexangin, Hexidin "Nyromed", Hexidin(e), Hexifoam, Hexizac, Hexol "Sigma", Hexophene, Hibicare, Hibiclens, Hibicol, Hibicrick, Hibideks DAP, Hibident, Hibidil, Hibigel, Hibiguard, Hibimax, Hibiscrub, Hibisel, Hibisol, Hibisorb, Hibisprint, Hibistat, Hibital "Zeneca", Hibitane, Hibitex, Hybiscrab, Hydrex "Adams; Unitech", ICI 10040, Kleenocid, Klorheksidos, Klorheksol, Klorhex, *Klorhexidin*, Klorhexol, K-Y Jelly, Lady Douche Plus, Larilon, Larylin-Gurgellsg., Lauvir, Lemocin CX, Lenil, Lenixil, Lifo-Scrub, Lisium, Lyasin, Lysofon "Lafon", Macrocide, Manusan, Maskin, Medicanol, Mediscrub, Mefren, Menalmina, Mentopin Gurgellösung, Merfène (new form), Microderm, Microlol, Microshield, Molfina, N 32, Naseptin, Neomercurocromo bianco, Neoxene, Neoxinal, Nolvasan, Normol, Novaclens, Novacol, Nur 1 Tropfen-Chlorhexidin, Odol-med dental, Odontoxina, Orahexal, Oralgene, Oro-Clense CHX, Oronine H, Orosept "Triomed", Pabron-Gargle, Peridex "Procter & Gamble", Periochip, PerioGard Chlorohex, pHiso-MED, Pixidin, Plac-Out, Plak-Out, Plaqacide, Plaqacide Mouthrinse, Plivasept, Plurexid, Präparat 10040, Prexidine, Primahex, Rally-A-S, Rapotec, Rexifoam, Rhino-Blache, Roter Keel, Rotersept, Rouhex-G, Salvequick, Salvesept, Sanoral "Bioprogress", Savacol, Savloclens, Savlodil, Savlon Medicated Powder, Scrub Care, Sebidin, Secalan, Septadin, Septalone, Septal-Scrub, Septéal, Septofort, Serotulle, Softaderm, Soludril sans sucre, Sorbifen-Wundpuder, Soretol, Spectro-Gram, Spotoway, Stanhexidine, Sterets Healthwipes, Sterexidine, Stericlon, Sterilon(e), Steripod, Steripod Pink, Steri-Sat, Steri-Stat, Syntasept, Tesan, Tisept, Travahex, Travasept, Triseptil, Uniscrub, Unisept "Seton", Urgospray, Uriflex C, Uroflex, Urogliss, Urogliss-S, Uterine, Vaxidina, Ventrosteril, Vetasept, Videorelax, Vigravit, Vitacontact, White Skodokugauze, Wick-Sulagil-Gurgellösung, Wound Wash "Zyma", Yavatox plus

U Topical antiseptic

11402-01 (9168-01)   77146-42-0
R (4-Aminophenyl)phosphate (1:2)
S BMY-30120, *Chlorhexidine phosphanilate*, CHP, WP 973

U Antiseptic

11403 (9169) $C_{22}H_{30}FNO_4$
84057-96-5

(*S*)-1-[*p*-[2-[(*p*-Fluorophenethyl)oxy]ethoxy]phenoxy]-3-(isopropylamino)-2-propanol = (*S*)-1-[4-[2-[2-(4-Fluorophenyl)ethoxy]ethoxy]phenoxy]-3-[(1-methylethyl)amino]-2-propanol (●)
S *Flusoxolol***, Ro 31-1411
Y β-Adrenergic blocker

11404 (9170) $C_{22}H_{30}FN_3OS_2$
131540-59-5

3-[4-[4-(6-Fluorobenzo[*b*]thiophen-3-yl)-1-piperazinyl]butyl]-2,5,5-trimethyl-4-thiazolidinone (●)
R Maleate (1:1) (131540-60-8)
S HP 236, P 9236
U Antipsychotic

11405   (9171) $C_{22}H_{30}FO_8P$
16562-18-8

6α-Fluoro-11β,17,21-trihydroxy-16α-methylpregna-1,4-diene-3,20-dione 21-(dihydrogen phosphate) = 6α-Fluoro-16α-methylprednisolone 21-phosphate = (6α,11β,16α)-6-Fluoro-11,17-dihydroxy-16-methyl-21-(phosphonooxy)pregna-1,4-diene-3,20-dione (●)
R Disodium salt (2145-14-4)
S Cortidene soluble, Monocortin-S, *Paramethasondinatriumphosphat*, *Paramethasone disodium phosphate*, RS-404, Solu-Dilar, Soludillar
U Glucocorticoid (anti-inflammatory, anti-allergic)

11406   (9172) $C_{22}H_{30}FO_8P$
312-93-6

9-Fluoro-11β,17,21-trihydroxy-16α-methylpregna-1,4-diene-3,20-dione 21-(dihydrogen phosphate) = 9-Fluoro-16α-methylprednisolone 21-phosphate = (11β,16α)-9-Fluoro-11,17-dihydroxy-16-methyl-21-(phosphonooxy)pregna-1,4-diene-3,20-dione (●)
S Calonat-liofilizado, Decadran-inyect., Dekasol, Dexafar "Faran", Dexagrin inyect., *Dexamethasone phosphate*, Dexona-Inj., Ificort-Inj., Juvadexon-Amp., Neodecadron, Trofinan, Wymesone-Inj.
U Glucocorticoid (anti-inflammatory, anti-allergic)

11406-01   (9172-01)   2392-39-4
R Disodium salt
S Aacidexam, Ak-Dex, Alba-Dex, Alin (inyect., crema simple, forte), Antillerg, Baldex, Betameth, Brulin, Cebedex, Colvasone, Corson-Amp., Dalalone, Dala-

ron, Decaderm "Frosst", Decadran inj., Decadrol, Decadron fosfato, Decadron phosphate, Decaject, Decameth, Decasone-Inj., Dekadex-5 inyect., Dekort inj., Deron S, Desalark (collirio + pomata), Desashok, Desocort (auriculaire), Desone, Detason, Dex-4, Dexa, Dexa 4/-8 von ct, Dexa "Vana", Dexa-Allvoran, Dexa-Augentropfen, Dexabene, Dexabeta, Dexa-Brachialin N, Dexacen-4, Dexa-clinit, Dexacollyre, Dexacom, Dexacortin Amp., Dexacort Inj. "Teva", Dexacortyl, Dexadreson, Dexa EDO, Dexa-Effekton, Dexa-F, Dexafar "Fournier", Dexafarm, Dexaflam N Lichtenstein, Dexafort, Dexagel, Dexagro, Dexa-Helvacort-Amp., Dexahexal, Dexa in der Ophtiole, Dexa-Inject, Dexair, Dexa-Jenapharm, Dexalavis, Dexalergin, Dexalone "Coophavet.", Dexamed "Medochemie", Dexamed-Monoampulle, Dexamedron, *Dexamethasone sodium phosphate*, Dexa-ratiopharm, Dexa-sine SE, Dexason "Indus", Dexasone "Ferndale; Hauck", Dexa TAD, Dexaven(e), Dex-A-vet, Dexawal, Dexawieb, Dexcorvis, Dexject, Dexon "Kay", Dexone-Amp., Dexsone, Dexzone, Dezone-Inj., Dibasona, Diodex, Duphacort, Etacortilen, F-Corten, Fortecortin Inject, Fortecortin-Mono-Amp., Fosfodexa, Grosodexon-Amp., Hexadreson, Hexadrol Phosphate Inj., I-Methasone, Izometazone, Lormine, Maxidex Ointment, Megacort, Mephameson inj., Novadex "Nova", NR 21 P, Ocu-Dex, Ocu-Dexa-Oculogutt, Oftan Dexa, Onadron-Inj., Opidexol, Oradexon-Amp., Orgadrone-Inj., Osatron-Solución, Penthasone, Rapison, Reusan, R.O.-Dexsone, Rupedex-inyect., Savacort-D, Soldesam, Soldesanil, Solone inyect., Solu-Decadron, Solurex, Spersadex, Spondy-Dexa, Storz-Dexa, Tendron, Totocortin, Triamcort "Faran", Turbinaire (-Decadron)
U Glucocorticoid (anti-inflammatory, anti-allergic)

11407   (9173) $C_{22}H_{30}FO_8P$
360-63-4

9-Fluoro-11β,17,21-trihydroxy-16β-methylpregna-1,4-diene-3,2-dione 21-(dihydrogen phosphate) = 9-Fluoro-16β-methylprednisolone 21-phosphate = (11β,16β)-9-Fluoro-11,17-dihydroxy-16-methyl-21-(phosphono-oxy)pregna-1,4-diene-3,20-dione (●)

R Disodium salt (151-73-5)

S Adbeon, Bedifos, Bentelan, Beta-corlan, Betafluorene-Gocce, Betallorens, Betameson, Betametasona Inyect "L. CW.", *Betamethasone sodium phosphate*, Betam-Ophtal, Betapred "Glaxo", Betasone, Beta-Stulln, Betnasol, Betnesol, Betnesol WL, Betsolan solubile, Bifosona, B-S-P, Celesdepot, Celestan Depot, Celestan solubile, Célestène injectable, Celestona-Amp., Celestone Repetabs, Cel-U-Jec., C-Lest, Corteroid inyect., Diprofos, Durabetason, Emilan, Flosteron solubile, *Flubenisolone disodium phosphate*, Hicort, Lenasone, Linolosal, Minisone, Paucisone, Pianet, Probeta "Ato", Sanbetason, Sclane-Amp., Selestoject, Solu-Celestan, Solu-Celeston, Solusone, Steronema, Tamezon, Topicorten "Chemica", Veticort, Vista-Methasone

U Glucocorticoid (anti-inflammatory, anti-allergic)

11408 (9174)

$C_{22}H_{30}F_2N_{10}$
53416-52-7

1,1'-Hexamethylenebis[5-(*p*-fluorophenyl)biguanide] = 1,6-Bis[5-(*p*-fluorophenyl)diguanido]hexane = *N,N"*-Bis(4-fluorophenyl)-3,12-diimino-2,4,11,13-tetraazatetradecanediimidamide (●)

S *Fluorhexidine*

U Dental caries prophylactic

11409

$C_{22}H_{30}MnN_4O_{14}P_2$
155319-91-8

Hexahydrogen (*OC*-6-13)-[[*N,N'*-ethylenebis[*N*-[[3-hydroxy-5-(hydroxymethyl)-2-methyl-4-pyridyl]methyl]glycine]-5,5'-bis(phosphato)](8–)]manganate(6–) = Hexahydrogen (*OC*-6-13)-[[*N,N'*-1,2-ethanediylbis[*N*-[[3-(hydroxy-κ*O*)-2-methyl-5-[(phosphonooxy)methyl]-4-pyridinyl]methyl]glycinato-κ*N*,κ*O*]](8–)]manganate(6–) (●)

R Trisodium salt (140678-14-4)

S *Mangafodipir trisodium***, MnDPDP, S-095, Teslasean, Win 59010

U Diagnostic (paramagnetic contrast agent)

11410 (9175)

$C_{22}H_{30}NO$

Triethyl[2-[(*E*)-(*p*-styrylphenoxy)ethyl]]ammonium = (*E*)-Triethyl(β-4-stilbenoxyethyl)ammonium = *N,N,N*-Triethyl-2-[4-[(1*E*)-2-phenylethenyl]phenoxy]ethanaminium (●)

R Iodide (77257-42-2) [also without (*E*)-definition (2551-76-0)]

S Elvetil, Elveton, M.G. 624, *Stilbethonii iodidum*, *Stilonium iodide***

U Ganglion blocking agent, spasmolytic

11411   (9176) $C_{22}H_{30}NO$

N,N-Diallyl-3-hydroxymorphinanium = 3-Hydroxy-
17,17-di-2-propenylmorphinanium (●)
R   Bromide (63868-47-3)
S   CM 32191, *Levallorphan allyl bromide*
U   Peripheral narcotic antagonist

11412   (9177) $C_{22}H_{30}NO$

1,1-Dimethyl-2-[[(*p*-methyl-α-phenylbenzyl)oxy]me-
thyl]piperidinium = 1,1-Dimethyl-2-[[(4-methylphe-
nyl)phenylmethoxy]methyl]piperidinium (●)
R   Bromide (35620-67-8)
S   *Pirdonium bromide***
U   Antihistaminic

11413   (9178) $C_{22}H_{30}NO_2$
                                                  7590-99-0

Diethyl(3-hydroxypropyl)methylammonium diphenylacetate = [3-(Diphenylacetoxy)propyl]diethylmethylam-
monium = 3-[(Diphenylacetyl)oxy]-N,N-diethyl-N-me-
thyl-1-propanaminium (●)
R   Methyl sulfate (3811-12-9)

S   Arpenal methyl methosulfate, Mesfenal, Mespenal,
        Mesphenal
U   Anticholinergic

11414   (9179) $C_{22}H_{30}N_2$
                                                  13838-14-7

1,4-Bis(α-methylphenethyl)piperazine = 1,4-Bis(1-phe-
nylisopropyl)piperazine = 1,4-Bis(1-methyl-2-phenyl-
ethyl)piperazine (●)
R   Dihydrochloride (13754-23-9)
S   Diphenazin, F 200, Quietidin
U   Neuroleptic

11415   (9180) $C_{22}H_{30}N_2$
                                                  37640-71-4

N,N-Diethyl-N'-2-indanyl-N'-phenyl-1,3-propanedi-
amine = N-[3-(Diethylamino)propyl]-N-indan-2-ylaniline
= N-[3-(Diethylamino)propyl]-N-phenyl-2-indanamine =
3-(N-Indan-2-yl-N-phenylamino)propyldiethylamine =
N-(2,3-Dihydro-1H-inden-2-yl)-N',N'-diethyl-N-phenyl-
1,3-propanediamine (●)
R   Monohydrochloride (33237-74-0)
S   AC 1802, Amidonal, *Aprindine hydrochloride***,
        Aspenon, Elpen, Fibocil, Fiboran, Lilly 83846,
        MS 5075, Ritmusin
U   Anti-arrhythmic

11416 (9181) $C_{22}H_{30}N_2O$
                                                            68252-19-7

(±)-*cis*-2,6-Dimethyl-α-phenyl-α-2-pyridyl-1-piperidine-
butanol = *cis*-(±)-α-[3-(2,6-Dimethyl-1-piperidinyl)pro-
pyl]-α-phenyl-2-pyridinemethanol (●)
R Monohydrochloride (61477-94-9)
S CCRIS 5243, CI-845, Pimenol, Pirmavar, *Pirmenol
      hydrochloride***
U Anti-arrhythmic

11417 (9183) $C_{22}H_{30}N_2O_2$
                                                            466-49-9

1-Acetyl-17-methoxyaspidospermidine (●)
S *Aspidospermine*
U Respiratory stimulant (alkaloid from *Aspidosperma
      quebracho-blanco* and other *Apocynaceae*)

11418 $C_{22}H_{30}N_2O_2$
                                                           174643-98-2

8-[4-(3,4-Dihydro-2(1*H*)-isoquinolinyl)butyl]-8-azaspi-
ro[4.5]decane-7,9-dione (●)

S MM-199
U Anxiolytic

11419 (9184) $C_{22}H_{30}N_2O_2$
                                                            79784-22-8

4-Benzyl-1,3-dihydro-7-[4-(isopropylamino)butoxy]-6-
methylfuro[3,4-*c*]pyridine = 4-[(4-Benzyl-1,3-dihydro-6-
methylfuro[3,4-*c*]pyridin-7-yl)oxy]-*N*-isopropylbutyl-
amine = 4-[[1,3-Dihydro-6-methyl-4-(phenylmethyl)fu-
ro[3,4-*c*]pyridin-7-yl]oxy]-*N*-(1-methylethyl)-1-butan-
amine (●)
S *Barucainide***
U Anti-arrhythmic

11420 (9182) $C_{22}H_{30}N_2O_2$
                                                            32665-36-4

4-(β-Methoxyphenethyl)-α-phenyl-1-piperazinepropa-
nol = 3-[4-(β-Methoxyphenethyl)-1-piperazinyl]-1-phe-
nyl-1-propanol = 4-(2-Methoxy-2-phenylethyl)-α-phe-
nyl-1-piperazinepropanol (●)
R Dihydrochloride (27588-43-8)
S Alecor, Asmisul, Brovel, *Eprozinol dihydrochlo-
      ride***, Eupneron, Nadyl
U Anti-asthmatic, bronchodilator

11421　(9185)　　　　　　　　　　　$C_{22}H_{30}N_2O_2S$
56030-54-7

N-[4-(Methoxymethyl)-1-[2-(2-thienyl)ethyl]-4-piperi-dyl]propionanilide = *N*-[4-(Methoxymethyl)-1-[2-(2-thie-nyl)ethyl]-4-piperidinyl]-*N*-phenylpropanamide (●)
R　also citrate (1:1) (60561-17-3)
S　Fentamorf, Fentatienil "Angelini", R 30730, R 33800, Sufenil, Sufenta, Sufentan, *Sufentanil**,* *Sulfentanyl*
U　Narcotic analgesic

11422　　　　　　　　　　　　　　　$C_{22}H_{30}N_2O_3$
147497-64-1

4-(2-Aminoethoxy)-*N*-[3-(3,4-dimethylphenyl)propyl]-3-methoxybenzeneacetamide (●)
R　Monohydrochloride (174661-97-3)
S　*Capsavanil hydrochloride*, DA 5018, KR-25018
U　Topical analgesic

11423　(9186)　　　　　　　　　　　$C_{22}H_{30}N_2O_4$
102908-55-4

(*S*)-(−)-8-[4-[[(1,4-Benzodioxan-2-ylmethyl)amino]bu-tyl]-8-azaspiro[4.5]decane-7,9-dione = (*S*)-8-[4-[[(2,3-Dihydro-1,4-benzodioxin-2-yl)methyl]amino]butyl]-8-azaspiro[4.5]decane-7,9-dione (●)
R　also with (±)-definition (113777-33-6)
S　MDL 72832
U　$5HT_{1\alpha}$-receptor and α_1-adrenoceptor antagonist

11424　(9187)　　　　　　　　　　　$C_{22}H_{30}N_2O_4S$
108498-50-6

2-[(2-Furanylmethyl)sulfinyl]-*N*-[3-[3-(1-piperidinylme-thyl)phenoxy]propyl]acetamide (●)
S　FRG-8701
U　Histamine H_2-receptor antagonist (anti-ulcer agent)

11425　(9188)　　　　　　　　　　　$C_{22}H_{30}N_2O_5$
87679-71-8

(2*S*,3a*S*,7a*S*)-1-[(*S*)-*N*-[(*S*)-1-Carboxy-3-phenylpro-pyl]alanyl]hexahydro-2-indolinecarboxylic acid = [2*S*-[1[*R**(*R**)],2α,3aα,7aβ]]-1-[2-[(1-Carboxy-3-phenylpro-pyl)amino]-1-oxopropyl]octahydro-1*H*-indole-2-carb-oxylic acid (●)
R　see also no. 12561
S　RU 44403, *Trandolaprilat**
U　Antihypertensive (ACE inhibitor)

11426　(9189)　　　　　　　　　　　$C_{22}H_{30}N_2O_5S_2$
83647-97-6

(8S)-7-[(S)-N-[(S)-1-Carboxy-3-phenylpropyl]alanyl]-1,4-dithia-7-azaspiro[4.4]nonane-8-carboxylic acid 1-ethyl ester = [8S-[7[R*(R*)],8R*]]-7-[2-[[1-(Ethoxycarbonyl)-3-phenylpropyl]amino]-1-oxopropyl]-1,4-dithia-7-azaspiro[4.4]nonane-8-carboxylic acid (●)
R Monohydrochloride (94841-17-5)
S Cardiopril "Schwarz Pharma; Searle", Quadropril, Renormax, Renpress, Sandopril, Sch 33844, Setrilan, *Spirapril hydrochloride***
U Antihypertensive (ACE inhibitor)

11427 (9190) $C_{22}H_{30}N_2O_6$
                       129689-30-1

(S)-2-[4-[2-[(2-Hydroxy-3-phenoxypropyl)amino]ethoxy]phenoxy]-N-(2-methoxyethyl)acetamide (●)
S ICI D 7114, ZD 7114
U β-Adrenoceptor agonist (anti-obesity)

11428 (9191) $C_{22}H_{30}N_2O_6$
                       17692-56-7

4-Methyl-5,7-bis(2-morpholinoethoxy)coumarin = 4-Methyl-5,7-bis[2-(4-morpholinyl)ethoxy]-2H-1-benzo-pyran-2-one (●)
R Dihydrochloride (94110-07-3)
S Fleboxil, *Moxicoumone dihydrochloride***, *Moxicumoni chloridum*, Moxile, Rec 15/0019
U Capillary protectant

11429 (9192) $C_{22}H_{30}N_4OS$
                       34349-88-7

4-Butoxy-4'-(4-methyl-1-piperazinyl)thiocarbanilide = N-(4-Butoxyphenyl)-N'-[4-(4-methyl-1-piperazinyl)phenyl]thiourea (●)
S Butomelid
U Tuberculostatic

11430 (9193) $C_{22}H_{30}N_4O_2$
                       126868-88-0

(Z)-5,6-Dimethyl-2-[[4-[3-(1-piperidinylmethyl)phenoxy]-2-butenyl]amino]-4(1H)-pyrimidinone (●)
R Dihydrochloride (126869-04-3)
S IGN-2098
U Histamine H_2-receptor antagonist (anti-ulcer agent)

11431 (9194) $C_{22}H_{30}N_4O_2S_2$
                       316-81-4

N,N-Dimethyl-10-[3-(4-methyl-1-piperazinyl)propyl]-2-phenothiazinesulfonamide = 2-(Dimethylsulfamoyl)-10-[3-(4-methyl-1-piperazinyl)propyl]phenothiazine = N,N-Dimethyl-10-[3-(4-methyl-1-piperazinyl)propyl]-10H-phenothiazine-2-sulfonamide (●)
R Dimethanesulfonate (2347-80-0)
S Cephalmin, IC 5911, Majeptil, Mayeptil, 7843 R.P., SKF 5883, Sulfenazin, Thioperazine, *Thioproperazine mesilate***, Vontil
U Neuroleptic, anti-emetic

11432 (9195) $[C_{22}H_{30}N_4O_5S]_x$
27307-30-8

Poly[iminocarbonyl[5-[4-carboxy-5,5-dimethyl-α-(2-phenylacetamido)-2-thiazolidineacetamido]pentylide-ne]] = N^6-[D-2-[(2R,4S)-4-Carboxy-5,5-dimethyl-2-thia-zolidinyl]-N-(phenylacetyl)glycyl]-L-lysine dodecapep-tide = [2R-[2α[R*(S*)],4β]]]-Poly[imino[1-[4-[[2-(4-carb-oxy-5,5-dimethyl-2-thiazolidinyl)-1-oxo-2-phenylace-tyl]amino]ethyl]amino]butyl]-2-oxo-1,2-ethanediyl (●)

S   *Benzylpenicilloyl Polylysine (Concentrate)*, Cilligen, *Penicilloyl-Polylysin*, *Penlysinum*, PPL, Pre-Pen, Te-starpen

U   Diagnostic aid (penicillin sensitivity)

11432-01 (9195-01)   61990-92-9
R   Dodecapotassium salt hydrate
S   *Benpenolisin***
U   Diagnostic aid (penicillin sensitivity)

11433 $C_{22}H_{30}N_6O_3S$
151140-96-4

3-[3-[4-(5-Methoxy-4-pyrimidinyl)-1-piperazinyl]pro-pyl]-N-methyl-1H-indole-5-methanesulfonamide (●)
R   Fumarate (1:1) (171171-42-9)
S   *Avitriptan fumarate***, BMS 180048-02
U   Serotonin agonist (migraine therapeutic)

11434 $C_{22}H_{30}N_6O_4S$
139755-83-2

5-[2-Ethoxy-5-[(4-methyl-1-piperazinyl)sulfonyl]phe-nyl]-6,7-dihydro-1-methyl-3-propylpyrazolo[4,3-d]pyri-midin-7-one = 1-[[3-(4,7-Dihydro-1-methyl-7-oxo-3-pro-pyl-1H-pyrazolo[4,3-d]pyrimidin-5-yl)-4-ethoxyphe-nyl]sulfonyl]-4-methylpiperazine (●)

R   Citrate (1:1) (171599-83-0)
S   Androx, Caverta, Eroxim, Patrex, Penegra, Segurex, *Sildenafil citrate***, UK-92480-10, Viagra, Wan Ai Ke
U   Vasodilator (type 5 cGMP phosphodiesterase inhibi-tor)

11435 $C_{22}H_{30}N_6O_4S$
187162-33-0

N-[(1-Amidino-4-piperidinyl)methyl]-3-[[(phenylme-thyl)sulfonyl]amino]-6-methyl-2-oxo-1(2H)pyridineacet-amide = N-[[1-(Aminoiminomethyl)-4-piperidinyl]me-thyl]-6-methyl-2-oxo-3-[[(phenylmethyl)sulfonyl]ami-no]-1(2H)-pyridineacetamide (●)
S   L 373890
U   Anticoagulant

11436 (9196)

$C_{22}H_{30}O$
54024-22-5

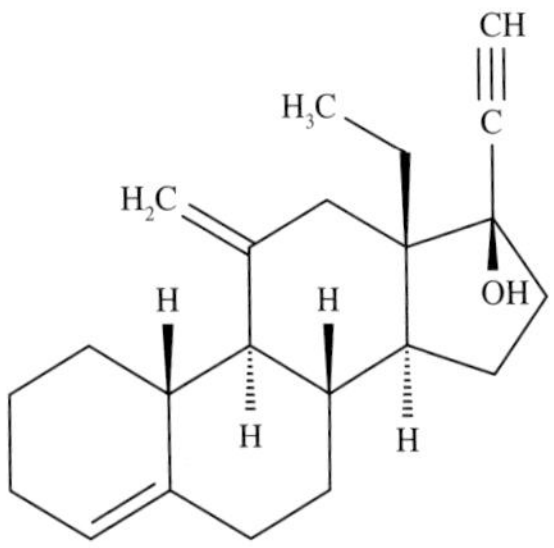

13-Ethyl-11-methylene-18,19-dinor-17α-pregn-4-en-20-yn-17-ol = (17α)-13-Ethyl-11-methylene-18,19-dinor-pregn-4-en-20-yn-17-ol (●)
S Ceracette, Cerazette, *Desogestrel***, Infar, Org 2969
U Progestin

11437 (9197)

$C_{22}H_{30}O_2$
34184-77-5

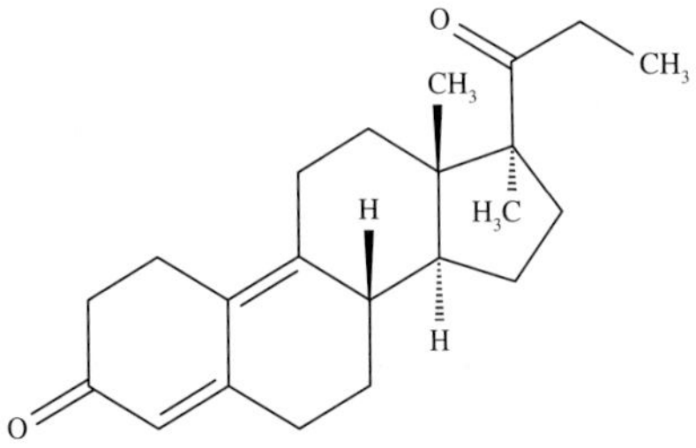

17α-Methyl-17-propionylestra-4,9-dien-3-one = 17,21-Dimethyl-19-norpregna-4,9-diene-3,20-dione = (17β)-17-Methyl-17-(1-oxopropyl)estra-4,9-dien-3-one (●)
S *Promegestone***, R 5020, Surgestone
U Progestin

11438 (9198)

$C_{22}H_{30}O_2$
1045-29-0

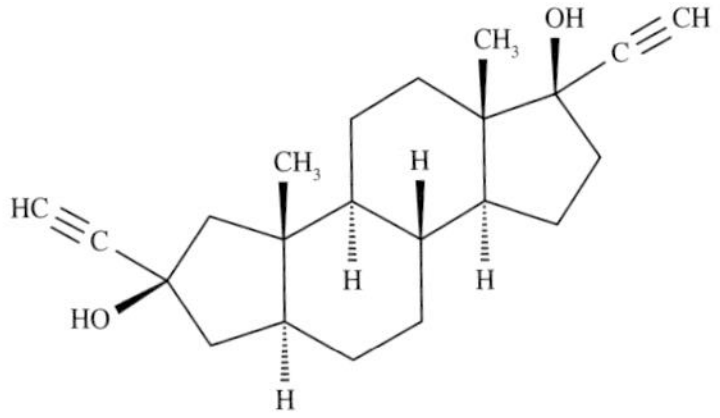

2α,17α-Diethynyl-A-nor-5α-androstane-2,17-diol = (2β,5α,17α)-2-Ethynyl-A-norpregn-20-yne-2,17-diol (●)
S AF 45, Anordiol

U Anti-estrogen

11439

$C_{22}H_{30}O_2$

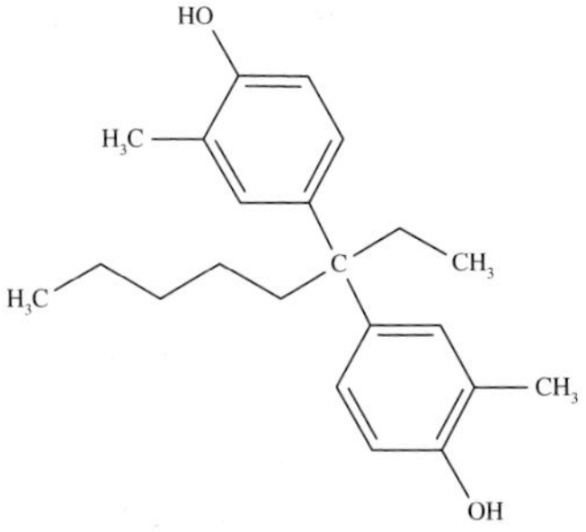

3,3-Bis(4-hydroxy-3-methylphenyl)octane = 4,4'-(1-Ethylhexylidene)bis[2-methylphenol] (●)
S PD 163892
U Antibacterial

11440 (9199)

$C_{22}H_{30}O_2$
19457-57-9

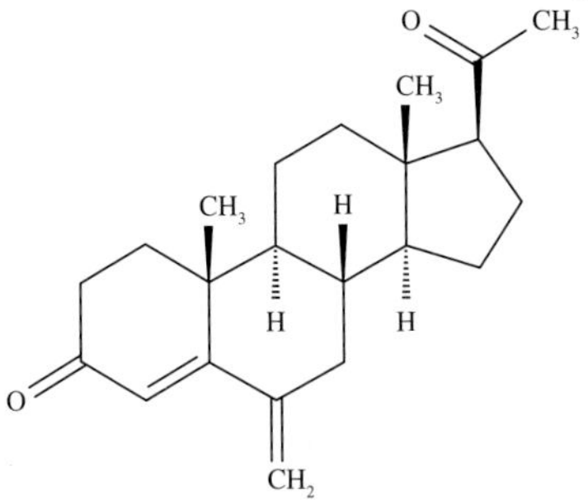

6-Methylenepregn-4-ene-3,20-dione (●)
S LY 207320, *Methyleneprogesterone*
U Anti-androgen

11441 (9203)

$C_{22}H_{30}O_3$
22733-60-4

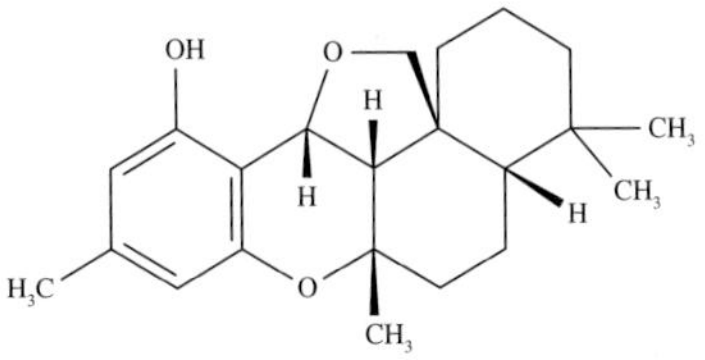

(13aS)-1,2,3,4,4aβ,5,6,6a,11bβ,13bβ-Decahydro-4,4,6aβ,9-tetramethyl-13H-benzo[a]furo[2,3,4-mn]xanthen-11-ol = [4aS-(4aα,6aα,11bα,13aR*,13bα)]-

1,2,3,4,4a,5,6,6a,11b,13b-Decahydro-4,4,6a,9-tetrame-
thyl-13H-benzo[a]furo[2,3,4-mn]xanthen-11-ol (●)
S CS 280, Piadar (active substance), SI 23548, *Sicca-
nin**, Tackle
U Antifungal (antibiotic from *Helminthosporium sic-
cans*)

11442 (9200) C_{22}H_{30}O_3
74513-62-5

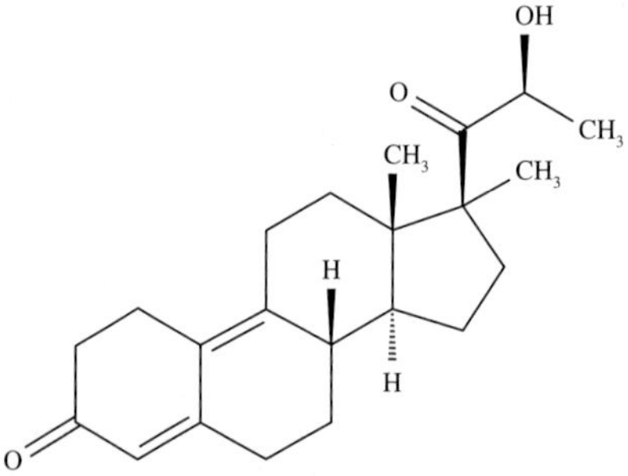

17β-[(S)-2-Hydroxypropanoyl]-17-methylestra-4,9-dien-
3-one = 17β-(S)-Lactoyl-17-methylestra-4,9-dien-3-one =
[17β(S)]-17-(2-Hydroxy-1-oxopropyl)-17-methylestra-
4,9-dien-3-one (●)
S Ondeva, RU 27987, *Trimegestone***
U Progestogen

11443 (9201) C_{22}H_{30}O_3
35100-44-8

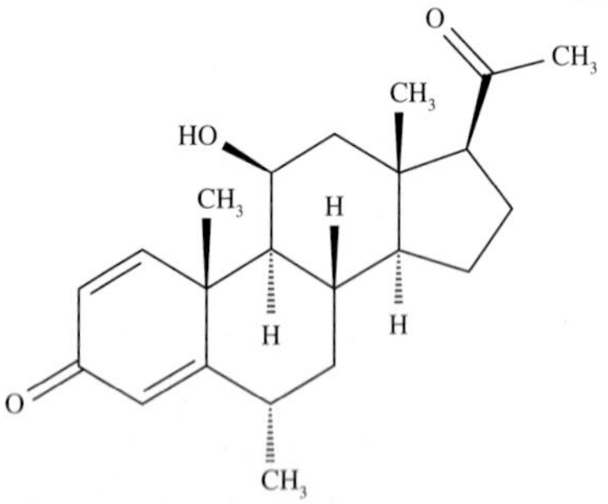

11β-Hydroxy-6α-methylpregna-1,4-diene-3,20-dione =
(6α,11β)-11-Hydroxy-6-methylpregna-1,4-diene-3,20-
dione (●)
S Aldrisone, Delta-Medryson, *Endrisone***, *Endrysone*
U Glucocorticoid (anti-inflammatory, anti-allergic)

11444 (9202) C_{22}H_{30}O_3
3562-63-8

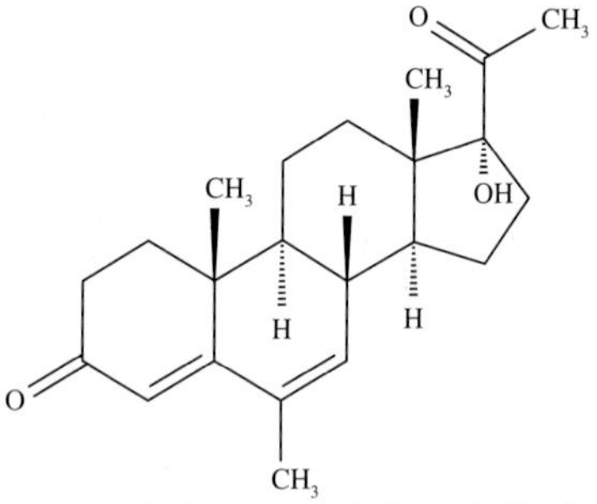

17-Hydroxy-6-methylpregna-1,4-diene-3,20-dione (●)
R see also no. 12515
S *Megestrol***
U Progestin, antineoplastic

11445 C_{22}H_{30}O_4
172923-88-5

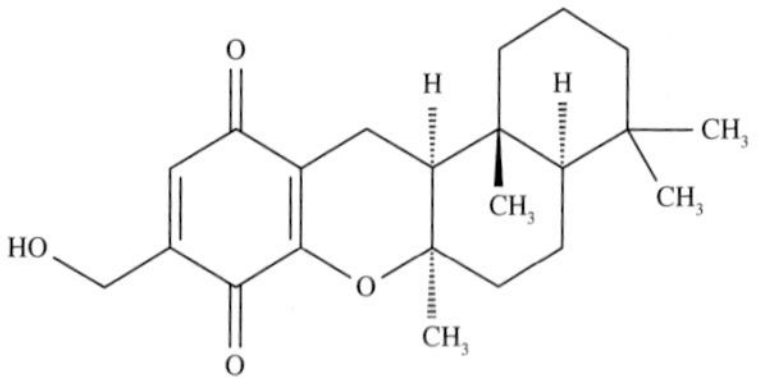

(6aα12bβ)-1,2,3,4,4a,5,6,6a,8,11,12,12a-Dodecahydro-
9-(hydroxymethyl)-4,4,6a,12b-tetramethylnaph-
tho[2,1-b][1]benzopyran-8,11-dione =
(4aα,6aα,12aα,12bβ)-1,3,4,4a,5,6,6a,12,12a,12b-Deca-
hydro-9-(hydroxymethyl)-4,4,6a,12b-tetramethyl-2H-
benzo[a]xanthene-8,11-dione (●)
S BE 40644
U Cytotoxic from *Actinoplanes* str. A 40644

11446 (9204) C_{22}H_{30}O_4
4138-96-9

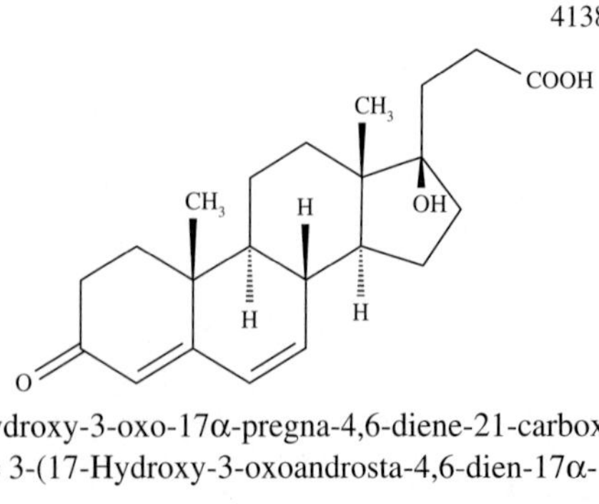

17-Hydroxy-3-oxo-17α-pregna-4,6-diene-21-carboxylic
acid = 3-(17-Hydroxy-3-oxoandrosta-4,6-dien-17α-

yl)propionic acid = (17α)-17-Hydroxy-3-oxopregna-4,6-diene-21-carboxylic acid (●)
S *Acide canrénoique***, *Acidum canrenoicum***, *Canrenoic acid***
U Aldosterone antagonist, diuretic

11446-01   (9204-01)   2181-04-6
R Potassium salt
S Aldactone pro inj., *Aldadiene Potassium*, *Canrenoate Potassium*, Canrenol, 8109 C.B., Doitolon, Gascool, Higroflux, *Kalii canrenoas***, Kalioral, Kalium-Can.-ratiopharm, Kanrenol, Luvion vena, MF 465a, Narmylon, Osiren pro inj., Osyrol pro inj., *Potassium canrenoate***, SC-14266, Sincomen pro inj., Soldactona, Soldactone, Soludactone, Spiroctan-m, Spiroctan pro inj., Venactone, Venectomin
U Aldosterone antagonist, diuretic

11447   (9205) $C_{22}H_{30}O_4S$
                                                       40845-00-9

3-[(3,17-Dioxoandrost-4-en-7α-yl)thio]propionic acid = 3-[[(7α)-3,17-Dioxoandrost-4-en-7-yl]thio]propanoic acid (●)
R Serum albumin conjugate
S GR 33207, *Ovandrotone albumin***
U Ovine fecundity

11448   (9208) $C_{22}H_{30}O_5$
                                                       95722-07-9

[2-[(2E,3aS,4S,5R,6aS)-Hexahydro-5-hydroxy-4-[(3S,4S)-3-hydroxy-4-methyl-1,6-nonadiynyl]-2-(1H)-pentalenylidene]ethoxy]acetic acid = [3aS-[2E,3aα,4α(3R*,4R*),5β,6aα]]-[2-[Hexahydro-5-hydroxy-4-(3-hydroxy-4-methyl-1,6-nonadiynyl)-2(1H)-pentalenylidene]ethoxy]acetic acid (●)
S *Cicaprost***, ZK 96480
U Prostaglandin

11449 $C_{22}H_{30}O_5$
                                                       60972-43-2

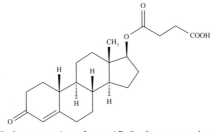

(1S,2S)-*rel*-2-[(1E,3R)-3-Hydroxy-4-phenoxy-1-butenyl]-5-oxocyclopentaneheptanoic acid (●)
S MB 28767
U Prostanoid EP$_3$ receptor agonist

11450   (9207) $C_{22}H_{30}O_5$
                                                       6785-62-2

17β-Hydroxyestr-4-en-3-one 17-(hydrogen succinate) = 19-Nortestosterone hemisuccinate = (17β)-17-(3-Carboxy-1-oxopropoxy)estr-4-en-3-one (●)
S Anabolico "Master", Menidrabol, *Nandrolone hydrogen succinate*, *Nortestosterone hemisuccinate*
U Anabolic

11451   (9206) $C_{22}H_{30}O_5$
                                                              83-43-2

11β,17,21-Trihydroxy-6α-methylpregna-1,4-diene-3,20-
dione = 6α-Methylprednisolone = (6α,11β)-11,17,21-Tri-
hydroxy-6-methylpregna-1,4-diene-3,20-dione (●)
R   see also nos. 11484, 12523, 13232, 13588, 14380,
       14803
S   Besônia, Bioprednon, Caberdelta-M, Cortalfa, Demi-
       cort, Dimedrol "Upjohn", Dura-Meth, Esametone,
       Eutisone, Firmacort, Genotan, Horusona, Lemod,
       Medesone, Medirona, Medixon, Medrate, Medrelon,
       Médrocortisone, Medrol, Medrone, Mega-Star, Me-
       prolone, Meristolone, Mesopren, Metastab, *Methyl-
       prednisolone***, Metilbetasone, Metilcort, Metilpred-
       nilone, *Metilprednisolone*, Metilstendiolo, Metipred,
       Metrisone, Metycortin, Metypred-Tabl., Metysolon,
       Moderin, Nirypan, Nixolan, Noretona, NSC-19987,
       Oro-Médrol, Prednilen, Predni-M-Tablinen, Pred-
       nol-L, Prednox, Promacortine, Radiosone, Reacte-
       nol, Sieropresol, Solomet, Summicort, Suprametil,
       U-7532, Urbason, Wyacort
U   Glucocorticoid (anti-inflammatory, anti-allergic)

11452   (9209) $C_{22}H_{30}O_5S_2$
                                                            119640-53-8

11β,17,21-Trihydroxypregn-4-ene-3,20-dione 21-(*O*-hy-
drogen dithiocarbonate) = Hydrocortisone 21-xanthoge-
nic acid = (11β)-21-[(Mercaptocarbonothioyl)oxy]-
11,17-dihydroxypregn-4-ene-3,20-dione = 11β-21-

[(Dithiocarboxy)oxy]-11,17-dihydroxypregn-4-ene-3,20-
dione (●)
R   also 2,2'-iminodiethanol salt (diethanolamine salt)
S   *Hydrocortisone xanthogenate*, Solvisat
U   Glucocorticoid (topical anti-allergic)

11453   (9210) $C_{22}H_{30}O_7$
                                                              82460-75-1

(5α)-8-(Acetyloxy)-8,10-deepoxy-13-deoxy-5-hydroxy-
enmein (●)
S   *Rabdophyllin G*
U   Antineoplastic from *Rabdosia macrophylla*

11454   (9211) $C_{22}H_{30}O_8$
                                                              18296-44-1

1,7a-Dihydro-1,6-dihydroxyspiro[cyclopenta[*c*]pyran-
7(6*H*),2'-oxirane]-4-methanol 4-acetate 1,6-diisovalerate
= 4-(Acetoxymethyl)-1,6-diisovaleryloxy-1,6,7,7a-tetra-
hydrocyclopenta[*c*]pyran-7-spiro-2'-oxirane = 3-Methyl-
butanoic acid (1*S*,2'*R*,6*S*,7a*S*)-4-[(acetyloxy)methyl]-
6,7a-dihydrospiro[cyclopenta[*c*]pyran-7(1*H*),2'-oxirane]-
1,6-diyl ester (●)
S   *Valepotriate, Valtrate**
U   Sedative (active principle from the roots of *Valeriana
       officinalis*)

11455 (9212) $C_{22}H_{31}ClO_2$
5591-27-5

6α-Chloro-16α-methylpregn-4-ene-3,20-dione =
(6α,16α)-6-Chloro-16-methylpregn-4-ene-3,20-dione
(●)
S *Clometerone***, *Clometherone*, L 38000
U Anti-estrogen

11456 (9213) $C_{22}H_{31}ClO_3$
2162-44-9

4-Chloro-17β-hydroxyandrost-4-en-3-one propionate =
4-Chlorotestosterone propionate = (17β)-4-Chloro-17-(1-
oxopropoxy)androst-4-en-3-one (●)
S *Clostebol propionate***, Yonchlon
U Anabolic

11457 (9214) $C_{22}H_{31}FO_2S_2$
85197-77-9

9-Fluoro-11β-hydroxyandrosta-1,4-diene-3,17-dione
17R)-17-(ethyl methyl mercaptole) = (11β,17α)-17-

(Ethylthio)-9-fluoro-11-hydroxy-17-(methylthio)andros-
ta-1,4-dien-3-one (●)
S Acti-Derm I (active substance), SQ 27239, *Tipre-
dane**
U Glucocorticoid (topical anti-inflammatory, anti-allergic)

11458 (9215) $C_{22}H_{31}NO$
13988-32-4

1-(p-Ethoxyphenyl)-N,N-diethyl-3-phenylbutylamine =
1-(Diethylamino)-1-(p-ethoxyphenyl)-3-methyl-3-phe-
nylpropane = p-Ethoxy-N,N-diethyl-α-(β-methylphen-
ethyl)benzylamine = α-(4-Ethoxyphenyl)-N,N-diethyl-γ-
methylbenzenepropanamine (●)
R Hydrochloride (10535-87-2)
S Sp 725, Spasmalfher (spasmolyt. component)
U Spasmolytic

11459 (9216) $C_{22}H_{31}NO$
124937-51-5

(+)-(R)-2-[α-[2-(Diisopropylamino)ethyl]benzyl]-p-cre-
sol = (R)-2-[3-[Bis(1-methylethyl)amino]-1-phenylpro-
pyl]-4-methylphenol (●)
R also tartrate (1:1) (124937-52-6)
S Detrol, Detrusitol, KABI-2234, PNU-200583, *Tol-
terodine***, Urotrol "Almirall"
U Muscarinic antagonist (for urinary incontinence)

11460

$C_{22}H_{31}NO_2$
207679-81-0

(*R*)-*N,N*-Diisopropyl-3-[2-hydroxy-5-(hydroxymethyl)phenyl]-3-phenylpropanamine = 3-[(1*R*)-3-[Bis(1-
methylethyl)amino]-1-phenylpropyl]-4-hydroxybenzene-
methanol (●)
S DD 01, PNU-200577
U Muscarinic antagonist

11461 (9220)

$C_{22}H_{31}NO_3$
80471-63-2

4α,5-Epoxy-3,17β-dihydroxy-4,17-dimethyl-5α-androst-
2-ene-2-carbonitrile = (4α,5α,17β)-4,5-Epoxy-3,17-di-
hydroxy-4,17-dimethylandrost-2-ene-2-carbonitrile (●)
S *Epostane***, Win 32729
U Contraceptive

11462 (9217)

$C_{22}H_{31}NO_3$
5633-20-5

4-(Diethylamino)-2-butynyl α-phenylcyclohexaneglycolate = α-Phenylcyclohexylglycolic acid 4-(diethylamino)-

2340

2-butynyl ester = α-Cyclohexyl-α-hydroxybenzeneacetic
acid 4-(diethylamino)-2-butynyl ester (●)
R also hydrochloride (1508-65-2)
S Apo-Oxybutynin, Cystonorm, Cystrin, Delak, Deli-
fon, Ditropan "Aventis; Mation Merrell Dow", Dri-
dase, Driptane, MJ 4309-1, Novitropan, Oros, *Oxibu-
tyninum*, Oxyb AbZ, Oxybase, Oxybubene, Oxybug-
amma, Oxybuton "Kanoldt", Oxybutyn, *Oxybu-
tynin***, Oxymedin, Oxynin, Oyrobin, Pasmonul,
Pollakisu, Retebem, Retemic, Ryol, Spasyt "TAD",
Tavor "Asofarma", Tropax, Urequim, Urequin, Uro-
butin, Uropan "Kocac Ilac San", Urotrol "Baker
Norton", Uroxal
U Anticholinergic, spasmolytic

11463

$C_{22}H_{31}NO_3$
146205-91-6

α-Cyclohexyl-α-hydroxybenzeneacetic acid 4-(ethylami-
no)-1,1-dimethyl-2-butynyl ester (●)
R Hydrochloride (129927-37-3)
S RCC-36
U Spasmolytic

11464 (9218)

$C_{22}H_{31}NO_3$
23271-63-8

2-Oxo-1-[β-(perhydro-1-azepinyl)ethyl]cyclohexane-
carboxylic acid benzyl ester = 2-(β-Hexamethyleneimi-
noethyl)cyclohexanone-2-carboxylic acid benzyl ester =
1-[2-(Hexahydro-1*H*-azepin-1-yl)ethyl]-2-oxocyclohe-
xanecarboxylic acid phenylmethyl ester (●)
S 580, *Amicibone***, Biotussal, Pectipront (old form)

U Antitussive

11465 (9219)

$C_{22}H_{31}NO_3$
118789-77-8

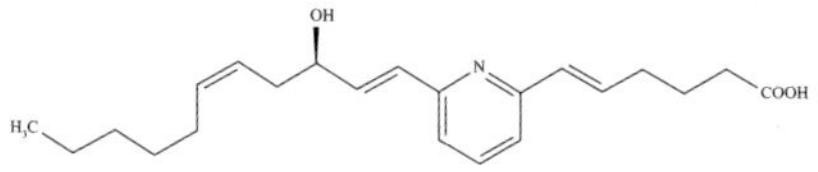

[R-(E,E,Z)]-6-[6-(3-Hydroxy-1,5-undecadienyl)-2-pyri-
dinyl]-5-hexenoic acid (●)
S U-77860
U Anti-allergic

11466 (9221)

$C_{22}H_{31}NO_4$
123618-00-8

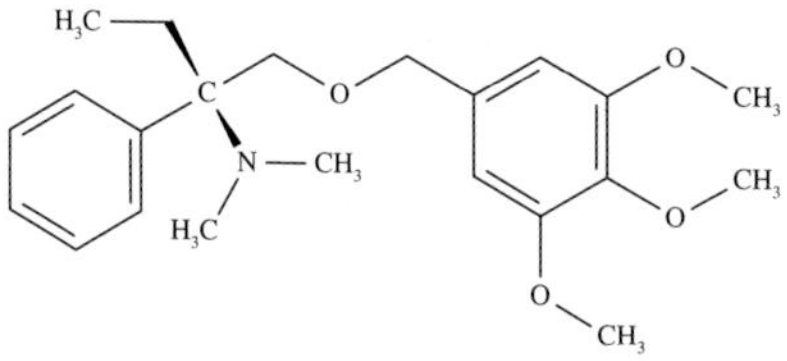

(+)-(R)-α-Ethyl-N,N-dimethyl-α-[[(3,4,5-trimethoxyben-
zyl)oxy]methyl]benzylamine = (+)-N,N-Dimethyl-1-phe-
nyl-1-[[(3,4,5-trimethoxybenzyl)oxy]methyl]propyl-
amine = (R)-α-Ethyl-N,N-dimethyl-α-[[(3,4,5-trimeth-
oxyphenyl)methoxy]methyl]benzenemethanamine (●)
R Tartrate (1:1) (133267-27-3)
S *Fedotozine tartrate**, JO-1196
U Gastrointestinal agent

11467 (9222)

$C_{22}H_{31}NO_5$
116290-93-8

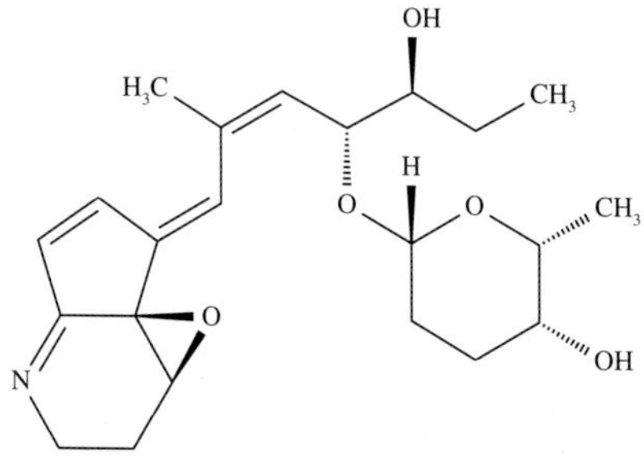

2R-[2α,3α,6α[1R*(S*),2Z,4E(1aS*,7aR*)]]]-6-[[4-(2,3-
Dihydrocyclopent[b]oxireno[c]pyridin-7-(1aH)-ylidene)-
-(1-hydroxypropyl)-3-methyl-2-butenyl]oxy]tetrahy-
dro-2-methyl-2H-pyran-3-ol (●)
S *Hatomamicin*, YL-0358M-A

U Antineoplastic antibiotic from a *Saccharopolyspora*
strain

11468 (9223)

$C_{22}H_{31}N_2O_2$

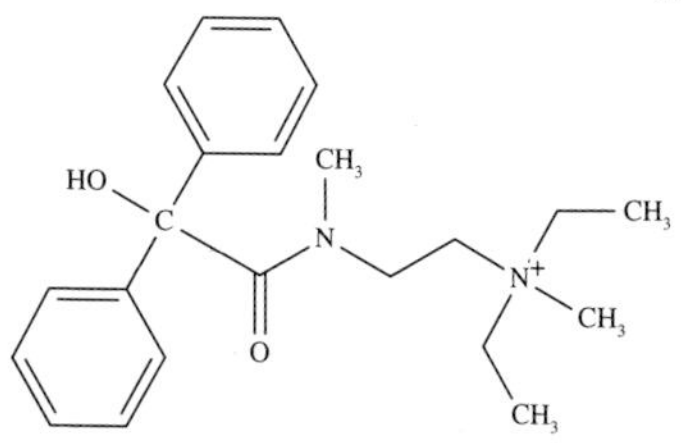

Diethylmethyl[2-(N-methylbenzilamido)ethyl]ammo-
nium = N,N-Diethyl-2-[(hydroxydiphenylacetyl)methyl-
amino]-N-methylethanaminium (●)
R Chloride (510-08-7)
S *Benzomethamine chloride*, Cotranul, MC 3743
U Anticholinergic, spasmolytic

11468-01 (9223-01)

3811-10-7

R Bromide
S *Benzomethamine bromide*, MC 3199
U Anticholinergic, spasmolytic

11469 (9224)

$C_{22}H_{31}N_3$

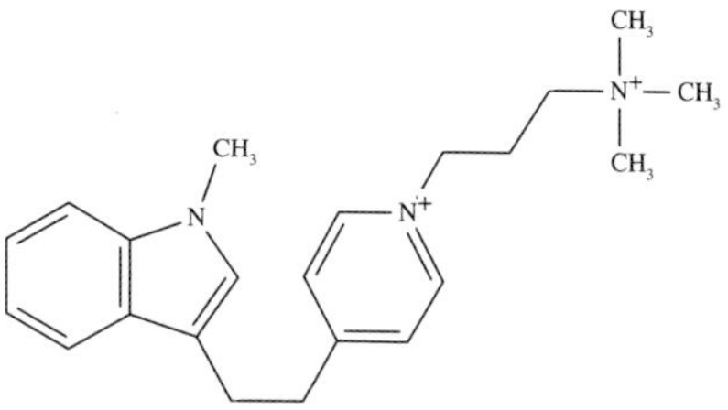

4-[2-(1-Methyl-1H-indol-3-yl)ethyl]-1-[3-(trimethylam-
monio)propyl]pyridinium (●)
R Dichloride (3426-58-2)
S IN 391, Isotensan, *Methindethyrium chloride*
U Antihypertensive

11470

$C_{22}H_{31}N_3O_2$
152811-62-6

N-[(1-Butyl-4-piperidinyl)methyl]-3,4-dihydro-2*H*-[1,3]oxazino[3,2-*a*]indole-10-carboxamide (●)
R   Monohydrochloride (178273-87-5)
S   *Piboserod hydrochloride**, SB 207266A
U   5-HT$_4$ receptor antagonist

11471   (9225)

$C_{22}H_{31}N_3O_2$
119375-01-8

[2*S*-(2*R**,5*R**)]-9-(1,1-Dimethyl-2-propenyl)-1,2,4,5,6,8-hexahydro-5-(hydroxymethyl)-1-methyl-2-(1-methyl-ethyl)-3*H*-pyrrolo[4,3,2-*gh*]-1,4-benzodiazonin-3-one (●)
S   *Pendolmycin*
U   Tumor promoting alkaloid from a *Nocardiopsis* strain

11472   (9226)

$C_{22}H_{31}N_3O_3$
21102-94-3

8-[2-[4-(2-Methoxyphenyl)-1-piperazinyl]ethyl]-8-aza-spiro[4.5]decane-7,9-dione (●)

R   Dihydrochloride (21102-95-4)
S   BMY-7378
U   Tranquilizer

11473   (9227)

$C_{22}H_{31}N_3O_3$

2,4'-Bis[2-(dimethylamino)-1-methylethoxy]-3-benzoyl-pyridine = [4-[2-(Dimethylamino)-1-methylethoxy]phe-nyl][2-[2-(dimethylamino)-1-methylethoxy]-3-pyridi-nyl]methanone (●)
R   Maleate (1:2) (105568-35-2)
S   Y-19995
U   Phagocytosis potentiator

11474   (9228)

$C_{22}H_{31}N_3O_3$
79253-92-2

(±)-N-[(4*R**,4a*R**,9b*S**)-1,2,3,4,4a,9b-Hexahydro-8,9b-dimethyl-3-oxo-4-dibenzofuranyl]-4-methyl-1-pipera-zinepropionamide = 1,2,3,4,4a,9b-Hexahydro-8,9b-dime-thyl-4-[3-(4-methyl-1-piperazinyl)propionamido]diben-zofuran-3-one = (4α,4aβ,9bβ)-(±)-N-(1,2,3,4,4a,9b-He-xahydro-8,9b-dimethyl-3-oxo-4-dibenzofuranyl)-4-me-thyl-1-piperazinepropanamide (●)
R   Dihydrochloride (68491-57-6)
S   Azipranone, Drazifon "Cassenne", RU 20201, *Tazi-prinone hydrochloride***
U   Antitussive

11475 $C_{22}H_{31}N_3O_4$
                                             227940-00-3

tert-Butyl 7-[(*S*)-3-(*p*-cyanophenoxy)-2-hydroxypropyl]-3,7-diazabicyclo[3.3.1]nonane-3-carboxylate = 7-[(2*S*)-3-(4-Cyanophenoxy)-2-hydroxypropyl]-3,7-diazabicyclo[3.3.1]nonane-3-carboxylic acid 1,1-dimethylethyl ester (●)

S *Adekalant**, H 345/52

U Anti-arrhythmic (class III), potassium channel blocker

11476   (9229) $C_{22}H_{31}N_3O_4S$
                                             3689-73-4

2*S*,5*R*,6*R*)-3,3-Dimethyl-7-oxo-6-(phenylacetamido)-4-hia-1-azabicyclo[3.2.0]heptane-2-carboxylic acid 2-(di-:thylamino)ethyl ester = Benzylpenicillin β-(diethylami-10)ethyl ester = [2*S*-(2α,5α,6β)]-3,3-Dimethyl-7-oxo-6-(phenylacetyl)amino]-4-thia-1-azabicyclo[3.2.0]heptane-2-carboxylic acid 2-(diethylamino)ethyl ester (●)

R Monohydriodide (808-71-9)

S Aephycillinum, Alivin, Bronchocillin, Bronchopen, Broncociline, Broncopen, Broncopenil, DAP, Deripen, Ephicillinum, Esterloven, Estopen, Estopenil, Iodocillina, Leocillin, Mamyzin, Neo-Penil, Pemacillin, Penester, Penetavet, *Pénéthacilline*, *Penethamate hydroiodide*, Penibronchial, Peni-Pulmo 500, Pneumocilina, Pneuropenil, Pulmaxil N, Pulmo 500, Toraxilina

U Antibiotic

11477   (9230) $C_{22}H_{31}N_3O_5$
                                             23887-46-9

1-[(1-Pyrrolidinylcarbonyl)methyl]-4-(3,4,5-trimethoxy-cinnamoyl)piperazine = 4-(3,4,5-Trimethoxycinnamoyl)-1-piperazineacetic acid pyrrolidide = 1-[2-Oxo-2-(1-pyrrolidinyl)ethyl]-4-[1-oxo-3-(3,4,5-trimethoxyphenyl)-2-propenyl]-piperazine (●)

R Maleate (1:1) (26328-04-1)

S Anapazin, Angiodisten, Arteripax, Brendil, Brentomine, Brepaneal, Cinema, Cinepadil, *Cinepazide maleate**, Ekarusin, Enpural, Flucetnal, Hishiline, Madesol, MD 67350, Neubeat, Prosmet, Scorojile, Sebdeel, Shulanderl, Sylpinale, Tatsumedil, Tineup, Vasodeniel, Vasodistal, Vasolande, Willbon

U Peripheral vasodilator

11478   (9231) $C_{22}H_{31}N_3O_5$
                                             88768-40-5

(1*S*,9*S*)-9-[[(*S*)-1-Carboxy-3-phenylpropyl]amino]octa-hydro-10-oxo-6*H*-pyridazino[1,2-*a*][1,2]diazepine-1-carboxylic acid 9-ethyl ester = [1*S*-[1α,9α(*R**)]]-9-[[1-(Ethoxycarbonyl)-3-phenylpropyl]amino]octahydro-10-oxo-6*H*-pyridazino[1,2-*a*][1,2]diazepine-1-carboxylic acid (●)

R also monohydrate (92077-78-6)

S *Cilazapril***, Cilazil, Dynorm, Inhibace "Merck; Roche", Inibace, Initiss, Inobes, Inocar, Justor, Prilazid, Ro 31-2848, Roxor, Vascace, Vascase, Zilacor

U Antihypertensive (ACE inhibitor)

11479 (9232) $C_{22}H_{31}N_3O_5S$
                                                  100277-62-1

(R)-3-[[(S)-1-Carboxy-5-(4-piperidyl)pentyl]amino]-
2,3,4,5-tetrahydro-4-oxo-1,5-benzothiazepine-5-acetic
acid = [S-(R*,S*)]-3-[[1-Carboxy-5-(4-piperidinyl)pen-
tyl]amino]-3,4-dihydro-4-oxo-1,5-benzothiazepine-
5(2H)-acetic acid (●)
S CV-5975
U Antihypertensive (ACE inhibitor)

11480 (9233) $C_{22}H_{31}N_3O_5S$
                                                  109683-61-6

(S)-2-tert-Butyl-4-[(S)-N-[(S)-1-carboxy-3-phenylpro-
pyl]alanyl]-Δ^2-1,3,4-thiadiazoline-5-carboxylic acid, 4-
ethyl ester = [2S-[2R*,3[R*(R*)]]]-5-(1,1-Dimethyl-
ethyl)-3-[2-[[1-(ethoxycarbonyl)-3-phenylpropyl]ami-
no]-1-oxopropyl]-2,3-dihydro-1,3,4-thiadiazole-2-carb-
oxylic acid (●)
S FPL 63547, Utibapril**
U Antihypertensive (ACE inhibitor)

11481 $C_{22}H_{31}N_3O_6S_2$
                                                  161715-24-8

(+)-Pivaloyloxymethyl (4R,5S,6S)-6-[(R)-1-hydroxy-
ethyl]-4-methyl-7-oxo-3-[[1-(1,3-thiazolin-2-yl)azetidin-
3-yl]thio]-1-azabicyclo[3.2.0]hept-2-ene-2-carboxylate =
(4R,5S,6S)-3-[[1-(4,5-Dihydro-2-thiazolyl)-3-azetidi-
nyl]thio]-6-[(1R)-1-hydroxyethyl]-4-methyl-7-oxo-1-
azabicyclo[3.2.0]hept-2-ene-2-carboxylic acid (2,2-dime-
thyl-1-oxopropoxy)methyl ester (●)
S Ebipenem pivoxil, Tebipenem**
U Antibiotic

11482 $C_{22}H_{31}N_5$
                                                  189152-50-9

trans-N-[4-[2-(4-Phenyl-1-piperazinyl)ethyl]cyclohe-
xyl]-2-pyrimidinamine (●)
S PD 158771
U Anxiolytic, antipsychotic

11483 $C_{22}H_{31}N_5O_4$
                                                  159776-70-2

N-[(R)-[[(2S)-2-[(p-Amidinobenzyl)carbmoyl]-1-azetidi-
nyl]carbonyl]cyclohexylmethyl]glycine = N-[(1R)-2-
[(2S)-2-[[[[4-(Aminoiminomethyl)phenyl]methyl]ami-

no]carbonyl]-1-azetidinyl]-1-cyclohexyl-2-oxoethyl]gly-
cine (●)

S H 319/68, *Melagatran***

U Thrombin inhibitor

11484 (9234)

$C_{22}H_{31}O_8P$
22252-38-6

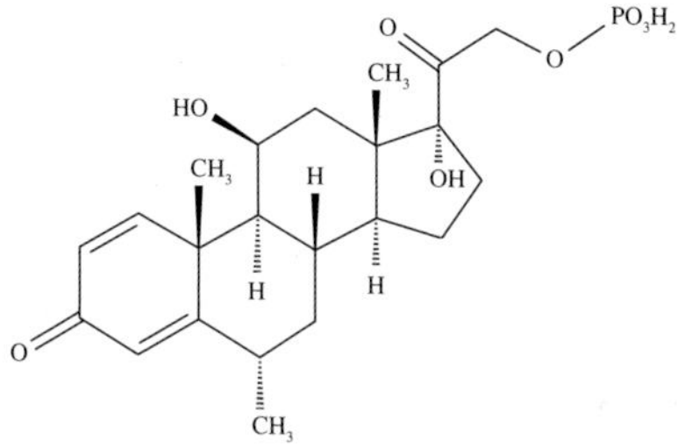

11β,17,21-Trihydroxy-6α-methylpregna-1,4-diene-3,20-
dione 21-(dihydrogen phosphate) = (6α,11β)-11,17-Dihy-
droxy-6-methyl-21-(phosphonooxy)pregna-1,4-diene-
3,20-dione (●)

R Disodium salt (5015-36-1)

S Medrol Stabisol, *Methylprednisolone sodium phosphate*, U-12019 E

U Glucocorticoid (anti-inflammatory, anti-allergic)

11485 (9235)

$C_{22}H_{32}Cl_2N_2O_4$
97964-56-2

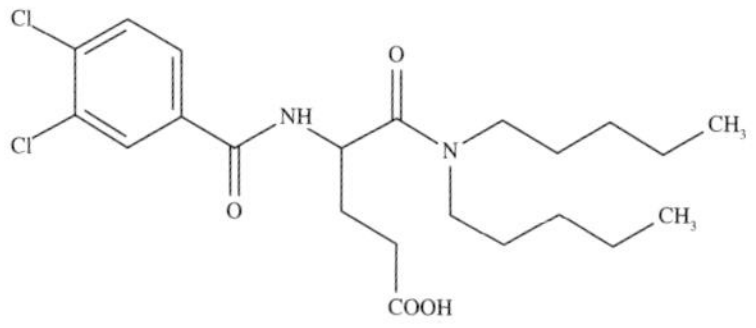

(±)-4-(3,4-Dichlorobenzamido)-*N,N*-dipentylglutaramic
acid = 4-[(3,4-Dichlorobenzoyl)amino]-5-(dipentylami-
no)-5-oxopentanoic acid (●)

S CR 1409, *Lorglumide***

U Cholecystokinin antagonist

11486 (9236)

$C_{22}H_{32}NO_3$

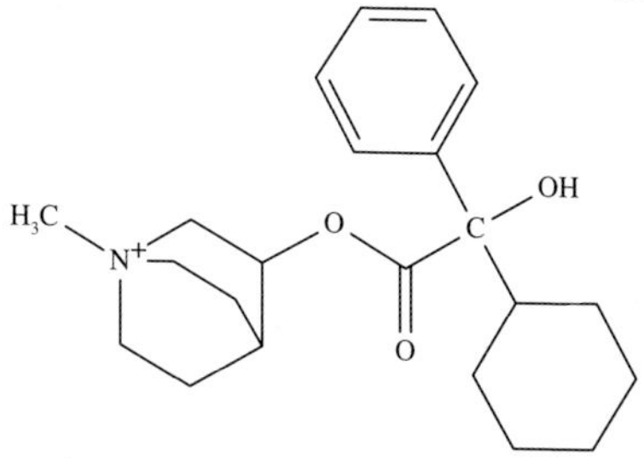

3-Hydroxy-1-methylquinuclidinium α-phenylcyclohexaneglycolate = 3-[(Cyclohexylphenylglycoloyl)oxy]-1-
methylquinuclidinium = 3-[(Cyclohexylhydroxyphenylacetyl)oxy]-1-methyl-1-azoniabicyclo[2.2.2]octane (●)

R Bromide (29125-56-2)

S *Droclidinium bromide***, LM 204

U Anticholinergic, spasmolytic

11487 (9237)

$C_{22}H_{32}N_2O_2$
96744-75-1

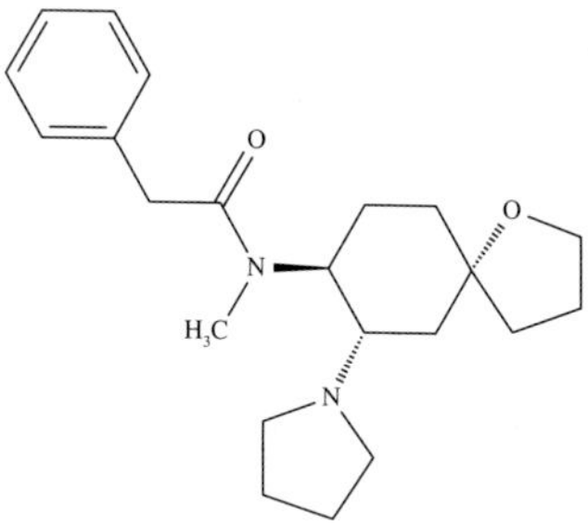

(5α,7α,8β)-(–)-*N*-Methyl-*N*-[7-(1-pyrrolidinyl)-1-oxa-
spiro[4.5]dec-8-yl]benzeneacetamide (●)

S U-69593

U Analgesic (kappa-opioid agonist)

11488 (9238)

$C_{22}H_{32}N_2O_2$
86197-47-9

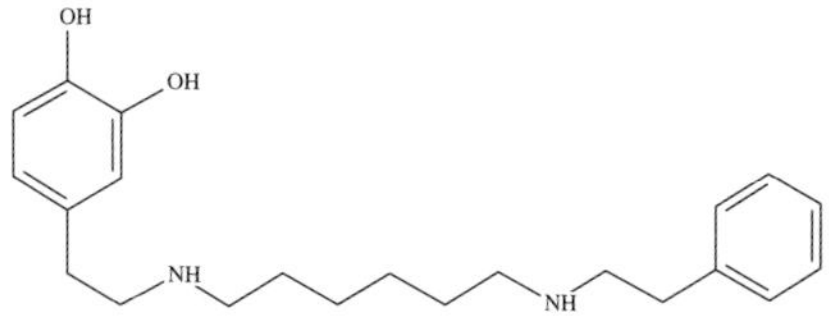

4-[2-[[6-(Phenethylamino)hexyl]amino]ethyl]pyroca-
techol = 4-[2-[[6-[(2-phenylethyl)amino]hexyl]ami-
no]ethyl]-1,2-benzenediol (●)

R Dihydrochloride (86484-91-5)

S Dopacard, *Dopexamine hydrochloride***, FPL 60278 AR
U Cardiovascular agent

11489 (9239)

$C_{22}H_{32}N_2O_3$
56693-13-1

1-[(1-Ethynylcyclohexyl)oxy]-3-[4-(2-methoxyphenyl)-1-piperazinyl]-2-propanol = α-[[(1-Ethynylcyclohexyl)oxy]methyl]-4-(2-methoxyphenyl)-1-piperazineethanol (●)
S CERM 3517, *Mociprazine***
U Anti-emetic

11490 (9240)

$C_{22}H_{32}N_2O_5$
63-12-7

N,N-Diethyl-1,3,4,6,7,11b-hexahydro-2-hydroxy-9,10-dimethoxy-2*H*-benzo[*a*]quinolizine-3-carboxamide acetate (ester) = 2-Acetoxy-1,3,4,6,7,11b-hexahydro-9,10-dimethoxy-2*H*-benzo[*a*]quinolizine-3-carboxylic acid diethylamide = 3-(Diethylcarbamoyl)-1,3,4,6,7,11b-hexahydro-9,10-dimethoxy-2*H*-benzo[*a*]quinolizin-2-yl acetate = 2-(Acetyloxy)-*N,N*-diethyl-1,3,4,6,7,11b-hexahydro-9,10-dimethoxy-2*H*-benzo[*a*]quinolizine-3-carboxamide (●)
R also monohydrochloride (30046-34-5)
S *Benzchinamidum, Benzoquinamide, Benzquinamide***, BZQ, Emete-Con, Emeticon, NSC-64375, P-2647, Promecon, Quantril
U Anti-emetic, tranquilizer

11491 (9241)

$C_{22}H_{32}N_2O_5$
91304-04-0

6-[1-(1-Imidazolylmethyl)-2-[(*p*-methoxybenzyl)oxy]ethoxy]-2,2-dimethylhexanoic acid = 6-[2-(1*H*-Imidazol-1-yl)-1-[[(4-methoxyphenyl)methoxy]methyl]ethoxy]-2,2-dimethylhexanoic acid (●)
S *Dimetagrel*, SC-41156
U Platelet aggregation inhibitor

11492 (9242)

$C_{22}H_{32}N_2O_6$
3215-70-1

α,α'-[Hexamethylenebis(iminomethylene)]bis(3,4-dihydroxybenzyl alcohol) = *N,N'*-Hexamethylenebis[4-(2-amino-1-hydroxyethyl)pyrocatechol] = *N,N'*-Hexamethylenebis[2-amino-1-(3,4-dihydroxyphenyl)ethanol] = 4,4'-[1,6-Hexanediylbis[imino(1-hydroxy-2,1-ethanediyl)]]bis[1,2-benzenediol] (●)
R also dihydrochloride (4323-43-7) or sulfate (1:1) (32266-10-7)
S Argocian, Bronalin, Broncholysin "Linz", Byk 1512, Delaprem, Etoscol, Gynipral, *Hexoprenaline***, Ipradol "Hafslund Nycomed; Lacer; Linz", Leanol, Prelin, ST 1512, Tocolysan, Tokolysan, Zacline
U Bronchodilator, uterine relaxant

11493 (9243) $C_{22}H_{32}N_4O$
77650-95-4

1,1-Diethyl-3-(6-propylergolin-8α-yl)urea = N,N-Di-ethyl-N'-[(8α)-6-propylergolin-8-yl]urea (●)
R Tartrate (2:1)
S *Proterguride tartrate***, ZK 39437
U Dopamine agonist

11493-01 (9243-01) 77650-96-5
R Phosphate (3:1)
S Propyldironyl, *Proterguride phosphate***,
VÚFB-13416
U Prolactin inhibitor

11494 $C_{22}H_{32}N_4O_4$
198958-88-2

(S)-β-[(R)-1-[3-(4-Piperidinyl)propionyl]nipecotamido]-3-pyridinepropionic acid = (βS)-β-[[[(3R)-1-[1-Oxo-3-(4-piperidinyl)propyl]-3-piperidinyl]carbonyl]amino]-3-pyridinepropanoic acid (●)
R also monohydrate (221005-96-5)
S *Elarofiban*, RWJ 53308
U Fibrinogen receptor (GPIIb/IIIa) antagonist

11495 (9244) $C_{22}H_{32}N_4O_4$
17299-00-2

3-Hydroxy-1-methylpyridinium hexamethylenebis[N-methylcarbamate] = 3,3'-[N,N'-Hexamethylenebis(methylcarbamoyloxy)]bis[1-methylpyridinium] = 3,3'-[1,6-Hexanediylbis[(methylimino)carbonyloxy]]bis[1-methylpyridinium] (●)
R Dibromide (15876-67-2)
S BC 51, *Distigmine bromide***, *Hexamarium bromide*, Tonus-Lab, Ubretid, Ubritil
U Parasympathomimetic (treatment of myasthenia gravis)

11496 $C_{22}H_{32}N_4O_{14}P_2$
118248-91-2

N,N'-Ethylenebis[N-[[3-hydroxy-5-(hydroxymethyl)-2-methyl-4-pyridyl]methyl]glycine] 5,5'-bis(dihydrogen phosphate) = N,N'-1,2-Ethanediylbis[N-[[3-hydroxy-2-methyl-5-[(phosphonooxy)methyl]-4-pyridinyl]methyl]glycine] (●)
S DPDP, *Fodipir***
U Diagnostic agent

11497

$C_{22}H_{32}N_6O_4S$
107478-35-3

N-[2-(Dimethylamino)ethyl]-N-methyl-4-(2,3,6,7-tetra-
hydro-2,6-dioxo-1,3-dipropyl-1H-purin-8-yl)benzenesul-
fonamide (●)
S   PD 115199
U   Adenosine receptor antagonist

11498   (9245)

$C_{22}H_{32}N_6O_4S$
96445-35-1

N-[3-(Dimethylamino)propyl]-4-(2,3,6,7-tetrahydro-2,6-
dioxo-1,3-dipropyl-1H-purin-8-yl)benzenesulfonamide
(●)
S   PD 113297
U   Adenosine receptor antagonist

11499   (9247)

$C_{22}H_{32}O_2$
6903-07-7

2-trans,trans-Farnesyl-5-methylresorcinol = (E,E)-5-
Methyl-2-(3,7,11-trimethyl-2,6,10-dodecatrienyl)-1,3-
benzenediol (●)
S   Grifolin
U   Antibiotic

11500   (9248)

$C_{22}H_{32}O_2$
117-51-1

3-Hexyl-6,6,9-trimethyl-7,8,9,10-tetrahydro-6H-diben-
zo[b,d]pyran-1-ol (●)
S   Parahexyl, Pyrahexyl, Synhexyl
U   Euphoric (formerly in thalamic dysfunctions)

11501   (9249)

$C_{22}H_{32}O_2$
6217-54-5

(all-Z)-4,7,10,13,16,19-Docosahexaenoic acid (●)
S   Cervonic acid, Doconexent**
U   Platelet aggregation inhibitor

11502   (9250)

$C_{22}H_{32}O_2$
39219-28-8

17β-Methoxy-3-propoxyestra-1,3,5(10)-triene = (17β)-
17-Methoxy-3-propoxyestra-3,5(10)-triene (●)
S   Colpotrofin(e), Colpotrophine, Delipoderm, Histo-
    trophyl, Promestriene**, Trophigyl
U   Glucocorticoid (topical anti-inflammatory, anti-aller-
    gic)

11503   (9246)

$C_{22}H_{32}O_2$
127-47-9

(all-E)-3,7-Dimethyl-9-(2,6,6-trimethyl-1-cyclohexen-1-
yl)-2,4,6,8-nonatetraenyl acetate = Retinol acetate (●)
S   A "Servet", Acaren, Afortin, Ahan, Akeral, Alphavi-
    san, Amirale, Amplex-A, A-Norman, Arcavit A,
    Avigan forte, A-Vite, Axerophthol acetate, Carovi-
    gen, Dagravit A, Forceval, Gerbera A, Hidrovitami-

na A, Homagenets-A, Nalfan, Neo Cystine,
NiO-A-Let, Palvit-A, Primavit, Reti-Nit, Retinol
"Ursapharm", *Retinol acetate*, Sehkraft A, Stabil A,
Viatate, Vitamin A "Spofa"-Caps, *Vitamin A aceta-
te*, Vit-A-N, Win-Vite A
U   Antixerophthalmic vitamin

11504   (9255) $C_{22}H_{32}O_3$
                                      434-05-9

17β-Hydroxy-1-methyl-5α-androst-1-en-3-one acetate =
1-Methyl-Δ^1-androstenolone acetate = (5α,17β)-17-(Ace-
tyloxy)-1-methylandrost-1-en-3-one (●)
R   see also no. 13652
S   *Metenolone acetate***, *Methenolone acetate*,
     Nibal Tablets, NSC-74226, Primobolan S, Primona-
     bol, SH 567, SQ 16496
U   Anabolic

11505   (9256) $C_{22}H_{32}O_3$
                                        1242-56-4

17β-Hydroxy-2-methyl-5α-androst-1-en-3-one acetate =
(5α,17β)-17-(Acetyloxy)-2-methylandrost-1-en-3-one
(●)
S   Anatrofin, RS-2106, S 3760, *Stenbolone acetate***,
     Stenbolone inyectable
U   Anabolic

11506   (9254) $C_{22}H_{32}O_3$
                                          57-85-2

17β-Hydroxyandrost-4-en-3-one propionate = Δ^4-Andro-
sten-17β-ol-3-one propionate = (17β)-17-(1-Oxopro-
poxy)androst-4-en-3-one (●)
S   Agovirin, Akroteston, Anderone, Andrest-Inj.,
     Andrhormone-Inj., Andro "Richter", Androfort, An-
     dro-Hart, Androlan in oil, Androlin Oil, Androlon,
     Andronex-Amp., Androsan-Amp., Androteston,
     Androtest P., Androxid, Androxil-intramusc., Andru-
     sol-P, Anertan, Arenorm, Biomestrone, Bio-Testiculi-
     na, Bio-Teston, Cosex TP 50, Enarmon-Inj., Endote-
     stina, Entestil, Espronate, Forten, Gonadrone, Her-
     mo M, Homandren-Amp., Homaninj., Homo-
     gene-P., Homosteron "Teva", Hormale-inj., Hor-
     mondrine-Inj., Hormoteston, Jebasteron, Malogen
     inj. in oil, Malotrone-P, Masenate, Naphateston, Na-
     pionate, Nasdol, Neo-Hombreol, Neo-Testosterona,
     Neoviron-Amp., NSC-9166, Opotestan, Orchic-Ol,
     Orchiol, Orchiormon, Orchisteron, Orchisterone-P.,
     Orchistin, Oreton, Orquisteron-P, Pantestin, Pe-
     randren-Amp., Perandrone-Amp., Pertestis, Primoni-
     at, Primotest, Primoteston, Propiokan, *Propionylte-
     stosterone*, Propiosteron, Proviron, Sanormon, Sig-
     matestrone, Sintest, Solvotest, Sterandril,
     Stérandryl-Amp., Sterotest, Sunmestrone, Supraste-
     ron-Amp., Synandrol, Syndren-Amp., Synerone, Te-
     lipex, Tesendocrin, Testadenos in oil, Testaform
     Aqueous, Testahomen, Testanon 25, Testarmon, Te-
     starona, Testavirol, Testeplex in oil, Testeubril,
     Testex, Testhormon, Testin-Amp., Testine, Testi-
     non, Testirene, Testo 100, Testobios, Testodet,
     Testodrin-Amp., Testo-Endo, Testoferol, Testogen,
     Testoici, Testoidral, Testokab, Testolets, Teston 25,
     Teston-Amp., Testonate, Testone, Testonil, Testo-
     nique, Testonorpon, Testopronil, Testormol, Testor-
     mon, Testorona-Amp., Testosaf, Testoselecta, Testo-
     sid, Testosteron "Vitis", Testosterona Dexter Inyect.,
     Testosteron-Amp. "Jenapharm", Testosterone Farma-

selecta, Testosterone Farmitalia, *Testosterone pro-pionate*, Testosteron forte Amp. "Orion", Testosteron Grossmann Amp., Testosteron ZZ, Testoviron, Testovis Fiale, Testovister, Testovitis, Testoxyl, Testrex, Testrol, Testron "Benzon", Testrone P, Tevirol, Tiverol, Tostrina, Uniteston, Uni-Testron, Virex-inj., Virormolofiale, Virormone-inj., Vulvan

U   Androgen

11507   (9251) $C_{22}H_{32}O_3$
                                                   33765-80-9

16β-Ethyl-17β-hydroxyestr-4-en-3-one acetate = 17β-(Acetyloxy)-16β-ethylestr-4-en-3-one = (16β,17β)-17-(Acetyloxy)-16-ethylestr-4-en-3-one (●)

S   TSAA 328
U   Anti-androgen

11508 $C_{22}H_{32}O_3$
                                                   24320-06-7

(16α)-16-Ethyl-21-hydroxy-19-norpregn-4-ene-3,20-dione (●)

S   Org 2058
U   Progestin

11509   (9252) $C_{22}H_{32}O_3$
                                                   2668-66-8

11β-Hydroxy-6α-methylpregn-4-ene-3,20-dione = (6α,11β)-11-Hydroxy-6-methylpregn-4-ene-3,20-dione (●)

S   Drixona, Episona, GSH-1043, HMS, *Hydroxymesterone, Hydroxymethylprogesterone*, Ipoflogin, Liquipom, Mecorvis, Medribioptal, Medrifar, Medriftalmina, Medrisocil, *Medrisona*, Medritonic, Medriusar, Medrixon, Medrocort, Medroftal, Medroptil, *Medrysone**, NSC-63278, Ophtocortin, Orion "North Medicamenta", Po-Medrisone, Sedesterol, Spectramedryn, U-8471, Visudrisone

U   Glucocorticoid (in treatment of ocular inflammation)

11510   (9253) $C_{22}H_{32}O_3$
                                                   520-85-4

17-Hydroxy-6α-methylpregn-4-ene-3,20-dione = 6α-Methyl-17-hydroxyprogesterone = (6α)-17-Hydroxy-6-methylpregn-4-ene-3,20-dione (●)

R   see also no. 12573
S   *Medrossiprogesterone, Medroxiprogesteronum, Medroxyprogesterone***
U   Progestin

11511 $C_{22}H_{32}O_4$
76675-97-3

11β,17α-Dihydroxy-17-propionylandrost-4-en-3-one =
(11β,17α)-11,17-Dihydroxy-17-(1-oxopropyl)androst-4-
en-3-one (●)
R   see also no. 13305
S   *Resocortol***
U   Glucocorticoid (topical anti-inflammatory)

11512   (9257) $C_{22}H_{32}O_4$
69648-40-4

trans-1-Hydroxy-16-phenoxy-ω-tetranorprostane-9,15-
dione = *trans*-(±)-2-(7-Hydroxyheptyl)-3-(3-oxo-4-phen-
oxybutyl)cyclopentanone (●)
S   M. & B. 33153, *Oxoprostol***
U   Gastric antisecretory

11513   (9258) $C_{22}H_{32}O_4$

5-Hydroxy-7-[4-(1-hydroxy-3-nonenyl)phenyl]-6-hepte-
noic acid (●)
R   Monolithium salt (120772-66-9)
S   SC-45694

U   Leukotriene B_4-receptor agonist

11514   (9259) $C_{22}H_{32}O_4$
78919-13-8

(*E*)-(3a*S*,4*R*,5*R*,6a*S*)-Hexahydro-5-hydroxy-4-[(*E*)-
(3*S*,4*RS*)-3-hydroxy-4-methyl-1-octen-6-ynyl]-$\Delta^{2(1H),\delta}$-
pentalenevaleric acid = 5-[(*E*)-(1*S*,5*S*,6*R*,7*R*)-7-Hydroxy-
6-[(*E*)-(3*S*,4*RS*)-3-hydroxy-4-methyl-1-octen-6-ynyl]bi-
cyclo[3,3,0]octen-3-ylidene]pentanoic acid = 5-[Hexahy-
dro-5-hydroxy-4-(3-hydroxy-4-methyl-1-octen-6-ynyl)-
2(1*H*)-pentalenylidene]pentanoic acid (●)
R   also trometamol (no. 314) salt
S   *Ciloprost*, Disoban, E-1030, Endoprost, Ilomed, Ilo-
medin(e), *Iloprost***, *Iloprost-Trometamol*, SH 401,
ZK 36374
U   Antihypertensive, vasodilator, anticoagulant

11515   (9260) $C_{22}H_{32}O_8$
18296-45-2

1,4a,5,7a-Tetrahydro-1,6-dihydroxyspiro[cyclopen-
ta[*c*]pyran-7(6*H*),2'-oxirane]-4-methanol 6-acetate 1,4-di-
isovalerate = [1*S*-(1α,4aα,6α,7β,7aα)]-3-Methylbutanoic
acid 6-(acetyloxy)-4a,5,6,7a-tetrahydro-4-[(3-methyl-1-
oxobutoxy)methyl]spiro[cyclopenta[*c*]pyran-7(1*H*),2'-
oxiran]-1-yl ester (●)
S   *Didrovaltrate***, *5,6-Dihydrovalepotriate*
U   Sedative (active principle from the roots of *Valeriana
officinalis*)

11516 (9261) $\qquad$ $C_{22}H_{33}ClN_2O$
85750-38-5

(*E*)-2-(*p*-Chlorobenzylidene)cyclohexanone (*E*)-*O*-[3-(diisopropylamino)propyl]oxime = 1-[(*E*)-3-(Diisopropylamino)propoxyimino]-2-(*p*-chlorophenylmethylene)cyclohexane = (*E*,*E*)-2-[(4-Chlorophenyl)methylene]cyclohexanone *O*-[3-[bis(1-methylethyl)amino]propyl]oxime (●)
S *Erocainide**
U Anti-arrhythmic, local anesthetic

11517 (9262) $\qquad$ $C_{22}H_{33}Cl_2N_3O_4$
13425-94-0

N-Acetylsarcolysyl-L-valine ethyl ester = *N*-[*N*-Acetyl-4-[bis(2-chloroethyl)amino]phenylalanyl]-L-valine ethyl ester (●)
S Asaline
U Antineoplastic

11518 (9263) $\qquad$ $C_{22}H_{33}Cl_2N_3O_4S$
3564-99-6

N-[*N*-Acetyl-4-[bis(2-chloroethyl)amino]phenylalanyl]methionine ethyl ester (●)
S Asamet
U Antineoplastic

11519 $\qquad$ $C_{22}H_{33}F_3O_2$
177080-77-2

3α-Hydroxy-3β-(trifluoromethyl)-5α-pregnan-20-one = (3α,5α)-3-Hydroxy-3-(trifluoromethyl)pregnan-20-one (●)
S Co 2-1970
U partial GABA(A) receptor agonist

11520 $\qquad$ $C_{22}H_{33}NO_2$
150748-23-5

(17β)-17-[(4-Hydroxybutyl)amino]estra-1,3,5(10)-trien-3-ol (●)
S *Butolame*
U Anticoagulant, estrogenic

11521 (9264) $\qquad$ $C_{22}H_{33}NO_2$
510-31-6

3β-(Methylamino)-20α-hydroxypregn-5-en-18-oic acid lactone(18→20) = (3β,20S)-20-Hydroxy-3-(methylamino)pregn-5-en-18-oic acid γ-lactone (●)
S *Paravallarine*

U Steroid alkaloid from *Paravallis microphylla* (primary product for drugs)

11522 (9265) $C_{22}H_{33}NO_3$
139-62-8

3-(2-Methylpiperidino)propyl *p*-(cyclohexyloxy)benzoate = *p*-(Cyclohexyloxy)benzoic acid 3-(2-methylpiperidino)propylester = 4-(Cyclohexyloxy)benzoic acid 3-(2-methyl-1-piperidinyl)propyl ester (●)
R also sulfate (1:1) (50978-10-4)
S Cainasurfa, Cyclocaine, *Cyclomethycaine***, Surfacaine, Surfathesin, Topocaine
U Local anesthetic

11523 (9266) $C_{22}H_{33}NO_5$
71097-83-1

(*E*)-(3a*R*,4*R*,5*R*,6a*S*)-δ-Cyano-3,3a,4,5,6,6a-hexahydro-5-hydroxy-4-[(*E*)-(3*S*,4*RS*)-3-hydroxy-4-methyl-1-octenyl]-2*H*-cyclopenta[*b*]furan-Δ$^{2,\delta}$-valeric acid = 5-Cyano-5-[(1*S*,5*R*,6*R*)-7-hydroxy-6-[(*E*)-(3*S*,4*RS*)-3-hydroxy-4-methyl-1-octenyl]-2-oxabicyclo[3.3.0]octan-3-ylidene]pentanoic acid = (5*E*,9α,11α,13*E*,15*S*)-5-Cyano-6,9-epoxy-11,15-dihydroxy-16-methylprosta-5,13-dien-1-oic acid (●)
S NIL, *Nileprost***, SH 427, ZK 34798
U Antihypertensive

11524 $C_{22}H_{33}NO_6$
124035-41-2

[3*R*-(3α,4aβ,5β,10α,10aβ,10bα)]-3-Ethenyl-2,3,4a,5,7,8,9,10,10a,10b-decahydro-10,10b-dihydroxy-3,4a,7,7,10a-pentamethyl-5-[[(methylamino)carbonyl]oxy]-1*H*-naphtho[2,1-*b*]pyran-1-one (●)
S HIL 568
U Ophthalmic (antiglaucoma)

11525 $C_{22}H_{33}N_5O_4S$
221452-76-2

N-[3-[4-[4-(Tetrahydro-1,3-dioxo-1*H*-pyrrolo[1,2-*c*]imidazol-2(3*H*)-yl)butyl]-1-piperazinyl]phenyl]ethanesulfonamide (●)
R Dihydrochloride (178482-61-6)
S EF-7412
U 5-HT$_{1A}$/D$_2$-receptor antagonist

11526 $C_{22}H_{33}N_5O_6$
144075-77-4

N-8-Guanidinooctanoyl-L-α-aspartyl-L-phenylalanine = *N*-[8-[(Aminoiminomethyl)amino]-1-oxooctyl]-L-α-aspartyl-L-phenylalanine (●)
S SC-49992
U Platelet aggregation inhibitor

11527 (9267)   C$_{22}$H$_{34}$ClN$_3$O$_2$
                                    116078-65-0

(±)-α-(o-Chlorophenyl)-α-[2-(N-isopropylacetami-
do)ethyl]-1-piperidinebutyramide = (±)-α-[2-[Acetyl(1-
methylethyl)amino]ethyl]-α-(2-chlorophenyl)-1-piperi-
dinebutanamide (●)
R   also without (±)-definition (103810-45-3)
S   *Bidisomide**, SC-40230
U   Anti-arrhythmic

11528 (9268)   C$_{22}$H$_{34}$ClN$_3$O$_3$
                                    81912-55-2

1-(2-Chloroethyl)-3-(3β-hydroxyandrost-5-en-17β-yl)-1-
nitrosourea = N-(2-Chloroethyl)-N'-[(3β,17β)-3-hydro-
xyandrost-5-en-17-yl]-N-nitrosourea (●)
S   *Sturamustine*
U   Antineoplastic

11529 (9269)   C$_{22}$H$_{34}$GdN$_5$O$_{10}$
                                    117827-80-2

[N,N-Bis[2-[(carboxymethyl)[(morpholinocarbonyl)me-
thyl]amino]ethyl]glycinato(3–)]gadolinium = [N,N-
Bis[2-[(carboxymethyl)[2-(4-morpholinyl)-2-oxo-
ethyl]amino]ethyl]glycinato(3–)]gadolinium (●)
S   *Gadopenamide**, ZK 117610
U   Diagnostic aid (paramagnetic contrast medium)

11530 (9270)   C$_{22}$H$_{34}$NO

Diethylmethyl[2-[(α-methyl-α-5-norbornen-2-ylben-
zyl)oxy]ethyl]ammonium = [2-[(α-Bicyclo[2.2.1]hept-5-
en-2-yl-α-methylbenzyl)oxy]ethyl]diethylmethylammo-
nium = 2-(1-Bicyclo[2.2.1]hept-5-en-2-yl-1-phenyleth-
oxy)-N,N-diethyl-N-methylethanaminium (●)
R   Bromide (29546-59-6)
S   Adamon "Asta", Asta 3746, *Ciclonium bromide**
U   Anticholinergic, spasmolytic

11531 (9271)   C$_{22}$H$_{34}$NO$_2$
                                                    17834-29-6

1-[(2-Cyclohexyl-2-phenyl-1,3-dioxolan-4-yl)methyl]-1-
methylpiperidinium (●)
R   Iodide (6577-41-9)
S   Allyproid, *Cyclonium iodide*, Espalexan, Esperan,
        Lynearmol, Oxaperan, *Oxapium iodide**, SH 100,
        Toiperan
U   Spasmolytic

11532   (9272) $C_{22}H_{34}N_2O$
                                                        78372-27-7

(*E*)-2-Benzylidenecycloheptanone (*E*)-*O*-[2-(diisopropyl-amino)ethyl]oxime = 1-[(*E*)-2-(Diisopropylamino)eth-oxyimino]-2-(*E*)-benzylidenecycloheptane = (*E,E*)-2-(Phenylmethylene)cycloheptanone *O*-[2-[bis(1-methyl-ethyl)amino]ethyl]oxime (●)
S   Egyt-1855, *Stirocainide***, Th 494
U   Anti-arrhythmic

11533   (9274) $C_{22}H_{34}N_2O_2$
                                                        104454-71-9

(+)-(4*S*,5*R*)-3-[(Hexahydro-1*H*-azepin-1-yl)propyl]-4-isobutyl-5-phenyl-2-oxazolidinone = (4*S*-*cis*)-3-[3-(He-xahydro-1*H*-azepin-1-yl)propyl]-4-(2-methylpropyl)-5-phenyl-2-oxazolidinone (●)
S   *Ipenoglan, Ipenoxazone***, MLV-6976, NC 1200
U   Muscle relaxant, NMDA receptor antagonist

11534   (9273) $C_{22}H_{34}N_2O_2$
                                                        135338-30-6

(*E*)-2-Benzylidenecyclohexanone (±)-(*E*)-*O*-[3-(diisopro-pylamino)-2-hydroxypropyl]oxime = (±)-1-(Diisopropyl-amino)-3-[[[(*E*)-2-(phenylmethylene)cyclohexylide-

ne]imino]oxy]-2-propanol = (*E,E*)-(±)-2-(Phenylmethyle-ne)cyclohexanone *O*-[3-[bis(1-methylethyl)amino]-2-hydroxypropyl]oxime (●)
R   Fumarate (1:1) (135338-31-7)
S   *Dridocainide fumarate*, Egis-3966
U   Anti-arrhythmic

11535   (9275) $C_{22}H_{34}N_2O_3$
                                                        104485-01-0

(±)-*trans*-2-(1-Pyrrolidinyl)cyclohexyl *m*-(pentyl-oxy)carbanilate = *trans*-(±)-[3-(Pentyloxy)phenyl]carb-amic acid 2-(1-pyrrolidinyl)cyclohexyl ester (●)
R   Monohydrochloride (77656-21-4)
S   Anulcen, BK-23, K 1902, *Pentacaine, Trapencaine hydrochloride**
U   Local anesthetic, anti-ulcer agent

11536 $C_{22}H_{34}N_2O_3$
                                                        147951-45-9

(±)-*trans*-2-(1-Pyrrolidinyl)cyclohexyl *p*-(pentyl-oxy)carbanilate = *trans*-[4-(Pentyloxy)phenyl]carbamic acid 2-(1-pyrrolidinyl)cyclohexyl ester (●)
S   K-2002
U   Anti-ulcer agent

11537 (9276) $C_{22}H_{34}N_2O_4$
15301-80-1

6,7-Bis[2-(diethylamino)ethoxy]-4-methylcoumarin =
6,7-Bis[2-(diethylamino)ethoxy]-4-methyl-2*H*-1-benzo-
pyran-2-one (●)
R also dihydrochloride (6830-17-7)
S *Ethoxarine*, Idro P$_3$, M.G. 652, *Oxamarin**,
SKF 17035-A
U Hemostatic

11538 $C_{22}H_{34}N_2O_6$
182292-49-5

4-Amino-4,5,6,7-tetradeoxy-*N*-[1-(3,4-dihydro-8-hydr-
oxy-1-oxo-1*H*-2-benzopyran-3-yl)-3-methylbutyl]-6-me-
thylheptonamide (●)
S PM 94128
U Cytotoxic

11539 (9277) $C_{22}H_{34}N_2S_2$
47510-96-3

4,4'-(Decamethylenedithio)bis(1-methylpyridinium) =
4,4'-[1,10-Decanediylbis(thio)]bis[1-methylpyridinium]
(●)
R Ditosylate (di-*p*-toluenesulfonate) (3785-02-2)

S BA 9-UTH 31, Brintobal, UTH 31
U Topical antibacterial, antifungal

11540 $C_{22}H_{34}N_4O_2$
177697-72-2

N-Methyl-D-phenylalanyl-*N*-[(*trans*-4-aminocyclohe-
xyl)methyl]-L-prolinamide (●)
S L 371912
U Antithrombotic

11541 (9278) $C_{22}H_{34}N_4O_{10}$
98631-95-9

4,4'-Ethylenebis[1-(hydroxymethyl)-2,6-piperazine-
dione] bis(isobutyl carbonate) (ester) = 1,1'-Ethylene-
bis[4-(isobutoxycarbonyloxymethyl)-3,5-dioxopipera-
zine] = Carbonic acid 1,2-ethanediylbis[(2,6-dioxo-4,1-
piperazinediyl)methylene] bis(2-methylpropyl) ester (●)
S MST-16, Perazolin, Perazoline, *Sobuzoxane**
U Antineoplastic

11542 (9279) $C_{22}H_{34}O_2$
60607-35-4

17β-Hydroxy-17-propylandrost-4-en-3-one = (17β)-17-
Hydroxy-17-propylandrost-4-en-3-one (●)
S *Topterone**, Win 17665
U Anti-androgen (anti-acne)

11543   (9280) $C_{22}H_{34}O_2$
2548-85-8

Clupanodonic acid = 4,8,12,15,19-Docosapentaenoic acid (●)

S   *Acidum clupadonicum*, Clupadene, Clupanina, Duodenale

U   Roborant, anti-ulcer agent

11543-01   (9280-01)

R   Calcium salt

S   Clupocalcium

U   Roborant

11544   (9281) $C_{22}H_{34}O_2$
86227-47-6

Ethyl (*all-Z*)-5,8,11,14,17-eicosapentaenoate = (*all-Z*)-5,8,11,14,17-Eicosapentaenoic acid ethyl ester (●)

S   Epadel, EPA-E, Eskima, *Ethyl icosapentate*, *Icosapent ethyl*, Solmiran

U   Platelet aggregation inhibitor, antihyperlipidemic

11545   (9282) $C_{22}H_{34}O_2$
2740-52-5

17-Hydroxy-6α-methylpregn-4-en-20-one = (6α)-17-Hydroxy-6-methylpregn-4-en-20-one (●)

R   see also no. 12604

S   *Anagestone***

U   Progestin

11546   (9283) $C_{22}H_{34}O_3$
855-22-1

17β-Hydroxy-5α-androstan-3-one propionate = (5α,17β)-17-(1-Oxopropoxy)androstan-3-one (●)

S   Androlone "Orma", *Androstanolone propionate*

U   Anabolic

11547 $C_{22}H_{34}O_4$
156979-72-5

(Z)-2,5-Dihydroxy-3-methyl-6-(10-pentadecenyl)-*p*-benzoquinone = 2,5-Dihydroxy-3-methyl-6-(10Z)-10-pentadecenyl-2,5-cyclohexadiene-1,4-dione (●)

S   *Maesanol*

U   Cytotoxic from *Maesa lanceolata*

11548   (9284) $C_{22}H_{34}O_4$
82380-21-0

(Z)-2-Hydroxy-5-methoxy-3-(10-pentadecenyl)-*p*-benzoquinone = (Z)-2-Hydroxy-5-methoxy-3-(10-pentadecenyl)-2,5-cyclohexadiene-1,4-dione (●)

S   *Maesanin*

U   Immunostimulant from fruit of *Maesa lanceolata*

11549　(9285)　　　　　　　　　　　　$C_{22}H_{34}O_5$
125-65-5

Glycolic acid 8-ester with octahydro-5,8-dihydroxy-4,6,9,10-tetramethyl-6-vinyl-3a,9-propano-3a*H*-cyclopentacyclooctan-1(4*H*)-one = Hydroxyacetic acid [3a*S*-(3aα,4β,5α,6α,8β,9α,9aβ,10*S**)]-6-ethenyldecahydro-5-hydroxy-4,6,9,10-tetramethyl-1-oxo-3a,9-propano-3a*H*-cyclopentacycloocten-8-yl ester (●)
S　*Drosophilin B, Pleuromulin**, Pleuromutilin*
U　Antibiotic from *Pleurotus mutilis* and other *Pleurotus* sp.

11550　(9286)　　　　　　　　　　　　$C_{22}H_{34}O_7$
132340-44-4

4,5-Dihydroxy-2-hexenoic acid 5-methoxy-4-[2-methyl-3-(3-methyl-2-butenyl)oxiranyl]-1-oxaspiro[2.5]oct-6-yl ester (●)
S　FR 111142, WF-2015 A
U　Antineoplastic (angiogenesis inhibitor from *Scole-cobasidium arenarium* F-2015)

11551　(9287)　　　　　　　　　　　　$C_{22}H_{34}O_7$
66575-29-9

(3*R*,4a*R*,5*S*,6*S*,6a*S*,10*S*,10a*R*,10b*S*)-Dodecahydro-5,6,10,10b-tetrahydroxy-3,4a,7,7,10a-pentamethyl-3-vinyl-1*H*-naphtho[2,1-*b*]pyran-1-one 5-acetate = [3*R*-(3α,4aβ,5β,6β,6aα,10α,10aβ,10bα)]-5-(Acetyloxy)-3-ethenyldodecahydro-6,10,10b-trihydroxy-3,4a,7,7,10a-pentamethyl-1*H*-naphtho[2,1-*b*]pyran-1-one (●)
R　see also no. 13660
S　*Boforsin, Coleonol, Colforsin**, Forskolin, HL 362, L 751362 B, M 410*
U　Cardiotonic, antiglaucoma agent (adenylatecyclase-activator) from *Coleus forskohli*

11552　(9288)　　　　　　　　　　　　$C_{22}H_{34}O_{19}$
77752-20-6

Homocitric acid tririboside = (*S*)-2-[(*O*-α-D-Ribofuranosyl-(1→2)-*O*-α-D-ribofuranosyl-(1→2)-α-D-ribofuranosyl)oxy]-1,2,4-butanetricarboxylic acid (●)
S　*Ribocitrin*
U　Dextransucrase inhibitor from *Streptomyces* sp. MF 980-CF 1

11553　　　　　　　　　　　　　　　　$C_{22}H_{35}^{123}IO_2$
123748-56-1

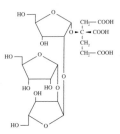

ω-(p-[123*I*]Iodophenyl)-β-methylpentadecanoic acid = 4-(Iodo-^{123}I)-β-methylbenzenepentadecanoic acid (●)
S Cardiodine, 123I-BMIPP
U Myocardial imaging agent

11554 (9289) $C_{22}H_{35}NO$
562-02-7

3β-(Methylamino)pregn-5-en-20-one = (3β)-3-(Methylamino)pregn-5-en-20-one (●)
S *Holaphylline*
U Anti-inflammatory (steroid alkaloid from *Holarrhenia floribunda*)

11555 (9290) $C_{22}H_{35}NO$
18841-58-2

4'-Octyl-3-piperidinopropiophenone = 1-(4-Octylphenyl)-3-(1-piperidinyl)-1-propanone (●)
R Hydrochloride (18787-40-1)
S N-1113, Nioben, *Pipoctanone hydrochloride***
U Antihypertensive

11556 (9291) $C_{22}H_{35}NO_2$
57558-44-8

1-Cyclohexyl-4-[ethyl(p-methoxy-α-methylphenethyl)amino]-1-butanone = 1-Cyclohexyl-4-[ethyl-[2-(4-methoxyphenyl)-1-methylethyl]amino]-1-butanone (●)
R Hydrochloride (57558-46-0)
S DU 23849, *Secoverine hydrochloride***

U Spasmolytic

11557 (9293) $C_{22}H_{35}NO_2$
90237-04-0

(+)-(S)-1-Cyclohexyl-4-[ethyl(p-methoxy-α-methylphenethyl)amino]-1-butanone = (+)-1-Cyclohexyl-4-[ethyl[2-(4-methoxyphenyl)-1-methylethyl]amino]-1-butanone (●)
R Hydrochloride (90237-06-2)
S *Dexsecoverine hydrochloride***, DU 23903
U Spasmolytic

11558 $C_{22}H_{35}NO_2$
6879-74-9

(3S,3aR,4R,4aS,8aR,9aS)-4-[(1E)-2-[(2R,6S)-1,6-Dimethyl-2-piperidinyl]ethenyl]decahydro-3-methylnaphtho[2,3-c]furan-1(3H)-one (●)
S *Himbacine*
U Muscarinic receptor antagonist (cognition enhancer) from *Galbulimina* sp.

11559 (9292) $C_{22}H_{35}NO_2$
119193-09-8

(αS,3R,4R)-α-Cyclohexyl-4-(3-hydroxyphenyl)-3,4-dimethyl-1-piperidinepropanol (●)

S LY 255582
U Appetite suppressant

11560   (9294)   C$_{22}$H$_{35}$NO$_3$
                                                    119477-85-9

[R-[R*,S*-(E,Z)]]-6-[6-(3-Hydroxy-1,5-undecadienyl)-2-
pyridinyl]-1,5-hexanediol (●)
S U-75302
U Anti-allergic

11561   (9295)   C$_{22}$H$_{35}$NO$_5$
                                                    54592-27-7

(±)-5-[2-(*tert*-Butylamino)-1-hydroxyethyl]-*m*-pheny-
lene dipivalate = 2,2-Dimethylpropanoic acid 5-[2-[(1,1-
dimethylethyl)amino]-1-hydroxyethyl]-1,3-phenylene ester (●)
S *Divabuterol***, *Terbutaline dipivalate*
U Bronchospasmolytic, uterine relaxant

11562   (9296)   C$_{22}$H$_{35}$NO$_7$
                                                    22661-76-3

α-[(Isopentyloxy)methyl]-4-morpholineethanol 3,4,5-tri-
methoxybenzoate (ester) = 1-[(Isopentyloxy)methyl]-2-
morpholinoethyl 3,4,5-trimethoxybenzoate = 3,4,5-Tri-
methoxybenzoic acid 1-[(3-methylbutoxy)methyl]-2-(4-
morpholinyl)ethyl ester (●)
R Hydrochloride (22661-96-7)

S *Amoproxan hydrochloride***, Aproxim, CERM 730,
       Médérel
U Anti-anginal, anti-arrhythmic

11563   (9297)   C$_{22}$H$_{35}$N$_2$
                                                    6803-62-9

4-Amino-1-dodecylquinaldinium = 4-Amino-1-dodecyl-
2-methylquinolinium (●)
R Acetate (146-37-2)
S Laurodin, *Laurolinium acetate***
U Antiseptic

11564   C$_{22}$H$_{35}$N$_3$O
                                                    147138-01-0

N-[(Z)-[3-Methoxy-5-(1H-pyrrol-2-yl)-2H-pyrrol-2-yli-
dene]methyl]-1-dodecanamine (●)
S BE 18591
U Antineoplastic, antibacterial from *Streptomyces* A
       18591

11565 (9298) $C_{22}H_{35}N_3O_2$
88296-62-2

(±)-*trans*-4-(Dimethylamino)-1-(2-hydroxycyclohexyl)-2',6'-isonipecotoxylidide = *trans*-(±)-4-(Dimethylamino)-*N*-(2,6-dimethylphenyl)-1-(2-hydroxycyclohexyl)-4-piperidinecarboxamide (●)
S   R 54718, *Transcainide***
U   Anti-arrhythmic

11566 (9299) $C_{22}H_{35}N_5O_5$
126631-86-5

N^2-(8-Guanidinooctanoyl)-N^1-(*p*-methoxyphenethyl)-α-asparagine = (*S*)-3-[[8-[(Aminoiminomethyl)amino]-1-oxooctyl]amino]-4-[[2-(4-methoxyphenyl)ethyl]amino]-4-oxobutanoic acid (●)
S   SC-47643
U   Antithrombotic

11567 (9300) $C_{22}H_{36}N_2O_3$
54063-39-7

1-(Isobutoxymethyl)-2-(4-methyl-1-piperazinyl)ethyl 2-phenylbutyrate = α-Ethylbenzeneacetic acid 1-[(4-methyl-1-piperazinyl)methyl]-2-(2-methylpropoxy)ethyl ester (●)
S   *Fenetradil***

U   Coronary vasodilator, anti-arrhythmic

11568 (9301) $C_{22}H_{36}N_2O_4$
105919-73-1

2-(Hexahydro-1*H*-azepin-1-yl)-1-(methoxymethyl)ethyl 2-(pentyloxy)carbanilate = [2-(Pentyloxy)phenyl]carbamic acid 2-(hexahydro-1*H*-azepin-1-yl)-1-(methoxymethyl)ethyl ester (●)
R   Monohydrochloride (112726-11-1)
S   BK-129
U   Anti-arrhythmic

11569 $C_{22}H_{36}N_2O_5S$
144494-65-5

N-(Butylsulfonyl)-4-[4-(4-piperidinyl)butoxy]-L-phenylalanine = *N*-(Butylsulfonyl)-*O*-[4-(4-piperidinyl)butyl]-L-tyrosine (●)
R   Monohydrochloride (142373-60-2) or monohydrochloride monohydrate (150915-40-5)
S   Aggrastat, L 700462, MK-383, *Tirofiban hydrochloride***
U   Fibrinogen receptor antagonist (GP IIb/IIIa antagonist)

11570 (9302) $C_{22}H_{36}N_4O_4S_2$
10038-83-2

[4,4'-Biphenylenebis(sulfonyliminoethylene)]bis[trimethylammonium] = *N,N'*-[2,2'-(4,4'-Biphenyldisulfonamido)diethylene]bis(trimethylammonium) = 2,2'-[[1,1'-Biphenyl]-4,4'-diylbis(sulfonylimino)]bis[*N,N,N*-trimethylethanaminium] (●)

R Diiodide (5184-79-2)
S Disufene
U Muscle relaxant

11570-01 (9302-01)   23131-02-4
R Dibenzenesulfonate
S Benzodisufene, Benzodisuphen
U Muscle relaxant

11571 $C_{22}H_{36}N_4O_5$
190648-49-8

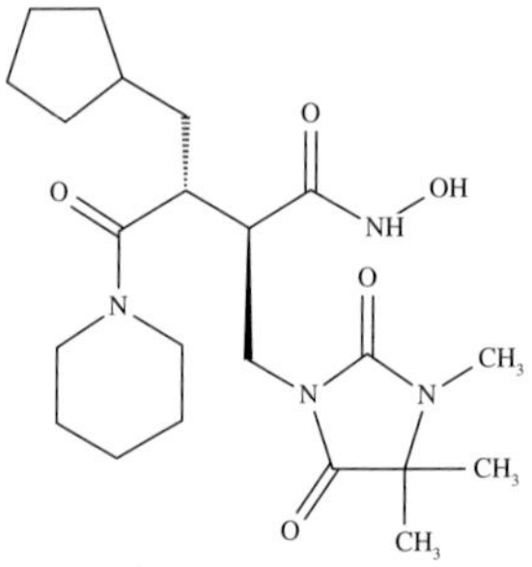

3(*R*)-(Cyclopentylmethyl)-2(*R*)-[(3,4,4-trimethyl-2,5-di-oxo-1-imidazolidinyl)methyl]-4-oxo-4-piperidinobutyro-hydroxamic acid = (α*R*,β*R*)-β-(Cyclopentylmethyl)-*N*-hydroxy-γ-oxo-α-[(3,4,4-trimethyl-2,5-dioxo-1-imidazo-lidinyl)methyl]-1-piperidinebutanamide (●)
S *Cipemastat***, CPA, Ro 32-3555, Trocade
U Collagenase inhibitor

11572 (9303) $C_{22}H_{36}N_8O_{11}$
62087-72-3

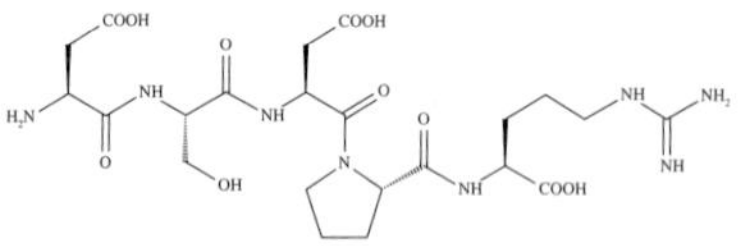

N^2-[1-[*N*-(*N*-L-α-Aspartyl-L-seryl)-L-α-aspartyl]-L-pro-lyl]-L-arginine (●)
S HEPP, *IgE pentapeptide*, Igevac,
*Pentapeptide DSDPR, Pentigetide***, Pentyde
U Immunomodulator, anti-allergic

11573 $C_{22}H_{36}O_2$
38398-32-2

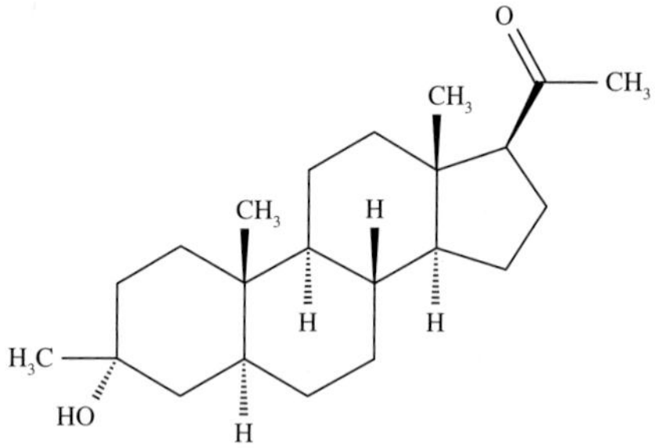

(3α,5α)-3-Hydroxy-3-methylpregnan-20-one (●)
S CCD 1042, *Ganaxolone***
U Anticonvulsant

11574 (9304) $C_{22}H_{36}O_2Si$
5055-42-5

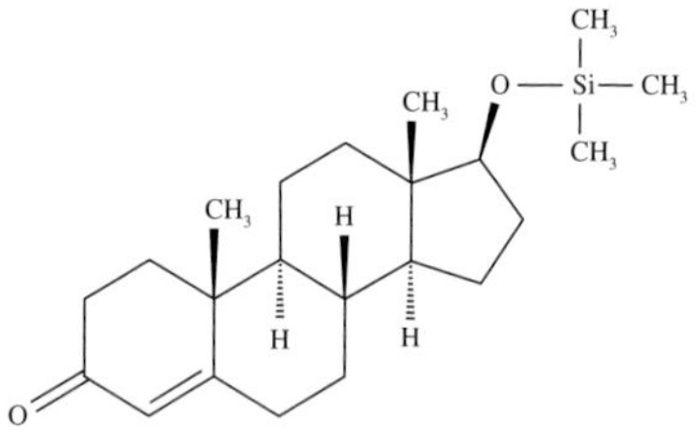

17β-(Trimethylsiloxy)androst-4-en-3-one = (17β)-17-[(Trimethylsilyl)oxy]androst-4-en-3-one (●)
S NSC-95147, SC-16148, *Silandrone***
U Androgen

11575 $C_{22}H_{36}O_4$
133906-74-8

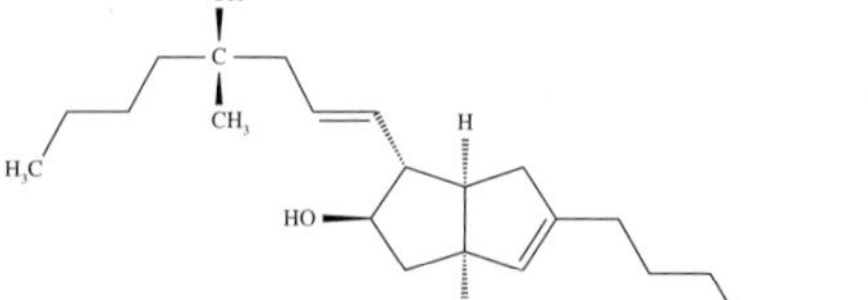

(3a*S*,5*R*,6*R*,6a*S*)-1,3a,4,5,6,6a-Hexahydro-5-hydroxy-6-[(1*E*,4*S*)-4-hydroxy-4-methyl-1-octenyl]-2-pentalene-pentanoic acid (●)
S TEI-3356
U Hepatoprotectant (Prostanoid EP$_3$ receptor agonist)

11576   (9305) $C_{22}H_{36}O_4$
                                                      88931-51-5

(+)-Methyl (3aS,5R,6R,6aS)-1,3a,4,5,6,6a-hexahydro-5-
hydroxy-6-[(E)-(3S)-3-hydroxy-1-octenyl]-2-pentalene-
valerate = (+)-(1S,5S,6R,7R)-Methyl 5-[7-hydroxy-6-
[(E)-(S)-3-hydroxy-1-octenyl]bicyclo[3.3.0]oct-2-en-3-
yl]-pentanoate = [3aS-[3aα,5β,6α-(1E,3R*),6aα]]-
1,3a,4,5,6,6a-Hexahydro-5-hydroxy-6-(3-hydroxy-1-oc-
tenyl)-2-pentalenepentanoic acid methyl ester (●)
S   Alteon, Arteon, *Clinprost***, *Isocarbacyclin methyl
      ester*, Lipocurren, Lipo PGL_2, TEI-9090, TTC-909
U   Antithrombotic, platelet aggregation inhibitor

11577   (9306) $C_{22}H_{36}O_4$
                                                      81845-44-5

(Z)-(3aS,5R,6R,6aR)-Hexahydro-5-hydroxy-6-[(E)-(3S)-
3-hydroxy-1-octenyl]-3a-methyl-$\Delta^{2(1H),\delta}$-pentalenevale-
ric acid = [3aS-[2Z,3aα,5β,6α(1E,3R*),6aα]]-5-[Hexahy-
dro-5-hydroxy-6-(3-hydroxy-1-octenyl)-3a-methyl-
2(1H)-pentalenylidene]pentanoic acid (●)
R   Calcium salt (2:1) (81703-55-1)
S   *Ciprostene calcium**, U-61431 F
U   Platelet aggregation inhibitor

11578   (9307) $C_{22}H_{36}O_5$
                                                      74397-12-9

(E)-7-[(1R,2R,3R)-3-Hydroxy-2-[(E)-(3S,5S)-3-hydroxy-
5-methyl-1-nonenyl]-5-oxocyclopentyl]-2-heptenoic acid

= (2E,11α,13E,15S,17S)-11,15-Dihydroxy-17,20-dime-
thyl-9-oxoprosta-2,13-dien-1-oic acid (●)
R   also compd. with α-cyclodextrin (88852-12-4)
S   *Limaprost***, ONO 1206, OP-1206, Opalmon, Pro-
      penal, Prorenal
U   Vasodilator, platelet aggregation inhibitor

11579   (9308) $C_{22}H_{36}O_5$
                                                      81026-63-3

(±)-Methyl (Z)-7-[(1R,2R,3R)-3-hydroxy-2-[(E)-(4RS)-4-
hydroxy-4-methyl-1-octenyl]-5-oxocyclopentyl]-4-hep-
tenoate = (4Z,11α,13E)-(±)-11,16-Dihydroxy-16-methyl-
9-oxoprosta-4,13-dien-1-oic acid methyl ester (●)
S   *Enisoprost***, SC-34301
U   Anti-ulcer agent

11580   (9309) $C_{22}H_{36}O_5$
                                                      54120-61-5

(±)-Methyl 7-[(1R*,2R*,3R*,5S*)-3,5-dihydroxy-2-[(E)-
3-hydroxy-3-methyl-1-octenyl]cyclopentyl]-4,5-hepta-
dienoate = (±)-Methyl (13E)-(9S,11R,15R)-9,11,15-tri-
hydroxy-15-methylprosta-4,5,13-trienoate =
(9α,11α,13E,15R)-(±)-9,11,15-Trihydroxy-15-methyl-
prosta-4,5,13-trien-1-oic acid methyl ester (●)
S   *Prostalene**, RS-9390, Synchrocept
U   Luteolytic

11581 (9310) $C_{22}H_{36}O_7$
4720-09-6

(3β,6β,14R)-Grayanotoxane-3,5,6,10,14,16-hexol 14-acetate (●)
S *Acetylandromedol*, Andromedotoxin, Andrometon, *Asebotoxin*, *Grayanotoxin I*, *Rhodotoxin*
U Antihypertensive from *Ericaceae*

11582 (9311) $C_{22}H_{37}ClO_4$
79360-43-3

(Z)-7-[(1R,2R,3R,5R)-5-Chloro-3-hydroxy-2-[(E)-(3R)-3-hydroxy-4,4-dimethyl-1-octenyl]cyclopentyl]-5-heptenoic acid = (5Z,9β,11α,13E,15R)-9-Chloro-11,15-dihydroxy-16,16-dimethylprosta-5,13-dien-1-oic acid (●)
S *Nocloprost***, SH 475, ZK 94726
U Cytoprotective, gastric antisecretory

11583 $C_{22}H_{37}NO_2$
94421-68-8

Arachidonylethanolamide = (5Z,8Z,11Z,14Z)-N-(2-Hydroxyethyl)-5,8,11,14-eicosatetraenamide (●)
S ANA, *Anandamide*

U Cannabinoid receptor agonist

11584 (9312) $C_{22}H_{37}NO_3$
103081-25-0

15-Dehydroxy-13,14-desethylene-11a,11a-dimethyl-15-oxo-13-aza-11a,12-dicarbathromboxane A$_2$ = [1S-[1α,2β(Z),3α,5α]]-7-[6,6-Dimethyl-3-[(1-oxohexyl)amino]bicyclo[3.1.1]hept-2-yl]-5-heptenoic acid (●)
S ONO 1217
U Thromboxane antagonist

11585 $C_{22}H_{37}NO_5$
57818-92-5

3,4,5-Trimethoxybenzoic acid 8-(diethylamino)octyl ester (●)
S TMB 8
U Multifunctional therapeutic, calcium antagonist

11586 (9313) $C_{22}H_{38}NO_3$
16758-41-1

[3-[(p-Butoxybenzoyl)oxy]-2-methylbutyl]triethylammonium = 3-[(4-Butoxybenzoyl)oxy]-N,N,N-triethyl-2-methyl-1-butanaminium (●)

R Iodide (3818-40-4)
S Cvaterone, Kvateleron, Quateleron, Quateron
U Ganglion blocking agent

11587 (9314) $C_{22}H_{38}N_4O_9$
79335-75-4

N-[(*R*)-6-Carboxy-*N*²-[*N*-(1-oxoheptyl)-D-γ-glutamyl]-L-lysyl]-D-alanine (●)
S FK 565, FR 41565
U Antineoplastic, immunomodulator

11588 $C_{22}H_{38}O_5$
157318-92-8

[2*R*-[2α,3aα,4α(1*E*,3*S**,5*S**),5β,6aα]]-[2-[Octahydro-5-hydroxy-4-(3-hydroxy-5-methyl-1-nonenyl)-2-pentalenyl]ethoxy]acetic acid (●)
S SM 10906
U Platelet aggregation inhibitor

11589 (9315) $C_{22}H_{38}O_5$
36950-85-3

[1*R*-[1α(*Z*),2β(1*E*,3*S**),3α,5α]]-7-[3,5-Dihydroxy-2-(3-hydroxy-1-decenyl)cyclopentyl]-5-heptenoic acid (●)
S ICI 74205
U Antifertility (synthetic prostaglandin)

11590 (9316) $C_{22}H_{38}O_5$
120373-36-6

(+)-(*Z*)-7-[(1*R*,2*R*,3*R*,5*S*)-3,5-Dihydroxy-2-(3-oxodecyl)cyclopentyl]-5-heptenoic acid = [1*R*-[1α(*Z*),2β,3α,5α]]-7-[3,5-Dihydroxy-2-(3-oxodecyl)cyclopentyl]-5-heptenoic acid (●)
R see also no. 13014
S *Unoprostone***
U Antiglaucoma agent

11591 (9317) $C_{22}H_{38}O_5$
35700-21-1

Methyl (*E*,*Z*)-(1*R*,2*R*,3*R*,5*R*)-7-[3,5-dihydroxy-2-[(3*S*)-3-hydroxy-3-methyl-1-octenyl]cyclopentyl]-5-heptenoate =
(5*Z*,9α,11α,13*E*,15*S*)-9,11,15-Trihydroxy-15-methyl-prosta-5,13-dien-1-oic acid methyl ester (●)
S *Carboprost methyl*, U-36384
U Oxytocic

11592 (9318) $C_{22}H_{38}O_5$
74159-84-5

(±)-11α,16-Dihydroxy-1,9-dioxo-1-(hydroxymethyl)-16-methyl-13-*trans*-prostene = (11α,13*E*)-(±)-11,16-Dihy-

droxy-1-(hydroxymethyl)-16-methylprost-13-ene-1,9-dione (●)
S   CL 115574
U   Gastric antisecretory, cytoprotective

11593   (9319) $C_{22}H_{38}O_5$
                                                            59122-46-2

(±)-Methyl (1*R*,2*R*,3*R*)-3-hydroxy-2-[(*E*)-(4*RS*)-4-hydroxy-4-methyl-1-octenyl]-5-oxocyclopentaneheptanoate = (11α,13*E*)-11,16-Dihydroxy-16-methyl-9-oxoprost-13-en-1-oic acid methyl ester (●)
S   Acetate, Cyprostil, Cyprostol, Cytotec, Dazitum, Menpros, Misodex, *Misoprostol***, Oxaprost, Procyt, Prostalgin, SC-29333, Symbol
U   Anti-ulcer agent

11594 $C_{22}H_{38}O_5S$
                                                            83058-69-9

6-[[(1*R*,2*S*,3*R*)-3-Hydroxy-2-[(1*E*,3*S*,5*R*)-3-hydroxy-5-methyl-1-nonenyl]-5-oxocyclopentyl]thio]hexanoic acid methyl ester (●)
S   TFC-612
U   Platelet aggregation inhibitor

11595   (9320) $C_{22}H_{38}O_7$
                                                            137-66-6

Palmitoylascorbic acid = (*S*)-2-[(2*R*)-2,5-Dihydro-3,4-dihydroxy-5-oxo-2-furyl]-2-hydroxyethyl hexadecanoate = L-Ascorbic acid 6-hexadecanoate (●)
S   *Acidum palmitoylascorbicum, Ascorbyl palmitate*, Ondascora, Quicifal
U   Dermatologic, anti-oxidant

11596   (9321) $C_{22}H_{39}ClNO$
                                                            123247-90-5

[2-(*p*-Chlorophenoxy)ethyl]dodecyldimethylammonium = *N*-[2-(4-Chlorophenoxy)ethyl]-*N*,*N*-dimethyl-1-dodecanaminium (●)
R   Bromide (15687-13-5)
S   *Dodeclonium bromide***, GR 412, Otinyl
U   Antiseptic, detergent

11597   (9322) $C_{22}H_{40}AsClO_3$

or

13(or 12)-Chloro-12(or 13)-arsenoso-12-henicosene-1-carboxylic acid = Chloroarsinobehenolic acid = 14(or 13)-Chloro-13(or 14)-arsenoso-13-docosenoic acid (●)
R   Strontium salt

S Elarson
U Antibacterial

11598 (9323) $C_{22}H_{40}BrN_3O_2$
60833-68-3

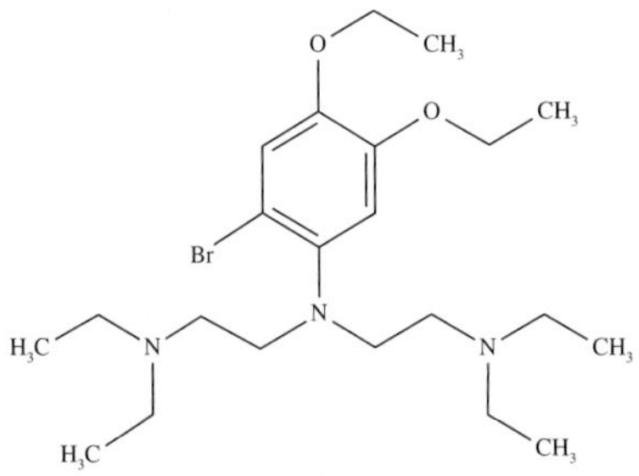

2-Bromo-*N,N*-bis[2-(diethylamino)ethyl]-4,5-diethoxy-
aniline = 5-Bromo-4-[bis(2-diethylaminoethyl)amino]-
1,2-diethoxybenzene = *N*-(2-Bromo-4,5-diethoxyphe-
nyl)-*N*-[2-(diethylamino)ethyl]-*N'*,*N'*-diethyl-1,2-ethane-
diamine (●)
S Diapromin
U Antimalarial

11599 (9324) $C_{22}H_{40}N$
47312-91-4

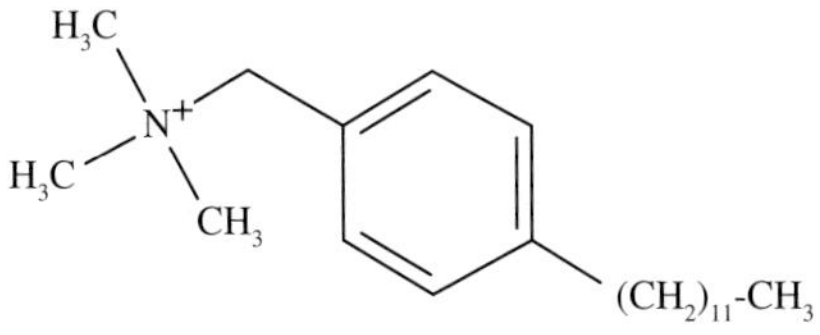

(*p*-Dodecylbenzyl)trimethylammonium = 4-Dodecyl-
N,N,N-trimethylbenzenemethanaminium (●)
R Chloride (19014-05-2)
S Halimide
U Antiseptic, detergent

11600 (9325) $C_{22}H_{40}N$
14054-12-7

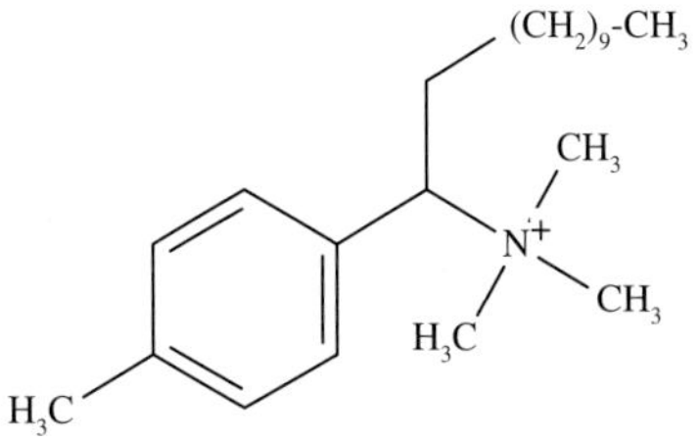

Trimethyl(1-*p*-tolyldodecyl)ammonium = *N,N,N*,4-Tetra-
methyl-α-undecylbenzenemethanaminium (●)
R Methyl sulfate (552-92-1)
S Amoseptic, Desogene, Desogen-flüssig, Stomato-
san, *Tolocinium methylsulfate*, *Toloconium metilsulfate**, *Tolytrimonium methylsulfate*
U Antiseptic

11601 (9326) $C_{22}H_{40}NO$
13900-14-6

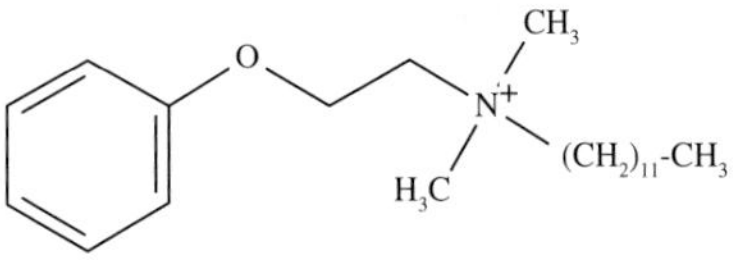

Dodecyldimethyl(2-phenoxyethyl)ammonium = *N,N*-Di-
methyl-*N*-(2-phenoxyethyl)-1-dodecanaminium (●)
R Bromide (538-71-6)
S Antiseptic "Confab", Antiseptique "Prodemdis", Bra-
donit, Bradoral, Bradosol, Bronchodex Pastilles,
Desept, *Domifène bromide*, *Domiphen bromide**,
Domittol, JRP 15/22, Liebelo, Modicare, Neo-Bra-
doral, Neo-Tysal, NSC-39415, Oradol, Oraseptic
"Prodemdis", PDDB, *Phenododecinium bromide*,
Servisept, Terrafine
U Antiseptic, detergent

11602 (9327) $C_{22}H_{40}N_4O_3S$
121100-28-5

N'-[2-(Diethylamino)ethyl]-N-[3-(diethylamino)propyl]-N-[2-(phenylsulfonyl)ethyl]urea (●)
R Citrate (1:2) (125651-31-2)
S AHR-12234
U Anti-arrhythmic

11603 (9328) $C_{22}H_{40}N_4O_5$
140923-32-6

[6S-[1(S*),6R*(1R*,2R*)]]-Tetrahydro-N-hydroxy-6-[[[2-methyl-1-(1-oxopropyl)butyl]amino]carbonyl]-γ-oxo-β-pentyl-1(2H)-pyridazinebutanamide (●)
S *Matlystatin B*, SF 2197
U Antibiotic from *Actinomadura atramentaria*

11604 (9329) $[C_{22}H_{40}O_4]_x$
71251-04-2

Poly(1,2-dicarboxy-3-hexadecyltetramethylene) = Poly(1,2-dicarboxy-3-hexadecyl-1,4-butanediyl) (●)
S AOMA, *Surfomer***
U Antihyperlipidemic

11605 (9330) $C_{22}H_{40}O_7$
666-99-9

α-Hexadecylcitric acid = 2-Hydroxy-1,2,3-nonadecane-tricarboxylic acid (●)
S *Acidum agaricinicum, Acidum agaricum, Acidum laricinicum, Agaricic acid, Agaricin, Agarizinsäure, Laricic acid, Laricinsäure*
U Anhidrotic, antibacterial, antifungal from *Polyporus officinalis* or *Laricifomus officinalis*

11606 (9331) $C_{22}H_{41}NO_7$
110231-33-9

4-Acetoxy-2-amino-3,5,14-trihydroxyeicos-6-enoic acid = 4-(Acetyloxy)-2-amino-3,5,14-trihydroxy-6-eicosenoic acid (●)
S *Fumifungin*
U Antifungal antibiotic from *Aspergillus fumigatus*

11607 (9332) $C_{22}H_{42}Br_2O_2$
95806-39-6

13,14-Dibromobehenic acid = 13,14-Dibromodocosanoic acid (●)
R Calcium salt (2:1) (1301-35-5)

S Asabromin, Calbroben, Sabromin (old form)
U Sedative

11608 (9333) $C_{22}H_{42}N_4O_8S_2$
16816-67-4

(+)-*N,N'*-[Dithiobis(ethyleneiminocarbonylethylene)]
bis[2,4-dihydroxy-3,3-dimethylbutyramide] = *N,N'*-
(Dithiodiethylene)bis[(+)-pantothenamide] = Bis[*N*-(+)-
pantothenoyl-2-aminoethyl] disulfide = [*R*-(*R**,*R**)]-
N,N'-[Dithiobis[2,1-ethanediylimino(3-oxo-3,1-propane-
diyl)]]bis[2,4-dihydroxy-3,3-dimethylbutanamide] (●)
S Analip, Atarone, Christamin, Dermorizin, DF 72, Ei-
pante, Freelon, Kohlchin, *LBF disulfide form*, Lipo-
del, Lipol, Liponet, Lubtethine, Lumiclin, Mitapan,
Obliterol "Faes", Paddy-Pow, Palfadin, Palmitocin,
Pancerol, Panfull, Pangare, Pansatol, Pantejust, *Pan-
tethine*, *Pantetina*, Pantetran, Panthecin S, Pantline,
Pantogen "Maruko", Pantomin, Pantosin, Pantozy-
me, Pantramin, Papaletin, Parutox, Raddy, Rape-
sten, Retten, Sekten, Silenitin, Taropatine, Tricedi-
ne, Vitamel, Youtetin
U Growth factor, antilipidemic

11609 $C_{22}H_{42}O_4$
103-23-1

Di(2-ethylhexyl)adipate = Adipic acid bis(2-ethylhexyl)
ester = Hexanedioic acid bis(2-ethylhexyl) ester (●)
S ADO, DEHA, *Dioctyl adipate*, DOA, Protec
U Insect repellent

11610 (9334) $C_{22}H_{43}IO_2$

13(or 14)-Iodobehenic acid = 13(or 14)-Iododocosanoic
acid (●)
R Calcium salt (2:1) (25498-47-9)
S *Calcium iodbehenicum, Calcium monojodbeheni-
cum*, Calioben, *Iodobehenate Calcium*, Sajodin
U Iodine source

11610-01 (9334-01)
R Iron salt
S Eisen-Sajodin, Picural
U Iodine and iron source

11611 (9335) $C_{22}H_{43}N_5O_{12}$
58152-03-7

O-6-Amino-6-deoxy-α-D-glucopyranosyl-(1→4)-*O*-[3-
deoxy-4-*C*-methyl-3-(methylamino)-β-L-arabinopyrano-
syl-(1→6)]-2-deoxy-*N*^{1}-[(*S*)-isoseryl]-D-streptamine =
(*S*)-*O*-6-Amino-6-deoxy-α-D-glucopyranosyl-(1→4)-*O*-
[3-deoxy-4-*C*-methyl-3-(methylamino)-β-L-arabinopyra-
nosyl-(1→6)]-*N*^{1}-(3-amino-2-hydroxy-1-oxopropyl)-2-
deoxy-D-streptamine (●)
R also sulfate (1:2)(salt) (68000-78-2)
S Exacin "Toyo Jozo", HAPA-5-epi-Gmb, HAPA-B,
Isepacin, Isépalline, *Isepamicin***, NEC 9801,
Sch 21420, Vizax
U Antibiotic

11612   (9336) $C_{22}H_{43}N_5O_{13}$
37517-28-5

O-3-Amino-3-deoxy-α-D-glucopyranosyl-(1→4)-O-[6-amino-6-deoxy-α-D-glucopyranosyl-(1→6)]-N^3-(4-amino-L-2-hydroxybutyryl)-2-deoxy-L-streptamine = 1-N-[L(−)-4-Amino-2-hydroxybutyryl]kanamycin A = (S)-O-3-Amino-3-deoxy-α-D-glucopyranosyl-(1→6)-O-[6-amino-6-deoxy-α-D-glucopyranosyl-(1→4)]-N^1-(4-amino-2-hydroxy-1-oxobutyl)-2-deoxy-D-streptamine (●)

R   most sulfate (1:2) (39831-55-5)

S   Akacin, Akicin, Akim, Amic "Jaba", Amicacide, Amicacina medical, Amicacina Northia, Amicasil, Amicin "Biochem", Amiglicin, Amiglyde-V, Amikabiot, *Amikacin sulfate***, Amikacit, Amikafur, Amikalem, Amikan, Amikasin, Amikason's, Amikaver, Amikavet, Amikayect, Amikazit, Amikhaver, Amikin, Amiklin, Amikozit, Amisin, Amitrex, A-M-K, Amukin, Anicamil, BB-K 8, Belmaton, Biclim, Biclin, Biklin, Biodacyna, Briclin, Briklin, Burecasin, Cashimy, Chemacin, Consumonit, Dramigel, Fabianol, Farcyclin, Flexelite, Fromentyl, Gamikal, Ivimycin, Kacin, Kacinth-A, Kamina, Kaminax, Kanbine, Kancin "Gap", Karuamine, Lanomycin "Pharmathen", Likacin, Lukadin, Medilen, Migracin, Mikacin, Mikamic, Mikan, Mikasin, Mikasome, Mikavir, Negasin, Novamin "Bristol", NSC-177001, Oprad, Orlobin, Pediakin, Pierami, Prutetucin, Romikacin, Selaxa, Selemycin, Sifamic, Slamik, Tipkin, Uzix, Yectamid

U   Antibiotic

11613 $C_{22}H_{44}N_2$
123018-47-3

2-[3-(Diethylamino)propyl]-8,8-dipropyl-2-azaspiro[4.5]decane = N,N-Diethyl-8,8-dipropyl-2-azaspiro[4.5]decane-2-propanamine (●)

R   Dihydrochloride (130065-61-1)

S   *Atiprimod dihydrochloride***, SKF 106615-A_2

U   Immunomodulator (anti-inflammatory, anti-arthritic)

11613-01   183063-72-1

R   (Z)-2-Butenedioate (1:2)

S   *Atriprimod dimaleate***, SKF 106615-I_2

U   Immunomodulator (anti-inflammatory, anti-arthritic)

11614   (9337) $C_{22}H_{44}N_4$
47473-71-2

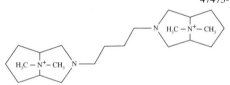

3,3'-Tetramethylenebis(9,9-dimethyl-3-aza-9-azoniabicyclo[3.3.1]nonane) = 3,3'-(1,4-Butanediyl)bis-[9,9-dimethyl-3-aza-9-azoniabicyclo[3.3.1]nonane] (●)

R   Diiodide (3406-44-8)

S   Nobutan, OF 1478

U   Curaremimetic, muscle relaxant

11614-01   (9337-01)   3841-25-6

R   Bis(methyl sulfate)

S   Mebutan

U   Curaremimetic

11615　(9338)　　　　　　　　　　　　$C_{22}H_{44}N_6O_{10}$
　　　　　　　　　　　　　　　　　　　　　51025-85-5

O-3-Amino-3-deoxy-α-D-glucopyranosyl-(1→4)-*O*-[2,6-
diamino-2,3,4,6-tetradeoxy-α-D-*erythro*-hexopyranosyl-
(1→6)]-*N'*-[(2*S*)-4-amino-2-hydroxybutyryl]-2-deoxy-L-
streptamine = 1-*N*-[(*S*)-4-Amino-2-hydroxybutyryl]dibe-
kacin = (*S*)-*O*-3-Amino-3-deoxy-α-D-glucopyranosyl-
(1→6)-*O*-[2,6-diamino-2,3,4,6-tetradeoxy-α-D-*erythro*-
hexopyranosyl-(1→4)]-*N¹*-(4-amino-2-hydroxy-1-oxo-
butyl)-2-deoxy-D-streptamine (●)
R　also sulfate (104931-87-5)
S　*Arbekacin***, HABA-Dibekacin, Habekacin, HBK,
　　1665 RB
U　Antibiotic

11616　　　　　　　　　　　　　　　$C_{22}H_{45}NO_2$
　　　　　　　　　　　　　　　　　　　　13127-82-7

N-Oleyldiethanolamine = 2,2'-[(9*Z*)-9-Octadecenylimi-
no]bis[ethanol] (●)
R　Hydrofluoride (207916-33-4)
S　Oleaflur
U　Dental caries prophylactic

11617　(9339)　　　　　　　　　　　　　$C_{22}H_{45}N_3$
　　　　　　　　　　　　　　　　　　　　5980-31-4

**2,6-Bis(2-ethylhexyl)hexahydro-7a-methyl-1*H*-imida-
zo[1,5-*c*]imidazole (●)
S　*Hexedine***, Sterisol, W 4701
U　Antiseptic

11618　(9340)　　　　　　　　　　　　$C_{22}H_{45}N_5O_{12}$
　　　　　　　　　　　　　　　　　　　　59733-86-7

(*S*)-*O*-3-Amino-3-deoxy-α-D-glucopyranosyl-(1→6)-*O*-
[6-amino-6-deoxy-α-D-glucopyranosyl-(1→4)]-*N¹*-(4-
amino-2-hydroxybutyl)-2-deoxy-D-streptamine (●)
S　*Butikacin***, UK 18892
U　Antibiotic

11619　(9341)　　　　　　　　　　$C_{22}H_{46}N_4O_3S_2V$
　　　　　　　　　　　　　　　　　　　　122575-28-4

Bis[2-amino-3-mercapto-*N*-octylpropionamidato(1–)-
S]oxovanadium = *S,S'*-(Vanadium(IV) oxide) of bis(*N*-oc-
tylcysteinamide) = Bis[2-amino-3-(mercapto-κ*S*)-*N*-octy-
lpropanamidato-κ*O*]oxovanadium (●)
R　also trihydrate
S　*Naglivan***
U　Antidiabetic

11620　(9342)　　　　　　　　　　　　　$C_{22}H_{46}O$
　　　　　　　　　　　　　　　　　　　　661-19-8

Behenyl alcohol = 1-Docosanol (●)
S　Abreva, Abreve, *Behenyl alcohol*, Docosanol, IK. 2,
　　Lidakol, V-1326
U　Antiviral, treatment of prostate disorders

11621   (9343) $C_{22}H_{47}N_3O$
60209-20-3

L-2,6-Diamino-*N*-hexadecylcapronamide = (*S*)-2,6-Di-
amino-*N*-hexadecylhexanamide (●)
S   *Lycetamine*, P-71
U   Antibacterial

11622   (9344) $C_{22}H_{48}N$
20256-56-8

Didecyldimethylammonium = *N*-Decyl-*N*,*N*-dimethyl-1-
decanaminium (●)
R   Chloride (7173-51-5)
S   Adrocid neue Formulierung, Amosept N, Bardac 22,
Desamon, *Didecyldimonium chloride*, Du-Muc,
Hoe S 2922, Orosept, Pedicid
U   Topical antibacterial

11623   (9345) $C_{22}H_{48}N$
45273-64-1

Ethyldimethyloctadecylammonium = *N*-Ethyl-*N*,*N*-dime-
thyl-1-octadecanaminium (●)
R   Ethyl sulfate (110-07-6)
S   *Aethalkoni aethylsulfas, Ethalkonium ethylsulfate*,
Querton
U   Antiseptic

11624 $C_{23}H_{14}F_9NO_4S$
169527-42-8

2-[[2-[[[3,5-Bis(trifluoromethyl)phenyl]sulfonyl]amino]-
4-(trifluoromethyl)phenyl]methyl]benzoic acid (●)
S   SB 203347
U   Phospholipase A$_2$ inhibitor

11625   (9346) $C_{23}H_{15}F_2NO_2$
96187-53-0

6-Fluoro-2-(2'-fluoro-4-biphenylyl)-3-methyl-4-quino-
linecarboxylic acid = 6-Fluoro-2-(2'-fluoro[1,1'-biphe-
nyl]-4-yl)-3-methyl-4-quinolinecarboxylic acid (●)
R   Sodium salt (96201-88-6)
S   *Bipenquinate sodium, Brequinar sodium***,
DuP 785, NSC-368390
U   Antineoplastic

11626 $C_{23}H_{16}F_4N_2O_2$
147838-04-8

8-Fluoro-4-[2-[4-[4-(trifluoromethyl)phenoxy]phe-
nyl]ethoxy]quinazoline (●)
S   XR-100

U Nematocide, insecticide

11627 $C_{23}H_{16}N_2O_5$
                                                            151360-46-2

Methyl 4-[2-[4-[2-(2-methoxyphenyl)benzoxazolyl]ben-
zoxazolyl]]carboxylate = 2'-(2-Methoxyphenyl)-[2,4'-bi-
benzoxazole]-4-carboxylic acid methyl ester (●)
S Methyl UK-1, MUK-1
U Antibacterial, antifungal, cytotoxic

11628   (9347) $C_{23}H_{16}O_3$
                                                            82-66-6

2-(Diphenylacetyl)-1,3-indandione = 2-(Diphenylacetyl)-
1H-indene-1,3(2H)-dione (●)
S Didandin, Didion, *Difenadione*, Difenatsin, Difexan,
Dipaxin, *Diphacinone*, Diphenacin, *Diphenadione**,
Oragulant, Solvan, U-1363, URI 788
U Anticoagulant, rodenticide

11629   (9348) $C_{23}H_{16}O_{11}$
                                                            16110-51-3

5,5'-[(2-Hydroxytrimethylene)dioxy]bis[4-oxo-4H-1-
benzopyran-2-carboxylic acid] = 5,5'-(2-Hydroxytrime-
thylenedioxy)bis(4-oxo-4H-chromene-2-carboxylic acid)

= 1,3-Bis[(2-carboxy-4-oxochromen-5-yl)oxy]-2-propa-
nol = 5,5'-[(2-Hydroxy-1,3-propanediyl)bis(oxy)]bis[4-
oxo-4H-1-benzopyran-2-carboxylic acid] (●)
R see also no. 14740
S *Acide cromoglicique***, Acidum cromoglicicum**,*
    *Cromoglicic acid**, Cromoglycic acid*
U Anti-allergic, anti-asthmatic (prophylactic)

11629-01   (9348-01)   15826-37-6
R Disodium salt
S Aarane "Syntex", Acecromol, Acromax, Aerocrom,
    Aeropaxyn, Alerbol, Alercrom, Alerg AT, Alerg-Na-
    senspray, Alergosovis, Alérion, Alersiris, Aller-
    crom, Allergo-COMOD, Allergocrom, Allergojovis,
    Allergotin, Allergoval, Allersol, Astmocupin, Be-
    nasol, Bicromat, Bikromat, Blacil, Botastin,
    Brol-Eze, Clariteyes, Claroftal, Clesin, CLO-5, Coli-
    mune, Coolway, Croglicina, Croglina, Crolom, Cro-
    mabak, Cromadura, Cromal, Croman, Cromantal,
    Cromedil, Cromese Sterinebs, Cromex "Davi",
    Cromeze, Cromisol, Cromo-Asma, Cromobene,
    Cromodyn, Cromoftal, Cromogen, *Cromoglicat Di-
    natrium, Cromoglicato di Sodico, Cromoglicin,*
    Cromoglin, Cromogloz, Cromoglyn, Cromohexal,
    Cromol, Cromolerg, Cromolergin, Cromolin, Crom-
    olind, Cromolyn-Fatol, Cromolyn Orion, *Cromolyn
    Sodium*, Crom-Ophtal, Cromopp, Cromoptic, Cromo
    pur von ct, Cromoral, Cromo-ratiopharm, Cromo-
    reak, Cromosan, Cromosol, Cromo-Stulln UD, Cro-
    moturmant, Cromovet, Cromovist, Cromo von ct,
    Cronacol, Cropoz, Cryl ofteno, Cumorol, Cusicrom,
    Cusilyn, Diffusyl, Dilospir, *Dinatrium cromoglici-
    cum, Disodium cromoglycate*, Dispacromil, DNCG,
    Dolmin, DSCG, Duobetic, Duracroman, Erysta-
    min K, Esirhinol, Ethicrom, Eye-crom, Farmacrom,
    Fenistil Augentropfen Cromoglicin, Fenolip, Fintal,
    Fivent, Flenid, Flui-DNCG, Fluvet "Vianex",
    FPL 670, Frenal, Frenasma, Gaster "So. Se. Pharm",
    Gastrocom, Gastrocrom, Gastrofrenal, Gelodrin, Gli-
    cacil, Glicin "Hexal", Glicinal, Glicrom, Glinor,
    Hay-Crom, Heusnif, Hexacroman, Histalyn, Hyper-
    san, Ifiral, Indoprex, Inhibin "Thiemann", Inostral,
    Intal, Intercron, Introl, Iopanchol, Irtan (old form),
    Joglicin, Kaosyl, Kiddicrom, Kromolin, Lecrolyn,
    Logomed-Heuschnupfen-Spray, Lomudal, Lomufer-
    te, Lomupren, Lomusol, Lomuspray, Lotal, Mainter,
    Maxicrom, Maxirom, Mitayaku, Nalcrom, Nalcron,
    Nasalcrom, Nasivin gegen Heuschnupfen, Nasmil,

Nasmum, Nasmun, *Natrii cromoglicas*, Nazotral, Nebulasma, Nebulcrom, Noaler, Noslan, Novacrom, Novo-Cromolyn, Oculos cromoglicato, Oftakrom, Oftalmobifan, Opticrom, Opticron, Optrex Hayfever, Oralcrom, Otriven H, Pädiacrom, Pentacrom UD, Pentatop, Polcrom, Poledin, Pollyferm, Prevalin, Primover, Prothanon cromo, Proventol, Pulbil, Pulmosin, Quimbar, Rhinocrome, Rhinol H, Rilan, Rinil "ACO; Pharmacia", Rinobifan, Rinofrenal, Rinoglin, Rugeon, Rynacrom, Sificrom, Smarodax, *Sodium cromoglicate*, Sofro, Spaziron, Stadaglicin, Steri-Neb Cromogen, Taleum, Talium, Tional, Ufocollyre, Urocollyre, Vekfanol, Vicrom, Vistacrom, Visuphrine, Vividrin, Viz-On, Wick Contrallerg, Zineli

U Anti-allergic, anti-asthmatic (prophylactic)

11630 (9349) $C_{23}H_{16}O_{11}$
                                                           16139-47-2

7,7'-(2-Hydroxytrimethylenedioxy)bis(4-oxo-4*H*-chromene-2-carboxylic acid) = 7,7'-[(2-Hydroxy-1,3-propanediyl)bis(oxy)]bis[4-oxo-4*H*-1-benzopyran-2-carboxylic acid] (●)

R Disodium salt (37092-38-9)
S *Disodium crompoxate*
U Anti-allergic

11631 $C_{23}H_{17}Cl_2NO_4$
                                                   147960-65-4

p-[3-(2,7-Dichloro-9-fluorenyloxy)propionamido]benzoic acid = 4-[[3-[(2,7-Dichloro-9*H*-fluoren-9-yl)oxy]-1-oxopropyl]amino]benzoic acid (●)

S NPC 17923
U Anti-inflammatory

11632 $C_{23}H_{17}NO_2$
                                                   194606-69-4

10-(4-Acetamidobenzylidene)-9-anthrone = *N*-[4-[(10-Oxo-9(10*H*)-anthracenylidene)methyl]phenyl]acetamide (●)

S DK-V-47
U Tyrosine kinase inhibitor

11633 (9350) $C_{23}H_{17}NO_3$
                                                           60883-69-4

o-Methoxyphenyl 2-phenyl-4-quinolinecarboxylate = 2-Phenyl-4-quinolinecarboxylic acid guaiacol ester = 2-Phenyl-4-quinolinecarboxylic acid 2-methoxyphenyl ester (●)

S *Guaiacol cinchophenate*, Guphen
U Bronchotherapeutic

11634 (9351) $C_{23}H_{18}F_2N_2O_2$
                                                           73445-46-2

2-(2,4-Difluorophenyl)-4,5-bis(*p*-methoxyphenyl)imida-zole = 2-(2,4-Difluorophenyl)-4,5-bis(4-methoxyphenyl)-1*H*-imidazole (●)
S A-214, *Fenflumizole***
U Anti-inflammatory

11635

$C_{23}H_{18}F_2N_4O$
124669-93-8

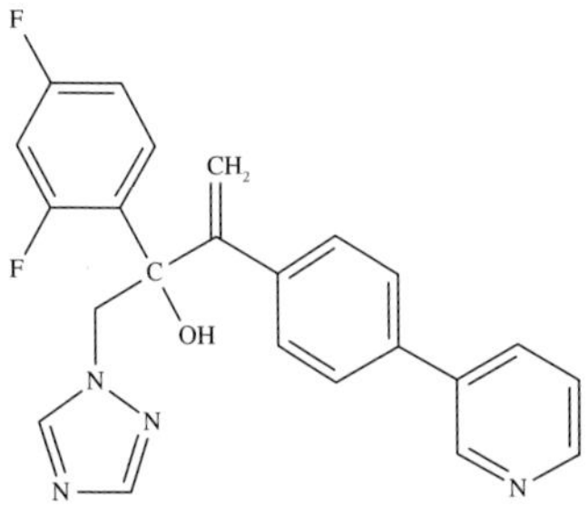

α-(2,4-Difluorophenyl)-α-[1-[4-(3-pyridinyl)phe-nyl]ethenyl]-1*H*-1,2,4-triazole-1-ethanol (●)
R Dimethanesulfonate (salt) (150525-63-6)
S XD 405
U Antifungal

11636 (9352)

$C_{23}H_{18}N_2O_2$
32710-91-1

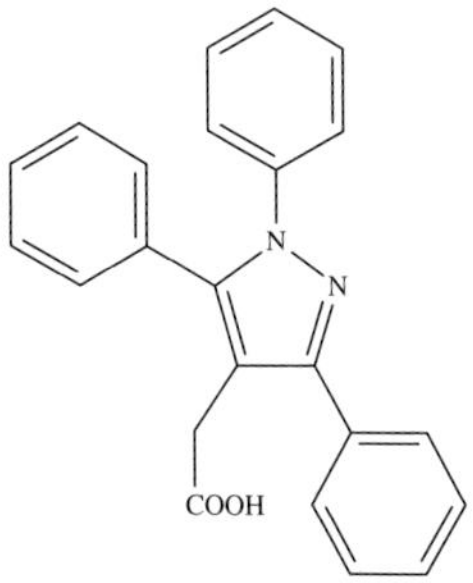

1,3,5-Triphenyl-1*H*-pyrazole-4-acetic acid (●)
S *Trifezolac***
U Analgesic, anti-inflammatory

11637 (9353)

$C_{23}H_{18}N_2O_2$
50270-33-2

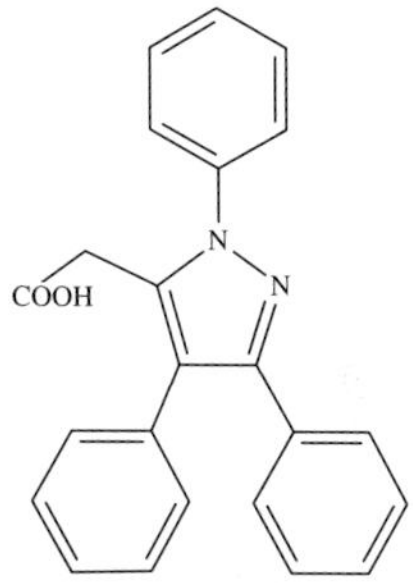

1,3,4-Triphenyl-1*H*-pyrazole-5-acetic acid (●)
S *Isofezolac***, LM 22102, Sofenac
U Analgesic, anti-inflammatory

11638 (9354)

$C_{23}H_{18}N_2O_3$
105350-26-3

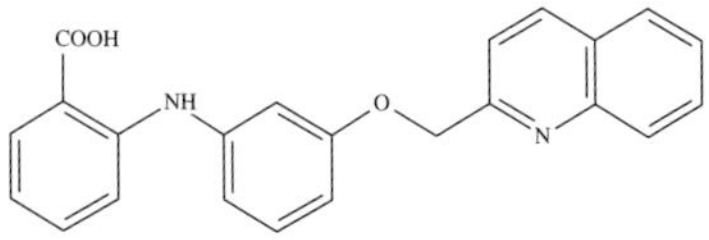

o-[*m*-(2-Quinolylmethoxy)anilino]benzoic acid = 2-[[3-(2-Quinolinylmethoxy)phenyl]amino]benzoic acid (●)
S SR 2640
U Anti-allergic

11639 (9355)

$C_{23}H_{18}N_2O_5$
83198-27-0

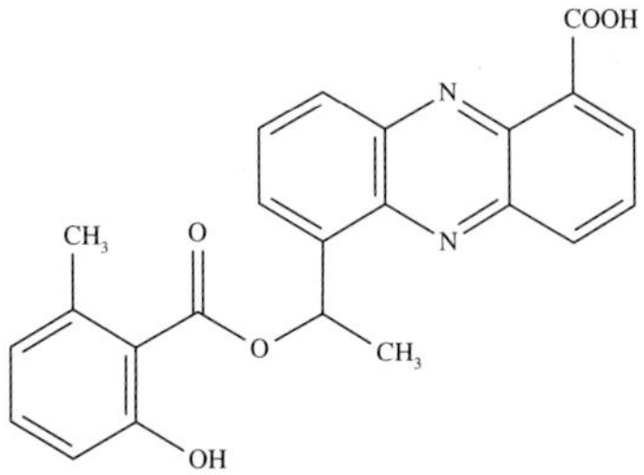

6-(1-Hydroxyethyl)-1-phenazinecarboxylic acid 6-me-thylsalicylate = 6-[1-[(2-Hydroxy-6-methylben-zoyl)oxy]ethyl]-1-phenazinecarboxylic acid (●)
S A 32256, *Saphenamycin*
U Antibiotic from *Streptomyces canarius*

11640   (9356) $C_{23}H_{18}N_4$
122955-18-4

5,5-Bis(4-pyridinylmethyl)-5*H*-cyclopenta[2,1-*b*:3,4-*b*']dipyridine (●)
R   also monohydrate (139781-09-2)
S   DuP 921, EXP-921, *Sibopirdine***
U   Cognition enhancer

11641   (9357) $C_{23}H_{18}O_6$
11029-70-2

S   *Heliomycin***
U   Antibiotic from *Actinomyces flavochromogenes* var. *heliomycini*

11642   (9358) $C_{23}H_{19}ClF_3NO_3$
68085-85-8

(*RS*)-α-Cyano-3-phenoxybenzyl (*Z*)-(1*RS*,3*RS*)-3-(2-chloro-3,3,3-trifluoropropenyl)-2,2-dimethylcyclopro-panecarboxylate = 3-(2-Chloro-3,3,3-trifluoro-1-prope-nyl)-2,2-dimethylcyclopropanecarboxylic acid cyano(3-phenoxyphenyl)methyl ester (●)
S   *Cyhalothrin*, PP 563
U   Insecticide

11643 $C_{23}H_{19}ClO_3S$
179545-77-8

11640   (9356) [right column]

4-(4'-Chlorobiphenyl-4-yl)-4-oxo-2(*S*)-(phenylthiome-thyl)butyric acid = (*S*)-3-[(4'-Chloro-4-biphenylyl)carbo-nyl]-2-[(phenylthio)methyl]propionic acid = (α*S*)-4'-Chloro-γ-oxo-α-[(phenylthio)methyl]-[1,1'-biphenyl]-4-butanoic acid (●)
S   Bay 12-9566, *Tanomastat***
U   Oncolytic, anti-arthritic (matrix metalloproteinase in-hibitor)

11644 $C_{23}H_{19}Cl_2NO_3S$
154413-61-3

(*E*)-6-[[(2,6-Dichlorophenyl)thio]methyl]-3-(phenethyl-oxy)-2-pyridineacrylic acid = (2*E*)-3-[6-[[(2,6-Dichloro-phenyl)thio]methyl]-3-(2-phenylethoxy)-2-pyridinyl]-2-propenoic acid (●)
S   SB 209247, *Ticolubant***
U   Anti-inflammatory, antipsoriatic

11645   (9359) $C_{23}H_{19}F_5N_6O_2$
124750-95-4

4-(Pentafluoroethyl)-2-propyl-1-[[2'-(1*H*-tetrazol-5-yl)[1,1'-biphenyl]-4-yl]methyl]-1*H*-imidazole-5-carb-oxylic acid (●)
S   DuP 532
U   Antihypertensive (angiotensin II receptor antagonist)

11646 $C_{23}H_{19}NO_3$
193620-69-8

3-[Bis(4-methoxyphenyl)methylene]-2-indolinone = 3-[Bis(4-methoxyphenyl)methylene]-1,3-dihydro-2*H*-indol-2-one (●)
S TAS-301
U Cytotoxic (granulation inhibitor)

11647 (9360) $C_{23}H_{19}NO_3$
123016-21-7

(*S*)-2-[6-(2-Quinolylmethoxy)-2-naphthyl]propionic acid = (*S*)-α-Methyl-6-(2-quinolinylmethoxy)-2-naphthaleneacetic acid (●)
S Wy-50295
U Anti-inflammatory

11648 $C_{23}H_{19}NO_4$
150256-47-6

4-[[[2-(9*H*-Fluoren-9-yl)ethoxy]carbonyl]amino]benzoic acid (●)
S NPC 16570
U Anti-allergic, immunosuppressive

11649 $C_{23}H_{19}N_5O$
152459-94-4

N-[4-Methyl-3-[[4-(3-pyridinyl)-2-pyrimidinyl]amino]phenyl]benzamide (●)
S CGP 53716
U antineoplastic, antiproliferative

11650 $C_{23}H_{20}F_2N_2O_4$
194804-75-6

1-Cyclopropyl-8-(difluoromethoxy)-7-[(1*R*)-2,3-dihydro-1-methyl-1*H*-isoindol-5-yl]-1,4-dihydro-4-oxo-3-quinolinecarboxylic acid (●)
R Monomethanesulfonate monohydrate (223652-90-2)
S BMS 284756, T 3811ME
U Antibacterial (gram-positive)

11651 (9361) $C_{23}H_{20}F_2N_4O_3$
118845-49-1

trans-6-[4,4-Bis(4-fluorophenyl)-3-(1-methyl-1*H*-tetrazol-5-yl)-1,3-butadienyl]tetrahydro-4-hydroxy-2*H*-pyran-2-one (●)
S BMY-22089

U Antihypercholesteremic (HMGCoA reductase inhibitor)

11652 $C_{23}H_{20}F_3N_7$
149285-55-2

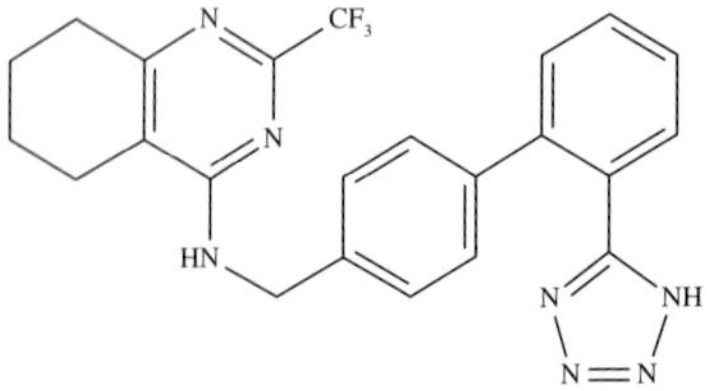

5,6,7,8-Tetrahydro-N-[[2'-(1H-tetrazol-5-yl)[1,1'-biphenyl]-4-yl]methyl]-2-(trifluoromethyl)-4-quinazolinamine (●)

S WAY-126227
U Antihypertensive (angiotensin II antagonist)

11653 $C_{23}H_{20}F_6N_6O_3$
155432-64-7

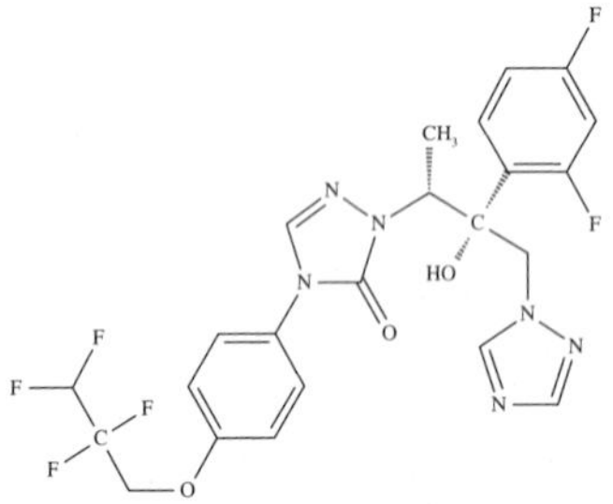

2-[(1R,2R)-2-(2,4-Difluorophenyl)-2-hydroxy-1-methyl-3-(1H-1,2,4-triazol-1-yl)propyl]-2,4-dihydro-4-[4-(2,2,3,3-tetrafluoropropoxy)phenyl]-3H-1,2,4-triazol-3-one (●)

S TAK-187
U Antifungal

11654   (9362) $C_{23}H_{20}N_2O_2S$
3736-92-3

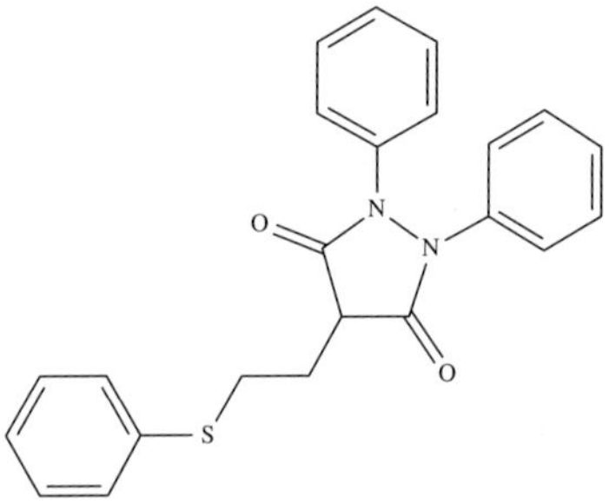

1,2-Diphenyl-4-[2-(phenylthio)ethyl]-3,5-pyrazolidinedione (●)
S G 25671, *Thiophenylpyrazolidine*
U Antirheumatic, anti-arthritic

11655   (9363) $C_{23}H_{20}N_2O_3S$
57-96-5

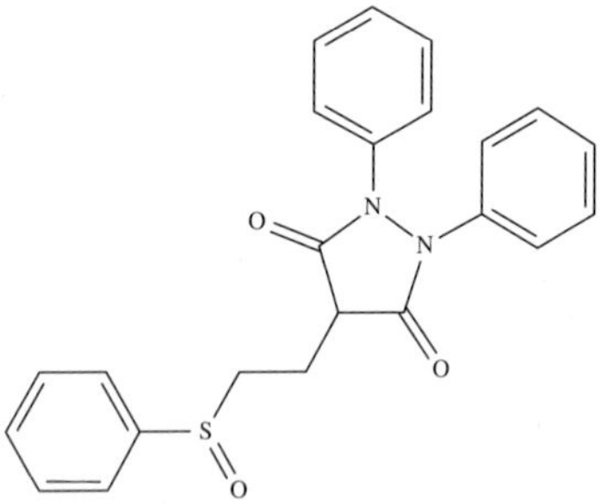

1,2-Diphenyl-4-[2-(phenylsulfinyl)ethyl]-3,5-pyrazolidinedione (●)
S Antazone, Anturan(e), Anturanil, Anturano, Anturen, Anturidin, Apo-Sulfinpyrazone, Aprazone, Enturan, Enturen, Eryfrace, Falizal, G 28315, Novopyrazone, Pyrocard, Rabenid, Sulfazone, Sulfinona, *Sulfinpyrazone***, Sulfizone, *Sulfoxyphenylpyrazolidine*, *Sulphinpyrazone*, TX 1320, Zynol
U Uricosuric, anti-arthritic, platelet aggregation inhibitor

11656   (9364) $C_{23}H_{20}N_2O_4S$
141200-24-0

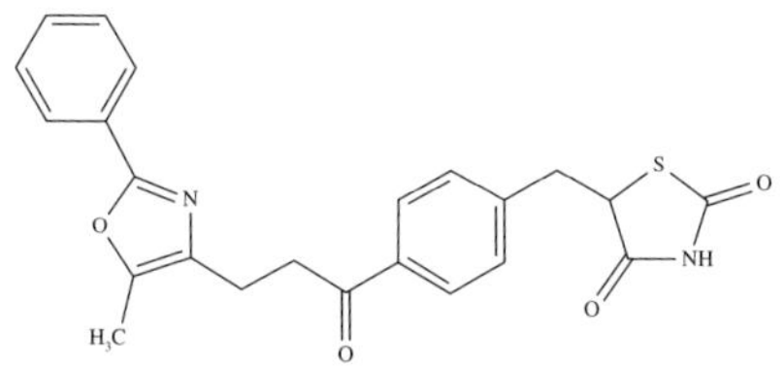

(±)-5-[*p*-[3-(5-Methyl-2-phenyl-4-oxazolyl)propio-nyl]benzyl]-2,4-thiazolidinedione = (±)-5-[[4-[3-(5-Methyl-2-phenyl-4-oxazolyl)-1-oxopropyl]phenyl]me-thyl]-2,4-thiazolidinedione (●)
R   Sodium salt (141683-98-9)
S   CP-86325-2, *Darglitazone sodium***
U   Oral hypoglycemic

11657   (9365) $C_{23}H_{20}N_2O_5$
37106-97-1

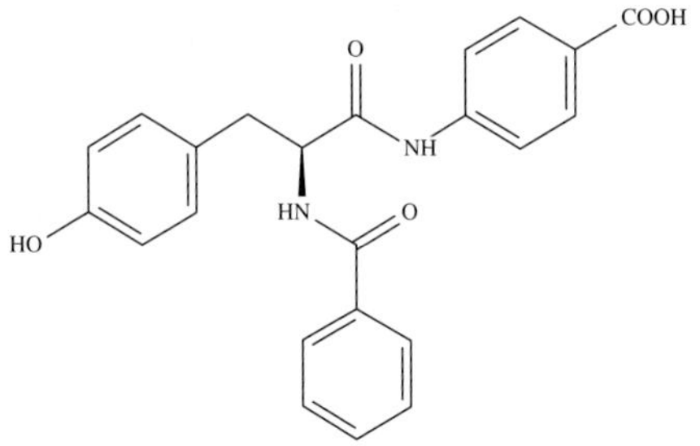

(*S*)-*p*-(α-Benzamido-*p*-hydroxyhydrocinnamamido)ben-zoic acid = *p*-(*N*-Benzoyl-L-tyrosinamido)benzoic acid = (*S*)-4-[[2-(Benzoylamino)-3-(4-hydroxyphenyl)-1-oxo-propyl]amino]benzoic acid (●)
S   *Bentiromide***, *Bentirosin*, BTM, BTPABA, Chymex, E-2663, Exocrine, Pankreas-Funktionstest "Roche", PFD, PFT "Roche", Ro 11-7891
U   Diagnostic aid (pancreas function determination)

11658   (9366) $C_{23}H_{20}O_7$
24340-62-3

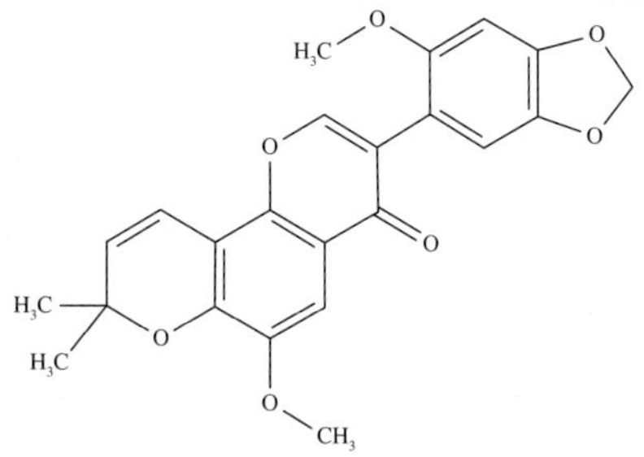

6-Methoxy-3-(6-methoxy-1,3-benzodioxol-5-yl)-8,8-di-methyl-4*H*,8*H*-benzo[1,2-*b*:3,4-*b'*]dipyran-4-one (●)
S   *Ichthynone*
U   Isoflavone from *Piscida erythrina*

11659   (9367) $C_{23}H_{20}O_8$
129586-20-5

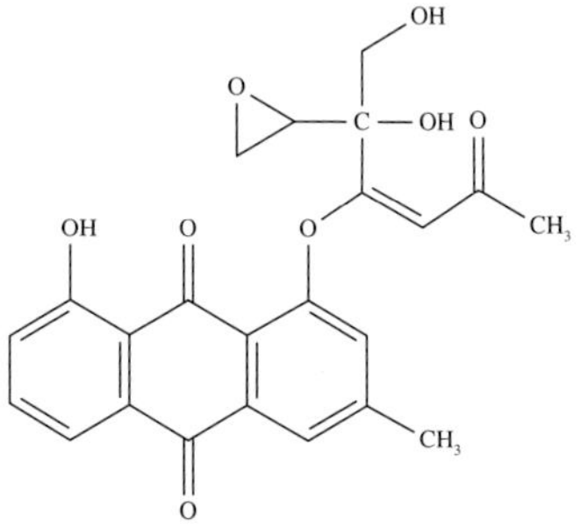

1-[3,4-Epoxy-2-hydroxy-2-(hydroxymethyl)-1-(2-oxo-propylidene)butoxy]-8-hydroxy-3-methylanthraquinone = 1-[[1-(1,2-Dihydroxy-1-oxiranylethyl)-3-oxo-1-bute-nyl]oxy]-8-hydroxy-3-methyl-9,10-anthracenedione (●)
S   BU-3839 T
U   Antineoplastic, antibacterial from *Streptomyces viola-ceus*

11660   (9368) $C_{23}H_{21}BrClN$
49857-37-6

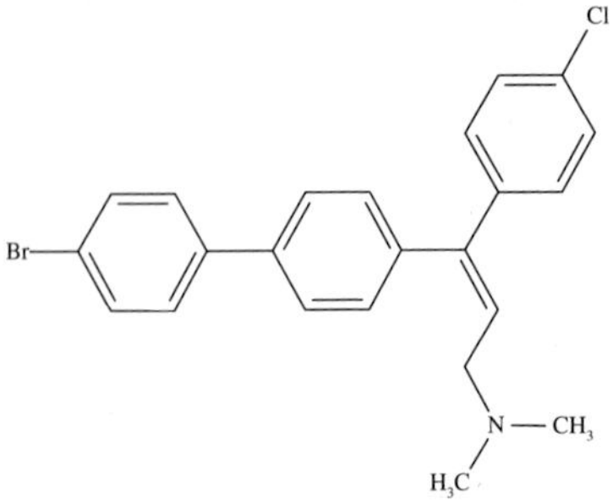

(*E*)-1-(4'-Bromo-4-biphenylyl)-1-(4-chlorphenyl)-3-(dimethylamino)-1-propene = (*E*)-3-(4'-Bromo[1,1'-biphenyl]-4-yl)-3-(4-chlorophenyl)-*N*,*N*-dimethyl-2-propen-1-amine (●)

S   353 C

U   Antiprotozoal

11661   (9369) $C_{23}H_{21}ClN_2O_4$
109623-97-4

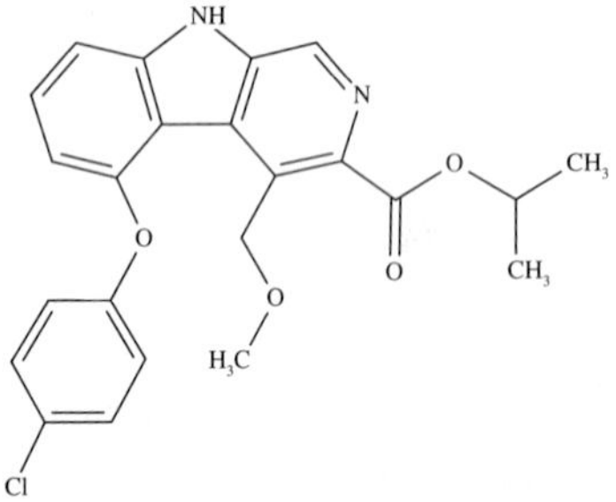

Isopropyl 5-(*p*-chlorophenoxy)-4-(methoxymethyl)-9*H*-pyrido[3,4-*b*]indole-3-carboxylate = 5-(4-Chlorophenoxy)-4-(methoxymethyl)-9*H*-pyrido[3,4-*b*]indole-3-carboxylic acid 1-methylethyl ester (●)

S   *Gedocarnil***, ZK 113315

U   Anticonvulsant, anxiolytic (benzodiazepine partial agonist)

11662   (9370) $C_{23}H_{21}ClN_6O_3$
61197-73-7

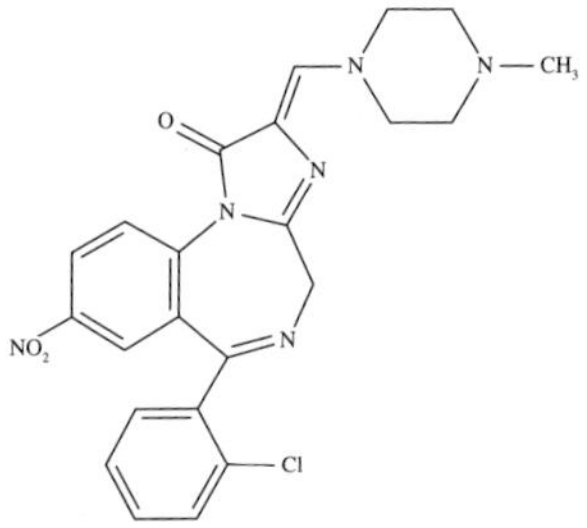

(*Z*)-6-(2-Chlorophenyl)-2,4-dihydro-2-[(4-methyl-1-piperazinyl)methylene]-8-nitro-1*H*-imidazo[1,2-*a*][1,4]benzodiazepin-1-one (●)

R   Monomethanesulfonate (70111-54-5) [or monomethanesulfonate monohydrate]

S   Avlane, Dormonoct, Halvane, Havlane, HR 158, *Loprazolam mesilate***, RU 31158, Somnovit, Sonin "Lipha"

U   Hypnotic

11663   (9371) $C_{23}H_{21}ClO_3$
569-57-3

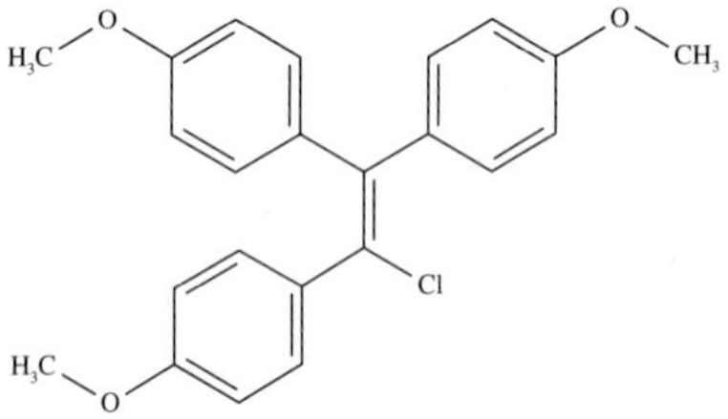

Chlorotris(*p*-methoxyphenyl)ethylene = Tri-*p*-anisylchloroethylene = 1,1',1''-(1-Chloro-1-ethenyl-2-ylidene)tris[4-methoxybenzene] (●)

S   Anisene, *Chlorotrianisene***, *Chlortrianisenum*, *Chlortrianisoestrolum*, Clorestrolo, *Clorotrianisen*, Clorotrisin, Hormonisene, Merbentul, Metace, NSC-10108, Restrol "Inibsa", Rianil, TACE, Triagen, *Trianisoestrol*

U   Synthetic estrogen

11664   (9372) $C_{23}H_{21}ClO_3$
80565-35-1

2-[[4-(4-Chlorophenyl)benzyl]oxy]-2-phenylbutyric acid
= α-[(4'-Chloro[1,1'-biphenyl]-4-yl)methoxy]-α-ethyl-
benzeneacetic acid (●)
S   ICI 115432
U   Anti-inflammatory

11665   (9373) $C_{23}H_{21}F_6NO_3$
110283-79-9

1-[2-[Bis-[4-(trifluoromethyl)phenyl]methoxy]ethyl]-
1,2,5,6-tetrahydro-3-pyridinecarboxylic acid (●)
R   Hydrochloride (110283-66-4)
S   CI-966
U   Anticonvulsant

11666 $C_{23}H_{21}F_7N_4O_3$
170729-80-3

3-[[(2R,3S)-3-(p-Fluorophenyl)-2-[[(αR)-α-methyl-3,5-
bis(trifluoromethyl)benzyl]oxy]morpholino]methyl]-Δ²-

1,2,4-triazolin-5-one = 5-[[(2R,3S)-2-[(1R)-1-[3,5-Bis(trifluoromethyl)phenyl]ethoxy]-3-(4-fluorophenyl)-4-mor-
pholinyl]methyl]-1,2-dihydro-3H-1,2,4-triazol-3-one (●)
S   Aprepitant*, L 754030, MK-869
U   Anti-emetic, antidepressant (neurokinin NK_1 receptor
antagonist)

11667 $C_{23}H_{21}NO$
155471-08-2

1-Propyl-2-methyl-3-(1-naphthoyl)indole = (2-Methyl-1-
propyl-1H-indol-3-yl)-1-naphthalenylmethanone (●)
S   JWH-015
U   Cannabinoid CB_2 receptor antagonist

11668   (9374) $C_{23}H_{21}NO_4$
1174-11-4

p-[(α-Ethoxy-p-phenylphenacyl)amino]benzoic acid = 4-
[(2-[1,1'-Biphenyl]-4-yl-1-ethoxy-2-oxoethyl)ami-
no]benzoic acid (●)
S   Acidum xenazoicum**, Antiviral "Usar",
C.V. 58903, LG 278, SKF 8318, Xenalamina, Xena-
zoic acid**, Xenovis, Xenovister
U   Antiviral

11669 (9375) $C_{23}H_{21}NO_4S$
127683-04-9

Methyl (±)-1,4,5,7-tetrahydro-2-methyl-5-oxo-4-[2-(phe-nylmethoxy)phenyl]thieno[3,4-b]pyridine-3-carboxylate = (±)-1,4,5,7-Tetrahydro-2-methyl-5-oxo-4-[2-(phenyl-methoxy)phenyl]thieno[3,4-b]pyridine-3-carboxylic acid methyl ester (●)
S   RS-30026
U   Cardiotonic (positive inotropic calcium agonist)

11670 (9376) $C_{23}H_{21}N_3O_4S$
123250-77-1

1-Phenylethyl (R)-1,4-dihydro-2,6-dimethyl-5-nitro-4-thieno[3,2-c]pyridin-3-yl-3-pyridinecarboxylate = 1,4-Dihydro-2,6-dimethyl-5-nitro-4-thieno[3,2-c]pyridin-3-yl-3-pyridinecarboxylic acid 1-phenylethyl ester (●)
S   LY 249933
U   Cardiotonic (positive inotropic calcium agonist)

11671 $C_{23}H_{21}N_3O_5$
173315-47-4

1-Cyclopropyl-7-(3,4-dihydro-2(1H)-isoquinolinyl)-1,4-dihydro-8-methyl-6-nitro-4-oxo-3-quinolinecarboxylic acid (●)
S   VG 6/1
U   Antibacterial

11672 $C_{23}H_{21}N_7O$
145733-36-4

5,8-Dihydro-2,4-dimethyl-8-[p-(o-1H-tetrazol-5-ylphe-nyl)benzyl]pyrido[2,3-d]pyrimidin-7(6H)-one = 5,8-Di-hydro-2,4-dimethyl-8-[[2'-(1H-tetrazol-5-yl)[1,1'-biphe-nyl]-4-yl]methyl]pyrido[2,3-d]pyrimidin-7(6H)-one (●)
S   ANA-756, *Apasartan*, *Tasosartan***, Verdia, WAY-ANA-756
U   Antihypertensive (angiotensin II receptor antagonist)

11673 $C_{23}H_{22}ClNO_2$
156897-06-2

6-(4-Chlorophenyl)-2,3-dihydro-2,2-dimethyl-7-phenyl-1H-pyrrolizine-5-acetic acid (●)
S   *Licocinac*, ML 3000, *Plicoxinac*
U   Anti-inflammatory (cyclooxygenase and 5-lipoxy-genase inhibitor)

11674   (9377) $C_{23}H_{22}ClNO_6$
76812-43-6

1-(4-Chlorobenzoyl)-5-methoxy-2-methyl-1H-indole-3-acetic acid 2-ethoxy-2-oxoethyl ester (●)
S   *Indometacin ethoxycarbonylmethyl ester, Lipoindomethacin*
U   Anti-inflammatory

11675 $C_{23}H_{22}ClN_3O_2$
133737-32-3

(+)-2-(7-Chloro-1,8-naphthyridin-2-yl)-3-(5-methyl-2-oxohexyl)phthalimidine = (+)-2-(7-Chloro-1,8-naphthyridin-2-yl)-2,3-dihydro-3-(5-methyl-2-oxohexyl)-1H-isoindol-1-one (●)
S   Anxiolyt, Bextra "Interneuron", IP 456, *Pagoclone***, RP 62955
U   Anxiolytic (GABA modulator)

11676   (9378) $C_{23}H_{22}ClN_5OS$
131614-02-3

(S)-6-(2-Chlorophenyl)-9-(cyclopropylcarbonyl)-7,8,9,10-tetrahydro-1,4-dimethyl-4H-pyrido[4',3':4,5]thieno[3,2-f][1,2,4]triazolo[4,3-a][1,4]diazepine (●)
S   E-6123
U   PAF antagonist

11677   (9379) $C_{23}H_{22}ClN_5O_2S$
114776-28-2

4-[[6-(o-Chlorophenyl)-8,9-dihydro-1-methyl-4H,7H-cyclopenta[4,5]thieno[3,2-f]-s-triazolo[4,3-a][1,4]diazepin-8-yl]carbonyl]morpholine = 6-(2-Chlorophenyl)-8,9-dihydro-1-methyl-8-(4-morpholinylcarbonyl)-4H,7H-cyclopenta[4,5]thieno[3,2-f][1,2,4]triazolo[4,3-a][1,4]diazepine = 4-[[6-(2-Chlorophenyl)-8,9-dihydro-1-methyl-4H,7H-cyclopenta[4,5]thieno[3,2-f][1,2,4]triazolo[4,3-a][1,4]diazepin-8-yl]carbonyl]morpholine (●)
S   *Bepafant***, WEB 2170 BS
U   PAF antagonist (anti-allergic, anti-inflammatory, hypertensive)

11678 $C_{23}H_{22}FNO_4$
                                                    220551-92-8

N-(2,5-Dimethoxybenzyl)-*N*-(5-fluoro-2-phenoxyphe-
nyl)acetamide = *N*-[(2,5-Dimethoxyphenyl)methyl]-*N*-
(5-fluoro-2-phenoxyphenyl)acetamide (●)
S   DAA 1106
U   Benzodiazepine receptor agonist

11679   (9380) $C_{23}H_{22}FN_3O$
                                                    115464-77-2

2-[[4-(7-Benzofuranyl)-1-piperazinyl]methyl]-5-(4-
fluorophenyl)pyrrole = 1-(7-Benzofuranyl)-4-[[5-(4-
fluorophenyl)-1*H*-pyrrol-2-yl]methyl]piperazine (●)
S   DU 29894, *Elopiprazole***, Silperidone*
U   Antipsychotic

11680   (9381) $C_{23}H_{22}F_3N_3O$
                                                    23419-43-4

4-Phenyl-2-[2-(1-pyrrolidinyl)ethyl]-6-(α,α,α-trifluoro-
m-tolyl)-3(2*H*)-pyridazinone = 4-Phenyl-2-[2-(1-pyrroli-

dinyl)ethyl]-6-[3-(trifluoromethyl)phenyl]-3(2*H*)-pyrida-
zinone (●)
S   Delibryl, *Ridaflone, Ridiflone*
U   Tranquilizer

11681   (9382) $C_{23}H_{22}F_3N_3O_2$
                                                    83863-79-0

2-(1-Pyrrolidinyl)ethyl *N*-[7-(trifluoromethyl)-4-quino-
lyl]anthranilate = 2-[[7-(Trifluoromethyl)-4-quinoli-
nyl]amino]benzoic acid 2-(1-pyrrolidinyl)ethyl ester (●)
S   FI-2522, *Florifenine***
U   Analgesic, anti-inflammatory

11682 $C_{23}H_{22}N_2O_3$
                                                    111469-88-6

(4b*S*,8*R*,8a*S*,14b*R*)-5,6,7,8,14,14b-Hexahydro-7-methyl-
4,8-methanobenzofuro[2,3-*a*]pyrido[4,3-*b*]carbazole-
1,8a(9*H*)-diol (●)
S   *Oxymorphindole*
U   δ-Opioid agonist

11683 (9385) $C_{23}H_{22}N_2O_4$
83910-34-3

Ethyl 5-(benzyloxy)-4-(methoxymethyl)-β-carboline-3-carboxylate = 4-(Methoxymethyl)-5-(phenylmethoxy)-9H-pyrido[3,4-b]indole-3-carboxylic acid ethyl ester (●)
S ZK 91296
U Anticonvulsant

11684 (9383) $C_{23}H_{22}N_2O_4$
83910-44-5

Ethyl 6-(benzyloxy)-4-(methoxymethyl)-β-carboline-3-carboxylate = 4-(Methoxymethyl)-6-(phenylmethoxy)-9H-pyrido[3,4-b]indole-3-carboxylic acid ethyl ester (●)
S ZK 93423
U Anticonvulsant, anxiolytic (benzodiazepine partial agonist)

11685 (9384) $C_{23}H_{22}N_2O_4$
82239-77-8

4-Acetamidophenyl 2-(5-p-toluoyl-1-methyl-2-pyrrolyl)acetate = 1-Methyl-5-(4-methylbenzoyl)-1H-pyrrole-2-acetic acid 4-(acetylamino)phenyl ester (●)
S AU-8001, Tolmetin Paracetamol ester
U Anti-inflammatory, analgesic

11686 (9386) $C_{23}H_{22}N_2O_4S$
27450-21-1

[2-(Phenylsulfinyl)ethyl]malonic acid mono(1,2-diphenylhydrazide) = [2-(Phenylsulfinyl)ethyl]propanedioic acid mono(1,2-diphenylhydrazide) (●)
S Osmadizone**
U Uricosuric

11687 (9387) $C_{23}H_{22}N_2O_6S$
27025-49-6

(2S,5R,6R)-N-(2-Carboxy-3,3-dimethyl-7-oxo-4-thia-1-azabicyclo[3.2.0]hept-6-yl)-2-phenylmalonamic acid 1-phenyl ester = (2S,5R,6R)-3,3-Dimethyl-7-oxo-6-[α-(phenoxycarbonyl)phenylacetamido]-4-thia-1-azabicyclo[3.2.0]heptane-2-carboxylic acid = (6R)-6-(2-Phenoxycarbonyl-2-phenylacetamido)penicillanic acid = [2S-(2α,5α,6β)]-6-[(1,3-Dioxo-3-phenoxy-2-phenylpropyl)amino]-3,3-dimethyl-7-oxo-4-thia-1-azabicyclo[3.2.0]heptane-2-carboxylic acid (●)
R Monosodium salt (21649-57-0)
S BRL 3475, Carbenicillin phenyl sodium, Carbiphen, Carfecillin Sodium*, Carfexil, Gripenin-O, Pencina, Pionin, Purapen, Safepen, Uricillina-Ibi, Urocarf, Uticillin, Vexyl
U Antibiotic

11688   (9388) $C_{23}H_{22}N_3O_2$
                                            47570-45-6

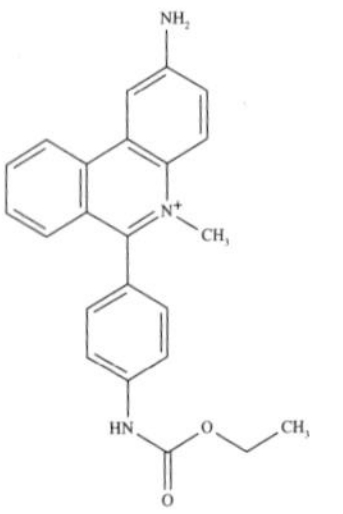

2-Amino-6-[4-[(ethoxycarbonyl)amino]phenyl]-5-me-
thylphenanthridinium (●)
R   Ethanesulfonate (3811-06-1)
S   *Carbidium(ethane sulfonate)*
U   Antibacterial

11689   (9389) $C_{23}H_{22}N_4O_3$
                                            132523-92-3

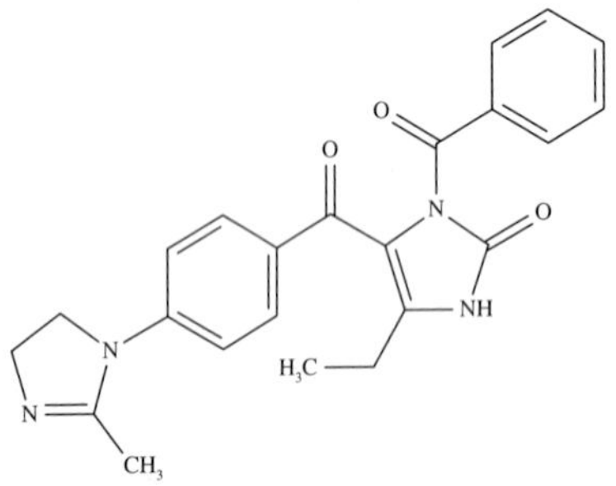

1-Benzoyl-5-[4-(4,5-dihydro-2-methyl-1H-imidazol-1-
yl)benzoyl]-4-ethyl-1,3-dihydro-2H-imidazol-2-one (●)
S   CK 3197
U   Cardiotonic (positive inotropic)

11690 $C_{23}H_{22}N_4O_3$
                                            144928-17-6

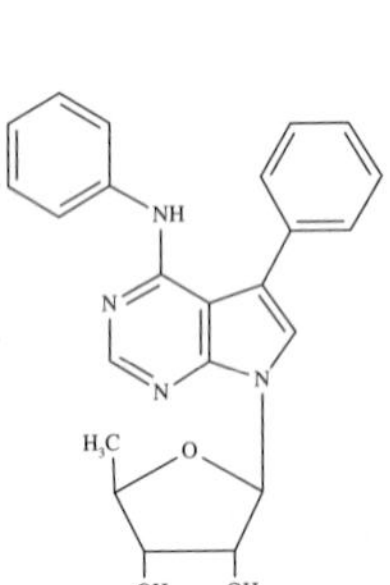

4-(Phenylamino)-5-phenyl-7-(5-deoxy-β-D-ribofurano-
syl)pyrrolo[2,3-d]pyrimidine = 7-(5-Deoxy-β-D-ribofura-
nosyl)-N,5-diphenyl-7H-pyrrolo[2,3-d]pyrimidin-4-
amine (●)
S   GP-1-683, GP 683
U   Adenosine kinase inhibitor

11691   (9390) $C_{23}H_{22}N_4O_5$
                                            116448-87-4

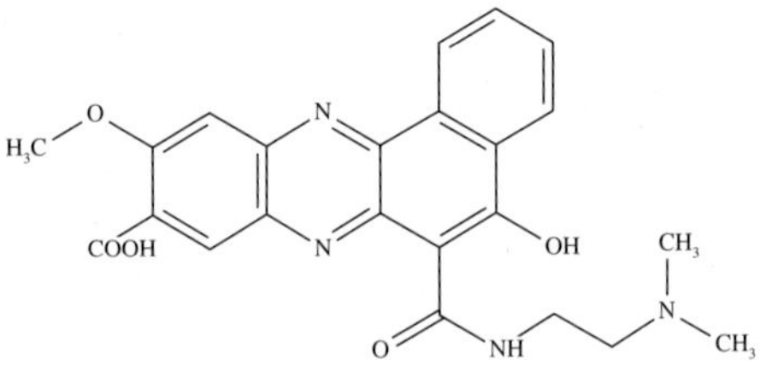

6-[[[2-(Dimethylamino)ethyl]amino]carbonyl]-5-hydr-
oxy-10-methoxybenzo[a]phenazine-9-carboxylic acid (●)
R   Monosodium salt (120602-99-5)
S   NC 190, NCU 190 Na
U   Antineoplastic

11692 $C_{23}H_{22}N_6O_2$
                                            141887-34-5

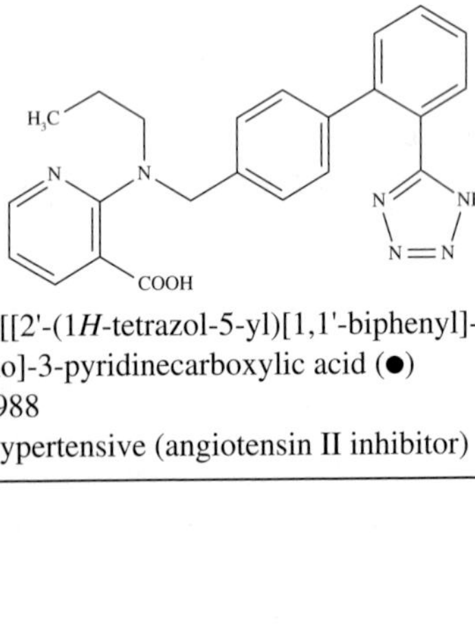

2-[Propyl[[2'-(1H-tetrazol-5-yl)[1,1'-biphenyl]-4-yl]me-
thyl]amino]-3-pyridinecarboxylic acid (●)
S   A-81988
U   Antihypertensive (angiotensin II inhibitor)

11693

$C_{23}H_{22}N_8O$
148504-51-2

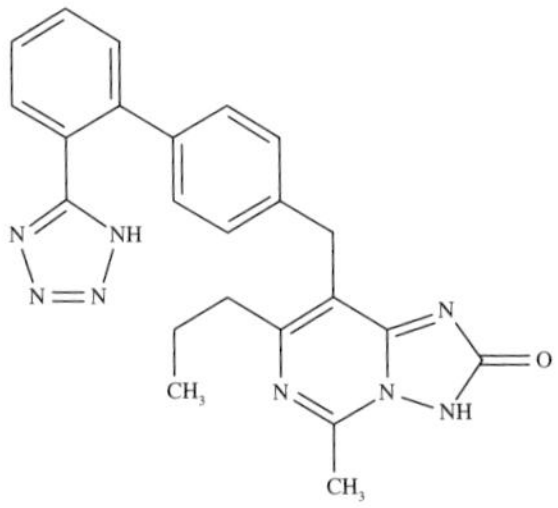

5-Methyl-7-propyl-8-[4-[2-(1*H*-tetrazol-5-yl)phe-
nyl]benzyl][1,2,4]triazolo[1,5-*c*]pyrimidin-2(3*H*)-one =
5-Methyl-7-propyl-8-[[2'-(1*H*-tetrazol-5-yl)[1,1'-biphe-
nyl]-4-yl]methyl]-[1,2,4]triazolo[1,5-*c*]pyrimidin-2(3*H*)-
one (●)
S   *Ripisartan***, UP 269-06
U   Antihypertensive (angiotensin II antagonist)

11694   (9391)

$C_{23}H_{22}N_{10}$

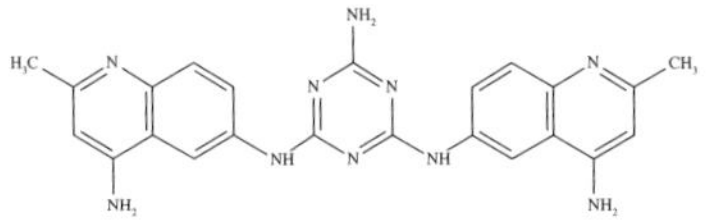

N^1,N^2-Bis(4-amino-2-methyl-6-quinolyl)melamine =
N,*N*'-Bis(4-amino-2-methyl-6-quinolinyl)-1,3,5-triazine-
2,4,6-triamine (●)
R   Trihydrochloride (33608-18-3)
S   Congasin, Surfen C
U   Antibacterial

11695   (9392)

$C_{23}H_{22}O_5$
58449-06-2

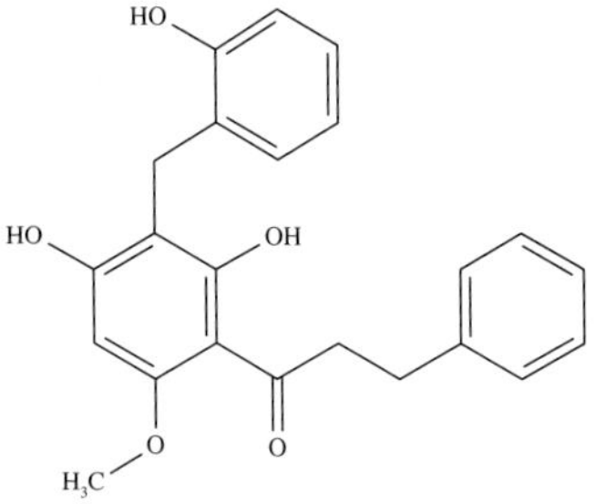

2',4'-Dihydroxy-3'-(*o*-hydroxybenzyl)-6'-methoxy-3-phe-
nylpropiophenone = 1-[2,4-Dihydroxy-3-[(2-hydroxy-
phenyl)methyl]-6-methoxyphenyl]-3-phenyl-1-propa-
none (●)

S   *Uvaretin*
U   Antineoplastic, antibacterial

11696   (9393)

$C_{23}H_{22}O_6$
83-79-4

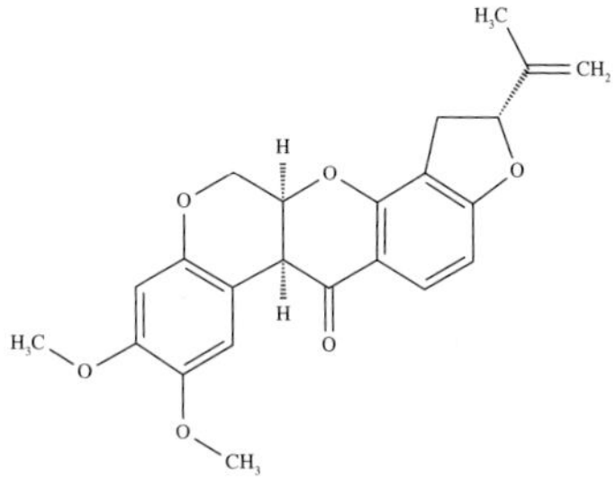

[2*R*-(2α,6aα,12aα)]-1,2,12,12a-Tetrahydro-8,9-dimeth-
oxy-2-(1-methylethenyl)[1]benzopyrano[3,4-*b*]fu-
ro[2,3-*h*][1]benzopyran-6(6a*H*)-one (●)
S   *Derrin, Rotenone, Tubatoxin*
U   Insecticide, dermatic (from the roots of *Derris* sp.)

11697

$C_{23}H_{22}O_6$
522-17-8

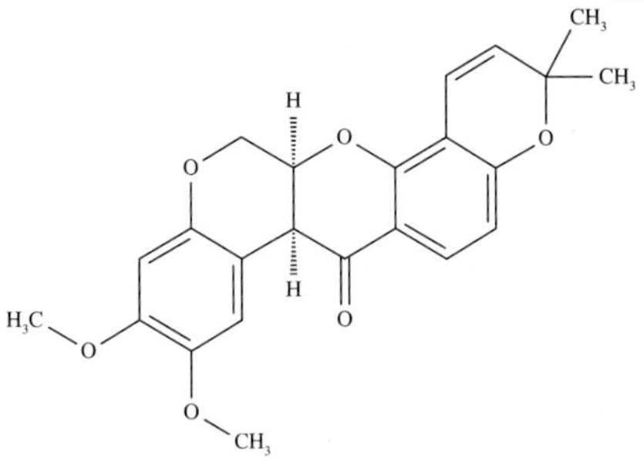

(7a*S*,13a*S*)-13,13a-Dihydro-9,10-dimethoxy-3,3-dime-
thyl-3*H*-bis[1]benzopyrano[3,4-*b*:6',5'-*e*]pyran-7(7a*H*)-
one (●)
S   *Deguelin*
U   Ornithine decarboxylase inhibitor from many plants

11698 (9394) $C_{23}H_{22}O_7$
65725-11-3

[3aR-(3aα,4α,9aα,9bβ)]-4-Hydroxybenzeneacetic acid
2,3,3a,4,5,7,9a,9b-octahydro-9-(hydroxymethyl)-6-me-
thyl-3-methylene-2,7-dioxoazuleno[4,5-b]furan-4-yl ester (●)
S *Lactucopicrin, Lactupikrin*
U Antitussive, sedative (from *Cichorium intybus* and
Lactuca sp.)

11699 (9395) $C_{23}H_{23}ClN_4O_9$
65635-35-0

[3S-[1(S*),3R*[Z(S*)]]]-3-[[[4-(3-Amino-3-carboxy-
propoxy)-3-chlorophenyl](hydroxyimino)acetyl]amino]-
α-(4-hydroxyphenyl)-2-oxo-1-azetidineacetic acid (●)
R also definition as compound with chloro in 3-position
on the other phenyl ring (95927-71-2)
S *Chlorocardicin, 3″-Chloronocardicin A*
U Antibiotic from a *Streptomyces* sp.

11700 (9396) $C_{23}H_{23}Cl_2N_3O_4S$
84697-22-3

Ethyl (±)-cis-p-[[[2-(2,4-dichlorophenyl)-2-(imidazol-1-
ylmethyl)-1,3-dioxolan-4-yl]methyl]thio]carbanilate =
cis-(±)-[4-[[[2-(2,4-Dichlorophenyl)-2-(1H-imidazol-1-
ylmethyl)-1,3-dioxolan-4-yl]methyl]thio]phenyl]carb-
amic acid ethyl ester (●)
R Monohydrochloride (83529-08-2)

S R 46846, *Tubulozole hydrochloride***
U Antifungal, microtubule inhibitor

11701 $C_{23}H_{23}Cl_3N_2O_3$
184300-67-2

1,5-Diamino-2,3,4-tris-O-(4-chlorophenyl)-1,5-dideoxy-
D-arabinitol (●)
S ZM 240304
U Antibacterial

11702 (9397) $C_{23}H_{23}FNO_5P$
119901-68-7

(S)-4-[[[1-(4-Fluorophenyl)-3-(1-methylethyl)-1H-indol-
2-yl]ethynyl]hydroxyphosphinyl]-3-hydroxybutanoic
acid (●)
R Disodium salt (133983-25-2)
S SQ 33600
U Antihyperlipidemic (HMGCoA reductase inhibitor)

11703 $C_{23}H_{23}FN_4O_3S$
150396-83-1

1-[3-[(6-Benzothiazolylcarbamoyl)oxy]propyl]-4-(6-fluoro-1,2-benzisoxazol-3-yl)piperidine = 6-Benzothia-zolylcarbamic acid 3-[4-(6-fluoro-1,2-benzisoxazol-3-yl)-1-piperidinyl]propyl ester (●)
S NNC 19-1228
U Neuroleptic, antipsychotic

11704 $C_{23}H_{23}FN_4O_5S$
154269-12-2

1-Cyclopropyl-6-fluoro-7-(4-sulfanilyl-1-piperazinyl)-4-quinolone-3-carboxylic acid = 7-[4-[(4-Amino-phenyl)sulfonyl]-1-piperazinyl]-1-cyclopropyl-6-fluoro-1,4-dihydro-4-oxo-3-quinolinecarboxylic acid (●)
S NSFQ 105
U Antibacterial

11705 $C_{23}H_{23}F_2NO$
140890-70-6

N-[2-[Bis(4-fluorophenyl)methoxy]ethyl]benzeneethan-amine (●)

R Maleate (1:1) (140890-71-7)
S VUF 8929
U Calcium antagonist (cardioprotective)

11706 $C_{23}H_{23}F_2N_3OS$
183949-64-6

(*R*)-(+)-2-Amino-4-(4-fluorophenyl)-5-[1-[4-(4-fluoro-phenyl)-4-oxobutyl]-3-pyrrolidinyl]thiazole = (*R*)-4-[3-[2-Amino-4-(4-fluorophenyl)-5-thiazolyl]-1-pyrrolidi-nyl]-1-(4-fluorophenyl)-1-butanone (●)
S NRA 0045
U Antipsychotic (dopamine D_4 and 5-HT$_{2A}$ receptor antagonist)

11707 (9398) $C_{23}H_{23}NO_2$
96389-68-3

2-[(6-Chrysenylmethyl)amino]-2-methyl-1,3-propane-diol (●)
R Methanesulfonate (salt) (96389-69-4)
S BW A 770 U mesylate, *Crisnatol mesilate**, 770 U 82
U Antineoplastic

11708 (9399) $C_{23}H_{23}NO_3$
120822-71-1

2-[p-(Benzyloxy)phenyl]-N-phenethylacetohydroxamic
acid = N-Hydroxy-N-(2-phenylethyl)-4-(phenylmethoxy)benzeneacetamide (●)
S   RG 6820
U   Anti-inflammatory

11709 $C_{23}H_{23}NO_3$
128253-31-6

(R)-2-Cyclopentyl-2-[4-(2-quinolinylmethoxy)phe-
nyl]acetic acid = (αR)-α-Cyclopentyl-4-(2-quinolinyl-
methoxy)benzeneacetic acid (●)
S   BAY x 1005
U   Leukotriene inhibitor (anti-allergic, cardioprotectant)

11710 (9400) $C_{23}H_{23}N_2S_2$
7187-55-5

3-Ethyl-2-[5-(3-ethyl-2-benzothiazolinylidene)-1,3-pen-
tadienyl]benzothiazolium = 3,3'-Diethyldithiadicarbo-
cyanine = 3-Ethyl-2-[5-(3-ethyl-2(3H)-benzothiazolyli-
dene)-1,3-pentadienyl]benzothiazolium (●)
R   Iodide (514-73-8)
S   Abminthic, Anelmid, Anguifugan, Antelmint, Blu-
cianina, Ceselmint, Compound 01748, Dejo, Del-
vex, Déselmine, Dilbrin, Dilombrin, D.I.M., *Dithia-
zanine iodide**, Ditialmin, Ditiazan, Dizan, Duxen
"Gedeon Richter", Elmetina, Elmizin, Fedal-Ditia-
zin, Gucifor, Helmisin, Helmiveril, Lombrizanina,
Nectocyd, Nekel, Nematan, Netocyd, Norivan, Nula-
verm, Omni-Passin, Ossiurene, Pankiller, Partel, Sa-

nivermin, Telmicid, Telmid, Totelmin, Tulbonal,
Vercidon, Vermicina
U   Anthelmintic

11711 (9401) $C_{23}H_{23}N_3O_2$
84260-62-8

2-(1,2-Benzisoxazol-3-yl)-3-[2-(2-piperidinoethoxy)phe-
nyl]acrylonitrile = α-[[2-[2-(1-Piperidinyl)ethoxy]phe-
nyl]methylene]-1,2-benzisoxazole-3-acetonitrile (●)
R   Monohydrochloride (81813-95-8)
S   SX-284
U   Antispasmodic (acetylcholine inhibitor)

11712 $C_{23}H_{23}N_3O_3$
143881-08-7

(2R)-1-[(2E)-1-Oxo-3-(2-phenylpyrazolo[1,5-a]pyridin-
3-yl)-2-propenyl]-2-piperidineacetic acid (●)
S   FK 352
U   Diuretic, antihypertensive (adenosine A_1 receptor antagonist)

11713   (9402) $C_{23}H_{23}N_3O_5$
123948-87-8

(*S*)-10-[(Dimethylamino)methyl]-4-ethyl-4,9-dihydroxy-1*H*-pyrano[3',4':6,7]indolizino[1,2-*b*]quinoline-3,14(4*H*,12*H*)-dione (●)
R   Monohydrochloride (119413-54-6)
S   Cantop, E-89/001, Evotopin, Hycamtin,
NSC-609699, SKF 104864-A, *Topotecan hydrochloride***
U   Antineoplastic

11714   (9403) $C_{23}H_{23}N_3O_5S$
50894-67-2

(2*S*,5*R*,6*R*)-6-[α-[[3-(2-Furyl)allylidene]amino]phenyl-acetamido]-3,3-dimethyl-7-oxo-4-thia-1-azabicy-clo[3.2.0]heptane-2-carboxylic acid = (6*R*)-6-[α-[[3-(2-Furyl)allylidene]amino]phenylacetamido]penicillanic acid = [2*S*-[2α,5α,6β(*S**)]]-6-[[[[3-(2-Furanyl)-2-prope-nylidene]amino]phenylacetyl]amino]-3,3-dimethyl-7-oxo-4-thia-1-azabicyclo[3.2.0]heptane-2-carboxylic acid (●)
R   Sodium salt (41136-10-1)
S   *Propampicillin*
U   Antibiotic

11715 $C_{23}H_{23}N_5O_2S$
150893-78-0

N^6-Ethyl-N^6-[[4-(phenylsulfonyl)phenyl]methyl]-2,4,6-quinazolinetriamine (●)
S   AG-377
U   Antineoplastic (dihydrofolate reductase inhibitor)

11716 $C_{23}H_{23}N_5O_4$
110189-06-5

4-[[5-[(4-Hydroxyphenyl)methyl]-4-[(4-methoxyphe-nyl)methyl]-1-methyl-1*H*-imidazol-2-yl]amino]-1-me-thyl-1*H*-imidazole-2,5-dione (●)
S   *Naamidine A*
U   Epidermal growth factor receptor antagonist, antineo-plastic (alkaloid from Fijian *Leucetta* sp.)

11717 $C_{23}H_{23}N_7O_2$
141872-46-0

4-[Butyl[[2'-(1*H*-tetrazol-5-yl)[1,1'-biphenyl]-4-yl]me-thyl]amino]-5-pyrimidinecarboxylic acid (●)
S   A-81282, Abbott-81282
U   Antihypertensive (angiotensin II antagonist)

11718 $C_{23}H_{24}ClNO_4$
160492-05-7

4-[3-[(4-Chlorophenyl)methyl]-1-(phenylmethyl)-3-pi-
peridinyl]-2-hydroxy-4-oxo-2-butenoic acid (●)
R Hydrochloride (174605-75-5)
S L 735882
U Antiviral

11719 $C_{23}H_{24}ClN_3O_5$
186392-65-4

5-Chloro-N-[(1S,2R)-3-[(3R,4S)-3,4-dihydroxy-1-pyrro-
lidinyl]-2-hydroxy-3-oxo-1-(phenylmethyl)propyl]-1H-
indole-2-carboxamide (●)
S CP-368296, *Ingliforin*
U Antidiabetic (glycogen phosphorylase inhibitor)

11720 (9404) $C_{23}H_{24}ClN_5OS$
114776-47-5

6-(2-Chlorophenyl)-8,9-dihydro-1-methyl-8-(4-morpho-
linylmethyl)-4H,7H-cyclopenta[4,5]thie-
no[3,2-f][1,2,4]triazolo[4,3-a][1,4]diazepine (●)
S STY 2108
U Hypertensive

11721 (9405) $C_{23}H_{24}FNO_4$
140841-32-3

6-[[3-Fluoro-5-(tetrahydro-4-methoxy-2H-pyran-4-
yl)phenoxy]methyl]-1-methyl-2(1H)-quinolinone (●)
S ICI D 2138, ZD 2138
U Anti-inflammatory

11722 (9406) $C_{23}H_{24}FN_3O$
82117-51-9

4'-Fluoro-4-[4-(3-isoquinolyl)-1-piperazinyl]butyrophe-
none = 3-[4-[3-(4-Fluorobenzoyl)propyl]-1-piperazi-
nyl]isoquinoline = 1-(4-Fluorophenyl)-4-[4-(3-isoquino-
linyl)-1-piperazinyl]-1-butanone (●)
R also monohydrochloride (82117-52-0)
S *Cinuperone**, HR 375
U Neuroleptic

11723 (9407) $C_{23}H_{24}FN_3O_2$
75444-65-4

3-[2-[4-(p-Fluorobenzoyl)piperidino]ethyl]-2-methyl-
4H-pyrido[1,2-a]pyrimidin-4-one = 3-[2-[4-(4-Fluoro-

benzoyl)-1-piperidinyl]ethyl]-2-methyl-4*H*-pyrido[1,2-*a*]pyrimidin-4-one (●)
S *Pirenperone***, R 47465
U Antipsychotic, neuroleptic (lysergic acid antagonist)

11724 (9408) $C_{23}H_{24}FN_3O_2S$
                                            127625-29-0

2-[3-[4-(4-Fluorophenyl)-1-piperazinyl]propyl]-2*H*-naphth[1,8-*cd*]isothiazole 1,1-dioxide (●)
S *Fananserin***, RP 62203
U 5-HT$_2$-receptor antagonist

11725 (9409) $C_{23}H_{24}FN_3O_3$
                                            95374-52-0

5-Cyano-*N*-[2-[4-(*p*-fluorobenzoyl)piperidino]ethyl]-*o*-anisamide = 5-Cyano-*N*-[2-[4-(4-fluorobenzoyl)-1-piperidinyl]ethyl]-2-methoxybenzamide (●)
S *Prideperone***
U Neuroleptic

11726 $C_{23}H_{24}FN_3O_5$
                                            150396-89-7

4-(6-Fluoro-1,2-benzisoxazol-3-yl)-1-[3-[[(3,4-methylenedioxyphenyl)carbamoyl]oxy]propyl]piperidine = 1,3-Benzodioxol-5-ylcarbamic acid 3-[4-(6-fluoro-1,2-benzisoxazol-3-yl)-1-piperidinyl]propyl ester (●)
R Monohydrochloride (160436-12-4)

S NNC 22-0031
U Neuroleptic, antipsychotic

11727 (9410) $C_{23}H_{24}F_3N_3OS$
                                            54063-26-2

10-[3-(Hexahydropyrrolo[1,2-*a*]pyrazin-2(1*H*)-yl)-propionyl]-2-(trifluoromethyl)phenothiazine = 10-[3-(Hexahydropyrrolo[1,2-*a*]pyrazin-2(1*H*)-yl)-1-oxopropyl]-2-(trifluoromethyl)-10*H*-phenothiazine (●)
S *Azaftozine***, Nonaftazin, Nonaphthazine
U Coronary vasodilator

11728 (9411) $C_{23}H_{24}NO_4$
                                            51865-94-2

8-Ethyl-2,3,10,11-tetramethoxydibenzo[*a,g*]quinolizinium (●)
R 2-Oxopropyl sulfate (51865-95-3)
S *Homocoralyne*, NSC-156625
U Antineoplastic, antileukemic

11729   (9412) $C_{23}H_{24}N_2$
                                                   115717-42-5

(Z)-5-Methyl-2-[2-(1-naphthyl)vinyl]-4-piperidinopyri-
dine = (Z)-5-Methyl-2-[2-(1-naphthalenyl)ethenyl]-4-(1-
piperidinyl)pyridine (●)
R   Monohydrochloride (115717-83-4)
S   AU-1421
U   Anti-ulcer agent

11730   (9413) $C_{23}H_{24}N_2$
                                                   80038-49-9

1,1'-Pentamethylenebis(quinolinium) = 1,1'-(1,5-Pentane-
diyl)bis[quinolinium] (●)
R   Diiodide (99218-67-4)
S   *Pentaquinomethonium*
U   Ganglion blocking agent

11731 $C_{23}H_{24}N_2O$
                                                   148545-09-9

2-Methyl-4aα-(3-hydroxyphenyl)-1,2,3,4,4a,5,12,12aα-
octahydroquino[2,3-g]isoquinoline = *rel*-3-[(4aR,12aS)-
1,3,4,5,12,12a-Hexahydro-2-methylpyrido[3,4-b]acridin-
4a(2H)-yl]phenol (●)
S   TAN 67

2394

U   δ-Opioid receptor agonist

11732 $C_{23}H_{24}N_2O_4$
                                                   150907-43-0

8-Methoxy-4-[(2-methylphenyl)amino]-γ-oxo-3-quino-
linebutanoic acid ethyl ester (●)
R   Monohydrochloride (150620-63-6)
S   CP-133411
U H^+/K^+-ATPase inhibitor

11733   (9414) $C_{23}H_{24}N_2O_4S$
                                                   133040-01-4

(E)-3-[1-[(4-Carboxyphenyl)methyl]-2-butyl-1H-imida-
zol-5-yl]-2-(2-thienylmethyl)-2-propenoic acid = (E)-2-
Butyl-1-(p-carboxybenzyl)-α-2-thenylimidazole-5-acry-
lic acid = (E)-α-[2-Butyl-5-[2-carboxy-3-(2-thienyl)prop-
1-enyl]-1H-imidazol-1-yl]-p-toluic acid = (E)-α-[[2-
Butyl-1-[(4-carboxyphenyl)methyl]-1H-imidazol-5-
yl]methylene]-2-thiophenepropanoic acid (●)
S   *Eprosartan***, SKF 108566
U   Antihypertensive (angiotensin II receptor antagonist)

11733-01   144143-96-4
R   Monomethanesulfonate
S   *Eprosartan mesilate***, SKF 108566J, Teveten, Te-
       vetens, Tevetenz
U   Antihypertensive (angiotensin II receptor antagonist)

11734   (9415) $C_{23}H_{24}N_2O_6$
                                                          75617-77-5

6-(Dimethylamino)-5,6-dihydro-9,10,14-trimethoxy-8*H*-1,3-dioxolo[4,5-*h*]isoindolo[1,2-*b*][3]benzazepin-8-one (●)

S   QF 1
U   Antineoplastic

11735   (9416) $C_{23}H_{24}N_4$
                                                          83166-17-0

11-[3-(Dimethylamino)propyl]-6-phenyl-11*H*-pyrido[2,3-*b*][1,4]benzodiazepine = *N,N*-Dimethyl-6-phenyl-11*H*-pyrido[2,3-*b*][1,4]benzodiazepine-11-propanamine (●)

R   Fumarate (1:1) (83166-18-1)
S   AHR-9377, *Tampramine fumarate***
U   Antidepressant

11736 $C_{23}H_{24}N_4O_3S$
                                                          138384-68-6

4-[[α-[(2-Aminobenz[*cd*]indol-6-yl)methylamino]-*p*-tolyl]sulfonyl]morpholine = N^6-[4-(Morpholinosulfonyl)benzyl]-N^6-methylbenz[*cd*]indole-2,6-diamine = 4-

[[4-[[(2-Aminobenz[*cd*]indol-6-yl)methylamino]methyl]phenyl]sulfonyl]morpholine (●)
R   Mono-D-glucuronate (157182-23-5)
S   AG-331, *Metesind glucuronate***
U   Antineoplastic (thymidylate synthase inhibitor)

11737 $C_{23}H_{24}N_4O_6$
                                                          198821-22-6

N-[3-[3-[3-Methoxy-4-(5-oxazolyl)phenyl]ureido]benzyl]carbamic acid tetrahydrofuran-3(*S*)-yl ester = [[3-[[[[3-Methoxy-4-(5-oxazolyl)phenyl]amino]carbonyl]amino]phenyl]methyl]carbamic acid (3*S*)-tetrahydro-3-furanyl ester (●)
S   *Merimpodib*, VI-21497, VX-497
U   Antipsoriatic (IMPDH inhibitor)

11738   (9417) $C_{23}H_{24}N_4O_9$
                                                          39391-39-4

[3*S*-[1(*S**),3*R**[*Z*(*S**)]]]-3-[[[4-(3-Amino-3-carboxypropoxy)phenyl](hydroxyimino)acetyl]amino]-α-(4-hydroxyphenyl)-2-oxo-1-azetidineacetic acid (●)
S   FR 1923, NCA, *Nocardin A*
U   Antibiotic from *Nocardia uniformis* var. *tsuyamanensis* ATCC 21806

11739 $C_{23}H_{24}N_6O$
139964-18-4

4-Ethyl-2-propyl-1-[[2'-(1H-tetrazol-5-yl)[1,1'-biphe-nyl]-4-yl]methyl]-1H-imidazole-5-carboxaldehyde (●)
S   DMP 581
U   Antihypertensive (angiotensin II receptor antagonist)

11740 $C_{23}H_{24}N_6O_2$
139964-19-5

4-Ethyl-2-propyl-1-[[2'-(1H-tetrazol-5-yl)[1,1'-biphe-nyl]-4-yl]methyl]-1H-imidazole-5-carboxylic acid (●)
S   DMP 811, DuP 811, L 908404
U   Antihypertensive (Angiotensin II receptor antagonist)

11741   (9418) $C_{23}H_{24}N_6O_5S_2$
84957-30-2

1-[[(6R,7R)-7-[2-(2-Amino-4-thiazolyl)glyoxylamido]-2-carboxy-8-oxo-5-thia-1-azabicyclo[4.2.0]oct-2-en-3-yl]methyl]-5,6,7,8-tetrahydroquinolinium hydroxide inner salt 7^2-(Z)-(O-methyloxime) = (Z)-(7R)-7-[2-(2-Amino-4-thiazolyl)-2-(methoxyimino)acetamido]-3-

[(5,6,7,8-tetrahydro-1-quinolinio)methyl]-3-cephem-4-carboxylate = [6R-[6α,7β(Z)]]-1-[[7-[[(2-Amino-4-thia-zolyl)(methoxyimino)acetyl]amino]-2-carboxy-8-oxo-5-thia-1-azabicyclo[4.2.0]oct-2-en-3-yl]methyl]-5,6,7,8-te-trahydroquinolinium hydroxide inner salt (●)
R   also sulfate (2:1) (123766-80-3)
S   *Cefquinome***, Cephaguard, Cobactan, HR 111 V
U   Antibiotic

11742   (9419) $C_{23}H_{24}N_6O_7S_2$
65761-24-2

α-[p-[(6-Methoxy-3-pyridazinyl)sulfamoyl]anilino]-2,3-dimethyl-5-oxo-1-phenyl-3-pyrazoline-4-methanesulfo-nic acid = α-(2,3-Dimethyl-5-oxo-1-phenyl-3-pyrazolin-4-yl)-p-[(6-methoxy-3-pyridazinyl)sulfamoyl]anilinome-thanesulfonic acid = 2,3-Dihydro-α-[[4-[[(6-methoxy-3-pyridazinyl)amino]sulfonyl]phenyl]amino]-1,5-dime-thyl-3-oxo-2-phenyl-1H-pyrazole-4-methanesulfonic acid (●)
R   Sodium salt (13061-27-3)
S   Genpiral, Marespin, *Sulfamazone sodium**, *Sulfena-zone*
U   Antibacterial, antipyretic

11743   (9420) $C_{23}H_{24}N_6O_7S_3$
115369-52-3

(6R,7R)-4-[[[7-[2-(2-Amino-4-thiazolyl)glyoxylamido]-2-carboxy-8-oxo-5-thia-1-azabicyclo[4.2.0]oct-2-en-3-yl]methyl]thio]-1-ethylpyridinium hydroxide inner salt 7^2-(Z)-[O-[(S)-1-carboxyethyl]oxime] = [6R-[6α,7β[Z(S*)]]]-4-[[[7-[[(2-Amino-4-thiazolyl)][(1-carb-oxyethoxy)imino]acetyl]amino]-2-carboxy-8-oxo-5-thia-

1-azabicyclo[4.2.0]oct-2-en-3-yl]methyl]thio]-1-ethylpyridinium hydroxide inner salt (●)
S ME-1228
U Antibiotic

11744 (9421)

$C_{23}H_{24}O_4$
2624-43-3

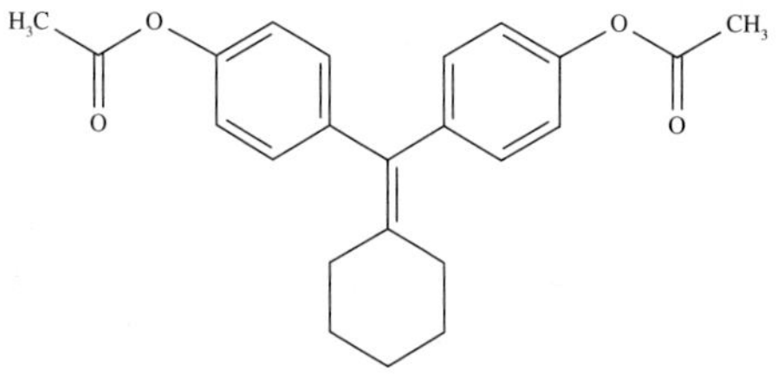

4,4'-(Cyclohexylidenemethylene)diphenol diacetate (ester) = 4,4'-(Cyclohexylidenemethylene)bis(phenyl acetate) = 4,4'-Diacetoxybenzhydrylidenecyclohexane = α-Cyclohexylidene-α-(p-hydroxyphenyl)-p-cresol diacetate = 4-[[4-(Acetyloxy)phenyl]cyclohexylidenemethyl]phenol acetate (●)
S Ciclifen, *Cyclofenil***, F 6066, Fertodur, Gyneuro, H 3452, ICI 48213, Klofenil, Menoferil, Menopax, Neoclym, Oginex, Ondogyne, Ondonid, Rehibin "Serono; Thomas", Sexadieno, Sexovar, Sexovid
U Ovulation stimulant (treatment of infertility)

11745

$C_{23}H_{24}O_6$
31271-07-5

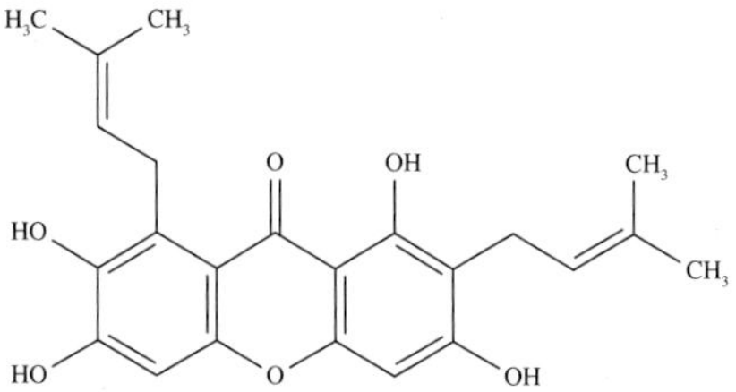

1,3,6,7-Tetrahydroxy-2,8-bis(3-methyl-2-butenyl)-9H-xanthen-9-one (●)
S γ-Mangostin, Normangostin
U 5-HT$_{2A}$ receptor antagonist

11746 (9422)

$C_{23}H_{24}O_7$
133806-61-8

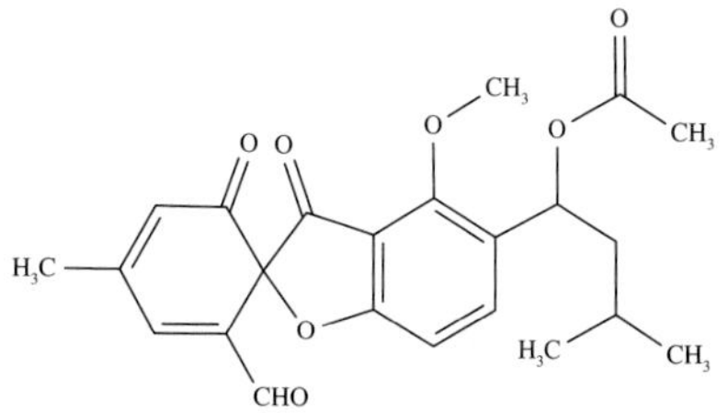

5-[1-(Acetyloxy)-3-methylbutyl]-2'-formyl-4-methoxy-4'-methylspiro[benzofuran-2(3H),1'-[2,4]cyclohexadiene]-3,6'-dione (●)
S FO-608 C, PI 885, *Purpactin C*
U Anticoagulant, antihyperlipidemic from *Penicillium* sp.

11747

$C_{23}H_{24}O_8$
19545-26-7

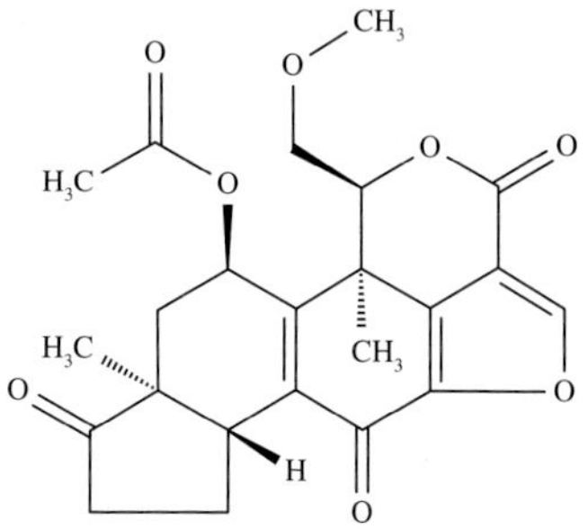

(1S,6bR,9aS,11R,11bR)-11-(Acetyloxy)-1,6b,7,8,9a,10,11,11b-octahydro-1-(methoxymethyl)-9a,11b-dimethyl-3H-furo[4,3,2-de]indeno[4,5-h]-2-benzopyran-3,6,9-trione (●)
S *Wortmannin*
U Antiviral from *Fusarium* sp.

11748

$C_{23}H_{24}O_{11}$
13020-19-4

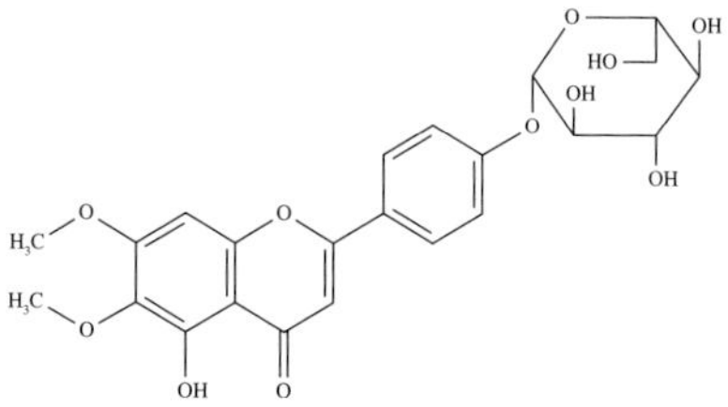

Cirsimaritin 4'-*O*-glucoside = 2-[4-(β-D-Glucopyranosy-loxy)phenyl]-5-hydroxy-6,7-dimethoxy-4*H*-1-benzopy-ran-4-one (●)
S *Cirsimarin*
U Adenosine A$_1$ receptor antagonist from *Microtea debilis*

11749 C$_{23}$H$_{25}$ClN$_2$O
156337-32-5

5-(4-Chlorophenyl)-4-methyl-3-[1-(2-phenylethyl)-4-pi-peridinyl]isoxazole = 4-[5-(4-Chlorophenyl)-4-methyl-3-isoxazolyl]-1-(2-phenylethyl)piperidine (●)
S L 741742
U Dopamine D$_4$ receptor antagonist

11750 (9423) C$_{23}$H$_{25}$ClN$_2$O$_9$
1181-54-0

7-Chloro-4-(dimethylamino)-1,4,4a,5,5a,6,11,12a-octa-hydro-3,6,10,12,12a-pentahydroxy-*N*-(hydroxymethyl)-6-methyl-1,11-dioxo-2-naphthacenecarboxamide = 7-chloro-2-*N*-(hydroxymethyl)tetracycline = [4*S*-(4α,4aα,5aα,6β,12aα)]-7-Chloro-4-(dimethylamino)-1,4,4a,5,5a,6,11,12a-octahydro-3,6,10,12,12a-pentahydr-oxy-*N*-(hydroxymethyl)-6-methyl-1,11-dioxo-2-naphtha-cenecarboxamide (●)
R also sodium salt (14448-68-1), monosodium salt (68-20-2)
S *Chlormethylencycline, Clomociclina, Clomocycli-ne**, Megaclor, Methylolchlortetracycline*
U Antibiotic

11751 (9424) C$_{23}$H$_{25}$ClN$_2$O$_{10}$
110298-64-1

4a-Hydroxy-8-methoxychlorotetracycline = [4*R*-(4α,4aα,5aα,6β,12aα)]-7-Chloro-4-(dimethylamino)-1,4,4a,5,5a,6,11,12a-octahydro-3,4a,6,10,12,12a-hexahy-droxy-8-methoxy-6-methyl-1,11-dioxo-2-naphthacene-carboxamide (●)
S *Dactylocyclinone*, Sch 34164
U Antibiotic from *Dactylosporangium vescum*

11752 (9425) C$_{23}$H$_{25}$ClN$_4$O
72593-09-0

2-Chloro-*N*-[2-(3,4,6,7,12,12a-hexahydropyrazi-no[1',2':1,6]pyrido[3,4-*b*]indol-2(1*H*)-yl)ethyl]benz-amide (●)
S *Chlorethindole*
U Antihistaminic

11753 (9426) C$_{23}$H$_{25}$ClN$_4$O$_4$S
24455-58-1

5'-Chloro-2-[*p*-[(5-isobutyl-2-pyrimidinyl)sulfamoyl]phenyl]-*o*-acetanisidide = *N*-(5-Chloro-2-methoxyphe-nyl)-4-[[[5-(2-methylpropyl)-2-pyrimidinyl]amino]sulfo-nyl]benzeneacetamide (●)
R Monosodium salt (24428-71-5)
S BS 1051, *Glicetanile sodium**, Glidanil-Natrium, Glydanile sodium*, SH 1051
U Oral hypoglycemic

11754　　　　　　　　　　　　　　$C_{23}H_{25}ClO_2$
153559-57-0

4-[1-(3-Chloro-5,6,7,8-tetrahydro-5,5,8,8-tetramethyl-2-naphthalenyl)ethenyl]benzoic acid (●)
S　LG-100153
U　Antineoplastic (retinoid x receptor agonist)

11755　　　　　　　　　　　　　　$C_{23}H_{25}FN_4O_2$
200398-40-9

1-[2-[4-(6-Fluoro-1,2-benzisoxazol-3-yl)-1-piperidinyl]ethyl]-3-phenyl-2-imidazolidinone (●)
S　S 18327
U　Antipsychotic

11756　(9427)　　　　　　　　　　$C_{23}H_{25}F_2N_3O_2$
54965-22-9

8-[3-(p-Fluorobenzoyl)propyl]-1-(p-fluorophenyl)-1,3,8-triazaspiro[4.5]decan-4-one = 1-(4-Fluorophenyl)-8-[4-(4-fluorophenyl)-4-oxobutyl]-1,3,8-triazaspiro[4.5]decan-4-one (●)
S　*Fluspiperone***, R 28930

U　Neuroleptic, antipsychotic

11757　(9428)　　　　　　　　　　$C_{23}H_{25}F_3N_2OS$
2709-56-0

2-[4-[3-[2-(Trifluoromethyl)-9-thioxanthenylidene]propyl]-1-piperazinyl]ethanol = 2-(Trifluoromethyl)-9-[3-[4-(2-hydroxyethyl)-1-piperazinyl]propylidene]thioxanthene = 4-[3-[2-(Trifluoromethyl)-9H-thioxanthen-9-ylidene]propyl]-1-piperazineethanol (●)
R　see also no. 14779;
　　also dihydrochloride (2413-38-9)
S　Depixol-Tabl., Émergil, Fluancsol, Fluanxol, Fluanzol, *Flupenthixol, Flupentixol***, FX-703, LC 44, Lu 5-110, Metamin "Takeda", N 7009, Phrenixol, Siplaril, Siplarol, SKF 10812, Umbralina
U　Neuroleptic

11758　　　　　　　　　　　　　　$C_{23}H_{25}F_3N_4O_5$
137888-49-4

(±)-1-[2-(3-Carbamoyl-4-hydroxyphenoxy)ethylamino]-3-[4-[1-methyl-4-(trifluoromethyl)-2-imidazolyl]phenoxy]-2-propanol = 2-Hydroxy-5-[2-[[2-hydroxy-3-[4-[1-methyl-4-(trifluoromethyl)-1H-imidazol-2-yl]phenoxy]propyl]amino]ethoxy]benzamide (●)
R　Monomethanesulfonate (salt) (105737-62-0)
S　CGP 20712A
U　β_1- and β_2-adrenoceptor antagonist

11759 $C_{23}H_{25}IN_2O_3$
164178-33-0

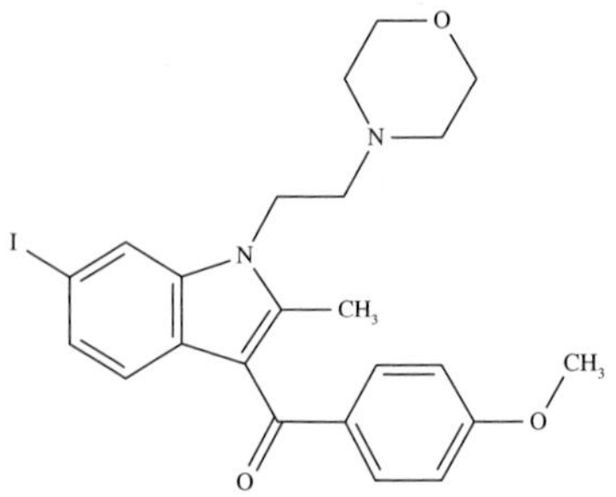

[6-Iodo-2-methyl-1-[2-(4-morpholinyl)ethyl]-1H-indol-3-yl](4-methoxyphenyl)methanone (●)
S AM-630, *Iodopravadoline*
U Cannabinoid receptor antagonist

11760 (9429) $C_{23}H_{25}N$
13042-18-7

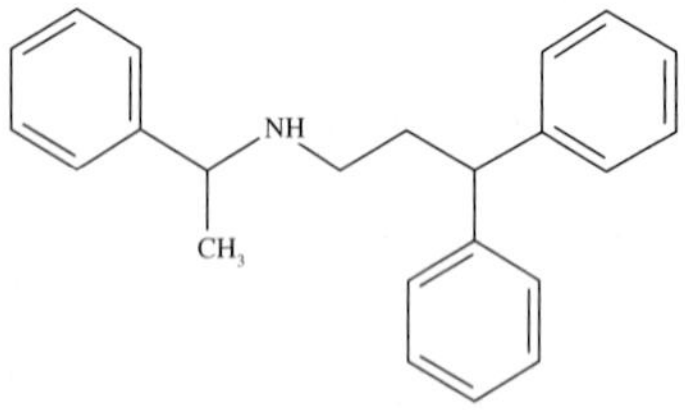

N-(3,3-Diphenylpropyl)-α-methylbenzylamine = 2,6,6-Triphenyl-3-azahexane = γ-Phenyl-N-(1-phenylethyl)benzenepropanamine (●)
R Hydrochloride (13636-18-5)
S Cordan "IBI", Difmecor, Fendilar, *Fendiline hydrochloride**, HK 137, Olbiacor, *Phenaxazan*, *Phendiline hydrochloride*, *Phenoxan*, Plenum, Sensit, Tifendil
U Coronary vasodilator (calcium antagonist)

11761 (9431) $C_{23}H_{25}NO$
1242-69-9

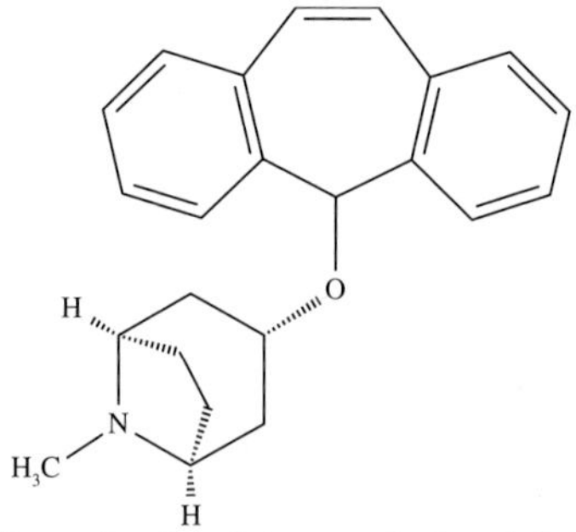

3α-(5H-Dibenzo[a,d]cyclohepten-5-yloxy)-1αH,5αH-tropane = endo-3-(5H-Dibenzo[a,d]cyclohepten-5-yloxy)-8-methyl-8-azabicyclo[3.2.1]octane (●)
R Maleate (1:1) (10139-99-8)
S B.S. 7039, *Decitropine maleate***
U Anticholinergic, anti-ulcer agent

11762 (9432) $C_{23}H_{25}NO$
84388-96-5

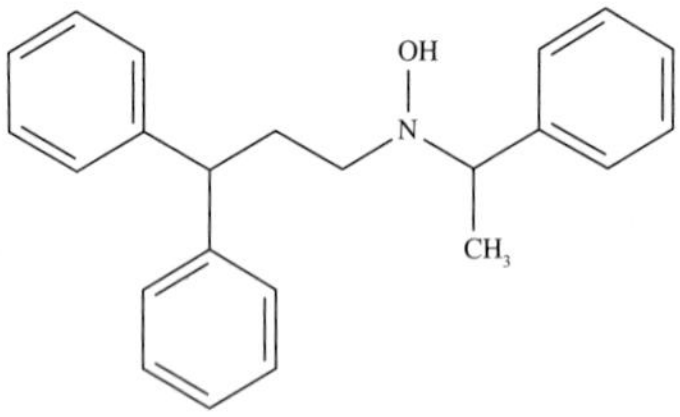

N-(3,3-Diphenylpropyl)-N-(1-phenylethyl)hydroxylamine = N-Hydroxy-γ-phenyl-N-(1-phenylethyl)benzenepropanamine (●)
R Hydrochloride (84388-97-6)
S Oxydilin
U Antihypertensive

11763 (9430) $C_{23}H_{25}NO$
14089-87-3

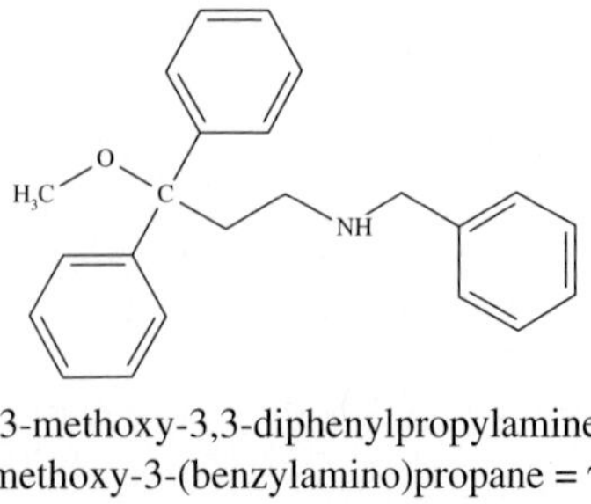

N-Benzyl-3-methoxy-3,3-diphenylpropylamine = 1,1-Diphenyl-1-methoxy-3-(benzylamino)propane = γ-Me-

thoxy-γ-phenyl-*N*-(phenylmethyl)benzenepropanamine
(●)
R Hydrochloride (14089-88-4)
S ST 4250
U Analgesic

11764 (9433)

$C_{23}H_{25}NO_4$
87626-56-0

2-(Diethylamino)ethyl 2-(8-flavonyl)acetate = 4-Oxo-2-phenyl-4*H*-1-benzopyran-8-acetic acid 2-(diethylamino)ethyl ester (●)
R Hydrochloride (87626-57-1)
S LM 985, NSC-293015
U Antineoplastic

11765 (9434)

$C_{23}H_{25}NO_4$
134332-63-1

(*E*)-3-Ethyl-6-methoxy-5-methyl-2-[2-(3-methyl-4-phenyl-3-butenyl)-4-oxazolyl]-4*H*-pyran-4-one (●)
S *Phenoxan*
U Antiviral from *Polyangium* sp.

11766 (9435)

$C_{23}H_{25}NO_5$
2748-74-5

17-Allyl-4,5α-epoxymorphin-7-ene-3,6α-diyl diacetate = Diacetyl-*N*-allylnormorphine = (5α,6α)-7,8-Didehydro-

4,5-epoxy-17-(2-propenyl)morphinan-3,6-diol diacetate (ester) (●)
S *Diacetylnalorphine*
U Antidote

11767

$C_{23}H_{25}NO_6S$
135038-57-2

N-[(*S*)-α-(Mercaptomethyl)-3,4-(methylenedioxy)hydrocinnamoyl]-L-alanine benzyl ester acetate (ester) = Benzyl (*S*)-2-[[(*S*)-2-[(acetylthio)methyl]-3-[3,4-(methylenedioxy)phenyl]propanoyl]amino]propanoate = *N*-[(2*S*)-3-(Acetylthio)-2-(1,3-benzodioxol-5-ylmethyl)-1-oxopropyl]-L-alanine phenylmethyl ester (●)
S *Alatriopril*, BP 1.137, *Fasidotril***
U Cardiotonic (endopeptiase inhibitor), antihypertensive

11768

$C_{23}H_{25}NO_8S_2$
144967-66-8

3(*S*)-(α-L-Daunosaminyloxy)-1,10-dihydro-4,11-dimethoxy-3*H*-thieno[3,4-*b*]thioxanthen-10-one 2,2-dioxide = (*S*)-3-[(3-Amino-2,3,6-trideoxy-α-L-*lyxo*-hexopyranosyl)oxy]-1,3-dihydro-4,11-dimethoxy-10*H*-thieno[3,4-*b*]thioxanthen-10-one 2,2-dioxide (●)
S BCH 255
U Cytotoxic

11769 (9436)

$C_{23}H_{25}N_2$
10309-95-2

[4-[*p*-(Dimethylamino)diphenylmethylene]-2,5-cyclohe-xadienylidene]dimethylammonium = [4-(Dimethylami-no)phenyl]phenyl[4-(dimethylimino)-2,5-cyclohexadi-enylidene]methane = *N*-[4-[[4-(Dimethylamino)phe-nyl]phenylmethylene]-2,5-cyclohexadien-1-ylidene]-*N*-methylmethanaminium (●)

R Chloride (569-64-2) [or hydrogen sulfate (16044-24-9)]

S *Anilin Green, Anilingrün, Bittermandelölgrün, China Green, Chinagrün,* C.I. Basic Green 4, *Diamantgrün B, Malachite Green, Malachitgrün,* Vermala, *Vert Malachite,* Victoria green, *Viktoria-grün, Viride malachitum*

U Topical antiseptic, anthelmintic, dye

11770

$C_{23}H_{25}N_3$
115621-81-3

2,3-Dihydro-5-[4-(1-piperidinylmethyl)phenyl]imida-zo[2,1-*a*]isoquinoline (●)

R Dihydrochloride (115621-95-9)

S SDZ 62-434

U Antineoplastic

11771

$C_{23}H_{25}N_3O_2$
159533-26-3

3-[3-(Dimethylamino)propyl]-4-hydroxy-*N*-[4-(4-pyridi-nyl)phenyl]benzamide (●)

R Monohydrochloride (172854-55-6)

S GR 55562

U 5-HT$_{1D}$ receptor antagonist

11772 (9437)

$C_{23}H_{25}N_3O_2$
121524-18-3

(+)-2(*R*)-(2-Hydroxyethyl)-1-[(*E*)-3-(2-phenylpyrazo-lo[1,5-*a*]pyridin-3-yl)acryloyl]piperidine = [*R*-(*E*)]-1-[1-Oxo-3-(2-phenylpyrazolo[1,5-*a*]pyridin-3-yl)-2-prope-nyl]-2-piperidineethanol (●)

S FK 453, FR 453, FR 113453

U Antihypertensive, diuretic

11773

$C_{23}H_{25}N_3O_2$
145508-78-7

3-[2-(1-Benzyl-4-piperidyl)ethyl]-5,7-dihydro-6*H*-pyrro-lo[3,2-*f*]-1,2-benzisoxazol-6-one = 5,7-Dihydro-3-[2-[1-(phenylmethyl)-4-piperidinyl]ethyl]-6*H*-pyrrolo[3,2-*f*]-1,2-benzisoxazol-6-one (●)

R Maleate (1:1) (145815-98-1)

S CP-118954-11, *Icopezil maleate***
U Acetylcholinesterase inhibitor (cognition activator)

11774 (9438)

$C_{23}H_{25}N_3O_3$
13715-74-7

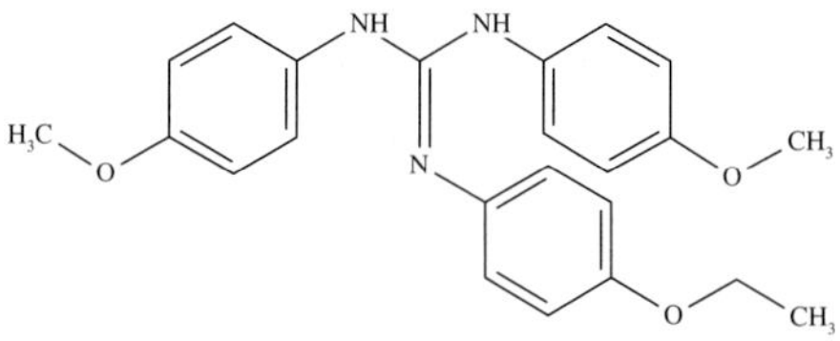

2-(*p*-Ethoxyphenyl)-1,3-bis(*p*-methoxyphenyl)guanidine
= *N*''-(4-Ethoxyphenyl)-*N*,*N*'-bis(4-methoxyphenyl)gua-
nidine (●)
R Monohydrochloride (537-05-3)
S Acoin(e) (old form), Guanicaine, *Phenodianisyl*
U Local anesthetic

11775

$C_{23}H_{25}N_3O_3S$
130717-22-5

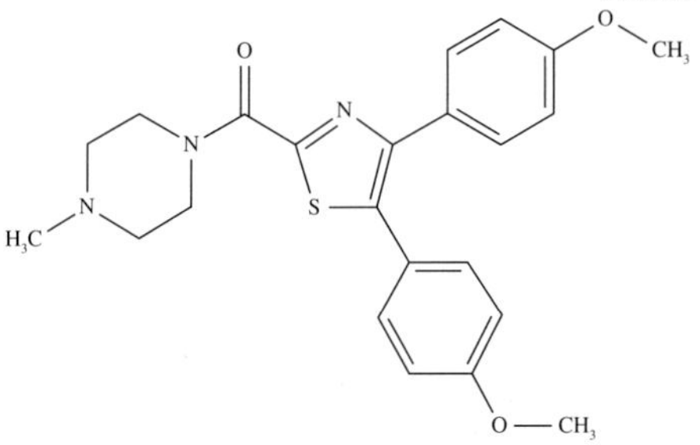

1-[[4,5-Bis(4-methoxyphenyl)-2-thiazolyl]carbonyl]-4-
methylpiperazine (●)
R Monohydrochloride (130717-51-0)
S FR 122047
U Cyclooxygenase inhibitor

11776 (9439)

$C_{23}H_{25}N_3O_4$
74512-62-2

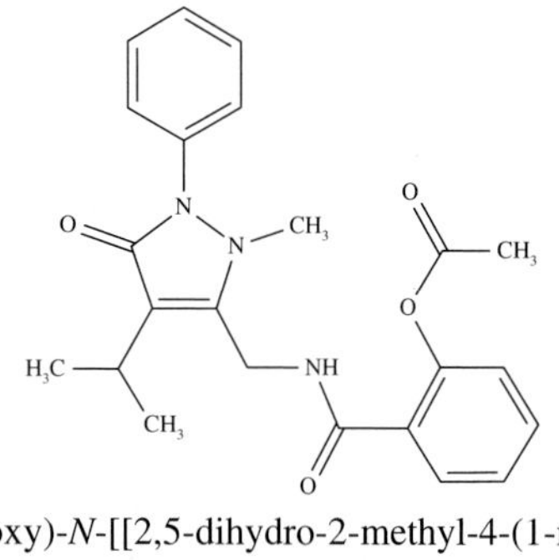

2-(Acetyloxy)-*N*-[[2,5-dihydro-2-methyl-4-(1-methyl-
ethyl)-5-oxo-1-phenyl-1*H*-pyrazol-3-yl]methyl]benz-
amide (●)
S AIA, Aspirin-isopropylantipyrine
U Analgesic, antipyretic, anti-inflammatory, platelet aggregation inhibitor

---

11777

$C_{23}H_{25}N_3O_6$
132877-29-3

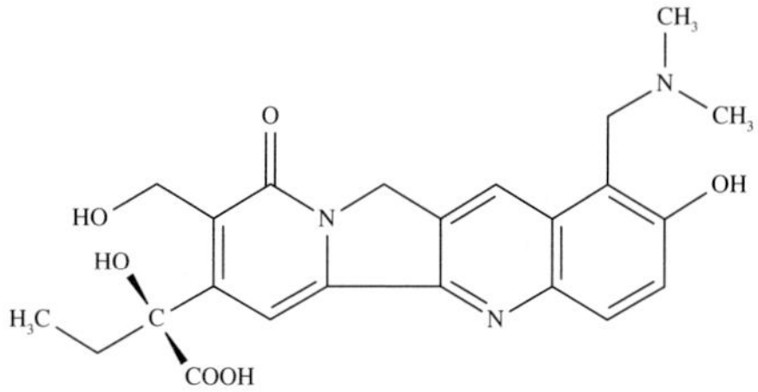

(α*S*)-1-[(Dimethylamino)methyl]-α-ethyl-9,11-dihydro-
α,2-dihydroxy-8-(hydroxymethyl)-9-oxoindolizi-
no[1,2-*b*]quinoline-7-acetic acid (●)
S SKF 105992
U Antineoplastic

11778

$C_{23}H_{25}N_3O_6$
149503-79-7

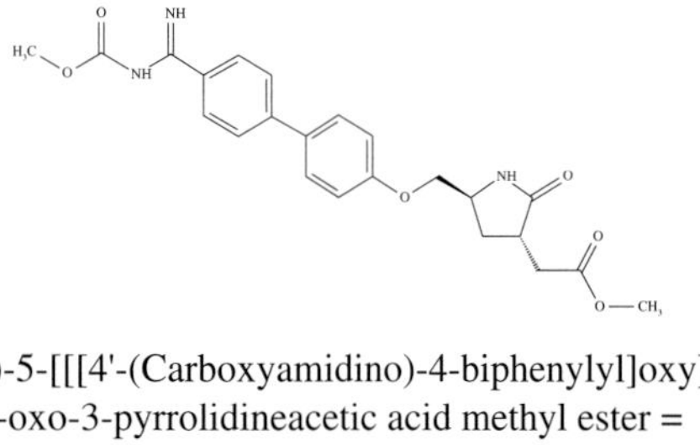

(3*S*,5*S*)-5-[[[4'-(Carboxyamidino)-4-biphenylyl]oxy]me-
thyl]-2-oxo-3-pyrrolidineacetic acid methyl ester =
(3*S*,5*S*)-5-[[[4'-[Imino[(methoxycarbonyl)amino]me-

thyl][1,1'-biphenyl]-4-yl]oxy]methyl]-2-oxo-3-pyrroli-
dineacetic acid methyl ester (●)
S BIBU 104XX, *Fradafiban mecoxil, Lefradafiban***
U Fibrinogen receptor (α_{2B}/β_3)-antagonist

11779 $C_{23}H_{25}N_5O_2$
                                          170912-52-4

4-[4-[[[3-(2-Aminoethyl)-1*H*-indol-5-yl]oxy]acetyl]-1-
piperazinyl]benzonitrile = 1-[[[3-(2-Aminoethyl)-1*H*-in-
dol-5-yl]oxy]acetyl]-4-(4-cyanophenyl)piperazine (●)
R Monohydrochloride (170911-68-9)
S *Doniptan hydrochloride, Donitriptan hydrochloride***, F 1135G
U Serotonin agonist (migraine therapeutic)

11780 (9440) $C_{23}H_{25}N_5O_5$
                                          74191-85-8

1-(4-Amino-6,7-dimethoxy-2-quinazolinyl)-4-(1,4-ben-
zodioxan-2-ylcarbonyl)piperazine = 1-(4-Amino-6,7-di-
methoxy-2-quinazolinyl)-4-[(2,3-dihydro-1,4-benzodi-
oxin-2-yl)carbonyl]piperazine (●)
R Monomethanesulfonate (77883-43-3)
S Alfadil, Benur, Cardenaline, Cardular, Cardular Uro,
Cardura, Carduran, Dedralen, Diblocin, Diblocin
Uro, Doksura, Doxacor, Doxagamma, Doxa-Puren,
Doxazomerck, *Doxazosin mesilate***, Doxolbran,
Jutalar, Normothen, Prodil "Farmasa", Progandol,
Prostadilat, Supress, Supressin "Mack; Pfizer", Ten-
siobas, Tonocardin "Pliva", Tonokardin,
UK-33274-27, Unoprost "Apsen", Uriduct, Zoxan
U Antihypertensive, vasodilator

11781 (9441) $C_{23}H_{25}N_5O_6$
                                          37739-04-1

N-[2-Hydroxy-3-(1-naphthalenyloxy)propyl]adenosine
(●)
S SC-46256
U Coronary vasodilator

11782 (9442) $C_{23}H_{25}N_5O_7S_2$
                                          95056-36-3

7-*N*-[2-[(4-Nitrophenyl)dithio]ethyl]mitomycin C = [1a*S*-
(1aα,8β,8aα,8bα)]-8-[[(Aminocarbonyl)oxy]methyl]-
1,1a,2,8,8a,8b-hexahydro-8a-methoxy-5-methyl-6-[[2-
[(4-nitrophenyl)dithio]ethyl]amino]azirino[2',3':3,4]pyr-
rolo[1,2-*a*]indole-4,7-dione (●)
S BMS-181174, BMY-25067
U Antineoplastic

11783 (9443) $C_{23}H_{25}N_6O_8S_3$

4-[[[(6*R*,7*R*)-7-[2-(2-Amino-4-thiazolyl)glyoxylamido]-
2-carboxy-8-oxo-5-thia-1-azabicyclo[4.2.0]oct-2-en-3-
yl]methyl]thio]-1-methylpyridinium 7^2-(*Z*)-[*O*-(1-carb-

oxy-1-methylethyl)oxime] S-oxide = [5S-[5α,6β,7α(Z)]]-4-[[[7-[[(2-Amino-4-thiazolyl)][(1-carboxy-1-methylethoxy)imino]acetyl]amino]-2-carboxy-8-oxo-5-thia-1-azabicyclo[4.2.0]oct-2-en-3-yl]methyl]thio]-1-methylpyridinium S-oxide (●)

R Chloride (107452-79-9)
S *Cefmepidium chloride***
U Antibiotic

11784 (9444) $C_{23}H_{26}As_2N_6O_4S$

p,p′-Bis(4-amino-2,3-dimethyl-5-oxopyrazolin-1-yl)arsenobenzene-*N*-methylenesulfoxylic acid = [[2-[4-[[4-(4-Amino-2,5-dihydro-2,3-dimethyl-5-oxo-1*H*-pyrazol-1-yl)phenyl]diarsenyl]phenyl]-2,3-dihydro-1,5-dimethyl-3-oxo-1*H*-pyrazol-4-yl]amino]methanesulfonic acid (●)

R Sodium salt
S Sulfoxyl-Salvarsan
U Antibacterial

11785 (9445) $C_{23}H_{26}ClNO_5S_2$
                                               129648-97-1

3-[4-(4-Chlorobenzenesulfonamido)butyl]-6-isopropyl-1-azulenesulfonic acid = 3-[4-[[(4-Chlorophenyl)sulfonyl]amino]butyl]-6-(1-methylethyl)-1-azulenesulfonic acid (●)

R Monosodium salt (129648-96-0)
S KT 2-962
U Anti-anaphylactic, vasodilator

11786 $C_{23}H_{26}ClN_3O_3$
                                               169451-66-5

(*S*)-(+)-*N*-(1-Benzyl-3-pyrrolidinyl)-5-chloro-4-[(cyclopropylcarbonyl)amino]-2-methoxybenzamide = (*S*)-5-Chloro-4-[(cyclopropylcarbonyl)amino]-2-methoxy-*N*-[1-(phenylmethyl)-3-pyrrolidinyl]benzamide (●)

S YM-43611
U Dopamine D_3 and D_4 receptor antagonist

11787 $C_{23}H_{26}ClN_3O_5S$
                                               161611-99-0

2-[4-[4-(7-Chloro-2,3-dihydro-1,4-benzodioxin-5-yl)-1-piperazinyl]butyl]-1,2-benzisothiazol-3(2*H*)-one 1,1-dioxide (●)

S DU 125530
U 5-HT$_{1A}$ receptor antagonist

11788 $C_{23}H_{26}Cl_2N_2O_2$
                                               115904-92-2

trans-3,4-Dichloro-*N*-methyl-*N*-[1,2,3,4-tetrahydro-6-hydroxy-2-(1-pyrrolidinyl)-1-naphthalenyl]benzeneacetamide (●)

R Phosphate (1:1) (salt) (150433-16-2)
S E 3800
Y κ-Agonist analgesic

11789 (9446) $C_{23}H_{26}Cl_2N_2O_4$
125190-72-9

(±)-[3-[1-[[(3,4-Dichlorophenyl)acetyl]methylamino]-2-
(1-pyrrolidinyl)ethyl]phenoxy]acetic acid (●)
R Monohydrochloride (121264-04-8)
S ICI 204448
U Analgesic (κ-opioid)

11790 (9447) $C_{23}H_{26}Cl_2N_4O_6$
97290-20-5

3-Ethyl 5-methyl 2-[[2-(2-amino-4,5-dihydro-4-oxo-1-
imidazolyl)ethoxy]methyl]-4-(2,3-dichlorophenyl)-1,4-
dihydro-6-methyl-3,5-pyridinedicarboxylate = 2-[[2-(2-
Amino-4,5-dihydro-4-oxo-1*H*-imidazol-1-yl)ethoxy]me-
thyl]-4-(2,3-dichlorophenyl)-1,4-dihydro-6-methyl-3,5-
pyridinedicarboxylic acid 3-ethyl 5-methyl ester (●)
S UK-55444
U Coronary vasodilator (calcium antagonist)

11791 (9448) $C_{23}H_{26}Cl_2O_6$
14929-11-4

2-(*p*-Chlorophenoxy)-2-methylpropionic acid trimethylene ester = Trimethylene bis[2-(*p*-chlorophenoxy)-2-me-
thylpropionate] = 2-(4-Chlorophenoxy)-2-methylpropa-
noic acid 1,3-propanediyl ester (●)
S Cholesolvin, CLY-503, Colesolvin, *Diclofibrate*, Fi-
brafit, Halofide, Kaseolin, Liposolvin, *Simfibrate***,
Sinfibrex, Taisterolin, *Trimethylenebis(clofibrate)*
U Antihyperlipidemic

11792 (9449) $C_{23}H_{26}FN_3O_2$
749-02-0

8-[3-(*p*-Fluorobenzoyl)propyl]-1-phenyl-1,3,8-triazaspi-
ro[4.5]decan-4-one = 8-[4-(4-Fluorophenyl)-4-oxobutyl]-
1-phenyl-1,3,8-triazaspiro[4.5]decan-4-one (●)
S E-525, *Espiperona*, R 5147, *Spiperone***, *Spiro-
decanone*, *Spiroperidol*, Spiropitan
U Neuroleptic

11793 (9450) $C_{23}H_{26}F_3N_3S$
17692-26-1

10-[3-(4-Cyclopropyl-1-piperazinyl)propyl]-2-(trifluoro-
methyl)phenothiazine = 10-[3-(4-Cyclopropyl-1-pipera-
zinyl)propyl]-2-(trifluoromethyl)-10*H*-phenothiazine (●)
R also dihydrochloride (15686-74-5)
S *Ciclofenazine***, Cyclophenazine, Lilly 60284
U Neuroleptic

11794　(9451)　　　　　　　　　　　　$C_{23}H_{26}F_4N_2OS$
　　　　　　　　　　　　　　　　　　　　　55837-23-5

4-[3-[6-Fluoro-2-(trifluoromethyl)-9-thioxanthenyl]pro-
pyl]-1-piperazineethanol = 4-[3-[6-Fluoro-2-(trifluoromethyl)-9*H*-thioxanthen-9-yl]propyl]-1-piperazineetha-
nol (●)
R　Dihydrochloride (53542-42-0)
S　Flutixan, Lu 10-022, *Teflutixol dihydrochloride***
U　Neuroleptic

11795　(9452)　　　　　　　　　　　　$C_{23}H_{26}NO_2S$

(±)-3α-[(6,11-Dihydrodibenzo[*b,e*]thiepin-11-yl)oxy]-
6β,7β-epoxy-8-methyl-1α*H*,5α*H*-tropanium = 7-[(6,11-
Dihydrodibenzo[*b,e*]thiepin-11-yl)oxy]-9,9-dimethyl-3-
oxa-9-azoniatricyclo[3.3.1.0²,⁴]nonane (●)
R　Methanesulfonate (88199-75-1)
S　BEA 1306 MS, *Sevitropium mesilate***
U　Bronchospasmolytic

11796　(9453)　　　　　　　　　　　　$C_{23}H_{26}NO_3$
　　　　　　　　　　　　　　　　　　　　　22059-05-8

8-Methyltropinium xanthene-9-carboxylate = *N*-Methyl-
O-(xanthene-9-carbonyl)tropinium = (3-*endo*)-8,8-Di-

methyl-3-[(9*H*-xanthen-9-ylcarbonyl)oxy]-8-azoniabicy-
clo[3.2.1]octane (●)
R　Bromide (4047-34-1)
S　Dexotropin, Gastrixon, N-640, *Trantelinium bromi-
de***
U　Anticholinergic, spasmolytic

11797　(9454)　　　　　　　　　　　　$C_{23}H_{26}N_2$
　　　　　　　　　　　　　　　　　　　　　73966-53-7

3-[4-(1,2,3,6-Tetrahydro-4-phenyl-1-pyridyl)butyl]in-
dole = 3-[4-(3,6-Dihydro-4-phenyl-1(2*H*)-pyridinyl)bu-
tyl]-1*H*-indole (●)
R　Hydrochloride (73966-59-3)
S　EMD 23448
U　Antidepressant (dopamine agonist)

11798　(9455)　　　　　　　　　　　　$C_{23}H_{26}N_2O$
　　　　　　　　　　　　　　　　　　　　　112192-04-8

3-[4-(3,6-Dihydro-4-phenyl-1(2*H*)-pyridyl)butyl]indol-
5-ol = 3-[4-(3,6-Dihydro-4-phenyl-1(2*H*)-pyridinyl)bu-
tyl]-1*H*-indol-5-ol (●)
R　Monohydrochloride (108050-82-4)
S　EMD 38362, *Roxindole hydrochloride***
U　Neuroleptic (presynaptic dopamine agonist), antidepressant

11798-01　(9455-01)　　　　　　　　　　119742-13-1
R　Monomethanesulfonate (salt)
S　EMD 49980, *Roxindole mesilate***
U　Neuroleptic, antidepressant

11799 $C_{23}H_{26}N_2O_2$
242478-37-1

1(*S*)-Phenyl-1,2,3,4-tetrahydroisoquinoline-2-carboxylic acid 3(*R*)-quinuclidinyl ester = (1*S*)-3,4-Dihydro-1-phenyl-2(1*H*)-isoquinolinecarboxylic acid (3*R*)-1-azabicyclo[2.2.2]oct-3-yl ester (●)

R   Succinate (1:1) (242478-38-2)
S   *Solifenacin succinate*, YM-905, YM-67905
U   Muscarinic M_3 antagonist

11800   (9456) $C_{23}H_{26}N_2O_2$
14051-33-3

2-(1-Benzyl-4-piperidyl)-2-phenylglutarimide = 3-(1-Benzyl-4-piperidyl)-3-phenyl-2,6-piperidinedione = 3-Phenyl-1'-(phenylmethyl)[3,4'-bipiperidine]-2,6-dione (●)

R   also with (±)-definition (119391-55-8); Monohydrochloride (5633-14-7)
S   *Benzetimide hydrochloride***, Dioxatrine, McN-JR 4929-11, R 4929, Spasmentral
U   Anticholinergic, spasmolytic, antidiarrhoic

11801   (9457) $C_{23}H_{26}N_2O_2$
21888-99-3

(−)-2-(1-Benzyl-4-piperidyl)-2-glutarimide = (*R*)-3-Phenyl-1'-(phenylmethyl)[3,4'-bipiperidine]-2,6-dione (●)
S   *Levetimide*
U   Muscarinic antagonist

11802   (9458) $C_{23}H_{26}N_2O_2$
21888-98-2

(+)-2-(1-Benzyl-4-piperidyl)-2-phenylglutarimide = (+)-3-(1-Benzyl-4-piperidyl)-3-phenyl-2,6-piperidinedione = (*S*)-3-Phenyl-1'-(phenylmethyl)[3,4'-bipiperidine]-2,6-dione (●)

R   also monohydrochloride (21888-96-0)
S   *Dexbenzetimide*, *Dexetimide***, R 16470, Serenone, Tremblex, Trembley
U   Anticholinergic, antiparkinsonian

11803   (9459) $C_{23}H_{26}N_2O_3$
92623-83-1

p-Methoxyphenyl 2-methyl-1-(2-morpholinoethyl)-3-indolyl ketone = (4-Methoxyphenyl)[2-methyl-1-[2-(4-morpholinyl)ethyl]-1*H*-indol-3-yl]methanone (●)
R   Maleate (1:1) (92623-84-2)
S   *Pravadoline maleate***, Win 48098-6
U   Analgesic

11804　(9460)

$C_{23}H_{26}N_2O_4$
357-57-3

2,3-Dimethoxystrychnine = 2,3-Dimethoxystrychnidin-10-one (●)

S *Brucine, Bruzin, Caniramin, Dimethoxystrychnine, Vomicinum*

U CNS stimulant, cholinesterase inhibitor, reagent (Strychnos alkaloid)

11805　(9461)

$C_{23}H_{26}N_2O_5$
82768-85-2

(3S)-2-[(S)-N-[(S)-1-Carboxy-3-phenylpropyl]alanyl]-1,2,3,4-tetrahydro-3-isoquinolinecarboxylic acid = [3S-[2[R*(R*)],3R*]]-2-[2-[(1-Carboxy-3-phenylpropyl)amino]-1-oxopropyl]-1,2,3,4-tetrahydro-3-isoquinolinecarboxylic acid (●)

R see also no. 12820;
also monohydrate

S Accupro i.v., Accupro-parenteral, CI-928, *Quinaprilat***, Tonovascon parenteral

U Antihypertensive (ACE inhibitor)

11806

$C_{23}H_{26}N_2O_5S$
159182-14-6

N-[4-[2-[[(2S)-2-Hydroxy-3-(4-hydroxyphenoxy)propyl]amino]ethyl]phenyl]benzenesulfonamide (●)

S L 739574

U Human β_3 receptor agonist

11807

$C_{23}H_{26}N_2O_6$
139225-22-2

(5S)-5-[[4-Hydroxy-4-[3,4-(methylenedioxy)phenyl]piperidino]methyl]-3-(p-methoxyphenyl)-2-oxazolidinone = (5S)-5-[[4-(1,3-Benzodioxol-5-yl)-4-hydroxy-1-piperidinyl]methyl]-3-(4-methoxyphenyl)-2-oxazolidinone (●)

R Monohydrochloride (181520-55-8)

S EMD 57445, *Panamesine hydrochloride***

U Neuroleptic (σ-ligand)

11808　(9462)

$C_{23}H_{26}N_4OS$
79449-97-1

6-[2-(4-Indol-3-ylpiperidino)ethyl]-3,7-dimethyl-5H-thiazolo[3,2-a]pyrimidin-5-one = 6-[2-[4-(1H-Indol-3-yl)-1-piperidinyl]ethyl]-3,7-dimethyl-5H-thiazolo[3,2-a]pyrimidin-5-one (●)

S R 51703

U Neuroleptic

11809

$C_{23}H_{26}N_4O_2$
133510-11-9

1-(Phenylacetyl)-4-[4-(1,4,5,6-tetrahydro-4-methyl-6-oxo-3-pyridazinyl)phenyl]piperazine (●)

S J-894

U Calcium antagonist

11810 (9463) $C_{23}H_{26}O_2$
                                                             51018-08-7

19β-*cis*-Ethylidene-17α-ethynyl-17-hydroxyandrosta-1,4,6-trien-3-one = [10(Z),17α]-17-Hydroxy-10-(1-propenyl)-19-norpregna-1,4,6-trien-20-yn-3-one (●)
S Ro 6-5403
U Gonadotropin secretion inhibitor

11811 (9465) $C_{23}H_{26}O_3$
                                                             26002-80-2

m-Phenoxybenzyl (±)-*cis,trans*-2,2-dimethyl-3-(2-methylpropenyl)cyclopropanecarboxylate = 3-Phenoxybenzyl (±)-*cis,trans*-chrysanthemate = 2,2-Dimethyl-3-(2-methyl-1-propenyl)cyclopropanecarboxylic acid (3-phenoxyphenyl)methyl ester (●)
R also stereoisomers
S Anti Bit, Ch. Antipiojos Cruz, Cif, Cusitrim, *Fenotrin*, Full Marks, Head Master, Hégor Anti-Poux, Itax, Item Lot Antipoux, Ivaliten, Mediker, MOM "Celi", Parasidone, Parasidose shampooing, Phenoderm, *Phenothrin**, S-2539, Sifax, Sumithrin, Venatren
U Insecticide, pediculicide

11812 (9464) $C_{23}H_{26}O_3$
                                                             26046-85-5

(1*R*-*trans*)-2,2-Dimethyl-3-(2-methyl-1-propenyl)cyclopropanecarboxylic acid (3-phenoxyphenyl)methyl ester (●)
S *Dexphenothrin*, Zinkan
U Insecticide, pediculicide

11813 (9466) $C_{23}H_{26}O_4$

Estra-1,3,5(10)-triene-3,17β-diol 17-furoate = (17β)-Estra-1,3,5(10)-triene-3,17-diol 17-(2-furancarboxylate) (●)
S Di-folliculine, *Estradiol furoate*, Östradiolfuroat
U Estrogen

11814 $C_{23}H_{26}O_7$
                                                             133806-59-4

3-[(1*S*)-1-(Acetyloxy)-3-methylbutyl]-11-hydroxy-4-methoxy-9-methyl-5*H*,7*H*-dibenzo[*b*,*g*][1,5]dioxocin-5-one (●)
S α-*Acetylpenicillide*, AS 186b, *Purpactin A*
U Enzyme inhibitor (hypocholesteremic) from *Penicillium asperosporum* KY 1635

11815

$C_{23}H_{26}O_8$
58053-83-1

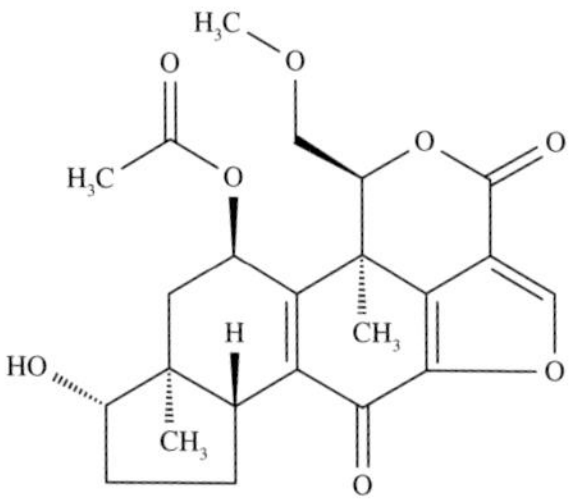

(1*S*,6b*R*,9*S*,9a*S*,11*R*,11b*R*)-11-(Acetyloxy)-
6b,7,8,9,9a,10,11,11b-octahydro-9-hydroxy-1-(methoxymethyl)-9a,11b-dimethyl-3*H*-furo[4,3,2-*de*]inde-
no[4,5-*h*]-2-benzopyran-3,6(1*H*)-dione (●)
S LY 301497
U Antiviral (PI 3-kinase inhibitor)

11816

$C_{23}H_{27}BrFNO_2$
161582-11-2

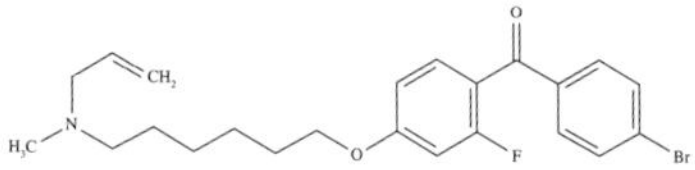

[4-[[6-(Allylmethylamino)hexyl]oxy]-2-fluorophenyl](4-
bromophenyl)methanone = (4-Bromophenyl)[2-fluoro-4-
[[6-(methyl-2-propenylamino)hexyl]oxy]phenyl]metha-
none (●)
R Fumarate (1:1) (189197-69-1)
S Ro 48-8071
U Oxidosqualene cyclase inhibitor

11817 (9467)

$C_{23}H_{27}ClFN_3O_2$
86580-77-0

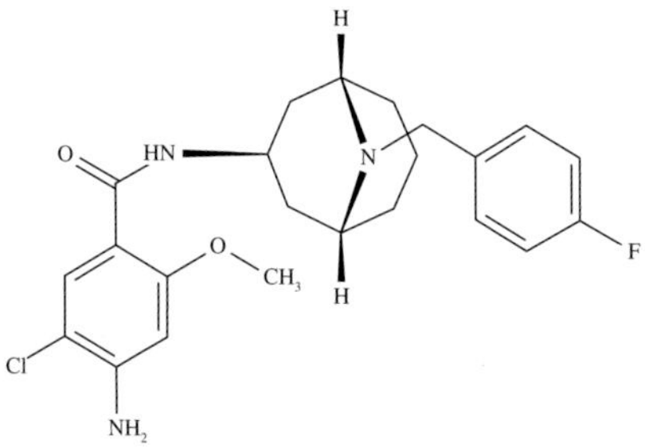

4-Amino-5-chloro-*N*-[9-(*p*-fluorobenzyl)-9-azabicy-
clo[3.3.1]non-3-yl]-*o*-anisamide = *exo*-4-Amino-5-chlo-
ro-*N*-[9-[(4-fluorophenyl)methyl]-9-azabicy-
clo[3.3.1]non-3-yl]-2-methoxybenzamide (●)

S BRL 34778
U Antipsychotic

11818 (9468)

$C_{23}H_{27}ClN_2O_2$
14008-66-3

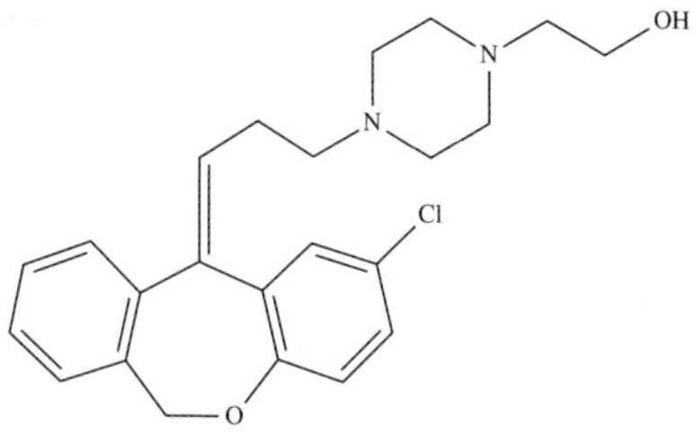

cis-4-[3-(2-Chlorodibenz[*b,e*]oxepin-11(6*H*)-yli-
dene)propyl]-1-piperazineethanol = (*Z*)-4-[3-(2-Chlorodi-
benz[*b,e*]oxepin-11(6*H*)-ylidene)propyl]-1-piperazine-
ethanol (●)
R Dihydrochloride (14008-46-9)
S P-5227, *Pinoxepin hydrochloride***
U Neuroleptic, antipsychotic

11819

$C_{23}H_{27}ClN_4O_3$
170568-47-5

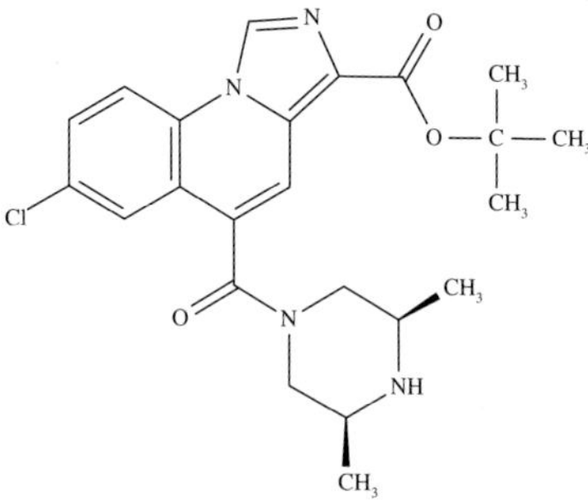

rel-7-Chloro-5-[[(3*R*,5*S*)-3,5-dimethyl-1-piperazinyl]car-
bonyl]imidazo[1,5-*a*]quinoline-3-carboxylic acid 1,1-di-
methylethyl ester (●)
S U 101017
U Partial benzodiazepine agonist (anxiolytic)

11820 (9469)

$C_{23}H_{27}ClN_4O_4S_3$
22561-27-9

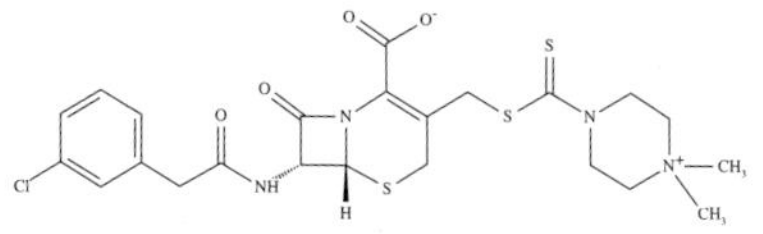

(6R,7S)-7-(3-Chlorophenylacetamido)-3-[[(4,4-dime-thylpiperazinium-1-thiocarbonyl)thio]methyl]-8-oxo-5-thia-1-azabicyclo[4.2.0]oct-2-ene-2-carboxylic acid be-taine = (6R-trans)-4-[[[[2-Carboxy-7-[[(3-chlorophe-nyl)acetyl]amino]-8-oxo-5-thia-1-azabicyclo[4.2.0]oct-2-en-3-yl]methyl]thio]thioxomethyl]-1,1-dimethylpipe-razinium hydroxide inner salt (●)
S *Cefaclomezine, Cephachlomezine*
U Antibiotic

11821 (9470) $C_{23}H_{27}ClO_2$
19291-69-1

6-Chloro-1α,2α:16α,17-dimethylenepregna-4,6-diene-3,20-dione = 17β-Acetyl-6-chloro-1β,1a,2β,8β,9α,10,11,12,13,14α,15,16β,16a,17-tetrade-cahydro-10β,13β-dimethyl-3H-dicyclopro-pa[1,2:16,17]cyclopenta[a]phenanthren-3-one = (1α,2α,8α,9β,10α,13α,14β,16α,17α)-17-Acetyl-6-chlo-ro-1,2,8,9,10,11,12,13,14,15,16,17,20,21-tetradecahy-dro-10,13-dimethyl-3H-dicyclopropa[1,2:16,17]cyclo-penta[a]phenanthren-3-one (●)
S *Gestaclone**, SH 1040
U Progestin

11822 (9471) $C_{23}H_{27}ClO_4$
13698-49-2

6-Chloro-17-hydroxypregna-1,4,6-triene-3,20-dione ace-tate = 17-(Acetyloxy)-6-chloropregna-1,4,6-triene-3,20-dione (●)

S *Delmadinone acetate**, Delmate, Estrex, RS-1301, Tardak, Tardastren, Tardastrex, Tarden, Zenadrex
U Progestin, anti-androgen, anti-estrogen

11823 (9472) $C_{23}H_{27}ClO_6$
14066-79-6

6α-Chloro-17,21-dihydroxypregna-1,4-diene-3,11,20-tri-one 21-acetate = (6α)-21-(Acetyloxy)-6-chloro-17-hydr-oxypregna-1,4-diene-3,11,20-trione (●)
S Adremycine, *Chloroprednisone acetate**, Chlor-prednisoni acetas*, Localyn, Topilan
U Glucocorticoid (topical anti-inflammatory, anti-aller-gic)

11824 $C_{23}H_{27}Cl_2N_3O_2$
129722-12-9

7-[4-[4-(2,3-Dichlorophenyl)-1-piperazinyl]butoxy]-3,4-dihydrocarbostyril = 7-[4-[4-(2,3-Dichlorophenyl)-1-pi-perazinyl]butoxy]-3,4-dihydro-2(1H)-quinolinone (●)
S *Aripiprazole**, OPC-31, OPC-14597
U Antipsychotic

11825 (9473) $C_{23}H_{27}Cl_2N_3O_4$
5377-74-2

p-[p-[Bis(2-chloroethyl)amino]-α-formamidohydrocin-namamido]benzoic acid ethyl ester = 4-[[3-[4-[Bis(2-

chloroethyl)amino]phenyl]-2-(formylamino)-1-oxopropyl]amino]benzoic acid ethyl ester (●)
S Pharsazin, Phosarzin
U Antineoplastic

11826

$C_{23}H_{27}FN_2O_2S$
158848-32-9

5-Fluoro-3-[2-[4-methoxy-4-[[(R)-phenylsulfinyl]methyl]-1-piperidinyl]ethyl]-1H-indole (●)
S GR 159897
U Bronchodilator (neurokinin NK$_2$ receptor antagonist)

11827

$C_{23}H_{27}FN_2O_3S$

(+)-(8aR,12aS,13aS)-12-[(4-Fluorophenyl)sulfonyl]-5,8,8a,9,10,11,12,12a,13,13a-decahydro-3-methoxy-6H-isoquino[2,1-g][1,6]naphthyridine
S RS-15385-FPh
U Antihypertensive, antidepressant (α$_2$-adrenoceptor antagonist)

11828 (9474)

$C_{23}H_{27}FN_4O_2$
106266-06-2

3-[2-[4-(6-Fluoro-1,2-benzisoxazol-3-yl)piperidino]ethyl]-6,7,8,9-tetrahydro-2-methyl-4H-pyrido[1,2-a]pyrimidin-4-one = 3-[2-[4-(6-Fluoro-1,2-benzi-

soxazol-3-yl)-1-piperidinyl]ethyl]-6,7,8,9-tetrahydro-2-methyl-4H-pyrido[1,2-a]pyrimidin-4-one (●)
S Belivon, R 64766, Rehablit, Risolept, Risperdal, Risperidal, *Risperidone***, Risperin, Rispolept, Rispolin, Rispon, Sequinan, Zargus
U Neuroleptic

11829

$C_{23}H_{27}FN_4O_3$
144598-75-4

3-[2-[4-(6-Fluoro-1,2-benzisoxazol-3-yl)-1-piperidinyl]ethyl]-6,7,8,9-tetrahydro-9-hydroxy-2-methyl-4H-pyrido[1,2-a]pyrimidin-4-one (●)
R see also no. 15240
S *9-Hydroxyrisperidone, Paliperidone**, R 76477
U Antipsychotic (D$_2$/5HT$_{2A}$ antagonist)

11830 (9475)

$C_{23}H_{27}F_3N_2O_2$
138335-21-4

cis-(±)-1-[2-(Dimethylamino)ethyl]-1,3,4,5-tetrahydro-4-(4-methoxyphenyl)-3-methyl-6-(trifluoromethyl)-2H-1-benzazepin-2-one (●)
S SQ 32910
U Calcium antagonist

11831　(9476)　　　　　　　　　　　　$C_{23}H_{27}F_3N_2O_2S$
　　　　　　　　　　　　　　　　　　　47682-41-7

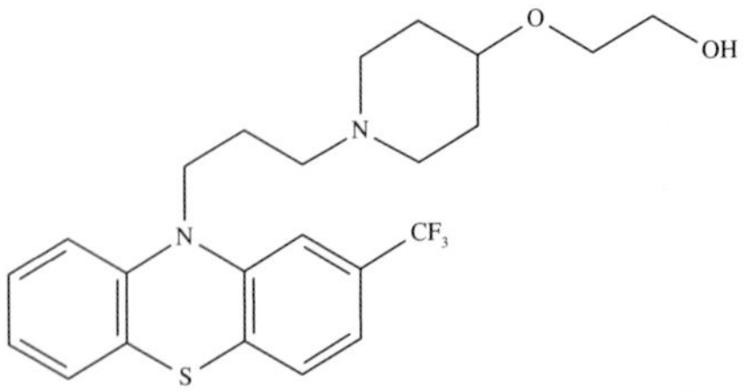

2-[[1-[3-[2-(Trifluoromethyl)phenothiazin-10-yl]pro-
pyl]-4-piperidyl]oxy]ethanol = 2-[[1-[3-[2-(Trifluoromethyl)-10H-phenothiazin-10-yl]propyl]-4-piperidi-
nyl]oxy]ethanol (●)
S　*Flupimazine***, RU 17033
U　Neuroleptic

11832　(9477)　　　　　　　　　　　　　　$C_{23}H_{27}N$
　　　　　　　　　　　　　　　　　　　101828-21-1

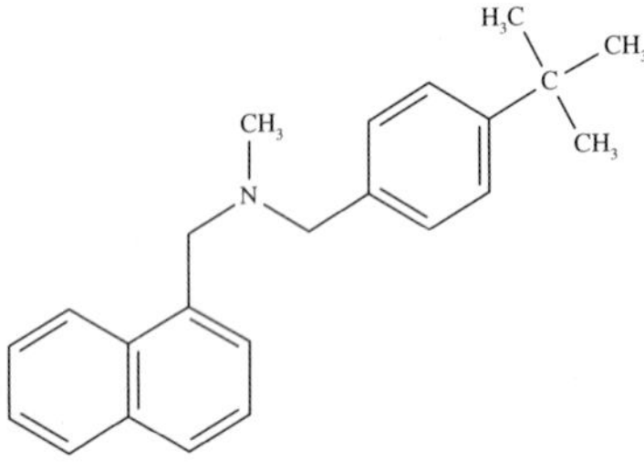

N-(p-tert-Butylbenzyl)-N-methyl-1-naphthalenemethyl-
amine = N-(4-tert-Butylbenzyl)-N-(1-naphthylme-
thyl)methylamine = N-[[4-(1,1-Dimethylethyl)phe-
nyl]methyl]-N-methyl-1-naphthalenemethanamine (●)
R　Hydrochloride (101827-46-7)
S　*Butenafine hydrochloride***, Dermax, KP 363, Men-
tax, Volley
U　Antifungal (topical)

11833　(9478)　　　　　　　　　　　　　　$C_{23}H_{27}NO$
　　　　　　　　　　　　　　　　　　　604-51-3

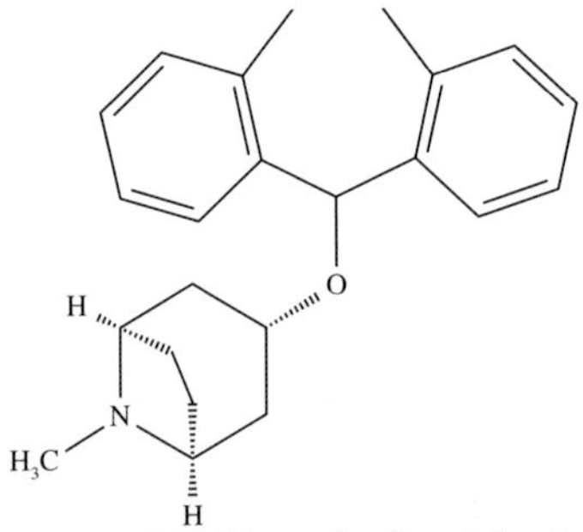

3-[(10,11-Dihydro-5H-dibenzo[a,d]cyclohepten-5-
yl)oxy]-1αH,5αH-tropane = (1R,3r,5S)-3-(10,11-Dihy-
drodibenzo[a,d]cycloheptadien-5-yloxy)tropane = *endo*-
3-[(10,11-Dihydro-5H-dibenzo[a,d]cyclohepten-5-
yl)oxy]-8-methyl-8-azabicyclo[3.2.1]octane (●)
R　see also no. 12417;
　　Citrate (1:1) (2169-75-7)
S　Brontin(a), Brontine, B.S. 6987, *Deptropine citrate***, *Dibenzheptropine citrate*, Su-Brontine
U　Anticholinergic, antihistaminic, bronchodilator

11834　　　　　　　　　　　　　　　　　$C_{23}H_{27}NO_2$
　　　　　　　　　　　　　　　　　　　152148-63-5

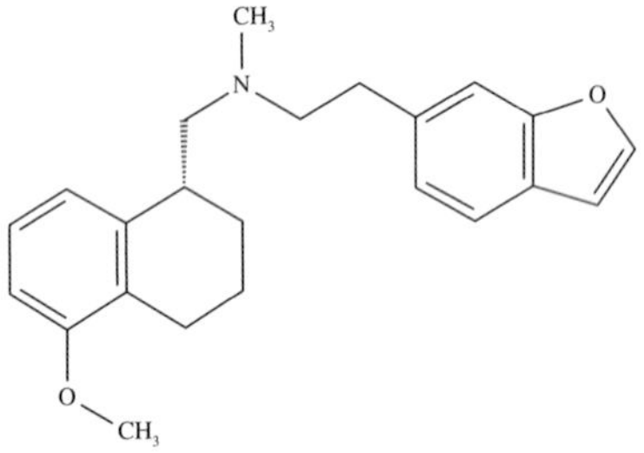

(R)-N-Methyl-N-[(1,2,3,4-tetrahydro-5-methoxy-1-naph-
thalenyl)methyl]-6-benzofuranethanamine (●)
S　A-80426
U　Antidepressant (α_2-adrenoceptor antagonist)

11835 (9479)

$C_{23}H_{27}NO_2$
107010-27-5

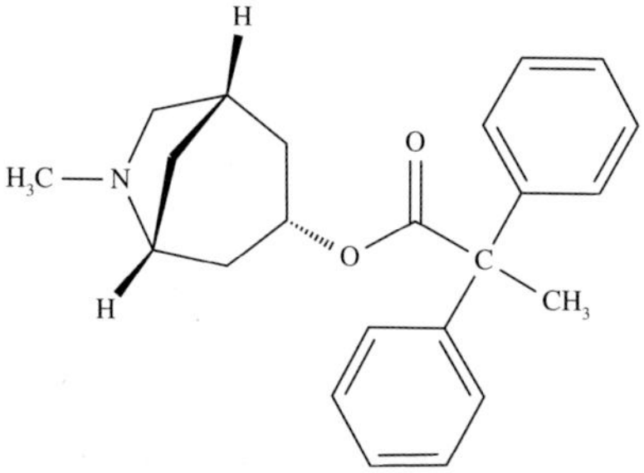

6-Methyl-6-azabicyclo[3.2.1]octan-3α-ol 2,2-diphenyl-
propionate = endo-α-Methyl-α-phenylbenzeneacetic acid
6-methyl-6-azabicyclo[3.2.1]oct-3-yl ester (●)
S *Azaprophen*
U Muscarinic antagonist

11836 (9480)

$C_{23}H_{27}NO_2$
39123-11-0

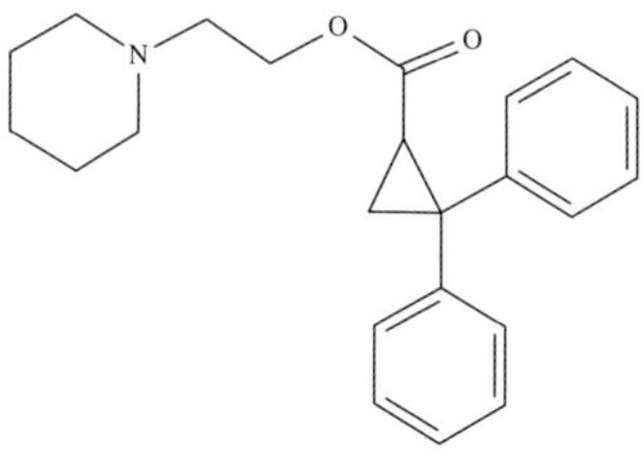

2-Piperidinoethyl 2,2-diphenylcyclopropanecarboxylate
= 2,2-Diphenylcyclopropanecarboxylic acid 2-(1-piperi-
dinyl)ethyl ester (●)
S *Pituxate**,* Upsatux
U Bronchospasmolytic, antitussive

11837

$C_{23}H_{27}NO_2$
184433-64-5

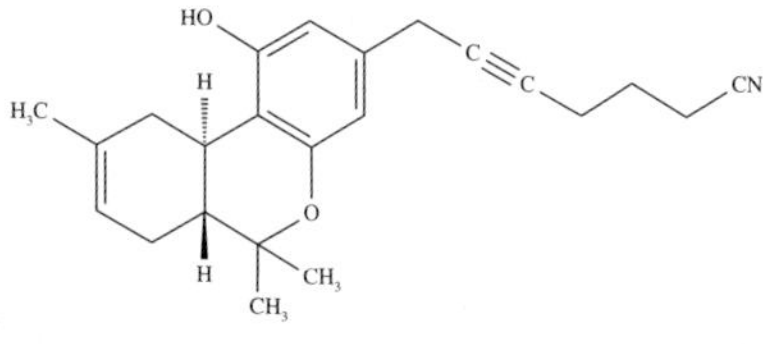

(6aR-trans)-7-(6a,7,10,10a-Tetrahydro-1-hydroxy-6,6,9-
trimethyl-6H-dibenzo[b,d]pyran-3-yl)-5-heptynenitrile
(●)
S O-823
U Cannabinoid CB_1 receptor agonist

11838 (9481)

$C_{23}H_{27}NO_3$
76727-03-2

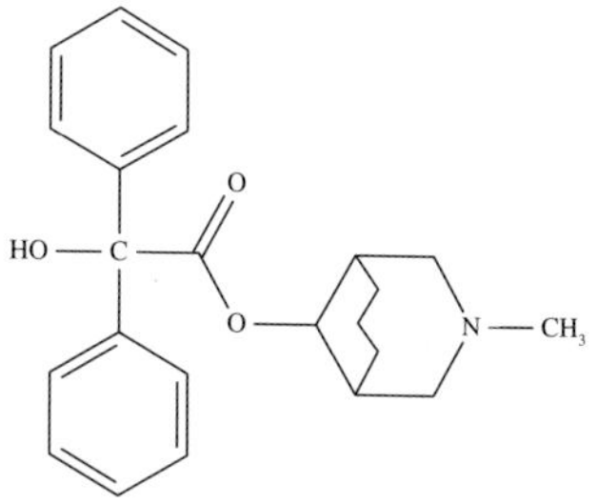

3-Methyl-3-azabicyclo[3.3.1]non-9-yl benzilate = α-Hydroxy-α-phenylbenzeneacetic acid 3-methyl-3-azabicy-
clo[3.3.1]non-9-yl ester (●)
S *Enpiperate*
U Calcium antagonist

11839 (9482)

$C_{23}H_{27}NO_3$
15686-63-2

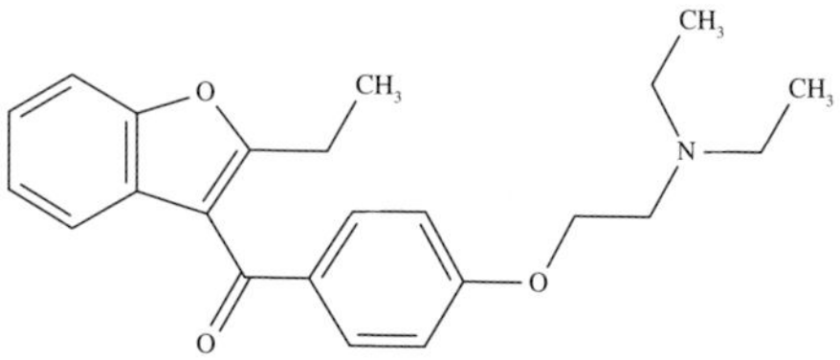

p-[2-(Diethylamino)ethoxy]phenyl 2-ethyl-3-benzofura-
nyl ketone = [4-[2-(Diethylamino)ethoxy]phenyl](2-
ethyl-3-benzofuranyl)methanone (●)
R also hydrochloride (73416-48-5)
S *Etabenzarone**,* L 2642-Labaz, Sedanyl "Labaz"
U Anticoagulant, anti-inflammatory

11840 (9483)

$C_{23}H_{27}NO_4$
132684-62-9

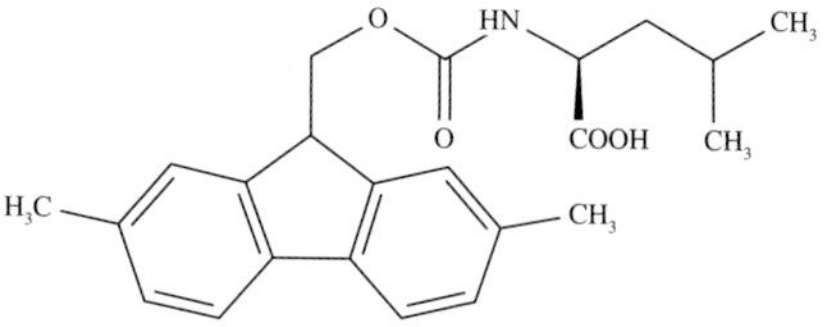

N-[[(2,7-Dimethyl-9H-fluoren-9-yl)methoxy]carbonyl]-
L-leucine (●)
S NPC 15669
U Treatment of septic shock

11841 (9484) $C_{23}H_{27}NO_4$
39537-99-0

Nicotinic acid ester with *cis*-3,3,5-trimethylcyclohexyl (±)-mandelate = (±)-α-(Nicotinoyloxy)phenylacetic acid *cis*-3,3,5-trimethylcyclohexyl ester = (±)-*cis*-3-Pyridinecarboxylic acid 2-oxo-1-phenyl-2-[(3,3,5-trimethylcyclohexyl) oxy]ethyl ester (●)
S *Cyclandelate nicotinate, Micinicate***, Micivas, RV 12128
U Vasodilator

11842 (9485) $C_{23}H_{27}NO_5$
549-68-8

6,7-Dimethoxy-1-(3,4,5-triethoxyphenyl)isoquinoline (●)
R Hydrochloride (6775-26-4)
S Gastrolena, *Octaverine hydrochloride***
U Spasmolytic

11843 (9486) $C_{23}H_{27}NO_8$
131-28-2

6-[[6-[2-(Dimethylamino)ethyl]-2-methoxy-3,4-(methylenedioxy)phenyl]acetyl]-*o*-veratric acid = 3,4,6'-Trimethoxy-4',5'-(methylenedioxy)deoxybenzoin-2-carboxylic acid = 6-[[6-[2-(Dimethylamino)ethyl]-4-methoxy-1,3-benzodioxol-5-yl]acetyl]-2,3-dimethoxybenzoic acid (●)
S *Narceine*
U Spasmolytic, sedative, analgesic (opium alkaloid)

11844 $C_{23}H_{27}NO_9$
20290-10-2

Morphine 6-glucuronide = (5α,6α)-7,8-Didehydro-4,5-epoxy-3-hydroxy-17-methylmorphinan-6-yl β-D-glucopyranosiduronic acid (●)
S CEE 04-410, M6G
U Narcotic analgesic

11845 (9488)

$C_{23}H_{27}N_3O$
90828-99-2

N-[2-(Diethylamino)ethyl]-1-o-tolyl-4-isoquinolinecarb-oxamide = N-[2-(Diethylamino)ethyl]-1-(2-methylphe-nyl)-4-isoquinolinecarboxamide (●)
S Itrocainide**
U Anti-arrhythmic

11846 (9487)

$C_{23}H_{27}N_3O$
47543-65-7

3-(2,2-Diphenylethyl)-5-(2-piperidinoethyl)-1,2,4-oxa-diazole = 1-[2-[3-(2,2-Diphenylethyl)-1,2,4-oxadiazol-5-yl]ethyl]piperidine (●)
R Monohydrochloride (982-43-4) [or mono[2-(4-hydro-xybenzoyl)benzoate] = monohibenzate (37671-82-2)]
S Beclodin, Glibexin, HK 256, Libexin(e),
    Lomapect-Hustensirup (-Tropfen), Mephaxine,
    Prenoxdiazine hydrochloride**, Sedel jarabe, Tibe-xin, Toparten, Varoxil
U Antitussive

11847

$C_{23}H_{27}N_3O_2$
139314-01-5

1,2,3,4-Tetrahydro-2-isoquinolinecarboxylic acid (3aS-cis)-1,2,3,3a,8,8a-hexahydro-1,3a,8-trimethylpyrro-lo[2,3-b]indol-5-yl ester = (3aS,8aR)-3,4-Dihydro-2(1H)-isoquinolinecarboxylic acid 1,2,3,3a,8,8a-hexahydro-1,3a,8-trimethylpyrrolo[2,3-b]indol-5-yl ester (●)
S HP 290, NXX 066, Quilostigmine**
U Acetylcholinesterase inhibitor, cognition activator

11848 (9489)

$C_{23}H_{27}N_3O_2$
103238-56-8

(±)-1-[(2-Indol-3-yl-1,1-dimethylethyl)amino]-3-(indol-4-yloxy)-2-propanol = 1-[[2-(1H-Indol-3-yl)-1,1-dime-thylethyl]amino]-3-(1H-indol-4-yloxy)-2-propanol (●)
S Parodilol*
U Antihypertensive, anti-arrhythmic (β-adrenergic
    blocker)

11849

$C_{23}H_{27}N_3O_2$
185259-85-2

3-[3-[2-(Dimethylamino)ethyl]-1H-indol-5-yl]-N-(4-methoxybenzyl)acrylamide = 3-[3-[2-(Dimethylami-no)ethyl]-1H-indol-5-yl]-N-[(4-methoxyphenyl)methyl]-2-propenamide (●)
S GR 46611
U 5-HT$_{1A,1B,1D}$ receptor agonist

11850   (9490) $C_{23}H_{27}N_3O_2$
6536-18-1

2,3-Dimethyl-1-phenyl-4-[(3-methyl-2-phenylmorpho-
lino)methyl]-5-pyrazolone = 4-[(3-Methyl-2-phenylmor-
pholino)methyl]antipyrine = 1,2-Dihydro-1,5-dimethyl-
4-[(3-methyl-2-phenyl-4-morpholinyl)methyl]-2-phenyl-
3H-pyrazol-3-one (●)
R   also monohydrochloride (50321-35-2)
S   *Morazone***, Novartrina, Orsimon, R 445, *Tarcuza-
te*, Tarugan (old form)
U   Analgesic, anti-inflammatory

11850-01   (9490-01)
R   Lactate
S   Rosimon-Neu-Amp.
U   Analgesic, anti-inflammatory

11851   (9491) $C_{23}H_{27}N_3O_3$
35135-01-4

cis-4'-(1,2,3,4,4a,10b-Hexahydro-8,9-dimethoxy-2-me-
thylbenzo[*c*][1,6]naphthyridin-6-yl)acetanilide = *cis-N*-
[4-(1,2,3,4,4a,10b-Hexahydro-8,9-dimethoxy-2-methyl-
benzo[*c*][1,6]naphthyridin-6-yl)phenyl]acetamide (●)
R   also maleate (1:2) (76166-55-7)
S   AH 21-132, *Benafentrine***
U   Cardiotonic

11852   (9492) $C_{23}H_{27}N_3O_3$
102392-05-2

N-[4-[4-(*o*-Methoxyphenyl)-1-piperazinyl]butyl]phthal-
imide = 2-[4-[4-(2-Methoxyphenyl)-1-piperazinyl]butyl]-
1H-isoindole-1,3(2H)-dione (●)
R   Monohydrobromide (115338-32-4)
S   NAN-190
U   5-HT$_{1A}$-antagonist

11853   (9493) $C_{23}H_{27}N_3O_4$
78718-52-2

Benzyl salicylate *trans*-4-(guanidinomethyl)cyclohexane-
carboxylate = *trans*-2-[[[4-[[(Aminoiminomethyl)ami-
no]methyl]cyclohexyl]carbonyl]oxy]benzoic acid phenylmethyl ester (●)
R   Monohydrochloride (78718-25-9)
S   *Beguhexate hydrochloride, Benexate hydrochloride***, CM 125, TKG 01
U   Anti-ulcer agent

11853-01   (9493-01)   91574-91-3
R   Clathrate of the monohydrochloride with β-cyclodex-
trin (1:1)
S   *Beguhexate CD, Benexale, Benexate CD*, BGH-CD,
Lonmiel, TA-903, Ulgut
U   Anti-ulcer agent

11854 $C_{23}H_{27}N_3O_4S$
                                   233254-24-5

N-[4-[5-(Dimethylamino)-1-naphthylsulfonamido]phe-
nyl]-3-hydroxy-2,2-dimethylpropionamide = N-[4-[[[5-
(Dimethylamino)-1-naphthalenyl]sulfonyl]amino]phe-
nyl]-3-hydroxy-2,2-dimethylpropanamide (●)
S   Bay 38-4766, *Tomeglovir**
U   Anticytomegalovirus drug

11855   (9494) $C_{23}H_{27}N_3O_6$
                                   122328-38-5

trans-2-[3-Azido-5-methoxy-4-(2-propenyloxy)phe-
nyl]tetrahydro-5-(3,4,5-trimethoxyphenyl)furan (●)
S   L 662025
U   PAF receptor antagonist

11856 $C_{23}H_{27}N_3O_6S$
                                   150725-87-4

4-(1,1-Dimethylethyl)-N-[6-(2-hydroxyethoxy)-5-(3-
methoxyphenoxy)-4-pyrimidinyl]benzenesulfonamide
(●)
S   Ro 46-2005
U   Antihypotensive (endothelin receptor antagonist)

11857 $C_{23}H_{27}N_3O_6S$
                                   152684-53-2

(2R)-L-γ-Glutamyl-S-(phenylmethyl)-L-cysteinyl-2-phe-
nylglycine (●)
S   TER 117
U   Potentiator of chemotherapeutics

11858   (9495) $C_{23}H_{27}N_3O_7$
                                   10118-90-8

4,7-Bis(dimethylamino)-1,4,4a,5,5a,6,11,12a-octahydro-
3,10,12,12a-tetrahydroxy-1,11-dioxo-2-naphthacene-
carboxamide = 6-Demethyl-6-deoxy-7-(dimethylamino)tetracycline = [4S-(4α,4aα,5aα,12aα)]-4,7-Bis(dimethylamino)-1,4,4a,5,5a,6,11,12a-octahydro-3,10,12,12a-
tetrahydroxy-1,11-dioxo-2-naphthacenecarboxamide (●)
R   also monohydrochloride (13614-98-7) or monohydro-
     chloride dihydrate (128420-71-3)
S   Acneline, Akamin, Aknemin, Akne-Puren, Aknere-
     duct, Aknin-Mino, Aknin-N, Aknoral, Aknosan,
     Aliucin, Alti-Minocycline, Apo-Minocycline, Aura-
     min, Blemix, Chemocycline, Cipancin, Coupelacin,
     Cychlocin, Cyclimycin, Cyclomin "APS; Berk", Cy-
     clops, Cynomycin, Dentomycin "Lederle; Wyeth",
     Durakne, Dynacin, Evitil "Luso-Farmaco", Gonocin,
     Icht-Oral, Klinoc, Klinomycin, Lederderm, Logryx,
     LS 007, Mestacine, Methiocil "Euro Generics", Mi-
     nac, Minakne, Minitetra "Luso-Farmaco", Mino-50,
     Minocal, Minocin, Minoclir, *Minocycline***, Mino-
     cyn, Minogal, Minogalen, Minolis, Minomax, Mino-
     mycin, Minopen, Minoplus, Minosacin, Minostacin,
     Mino T, Minotab, Minotowa, Minotrex, Minotyrol,

Mino-Wolff, Minox "Rowex", Mynocine, Namimy-cin, Novo-Minocycline, Oracyclin, Pardoclin, Peri-ocline, Robafirin, Romin "Rolab", Skid, Skinocyc-lin, Spicline, Syn-Minocycline, Triomin "Triomed", Udima (Kapseln + Tabl.), Ultramycin "Parke-Da-vis", Vectrin, Zacnan
U Antibiotic

11859 $C_{23}H_{27}N_5OS$
149847-79-0

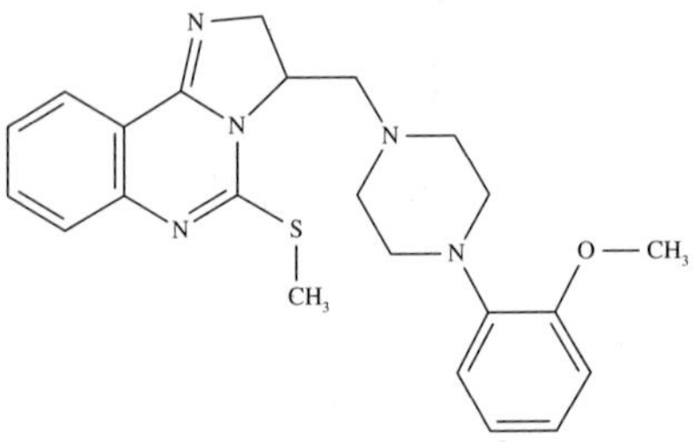

2,3-Dihydro-3-[[4-(2-methoxyphenyl)-1-piperazinyl]me-thyl]-5-(methylthio)imidazo[1,2-c]quinazoline (●)
S DL-017
U Antihypertensive

11860 $C_{23}H_{27}N_5O_2$
149979-74-8

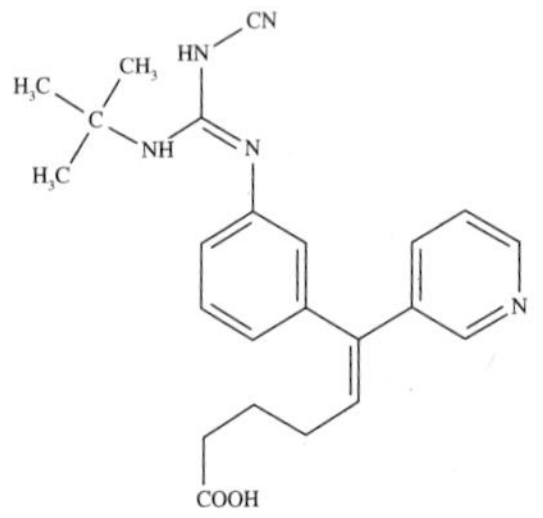

(5E)-6-[m-(3-tert-Butyl-2-cyanoguanidino)phenyl]-6-(3-pyridyl)-5-hexenoic acid = (5E)-6-[3-[[(Cyanoami-no)[(1,1-dimethylethyl)amino]methylene]amino]phe-nyl]-6-(3-pyridinyl)-5-hexenoic acid (●)
S BIBV 308SE, *Terbogrel***
U Platelet aggregation inhibitor

11861 $C_{23}H_{27}N_5O_2$
187162-67-0

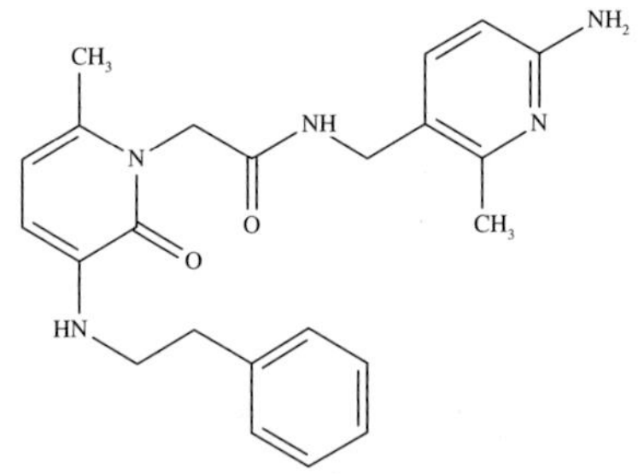

N-[(6-Amino-2-methyl-3-pyridinyl)methyl]-6-methyl-2-oxo-3-[(2-phenylethyl)amino]-1(2H)-pyridineacetamide (●)
S L 375378
U Thrombin inhibitor

11862 $C_{23}H_{27}N_5O_3$
107021-36-3

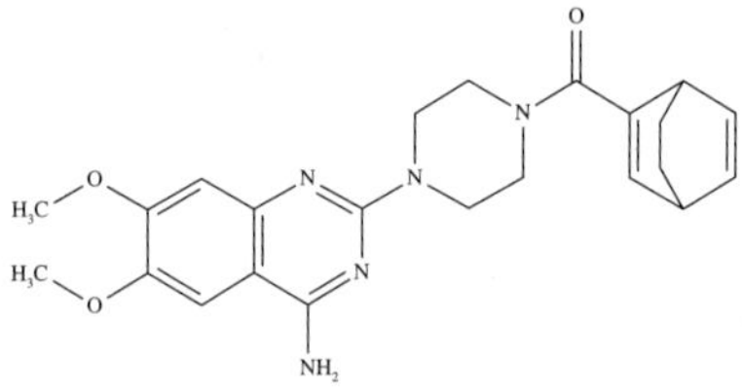

1-(4-Amino-6,7-dimethoxy-2-quinazolinyl)-4-(bicy-clo[2.2.2]octa-2,5-dien-2-ylcarbonyl)piperazine (●)
S SZL 49
U α_1-Adrenoceptor antagonist

11863 $C_{23}H_{27}N_5O_4$
139953-73-4

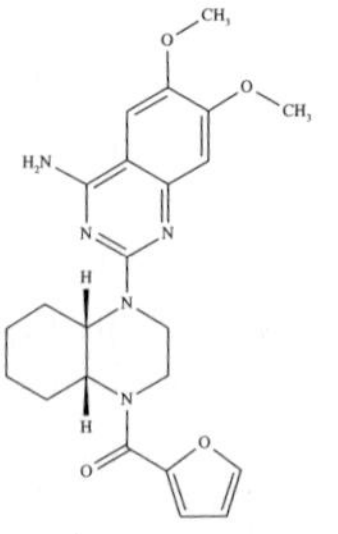

cis-1-(4-Amino-6,7-dimethoxy-2-quinazolinyl)-4-(2-fu-ranylcarbonyl)decahydroquinoxaline (●)
R Monohydrochloride (146929-33-1)
S *Cyclazosin*

U α_{1B}-Adrenoceptor antagonist

11864 $C_{23}H_{27}N_5O_5$
133790-13-3

N-[[[3-(2-Aminoethyl)-1*H*-indol-5-yl]oxy]acetyl]-L-tyrosylglycinamide (●)
S IS-159
U 5-HT$_{1B/1D}$ receptor agonist

11865   (9496) $C_{23}H_{27}N_5O_7S$
61477-96-1

(2*S*,5*R*,6*R*)-6-[(*R*)-2-(4-Ethyl-2,3-dioxo-1-piperazine-carboxamido)-2-phenylacetamido]-3,3-dimethyl-7-oxo-4-thia-1-azabicyclo[3.2.0]heptane-2-carboxylic acid = (6*R*)-6-[(*R*)-2-(4-Ethyl-2,3-dioxopiperazine-1-carboxamido)-2-phenylacetamido]penicillanic acid = [2*S*-[2α,5α,6β(*S**)]]-6-[[[[(4-Ethyl-2,3-dioxo-1-piperazinyl)carbonyl]amino]phenylacetyl]amino]-3,3-dimethyl-7-oxo-4-thia-1-azabicyclo[3.2.0]heptane-2-carboxylic acid (●)
R Monosodium salt (59703-84-3)
S AB-Piperacillin, Akocil, Alukabemin, Avocin, CL 227193, Cypercil, Eril "Foletto; Savio I.B.N.", Fidelbiotico, Hishiyaclorin, Isipen, Ivacin, Ledercil, Penmalin, Pentcillin, Peracil "Bouiscontro & Gazzone", Peracillin, Percilin, Picillin "CT", Pipcil, *Piperacillin sodium***, Pipercillin, Pipérilline, Piperital, Piperzam, Pipracil, Pipracin, Pipraks, Piprasin, Piprax "Eczacibasi", Pipril "Cyanamid-Lederle", Piprilin, Pirasillin, Semipen, T-1220, Taiperacillin
U Antibiotic

11865-01   (9496-01)   123683-33-0
R Mixture with Tazobactam (no. 2172)

S Penmode, *Piperacillin-Tazobactan*, Tasocine, Tazobac, Tazocel, Tazocilline, Tazocin, Tazonam, Tazosyn, TAZ/PIPC, YP-14, Zocyn, Zosin, Zosyn
U Antibiotic

11866   (9497) $C_{23}H_{28}$
75078-91-0

1,2,3,4-Tetrahydro-1,1,4,4-tetramethyl-6-[(*E*)-α-methylstyryl]naphthalene = (*E*)-1,2,3,4-Tetrahydro-1,1,4,4-tetramethyl-6-(1-methyl-2-phenylethenyl)naphthalene (●)
S Ro 15-778, *Temarotene***
U Dermatologic, antineoplastic

11867   (9498) $C_{23}H_{28}Br_2N_2O_3$
84461-99-4

N-[2,4-Dibromo-6-[(cyclohexylmethylamino)methyl]phenyl]-3,4-dimethoxybenzamide (●)
S BR 227
U Mucolytic

11868 $C_{23}H_{28}ClNO_4$
152357-14-7

Ethyl [[(*S*)-8-[[(*R*)-2-(3-chlorophenyl)-2-hydroxyethyl]amino]-6,7,8,9-tetrahydro-5*H*-benzocyclohepten-2-

yl]oxy]acetate = [R-(R*,S*)]-[[8-[[2-(3-Chlorophenyl)-2-hydroxyethyl]amino]-6,7,8,9-tetrahydro-5H-benzocyclo-hepten-2-yl]oxy]acetic acid ethyl ester (●)

R Hydrochloride (152357-12-5)

S FK 175, FR 149175

U β_3-Adrenoceptor agonist

11869   (9499) $C_{23}H_{28}ClN_3O_2S$
84-06-0

2-[4-[3-(2-Chloro-10-phenothiazinyl)propyl]-1-piperazi-nyl]ethyl acetate = 10-[3-[4-(2-Acetoxyethyl)-1-piperazi-nyl]propyl]-2-chlorophenothiazine = 4-[3-(2-Chloro-10H-phenothiazin-10-yl)propyl]-1-piperazineethanol acetate (ester) (●)

R also dihydrochloride (146-28-1)

S Artalan, BS 14, Dartal, Dartalan, Dartan, *Perphena-zine acetate*, SC-7105, *Thiopropazate***

U Neuroleptic

11870   (9500) $C_{23}H_{28}ClN_3O_5S$
10238-21-8

1-[[p-[2-(5-Chloro-o-anisamido)ethyl]phenyl]sulfonyl]-3-cyclohexylurea = 1-[[4-[2-(5-Chloro-2-methoxybenz-amido)ethyl]phenyl]sulfonyl]-3-cyclohexylurea = 5-Chloro-N-[2-[4-[[[(cyclohexylamino)carbonyl]ami-no]sulfonyl]phenyl]ethyl]-2-methoxybenzamide (●)

S Abbenclamide, Adiab, Adiabet, Aglucil, Albert Gly-buride, Apo-Gliburid, Apo-Glyburide, Armoniol, Asangon, Azuglucon, Bastiverit, Benclart, Bengla-mid, Benil "Reko", Betanase, Bevoren "Sintesa", Bratogen, Calabren, Ciana, Clamid, Cliniben, Dao-nil, Deklara, Deptan, Deroctyl, Dia-basan, Dia-BASF, Diabefar, Diabenium, Diabeta, Diabet-amide "Ashbourne", Dia-Eptal, Dianorm, Dibelet, Digaben, Dimel, Dioben, Duraglucon, Duvit, Entre-zon, Euclamin, Euglucan, Euglucon, Euglucon 5, Eu-

glusid, Euglyben, Euglykon, Eunil, Gardoton, Gen-Glib, Gen-Glybe, Gewaglucon, Gilemal, Gile-mid, Gliamide, Glib, Glibamid, Gliban, Gli-basan, Glibedal, Glibelam, Glibemex, Gliben, Glibenase, Gliben-b, Glibenbeta, Glibencil, *Glibenclamide***, Glibendoc, Glibenhexal, Glibenil, *Glibenklamid*, Gli-ben-Puren N, Glibens, Gliben von ct, Glibesifar, Gli-betic, Glibil, Gliboral, Glib-ratiopharm, Glicem, Gli-cemin, Glicon, Glidanil "Montpellier", Glidiabet, Gliformin, Glimel, Glimidstada, Glitisol "Remedi-ca", Glubate, Gluben, Glucal "Chemia", Glucobene, Glucolon, Glucomid, Gluconorm "Wolff", Glucore-med, Gluco-Tablinen, Glucoven, Glukofug, Glukore-duct, Glukovital, Glyben, *Glybenzyclamide, Glyburi-de*, Glyburon, Glycolande, Glycomin, Glynase, Gly-nil, Glyrin "Pharmafrid", HB 419, Hemi-Daonil, He-xaglucon, Humedia, Idimide, Ikaton, Lederglib, Li-banil, Lisaglucon, Lufranclamida, Maclamide, Ma-lix, Maninil, Manirom, Margleid, Melbetese, Melix, Micronase, Miglucan, Nadib, Neogluconin, Nor-boral "Silanes", Norglicem, Norminsul, Normodia-bet, Normoglucon, Novo-Glyburide, Nu-Glyburide, Opeamin, Orabetic "Dorsch; Mibe; YSS", Ozepal, Pamilcon, Pira, Praeciglucon, Renabetic, Renob, Se-mi-Daonil, Semi-Euglucon, Semi-Gliben-Puren N, Seogrumin, Somangwhan, Sugril, Switlus, Syndigin, Tao "Nemi", U-26452, UR-606

U Oral hypoglycemic

11870-01   155021-30-0

R Compd. with β-cyclodextrin (no. 15419) (1:1)

S CHF-1522

U Oral hypoglycemic

11871 $C_{23}H_{28}ClN_5O_3$
149908-53-2

(E)-1-[[5-(p-Chlorophenyl)furfurylidene]amino]-3-[4-(4-methyl-1-piperazinyl)butyl]hydantoin = 1-[[[5-(4-Chlo-rophenyl)-2-furanyl]methylene]amino]-3-[4-(4-methyl-1-piperazinyl)butyl]-2,4-imidazolidinedione (●)

R Dihydrochloride (149888-94-8)
S *Azimilide hydrochloride***, NE 10064, Stedicor "Procter & Gamble"
U Anti-arrhythmic

11872 (9501) $C_{23}H_{28}Cl_2N_2O_3$
                    54301-18-7

p-[[4-[*p*-[Bis(2-chloroethyl)amino]phenyl]butyryl]amino]benzoic acid ethyl ester = 4-[[4-[4-[Bis(2-chloroethyl)amino]phenyl]-1-oxobutyl]amino]benzoic acid ethyl ester (●)
S Butastezine
U Antineoplastic

11873 (9502) $C_{23}H_{28}Cl_2N_2O_3$
                    13382-10-0

N-[[4-[Bis(2-chloroethyl)amino]phenyl]acetyl]-DL-phenylalanine ethyl ester (●)
S Fenafan, Phenaphan
U Antineoplastic

11874 (9503) $C_{23}H_{28}Cl_2O_5$
                    79-61-8

9,11β-Dichloro-17,21-dihydroxypregna-1,4-diene-3,20-dione 21-acetate = (11β)-21-(Acetyloxy)-9,11-dichloro-17-hydroxypregna-1,4-diene-3,20-dione (●)
S Astroderm, Cloriderm, Dermaren, Dermocid, *Dichlorisone acetate*, Diclasone, Dicloderm, Diclor "Cilag", Diclorisona, *Diclorisone acetate*, Diloderm, Disoderm, Epsibom, Pelfar, Sch 5005, Visoderm
U Glucocorticoid (topical anti-inflammatory, anti-allergic)

11875 (9504) $C_{23}H_{28}Cl_3N_3O$
                    64046-79-3

9-[[4-[Bis(2-chloroethyl)amino]-1-methylbutyl]amino]-6-chloro-2-methoxyacridine = N^1,N^1-Bis(2-chloroethyl)-N^4-(6-chloro-2-methoxy-9-acridinyl)-1,4-pentanediamine (●)
R Dihydrochloride (4213-45-0)
S ICR 10, *Quinacrine mustard*
U Antineoplastic

11876 (9505) $C_{23}H_{28}FN_3OS$
                    104606-13-5

N-[1-[4-(4-Fluorophenoxy)butyl]-4-piperidinyl]-*N*-methyl-2-benzothiazolamine (●)
S R 56865
U Vasodilator, anti-arrhythmic

11877 (9506) $C_{23}H_{28}FN_5O_3$
110390-84-6

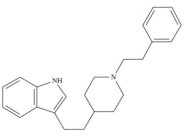

7-[4-[4-(*p*-Fluorobenzoyl)piperidino]butyl]theophylline = 7-[4-[4-(4-Fluorobenzoyl)-1-piperidinyl]butyl]-3,7-dihydro-1,3-dimethyl-1*H*-purine-2,6-dione (●)
S *Perbufylline***
U Coronary vasodilator

11878 (9507) $C_{23}H_{28}F_3N_3OS$
3833-99-6

2-[4-[3-[2-(Trifluoromethyl)-10-phenothiazinyl]propyl]perhydro-1,4-diazepin-1-yl]ethanol = 1-(2-Hydroxyethyl)-4-[3-[2-(trifluoromethyl)-10-phenothiazinyl]propyl]homopiperazine = Hexahydro-4-[3-[2-(trifluoromethyl)-10*H*-phenothiazin-10-yl]propyl]-1*H*-1,4-diazepine-1-ethanol (●)
R Dihydrochloride (1256-01-5)
S D 775, HFZ, *Homofenazine hydrochloride***, Irexan, Oldagen, Pasaden "Homburg"
U Neuroleptic

11879 (9508) $C_{23}H_{28}I_2O_4$
551-90-6

3-[4-(3,5-Di-*tert*-butyl-4-hydroxyphenoxy)-3,5-diiodophenyl]propionic acid = 4-[3,5-Bis(1,1-dimethylethyl)-4-hydroxyphenoxy]-3,5-diiodobenzenepropanoic acid (●)
S *Hinderin*
U Thyromimetic

11880 (9509) $C_{23}H_{28}N_2$
3569-26-4

3-[2-(1-Phenethyl-4-piperidyl)ethyl]indole = 3-[2-[1-(2-Phenylethyl)-4-piperidinyl]ethyl]-1*H*-indole (●)
S *Indopine***
U Analgesic, sedative

11881 (9510) $C_{23}H_{28}N_2O$
5633-16-9

1-[*o*-[2-(Diethylamino)ethoxy]phenyl]-2-methyl-5-phenylpyrrole = *N,N*-Diethyl-2-[2-(2-methyl-5-phenyl-1*H*-pyrrol-1-yl)phenoxy]ethanamine (●)
R Monohydrochloride (14435-78-0)
S DV 714, Léioplégil, *Leiopyrrole hydrochloride***, Leplegil

U   Spasmolytic

11882   (9511) $C_{23}H_{28}N_2O_2$
127805-70-3

(R^*,R^*)-$(\pm)$-1,3-Dihydro-5-[1-hydroxy-2-[4-(phenylme-thyl)-1-piperidinyl]propyl]-2H-indol-2-one ($\bullet$)
S   CP-112116
U   Vasodilator (NMDA antagonist)

11883   (9512) $C_{23}H_{28}N_2O_3$
748-44-7

$(\pm)$-N-[[1-(1,4-Benzodioxan-2-ylmethyl)-4-phenyl-4-pi-peridyl]methyl]acetamide = DL-4-(N-Acetylaminome-thyl)-1-[2-(1,4-benzodioxanyl)methyl]-4-phenylpiperi-dine = $(\pm)$-N-[[1-[(2,3-Dihydro-1,4-benzodioxin-2-yl)me-thyl]-4-phenyl-4-piperidinyl]methyl]acetamide ($\bullet$)
S   Acetoxatrine, *Acoxatrine***, R 5385
U   Vasodilator, antihypertensive

11884   (9513) $C_{23}H_{28}N_2O_3$
76352-13-1

N-(8-Benzyl-1αH,5αH-nortropan-3β-yl)-o-veratramide = *exo*-2,3-Dimethoxy-N-[8-(phenylmethyl)-8-azabicy-clo[3.2.1]oct-3-yl]benzamide ($\bullet$)
R   also monohydrochloride
S   MD 790501, *Tropapride***
U   Antipsychotic

11885 $C_{23}H_{28}N_2O_3$
173273-41-1

$(5RS,1'SR)$-5-Benzyl-3-(3-morpholino-1-phenylpro-pyl)oxazolidine-2-one = (R^*,S^*)-3-[3-(4-Morpholinyl)-1-phenylpropyl]-5-(phenylmethyl)-2-oxazolidinone ($\bullet$)
R   Fumarate (1:1) (173273-43-3)
S   NC 1800
U   Treatment of dysuria

11886   (9514) $C_{23}H_{28}N_2O_3$
70260-53-6

1-(4-Indolyloxy)-3-[4-(phenoxymethyl)piperidino]-2-propanol = 4-[2-Hydroxy-3-[4-(phenoxymethyl)piperidi-no]propoxy]indole = α-[[(1H-Indol-4-yloxy)methyl]-4-(phenoxymethyl)-1-piperidineethanol ($\bullet$)
S   BM 12434, *Mindodilol***
Y β-Adrenergic blocker, vasodilator

11887   (9515) $C_{23}H_{28}N_2O_3$
62658-63-3

$(\pm)$-1-(*tert*-Butylamino)-3-[(2-methyl-4-indolyl)oxy]-2-propanol benzoate (ester) = $(\pm)$-1-[(1,1-Dimethyl-ethyl)amino]-3-[(2-methyl-1H-indol-4-yl)oxy]-2-propa-nol benzoate (ester) ($\bullet$)
R   also maleate (1:1) or malonate (1:1) (82857-38-3)
S   *Bopindolol***, LT 31-200, Sandonorm, Wandonorm

U β-Adrenergic blocker

11888 $C_{23}H_{28}N_2O_3S$
                                 122024-96-8

cis-(−)-3-(Acetyloxy)-5-[2-(dimethylamino)ethyl]-2,3-dihydro-8-methyl-2-(4-methylphenyl)-1,5-benzothiazepin-4(5*H*)-one (●)
R Maleate (122024-98-0)
S TA-993
U Antithrombotic

11889   (9516) $C_{23}H_{28}N_2O_3S$
                                 128312-51-6

(*E*)-3-[[[3-[2-(4-Cyclobutyl-2-thiazolyl)vinyl]phenyl]amino]carbonyl]-2,2-diethylpropionic acid = 3'-[(*E*)-2-(4-Cyclobutyl-2-thiazolyl)vinyl]-2,2-diethylsuccinanilic acid = (*E*)-4-[[3-[2-(4-Cyclobutyl-2-thiazolyl)ethenyl]phenyl]amino]-2,2-diethyl-4-oxobutanoic acid (●)
S *Cinalukast***, Ro 24-5913
U Smooth muscle relaxant (anti-asthmatic, anti-allergic)

11890 $C_{23}H_{28}N_2O_3S$
                                 133646-11-4

(1*S*,6*S*)-1-Benzyl-10-(3-pyridinylmethyl)-7-thia-10-azaspiro[5.6]dodecan-11-one 7,7-dioxide = (1*S*-*trans*)-1-(Phenylmethyl)-10-(3-pyridinylmethyl)-7-thia-10-azaspiro[5.6]dodecan-11-one 7,7-dioxide (●)
R Monohydrochloride (133321-57-0)
S FR 128998
U PAF receptor antagonist

11891   (9517) $C_{23}H_{28}N_2O_4$
                                 65655-59-6

(−)-*p*-[3-[(3,4-Dimethoxyphenethyl)amino]-2-hydroxypropoxy]-β-methylcinnamonitrile = (−)-3-[4-[3-[[2-(3,4-Dimethoxyphenyl)ethyl]amino]-2-hydroxypropoxy]phenyl]crotononitrile = (−)-3-[4-[3-[[2-(3,4-Dimethoxyphenyl)ethyl]amino]-2-hydroxypropoxy]phenyl]-2-butenenitrile (●)
S *Crinolol*, Hoechst 224, *Pacrinolol***
U Antihypertensive

11892   (9518) $C_{23}H_{28}N_2O_4$
                                 83-75-0

Ethyl (6-methoxy-4-quinolyl)(5-vinyl-2-quinuclidinyl)methyl carbonate = Quinine-*O*-carboxylic acid ethyl

ester = (8α,9R)-6'-Methoxycinchonan-9-ol ethyl carbonate (ester) (●)
S *Aecachininum, Chininethylcarbonat, Chinin (geschmackloses), Chininum aethylcarbonicum,* Euchinin, Euquinine, *Quinine ethylcarbonate, Quinini aethylcarbonas, Tasteless Quinine*
U Antimalarial, influenca prophylactic, muscle relaxant

11893 (9519) $C_{23}H_{28}N_2O_5$
131-02-2

(3β,19α,20α)-16,17-Didehydro-11,11-dimethoxy-19-methyloxayohimban-16-carboxylic acid methyl ester (●)
S *Elliptamine,* Redouline, *Reserpiline*
U Antihypertensive (alkaloid from *Rauwolfia serpentina*)

11893-01 (9519-01) 63647-55-2
R Monohydrochloride
S Grona
U Antihypertensive

11893-02 (9519-02) 21086-99-7
R Nicotinate (1:1)
S *Reserpiline nicotinate*
U Antihypertensive

11894 (9520) $C_{23}H_{28}N_2O_5S_2$
111902-57-9

(+)-(2S,6R)-6-[[(1S)-1-Carboxy-3-phenylpropyl]amino]tetrahydro-5-oxo-2-(2-thienyl)-1,4-thiazepine-4(5H)-acetic acid 6-ethyl ester = [2S-[2α,6β(R*)]]-6-[[1-(Eth-

oxycarbonyl)-3-phenylpropyl]amino]tetrahydro-5-oxo-2-(2-thienyl)-1,4-thiazepine-4(5H)-acetic acid (●)
R Monohydrochloride (110221-44-8)
S Acecol, Acecor "Sankyo", CS 622, Escol, *Locapril hydrochloride,* RS-5142, *Temocapril hydrochloride**
U Antihypertensive (ACE inhibitor)

11895 (9521) $C_{23}H_{28}N_3O$
81531-57-9

2-[2-(Diethylamino)ethyl]-9-hydroxy-5,11-dimethyl-6H-pyrido[4,3-b]carbazolium (●)
R Chloride (105118-14-7)
S *Datelliptium chloride**, Detalliptium chloride,* DHE, NSC-311152, SR 95156 B
U Antineoplastic

11896 (9522) $C_{23}H_{28}N_4O_2$
105150-87-6

2-[[3-[3-(1-Piperidinylmethyl)phenoxy]propyl]amino]-4(3H)-quinazolinone (●)
S NO-794
U Histamine H_2-receptor antagonist

11897 (9523) $C_{23}H_{28}N_4O_2S$
125235-15-6

1-Ethyl-3-[4-[4-(4-hydroxyphenyl)-1-piperazinyl]phenyl]-5,5-dimethyl-2-thioxo-4-imidazolidinone (●)

S R 68151
U Anti-inflammatory, antipsoriatic

11898 (9525) $C_{23}H_{28}N_4O_4$
                                                          90326-85-5

(±)-1-[4-(*o*-Methoxyphenyl)-1-piperazinyl]-3-[*m*-(5-me-
thyl-1,3,4-oxadiazol-2-yl)phenoxy]-2-propanol = (±)-4-
(2-Methoxyphenyl)-α-[[3-(5-methyl-1,3,4-oxadiazol-2-
yl)phenoxy]methyl]-1-piperazineethanol (●)
S LU 45966, *Nesapidil***
U Antihypertensive

11899 (9524) $C_{23}H_{28}N_4O_4$
                                                          35265-50-0

6,7-Dimethoxy-2-[2-[4-(*o*-methoxyphenyl)-1-piperazi-
nyl]ethyl]-4(3*H*)-quinazolinone = 6,7-Dimethoxy-2-[2-
[4-(2-methoxyphenyl)-1-piperazinyl]ethyl]-4(1*H*)-quina-
zolinone (●)
S *Peraquinsin**
U Antihypertensive

11900 (9526) $C_{23}H_{28}N_4O_8S$
                                                          81523-49-1

(±)-α-[[4-Amino-5-(3,4,5-trimethoxybenzyl)-2-pyrimidi-
nyl]amino]-3-ethoxy-4-hydroxy-α-toluenesulfonic acid =
(±)-α-[[4-Amino-5-[(3,4,5-trimethoxyphenyl)methyl]-2-

---

pyrimidinyl]amino]-3-ethoxy-4-hydroxybenzeneme-
thanesulfonic acid (●)
S MD-102, *Vaneprim***
U Antibacterial

11901 (9527) $C_{23}H_{28}N_4O_{11}S$
                                                          100680-33-9

1-[(2-Methoxy-2-methylpropionyl)oxy]ethyl (6*R*,7*R*)-7-
[2-(2-furyl)glyoxylamido]-3-(hydroxymethyl)-8-oxo-5-
thia-1-azabicyclo[4.2.0]oct-2-ene-2-carboxylate, 7^2-(*Z*)-
(*O*-methyloxime) carbamate (ester) = (*Z*)-(7*R*)-3-[(Carbamoyloxy)methyl]-7-[2-(2-furyl)-2-(methoxyimino)acet-
amido]-3-cephem-4-carboxylic acid 1-[(2-methoxy-2-
methylpropionyl)oxy]ethyl ester = [6*R*-[6α,7β(*Z*)]]-3-
[[(Aminocarbonyl)oxy]methyl]-7-[[2-furanyl(methoxyimino)acetyl]amino]-8-oxo-5-thia-1-azabicy-
clo[4.2.0]oct-2-ene-2-carboxylic acid 1-(2-methoxy-2-
methyl-1-oxopropoxy)ethyl ester (●)
S CCI 23628, *Cefuroxime pivoxetil***, E 243
U Antibiotic

11902 $C_{23}H_{28}N_6O_6$
                                                          164178-54-5

Ethyl 5-[(3*S*,4*R*)-4-[(1,6-dihydro-6-oxo-3-pyridazi-
nyl)oxy]-3-hydroxy-2,2,3-trimethyl-6-chromanyl]-1*H*-
tetrazole-1-butyrate = Ethyl (3*S*,4*R*)-4-[5-[3-hydroxy-
2,2,3-trimethyl-4-[(3-oxo-2,3-dihydro-6-pyridazi-
nyl)oxy]-6-chromanyl]-1-tetrazolyl]butyrate = 5-
[(3*S*,4*R*)-4-[(1,6-Dihydro-6-oxo-3-pyridazinyl)oxy]-3,4-
dihydro-3-hydroxy-2,2,3-trimethyl-2*H*-1-benzopyran-6-
yl]-1*H*-tetrazole-1-butanoic acid ethyl ester (●)

S *Mazokalim***, UK-143220
U Potassium channel activator (antihypertensive, bronchodilator)

11903 $C_{23}H_{28}N_8$
145216-43-9

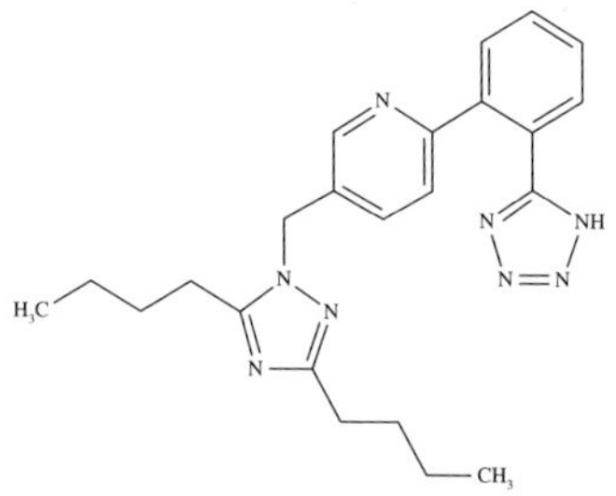

5-[(3,5-Dibutyl-1*H*-1,2,4-triazol-1-yl)methyl]-2-[2-(1*H*-tetrazol-5-yl)phenyl]pyridine (●)
S *Forasartan***, SC-52458
U Angiotensin AT$_1$ antagonist (antihypertensive)

11904 (9528) $C_{23}H_{28}O_2$
91587-01-8

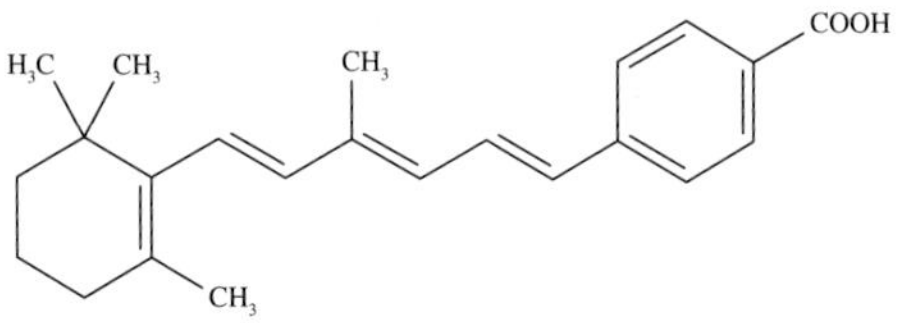

(*E,E,E*)-4-[4-Methyl-6-(2,6,6-trimethyl-1-cyclohexen-1-yl)-1,3,5-hexatrienyl]benzoic acid (●)
S BASF 43915, ORF 18704, *Pelretin***, RWJ 18704
U Keratolytic

11905 $C_{23}H_{28}O_2S$
156691-84-8

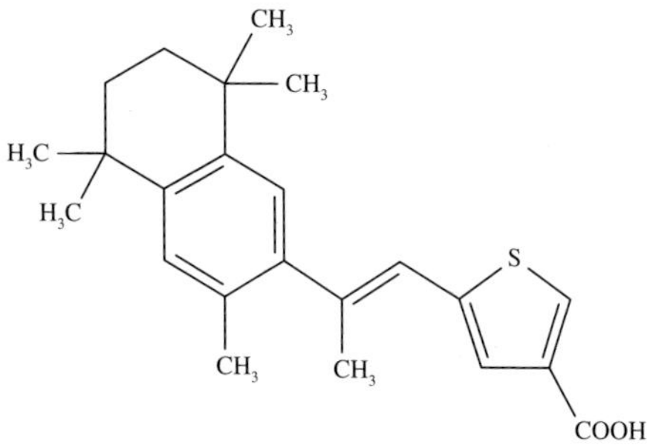

5-[(1*E*)-2-(5,6,7,8-Tetrahydro-3,5,5,8,8-pentamethyl-2-naphthalenyl)-1-propenyl]-3-thiophenecarboxylic acid
(●)
S AGN 191701
U Retinoic X receptor agonist

11906 $C_{23}H_{28}O_3$
74915-64-3

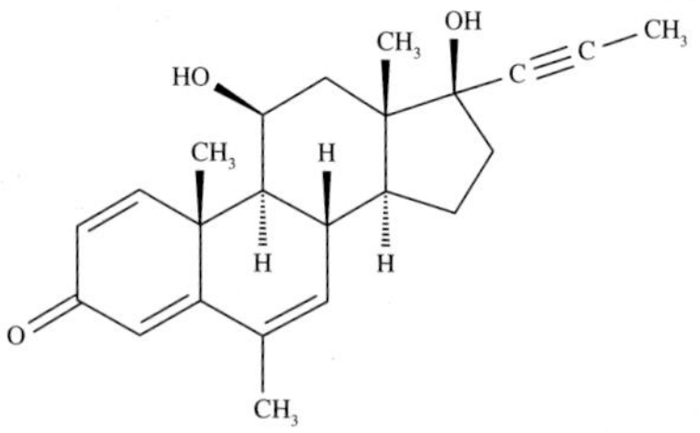

(11β,17β)-11,17-Dihydroxy-6-methyl-17-(1-propynyl)androsta-1,4,6-trien-3-one (●)
S RU 28362
U Glucocorticoid receptor agonist

11907 (9529) $C_{23}H_{28}O_5$
121571-14-0

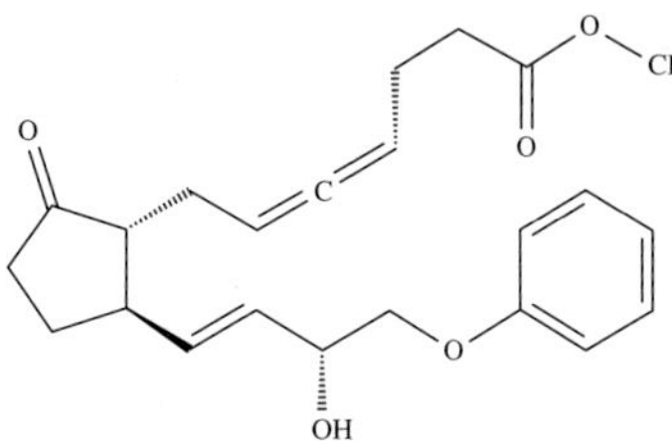

[1*R*-[1α(*R**),2β(1*E*,3*R**)]]-7-[2-(3-Hydroxy-4-phenoxy-1-butenyl)-5-oxocyclopentyl]-4,5-heptadienoic acid methyl ester (●)
S RS-61756-007
U Thromboxane receptor antagonist

11908 (9530) $C_{23}H_{28}O_6$
61281-37-6

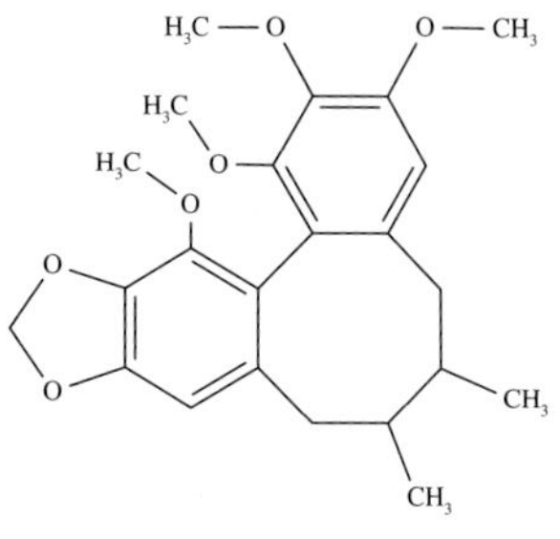

Stereoisomer of 5,6,7,8-tetrahydro-1,2,3,13-tetrameth-
oxy-6,7-dimethylbenzo[3,4]cycloocta[1,2-*f*][1,3]benzodi-
oxole (●)
S   *Schisandrin B*, *Schizandrin B*
U   Hepatoprotectant from *Fructus schizandrae*

11909   (9531) $C_{23}H_{28}O_6$
73121-56-9

Methyl 7-[(1*R**,2*R**,3*R**)-3-hydroxy-2-[(*E*)-(3*R**)-3-
hydroxy-4-phenoxy-1-butenyl]-5-oxocyclopentyl]-4,5-
heptadienoate = [1α,2β(1*E*,3*R**),3α]-7-[3-Hydroxy-2-(3-
hydroxy-4-phenoxy-1-butenyl)-5-oxocyclopentyl]-4,5-
heptadienoic acid methyl ester (●)
S   Camleed, *Enprostil***, Fundyl, Gardrin(e),
        RS-84135, Syngard, TA-84135
U   Gastric antisecretory, anti-ulcer agent

11910   (9532) $C_{23}H_{28}O_7$
58546-54-6

(+)-(6*S*,7*S*,*R-biar*)-5,6,7,8-Tetrahydro-1,2,3,12-tetra-
methoxy-6,7-dimethyl-10,11-methylenedioxydiben-
zo[*a*,*c*]cycloocten-6-ol = (+)-(6*S*,7*S-biar-R*)-5,6,7,8-Tet-
rahydro-1,2,3,13-tetramethoxy-6,7-dimethylben-
zo[3,4]cycloocta[1,2-*f*][1,3]benzodioxol-6-ol (●)
S   *Besigomsin***, *Gomibecin*, *Gomisin A*,
        Schizandrol B, TJN-101
U   Hepatoprotectant, cytoprotective from *Schizandra*
        fruits

11911 $C_{23}H_{28}O_{11}$
25514-29-8

(11β,12α,15β)-15-(Acetyloxy)-13,20-epoxy-3,11,12-tri-
hydroxy-2,16-dioxopicras-3-en-21-oic acid methyl ester
(●)
S   *Bruceine B*
U   Leucocyte-endothelia cell adhesion inhibitor

11912   (9533) $C_{23}H_{29}ClFN_3O_4$
81098-60-4

cis-4-Amino-5-chloro-*N*-[1-[3-(*p*-fluorophenoxy)pro-
pyl]-3-methoxy-4-piperidyl]-*o*-anisamide = *rel*-4-Amino-
5-chloro-*N*-[1-[(3*R*,4*S*)-3-(4-fluorophenoxy)propyl]-3-
methoxy-4-piperidinyl]-2-methoxybenzamide (●)
R   also monohydrate (260779-88-2)
S   Aceleran, Acenalin, Acpulsif, Adamin, Adiparsic,
        Alimex, Alimix(a), Arcasin "Esteve", Bondigest,
        Calmax "Biogen", Cefanyl, Cinacol, Cipramax, Ci-
        pril "Fisons; Italchimici", Circocric, Cisap, Cisapid,
        Cisapride**, Cisapron, Cisawal, Ciside, Cismotil,
        Cispride, Clotioride, Conacid, Coordinax, Cordinax,
        Cyprid, Desaprid, Digenol, Digenormotil, Dispep,
        Dispepci, Dispeprid, Enteropride, Esorid, Etacril
        "Beta", Fisiogastrol, Gastrenol, Gastrokin, Gastro-
        met "Pharma Investi; Recalcine", Gastrum, Gutpro,
        Hagasical, Isaprid, Kaudalit, Kelosal, Kinestase,
        Kinet, Kinetizine, Kinussen, Lamafer, Lavinol, Libe-
        rin, Minsk, Motilar, Norpride, Ondax, Pangest "Far-
        masa", Peristal "Mustafa Nevzat", Prepulsid, Prepul-
        sio, Procinet, Properistal, Propulsid, Propulsin, Pul-
        sar "Phoenix", Pulsitil, R 51619, Rapulid, Risamal,

Risamol, Ruvetine, Saprid, Sedolax "California", Sisapro, Sisarid, Stimulit, Syspride, Tonocis, Trautil, Unamol, Unipride
U　Peristaltic stimulant

11913　(9534)　　　　　　　　　　　　$C_{23}H_{29}ClN_2OS$
14008-71-0

4-[3-(2-Chlorothioxanthen-9-yl)propyl]-1-piperazinepropanol = 2-Chloro-9-[3-[4-(3-hydroxypropyl)-1-piperazinyl]propyl]thioxanthene = 4-[3-(2-Chloro-9*H*-thioxanthen-9-yl)propyl]-1-piperazinepropanol (●)
R　Dihydrochloride (17162-32-2)
S　Daxid, S 9-888, *Xanthiol hydrochloride**
U　Anti-emetic, psychosedative

11914　(9535)　　　　　　　　　　　$C_{23}H_{29}ClN_2O_5S$
119636-74-7

(±)-2-Chloro-4'-hydroxy-5-(2-hydroxy-1-methyl-5-oxo-2-pyrrolidinyl)-3',5'-diisopropylbenzenesulfonanilide = (±)-5-[4-Chloro-3-[(4-hydroxy-3,5-diisopropylanilino)sulfonyl]phenyl]-5-hydroxy-1-methyl-2-pyrrolidinone = 2-Chloro-*N*-[4-hydroxy-3,5-bis(1-methylethyl)phenyl]-5-(2-hydroxy-1-methyl-5-oxo-2-pyrrolidinyl)benzenesulfonamide (●)
S　*Sitalidone**
U　Diuretic, antihyperlipidemic

11915　(9536)　　　　　　　　　　　　$C_{23}H_{29}ClN_4O_3$
54063-30-8

1-(5-Chloro-2-methoxybenzoyl)-3-[3-(4-*m*-tolyl-1-piperazinyl)propyl]urea = 5-Chloro-2-methoxy-*N*-[[[3-[4-(3-methylphenyl)-1-piperazinyl]propyl]amino]carbonyl]benzamide (●)
S　*Ciltoprazine**
U　Muscle relaxant

11916　　　　　　　　　　　　　　　$C_{23}H_{29}ClO_3$
145096-02-2

(−)-2-(4-*tert*-Butylphenoxy)-7-(4-chlorophenyl)heptanoic acid = (−)-4-Chloro-α-[4-(1,1-dimethylethyl)phenoxy]benzeneheptanoic acid (●)
R　Sodium salt (145521-48-8)
S　BM 131074-Na
U　Hypoglycemic

11917　(9539)　　　　　　　　　　　　$C_{23}H_{29}ClO_4$
84782-20-7

1-(4-Carboxyphenoxy)-10-(4-chlorophenoxy)decane = 4-[[10-(4-Chlorophenoxy)decyl]oxy]benzoic acid (●)
S　BRL 24139
U　Antihyperlipidemic

11918   (9537) $C_{23}H_{29}ClO_4$
                                                                            151-69-9

6α-Chloro-17-hydroxypregna-1,4-diene-3,20-dione acetate = (6α)-17-(Acetyloxy)-6-chloropregna-1,4-diene-3,20-dione (●)
S   *Cismadinone acetate***
U   Progestin

11919   (9538) $C_{23}H_{29}ClO_4$
                                                                            302-22-7

6-Chloro-17-hydroxypregna-4,6-diene-3,20-dione acetate = 6-Chloro-6-dehydro-17-acetoxyprogesterone = 17-(Acetyloxy)-6-chloropregna-4,6-diene-3,20-dione (●)
S   Alternyl, Anifertil, Bovisynchron, Cero, CGD 4 I, Chlokinan, *Chlormadinone acetate***, Chlormadinon-Jenapharm, Chlormadinon-Polfa, Chronosyn, Clordion, *Clormadinone acetato*, Cyclonorm, Efumin, Fertiletten, Gesin, Gestafortin, Gestogan, Hypostat "Kyowa", ICI 39575, Kishilinon, Lecoruc, Lormin, Lunapyran, Lutéran, Lutoral "Shionogi", Lutoral "Syntex", Madinon-S, Matrol, Medonsan, Menstridyl, Minipill, Nacenyl, Non-stop, Normenon, NSC-92338, Ovamizole, Papachol, Praxan, Prenival, Prestron, Prococyd, Progestormon, Promadinal, Prostal "Teikoku", Prostat, Protagen S, Retex, Ronsteron, RS-1280, Saiteras, Sakiodil, Skedule, STG 155, Synchrogest, Synchrosyn, Tolecalmo A, Traslán, Urbal, Verton
U   Progestin

11920   (9540) $C_{23}H_{29}ClO_5$
                                                                            38462-04-3

[*S*-(*E,E*)]-3-Chloro-4,6-dihydroxy-2-methyl-5-[3-methyl-7-(tetrahydro-5,5-dimethyl-4-oxo-2-furanyl)-2,6-octadienyl]benzaldehyde (●)
S   *Ascofuranone*
U   Antihyperlipidemic, antineoplastic (antibiotic from *Ascochyta viciae*)

11921   (9541) $C_{23}H_{29}ClO_6$
                                                                            62524-99-6

Methyl (2*E*,5*Z*)-7-[(1*R*,2*R*,3*R*,5*S*)-2-[(*E*)-(3*R*)-4-(*m*-chlorophenoxy)-3-hydroxy-1-butenyl]-3,5-dihydroxycyclopentyl]-2,5-heptadienoate = Methyl (2*E*,5*Z*,13*E*)-(9*S*,11*R*,15*R*)-16-(3-chlorophenoxy)-9,11,15-trihydroxy-ω-tetranorprosta-2,5,13-trienoate = [1*R*-[1α(2*E*,5*Z*),2β(1*E*,3*R**),3α,5α]]-7-[2-[4-(3-Chlorophenoxy)-3-hydroxy-1-butenyl]-3,5-dihydroxycyclopentyl]-2,5-heptadienoic acid methyl ester (●)
S   A 2774, *Delprostenate***, ONO 1052
U   Luteolytic (veterinary)

11922   (9542) $C_{23}H_{29}Cl_2N_3O_4$
                                                                            74296-42-7

2,3-Dihydro-5-(3,4-dichlorophenyl)-6,7-bis(hydroxyme-thyl)-1*H*-pyrrolizine bis(isopropylcarbamate) = (1-Meth-ylethyl)carbamic acid [5-(3,4-dichlorophenyl)-2,3-dihy-dro-1*H*-pyrrolizine-6,7-diyl]bis(methylene) ester (●)
S *Isopropylcad*, NSC-278214
U Antineoplastic

11923 (9543)   C₂₃H₂₉FN₂OS
70931-18-9

3-Fluoro-10,11-dihydro-10-[4-(2-hydroxyethyl)-1-pipe-razinyl]-8-isopropyldibenzo[*b,f*]thiepin = 4-[3-Fluoro-10,11-dihydro-8-(1-methylethyl)dibenzo[*b,f*]thiepin-10-yl]-1-piperazineethanol (●)
S *Isofloxythepin*, VÚFB-10662
U Neuroleptic

11924 (9544)   C₂₃H₂₉FO₆
570-36-5

6α-Fluoro-11β,17,21-trihydroxypregna-1,4-diene-3,20-dione 21-acetate = 6α-Fluoroprednisolone 21-acetate = (6α,11β)-21-(Acetyloxy)-6-fluoro-11,17-dihydroxypreg-na-1,4-diene-3,20-dione (●)
S *Fluprednisolone acetate*, Isopredon-Kristallsusp., Selectren retard inyect.
U Glucocorticoid (anti-inflammatory, anti-allergic)

11925 (9545)   C₂₃H₂₉FO₆
338-98-7

9-Fluoro-11β,17,21-trihydroxypregna-1,4-diene-3,20-dione 21-acetate = 9-Fluoroprednisolone 21-acetate = (11β)-21-(Acetyloxy)-9-fluoro-11,17-dihydroxypregna-1,4-diene-3,20-dione (●)
S Abicorten R, *9-Fluorprednisolone acetate*, *Isoflupre-done acetate*, Menaderm simplex (old form), Predef, U-6013
U Glucocorticoid (anti-inflammatory, anti-allergic)

11926 (9546)   C₂₃H₂₉F₂N₃O
75558-90-6

4-[4,4-Bis(4-fluorophenyl)butyl]-*N*-ethyl-1-piperazine-carboxamide (●)
R also hydrochloride (75529-73-6)
S *Amperozide***, FG 5606, Hogpax
U Anti-aggressive, anti-arrhythmic

11927 (9547)   C₂₃H₂₉F₃O₆
40666-16-8

(±)-(Z)-7-[(1*R**,2*R**,3*R**,5*S**)-3,5-Dihydroxy-2-[(*E*)-(3*R**)-3-hydroxy-4-[(α,α,α-trifluoro-*m*-tolyl)oxy]-1-bu-tenyl]cyclopentyl]-5-heptenoic acid = (±)-(Z)-7-

[(1R,2R,3R,5S)-2-[(E)-(3R)-3-Hydroxy-4-[3-(trifluoro-methyl)phenoxy]-1-butenyl]-3,5-dihydroxycyclopentyl]-5-heptenoic acid = (±)-(5Z,13E)-(9S,11R,15R)-9,11,15-Trihydroxy-16-[3-(trifluoromethyl)phenoxy]-ω-tetranor-prosta-5,13-dienoic acid = [1α(Z),2β(1E,3R*),3α,5α]-(±)-7-[3,5-Dihydroxy-2-[3-hydroxy-4-[3-(trifluorome-thyl)phenoxy]-1-butenyl]cyclopentyl]-5-heptenoic acid (●)

R Sodium salt (55028-71-2)
S Alestrum, Equimate, *Fluprostenol Sodium***, ICI 81008
U Synthetic prostaglandin (treatment of infertility in ma-res)

11928 $C_{23}H_{29}IN_2O$
                                                                               147590-37-2

(*m*-Iodobenzyl)trozamicol = (3'R,4'R)-*rel*-1'-[(3-Iodophe-nyl)methyl]-4-phenyl-[1,3'-bipiperidin]-4'-ol (●)
R Dihydrochloride (153969-57-4)
S MIBT
U Coronary vasodilator, anticholinergic

11929   (9548) $C_{23}H_{29}N$
                                                                               140850-73-3

(+)-α-[(E)-Cinnamyl]-N-(cyclopropylmethyl)-α-ethyl-N-methylbenzylamine = (+)-(E)-(Cyclopropylmethyl)(1-ethyl-1,4-diphenylbut-3-enyl)(methyl)amine = (+)-N-(Cyclopropylmethyl)-α-ethyl-N-methyl-α-[(2E)-(3-phe-nyl-2-propenyl)]benzenemethanamine (●)
R also hydrochloride (130152-35-1)
S CI-1019, *Igmesine***, JO-1784

U Psychotropic (σ opiate receptor antagonist)

11930   (9549) $C_{23}H_{29}NO$
                                                                               15386-01-3

3α-(Diphenylmethoxy)-8-isopropyl-1αH,5αH-nortro-pane = *endo*-3-(Diphenylmethoxy)-8-(1-methylethyl)-8-azabicyclo[3.2.1]octane (●)
R Methanesulfonate (17616-19-2)
S Sch 221
U Anticholinergic, antihistaminic

11931   (9551) $C_{23}H_{29}NO$
                                                                               60996-94-3

4,4-Diphenyl-6-pyrrolidino-3-heptanone = 4,4-Diphenyl-6-(1-pyrrolidinyl)-3-heptanone (●)
S Hoechst 10819
U Narcotic analgesic

11932   (9550) $C_{23}H_{29}NO$
                                                                               561-48-8

4,4-Diphenyl-6-piperidino-3-hexanone = 4,4-Diphenyl-6-(1-piperidinyl)-3-hexanone (●)
R also hydrochloride (6033-41-6) or hydrobromide (6033-42-7)
S Hexalgon, Hoechst 10495, NIH 7557, *Norpipanone***, Ortenso
U Narcotic analgesic

11933 (9552) $C_{23}H_{29}NO_2$
                                                                85213-92-9

4'-(*cis-p*-Menthan-8-yloxy)benzanilide = *cis-N*-[4-[1-Methyl-1-(4-methylcyclohexyl)ethoxy]phenyl]benzamide (●)
S YM-95831
U Antihyperlipidemic

11934 (9553) $C_{23}H_{29}NO_2$
                                                                28240-18-8

D-(+)-1-Methyl-1-(1-methyl-2-piperidyl)ethyl diphenylacetate = 1,α,α-Trimethyl-2-piperidinemethanol diphenylacetate = (+)-α-Phenylbenzeneacetic acid 1-methyl-1-(1-methyl-2-piperidinyl)ethyl ester (●)
S *Pinolcaine***, Seristan
U Local anesthetic

11935 (9554) $C_{23}H_{29}NO_2$
                                                                467-84-5

6-Morpholino-4,4-diphenyl-3-heptanone = 6-(4-Morpholinyl)-4,4-diphenyl-3-heptanone (●)
R also hydrochloride (545-91-5)
S C.B. 11, *Fenadossone*, *Fenadoxone*, Flogodin "Volpino", Hepagin, Heptalgin, Heptalin, *Heptazone*, Heptone, Hoechst 10600, *Morphodone*, *Phenadoxone***, Supralgin "Gerot"
U Narcotic analgesic, hypnotic

11936 (9555) $C_{23}H_{29}NO_2$
                                                                15686-97-2

(+)-α-Benzyl-β-methyl-α-phenyl-1-pyrrolidinepropanol acetate (ester) = D-1-Benzyl-2-methyl-1-phenyl-3-(1-pyrrolidinyl)propyl acetate = α-D-2-Acetoxy-1,2-diphenyl-3-methyl-4-pyrrolidinobutane = β-Methyl-α-phenyl-α-(phenylmethyl)-1-pyrrolidinepropanol acetate (ester) (●)
R Hydrochloride (5591-44-6)
S Lilly 31518, *Pyrrolifene hydrochloride***, *Pyrroliphene hydrochloride*
U Analgesic

11937 (9558) $C_{23}H_{29}NO_3$
4378-36-3

2-(3-Methyl-2-phenylmorpholino)ethyl 2-phenylbutyrate = α-Phenylbutyric acid 2-(3-methyl-2-phenylmorpholino)ethyl ester = α-Ethylbenzeneacetic acid 2-(3-methyl-2-phenyl-4-morpholinyl)ethyl ester (●)
S *Fenbutrazate**, Phenbutrazate*
U Anorexic, central stimulant

11937-01 (9558-01) 8004-38-4
R Mixture of the hydrochloride with phenmetrazine teoclate (no. 2872-01)
S Abstin, Cafilon, Filon, Sabacid
U Anorexic

11938 (9560) $C_{23}H_{29}NO_3$
60569-19-9

1-Methyl-4-piperidyl diphenylpropoxyacetate = Diphenylpropoxyacetic acid 1-methyl-4-piperidyl ester = O-Propylbenzilic acid 1-methyl-4-piperidyl ester = α-Phenyl-α-propoxybenzeneacetic acid 1-methyl-4-piperidinyl ester (●)
R Hydrochloride (54556-98-8)
S Apomed, Bup 4, Detrunorm, Mictonetten, Mictonorm, P 4, Papfo, *Propiverine hydrochloride**
U Anticholinergic, spasmolytic, antiparkinsonian

11939 $C_{23}H_{29}NO_3$
86383-88-2

2'-[2-Hydroxy-3-(1-piperidinyl)propoxy]-3-phenylpropiophenone = 1-[2-[2-Hydroxy-3-(1-piperidinyl)propoxy]phenyl]-3-phenyl-1-propanone (●)
S AM 05
U Multidrug resistance modulator

11940 (9559) $C_{23}H_{29}NO_3$
72060-05-0

17-(Cyclopropylmethyl)-4,5α-epoxy-8β-ethyl-3-methoxymorphinan-6-one = (5α,8β)-17-(Cyclopropylmethyl)-4,5-epoxy-8-ethyl-3-methoxymorphinan-6-one (●)
R Hydrochloride (70865-14-4)
S *Codorphone hydrochloride, Conorfone hydrochloride**, Conorphone hydrochloride,* TR-5109
U Analgesic, antagonist to narcotics

11941 (9557) $C_{23}H_{29}NO_3$
562-26-5

Ethyl 1-(3-hydroxy-3-phenylpropyl)-4-phenylpiperidine-4-carboxylate = 1-(3-Hydroxy-3-phenylpropyl)-4-phenylisonipecotic acid ethyl ester = 1-(3-Hydroxy-3-phenyl-

propyl)-4-phenyl-4-piperidinecarboxylic acid ethyl ester (●)
R also hydrochloride (3627-49-4)
S *Fenoperidine*, Lealgin, NIH 7591, Operidine "Janssen-Cilag", *Pheniperidinum*, *Phenoperidine***, R 1406, SC-9369
U Narcotic analgesic

11942 (9556) $C_{23}H_{29}NO_3$
3691-78-9

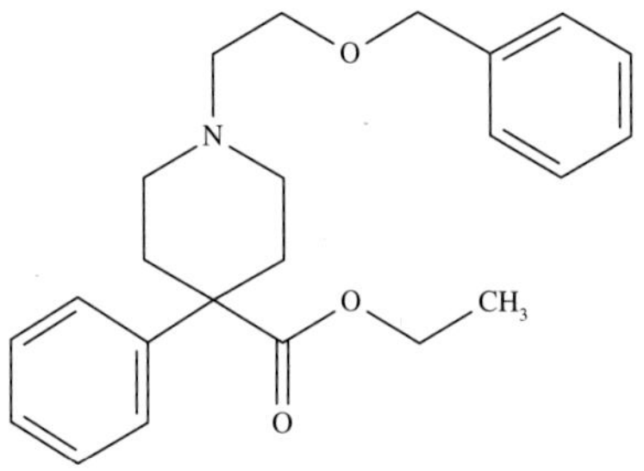

1-[2-(Benzyloxy)ethyl]-4-phenyl-4-piperidinecarboxylic acid ethyl ester = Ethyl 1-(2-benzyloxyethyl)-4-phenylpiperidine-4-carboxylate = 4-Phenyl-1-[2-(phenylmethoxy)ethyl]-4-piperidinecarboxylic acid ethyl ester (●)
S *Benzethidine***, *Benzetidina*, NIH 7574, TA 28
U Narcotic analgesic

11943 $C_{23}H_{29}NO_3S$
128995-52-8

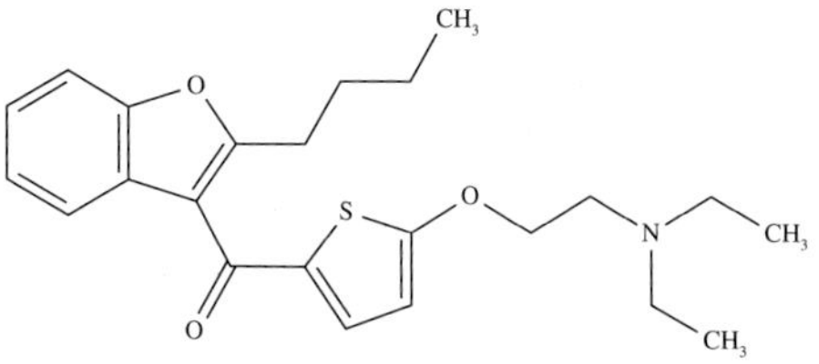

(2-Butyl-3-benzofuranyl)[5-[2-(diethylamino)ethoxy]-2-thienyl]methanone (●)
R Hydrochloride (128995-51-7)
S E 047/1
U Anti-arrhythmic

11944 $C_{23}H_{29}NO_3S$
212329-37-8

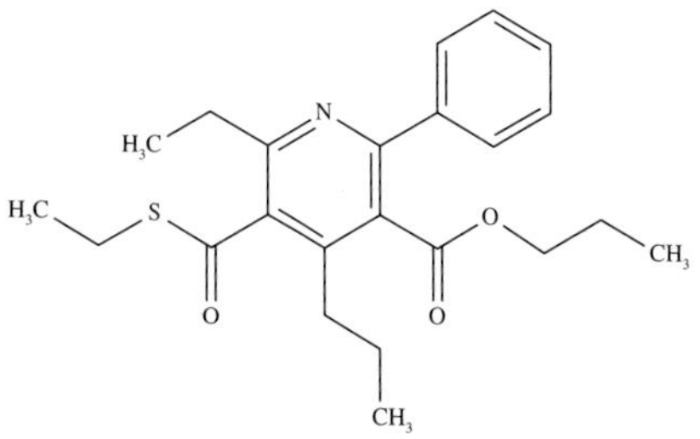

Propyl 2-ethyl-4-propyl-3-(ethylthiocarbonyl)-6-phenylpyridine-5-carboxylate = 6-Ethyl-5-[(ethylthio)carbonyl]-2-phenyl-4-propyl-3-pyridinecarboxylic acid propyl ester (●)
S MRS 1523
U Adenosine A_3 receptor antagonist

11945 (9561) $C_{23}H_{29}NO_{12}$
6379-56-2

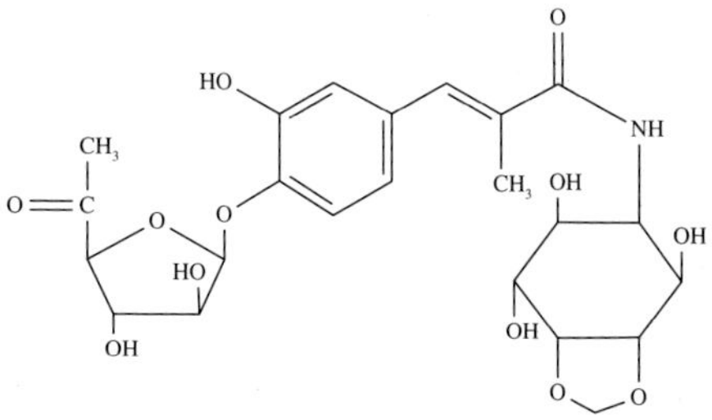

5-Deoxy-5-[[(2E)-3-[4-[(6-deoxy-β-D-*arabino*-hexofuranos-5-ulos-1-yl)oxy]-3-hydroxyphenyl]-2-methyl-1-oxo-2-propenyl]amino]-1,2-O-methylene-D-*neo*-inositol (●)
S Homomycin, Hygromix, *Hygromycin A*, XK 43-2
U Antibiotic from *Streptomyces hygroscopicus*

11946 (9562) $C_{23}H_{29}N_3O$
103844-77-5

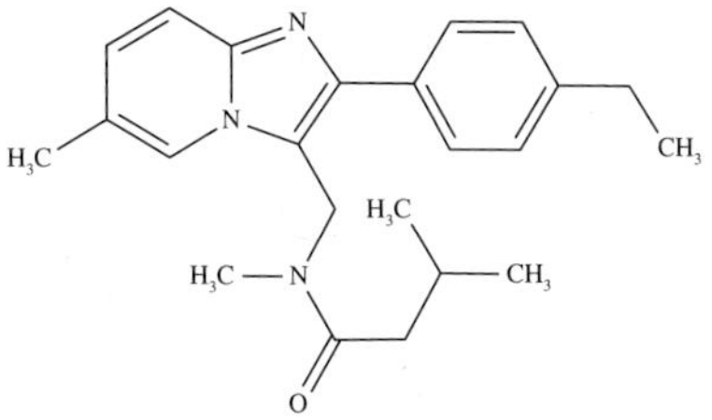

N-[[2-(p-Ethylphenyl)-6-methylimidazo[1,2-a]pyridin-3-yl]methyl]-N,3-dimethylbutyramide = N-[[2-(4-Ethylphe-

nyl)-6-methylimidazo[1,2-*a*]pyridin-3-yl]methyl]-*N*,3-
dimethylbutanamide (●)
S *Necopidem***, SL 850355
U Anesthetic (general and local)

11947 (9563) $C_{23}H_{29}N_3O$
315-72-0

2-[4-[3-(5*H*-Dibenz[*b,f*]azepin-5-yl)propyl]-1-piperazi-
nyl]ethanol = 5-[3-(*N'*-β-Hydroxyethylpiperazino)pro-
pyl]-5*H*-dibenz[*b,f*]azepine = 4-[3-(5*H*-Dibenz[*b,f*]aze-
pin-5-yl)propyl]-1-piperazineethanol (●)
R Dihydrochloride (909-39-7)
S Deprenil "Yurtoglu", Dinsidon, Ensidon, G 33040,
Insidon, Insifen, Insomin "Ibrahim", Inzeton, Nisida-
na, *Opipramol hydrochloride***, Opramol, Opridon,
Oprimol, Pramolan, 8307 R.P.
U Antidepressant

11948 (9564) $C_{23}H_{29}N_3O$
39186-49-7

Hexahydro-α,α-diphenylpyrrolo[1,2-*a*]pyrazine-2(1*H*)-
butyramide = Hexahydro-α,α-diphenylpyrrolo[1,2-*a*]py-
razine-2(1*H*)-butanamide (●)
S *Pirolazamide***, SC-26438
U Anti-arrhythmic

11949 $C_{23}H_{29}N_3OS$
135003-30-4

10-[(1*R*)-1-Methyl-2-(1-pyrrolidinyl)ethyl]-*N*-propyl-
10*H*-phenothiazine-2-carboxamide (●)
R Monohydrochloride (135092-09-0)
S *Apadoline hydrochloride**, RP 60180
U Analgesic (κ-opioid agonist)

11950 (9568) $C_{23}H_{29}N_3O_2$
81079-97-2

4'-Hydroxy-3',5'-bis(1-pyrrolidinylmethyl)benzanilide =
N-[4-Hydroxy-3,5-bis(1-pyrrolidinylmethyl)phe-
nyl]benzamide (●)
S ACC-9164
U Anti-arrhythmic

11951 (9565) $C_{23}H_{29}N_3O_2$
90446-66-5

4-Hydroxy-3,5-bis(1-pyrrolidinylmethyl)benzanilide = 4-
Hydroxy-*N*-phenyl-3,5-bis(1-pyrrolidinylmethyl)benz-
amide (●)
R   also dihydrochloride (90446-69-8)
S   ACC-9358, DuP 923
U   Anti-arrhythmic

11952   (9567) $C_{23}H_{29}N_3O_2$
                                                       153-87-7

5,6-Dimethoxy-2-methyl-3-[2-(4-phenyl-1-piperazi-
nyl)ethyl]indole = 5,6-Dimethoxy-2-methyl-3-[2-(4-phe-
nyl-1-piperazinyl)ethyl]-1*H*-indole (●)
R   also hydrochloride (40523-01-1)
S   DO-180, DP 181, Equipertine, Forit, Integrin, Lantu-
      ril, Lotawin, Opertil, *Oxipertinum*, *Oxypertine***,
      Win 18501
U   Antipsychotic, neuroleptic

11953   (9566) $C_{23}H_{29}N_3O_2$
                                                       21590-92-1

2-Ethyl-3-(β-piperidino-*p*-phenetidino)phthalimidine =
2-Ethyl-3-[*p*-(β-piperidinoethoxy)anilino]isoindolin-1-
one = 2-Ethyl-2,3-dihydro-3-[[4-[2-(1-piperidinyl)eth-
oxy]phenyl]amino]-1*H*-isoindol-1-one (●)
S   *Amidoline*, Dolispasmo, *Etomidoline***, K 2680, Mi-
      antor-S, Smedolin
U   Muscle relaxant, spasmolytic

11954   (9569) $C_{23}H_{29}N_3O_2S$
                                                       2751-68-0

10-[3-[4-(2-Hydroxyethyl)-1-piperazinyl]propyl]pheno-
thiazin-2-yl methyl ketone = 2-Acetyl-10-[3-[4-(β-hydroxyethyl)-1-piperazinyl]propyl]phenothiazine = 1-[10-[3-
[4-(2-Hydroxyethyl)-1-piperazinyl]propyl]-10*H*-pheno-
thiazin-2-yl]ethanone (●)
R   also maleate (1:2) (5714-00-1)
S   *Acephenazini maleas*, *Acetophenazine maleate***,
      NSC-70600, Phenthoxate, Sch 6673, Tindal(a)
U   Antipsychotic, neuroleptic

11955   (9570) $C_{23}H_{29}N_3O_2S_2$
                                                       3313-26-6

cis-*N*,*N*-Dimethyl-9-[3-(4-methyl-1-piperazinyl)propyli-
dene]thioxanthene-2-sulfonamide = (*Z*)-*N*,*N*-Dimethyl-9-
[3-(4-methyl-1-piperazinyl)propylidene]-9*H*-thio-
xanthene-2-sulfonamide (●)
R   also dihydrochloride dihydrate (22189-31-7)
S   CP 12252-1, Navan(a), Navane, Navaron,
      NSC-108165, Orbinamon, P-4657 B, *Thiothixene*,
      *Tiotixene***
U   Antipsychotic, neuroleptic

11956 (9571) $C_{23}H_{29}N_3O_3$
83200-08-2

(±)-2-[[[3-(Diethylamino)-2-hydroxypropyl]amino]car-
bonyl]-3-methoxy-1-phenylindole = (±)-*N*-[3-(Diethylamino)-2-hydroxypropyl]-3-methoxy-1-phenyl-1*H*-in-
dole-2-carboxamide (●)
R   also without (±)-definition (85793-29-9); Monohydrochloride (85793-72-2)
S   *Eproxindine hydrochloride***, KC 3791
U   Anti-arrhythmic

11957 (9572) $C_{23}H_{29}N_3O_4$
74178-99-7

1-[1-(β-Hydroxy-3,4-dimethoxy-α-methylphenethyl)-4-
piperidyl]-2-benzimidazolinone = 1-[1-[2-(3,4-Dime-
thoxyphenyl)-2-hydroxy-1-methylethyl]-4-piperidinyl]-
1,3-dihydro-2*H*-benzimidazol-2-one (●)
R   Succinate (85984-40-3)
S   KF 4307
U   Antihypertensive

11958 (9573) $C_{23}H_{29}N_3O_4$
95893-19-9

(±)-2-[4-[3-(*tert*-Butylamino)-2-hydroxypropoxy]phe-
nyl]-6-methoxy-3-methyl-4(3*H*)-quinazolinone = 2-[4-

2440

[3-[(1,1-Dimethylethyl)amino]-2-hydroxypropoxy]phe-
nyl]-6-methoxy-3-methyl-4-(3*H*)-quinazolinone (●)
S   (±)-HX-CH 44 BS
U   β$_1$-selective adrenoceptor antagonist

11959 $C_{23}H_{29}N_3O_6S_2$
161162-21-6

(2*R*)-*N*-[3-[[2-[(2,3-Dihydro-6-hydroxy-2-oxo-1*H*-naph-
tho[2,1-*b*][1,4]thiazin-5-yl)thio]ethyl]amino]-3-oxopro-
pyl]-2,4-dihydroxy-3,3-dimethylbutanamide (●)
S   FR 901537
U   Aromatase inhibitor from *Bacillus* sp. 3072

11960 (9574) $C_{23}H_{29}N_5O$
7125-71-5

N-(4-Ethyl-4,6,6a,7,8,9,10,10a-octahydro-7-methylindo-
lo[4,3-*fg*]quinolin-9-yl)-3,5-dimethyl-1*H*-pyrazole-1-
carboxamide = 4-Ethyl-7-methyl-9-(3,5-dimethylpyra-
zole-1-carboxamido)-4,6,6a,7,8,9,10,10a-octahydroindo-
lo[4,3-*fg*]quinoline = *N*-[(8β)-1-Ethyl-6-methylergolin-8-
yl]-3,5-dimethyl-1*H*-pyrazole-1-carboxamide (●)
S   Lilly 44106, *Toquizine**
U   Anticholinergic, anti-ulcer agent

11961 (9575) $C_{23}H_{29}N_5O_3$

1-(4-Amino-6,7-dimethoxy-2-quinazolinyl)-4-(bicyclo[2.2.2]oct-5-en-2-ylcarbonyl)piperazine (●)

R Monohydrochloride (99899-45-3)
S SM 2470
U Antihypertensive, antilipidemic

11962 (9576)

$C_{23}H_{29}N_5O_8S_2$
105889-45-0

(Pivaloyloxy)methyl (+)-(6*R*,7*R*)-7-[(*Z*)-2-(2-amino-4-
thiazolyl)-2-pentenamido]-3-[(carbamoyloxy)methyl]-8-
oxo-5-thia-1-azabicyclo[4.2.0]oct-2-ene-2-carboxylate =
[6*R*-[6α,7β(*Z*)]]-3-[[(Aminocarbonyl)oxy]methyl]-7-[[2-
(2-amino-4-thiazolyl)-1-oxo-2-pentenyl]amino]-8-oxo-5-
thia-1-azabicyclo[4.2.0]oct-2-ene-2-carboxylic acid (2,2-
dimethyl-1-oxopropoxy)methyl ester (●)
R Monohydrochloride monohydrate (147816-24-8)
S *Cefcamate pivoxil hydrochloride, Cefcapene pivoxil
hydrochloride**, Flomox, Flumax, S 1108
U Antibiotic

11963 (9577)

$C_{23}H_{29}N_7O_6$
120225-54-9

2-[[4-(2-Carboxyethyl)phenethyl]amino]adenosine-5'-*N*-
ethylcarboxamide = 4-[2-[[6-Amino-9-(*N*-ethyl-β-D-ribo-
furanuronamidosyl)-9*H*-purin-2-yl]amino]ethyl]ben-
zenepropanoic acid (●)
R Monosodium salt (120225-64-1)
S CGS 21680 C
U Antihypertensive, coronary vasodilator

11964 (9578)

$C_{23}H_{30}BrN_3O_2$
60019-20-7

2-Bromo-6-methylergoline-8β-methanol hexahydro-1*H*-
azepine-1-carboxylate (ester) = (8β)-2-Bromo-6-methyl-
ergoline-8-methanol hexahydro-1*H*-azepine-1-carb-
oxylate (ester) (●)
R Monomethanesulfonate (60019-21-8)
S *Brazergoline mesilate**, Luvion (old form)
U Vasodilator

11965 (9579)

$C_{23}H_{30}ClFO_3$
111157-06-3

21-Chloro-9-fluoro-11β-hydroxy-16α,17-dimethylpreg-
na-1,4-diene-3,20-dione = (11β,16α)-21-Chloro-9-
fluoro-11-hydroxy-16,17-dimethylpregna-1,4-diene-
3,20-dione (●)
S Org 6632
U Immunosuppressive, antineoplastic

11966 (9580)

$C_{23}H_{30}ClNO$
59767-12-3

1-[2-[(*p*-Chloro-α-methyl-α-phenylbenzyl)oxy]ethyl]oc-
tahydroazocine = 1-[2-[1-(4-Chlorophenyl)-1-phenyleth-
oxy]ethyl]octahydroazocine (●)
S *Octastine***
U Antihistaminic

11967 (9581)   C$_{23}$H$_{30}$ClN$_3$O
                                                        83-89-6

6-Chloro-9-[[4-(diethylamino)-1-methylbutyl]amino]-2-
methoxyacridine = *N*^{4}-(6-Chloro-2-methoxy-9-acridi-
nyl)-*N*^{1},*N*^{1}-diethyl-1,4-pentanediamine (●)
R most dihydrochloride (69-05-6) [or dihydrochloride
       dihydrate (6151-30-0)]
S *Acrichin*, Acrihin, *Acrinaminum*, Acriquine, Akri-
       chin, Anofelin, Antimalarina, Arichin, Atabrin(e),
       Atatrin, Atebrin, Chemiochin, *Chinacrin*, Crinodora,
       Erion, Fedal-Lamb, Golo, Haffkinine, Hepacrine,
       Italchina, Maladin, Malaricida, *Mepacrine***, Méta-
       quine, Methoquine, Metochin, Metocrine, Metoqui-
       ne, Palacrin, Palusan, Pentilen, *Quinacrine*,
       866 R.P., S.N. 390, Ténicridine
U Antimalarial, anthelmintic

11967-01 (9581-01)   316-05-2
R Dimethanesulfonate
S Musonal, *Quinacrine Soluble*
U Antimalarial, anthelmintic

11968   C$_{23}$H$_{30}$ClN$_5$O$_3$
                                                        148913-08-0

tert-Butyl 7-chloro-4,5-dihydro-5-[(3,4,5-trimethylpipe-
razino)carbonyl]imidazo[1,5-*a*]quinoxaline-3-carb-
oxylate = 7-Chloro-4,5-dihydro-5-[(3,4,5-trimethyl-1-pi-
perazinyl)carbonyl]imidazo[1,5-*a*]quinoxaline-3-carb-
oxylic acid 1,1-dimethylethyl ester (●)
S U-97775
U Anxiolytic (GABA$_A$ receptor ligand)

11969 (9582)   C$_{23}$H$_{30}$FN$_3$
                                                        132810-10-7

2-(4-Ethyl-1-piperazinyl)-4-(4-fluorophenyl)-
5,6,7,8,9,10-hexahydrocycloocta[*b*]pyridine (●)
S AD 5423, Bionanserin, *Blonanserin*
U Antipsychotic

11970 (9583)   C$_{23}$H$_{30}$GdN$_3$O$_{11}$
                                                        135326-11-3

(4*S*)-4-(4-Ethoxybenzyl)-3,6,9-tris(carboxylatomethyl)-
3,6,9-triazaundecanedioic acid gadolinium complex =
(*SA*-8-11252634)-Dihydrogen [*N*-[(2*S*)-2-[bis[(carboxy-
κ*O*)methyl]amino-κ*N*]-3-(4-ethoxyphenyl)propyl]-*N*-[2-
[bis[(carboxy-κ*O*)methyl]amino-κ*N*]ethyl]glycinato(5–)-
κ*N*,κ*O*]gadolinate(2–) (●)
S *Acidum gadoxeticum***, *Gadoxetic acid***,
       Gd-EOB-DTPA
U Diagnostic aid

11970-01   135326-22-6
R Disodium salt

S ZK 139834
U Diagnostic aid

11971 (9584) $C_{23}H_{30}NO_3$
23795-19-9

1-Ethyl-1-(2-hydroxyethyl)piperidinium benzilate = 1-[2-(Benziloyloxy)ethyl]-1-ethylpiperidinium = 1-Ethyl-1-[2-[(hydroxydiphenylacetyl)oxy]ethyl]piperidinium (●)
R Bromide (23182-46-9)
S *Ethylpipethanate bromide*, Flespan, Panpurol, PB 106, *Pipethanate ethobromide*, Spalgin, Spasmodil "ABC; Bioprogress"
U Anticholinergic

11972 (9585) $C_{23}H_{30}NO_3$
298-50-0

(2-Hydroxyethyl)diisopropylmethylammonium xanthene-9-carboxylate = Diisopropylmethyl-[(xanthene-9-carbonyloxy)ethyl]ammonium = *N*-Methyl-*N*-(1-methylethyl)-*N*-[2-[(9*H*-xanthen-9-ylcarbonyl)oxy]ethyl]-2-propanaminium (●)
R Bromide (50-34-0)
S Aclobrom, Apopant, Astrospas, Banlin, Bantinova, Bantinyl-duplex, Bropantil, Clontheline, Corrigast, Ercorax, Ercoril, Ercotina, Giquel, Ketaman, Kivatin, Lenigastril, Mephathelin, Neo-Banex, Neo-dexabine, Neo-Gastrosedan, Neo-Metantyl, Neopepulsan, Neo-Ulkophob, Neo-ventrisan, Norpanth, Novopropanthil, Pantheline, Panthene, Pervagal, Pro-Alusine, Probamide, Pro-Banthin(e), Prodixamon, Pro-Gastron, Propantel, *Propantelina*, Propanthel, *Propantheline bromide***, Robantaline, Ropanth, SC 3171, Sedalcer, Spastil, Suprantil, Uro-Gastrosedan
U Anticholinergic, spasmolytic

11972-01 72442-47-8
R Bromide mixt. with magnesium silicon oxide $(Mg_2Si_3O_8)$ and sodium dihydrogen (*SP*-4-2)-[(2*S*,3*S*)-18-carboxy-20-(carboxymethyl)-8-ethenyl-13-ethyl-2,3-dihydro-3,7,12,17-tetramethyl-21*H*,23*H*-porphine-2-propanoato(5−)-$\kappa N^{21},\kappa N^{22},\kappa N^{23},\kappa N^{24}$]cuprate(3−)
S Methaphyllin
U Anti-ulcer agent

11973 (9586) $C_{23}H_{30}NO_3$
59270-39-2

1-(2-Hydroxyethyl)-1,2,5-trimethylpyrrolidinium benzilate = 1-[(2-Benziloyloxy)ethyl]-1,2,5-trimethylpyrrolidinium = 1-[2-[(Hydroxydiphenylacetyl)oxy]ethyl]-1,2,5-trimethylpyrrolidinium (●)
R Bromide (51047-24-6)
S *Dimetipirium bromide***
U Anticholinergic, spasmolytic

11974 (9588) $C_{23}H_{30}N_2$
78370-13-5

2-Isopropyl-5-(methylphenethylamino)-2-phenylvaleronitrile = 1,7-Diphenyl-3-(methylaza)-7-cyano-8-methyl-

nonane = α-(1-Methylethyl)-α-[3-[methyl(2-phenyl-ethyl)amino]propyl]benzeneacetonitrile (●)
S *Emopamil**, Sz 45
U Coronary vasodilator

11975 (9587)

$C_{23}H_{30}N_2$
101238-51-1

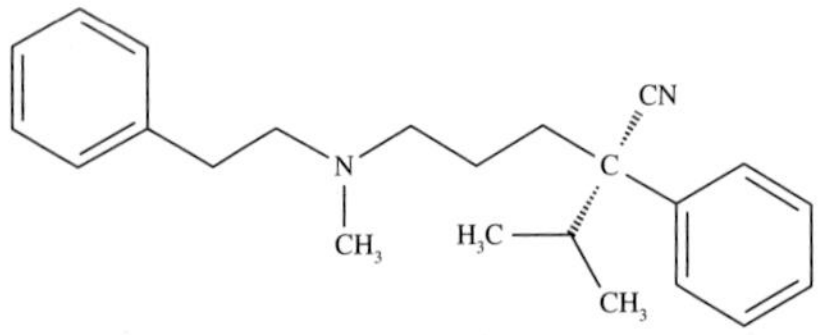

(−)-(S)-2-Isopropyl-5-(methylphenethylamino)-2-phenyl-valeronitrile = (S)-α-(1-Methylethyl)-α-[3-[methyl(2-phenylethyl)amino]propyl]benzeneacetonitrile (●)
S *Levemopamil***, (S)-*Emopamil*
U Calcium antagonist, antihypoxic

11976

$C_{23}H_{30}N_2O$

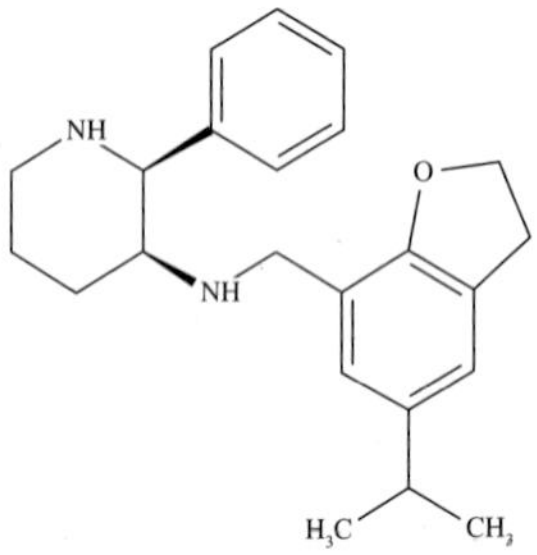

(2S,3S)-3-[[(5-Isopropyl-2,3-dihydrobenzofuran-7-yl)methyl]amino]-2-phenylpiperidine = (2S,3S)-N-[[2,3-Dihydro-5-(1-methylethyl)-7-benzofuranyl]methyl]-2-phenyl-3-piperidinamine (●)
R Dihydrochloride (183673-27-0)
S HSP-117
U Anti-emetic (tachykinin NK₁ receptor antagonist)

11977 (9589)

$C_{23}H_{30}N_2O$
42045-86-3

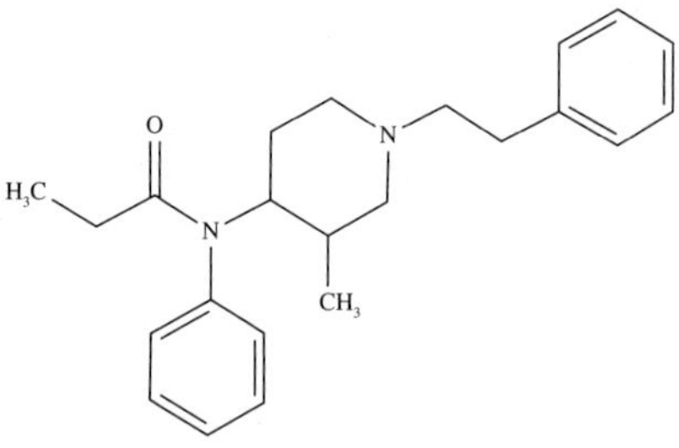

N-(3-Methyl-1-phenethyl-4-piperidyl)propionanilide = N-[3-Methyl-1-(2-phenylethyl)-4-piperidinyl]-N-phenyl-propanamide (●)
S *Mefentanyl*
U Narcotic analgesic

11978 (9590)

$C_{23}H_{30}N_2O_2$
13495-09-5

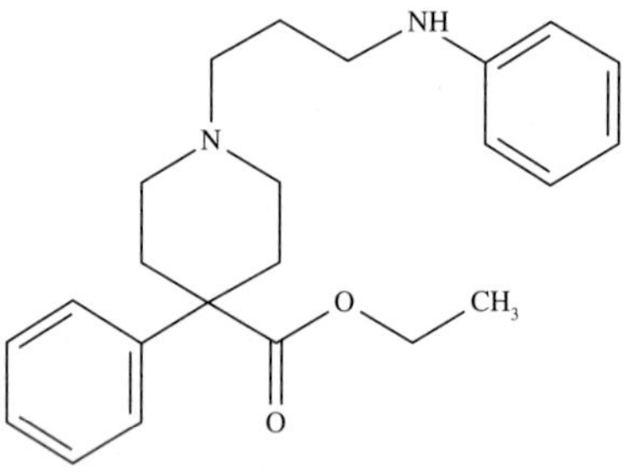

Ethyl 4-phenyl-1-(3-anilinopropyl)piperidine-4-carb-oxylate = 1-(3-Anilinopropyl)-4-phenylisonipecotic acid ethyl ester = 4-Phenyl-1-[3-(phenylamino)propyl]-4-pi-peridinecarboxylic acid ethyl ester (●)
R also ethanesulfonate (1:1) (7081-52-9)
S Alvodine, Anopridine, Cimadon, NIH 7590, Pima-din "Winthrop", *Piminodine***, Win 14098
U Narcotic analgesic

11979 (9591) $C_{23}H_{30}N_2O_2$
                                                             78995-14-9

N-[1-(β-Hydroxyphenethyl)-3-methyl-4-piperidyl]pro-
pionanilide = *N*-[1-(2-Hydroxy-2-phenylethyl)-3-methyl-
4-piperidinyl]-*N*-phenylpropanamide (●)
R also monohydrochloride (135159-44-3)
S *Ohmefentanyl*, OMF, RTI-4614-4
U Narcotic analgesic

11980 (9592) $C_{23}H_{30}N_2O_2S$
                                                             127304-28-3

5',6',7',8'-Tetrahydro-5',5',8',8'-tetramethyl-2'-acetonaph-
thone (*E*)-[*p*-(methylsulfonyl)phenyl]hydrazone = (*E*)-1-
(5,6,7,8-Tetrahydro-5,5,8,8-tetramethyl-2-naphthale-
nyl)ethanone [4-(methylsulfonyl)phenyl]hydrazone (●)
S BASF 52404, *Linarotene***, RWJ 24834
U Antikeratolytic

11981 (9593) $C_{23}H_{30}N_2O_3$

2-(Diethylamino)ethyl 2-[(dimethylamino)carbonyl]-2,2-
diphenylacetate = α-[(Dimethylamino)carbonyl]-α-phe-
nylbenzeneacetic acid 2-(diethylamino)ethyl ester (●)
S *Carbactyzin*

U Antidote

11982 (9594) $C_{23}H_{30}N_2O_3$
                                                             57694-27-6

2-Hydroxypropyl 14-deoxyvincaminate = 14-Deoxyvin-
caminic acid 2-hydroxypropyl ester = (3α,14α,16α)-
14,15-Dihydroeburnamenine-14-carboxylic acid 2-hydr-
oxypropyl ester (●)
S *Vinpoline***
U Cerebrotonic

11983 (9595) $C_{23}H_{30}N_2O_3$
                                                             49646-12-0

(±)-1-(*o*-Acetylphenoxy)-3-[4-(3,4-dimethylphenyl)-1-
piperazinyl]-2-propanol = (±)-1-[2-[3-[4-(3,4-Dimethyl-
phenyl)-1-piperazinyl]-2-hydroxypropoxy]phenyl]etha-
none (●)
S *Centxylazine*
U Local anesthetic

11984   (9596) $C_{23}H_{30}N_2O_3$
                                       136471-32-4

3a,4,9,9a-Tetrahydro-2,9-dimethylfuro[2,3-*b*]quinoxaline-3-carboxylic acid 1,7,7-trimethylbicyclo[2.2.1]hept-2-yl ester (●)
S   Encyclan
U   Antiviral

11985   (9597) $C_{23}H_{30}N_2O_3$
                                       113696-56-3

1-Cinnamyl-4-(3,4,5-trimethoxybenzyl)piperazine = 1-(3-Phenyl-2-propenyl)-4-[(2,3,4-trimethoxyphenyl)methyl]piperazine (●)
S   8205, Corocinnarine
U   Anti-arrhythmic (calcium antagonist)

11986   (9598) $C_{23}H_{30}N_2O_4$
                                       4098-40-2

(*E*)-16,17-Didehydro-9,17-dimethoxy-17,18-seco-20α-yohimban-16-carboxylic acid methyl ester = 9-Methoxycorynantheidine = (16*E*,20β)-16,17-Didehydro-9,17-dimethoxycorynan-16-carboxylic acid methyl ester (●)
S   *Mitragynine*, SKF 12711
U   Analgesic, antitussive (alkaloid from *Mitragyna speciosa*)

11987   (9599) $C_{23}H_{30}N_2O_4$
                                       509-67-1

4,5α-Epoxy-17-methyl-3-(2-morpholinoethoxy)morphin-7-en-6α-ol = 3-*O*-(2-Morpholinoethyl)morphine = Morphine 3-(2-morpholinoethyl) ether = (5α,6α)-7,8-Didehydro-4,5-epoxy-17-methyl-3-[2-(4-morpholinyl)ethoxy]morphinan-6-ol (●)
R   also monohydrate (6254-99-5)
S   Actuss, Adaphol, Adaphol Linetus, Benylin Childrens Dry Coughs, Caltoson, Codisol, Codotussyltoux sèche, Codylin, Contrapect infant, Dia-Tuss, Duro-Tuss, Ethnin(e), Evaphol, Expulin Dry Cough, Folcodan, *Folcodina*, Folco-Retard, Folcovin, *Folkodin*, Galenphol, Galphol, Glycodine, Hibernyl, Homocodeine, Infantussin, Lantuss, Linctus Tussinol, M.E.M., Memine, Neocodin, Ordov Dry Tickly Cough, Paediatric, Pavacol D, Pectolin, Phocil, *Pholcodine***, Pholcolin, Pholcolinet, Pholcomed, Pholcotussin, Pholdine, Pholevan, Pholtrate, Pholtussin, Prentosse, Prodromine, Respilene, Sedlingtus, Sednaco, Sednine, Sirop des Vosges, Tixylix Daytime "Intercare", Triopaed, Trophires (jarabe + supos.), Tussinol "G.P.", Tussokon, Tuxi forte, Weifacodin
U   Antitussive, narcotic analgesic

11988 (9600) $C_{23}H_{30}N_2O_5$
54063-41-1

3,4,5-Trimethoxy-*N*-[1-(phenoxymethyl)-2-(1-pyrrolidi-nyl)ethyl]benzamide (●)
S CERM 1875, *Fepromide***
U Anti-arrhythmic

11989 (9601) $C_{23}H_{30}N_2O_6$
113594-64-2

p-Methoxyphenyl 4-(3,4,5-trimethoxybenzyl)-1-pipera-zineacetate = 1-(3,4,5-Trimethoxybenzyl)-4-[[(4-meth-oxyphenoxy)carbonyl]methyl]piperazine = 4-[(3,4,5-Tri-methoxyphenyl)methyl]-1-piperazineacetic acid 4-meth-oxyphenyl ester (●)
R Fumarate (1:1) (129200-10-8)
S KB 5492
U Anti-ulcer agent, cytoprotective

11990 (9602) $C_{23}H_{30}N_4O_2S$
150915-41-6

cis-*N*-[[4-(1,2-Benzisothiazol-3-yl)-1-piperazinyl]butyl]-1,2-cyclohexanedicarboximide = *cis*-2-[4-[4-(1,2-Benzi-sothiazol-3-yl)-1-piperazinyl]butyl]hexahydro-1*H*-iso-indole-1,3(2*H*)-dione (●)
R Monohydrochloride (129273-38-7)
S Lullan, *Peraspirone hydrochloride, Perospirone hy-drochloride***, SM 9018
U Antipsychotic

11991 (9603) $C_{23}H_{30}N_4O_4S$
139133-26-9

Ethyl *N*-methyl-*N*-[α-(2-methylimidazo[4,5-*c*]pyridin-1-yl)tosyl]-L-leucinate = *N*-Methyl-*N*-[[α-(2-methyl-1*H*-imidazo[4,5-*c*]pyridin-1-yl)-*p*-tolyl]sulfonyl]-L-leucine ethyl ester = *N*-Methyl-*N*-[[4-[(2-methyl-1*H*-imida-zo[4,5-*c*]pyridin-1-yl)methyl]phenyl]sulfonyl]-L-leucine ethyl ester (●)
S BB 882, DO 6, *Lexipafant***, Zacutex
U PAF antagonist (in pancreatitis)

11992 (9604) $C_{23}H_{30}N_6O_4$
65184-10-3

7-[3-[[2-Hydroxy-3-[(2-methyl-4-indolyl)oxy]pro-pyl]amino]butyl]theophylline = 3,7-Dihydro-7-[3-[[2-hydroxy-3-[(2-methyl-1*H*-indol-4-yl)oxy]propyl]ami-no]butyl]-1,3-dimethyl-1*H*-purine-2,6-dione (●)
S D-13312, *Teoprolol***
U β-Adrenergic blocker

11993 $C_{23}H_{30}N_7O_{12}P$
                                         158275-32-2

4-Amino-1-[5-O-[[[2,7-diamino-9-[[(aminocarbo-
nyl)oxy]methyl]-2,3-dihydro-5,8-dihydroxy-6-methyl-
1H-pyrrolo[1,2-a]indol-1-yl]oxy]hydroxyphosphinyl]-β-
D-arabinofuranosyl]-2(1H)-pyrimidinone (●)
S   *Cytaramycin*
U   Antileukemic

11994 $C_{23}H_{30}O_2$
                                         79073-31-7

(2E,4E,6E)-3-Methyl-7-(5,6,7,8-tetrahydro-5,5,8,8-tetra-
methyl-2-naphthalenyl)-2,4,6-octatrienoic acid (●)
S   Ro 13-6307
U   Antileukemic

11995   (9605) $C_{23}H_{30}O_3$
                                         40574-52-5

6,7-Dihydro-17β-hydroxy-3-oxo-(6α,7α)-3'H-cyclopro-
pa[6,7]-17α-pregna-4,6-diene-21-carboxylic acid γ-lactone = (6α,7α,17α)-6,7-Dihydro-17-hydroxy-3-oxo-3'H-
cyclopropa[6,7]pregna-4,6-diene-21-carboxylic acid γ-
lactone (●)

S   *Prorenone*, SC 23233
U   Diuretic (antimineralocorticoid)

11996   (9606) $C_{23}H_{30}O_3$
                                         54350-48-0

Ethyl (*all-E*)-9-(4-methoxy-2,3,6-trimethylphenyl)-3,7-
dimethyl-2,4,6,8-nonatetraenoate = Ethyl 3-methoxy-15-
apo-ψ-caroten-15-oate = (*all-E*)-9-(4-Methoxy-2,3,6-tri-
methylphenyl)-3,7-dimethyl-2,4,6,8-nonatetraenoic acid
ethyl ester (●)
S   *Etretinate***, Ro 10-9359, Tegison, Tigasan, Tigason
U   Antipsoriatic, antineoplastic

11997   (9607) $C_{23}H_{30}O_3$
                                         5633-18-1

6-Methyl-16-methylene-17-hydroxy-Δ^6-progesterone =
17-Hydroxy-6-methyl-16-methylenepregna-4,6-diene-
3,20-dione (●)
R   see also no. 12892
S   *Melengestrol***
U   Progestin, antineoplastic

11998   (9609) $C_{23}H_{30}O_4$
                                         58652-20-3

17-(Acetyloxy)-6-methyl-19-norpregna-4,6-diene-3,20-dione (●)
S Lutenil, Lutenyl, NOM-Ac, *Nomegestrol acetate**,* TX-066, Uniplant
U Progestin

11999 (9608)

$C_{23}H_{30}O_4$
7759-35-5

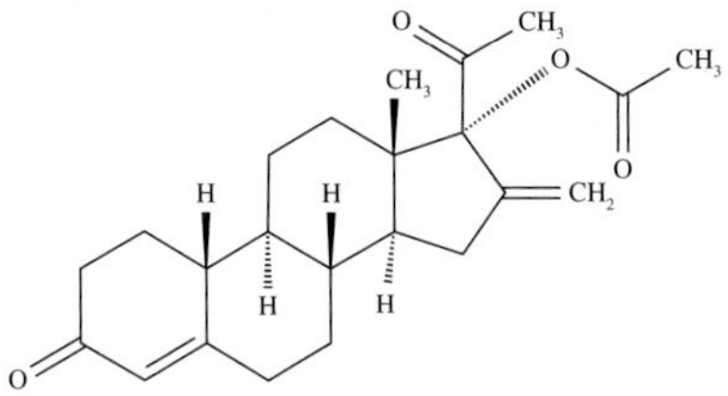

17-(Acetyloxy)-16-methylene-19-norpregn-4-ene-3,20-dione (●)
S *Elcometrine,* Nestorone, ST-1435
U Progestin (contraceptive)

12000 (9610)

$C_{23}H_{30}O_4S$
28913-23-7

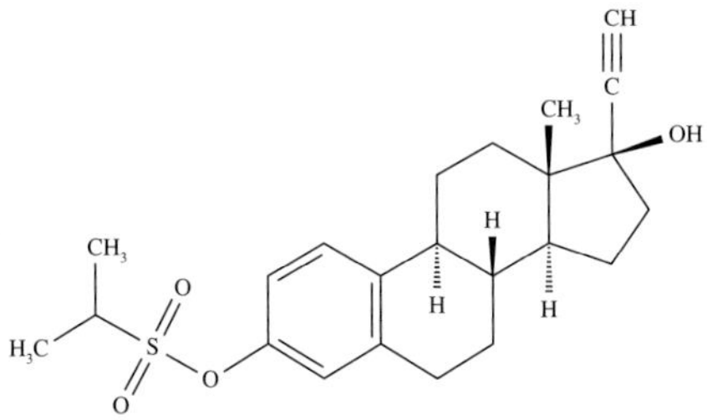

17-Ethynylestra-1,3,5(10)-triene-3,17β-diol 3-isopropyl-sulfonate = 17α-Ethynyl-3-(isopropylsulfonyloxy)estra-diol = (17α)-19-Norpregna-1,3,5(10)-trien-20-yne-3,17-diol 3-(2-propanesulfonate) (●)
S *Äthinylöstradiol-3-isopropylsulfonat,* Deposi-ston-Östrogen,
Ethinylestradiol 3-isopropylsulfonate, Ethinylestra-diolum propansulfonicum, J 96, Turisteron
U Depot-estrogen

12001

$C_{23}H_{30}O_5$
160219-86-3

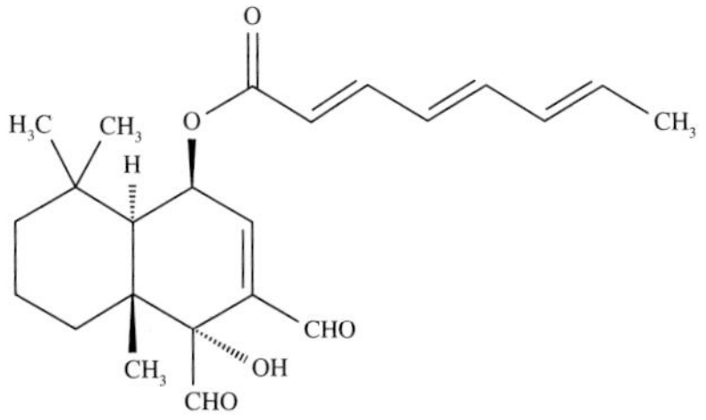

[1S-[1α(2E,4E,6E),4β,4aα,8aβ]]-2,4,6-Octatrienoic acid 3,4-diformyl-1,4,4a,5,6,7,8,8a-octahydro-4-hydroxy-4a,8,8-trimethyl-1-naphthalenyl ester (●)
S RES 1149-1
U Antibacterial (gram-positive), endothelin antagonist

12002

$C_{23}H_{30}O_5$
7753-60-8

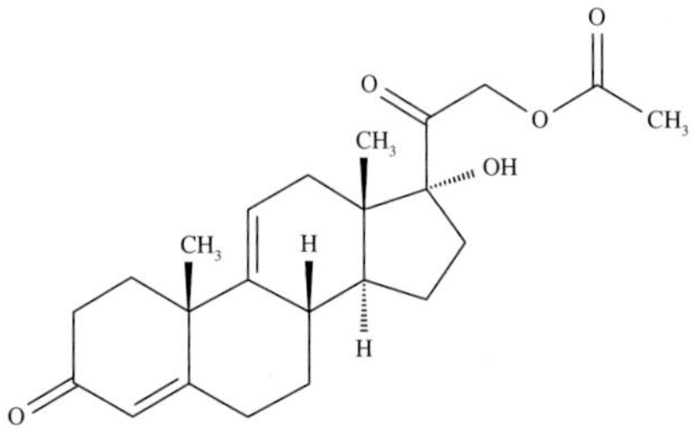

17,21-Dihydroxypregna-4,9(11)-diene-3,20-dione 21-acetate = 21-(Acetyloxy)-17-hydroxypregna-4,9(11)-di-ene-3,20-dione (●)
S AL-3789, *Anecortave**, Anecortave acetate*
U Angiostatic steroid

12003 (9611)

$C_{23}H_{30}O_6$
79378-27-1

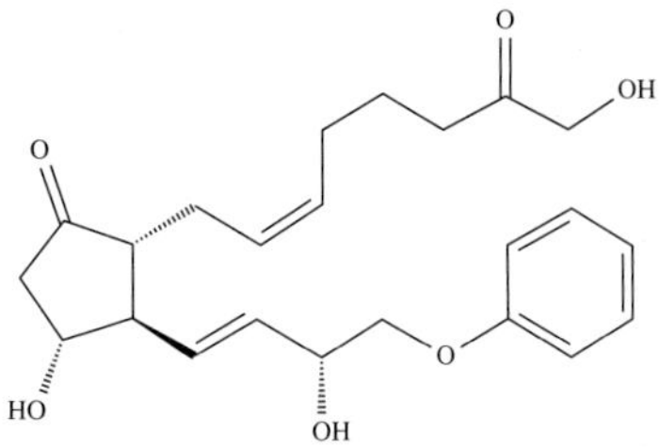

[2α(Z),3β(1E,3R*),4α]-(±)-4-Hydroxy-2-(8-hydroxy-7-oxo-2-octenyl)-3-(3-hydroxy-4-phenoxy-1-butenyl)cy-clopentanone (●)
S CL 116069

U   Nasal decongestant

12004   (9612)

$C_{23}H_{30}O_6$
69363-14-0

Stereoisomer of 5,6,7,8-tetrahydro-2,3,10,11,12-penta-methoxy-6,7-dimethyldibenzo[a,c]cycloocten-1-ol (●)
S   *Schisanhenol*
U   Hepatoprotectant from the kernels of *Schisandia rubiflora*

12005   (9613)

$C_{23}H_{30}O_6$
69381-94-8

Methyl (±)-7-[(1R*,2R*,3R*,5S*)-3,5-dihydroxy-2-[(E)-(3R*)-3-hydroxy-4-phenoxy-1-butenyl]cyclopentyl]-4,5-heptadienoate = Methyl (13E)-(9S,11R,15R)-16-phen-oxy-9,11,15-trihydroxy-ω-tetranorprosta-4,5,13-trienoate = [1α,2β(1E,3R*),3α,5α]-(±)-7-[3,5-Dihydroxy-2-(3-hydroxy-4-phenoxy-1-butenyl)cyclopentyl]-4,5-heptadienoic acid methyl ester (●)
S   Bovilene, *Fenprostalene***, Porcilene, RS-84043, Synchrocept B
U   Luteolytic, abortifacient (veterinary)

12006   (9615)

$C_{23}H_{30}O_6$
52-21-1

11β,17,21-Trihydroxypregna-1,4-diene-3,20-dione 21-acetate = (11β)-21-(Acetyloxy)-11,17-dihydroxypregna-1,4-diene-3,20-dione (●)
S   Ak-Tate, Alto-Pred LA, Articulose-50, Balpred, Benisolon-Injection, Biocortin, Bio-Pred, Cormalone, Cortipred, Dacortin-Kristallsuspension, Davisolona, Decaprednil H inj., Decortin-H-Kristallsuspension, Deltacortenolo i.m., Delta-Cortilen, Deltacortril-Suspension, Delta-Efcorlin, Deltasol "Lyka", Deltastab-Injection, Deltilen, Depo-Pred 100, depPredalone, Dermo-Nydol, Di-Adreson-F inj., Diopred, Di-Pres, Dontisolon D-Mundheilpaste, Durapred, Duraprednisolon-Kristallsusp., Econopred, Esidolene, Frisolona, Hexacorton "Spirig", Hidrodecortancyl, Hostacortin-H-Kristallsuspension, Hydrocortancyl Suspension injectable, Ibisterolon, Inflanefran, Inflastat, I-Prednicet, Key-Pred, Lepicortinolo i.m., Meticortelone aqueous, Neocortil, Nisolone i.m., Novosterol, Ocu-Pred-A, Oftalmol, Ophtho-Tate, Optalan, Precortalon-inj., PreCortisyl-injectable, Prectal, Pred-A, Predair-A, Predaject, Predalone 50, Predcor, Predcortol, Pred-Forte, Prediacortine, Predicort, Pred-Mild, Prednefrin SF, Prednelan-N, Prednidoren, Prednifor-Tropfen, Prednigalen, PredniHexal, Predni-H-injekt, Predni-Lichtenstein, Predniocil, Predni-Ophtal, Predni-POS, *Prednisolone acetate***, Predonine-Inj., Predsol "Acromax", Predsolets, Premandol, Prenema, Pricortin, PSA "SS", Savacort, Scherisolon-Kristallsuspension, Sigpred, Solona Plus, Sophipren, Spolotan-Kristallsuspension, Steraject, Sterane intramuscular, Sterofrin, Tarocortelone Injections, Ulacort-Inj., Ultracortenol-Augensalbe (-Tropfen), Ultrapred, Wedecort
U   Glucocorticoid (anti-inflammatory, anti-allergic)

12007   (9614) $C_{23}H_{30}O_6$
                                                              50-04-4

17,21-Dihydroxypregn-4-ene-3,11,20-trione 21-acetate =
21-(Acetyloxy)-17-hydroxypregn-4-ene-3,11,20-trione
(●)
R   see also no. 10653
S   Acetisone, Adreson, Adricort, Altesona, Artriona,
       Bencorten, Berlison (old form), Colirio Collado Cor-
       tioftal, Corace, Corlin, Cortadren, Cortal "Organon",
       Cortasson, Cortate "Protea", Cortelan, Corticil, Cor-
       ticlin, Cortic (old form), Cortifor, Cortilen, Corti-
       pon, Cortisat(e), Cortisol "Vis", Cortison "Ciba; Dr.
       Winzer; Jenapharm; Spofa", Cortisone "Lepetit;
       Roussell", *Cortisone acetate***, Cortistab, Cortistal,
       Cortisyl, Cortodrin, Cortogen Acetate, Cortone (Ace-
       tate), Cortosal, Enticort, Incortin, Novocort "Novo-
       pharm", Oftasone, Oftavisan, Ophthalmo-Cortisone,
       Orgasona, Pan-Cort, Pantisone, Ricortex, Scheral-
       son, Scheroson
U   Glucocorticoid (anti-inflammatory, anti-allergic)

12008   (9616) $C_{23}H_{30}O_7$
                                                              109637-83-4

4-[[(10-Dihydroartemisin-10-yl)oxy]methyl]benzoic acid
= 4-[[(Decahydro-3,6,9-trimethyl-3,12-epoxy-12*H*-pyra-
no[4,3-*j*]-1,2-benzodioxepin-10-yl)oxy]methyl]benzoic
acid (●)
R   also with [3*R*-(3α,5aβ,6β,8aβ,9α,10α,12β,12a*R**)]-
       definition (120020-26-0) (see structural formula)
S   *Artelinic acid*, ARTZ, WR 155663

U   Antimalarial, antiviral

12009   (9617) $C_{23}H_{30}O_7$
                                                              111802-47-2

Methyl 11β,17,21-trihydroxy-3,20-dioxopregna-1,4-di-
ene-16α-carboxylate = (11β,16α)-11,17,21-Trihydroxy-
3,20-dioxopregna-1,4-diene-16-carboxylic acid methyl
ester (●)
S   P 16 CM
U   Glucocorticoid (anti-inflammatory, anti-allergic)

12010 $C_{23}H_{31}CaN_3O_{11}$

Trihydrogen [*N*-[(2*S*)-2-[bis(carboxymethyl)amino]-3-
(4-ethoxyphenyl)propyl]-*N*-[2-[bis(carboxymethyl)ami-
no]ethyl]glycinato(5–)]calciate(3–) (●)
R   see also no. 12101
S   *Acidum caloxeticum***, *Caloxetic acid***
U   Pharmaceutical aid

12011   (9618) $C_{23}H_{31}ClN_2O_3$
                                                              17692-34-1

2-[2-[2-[4-(*p*-Chloro-α-phenylbenzyl)-1-piperazinyl]eth-oxy]ethoxy]ethanol = 8-[4-(*p*-Chlorobenzhydryl)-1-pipe-razinyl]-3,6-dioxaoctan-1-ol = 1-(*p*-Chlorobenzhydryl)-4-[2-[2-(2-hydroxyethoxy)ethoxy]ethyl]piperazine = 2-[2-[2-[4-[(4-Chlorophenyl)phenylmethyl]-1-piperazi-nyl]ethoxy]ethoxy]ethanol (●)
S *Äthoxyhydroxyzin, Etodroxizine**, Hydrochlorben-zäthylamin, Hydrochlorbenzethylamine*
U Tranquilizer, hypnotic

12011-01 (9618-01)   53859-10-2
R Maleate (1:2)
S Drimyl, Indunox "UCB", Noxadron, U.C.B. 1414
U Tranquilizer, hypnotic

12012 (9619) $C_{23}H_{31}ClN_4O$
                                    3734-75-6

7-Amino-6-chloro-9-[[4-(diethylamino)-1-methylbu-tyl]amino]-2-methoxyacridine = 3-Chloro-N^9-[4-(di-ethylamino)-methylbutyl]-7-methoxy-2,9-acridinedi-amine (●)
R also dihydrochloride dihydrate
S *Aminoacrichine, Aminoacriquine*
U Antibacterial

12013 (9620) $C_{23}H_{31}ClO_4$
                                    2477-73-8

6α-Chloro-17-hydroxypregn-4-ene-3,20-dione acetate = 6α-Chloro-17-hydroxyprogesterone acetate = (6α)-17-(Acetyloxy)-6-chloropregn-4-ene-3,20-dione (●)
S CAP, *Hydromadinone acetate***

U Progestin

12014 (9621) $C_{23}H_{31}Cl_2NO_3$
                                    2998-57-4

Estra-1,3,5(10)-triene-3,17β-diol 3-[bis(2-chloro-ethyl)carbamate] = Estradiol 3-[bis(2-chloroethyl)carb-amate] = (17β)-Estra-1,3,5(10)-triene-3,17-diol 3-[bis(2-chloroethyl)carbamate] (●)
R see also nos. 12043, 13256
S *Estramustine***, Leo 275, Ro 21-8837
U Antineoplastic

12015 (9622) $C_{23}H_{31}Cl_3O_4$
                                    13867-82-8

17β-(2,2,2-Trichloro-1-hydroxyethoxy)androst-4-en-3-one acetate = Testosterone 17-chloral hemiacetal acetate = (17β)-17-[1-(Acetyloxy)-2,2,2-trichloroethoxy]androst-4-en-3-one (●)
S Caprosem, *Cloxotestosterone acetate***, *Testoster-one chloral hemiacetal acetate*
U Androgen

12016 (9623) $C_{23}H_{31}FO_4$
25092-07-3

9-Fluoro-11β,21-dihydroxy-16α,17-dimethylpregna-1,4-diene-3,20-dione = (11β,16α)-9-Fluoro-11,21-dihydroxy-16,17-dimethylpregna-1,4-diene-3,20-dione (●)
S *Dimesone**
U Glucocorticoid (anti-inflammatory, anti-allergic)

12017 (9624) $C_{23}H_{31}FO_5$
2529-45-5

9-Fluoro-11β,17-dihydroxypregn-4-ene-3,20-dione 17-acetate = (11β)-17-(Acetyloxy)-9-fluoro-11-hydroxy-pregn-4-ene-3,20-dione (●)
S Chronogest, Cronolone, FGA, *Flugestone acetate**,
Flurogestone acetate, NSC-65411, SC-9880
U Progestin

12018 (9625) $C_{23}H_{31}FO_6$
514-36-3

9-Fluoro-11β,17,21-trihydroxypregn-4-ene-3,20-dione 21-acetate = 9-Fluorohydrocortisone 21-acetate = (11β)-21-(Acetyloxy)-9-fluoro-11,17-dihydroxypregn-4-ene-3,20-dione (●)

R see also no. 10664
S Alfa-Fluorone (old form), Alfanonidrone, Alflorone Acetate, Cortineff, F-Cortef, Florinef (acetate), Fludrocortisat, Fludrocortison "Heyden", *Fludrocortisone acetate**, Fludrocortone(acetate), Fludronil, *Fluorhydrocortisonum aceticum*, Scheroflu-ron
U Corticoid (mainly mineralocorticoid)

12019 (9626) $C_{23}H_{31}NO$
77287-89-9

17-(Cyclobutylmethyl)-8β-methyl-6-methylenemor-phinan-3-ol = (8β)-17-(Cyclobutylmethyl)-8-methyl-6-methylenemorphinan-3-ol (●)
R Methanesulfonate (77287-90-2)
S 4 Bb, TR-5379 M, *Xorphanol mesilate**
U Narcotic analgesic

12020 (9631) $C_{23}H_{31}NO_2$
302-33-0

2-(Diethylamino)ethyl 2,2-diphenylvalerate = 2,2-Diphe-nylvaleric acid 2-(diethylamino)ethyl ester = α,α-Di-phenyl-α-propylacetic acid β'-diethylaminoethyl ester = α-Phenyl-α-propylbenzeneacetic acid 2-(diethylami-no)ethyl ester (●)
R Hydrochloride (62-68-0)
S NSC-39690, *Proadifen hydrochloride**, *Propyla-diphenine*, 5171 R.P., SKF-525-A, U-5446
U Drug potentiator (inhibitor of drug metabolism)

12021   (9627) $C_{23}H_{31}NO_2$
509-74-0

6-(Dimethylamino)-4,4-diphenyl-3-heptanol acetate (ester) = 4-(Dimethylamino)-1-ethyl-2,2-diphenylpentyl acetate = 3-Acetoxy-6-(dimethylamino)-4,4-diphenyl-heptane = β-[2-(Dimethylamino)propyl]-α-ethyl-β-phenylbenzeneethanol acetate (ester) (●)
S   *Acemethadone, Acetyldimepheptanol, Acetylmethadol**, Amidolacetat, Dimepheptanolacetat, Methadyl acetate*, NIH 2953, *Race-Acetylmethadol*
U   Narcotic analgesic

12022   (9628) $C_{23}H_{31}NO_2$
34433-66-4

(–)-6-(Dimethylamino)-4,4-diphenyl-3-heptanol acetate (ester) = (–)-4-(Dimethylamino)-1-ethyl-2,2-diphenylpentyl acetate = (–)-3-Acetoxy-6-(dimethylamino)-4,4-phenylheptane = (–)-β-[2-(Dimethylamino)propyl]-α-ethyl-β-phenylbenzeneethanol acetate (ester) (●)
R   also [S-(R^*,R^*)]-compd. (1477-40-3) (see structural formula) and [S-(R^*,R^*)]-hydrochloride (43033-72-3)
S   LAAM, *Levacetylmethadol**, Levomethadyl acetate*, MK-790, Orlaam
U   Narcotic analgesic

12023   (9629) $C_{23}H_{31}NO_2$
17199-58-5

(3R,6R)-6-(Dimethylamino)-4,4-diphenyl-3-heptanol acetate (ester) = α-4-(Dimethylamino)-1-ethyl-2,2-diphenylpentyl acetate = α-3-Acetoxy-6-(dimethylamino)-4,4-diphenylheptane = [R-(R^*,R^*)]-β-[2-(Dimethylamino)propyl]-α-ethyl-β-phenylbenzeneethanol acetate (ester) (●)
S   *Alfacetilmetadolo, Alphacemethadone, Alphacetylmethadol***, DAAM
U   Narcotic analgesic

12024   (9630) $C_{23}H_{31}NO_2$
17199-59-6

(3S,6R)-6-(Dimethylamino)-4,4-diphenyl-3-heptanol acetate (ester) = β-4-(Dimethylamino)-1-ethyl-2,2-diphenylpentyl acetate = β-3-Acetoxy-6-(dimethylamino)-4,4-diphenylheptane = [S-(R^*,S^*)]-β-[2-(Dimethylamino)propyl]-α-ethyl-β-phenylbenzeneethanol acetate (ester) (●)
S   *Betacemethadone, Betacetilmetadolo, Betacetylmethadol***
U   Narcotic analgesic

12025   (9632) $C_{23}H_{31}NO_2$
56281-36-8

(*all-E*)-*N*-Ethyl-9-(4-methoxy-2,3,6-trimethylphenyl)-
3,7-dimethyl-2,4,6,8-nonatetraenamide (●)
S   Moncler "Sauter", *Motretinide***, Ro 11-1430, Tas-
maderm
U   Keratolytic, anti-acne, antineoplastic (in epithelial tu-
mors)

12026 $C_{23}H_{31}NO_2$
204336-80-1

(4*R*,6*S*)-*rel*-1-Hydroxy-4-(1-naphthalenyl)-6-octyl-2-pi-
peridinone (●)
S   BMD 188
U   Antineoplastic

12027   (9633) $C_{23}H_{31}NO_3$
3626-03-7

α-Ethoxydiphenylacetic acid 3-(diethylamino)propyl ester = 3-(Diethylamino)propyl *O*-ethylbenzilate = α-Ethoxy-α-phenylbenzeneacetic acid 3-(diethylamino)propyl
ester (●)
R   also hydrochloride (2740-17-2)
S   Aethpenalum, Ethpenal, Etpenal
U   Anticholinergic

12028   (9634) $C_{23}H_{31}NO_3$
35189-28-7

(+)-13-Ethyl-17-hydroxy-18,19-dinor-17α-pregn-4-en-
20-yn-3-one oxime acetate (ester) = (+)-17-(Acetyloxy)-
13-ethyl-17α-ethynylgon-4-en-3-one oxime = 13β-Ethyl-
3-hydroxyimino-18,19-dinor-17-pregn-4-en-20-yn-17β-
yl acetate = (17α)-17-(Acetyloxy)-13-ethyl-18,19-dinor-
pregn-4-en-20-yn-3-one oxime (●)
S   D-138, *Dexnorgestrel acetime, Norgestimate***,
ORF 10131, Ortrel, RWJ 10131
U   Progestin

12029 $C_{23}H_{31}NO_3$
94560-98-2

α-[[2-(1-Hydroxy-3-phenylpropyl)phenoxy]methyl]-1-
piperidineethanol (●)
R   also hydrochloride (179319-76-7)
S   GP-88
U   Multidrug resistance modulator

12030   (9635) $C_{23}H_{31}NO_3$
81447-80-5

(±)-2'-[2-Hydroxy-3-(*tert*-pentylamino)propoxy]-3-phe-
nylpropiophenone = (±)-2'-[3-[(1,1-Dimethylpropyl)ami-
no]-2-hydroxypropoxy]-3-phenylpropiophenone = (±)-1-
[2-[3-[(1,1-Dimethylpropyl)amino]-2-hydroxyprop-
oxy]phenyl]-3-phenyl-1-propanone (●)
S   *Diprafenone***
U   Anti-arrhythmic

12031   C$_{23}$H$_{31}$NO$_4$S
                                              178961-24-5

trans-3-Butyl-3-ethyl-2,3,4,5-tetrahydro-7,8-dimethoxy-
5-phenyl-1,4-benzothiazepine 1,1-dioxide (●)
S   GW 264, 264 W 94
U   Hypocholesterolemic (IBAT inhibitor)

12032   (9636)   C$_{23}$H$_{31}$NO$_7$
                                              128794-94-5

2-Morpholinoethyl (*E*)-6-(4-hydroxy-6-methoxy-7-me-
thyl-3-oxo-5-phthalanyl)-4-methyl-4-hexenoate = (*E*)-6-
(1,3-Dihydro-4-hydroxy-6-methoxy-7-methyl-3-oxo-5-
isobenzofuranyl)-4-methyl-4-hexenoic acid 2-(4-morpho-
linyl)ethyl ester (●)
R   also hydrochloride (116680-01-4)
S   Cellcept, ME-MPA, MMF, Munoloc, *Mycophenola-
        te mofetil*, RS-61443
U   Anti-arthritic, immunosuppressive

12033   (9637)   C$_{23}$H$_{31}$NO$_7$S
                                              60325-46-4

(*Z*)-7-[(1*R*,2*R*,3*R*)-3-Hydroxy-2-[(*E*)-(3*R*)-(3-hydroxy-4-
phenoxy-1-butenyl)]-5-oxocyclopentyl]-*N*-(methylsulfonyl)-5-heptenamide = (5*Z*,13*E*)-(8*R*,11*R*,12*R*,15*R*)-
11,15-Dihydroxy-*N*-(methylsulfonyl)-9-oxo-16-phen-
oxy-17,18,19,20-tetranor-5,13-prostadienoic acid amide
= [1*R*-[1α(*Z*),2β(1*E*,3*R**),3α]]-7-[3-Hydroxy-2-(3-hydr-
oxy-4-phenoxy-1-butenyl)-5-oxocyclopentyl]-*N*-(methylsulfonyl)-5-heptenamide (●)
S   CP-34089, Nalador, SHB 286, *Sulprostone***,
        ZK 57671
U   Prostaglandin (abortifacient)

12034   (9638)   C$_{23}$H$_{31}$N$_2$O
                                              23724-95-0

1-(3-Carbamoyl-3,3-diphenylpropyl)-1-methylperhy-
droazepinium = 1-(4-Amino-4-oxo-3,3-diphenylbutyl)he-
xahydro-1-methyl-1*H*-azepinium (●)
R   Iodide (15351-05-0)
S   *Buzepide metiodide***, *Diphexamide iodomethylate*,
        F.I. 6146, *Métazépium iodide*, R 661, Spactin
U   Anticholinergic

12035 $C_{23}H_{31}N_3O$
142873-40-3

2,3-Dihydro-2,4,6,7-tetramethyl-2-[(4-phenyl-1-pipera-
zinyl)methyl]-5-benzofuranamine (●)
S   YT-18
U   5-Lipogenase inhibitor, anti-atherosclerotic

12036   (9639) $C_{23}H_{31}N_3O_2$
67449-00-7

cis-5,6-Dimethoxy-2-methyl-3-[2-(4-phenyl-1-piperazi-
nyl)ethyl]indoline = cis-2,3-Dihydro-5,6-dimethoxy-2-
methyl-3-[2-(4-phenyl-1-piperazinyl)ethyl]-1H-indole
(●)
S   CL 77328, DHO
U   Antipsychotic

12037 $C_{23}H_{31}N_3O_3$
142223-40-3

5-[[[1-Cyclohexyl-2-(1H-imidazol-1-yl)-3-phenylpropy-
lidene]amino]oxy]pentanoic acid (●)
S   FCE 27262

U   Antithrombotic

12038   (9640) $C_{23}H_{31}N_3O_4S_2$
130370-60-4

(2S,3R)-N^1-Hydroxy-3-isobutyl-N^4-[(S)-α-(methylcarbamoyl)phenethyl]-2-[(2-thienylthio)methyl]succinamide
= (2S,3R)-5-Methyl-3-[[(αS)-α-(methylcarbamoyl)phen-
ethyl]carbamoyl]-2-[(2-thienylthio)methyl]hexanohydro-
xamic acid = [2R-[2(S*),2R*,3S*]]-N^4-Hydroxy-N^1-[2-
(methylamino)-2-oxo-1-(phenylmethyl)ethyl]-2-(2-me-
thylpropyl)-3-[(2-thienylthio)methyl]butanediamide (●)
S   Barinatrix, Batimastat**, BB-94, ISV-120
U   Antineoplastic (matrix metalloproteinase inhibitor)

12039   (9641) $C_{23}H_{31}N_3O_5$
117782-84-0

(E)-7-[[4,6-Dideoxy-3-C-methyl-4-(methylamino)-α-L-
mannopyranosyl]oxy]-1,2,3,11a-tetrahydro-2-propyli-
dene-5H-pyrrolo[2,1-c][1,4]benzodiazepin-5-one (●)
S   Sibanomicin
U   Antileukemic from Micromonospora sp. SF 2364

12040   (9642) $C_{23}H_{31}N_5O_4S$
52430-65-6

1-Methyl-3-[p-[[3-(4-methylcyclohexyl)ureido]sulfo-
nyl]phenethyl]-1-(2-pyridyl)urea = N-(4-Methylcyclohe-
xyl)-N'-[4-[2-(N'-methyl-N'-2-pyridylureido)ethyl]phe-

nylsulfonyl]urea = N-[[(4-Methylcyclohexyl)amino]car-
bonyl]-4-[2-[[(methyl-2-pyridinylamino)carbonyl]ami-
no]ethyl]benzenesulfonamide (●)
S *Glisamuride***, HB 180
U Oral hypoglycemic

12041

$C_{23}H_{31}N_5O_{12}$
102770-00-3

1-Deoxy-5-O-[4-deoxy-N-[(3S)-hexahydro-2-oxo-1H-
azepin-3-yl]-β-L-*erythro*-hex-4-enopyranuronamidosyl]-
1-(3,4-dihydro-2,4-dioxo-1(2H)-pyrimidinyl)-3-O-me-
thyl-β-D-allofuranuronamide (●)
S 446-53-1, *Capuramycin*
U Antibacterial from *Streptomyces griseus* strain 446-53

12042

$C_{23}H_{32}ClN_2O_5$
161729-47-1

2-[[4-(2-Chlorophenyl)-3-(ethoxycarbonyl)-1,4-dihydro-
5-(methoxycarbonyl)-6-methyl-2-pyridinyl]methoxy]-
N,N,N-trimethylethanaminium (●)
R Chloride
S UK-118434-05
U Antihypertensive

12043   (9643)

$C_{23}H_{32}Cl_2NO_6P$
4891-15-0

Estra-1,3,5(10)-triene-3,17β-diol 3-[bis(2-chloro-
ethyl)carbamate] 17-(dihydrogen phosphate) = Estradiol
3-[bis(2-chloroethyl)carbamate] 17-(dihydrogen phos-
phate) = (17β)-Estra-1,3,5(10)-triene-3,17-diol 3-[bis(2-
chloroethyl)carbamate] 17-(dihydrogen phosphate) (●)
R also disodium salt (52205-73-9) or meglumine salt
S Amsupros, Biasetyl, Cellmustin, Emcyt "Kabi
Pharmacia; Pharmacia-Upjohn; Roche", Estracyt,
*Estramustine phosphate***, Extramustin, Leo 299,
Multosin, NSC-89199, Proesta, Prostamustin,
Ro 21-8837/001
U Antineoplastic, radioprotector

12044   (9644)

$C_{23}H_{32}Cl_2N_2O_4S$
100667-31-0

3,4-Dichloro-N-[3-[[2-(3,4-dimethoxyphenyl)ethyl]me-
thylamino]propyl]-N-(1-methylethyl)benzenesulfon-
amide (●)
S Wy-46622
U Coronary vasodilator (calcium antagonist)

12045   (9645)

$C_{23}H_{32}N_2O$
50794-02-0

p-(Dipropylamino)-N-(α-methylphenethyl)phenylacet-
amide = 4-(Dipropylamino)-N-(1-methyl-2-phenyl-
ethyl)benzeneacetamide (●)
S IEM-611, Propylphepracet
U β-Adrenolytic

12046 (9646) $C_{23}H_{32}N_2O$
53076-26-9

N,N-Diethyl-*N'*-(1-methoxy-2-indanyl)-*N'*-phenyl-1,3-propanediamine = *N*-[3-(Diethylamino)propyl]-*N*-phenyl-1-methoxy-2-indanamine = *N*-(2,3-Dihydro-1-methoxy-1*H*-inden-2-yl)-*N',N'*-diethyl-*N*-phenyl-1,3-propanediamine (●)
S *Methoxyaprindine, Moxaprindine**
U Anti-arrhythmic

12047 (9647) $C_{23}H_{32}N_2O_2$
13071-27-7

1,5-Dimorpholino-3-(1-naphthyl)pentane = 4,4'-[3-(1-Naphthalenyl)-1,5-pentanediyl]bis[morpholine] (●)
S DA 1686
U Anti-arrhythmic

12048 $C_{23}H_{32}N_2O_2$
165377-43-5

1-[2-(3,4-Dimethoxyphenyl)ethyl]-4-(3-phenylpropyl)piperazine (●)
R Dihydrochloride (165377-44-6)

S SA 4503
U σ_1 Receptor agonist, cognition enhancer

12049 (9648) $C_{23}H_{32}N_2O_2S$
910-86-1

4,4'-Bis(isopentyloxy)thiocarbanilide = 1,3-Bis[4-(isopentyloxy)phenyl]thiourea = 4,4'-Diisoamyloxythiocarbanilide = *N,N'*-Bis[4-(3-methylbutoxy)phenyl]thiourea (●)
S Amixyl, B 27, CP 919, DAT, Datanil, DATC, Dat-Orion, Dat-Wander, Disocarban, Disoxyl, Epomon (active substance), Isofetil, Isoxyl, Purazona, Sanoxyl, Sarbamyl, *Thiocarlide, Tiocarlide**
U Tuberculostatic, leprostatic

12050 $C_{23}H_{32}N_2O_3$
32896-53-0

(8β)-6-Methyl-1-(1-methylethyl)ergoline-8-carboxylic acid 2-hydroxy-1-methylpropyl ester (●)
R Maleate (1:1) (60634-51-7)
S LY 53857
U Serotonin antagonist

12051 (9649) $C_{23}H_{32}N_2O_3$
34758-83-3

α-(α-Methoxybenzyl)-4-(β-methoxyphenethyl)-1-piperazineethanol = 1-Methoxy-3-[4-(β-methoxyphenethyl)-1-piperazinyl]-1-phenyl-2-propanol = 4-(2-Methoxy-2-

phenylethyl)-α-(methoxyphenylmethyl)-1-piperazine-
ethanol (●)
R Dihydrochloride (34758-84-4)
S Antituxil-Z, Athos "Silesia", Balutox, Bechizolo,
Broncovis, Broncozina, Bronx (old form), Centrum
"Palenzona", CERM 3024, Chilvax (old form), Citi-
zeta, Coloplex "Andromaco", Delaviral, Demetovix,
Devixil, Dovavixin, Duoextolen, Eritós, Frenotos
"Lafi", Jactuss, Jiperol, Mirsol "Permamed", Nan-
tux, Oxiladin, Respilène, Respiral, Respirase, Respir-
ex "Inibsa; Sterling-Winthrop", Restrin, Sanotus, Si-
lentos, Sousibim, Talasa, Tusigen, Tusipriv, Tussi-
flex "Abbott", *Zipeprol dihydrochloride***, Zipertos,
Zipetoss, Ziprol, Zitoxil "Italfarmaco"
U Antitussive

12052 $C_{23}H_{32}N_2O_3$
                                                150710-80-8

(−)-*N*-[2-[[(2,3-Dihydro-8-methyl-1,4-benzodioxin-2-
yl)methyl]amino]ethyl]tricyclo[3.3.1.1^{3,7}]decane-1-carb-
oxamide (●)
S HT-90B
U Anxiolytic (5-HT$_{1A}$ receptor agonist/5-HT$_2$ receptor
antagonist)

12053   (9650) $C_{23}H_{32}N_2O_3S$
                                                107000-34-0

1'-(Methylsulfonyl)-1'*H*-5α,17α-pregn-20-yno[3,2-*c*]py-
razol-17-ol = (5α,17α)-1'-(Methylsulfonyl)-1'*H*-pregn-
20-yno[3,2-*c*]pyrazol-17-ol (●)
S Win 49596, *Zanoterone***
U Anti-androgen

12054   (9651) $C_{23}H_{32}N_2O_4$
                                                76644-53-6

N-[2-(Diethylamino)ethyl]-3,4,5-trimethoxy-2'-methyl-
benzanilide = *N*-[2-(Diethylamino)ethyl]-3,4,5-trimeth-
oxy-*N*-(2-methylphenyl)benzamide (●)
R also maleate (1:1) (79796-08-0)
S *Bernzamide, Gallanilide 603*
U Anti-arrhythmic

12055   (9652) $C_{23}H_{32}N_2O_5$
                                                83059-56-7

(3*S*)-2-[(2*S*)-*N*-[(1*S*)-1-Carboxy-3-phenylpropyl]alanyl]-
2-azabicyclo[2.2.2]octane-3-carboxylic acid 1-ethyl ester
= [3*S*-[2[*R**(*R**)],3*R**]]-2-[2-[[1-(Ethoxycarbonyl)-3-
phenylpropyl]amino]-1-oxopropyl]-2-azabicy-
clo[2.2.2]octane-3-carboxylic acid (●)
S *Zabicipril***
U Antihypertensive (ACE inhibitor)

12056   (9653) $C_{23}H_{32}N_2O_5$
                                                87333-19-5

(2*S*,3a*S*,6a*S*)-1-[(*S*)-*N*-[(*S*)-1-Carboxy-3-phenylpro-
pyl]alanyl]octahydrocyclopenta[*b*]pyrrole-2-carboxylic

acid 1-ethyl ester = 2-[*N*-[(*S*)-1-(Ethoxycarbonyl)-3-phe-
nylpropyl]-L-alanyl]-(1*S*,3*S*,5*S*)-2-azabicyclo[3.3.0]oc-
tane-3-carboxylic acid = [2*S*-[1[*R**(*R**)],2α,3aβ,6aβ]]-1-
[2-[[1-(Ethoxycarbonyl)-3-phenylpropyl]amino]-1-oxo-
propyl]octahydrocyclopenta[*b*]pyrrole-2-carboxylic acid
(●)

R see also no. 14558

S Acovil, Altace, Carasel, Cardace "Hoechst", Delix
"Hoechst", Hoechst 498, Hypren, Lostapres, Prama-
ce, Quark, Ramace, Ramicor, *Ramipril***, Stibenyl,
Triatec, Tritace, Unipril "Astra Simes", Vasotop
"Provet AG", Vesdil

U Antihypertensive (ACE inhibitor)

12057 (9655) $C_{23}H_{32}N_2O_6$
84768-09-2

1-[*N*-[4-(2,3-Dihydro-2-benzofuranyl)-1-(ethoxycarbonyl)butyl]-L-alanyl]-L-proline (●)

S BRL 36378

U Antihypertensive (ACE inhibitor)

12058 (9654) $C_{23}H_{32}N_2O_6$
68576-86-3

(*RS*)-1-[4-(2-Methoxyphenyl)-1-piperazinyl]-3-(3,4,5-
trimethoxyphenoxy)-2-propanol = (±)-4-(2-Methoxyphe-
nyl)-α-[(3,4,5-trimethoxyphenoxy)methyl]-1-piperazine-
ethanol (●)

R also dihydrochloride (68576-88-5)

S D-13112, *Enciprazine***, Wy-48624

U Anxiolytic, anti-aggressive

12059 $C_{23}H_{32}N_4$
157286-86-7

N-Butyl-*N*-ethyl-2,5-dimethyl-7-(2,4,6-trimethylphenyl)-
7*H*-pyrrolo[2,3-*d*]pyrimidin-4-amine (●)

S CP-154526

U Antidepressant (CRF$_1$ receptor antagonist)

12060 (9656) $C_{23}H_{32}N_4O_3S$
60996-95-4

9-[[3-(Diethylamino)-2-hydroxypropyl]amino]-*N*,*N*,7-
trimethyl-2-acridinesulfonamide (●)

S Domigon

U Antibacterial

12061 (9657) $C_{23}H_{32}N_4O_3S$
139133-27-0

(*S*)-*N*-[1-(Ethoxymethyl)-3-methylbutyl]-*N*-methyl-α-(2-
methyl-1*H*-imidazo[4,5-*c*]pyridin-1-yl)-*p*-toluenesulfon-
amide = (*S*)-*N*-[1-(Ethoxymethyl)-3-methylbutyl]-*N*-me-
thyl-4-[(2-methyl-1*H*-imidazo[4,5-*c*]pyridin-1-yl)me-
thyl]benzenesulfonamide (●)

R also monohydrochloride (144736-31-2)

S　BB-2113, *Nupafant***

U　PAF antagonist

12062　　　　　　　　　　　　　　　　$C_{23}H_{32}N_4O_4$
　　　　　　　　　　　　　　　　　　171049-14-2

(*S*)-2,3,4,5-Tetrahydro-4-methyl-3-oxo-7-[[4-(4-piperi-dyl)piperidino]carbonyl]-1*H*-1,4-benzodiazepine-2-ace-tic acid = (2*S*)-7-([4,4'-Bipiperidin]-1-ylcarbonyl)-2,3,4,5-tetrahydro-4-methyl-3-oxo-1*H*-1,4-benzodiaze-pine-2-acetic acid (●)

R　Monohydrochloride (179599-82-7)

S　*Lotrafiban hydrochloride***, SB 214857-A

U　Platelet aggregation inhibitor (GP IIB/IIIa fibrinogen receptor antagonist)

12063　(9658)　　　　　　　　　　　$C_{23}H_{32}N_4O_7S$
　　　　　　　　　　　　　　　　　　27826-45-5

2-[[(5-Carboxy-5-formamidopentyl)carbamoyl](2-phe-nylacetamido)methyl]-5,5-dimethyl-4-thiazolidinecarb-oxylic acid = (2*R-trans*)-N^6-[L-2-(4-Carboxy-5,5-dime-thyl-2-thiazolidinyl)-*N*-(phenylacetyl)glycyl]-N^2-formyl-L-lysine (●)

S　BPO-FLYS, *Libecillide***, *Lisocillide*, Ro 6-0787

U　Anti-allergic (inhibits penicillin-induced antibodies)

12064　　　　　　　　　　　　　　　$C_{23}H_{32}N_4O_8$
　　　　　　　　　　　　　　　　　　143313-51-3

Acetyl-tyrosinyl-valinyl-alanyl-aspartylaldehyde = *N*-Acetyl-L-tyrosyl-L-valyl-*N*-[(1*S*)-2-carboxy-1-formyl-ethyl]-L-alaninamide (●)

S　L 709049

U　Interleukin-1β converting enzyme inhibitor

12065　(9659)　　　　　　　　　　　$C_{23}H_{32}N_6O_3S$
　　　　　　　　　　　　　　　　　　79712-55-3

7-[2-Hydroxy-3-[4-[3-(phenylthio)propyl]-1-piperazi-nyl]propyl]theophylline = 3,7-Dihydro-7-[2-hydroxy-3-[4-[3-(phenylthio)propyl]-1-piperazinyl]propyl]-1,3-di-methyl-1*H*-purine-2,6-dione (●)

R　Dihydrochloride (79712-53-1)

S　LN 2974, RS-49014, *Tazifylline hydrochloride***

U　Antihistaminic, bronchodilator

12066　　　　　　　　　　　　　　　$C_{23}H_{32}N_6O_4S$
　　　　　　　　　　　　　　　　　　224785-90-4

2-[2-Ethoxy-5-[(4-ethyl-1-piperazinyl)sulfonyl]phenyl]-5-methyl-7-propyl-3*H*-imidazo[5,1-*f*][1,2,4]triazin-4-one = 1-[[3-(1,4-Dihydro-5-methyl-4-oxo-7-propylimida-zo[5,1-*f*][1,2,4]triazin-2-yl)-4-ethoxyphenyl]sulfonyl]-4-ethylpiperazine (●)

S　Bay 38-7268, *Vardenafil***

U　Vasodilator (PDE 5 inhibitor)

12066-01　　　　　　　　　　　　　　　224785-91-5

R　Hydrochloride

S　Bay 38-9456

U　Vasodilator (PDE 5 inhibitor)

12067　(9660)　　　$C_{23}H_{32}N_8O_{14}P_2$
72627-94-2

N-(2-Ammoniumethyl)nicotinamide-adenine dinucleo-
tide = Adenosine 5'-(trihydrogen diphosphate) 5'→5'-es-
ter with 3-[[(2-aminoethyl)amino]carbonyl]-1-β-D-ribo-
furanosylpyridinium hydroxide inner salt (●)
S　N^3-*(2-Ammoniumethyl)nadide*, Riboksin, Riboxine
U　Antihypoxic

12068　(9661)　　　$C_{23}H_{32}O_2$
79-64-1

17β-Hydroxy-6α-methyl-17-(1-propynyl)androst-4-en-
3-one = 17β-Hydroxy-6α,21-dimethylpregn-4-en-20-yn-
3-one = 6α,21-Dimethylethisterone = (6α,17β)-17-Hy-
droxy-6-methyl-17-(1-propynyl)androst-4-en-3-one (●)
R　also monohydrate (41354-30-7)
S　*Dimethisterone***, *Dimethylethisterone*, Lutogan,
　　MJ 5048, P-5048, Secrosteron
U　Progestin

12069　　　　$C_{23}H_{32}O_2$
178600-20-9

(2E,4E,6E)-7-(3,5-Di-*tert*-butylphenyl)-3-methylocta-
2,4,6-trienoic acid = (*all-E*)-7-[3,5-Bis(1,1-dimethyl-
ethyl)phenyl]-3-methyl-2,4,6-octatrienoic acid (●)
S　ALRT-1550
U　Antiproliferative

12070　　　　$C_{23}H_{32}O_2$
178600-19-6

(2E,4E,6Z)-7-(3,5-Di-*tert*-butylphenyl)-3-methylocta-
2,4,6-trienoic acid = (2E,4E,6Z)-7-[3,5-Bis(1,1-dimethyl-
ethyl)phenyl]-3-methyl-2,4,6-octatrienoic acid (●)
S　LG 100567
U　Antiproliferative

12071　　　　$C_{23}H_{32}O_2$
119-47-1

2,2'-Methylenebis[6-*tert*-butyl-p-cresol] = Bis(2-hydr-
oxy-3-*tert*-butyl-5-methylphenyl)methane = 2,2'-Methy-
lenebis[6-(1,1-dimethylethyl)-4-methylphenol] (●)
S　GERI-BP 002-A
U　Cytotoxic, anti-oxidant

12072   (9662) $C_{23}H_{32}O_2$
                                                                 977-79-7

6,17-Dimethylpregna-4,6-diene-3,20-dione (●)
S   AY-62022, Ayerluton, Colpro, Colpron(e), Etogyn,
     *Medrogestone**, Metrogestone*, NSC-123018, Prot-
     hil, R 13615
U   Progestin

12073   (9663) $C_{23}H_{32}O_2$
                                                                 124-85-6

5α,17α-Pregn-2-en-20-yn-17-ol acetate = (5α,17α)-
Pregn-2-en-20-yn-17-ol acetate (●)
S   Regonyl, TX-380
U   Antiprolactin

12074   (9665) $C_{23}H_{32}O_3$
                                                                 24894-50-6

Estra-1,3,5(10)-triene-3,17β-diol 17-pivalate = Estradiol
17-trimethylacetate = (17β)-Estra-1,3,5(10)-triene-3,17-
diol 17-(2,2-dimethylpropanoate) (●)
S   *Estradiol pivalate*, Estrotate, *Östradioltrimethylace-*
     tat
U   Depot estrogen

12075   (9664) $C_{23}H_{32}O_3$
                                                                 979-32-8

Estra-1,3,5(10)-triene-3,17β-diol 17-valerate = Estradiol
17-valerate = (17β)-Estra-1,3,5(10)-triene-3,17-diol 17-
pentanoate (●)
S   Ardefem, Atladiol, Climaval, Climen, Cyclabil, Cy-
     clocur, Deladiol, Deladumone, Delestrogen, Depe-
     striol, Depodiol, Depo-Estro-Med, Depogen "Hyrex;
     Sig", Depostradol, Depot-Estiol, Diol-20, Dioval,
     Dura-Chorion Plus, Dura-Estate, Dura-Estradiol,
     Duragen, Duratrad, Enadiol, Estate "Savage", Estra-
     din Depot, Estradiol "Jenapharm", *Estradiol valer-*
     *ate**, Estra-L, Estra-V, Estraval, Estra-Vate,
     Estro-Pause, Estroval-10, Exten Strone, Femo-
     gen-L.A., Femogex, Filena, Gynogen L.A., Gynoka-
     din, L.A.E. 20 (40), Lastrogen, Lenadiol, Medor-
     mon, Menaval, Merimono, Mirion "Erba",
     Neofollin, Neo-Östrogynal, NSC-17590, Oestradiol
     "Jenapharm" (Depot-Amp. + Tabl.), Oestradiol-re-
     tard Pharlon, *Oestradiol valerate, Östradiolvaleria-*
     nat, Östrin-Depo, Östrogynal sine, Ovormol-Depot,
     Pan-Estra, Pelanin Depot, Postoval, Primofol-Depot,
     Primogyna oral (-depot), Primogyn-Depot, Progy-
     non-Depot, Progynova, Progynovum, Rep-Estra, Re-
     pestrogen, Repo-Estro Med, Reposo-E, Retestrin,
     Ronfase, SH 743, Span-Est, Suitest, Valergen, Vale-
     riol "Klinos", Valestra
U   Depot estrogen

12076   (9666) $\qquad$ $C_{23}H_{32}O_3$
1169-79-5

3-(Cyclopentyloxy)estra-1,3,5(10)-triene-16α,17β-diol =
Estriol 3-cyclopentyl ether = (16α,17β)-3-(Cyclopentyloxy)estra-1,3,5(10)-triene-16,17-diol (●)
S   *Chinoestradiol*, Colpovis, Colpovister, *Estriol cyclopentyl ether*, *Östriolcyclopentyläther*, Pentovis, *Qui-
nestradiol*, *Quinestradol***
U   Estrogen

12077   (9667) $\qquad$ $C_{23}H_{32}O_4$
49848-01-3

6,7-Dihydro-17-hydroxy-3-oxo-3'H-cyclopropa[6,7]-
17α-pregna-4,6-diene-21-carboxylic acid = 6α,7α-Dihy-
dro-17β-hydroxy-3-oxo-3'H-cyclopropa[6,7]-21,24-di-
nor-17α-chol-4-en-23-oic acid = 3-(3-Oxo-17β-hydroxy-
6β,7β-methyleneandrost-4-en-17-yl)propionic acid =
(6α,7α,17α)-6,7-Dihydro-17-hydroxy-3-oxo-3'H-cyclopropa[6,7]pregna-4,6-diene-21-carboxylic acid (●)
R   Potassium salt (49847-97-4)
S   *Prorenoate potassium***, *Prorenoatum kalicum***,
SC-23992
U   Aldosterone antagonist

12078   (9668) $\qquad$ $C_{23}H_{32}O_4$
25092-41-5

17-Hydroxy-11β-methyl-19-norpregn-4-ene-3,20-dione
acetate = (11β)-17-(Acetyloxy)-11-methyl-19-norpregn-
4-ene-3,20-dione (●)
S   *Norgestomet***, SC-21009
U   Progestin

12079   (9669) $\qquad$ $C_{23}H_{32}O_4$
130273-99-3

[3aS-[2(Z),3aα,5β,6α(1E,3R*,4R*),6aα]]-5-
[1,3a,4,5,6,6a-Hexahydro-5-hydroxy-6-(3-hydroxy-4-
methyl-1-nonen-6-ynyl)-2-pentalenyl]-4-pentenoic acid
(●)
S   KP 10614
U   Platelet aggregation inhibitor

12080   (9671) $\qquad$ $C_{23}H_{32}O_4$
302-23-8

17-Hydroxypregn-4-ene-3,20-dione acetate = 17-Hy-
droxyprogesterone acetate = 17-(Acetyloxy)pregn-4-ene-
3,20-dione (●)

S *Acetoxyprogesterone*, Gestageno Gador, *Hydroxy-progesterone acetate***, Kyormon, Lutate-Inj., *Oxi-progesteroni acetas*, Prodix, Prodox, Prokan

U Progestin

12081   (9670) $C_{23}H_{32}O_4$
                                                           56-47-3

21-Hydroxypregn-4-ene-3,20-dione acetate = 11-Deoxy-corticosterone acetate = 21-(Acetyloxy)pregn-4-ene-3,20-dione (●)

R see also nos. 10723, 13304, 13642, 13958, 14197

S A.D.C., Altecortian, Arcort, Articortal, Artrisone, Bio-Corten, Cecostrate, Coralets, Corendocrin, Cortacet, Cortarmour, Cortate (old form), Cortenil(etten), Cortesan, Cortical (old form), Corticoici, Corticona, Corticosir, Corticosterole, Corticosteron "Vitis", Corticosterone "Eupharma; Taricco", Cortifar, Cortigan, Cortigen, Cortin "Galenika", Cortina, Cortinaq, Cortiron, Cortisaf, Cortisal, Cortisteril, Cortisteron, Cortivis, Cortivister, Cortixyl, Cortolipex, Cortormon, Costrex, DCA, Decort, Decortacete, Decorten, Decorterone, Decortic, Decortin "Schieffelin", Decorton "Léciva", Decosone, Decosteron, Decostrate, Dekorton, *Deoxycortone Acetate*, Descornaq, Descort 10, Descorterone, Descotone, Desocort, *Desoxicortoni acetas*, *Desoxycorticosterone acetate*, *Desoxycortone acetate*, Desoxycortonum-Polfa, Dexon "Pasadena", *DOCA*, Doca Acetate, Dohycamon, Dorcostrin, Doxo, Doxotone, Doxycamon, Endoxicortinal, Glosso-Syncortyl, Krinocort, Leocortex, Medicosteron, Mephadecortyl, Mincortid, Neocortin, Ocritan, Ocritena, Pancortyl, Percorten (acetate), Percotol, Primocort, Primocortan, Prodecort, Sincortex, Steraq, Sterone, Supracortex, Surrenon, Surrenosterone, Syncort(a), Syncortin, Syncortyl, Unidocan

U Mineralocorticoid (salt-regulating)

12082 $C_{23}H_{32}O_5$
                                                         159121-98-9

[1'R-(1'α,2'α,4'aα,6'α,8'aβ)]-3',4',4'a,5',6',7',8',8'a-Octahydro-4,6'-dihydroxy-6-(hydroxymethyl)-2',5',5',8'a-tetramethylspiro[benzofuran-2(3H),1'(2'H)-naphthalene]-7-carboxaldehyde (●)

S Mer-NF 5003E

U Antiviral

12083 $C_{23}H_{32}O_6$
                                                         126382-01-2

5-Hydroxy-7-methoxy-3-[5-(2-methyl-1,6-dioxaspiro[4.5]dec-7-yl)pentyl]-1(3H)-isobenzofuranone (●)

S *Spirolaxin*

U Antibacterial, antihyperlipidemic

12084   (9672) $C_{23}H_{32}O_6$
                                                           50-03-3

11β,17,21-Trihydroxypregn-4-ene-3,20-dione 21-acetate = 17-Hydroxycorticosterone 21-acetate = (11β)-21-(Acetyloxy)-11,17-dihydroxypregn-4-ene-3,20-dione (●)

R see also nos. 10737, 10756, 11452, 12933, 13640, 13643, 14199

S Abbocort, Acepolcort H, Aceto-Cort, *Acetylhydrocortisone*, Adacor, Ak-cort, Alfacorton, Allocort, Anucort HC, Anu-Med HC, Anuprep HC, Anusol HC, Apocort, Apocortal, Bambicort, Behrederm H, Berlison (-F), Biocortar, Bio-F, Biosone "Kay", Calacort, Caldecort, Carmol-HC, Chemysone, Clearaid, Colifoam, Colofoam, Cordes H, Coreton "Taisho", Cortacet "Ayerst", Cortacream, Cortagel, Cortaid, Cortamed, Cort-Dome High Potency, Cortef acetate, Cortell, Cortes, Corti-Basileos, Cortibel, Cortic, Cortic "Sigma", Cortical "Anacel", Corticina, Corticorenol, Corticreme, Corti-Creme Lichtenstein, Cortiderm, Corti-dermosina, Cortidro, Cortifoam, Cortiform, Cortiment, Cortimycine, Cortioftal, Cortiprel, *Cortisol acetate*, Cortocaps, Cortocin Ointm., Cortoderm "Crookes; Taro", Cortomister, Cortosterone F, Cortril acetate, crema transcutan "Semar", Cremocort, Dermacalm crème, Dermacort "Parke-Davis", Dermacortin F, Derminovag, Dermo-Cortison, Dilucort, Dortizon, Ebenol, Efzem, Ekzemsalbe "F", Ekzemsalbe Agepha, Ekzesin, Entybase A, Fenistil-Hydrocortison, Fenitral, Ferncort, Fernisone, Ficortrit Augensalbe, Foillecort, Genocortison, Glycocortison, Glycocortison H, Gynecort, HC 45, HCA, Hemorrhoid-HC, Hemril-HC, Hemsol, Hidalone, Hidrocet, Hidrocorticil, Hidrocorticlin, Hidro-Corti-Ofteno, Hidrocortisan, Hidro-Scheralson, Hipoge, Hipokort, Hycin, Hycor Eye Ointment, Hyderm, Hydrin-2, Hydrison, Hydro-Adreson, Hydro-Bencorton, Hydrocal, Hydro-Can, Hydrocort "Dunhall", Hydrocortifor-Susp., Hydrocortisate, Hydrocortison "Ciba", *Hydrocortisone acetate***, Hydrocortison POS N, Hydrocortistab, Hydrocortone acetate, Hydrocort von ct, Hydrocutan-Creme, Hydroftal, Hydroricortex, Hydrosone, Hysone-A, Idracetisone, Idrocort, Idrocortigamma, *Idrocortisone acetato*, Incortin H, Komed-HC, Korti, Kortibalsam, Labocort, Lanacort, Latimit, Lenirit, Litraderm, Lycortin, Maximum Strength Corticaine "Whitby", Maximum Strength Dermatest Dricort, Medithane, My-Cort Lotion, Mylocort, Mysone, Neo-HC, Neo-Hycortole, Nericort, Novohydrocort, NSC-741, Nutracort "Lääkefarmos", Ofta-Hydrone, Ophticor H, Optacort, Orabase HCA, Otozon-Base, Pabracort(in), Pancortisona, Pannocort, Pantison, Paro "Hoechst Marion Roussel; Milanfarma", Phar-

mac-Cort, Pharma-Cort, Pharmecortisone, Posterine-Corte, Pramosone, Proctocort "Biotherax", Proctosol-HC, Promecort, Rectocort, Rectoparin H, Resicort, Rhulicort, Sagittacortin-Salbe, Sanaderm "Spirig", Sanadermil, Scheroson F, Sebamed Lotion C, Servicort, Siguent Hycor, Sintotrat, Sohydrone, Soventol-Hydrocortison, Span-Ster, Squibb-HC, Ster-Jet 50, Steroderm "Medikon Prima", Stopitch, Supralef, Surfa-Cort, Synthacort, U-Cort, Ultimacort, Urecortyn, Velopural, Velopural-OPT, Wellcortin, Wincort Acetate, Wycort

U Glucocorticoid (anti-inflammatory, anti-allergic)

12085 $C_{23}H_{32}O_6$
                                                       134366-08-8

Estradiol 3-β-D-xyloside = (17β)-17-Hydroxyestra-1,3,5(10)-trien-3-yl β-D-xylopyranoside (●)

S *Estradiol xyloside*

U Anticoagulant

12086 (9673) $C_{23}H_{33}Cl_2NO_3$
                                                       3096-15-9

17β-Hydroxyestr-4-en-3-one 17-[bis(2-chloroethyl)carbamate] = (17β)-17-[[[Bis(2-chloroethyl)amino]carbonyl]oxy]estr-4-en-3-one (●)

S LS 1727,
      Nandrolone 17-bis(2-chloroethyl)carbamate

U Antineoplastic

12087   (9674) $C_{23}H_{33}FN_2O_2$
3781-28-0

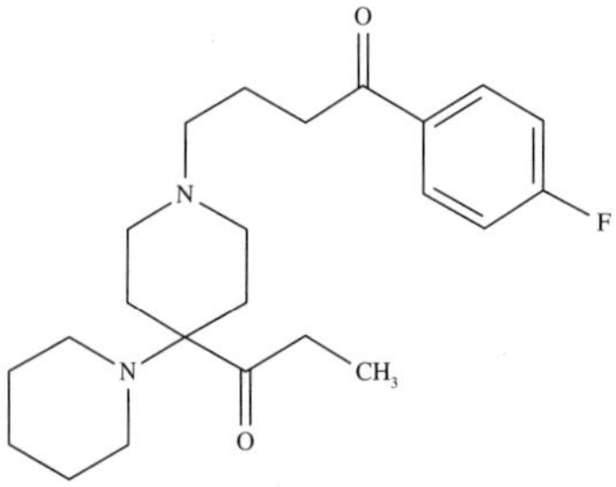

4'-Fluoro-4-(4-piperidino-4-propionylpiperidino)butyro-
phenone = 1-[3-(4-Fluorobenzoyl)propyl]-4-piperidino-
4-propionylpiperidine = 1-(4-Fluorophenyl)-4-[4'-(1-oxo-
propyl)[1,4'-bipiperidin]-1'-yl]-1-butanone (●)
R   Dihydrochloride (2024-11-5)
S   *Floropipetone, Propyperone dihydrochloride**,*
R 4082
U   Neuroleptic

12088   (9675) $C_{23}H_{33}NO$
15266-38-3

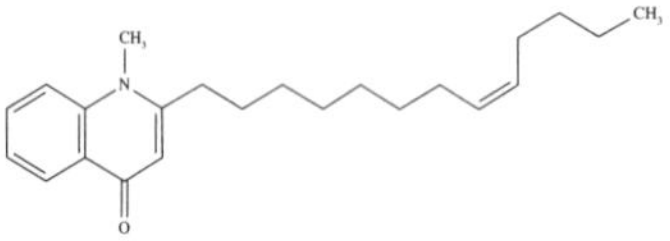

(Z)-1-Methyl-2-(8-tridecenyl)-4(1H)-quinolinone (●)
S   *Evocarpine*
U   Vasorelaxant, calcium antagonist from fruits of
Evodia sp.

12089   (9676) $C_{23}H_{33}NO_2$
13074-00-5

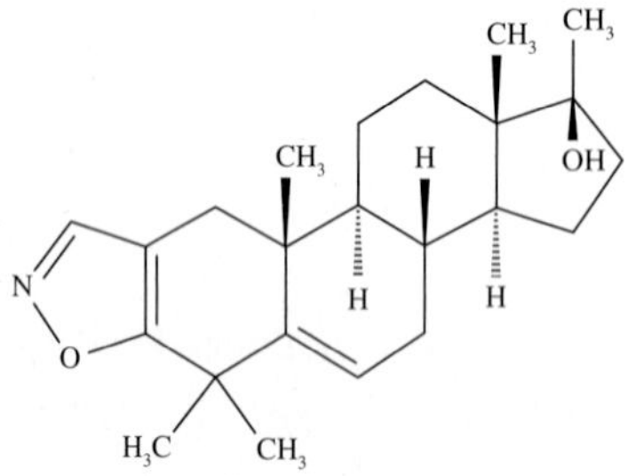

4,4,17-Trimethylandrosta-2,5-dieno[2,3-d]isoxazol-17β-
ol = (17β)-4,4,17-Trimethylandrosta-2,5-di-
eno[2,3-d]isoxazol-17-ol (●)
S   *Azastene*, Win 17625

U   Contraceptive

12090 $C_{23}H_{33}NO_2$
149860-29-7

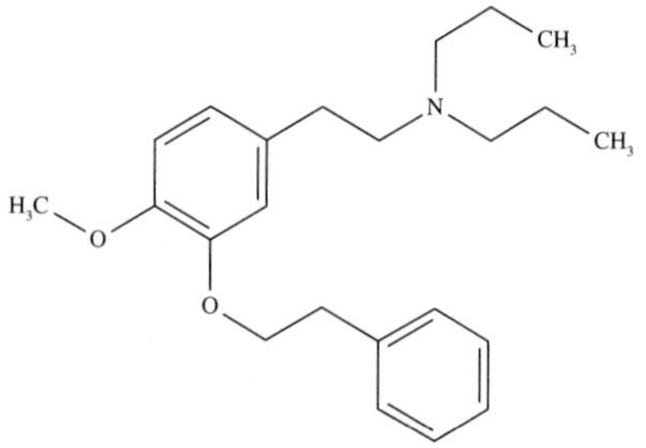

N-[2-[4-Methoxy-3-(2-phenylethoxy)phenyl]ethyl]-N,N-
dipropylamine = 4-Methoxy-3-(2-phenylethoxy)-N,N-di-
propylbenzeneethanamine (●)
R   Hydrochloride (149409-57-4)
S   NE 100
U   Narcotic antagonist, antipsychotic

12091   (9677) $C_{23}H_{33}NO_2$
150443-71-3

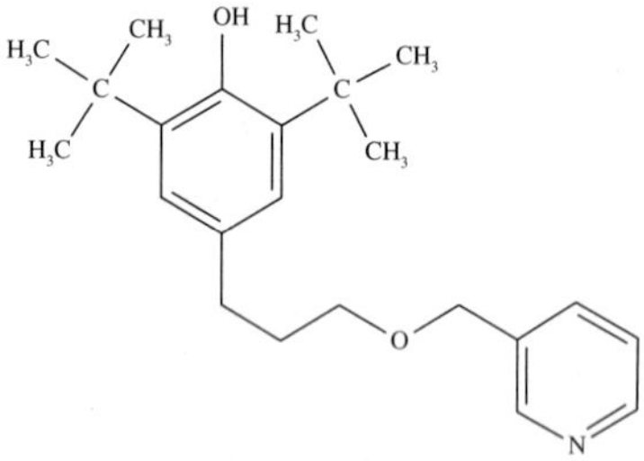

2,6-Di-*tert*-butyl-4-[2-(3-pyridylmethoxy)propyl]phenol
= 2,6-Bis(1,1-dimethylethyl)-4-[3-(3-pyridinylmeth-
oxy)propyl]phenol (●)
S   Mrz 3/124, *Nicanartine**
U   Anti-oxidant, antihyperlipidemic, antiproliferative

12092   (9679) $C_{23}H_{33}NO_2$
1600-19-7

2-[2-(Di-2,5-xylylmethoxy)ethoxy]-N,N-dimethylamine
= 2-[2-[Bis(2,6-dimethylphenyl)methoxy]ethoxy]-N,N-
dimethylethanamine (●)
R Hydrochloride (2827-06-7)
S B.S. 6748, Xyloxemine hydrochloride*
U Antitussive

12093   (9678) $C_{23}H_{33}NO_2$
19179-78-3

1-(Di-2,6-xylylmethoxy)-3-(isopropylamino)-2-propanol
= 1-[Bis(2,6-dimethylphenyl)methoxy]-3-[(1-methyl-
ethyl)amino]-2-propanol (●)
R Hydrochloride (19179-88-5)
S B.S. 7977-D, Xipranolol hydrochloride**
U β-Adrenergic blocker, anti-arrhythmic

12094   (9680) $C_{23}H_{33}NO_6$
117184-53-9

N-(6,7:9,10-Diepoxy-8-oxo-1-oxaspiro[4.5]dec-3-yl)-
4,6-dimethyl-2,4-dodecadienamide = N-(Dihydro-5'-
hydroxy-6-oxospiro[4,8-dioxatricyclo[5.1.0.0^{3,5}]octane-

2,2'(3'H)-furan]-4'-yl)-4,6-dimethyl-2,4-dodecadien-
amide (●)
S Aranorosin
U Antibacterial, antifungal from Pseudoarachniotus roseus

12095 $C_{23}H_{33}NO_8$
141646-08-4

1-[[(Cyclohexyloxy)carbonyl]oxy]ethyl (1S,5S,8aS,8bR)-
1,2,5,6,7,8,8a,8b-octahydro-1-[(R)-1-hydroxyethyl]-5-
methoxy-2-oxoazeto[2,1-a]isoindole-4-carboxylate =
(1S,5S,8aS,8bR)-1,2,5,6,7,8,8a,8b-Octahydro-1-[(1R)-1-
hydroxyethyl]-5-methoxy-2-oxoazeto[2,1-a]isoindole-4-
carboxylic acid 1-[[(cyclohexyloxy)carbonyl]oxy]ethyl
ester (●)
S GV 118819X, Sanfetrinem cilexetil**
U Antibiotic

12096   (9681) $C_{23}H_{33}N_2O$
7492-32-2

(3-Carbamoyl-3,3-diphenylpropyl)diisopropylmethylam-
monium = γ-(Aminocarbonyl)-N-methyl-N,N-bis(1-me-
thylethyl)-γ-phenylbenzenepropanaminium (●)
R Iodide (71-81-8)
S Antidiarroico Gellini, Darbid, Dipramid, Embamida,
Isamid, Isopropamide iodide**, Isoproponi jodidum,
Marygin-M, 5579 MD, Piaccamide, Priamide, Priazi-
mide, R 79, Raspon, Sanulcin, SKF 4740, Tyrimid(e)

U   Anticholinergic, spasmolytic

12096-01   (9681-01)   16564-41-3
R   Bromide
S   Diapantin, Neopant
U   Anticholinergic, spasmolytic

12097   (9682) $C_{23}H_{33}N_2O_2$
                                                              35080-11-6

N-Propylajmalinium = (17*R*,21α)-17,21-Dihydroxy-4-propylajmalanium (●)
R   Tartrate (1:1) (2589-47-1)
S   GT 1012, Neoaritmina, Neo-Gilurytmal, Neorythmin, NPAB, Prajmalin, *Prajmalium bitartrate**
U   Anti-arrhythmic

12098 $C_{23}H_{33}N_3O$
                                                              121879-58-1

trans-*N*-[(8α)-2,6-Dimethylergolin-8-yl]-2-ethyl-2-methylbutyramide = *N*-[(8α)-2,6-Dimethylergolin-8-yl]-2-ethyl-2-methylbutanamide (●)
R   Maleate (1:1) (121879-59-2)
S   MAR 327, SDZ-MAR 327
U   Antipsychotic (partial dopamine D_2-receptor agonist)

12099 $C_{23}H_{33}N_3O_5$
                                                              181372-99-6

N^2-[4-(2-Hexyl-2,3,4,4a-tetrahydrocyclopent[*b*]oxireno[*c*]pyridin-7(1a*H*)-ylidene)-1-oxo-2-butenyl]-L-glutamine (●)
S   *Epostatin*
U   Antirheumatic (dipeptidyl peptidase II inhibitor)

12100   (9683) $C_{23}H_{33}N_3O_6S$
                                                              56211-43-9

2-(Diethylamino)ethyl (2*S*,5*R*,6*R*)-6-(2,6-dimethoxybenzamido)-3,3-dimethyl-7-oxo-4-thia-1-azabicyclo[3.2.0]heptane-2-carboxylate = 2-(Diethylamino)ethyl (6*R*)-6-(2,6-dimethoxybenzamido)penicillanate = [2*S*-(2α,5α,6β)]-6-[(2,6-Dimethoxybenzoyl)amino]-3,3-dimethyl-7-oxo-4-thia-1-azabicyclo[3.2.0]heptane-2-carboxylic acid 2-(diethylamino)ethyl ester (●)
S   DAN-523, *Tameticillin***
U   Antibiotic

12101 $C_{23}H_{33}N_3O_{11}$
                                                              158599-72-5

(4*S*)-4-(4-Ethoxybenzyl)-3,6,9-tris(carboxymethyl)-3,6,9-triazaundecanedioic acid = *N*-[(2*S*)-2-[Bis(carboxymethyl)amino]-3-(4-ethoxyphenyl)propyl]-*N*-[2-[bis(carboxymethyl)amino]ethyl]glycine (●)
R   Calcium sodium salt (1:1:3) (153924-80-2)
S   Ca-EOB-DTPA, ZK 155116

U Pharmaceutical aid

12102   C$_{23}$H$_{33}$N$_5$O$_5$
                                         160470-73-5

(αS,βR)-β-Amino-α-hydroxybenzenebutanoyl-L-prolyl-L-prolyl-L-alaninamide = 1-[(2S,3R)-3-Amino-2-hydroxy-1-oxo-4-phenylbutyl]-L-prolyl-L-prolyl-L-alaninamide (●)

S *Apstatin*
U Aminopeptidase P inhibitor

12103 (9684)   C$_{23}$H$_{33}$N$_5$O$_5$S
                                         51876-98-3

endo-1-[[4-[2-(2-Methoxynicotinamido)ethyl]piperidino]sulfonyl]-3-(5-norbornen-2-ylmethyl)urea = *endo-N*-[2-[1-[[[[(Bicyclo[2.2.1]hept-5-en-2-ylmethyl)amino]carbonyl]amino]sulfonyl]-4-piperidinyl]ethyl]-2-methoxy-3-pyridinecarboxamide (●)

S CP-27634, *Gliamilide***
U Oral hypoglycemic

12104   C$_{23}$H$_{34}$ClN
                                         132173-06-9

(Z)-N-[3-(3-Chloro-4-cyclohexylphenyl)-2-propenyl]-N-ethylcyclohexanamine (●)

R Hydrochloride (132173-07-0)
S SR 31747
U Immunomodulator

12105 (9685)   C$_{23}$H$_{34}$NO$_5$P
                                         95399-71-6

(4S)-4-Cyclohexyl-1-[[hydroxy(4-phenylbutyl)phosphinyl]acetyl]-L-proline = *trans*-4-Cyclohexyl-1-[[hydroxy(4-phenylbutyl)phosphinyl]acetyl]-L-proline (●)

R see also no. 14399
S *Fosinoprilat***, SQ 27519
U Antihypertensive (ACE inhibitor)

12106 (9686)   C$_{23}$H$_{34}$N$_2$OS$_2$
                                         129184-48-1

2-[[8-(Dimethylamino)octyl]thio]-6-isopropyl-3-(2-thenoyl)pyridine = [2-[[8-(Dimethylamino)octyl]thio]-6-(1-methylethyl)-3-pyridinyl]-2-thienylmethanone (●)

R Citrate (1:1) (143984-30-9)
S Y-29794
U Prolyl endopeptidase inhibitor

12107 $C_{23}H_{34}N_2O_2$
                                              158182-74-2

Dimethylcarbamic acid 3-(2-cyclohexylethyl)-
2,3,3a,4,5,9b-hexahydro-1*H*-benz[*e*]indol-6-yl ester (●)
S   Ro 46-5934
U   Cognition enhancer (acetylcholinesterase inhibitor)

12108   (9688) $C_{23}H_{34}N_3O_3$
                                              86434-70-0

cis-trans Mixture (1:1) of 1-ethyl-4-hydroxy-1-methylpi-
peridinium (±)-α-(hexahydro-1*H*-azepin-1-yl)-1,2-benz-
isoxazole-3-acetate = 4-[2-(1,2-Benzisoxazol-3-yl)-2-
(hexahydro-1*H*-azepin-1-yl)acetoxy]-1-ethyl-1-methyl-
piperidinium = 4-[[1,2-Benzisoxazol-3-yl(hexahydro-1*H*-
azepin-1-yl)acetyl]oxy]-1-ethyl-1-methylpiperidinium
(●)
R   Iodide (86434-57-3)
S   *Beperadium iodide, Beperidium iodide**, SX-810
U   Anticholinergic, gastric cytoprotective, spasmolytic,
       mydriatic

12109   (9689) $C_{23}H_{34}N_3O_{10}P$
                                              36357-77-4

N-[*N*-[[(6-Deoxy-α-L-mannopyranosyl)oxy]hydroxy-
phosphinyl]-L-leucyl]-L-tryptophan (●)
S   *Phosphoramidon*
U   Protection against corneal ulcer (thermolysin inhibitor
       from *Streptomyces tanashiensis*)

12110 $C_{23}H_{34}N_6O$
                                              140945-32-0

1-(2-Ethoxyethyl)-2-[[4-[4-(1*H*-pyrazol-1-yl)butyl]-1-pi-
perazinyl]methyl]-1*H*-benzimidazole (●)
R   Maleate (1:2) (179188-69-3)
S   E-4716, *Mapinastine maleate**, Miostina
U   Antihistaminic, anti-allergic

12111   (9690) $C_{23}H_{34}O_2$
                                              7069-42-3

(*all-E*)-3,7-Dimethyl-9-(2,6,6-trimethyl-1-cyclohexen-1-
yl)-2,4,6,8-nonatetraenyl propionate = Retinol propanoate
(●)
S   Axerofluid, *Axerophthol propionate, Retinol propio-
       nate, Vitamin A propionate*
U   Antixerophthalmic vitamin

12112   (9691) $C_{23}H_{34}O_2S$
                                                     50708-95-7

8-(1,2-Dimethylheptyl)-1,2,3,5-tetrahydro-5,5-dime-
thylthiopyrano[2,3-*b*][1]benzopyran-10-ol (●)
S   SP-119, *Tinabinol***
U   Antihypertensive

12113   (9693) $C_{23}H_{34}O_3$
                                                     1169-49-9

17β-Hydroxyandrost-4-en-3-one isobutyrate = (17β)-17-
(2-Methyl-1-oxopropoxy)androst-4-en-3-one (●)
S   Agovirin-Depot, Perandren M, Perandrone-Ampou-
      les cristallines, Testex Leo krystalsuspension, Testo-
      cryst, *Testosterone isobutyrate*, Virex-cryst
U   Depot-androgen

12114   (9692) $C_{23}H_{34}O_3$
                                                     1778-02-5

3β-Hydroxypregn-5-en-20-one acetate = (3β)-3-(Acetyl-
oxy)pregn-5-en-20-one (●)
S   Antofin, Arthenolone, Eterna 27, *Pregnenolone ace-
      tate***, Pregno-Pan, Prévisone
U   Glucocorticoid (anti-inflammatory, anti-allergic)

12115 $C_{23}H_{34}O_4$
                                                     156722-18-8

17β-(3-furyl)-5β-androstane-3β,14β,17α-triol =
(3β,5β,14β)-21,23-Epoxy-24-norchola-20,22-diene-
3,14,17-triol (●)
S   PST 2238
U   Antihypertensive

12116   (9694) $C_{23}H_{34}O_4$
                                                     566-78-9

3β,21-Dihydroxypregn-5-en-20-one 21-acetate = 21-Ace-
toxypregn-5-en-3-ol-20-one = (3β)-21-(Acetyloxy)-3-
hydroxypregn-5-en-20-one (●)
S   Acetoxanon, *21-Acetoxypregnenolone*, Acetoxy-Pre-
      nolon, A.O.P., Artisone, Artivis, *Prebediolone aceta-
      te*, Pregnartrone, Sterosone
U   Glucocorticoid (anti-inflammatory, anti-allergic)

12117 $C_{23}H_{34}O_5$
                                                     81846-19-7

2473

[[(1R,2R,3aS,9aS)-2,3,3a,4,9,9a-Hexahydro-2-hydroxy-1-[(3S)-3-hydroxyoctyl]-1H-benz[f]inden-5-yl]oxy]acetic acid (●)
S 15AU81, *Treprostinol*, U-62840, Uniprost
U Treatment for congestive heart failure

12118 (9695) $C_{23}H_{34}O_5$
73573-88-3

(1S,7S,8S,8aR)-1,2,3,7,8,8a-Hexahydro-7-methyl-8-[2-[(2R,4R)-tetrahydro-4-hydroxy-6-oxo-2H-pyran-2-yl]ethyl]-1-naphthyl (S)-2-methylbutyrate = [1S-[1α(R*),7β,8β(2S*,4S*),8aβ]]-2-Methylbutanoic acid 1,2,3,7,8,8a-hexahydro-7-methyl-8-[2-(tetrahydro-4-hydroxy-6-oxo-2H-pyran-2-yl)ethyl]-1-naphthalenyl ester (●)
S *Compactin*, CS 500, *6-Demethylmevinolin, Mevastatin**, Mevastin*, ML 236B
U Antihyperlipidemic (HMGCoA reductase inhibitor), antifungal (from *Penicillium brevicompactum*)

12119 $C_{23}H_{34}O_5$
41639-83-2

(5Z)-7-[(1R,2R,3R,5S)-3,5-Dihydroxy-2-[(3R)-3-hydroxy-5-phenylpentyl]cyclopentyl]-5-heptenoic acid (●)
S PhXA 85
U Synthetic prostaglandin

12120 (9696) $C_{23}H_{34}O_5$
23930-37-2

3α,21-Dihydroxy-5α-pregnane-11,20-dione 21-acetate = (3α,5α)-21-(Acetyloxy)-3-hydroxypregnane-11,20-dione (●)
R see also no. 10777-01
S *Alfadolone acetate**, Alphadolone acetate*
U Anesthetic

12121 (9697) $C_{23}H_{34}O_8S$

5,6-Dihydro-5-hydroxy-6-[6-hydroxy-5-methyl-4-(sulfooxy)-1,7,9,11-heptadecatetraenyl]-2H-pyran-2-one (●)
R Monosodium salt (131774-59-9)
S BU-3285T, *Sultriecin*
U Antineoplastic antibiotic from *Chainia rosea*

12122 (9698) $C_{23}H_{35}ClN_2O_2$
55313-67-2

(±)-1-(*o*-Chlorophenethyl)-*N*-cyclohexyl-4-hydroxy-*N*,α-dimethyl-4-piperidineacetamide = (±)-2-[1-(*o*-Chlorophenethyl)-4-hydroxy-4-piperidyl]-*N*-cyclohexyl-*N*-methylpropionamide = 1-[2-(2-Chlorophenyl)ethyl]-*N*-cyclohexyl-4-hydroxy-*N*,α-dimethyl-4-piperidineacetamide (●)

S FQ 27-096, *Pipramadol***

U Analgesic

12123 (9699)

$C_{23}H_{35}Cl_2N_3O_4$
3577-89-7

α-*N*-Acetyl-DL-sarcolysyl-L-leucine ethyl ester = *N*-[*N*-Acetyl-4-[bis(2-chloroethyl)amino]-DL-phenylalanyl]-L-leucine ethyl ester (●)

S Asalei, Asaley, NSC-167780

U Antineoplastic

12124

$C_{23}H_{35}NOS$
155233-30-0

4-[11-Methoxy-8-methyl-1(*Z*),5(*E*),7(*E*)-13-tetradecatetraenyl]-2-(2-methylcyclopropyl)-4,5-dihydrothiazole = (4*R*)-4,5-Dihydro-4-[(1*Z*,5*E*,7*E*,11*R*)-11-methoxy-8-methyl-1,5,7,13-tetradecatetraenyl]-2-[(1*R*,2*S*)-2-methylcyclopropyl]thiazole (●)

S *Curacin A*

U Cytotoxic from marine cyanobacterium *Lyngbya majuscula*

12125

$C_{23}H_{35}NO_2$
103497-68-3

(5α)-23-Methyl-4-aza-21-norchol-1-ene-3,20-dione = (4a*R*,4b*S*,6a*S*,7*S*,9a*S*,9b*S*,11a*R*)-1,4a,4b,5,6,6a,7,8,9,9a,9b,10,11,11a-Tetradecahydro-4a,6a-dimethyl-7-(3-methyl-1-oxobutyl)-2*H*-indeno[5,4-*f*]quinolin-2-one (●)

S L 654066, MK-963

U 5α-Reductase inhibitor

12126

$C_{23}H_{35}NO_2$
150748-24-6

(17β)-17-[(5-Hydroxypentyl)amino]estra-1,3,5(10)-trien-3-ol (●)

S *Pentolame*

U Anticoagulant, estrogenic

12127 (9700)

$C_{23}H_{35}NO_2$
68681-43-6

(±)-1-[(2*R**,6*S**,11*S**)-1,2,3,4,5,6-Hexahydro-8-hydroxy-3,6,11-trimethyl-2,6-methano-3-benzazocin-11-yl]-6-methyl-3-heptanone = (2α,6α,11*S**)-(±)-1-(1,2,3,4,5,6-

Hexahydro-8-hydroxy-3,6,11-trimethyl-2,6-methano-3-benzazocin-11-yl)-6-methyl-3-heptanone (●)
R Methanesulfonate (salt) (74559-85-6)
S Win 42964-4, *Zenazocine mesylate*
U Analgesic

12128 (9701) $C_{23}H_{35}NO_2$
                                                           71461-18-2

(±)-1-[(2R*,6S*,11S*)-1,2,3,4,5,6-Hexahydro-8-hydroxy-3,6,11-trimethyl-2,6-methano-3-benzazocin-11-yl]-3-octanone = (2α,6α,11S*)-(±)-1-(1,2,3,4,5,6-Hexahydro-8-hydroxy-3,6,11-trimethyl-2,6-methano-3-benzazocin-11-yl)-3-octanone (●)
R Methanesulfonate (salt) (73789-00-1)
S *Tonazocine mesilate**, Win 42156-2
U Analgesic

12129 $C_{23}H_{35}NO_6$
                                                           145147-04-2

N-(6,7:9,10-Diepoxy-2,8-dihydroxy-1-oxaspiro[4.5]dec-3-yl)-3,4-dimethyl-2,4-dodecadienamide = N-(Dihydro-5',6-dihydroxyspiro[4,8-dioxatricyclo[5.1.0.0³,⁵]octane-2,2'(3'H)-furan]-4'-yl)-4,6-dimethyl-2,4-dodecadienamide (●)
S *Aranorosinol-A*
U Antibacterial, antifungal from *Pseudoarachniotus roseus*

12130 (9702) $C_{23}H_{35}N_3O$
                                                           96914-39-5

(±)-*cis*-4-[2-(Diisopropylamino)ethyl]-4,4a,5,6,7,8-hexahydro-1-methyl-4-phenyl-3H-pyrido[1,2-c]pyrimidin-3-one = *cis*-(±)-4-[2-[Bis(1-methylethyl)amino]ethyl]-4,4a,5,6,7,8-hexahydro-1-methyl-4-phenyl-3H-pyrido[1,2-c]pyrimidin-3-one (●)
S *Actisomide**, *Dizactamide*, SC-36602
U Cardiac depressant (anti-arrhythmic)

12131 (9703) $C_{23}H_{35}N_3O_2$
                                                           81947-79-7

8-[[4-(Octanoylamino)-1-methylbutyl]amino]-6-methoxyquinoline = N-[4-[(6-Methoxy-8-quinolinyl)amino]pentyl]octanamide (●)
S *Octanoylprimaquine*
U Antimalarial

12132 (9704) $C_{23}H_{36}N_2$
                                                           82985-31-7

cis-N-(2-Phenylcyclopentyl)azacyclotridec-1-en-2-amine (●)
R Monohydrochloride (40297-09-4)
S MDL 12330 A, RMI 12330 A

U   Antisecretory (adenylate cyclase inhibitor)

12133   (9705)

$C_{23}H_{36}N_2O_2$
98319-26-7

N-tert-Butyl-3-oxo-4-aza-5α-androst-1-ene-17β-carbox-
amide = (5α,17β)-*N*-(1,1-Dimethylethyl)-3-oxo-4-
azaandrost-1-ene-17-carboxamide (●)
S   Alfasin, Anatine, Andozac, Avertex, Beneprost,
Capiro, Chibro-Proscar, Dinaprost, Eutiz, Fena-
sten, Finaspros, Finast, *Finasteride***, Finasterin, Fi-
nastid, MK-906, Nasterid-A, Nasteril, Poruxin, Pro-
cure, Propecia, Proscar, Prosh, Prostene, Prosterid,
Prostide, Reprostom, Tealep, Urprosan, YM-152
U   Antineoplastic (5α-reductase inhibitor)

12134   (9706)

$C_{23}H_{36}N_2O_2$
77582-30-0

2-Pentadecyl-1*H*-benzimidazole-5-carboxylic acid (●)
S   M. & B. 35347 B
U   Hypolipidemic, hypoglycemic, anti-obesity

12135

$C_{23}H_{36}N_4O_5$
106314-87-8

N-[2-[2-(Hydroxyamino)-2-oxoethyl]-4-methyl-1-oxo-
pentyl]-L-leucyl-L-phenylalaninamide (●)
S   U-24522
U   Chondroprotective (metalloproteinase inhibitor)

12136   (9707)

$C_{23}H_{36}N_4O_5S_3$
137-86-0

8-[[2-[*N*-[(4-Amino-2-methyl-5-pyrimidinyl)me-
thyl]formamido]-1-(2-hydroxyethyl)propenyl]dithio]-6-
mercaptooctanoic acid methyl ester *S*-acetate = 6-(Acetylthio)-8-[[2-[[(4-amino-2-methyl-5-pyrimidinyl)me-
thyl]formylamino]-1-(2-hydroxyethyl)-1-prope-
nyl]dithio]octanoic acid methyl ester (●)
S   Neuvita(n), *Octotiamine***, TATD
U   Analgesic (Vitamin B$_1$ source)

12137   (9708)

$C_{23}H_{36}N_6O_5S$
74863-84-6

(2*R*,4*R*)-4-Methyl-1-[(*S*)-*N*^{2}-[[(*RS*)-1,2,3,4-tetrahydro-3-
methyl-8-quinolyl]sulfonyl]arginyl] pipecolic acid =
(2*R*,4*R*)-4-Methyl-1-[*N*^{2}-[(1,2,3,4-tetrahydro-3-methyl-
8-quinolyl)sulfonyl]-L-arginyl]-2-piperidinecarboxylic
acid = [2*R*-[1(2*S**),2α,4β]]-[*partial*]-1-[5-[(Aminoiminomethyl)amino]-1-oxo-2-[[(1,2,3,4-tetrahydro-3-methyl-
8-quinolinyl)sulfonyl] amino]pentyl]-4-methyl-2-piperi-
dinecarboxylic acid (●)
R   also monohydrate (141396-28-3)

S *Argatroban***, *Argipidine*, DK 7419, GN 1600, MCI-9038, MD 805, MQPA, NA, Novastan, OM-805, Slonnon
U Antithrombotic

12138 $\qquad$ $C_{23}H_{36}O_2$
171063-54-0

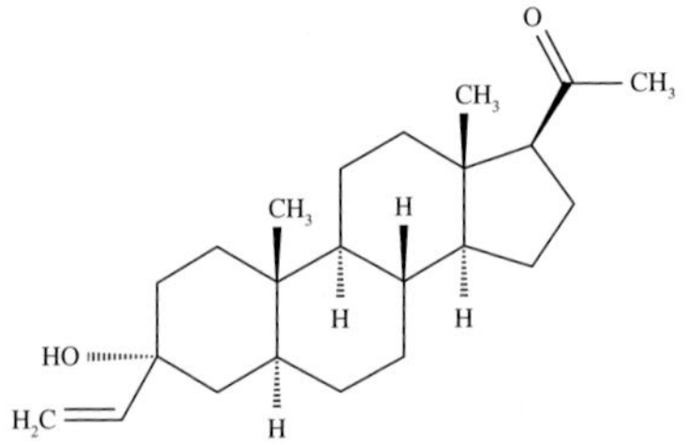

3β-Ethenyl-3α-hydroxy-5α-pregnan-20-one = (3α,5α)-3-Ethenyl-3-hydroxypregnan-20-one (●)
S Co 3-0593
U Anxiolytic, anticonvulsant (GABA$_A$ receptor modulator)

12139 (9709) $\qquad$ $C_{23}H_{36}O_2$
21208-26-4

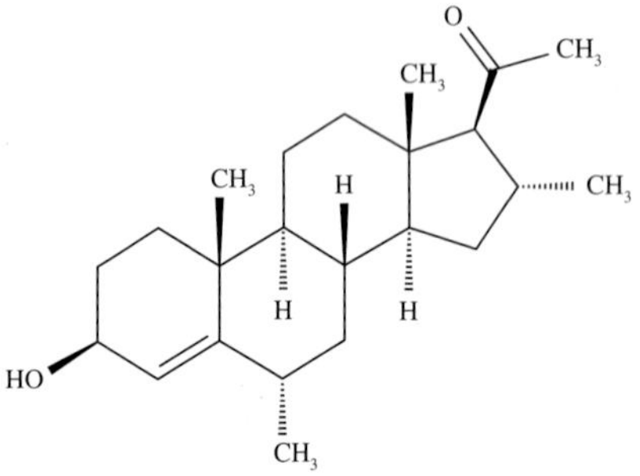

3β-Hydroxy-6α,16α-dimethylpregn-4-en-20-one = (3β,6α,16α)-3-Hydroxy-6,16-dimethylpregn-4-en-20-one (●)
S *Dimepregnen***, St 1411
U Anti-estrogen

12140 (9712) $\qquad$ $C_{23}H_{36}O_3$
521-12-0

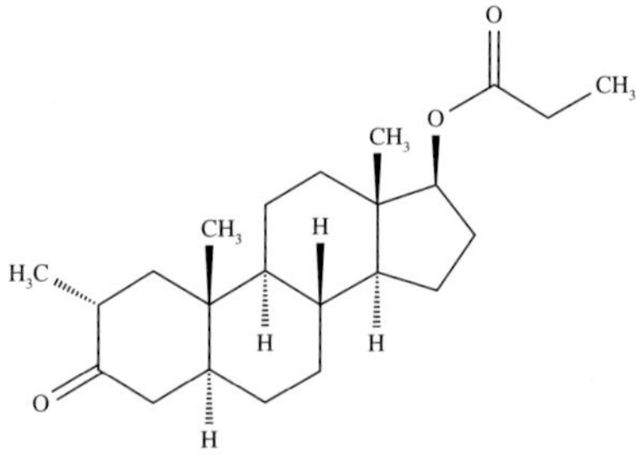

17β-Hydroxy-2α-methyl-5α-androstan-3-one propionate = 2α-Methyldihydrotestesterone propionate = (2α,5α,17β)-2-Methyl-17-(1-oxopropoxy)androstan-3-one (●)
S CS 1507, Drolban, *Dromostanolone propionate*, *Drostanolone propionate***, Emdisterone, Lilly 32379, Masterid, Masteril, Masteron, Mastisol, Medrotestron propionate, *Metalona propionato*, Metormon, NSC-12198, Permastril, *Prometholone*, RS-877
U Antineoplastic (treatment of advanced or metastatic breast cancer)

12141 (9711) $\qquad$ $C_{23}H_{36}O_3$
60883-73-0

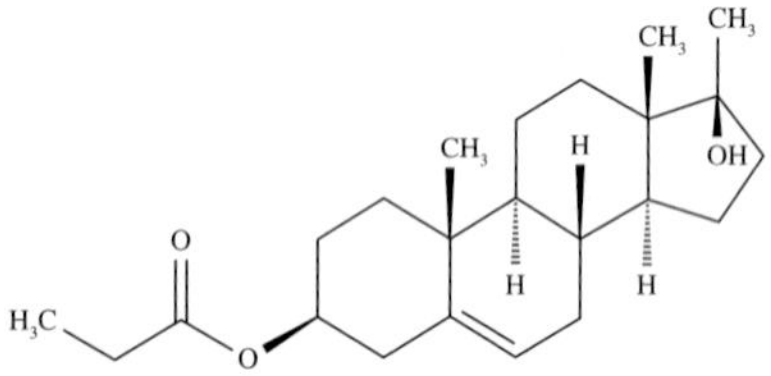

17-Methylandrost-5-ene-3β,17β-diol 3-propionate = (3β,17β)-17-Methylandrost-5-ene-3,17-diol 3-propanoate (●)
S *Methandriol propionate*, *Methylandrostendiol propionate*, Metilbisexovis-fiale, Metil-Bisexovister, Metildiolo-fiale, Stenosterone-fiale
U Anabolic

12142   (9710) $C_{23}H_{36}O_3$
3638-82-2

19-Nor-17α-pregn-4-ene-3β,17-diol 3-propionate = 17α-
Ethylestr-4-ene-3β,17-diol 3-propionate = (3β,17α)-19-
Norpregn-4-ene-3,17-diol 3-propanoate (●)
S   *Propetandrol***, *Propethandrol*, SC-7294, Solevar
U   Anabolic, androgen

12143   (9713) $C_{23}H_{36}O_5$
59982-03-5

20-Isopropylidene-PGE$_2$ = [1R-[1α(Z),2β(1E,3S*),3α]]-
7-[3-Hydroxy-2-(3-hydroxy-9-methyl-1,8-decadienyl)-5-
oxocyclopentyl]-5-heptenoic acid (●)
S   CS 412
U   Bronchodilator

12144   (9714) $C_{23}H_{36}O_5$
73647-73-1

(±)-Methyl (Z)-7-[(1R,2R,3R)-2-[(E)-(4RS)-4-butyl-1,5-
hexadienyl]-3-hydroxy-5-oxocyclopentyl]-5-heptenoate
= (±)-15-Deoxy-16-hydroxy-16-vinylprostaglandin E$_2$
methyl ester = Methyl (5Z,13E)-(11R,16RS)-7-[3-hydr-
oxy-2-(4-hydroxy-4-vinyloct-1-enyl)-5-oxocyclopen-
tyl]hept-5-enoate = (5Z,11α,13E)-(±)-16-Ethenyl-11,16-

dihydroxy-9-oxoprosta-5,13-dien-1-oic acid methyl ester
(●)
S   CL 115347, DHV-PGE$_2$ME, *Viprostol***
U   Vasodilator, antihypertensive

12145 $C_{23}H_{36}O_6$
155468-85-2

(4S,4aR,5S,7R,8R,8aS)-*rel*-(+)-4a,5,6,7,8,8a-Hexahydro-
2,5,8-trihydroxy-4-[(2Z)-3-hydroxy-1-oxo-2-propenyl]-
4,7-dimethyl-3-[(1R)-1-methylheptyl]-1(4H)-naphthale-
none (●)
S   *Australifungin*
U   Sphingolipid synthesis inhibitor fro *Sporormiella au-
stralis*

12146   (9715) $C_{23}H_{36}O_7$
81093-37-0

(+)-(βR,δR,1S,2S,6S,8S,8aR)-1,2,6,7,8,8a-Hexahydro-
β,δ,6,8-tetrahydroxy-2-methyl-1-naphthaleneheptanoic
acid 8-[(2S)-2-methylbutyrate] = (3R,5R)-7-
[(1S,2S,6S,8S,8aR)-1,2,6,7,8,8a-Hexahydro-6-hydroxy-
2-methyl-8-[(S)-2-methylbutyryloxy]-1-naphthyl]-3,5-di-
hydroxyheptanoic acid = [1S-[1α(βS*,δS*),2α,6α,8β-
(R*),8aα]]-1,2,6,7,8,8a-Hexahydro-β,δ,6-trihydroxy-2-
methyl-8-(2-methyl-1-oxobutoxy)-1-naphthalenehepta-
noic acid (●)
R   Monosodium salt (81131-70-6)
S   Aplactin, Bristacol, CS 514, Elisor, *Eptastatin sodium*, Lipemol, Lipidal, Liplat, Lipostat "Bristol-Myers
Squibb", Liprevil, Maxudin, Mevalotin, Oliprevin,
Pralidon, Prareduct, Prasterol, Prava "B.-M. Sq./

South Africa", Pravach L, Pravachol, Pravacol, Pravaselect, Pravasin, *Pravastatin sodium***, Prevacol, Privil, Sanaprav, Selectin, Selektine, Selipran, Seliprem, SQ 31000, Vasten

U Antihyperlipidemic (HMGCoA reductase inhibitor)

12147　　　　　　　　　　　　　　　　　　$C_{23}H_{37}BN_6O_5$
　　　　　　　　　　　　　　　　　　　　167843-21-2

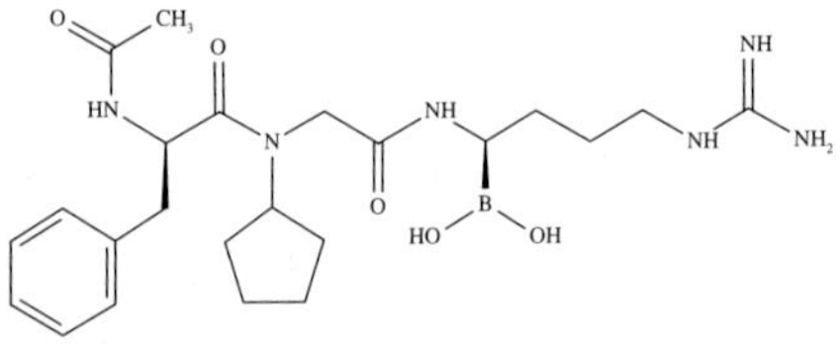

Acetyl-D-phenylalanyl-N-(cyclopentyl)glycyl-L-1-boroarginine = N-Acetyl-D-phenylalanyl-N-[(1R)-4-[(aminoiminomethyl)amino]-1-boronobutyl]-N^2-cyclopentylglycinamide (●)

S S 18326

U Antithrombotic

12148　(9716)　　　　　　　　　　　　　　$C_{23}H_{37}NO_4$
　　　　　　　　　　　　　　　　　　　　102191-05-9

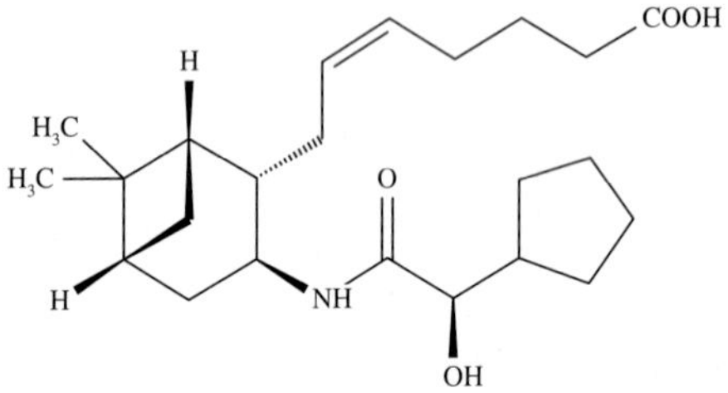

[1S-[1α,2β(Z),3α(S*),5α]]-7-[3-[(Cyclopentylhydroxyacetyl)amino]-6,6-dimethylbicyclo[3.1.1]hept-2-yl]-5-heptenoic acid (●)

S ONO 3708

U Thromboxane A_2- and prostaglandin H_2-receptor antagonist

12149　(9717)　　　　　　　　　　　　　　$C_{23}H_{38}ClN_3O$
　　　　　　　　　　　　　　　　　　　　68284-69-5

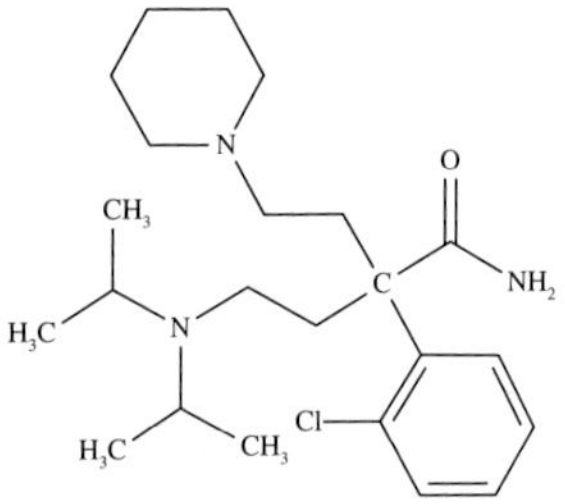

α-(o-Chlorophenyl)-α-[2-(diisopropylamino)ethyl]-1-piperidinebutyramide = α-[2-[Bis(1-methylethyl)amino]ethyl]-α-(2-chlorophenyl)-1-piperidinebutanamide (●)

S *Disobutamide***, SC-31828

U Anti-arrhythmic

12150　　　　　　　　　　　　　　　　　　$C_{23}H_{38}FNO$
　　　　　　　　　　　　　　　　　　　　166100-39-6

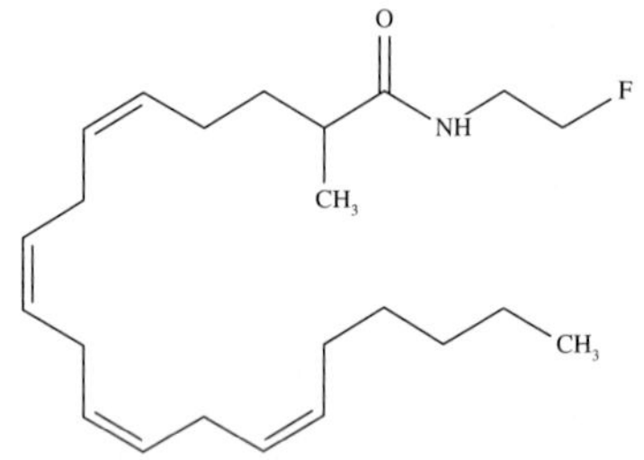

(−)-N-(2-Fluoroethyl)-2-methylarachidonamide = (5Z,8Z,11Z,14Z)-N-(2-Fluoroethyl)-2-methyl-5,8,11,14-eicosatetraenamide (●)

S O-689

U Anti-inflammatory

12151　　　　　　　　　　　　　　　　　　$C_{23}H_{38}N_2O_3$
　　　　　　　　　　　　　　　　　　　　109836-81-9

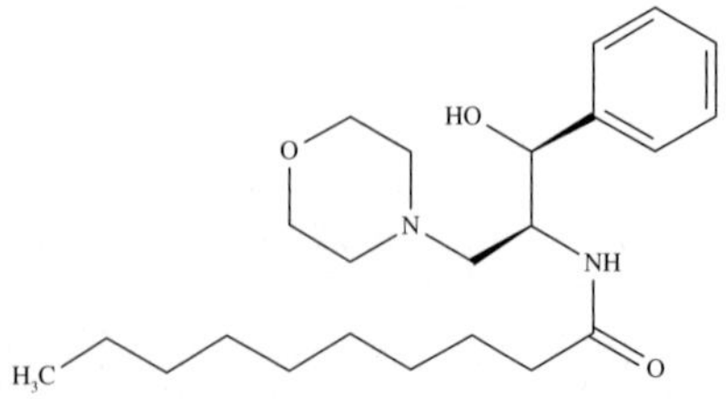

L-*threo*-1-Phenyl-2-decanoylamino-3-morpholino-1-propanol = N-[(1S,2S)-2-Hydroxy-1-(4-morpholinylmethyl)-2-phenylethyl]decanamide (●)

S   L-PDMP
U   Cognition enhancer

12152   (9718) $C_{23}H_{38}N_2O_5$
                                                              107332-47-8

[1S-[1α,2α(Z),3α,4α]]-7-[3-[[[[(1-Oxoheptyl)ami-
no]acetyl]amino]methyl]-7-oxabicyclo[2.2.1]hept-2-yl]-
5-heptenoic acid (●)
S   SQ 30741
U   Bronchospasmolytic (thromboxane A_2-receptor antagonist)

12153   (9719) $C_{23}H_{38}O$
                                                              6809-52-5

Geranylgeranylacetone = 6,10,14,18-Tetramethyl-
5,9,13,17-nonadecatetraen-2-one (●)
R   Mixture (3:2) of (5E,9E,13E)- and (5Z,9E,13E)-iso-
mers
S   Celoop, Cerbex, E 671, GGA, Selbex, *Teprenone***,
Tetprenone
U   Anti-ulcer agent

12154   (9720) $C_{23}H_{38}O_2$
                                                              79243-67-7

17β-Hydroxy-1α-methyl-17-propyl-5α-androstan-3-one
= (1α,5α,17β)-17-Hydroxy-1-methyl-17-propylandros-
tan-3-one (●)
S   17α-*Propylmesterolone, Rosterelone, Rostero-
lone***, SH 434, ZK 36868
U   Anti-androgen, sebosuppressive

12155   (9721) $C_{23}H_{38}O_4$
                                                              61263-35-2

(Z)-7-[(1R,2R,3R)-3-Hydroxy-2-[(E)-(3R)-3-hydroxy-
4,4-dimethyl-1-octenyl]-5-methylenecyclopentyl]-5-hep-
tenoic acid = (5Z,11α,13E,15R)-11,15-Dihydroxy-16,16-
dimethyl-9-methyleneprosta-5,13-dien-1-oic acid (●)
S   *Meteneprost***, U-46785
U   Oxytocic (abortifacient)

12156   (9722) $C_{23}H_{38}O_4$
                                                              69900-72-7

(Z)-7-[(1R,2R,3R)-2-[(E)-(3R)-3-Hydroxy-4,4-dimethyl-
1-octenyl]-3-methyl-5-oxocyclopentyl]-5-heptenoic acid
= (5Z,11α,13E,15R)-15-Hydroxy-11,16,16-trimethyl-9-
oxoprosta-5,13-dien-1-oic acid (●)
S   Ro 21-6937/000, *Trimoprostil***, Ulpax "Roche",
Ulstar
U   Gastric antisecretory

12157   (9723) $C_{23}H_{38}O_5$
64318-79-2

Methyl (*E*)-7-[(1*R*,2*R*,3*R*)-3-Hydroxy-2-[(*E*)-(3*R*)-3-
hydroxy-4,4-dimethyl-1-octenyl]-5-oxocyclopentyl]-2-
heptenoate = (2*E*,11α,13*E*,15*R*)-11,15-Dihydroxy-16,16-
dimethyl-9-oxoprosta-2,13-dien-1-oic acid methyl ester
(●)
S   Cergem, Cervagem(e), Cervegem, Cervidil "Serono", *Gemeprost***, ONO 802, Preglandin, SC-37681
U   Inducer of uterine muscle contraction

12158   (9724) $C_{23}H_{38}O_6$
70667-26-4

Methyl (−)-(1*R*,2*R*,3*R*)-3-hydroxy-2-[(*E*)-(3*S*,5*S*)-3-hydr-
oxy-5-methyl-1-nonenyl]-ε,5-dioxocyclopentanehepta-
noate = (11α,13*E*,15*S*,17*S*)-11,15-Dihydroxy-17,20-di-
methyl-6,9-dioxoprost-13-en-1-oic acid methyl ester (●)
S   Alloca, ONO 1308, *Ornoprostil***, OU-1308,
Ronok, *Ronoprost*
U   Anti-ulcer agent, cytoprotective

12159   (9725) $C_{23}H_{39}NO_2$
55986-43-1

4-(Hexadecylamino)benzoic acid (●)
R   Sodium salt (64059-66-1)

S   *Cetaben sodium***, CL 203821, PHB
U   Antihyperlipidemic

12160   (9726) $C_{23}H_{39}NO_4$
114289-47-3

2,2-Dimethyl-*N*-(2,4,6-trimethoxyphenyl)dodecanamide
(●)
S   CI-976, PD 128042
U   Antihyperlipidemic (HMGCoA reductase inhibitor)

12161 $C_{23}H_{39}N_3O_6S_2$
160141-08-2

(2*S*)-2-[[(2*S*)-2-[[(2*S*,3*S*)-2-[[(2*R*)-2-Amino-3-mercapto-
propyl]amino]-3-methylpentyl]oxy]-1-oxo-3-phenylpro-
pyl]amino]-4-(methylsulfonyl)butanoic acid (●)
S   L 739750
U   Farnesyltransferase inhibitor

12162 $C_{23}H_{39}N_9O_8S_2$
126053-71-2

N-Acetyl-L-cysteinyl-*N*²-methyl-L-arginylglycyl-L-α-as-
partyl-3-mercapto-L-valinamide cyclic (1→5)-disulfide
(●)
S   SKF 106760
U   Fibrinogen receptor antagonist

12163   (9727) $C_{23}H_{40}FN_2O_8P$
                                                        86976-77-4

Tetradecyl 2'-deoxy-5-fluoro-5'-uridylate = 2'-Deoxy-5-
fluoro-5'-uridylic acid monotetradecyl ester (●)
S   *Floxuridine fosdecate*, *Floxuridine fostedate***,
      TEI-6170
U   Antineoplastic, antiviral

12164   (9728) $C_{23}H_{40}N_2O_3$

1-(α-Carboxybenzyl)-1-methylpiperidinium diethyl[2-(2-
hydroxyethoxy)ethyl]methylammonium ester = 1-[α-[[2-
[2-(Diethylmethylammonio)ethoxy]ethoxy]carbo-
nyl]benzyl]-1-methylpiperidinium = 1-[2-[2-[2-(Diethylmethylammonio)ethoxy]ethoxy]-2-oxo-1-phenylethyl]-
1-methylpiperidinium (●)
R   Diiodide (3562-55-8)
S   Brévicurarine, L.D. 2480, *Piprocurarium iodide***
U   Muscle relaxant

12165 $C_{23}H_{40}N_2O_4$
                                                        153601-03-7

N-[4-[2-Hydroxy-3-(isopropylamino)propoxy]-3-meth-
oxybenzyl]nonanamide = N-[[4-[2-Hydroxy-3-[(1-me-

thylethyl)amino]propoxy]-3-methoxyphenyl]methyl]no-
nanamide (●)
S   *Capsinolol*
U   β-Adrenergic blocker, CGRC releasing drug

12166   (9729) $C_{23}H_{40}N_4O_{11}$
                                                        74817-61-1

2-Acetamido-3-O-[(R)-1-[[(S)-1-[[(R)-3-carbamoyl-1-
carboxypropyl]carbamoyl]ethyl]carbamoyl]ethyl]-2-de-
oxy-D-glucopyranose butyl ester = N^2-[N-(N-Acetylmuramoyl)-L-alanyl]-D-glutamine butyl ester (●)
S   *Murabutide***
U   Immunomodulator

12167 $C_{23}H_{40}O_2$
                                                        69505-74-4

2-(9-Methyldecyl)-5-(4-methylpentyl)-1,3-benzenediol
(●)
S   *Resorcinin*
U   Mitogenic

12168   (9731) $C_{23}H_{40}O_5$
                                                        139403-31-9

(+)-Methyl [2-[(2R,3aS,4R,5R,6aS)-octahydro-5-hydr-
oxy-4-[(1E,3S,5S)-3-hydroxy-5-methyl-1-nonenyl]-2-
pentalenyl]ethoxy]acetate = [2R-
[2α,3aα,4α(1E,3S*,5S*),5β,6aα]]-2-[Octahydro-5-
hydroxy-4-(3-hydroxy-5-methyl-1-nonenyl)-2-pentale-
nyl]ethoxy]acetic acid methyl ester (●)

S *Pidilprost, Pimilprost***, SM 10902
U Peripheral vasodilator, platelet aggregation inhibitor

12169   (9730) $C_{23}H_{40}O_5$
7311-27-5

α-(*p*-Nonylphenyl)-ω-hydroxytetra(oxyethylene) = 11-
(*p*-Nonylphenoxy)-3,6,9-trioxa-1-undecanol = 2-[2-[2-[2-
(4-Nonylphenoxy)ethoxy]ethoxy]ethoxy]ethanol (●)
R Average for a *p*-nonylphenoxypolyethoxyethanol
S Igepal CO-430, *Nonoxinol 4***, *Nonoxynol 4*
U Antiseptic, surfactant

12170   (9732) $C_{23}H_{40}O_6$
88980-20-5

Methyl (1*R*,2*R*,3*R*)-3-hydroxy-2-[(*E*)-3-hydroxy-4-meth-
oxy-4-methyloctyl]-5-oxocyclopentaneheptanoate =
(11α,13*E*,15*R*)-11,15-Dihydroxy-16-methoxy-16-me-
thyl-9-oxoprost-13-en-1-oic acid methyl ester (●)
S DL 646, MDL 646, *Mexiprostil***
U Gastroprotective

12171   (9733) $C_{23}H_{41}N_2O$
73091-68-6

Benzyl[(dodecylcarbamoyl)methyl]dimethylammonium
= *N*-[2-(Dodecylamino)-2-oxoethyl]-*N,N*-dimethylben-
zenemethanaminium (●)
R Chloride (100-95-8)

S *Dodecarbonium chloride*, Hydramon, *Metalkonium
chloride***, Nopcocide, Straminol, Theotex, Uroloci-
de
U Antiseptic

12172 $C_{23}H_{41}N_5O_3$
115976-91-5

(α*S*)-*N*-[4-[[3-[(3-Aminopropyl)amino]propyl]amino]bu-
tyl]-4-hydroxy-α-[(1-oxobutyl)amino]benzenepro-
panamide (●)
S *Philanthotoxin 433*, PhTX-433
U Noncompetitive inotropic receptor antagonist

12173 $C_{23}H_{41}N_5O_5S$
259188-38-0

N-[(2*S*)-2-Mercapto-1-oxo-4-(3,4,4-trimethyl-2,5-dioxo-
1-imidazolidinyl)butyl]-L-leucyl-*N*,3-dimethyl-L-valin-
amide (●)
S BMS-275291-01, *Crebizmastat*, D 2163
U Matrix metalloproteinase inhibitor, anti-angiogenic

12174   (9734) $C_{23}H_{41}N_7O_{14}$
123482-12-2

2″-*N*-Glycoloylstreptomycin = *O*-2-Deoxy-2-[(hydroxyacetyl)methylamino]-α-L-glucopyranosyl-(1→2)-*O*-5-
deoxy-3-*C*-formyl-α-L-lyxofuranosyl-(1→4)-*N*,*N'*-
bis(aminoiminomethyl)-D-streptamine (●)
S *Ashimycin B*

U   Antibacterial (broad spectrum)

12175   (9735) $C_{23}H_{42}N$
16287-71-1

Benzyldimethyltetradecylammonium = *N,N*-Dimethyl-*N*-tetradecylbenzenemethanaminium (●)
R   Chloride (139-08-2)
S   Alpagelle, Faringets (active substance), *Miristalkonium chloride***, *Miristylbenzalkonium chloride*
U   Antiseptic

12175-01   61134-95-0
R   Fluoride
S   TDBAF
U   Tuberculostatic

12176   (9736) $C_{23}H_{42}NO_2$
23884-64-2

Benzyldodecylbis(2-hydroxyethyl)ammonium = *N*-Dodecyl-*N,N*-bis(2-hydroxyethyl)benzenemethanaminium (●)
R   Chloride (19379-90-9)
S   Absonal, Bactofen, Bactofen-blu, *Benzoxonium chloride***, Bialcol, Bradophen, D 301, Desinfektionsspray "Atarost", *Disinfettante D. 77*, *Dodetonii chloridum*, Lomades, Orocil, Orofar, Sinecod Bocca
U   Antiseptic, surfactant

12177   (9737) $C_{23}H_{42}N_2$
4282-07-9

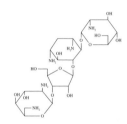

3β-Amino-20(*S*)-(dimethylamino)-5α-pregnane =
(3β,5α,20*S*)-N^{20},N^{20}-Dimethylpregnane-3,20-diamine
(●)
S   *Chonemorphine*
U   Anti-amebic

12178   (9738) $C_{23}H_{45}N_5O_{14}$
7542-37-2

O-2-Amino-2-deoxy-α-D-glucopyranosyl-(1→4)-*O*-[*O*-2,6-diamino-2,6-dideoxy-β-L-idopyranosyl-(1→3)-β-D-ribofuranosyl-(1→5)]-2-deoxy-D-streptamine (●)
R   also sulfate (1263-89-4)
S   *Aminosidine*, Aminôxidin, Amminofarma, *Amminosidina*, Amoxoral, *Antibiotic 1600*, C 1488, *Catenulin*, *Crestomicina*, *Crestomycin*, *Estomycin*, Farmiglucina, Farminosidin, F.I. 5853, Gabbrocol, Gabbromicina, Gabbromycin, Gabbroral, Gabromicina, Gabroral, Glusidin, Gluxidin, Hucyn, Humagel, Humatin, Humycin, *Hydroxymycin*, *Monomycin A*, *Neomycin E*, Pangonyl, Paramicina, Pargonyl, Paricina, *Paromomycin***, *Paucimycin*, R 400, Sinosid, Utramicina "Ravasi", Vetgabbromicina, *Zygomycin A$_1$*
U   Anti-amebic (antibiotic from *Streptomyces rimosus* var. *paromomycinus* or *Streptomyces chrestomyceticus*)

12179   (9739) $C_{23}H_{46}N_2O_3$
32954-43-1

(Carboxymethyl)dimethyl(3-palmitamidopropyl)ammo-
nium hydroxide inner salt = Dimethyl(3-palmitamidopro-
pyl)ammonioacetate = *N,N*-Dimethyl(3-palmitamidopro-
pyl)aminoacetic acid betaine = *N*-(Carboxymethyl)-*N,N*-
dimethyl-3-[(1-oxohexadecyl)amino]-1-propanaminium
hydroxide inner salt (●)
S   *Pendecamaine***, Tego-Betaines
U   Surfactant

12180   (9741) $C_{23}H_{46}N_6O_{13}$
66-86-4

O-2,6-Diamino-2,6-dideoxy-α-D-glucopyranosyl-(1→4)-
O-[*O*-2,6-diamino-2,6-dideoxy-α-D-glucopyranosyl-
(1→3)-β-D-ribofuranosyl-(1→5)]-2-deoxy-D-strept-
amine (●)
R   see also no. 12181-01
S   *Neomycin C, Streptothricin B-I*
U   Antibiotic from *Streptomyces* sp.

12181   (9740) $C_{23}H_{46}N_6O_{13}$
119-04-0

O-2,6-Diamino-2,6-dideoxy-α-D-glucopyranosyl-(1→4)-
O-[*O*-2,6-diamino-2,6-dideoxy-β-L-idopyranosyl-
(1→3)-β-D-ribofuranosyl-(1→5)]-2-deoxy-D-strept-
amine (●)
R   also sulfate (1:3) (4146-30-9)
S   Actilin, Anti-Rhinyl, Carident (active substance),
Daryant-Tulle, EF-185, Enevis, Enterfram, Frakitaci-
ne, *Framicetina*, Framidal, Framitulle, Framoccid,
Framomycin, Framybiotal, *Framycetin***, Framy-
cin, Framycoccid, Framygen, Francetin, Fraquinol,
Isofra, Isoframicol, Leukase N, *Neomycin B*, Nisoc-
lyn, Rhinalène, Rhinobioptal, Rhinobiotal, Rhinyl
(active substance), Sofracaps, Soframycin(e), So-
fra-Tulle (-Tüll), Solflen, *Streptothricin B-II*, Tutto-
mycin
U   Antibiotic from *Streptomyces* sp.

12181-01   (9740-01)   1404-04-2
R   Mixture with Neomycin A (no. 3739) and Neomycin
C (no. 12180) [also as sulfate (1405-10-3)]
S   Ani-Neopre, Apobacyn, Apokalin, Baneocin, Biofra-
din, Biosol, Burn-Gel, Bykomycin, Colicet, Colino-
vina, Colivet "A.L.", Concentrat VO 59, Cysto-Mya-
cine N, Dentargle F, Dermonalef, Dexmy,
Emorex N, Endomixin, Endomycin, Enteromicina,
Fradio, *Fradiomycin*, Fradyl, Fungicina, Gastromy-
cin, Glycomycin, Herisan antibiotic, Izoneocol, Lar-
micin, Micina, Modocel, Myacyne, Mycerin, Myci-
fradin, Myciguent, Nebacetin N, Negamicin B,
Neo 200, Neobicin vet., Neobiotic, Neobitiol, Neob-
rettin, Neocin, Neodermil, Neofracin, Neo-Fradin,
Neo-IM, Neointestin, Neolate, Neomas, *Neomicina*,
Neomin, Neomix, Neomy, Neomycane, *Neomy-
cin***, Néomydiar, Neopan, Neopt, Neo-RX,
Neo-Sol 50, Neosulf, Neo-Tabs, Neovet, Nisocla,
Nivemycin, Nokamycin, Océmycine, Ophtalkan,
Oranecin, Orojet N, Peniderm, Piodercina, P.M. 15,

Rovicine, Sogémycine, Stol, Uro-Beniktol N,
Uro-Nebacetin N, Uvanovine, Vagicillin,
Viro Neo M, Vonamycin
U Antibiotic from *Streptomyces* sp.

12181-02 (9740-02) 55298-68-5
R Palmitate of no. 12181-01
S *Neomycin palmitate*
U Antibacterial

12181-03 (9740-03) 1406-04-8
R Undecylenate of no. 12181-01
S Neodecyllin, *Neomycin undecylenate*
U Antibacterial, antifungal

12182 $C_{23}H_{50}AsO_4P$
156825-86-4

Octadecyl [2-(trimethylarsonio)ethyl] phosphate = [2-
[[Hydroxy(octadecyloxy)phosphinyl]oxy]ethyl]trime-
thylarsonium inner salt (●)
S D-21805
U Antineoplastic

12183 $C_{23}H_{50}NO_4P$
65956-63-0

Choline hydroxide octadecyl hydrogen phosphate inner
salt = 2-[[Hydroxy(octadecyloxy)phosphinyl]oxy]-
N,N,N-trimethylethanaminium inner salt (●)
S D-19391, *Octadecylphosphocholine*
U Antineoplastic

12184 $C_{23}H_{52}N_8$
110078-40-5

1,1'-(1,3-Propanediyl)bis-[1,4,8,11-tetraazacyclotetrade-
cane] (●)
S AMD 2763, JM-2763
U Antiviral

12185 $C_{24}H_{12}N_4$
120154-96-3

Dibenzo[*b,j*]dipyrido[4,3,2-*de*:2',3',4'-*gh*][1,10]phenan-
throline (●)
S *Eilatine*
U Antileukemic from the Red Sea purple tunicate *Eudis-
toma* sp.

12186 $C_{24}H_{17}NO_4$
176977-56-3

4-[[6-Methoxy-2-(4-methoxyphenyl)-3-benzofura-
nyl]carbonyl]benzonitrile (●)
S LY 320135
U Cannabinoid CB_1 receptor antagonist

12187 $C_{24}H_{18}ClF_2N_3O$
128831-46-9

N-[4-(2-Chlorophenyl)-6,7-dimethyl-3-quinolinyl]-N'-(2,4-difluorophenyl)urea (●)
S TMP-153
U ACAT inhibitor

12188 (9742) $C_{24}H_{18}FNO_3$
136326-31-3

2-Fluoro-4'-(2-quinolinylmethoxy)[1,1'-biphenyl]-4-acetic acid (●)
S WAY-121006
U Anti-allergic, anti-inflammatory

12189 $C_{24}H_{18}FN_5O_2$
158876-65-4

1-[7-(4-Fluorophenyl)-1,2,3,4-tetrahydro-8-(4-pyridinyl)pyrazolo[5,1-c][1,2,4]triazin-2-yl]-2-phenylethanedione = 7-(4-Fluorophenyl)-1,2,3,4-tetrahydro-2-(oxophenylacetyl)-8-(4-pyridinyl)pyrazolo[5,1-c][1,2,4]triazine (●)
R Sulfate (1:1) (158876-66-5)
S FR 167653
U Interleukin-1 and tumor necrosis factor inhibitor

12190 $C_{24}H_{18}N_2O_4$
26612-48-6

5,12-Dihydrocycloocta[1,2-b:5,6-b']diindole-6,13-dicarboxylic acid dimethyl ester (●)
S *Caulerpin*
U Pigment from *Caulerpa racemosa* with antineoplastic activity

12191 $C_{24}H_{18}N_4O$
136194-77-9

5,6,7,13-Tetrahydro-13-methyl-5-oxo-12H-indolo[2,3-a]pyrrolo[3,4-c]carbazole-12-propanenitrile (●)
S Go 6976, Goe 6976
U Protein kinase C inhibitor

12192 $C_{24}H_{18}N_4O_2Se_2$
115369-67-0

Bis[2-(2-pyridylcarbamoyl)phenyl]diselenide = 2,2'-Diselenobis[N-2-pyridinylbenzamide] (●)
S AE 22
U Immunomodulator

12193 (9743)

$C_{24}H_{18}N_4O_3$
101388-47-0

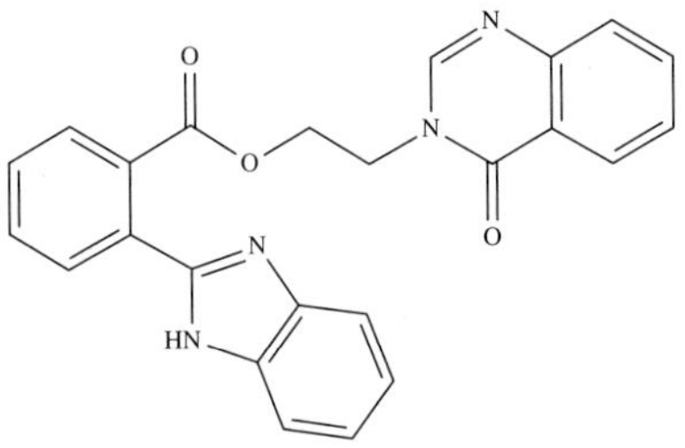

2-(3,4-Dihydro-4-oxo-3-quinazolinyl)ethyl 2-(2-benzimidazolyl)benzoate = 2-(1H-Benzimidazol-2-yl)benzoic acid 2-(3,4-dihydro-4-oxo-3-quinazolinyl)ethyl ester (●)
S VÚFB-12987
U Analgesic

12194 (9744)

$C_{24}H_{18}O_6$
5449-84-3

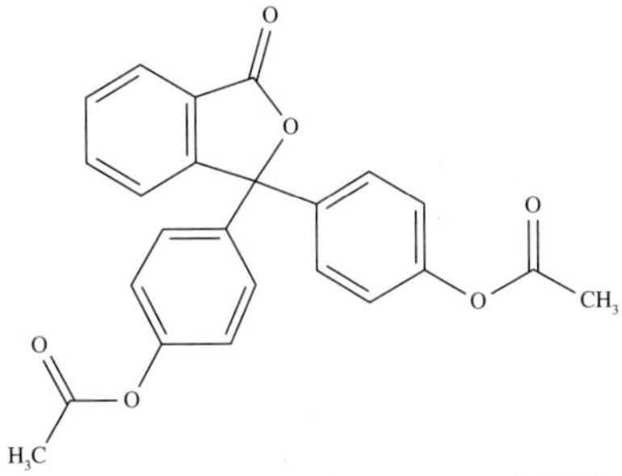

p,p'-Phthalidylidenebis(phenyl acetate) = 3,3-Bis(p-acetoxyphenyl)phthalide = Phenolphthalein diacetate = 3,3-Bis[4-(acetyloxy)phenyl]-1(3H)-isobenzofuranone (●)
R see also no. 9171-01
S Diacetophthalein, Laxoral
U Laxative

12195 (9745)

$C_{24}H_{19}BrO_4$
6192-24-1

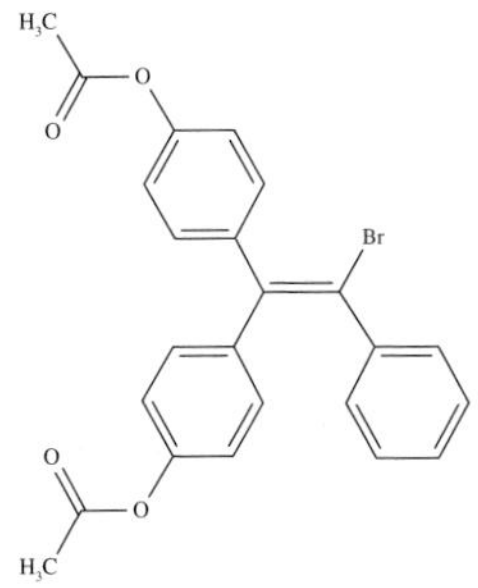

p,p'-(β-Bromostyrylidene)bis(phenyl acetate) = α,α-Bis(p-acetoxyphenyl)-β-bromo-β-phenylethene = 4,4'-(Bromophenylethylidene)bis[phenol] diacetate (ester) (●)
S Ovobrene
U Synthetic estrogen

12196 (9746)

$C_{24}H_{19}ClN_4O_3$
64039-88-9

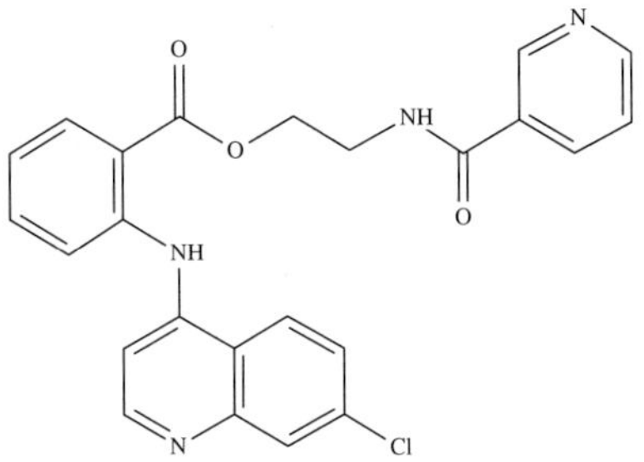

N-(7-Chloro-4-quinolyl)anthranilic acid ester with N-(2-hydroxyethyl)nicotinamide = N-(7-Chloro-4-quinolyl)anthranilic acid 2-nicotinamidoethyl ester = 2-[(7-Chloro-4-quinolinyl)amino]benzoic acid 2-[(3-pyridinylcarbonyl)amino]ethyl ester (●)
S Nicafenine**
U Analgesic

12197

$C_{24}H_{19}F_3O_4$
204981-48-6

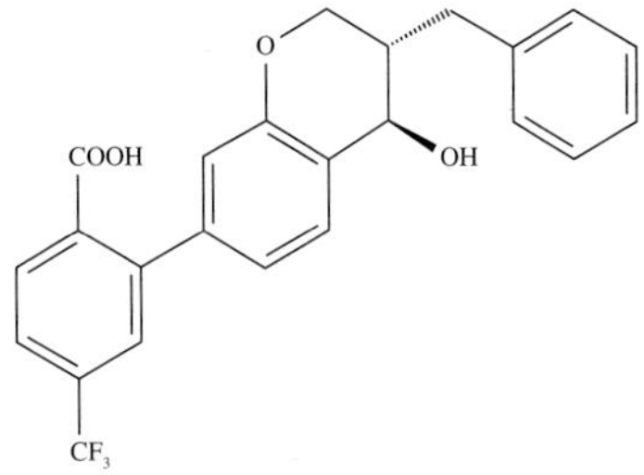

(+)-2-[(3S,4R)-3-Benzyl-4-hydroxychroman-7-yl]-4-(trifluoromethyl)benzoic acid = 2-[(3S,4R)-3,4-Dihydro-4-hydroxy-3-(phenylmethyl)-2H-1-benzopyran-7-yl]-4-(trifluoromethyl)benzoic acid (●)
S CP-195543
U Leukotriene B$_4$ receptor antagonist

12198 (9747) $C_{24}H_{19}NO_5$
                                                       115-33-3

p,p'-(2-Oxoindolin-3-ylidene)bis(phenyl acetate) = 3,3-Bis(p-acetoxyphenyl)oxindole = Diacetylbis(p-hydroxyphenyl)isatin = 3,3-Bis(p-acetoxyphenyl)indolin-2-one = 3,3-Bis[4-(acetyloxy)phenyl]-1,3-dihydro-2H-indol-2-one (●)

S Acelax, Acetalax, *Acetfenolisatin*, *Acetphenolisatin*, *Acetylphenylisatin*, Apitin, Apperisatin, Asitin, *Bisatin*, Bisco-Zitron-Dragees, Brocatine, Bydolax, Ciraceen, Cirotex, Cirotyl, Contax, Curalax, Darmoletten (old form), *Diacetyldiphenolisatin*, *Diasatin*, *Dioxindol*, *Diphésatine*, Disacetine, Distalene, Ditin, Emolax, *Endophenolphthalein*, Eulaxin, Fenisan, Fenlaxin, Fiel-Lax, Gardalaxan, Granulax, Inlax, Isacen, Isalax, Isaphen, Isaphenin, Isocrin, Izafenin, Izaman, Laxaseptol, Laxatan forte, Laxen, Laxigen, Laxnormal, Laxobilina, Laxocol, Laxo-Isatin, Laxotinal, Laxyl, Leacen, Lenavac, Lisagal, Mileval, Neda-Dragees, Neolax "Interpharm", Neo-Prunex, Normacen, Nourilax, Novolax, NSC-59687, Nurilaksi, Obstipan, *Oxyphenisatin acetate*, *Oxyphenisatinum diaceticum*, Phenlaxinum, Phenylisatin, Promassolax, Prudents, Prulax, Prulet (old form), Prutab, Purgaceen, Purgazen-Tabl., Purgetten, Purgophen "Glutan", Regal "Ferrosan", Sanapert, Schokolax, Surlax, Sur-Laxante, Taxin, Tete-Lax, Vinco-Abführperlen (old form)
U Laxative

12199 (9748) $C_{24}H_{19}NO_6$
                                                       14008-48-1

p,p'-(2,3-Dihydro-3-oxo-1,4-benzoxazin-2-ylidene)bis(phenyl acetate) = 2,2-Bis(p-acetoxyphenyl)-2,3-dihydro-3-oxobenz-1,4-oxazine = 2,2-Bis[4-(acetyloxy)phenyl]-2H-1,4-benzoxazin-3(4H)-one (●)
S *Bisoxatin acetate***, Exodol, Kritel, La 271a, Lasotin, Laxonalin, Maratan, Metrolax, Regoxal, Talsis, Wy-8138, Wylaxine
U Laxative

12200 $C_{24}H_{19}N_3O$
                                                       174232-22-5

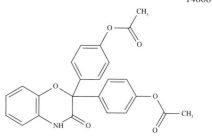

N-5-Acenaphthylenyl-N'-(4-methoxy-1-naphthalenyl)guanidine (●)
S CNS 1237
U Ca^{2+} and Na^+ channel inhibitor

12201 (9749) $C_{24}H_{19}N_3O_5S$
                                                       87234-24-0

2-Methyl-3-(2-pyridylcarbamoyl)-2*H*-1,2-benzothiazin-4-yl cinnamate *S,S*-dioxide = 3-Phenyl-2-propenoic acid 2-methyl-3-[(2-pyridinylamino)-carbonyl]-2*H*-1,2-benzothiazin-4-yl ester *S,S*-dioxide (●)

S *Cinnoxicam*, *Piroxicam cinnamate*, Sinartrol "SPA", SPA-S 510, Zelis, Zen

U Anti-inflammatory

12202 $C_{24}H_{20}BrClN_2O$
                                           191034-25-0

2-(4-Pyridyl)-5-(4-chlorophenyl)-3-(5-bromo-2-propoxyphenyl)pyrrole = 4-[3-(5-Bromo-2-propoxyphenyl)-5-(4-chlorophenyl)-1*H*-pyrrol-2-yl]pyridine (●)

S L 168049

U Human glucagon receptor antagonist

12203   (9750) $C_{24}H_{20}BrClN_6O_3$
                                           145781-32-4

1-[[3-Bromo-2-[2-(1*H*-tetrazol-5-yl)phenyl]-5-benzofuranyl]methyl]-2-butyl-4-chloro-1*H*-imidazole-5-carboxylic acid (●)

S GR 117289, GR 117289X, *Zolasartan***

U Antihypertensive (angiotensin II antagonist)

12204   (9751) $C_{24}H_{20}Cl_2N_2OS$
                                           72479-26-6

1-[2,4-Dichloro-β-[[*p*-(phenylthio)benzyl]oxy]phenethyl]imidazole = 1-(2,4-Dichlorophenyl)-2-imidazol-1-ylethyl 4-(phenylthio)benzyl ether = 1-[2-(2,4-Dichlorophenyl)-2-[[4-(phenylthio)phenyl]methoxy]ethyl]-1*H*-imidazole (●)

R Mononitrate (73151-29-8)

S Falvin, Fenizolan, *Fenticonazole nitrate***, Fentiderm, Fentigyn, Fentikol, Gyno-Lomexin, Gyno-Mycodermil, Gynoxin, Lomexin, Micofentin, Mycodermil, Rec 15/1476, Terlomexin

U Antifungal

12205   (9752) $C_{24}H_{20}Cl_2N_2O_3S$
                                           86273-92-9

7-Chloro-5-(*o*-chlorophenyl)-1,3-dihydro-1-[2-(*p*-tolylsulfonyl)ethyl]-2*H*-1,4-benzodiazepin-2-one = 7-Chloro-5-(2-chlorophenyl)-1,3-dihydro-1-[2-[(4-methylphenyl)sulfonyl]ethyl]-2*H*-1,4-benzodiazepin-2-one (●)

S *Tolufazepam***

U Anticonvulsant, tranquilizer

12206 (9753)

$C_{24}H_{20}F_6N_6O_2$
103961-78-0

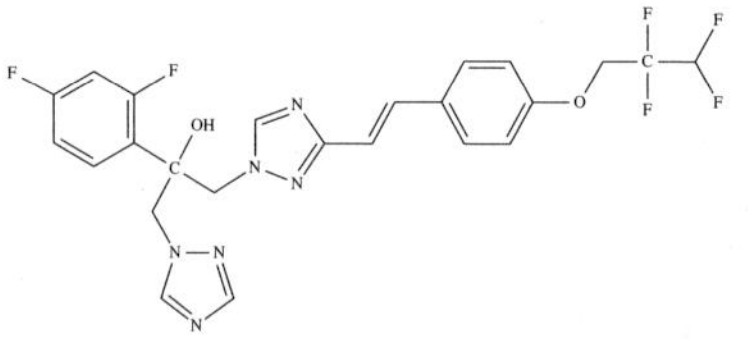

α-(2,4-Difluorophenyl)-3-[2-[4-(2,2,3,3-tetrafluoroprop-oxy)phenyl]ethenyl]-α-(1H-1,2,4-triazol-1-ylmethyl)-1H-1,2,4-triazole-1-ethanol (●)
S ICI 195739
U Antifungal

12207

$C_{24}H_{20}F_6N_6O_2$
149715-95-7

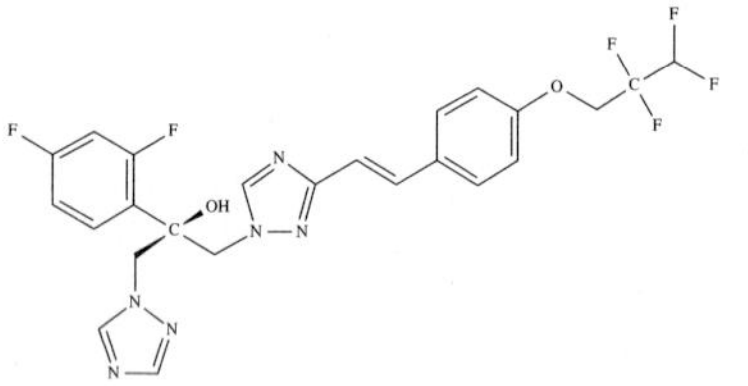

(αR)-α-(2,4-Difluorophenyl)-3-[(1E)-2-[4-(2,2,3,3-tetra-fluoropropoxy)phenyl]ethenyl]-α-(1H-1,2,4-triazol-1-yl-methyl)-1H-1,2,4-triazole-1-ethanol (●)
R Sulfate (1:1) (salt) (141113-29-3)
S D-0870, ICID 870, M-16354, ZD 0870
U Antifungal

12208 (9754)

$C_{24}H_{20}I_6N_4O_8$
10397-75-8

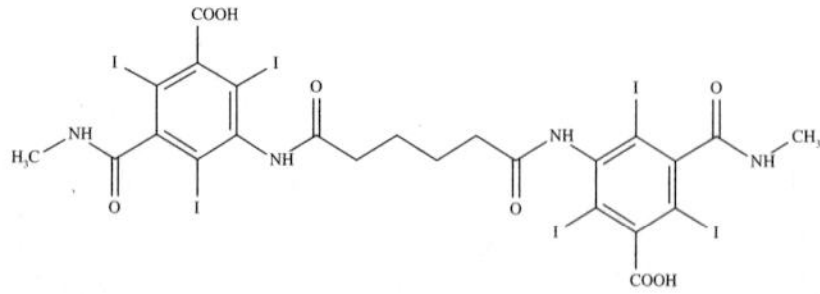

5,5'-(Adipoiyldiimino)bis[2,4,6-triiodo-N-methyliso-phthalamic acid] = 2,2',4,4',6,6'-Hexaiodo-5,5'-adipoyl-diamidobis(N-methylisophthalamic acid) = 3,3'-[(1,6-Di-oxo-1,6-hexanediyl)diimino]bis[2,4,6-triiodo-5-[(me-thylamino)carbonyl]benzoic acid] (●)
R also N-methylglucamine salt (1:2) (54605-45-7)
S Acide iocarmique**, Acidum iocarmicum**, Acidum jocarmicum, DB-2041, Dimeray, Dimerex, Dimer-X, Dirax, Iocarmate meglumine, Iocarmic

acid**, Meglumine iocarmate, MP-2032, Myelotrast "Guerbet"
U Diagnostic aid (radiopaque medium)

12209 (9755)

$C_{24}H_{20}N_2O_3$
3878-14-6

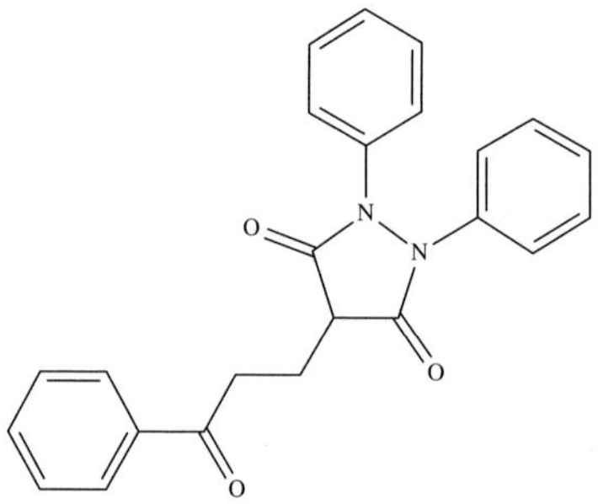

4-(3-Oxo-3-phenylpropyl)-1,2-diphenyl-3,5-pyrazoli-dinedione (●)
S Benzopyrazone
U Anti-inflammatory, antirheumatic

12210

$C_{24}H_{20}N_2O_4S_2$
213411-83-7

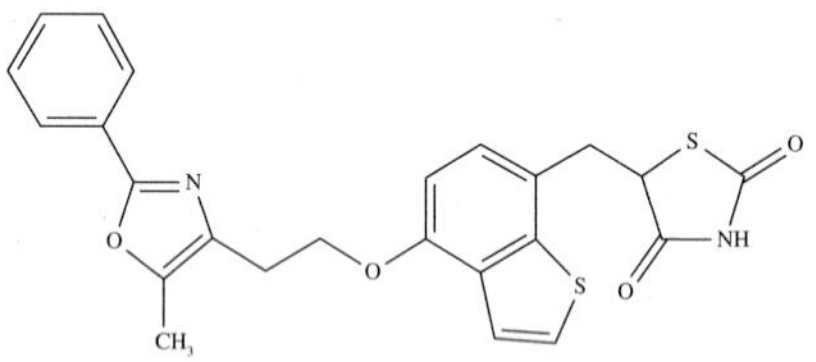

5-[[4-[2-(5-Methyl-2-phenyl-4-oxazolyl)ethoxy]ben-zo[b]thien-7-yl]methyl]-2,4-thiazolidinedione (●)
S BM 131258
U Hypoglycemic

12211 (9756)

$C_{24}H_{20}N_4O$
85-83-6

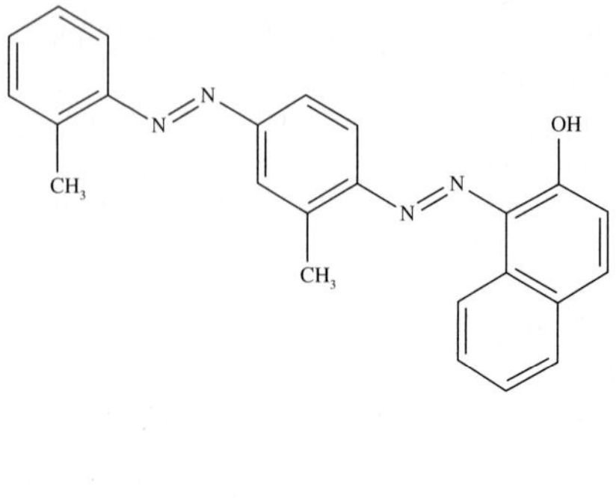

1-[4-(*o*-Tolylazo)-*o*-tolylazo]-2-naphthol = 1-[[2-Methyl-4-[(2-methylphenyl)azo]phenyl]azo]-2-naphthalenol (●)
S   *Biebricher Scharlachrot*, Biebrich Scarlet Red,
C.I. 26105, C.I. Solvent Red 24, *Fat Ponceau R*,
Fettponceau R, Rubren, *Scarlet Red, Scharlachrot,
Sudan IV*
U   Promote wound healing, dye

12212 $C_{24}H_{20}N_6O_3$
                                        139481-59-7

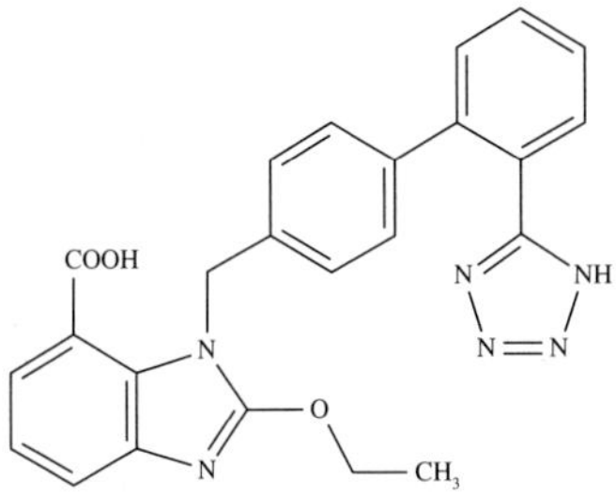

2-Ethoxy-1-[*p*-(*o*-1*H*-tetrazol-5-ylphenyl)benzyl]-7-ben-
zimidazolecarboxylic acid = 2-Ethoxy-1-[[2'-(1*H*-tetra-
zol-5-yl)[1,1'-biphenyl]-4-yl]methyl]-1*H*-benzimidazole-
7-carboxylic acid (●)
R   see also no. 14733
S   *Candesartan***, CV-11974
U   Antihypertensive (angiotensin II receptor antagonist)

12213 $C_{24}H_{20}O_8$
                                        150829-94-0

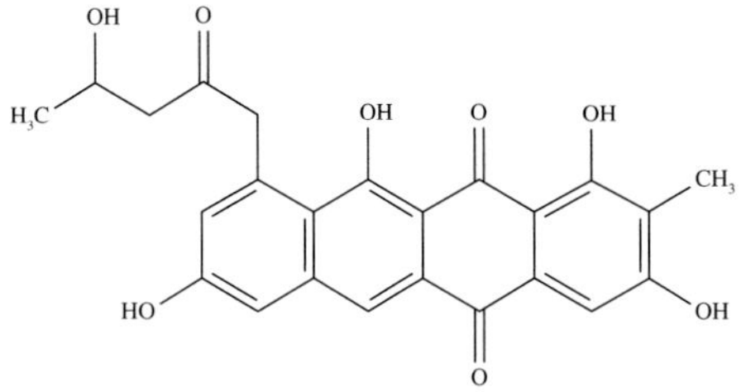

(+)-1,3,8,11-Tetrahydroxy-10-(4-hydroxy-2-oxopentyl)-
2-methyl-5,12-naphthacenedione (●)
S   UCE 6
U   Cytotoxic